Metalloenzymes

Metalloenzymes

From Bench to Beside

Edited by

Claudiu T. Supuran
William A. Donald

ELSEVIER

ACADEMIC PRESS
An imprint of Elsevier

Academic Press is an imprint of Elsevier
125 London Wall, London EC2Y 5AS, United Kingdom
525 B Street, Suite 1650, San Diego, CA 92101, United States
50 Hampshire Street, 5th Floor, Cambridge, MA 02139, United States
The Boulevard, Langford Lane, Kidlington, Oxford OX5 1GB, United Kingdom

ISBN 978-0-12-823974-2

For information on all Academic Press publications
visit our website at https://www.elsevier.com/books-and-journals

Publisher: Stacy Masucci
Acquisitions Editor: Peter B. Linsley
Editorial Project Manager: Susan E. Ikeda
Production Project Manager: Omer Mukthar
Cover Designer: Matthew Limbert

Typeset by STRAIVE, India

Contents

7.2 Indoleamine 2,3-dioxygenase

Michele Coluccia, Daniela Secci, and Paolo Guglielmi

Section H
Copper enzymes

8.1 Superoxide dismutases inhibitors

Azadeh Hekmat, Ali Akbar Saboury, and Luciano Saso

8.2 Tyrosinase enzyme and its inhibitors: An update of the literature

Simone Carradori, Francesco Melfi, Josip Rešetar, and Rahime Şimşek

Section I
Cadmium enzymes CAs

9. CDCA1, a versatile member of the ζ-class of carbonic anhydrase family

Vincenzo Alterio, Emma Langella, Davide Esposito, Martina Buonanno, Simona Maria Monti, and Giuseppina De Simone

Section J Molybdenum enzymes

10. Molybdenum enzymes

Simone Giovannuzzi

Section K Tungsten-containing enzymes

11. Tungsten-containing enzymes

Niccolò Paoletti

Contributors

Numbers in parentheses indicate the pages on which the authors' contributions begin.

Atilla Akdemir (449), Istinye University, Faculty of Pharmacy, Istanbul, Turkey

Vincenzo Alterio (549), Institute of Biostructures and Bioimaging-CNR, Naples, Italy

Andrea Angeli (23), NEUROFARBA Department, Pharmaceutical Sciences Section, University of Florence, Florence, Italy

Guillaume Arlet (157), Sorbonne Université, U1135, CIMI-Paris, Paris, France

Francesca Arrighi (239), Department of Drug Chemistry and Technologies, Sapienza University of Rome, Rome, Italy

Giulia Barchielli (305), Department of Chemistry "Ugo Schiff", University of Florence, Florence, Italy

Emanuela Berrino (239), Department of Drug Chemistry and Technologies, Sapienza University of Rome, Rome, Italy

Liberato Berrino (321), Department of Experimental Medicine—Section of Pharmacology "L. Donatelli", University of Campania "Luigi Vanvitelli", Naples, Italy

Jörg Bomke (343), Merck Healthcare KGaA, Darmstadt, Germany

Alessandro Bonardi (411), Department of NEUROFARBA, Section of Pharmaceutical and Nutraceutical Sciences, Pharmaceutical and Nutraceutical Section, University of Florence, Firenze, Italy

Martina Buonanno (549), Institute of Biostructures and Bioimaging-CNR, Naples, Italy

Matteo Calligaris (207), Department of Pharmacy, University of Pisa, Pisa; Proteomics Group of Fondazione Ri.MED, Research Department IRCCS ISMETT (Istituto Mediterraneo per i Trapianti e Terapie ad Alta Specializzazione), Palermo, Italy

Clemente Capasso (185,283), Department of Biology, Agriculture and Food Sciences, CNR, Institute of Biosciences and Bioresources, Napoli, Italy

Antonella Capperucci (305), Department of Chemistry "Ugo Schiff", University of Florence, Florence, Italy

Annalisa Capuano (321), Department of Experimental Medicine—Section of Pharmacology "L. Donatelli", University of Campania "Luigi Vanvitelli"; Campania Regional Centre for Pharmacovigilance and Pharmacoepidemiology, Naples, Italy

Simone Carradori (63,393,533), Department of Pharmacy, University "G. d'Annunzio" of Chieti-Pescara, Chieti, Italy

Fabrizio Carta (35,265,293), NEUROFARBA Department, Section of Pharmaceutical and Nutraceutical Sciences, University of Florence, Florence, Italy

Michele Coluccia (485), Department of Drug Chemistry and Technologies, Sapienza University of Rome, Rome, Italy

Doretta Cuffaro (207), Department of Pharmacy, University of Pisa, Pisa, Italy

Ilaria D'Agostino (63,393), Department of Pharmacy, University "G. d'Annunzio" of Chieti-Pescara, Chieti, Italy

Antonella De Angelis (321), Department of Experimental Medicine—Section of Pharmacology "L. Donatelli", University of Campania "Luigi Vanvitelli", Naples, Italy

Giuseppina De Simone (549), Institute of Biostructures and Bioimaging-CNR, Naples, Italy

Elsa Denakpo (157), Université Paris-Saclay, CNRS, Institut de Chimie des Substances Naturelles, Gif-sur-Yvette, France

Luigi F. Di Costanzo (333), Department of Agriculture - Department of Excellence - University of Naples Federico II - Palace of Portici - Piazza Carlo di Borbone, Portici (NA), Italy

William A. Donald (3), School of Chemistry, University of New South Wales, Sydney, NSW, Australia

Ghada F. Elmasry (51), Department of Pharmaceutical Chemistry, Faculty of Pharmacy, Cairo University, Cairo, Egypt

Davide Esposito (549), Institute of Biostructures and Bioimaging-CNR, Naples, Italy

Marta Ferraroni (467), Dipartimento di Chimica "Ugo Schiff", Università di Firenze, Firenze, Italy

Manja Friese-Hamim (343), Merck Healthcare KGaA, Darmstadt, Germany

Wen-Yun Gao (375), College of Life Sciences, Northwest University, Xi'an, PR China

Simone Giovannuzzi (557), NEUROFARBA Department, Pharmaceutical and Nutraceutical Section, University of Florence, Firenze, Italy

Paolo Guglielmi (485), Department of Drug Chemistry and Technologies, Sapienza University of Rome, Rome, Italy

Jakub Gunera (343), Merck Healthcare KGaA, Darmstadt, Germany

Özlen Güzel-Akdemir (459), Istanbul University, Faculty of Pharmacy, Department of Pharmaceutical Chemistry, Istanbul, Turkey

Timo Heinrich (343), Merck Healthcare KGaA, Darmstadt, Germany

Azadeh Hekmat (523), Department of Biology, Science and Research Branch, Islamic Azad University, Tehran, Iran

Philip Hewitt (343), Merck Healthcare KGaA, Darmstadt, Germany

Marc A. Ilies (101), Department of Pharmaceutical Sciences and Moulder Center for Drug Discovery Research, Temple University School of Pharmacy, Philadelphia, PA, United States

Bogdan I. Iorga (157), Université Paris-Saclay, CNRS, Institut de Chimie des Substances Naturelles, Gif-sur-Yvette, France

Shibbir Ahmed Khan (101), Department of Pharmaceutical Sciences and Moulder Center for Drug Discovery Research, Temple University School of Pharmacy, Philadelphia, PA, United States

Emma Langella (549), Institute of Biostructures and Bioimaging-CNR, Naples, Italy

Elena Lenci (197), Department of Chemistry "Ugo Schiff", University of Florence, Sesto Fiorentino, Italy

Heng Li (375), College of Life Sciences, Northwest University, Xi'an, PR China

Samuele Maramai (63), Department of Biotechnology, Chemistry and Pharmacy, University of Siena, Siena, Italy

Annamaria Mascolo (321), Department of Experimental Medicine—Section of Pharmacology "L. Donatelli", University of Campania "Luigi Vanvitelli"; Campania Regional Centre for Pharmacovigilance and Pharmacoepidemiology, Naples, Italy

Francesco Melfi (533), Department of Pharmacy, University "G. d'Annunzio" of Chieti-Pescara, Chieti, Italy

Simona Maria Monti (549), Institute of Biostructures and Bioimaging-CNR, Naples, Italy

Mattia Mori (9), Department of Biotechnology, Chemistry and Pharmacy, University of Siena, Siena, Italy

Djordje Musil (343), Merck Healthcare KGaA, Darmstadt, Germany

Alessio Nocentini (83), NEUROFARBA Department, Pharmaceutical and Nutraceutical Section, University of Florence, Firenze, Italy

Elisa Nuti (207), Department of Pharmacy, University of Pisa, Pisa, Italy

Niccolò Paoletti (583), Department of NEUROFARBA, Section of Pharmaceutical and Nutraceutical Sciences, Pharmaceutical and Nutraceutical Section, University of Florence, Firenze, Italy

Alain Philippon (157), Faculté de Médecine, Bactériologie, Université de Paris-Cité, Paris, France

Francesca Picarazzi (9), Department of Biotechnology, Chemistry and Pharmacy, University of Siena, Siena, Italy

Josip Rešetar (533), Faculty of Pharmacy and Biochemistry, University of Zagreb, Zagreb, Croatia

Maria Antonietta Riemma (321), Department of Experimental Medicine—Section of Pharmacology "L. Donatelli", University of Campania "Luigi Vanvitelli", Naples, Italy

Felix Rohdich (343), Merck Healthcare KGaA, Darmstadt, Germany

Maria Novella Romanelli (431), NEUROFARBA—Department of Neurosciences, Psychology, Drug Research and Child Health, Section of Pharmaceutical and Nutraceutical Sciences, University of Florence, Italy

Armando Rossello (207), Department of Pharmacy, University of Pisa, Pisa, Italy

Ali Akbar Saboury (523), Institute of Biochemistry and Biophysics, University of Tehran, Tehran, Iran

Luciano Saso (523), Department of Physiology and Pharmacology "vittorio erspamer", Sapienza University, Rome, Italy

Simone D. Scilabra (207), Proteomics Group of Fondazione Ri.MED, Research Department IRCCS ISMETT (Istituto Mediterraneo per i Trapianti e Terapie ad Alta Specializzazione), Palermo, Italy

Daniela Secci (239,485), Department of Drug Chemistry and Technologies, Sapienza University of Rome, Rome, Italy

Mario Sechi (35), Department of Medicine, Surgery and Pharmacy, Laboratory of Drug Design and Nanomedicine, University of Sassari, Sassari, Italy

Rahime Şimşek (533), Faculty of Pharmacy, Department of Pharmaceutical Chemistry, Hacettepe University, Ankara, Turkey

Donatella P. Spanò (207), Proteomics Group of Fondazione Ri.MED, Research Department IRCCS ISMETT (Istituto Mediterraneo per i Trapianti e Terapie ad Alta Specializzazione); STEBICEF (Dipartimento di Scienze e Tecnologie Biologiche Chimiche e Farmaceutiche), University of Palermo, Palermo, Italy

Liberata Sportiello (321), Department of Experimental Medicine—Section of Pharmacology "L. Donatelli", University of Campania "Luigi Vanvitelli"; Campania Regional Centre for Pharmacovigilance and Pharmacoepidemiology, Naples, Italy

Claudiu T. Supuran (3,93,139,185,283), Neurofarba Department, Section of Pharmaceutical and Nutraceutical Sciences, University of Florence, Florence, Italy

Damiano Tanini (305), Department of Chemistry "Ugo Schiff", University of Florence, Florence, Italy

Giusy Tassone (63), Department of Biotechnology, Chemistry and Pharmacy, University of Siena, Siena, Italy

Andrea Trabocchi (197), Department of Chemistry "Ugo Schiff", University of Florence, Sesto Fiorentino, Italy

Ansgar Wegener (343), Merck Healthcare KGaA, Darmstadt, Germany

Jean-Yves Winum (255), IBMM, Univ Montpellier, CNRS, ENSCM, Montpellier, France

Frank T. Zenke (343), Merck Healthcare KGaA, Darmstadt, Germany

Preface

The idea for metalloenzymes arose during a visit in February 2020 by one of the editors (Supuran) to the University of New South Wales in Sydney, where the other editor (Donald) is active. With both of our research groups having a long history of collaboration on metalloenzymes, we noticed a lack of a comprehensive monograph in the field, which inspired us to undertake this book project. Unfortunately, this coincided with the onset of the COVID-19 pandemic, which officially started a month later and that we knew would pose many challenges and limitations, resulting in significant delays. Despite these challenges, we were ultimately able to complete the book as envisioned with gratitude to the many dedicated subject-expert authors, and Susan Ikeda from the publisher who facilitated the project.

Metalloenzymes: From Bench to Bedside is a comprehensive resource featuring more than 35 chapters on validated or promising metalloenzyme drug targets. The book has a clear structure, with each chapter reviewing an enzyme's structure, function, physiological/pathological role, known inhibitors/activators, and drug design as well as any clinically used agents or compounds in (pre)clinical development. The book is organized according to the nature of the metal ion present in the enzyme's active site, beginning with eight chapters on magnesium- and calcium-based enzymes. These chapters cover a range of enzymes including nucleic acid processing enzymes, kinases such as CDK2 and enzymes involved in the design of antibacterials. The latter includes D-Ala-D-Ala ligase and the calcium enzyme paraoxonase, whose role in a variety of diseases remains controversial. Inhibitors of some of these enzymes are widely used clinically to manage viral (DNA polymerase, reverse transcriptase, and integrase) and bacterial (D-Ala-D-Ala ligase) infections.

The next section of the book covers zinc enzymes, which are the most abundant known metalloenzymes, with 12 different chapters. This section discusses many pharmacological and medical applications for their inhibitors and activators for well-known and clinically relevant enzymes such as carbonic anhydrases, beta-lactamases, matrix metalloproteases, angiotensin-converting enzyme, and histone deacetylases. In the book, additional enzymes are examined that were only recently considered possible drug targets, including histidinol dehydrogenases, bacterial proteases, CD73, glyoxalase II, and neprilysin, and compounds are detailed that are in (pre)clinical development for targeting these enzymes and may lead to therapeutic drugs.

The book contains additional sections that focus on metalloenzymes containing functionally important cofactors such as manganese, nickel, iron, copper, cadmium, molybdenum, and tungsten. For instance, the book dedicates five chapters to discussing established and potential targets of manganese- and nickel-containing metalloenzymes, such as arginase, methionine aminopeptidase, and urease. The book also reviews both heme- and non-heme-containing iron enzymes, including clinically important cyclooxygenases, cytochrome P450s, and aromatase and emerging drug targets indoleamine 2,3-dioxyganese and mono- and dioxygenases. The book also covers copper-containing enzymes, such as superoxide dismutase and tyrosinases and cadmium-, molybdenum-, and tungsten-containing enzymes in separate chapters.

The book will be a valuable resource for researchers in academia, government, and industry engaged in medicinal chemistry, pharmacology, and molecular biology of metalloenzymes as well as PhD students in these fields. Given the constant influx of pertinent discoveries in the field, we are confident that this comprehensive book will make a significant contribution to our understanding of these fascinating enzymes.

William Alex Donald
Claudiu T. Supuran

Section A

Metalloenzymes

Chapter 1

Introduction to metalloenzymes: From bench to bedside

William A. Donald[a] and Claudiu T. Supuran[b]

[a]*School of Chemistry, University of New South Wales, Sydney, NSW, Australia,* [b]*NEUROFARBA Department, Section of Pharmaceutical and Nutraceutical Sciences, University of Florence, Florence, Italy*

1 Introduction

In the postgenomic era, after the discovery that the human genome encodes only for ~30,000 genes, there is still much debate regarding how many can be considered druggable [1]. Although there is considerable uncertainty in different approximations [1–4], reports indicate that all drugs that are currently used clinically act on ~400 targets that are typically proteins, but also include nucleic acids and more rarely sugars or lipids. However, there are 5000–10,000 potentially druggable targets that might be encoded by the human genome in addition to the genomes of parasites (bacteria, fungi, protozoan, worms, etc.) that can infect humans. Furthermore, ~50% of all drugs act as enzyme inhibitors, making these proteins among the most relevant drug targets [1–4]. Of the known enzymes, a rather large number incorporate metal ions that are essential either for the catalytic activity of the enzyme or for structural reasons (e.g., for stabilizing tertiary/quaternary structures of the enzyme or for orientating substrates/modulators of activity when bound within the enzyme active site) [5–9].

The presence of metal ions in metalloproteins is highly relevant for their structural and functional roles. The following metal ions are frequently found in many classes of enzymes: Mg(II), Ca(II), Zn(II), Mn(II), Ni(II), Fe(II), and Fe(III), including heme and non-heme proteins (although the iron may be present also in higher oxidation states during the catalytic cycle), Cu(I) and Cu(II), Cd(II), Mo(IV), Mo(VI), and W(VI) [5–11]. For the latter two metal ions, tetra- and pentavalent species can also be involved during catalysis. In most metalloenzyme active sites, one or more of the metal ions are present (most of the time the same ions, but in some cases also in various combinations, for example, Cu(II) and Zn(II) in superoxide dismutase [12]) and are coordinated by amino acid residues and water molecules, many of which thereafter take part in the catalytic cycle [5–9]. Furthermore, in many cases the metal ion may also directly interact with fragments of the substrate(s), activating it for the catalysis [13]. In other cases, the metal ions are present in cofactors, such as heme or iron-sulfur clusters [14,15], which are tightly bound in the active site and participate in various steps of the catalytic cycle.

Non-catalytic metal ions are often essential for the stabilization of protein structures by creating or maintaining secondary/tertiary structural elements similarly to disulfide bridges, and this is mostly achieved by calcium and zinc ions [16]. The metal ions induce the correct folding of protein sequences, for example as zinc-fingers, zinc-twists, or zinc-clusters in numerous regulatory proteins and hormone receptors, contributing to the overall stability of these domains [17]. Zinc fingers are structurally diverse and are present in proteins that perform a broad range of functions in various cellular processes. Such structurally stabilizing motifs are as diverse as their functions, being associated also with protein-nucleic acid recognition as well as protein-protein interactions [17,18]. The zinc ion can be also involved in the structural maintenance of chromatin and biomembranes with crucial roles in the regulation of their functions [18]. In paraoxonase, which contains two calcium ions, one of the metal ions is involved in the catalytic cycle, whereas the second one has a structural role, and its removal can cause irreversible structural disruption of the protein [19].

In catalytic sites, the metal ions participate directly in the catalytic process and may exhibit rather diverse geometries. The distorted-tetrahedral geometry with three O/N/S ligands bound to the metal ion and the fourth ligand being a water molecule, acting as an activated nucleophile for the catalytic process, is one of the most common motifs [5–9]. However, the coordination number may be higher (5 or 6), with trigonal-bipyramidal or octahedral geometries of the metal center possessing a highly relevant catalytic function in many copper and iron-containing enzymes in addition to other metal ions [9,12]. For example, zinc is essential for the catalytic activity of more than 300 enzymes belonging to all six classes of enzymes. Zinc ions are often located at the core of the enzyme active sites and participate directly in

Metalloenzymes. https://doi.org/10.1016/B978-0-12-823974-2.00038-3

the catalytic mechanism through interactions with substrate molecules undergoing the catalyzed chemical transformation [5–9]. Typical reactions that are catalyzed by metalloenzymes and are relevant to physiology include carbonic anhydrases, alcohol dehydrogenase, metalloproteinases (matrix metalloproteinases and many related bacterial/parasite proteases), oxidoreductases, cyclooxygenase, lipoxygenase, tyrosinase, catechol-*O*-methyl-transferase, phosphodiesterases, integrase, paraoxonase, a number of peptidases (angiotensin-converting enzyme, a disintegrin and metalloprotease enzyme, ADAM, elastase, etc.), arginase, histone deacetylases, metallo-β-lactamases, and indoleamine 2,3-dioxygenase [5–9]. Thus, metalloenzymes play a massive role of in the proliferation of human diseases, and their druggability has been explored extensively in the past 50 years. There are many classes of metalloenzyme inhibitors in clinical use for decades, and they will be presented in detail in multiple chapters of this book. Indeed, the scope of this book is to cover the major metalloenzyme drug targets that are either validated or for which evidence is building that they may be promising targets in the future. This is also why chapters were also included on less well-investigated metalloenzymes, such as those containing Mo(IV/VI) and W(VI) [10,11].

The goal of this book is to provide a reference for years to come, written by world renowned expert investigators studying key metalloenzymes that have key biological roles in many different biological functions and diseases such as obesity, diabetes, fatty liver disease, inflammation, cancer, cardiovascular and mood-related manifestations, infections, and more, which are being uncovered at an expanding rate. Increasing our understanding of metalloenzymes and modulating their function with inhibitors and activators affords in health and disease the opportunity for novel therapeutics. This book will offer a thorough overview of metalloenzymes, spanning biochemical and structural features, pharmacology, and biotechnological applications. Each chapter follows the general outline: (i) structure and function, (ii) physiological/pathological role, (iii) classes of inhibitors and activators and their design, and (iv) clinically applied agents and/or compounds that are in clinical development.

2 Druggability of metalloenzymes: Challenges and opportunities

The advantages of targeting metalloenzymes reside in the following factors, which are detailed in this book:

- **(i)** the enzyme function can often be disrupted by targeting the metal site,
- **(ii)** metal sites can often promote ligand-protein interactions,
- **(iii)** metalloenzymes are often amenable to high-throughput, optical-based assays by tracking the kinetics of chemical transformations, and
- **(iv)** the normal function of such enzymes is typically involved in disease proliferation.

In addition, there are also many challenges, including:

- **(i)** a need for selectivity, as metal sites can be "sticky" owing to highly conserved active sites and strong Coloumbic interactions between anionic inhibitors with cationic metal sites, can lead to off-target effects [5–9],
- **(ii)** many metalloenzymes have a large number of isoforms, some of which possess very diverse functions and physiological/pathological roles [20]. As a consequence, developing isoform-selective inhibitors is crucial [21],
- **(iii)** delivering drugs to their site of action often within cells, tissues, or dense tumor microenvironments [22], and
- **(iv)** high-throughput target-based screening of mixtures of chemicals should be further developed [23].

In summary, there are plenty of opportunities for positively impacting human health by the development of metalloenzyme inhibitors as demonstrated by those already available and the many more that are anticipated to be established in the future. This book details more than 36 metalloenzymes, their structure, druggability, and strategies for modulating their function for therapeutic goals. The future is bright for metalloenzyme research in terms of new approaches for more efficiently discovering bioactive molecules involving high-throughput screening, computational chemistry, artificial intelligence, new methods for rapidly profiling off target effects, and mining natural products [24,25], which should also be taken into consideration when dealing with metalloenzyme drug design, and some of these aspects are dealt with in this book.

References

[1] Hopkins AL, Groom CR. The druggable genome. Nat Rev Drug Discov 2002;1(9):727–30.

[2] Kramer R, Cohen D. Functional genomics to new drug targets. Nat Rev Drug Discov 2004;3(11):965–72.

[3] Overington JP, Al-Lazikani B, Hopkins AL. How many drug targets are there? Nat Rev Drug Discov 2006;5(12):993–6.

[4] Imming P, Sinning C, Meyer A. Drugs, their targets and the nature and number of drug targets. Nat Rev Drug Discov 2006;5(10):821–34.

[5] Supuran CT, Winum JY. Introduction to zinc enzymes as drug targets. In: Supuran CT, Winum JY, editors. Drug design of zinc-enzyme inhibitors: Functional, structural, and disease applications. Hoboken: Wiley; 2009. p. 3–12.

[6] Nocentini A, Supuran CT. Carbonic anhydrases: an overview. In: Supuran CT, Nocentini A, editors. Carbonic anhydrases – Biochemistry and pharmacology of an evergreen pharmaceutical target. London, UK: Elsevier – Academic Press; 2019. p. 3–16.

[7] Capasso C, Supuran CT. Protozoan, fungal and bacterial carbonic anhydrases targeting for obtaining antiinfectives. In: Supuran CT,

Capasso C, editors. Targeting carbonic anhydrases. London: Future Science Ltd.; 2014. p. 133–41.

[8] Supuran CT, Scozzafava A. Matrix metalloproteinases (MMPs). In: Smith HJ, Simons C, editors. Proteinase and peptidase inhibition: Recent potential targets for drug development. London & New York: Taylor & Francis; 2002. p. 35–61.

[9] Chen AY, Adamek RN, Dick BL, Credille CV, Morrison CN, Cohen SM. Targeting metalloenzymes for therapeutic intervention. Chem Rev 2019;119(2):1323–455.

[10] Ott G, Havemeyer A, Clement B. The mammalian molybdenum enzymes of mARC. J Biol Inorg Chem 2015;20(2):265–75.

[11] L'vov NP, Nosikov AN, Antipov AN. Tungsten-containing enzymes. Biochemistry (Mosc) 2002;67(2):196–200.

[12] Ferraroni M, Rypniewski W, Wilson KS, Viezzoli MS, Banci L, Bertini I, Mangani S. The crystal structure of the monomeric human SOD mutant F50E/G51E/E133Q at atomic resolution. The enzyme mechanism revisited. J Mol Biol 1999;288(3):413–26.

[13] Bigley AN, Raushel FM. Catalytic mechanisms for phosphotriesterases. Biochim Biophys Acta 2013;1834(1):443–53.

[14] Crielaard BJ, Lammers T, Rivella S. Targeting iron metabolism in drug discovery and delivery. Nat Rev Drug Discov 2017;16 (6):400–23.

[15] Schulz V, Freibert SA, Boss L, Mühlenhoff U, Stehling O, Lill R. Mitochondrial [2Fe-2S] ferredoxins: new functions for old dogs. FEBS Lett 2022. https://doi.org/10.1002/1873-3468.14546 [in press].

[16] Lee YM, Lim C. Physical basis of structural and catalytic Zn-binding sites in proteins. J Mol Biol 2008;379:545–53.

[17] Vallee BL, Coleman JE, Auld DS. Zinc fingers, zinc clusters, and zinc twists in DNA-binding protein domains. Proc Natl Acad Sci U S A 1991;88:999–1003.

[18] Cox EH, McLendon GL. Zinc-dependent protein folding. Curr Opin Chem Biol 2000;4:162–5.

[19] Harel M, Aharoni A, Gaidukov L, Brumshtein B, Khersonsky O, Meged R, Dvir H, Ravelli RB, McCarthy A, Toker L, Silman I, Sussman JL, Tawfik DS. Structure and evolution of the serum paraoxonase family of detoxifying and anti-atherosclerotic enzymes. Nat Struct Mol Biol 2004;11(5):412–9.

[20] Supuran CT. Carbonic anhydrases: novel therapeutic applications for inhibitors and activators. Nat Rev Drug Discov 2008;7(2):168–81.

[21] Supuran CT. Emerging role of carbonic anhydrase inhibitors. Clin Sci (Lond) 2021;135(10):1233–49.

[22] McDonald PC, Chafe SC, Supuran CT, Dedhar S. Cancer therapeutic targeting of hypoxia induced carbonic anhydrase IX: from bench to bedside. Cancers (Basel) 2022;14(14):3297.

[23] Bennett JL, Nguyen GTH, Donald WA. Protein-small molecule interactions in native mass spectrometry. Chem Rev 2022;122(8): 7327–85.

[24] Atanasov AG, Zotchev SB, Dirsch VM, International Natural Product Sciences Taskforce, Supuran CT. Natural products in drug discovery: advances and opportunities. Nat Rev Drug Discov 2021;20(3):200–16.

[25] Nguyen GTH, Bennett JL, Liu S, Hancock SE, Winter DL, Glover DJ, Donald WA. Multiplexed screening of thousands of natural products for protein-ligand binding in native mass spectrometry. J Am Chem Soc 2021;143(50):21379–87.

Section B

Magnesium and calcium-containing enzymes

Chapter 2.1

DNA and RNA polymerases

Francesca Picarazzi and Mattia Mori
Department of Biotechnology, Chemistry and Pharmacy, University of Siena, Siena, Italy

1 Structure and function of the enzyme(s)

1.1 DNA polymerase

DNA polymerases (DNAPs) are a family of enzymes and multiprotein complexes classified as transferases (EC 2.7.7.6), which catalyze the synthesis of long polymer chains of DNA based on the sequence of a complementary template strand. The template used for the synthesis of the new DNA strand can be composed of DNA nucleotides, in the case of DNA-dependent DNA polymerase, or RNA nucleotides, in the case of RNA-dependent DNA polymerase [commonly referred as reverse transcriptase (RT)], this latter being characteristic of some viruses.

DNAPs play a central role in the processes of life by duplicating genetic information of cells during the cell division process and passing it to daughter cells. A DNAP copies the DNA template strand in the 5′-3′ direction, generating two newly synthesized DNA molecules using a semiconservative process [1,2]. In particular, the leading strand is continuously synthetized in the same direction as the replicative fork (5′-3′) while the lagging strand is synthesized in short fragments (called Okazaki fragments) in the opposite direction (3′-5′) with respect to the replication fork [3–5]. Another important feature of polymerases is their inability to initiate the synthesis of nucleotide chains de novo; in fact, to start the elongation of a strand, a DNAP needs a short preexisting RNA or DNA segment often referred as a primer. When the bases of the primer are paired with those of the template, the DNAP adds nucleotides to the free 3′-OH hydroxyl group of the primer [6]. The DNA synthesis process involves a nucleophilic attack from the 3′-OH hydroxyl group of the terminal nucleotide in the nascent strand to the α-phosphate group of the subsequent 5′-triphosphate deoxynucleoside (dNTP). The catalyzed reaction is: $(dNMP)_n + dNTP \rightarrow (dNMP)_{n+1} + $pyrophosphate, in which each dNTP is added according to strand complementarity, which contributes to DNA polynucleotide chain elongation in the 5′-3′ direction. This reaction requires the presence of a single unpaired chain as a template and a primer chain having a free 3′-OH hydroxyl group to which new nucleotide units are added. Incoming nucleotides are selected based on base-pair complementarity with the DNA template strand, according to the Watson-Crick rule associating A-T and C-G. The reaction product has a new free hydroxyl group at the 3′ terminus (3′-OH) that permits chain elongation through the addition of the subsequent nucleotide [7].

In the catalytic process, the enzyme first binds to the already synthesized template and primer. Then, based on the template, the complementary dNTP binds to the polymerase-DNA complex, giving rise to the nucleophilic attack toward the phosphodiester bond of the dNTP, which leads to the incorporation of the nucleotide into the nascent DNA molecule. During this reaction, an inorganic pyrophosphate (PPi) molecule is released. The polymerase active site is characterized by the presence of highly conserved aspartate residues that coordinate two Mg^{2+} ions. These cations catalyze the enzymatic reaction with specific and differential roles: (i) an Mg^{2+} ion promotes the nucleophilic attack of the free 3′-OH hydroxyl group to the α-phosphate of the incoming dNTP, while (ii) the second Mg^{2+} ion facilitates the removal of the PPi from the reaction environment. Given the peculiar role of the Mg^{2+} ions, this reaction mechanism is also referred as "two-metal-ion catalysis", and it is shared by all polymerases, including RNA polymerases and RT [7]. Once the catalytic process is terminated, i.e., all nucleotides complementary to the templates are included in the nascent nucleic acid molecule, two identical DNA chains are generated, and the DNAP detaches from the template, leading to the disassembly of the replication machinery (Fig. 1) [8,9].

The replication process is highly accurate, and a base insertion error can generally occur with an approximate frequency of one wrong nucleotide incorporated every 10^8–10^{10} copied nucleotides. Presynthesis error control, called "base selection" or insertion fidelity, contributes by a factor of 10^4–10^5 to replication fidelity. A final contribution of about 10^2–10^3 to replication fidelity is provided by the post replicative repair system [10]. Some DNAPs have a 3′-5′ exonuclease activity, also known as proofreading activity, which can increase replication fidelity by checking, and

Metalloenzymes. https://doi.org/10.1016/B978-0-12-823974-2.00011-5

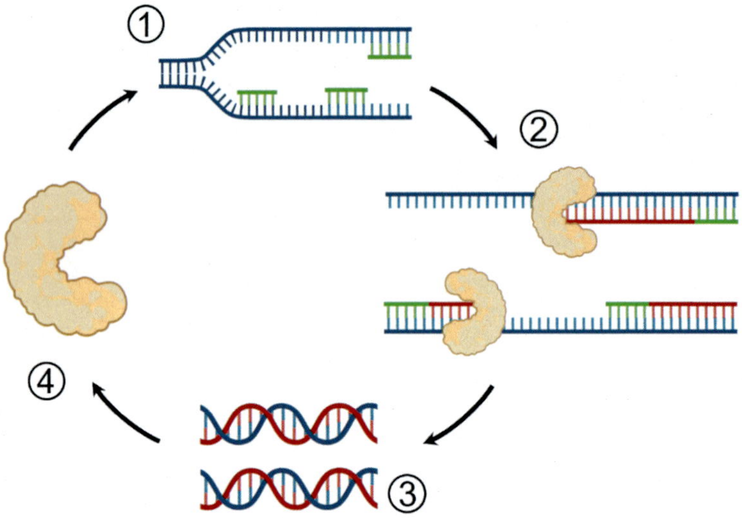

FIG. 1 Schematic representation of DNA replication process. (1) DNAP recognizes and binds the template. (2) Elongation of the complementary chains using the "two metal ion catalysis" mechanism. (3) The two identical DNA molecules generated by the catalytic processes are resolved. (4) The replication machinery is disassembled, and the free DNAP is ready to start another replication cycle.

possibly correcting, the newly synthesized DNA, which is estimated to contribute by a factor of 10^2–10^3 to the fidelity of polymerization [11,12]. The proofreading activity allows the enzyme to remove the newly inserted nucleotide and is highly specific for mismatches [13]. If a wrong (i.e., non-complementary) nucleotide is inserted in the nascent polynucleotide chain, translocation of the polymerase to the next position of the template is inhibited. The delay induced by this partial inhibition allows the enzyme to correct the error by switching the DNA from the polymerase site to the exonuclease site. DNA repair mechanisms are essential to preserve the integrity of genetic information and to prevent the formation of dysfunctional proteins as well as the onset of diseases [14]. After the addition of a nucleotide to the nascent DNA chain, the polymerase must either dissociate or move along the template to add another nucleotide. The association and dissociation of the polymerase can limit the overall speed of the reaction; therefore, DNA synthesis is faster when the polymerase continuously adds nucleotides without dissociating from the template. The average number of nucleotides added before the polymerase dissociates is called processivity, and it is a unique feature of DNAPs. In fact, these enzymes can remain associated with the primer-template substrate for various catalytic cycles. DNAPs are characterized by highly variable processivities as some members of the enzyme family can only add a few nucleotides, whereas other members may add thousands of them before dissociating from DNA. Association with the DNA template during the replication of the genome may be facilitated by additional protein factors, such as in the case of many DNAPs, although these proteins are not directly involved in the catalytic process [15].

From a structural standpoint, all polymerases share a common architecture. The structure is organized into three domains resembling a human right hand with fingers, palm, and thumb (Fig. 2) [17–20].

Each domain is designed to perform a specific function: the fingers interact with nucleotides that will be inserted in the nascent chain, the palm accommodates the active site with divalent Mg^{2+} cations and conserved catalytic residues that interact with the incoming nucleotides, while the thumb may bind the newly formed double strand of DNA. In general, the sequence of the palm domain is extremely conserved between different species, while other regions of the protein may exert a higher degree of sequence variation [21].

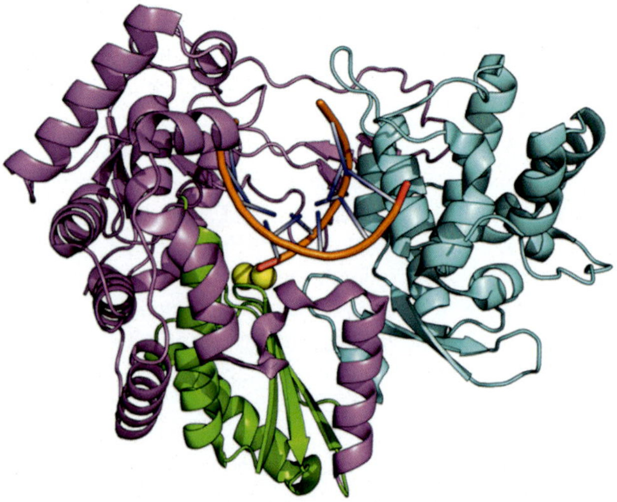

FIG. 2 Graphical representations of palm (*green*), thumb (*cyan*), and fingers (*pink*) subdomains in RNA-bound RdRp (PDB-ID: 4WTG) [16]. Mg^{2+} ions within the catalytic site are represented as *yellow spheres*.

The first DNAP has been isolated from *E. coli* by Arthur Kornberg in 1956 [22]. From that milestone, several polymerases from eukaryotes, archaea, and viruses have been isolated and characterized. At the state of the art, polymerases can be divided into seven families based on their sequence homology: A, B, C, D, X, Y, and RT [23]. Members of families A, B, C, and D participate directly in DNA replication, while those of X and Y families are involved in DNA repair processes. However, with the only exception of the RT that replicates DNA from a RNA template, all other DNAP family members polymerize DNA molecules. In general, prokaryotes possess five different DNAPs indicated by roman numbers (I, II, III, IV, and V). Among them, DNAP III is the most complex and replicates the bacterial chromosome, while DNAP II, IV, and V play a central role in the repair of damaged DNA. Finally, the DNAP I is involved in both replication and repair processes [24]. Conversely, in eukaryotic cells, up to 15 different types of polymerases have been identified and are indicated by Greek letters (α, β, γ, δ, ε, η, ι, κ, ζ, θ, λ, φ, σ, and μ) except for the terminal transferase Rev1. Each of them has a specific task in the replication process. Some DNA viruses, either composed by double-stranded or single-stranded DNA (dsDNA and ssDNA, respectively), code for their own DNAP that generally acts as a single protein that carries out multiple important functions in viral replication [25].

1.2 RNA polymerases

RNA polymerases (RNAPs) are a class of enzymes that synthesize RNA molecules using a single strand of DNA as a template. The transmission of genetic information from DNA to RNA is a process called transcription, in which DNA-dependent RNA polymerase is the main character. Some RNAPs synthesize RNA molecules using an RNA strand as a template, in this case the enzyme is referred as RNA-dependent RNA polymerase (RdRp) and is typically encoded by many RNA viruses. In addition to the latter, RNAPs are found in all living organisms, albeit with some sequence differences between the species [26].

The transcription process is divided into three steps: (i) initiation, (ii) elongation, and (iii) termination. The synthesis of the RNA molecule initiates with the binding of RNAP to specific regions of template DNA that are referred as promoters [27,28], although different from DNAPs, the RNAP is also able to add the new nucleotides with a de novo mechanism, i.e., in the absence of a primer sequence [29]. Furthermore, unlike DNAP, RNAP also has a helicase activity that allows the double strand of DNA to be opened with no need for an additional enzyme [30]. The elongation process of the new RNA strand proceeds in the 5′-3′ direction, while the DNA strand is read in the antiparallel 3′-5′ direction. The catalysis mechanism of RNAP is identical to that of DNAP: the active site is characterized by the presence of highly conserved aspartate residues that coordinate Mg^{2+} ions, which facilitate the nucleophilic attack of the free 3′-OH hydroxyl group to the incoming nucleotide, and promote the release of the PPi molecule. The overall phosphodiester bond formation reaction is: $(NMP)_n + NTP \rightarrow (NMP)_{n+1} + PPi$, in analogy to what described in Fig. 1. The substrates of the reaction are paired according to the Watson-Crick rule, with A pairing to U (in RNA thymine is replaced by uracil) and C pairing to G [7]. When the RNAP recognizes specific DNA terminator sequences encoded at the end of the gene, the RNA transcript is released, and the process ends. RNAP also has a proofreading mechanism [31]. Differently from the DNAP, the mismatched nucleotide is removed in the same site as that of polymerization. However, the error rate is slightly higher than that of DNAP, and it is approximately 10^{-6}–10^{-5} [32].

In general, the structure of RNAP is composed of multiple subunits, whose size and number can vary from species to species. Bacteria and archaea have single RNAPs capable of synthesizing both messenger RNA and non-coding RNA. The bacterial RNAP consists of a total of five subunits: two small α subunits (36 kDa), a β and β' subunit (150 kDa and 155 kDa, respectively) and a small ω subunit. A sigma (σ) factor binds to the core, forming the holoenzyme [33]. Archaea RNAP is more complex than that of bacteria, and it is structurally and mechanistically related to the RNAP II of eukaryotic cells [34,35].

The transcription apparatus of eukaryotic cells is much more complex than that of bacteria. Eukaryotes have three different RNAPs (I, II, and III), which are distinct protein complexes, although sharing some subunit types. RNAP I is a 14-subunit protein responsible for the synthesis of preribosomal RNA, which contains precursors of 28S, 18S, and 5.8S rRNAs. RNAP II is composed of 10–12 subunits and synthesizes mRNA and some specialized RNAs. Finally, the 17-subunit RNAP III synthesizes tRNAs, 5S rRNAs, and other small RNAs with specialized functions [36].

2 Physiologic/pathologic role

2.1 Eukaryotic cells reproduce by mitosis

All living organisms are made of cells that reproduce to increase the size of organs and systems or to replace cellular elements that are destroyed by physiological or pathological processes. To cope with renewal and growth needs, cells replicate themselves giving rise to daughter cells. Cell replication occurs by the division of a "mother" cell into two daughter cells, a process called cell division or mitosis. In an ideal replication process, the two daughter cells are exact copies of the mother cell, which also guarantees the preservation of the original cell functions.

The cell cycle is composed of a series of ordered events that lead to cell growth and duplication. It is organized into two major phases: interphase and mitosis. The interphase is the period between one division and another. During the interphase, a cell grows and accumulates nutrients preparing to replicate its own DNA and to divide into two daughter cells. The interphase consists of three distinct phases: (i) G_1, in which the cell grows and prepares for DNA synthesis; (ii) S, in which the synthesis of DNA does occurs; and (iii) G_2, in which the cell continues to grow waiting for the initiation of the final cell division process. Finally, mitosis occurs. During this phase, the chromosomes separate and migrate to opposite poles of the cytoplasm, resulting giving rise to daughter cells. After cell division, each daughter cell restarts the interphase in a new cycle. Not all the cells continue to divide as under certain conditions cell cycle progression stops, entering the so-called G_0 phase. Indeed, many cells of multicellular eukaryotes are non-proliferative, and once cell differentiation is achieved, they enter a state of quiescence. Each step of the cell cycle is highly regulated by control systems called checkpoints. Each checkpoint prevents the cycle from progressing until the necessary requirements are met. Many types of cancer are due to mutations in genes responsible for controlling cell cycle regulation, causing abnormal and potentially invasive growth of the affected tissue [37].

The S phase is one of the most important and delicate phases of the cell cycle, as any errors at this stage can lead to the emergence of mutations, genetic dysfunctions, and diseases such as cancer. During this phase, the DNA duplex of the mother cell is unwound by an enzyme called gyrase, the two annealed strands are separated by the helicase and used as templates for polymerization. The DNAP reads each of the single strand of DNA and generates two new duplexes that are completely identical to the original nucleotide. Following this mechanism, the genetic information is faithfully transferred to daughter cells [37]. The correct execution of the reactions catalyzed by polymerases is crucial for the maintenance of life [38]. Indeed, if DNA is not copied correctly during cell division, the daughter cells could have abnormal shape and function. The deregulation of cell division processes plays a key role in tumorigenesis. Although the cell has several control mechanisms at different stages of cell division, as well as holds many repair mechanisms of damaged DNA, a small percentage of mutations in the DNA nucleotide sequence can be passed to the daughter cell, and they can strongly influence its structure and function. Over the time, accumulation of uncorrected errors can result in the production of malfunctioning proteins, leading to the loss of effectiveness of the control and repair systems. In this context, the uncontrolled proliferation of cells can result in cancer with mutations or overexpression of DNAP [39]. Indeed, although mechanisms of exonuclease-proofreading and DNA mismatch repair work together to ensure fidelity in the DNA duplication process, defects in these mechanisms are associated with the increased incidence of cancer. Specifically, mutations in the exonuclease site of DNAP cause an inactivation of the proofreading mechanism leading to hypermutated tumors [40–42]. It is worth noting that polymerase mutations do not always correlate with diseases. In simpler organisms, but also in the evolution of the human species, mutations have also had positive effects, such as the acquisition of structural characteristics and functions that have allowed adaptation to environmental changes, although these occur and consolidate in rather long periods. On the other hand, in the current phase of the evolution of the human species, gene mutations are attributed largely to negative effects because they are associated with the predisposition to development of many diseases, such as cancer.

2.2 Prokaryotic cells reproduce by binary fission

Prokaryotes are living organisms characterized by the absence of a cell nucleus. Their genome consists of a single DNA chromosome enclosed in a specific area within the cell called nucleoid, whereas processes such as DNA duplication, mRNA transcription, and protein synthesis occur in the cytoplasm. Prokaryotes reproduce by binary fission, a proliferative mechanism highly similar to that of eukaryotic cells, although simpler and faster. The separation of daughter cells occurs through the formation of a septum in the mother cell, which derives from the introflexion of the plasma membrane and the cell wall, extending toward the center of the cell from opposite directions. Each daughter cell receives various organic and inorganic compounds and macromolecules that are important nutrients and cofactors for survival. Shortly before binary cleavage occurs, and concomitant with the growth of the mother cell, the DNA duplicates, remaining anchored to the plasma membrane. Finally, the two DNA molecules are separated in the two daughter cells by the septum. The time required for bacterial division depends on various nutritional and genetic factors [43]. The replicative apparatus of the genome of prokaryotes is less efficient than that of eukaryotes, and during the synthesis of the DNA strand, there is a greater probability of errors (mutations). In most prokaryotes, one cell continues to grow until it is divided by binary fission, resulting in two daughter cells. In this sense, the final goal is the same as for eukaryotes, and it consists of the production of two identical individuals (i.e., clones), while the variability that can be observed from one generation to the other might derive from incorrect DNA duplication. However, as anticipated above, this might not be a disadvantage, because mutations are a source of genetic variability and therefore greater adaptability to the environment over the bacterial generations.

2.3 Viral replication

Viruses are obligate intracellular parasitic entities that have no cellular organization or metabolic processes and are thus entirely dependent on the host cell's reproductive apparatus to reproduce themselves. Indeed, viruses use the host's enzymes involved in the transcriptional and translational processes to produce many new viruses that will be released in the extracellular environment. For these reasons, they are not classified in any domain and are found on the border between living and non-living organisms. Each virus has an extracellular, metabolically inert form, called virion, which is responsible for recognizing the host cell into which it injects the nucleic acid (e.g., DNA or RNA), a process that initiates the intracellular and infective phase of viral life cycle. The structure of a virion is extremely simple: it consists of a protein capsid capable of recognizing specific structures on the surface of the host cells to be infected and a nucleic acid molecule inside it. Some virions have an additional external envelope made of protein or phospholipids from the host cell. Unlike other organisms, the viral genome can be either DNA or RNA. During the infection of a host cell, a virus first attaches to its surface, penetrates the cytoplasm, and sheds its outer envelopes to expose the genome to host's enzymes involved in nucleic acids replication and in the expression of encoded proteins. Ultimately, new viral particles assembled in the host cell will be released in the extracellular space where they can infect other cells and progress the infection process [44]. Exposure of viral nucleic acids to cellular enzymes initiates the replication phase in which numerous copies of the viral genome are reproduced, and the capsid proteins are synthesized. This stage differs among viruses depending on the nature of nucleic acids, their secondary structure (e.g., double- or single-stranded configuration), and the polarity of the RNA genome. Based on these properties, viruses can be classified into seven classes, according to the Baltimore scheme [45,46]. The outcomes of viral infections on cells can be different. Indeed, some viruses cause lytic infections, which lead to the death of the host cell. Otherwise, the newly assembled virions can be released slowly while the host cell remains viable for a long time, although this generally provokes a persistent infection. Latent infections are also possible when there is a delay between the time of infection and lysis (e.g., herpes simplex infection). Finally, some viruses can cause transformation of the host cell into a cancer cell because they interfere with the molecular mechanisms that are responsible for controlling cell cycle. Among them, some retroviruses are known to induce leukemia in humans and animals, in addition to DNA viruses such as Hepatitis B and Hepatitis C, and some herpesviruses and papillomaviruses that may cause different types of cancer. Viral polymerases are crucial enzymes for viral replication and transcription and, depending on the needs of each virus family, over time they have evolved to better adapt to host structures [47].

3 Classes of inhibitors/activators and their design

DNA and RNA polymerases play a key role in the replication of nucleic acids within the framework of cell division or microorganisms' proliferation, and their pharmacological inhibition is conceived as a good strategy for the treatment of hyperproliferative and infection diseases. Particularly, cancer cells can escape the normal mechanisms of replication, and some of them have a high rate of proliferation, spending most of their cell cycle in the S phase. The aim of pharmacological therapy is to target a process that is much more active in cancer cells than in healthy ones. In this context, the antiproliferative effect of DNAP inhibitors is mainly S-phase specific. Therefore, cells with a high rate of proliferation will be mostly affected by DNAP inhibitors compared to cells in the G_0 phase [48]. In bacteria, RNAP is an essential enzyme for transcription. Although it shares the reaction mechanism, substrates and products with the human orthologue, the low degree of sequence similarity between polymerases from different organisms makes bacterial RNAP an excellent target for the development of safe and specific antibacterial drugs [49]. In the case of viral infections, the quality of DNAP as a valuable therapeutic target is sustained by the required activation of specific prodrug inhibitors, a process that occurs only in infected cells through viral proteins [50]. Finally, RdRp is a peculiar viral protein, whose targeting by small molecular drugs represents an effective strategy that does not harm healthy human cells [51].

Pharmacological inhibition of DNAP and RNAP enzymes has been extensively explored through the development of small molecule acting with different mechanisms of action. One of the most effective strategies relies on the design of nucleos(*t*)ide inhibitors (NIs) that mimic the substrate without allowing the catalytic reaction to occur or to continue. This class of compounds includes most of the commercially available drugs that act against human, bacterial, and viral DNAPs, which bear purine or pyrimidine moieties structurally resembling the natural nucleotide substrates of DNAPs or RNAPs. Usually, NIs lack the free hydroxyl group in position 3′ that, in physiological substrates, allows the attack of the next NTP through the catalytic mechanism described above (e.g., Fig. 1). The lack of the free 3′-OH hydroxyl group in NIs prevents the subsequent insertion of NTP by terminating the polynucleotide chain. A notable member of this family of NIs is Acyclovir, a drug used in the treatment of infections by herpes simplex virus, which is highly selective for infected cells because it is a prodrug activated by specific viral enzymes. However, some NIs have the 3′-OH hydroxyl group, which allows the insertion of additional base through catalytic reaction. The mechanism of polymerase inhibition by these NIs consists of a steric clash following the insertion of two or three additional bases after the NI drug, with prevents chain elongation. For this reason, these NIs are also referred as delayed chain terminators. An example of this class is sofosbuvir, a drug used in the treatment of Hepatitis C virus infections. Following the insertion of two bases after sofosbuvir, the viral polymerase undergoes chain termination thanks to the steric hindrance of a methyl group of the drug that impairs the sliding of the nascent strand on the protein [52]. Overall, NIs act within the catalytic site of polymerases in the form of bioactive triphosphates that mimic NTP substrates. Unfortunately, the strong polar charge localized on the phosphate groups makes these molecules unsuitable to permeate the cell membrane. To enhance their bioavailability, several prodrugs that are activated in cells by chemical modifications have been developed [53].

Different from NIs, non-nucleos(*t*)idic inhibitors (NNIs) are compounds that inhibit polymerases by binding to these enzymes through a non-competitive mechanism. A notable example of NNIs is rifamycin, a broad-spectrum antibacterial drug that binds in a pocket adjacent to the active site, although 12 Å apart from it [54]. In addition, some antiviral drugs can bind in pockets that are in different subdomains of the RdRp, implementing different mechanisms of action based on the specific allosteric site targeted by the drugs [55,56].

Enzyme inhibition can also occur by covalent inhibitors, namely small molecules endowed with chemically reactive groups that are capable of binding covalently the enzyme at specific sites and to inhibit its catalytic functions. The widest group of covalent inhibitors of polymerases react with the functional groups of the enzyme's active site, generating a steric hindrance that prevents the substrate from accessing to catalytic residues or inhibiting its transformation. This inhibition mechanism can be either reversible or irreversible. In reversible covalent inhibition, there is an equilibrium between the bound and the unbound state of the drug, whereas in the case of irreversible covalent inhibition, the complex cannot be dissociated, and the recovery of enzymatic functions depends merely on the enzyme physiological turnover.

Recently, other strategies for polymerase inhibition such as the development of protein-protein interaction (PPI) inhibitors have been investigated [57]. Indeed, polymerases consist of an assembly between multiple subunits, which are usually associated with other proteins to exert their enzymatic function. At the state of the art, no PPI inhibitors of polymerases have been approved by regulatory authorities, but a phase I clinical trial for the evaluation of compounds inhibiting human RNAP I against advanced hematologic cancers (ClinicalTrial.gov Identifier: NCT02719977) is running [58].

Polymerase inhibition can also occur indirectly, such as in the case of the drug candidate NCT02719977 that is running in clinical phase I [59]. Besides, most indirect polymerase inhibitors intercalate the DNA duplex substrate of human RNAP with a mechanism that interferes with the catalytic activity of the enzyme and results in transcriptional block.

In recent years, drug repurposing is emerged as a valuable strategy in drug discovery, to identify new uses for approved or advanced clinical drugs that do not fall within the scope of the original medical indication. This strategy offers several advantages over developing a completely new drug, including the safety of the repurposed entity (that have been already tested in preclinical and clinical studies) and the relatively lower time and budget investments required in drug development, given that several assessments have been already completed. In addition to sildenafil, which is often referred as a popular and successful example of drug repurposing [60], the approved RdRp inhibitor Remdesivir has been recently repurposed for the treatment of SARS-CoV-2 infections, i.e., COVID-19 [61–64].

4 Polymerase inhibitors in clinical use or in advanced development stages

To date, more than 40 polymerase inhibitors have been approved for clinical use with different applications (Table 1). Here, polymerase inhibitor drugs are discussed, and the mechanisms of action of the most notable examples of each category are described in detail. Finally, drugs are grouped based on their target (i.e., DNAP inhibitors and RNAP inhibitors including RdRp inhibitors).

4.1 DNAP inhibitors

4.1.1 Fludarabine

One of the most widely used NIs in cancer therapy is Fludarabine, which has been approved by the Food and Drugs Administration (FDA) in 1991 for the treatment of hematological cancers including non-Hodgkin's lymphoma and B-cell chronic lymphocytic leukemia [65]. Fludarabine phosphate is a purine analog antimetabolite, corresponding to the 2-fluoro-5′-monophosphate derivative of Vidarabine (ara-A). Different from Vidarabine, Fludarabine is a poor substrate of the adenosine deaminase enzyme, which offers a competitive advantage. Upon administration, Fludarabine is rapidly dephosphorylated to 2-fluoro-ara-A and then undergoes intracellular metabolic conversion by deoxycytidine kinase to the active triphosphate form, i.e., 2-fluoro-ara-ATP also referred as Fludarabine triphosphate (Fig. 3).

Fludarabine is a cell cycle-specific drug that inhibits DNA synthesis at multiple levels by interfering with the catalytic activity of various enzymes, including the ribonucleotide reductase, DNA primase, and human DNAP α and ε. Specifically, these DNAPs are highly activated in the S phase of the cell cycle, and they are inhibited by Fludarabine in vitro with IC_{50} values of 1.6 and 1.3 μM, respectively. The bioactive triphosphate form of fludarabine competes

TABLE 1 List of approved polymerase inhibitor drugs.

Category	Drug name	Abbreviation	Clinical use	FDA first approval date
DNAP inhibitors				
NI	Cytarabine	Ara-C	Acute myeloid leukemia, acute lymphocytic leukemia, chronic myelogenous leukemia, and non-Hodgkin's lymphoma	1969
NI	Fludarabine	2-F-ara-A	Chronic lymphocytic leukemia, non-Hodgkin's lymphoma, acute myeloid leukemia, and acute lymphocytic leukemia	1991
NI	Gemcitabine	dFdC	Pancreatic cancer, non-small cell lung cancer, metastatic breast cancer	1996
NI	Clofarabine	CAFdA	Acute lymphoblastic leukemia	2004
NI	Nelarabine	Ara-G (prodrug)	T-cell acute lymphoblastic leukemia and T-cell lymphoblastic lymphoma	2005
NI	Acyclovir	ACV	Herpes simplex virus, Varicella zoster virus	1982
NI	Ganciclovir	GCV	Human Cytomegalovirus	1989
NI	Famciclovir	FCV	Herpes simplex virus, Varicella zoster virus	1994
NI	Valacyclovir	VACV	Herpes simplex virus, Varicella zoster virus	1995
NI	Penciclovir	PCV	Herpes simplex virus	1996
NI	Valganciclovir	VGCV	Human Cytomegalovirus	2001
NI	Vidarabine	VDR	Herpes simplex virus, Varicella zoster virus	1976
NI	Entecavir	ETV	Hepatitis B virus	2005
NI	Edoxudine	EDU	Herpes simplex virus	1969
NI	Telbivudine	LdT	Hepatitis B virus	2006
NI	Idoxuridine	IDU	Herpes simplex virus 1	1963
NI	Trifluridine	TFT	Herpes simplex virus	1980
NI	Brivudine	BVDU	Herpes simplex virus 1, Varicella zoster virus	2000
NI	Foscarnet	PFA	Human Cytomegalovirus, Herpes simplex virus	1991
NI	Cidofovir	CDV	Human Cytomegalovirus	1996
NI	Tenofovir	TDF	Human immunodeficiency virus, Hepatitis B virus	2001
NI	Adefovir dipivoxil	ADV	Hepatitis B virus	2002
RNAP inhibitors				
NNI	Rifampicin	RMP	Tuberculosis	1971
NNI	Rifaximin	RFX	Traveler's diarrhea, irritable bowel syndrome, and hepatic encephalopathy	2004
NNI	Rifapentine	RPT	Pulmonary tuberculosis	1998
NNI	Rifabutin	RFB	Pulmonary tuberculosis	1992
NNI	Actinomicyn D	AMD	Wilms' tumor, rhabdomyosarcoma, Ewing's sarcoma, trophoblastic neoplasm, testicular cancer, ovarian cancer	1964
NNI	Doxorubicin	DOX	Breast cancer, bladder cancer, Kaposi's sarcoma, lymphoma, and acute lymphocytic leukemia	1974
NNI	Daunorubicin	DNR	Acute myeloid leukemia, acute lymphoblastic leukemia, chronic myelogenous leukemia, and Kaposi's sarcoma	1979

Continued

TABLE 1 List of approved polymerase inhibitor drugs—cont'd

Category	Drug name	Abbreviation	Clinical use	FDA first approval date
NNI	Epirubicin	EPI	Breast cancer, ovarian cancer, gastric cancer, lung cancer and lymphomas	1999
NNI	Idarubicin	IDA	Acute lymphoblastic leukemia and chronic myelogenous leukemia	1990
NI	Remdesivir	RDV	Severe acute respiratory syndrome 2	2020
NI	Sofosbuvir	SOF	Hepatitis C virus 2 or 3	2013
NI	Dasabuvir	DSV	Hepatitis C virus	2014
NI	Ribavirin	RBV	Hepatitis C virus, respiratory syncytial virus, hemorrhagic fever	1985
NI	Favipiravir	FPV	Influenza viruses A, B, and C	2014

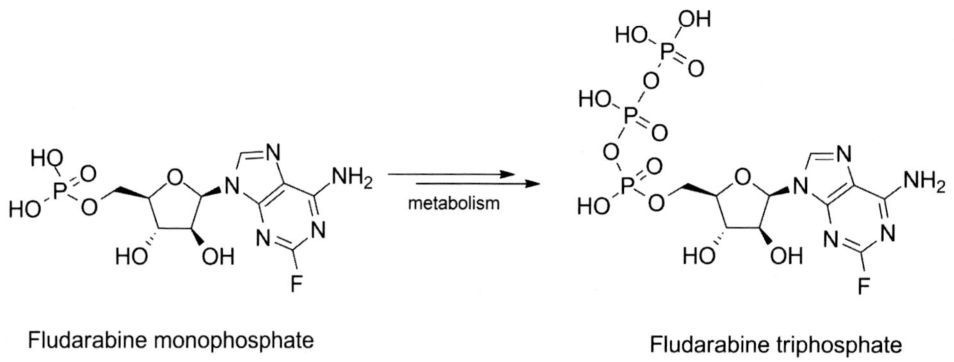

FIG. 3 Metabolic phosphorylation of Fludarabine monophosphate to its bioactive triphosphate form by deoxycytidine kinase.

with the natural substrate 2′-deoxyadenosine 5′-triphosphate (dATP) for the incorporation into nascent DNA. Through pairing with the opposite thymine of the template strand, Fludarabine directly blocks the further elongation of the nascent DNA strand [66]. At the same time, the concentration of dATP decreases due to the inhibition of ribonucleotide reductase, which enhances the cytotoxic activity of 2-fluoro-ara-ATP.

4.1.2 *Acyclovir*

Acyclovir (ACV) was discovered in 1977 [67] and represented the prototype of the second generation of antiviral nucleoside analogues characterized by lower toxicity and greater selectivity compared to first generation drugs. ACV is a nucleoside analogue of guanosine (9-[2-hydroxyethoxymethyl] guanine) approved by the FDA in 1982 for the treatment of the herpes simplex virus infection. The bioactive form of ACV is the triphosphate form, which is generated in infected cells by monophosphorylation operated by the viral thymidine kinase (TK) followed by conversion to the bioactive triphosphate form by cellular enzymes, which make it a suitable substrate for viral DNAP (Fig. 4).

ACV triphosphate acts as an inhibitor of viral DNAP by competing with 2′-deoxyguanosine-5-triphosphate (dGTP) for interaction within the catalytic site and blocking the elongation of the nascent viral DNA chain. Early termination of the nascent DNA chain occurs due to the lack of the 3′-OH hydroxyl group required for the attachment of the next NTP. ACV has a great selectivity of action with a good safety profile thanks to (i) the high specificity to act as a substrate of viral TK, which is about 200 times greater than for cellular TK; and (ii) an affinity for viral DNAP that is about 10 times higher than that for human DNAP. However, its selectivity and spectrum of action are limited

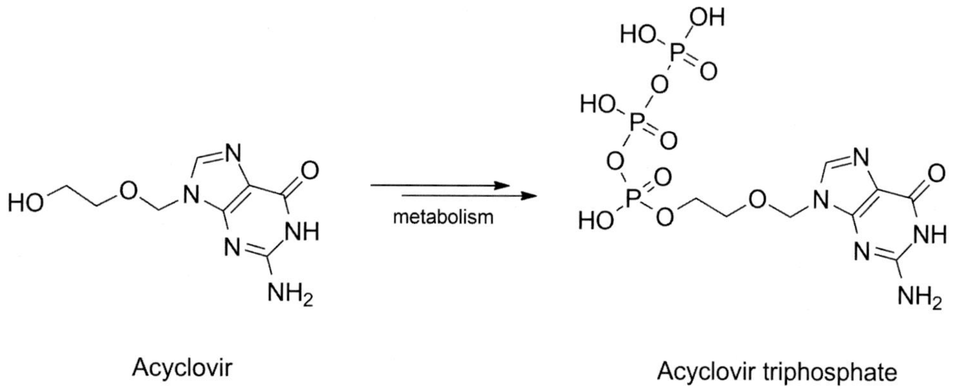

FIG. 4 Metabolic phosphorylation of acyclovir to its bioactive triphosphate form.

to viruses that code for their own TKs. Over the years, many second-generation nucleoside analogues of ACV have been developed, including Bromovinildeoxyuridine (BVDV), Ganciclovir (GCV), and Penciclovir (PCV). However, these molecules have experienced a limited oral bioavailability, which has led to the development of prodrugs (e.g., Famciclovir (FCV) as a prodrug of PCV, Valaciclovir (VACV) as a prodrug of ACV, and Vanganciclovir (VGCV) as a prodrug of GCV) that are converted in the corresponding parent compounds by cellular enzymes.

4.1.3 Ibezapolstat

In the last decades, the massive—and often deregulated—use of antibacterial drugs has exacerbated the antibiotic-resistance issue, raising the need for novel bioactive molecules able to prevent and treat bacterial infections and related antibiotic resistance. To date, there are no approved drugs acting on bacterial replication mechanisms. In this context, bacterial DNAP is a new and unexplored target that could pave the way to the discovery of antibacterial drugs endowed with novel mechanisms of action. Currently, the N7-substitute guanine analog Ibezapolstat (ACX-362E) (Fig. 5) is in clinical phase II for the treatment of *Clostridioides difficile* infection (ClinicalTrials.gov Identifier: NCT04247542), becoming the first NI with selectivity for bacterial DNAP IIIC undergoing clinical development as a drug candidate [68]. The mechanism of action of Ibezapolstat

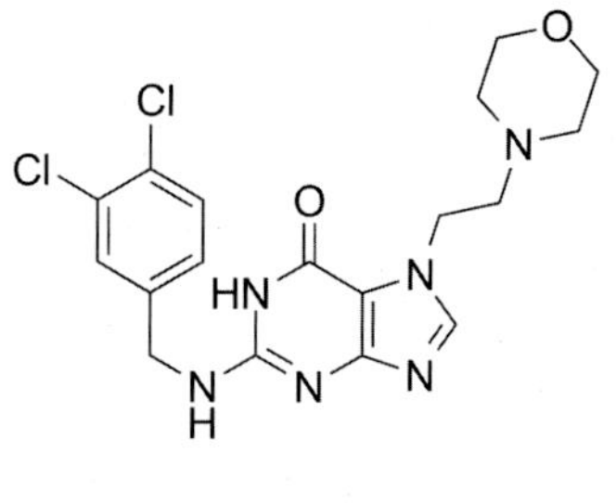

FIG. 5 Chemical structure of Ibezapolstat.

is common to 2-phenylguanines (PGs), 6-anilinouracils (AUs), and analogues of 2′-deoxyguanosine-5′-triphosphate (dGTP) that mimic the guanine moiety of dGTP and bind DNAP catalytic site forming an inactive ternary complex. Specifically, Ibezapolstat inhibits purified *C. difficile* DNAP IIIC with a K_i of 0.325 μM [68–70].

4.2 RNA polymerase inhibitors

4.2.1 Rifampicin

Discovered in 1965, firstly marketed in Italy in 1968, and finally approved by the FDA in 1971 for the treatment of tuberculosis caused by *Mycobacterium tuberculosis*, rifampicin is an NNI belonging to the rifamycin family [71]. Rifamycins are wide-spectrum natural or semisynthetic antibiotics used in the therapy of several bacterial infections [72]. Among them, rifampicin is one of the most powerful antituberculosis drugs, having an in vitro activity against human *M. tuberculosis* in the range of 0.005–0.5 μg/mL depending on the specific culture medium [73]. From a chemical standpoint, Rifampicin is a polyketide endowed with a naphthoquinone core, which is derived from rifampicin B that is synthesized by the actinomycete *Streptomyces mediterranei*. Rifampicin has also four hydroxyl groups that are crucial to establish hydrogen bonds with key residues of RNAP (Fig. 6) [54].

Allosteric modulation of RNA synthesis by rifampicin was assessed by X-ray crystallography studies, demonstrating that the drug inhibits bacterial RNAP by binding non-competitively to the β subunit of the enzyme at around 12 Å distance from the catalytic site (Fig. 6) [54]. The binding of the Rifampicin to RNAP causes a steric occlusion that physically blocks the transcription process, preventing the formation of the second or third phosphodiester bond [74,75]. Rifampicin has a high affinity for prokaryotic RNAP, not affecting the human ones [76], while mutations of key residues involved in drug-protein interaction decrease the drug's affinity and promote the emergence of drug resistance [74].

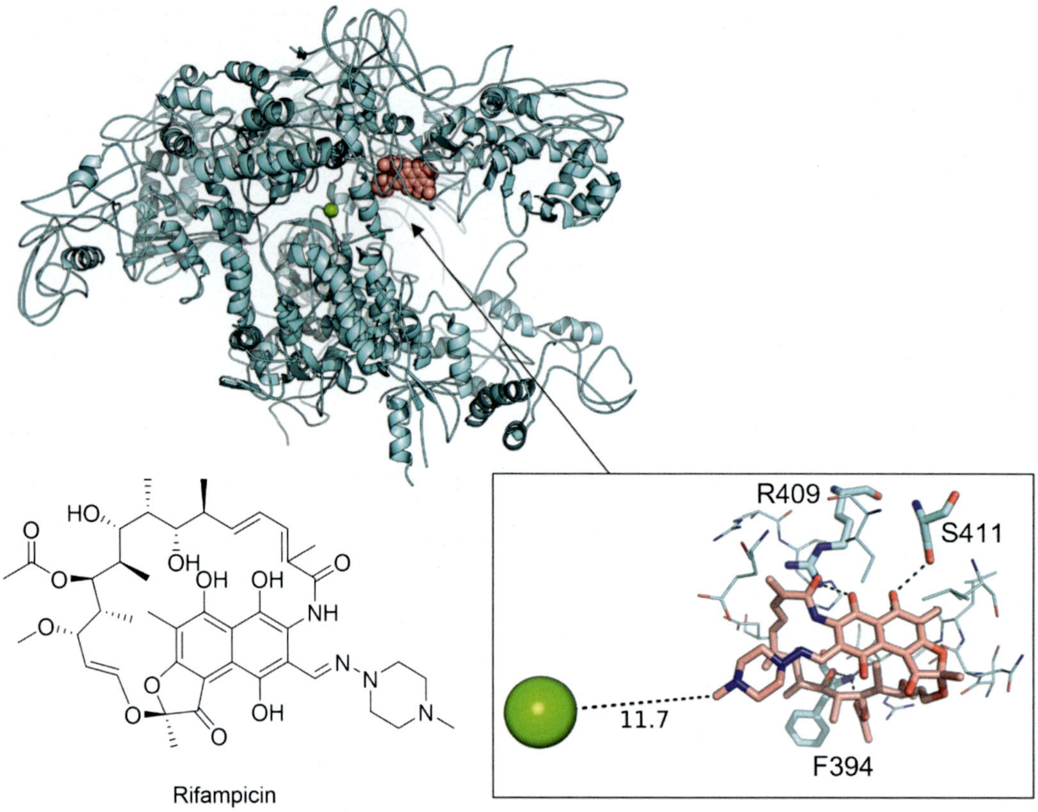

FIG. 6 Crystallographic structure of *Thermus aquaticus* RNAP in complex with rifampicin (PDB-ID: 1I6V) [54]. Residues involved in polar interactions are represented as sticks and are labeled in the bottom-right panel. Polar contacts and the distance between the molecule and the Mg^{2+} ion of the catalytic site are highlighted by *black dashed lines*.

4.2.2 Remdesivir

Originally selected from a screening of candidate anti-HCV drugs, remdesivir was later used against Ebola virus infections [77]. Given its broad-spectrum antiviral activity, it was tested on various organisms including SARS-CoV and MERS-CoV, demonstrating IC_{50} values in the sub-micromolar range in different in vitro systems. For this reason, in 2020, it was tested against SARS-CoV-2, becoming the first drug approved by the FDA for the treatment of hospitalized patients with COVID-19 [51,61,78,79].

Remdesivir is a phosphoramidate NI bearing a nitrile substituent in position 1′ of the sugar moiety, which acts as a prodrug of adenosine. Indeed, Remdesivir is converted to the bioactive triphosphate NTP form by cellular metabolism (Fig. 7).

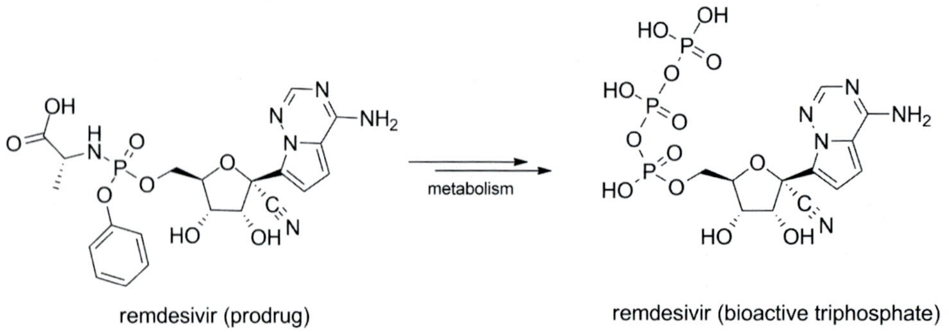

FIG. 7 Metabolic transformation of Remdesivir prodrug to its bioactive form by cellular enzymes.

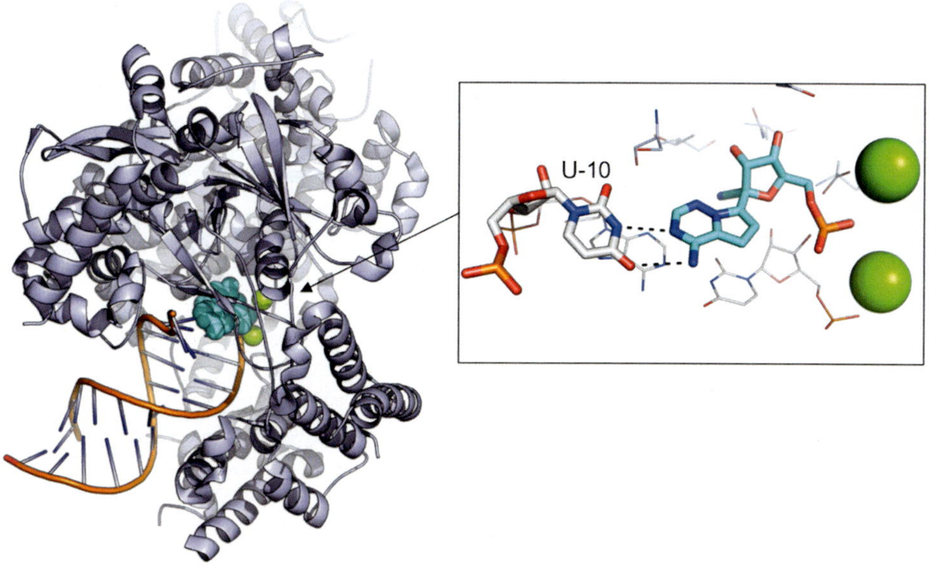

FIG. 8 Electron microscopy structure of the RdRp of SARS-CoV-2 in complex with bioactive triphosphate Remdesivir (*cyan*). The opposite base is represented by *sticks*, while polar interactions are highlighted by *black dashed lines*. Mg^{2+} ions are represented as *green spheres*.

The bioactive form of Remdesivir competes with the natural substrates of viral RdRp for incorporation into nascent RNA chain, causing delayed chain termination during viral RNA replication. Detailed analysis of Remdesivir mechanism of action has suggested that the drug can induce a delay in chain termination thanks to the presence of the 3′-OH hydroxyl group, to which three subsequent NTPs attach until a steric clash between the 1′-CN group of the drug and the S861 residue of RdRp blocks the sliding of the nascent RNA chain. Viral RdRp inhibition by Remdesivir impairs viral replication within infected cells, facilitating the recovery from infection by viruses such as the SARS-CoV-2 [80]. This mechanism of action has been recently corroborated by Cryo-EM studies, showing the structure of the drug within the catalytic site of the RdRp (PDB-ID 7BV2) (Fig. 8) [81].

4.2.3 Actinomycin D

Actinomycin D is a cytotoxic antibiotic isolated in 1940 from the *Streptomyces* species and is the main member of Actinomycins group. It was approved by the FDA in 1964 for the treatment of several tumors including Ewing's sarcoma, Wilms' tumor, rhabdomyosarcoma, testicular cancer, and certain types of ovarian cancer [82]. From a chemical standpoint, Actinomycin D is composed of a heterocyclic phenoxazine core containing a quinonimine portion, linked to two cyclic pentapeptide lactone rings. The heterocyclic core is responsible for the color of the compound, as well as its ability to intercalate nucleic acid duplexes thanks to its flat and aromatic nature (Fig. 9).

Several X-ray crystallographic structures describe the binding mode of Actinomycin D to DNA duplexes, notably clarifying its mechanism of action. The planar aromatic chromophore strongly and stably intercalates in the minor groove of dsDNA preferably in correspondence of G-C pairs. Moreover, threonine residues of the cyclic pentapeptides establish specific hydrogen bonds with the G-C pairs, while the other residues are involved in hydrophobic interactions with DNA nucleotides [84]. Actinomycin D intercalates in DNA duplex with a high association constant ($2.3 \times 10^6 M^{-1}$) [85], compromising the integrity of the double helix and inhibiting DNA-dependent RNA synthesis [86]. Notably, the drug is particularly selective for RNAP I over RNAP II, with an IC_{50} value of 0.05 μg/mL and 0.5 μg/mL, respectively [87]. Since RNAP inhibition induces apoptosis in cancer cells, the moderate cytotoxicity and broad-spectrum anticancer activity of the drug limit its extensive clinical use [85].

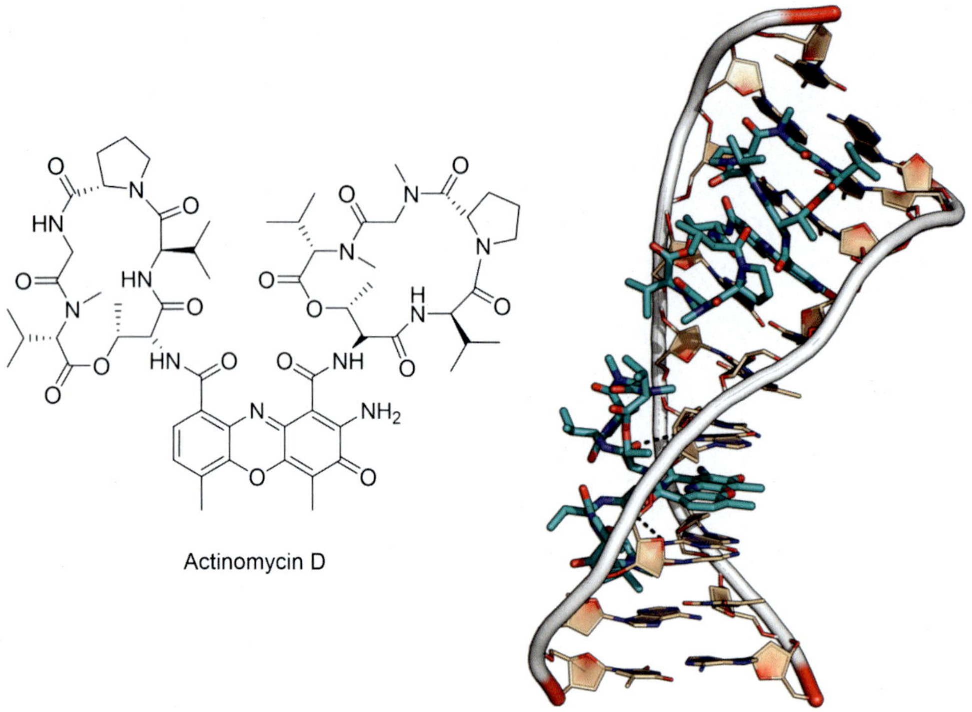

FIG. 9 *Left panel*: chemical structure of Actinomycin D. *Right panel*: X-ray crystallographic structure of actinomycin D bound to a DNA duplex (PDB-ID: 1MNV) [83]. Actinomycin D is represented as *blue sticks*.

References

[1] Cairns J. The bacterial chromosome and its manner of replication as seen by autoradiography. J Mol Biol 1963;6:208–13. https://doi.org/10.1016/s0022-2836(63)80070-4.

[2] Meselson M, Stahl FW. The replication of DNA in Escherichia coli. Proc Natl Acad Sci U S A 1958;44(7):671–82. https://doi.org/10.1073/pnas.44.7.671.

[3] Lee J, Chastain 2nd PD, Kusakabe T, Griffith JD, Richardson CC. Coordinated leading and lagging strand DNA synthesis on a minicircular template. Mol Cell 1998;1(7):1001–10. https://doi.org/10.1016/s1097-2765(00)80100-8.

[4] Ogawa T, Okazaki T. Discontinuous DNA replication. Annu Rev Biochem 1980;49:421–57. https://doi.org/10.1146/annurev.bi.49.070180.002225.

[5] Spiering MM, Benkovic SJ. DNA replication fork, bacterial. In: E. Inc, editor. Encyclopedia of biological chemistry. 2nd ed. Elsevier Inc.; 2013. p. 114–7.

[6] Geider K, Kornberg A. Conversion of the M13 viral single strand to the double-stranded replicative forms by purified proteins. J Biol Chem 1974;249(13):3999–4005. Retrieved from https://www.ncbi.nlm.nih.gov/pubmed/4604274.

[7] Steitz TA, Steitz JA. A general two-metal-ion mechanism for catalytic RNA. Proc Natl Acad Sci U S A 1993;90(14):6498–502. https://doi.org/10.1073/pnas.90.14.6498.

[8] Dewar JM, Budzowska M, Walter JC. The mechanism of DNA replication termination in vertebrates. Nature 2015;525(7569):345–50. https://doi.org/10.1038/nature14887.

[9] Dewar JM, Walter JC. Mechanisms of DNA replication termination. Nat Rev Mol Cell Biol 2017;18(8):507–16. https://doi.org/10.1038/nrm.2017.42.

[10] Bebenek A, Ziuzia-Graczyk I. Fidelity of DNA replication-a matter of proofreading. Curr Genet 2018;64(5):985–96. https://doi.org/10.1007/s00294-018-0820-1.

[11] Macao B, Uhler JP, Siibak T, Zhu X, Shi Y, Sheng W, Falkenberg M. The exonuclease activity of DNA polymerase gamma is required for ligation during mitochondrial DNA replication. Nat Commun 2015;6:7303. https://doi.org/10.1038/ncomms8303.

[12] Wood RD, Shivji MKK. Which DNA polymerases are used for DNA-repair in eukaryotes? Carcinogenesis 1997;18(4):605–10. https://doi.org/10.1093/carcin/18.4.605.

[13] Hunter WN, Brown T, Anand NN, Kennard O. Structure of an adenine. Cytosine base pair in DNA and its implications for mismatch repair. Nature 1986;320(6062):552–5. https://doi.org/10.1038/320552a0.

[14] Chatterjee N, Walker GC. Mechanisms of DNA damage, repair, and mutagenesis. Environ Mol Mutagen 2017;58(5):235–63. https://doi.org/10.1002/em.22087.

[15] Alberts B, Johnson A, Lewis J, Raff M, Roberts K, Walter P. DNA replication mechanisms. In: Molecular biology of the cell. 4th ed. New York: Garland Science; 2002.

[16] Appleby TC, Perry JK, Murakami E, Barauskas O, Feng J, Cho A, Edwards TE. Viral replication. Structural basis for RNA replication by the hepatitis C virus polymerase. Science 2015;347(6223):771–5. https://doi.org/10.1126/science.1259210.

[17] Kohlstaedt LA, Wang J, Friedman JM, Rice PA, Steitz TA. Crystal structure at 3.5 A resolution of HIV-1 reverse transcriptase complexed with an inhibitor. Science 1992;256(5065):1783–90. https://doi.org/10.1126/science.1377403.

[18] Ollis DL, Brick P, Hamlin R, Xuong NG, Steitz TA. Structure of large fragment of Escherichia coli DNA polymerase I complexed with dTMP. Nature 1985;313(6005):762–6. https://doi.org/10.1038/313762a0.

[19] Sawaya MR, Pelletier H, Kumar A, Wilson SH, Kraut J. Crystal structure of rat DNA polymerase beta: evidence for a common polymerase mechanism. Science 1994;264(5167):1930–5. https://doi.org/10.1126/science.7516581.
[20] Steitz TA. DNA polymerases: structural diversity and common mechanisms. J Biol Chem 1999;274(25):17395–8. https://doi.org/10.1074/jbc.274.25.17395.
[21] Wu S, Beard WA, Pedersen LG, Wilson SH. Structural comparison of DNA polymerase architecture suggests a nucleotide gateway to the polymerase active site. Chem Rev 2014;114(5):2759–74. https://doi.org/10.1021/cr3005179.
[22] Lehman IR, Bessman MJ, Simms ES, Kornberg A. Enzymatic synthesis of deoxyribonucleic acid. I. Preparation of substrates and partial purification of an enzyme from Escherichia coli. J Biol Chem 1958;233(1):163–70. Retrieved from https://www.ncbi.nlm.nih.gov/pubmed/13563462.
[23] Ohmori H, Friedberg EC, Fuchs RP, Goodman MF, Hanaoka F, Hinkle D, Woodgate R. The Y-family of DNA polymerases. Mol Cell 2001;8(1):7–8. https://doi.org/10.1016/s1097-2765(01)00278-7.
[24] O'Donnell M. Replisome architecture and dynamics in Escherichia coli. J Biol Chem 2006;281(16):10653–6. https://doi.org/10.1074/jbc.R500028200.
[25] Garcia-Diaz M, Bebenek K. Multiple functions of DNA polymerases. CRC Crit Rev Plant Sci 2007;26(2):105–22. https://doi.org/10.1080/07352680701252817.
[26] Werner F, Grohmann D. Evolution of multisubunit RNA polymerases in the three domains of life. Nat Rev Microbiol 2011;9(2):85–98. https://doi.org/10.1038/nrmicro2507.
[27] Murakami KS, Masuda S, Campbell EA, Muzzin O, Darst SA. Structural basis of transcription initiation: an RNA polymerase holoenzyme-DNA complex. Science 2002;296(5571):1285–90. https://doi.org/10.1126/science.1069595.
[28] Young BA, Gruber TM, Gross CA. Views of transcription initiation. Cell 2002;109(4):417–20. https://doi.org/10.1016/s0092-8674(02)00752-3.
[29] Kennedy WP, Momand JR, Yin YW. Mechanism for de novo RNA synthesis and initiating nucleotide specificity by t7 RNA polymerase. J Mol Biol 2007;370(2):256–68. https://doi.org/10.1016/j.jmb.2007.03.041.
[30] Ayoubi LE, Dumay-Odelot H, Chernev A, Boissier F, Minvielle-Sebastia L, Urlaub H, Teichmann M. The hRPC62 subunit of human RNA polymerase III displays helicase activity. Nucleic Acids Res 2019;47(19):10313–26. https://doi.org/10.1093/nar/gkz788.
[31] Thomas MJ, Platas AA, Hawley DK. Transcriptional fidelity and proofreading by RNA polymerase II. Cell 1998;93(4):627–37. https://doi.org/10.1016/s0092-8674(00)81191-5.
[32] Carey LB. RNA polymerase errors cause splicing defects and can be regulated by differential expression of RNA polymerase subunits. elife 2015;4, e09945. https://doi.org/10.7554/eLife.09945.
[33] Zhang G, Campbell EA, Minakhin L, Richter C, Severinov K, Darst SA. Crystal structure of Thermus aquaticus core RNA polymerase at 3.3 A resolution. Cell 1999;98(6):811–24. https://doi.org/10.1016/s0092-8674(00)81515-9.
[34] Ramsay EP, Abascal-Palacios G, Daiss JL, King H, Gouge J, Pilsl M, Vannini A. Structure of human RNA polymerase III. Nat Commun 2020;11(1):6409. https://doi.org/10.1038/s41467-020-20262-5.
[35] Werner F. Structure and function of archaeal RNA polymerases. Mol Microbiol 2007;65(6):1395–404. https://doi.org/10.1111/j.1365-2958.2007.05876.x.
[36] Carter R, Drouin G. Structural differentiation of the three eukaryotic RNA polymerases. Genomics 2009;94(6):388–96. https://doi.org/10.1016/j.ygeno.2009.08.011.
[37] Cooper G. Eukaryotic RNA polymerases and general transcription factors. In: The cell: A molecular approach. 2nd ed. Sunderland, MA: Sinauer Associates; 2000.
[38] Garcia-Diaz M, Bebenek K, Krahn JM, Pedersen LC, Kunkel TA. Role of the catalytic metal during polymerization by DNA polymerase lambda. DNA Repair (Amst) 2007;6(9):1333–40. https://doi.org/10.1016/j.dnarep.2007.03.005.
[39] Albertella MR, Lau A, O'Connor MJ. The overexpression of specialized DNA polymerases in cancer. DNA Repair (Amst) 2005;4(5):583–93. https://doi.org/10.1016/j.dnarep.2005.01.005.
[40] Barbari SR, Shcherbakova PV. Replicative DNA polymerase defects in human cancers: consequences, mechanisms, and implications for therapy. DNA Repair 2017;56:16–25. https://doi.org/10.1016/j.dnarep.2017.06.003.
[41] Herzog M, Alonso-Perez E, Salguero I, Warringer J, Adams DJ, Jackson SP, Puddu F. Mutagenic mechanisms of cancer-associated DNA polymerase epsilon alleles. Nucleic Acids Res 2021;49(7):3919–31. https://doi.org/10.1093/nar/gkab160.
[42] Kane DP, Shcherbakova PV. A common cancer-associated DNA polymerase epsilon mutation causes an exceptionally strong mutator phenotype, indicating fidelity defects distinct from loss of proofreading. Cancer Res 2014;74(7):1895–901. https://doi.org/10.1158/0008-5472.CAN-13-2892.
[43] Pham TM, Tan KW, Sakumura Y, Okumura K, Maki H, Akiyama MT. A single-molecule approach to DNA replication in Escherichia coli cells demonstrated that DNA polymerase III is a major determinant of fork speed. Mol Microbiol 2013;90(3):584–96. https://doi.org/10.1111/mmi.12386.
[44] Louten J. Virus replication. In: Louten J, editor. Essential human virology. Academic Press; 2016. p. 49–70.
[45] Baltimore D. Expression of animal virus genomes. Bacteriol Rev 1971;35(3):235–41. https://doi.org/10.1128/br.35.3.235-241.1971.
[46] Rampersad S, Tennant P. Replication and expression strategies of viruses. Viruses 2018;55–82.
[47] Choi KH. Viral polymerases. Adv Exp Med Biol 2012;726:267–304. https://doi.org/10.1007/978-1-4614-0980-9_12.
[48] Malhotra V, Perry MC. Classical chemotherapy: mechanisms, toxicities and the therapeutic window. Cancer Biol Ther 2003;2(4 Suppl 1):S2–4. Retrieved from https://www.ncbi.nlm.nih.gov/pubmed/14508075.
[49] Villain-Guillot P, Bastide L, Gualtieri M, Leonetti JP. Progress in targeting bacterial transcription. Drug Discov Today 2007;12(5–6):200–8. https://doi.org/10.1016/j.drudis.2007.01.005.
[50] Gnann Jr JW, Barton NH, Whitley RJ. Acyclovir: mechanism of action, pharmacokinetics, safety and clinical applications. Pharmacotherapy 1983;3(5):275–83. https://doi.org/10.1002/j.1875-9114.1983.tb03274.x.
[51] Picarazzi F, Vicenti I, Saladini F, Zazzi M, Mori M. Targeting the RdRp of emerging RNA viruses: the structure-based drug design challenge. Molecules 2020;25(23). https://doi.org/10.3390/molecules25235695.
[52] Xu HT, Colby-Germinario SP, Hassounah SA, Fogarty C, Osman N, Palanisamy N, Wainberg MA. Evaluation of Sofosbuvir (beta-D-2 '-deoxy-2 '-alpha-fluoro-2 '-beta-C-methyluridine) as an inhibitor of dengue virus replication. Sci Rep 2017;7:6345. https://doi.org/10.1038/s41598-017-06612-2.
[53] Slusarczyk M, Serpi M, Pertusati F. Phosphoramidates and phosphonamidates (ProTides) with antiviral activity. Antivir Chem Chemother 2018;26. https://doi.org/10.1177/2040206618775243.
[54] Campbell EA, Korzheva N, Mustaev A, Murakami K, Nair S, Goldfarb A, Darst SA. Structural mechanism for rifampicin inhibition of bacterial rna polymerase. Cell 2001;104(6):901–12. https://doi.org/10.1016/s0092-8674(01)00286-0.

[55] Mosley RT, Edwards TE, Murakami E, Lam AM, Grice RL, Du J, Otto MJ. Structure of hepatitis C virus polymerase in complex with primer-template RNA. J Virol 2012;86(12):6503–11. https://doi.org/10.1128/JVI.00386-12.

[56] Tomei L, Altamura S, Bartholomew L, Bisbocci M, Bailey C, Bosserman M, Migliaccio G. Characterization of the inhibition of hepatitis C virus RNA replication by nonnucleosides. J Virol 2004;78(2):938–46. https://doi.org/10.1128/jvi.78.2.938-946.2004.

[57] Takahashi H, Takahashi C, Moreland NJ, Chang YT, Sawasaki T, Ryo A, Yamamoto N. Establishment of a robust dengue virus NS3-NS5 binding assay for identification of protein-protein interaction inhibitors. Antivir Res 2012;96(3):305–14. https://doi.org/10.1016/j.antiviral.2012.09.023.

[58] Khot A, Brajanovski N, Cameron DP, Hein N, Maclachlan KH, Sanij E, Harrison SJ. First-in-human RNA polymerase I transcription inhibitor CX-5461 in patients with advanced hematologic cancers: results of a phase I dose-escalation study. Cancer Discov 2019;9(8):1036–49. https://doi.org/10.1158/2159-8290.CD-18-1455.

[59] Xu H, Di Antonio M, McKinney S, Mathew V, Ho B, O'Neil NJ, Aparicio S. CX-5461 is a DNA G-quadruplex stabilizer with selective lethality in BRCA1/2 deficient tumours. Nat Commun 2017;8:14432. https://doi.org/10.1038/ncomms14432.

[60] Ghofrani HA, Osterloh IH, Grimminger F. Sildenafil: from angina to erectile dysfunction to pulmonary hypertension and beyond. Nat Rev Drug Discov 2006;5(8):689–702. https://doi.org/10.1038/nrd2030.

[61] Cusinato J, Cau Y, Calvani AM, Mori M. Repurposing drugs for the management of COVID-19. Expert Opin Ther Pat 2021;31(4):295–307. https://doi.org/10.1080/13543776.2021.1861248.

[62] Elfiky AA. Anti-HCV, nucleotide inhibitors, repurposing against COVID-19. Life Sci 2020;248, 117477. https://doi.org/10.1016/j.lfs.2020.117477.

[63] Serafin MB, Bottega A, Foletto VS, da Rosa TF, Horner A, Horner R. Drug repositioning is an alternative for the treatment of coronavirus COVID-19. Int J Antimicrob Agents 2020;55(6) 105969. https://doi.org/10.1016/j.ijantimicag.2020.105969.

[64] Singh TU, Parida S, Lingaraju MC, Kesavan M, Kumar D, Singh RK. Drug repurposing approach to fight COVID-19. Pharmacol Rep 2020;72(6):1479–508. https://doi.org/10.1007/s43440-020-00155-6.

[65] Hallek M. State-of-the-art treatment of chronic lymphocytic leukemia. Hematology Am Soc Hematol Educ Program 2009;440–9. https://doi.org/10.1182/asheducation-2009.1.440.

[66] Spriggs D, Robbins G, Mitchell T, Kufe D. Incorporation of 9-beta-D-arabinofuranosyl-2-fluoroadenine into HL-60 cellular RNA and DNA. Biochem Pharmacol 1986;35(2):247–52. https://doi.org/10.1016/0006-2952(86)90521-6.

[67] Elion GB, Furman PA, Fyfe JA, de Miranda P, Beauchamp L, Schaeffer HJ. Selectivity of action of an antiherpetic agent, 9-(2-hydroxyethoxymethyl) guanine. Proc Natl Acad Sci U S A 1977;74(12):5716–20. https://doi.org/10.1073/pnas.74.12.5716.

[68] Xu WC, Silverman MH, Yu XY, Wright G, Brown N. Discovery and development of DNA polymerase IIIC inhibitors to treat gram-positive infections. Bioorg Med Chem 2019;27(15):3209–17. https://doi.org/10.1016/j.bmc.2019.06.017.

[69] Torti A, Lossani A, Savi L, Focher F, Wright GE, Brown NC, Xu WC. Clostridium difficile DNA polymerase IIIC: basis for activity of antibacterial compounds. Curr Enzym Inhib 2011;7(3):147–53. https://doi.org/10.2174/157340811798807597.

[70] van Eijk E, Wittekoek B, Kuijper EJ, Smits WK. DNA replication proteins as potential targets for antimicrobials in drug-resistant bacterial pathogens. J Antimicrob Chemother 2017;72(5):1275–84. https://doi.org/10.1093/jac/dkw548.

[71] Sensi P. History of the development of rifampin. Rev Infect Dis 1983;5(Suppl 3):S402–6. https://doi.org/10.1093/clinids/5.supplement_3.s402.

[72] Lester W. Rifampin: a semisynthetic derivative of rifamycin—a prototype for the future. Annu Rev Microbiol 1972;26:85–102. https://doi.org/10.1146/annurev.mi.26.100172.000505.

[73] Rifampicin: a review. Drugs 1971;1(5):354–98. https://doi.org/10.2165/00003495-197101050-00002.

[74] Feklistov A, Mekler V, Jiang Q, Westblade LF, Irschik H, Jansen R, Ebright RH. Rifamycins do not function by allosteric modulation of binding of Mg2+ to the RNA polymerase active center. Proc Natl Acad Sci U S A 2008;105(39):14820–5. https://doi.org/10.1073/pnas.0802822105.

[75] McClure WR, Cech CL. On the mechanism of rifampicin inhibition of RNA synthesis. J Biol Chem 1978;253(24):8949–56. Retrieved from https://www.ncbi.nlm.nih.gov/pubmed/363713.

[76] Hartmann G, Honikel KO, Knusel F, Nuesch J. The specific inhibition of the DNA-directed RNA synthesis by rifamycin. Biochim Biophys Acta 1967;145(3):843–4. https://doi.org/10.1016/0005-2787(67)90147-5.

[77] Pardo J, Shukla AM, Chamarthi G, Gupte A. The journey of remdesivir: from Ebola to COVID-19. Drugs Context 2020;9. https://doi.org/10.7573/dic.2020-4-14.

[78] Beigel JH, Tomashek KM, Dodd LE, Mehta AK, Zingman BS, Kalil AC, Members A-SG. Remdesivir for the treatment of Covid-19 - final report. N Engl J Med 2020;383(19):1813–26. https://doi.org/10.1056/NEJMoa2007764.

[79] Lamb YN. Remdesivir: first approval. Drugs 2020;80(13):1355–63. https://doi.org/10.1007/s40265-020-01378-w.

[80] Elsawah HK, Elsokary MA, Abdallah MS, ElShafie AH. Efficacy and safety of remdesivir in hospitalized Covid-19 patients: systematic review and meta-analysis including network meta-analysis. Rev Med Virol 2021;31(4) e2187. https://doi.org/10.1002/rmv.2187.

[81] Yin W, Mao C, Luan X, Shen DD, Shen Q, Su H, Xu HE. Structural basis for inhibition of the RNA-dependent RNA polymerase from SARS-CoV-2 by remdesivir. Science 2020;368(6498):1499–504. https://doi.org/10.1126/science.abc1560.

[82] Hollstein U. Actinomycin. Chemistry and mechanism of action. Chem Rev 1974;74(6):625–52.

[83] Hou MH, Robinson H, Gao YG, Wang AH. Crystal structure of actinomycin D bound to the CTG triplet repeat sequences linked to neurological diseases. Nucleic Acids Res 2002;30(22):4910–7. https://doi.org/10.1093/nar/gkf619.

[84] Bailey SA, Graves DE, Rill R. Binding of actinomycin D to the T(G)nT motif of double-stranded DNA: determination of the guanine requirement in nonclassical, non-GpC binding sites. Biochemistry 1994;33(38):11493–500. https://doi.org/10.1021/bi00204a011.

[85] Takusagawa F, Carlson RG, Weaver RF. Anti-leukemia selectivity in actinomycin analogues. Bioorg Med Chem 2001;9(3):719–25. https://doi.org/10.1016/s0968-0896(00)00293-5.

[86] Sobell HM. Actinomycin and DNA transcription. Proc Natl Acad Sci U S A 1985;82(16):5328–31. https://doi.org/10.1073/pnas.82.16.5328.

[87] Bensaude O. Inhibiting eukaryotic transcription: which compound to choose? How to evaluate its activity? Transcription 2011;2(3):103–8. https://doi.org/10.4161/trns.2.3.16172.

Chapter 2.2

Reverse transcriptase

Andrea Angeli
NEUROFARBA Department, Pharmaceutical Sciences Section, University of Florence, Florence, Italy

1 An overview of reverse transcriptase function

Reverse transcriptase (RT) enzyme family is responsible for the conversion of RNA first into complementary DNA (cDNA) and then into double-stranded DNA that integrates into the host genome [1]. These enzymes are conserved in all kingdom life based on the homology of their amino acid sequences playing a pivotal role in the evolution facilitating the transfer of the genetic information encoded in the RNA template to DNA [2]. RT was identified as part of the retroviral life cycle in 1970 by two researchers, Baltimore and Temin, showing, for the first time, a breakthrough of the "Central Dogma" of biology with an enormous impact on life sciences [3,4]. Before that, the genetic information flowed in an obligatory direction from DNA to RNA to protein as described by the "Central Dogma" of molecular biology coined by Crick [5].

RTs are nucleic acid-dependent polymerases, which use deoxyribonucleotide triphosphates as building blocks, to generate DNA polymers complementary to a nucleic acid template strand proving to be essential in the life cycle of several retroviruses, eukaryote organisms (including yeast, drosophila, and human cells), participating in the duplication of transposable genetic elements called retrotransposons, and although RTs are considered as eukaryotic enzymes, they are also abundant in bacteria known as retrons [6,7]. In 1997, this diverse superfamily of enzymes was joined by the telomerase reverse transcriptase (TERT), a specialized RT that maintains the ends of eukaryotic linear chromosomes by addition of short G-rich repeated DNA sequences that are copied multiple times via reverse transcription of a specific region of the associated RNA template constituting part of the holoenzyme becoming the unique single-copy eukaryotic RT genes that do not represent a component of any mobile element or virus [8]. In order to carry out all of the steps in the process, RTs exhibit two enzymatic activities that are both required to synthesize double-stranded DNA from a single-stranded RNA template. The first one is a DNA polymerase activity that can use both DNA and RNA, as a template with a primer ($tRNA^{Lys3}$ for HIV-1 RT), to synthesize a complementary DNA strand generating an RNA/DNA hybrid molecule. Subsequently, an RNase H endonuclease activity hydrolyzes the RNA strand when it is annealed to a DNA lattice. Without one of these processes, the reverse transcription cannot take place, since RNA degradation is absolutely required for synthesis of the second DNA strand (Fig. 1) [9].

All RTs follow an obligatory sequential ordered mechanism, whereby the nucleic acid binds first, followed by the nucleoside triphosphate. Nucleic acid binding induces a first conformational change in the p66 subunit, exposing the nucleotide-binding site. After nucleotide binding, a second conformational change occurs, in which the p66 fingers close down on the active site, properly positioning the 3′ hydroxyl primer end and the 5′-α-phosphate group of the incoming nucleotide for catalysis. After incorporation of the nucleotide into the growing DNA chain, a rapid conformational change causes the opening of fingers to release the pyrophosphate and allow the translocation of enzyme along the nucleic acid lattice. In the case of HIV RT, an unusual ability to reverse the polymerization reaction by removing the last incorporated nucleotide has been observed. This reaction, termed phosphorolysis, generates a DNA product shorter than one residue showing this phosphorolytic mechanism one of the major determinant of HIV RT resistance to drugs used to inhibit HIV replication [10]. The RNase H mechanism is an endonuclease that specifically hydrolyzes the RNA present in RNA/DNA hybrids and the position of the site to be cut is strictly dependent on the neighboring sequences [10].

Retrotransposons are genetic mobile elements that utilize reverse transcription to generate a DNA copy of themselves to be inserted into eukaryotic genomes and are divided into two groups, depending on the presence of flanking long-terminal repeat (LTR) sequences. Similarly to retroviruses, their genomes possess the gene for reverse transcriptase, RNase H, and protease, but they lack the genes for the proteins of the viral capsid, so they do not generate virions and therefore do not spread among individuals, but are genetically inherited by the descendants if present in the chromosomes of germline cells. Thus, the

Metalloenzymes. https://doi.org/10.1016/B978-0-12-823974-2.00032-2

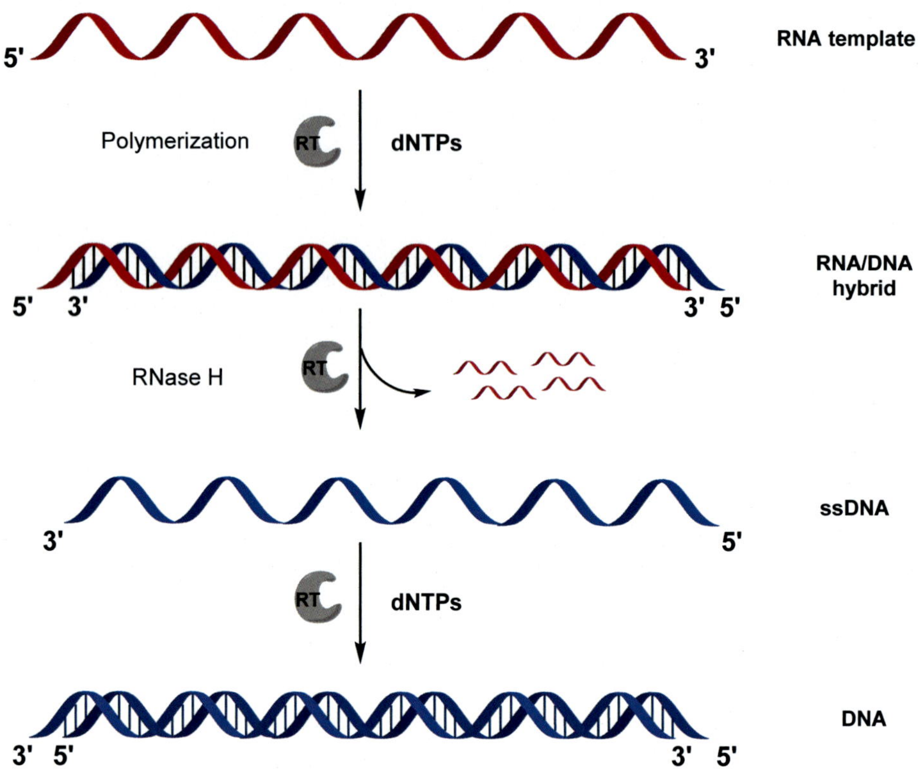

FIG. 1 Reverse transcriptase activities and mechanism of action.

retrotransposons were viewed as possible progenitors of the retroviruses or alternatively as descendants of the retroviruses by loss of their envelope gene [11]. The inability to leave an organism also means that retrotransposons must be more circumspect than a virus in how often they replicate due to the potential damage caused by their insertion into the host genome. The LTR retrotransposons can be divided into three groups with the name of Ty1/copia group, the Bel group, and the Ty3/gypsy group. The Ty1/copia and Ty3/gypsy groups of elements have extremely wide distributions in animals, plants, and fungi, while the Bel class of elements have, to date, only been reported in animals [12].

Telomerase is a complex ribonucleoprotein composed of multi-subunits containing in its holoenzyme core only an RNA template (TER) and catalytic telomerase reverse transcriptase (TERT) counterbalancing the progressive loss of chromosome terminal DNA by adding telomeric DNA (TTAGGG) that repeats at the terminal of linear chromosomes, which prevents chromosome attrition resulting from incomplete semiconservative DNA replication at chromosomal ends [13,14]. TER is ubiquitously expressed, whereas the TERT gene is stringently repressed in most human somatic cells, which consequently results in telomerase silencing [15]. When a telomerase impairment is present, the cells have the capacity to continuously replicate, which is often referred to as "cell immortality." In particular, TERT overexpressed together with simian virus 40 large T antigen and RAS oncoproteins in human cells, the cells become "immortal," are malignantly transformed, and can form tumors in vivo [16]. From the general point of view, TERTs are well conserved and share significant sequence homology with viral RT enzymes showing reverse transcription reaction to synthesize telomere DNA from its RNA template. However, telomerase has also the ability to synthesize multiple telomere DNA repeats during a single binding event becoming, thus, the rate-limiting component to control telomerase activity [17]. When telomerase activity is unable to maintain telomere length required for sustained cell proliferation, it leads to a permanent cell growth arrest, so-called replicative senescence, acting as a key driving force for aging and age-related degenerative diseases in humans [18].

However, there are some important differences in the RTs, and the process of reverse transcription among the different classes, anyway, the role of RTs in the replication of retroviruses attracted particular attention due to notorious relevance for understanding malignant transformation and as a main target of antiretroviral therapies.

2 Structure and function of HIV-1 reverse transcriptase

In 1983, human immunodeficiency virus (HIV) was isolated and identified as the causative agent of AIDS. To date, two type of HIV have been discovered: namely HIV-1 and HIV-2. The first type remains a threat to global public health and no vaccine available, while the related virus HIV-2 is capable of generating full-blown AIDS; however, infection rates are more localized and on the decline [19]. For this reason, the bulk of current investigative efforts are focused on the structural study of HIV-1 in order to understand its mechanism of action and develop different class of inhibitors. HIV-1 RT initially exists as homodimer and undergoes maturation, leading to a heterodimer consisting of a 560 residue chain (called p66) and a second chain comprising the initial 440 residues of p66 (called p51). p66 has an N-terminal RNA-dependent and DNA-dependent DNA polymerase domain folds into five domains named fingers (residues 1–85 and 118–155), palm (residues 86–117 and 156–236), thumb (237–318), connection (319–426), and RNase H domain that digests the RNA. On the other hand, the p51 subunit comprises the first four domains of p66, and both subunits contain the residues important for polymerase activity, but only in p66 do these forms an active site (Fig. 2) [20].

Both polymerase and RNase H active sites are located in different protein domains, separated by over 50 Å requiring Mg^{2+} for catalysis in vivo; however, Mn^{2+} can support polymerization in vitro. In particular, the polymerase active site is composed of three carboxylate residues in the palm subdomain of p66 (D110, D185, and D186) that bind divalent ion Mg^{2+} and RNase H active site is composed of four conserved residues D443, E478, D498, and D549 that bind a second divalent ion Mg^{2+} [21]. Both activities are essential for HIV replication anyway, all current drugs against RTs targeting only the RT DNA polymerase activity with, to date, 13 clinical inhibitors of RT divided in nucleoside or nucleotide inhibitors (NRTIs), which are incorporated into the growing DNA strand and act as chain terminators or nonnucleoside RT inhibitors (NNRTIs) that are allosteric inhibitors of DNA polymerization. Regarding the mechanism of inhibition with NRTIs, they act as chain terminators, since when incorporated into the nascent DNA primer strand, they lack a 3′ OH that terminates the RT polymerase function (Fig. 3A).

However, resistant RTs can discriminate against NRTIs by two mechanisms. The first is a mutation of M184V or M184I, which replaces a flexible side chain near the polymerase active sites with a β-branched amino acid that selectively discriminates against NRTIs. The second mechanism is a mutation of K65R, which forms a molecular platform with the conserved R72; allowing the reverse transcriptase to discriminate tenofovir diphosphate from dATP [22]. In contrast, NNRTIs have a destabilizing effect as the consequence of loss of thumb and fingers interactions with nucleic acid since NNRTIs bind in a deep pocket that lies between the β sheets of the palm and at the base of the thumb subdomains close to the expected primer terminus, but not overlapping the DNA-binding sites (Fig. 3B) [23]. The binding of this allosteric site by NNRTIs has a profound effect on RT-DNA conformation compared to the other class of inhibitors. Indeed, the "primer grip" shifted lifting the DNA primer terminus away from the P site, and key interactions of the conserved catalytic motif with the primer terminus are lost. The loss of contacts between the template strand and fingers has a profound impact on the positional stability of both; as a result, the fingers subdomain is also weakly ordered. The binding and incorporation of a dNTP require interactions with fingers base pairing with the first template overhang and chelation with the metal ion. Thus, the interactions of DNA with palm and fingers subdomains are substantially reduced with NNRTIs; however, the

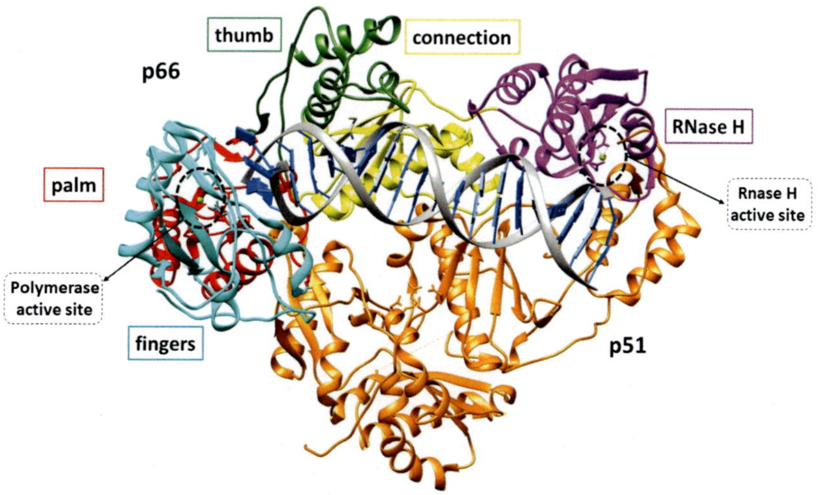

FIG. 2 Representation of HIV-1 RT in a complex with nucleic acid. The fingers, palm, thumb, connection, and RNase H subdomains of the p66 subunit are shown in *cyan*, *red*, *green*, *yellow*, and *magenta*, respectively. The p51 subunit is shown in *orange*. The template and primer DNA strands are shown in *dark gray*, and Mg^{2+} ion as yellow sphere, respectively.

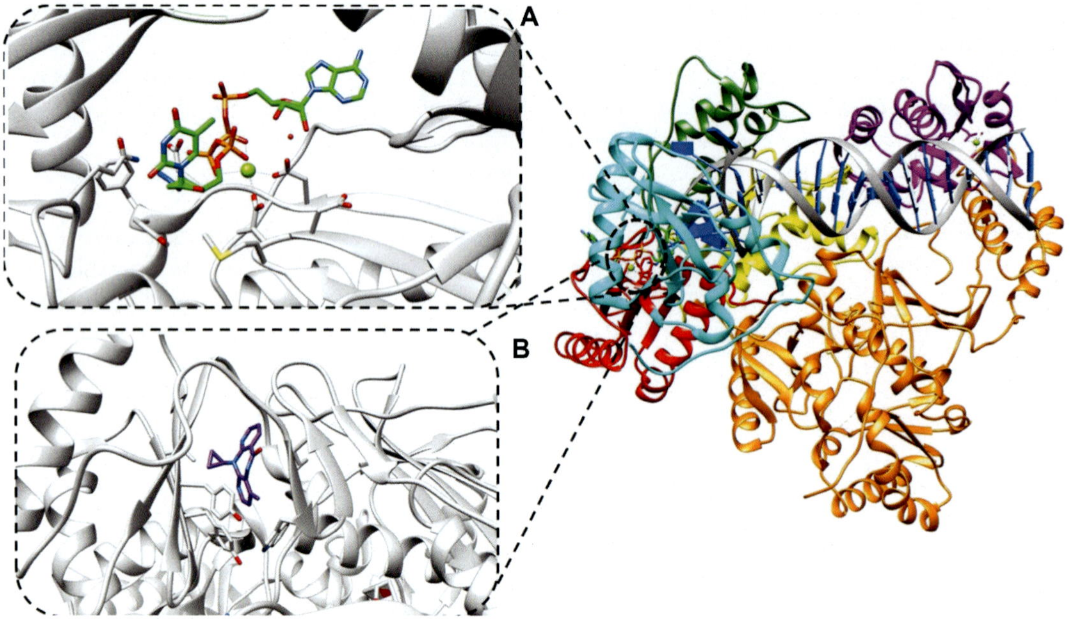

FIG. 3 (A) NRTI (zidovudine) triphosphate complex at the dNTP-binding site. (B) NNRTI nevirapine in complex at the allosteric binding site.

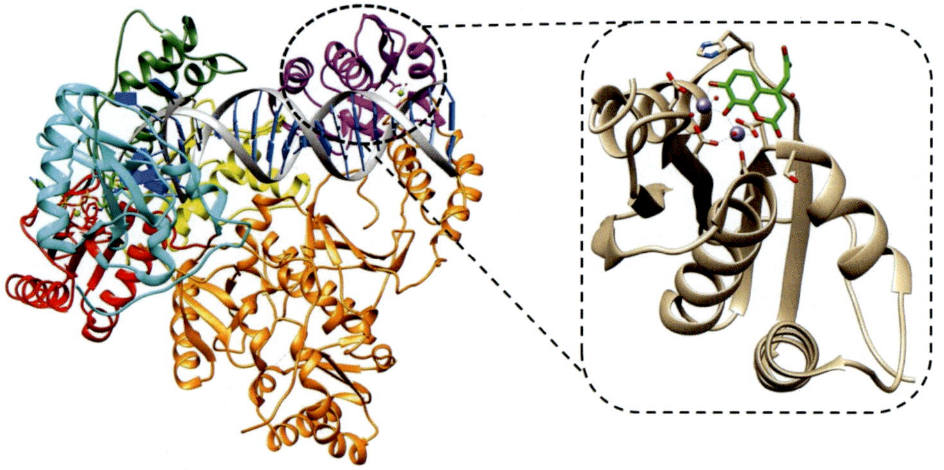

FIG. 4 F3284-8495 in complex with HIV-1 RT at the RNase H-binding site [29].

remaining DNA-protein interactions are less perturbed despite all subdomains being rearranged [24,25]. On the other hand, the active site of RT RNase H has proven to be a difficult target for drug discovery and development due to the shallow active site, which presents few binding opportunities for a ligand, apart from hydrophilic interactions, which has led, despite considerable effort, to no clinically useful drugs targeting RT RNase H [26–28]. In the last years, however, several inhibitors were discovered to bound the RNase H active site exhibiting very similar cation coordination geometry, although the inhibitory activities differed significantly. The active site of RT-RNase H contains two divalent cations with an approximately octahedral coordination geometry. These cations do not exhibit any significant changes in the conformation of the RNase H domain. The features described, including the role of the amino acid residue H539 in modulating activity, were observed in the structure of HIV-1 RT bound to an RNase H inhibitor (specifically F3284-8495), as reported by Himmel et al. [29] (Fig. 4).

3 Structure and function of Ty3 retrotransposon

Retrotransposons had a life cycle similar to a retrovirus that has either lost, or never acquired, the ability to be transmitted horizontally from one cell to another; therefore, they could be viewed as possible progenitors of the retroviruses or alternatively as descendants of the retroviruses by loss of

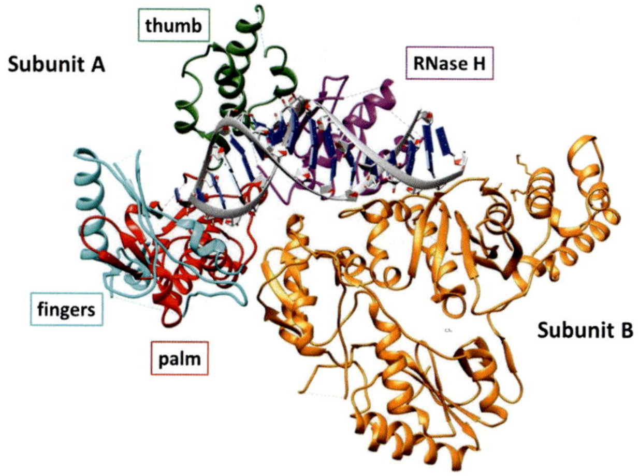

FIG. 5 Representation of Ty3 RT from *Saccharomyces cerevisiae* in a complex with nucleic acid. The fingers, palm, thumb and RNase H subdomains of the subunit A are shown in *cyan*, *red*, *green* and *magenta*, respectively. The subunit B is shown in *orange*. The template and primer DNA strands are shown in *light gray*.

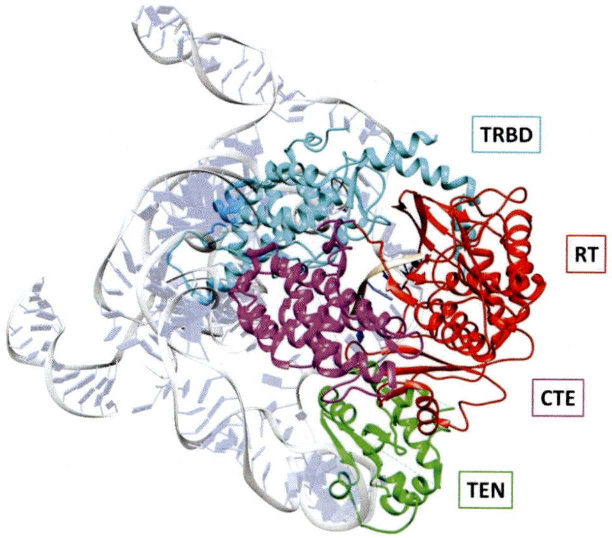

FIG. 6 Human telomerase reverse transcriptase (TERT) structure. The four subunits are labeled in *cyan* (TRBD), *red* (RT), *green* (TEN), and *magenta* (CTE).

their envelope gene [11]. However, there are some important differences in the RTs and the process of reverse transcription, among the different retroviruses and LTR retrotransposons. One of the most studied was LTR-retrotransposon Ty3 from *Saccharomyces cerevisiae* as asymmetric homodimer of 55-kDa. The overall conformations of Ty3 and HIV-1 RTs are quite similar, showed two subunits, one of them mimics p51 and the other p66 of HIV-1 RT, divided in four subdomains: fingers, palm, thumb, and RNase H (Fig. 5) [30].

On the other hand, the position of the Ty3 RNase H domain corresponds with the retroviral connection subdomain of HIV-1, supporting the hypothesis that evolution of retroviral RTs from LTR retrotransposon enzymes involved converting their RNase H domain to a "connector" with loss of catalytic function and recruitment of a new RNase H domain [31]. Although the structural motifs mediating substrate recognition and catalysis generally resemble those of vertebrate retroviral RTs, a notable difference between the Ty3 and retroviral enzymes is the separation of its DNA polymerase and active RNase H.

4 Structure and function of telomerase reverse transcriptase

In most eukaryotic cells, telomeres contain G-rich repeats cap and protect the ends of linear chromosomes from degradation, fusion, and shortening. As mentioned above, telomeres are composed of two essential components: telomerase reverse transcriptase and telomerase RNA along with a battery of cofactors [32]. The TERT protein is structurally similar to retroviral RTs and is composed of four conserved domains: the telomerase essential amino-terminal (TEN) domain, the RNA-binding domain (TRBD), the RT domain, and the C-terminal extension (CTE), which is also known as the thumb domain [33]. The TEN domain (Fig. 6) is critical for maintaining the repeat addition processivity (RAP), a unique property of telomerase to recycle its internal RNA template and synthesize multiple telomere DNA repeats without dissociating from the telomere, anyway, is not required during the addition of a single telomere repeat [34]. The TRBD (Fig. 6) contains conserved motifs and is responsible for establishing RNA-protein interactions between TERT and TR, it is absolutely essential for telomerase activity highlighting the existence of common mechanistic aspects of telomere replication across phylogenetic groups [35]. Indeed, telomerase is a high-fidelity enzyme able to add multiple identical repeats of DNA, a capability that is conferred in large part by its integral RNA-templating region and, after the completion of one round of telomere extension, leads to a transient dissociation of the RNA-DNA hybrid, followed by repositioning of the DNA end at the active site of TERT and RNA-DNA pairing at the other end of the template. Telomerase remains associated with the chromosome end until several telomeric repeats have been added, a process that is highly regulated by telomere-binding proteins, such as the TPIP1-POT1 complex in humans [36,37]. The RT is a 5′ pseudoknot structure that interacts with TERT and includes a region that serves as a template for telomere synthesis and with CTE domains being homologous to the palm and thumb domains of canonical polymerases, respectively (Fig. 6) [38].

Several mutagenesis studies have been conducted, in the last years, to understand which amino acid residues play a key role in the activity of TERT. In particular, K570E mutant was discovered as severely defective in nucleotide addition processivity, and that mutations affecting thumb residues L958, L980, and K981 lead to reductions in repeat addition processivity [39,40]. In addition, Y717 acts as a steric gate that discriminates rNTPs from dNTPs and allows only dNTPs to be incorporated [41]. It is striking, however, that despite this wealth of information and the compelling rationale for pursuing telomerase as a therapeutic target, only one telomerase inhibitor, imetelstat (also known as GRN163L), has been sufficiently developed to move into clinical trials [42].

5 Clinically used inhibitors

As stated above, RTs have been considered as the most attractive targets for antiviral purposes; therefore, they are extensively exploited as a target for antiviral drug discovery. Most of the Reverse Transcriptase inhibitors (RTIs) are target-specific viruses, and in particular, many efforts are being made against the causative agent of AIDS, the HIV-1 RT. To date, many different small molecules with distinct mechanisms of action target RT, resulting in inhibition of viral DNA synthesis and thus inhibiting HIV replication. All RTIs tend to divide into three classes:

The nucleoside reverse-transcriptase inhibitors (NRTIs) were the first class of antiretroviral drugs to be approved by the FDA and compete with natural deoxynucleotides for incorporation into growing viral DNA chains. However, NRTIs are taken as prodrugs and must be introduced into the host cell and phosphorylated before they become active. Once activated to their 5′-triphosphate forms by cell-derived kinases, they can be incorporated by HIV RT, but because they lack a 3′-hydroxyl group, the next incoming deoxynucleotide cannot form the following 5′-, 3′-phosphodiester bond needed to extend the DNA chain acting, in fact, as DNA chain terminators [43]. Usually, NRTIs are transported into cells by simple diffusion (mainly due to their hydrophobic characteristics) or diffusion via nucleoside carrier-mediated transporter belonging to the solute carrier transporter (SLC) families [44]. The first NRTI approved by FDA for the treatment of HIV was the thymidine analogue zidovudine (AZT, Fig. 7) in 1987 inhibiting the HIV-1, HIV-2, and other mammalian retroviruses [45,46].

As mentioned above, AZT is structurally similar to the endogenous nucleoside thymidine, differing only by the presence of an azido group (N_3) in the ribose ring instead of a hydroxyl group. The antiretroviral therapy requires the combined use of zidovudine with other antivirals even if AZT is also recommended for pregnant women infected with HIV, however, by itself, leads to the selection of resistant strains [46]. On the other hand, zidovudine combination with other HIV medications such as lamivudine and abacavir has proved to be more effective in suppressing viral replication decreasing in plasma levels of HIV-1 to below the detection limit [47,48]. After oral administration, AZT is rapidly and almost completely absorbed from the gastrointestinal tract and the antiviral effect of zidovudine is dependent, not only on dose and rate of elimination but also on the rate of intracellular phosphorylation. Indeed, phosphorylation reaction is a saturable process, and an increment in the plasma concentration of AZT does not lead to parallel increase in the concentration of phosphorylated AZT [49]. Finally, the adverse effects with zidovudine tend to be dose-related and reversible

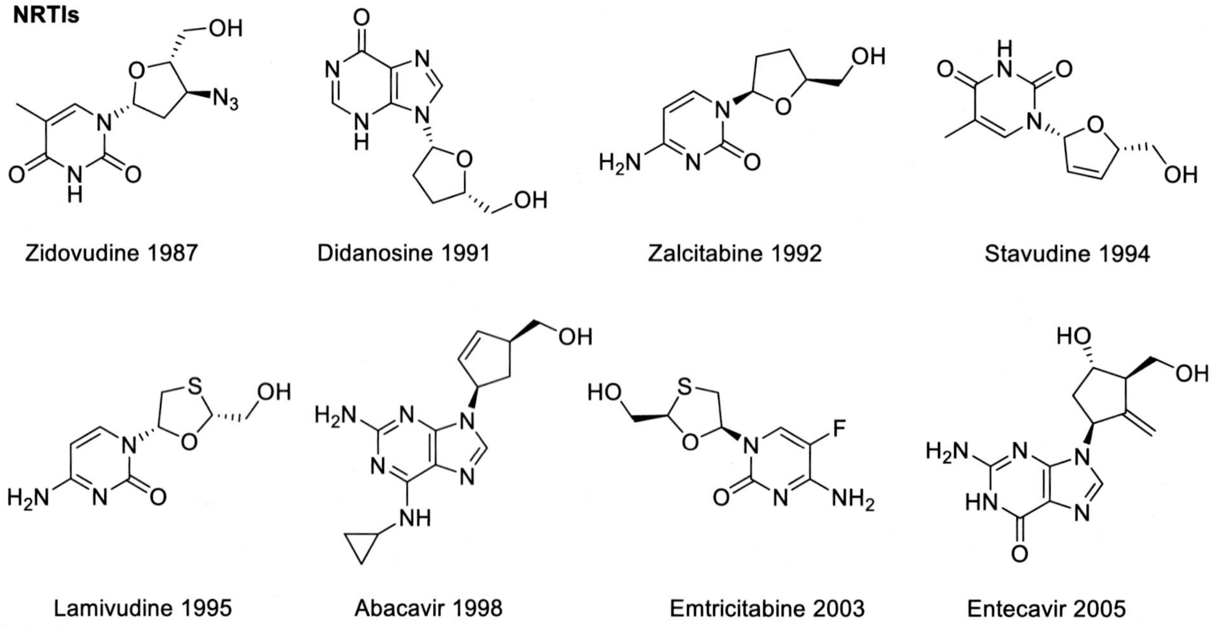

FIG. 7 Structures of clinically NRTI drugs.

with common side effects such as anemia, granulocytopenia, and less often cardiomyopathy [50]. Four years later, the FDA has been approved didanosine (Fig. 7), a synthetic purine nucleoside analogue of adenosine becoming a cornerstone of HIV management, originally introduced as alternative to intolerance to zidovudine but, to date, it is replaced by better tolerated agents [51]. Didanosine inhibits HIV replication by competing with naturally occurring adenosine for incorporation into the growing viral DNA chain, causing inhibition of the RT by chain termination. Combination therapy, as for zidovudine, delays the disease progression and prolongs survival in patients with intermediate or advanced HIV infection [52]. Didanosine is unstable under acidic conditions, so it is formulated for oral administration with a buffer and a recommendation that it be administered on an empty stomach [53]. The side effects are pancreatitis and peripheral neuropathy of the lower extremities, and both adverse effects seem to be dose-related in particular for the pancreatitis that may limit its use in certain HIV-infected populations [54]. To date, better and safer alternatives have led to didanosine no further being recommended for the treatment of HIV and being withdrawn from the market in 2020.

Zalcitabine (Fig. 7) has been approved 1 year later (1992) by FDA, begun the third compound in the antiretroviral armamentarium showing approximately 10-fold more potent than AZT on in vitro experiments. Furthermore, in 1996, the FDA approved zalcitabine as part of a combination regimen with AZT for the treatment of AIDS, since synergistic antiretroviral activity has been reported for zalcitabine in combination with several other antiretroviral agents [55,56]. Peripheral neuropathy is the major dose-limiting side effect and less frequent mouth ulcers, stomatitis, and pancreatitis; thus, zalcitabine was removed from the market in 2006 [57].

Stavudine (Fig. 7), a synthetic pyrimidine nucleoside analogue where the 3′-hydroxyl group is replaced by a double bond between the 2′- and 3′-carbons of the pentose ring (Fig. 7), was approved in 1994 by FDA. Its potency of inhibit HIV RT is comparable to zidovudine although, stavudine was found equally effective against HIV-1 and HIV-2 in MT-4 cells, but 20-fold less potent than zidovudine against these viruses and did not show significant activity against hepatitis B virus [58]. The drug is generally well tolerated ongoing prolonged therapy without viral resistance during the treatment and showed clinical benefit in zidovudine pretreated patients [59]. As for the case of didanosine, the primary dose-limiting event of stavudine observed was the peripheral neuropathy and, for this reason, should be used cautiously in combination with didanosine [60]. However, even though stavudine has a favorable viral resistance profile compared to the other NRTIs reported so far, to date, due to its toxicity and safer alternatives, stavudine was removed from the market in 2020.

Lamivudine (Fig. 7), the fifth approved cytosine nucleoside analogue, is used in combination with other drugs in the treatment of HIV-1 infection and, as monotherapy, in the treatment of hepatitis B virus (HBV) infection [61]. It is a negative enantiomer of 2′,3′-dideoxy-3′-thiacytidine, a dideoxypyridimine, in which the 3′ carbon of the ribose ring is replaced by a sulfur atom, and this substitution allows lamivudine to be resistant to cleavage from the 3′ terminals of the RNA/DNA complexes by 13′,5′-exonuclease [62,63]. Lamivudine is well absorbed after oral administration and one of the most well-tolerated agents in clinical studies among patients with HIV or HBV infection. Indeed, although lamivudine is part of the NRTIs class, where specific toxicities are present such as mitochondrial toxicity, lactic acidosis, pancreatitis, peripheral neuropathy, lipoatrophy, and bone marrow toxicity, the most commonly reported adverse reactions are headache, nausea, malaise and fatigue, nasal signs, diarrhea, and cough. Compared to other NRTIs, lamivudine is not associated with anemia (as is zidovudine) or nephrotoxicity (as is tenofovir). It has better gastrointestinal tolerability and less potential for decreased bone density, dyslipidemia, fat redistribution, or long-term mitochondrial toxicity resulting in lactic acidosis or hepatic steatosis compared to other NRTIs [64]. In this context, Lamivudine is considered one of the best-tolerated agents among all antiretroviral classes, and it is a main component of combination treatments and remained a backbone agent of choice since its initial approval in 1995 based on its established efficacy and favorable safety profile in a wide variety of patient populations.

Abacavir (Fig. 7), the first guanosine analogues approved by FDA (1998), is a carbocyclic nucleosides gained significant attention due to lacking a proper glycosidic linkage between the heterocyclic nucleobase and the sugar moiety, thereby increasing in vivo stability [65]. Its activation mechanism is quite different when compared to other nucleoside analogues mentioned above, because abacavir is first converted into its monophosphate form by adenosine phosphotransferase and further deaminated by cytosolic enzymes to form carbovir monophosphate. The monophosphate compound subsequently undergoes to two subsequent phosphorylation steps to form active metabolite carbovir triphosphate inhibiting the HIV-1 RT and acting as chain terminator in the same manner of the other NRTIs [66]. The major side effects observed by abacavir include hypersensitivity reactions, fever, maculopapular rash, fatigue, malaise, myalgias, arthralgias, lymphadenopathy, mucositis, pneumonitis, myocarditis, hepatitis, interstitial nephritis, atypical lymphocytosis, and eosinophilia [67].

The seventh NRTI approved is emtricitabine (Fig. 7), an analogue of cytidine and structurally related to lamivudine scaffold. Because of emergence of resistance, combination therapies with nucleoside inhibitors became the standard into the control of HIV replication; thus, in 2003, the FDA approved a combination of emtricitabine and tenofovir, which became the first-line therapy when used with a third antiretroviral agent [68]. Indeed, when used as monotherapy, it was observed that emtricitabine decreases its efficacy by more than 1000-fold due to the occurrence of several drug resentences even in patients with the M184V mutation [69].

Entecavir (Fig. 7), a guanosine nucleoside analogue, was originally developed for the treatment of herpes simplex virus infections, but its moderate activity against herpes virus led to approval by the FDA, for the treatment of chronic HBV infection in 2005 [70]. Although entacavir is not used clinically to treat HIV, the co-infection with HIV-1 is common, showing at doses used to treat HBV partial inhibition of HIV-1 replication, and interesting data have shown it as a useful tool against the M184V resistance mutation in HIV RT [71–73]. Unlike the other NRTIs, discussed above, entecavir retains the 3′ -hydroxyl group on the sugar moiety, and this characteristic displayed anti-HIV activity via a unique, delayed chain termination mechanism. Indeed, at the point of incorporation of entacavir by HIV-RT, the incorporation of three additional nucleotides showed 2–3 orders of magnitude more efficient than inhibition at n+1, indicating that delayed chain termination is an important mechanism of action and protection from excision [73]. The most common side effects include headache, fatigue, dizziness, and less common gastrointestinal symptoms [74].

The second category includes the nucleotide reverse-transcriptase inhibitors (NtRTIs).

This class of drugs has a similar mode of action to NRTIs, but the presence of the phosphonate group, which is analogous to a phosphate one, means that only two phosphorylation steps by cellular kinases are required for conversion to the active metabolite. NtRTIs are therefore able to bypass the nucleoside-kinase reaction, which can limit the activity of the dideoxynucleoside analogues against HIV [75]. The first NtRTIs approved by FDA is the prodrug of adenosine nucleotide analogue tenofovir in 2001, namely tenofovir disoproxil (Fig. 8). Indeed, the low bioavailability caused by the highly polar phosphonic acid groups required its modification into ester prodrug. Once tenofovir prodrug entered the cell, it is rapidly hydrolyzed to tenofovir and further phosphorylated diphosphate one, which inhibits the activity of HIV-1 RT by competing with the natural substrate (ATP) leading to chain termination [76].

Although tenofovir disoproxil is currently used in triple combination with other antiviral drugs, it is absorbed quickly and well distributed, the renal and bone toxicity limits its use. For this reason, the second prodrug of tenofovir was developed and approved as tenofovir alafenamide fumarate in 2016 (Fig. 8) to overcome these side effects [77]. The novel prodrug showed a better plasma stability than the previously one and, once it has entered cells passively, is selectively hydrolyzed intracellularly by the lysosomal carboxypeptidase cathepsin A requiring a lower dose of drug [78]. The second NtRTI approved by FDA is the

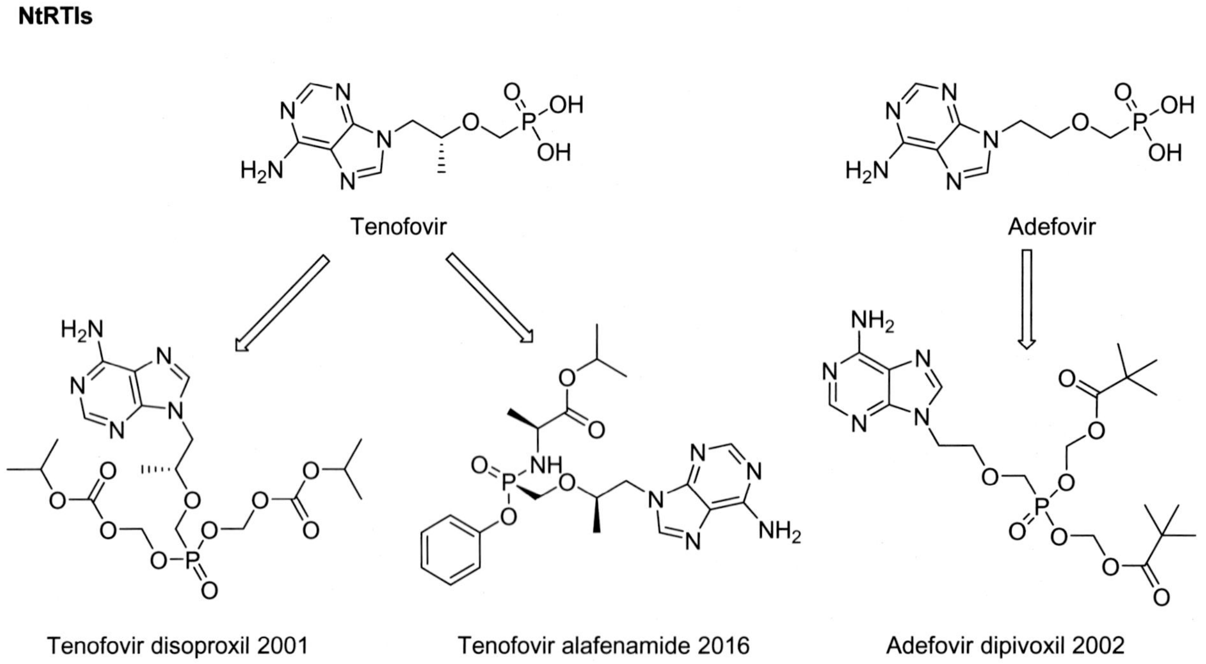

FIG. 8 Structures of clinically NtRTI drugs.

prodrug of adefovir, namely adefovir dipivoxil (Fig. 8) in 2002 as adenosine monophosphate analogue showed antiviral activity spectrum encompasses both retro- and hepadna-viruses and herpesviruses. However, when used in high doses, adefovir is associated with renal tubular dysfunction [79].

The third and last family of RT inhibitors are the nonnucleoside reverse-transcriptase inhibitors (NNRTIs, Fig. 9). This class of drugs was discovered for the first time in 1990 targeting a different binding pockets near the catalytic site of RT to block the viral transcription of a double-stranded viral DNA genome from a single-stranded viral RNA genome acting as allosteric inhibitors. To date, they have chemically highly diverse scaffolds and are a key part of the typical cocktail treatment (two NRTIs and one NNRTI) against HIV owing to their potency, favorable safety profile, and ease of dosing. However, the relatively rapid onset of resistance, resulting from mutations at amino acid residues at the rim of the NNRTI-binding site (in particular K103N and Y181C), generates a serious limitation of employment as monotherapy [80].

Nevirapine, a synthetic dipyridodiazepine derivative (Fig. 9), is the first member of NNRTIs approved by FDA in 1996 and used in synergistic cocktail with other NRTIs since nevirapine does not stop the elongation of the viral DNA complementary chain but modified only the conformation of RT slowing the enzymatic activity [81].

Nevirapine is highly specific for HIV-1 RT with respect HIV-2 or other retroviral RTs although the monotherapy was observed with the rapid selection of mutation leading to drug resistance, conferred by a single amino acid substitution in the HIV RT gene. For these reasons, nevirapine has been used synergistically with zidovudine, lamivudine, or stavudine against HIV-1 [82].

One year later, FDA approved the second NNRTIs delavirdine (Fig. 9) showing the same hydrophobic binding site, although the high diversity scaffold of nevirapine, close to the catalytic binding site stabilizing the polymerase site of the p66 subdomain [83]. Delavirdine demonstrated a high degree of selectivity for HIV-1 RT, with minimal effects against HIV-2, and unlike nevirapine, the binding of delavirdine to RT is stabilized by hydrogen bonding to lysine at position 103 and through several hydrophobic interactions with the proline residue at position 236 [84]. This particular interaction in the binding complex explains the pivotal role of these amino acids in the development of resistance mutations especially when delaviridine is used as monotherapy. For these reasons, NNRTIs in the current treatment guidelines are recommended as part of a combination antiretroviral regiment [85]. In 1998, the FDA approved efavirenz (Fig. 9) for the treatment of HIV-1 infection, after which it quickly became an important component of highly active antiretroviral therapy also due to better efficacy than nevirapine [86]. Efavirenz is generally well tolerated showing as major side effects such as rash, neuropsychiatric disturbances, teratogeny, and therefore contraindicated in pregnant women replaced in this condition by nevirapine [87]. Because of low genetic barrier

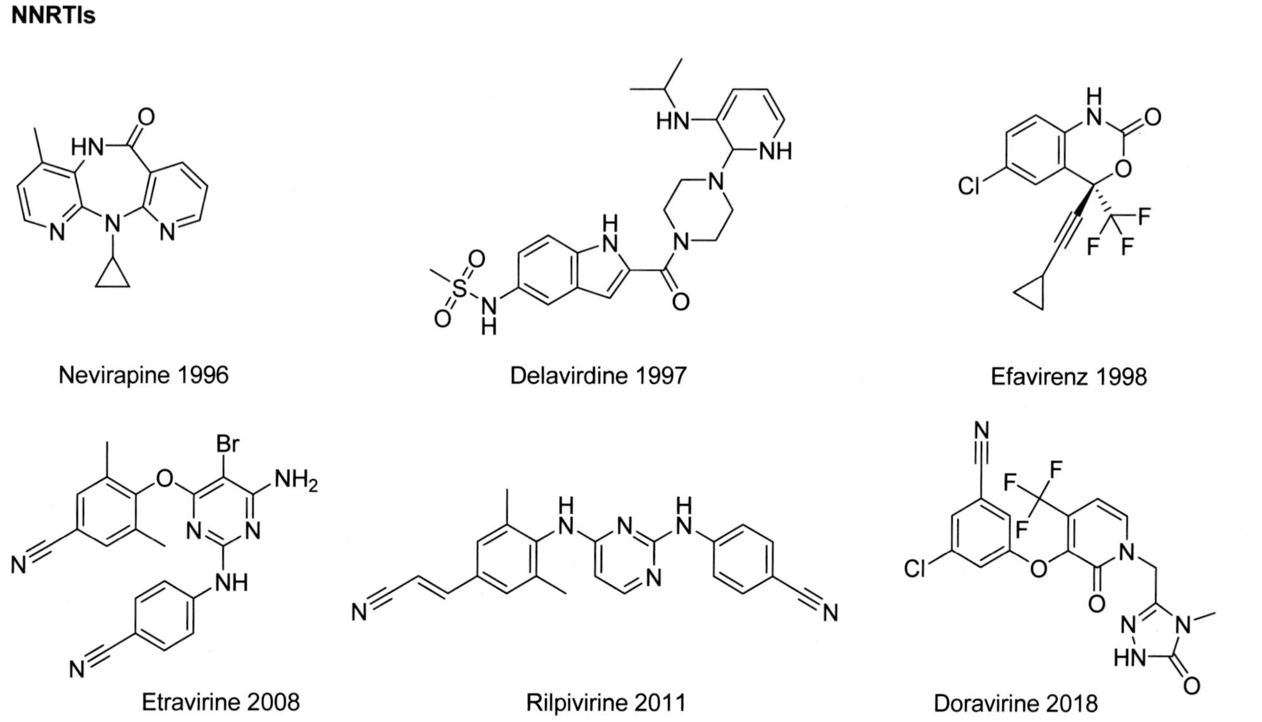

FIG. 9 Structures of clinically NNRTI drugs.

for resistance, to date, these first generation of NNRTIs (nevirapine, efavirenz and delavirdine) are rarely used beyond first- or second-line therapy in treatment-experienced patients. Therefore, there is a need for next-generation NNRTIs that present a higher genetic barrier to resistance while maintaining a good tolerability and pharmacokinetic profile [88].

In this context, etravirine (Fig. 9) was approved in 2008 by FDA (10 years later than efavirenz) becoming the first NNRTI of second generation. It is a di-arylpyrimidine analogue that acts principally by directly binding the allosteric site of RT but, this time, etravirine can adapt to changes in binding site caused by mutations, repositioning and reorienting itself explained the potent antiviral activity also in some NNRTI-resistant strains of HIV-1 [89]. In addition, etravirine also appears to affect post-integration steps, probably by enhancing the processing of gag and gag-pol precursor proteins in HIV-1 transfected cells, leading to a reduction in the viral particle formation [90]. Rilpivirine (Fig. 9) was the second diarylpyrimidine approved in 2011 by FDA, designed according to the strategic flexibility concept to overcome the resistance mutations in the NNRTIs-binding site. It is approved for use in combination with other antiretrovirals in treatment-experienced patients with HIV-1 infection [91]. The last NNRTI approved by the FDA was in 2018 with the name of doravirine (Fig. 9). It was a pyridone derivative, which has demonstrated a broad spectrum of antiviral activity against clinically relevant mutant viruses with improved pharmacokinetic parameters compared with preclinical compounds [92].

References

[1] Menéndez-Arias L, Sebastián-Martín A, Álvarez M. Viral reverse transcriptases. Virus Res 2017;234:153–76.

[2] Ellefson JW, Gollihar J, Shroff R, Shivram H, Iyer VR, Ellington AD. Synthetic evolutionary origin of a proofreading reverse transcriptase. Science 2016;352:1590–3.

[3] Baltimore D. RNA-dependent DNA polymerase in virions of RNA tumour viruses. Nature 1970;226:1209–11.

[4] Temin HM, Mizutani S. RNA-dependent DNA polymerase in virions of Rous sarcoma virus. Nature 1970;226:1211–3.

[5] Crick FH. The origin of the genetic code. J Mol Biol 1968;38:367–79.

[6] Dewannieux M, Heidmann T. Endogenous retroviruses: acquisition, amplification and taming of genome invaders. Curr Opin Virol 2013;3:646–56.

[7] Zimmerly S, Wu L. An unexplored diversity of reverse transcriptases in bacteria. Microbiol Spectr 2015;3. MDNA3-0058-2014.

[8] Lingner J, Hughes TR, Shevchenko A, Mann M, Lundblad V, Cech TR. Reverse transcriptase motifs in the catalytic subunit of telomerase. Science 1997;276:561–7.

[9] Mustafin RN, Khusnutdinova EK. The role of reverse transcriptase in the origin of life. Biochemistry (Mosc) 2019;84:870–83.

[10] Mas A, Vázquez-Alvarez BM, Domingo E, Menéndez-Arias L. - Multidrug-resistant HIV-1 reverse transcriptase: involvement of ribonucleotide-dependent phosphorolysis in cross-resistance to nucleoside analogue inhibitors. J Mol Biol 2002;323:181–97.

[11] Eickbush TH, Jamburuthugoda VK. The diversity of retrotransposons and the properties of their reverse transcriptases. Virus Res 2008;134:221–34.

[12] SanMiguel P, Gaut BS, Tikhonov A, Nakajima Y, Bennetzen JL. The paleontology of intergene retrotransposons of maize. Nat Genet 1998;20:43–5.

[13] Shippen-Lentz D, Blackburn EH. Functional evidence for an RNA template in telomerase. Science 1990;247:546–52.

[14] Blackburn EH. Telomerases. Annu Rev Biochem 1992;61:113–29.

[15] Shay JW, Wright WE. Telomeres and telomerase: three decades of progress. Nat Rev Genet 2019;20:299–309.

[16] Hahn WC, Counter CM, Lundberg AS, Beijersbergen RL, Brooks MW, Weinberg RA. Creation of human tumour cells with defined genetic elements. Nature 1999;400:464–8.

[17] Zhao Y, Abreu E, Kim J, Stadler G, Eskiocak U, Terns MP, Terns RM, Shay JW, Wright WE. Processive and distributive extension of human telomeres by telomerase under homeostatic and nonequilibrium conditions. Mol Cell 2011;42:297–307.

[18] Yuan X, Xu D. Telomerase reverse transcriptase (TERT) in action: cross-talking with epigenetics. Int J Mol Sci 2019;20:3338.

[19] Davenport YW, West Jr AP, Bjorkman PJ. Structure of an HIV-2 gp120 in complex with CD4. J Virol 2015;90:2112–8.

[20] Xavier Ruiz F, Arnold E. Evolving understanding of HIV-1 reverse transcriptase structure, function, inhibition, and resistance. Curr Opin Struct Biol 2020;61:113–23.

[21] Davies 2nd JF, Hostomska Z, Hostomsky Z, Jordan SR, Matthews DA. Crystal structure of the ribonuclease H domain of HIV-1 reverse transcriptase. Science 1991;252:88–95.

[22] Das K, Bandwar RP, White KL, Feng JY, Sarafianos SG, Tuske S, Tu X, Clark Jr AD, Boyer PL, Hou X, Gaffney BL, Jones RA, Miller MD, Hughes SH, Arnold E. Structural basis for the role of the K65R mutation in HIV-1 reverse transcriptase polymerization, excision antagonism, and tenofovir resistance. J Biol Chem 2009;284:35092–100.

[23] Liu S, Abbondanzieri EA, Rausch JW, Le Grice SF, Zhuang X. Slide into action: dynamic shuttling of HIV reverse transcriptase on nucleic acid substrates. Science 2008;322:1092–7.

[24] Das K, Martinez SE, Bauman JD, Arnold E. HIV-1 reverse transcriptase complex with DNA and nevirapine reveals non-nucleoside inhibition mechanism. Nat Struct Mol Biol 2012;19:253–9.

[25] Tu X, Das K, Han Q, Bauman JD, Clark Jr AD, Hou X, Frenkel YV, Gaffney BL, Jones RA, Boyer PL, Hughes SH, Sarafianos SG, Arnold E. Structural basis of HIV-1 resistance to AZT by excision. Nat Struct Mol Biol 2010;17:1202–9.

[26] Roquebert B, Marcelin AG. The involvement of HIV-1 RNAse H in resistance to nucleoside analogues. J Antimicrob Chemother 2008;61:973–5.

[27] Xi Z, Wang Z, Sarafianos SG, Myshakina NS, Ishima R. Determinants of active-site inhibitor interaction with HIV-1 RNase H. ACS Infect Dis 2019;5:1963–74.

[28] Madia VN, Messore A, De Leo A, Tudino V, Pindinello I, Saccoliti F, De Vita D, Scipione L, Costi R, Di Santo R. Small-molecule inhibitors of HIV-1 reverse transcriptase-associated ribonuclease H function: challenges and recent developments. Curr Med Chem 2021;28:6146–78.

[29] Himmel DM, Myshakina NS, Ilina T, Van Ry A, Ho WC, Parniak MA, Arnold E. Structure of a dihydroxycoumarin active-site inhibitor in complex with the RNase H domain of HIV-1 reverse transcriptase and structure-activity analysis of inhibitor analogs. J Mol Biol 2014;426:2617–31.

[30] Nowak E, Miller JT, Bona MK, Studnicka J, Szczepanowski RH, Jurkowski J, Le Grice SF, Nowotny M. Ty3 reverse transcriptase

complexed with an RNA-DNA hybrid shows structural and functional asymmetry. Nat Struct Mol Biol 2014;21:389–96.
[31] Malik HS, Eickbush TH. Phylogenetic analysis of ribonuclease H domains suggests a late, chimeric origin of LTR retrotransposable elements and retroviruses. Genome Res 2001;11:1187–97.
[32] Ghanim GE, Fountain AJ, van Roon AM, Rangan R, Das R, Collins K, Nguyen THD. Structure of human telomerase holoenzyme with bound telomeric DNA. Nature 2021;593:449–53.
[33] Mitchell M, Gillis A, Futahashi M, Fujiwara H, Skordalakes E. Structural basis for telomerase catalytic subunit TERT binding to RNA template and telomeric DNA. Nat Struct Mol Biol 2010;17:513–8.
[34] Robart AR, Collins K. Human telomerase domain interactions capture DNA for TEN domain-dependent processive elongation. Mol Cell 2011;42:308–18.
[35] Chen JL, Greider CW. An emerging consensus for telomerase RNA structure. Proc Natl Acad Sci U S A 2004;101:14683–4.
[36] Greider CW. Telomerase is processive. Mol Cell Biol 1991;11:4572–80.
[37] Wang F, Podell ER, Zaug AJ, Yang Y, Baciu P, Cech TR, Lei M. The POT1-TPP1 telomere complex is a telomerase processivity factor. Nature 2007;445:506–10.
[38] Bachand F, Autexier C. Functional regions of human telomerase reverse transcriptase and human telomerase RNA required for telomerase activity and RNA-protein interactions. Mol Cell Biol 2001;21:1888–97.
[39] Wu RA, Tam J, Collins K. DNA-binding determinants and cellular thresholds for human telomerase repeat addition processivity. EMBO J 2017;36:1908–27.
[40] Qi X, Xie M, Brown AF, Bley CJ, Podlevsky JD, Chen JJ. RNA/DNA hybrid binding affinity determines telomerase template-translocation efficiency. EMBO J 2012;31:150–61.
[41] Schaich MA, Sanford SL, Welfer GA, Johnson SA, Khoang TH, Opresko PL, Freudenthal BD. Mechanisms of nucleotide selection by telomerase. elife 2020;9, e55438.
[42] Arndt GM, MacKenzie KL. New prospects for targeting telomerase beyond the telomere. Nat Rev Cancer 2016;16:508–24.
[43] Squires KE. An introduction to nucleoside and nucleotide analogues. Antivir Ther 2001;6(Suppl 3):1–14.
[44] Takeda M, Khamdang S, Narikawa S, Kimura H, Kobayashi Y, Yamamoto T, Cha SH, Sekine T, Endou H. Human organic anion transporters and human organic cation transporters mediate renal antiviral transport. J Pharmacol Exp Ther 2002;300:918–24.
[45] Maga G, Radi M, Gerard MA, Botta M, Ennifar E. HIV-1 RT inhibitors with a novel mechanism of action: NNRTIs that compete with the nucleotide substrate. Viruses 2010;2:880–99.
[46] Soares KC, Rediguieri CF, Souza J, Serra CH, Abrahamsson B, Groot DW, Kopp S, Langguth P, Polli JE, Shah VP, Dressman J. Biowaiver monographs for immediate-release solid oral dosage forms: zidovudine (azidothymidine). J Pharm Sci 2013;102:2409–23.
[47] Montaner JS, Reiss P, Cooper D, Vella S, Harris M, Conway B, Wainberg MA, Smith D, Robinson P, Hall D, Myers M, Lange JM. A randomized, double-blind trial comparing combinations of nevirapine, didanosine, and zidovudine for HIV-infected patients: the INCAS Trial. Italy, The Netherlands, Canada and Australia Study. JAMA 1998;279:930–7.
[48] Henry K, Erice A, Tierney C, Balfour Jr HH, Fischl MA, Kmack A, Liou SH, Kenton A, Hirsch MS, Phair J, Martinez A, Kahn JO. A randomized, controlled, double-blind study comparing the survival benefit of four different reverse transcriptase inhibitor therapies (three-drug, two-drug, and alternating drug) for the treatment of advanced AIDS. AIDS Clinical Trial Group 193A Study Team. J Acquir Immune Defic Syndr Hum Retrovirol 1998;19:339–49.
[49] Barry M, Mulcahy F, Merry C, Gibbons S, Back D. Pharmacokinetics and potential interactions amongst antiretroviral agents used to treat patients with HIV infection. Clin Pharmacokinet 1999;36:289–304.
[50] Koduri PR, Parekh S. Zidovudine-related anemia with reticulocytosis. Ann Hematol 2003;82:184–5.
[51] Cooper DA. Update on didanosine. J Int Assoc Physicians AIDS Care (Chic) 2002;1:15–25.
[52] Perry CM, Noble S. Didanosine: an updated review of its use in HIV infection. Drugs 1999;58:1099–135.
[53] Shyu WC, Knupp CA, Pittman KA, Dunkle L, Barbhaiya RH. Food-induced reduction in bioavailability of didanosine. Clin Pharmacol Ther 1991;50:503–7.
[54] Perry CM, Balfour JA. Didanosine. An update on its antiviral activity, pharmacokinetic properties and therapeutic efficacy in the management of HIV disease. Drugs 1996;52:928–62.
[55] Adkins JC, Peters DH, Faulds D. Zalcitabine. An update of its pharmacodynamic and pharmacokinetic properties and clinical efficacy in the management of HIV infection. Drugs 1997;53:1054–80.
[56] Devineni D, Gallo JM. Zalcitabine. Clinical pharmacokinetics and efficacy. Clin Pharmacokinet 1995;28:351–60.
[57] Lipsky JJ. Zalcitabine and didanosine. Lancet 1993;341:30–2.
[58] Yokota T, Mochizuki S, Konno K, Mori S, Shigeta S, De Clercq E. Inhibitory effects of selected antiviral compounds on human hepatitis B virus DNA synthesis. Antimicrob Agents Chemother 1991;35:394–7.
[59] Riddler SA, Anderson RE, Mellors JW. Antiretroviral activity of stavudine (2′,3′-didehydro-3′-deoxythymidine, D4T). Antivir Res 1995;27:189–203.
[60] Hurst M, Noble S. Stavudine: an update of its use in the treatment of HIV infection. Drugs 1999;58:919–49.
[61] Johnson MA, Moore KH, Yuen GJ, Bye A, Pakes GE. Clinical pharmacokinetics of lamivudine. Clin Pharmacokinet 1999;36:41–66.
[62] Soudeyns H, Yao XI, Gao Q, Belleau B, Kraus JL, Nguyen-Ba N, Spira B, Wainberg MA. Anti-human immunodeficiency virus type 1 activity and in vitro toxicity of 2′-deoxy-3′-thiacytidine (BCH-189), a novel heterocyclic nucleoside analog. Antimicrob Agents Chemother 1991;35:1386–90.
[63] Skalski V, Chang CN, Dutschman G, Cheng YC. The biochemical basis for the differential anti-human immunodeficiency virus activity of two cis enantiomers of 2′,3′-dideoxy-3′-thiacytidine. J Biol Chem 1993;268:23234–8.
[64] Kumar PN, Patel P. Lamivudine for the treatment of HIV. Expert Opin Drug Metab Toxicol 2010;6:105–14.
[65] Amblard F, Patel D, Michailidis E, Coats SJ, Kasthuri M, Biteau N, Tber Z, Ehteshami M, Schinazi RF. HIV nucleoside reverse transcriptase inhibitors. Eur J Med Chem 2022;240, 114554.
[66] Faletto MB, Miller WH, Garvey EP, St Clair MH, Daluge SM, Good SS. Unique intracellular activation of the potent anti-human immunodeficiency virus agent 1592U89. Antimicrob Agents Chemother 1997;41:1099–107.
[67] Hewitt RG. Abacavir hypersensitivity reaction. Clin Infect Dis 2002;34:1137–42.
[68] Anderson PL, Kiser JJ, Gardner EM, Rower JE, Meditz A, Grant RM. Pharmacological considerations for tenofovir and emtricitabine to prevent HIV infection. J Antimicrob Chemother 2011;66:240–50.
[69] Schinazi RF, Lloyd Jr RM, Nguyen MH, Cannon DL, McMillan A, Ilksoy N, Chu CK, Liotta DC, Bazmi HZ, Mellors JW. Characterization of human immunodeficiency viruses resistant to oxathiolane-cytosine nucleosides. Antimicrob Agents Chemother 1993;37:875–81.

[70] Opio CK, Lee WM, Kirkpatrick P. Entecavir. Nat Rev Drug Discov 2005;4:535–6.

[71] McMahon MA, Jilek BL, Brennan TP, Shen L, Zhou Y, Wind-Rotolo M, Xing S, Bhat S, Hale B, Hegarty R, Chong CR, Liu JO, Siliciano RF, Thio CL. The HBV drug entecavir - effects on HIV-1 replication and resistance. N Engl J Med 2007;356:2614–21.

[72] Domaoal RA, McMahon M, Thio CL, Bailey CM, Tirado-Rives J, Obikhod A, Detorio M, Rapp KL, Siliciano RF, Schinazi RF, Anderson KS. Pre-steady-state kinetic studies establish entecavir 5′-triphosphate as a substrate for HIV-1 reverse transcriptase. J Biol Chem 2008;283:5452–9.

[73] Tchesnokov EP, Obikhod A, Schinazi RF, Götte M. Delayed chain termination protects the anti-hepatitis B virus drug entecavir from excision by HIV-1 reverse transcriptase. J Biol Chem 2008;283:34218–28.

[74] Roade L, Riveiro-Barciela M, Esteban R, Buti M. Long-term efficacy and safety of nucleos(t)ides analogues in patients with chronic hepatitis B. Ther Adv Infect Dis 2021;8.

[75] De Clercq E. The design of drugs for HIV and HCV. Nat Rev Drug Discov 2007;6:1001–18.

[76] Wonganan P, Limpanasithikul W, Jianmongkol S, Kerr SJ, Ruxrungtham K. Pharmacokinetics of nucleoside/nucleotide reverse transcriptase inhibitors for the treatment and prevention of HIV infection. Expert Opin Drug Metab Toxicol 2020;16:551–64.

[77] Ray AS, Fordyce MW, Hitchcock MJ. Tenofovir alafenamide: a novel prodrug of tenofovir for the treatment of human immunodeficiency virus. Antivir Res 2016;125:63–70.

[78] Gibson AK, Shah BM, Nambiar PH, Schafer JJ. Tenofovir Alafenamide. Ann Pharmacother 2016;50:942–52.

[79] Naesens L, Snoeck R, Andrei G, Balzarini J, Neyts J, De Clercq E. HPMPC (cidofovir), PMEA (adefovir) and related acyclic nucleoside phosphonate analogues: a review of their pharmacology and clinical potential in the treatment of viral infections. Antivir Chem Chemother 1997;8:1–23.

[80] Ha B, Larsen KP, Zhang J, Fu Z, Montabana E, Jackson LN, Chen DH, Puglisi EV. High-resolution view of HIV-1 reverse transcriptase initiation complexes and inhibition by NNRTI drugs. Nat Commun 2021;12:2500.

[81] Milinkovic A, Martínez E. Nevirapine in the treatment of HIV. Expert Rev Anti-Infect Ther 2004;2:367–73.

[82] Podzamczer D, Fumero E. The role of nevirapine in the treatment of HIV-1 disease. Expert Opin Pharmacother 2001;2:2065–78.

[83] Esnouf R, Ren J, Ross C, Jones Y, Stammers D, Stuart D. Mechanism of inhibition of HIV-1 reverse transcriptase by non-nucleoside inhibitors. Nat Struct Biol 1995;2:303–8.

[84] Dueweke TJ, Poppe SM, Romero DL, Swaney SM, So AG, Downey KM, Althaus IW, Reusser F, Busso M, Resnick L, et al. U-90152, a potent inhibitor of human immunodeficiency virus type 1 replication. Antimicrob Agents Chemother 1993;37:1127–31.

[85] Esnouf RM, Ren J, Hopkins AL, Ross CK, Jones EY, Stammers DK, Stuart DI. Unique features in the structure of the complex between HIV-1 reverse transcriptase and the bis(heteroaryl)piperazine (BHAP) U-90152 explain resistance mutations for this nonnucleoside inhibitor. Proc Natl Acad Sci U S A 1997;94:3984–9.

[86] Maggiolo F. Efavirenz: a decade of clinical experience in the treatment of HIV. J Antimicrob Chemother 2009;64:910–28.

[87] Vrouenraets SM, Wit FW, van Tongeren J, Lange JM. Efavirenz: a review. Expert Opin Pharmacother 2007;8:851–71.

[88] Gardner EM, Hullsiek KH, Telzak EE, Sharma S, Peng G, Burman WJ, MacArthur RD, Chesney M, Friedland G, Mannheimer SB, Terry Beirn Community Programs for Clinical Research on AIDS and the International Network for Strategic Initiatives in Global HIV Trials. Antiretroviral medication adherence and class- specific resistance in a large prospective clinical trial. AIDS 2010;24:395–403.

[89] Das K, Clark Jr AD, Lewi PJ, Heeres J, De Jonge MR, Koymans LM, Vinkers HM, Daeyaert F, Ludovici DW, Kukla MJ, De Corte B, Kavash RW, Ho CY, Ye H, Lichtenstein MA, Andries K, Pauwels R, De Béthune MP, Boyer PL, Clark P, Hughes SH, Janssen PA, Arnold E. Roles of conformational and positional adaptability in structure-based design of TMC125-R165335 (etravirine) and related non-nucleoside reverse transcriptase inhibitors that are highly potent and effective against wild-type and drug-resistant HIV-1 variants. J Med Chem 2004;47:2550–60.

[90] Figueiredo A, Moore KL, Mak J, Sluis-Cremer N, de Bethune MP, Tachedjian G. Potent nonnucleoside reverse transcriptase inhibitors target HIV-1 Gag-Pol. PLoS Pathog 2006;2, e119.

[91] Deeks ED, Keating GM. Etravirine. Drugs 2008;68:2357–72.

[92] Côté B, Burch JD, Asante-Appiah E, Bayly C, Bédard L, Blouin M, Campeau LC, Cauchon E, Chan M, Chefson A, Coulombe N, Cromlish W, Debnath S, Deschênes D, Dupont-Gaudet K, Falgueyret JP, Forget R, Gagné S, Gauvreau D, Girardin M, Guiral S, Langlois E, Li CS, Nguyen N, Papp R, Plamondon S, Roy A, Roy S, Seliniotakis R, St-Onge M, Ouellet S, Tawa P, Truchon JF, Vacca J, Wrona M, Yan Y, Ducharme Y. Discovery of MK-1439, an orally bioavailable non-nucleoside reverse transcriptase inhibitor potent against a wide range of resistant mutant HIV viruses. Bioorg Med Chem Lett 2014;24:917–22.

Chapter 2.3

Integrase

Fabrizio Carta[a] and Mario Sechi[b]

[a]NEUROFARBA Department, Section of Pharmaceutical and Nutraceutical Sciences, University of Florence, Florence, Italy, [b]Department of Medicine, Surgery and Pharmacy, Laboratory of Drug Design and Nanomedicine, University of Sassari, Sassari, Italy

1 Introduction

The Baltimore virus classification method, currently adopted from the International Committee on Taxonomy of Viruses (ICTV), organizes the ribonucleic acid (RNA) containing viruses within groups III–VI on the basis of genome pairing (i.e., single-stranded RNA (ssRNA) and double-stranded RNA (dsRNA)) and sign [1–3]. Additional taxonomic features are also considered in order to discriminate each virus among the others [1–3]. All RNA viruses do replicate their genomes using their viral encoded and highly conserved RNA-dependent RNA polymerase (RdRp) enzymes. The RdRps and associated proteins required for viral genome synthesis to take place are properly refereed as the "replicase complex." In some viral strains, the replicase complexes may also contain the RNA-helicase enzymes, which are necessary to unwind highly base-paired regions of the RNA genomes, and the NTPases as energy suppliers of the polymerization cycle processes [4].

RNA viruses include a wide variety of human- and animal-affecting pathogens such as influenza, coronaviruses, HIVs, measles, and rabies [1,4]. A key feature of all RNA viruses when compared to the DNA counterparts is represented by their higher mutation rates, which make them very quick in adaptation to changing environments [5,6]. Elevated mutation rates usually are associated with the emergence of unprecedently reported viral strains, and that may easily cause outbreaks or pandemics when uncontrolled diffusion of the virus among human/animal/plant population takes place [5–7]. RNA viruses also have relatively simple replication cycles when compared to the DNA-containing ones [6]. Along the RNA viruses, the common replicative process of the family of retroviruses (e.g., human immunodeficiency virus 1 (HIV-1), avian sarcoma leukosis virus (ASLV), and murine leukemia virus (MLV)) includes the reverse transcription of the virion RNA into double-stranded DNA and its subsequent integration into the genome of the host cell [6].

Fig. 1 reports schematically the main steps that constitute a retrovirus replication cycle [6].

Thus, all retroviruses are all bound by the need of integrating their genomes within the host cellular DNA in order to replicate. The integration process is efficiently mediated by the viral enzyme integrase (IN; EC 2.7.7.49), and this chapter discusses the structural aspects of it, the catalytic mechanisms along with its role in the management of RNA viral promoted infection diseases. Since IN enzymes are highly conserved among the retroviruses, herein we refer to the HIV-1 expressed IN enzyme, which to the best of our knowledge is the best characterized so far. Claims to different viral INs will be indicated as needed.

2 Structure of IN

Within the general context of retroviral RNA virus replication cycle (Fig. 1), the linear double-stranded viral DNA (v-DNA) obtained by transcription of the native v-RNA genome exists at cytoplasm level in the form of the preintegration complex (PIC). Although known, PIC entities are vaguely characterized as on acute viral infections just a single copy per cell is usually retrieved [8]. To the best of our knowledge, PICs are quite complex units, which vary among viral strains and have been reported to incorporate host cellular and viral proteins, which among others is the IN [9–17]. Once the PICs gain access into the cellular nucleus, the v-DNA is inserted into a cellular chromosome by means of the IN enzyme, and then the process is completed by the host cell DNA repairing machinery, thus allowing the viral genome to be fully integrated within the cell host [9–17]. Within the PIC, the Intasome (INT) is defined, which properly refers to the minimal PIC substructure that catalyzes viral genome integration in vitro [18,19]. To date, the level of complexity of various INTs has been deciphered also by means of X-ray crystallographic and single-particle cryogenic electron microscopy (cryo-EM) experiments [20]. For the sake of clarity, it is enough to report in this chapter that INTs organization is highly specific for each retrovirus and their related mutant strains [20].

Metalloenzymes. https://doi.org/10.1016/B978-0-12-823974-2.00006-1

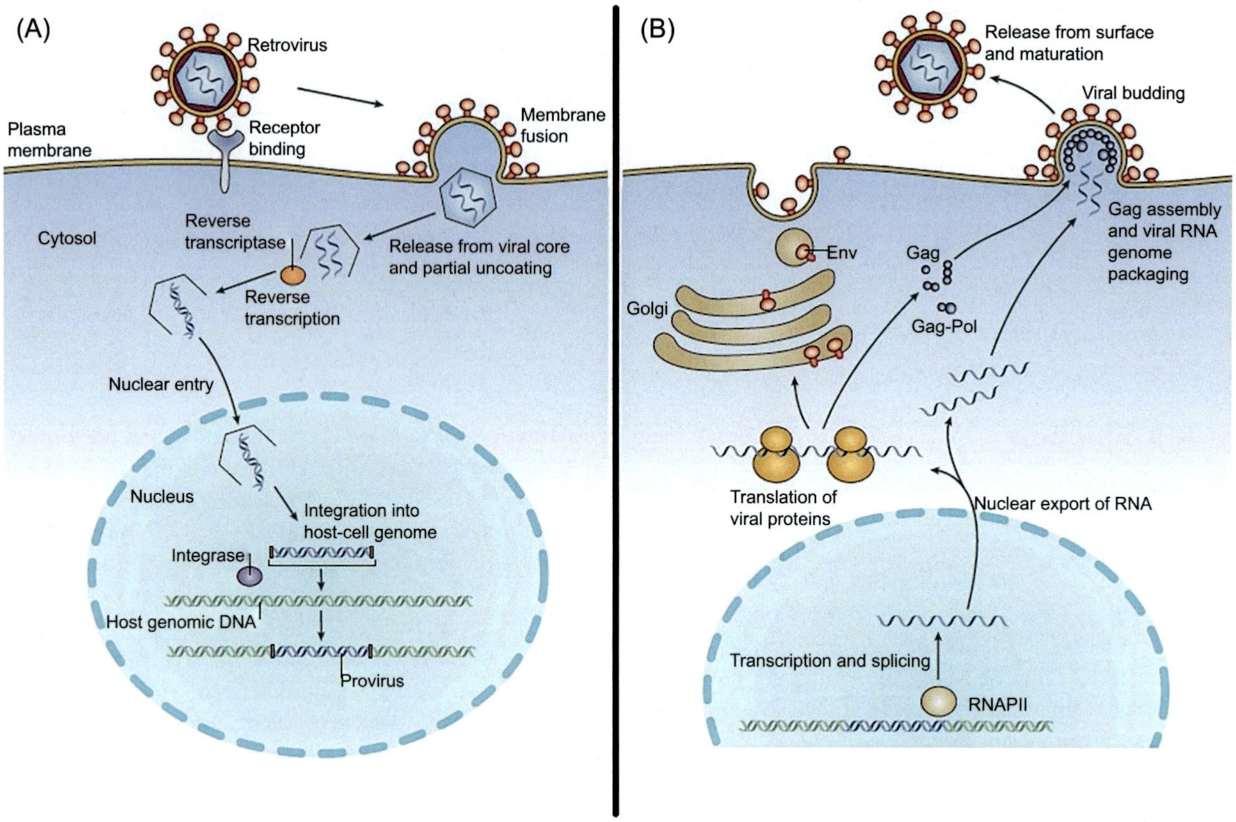

FIG. 1 Replication cycle of retroviruses: (A) Main steps for retrovirus entry within an eukaryotic cell and (B) for exit from it.

IN is a member of the tyrosine recombinase enzyme family and accounts for three highly conserved structural domains, which are schematically represented in Fig. 2 and named as the N-terminal domain (NTD), the catalytic core domain (CCD), and the C-terminal domain (CTD).

All three domains have been distinctively characterized by means of NMR and X-ray crystallographic experiments [21–27]. Specifically, the NTD-IN domain is constituted by three compact α-helices stabilized by coordination of a Zn(II) ion with the conserved HHCC motif [21,24]. The NTD domain appears extended up to ~40 amino acid residues in spumaviral ε- and γ-retroviral INs [28].

As for the CCD domain, it comprises the IN enzyme active site and reports the highly conserved DDE motif (Fig. 2), which is responsible in coordinating two divalent cations, almost certainly magnesium, to deprotonate the attacking oxygen nucleophiles and destabilize scissile phosphodiester bonds for the single-step trans-esterification process [29]. The crystal structures of the HIV-1 IN CCD section show a protein folding motif typical of the nucleotidyl transferase enzymes [22]. Most of the solved CCD-IN crystal structures reported in the literature showed this section being organized as a dimer and according to variable multimeric enzyme organizations (Fig. 3). The large spatial separation of the IN active sites within the CCD dimer are themselves indicative for high-order of protein multimeric organization, which is part of the functional INT being responsible for the v-DNA integration process.

Particularly challenging and yet to be clarified is the required IN multimeric functional organization within the INTs. To date, a tetrameric assembly for the HIV-1 IN has received the strongest experimental support [30,31], although superior aggregation states, such as the octameric, were also reported [32,33]. Quite of interest are the studies on variegate retroviral genera, which do account for different IN functional multimeric aggregations [34–37]. The purpose of this chapter is enough to state that the CCD domain dimerization is the basis for IN functional multimeric aggregations within the INT, and each dimer is interlocked to the other at the NTD-CCD level as reported in Fig. 3.

Among the three IN domains, CTD is the least conserved one. However, it still retains important features such as the β-folding of Src homology 3 (SH3) [23,25] and a closely related organization typically retrieved within the Tudor family domains, which are known to be implicated in chromatin-binding processes [23,25].

During the viral genome integration, both the NTD and the CTD are critically implicated in engaging and locking the v-DNA substrate at the INT complex [38,39].

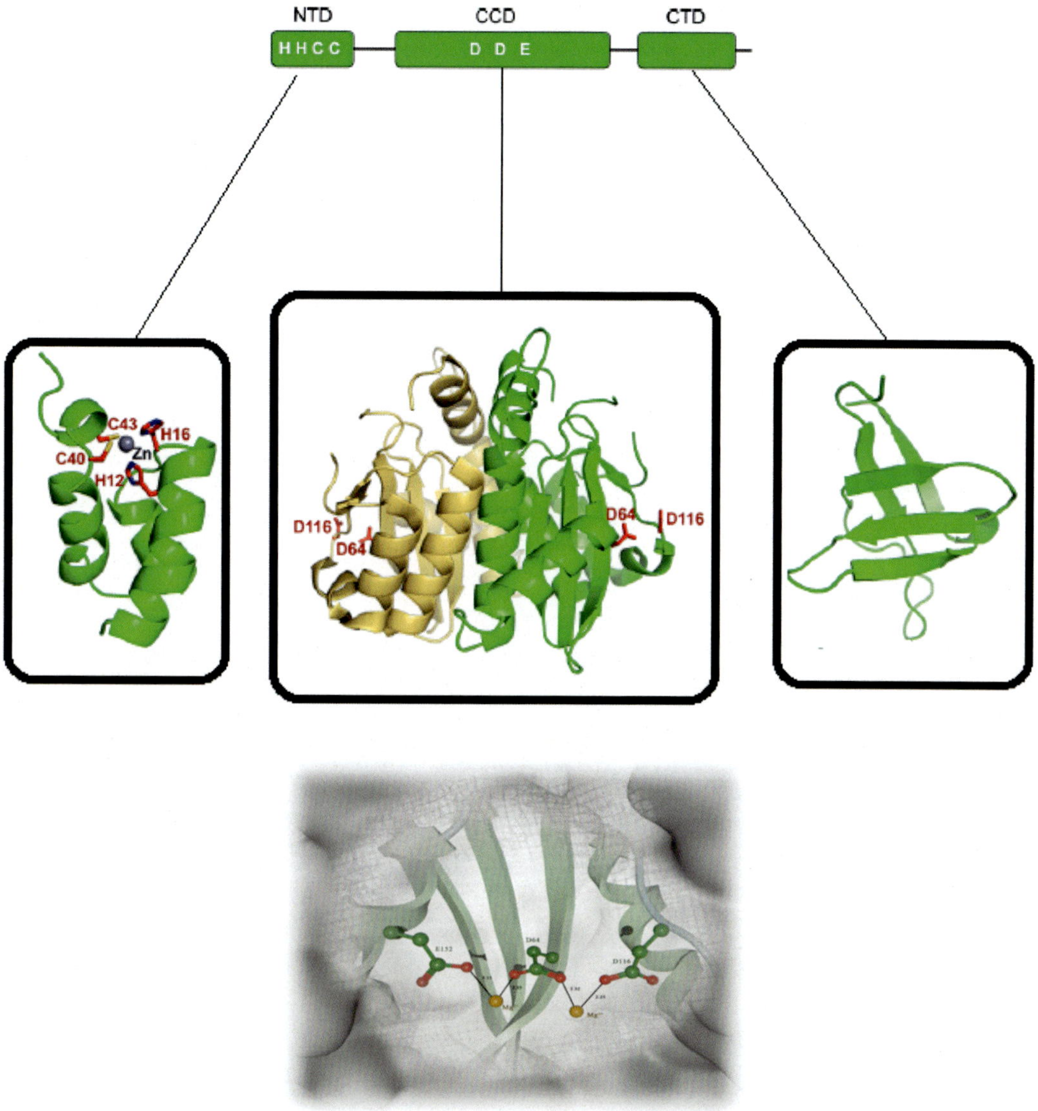

FIG. 2 A schematic diagram of the three domains of HIV1-IN (top). Structural representation of these domains of the enzyme (center), the catalytic core of HIV-1 IN containing the DDE motif (bottom).

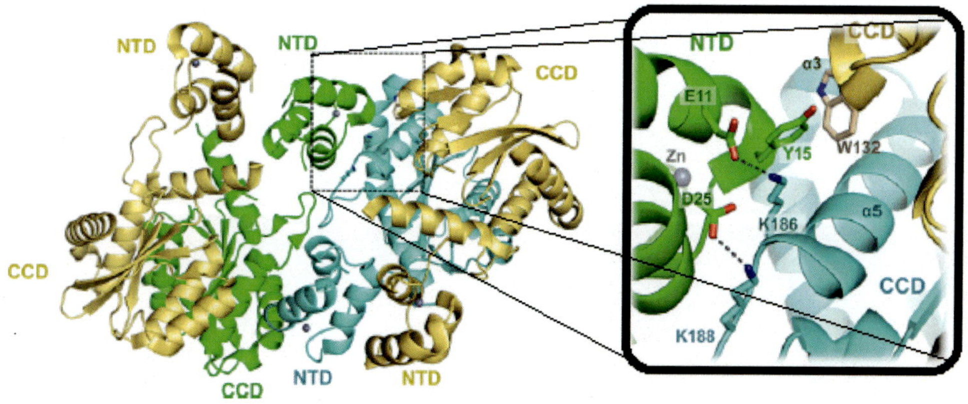

FIG. 3 Tetrameric organization of HIV1-INs based on CCD dimerization and details of the key contacts at the CCD-NTD interface (PDB ID **1K6Y**).

Particularly relevant in the v-DNA engagement is the high degree of flexibility of the CDD connecting linkers to the external domains (i.e., NTD and CDT), which also appear highly variable in length among retroviruses [28,40].

3 Reactions catalyzed by the IN enzyme

Two main reactions are associated with the IN enzyme within the PICs, namely, the 3′-OH processing and the strand transfer [41–44], and are schematically reproduced in Fig. 4.

During the 3′-OH processing step, the IN enzyme first recognizes and binds the v-DNA at specific recognition sites and then cleaves the 3′-OH ends of the v-DNA, thus creating a 3′-OH group at each strand that is therefore available for nucleophilic attack. These operations take place at the cytoplasm level. Next, the PIC translocates into the cellular nucleus, the IN enzyme binds to the host DNA at the attachment sites and uses the 3′-OH group on the v-DNA to attack the phosphodiester bond between the 3′-OH group on the host DNA strand and the 5′-phosphate group on the next nucleotide. This results in a covalent linkage between the v-DNA and the host cellular DNA, with effective integration of the v-DNA into the host genome. This step is therefore crucial for the replication of the virus, as it allows the viral genetic material to be passed onto the host cell's progeny. The specificity of the attachment site is determined by the virus and the associated IN enzyme, allowing integration to take place in a specific location of the host genome.

Since the 3′-processing of the v-DNA takes place in the cellular cytoplasm, the elevated local concentration of linear v-DNAs surrounding the PICs may represent a problem for retroviruses replication as they may act competitively as potential targets for the strand transfer process to give self-integration end-products. Such an effect is clearly detectable experimentally in PICs, and it was widely reported in cellular models of infection [45–47]. Quite interesting, variegate retrovirus classes seem to have developed highly efficient strategies to avoid autointegration of their genomes. An example is the highly conserved cellular protein named Barrier-to-Autointegration Factor (BAF). Such a molecule is a small DNA-binding protein able to bridge and to condense separate DNA molecules [48,49], thus suppressing the autointegration activity as observed for murine leukemia viruses (MLV) PICs in vitro [50]. Although BAF demonstrated excellent autointegration inhibition ability in vitro, still it has to be determined whether it

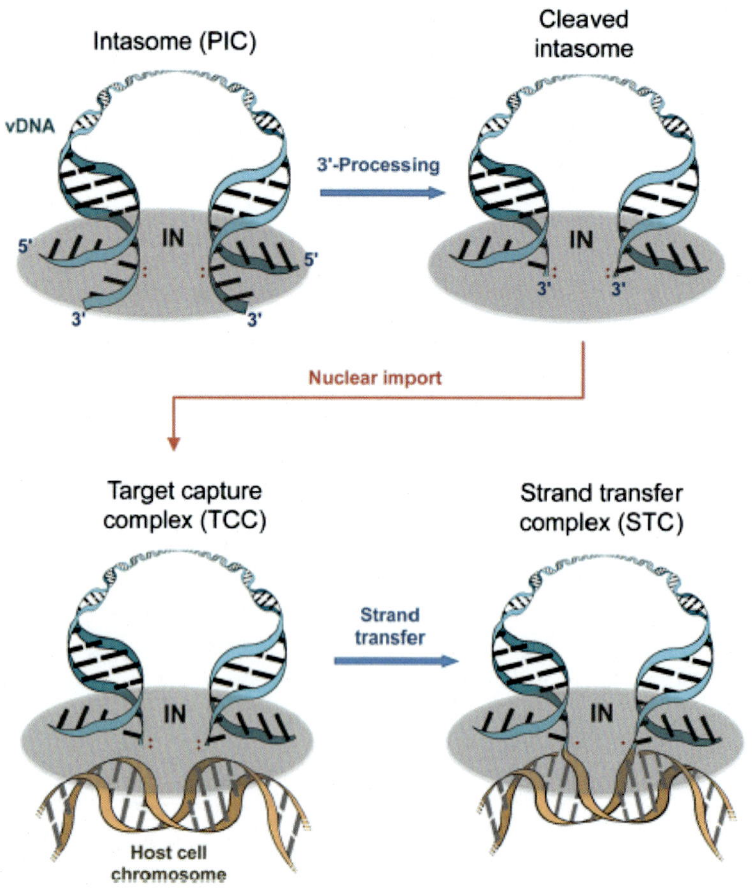

FIG. 4 Integrase mechanisms.

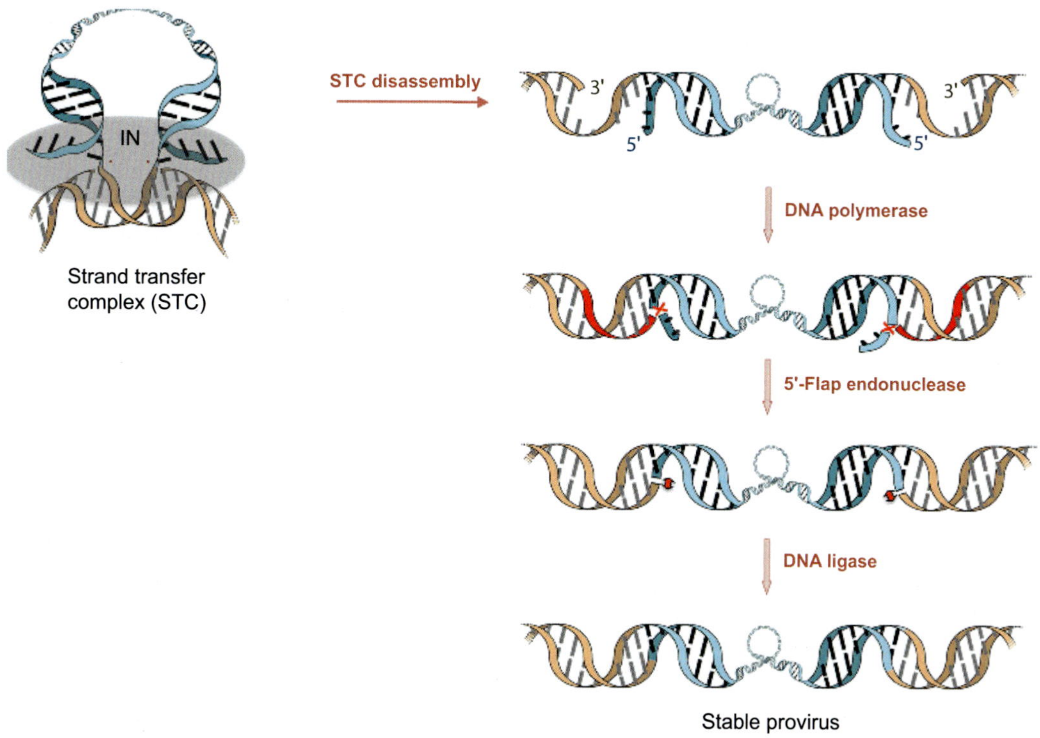

FIG. 5 Post integration DNA repairing processes.

efficiently acts during MLV promoted infections on cells [50]. Alternative autointegration preventive processes have been identified among retroviruses. The SET complex, which is an endoplasmic reticulum–associated complex containing three distinct DNase enzymes, was reported to suppress autointegration during HIV-1 infection [51,52]. Alternatively, the viral capsid protein was reported to play a key role in regulating simian immunodeficiency virus (SIV) autointegration processes [47]. Besides such specific examples, it is general opinion that different retroviral species have evolved unique ways to protect themselves from the common problem of suicidal genome integration.

The occurrence of aberrant strand transfer events (i.e., half-site integration processes) seems particularly relevant when in vitro functional IN PICs are considered, whereas they are scarcely associated with naturally occurring cellular PICs [53–60].

Finally, IN is also known to catalyze a reversal genome integration reaction properly referred as "disintegration" [61–70]. It is well supported that such an event is clearly associated only with experimentally reproduced PICs, whereas no activities as such were reported when in vivo infections take place [61–70].

Once the v-DNA is physically bound to the host genome, a series of DNA-repairing activities need take place in order to ensure complete merge. As a matter of fact, the strand transfer process catalyzed by the IN enzyme leaves a genome containing at the newly formed junction sites a pair of single-stranded gaps and short 5′ overhangs flanking the original v-DNA (Fig. 5).

A preliminary and necessary step prior to DNA repairing involves dismantle of the highly thermodynamic stable strand transfer complex (STC). Although marginally investigated, such an event is vital for ensuring successful viral replication. To the best of our knowledge, only one seminal study is reported on the subject [71]. Specifically the authors did demonstrate that the von Hippel-Lindau binding protein 1 is effectively implicated in proteasome-mediated HIV-1 IN degradation with subsequent exposure of the DNA to the activities of a DNA polymerase, a 5′ flap endonuclease, and a ligase enzyme, respectively (Fig. 5).

A putative role of IN in postintegration DNA repairing has been postulated [61,72,73], but evidence on its effectiveness in infection models is still missing.

4 IN-strand transfer inhibitors (INSTIs)

To date, the combination antiretroviral therapy (cART) accounts for the association of drugs, which belong to four distinct classes: (i) nucleoside reverse transcriptase (RT) inhibitors (NRTIs), (ii) nonnucleoside RT inhibitors (NNRTIs), (iii) protease (PR) inhibitors (PIs), (iv) and the last introduced, the IN-strand transfer inhibitors (INSTIs). The introduction into the clinics of INSTIs represented a true turning point in the cART therapy as such drugs are the only ones able to target a specific process, which is exclusive and essential for retroviruses (i.e., the viral genome integration) [74].

HIVs-directed cART in the timeframe of 1990s–early 2000s was represented by the NRTIs, NNRTIs, and PIs, whose clinical efficiency was drastically reduced by multi- and cross-drug resistances mainly ascribed to specific mutations on the target proteins. The discovery of the HIV-IN at first and the development of appropriate and reliable in vitro screening assays for small molecules came at a crucial time when the loss of cART effectiveness was seriously impacting HIV-affected clinical outcomes at global scale thus with a tangible risk of pandemic.

The chemical structures of currently used INSTIs are reported in Fig. 6.

Despite the classification in two generations, all clinically used INSTIs are structurally related to the experimental compounds 1-(5-chloroindol-3-yl)-3-hydroxy-3-(2*H*-tetrazol-5-yl)propenone **6** (**5-CITEP**) as do share the key diketoacid (DKA) moiety (Fig. 7) [75–77].

5-CITEP is an α,γ-DKA with the carboxylic acid functionality replaced by the acidic bioisosteric tetrazole ring. The ligand showed for the IN IC_{50} experimental values of 35 μM for the 3′-processing and 0.65 μM for the strand transfer respectively [77]. In addition, **5-CITEP** is the first prototype IN inhibitor successfully co-crystallized within the CCD domain [75]. Clear density maps accounted for the **5-CITEP** bound within the IN-CCD domain and more precisely allocated between D64, D116, and E152 residues. In addition, **5-CITEP** is engaged in several contacts with the CCD residues responsible in anchoring the host DNA.

Parallel research in the field allowed Merck to disclose in the same period derivatives **L-708,906** [78] and **L-731,988** [79], with the latter being particularly potent in inhibiting the HIV-1 IN in vitro with remarkable selectivity for the strand transfer process over the 3′-processing (i.e., IC_{50} of 80 nM and 6 μM respectively) [78,79]. Remarkable results on infected cells were obtained for the DKA compound series containing the 8-hydroxy-[1,6]-naphthyridine-7-carboxamide moiety, and among them the derivative **L-870,810** resulted potent IN strand transfer inhibitor (IC_{50} of 8 nM) and endowed with effective antiviral activity on cell-based assays (i.e., EC_{95} = 15 nM) [80]. Being the first INSTI that showed anti-HIV activity in experimental animal models, **L-870,810** entered clinical

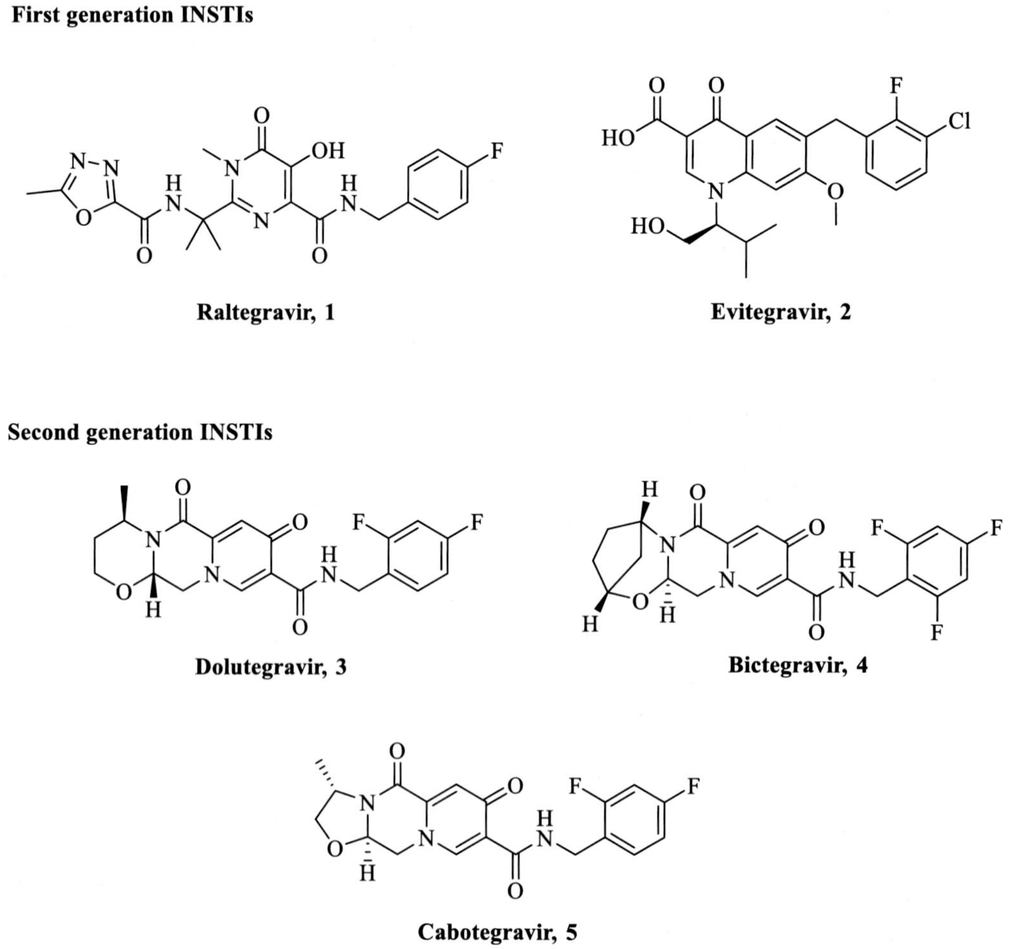

FIG. 6 Chemical structures of the experimental INSTI clinically used **Raltegravir 1**, **Elvitegravir 2**, **Dolutegravir 3**, **Bictegravir 4**, and **Cabotegravir 5**.

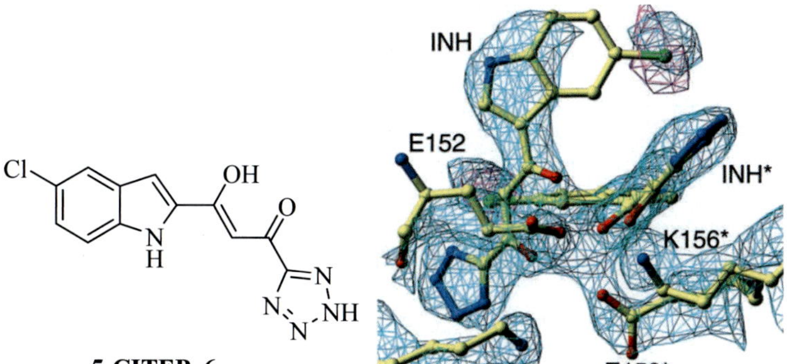

5-CITEP, 6

FIG. 7 Chemical structure of the experimental INSTI **5-CITEP 6** and electron density maps of the IN CCD active site region in complex with it [75].

trials, which however failed as elevated liver and kidney toxicity was reported in dogs [81,82]. Better results were obtained with the DKA derivative (*Z*)-1-[5-(4-fluorobenzyl)furan-2-yl]-3-hydroxy-3-(1*H*-1,2,4-triazol-3-yl)propenone **S-1360**, which was jointly developed by Shionogi (Osaka, Japan) and GlaxoSmithKline (London, United Kingdom). Compound **S-1360** exhibited in vitro an IC_{50} of 20nM for the strand transfer process, and it resulted highly effective in inhibiting the HIV viral replication on MTT assays with EC_{50} and CC_{50} values of 200nM and 12μM, respectively [83]. Although preclinical assays on **S-1360** gave satisfactory pharmacological, pharmacokinetic, safety, and toxicological outcomes in animal models, it suffered rapid metabolism and clearance via a non-CYP450-mediated pathway in humans [84]. The joint venture between Shionogi and GlaxoSmithKline was successful also in developing the 4-hydroxy-2-oxo-1,2-dihydro-1,5-naphthyridine GSK364735 [85], which resulted in a remarkable INSTI and effective in suppressing virus replication in cell assays [85]. Unfortunately, such a compound was discontinued from clinical development due to long-term safety studies in monkeys, which reported serious hepatotoxicity [86].

All the aforementioned compounds although failed to enter into the clinical armamentarium, resulted highly valuable within the research field as allowed scientists to assess in deep detail the intimate mechanisms of IN for the 3′-processing and the strand transfer catalyzed reactions. For the purposes of this chapter, it is enough to recapitulate that compounds in Figs. 7 and 8 strongly contributed to elucidate that: (i) the v-DNA integration process most likely relies on two magnesium(II) ions for the assembly of stable and functional PICs; (ii) one metal ion appears coordinated by D64 and D116 residues, whereas for the second, the D116 and E152 residues are involved in metal coordination [87,88]; (iii) the DKA-containing molecules being effective inhibitors on the IN enzyme are all characterized by a γ-ketone, an enolizable α-ketone and a carboxylic acid moiety. The latter can be substituted with acidic bioisosteric functional groups such as the tetrazole and triazole or basic ones (i.e., pyridines) [87–89]. Such chemical features endow DKAs with potent antiviral ability by means of divalent metal ion chelating features within the IN CCD active site [87–89]. Besides the well-known mono-DKA-containing molecules, large series of dimeric DKAs and triketoacids (TKAs) were explored with the intent to decipher their role of divalent metal ion chelation as potentially useful for the inhibition of the IN in various states of supramolecular organization [90]. In any case, dimeric-DKAs and TKAs reported moderate IN activities, much lower when compared to DKAs [91].

5 First-generation INSTIs

Raltegravir was the first INSTI approved by the US-FDA in 2007 for the treatment of HIV-sustained infections. [92–95]. The compound was discovered and developed at the Merck research laboratories in Pomezia (IT) as a result of important research investments directed to sustain a wide R&D program on HCV polymerases and based on small molecules bearing the DKA moiety. As a matter of fact, Raltegravir resulted ineffective on the HCV polymerase, instead it revealed a potent, reversible, and selective INSTI with an IC_{50} value of 0.085μM [96]. Raltegravir resulted highly effective on clinical uses when administered orally at the dosage of 400mg twice/day, showed good tolerability, safety profiles and was devoid of any significant drug interactions [92,95,97,98]. The wide use of Raltegravir after its approval determined rapidly the appearance of viral strains bearing mutations at the IN-CCD domain. The most important are the Q148/H/K/R, E138A/K, G140A, T66A, Q95K, Y143C/R, and the N155H [99]. Three signature resistant-associated mutations such as the Y143C/R±T97A, Q148H/K/R±G140S/A, and the N155H ±E92Q have been classified and defined as responsible of 10-fold reduced susceptibility of IN-CCD to Raltegravir [92], whereas the Q148K, the E138, and the G140A mutations were identified as responsible for the drug-reduced susceptibility up to several hundred-folds.

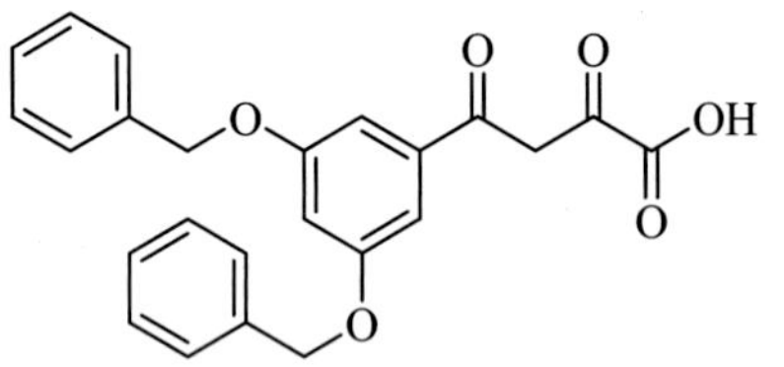

L-708,906, 7

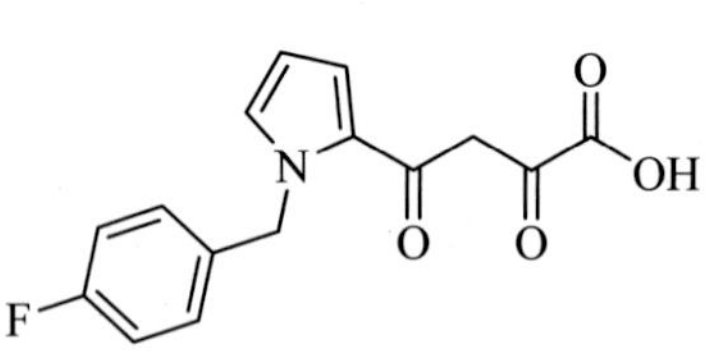

L-731,988, 8

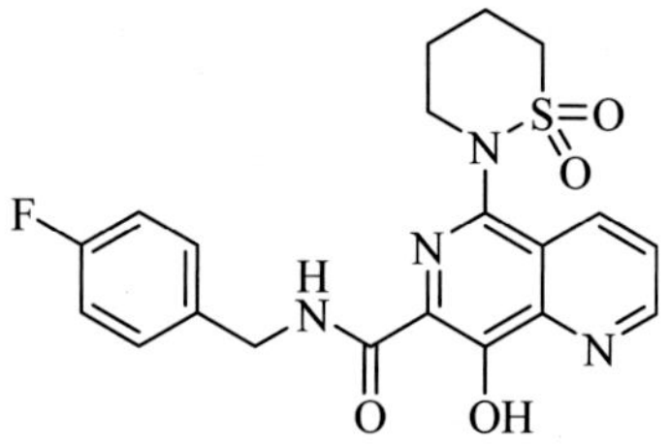

L-870,810, 9

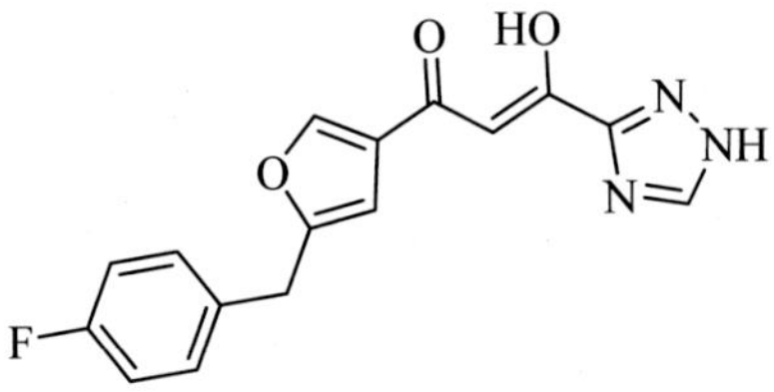

S-1360, 10

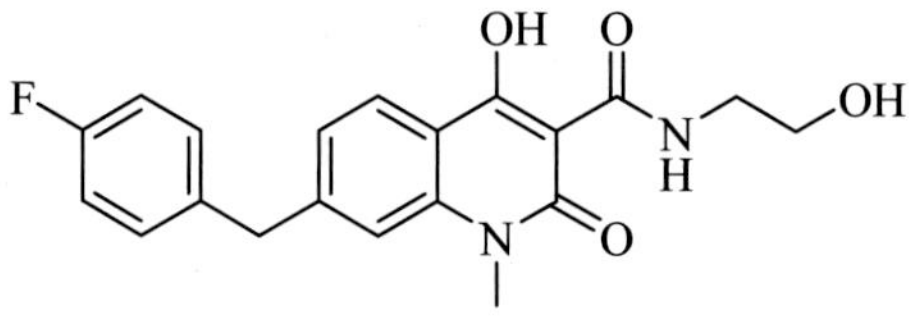

GSK364735, 11

FIG. 8 Chemical structures of DKA-containing compounds **7–11**.

Eviltegravir soon followed Raltegravir for approval by the FDA for the management of HIV-sustained infections. It was initially developed by the Central Pharmaceutical Research Institute of Japan Tobacco, Inc. (Osaka, Japan) and later licensed to Gilead Sciences (CA, United States) for clinical development. Eviltegravir is chemically derived from antibiotics of the quinolone type and makes use the 4-quinolone-3-carboxylic acid instead of the DKA moiety. Eviltegravir showed very potent IN inhibitory activity for the strand transfer with an IC_{50} value of 7.2nM, and even more interestingly, it resulted very effective on acute HIV-1 infection assay being active with an EC_{50} of 0.9nM [100,101]. Unfortunately, cross-resistance mutations between Raltegravir and Eviltegravir were reported, and this abolished the possibility to switch drugs when specific compound treatment failures take place. Eviltegravir was licensed in the United States and EU in a single table formulation named STRIBILD, which contains the pharmacokinetic enhancer cobicistat (COBI) able to inhibit the CYP3A4 enzyme, the NRTIs emtricitabine (FTC) and tenofovir disoproxil fumarate (TDF) [102]. Furthermore, a lower-dosed single-tablet regimen was approved (i.e., Genvoya) and showed significant reduced bone and renal side effects compared with STRIBILD [103]. The development of Eviltegravir and its formulation into a single-tablet dosage regimen represented a very step further the management of HIV-1-sustained infection.

6 Second-generation INSTIs

Compounds classified as second-generation INSTIs are all characterized by chemical structures able to fully occupy the IN active-site regions spanning from the DNA-binding on one side up to the connector sequence that links the IN structural elements β4 and α2 strands. INSTIs clinically licensed for therapeutic use are reported within this section. Experimental compounds actually under clinical trial investigation at different stages are not discussed since their chemical features and preliminary disclosed biomedical data are largely superimposable with known art.

Dolutegravir was jointly discovered and developed by Shionogi and GlaxoSmithKline [104,105] and marketed by the latter as a 50mg tablet under the brand name Tivicay R. The introduction of Dolutegravir into the market was a successful reply to genetic cross-resistances observed for the first-generation INSTIs Raltegravir and Eviltegravir. The chemical strategy behind Dolutegravir is connected to the optimization of a series of carbamoyl pyridone analogs, which do retain a two-metal chelation ability within the IN catalytic active site [104]. Important structure-activity relationships (SARs) account for the tricyclic carbamoyl pyridine as an essential group, which shares its oxygen-derived lone pairs to coordinate the two divalent metal ions within the IN active site [104]. Noteworthy is the carbonyl at the 5-position carboxamide, which is not directly involved in the IN metal coordination, whereas it endows the drug scaffold with enhanced structural flexibility, thus resulting in a better drug embedment into the enzyme active. Such a conformational freedom allows Dolutegravir to readjust its allocation within the enzymatic cleft in response to structural changes due to mutations typical of IN first-generation exposure [106].

Cabotegravir was developed cooperatively by Shionogi-ViiV Healthcare and GSK [107], and it was licensed by the US-FDA at the fall of 2021 [108] as preexposure prophylaxis agent to reduce the risk of sexually acquired HIV viruses. From the chemical viewpoint, Cabotegravir is structurally related to Dolutegravir [109,110] but endowed with superior genetic barrier to resistance, a better pharmacokinetic profile [110], and more importantly, it has half-life of 30h. Such features do allow Cabotegravir for low-dosage administrations up to once per month when parenteral nanosuspension formulation is used (i.e., Apretude) [111].

Bictegravir was first approved in 2018 as a potent INSTI (i.e., IC_{50} of 7.5nM) [112,113] and represented a good therapeutic achievement as it reported genetic barrier to resistances superior to first-generation INSTIs and Dolutegravir. More interestingly, Bictegravir exhibited synergistic in vitro antiviral effects when combined with the N/NRTIs tenofovir alafenamide (TAF), emtricitabine (FTC), or the PI Darunavir [112]. Gilead Sciences contributed by developing the single tablet Biktarvy, which associates Bictegravir with FTC and TAF [114].

7 New perspectives for IN inhibition

7.1 Allosteric IN inhibitors (ALLINIs)

Despite the great advantages to the cART therapy when INSTIs were introduced in clinic, state-of-the-art panorama accounts for continuous developing of cross-resistances, which are further facilitated since they all share the same mechanism of action. In this contest are the allosteric INIs (ALLINIs). ALLINIs bind to the IN in regions distinct from the catalytic site, and they determine disruption/inhibition of its enzymatic activities (i.e., 3′-processing and strand transfer) by altering the interactions occurring between IN and cellular cofactors necessary for the PIC/intasome to be functional [115–117]. To date, variegate chemical species have been identified as ALLINIs, and among them the majority have shown to act by tethering together the subunits of INs, thus promoting the formation of stable high-order enzyme multimers. Very general structural information is reported for such aberrant IN multimerization complexes, and any related knowledge on the mechanisms underpinning ALLINIs mode of action remains scarce. For the purposes of this chapter, it is relevant to consider that: (i) each specific aberrant enzymatic multimerization is consistently observed for determined chemical classes of ALLINIs and (ii) such an unnatural structural organization, although thermodynamically very effective, strongly affects the functional dynamic flexibility between IN subunits [118]. Quite recently were reported ALLINIs that bind to the interface of CCD IN domain and induce disruption of the IN activity by promoting aberrant multimerizations and by competing with the natural ligands. Below are reported the most important ALLINI classes.

7.2 LEDGF/p75

The majority of ALLINIs target the LEDGF/p75 cellular cofactor, which was firstly discovered as an essential cellular component able to improve the interaction of the host DNA with the PIC [119]. LEDGF/p75 binds to a specific site located at the IN protein C-terminus called integrase binding domain (IBD) by coordination to Ile365, Asp366, Phe406, and Val 408 residues. As for the host genome LEDGF/p75 binds with the N-terminal PWWP (Pro-Trp-Trp-Pro) with the help of A/T-hook elements [120,121].

Large series of compounds are reported with the aim to specifically target the LEDGF/p75 interaction site in order to affect the IN multimerization status and thus to allosterically affect its activity [121]. Such compounds are properly defined as LEDGINs [122] and besides the allosteric-induced INI are also responsible for affecting the enzyme catalytic activity with poor discrimination between the 3′-processing and the strand transfer [123]. Despite both LEDGINs and INSTIs being effective on the strand transfer IN catalytic activity, they act through diverse mechanisms of action and therefore no cross-resistances occur [124]. Notably LEDGINs and INSTIs were reported to act in an additive or synergistic way, and that paved the way for cART therapy including LEDGIN/INSTI. In Fig. 9 are reported the structures of the most promising LEDGIN compounds.

7.3 Multimeric INIs (MINIs)

A distinct class of ALLINIs are represented by the multimerization-selective INIs (MINIs), which are featured by inducing aberrant IN multimerizations only during the virion maturation steps, whereas resulted ineffective on the early steps of viral replication. The most advanced experimental compounds are **KF115** and **KF116**, which were obtained by switching the quinolone in **BI-1001** into the biaryl pyridine moiety (Fig. 10) [125,126].

8 Dual-acting inhibitors

New advancements in the field of IN modulators for the management of retrovirus promoted infections are the development of dual acting inhibitors, thus single molecules able to modulate at the same time multiple and different targets in order to induce therapeutic responses of the additive or synergistic type [127]. Since such a strategy is primarily oriented toward enzymes, which are critical for the viral replication cycle, their application in clinic is expected to afford far more efficient therapeutic results associated with fewer side effects along with an increased

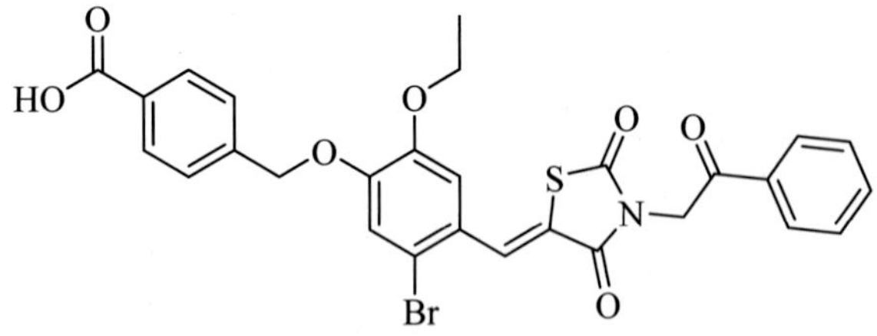

D77, 12

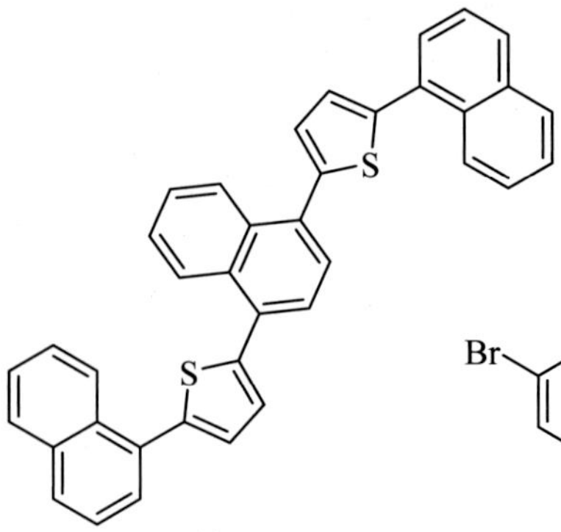

13

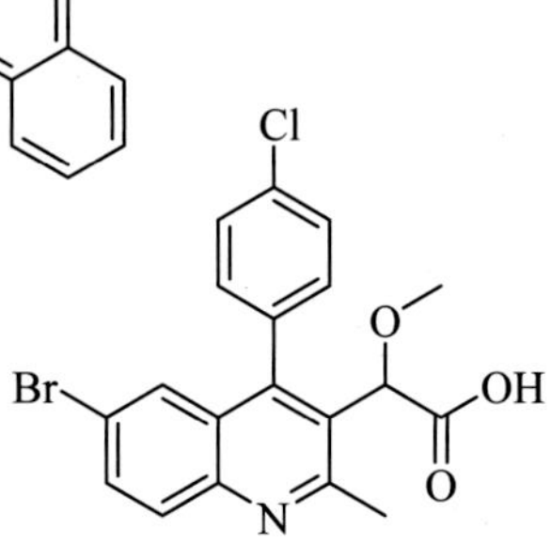

BI-1001, 14

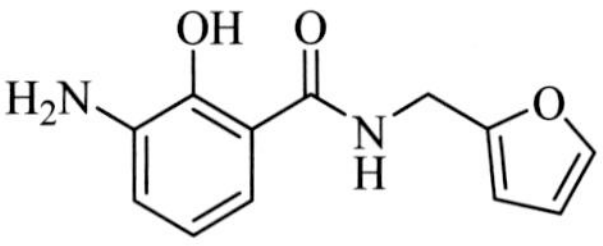

LZ-14, 15

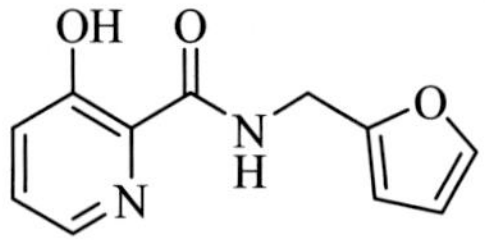

16

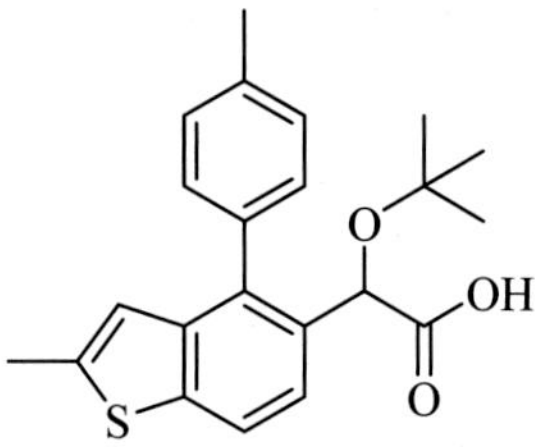

CX14442, 17

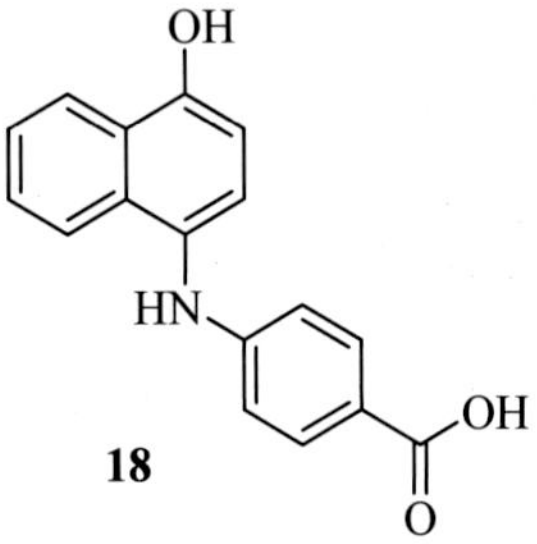

18

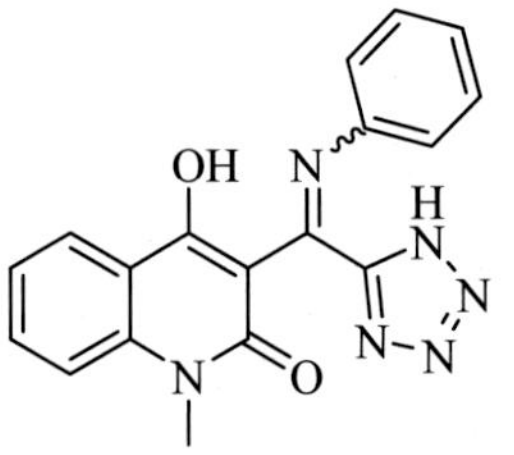

LEDGIN1, 19

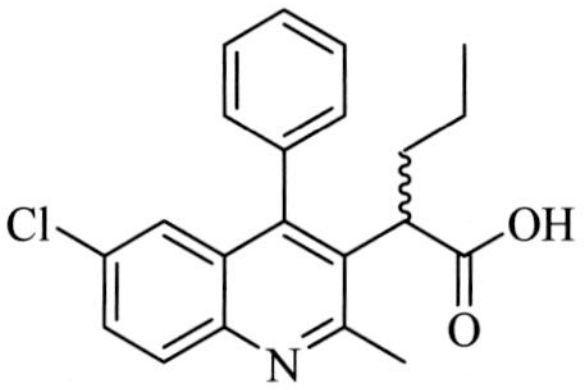

LEDGIN6, 20

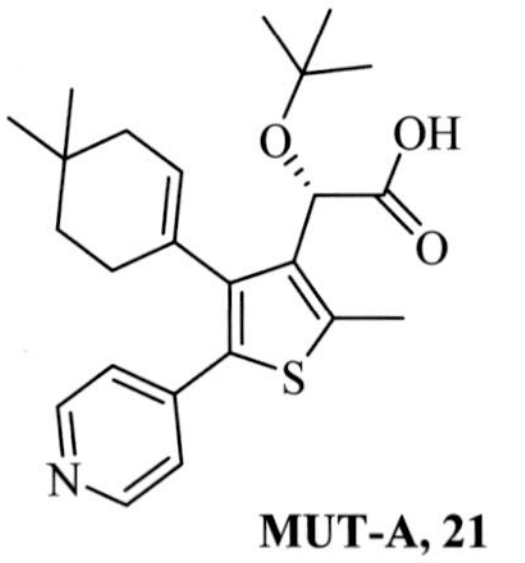

MUT-A, 21

FIG. 9 Chemical structures of the most representative LEDGINs.

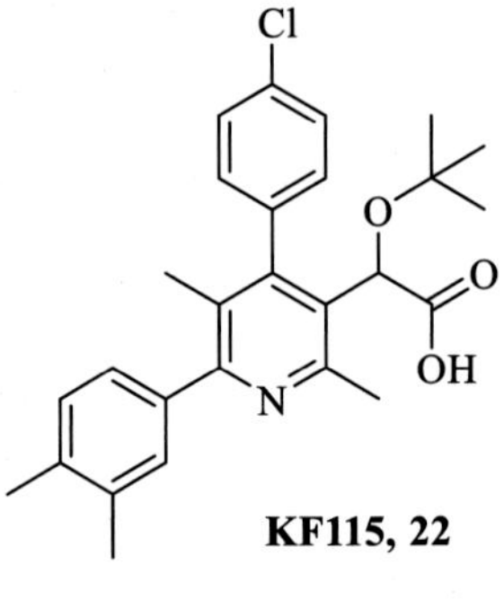

KF115, 22

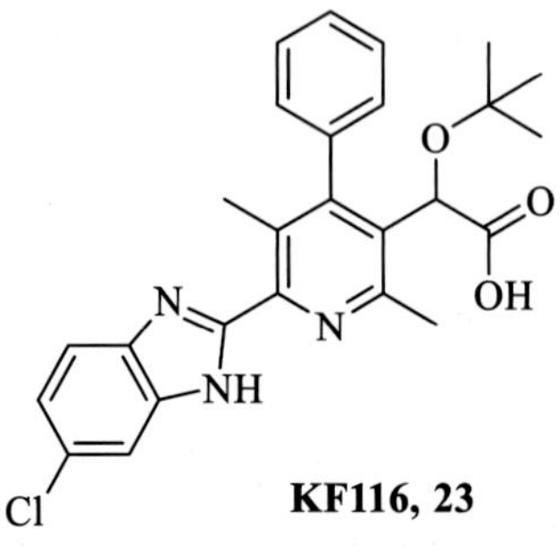

KF116, 23

FIG. 10 Chemical structures of the most representative MINIs.

patient compliance. Below are reported the most promising programs based on inhibition of the IN enzyme coupled to an additional viral target.

8.1 IN-RT RNase H inhibitors

Dual-acting inhibitors targeting the IN and RT RNase H enzymes represent the most immediate approach since both targets do share a common site [128–130]. Specifically, both enzymes present an αβ-fold that contains a central five-stranded mixed β-sheet next to α-helices on each side. In addition, the two enzymes have key acid amino acids (D64, D116, and E152 for HIV-1 IN; D443, E478, D498, and D549 for the RNase H domain) that could chelate two magnesium metal ions in their active sites [129]. Such a close structural similarities strongly suggested that DNA aptamers in analogy to DKAs may possess inhibitory activity for IN [131].

8.2 INI-LEDGF/p75-IN interaction disruptors

Specific experimental compounds endowed with INI activity and disruption of the LEDGF/p75-IN interaction are reported in Fig. 11.

By means of in silico studies, the compound **CHI-1043** was reported to bind the IN similarly to known INSTIs bearing the DKA moiety and thus by coordination of the two metal cofactors. In agreement, enzymatic assays on **CHI-1043** and its derivatives accounted for its metal complexes to inhibit the IN-associated strand transfer catalytic activity at concentrations spanning between nano- and micromolar, whereas the metal complexes and the free ligand counterparts were able to disrupt the interaction between IN and LEDGF/p75 when administered at low micromolar ranges. More interestingly were the data on infected cells, which proved such compounds as free ligands or in coordination with divalent metals such as Mg(II) were endowed with interesting antiviral features [132].

Derivatives of the type comprised in general structure of **28** in Fig. 11 were obtained by merging the scaffold of the CHIBA series with CHIs. The rationale behind such an approach relied on the experimentally reported activity of CHIBAs **3002**, **3003**, and **3053** to inhibit the association of LEDGF/p75 to IN at micromolar concentrations [133] and of **CHI-1043** to affect the IN strand transfer process and IN-LEDGF/p75 interaction [134].

GSK1264 is also endowed with dual inhibitor for the LEDGF/IN association as well as 3′-processing IN activity. More interestingly, such compound acts on late stages of viral replication (i.e., post integration), thus preventing new virions to infect host cells [135]. The binding mode of **GSK1264** within the IN was assessed by means of X-ray crystallization experiments and clearly showed it

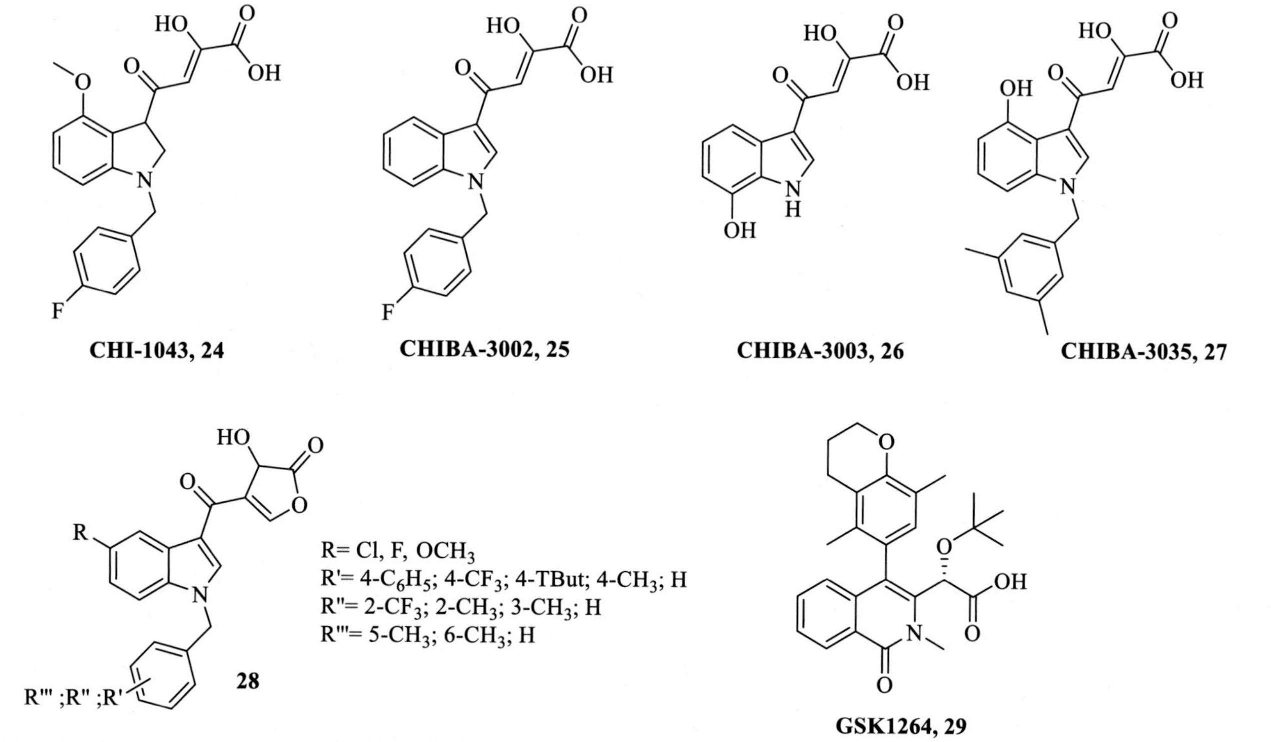

FIG. 11 Chemical structures of IN-LEDGF/p75-IN disruptors.

binds to the α1 and α3 helices of the first subunit IN monomer and to the α4 and α5 of the second monomer [135].

9 Conclusion

This chapter summarizes recent and major advances in Medicinal Chemistry within the field of IN modulators (i.e., HIV1-IN in particular) and thus potentially useful for the management of infections whose etiological agents are retroviruses. Additional experimental approaches, although on interest, were not discussed within this chapter as still at their infancy. Among others, such methods include: (i) the integrase-mediated nuclear import inhibitors, which bind to the host cell's nuclear import protein and prevent the transport of the PIC into the nucleus, thus inhibiting the integration of the viral DNA into the host genome [136]; and the (ii) the integrase-mediated chromatin remodeling inhibitors (ICRIs). In this case, the host cell's chromatin remodeling factors are targeted and thus do prevent the opening of the chromatin structure with inhibition of the integration of the v-DNA into the host genome [137]. Continuous research efforts to find alternative and efficient therapeutic approaches for the management of retrovirus infections are needed since the elevated replication cycles easily lead to resistances.

References

[1] https://ictv.global/taxonomy/about. [last access 15 March 2023].

[2] https://ictv.global/report. [last access 15 March 2023].

[3] http://www.virology.net/Big_Virology/BVHomePage.html. [last access 15 March 2023].

[4] Nisole S, Saïb A. Early steps of retrovirus replicative cycle. Retrovirology 2004;1:9.

[5] D'Souza V, Summers MF. How retroviruses select their genomes. Nat Rev Microbiol 2005;8:643–55.

[6] Stoye JP. Studies of endogenous retroviruses reveal a continuing evolutionary saga. Nat Rev Microbiol 2012;6:395–406.

[7] https://covid19.who.int/. [last access 15 March 2023].

[8] Lesbats P, Engelman AN, Cherepanov P. Retroviral DNA integration. Chem Rev 2016;116:12730–57.

[9] Farnet CM, Haseltine WA. Determination of viral proteins present in the human immunodeficiency virus type 1 preintegration complex. J Virol 1991;4:1910–5.

[10] Bukrinsky MI, Sharova N, McDonald TL, Pushkarskaya T, Tarpley WG, Stevenson M. Association of integrase, matrix, and reverse transcriptase antigens of human immunodeficiency virus type 1 with viral nucleic acids following acute infection. Proc Natl Acad Sci U S A 1993;13:6125–9.

[11] Heinzinger NK, Bukinsky MI, Haggerty SA, Ragland AM, Kewalramani V, Lee MA, Gendelman HE, Ratner L, Stevenson M, Emerman M. The Vpr protein of human immunodeficiency virus type 1 influences nuclear localization of viral nucleic acids in nondividing host cells. Proc Natl Acad Sci U S A 1994;91(15):7311–5.

[12] Miller MD, Farnet CM, Bushman FD. Human immunodeficiency virus type 1 preintegration complexes: studies of organization and composition. J Virol 1997;7:5382–90.

[13] Li L, Olvera JM, Yoder KE, Mitchell RS, Butler SL, Lieber M, Martin SL, Bushman FD. Role of the non-homologous DNA end joining pathway in the early steps of retroviral infection. EMBO J 2001;12:3272–81.

[14] Lin CW, Engelman A. The barrier-to-autointegration factor is a component of functional human immunodeficiency virus type 1 preintegration complexes. J Virol 2003;8:5030–6.

[15] Llano M, Vanegas M, Fregoso O, Saenz D, Chung S, Peretz M, Poeschla EM. LEDGF/p75 determines cellular trafficking of diverse lentiviral but not murine oncoretroviral integrase proteins and is a component of functional lentiviral preintegration complexes. J Virol 2004;17:9524–37.

[16] Raghavendra NK, Shkriabai N, Graham R, Hess S, Kvaratskhelia M, Wu L. Identification of host proteins associated with HIV-1 preintegration complexes isolated from infected CD4+ cells. Retrovirology 2010;7:66.

[17] Schweitzer CJ, Jagadish T, Haverland N, Ciborowski P, Belshan M. Proteomic analysis of early HIV-1 nucleoprotein complexes. J Proteome Res 2013;2:559–72.

[18] Wei S-Q, Mizuuchi K, Craigie R. A large nucleoprotein assembly at the ends of the viral DNA mediates retroviral DNA integration. EMBO J 1997;16:7511–20.

[19] Chen H, Wei S-Q, Engelman A. Multiple integrase functions are required to form the native structure of the human immunodeficiency virus type I intasome. J Biol Chem 1999;274:17358–64.

[20] Engelman AN, Cherepanov P. Retroviral intasomes arising. Curr Opin Struct Biol 2017;47:23–9.

[21] Cai M, Zheng R, Caffrey M, Craigie R, Clore GM, Gronenborn AM. Solution structure of the N-terminal zinc binding domain of HIV-1 integrase [Published Erratum Appears in Nat. Struct. Biol. 1997, 10, 839−840]. Nat Struct Biol 1997;7:567–77.

[22] Dyda F, Hickman AB, Jenkins TM, Engelman A, Craigie R, Davies DR. Crystal structure of the catalytic domain of HIV-1 integrase: similarity to other polynucleotidyl transferases. Science 1994;5193: 1981–6.

[23] Eijkelenboom AP, Puras Lutzke RA, Boelens R, Plasterk RH, Kaptein R, Hard K. The DNA-binding domain of HIV-1 integrase has an SH3-like fold. Nat Struct Biol 1995;9:807–10.

[24] Eijkelenboom AP, van den Ent FM, Vos A, Doreleijers JF, Hard K, Tullius TD, Plasterk RH, Kaptein R, Boelens R. The solution structure of the amino-terminal HHCC domain of HIV-2 integrase: a three-helix bundle stabilized by zinc. Curr Biol 1997;10: 739–46.

[25] Lodi PJ, Ernst JA, Kuszewski J, Hickman AB, Engelman A, Craigie R, Clore GM, Gronenborn AM. Solution structure of the DNA binding domain of HIV-1 integrase. Biochemistry 1995;31:9826–33.

[26] Bujacz G, Jaskolski M, Alexandratos J, Wlodawer A, Merkel G, Katz RA, Skalka AM. High-resolution structure of the catalytic domain of avian sarcoma virus integrase. J Mol Biol 1995;2: 333–46.

[27] Aiyer S, Rossi P, Malani N, Schneider WM, Chandar A, Bushman FD, Montelione GT, Roth MJ. Structural and sequencing analysis of local target DNA recognition by MLV integrase. Nucleic Acids Res 2015;11:5647–63.

[28] Ballandras-Colas A, Brown M, Cook NJ, Dewdney TG, Demeler B, Cherepanov P, Lyumkis D, Engelman AN. Cryo-EM reveals a novel octameric integrase structure for betaretroviral intasome function. Nature 2016;7590:358–61.

[29] Kulkosky J, Jones KS, Katz RA, Mack JP, Skalka AM. Residues critical for retroviral integrative recombination in a region that is

highly conserved among retroviral/retrotransposon integrases and bacterial insertion sequence transposases. Mol Cell Biol 1992;5: 2331–8.
[30] Li M, Mizuuchi M, Burke Jr TR, Craigie R. Retroviral DNA integration: reaction pathway and critical intermediates. EMBO J 2006;6:1295–304.
[31] Faure A, Calmels C, Desjobert C, Castroviejo M, Caumont-Sarcos A, Tarrago-Litvak L, Litvak S, Parissi V. HIV-1 integrase cross-linked oligomers are active in vitro. Nucleic Acids Res 2005;3: 977–86.
[32] Lee SP, Xiao J, Knutson JR, Lewis MS, Han MK. Zn2+ promotes the self-association of human immunodeficiency virus type-1 integrase in vitro. Biochemistry 1997;1:173–80.
[33] Heuer TS, Brown PO. Photo-cross-linking studies suggest a model for the architecture of an active human immunodeficiency virus type 1 integrase-DNA complex. Biochemistry 1998;19:6667–78.
[34] Hare S, Gupta SS, Valkov E, Engelman A, Cherepanov P. Retroviral intasome assembly and inhibition of DNA strand transfer. Nature 2010;7286:232–6.
[35] Valkov E, Gupta SS, Hare S, Helander A, Roversi P, McClure M, Cherepanov P. Functional and structural characterization of the integrase from the prototype foamy virus. Nucleic Acids Res 2009;1:243–55.
[36] Gupta K, Curtis JE, Krueger S, Hwang Y, Cherepanov P, Bushman FD, Van Duyne GD. Solution conformations of prototype foamy virus integrase and its stable synaptic complex with U5 viral DNA. Structure 2012;11:1918–28.
[37] Yin Z, Shi K, Banerjee S, Pandey KK, Bera S, Grandgenett DP, Aihara H. Crystal structure of the Rous sarcoma virus intasome. Nature 2016;7590:362–6.
[38] Hare S, Maertens GN, Cherepanov P. 3′-processing and strand transfer catalysed by retroviral integrase in crystallo. EMBO J 2012;13:3020–8.
[39] Maertens GN, Hare S, Cherepanov P. The mechanism of retroviral integration from X-ray structures of its key intermediates. Nature 2010;7321:326–9.
[40] Li X, Krishnan L, Cherepanov P, Engelman A. Structural biology of retroviral DNA integration. Virology 2011;411:194–205.
[41] Fujiwara T, Mizuuchi K. Retroviral DNA integration: structure of an integration intermediate. Cell 1988;4:497–504.
[42] Brown PO, Bowerman B, Varmus HE, Bishop JM. Retroviral integration: structure of the initial covalent product and its precursor, and a role for the viral in protein. Proc Natl Acad Sci U S A 1989;8:2525–9.
[43] Roth MJ, Schwartzberg PL, Goff SP. Structure of the termini of DNA intermediates in the integration of retroviral DNA: dependence on in function and terminal DNA sequence. Cell 1989;1: 47–54.
[44] Fujiwara T, Craigie R. Integration of mini-retroviral DNA: a cell-free reaction for biochemical analysis of retroviral integration. Proc Natl Acad Sci U S A 1989;9:3065–9.
[45] Shoemaker C, Hoffman J, Goff SP, Baltimore D. Intramolecular integration within moloney murine leukemia virus DNA. J Virol 1981;40:164–72.
[46] Li Y, Kappes JC, Conway JA, Price RW, Shaw GM, Hahn BH. Molecular characterization of human immunodeficiency virus type 1 cloned directly from uncultured human brain tissue: identification of replication-competent and -defective viral genomes. J Virol 1991;65:3973–85.
[47] Tipper C, Sodroski J. Enhanced autointegration in hyperstable simian immunodeficiency virus capsid mutants blocked after reverse transcription. J Virol 2013;7:3628–39.
[48] Zheng R, Ghirlando R, Lee MS, Mizuuchi K, Krause M, Craigie R. Barrier-to-autointegration factor (BAF) bridges DNA in a discrete, higher-order nucleoprotein complex. Proc Natl Acad Sci U S A 2000;16:8997–9002.
[49] Skoko D, Li M, Huang Y, Mizuuchi M, Cai M, Bradley CM, Pease PJ, Xiao B, Marko JF, Craigie R, et al. Barrier-to-autointegration factor (BAF) condenses DNA by looping. Proc Natl Acad Sci U S A 2009;39:16610–5.
[50] Lee MS, Craigie R. A previously unidentified host protein protects retroviral DNA from autointegration. Proc Natl Acad Sci U S A 1998;4:1528–33.
[51] Yan N, Cherepanov P, Daigle JE, Engelman A, Lieberman J. The SET complex acts as a barrier to autointegration of HIV-1. PLoS Pathog 2009;5:e1000327.
[52] Yan N, Cherepanov P, Daigle JE, Engelman A, Lieberman J. The SET complex acts as a barrier to autointegration of HIV-1. PLoS Pathog 2009;3:e1000327.
[53] Sinha S, Pursley MH, Grandgenett DP. Efficient concerted integration by recombinant human immunodeficiency virus type 1 integrase without cellular or viral cofactors. J Virol 2002;7:3105–13.
[54] Sinha S, Grandgenett DP. Recombinant human immunodeficiency virus type 1 integrase exhibits a capacity for full-site integration in vitro that is comparable to that of purified preintegration complexes from virus-infected cells. J Virol 2005;13:8208–16.
[55] Li M, Craigie R. Processing of viral DNA ends channels the HIV-1 integration reaction to concerted integration. J Biol Chem 2005;32:29334–9.
[56] Hindmarsh P, Ridky T, Reeves R, Andrake M, Skalka AM, Leis J. HMG protein family members stimulate human immunodeficiency virus type 1 and avian sarcoma virus concerted DNA integration in vitro. J Virol 1999;4:2994–3003.
[57] Hare S, Shun MC, Gupta SS, Valkov E, Engelman A, Cherepanov P. A novel co-crystal structure affords the design of gain-of-function lentiviral integrase mutants in the presence of modified PSIP1/LEDGF/p75. PLoS Pathog 2009;1:e1000259.
[58] Pandey KK, Bera S, Grandgenett DP. The HIV-1 integrase monomer induces a specific interaction with LTR DNA for concerted integration. Biochemistry 2011;45:9788–96.
[59] Li M, Ivanov V, Mizuuchi M, Mizuuchi K, Craigie R. DNA requirements for assembly and stability of HIV-1 intasomes. Protein Sci 2012;21:249–57.
[60] Li M, Jurado KA, Lin S, Engelman A, Craigie R. Engineered hyperactive integrase for concerted HIV-1 DNA integration. PLoS One 2014;9:e105078.
[61] Chow SA, Vincent KA, Ellison V, Brown PO. Reversal of integration and DNA splicing mediated by integrase of human immunodeficiency virus. Science 1992;5045:723–6.
[62] van Gent DC, Groeneger AA, Plasterk RH. Mutational analysis of the integrase protein of human immunodeficiency virus type 2. Proc Natl Acad Sci U S A 1992;20:9598–602.
[63] Engelman A, Craigie R. Identification of conserved amino acid residues critical for human immunodeficiency virus type 1 integrase function in vitro. J Virol 1992;11:6361–9.
[64] Leavitt AD, Shiue L, Varmus HE. Site-directed mutagenesis of HIV-1 integrase demonstrates differential effects on integrase functions in vitro. J Biol Chem 1993;3:2113–9.

[65] Vink C, Oude Groeneger AM, Plasterk RH. Identification of the catalytic and DNA-binding region of the human immunodeficiency virus type I integrase protein. Nucleic Acids Res 1993;6:1419–25.

[66] Gerton JL, Brown PO. The Core domain of HIV-1 integrase recognizes key features of its DNA substrates. J Biol Chem 1997;41:25809–15.

[67] Bushman FD, Engelman A, Palmer I, Wingfield P, Craigie R. Domains of the integrase protein of human immunodeficiency virus type 1 responsible for polynucleotidyl transfer and zinc binding. Proc Natl Acad Sci U S A 1993;90:3428–32.

[68] Bushman FD, Wang B. Rous sarcoma virus integrase protein: mapping functions for catalysis and substrate binding. J Virol 1994;68:2215–23.

[69] Kulkosky J, Katz RA, Merkel G, Skalka AM. Activities and substrate specificity of the evolutionarily conserved central domain of retroviral integrase. Virology 1995;206:448–56.

[70] Donzella GA, Jonsson CB, Roth MJ. Coordinated disintegration reactions mediated by moloney murine leukemia virus integrase. J Virol 1996;6:3909–21.

[71] Mousnier A, Kubat N, Massias-Simon A, Segeral E, Rain JC, Benarous R, Emiliani S, Dargemont C. Von Hippel Lindau binding protein 1-mediated degradation of integrase affects HIV-1 gene expression at a postintegration step. Proc Natl Acad Sci U S A 2007;34:13615–20.

[72] Studamire B, Goff SP. Host proteins interacting with the moloney murine leukemia virus integrase: multiple transcriptional regulators and chromatin binding factors. Retrovirology 2008;5:48.

[73] Faust EA, Triller H. Stimulation of human flap endonuclease 1 by human immunodeficiency virus type 1 integrase: possible role for flap endonuclease 1 in 5′-end processing of human immunodeficiency virus type 1 integration intermediates. J Biomed Sci 2002;3:273–87.

[74] Engelman AN. Multifaceted HIV integrase functionalities and therapeutic strategies for their inhibition. J Biol Chem 2019;41:15137–57.

[75] Sechi M, Derudas M, Dallocchio R, Dessì A, Bacchi A, Sannia L, Carta F, Palomba M, Ragab O, Chan C, Shoemaker R, Sei S, Dayam R, Neamati N. Design and synthesis of novel indole beta-diketo acid derivatives as HIV-1 integrase inhibitors. J Med Chem 2004;21:5298–310.

[76] Goldgur Y, Craigie R, Cohen GH, Fujiwara T, Yoshinaga T, Fujishita T, Sugimoto H, Endo T, Murai H, Davies DR. Structure of the HIV-1 integrase catalytic domain complexed with an inhibitor: a platform for antiviral drug design. Proc Natl Acad Sci U S A 1999;23:13040–3.

[77] Bacchi A, Biemmi M, Carcelli M, Carta F, Compari C, Fisicaro E, Rogolino D, Sechi M, Sippel M, Sotriffer CA, Sanchez TW, Neamati N. From ligand to complexes. Part 2. Remarks on human immunodeficiency virus type 1 integrase inhibition by beta-diketo acid metal complexes. J Med Chem 2008;22:7253–64.

[78] Hazuda DJ. Inhibitors of strand transfer that prevent integration and inhibit HIV-1 replication in cells. Science 2000;5453:646–50.

[79] Marchand C, Zhang X, Pais GC, Cowansage K, Neamati N, Burke TR, Pommier Y. Structural determinants for HIV-integrase inhibition by β-diketo acids. J Biol Chem 2002;15:12596–603.

[80] Hazuda DJ, Anthony NJ, Gomez RP, Jolly SM, Wai JS, Zhuang L, Fisher TE, Embrey M, Guare Jr JP, Egbertson MS, Vacca JP, Huff JR, Felock PJ, Witmer MV, Stillmock KA, Danovich R, Grobler J, Miller MD, Espeseth AS, Jin L, Chen IW, Lin JH, Kassahun K, Ellis JD, Wong BK, Xu W, Pearson PJ, Schleif WA, Cortese R, Emini E, Summa V, Holloway MK, Young SD. A naphthyridine carboxamide provides evidence for discordant resistance between mechanistically identical inhibitors of HIV-1 integrase. Proc Natl Acad Sci U S A 2004;31:11233–8.

[81] Semenova EA, Marchand C, Pommier Y. HIV-I integrase inhibitors: update and perspectives. Adv Pharmacol 2008;56:199.

[82] Nair V, Okello M. Integrase inhibitor prodrugs: approaches to enhancing the anti-HIV activity of beta-diketo acids. Molecules 2015;7:12623–51.

[83] Billich A. S-1360 Shionogi-GlaxoSmithKline. Curr Opin Investig Drugs 2003;2:206–9.

[84] Rosemond M, St John-Williams L, Yamaguchi T, Fujishita T, Walsh JS. Enzymology of a carbonyl reduction clearance pathway for the HIV integrase inhibitor, S-1360: role of human liver cytosolic aldo-keto reductases. Chem Biol Interact 2004;2:129–39.

[85] Garvey EP, Johns BA, Gartland MJ, Foster SA, Miller WH, Ferris RG, Hazen RJ, Underwood MR, Boros EE, Thompson JB, Weatherhead JG, Koble CS, Allen SH, Schaller LT, Sherrill RT, Yoshinaga T, Kobayashi M, Wakasa-Morimoto C, Miki S, Nakahara K, Noshi T, Sato A, Fujiwara T. The naphthyridinone GSK364735 is a novel, potent human immunodeficiency virus type 1 integrase inhibitor and antiretroviral. Antimicrob Agents Chemother 2008;3:901–8.

[86] Korolev S, Agapkina YY, Gottikh M. Clinical use of inhibitors of HIV-1 integration: problems and prospects. Acta Nat 2011;3:12.

[87] Grobler JA, Stillmock K, Hu B, Witmer M, Felock P, Espeseth AS, Wolfe A, Egbertson M, Bourgeois M, Melamed J, Wai JS, Young S, Vacca J, Hazuda DJ. Diketo acid inhibitor mechanism and HIV-1 integrase: implications for metal binding in the active site of phosphotransferase enzymes. Proc Natl Acad Sci U S A 2002;10:6661–6.

[88] Marchand C, Johnson AA, Karki RG, Pais GCG, Zhang X, Cowansage K, Patel TA, Nicklaus MC, Burke Jr TR, Pommier Y. Metal-dependent inhibition of HIV-1 integrase by β-diketo acids and resistance of the soluble double-mutant (F185K/C280S). Mol Pharmacol 2003;3:600–9.

[89] Pais GCG, Zhang X, Marchand C, Neamati N, Cowansage K, Svarovskaia ES, Pathak VK, Tang Y, Nicklaus M, Pommier Y, Burke Jr TR. Structure activity of 3-Aryl-1, 3-diketo-containing compounds as HIV-1 integrase inhibitors. J Med Chem 2002;15:3184–94.

[90] Long Y-Q, Jiang X-H, Dayam R, Sanchez T, Shoemaker R, Sei S, Neamati N. Rational design and synthesis of novel dimeric diketoacid-containing inhibitors of HIV-1 integrase: implication for binding to two metal ions on the active site of integrase. J Med Chem 2004;10:2561–73.

[91] Walker MA, Johnson T, Ma Z, Banville J, Remillard R, Kim O, Zhang Y, Staab A, Wong H, Torri A, Samanta H, Lin Z, Deminie C, Terry B, Krystal M, Meanwell N. Triketoacid inhibitors of HIV-integrase: a new chemotype useful for probing the integrase pharmacophore. Bioorg Med Chem Lett 2006;11:2920–4.

[92] Liedtke MD, Tomlin CR, Lockhart SM, Miller MM, Rathbun RC. Long-term efficacy and safety of raltegravir in the management of HIV infection. Infect Drug Resist 2014;7:73–84.

[93] Jaeckle M, Khaykin P, Haberl A, De Leuw P, Schüttfort G, Stephan C, Wolf T. Efficacy of raltegravir-containing regimens in antiretroviral-naive and -experienced individuals in routine clinical practice. Int J STD AIDS 2016;13:1170–9.

[94] Deeks ED. Raltegravir once-daily tablet: a review in HIV-1 infection. Drugs 2017;16:1789–95.

[95] De Miguel R, Montejano R, Stella-Ascariz N, Arribas JR. A safety evaluation of raltegravir for the treatment of HIV. Expert Opin Drug Saf 2018;2:217–23.

[96] Summa V, Petrocchi A, Bonelli F, Crescenzi B, Donghi M, Ferrara M, Fiore F, Gardelli C, Paz OG, Hazuda DJ, Jones P, Kinzel O, Laufer R, Monteagudo E, Muraglia E, Nizi E, Orvieto F, Pace P, Pescatore G, Scarpelli R, Stillmock K, Witmer MV, Rowley M. Discovery of raltegravir, a potent, selective orally bioavailable HIV-integrase inhibitor for the treatment of HIV-AIDS infection. J Med Chem 2008;18:5843–55.

[97] Steigbigel RT, Cooper DA, Kumar PN, Eron JE, Schechter M, Markowitz M, Loutfy MR, Lennox JL, Gatell JM, Rockstroh JR, Katlama C, Yeni P, Lazzarin A, Clotet B, Zhao J, Chen J, Ryan DM, Rhodes RR, Killar JA, Gilde LR, Strohmaier KM, Meibohm AR, Miller MD, Hazuda DJ, Nessly ML, DiNubile MJ, Isaacs RD, Nguyen B-Y, Teppler H, BENCHMRK Study Teams. Raltegravir with optimized background therapy for resistant HIV-1 infection. N Engl J Med 2008;4:339–54.

[98] Markowitz M, Nguyen B-Y, Gotuzzo E, Mendo F, Ratanasuwan W, Kovacs C, Prada G, Morales-Ramirez JO, Crumpacker CS, Isaacs RD, Gilde LR, Wan H, Miller MD, Wenning LA, Teppler H, Protocol 004 Part II Study Team. Rapid and durable antiretroviral effect of the HIV-1 integrase inhibitor raltegravir as part of combination therapy in treatment-naive patients with HIV-1 infection: results of a 48-week controlled study. J Acquir Immune Defic Syndr 2007;2:125–33.

[99] Jegede O, Babu J, Di Santo R, Mccoll DJ, Weber J, Quinones-Mateu M. HIV type 1 integrase inhibitors: from basic research to clinical implications. AIDS Rev 2008;3:172–89.

[100] Sato M, Motomura T, Aramaki H, Matsuda T, Yamashita M, Ito Y, Kawakami H, Matsuzaki Y, Watanabe W, Yamataka K, Ikeda S, Kodama E, Matsuoka M, Shinkai H. Novel HIV-1 integrase inhibitors derived from quinolone antibiotics. J Med Chem 2006;5:1506–8.

[101] Sato M, Kawakami H, Motomura T, Aramaki H, Matsuda T, Yamashita M, Ito Y, Matsuzaki Y, Yamataka K, Ikeda S, Shinkai H. Quinolone carboxylic acids as a novel monoketo acid class of human immunodeficiency virus type 1 integrase inhibitors. J Med Chem 2009;15:4869–82.

[102] Imaz A, Podzamczer D. Tenofovir alafenamide, emtricitabine, elvitegravir, and cobicistat combination therapy for the treatment of HIV. Expert Rev Anti Infect Ther 2017;3:195–209.

[103] Angione SA, Cherian SM, Ozdener AE. A review of the efficacy and safety of Genvoya(R) (elvitegravir, cobicistat, emtricitabine, and tenofovir alafenamide) in the management of HIV-1 infection. J Pharm Pract 2018;2:216–21.

[104] Johns BA, Kawasuji T, Weatherhead JG, Taishi T, Temelkoff DP, Yoshida H, Akiyama T, Taoda Y, Murai H, Kiyama R, Fuji M, Tanimoto N, Jeffrey J, Foster SA, Yoshinaga T, Seki T, Kobayashi M, Sato A, Johnson MN, Garvey EP, Fujiwara T. Carbamoyl pyridone HIV-1 integrase inhibitors 3. A diastereomeric approach to chiral nonracemic tricyclic ring systems and the discovery of dolutegravir (S/GSK1349572) and (S/GSK1265744). J Med Chem 2013;14:5901–16.

[105] Kobayashi M, Yoshinaga T, Seki T, Wakasa-Morimoto C, Brown KW, Ferris R, Foster SA, Hazen RJ, Miki S, Suyama-Kagitani A, Kawauchi-Miki S, Taishi T, Kawasuji T, Johns BA, Underwood MR, Garvey EP, Sato A, Fujiwara T. In vitro antiretroviral properties of S/GSK1349572, a next-generation HIV integrase inhibitor. Antimicrob Agents Chemother 2011;2:813–21.

[106] Hare S, Smith SJ, Métifiot M, Jaxa-Chamiec A, Pommier Y, Hughes SH, Cherepanov P. Structural and functional analyses of the second-generation integrase strand transfer inhibitor dolutegravir (S/GSK 1349572). Mol Pharmacol 2011;4:565–72.

[107] Karmon SL, Markowitz M. Next-generation integrase inhibitors: where to after raltegravir? Drugs 2013;3:213–28.

[108] https://www.fda.gov/news-events/press-announcements/fda-approves-first-injectable-treatment-hiv-pre-exposure-prevention?fbclid=IwAR3QFnXbPbzmA2SqSHOhewW7mmTluW7ilnpkqYqGqcXXjkpVv87OCzaQOMA.

[109] Johns BA, Kawasuji T, Weatherhead JG, Boros EE, Thompson JB, Koble CS, Garvey EP, Foster SA, Jeffrey JL, Fujiwara T. Naphthyridinone (NTD) integrase inhibitors: N1 protio and methyl combination substituent effects with C3 amide groups. Bioorg Med Chem Lett 2013;2:422–5.

[110] Ford SL, Gould E, Chen S, Lou Y, Dumont E, Spreen W, Piscitelli S. Effects of etravirine on the pharmacokinetics of the integrase inhibitor S/GSK1265744. Antimicrob Agents Chemother 2013;1: 277–80.

[111] Mcpherson TD, Sobieszczyk ME, Markowitz M. Cabotegravir in the treatment and prevention of Human Immunodeficiency Virus-1. Expert Opin Investig Drugs 2018;4:413–20.

[112] Tsiang M, Jones GS, Goldsmith J, Mulato A, Hansen D, Kan E, Tsai L, Bam RA, Stepan G, Stray KM, Niedziela-Majka A, Yant SR, Yu H, Kukolj G, Cihlar T, Lazerwith SE, White KL, Jin H. Antiviral activity of bictegravir (GS-9883), a novel potent HIV-1 integrase strand transfer inhibitor with an improved resistance profile. Antimicrob Agents Chemother 2016;12:7086–97.

[113] Gallant JE, Thompson M, DeJesus E, Voskuhl GW, Wei X, Zhang H, White K, Cheng A, Quirk E, Martin H. Antiviral activity, safety, and pharmacokinetics of bictegravir as 10-day monotherapy in HIV-1-infected adults. J Acquir Immune Defic Syndr 2017;1:61–6.

[114] Markham A. Bictegravir: first global approval. Drugs 2018;5:601–6.

[115] Burlein C, Wang C, Xu M, Bhatt T, Stahlhut M, Ou Y, Adam GC, Heath J, Klein DK, Sanders J, Narayan K, Abeywickrema P, Heo MR, Carroll SS, Grobler JA, Sharma S, Diamond TL, Converso A, Krosky DJ. Discovery of a distinct chemical and mechanistic class of allosteric HIV-1 integrase inhibitors with antiretroviral activity. ACS Chem Biol 2017;11:2858–65.

[116] Christ F, Shaw S, Demeulemeester J, Desimmie BA, Marchand A, Butler S, Smets W, Chaltin P, Westby M, Debyser Z, Pickford C. Small-molecule inhibitors of the LEDGF/p75 binding site of integrase block HIV replication and modulate integrase multimerization. Antimicrob Agents Chemother 2012;8:4365–74.

[117] Fenwick C, Amad M, Bailey MD, Bethell R, Bös M, Bonneau P, Cordingley M, Coulombe R, Duan J, Edwards P, Fader LD, Faucher A-M, Garneau M, Jakalian A, Kawai S, Lamorte L, LaPlante S, Luo L, Mason S, Poupart M-A, Rioux N, Schroeder P, Simoneau B, Tremblay S, Tsantrizos Y, Witvrouw M, Yoakim C. Preclinical profile of BI 224436, a novel HIV-1 non-catalytic-site integrase inhibitor. Antimicrob Agents Chemother 2014;6:3233–44.

[118] Bonnard D, Le Rouzic E, Eiler S, Amadori C, Orlov I, Bruneau J-M, Brias J, Barbion J, Chevreuil F, Spehner D, Chasset S, Ledoussal B, Moreau F, Saïb A, Klaholz BP, Emiliani S, Ruff M, Zamborlini A, Benarous A. Structure-function analyses unravel distinct effects of allosteric inhibitors of HIV-1 integrase on viral maturation and integration. J Biol Chem 2018;16:6172–86.

[119] Peat TS, Dolezal O, Newman J, Mobley D, Deadman JJ. Interrogating HIV integrase for compounds that bind—a SAMPL challenge. J Comput Aided Mol Des 2014;4:347–62.

[120] Gu W-G, Tsz-Ming Ip D, Liu S-J, Chan JH, Wang Y, Zhang X, Zheng Y-T, Chi-Cheong Wan D. 1,4-Bis(5-(naphthalen-1-yl) thiophen-2-yl)naphthalene, a small molecule, functions as a novel anti-HIV-1 inhibitor targeting the interaction between integrase

and cellular Lens epithelium-derived growth factor. Chem Biol Interact 2014;213:21–7.

[121] Blokken J, De Rijck J, Christ F, Debyser Z. Protein–protein and protein–chromatin interactions of LEDGF/p75 as novel drug targets. Drug Discov Today Technol 2017;24:25–31.

[122] Christ F, Voet A, Marchand A, Nicolet S, Desimmie BA, Marchand D, Bardiot D, Van der Veken NJ, Van Remoortel B, Strelkov SV, De Maeyer M, Chaltin P, Debyser Z. Rational design of small-molecule inhibitors of the LEDGF/p75-integrase interaction and HIV replication. Nat Chem Biol 2010;6:442–8.

[123] Christ F, Debyser Z. HIV-1 integrase inhibition: looking at cofactor interactions. Future Med Chem 2015;18:2407–10.

[124] Christ F, Debyser Z. The LEDGF/p75 integrase interaction, a novel target for anti-HIV therapy. Virology 2013;1:102–9.

[125] Sharma A, Slaughter A, Jena N, Feng L, Kessl JJ, Fadel HJ, Malani N, Male F, Wu L, Poeschla L, Bushman FD, Fuchs JR, Kvaratskhelia M. A new class of multimerization selective inhibitors of HIV-1 integrase. PLoS Pathog 2014;5:e1004171.

[126] Hoyte AC, Jamin AV, Koneru PC, Kobe MJ, Larue RC, Fuchs JR, Engelman AN, Kvaratskhelia M. Resistance to pyridine-based inhibitor KF116 reveals an unexpected role of integrase in HIV-1 Gag-Pol polyprotein proteolytic processing. J Biol Chem 2017;48: 19814–25.

[127] Soriano V, Fernandez-Montero JV, Benitez-Gutierrez L, Mendoza C, Arias A, Barreiro P, Peña JM, Labarga P. Dual antiretroviral therapy for HIV infection. Expert Opin Drug Saf 2017;8:923–32.

[128] Di Santo R. Inhibiting the HIV integration process: past, present, and the future. J Med Chem 2014;3:539–66.

[129] Desimmie BA, Schrijvers R, Demeulemeester J, Borrenberghs D, Weydert C, Thys W, Vets S, Van Remoortel B, Hofkens J, De Rijck J, Hendrix J, Bannert N, Gijsbers R, Christ F, Debyser Z. LEDGINs inhibit late stage HIV-1 replication by modulating integrase multimerization in the virions. Retrovirology 2013;10:57. Correction Retrovirology 2020, 1, 22.

[130] Gu SX, Xue P, Ju XL, Zhu YY. Advances in rationally designed dual inhibitors of HIV-1 reverse transcriptase and integrase. Bioorg Med Chem 2016;21:5007–16.

[131] Billamboz M, Bailly F, Barreca ML, De Luca L, Mouscadet J-F, Calmels C, Andréola M-L, Witvrouw M, Christ F, Debyser Z, Cotelle P. Design, synthesis, and biological evaluation of a series of 2-hydroxyisoquinoline-1,3(2H,4H)-diones as dual inhibitors of human immunodeficiency virus type 1 integrase and the reverse transcriptase RNase H domain. J Med Chem 2008;24: 7717–30.

[132] Rogolino D, Carcelli M, Compari C, De Luca L, Ferro S, Fisicaro E, Rispoli G, Neamati N, Debyser Z, Christ F, Chimirri A. Diketoacid chelating ligands as dual inhibitors of HIV-1 integration process. Eur J Med Chem 2014;78:425–30.

[133] De Luca L, Barreca ML, Ferro S, Christ F, Iraci N, Gitto R, Monforte AM, Debyser Z, Chimirri A. Pharmacophore-based discovery of small-molecule inhibitors of protein-protein interactions between HIV-1 integrase and cellular cofactor LEDGF/p75. ChemMedChem 2009;8:1311–6.

[134] Ferro S, De Luca L, Lo Surdo G, Morreale F, Christ F, Debyser Z, Gitto R, Chimirri A. A new potential approach to block HIV-1 replication via protein-protein interaction and strand-transfer inhibition. Bioorg Med Chem 2014;7:2269–79.

[135] Gupta K, Turkki V, Sherrill-Mix S, Hwang Y, Eilers G, Taylor L, McDanal C, Wang P, Temelkoff D, Nolte RT, Velthuisen E, Jeffrey J, Van Duyne GD, Bushman FD. Structural basis for inhibitor-induced aggregation of HIV integrase. PLoS Biol 2016;12: e1002584.

[136] Levin A, Armon-Omer A, Rosenbluh J, Melamed-Book N, Graessmann A, Waigmann E, Loyter A. Inhibition of HIV-1 integrase nuclear import and replication by a peptide bearing integrase putative nuclear localization signal. Retrovirology 2009;6:112.

[137] Matysiak J, Lesbats P, Mauro E, Lapaillerie D, Dupuy J-W, Lopez AP, Benleulmi MS, Calmels C, Andreola M-L, Ruff M, Llano M, Delelis O, Lavigne M, Parissi V. Modulation of chromatin structure by the FACT histone chaperone complex regulates HIV-1 integration. Retrovirology 2017;14:39.

Chapter 2.4

Cyclin-dependent kinase 2 (CDK2)

Ghada F. Elmasry
Department of Pharmaceutical Chemistry, Faculty of Pharmacy, Cairo University, Cairo, Egypt

1 Structure and function of CDK2

Protein kinases constitute a central component of cellular signaling pathways in living organisms [1,2]. The cell cycle regulation, transition from G1 to S phase, and G2 modulation are critically controlled by cyclin-dependent kinase-2 (CDK2), a mammalian serine/threonine protein kinase [3–5]. Other names for CDK2 include p33 protein kinase and cell division protein kinase 2 [4]. Like all CDKs, the activity of CDK2 needs binding with a protein subunit called cyclin [6]. Additional sequences for enzymatic activity are afforded by cyclin, the non-catalytic regulatory protein subunit (cyclin A or E). The outcome of the association of CDK2 and cyclin is a heterodimer CDK2-cyclin complex that is essential for controlling various cell processes [7,8].

Like other CDKs, CDK2 has the classical bilobal structure (Fig. 1), N-terminal domain made up of β-sheets, and C-terminal lobe composed of α-helices [9,10]. The N-terminal lobe consists of five antiparallel β-sheets with one αC-helix and also has an inhibitory component rich in glycine known as G-loop. The activation segment is accommodated by the larger C-terminal, which is known as T-loop and Thr-160 (activating phosphorylation site) [11–13]. The T-loop is the site for binding of the phosphorylation residues: serine or threonine. The flexible hinge region, which is situated in a deep cleft in the ATP-binding site, connects the N-terminal and C-terminal lobes. Both lobes are involved in ATP recognition. The ATP-γ-phosphate faces the Ser/Thr hydroxylated side chain on the surface of the substrate, while CDK2 provides neighboring binding sites for the substrate of the phosphor-acceptor protein and ATP [13]. It is worth to mention that small compounds acting as CDK2 inhibitors can target the ATP-binding region of CDK2-cyclin complexes either by allosteric inhibition or in a reversible and competitive manner (Fig. 2) [14–16].

Interestingly, Magnesium ion exists in the catalytic site of CDK2 [17] (Figs. 3 and 4). Protein Kinase A (PKA) research has shown that Mg^{2+} ions play a dynamic and intricate role at multiple phases of the catalytic cycle, with one Mg^{2+} ion being mentioned as "crucial" and the other as "inhibitory" [18]. As a result, it is indistinct if CDK2 uses one Mg^{2+} ion for the phosphoryl-transfer stage, as opposed to the two Mg^{2+} mechanism seen in PKA, or whether a 2 Mg CDK2 state simply has not been detected [17].

CDK2 is a key cell-cycle switch in dividing cells and is active from late G1 through the S-phase (Fig. 5). The CAK complex (CDK7, MAT1, cyclin H) phosphorylates CDK2 and removes CDC25A (cell division cycle 25 A) suppressing phosphorylations. Triggered CDK2–cyclin-E and CDK4/6–cyclin-D phosphorylate retinoblastoma protein (PRb) to release E2F from PRb during late G1, and this triggers the expression of genes essential to G1/S transition and S phase development [19]. CDK2 regulates the phosphorylation of additional regulatory proteins in addition to PRb, tying other activities to cell cycle development. The phosphorylation of SMAD3 (mothers against decapentaplegic homolog 3) by CDK2–cyclin E, for example, inhibits its transcriptional activity and thereby delays the advancement of the cell cycle [20].

CDK2 phosphorylates numerous constituents of the pre-replication complex, that is necessary for DNA synthesis to proceed. This comprises cell devision cycle 6 (CDC6), which controls the commencement of DNA replication by packing minichromosome maintenance (MCM) proteins onto DNA and MCM helicase proteins [21]. Finally, target centrosome proteins like nucleophosmin (NPM) and centriolar coiled-coil protein of 110 kDa (CCP110) should be phosphorylated by CDK2, in order to release centrioles and thus maintain centriole duplication [22,23].

2 Physiologic and pathologic role of CDK2

CDK2 physiologic and pathologic role has been investigated using a variety of approaches, including dominant-negative CDK2 mutant protein, CDK2 knockout, CDK2 siRNA (short interfering RNA) transfection, and CDK2 inhibition using small-molecular inhibitors [24–26]. When CDK2 is diminished, CDK1 interacts with and is triggered by cyclin E and cyclin A, making it challenging to characterize CDK2-explicit roles in CDK2 models or with short

Metalloenzymes. https://doi.org/10.1016/B978-0-12-823974-2.00030-9

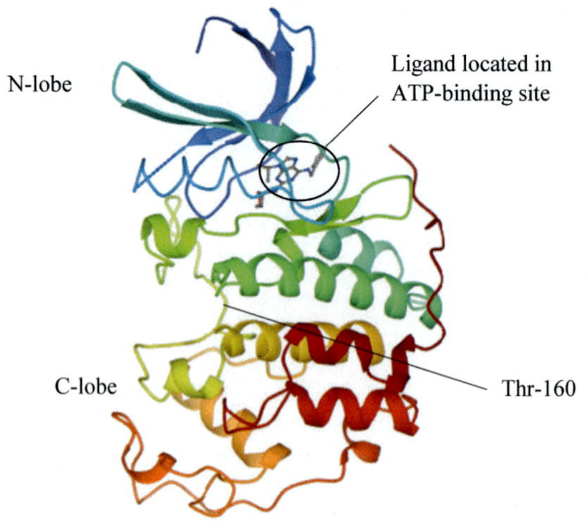

FIG. 1 Representation of CDK2 showing its bilobal structure (PDB ID is 2A4L).

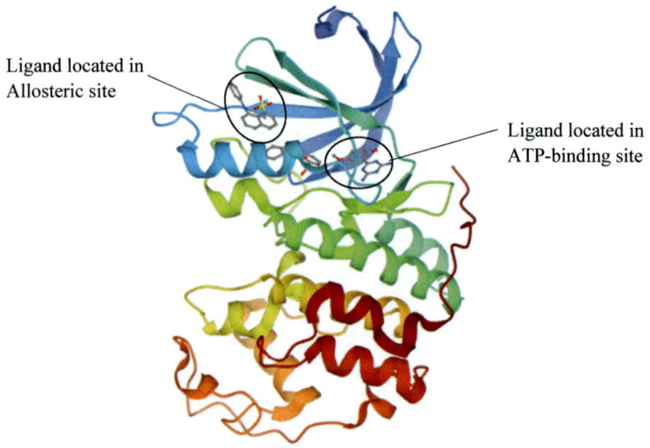

FIG. 2 Structure of CDK2 bound with JWS648 inhibitor in ATP-binding site and ANS in allosteric-binding site (PDB ID: 3PXZ).

interfering RNA (siRNA) or short hairpin RNA (shRNA) methods. Being promiscuous with other kinases, such as CDK1, CDK4/6, CDK7, and CDK9, the CDK2 inhibitors that are readily available now have restrictions [5].

Merrick and coworkers developed an innovative chemical genetic technique to solve these inadequacies, in which CDK2 was substituted with a form that expanded the ATP-binding pocket, permitting normal CDK2 activity while also allowing selective inhibition by immense adenine correspondents [27]. This revealed that CDK2 is necessary for S-phase entry, notably in the absence of growth factors, as well as anchorage-independent proliferation mediated by oncogenes. Accordingly, CDK2 pursuing has resurfaced as a therapeutic option for limiting cancer cell growth, particularly because the activities of CDK2 in apoptosis and DNA damage can be co-targeted [28].

Furthermore, CDK2/cyclins are frequently dysregulated in several malignancies and can promote uncontrolled cancer cell growth. Many recent investigations have found a link between cancer and transcription factors dysregulation as well. CDK2 may be associated with breast cancer [29], lymphoma [30], ovarian [31], endometrial [32], lung [23] carcinomas, melanoma [33,34], colorectal cancer [35,36], glioblastoma [37–39], prostate cancer [40], and BRCA-deficient cancers [26].

2.1 Role of CDK2 in apoptosis

CDK2 modulates the functional components and regulators of apoptosis. Forkhead box protein O1 (FOXO1), a CDK2 target protein, is involved in DNA-damage-persuaded apoptosis following dsDNA (double stranded DNA) breaks [41]. CDK2 does not phosphorylate FOXO1 at inhibitory sites after G1/S-phase arrest induced by DNA damage, enabling FOXO1 to transcript actively and initiate the apoptotic pathways through enhanced expression of multiple proapoptotic proteins, namely, B-cell lymphoma 2 (BCL-2), interacting mediator of cell death (Bim), and tumor necrosis factor (TNF)-related apoptosis-inducing ligand (TRAIL) [42].

Through activation of the pro-survival factor myeloid leukemia cell differentiation protein (MCL-1), CDK2 can guard cells against apoptosis. MCL-1 is phosphorylated on Thr92 and Thr163 by CDK2–cyclin-E to improve its stability, and CDK2 enables MCL-1 phosphorylation on Ser64 to confiscate the proapoptotic protein Bim. Inhibition of CDK2 by siRNA in human diffuse large B cell lymphoma cell lines (in which CDK2 is highly expressed) promotes apoptosis in combination with a drop in MCL-1 levels [30,43,44].

2.2 Role of CDK2 in DNA damage response (DDR)

Next to DNA damage, the DDR halts cell division at the G1/S transition to permit DNA repair and preserve genomic integrity in daughter cells. Through CDK2, two mechanisms combine to prevent proliferation at the G1/S DNA damage checkpoint. Cessation of cell-cycle is caused by p53 overexpression, which also increases the transcription of $p21^{Cip1/Waf1}$ and inhibits cyclin-D1–CDK4/6 and cyclin-E–CDK2 [45]. The second route destroys CDC25A, which causes the tyrosine kinase WEE1 to continue phosphorylating CDK2 at Thr14 and Tyr15 in an inhibitory manner and preventing S-phase entry [46].

Other DDR mechanisms, such as non-homologous end-joining repair, which involves the phosphorylation of Ku, the pathway's key protein, and p53 response, which involves the prompt phosphorylation of p53 and checkpoint kinase 1 (CHK1) after DNA damage, may involve CDK2 actions. These findings collectively point to a key role for CDK2 in DDR that works in conjunction with its cell-cycle function and having the capability of therapeutic treatment [47].

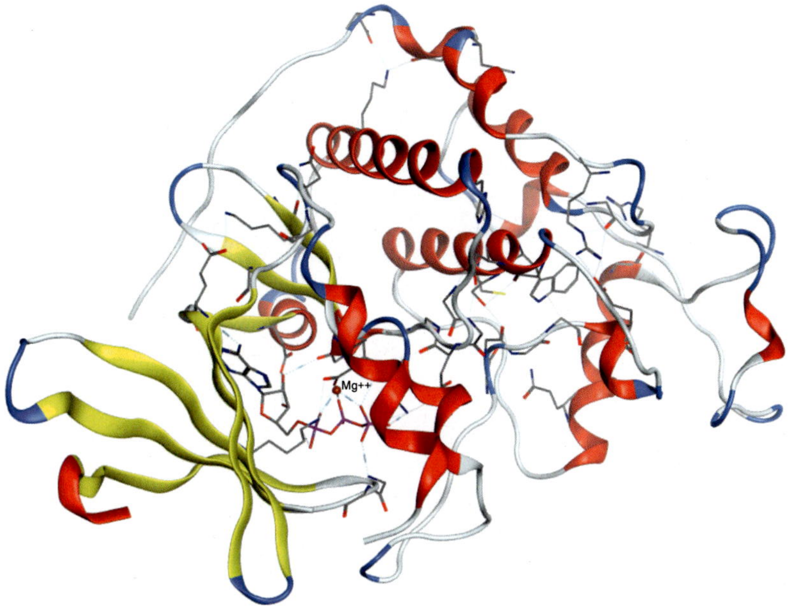

FIG. 3 Structure of CDK2 bound to ATP and Mg^{2+} ion (PDB ID: 1B38).

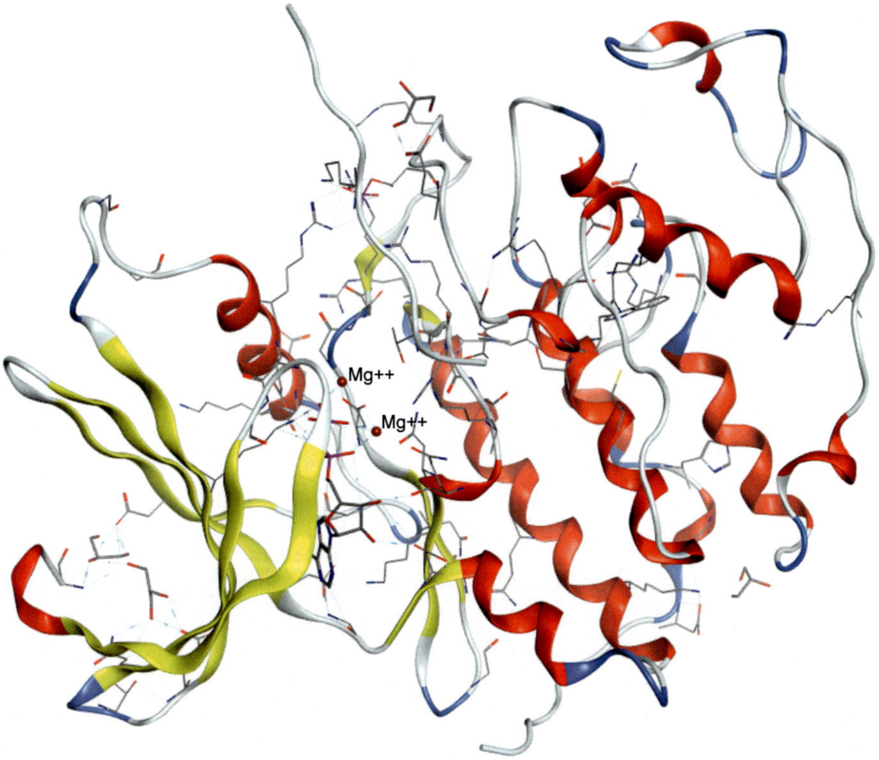

FIG. 4 The active form of CDK2 catalytic subunit, which contains ATP and illustrating the location of Mg^{2+} (1) and Mg^{2+} (2) (PDB ID: 3QHR).

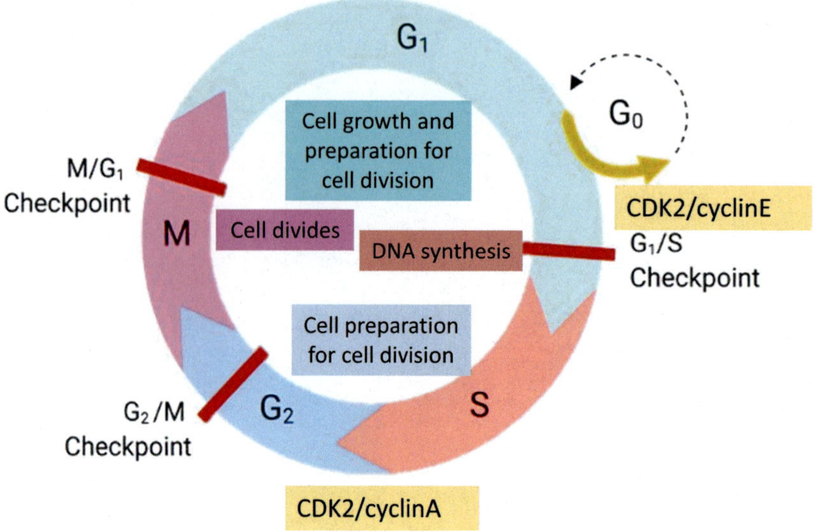

FIG. 5 Cell cycler regulation by CDK2: The descriptive diagram depicts the several phases of the cell division cycle, as well as the function of CDK2 and its cyclin counterparts.

3 Classes of inhibitors and their design

3.1 ATP-competitive CDK2 inhibitors

Identifying the protein-ligand interactions provides insight into the features of a specific drug. Co-crystallizing the ligand with the protein to examine CDK2-ligand interactions has revealed a lot of information about the interactions generated by the ligand in the ATP-binding pocket, CDK2/ligand crystal structures demonstrate that all reported binding modes entail at least one hydrogen bond with Leu83 backbone (Leu83-NH which exists in the hinge region). The ligand may also bind with Glu81-CO and/or Leu83-CO [48]. It is worth to mention that these inhibitors are type I kinase inhibitors. Type I inhibitors interact with the ATP-binding pocket directly and bind with the kinase DFG-in active form in addition to a phosphorylated activation loop. The hydrogen bonding interactions generated between the adenine moiety of ATP and the enzyme hinge region are mimicked by these inhibitors [49].

Literature review demonestrated that various scaffolds have CDK2 inhibitory activity, namely purines, pyrazolopyrimidines, pyrazolopyridines, and quinazolines and others [10,48,50].

3.1.1 Type I CDK2 inhibitors

Purine-based CDK2 inhibitors

Purine is a favored scaffold that can be found in the chemical framework of a variety of bioactive chemical compounds, such as ATP and diverse kinase inhibitors. 2,6,9-Trisubstituted purine-based CDK2 inhibitors were among the first to be established as anticancer drugs. Roscovitine **1** and H717 **2**, for example, are 2,6,9-trisubstituted purine CDK2 inhibitors and are also inhibitors of other CDKs (Fig. 6).

The structures of CDK2 co-crystallized with R-roscovitine ($IC_{50}=0.7\,\mu M$) **1** and H717 ($IC_{50}=48\,nM$) **2** revealed that both ligands achieved similar binding patterns in the ATP-binding site (Fig. 9), forming two preserved hydrogen bonds with the backbone amino and carbonyl groups of Leu83 where the crucial hetero-bicyclic ring structure was located in the ATP binding pocket, forming polar interactions to gain access to the phosphate-binding region of the ATP-binding site. Another hydrogen bond is exploited with the carbonyl of Glu-81. The N-9 substituent of H717 is buried in the Phe80 pocket. The benzylamino substituent of roscovitine exhibits π-π interaction with Ile10. Moreover, extra interactions with backbone atoms of Glu12, Asp145, and Asn132 in the glycine-rich loop are observed [51,52].

Inspired by CDK2 inhibitory activity of roscovitine **1**, Liang and coworkers designed and synthesized new 2-aminopurine derivatives depending on the fragment-centric pocket mapping analysis of CDK2 crystalline structure. The most active compound **3** displayed high CDK2 inhibition ($IC_{50}=0.019\,\mu M$) and possessed better anticancer activity in MDA-MB-231 cells than roscovitine (Fig. 6) [53].

Pyrazolopyrimidine-based CDK2 inhibitors

The co-crystal structure of CDK2 with dinaciclib **4** -(pyrazolopyrimidine-based CDK2 inhibitor with $IC_{50}=1\,nM$) demonstrated a crucial hydrogen bonding between dinaciclib, amino-pyrazolopyrimidine, and Leu83 (Fig. 9). The ethyl group filled a hydrophobic pocket and developed hydrophobic interactions with the gatekeeper residues, dinaciclib also expoited multiple hydrophobic interactions in the binding pocket [54].

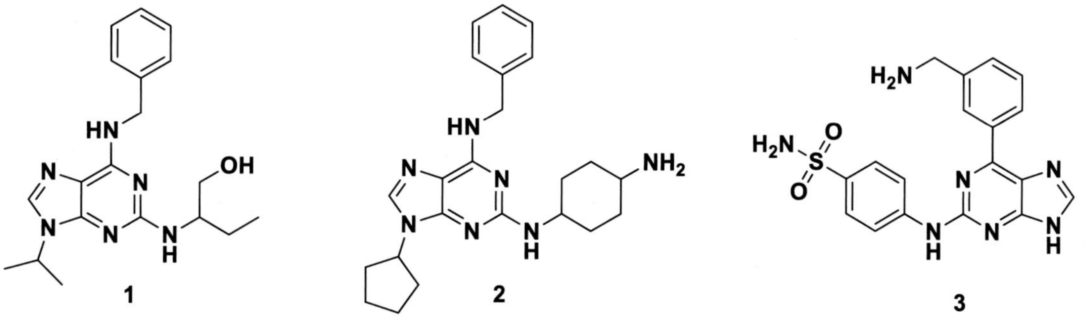

1 Roscovitine

2

3

FIG. 6 Chemical structures of some purine-based CDK2 inhibitors.

In search for new selective CDK2 inhibitors, Cherukupalli et al. synthesized a new series of of 4,6-disubstituted pyrazolo [3,4-*d*]pyrimidines (bioisosteres of purines) hybridized with various anilines at C-4 position and thiophenethyl or thiopentane moieties at C-6 position having the general formulas **5** and **6** (Fig. 7). Compounds **5a, 5b, 6a,** and **6b** displayed significant activity against CDK2/cyclin E with IC_{50} values ranging from 5.1 μM to 8.8 μM and succeeded to achieve strong H bond interactions with NH of Leu83 and C=O of Glu81, squentially. Moreover, the titled compounds exhibited cytotoxic activity against MCF-7 and K-562 cancer cell lines with IC_{50} values in a micromolar range [55].

Almehmadi and coworkers developed new derivatives bearing pyrazolo[1,5-α]pyrimidine core. Compunds **7a** and **7b** (Fig. 7) displayed comparable activity to that of the reference dinaciclib **4** against CDK2 (IC_{50} = 0.022 and 0.024 μM, respectively), MOLT-4, and HL-60 leukemia cell lines. It is worth to mention that the adopted design strategy relied on conserving the crucial binding interactions of dinaciclib with the ATP-binding pocket of CDK2 by preserving the planar pyrazolo[1,5-*a*]pyrimidine scaffold, which is attached to azo group to be able to bind to the essential residues of the hinge region via hydrogen bonds. The hydrophobic pocket inhabited by the ethyl group of dinaciclib

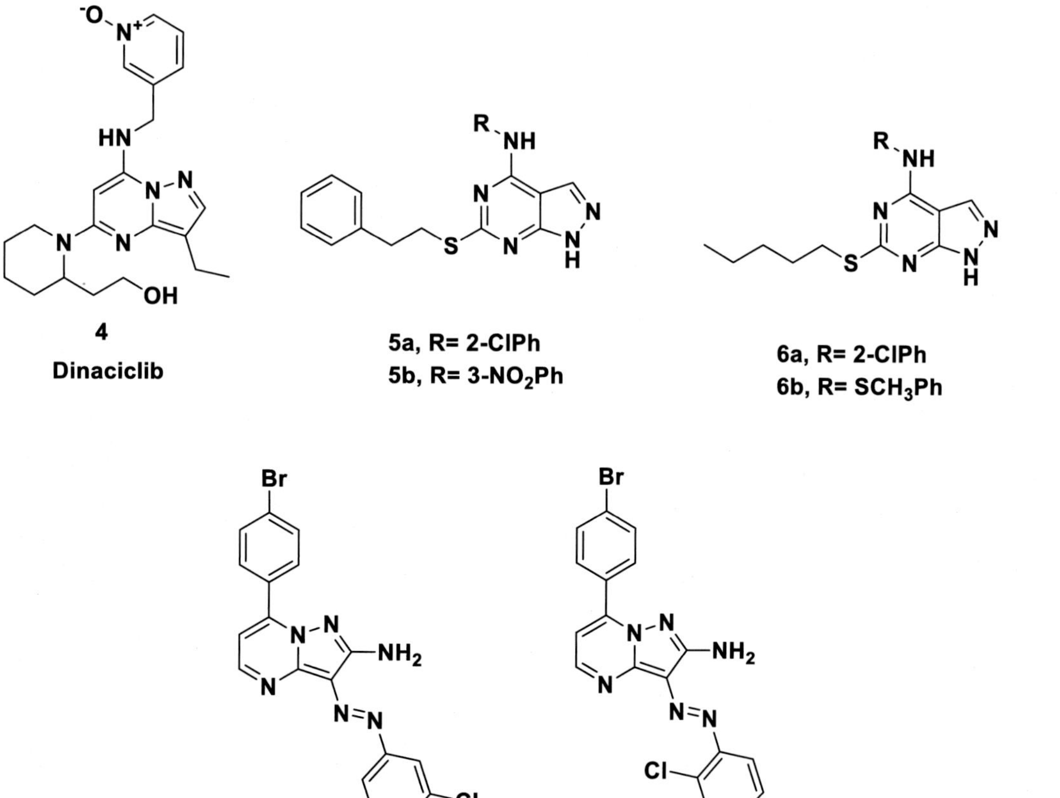

4 Dinaciclib

5a, R= 2-ClPh
5b, R= 3-NO₂Ph

6a, R= 2-ClPh
6b, R= SCH₃Ph

7a

7b

FIG. 7 Chemical structures of some pyrazolopyrimidine-based CDK2 inhibitors.

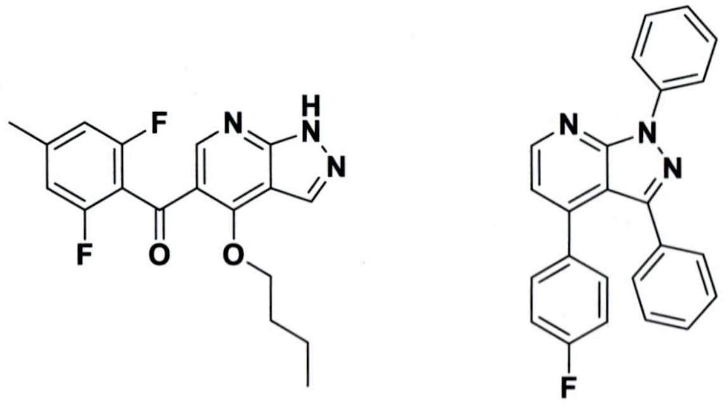

8 9

FIG. 8 Chemical structures of some pyridine-based CDK2 inhibitors.

was assumed to be occupied by the aryl fragment at the 7-position of the pyrazolo[1,5-α]pyrimidine. The aryl group linked to the azo moiety could also be directed toward the solvent-accessible area [56].

Pyridine-based CDK2 inhibitors

Bioisosteric replacement of the common pyrazolopyrimidine moiety of dinaciclib **4** with pyrazolopyridine resulted in potent CDK2 inhibition at a nanomolar level as in compound **8** (IC_{50} = 9 nM) [13,57]. Accordingly, new pyrazolo[3,4-*b*]pyridines were developed as CDK2 inhibitors and compound **9** (Fig. 8) displayed the highest CDK2 inhibitory activity among this series with IC_{50} of 0.0415 μM [58].

Quinazoline-based CDK2 inhibitors

Several quinazolines have been recognized as CDK2 inhibitors [59]. DIN-234325 (IC_{50} = 0.6 μM) **10** is a quinazoline-based CDk2 inhibitor. It interacts with the ATP-binding pocket of CDK2. The N1 of the quinazoline ring binds with the essential amino acids Glu81 and Leu83 via water-mediated hydrogen bond interactions as shown in Fig. 9. Val18 and Leu134 residues formed stacking interactions with the quinazoline ring (Fig. 9). Furthermore, the aniline moiety linked to the quinazoline ring at position 6 created a pi-staking contact with Gln85 [60].

Urged by the previous findings, some researchers characterized new series of quinazolin-4(3*H*)-ones as CDK2 inhibitors and apoptotic inducers, recently (Fig. 10). The most active CDK2 inhibitor among this series **11**

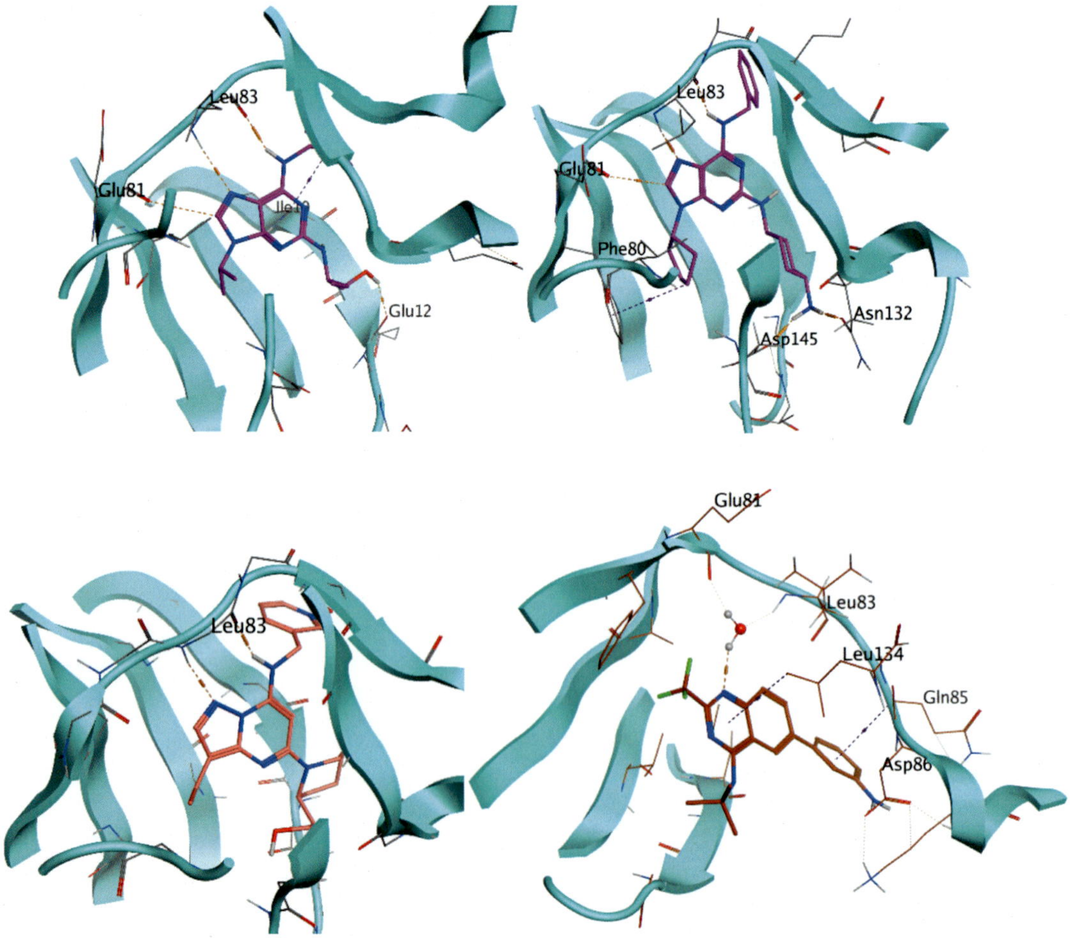

FIG. 9 3D diagrams of roscovitine (PDB ID:2A4L), H717 (PDB ID: 1G5S), dinaciclib (PDB ID: 5L2W), quinazoline CDK2 inhibitor (PDB ID:2B53) bound to CDK2 and exhibiting type I binding mode.

(IC_{50} = 0.63 μM against CDK2) attained promising activity in melanoma MDA-MB-435 and glioblastoma SNB-75 cell lines (GI% = 94.53 and 63.09, respectively) [61].

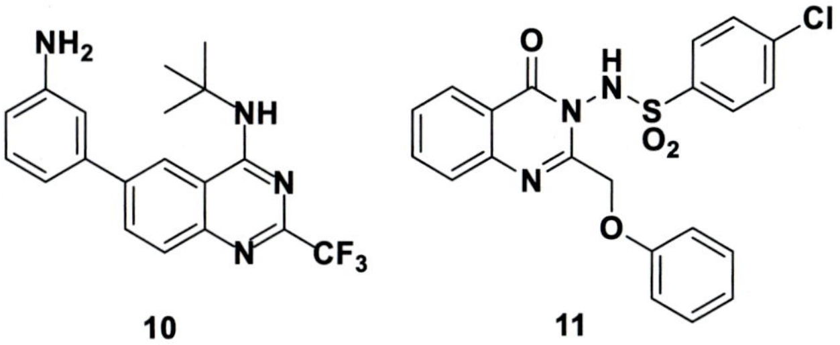

10

DIN-234325

11

FIG. 10 Chemical structures of some quinazoline-based CDK2 inhibitors.

3.2 Type II inhibitors

Type II inhibitors bind with the inactive DFG-out section of the enzyme and probe the hinge region and adenine binding pocket, similar to type I inhibitors [49,62].

K03861 (AUZ454) **12**, an aminopyrimidine-phenyl urea derivative (Fig. 11), is an example of an inhibitor that can assume the DFG-out conformation. This is a CDK2 inhibitor of class II, with a Kd of 50nM for CDK2. The co-crystal structure of K03861 bound to CDK2 adopts DFG-out conformation as illustrated in Fig. 12 (PDB ID: 5A14) [63].

Recently, novel series of CDK2 type II inhibitors were developed using a hybridization strategy involving the building units benzofuran and piperazine. Compounds **13a and 13b** (Fig. 11) revealed the most potent inhibitory activity among these series (IC_{50} of 40.91 and 41.70 nM, respectively) [64].

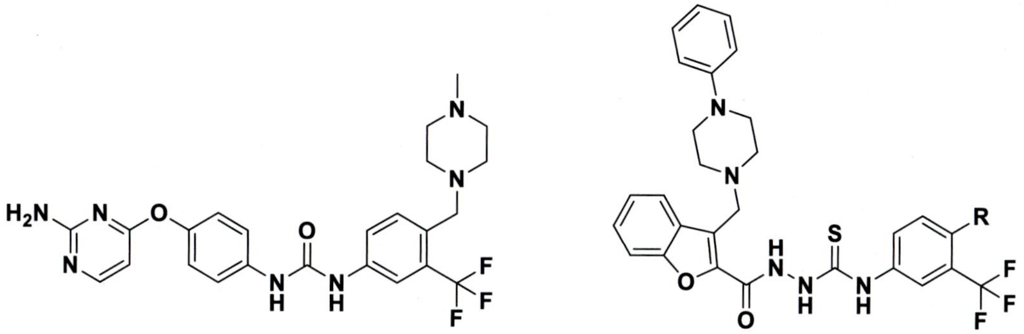

12

K03861

13a, R= H
13b, R= Cl

FIG. 11 Chemical structures of some type II CDK2 inhibitors.

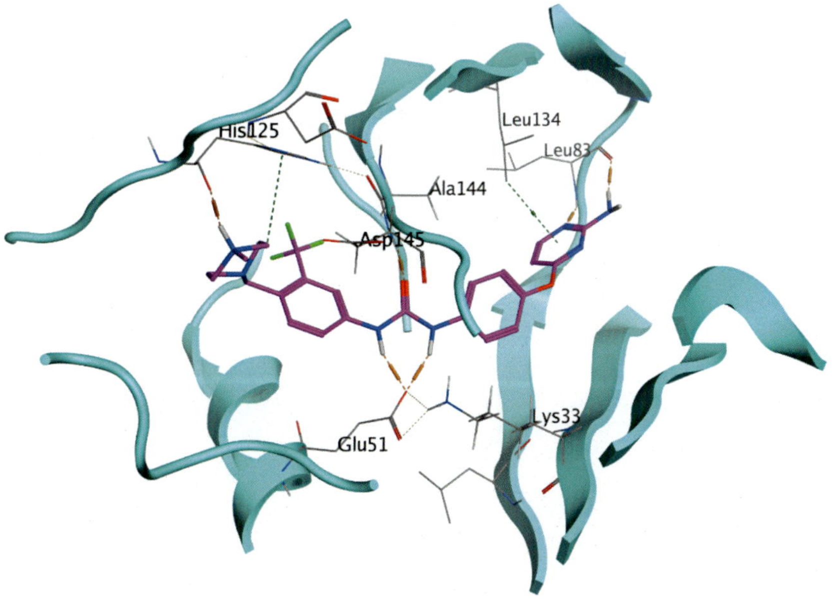

FIG. 12 3D diagrams of K03861 (PDB ID:5A14) bound to CDK2 and exhibiting type II binding mode.

3.3 Allosteric inhibitors (type IV CDK2 inhibitors)

Since CDK2 was identified as a potential cancer therapeutic target, various CDK2 inhibitors have been developed and thoroughly investigated for their preferential binding to CDK2 ATP-binding site. However, the highly conserved structure of the ATP-binding site relative to the other CDKs is the most significant impediment to constructing selective ATP competitive CDK2 inhibitors [65,66]. As a result, identifying additional binding pockets other than the catalytic

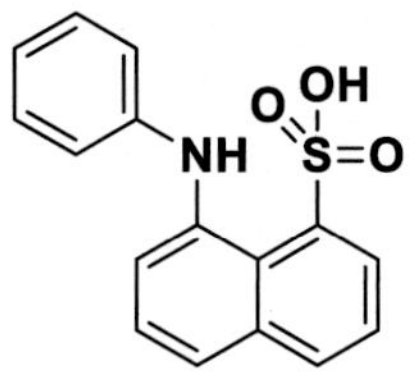

14

FIG. 13 Chemical structure of 8-anilino-1-naphthalene sulfonate (ANS), the type IV CDK2 inhibitor.

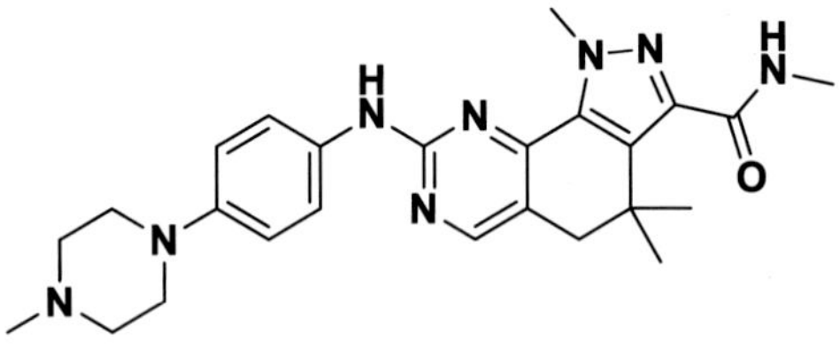

15
Milciclib

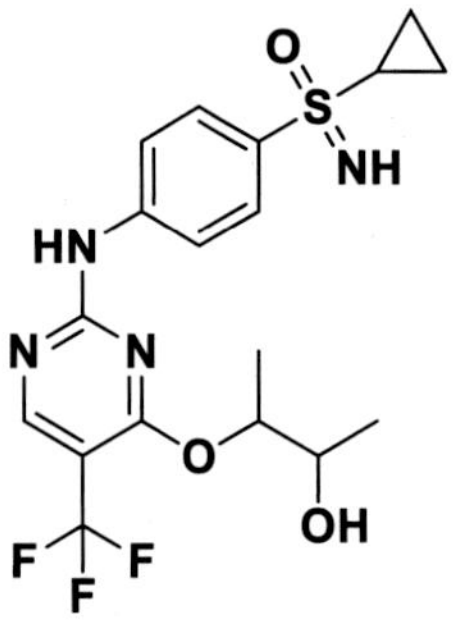

16
Roniciclib

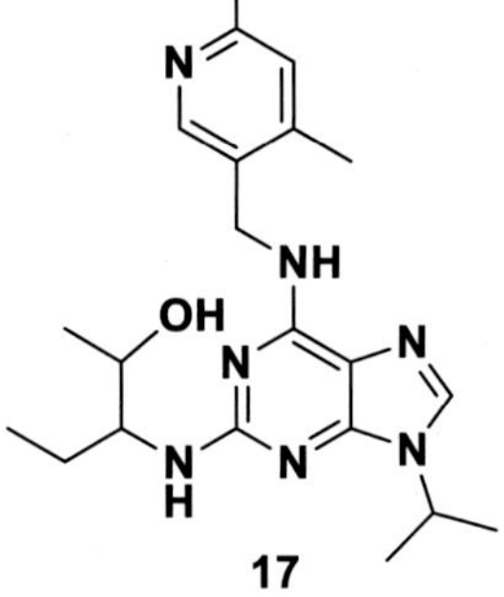

17
Fadraciclib

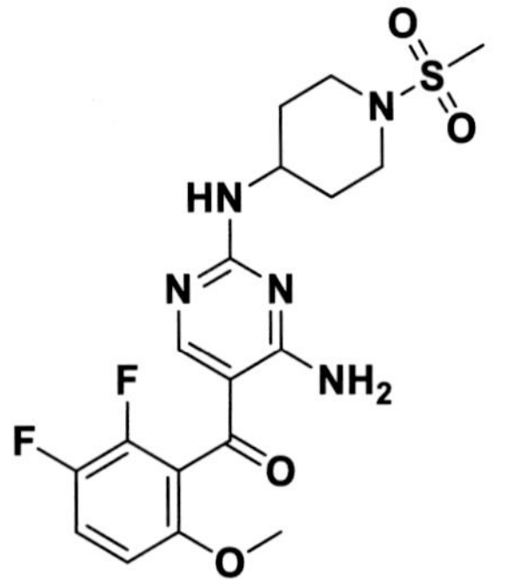

18
R-547

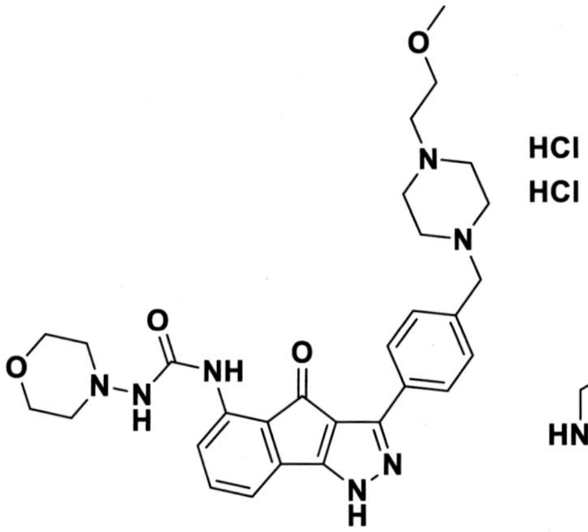

19
RGB-286638

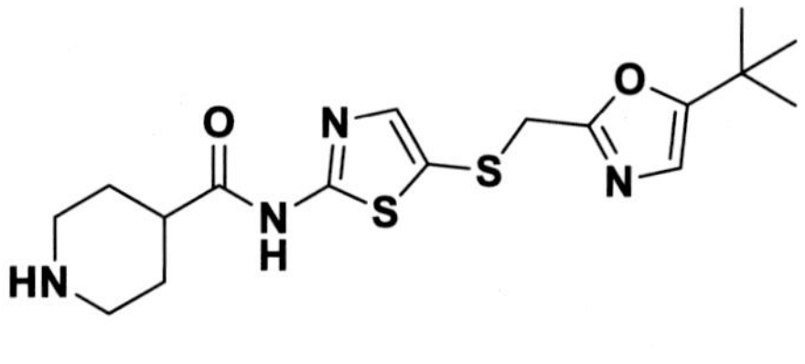

20
SNS-032

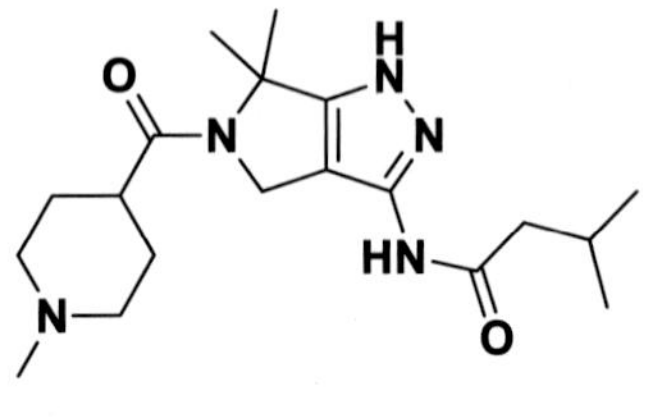

21
PHA-793887

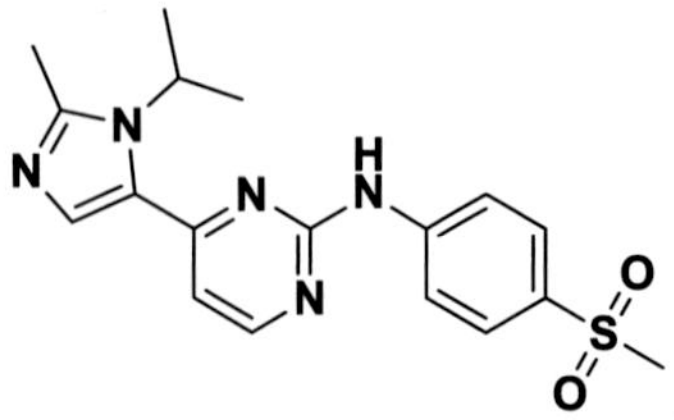

22
AZD5438

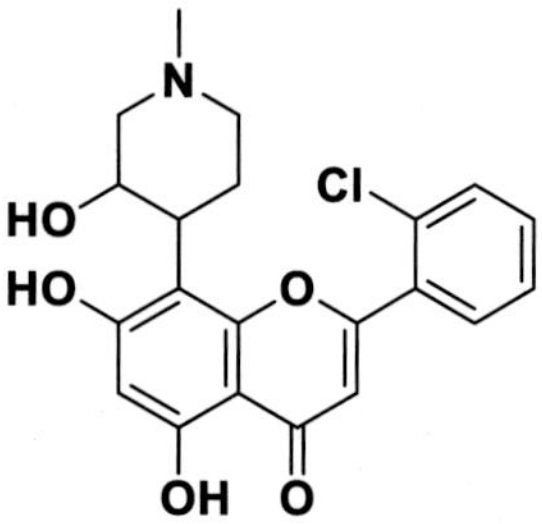

23
Flavopiridol

FIG. 14 Structures of some CDK2 inhibitors in clinical trials.

binding site is critical for the generation of a selective CDK2 inhibitor. Consequently, type IV inhibitors were identified as compounds that bind to distinctive structural pockets far from the ATP-binding pocket and interact with these allosteric areas by stabilizing inactive conformations. Except for the hydrophobic pocket around the ATP-binding site, which is the target of type III inhibitors, the allosteric pocket of type IV inhibitors can be found throughout the kinase [67]. In case of type III CDK2 inhibitors, no record of such a molecule exists till now.

Probing the extrinsic fluorophore 8-anilino-1-naphthalene sulfonate (ANS) **14** (Fig. 13) has revealed a novel allosteric ligand-binding site, which is nearly situated halfway between the ATP-binding site and C-helix as shown in Fig. 2. ANS **14** inhibits CDK2 (IC_{50} of 91 μM) by forming hydrogen bonds, hydrophobic contacts, and van der Waals interactions with the residues of the binding site. Despite its poor inhibitory effect, ANS binding to CDK2 causes remarkable conformational variations in the protein structure [14,68].

4 Clinically used agents or compounds in clinical development

As previously discussed, a variety of CDK2 inhibitors with various scaffolds have been developed over the past few decades. Some of these inhibitors have entered clinical trials as a result of their exceptional anticancer capabilities. However, not a single one has received commercial approval as an anticancer agent.

Clinical trials are being conducted on many promising CDK2 inhibitors, such as dinaciclib **4** (a multi-CDK inhibitor), milciclib, roniciclib, fadraciclib, and ribociclib. The most famous CDK2 inhibitors in clinical trials are illustrated in Fig. 14 and Table 1.

However, off-target effects, misregulated expression of CDK2 in different malignancies and the difficulty of designing selective CDK2 inhibitors proposed that more extensive studies will be necessary for developing new CDK2 inhibitors as chemotherapeutic agents.

TABLE 1 Some CDK2 inhibitors undergoing different phases of clinical and preclinical trials.

Compound	Condition	Clinical phase	Recruitment status
Roscovitine (Seliciclib) **1**	Breast cancer	I	Withdrawn
	Non-small-cell lung cancer (NSCLC)	II	Terminated
	Cushing disease	II	Terminated
	Advanced solid tumors	I	Completed
Dinaciclib **4**	Breast and lung cancer	II	Completed
	Acute myeloid leukemia	I	Active, not recruiting
	Solid tumors	I	Recruiting
Milciclib **15**	Malignant thymoma	II	Terminated
Roniciclib **16**	NSCLC	II	Withdrawn
	Solid tumors	II	Withdrawn
Fadraciclib **17**	Solid tumor and lymphoma	II	Recruiting
	Leukemia	I	Recruiting
	Myelodysplastic syndromes (MDS)	I	Recruiting
R547 **18**	Solid tumors	I	Completed
RGB-286638 **19**	Hematological malignancies	I	Withdrawn
SNS-032 **20**	Solid tumors	I	Completed
	Chronic lymphocytic leukemia	I	Completed
PHA-793887 **21**	Solid tumors	I	Terminated
AZD5438 **22**	Solid tumors	I	Completed

Continued

TABLE 1 Some CDK2 inhibitors undergoing different phases of clinical and preclinical trials—cont'd

Compound	Condition	Clinical phase	Recruitment status
Flavopiridol (Alvocidib) **23**	Esophageal cancer	II	Completed
	Breast cancer	I	Completed
	Pancreatic cancer	II	Completed
	Unspecified adult solid tumor	I	Terminated
	Liver cancer	II	Completed
	Prostate tumor	II	Completed

Information retrieved from https://clinicaltrials.gov/, accessed on 16 June 2022.

References

[1] Cormier KW, Woodgett JR. Protein kinases: Physiological roles in cell signalling. In: eLS. John Wiley & Sons, Ltd; 2016. p. 1–9.

[2] Ardito F, Giuliani M, Perrone D, Troiano G, Lo ML. The crucial role of protein phosphorylation in cell signaling and its use as targeted therapy (review). Int J Mol Med 2017;40(2):271–80.

[3] Besson A, Dowdy SF, Roberts JM. CDK inhibitors: cell cycle regulators and beyond. Dev Cell 2008;14(2):159–69.

[4] Chohan TA, Qian H, Pan Y, Chen J-Z. Cyclin-dependent kinase-2 as a target for cancer therapy: progress in the development of CDK2 inhibitors as anti-cancer agents. Curr Med Chem 2015;22(2):237–63.

[5] Tadesse S, Anshabo AT, Portman N, Lim E, Tilley W, Caldon CE, et al. Targeting CDK2 in cancer: challenges and opportunities for therapy. Drug Discov Today 2020;25(2):406–13.

[6] Zhang J, Yang PL, Gray NS. Targeting cancer with small molecule kinase inhibitors. Nat Rev Cancer 2009;9(1):28–39.

[7] Vivo MD, Bottegoni G, Berteotti A, Recanatini M, Gervasio FL, Cavalli A. Cyclin-dependent kinases: bridging their structure and function through computations. Future Med Chem 2011;3 (12):1551–9.

[8] Cheng W, Yang Z, Wang S, Li Y, Wei H, Tian X, et al. Recent development of CDK inhibitors: an overview of CDK/inhibitor co-crystal structures. Eur J Med Chem 2019;164:615–39.

[9] Floquet N, Costa MG, Batista PR, Renault P, Bisch PM, Raussin F, et al. Conformational equilibrium of CDK/cyclin complexes by molecular dynamics with excited Normal modes. Biophys J 2015;109(6):1179–89.

[10] Marak BN, Dowarah J, Khiangte L, Singh VP. A comprehensive insight on the recent development of cyclic dependent kinase inhibitors as anticancer agents. Eur J Med Chem 2020;203, 112571.

[11] Pavletich NP. Mechanisms of cyclin-dependent kinase regulation: structures of Cdks, their cyclin activators, and Cip and INK4 inhibitors. J Mol Biol 1999;287(5):821–8.

[12] Honda R, Lowe ED, Dubinina E, Skamnaki V, Cook A, Brown NR, et al. The structure of cyclin E1/CDK2: implications for CDK2 activation and CDK2-independent roles. EMBO J 2005;24(3):452–63.

[13] Tadesse S, Caldon EC, Tilley W, Wang S. Cyclin-dependent kinase 2 inhibitors in cancer therapy: an update. J Med Chem 2019;62 (9):4233–51.

[14] Betzi S, Alam R, Martin M, Lubbers DJ, Han H, Jakkaraj SR, et al. Discovery of a potential allosteric ligand binding site in CDK2. ACS Chem Biol 2011;6(5):492–501.

[15] Rastelli G, Anighoro A, Chripkova M, Carrassa L, Broggini M. Structure-based discovery of the first allosteric inhibitors of cyclin-dependent kinase 2. Cell Cycle 2014;13(14):2296–305.

[16] Asghar U, Witkiewicz AK, Turner NC, Knudsen ES. The history and future of targeting cyclin-dependent kinases in cancer therapy. Nat Rev Drug Discov 2015;14(2):130–46.

[17] Bao ZQ, Jacobsen DM, Young MA. Briefly bound to activate: transient binding of a second catalytic magnesium activates the structure and dynamics of CDK2 kinase for catalysis. Structure 2011;19 (5):675–90.

[18] Shaffer J, Adams JA. Detection of conformational changes along the kinetic pathway of protein kinase A using a catalytic trapping technique. Biochemistry 1999;38(37):12072–9.

[19] Lapenna S, Giordano A. Cell cycle kinases as therapeutic targets for cancer. Nat Rev Drug Discov 2009;8(7):547–66.

[20] Matsuura I, Denissova NG, Wang G, He D, Long J, Liu F. Cyclin-dependent kinases regulate the antiproliferative function of Smads. Nature 2004;430(6996):226–31.

[21] Chuang L-C, Teixeira LK, Wohlschlegel JA, Henze M, Yates JR, Méndez J, et al. Phosphorylation of Mcm2 by Cdc7 promotes pre-replication complex assembly during cell-cycle re-entry. Mol Cell 2009;35(2):206–16.

[22] Adon AM, Zeng X, Harrison MK, Sannem S, Kiyokawa H, Kaldis P, et al. Cdk2 and Cdk4 regulate the centrosome cycle and are critical mediators of centrosome amplification in p53-null cells. Mol Cell Biol 2010;30(3):694–710.

[23] Hu S, Danilov AV, Godek K, Orr B, Tafe LJ, Rodriguez-Canales J, et al. CDK2 inhibition causes anaphase catastrophe in lung cancer through the centrosomal protein CP110. Cancer Res 2015;75 (10):2029–38.

[24] Berthet C, Aleem E, Coppola V, Tessarollo L, Kaldis P. Cdk2 knockout mice are viable. Curr Biol 2003;13(20):1775–85.

[25] Long XE, Gong ZH, Pan L, Zhong ZW, Le YP, Liu Q, et al. Suppression of CDK2 expression by siRNA induces cell cycle arrest and cell proliferation inhibition in human cancer cells. BMB Rep 2010;43(4):291–6.

[26] Deans AJ, Khanna KK, McNees CJ, Mercurio C, Heierhorst J, McArthur GA. Cyclin-dependent kinase 2 functions in normal DNA repair and is a therapeutic target in BRCA1-deficient cancers. Cancer Res 2006;66(16):8219–26.

[27] Merrick KA, Wohlbold L, Zhang C, Allen JJ, Horiuchi D, Huskey NE, et al. Switching Cdk2 on or off with small molecules to reveal requirements in human cell proliferation. Mol Cell 2011;42(5):624–36.

[28] Horiuchi D, Huskey NE, Kusdra L, Wohlbold L, Merrick KA, Zhang C, et al. Chemical-genetic analysis of cyclin dependent kinase 2 function reveals an important role in cellular transformation by multiple oncogenic pathways. Proc Natl Acad Sci U S A 2012;109(17): E1019–27.
[29] Akli S, Van Pelt CS, Bui T, Meijer L, Keyomarsi K. Cdk2 is required for breast cancer mediated by the low-molecular-weight isoform of cyclin E. Cancer Res 2011;71(9):3377–86.
[30] Faber AC, Chiles TC. Inhibition of cyclin-dependent Kinase-2 induces apoptosis in human diffuse large B-cell lymphomas. Cell Cycle 2007;6(23):2982–9.
[31] Marone M, Scambia G, Giannitelli C, Ferrandina G, Masciullo V, Bellacosa A, et al. Analysis of cyclin E and CDK2 in ovarian cancer: gene amplification and RNA overexpression. Int J Cancer 1998;75 (1):34–9.
[32] Oshita T, Shigemasa K, Nagai N, Ohama K. p27, cyclin E, and CDK2 expression in normal and cancerous endometrium. Int J Oncol 2002;21(4):737–43.
[33] Desai BM, Villanueva J, Nguyen T-TK, Lioni M, Xiao M, Kong J, et al. The anti-melanoma activity of Dinaciclib, a cyclin-dependent kinase inhibitor, is dependent on p53 signaling. PLoS One 2013;8 (3), e59588.
[34] Du J, Widlund HR, Horstmann MA, Ramaswamy S, Ross K, Huber WE, et al. Critical role of CDK2 for melanoma growth linked to its melanocyte-specific transcriptional regulation by MITF. Cancer Cell 2004;6(6):565–76.
[35] Cam WR, Masaki T, Shiratori Y, Kato N, Okamoto M, Yamaji Y, et al. Activation of cyclin E-dependent kinase activity in colorectal cancer. Dig Dis Sci 2001;46(10):2187–98.
[36] Yamamoto H, Monden T, Miyoshi H, Izawa H, Ikeda K, Tsujie M, et al. Cdk2/cdc2 expression in colon carcinogenesis and effects of cdk2/cdc2 inhibitor in colon cancer cells. Int J Oncol 1998;13 (2):233–9.
[37] Juric V, Murphy B. Cyclin-dependent kinase inhibitors in brain cancer: current state and future directions. Cancer Drug Resist 2020;3(1):48–62.
[38] Wang J, Yang T, Xu G, Liu H, Ren C, Xie W, et al. Cyclin-dependent kinase 2 promotes tumor proliferation and induces radio resistance in glioblastoma. Transl Oncol 2016;9(6):548–56.
[39] Riess C, Irmscher N, Salewski I, Strüder D, Classen CF, Große-Thie C, et al. Cyclin-dependent kinase inhibitors in head and neck cancer and glioblastoma-backbone or add-on in immune-oncology? Cancer Metastasis Rev 2021;40(1):153–71.
[40] Yin X, Yu J, Zhou Y, Wang C, Jiao Z, Qian Z, et al. Identification of CDK2 as a novel target in treatment of prostate cancer. Future Oncol 2018;14(8):709–18.
[41] Huang H, Regan KM, Lou Z, Chen J, Tindall DJ. CDK2-dependent phosphorylation of FOXO1 as an apoptotic response to DNA damage. Science 2006;314(5797):294–7.
[42] Huang H, Tindall DJ. CDK2 and FOXO1: a fork in the road for cell fate decisions. Cell Cycle 2007;6(8):902–6.
[43] Choudhary GS, Al-Harbi S, Mazumder S, Hill BT, Smith MR, Bodo J, et al. MCL-1 and BCL-xL-dependent resistance to the BCL-2 inhibitor ABT-199 can be overcome by preventing PI3K/AKT/mTOR activation in lymphoid malignancies. Cell Death Dis 2015;6(1), e1593.
[44] Choudhary GS, Tat TT, Misra S, Hill BT, Smith MR, Almasan A, et al. Cyclin E/Cdk2-dependent phosphorylation of Mcl-1 determines its stability and cellular sensitivity to BH3 mimetics. Oncotarget 2015;6(19):16912–25.
[45] Shieh SY, Ahn J, Tamai K, Taya Y, Prives C. The human homologs of checkpoint kinases Chk1 and Cds1 (Chk2) phosphorylate p53 at multiple DNA damage-inducible sites. Genes Dev 2000;14(3):289–300.
[46] Mailand N, Falck J, Lukas C, Syljuâsen RG, Welcker M, Bartek J, et al. Rapid destruction of human Cdc25A in response to DNA damage. Science 2000;288(5470):1425–9.
[47] Müller-Tidow C, Ji P, Diederichs S, Potratz J, Bäumer N, Köhler G, et al. The cyclin A1-CDK2 complex regulates DNA double-strand break repair. Mol Cell Biol 2004;24(20):8917–28.
[48] Vulpetti A, Pevarello P. An analysis of the binding modes of ATP-competitive CDK2 inhibitors as revealed by X-ray structures of protein-inhibitor complexes. Curr Med Chem Anticancer Agents 2005;5(5):561–73.
[49] Li Y, Zhang J, Gao W, Zhang L, Pan Y, Zhang S, et al. Insights on structural characteristics and ligand binding mechanisms of CDK2. Int J Mol Sci 2015;16(5):9314–40.
[50] Łukasik P, Baranowska-Bosiacka I, Kulczycka K, Gutowska I. Inhibitors of cyclin-dependent kinases: types and their mechanism of action. Int J Mol Sci 2021;22(6).
[51] De Azevedo WF, Leclerc S, Meijer L, Havlicek L, Strnad M, Kim S-H. Inhibition of cyclin-dependent kinases by purine analogues. Eur J Biochem 1997;243(1–2):518–26.
[52] Dreyer MK, Borcherding DR, Dumont JA, Peet NP, Tsay JT, Wright PS, et al. Crystal structure of human cyclin-dependent kinase 2 in complex with the adenine-derived inhibitor H717. J Med Chem 2001;44(4):524–30.
[53] Liang H, Zhu Y, Zhao Z, Du J, Yang X, Fang H, et al. Structure-based design of 2-aminopurine derivatives as CDK2 inhibitors for triple-negative breast cancer. Front Pharmacol 2022;13.
[54] Chen P, Lee NV, Hu W, Xu M, Ferre RA, Lam H, et al. Spectrum and degree of CDK drug interactions predicts clinical performance. Mol Cancer Ther 2016;15(10):2273–81.
[55] Cherukupalli S, Chandrasekaran B, Kryštof V, Aleti RR, Sayyad N, Merugu SR, et al. Synthesis, anticancer evaluation, and molecular docking studies of some novel 4,6-disubstituted pyrazolo[3,4-d] pyrimidines as cyclin-dependent kinase 2 (CDK2) inhibitors. Bioorg Chem 2018;79:46–59.
[56] Almehmadi SJ, Alsaedi AMR, Harras MF, Farghaly TA. Synthesis of a new series of pyrazolo[1,5-a]pyrimidines as CDK2 inhibitors and anti-leukemia. Bioorg Chem 2021;117, 105431.
[57] Misra RN, Xiao H, Rawlins DB, Shan W, Kellar KA, Mulheron JG, et al. 1H-Pyrazolo[3,4-b]pyridine inhibitors of cyclin-dependent kinases: highly potent 2,6-Difluorophenacyl analogues. Bioorg Med Chem Lett 2003;13(14):2405–8.
[58] Hassan GS, Georgey HH, Mohammed EZ, George RF, Mahmoud WR, Omar FA. Mechanistic selectivity investigation and 2D-QSAR study of some new antiproliferative pyrazoles and pyrazolopyridines as potential CDK2 inhibitors. Eur J Med Chem 2021;218, 113389.
[59] Shewchuk L, Hassell A, Wisely B, Rocque W, Holmes W, Veal J, et al. Binding mode of the 4-anilinoquinazoline class of protein kinase inhibitor: X-ray crystallographic studies of 4-anilinoquinazolines bound to cyclin-dependent kinase 2 and p38 kinase. J Med Chem 2000;43(1):133–8.
[60] Sielecki TM, Johnson TL, Liu J, Muckelbauer JK, Grafstrom RH, Cox S, et al. Quinazolines as cyclin dependent kinase inhibitors. Bioorg Med Chem Lett 2001;11(9):1157–60.
[61] Mohammed ER, Elmasry GF. Development of newly synthesised quinazolinone-based CDK2 inhibitors with potent efficacy against melanoma. J Enzyme Inhib Med Chem 2022;37(1):686–700.

[62] Zhao Z, Wu H, Wang L, Liu Y, Knapp S, Liu Q, et al. Exploration of type II binding mode: a privileged approach for kinase inhibitor focused drug discovery? ACS Chem Biol 2014;9(6):1230–41.
[63] Alexander LT, Möbitz H, Drueckes P, Savitsky P, Fedorov O, Elkins JM, et al. Type II inhibitors targeting CDK2. ACS Chem Biol 2015;10 (9):2116–25.
[64] Eldehna WM, Maklad RM, Almahli H, Al-Warhi T, Elkaeed EB, Abourehab MAS, et al. Identification of 3-(piperazinylmethyl) benzofuran derivatives as novel type II CDK2 inhibitors: design, synthesis, biological evaluation, and in silico insights. J Enzyme Inhib Med Chem 2022;37(1):1227–40.
[65] Duca JS. Recent advances on structure-informed drug discovery of cyclin-dependent kinase-2 inhibitors. Future Med Chem 2009;1 (8):1453–66.
[66] Fischer PM, Lane DP. Inhibitors of cyclin-dependent kinases as anti-cancer therapeutics. Curr Med Chem 2000;7 (12):1213–45.
[67] Gavrin LK, Saiah E. Approaches to discover non-ATP site kinase inhibitors. Med Chem Commun 2013;4(1):41–51.
[68] Yueh C, Rettenmaier J, Xia B, Hall DR, Alekseenko A, Porter KA, et al. Kinase atlas: druggability analysis of potential allosteric sites in kinases. J Med Chem 2019;62(14):6512–24.

Chapter 2.5

Catechol-*O*-methyltransferase (COMT)

Giusy Tassone[a], Simone Carradori[b], Samuele Maramai[a], and Ilaria D'Agostino[b,*]

[a]*Department of Biotechnology, Chemistry and Pharmacy, University of Siena, Siena, Italy,* [b]*Department of Pharmacy, University "G. d'Annunzio" of Chieti-Pescara, Chieti, Italy*

1 Structure and function of the catechol-*O*-methyltransferase (COMT) enzyme

Methyltransferases (MTases, EC 2.1.1) are enzymes that catalyze the methylation reaction to their substrate, which consists of the transfer of a methyl group. The substrates of MTases are molecules involved in a wide variety of biological processes, such as biosynthesis, metabolism, and protein repair. MTases are grouped into five classes (Classes I–V), distinguished by different structural folds and substrate preferences [1,2]. The vast majority of MTases belong to Class I, including the catechol-*O*-methyltransferase (COMT, EC 2.1.1.6) enzyme, responsible for the metabolic *O*-methylation of endogenous catecholamines neurotransmitters and catechol estrogens (CEs) [3]. COMT is a ubiquitous magnesium-dependent enzyme that catalyzes the transfer of a methyl group from its cofactor *S*-adenosyl-L-methionine (SAM or AdoMet) to one of the phenolic hydroxyl groups in a catechol substrate, producing the corresponding mono-*O*-methoxyphenol and releasing *S*-adenosyl-L-homocysteine (SAH or AdoHcy) (Fig. 1A) [4–8]. The methylation reaction is regioselective and occurs via a sequentially second-ordered mechanism [9,10]. During the catalytic cycle, the binding sequence of SAM, the Mg(II) ion, and the substrate to the enzyme is strictly maintained. SAM binds first to the enzyme, followed by the Mg(II) ion and the substrate [6,11,12]. COMT is the primary enzyme that inactivates dopamine (DOPA), a catechol neurotransmitter that plays a key role in pleasure, motivation, and learning [13]. Other physiological substrates of COMT are catecholamine neurotransmitters such as noradrenaline, adrenaline, and their metabolites [13]. COMT also inactivates catechol hormones, such as CEs, and a plethora of other catechol compounds (Fig. 1B), as well as neuroactive drugs such as levodopa (L-DOPA), α-methyl-DOPA, and isoproterenol [10,14]. Indeed, previous studies have shown that the inhibition of COMT activity leads to a marked reduction in the inactivation of catechol-type neurotransmitters, resulting in the maintenance of their levels and efficacy in the brain [4,8,15].

In humans, there are two isoforms of this enzyme, the soluble cytoplasmatic COMT (S-COMT and henceforth referred to as COMT) and the membrane-bound form (MB-COMT). These two isoforms are coded by a single gene, using two separate promoters, assigned to chromosome 22 band q11.2 [16]. COMT consists of 221 residues, and it is expressed in the liver and kidneys, whereas MB-COMT is predominating present in the brain. MB-COMT differs from the soluble isoform by a 50 residue-long extension at the N-terminus, which is the signal sequence for membrane anchoring [10,16–19]. To date, in the Protein Data Bank (PDB), 25 crystal structures of human COMT are available, and among them, only one structure describes the apo state [20]. The ternary structure of human COMT is composed of a seven-stranded β-sheet core at the C-terminus sandwiched between two sets of α-helices (helices α1–α5 on one side and helices α6–α8 on the other side) at the N-terminus. The β-sheet contains five parallel β-strands and one antiparallel β-hairpin (Fig. 2). The active site is located on the outer surface of the enzyme and includes two distinct parts: the cofactor binding domain and the catalytic site situated in the proximity of the Mg(II). The binding motif of the cofactor site, called a Rossman fold, is based on a seven-strand twisted β-sheet structure, which is a common feature of many nucleotide-binding proteins [21]. In the active site, Lys144 is a key residue for the catalytic reaction since it acts as a general catalytic base in this base-catalyzed nucleophilic reaction. Indeed, it accepts a proton from one hydroxyl group of the catecholic substrate, and subsequently, the methyl group from SAM is transferred to the hydroxyl group [22]. In addition, the hydrophobic residues Trp38, Trp143, and Pro174 contribute significantly to the binding of the substrates (and inhibitors). These residues define the selectivity of COMT to the different side chains of the substrate and participate directly in the methylation reaction by maintaining the planar catechol ring in the correct

---

* Current Affiliation: Department of Pharmacy, University of Pisa, Pisa, Italy

Metalloenzymes. https://doi.org/10.1016/B978-0-12-823974-2.00029-2

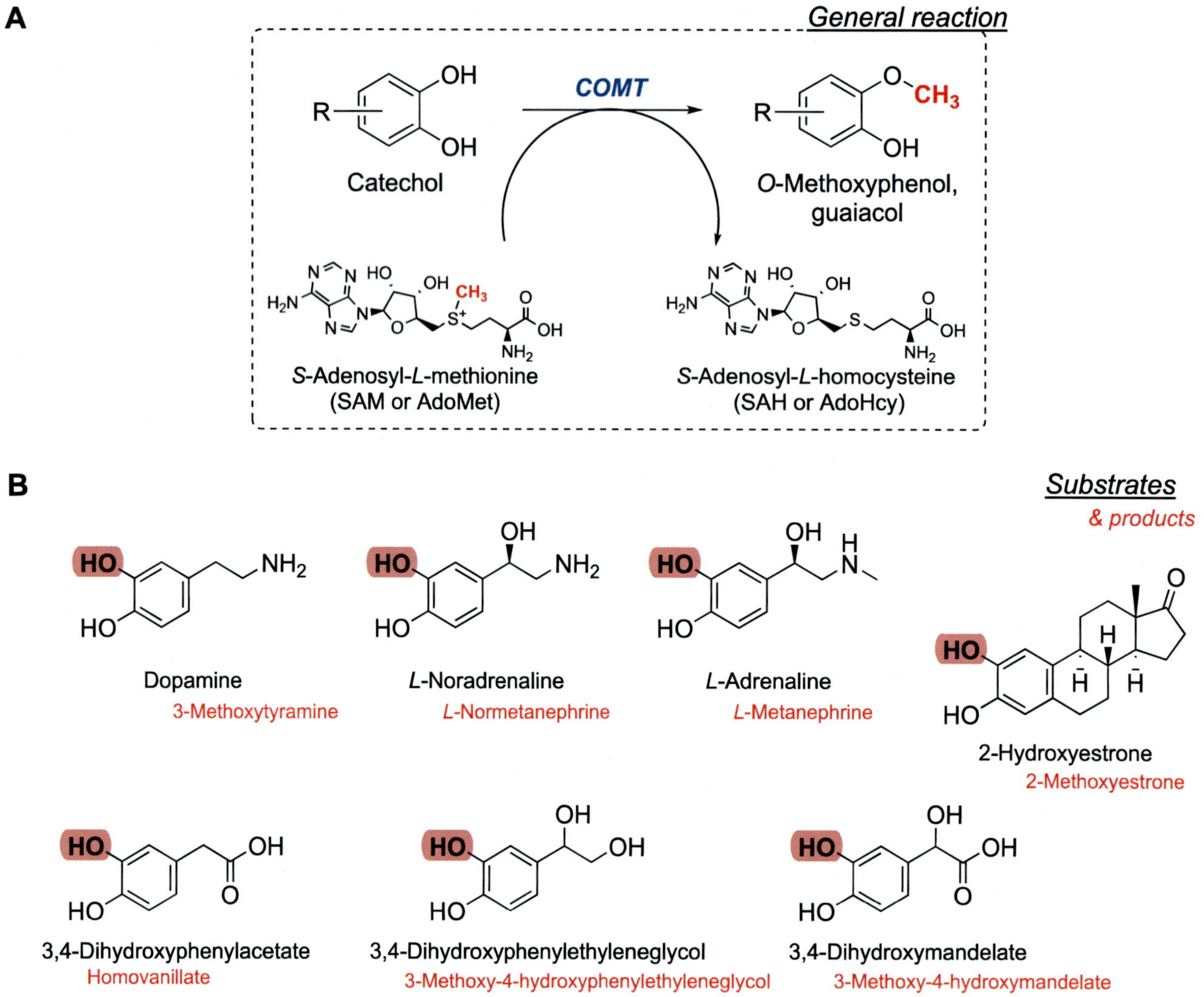

FIG. 1 COMT reactions and substrates. (A) General *O*-methylation of the catechol substrate catalyzed by COMT, along with the structures of *S*-adenosyl-L-methionine (SAM), and *S*-adenosyl-L-homocysteine (SAH); (B) Structure and name of the most recurrent substrates and reaction products (in *red*) of COMT, hydroxyl groups subjected to methylation are highlighted in *red*.

position [22]. The Mg(II) ion is also necessary for the methylation reaction since it makes the hydroxyl groups of the catechol substrate more easily ionizable [6]. In the structure of COMT in complex with inhibitors, this cation is octahedrally coordinated to two aspartic acid residues (Asp141 and Asp169), to one asparagine (Asn170), to both catechol hydroxyls, and a water molecule. Hence, Mg(II) controls the orientation of the catechol moiety.

2 Physiologic and pathologic roles of COMT

COMT has always attracted the scientific community for its great value in the neuroscience field, both for the modulation of brain functions and, in particular, as a druggable pharmacological target in DOPA-related neurobiological disorders. This is due to the expression of this enzyme in different compartments of the cell with a high degree of tissue specificity. As aforementioned, COMT is found as the soluble isoform S-COMT in the cytoplasm and bound to the membrane as MB-COMT on the extracellular side. MB-COMT is mainly expressed in the brain tissue, especially in the (pre)frontal cortex and hippocampus, in pre- and post-synaptic axons and neuron cells, and surrounding glia [23]. COMT is also present in the liver, placenta, lymphocytes, and erythrocytes. The interest in COMT is widely justified by the pivotal roles and functions of this enzyme, being implicated in several biological pathways. The *O*-methylation reaction catalyzed by COMT, as listed by KEGG databases (https://www.genome.jp/kegg/ Accessed on 05.12.2022), results in the degradation of catecholamines (they have an ~1–2 min half-life [24]) and the catabolism of a number of both endogenous substances and catechol-based neuroactive drugs, shortening their biological half-lives and impairing the effectiveness of the treatment.

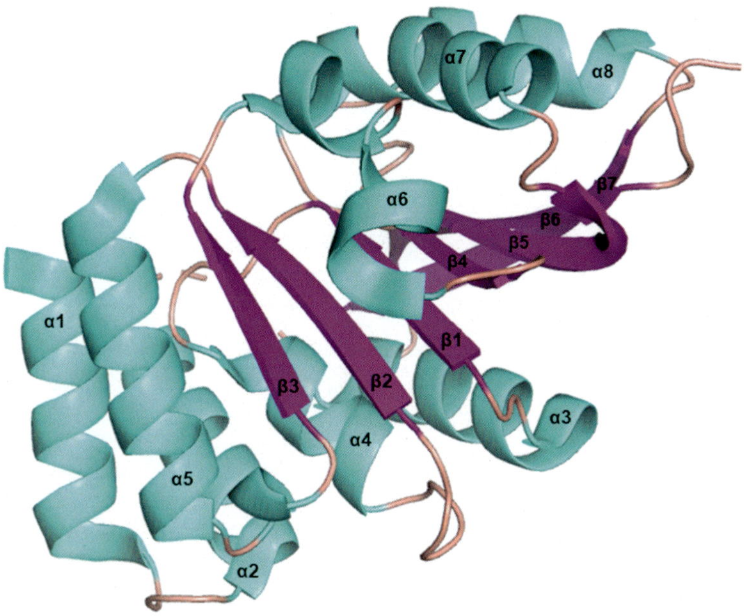

FIG. 2 Cartoon representation of the crystallographic structure of the human COMT (PDB ID 4PYI) [20] α-helices are colored cyan, β sheets magenta, and loops pink. *(The PyMOL Molecular Graphics System, version 2.0.4; Schrödinger LLC: New York, NY, United States, 2018.)*

COMT plays crucial and multifaceted roles in tyrosine metabolism, transferring a methyl group to neurotransmitters, and also producing homovanillate and the two 3-methoxy derivatives of 4-hydroxyphenylethyleneglycol and 4-hydroxymandelate (Fig. 1B). This methylation is also fundamental for the production of betalamic acid contained in betalains, nitrogen-containing pigments present in plants belonging to the order Caryophyllales (i.e., cactus and amaranth), and in higher fungi. More interestingly, for a MedChem purpose, COMT is also involved in DOPA-related activity in the dopaminergic synapses at two different levels: in the glial cells and at the synaptic cleft. In particular, in glial cells, COMT acts in a dual enzymatic system along with MAOs (monoamine oxidases), enzymes catalyzing the oxidative deamination of biogenic and xenobiotic amines, converting DOPA in 3-methoxytyramine (by COMT) or in 3,4-dihydroxyphenylacetate (by MAO) and subsequently producing from them homovanillate by MAO and COMT, respectively (Fig. 3).

The same pathway is present in presynaptic neurons. Moreover, COMTs are involved in the steroid hormone biosynthesis; COMTs catalyze the conversion of 2-hydroxyestrone into 2-methoxyestrone, an endogenous CE endowed with a very low affinity for the estrogen receptor and consequent low potency as an estrogenic agent. The multifaceted biological activity of COMT in humans is reflected by its wide protein-protein associations and interactions networks, as provided by String (https://string-db.org Accessed on 05.12.2022) and Stitch (http://stitch.embl.de Accessed on 05.12.2022) databases, respectively (https://string-db.org Accessed on 05.12.2022) (Fig. 4A and B). Besides MAOs, COMT establishes interactions with aromatic-L-amino-acid/L-tryptophan decarboxylase (DDC), which decarboxylates DOPA, L-5-hydroxytryptophan, and L-tryptophan, dopamine β-monooxygenase (DBH), which converts DOPA into noradrenaline, phenylethanolamine *N*-methyltransferase (PNMT), which converts noradrenaline to adrenaline, alcohol dehydrogenase 1B (ADH1B),

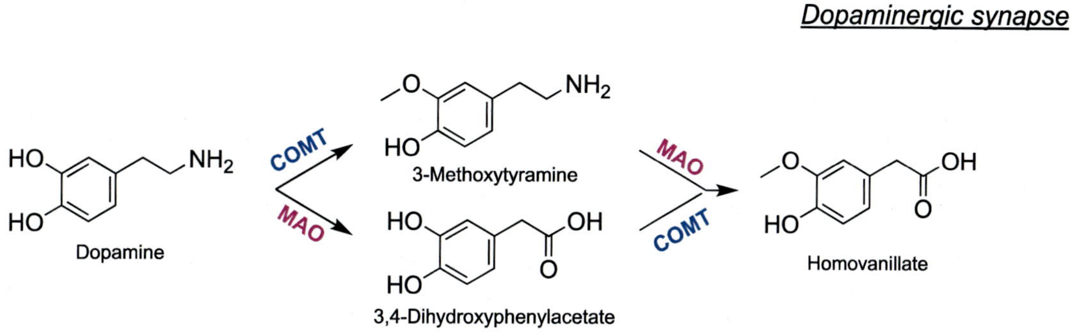

FIG. 3 The dual enzymatic system of COMT and MAO in the dopaminergic synapse.

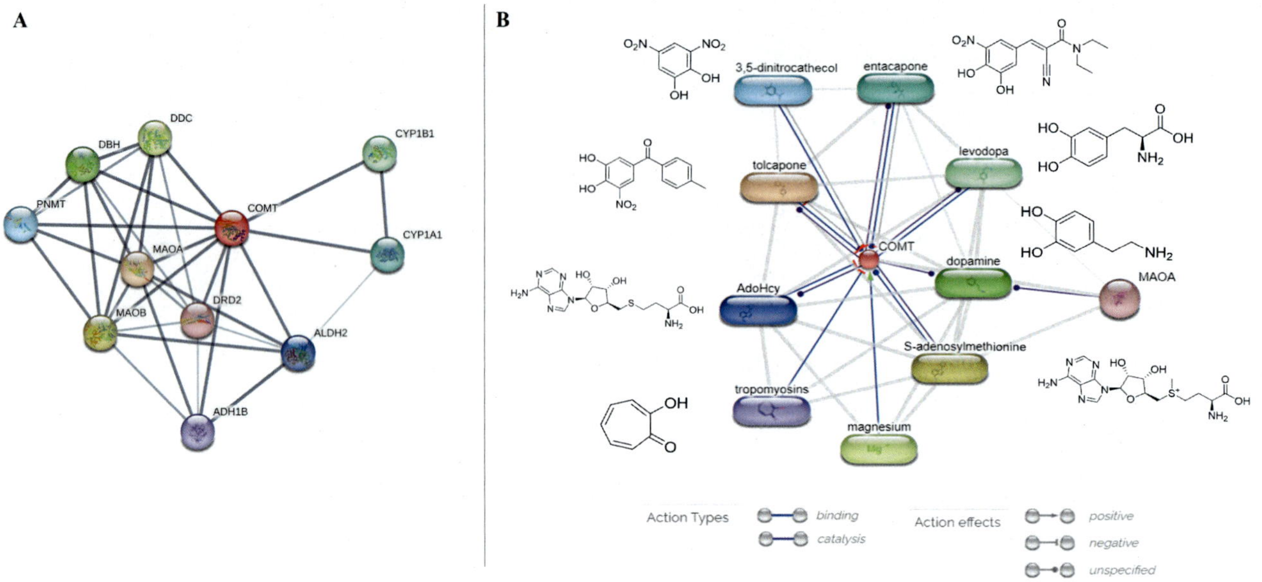

FIG. 4 Interaction networks for human COMT. (A) Protein-protein functional and physical association analysis performed on the STRING platform; lines are related to confidence view; stronger associations are represented by thicker lines; (B) Protein-chemical interaction network performed on STITCH web tool; line colors and symbols are related to the action mode view.

aldehyde dehydrogenase 2 (ALDH2), the dopamine receptor DRD2, and the cytochromes CYP1A1 and CYP1B1.

COMT is characterized by a dynamic regulation in physiopathological conditions, being differently expressed during healthy brain development and in response to environmental stimuli due to several recognized and unknown regulatory mechanisms. Several functional polymorphisms and haplotypes have been described for COMT, such as the Val158Met polymorphism in MB-COMT for the alleles COMT*1 (also named COMT*H or allele G) and COMT*2 (also known as COMT*L or allele A), which differ for the amino acid residue in position 158 (or 108 in S-COMT), valine or methionine, respectively. This residue change is responsible for a three- to fourfold lower enzymatic activity for COMT*2 and the resulting higher concentration of DOPA in the prefrontal cortex, thereby impacting performances in working memory and executive functions [25]. Also, it has been found to have a role in genetic susceptibility to alcoholism [26,27] and anorexia nervosa [28], due to the abundance of COMT in the frontal cortex, involved in eating disease behavior and the role of DOPA in mental disorders. Monarch Initiative reports 11 COMT phenotypes implicated in physiopathological conditions, such as body mass index, systolic blood pressure, abnormality of metabolism/homeostasis, attempted suicide, alcohol drinking, overall survival, and the measurement of urinary metabolites, low-density lipoprotein cholesterol, serum metabolites, hair color, and blood metabolites. (https://monarchinitiative.org/ Accessed on 05.12.2022)

Due to the key role played by COMT in the brain on the metabolism of catecholamines and other catechols, the inhibition of this enzyme is a validated strategy for the treatment of several diseases, mainly in central and peripheral nervous system disorders, such PD [29,30], schizophrenia [31–33], depression [34], Alzheimer's disease (AD) [35], restless leg syndrome [36], mood disorder [37], cognition improvement [38], bipolar disorders [25,39], panic [40,41] and obsessive-compulsive [42] disorders, attention-deficit and attention-deficit hyperactivity disorders [43]. Furthermore, recent publications unveiled the involvement of COMT also in osteosarcoma [44], adolescent idiopathic scoliosis [45], chronic kidney disease [46], cardiovascular diseases [47], gastrointestinal disturbances [48], DiGeorge syndrome [49], and other diseases [28,50].

Furthermore, COMT has been demonstrated to act as a physiological and protective barrier at several organs [8,51]. For example, it protects placenta and embryo from active hydroxyl compounds in the first months of pregnancy [8], it prevents the dissemination of catechols between compartments [52] in particular from injured tissues, at the neuronal level [53]. In addition, COMT has specific antiredox activity against free radicals generated by the DOPA oxidative metabolism, protecting the dopaminergic neuron [54]. Other pathological roles of COMT are obviously related to the involvement of catecholamines in vasoconstriction, sphincter contraction, hepatic glycogenolysis, and ionotropism and chronotropism in the heart [55]. Thereby, overactivity of catecholamines, due to COMT dysfunction or inhibition, can significantly affect the normal function of cardiovascular physiology [56]. Moreover, COMT is implicated in a complex pathway of polycyclic aromatic hydrocarbons (PAHs) and CEs metabolism, as illustrated in Fig. 5, associated with cancerogenesis and mutagenesis since their metabolites form DNA adducts and reactive oxygen species (ROS) [51].

In brief, PAHs are physiologically converted into several toxic metabolites due to the action of P450 peroxidase, that is, radical cations, or CYP450 and epoxide hydrolase, that is, PAH-(E)dihydrodiols (Fig. 5A). Besides their own toxicity, these latter are also substrates of Aldo-Keto Reductases (AKRs), which produce catechol derivatives and then, the more toxic *ortho*-quinones, able to form adducts with nucleic acids and glutathione (GSH). This pathway is partially inactivated by COMT by methylating the catechol derivatives in nontoxic methoxy phenols (Fig. 5A) [51]. Similarly, estrone (E1) and estradiol (E2), whose interconversion is mediated by 17β-hydroxysteroid dehydrogenases 1 and 2 (17β-HSD 1 and 2), are hydroxylated by CYP450, producing the CE estrons (2- or 4-OHE1 and 2), subsequently transformed into CE quinones, highly cytotoxic and mutagenic, and also COMT substrates (Fig. 5B) [51]. In this context, the role of COMT and its inhibitors in cancer and hormone-dependent diseases appears clear.

3 Inhibitors of COMT and their design

Due to its involvement in DOPA and other catecholamines metabolism, COMT began to attract increasing attention in the late 1970s as a promising target in CNS diseases associated with DOPA depletion, such as PD, schizophrenia, and depression [57], as outlined in the previous paragraph. PD is a chronic neurodegenerative disorder, principally affecting elderlies, caused by a nigrostriatal DOPA deficiency related to the gradual loss of nigral cells in the brain and characterized by tremors, muscle rigidity, and other motor symptoms [58]. The first-line treatment for PD is represented by L-DOPA, and since its discovery, it has always been used as an artificial substitute for DOPA to counteract this neurotransmitter deficit [59]. However, the major drawback of L-DOPA is the fast metabolism and, consequently, the short half-life. In addition to the modulation of endogenous neurotransmitters levels, COMT was found to be involved in the degradation of L-DOPA, which undergoes an *O*-methylation to the inert 3-*O*-methyl-L-DOPA (3-OMD) in the peripheral tissues before reaching the brain, resulting in a shortening of the drug half-life and limiting its efficacy. Thus, both academia and pharmaceutical companies directed their research efforts on optimizing L-DOPA therapy by targeting COMT [60]. This opened up the way to the discovery of novel agents able to interfere with COMT activity, and nowadays, COMT inhibitors can be classified based on their structures and pharmacokinetic properties, all of them sharing a similar, although not identical, mode of action. In the next paragraphs, we will discuss more in detail the different classes of clinically relevant COMT inhibitors, starting from the early compounds and moving on to the nitrocatechol-based drugs or agents under clinical investigation, highlighting the pros and cons of the different generations.

3.1 Early catechol- and pyrogallol-based competitive inhibitors

The first generation of COMT inhibitors is represented by simple and small molecules containing a vicinal-dihydroxy (catechol) or a trihydroxybenzene (pyrogallol) motif (**1–7**, Fig. 6). These compounds, displaying inhibition constants (K_I) in the low micromolar range, bind in a fully reversible manner to the enzyme, exploiting the 1,2-dihydroxy moiety or the 1,2,3-trihydroxybenzene pharmacophoric element for their interactions, and competitively inhibit COMT [61].

Catechol- and pyrogallol derivatives represent suitable substrates for COMT, as they are easily susceptible to *O*-methylation reactions leading to the formation of *O*-methylated metabolites. Nonetheless, this class of inhibitors is characterized by scarce in vivo efficacy, poor bioavailability, short time of action, and nonoptimal selectivity over other enzymatic systems, such as tyrosine hydroxylase, MAO, and dopamine-β-hydroxylase (DβH). In addition, they are accompanied by substantial toxic side effects, especially on the liver, and some of them can also induce convulsions, in correlation to the easy penetration into the brain [62]. The most potent compound of the catechol series was U-0521 (**3**, Fig. 6) [63], displaying a K_I of 7.8 μM, which was able to decrease plasma and brain levels of 3-OMD after i.p. injection in different animal species. However, it lacked selectivity toward COMT and was accompanied by high toxicity and unfavorable pharmacokinetic properties. Catechol itself (**1**, Fig. 6), displaying a $K_I = 30\,\mu M$, had to be repeatedly administered in vivo to achieve a noteworthy effect on COMT inhibition [64]. The naturally occurring caffeic acid (**4**, Fig. 6) and its esters have been the focus of different studies dealing with their ability to inhibit COMT, showing a pivotal effect on COMT-mediated *O*-methylation of CEs [65] or L-DOPA bioavailability in rabbit plasma [66], but their application is more related to the remarkable antioxidant activity rather than to COMT inhibition. Among the trihydroxybenzene series, pyrogallol (**5**, Fig. 6) showed one of the highest inhibition potencies, with a K_I value of 13 μM, and resulted even more potent than its catechol counterpart [67]. This has been partially attributed to the presence of an additional hydroxyl group able to establish stronger interactions within the enzyme catalytic pocket, even if the in vivo profile of pyrogallol derivatives traced out that of the catechol series, including high toxicity and inadequate efficacy. In mice, pyrogallol could reach the brain if repeatedly administered every half an hour at the dose of 50 mg/kg and showed a

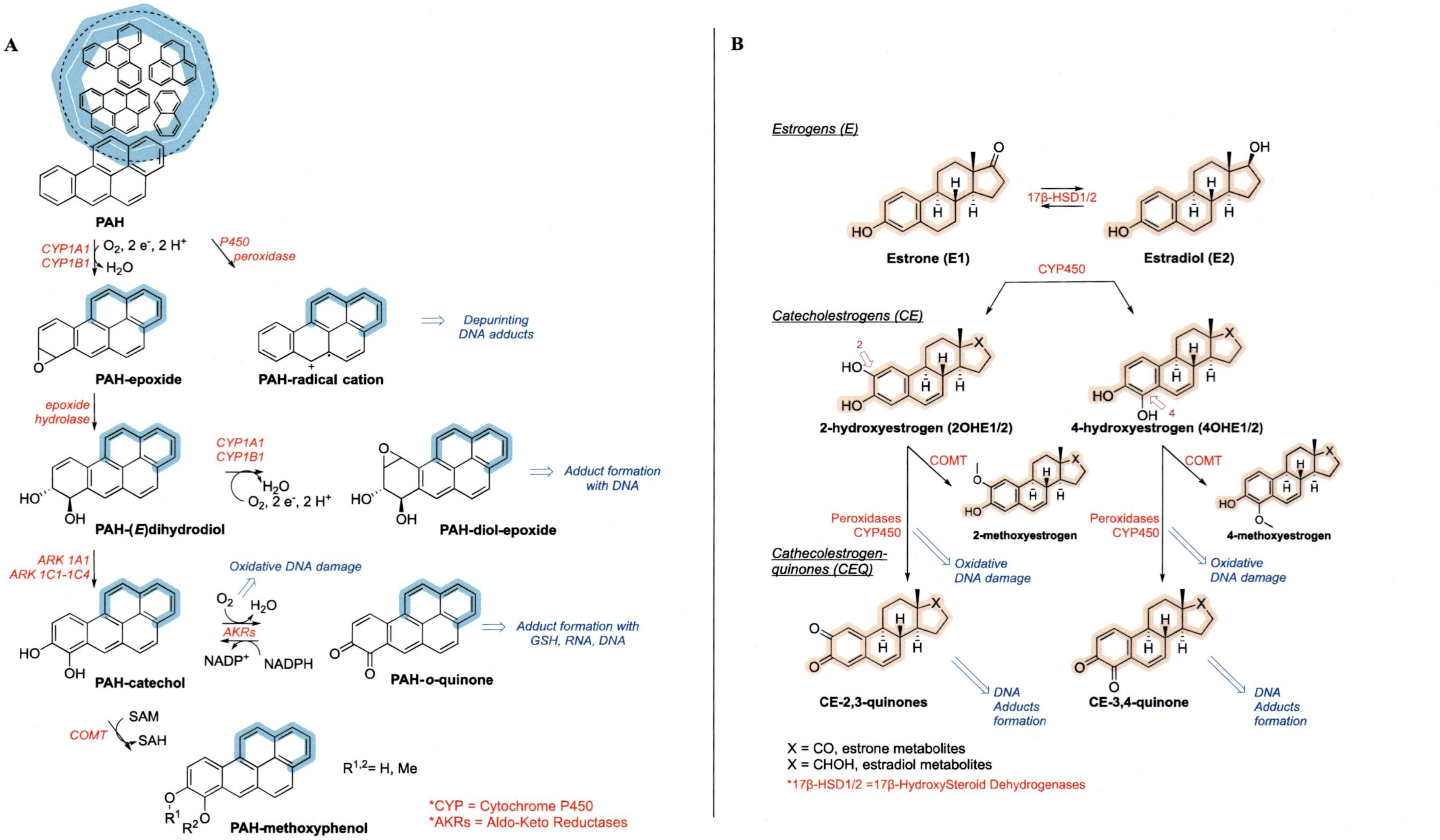

FIG. 5 Pathways involved in (A) PAH- and (B) CE-associated toxicity and their inactivation due to COMT activity.

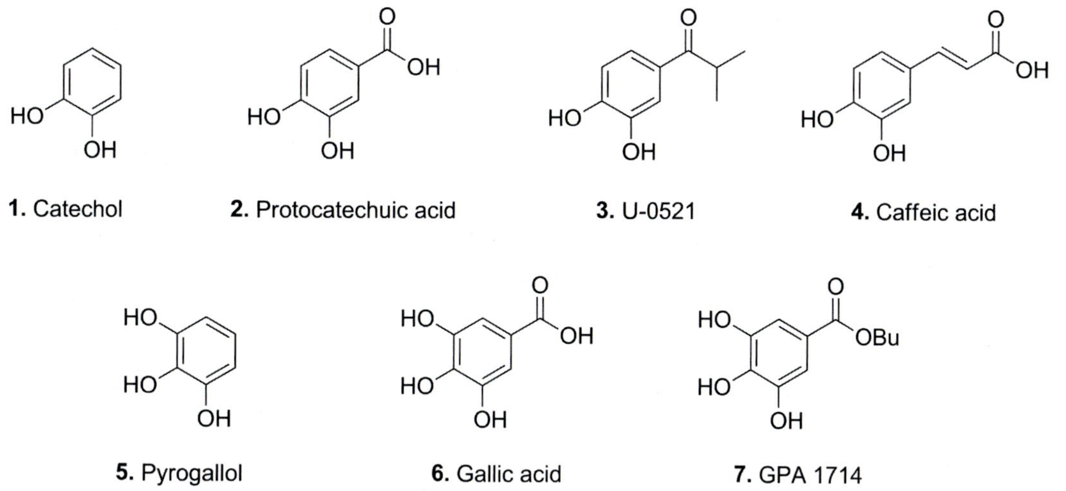

FIG. 6 Structures of early catechol- and pyrogallol-based COMT inhibitors.

short-lasting COMT inhibition at 200 mg/kg, but the effect on catecholamines levels was unsatisfactory. Gallic acid (**6**, Fig. 6), sharing the same 1,2,3-trihydroxybenzene scaffold, resulted even more toxic and less potent than pyrogallol [68], while its *n*-butyl ester GPA 1714 (**7**, Fig. 6) [69] exhibited some positive effects in patients affected by spasmodic torticollis and Huntington's chorea [70], counteracting the toxic dose-related side effects of L-DOPA, but failed in showing any real improvement if coadministered with L-DOPA or in patients affected by schizophrenia [71]. For all these reasons and for the uncertain efficacy against brain levels of catecholamines, no catechol- or pyrogallol-inspired compound has been advanced into clinical development.

3.2 Second-generation nitrated COMT inhibitors

Following the data on the early-discovered competitive COMT inhibitors, different research groups and companies found out that catechol could still be a valuable scaffold for building up efficient inhibitors, as long as this was substituted with strong electron-withdrawing groups (EWGs) in different positions of the aromatic ring with respect to the two hydroxyl groups (Fig. 7).

In particular, the substitution in the *ortho*-position to one of the hydroxyl groups could significantly lower the pK_a value of this latter, reducing the reactivity of the corresponding phenolate and thus making it a worse substrate for the COMT-mediated *O*-methylation. In contrast, the affinities of these new derivatives toward the catalytic site of the enzyme, and consequently the K_I values, were much higher, suggesting a stronger binding to COMT. In the general structure reported in Fig. 7, when $R^1 = R^3 = NO_2$ and $R^2 = H$, then the resulting compound was one of the most potent and selective COMT inhibitors ($IC_{50} = 12$ nM in rat brain homogenates) developed in the new series EWG-bearing catechols [72]. Coadministration of this inhibitor (50 mg/kg) with L-DOPA and an amino acid decarboxylase (AADC) inhibitor such as benserazide p.o. was able to strongly decrease the formation of 3-OMD [73]. Nonetheless, the dose responsible for acute toxicity in mice (312–625 mg/kg) was less than 10 times higher than the efficacy dose, and in addition to other pharmacological concerns, this hit compound was not selected as a clinical

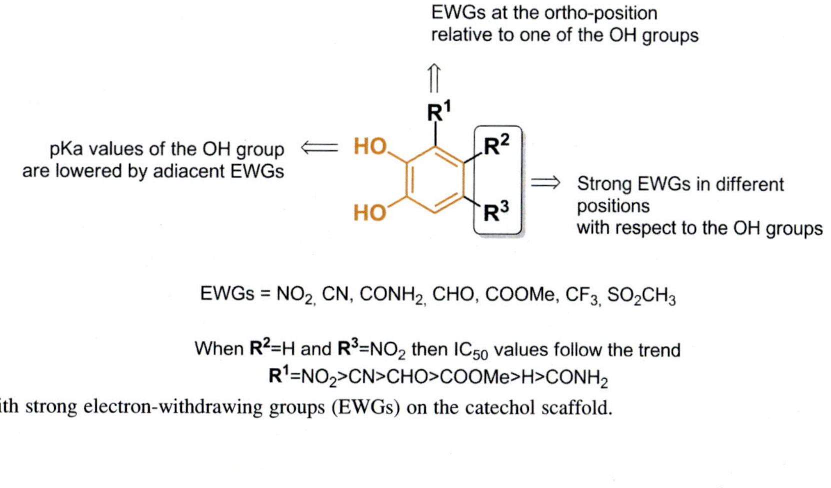

FIG. 7 Substitutions with strong electron-withdrawing groups (EWGs) on the catechol scaffold.

candidate for further development. The structure-activity relationship (SAR) study confirmed the strategic importance of the EWG in *ortho*-position to one of the OH groups and a small variety of EWGs other than NO_2 have been combined trying to improve the in vivo properties of the new analogues. Nothing but NO_2 substitution was increasing the binding affinity for COMT and different EWGs such as nitrile, aldehyde, ester, or carboxamide functionalities failed in this purpose. Therefore, the *ortho*-nitro substitution became a fundamental structural feature for all the following COMT inhibitors based on the catechol scaffold and resulted in the identification of a second generation of compounds, finally leading to a few promising candidates for clinical trials.

By substituting the 3-nitrocatechol moiety with vinylic structures, as in compounds **8** and **10** (Fig. 8) [74], the in vitro inhibition potencies toward COMT were extremely high and independent of the nature of the substituents on the vinylic portion. For instance, compound **8**, decorated with two nitrile groups, could reduce by 89% the plasma levels of 3-OMD at the dose of 30 mg/kg in rats and significantly increase the levels of L-DOPA 3 h after administration. Compound **10** showed similar effects, although with slightly lower potency. In addition, both analogues displayed an optimal selectivity profile over MAO, DβH, and also tyrosine hydroxylase. Closely related to the structures of compounds **8** and **10**, Entacapone and Nitecapone (**9** and **11** respectively, Fig. 8) were the two COMT inhibitors selected for clinical development. These two compounds were very effective inhibitors in vitro, with IC_{50} values ranging between 10 and 20 nM in brain tissues [75,76]. In vivo, they showed an important effect in COMT activity, causing almost complete inhibition in the duodenal, erythrocytic, or liver enzymes at the p.o. dose of 10 mg/kg [77]. Nitecapone was devoid of acute toxicity risk, and when administered in rats at the dose of 30 mg/kg along with L-DOPA and AADC inhibitors, it was able to decrease the formation of 3-OMD by 80% [78]. However, Nitecapone lacked activity on brain COMT, suggesting a scarce penetration in this compartment, and therefore identifying it as a peripherally acting agent. Entacapone was also found to be a purely peripheral COMT inhibitor at doses as high as 30 mg/kg, but its inhibitory activity was much longer and still significant 6 h postadministration [79]. When administered p.o. in rats, Entacapone could completely inhibit duodenal COMT, with a remarkable effect on 3-OMD plasma levels and L-DOPA pharmacokinetics. Of the two inhibitors, Entacapone only has progressed into late-stage clinical trials and finally became one of the most prescribed antiparkinsonian drugs worldwide. Nevertheless, the bioavailability of Entacapone is not optimal, and repeated daily administration is necessary to achieve a valuable effect

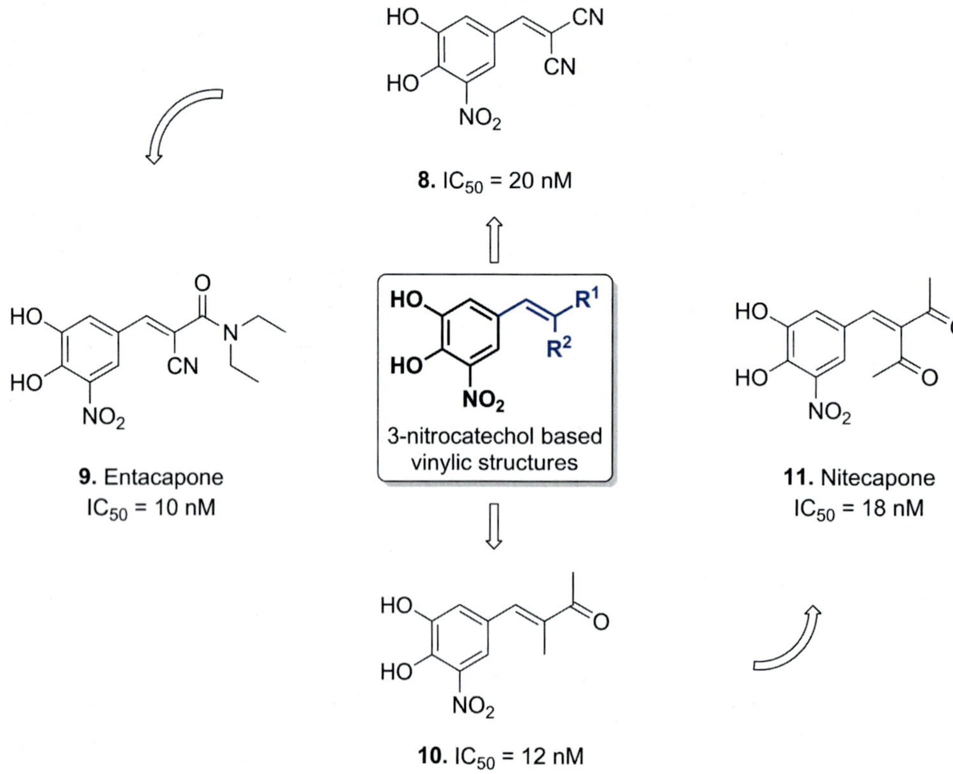

FIG. 8 Novel nitrated COMT inhibitors, including the approved drug Entacapone (**9**), and their IC_{50} values in rat brain homogenates.

during the therapy [80]. This has been partially attributed to the low aqueous solubility either at lower (<5) or at higher (>5) pH values, affecting its absorption in the stomach and its passage through membranes in the small intestine, respectively [81,82]. The scarce lipophilicity at physiological pH has also been correlated to the limited brain penetration while the glucuronidation process of the hydroxyl groups is responsible for the fast metabolism and relatively short half-life of Entacapone [83]. Although being developed as a coadjutant treatment for PD, this class of COMT inhibitors was also patented for the positive effects on ulcers and gastrointestinal lesions [84], cancer [85], and vascular dysfunctions [86]. In combination with L-DOPA and AADC inhibitors, they have been used to delay PD-associated motor dysfunctions [87] and to treat restless leg syndrome [88].

To find novel structures that could serve as orally bioavailable and brain-active COMT inhibitors, the 3-nitrocatechol scaffold was decorated with acetyl (or carbamoyl) groups in the place of the vinylic portion previously reported, achieving a new series of derivatives endowed with interesting in vivo properties (Fig. 9) [89].

The first good results came from benzoyl-substituted compounds **12a** and **12b** (Fig. 9), namely Ro41-0960 and Tolcapone, respectively. Ro41-0960 was a very potent inhibitor of COMT, with an IC_{50} value of 42nM, and possessed enhanced in vivo efficacy, blocking 3-OMD formation at the dose of 50mg/kg when coadministered with L-DOPA and benserazide [90]. Nonetheless, Tolcapone, bearing a 4-methyl substituent instead of a fluorine atom in the 2 position of the benzoyl portion, was the candidate selected for clinical evaluation in the anti-PD therapy. Tolcapone strongly bound to COMT and inhibited the enzyme in a reversible manner, resulting even more potent on brain COMT, with an IC_{50} value of 2.2nM. It was also characterized by a duration of action longer than other COMT inhibitors belonging to the same family, which was maintained up to 8 or 11h depending on the different animal species and body compartments (such as the liver, heart, or kidneys) [91]. When coadministered with L-DOPA and AADC inhibitors, Tolcapone affected the levels of 3-OMD by significantly decreasing them in a dose-dependent manner. What made Tolcapone a promising candidate was its ability to easily penetrate the blood-brain barrier (BBB), hence exerting its inhibitory effect with a different pharmacokinetic profile [92]. In fact, the presence of a more lipophilic 4-methyl substituent on the benzoyl system allowed a higher brain penetration, making Tolcapone an equipotent COMT inhibitor in both peripheral and central areas. The oral bioavailability resulted much higher than Entacapone, in correlation with the increased intestinal absorption and BBB permeation. Tolcapone entered different clinical trials until it received the first approval by the European Medicines Agency (EMA) and by the Food and Drug Administration (FDA) in 1997 and 1998, respectively, for the treatment of PD-related disorders [93]. Unfortunately,

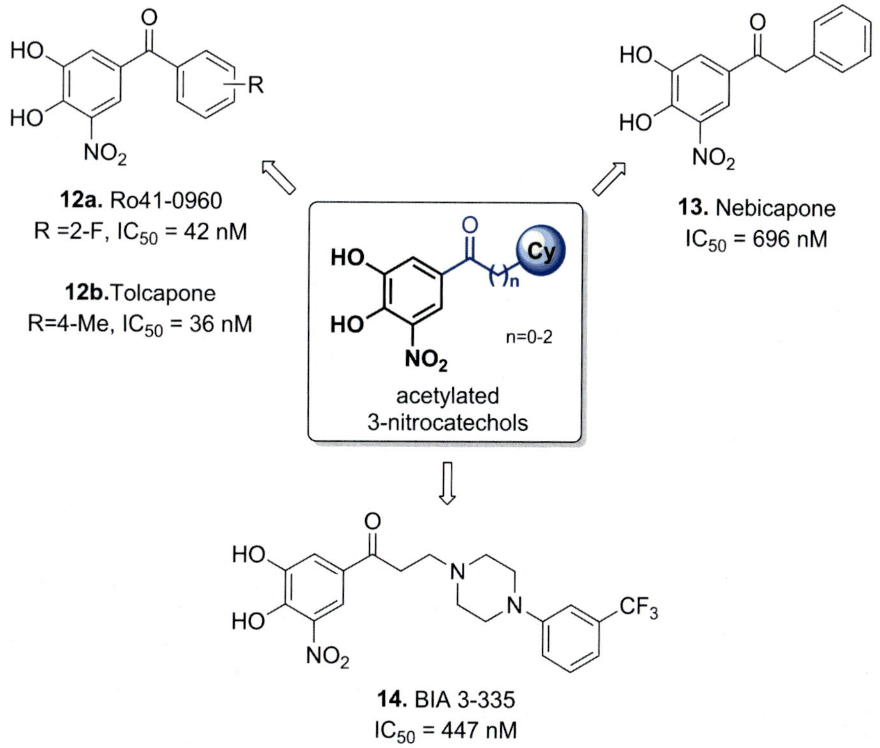

FIG. 9 Second-generation series of 3-nitrocatechol derivatives bearing acetylated portions and the related IC_{50} values in rat liver homogenates.

besides these optimal in vivo characteristics, Tolcapone was later found to have high toxicity on the liver, accompanied by the possibility to evoke episodes of fulminant hepatitis [94,95]. Accordingly, the nitro group of Tolcapone can easily be converted into the corresponding aniline prior to excretion, and further oxidation processes of this metabolite could be responsible for liver toxicity [96]. This has also been confirmed by in vitro studies on liver cells where these reactive metabolites, such as glutathione adducts, interacted with hepatic proteins causing damage to the liver [97]. Tolcapone marketing authorization was later suspended by the EMA in 1998 (and also by Canadian health national authority) while the FDA limited its use, with the addition of a black-box warning to its label. In 2004, EMA (and Canada) lifted the suspension but up to date, Tolcapone is not considered a first-choice L-DOPA add-on treatment in PD therapies. Nonetheless, it has been reevaluated by more recent clinical trials, suggesting that it could still be a valuable drug since liver enzyme elevation is often mild and aspecific in about 99% of cases and correct monitoring of PD patients would be sufficient to prevent severe side effects [93].

Due to the unfavorable toxicological profile of Tolcapone, several efforts have been dedicated to the discovery of different structures endowed with a safer profile and higher selectivity for COMT in peripheral areas. Still retaining an acetyl group on the nitrocatechol scaffold, the insertion of spacers between the carbonyl functionality and the aromatic substituent opened up the way for novel, potent, and peripherally acting inhibitors [98]. Aliphatic tethers with different lengths have been used to distance the aromatic portion from the carbonyl group (see general structure in Fig. 9), and the combination of a methylene spacer with an unsubstituted phenyl ring, namely a benzylic moiety, resulted in the discovery of Nebicapone (**13**, Fig. 9) [99]. Although having an IC_{50} value on rat liver COMT (696nM), which was higher than the value on the brain enzyme ($IC_{50} = 3.7$ nM), Nebicapone showed longer inhibitory activity on liver homogenates, still attaining 70% of COMT inhibition 9h postadministration, compared to the 22% on brain homogenates. In vivo, it was found to be equally potent to Tolcapone, with enhanced peripheral selectivity, and also exerted a prolonged peripheral COMT inhibition in comparison to Entacapone. The peripheral activity was also confirmed in rat amphetamine-induced hyperactivity tests, where Nebicapone did not alter any of the animal stereotypies at the dose of 30mg/kg. Nebicapone entered the clinical evaluation in combination with L-DOPA for treating PD, but it was discontinued during the trials for lacking an improvement in the toxicity profile in comparison to Tolcapone. Additional combinations of aliphatic chains and (hetero)cyclic substitutions led to the identification of another valuable series of analogues [100], including the alkylamino compound BIA 3-335 (**14**, Fig. 9) [101]. Here, the *N*-aryl piperazine moiety was distanced from the carbonyl group by a 2-carbon atoms chain, achieving a new inhibitor characterized by a long-lasting activity on liver COMT (74% inhibition 6h postdose at 30mg/kg) and enhanced peripheral selectivity. This latter was probably due to the presence of the piperazine ring, whose nitrogen atoms increased the hydrophilic surface of the molecule, hampering BBB penetration. Despite the promising features of BIA 3-335, no further information to date has been provided regarding the clinical evaluation and application of this COMT inhibitor.

3.3 Heterocyclic nitrocatechol derivatives

Moving away from the previously reported substitutions on the catechol pharmacophore, a different series of COMT inhibitors has been proposed by connecting heterocyclic structures to the 3-nitrocatechol scaffold in the *meta* position relative to the nitro group [102]. Several combinations have been proposed, bearing benzoxazinone (**15**), quinoxalinone (**16**), quinoxaline (**17**), and aminothiazole (**18**) moieties (Fig. 10), but despite the favorable IC_{50} values on rat brain tissues, they achieved no important improvement in vivo in comparison to Tolcapone, probably due to the poor bioavailability. Screening of compound libraries identified analogues **19** and **20** (Fig. 10) as promising COMT inhibitors in vitro, although **19** showed limited in vivo interference with COMT activity while **20**, which resulted equipotent to Tolcapone, displayed limited peripheral selectivity and considerable toxicity on cell viability [102]. A substantial step forward was taken by substituting the oxadiazole moiety of compound **20** with *meta*-pyridyl-*N*-oxides, obtaining derivatives **21a–c** (Fig. 10). *ortho*-Pyridyl- and *para*-pyridyl-*N*-oxides were generally less potent than the *meta*-regioisomers, and small lipophilic groups on the pyridyl ring, such as CF_3 or halogens, resulted in higher potency and reduced toxicity in vitro.

The best-performing compound of the series was **21c**, namely Opicapone (Fig. 10), which showed a remarkable affinity for COMT (with a K_d value in the low picomolar concentration) and a prolonged activity in vivo. In fact, at the dose of 3mg/kg p.o. in rats, it reached the maximum COMT inhibition 3h after administration, and it was still achieving 80%–90% inhibition 9h postdose, even longer than Tolcapone and Entacapone [103]. At the same dose, Opicapone could also elevate the plasma levels of L-DOPA and reduce the formation of 3-OMD by 43%–58% [104]. Its activity was limited to peripheral areas, due to the poor penetration across the BBB [102]. Opicapone entered different clinical trials aimed at evaluating its efficacy in anti-PD therapies [105]. It was approved in Europe in 2016 as the third COMT inhibitor prescribed in combination with L-DOPA and Carbidopa to treat

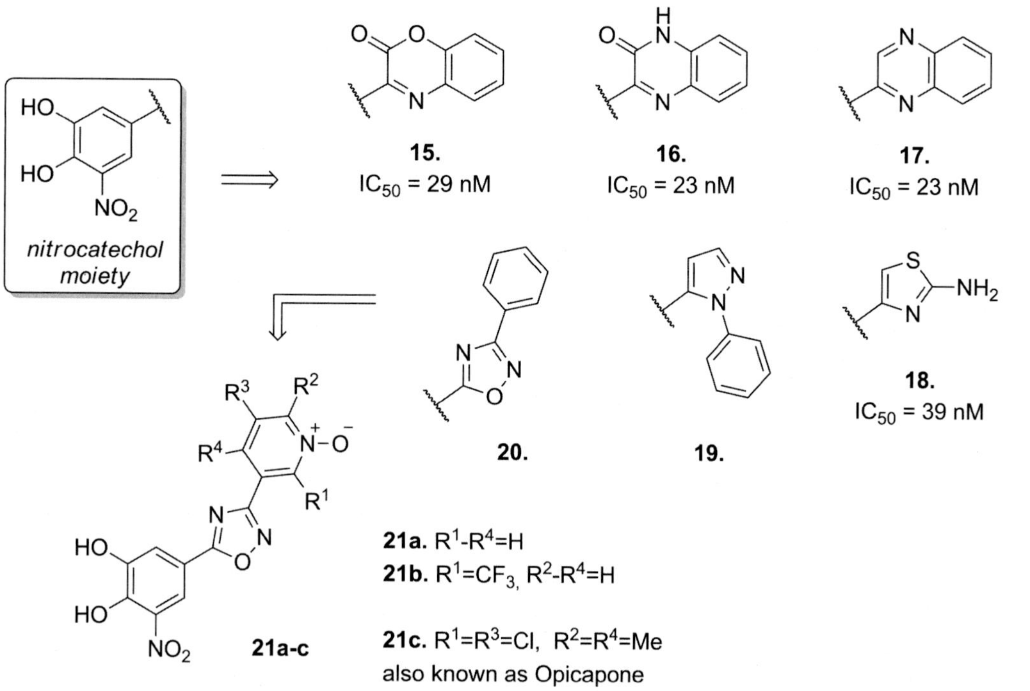

FIG. 10 SAR analysis around the heterocyclic substitutions for the 3-nitrocatechol scaffold and the last approved drug Opicapone (**21c**).

end-of-dose motor fluctuations in adult patients affected by PD. It also received FDA approval in 2020 for the US market, but its official release has been delayed due to the SARS-CoV-2 pandemic [106]. To date, no valuable safety concerns and no association with the risk of fulminant liver injuries have emerged.

3.4 Miscellaneous

Following the positive trend obtained with the catechol- and pyrogallol-based inhibitors previously described (Section 3.1), several other molecules bearing different pharmacophoric elements have been proposed as COMT inhibitors. These compounds were not exploiting the dihydroxy- or trihydroxybenzene motif for interacting with the enzyme but still retained remarkable activity in vitro and, for some of them, also in vivo, albeit being far from having optimal pharmacokinetics.

Tropolone (**27**, Fig. 11) was able to interact with COMT thanks to the α-keto-enol tautomerization of its structure, thus mimicking the dihydroxy moiety of catechol [107]. Tropolone and some of its 6-alky-substituted derivatives (here not shown) competitively inhibited COMT with K_I values in the low-to-medium micromolar range. Despite being able to concentrate in cerebral and peripheral areas, the relatively weak interaction with COMT resulted in a short in vivo effect and moderate efficacy on catecholamines levels. If we couple these aspects with the significant

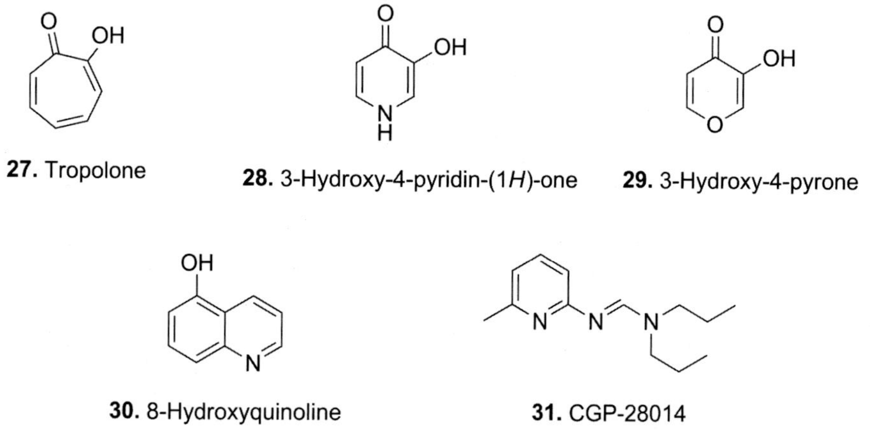

FIG. 11 Structures of noncatechol-based inhibitors of COMT.

toxicity at the tested doses, it is clear that no further clinical evaluation could be dedicated to tropolone and its analogues [108]. The same fate was reserved for pyridinone- and pyrone-analogues, here exemplified by compounds **28** and **29**, respectively (Fig. 11). These could be considered as isosteric replacements of catechol and entered in vitro evaluation for their ability to inhibit COMT. Nevertheless, the scarce activity, lack of selectivity, and significant toxicity led to the abandoning any further development [109]. 8-Hydroxyquinoline (**30**, Fig. 11) was also selected as a potential COMT inhibitor, even if it lacked favorable pharmacokinetic and pharmacodynamic features. However, it has been recently reinvestigated as a valuable scaffold for building up COMT inhibitors with enhanced properties. Small substituents at the 7-position of the quinoline ring demonstrated to increasing metabolic stability while retaining potency, and analogues with improved characteristics showed strong in vivo modulation of DOPA metabolites [110]. Worth of note is also compound CGP-28014 (**31**, Fig. 11), which is not structurally related to the previous classes of inhibitors. Interestingly, it is only believed to be a COMT inhibitor because of the lack of in vitro activity on the enzyme, but it possesses a promising in vivo efficacy [111]. In fact, administration of CGP-28014, alone or together with benserazide, decreased plasma concentrations of 3-OMD (by 43%–56%) and modulated the levels of homovanillic acid [112]. Despite being tested in healthy subjects, CGP-28014 did not enter any clinical development process.

Over the years, several other families of COMT inhibitors arose from differently decorated scaffolds, more or less related to the nitrocatechol moiety. Even if none of them completed the clinical evaluation and therefore failed in reaching the market, it is still worth mentioning some of the analogues that contributed to expanding the knowledge around COMT inhibition and their effectiveness. A quick overview is provided as follows.

- **Diphenyl sulfones**: the combination of the nitrocatechol pharmacophore with a benzenesulfonyl group at position 5 led to a series of inhibitors, exemplified by compound **32** (Fig. 12), with a remarkable activity in the low nanomolar range [113]. They have also been assessed in toxicity tests to determine the effect in hepatic cell lines, resulting to be less toxic than Tolcapone and Entacapone, but no further in vivo evaluation has been disclosed.
- ***Ortho*-nitrated catechols**: following the same SAR concept of the nitrated catechols bearing the nitro group in the *meta* position with respect to a vinylic substituent (see Figs. 7 and 8), also compounds with nitro groups adjacent to the vinylic portion (namely in the *ortho* position) have been evaluated as COMT inhibitors. As a representative analogue of this class, compound QO IIR2 (**33**, Fig. 12) showed the best results in terms of in vitro properties [114]. It was a selective, tight-binding, and reversible inhibitor [115] able to modulate the levels of 3-OMD when coadministered with L-DOPA and AADC inhibitors in a dose-dependent manner [116]. When the nitro group was in the *ortho* position with respect to a benzoyl group, inhibitors such as BIA 8-176 (**34**, Fig. 12) were obtained [117]. This benzophenone was a close analogue of Ro41-0960 (**12a**, Fig. 9), with in vitro IC_{50} values of 3 and

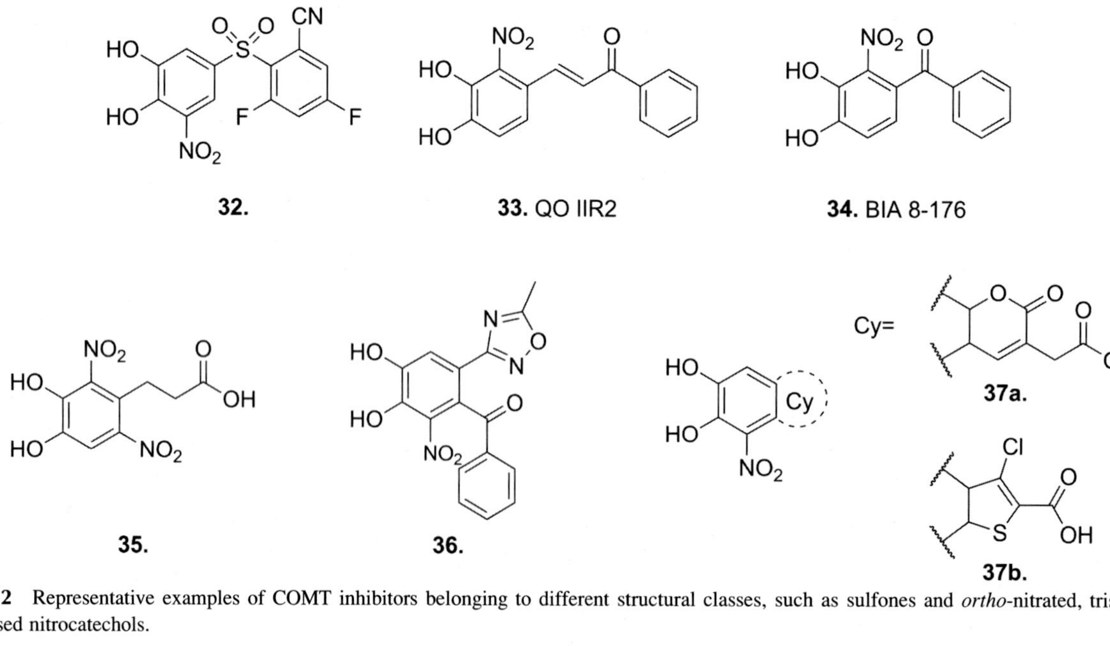

32. **33.** QO IIR2 **34.** BIA 8-176 **35.** **36.** **37a.** **37b.**

FIG. 12 Representative examples of COMT inhibitors belonging to different structural classes, such as sulfones and *ortho*-nitrated, trisubstituted, and fused nitrocatechols.

130nM on rat brain and liver COMT, respectively. In vivo, it displayed similar central and peripheral activities, with a rapid onset and a likewise fast termination. BIA 8-176 was advanced into preclinical evaluation, but the short-lasting effect, connected to the reduced metabolic stability, finally led to abandoning this and other *ortho*-nitrated inhibitors.

- **Trisubstituted catechols**: the addition of another substituent on the catechol scaffold led to a new series of trisubstituted analogues endowed with an inhibitory effect on COMT. Compound **35** (Fig. 12) showed better properties with respect to the mono-nitrated counterpart and exhibited improved bioavailability and was therefore proposed for the treatment of PD. [118] Also oxadiazole-substituted compound **36** (Fig. 12) [119], showed an IC_{50} value of 9.4nM on human COMT and a favorable toxicity profile. When coadministered with L-DOPA and AADC inhibitors at the dose of 3mg/kg, it doubled the plasma levels of L-DOPA within 4h after administration. Despite a few other preclinical evaluations, this compound probably lost interest over Opicapone and related analogues.
- **Fused nitrocatechols**: some fused bicyclic structures involving the nitrocatechol moiety have been proposed as COMT inhibitors, here represented by analogues **37a** and **37b** (Fig. 12). Compound **37a** possesses a coumarin-like scaffold and showed an IC_{50} value on COMT around 10nM, but no in vivo data are available [120]. In derivative **37b**, the nitrocatechol side was fused with a 5-membered ring, namely a substituted thiophene, reaching a K_I value of 0.5nM [121]. This compound showed a lack of severe toxicity and enhanced metabolic stability but possessed poor oral bioavailability. It was therefore discontinued, also in correlation to the scarce performance in comparison to Tolcapone and Entacapone [122].
- **Bifunctional analogues**: still inspired by di- and trihydroxybenzene structures, a series of bifunctional analogues bearing two catecholic groups were reported as COMT inhibitors [123]. The two portions were linked by a double carboxamide functionality and spaced with alkyl tethers of different lengths (**38a–c**, Fig. 13). No direct correlation between the bifunctional nature and the potency of COMT inhibition has ever been established, although the length of the alkyl chain could influence the in vitro activity. The best results were obtained by combining the *n*-propyl spacer with two catecholic groups (as in compound **38a**) or a catechol and pyrogallol moieties (exemplified by compound **38c**), thus obtaining inhibitors with K_I values in the sub- and low micromolar range, respectively. Other asymmetrical structures were synthesized and tested, such as phenol-nitrocatechol combinations (**39a–c**, Fig. 13) with improved in vitro potency [124], but no further in vivo evaluation has been reported.
- **Bisubstrate compounds**: to hit the catechol and the SAM binding sites in the catalytic pocket of COMT at the same time, bisubstrate inhibitors have been developed (**40a–c**, Fig. 13). Here, the selection and combination of catechol and a ribose-adenine moiety (mimicking the structure of SAM) were guided by the structural information provided on the quaternary complex between the enzyme, the cofactor, the Mg(II) ion, and the designed compounds [125]. The linkage of these two portions with short and conformationally constrained chains allowed the discovery of the potent inhibitor **40a** (Fig. 13) [126]

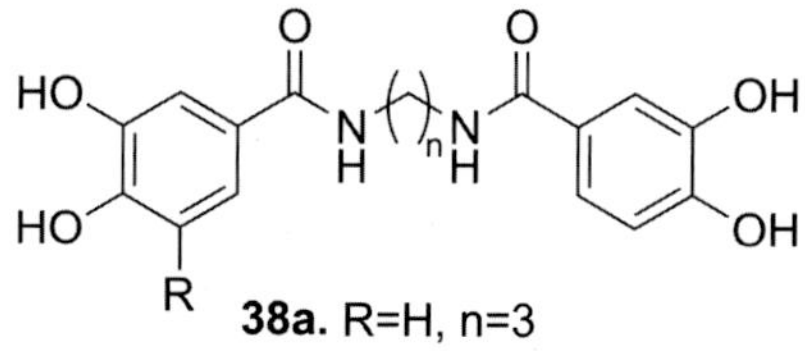

38a. R=H, n=3
38b. R=OH, n=2
38c. R=OH, n=3

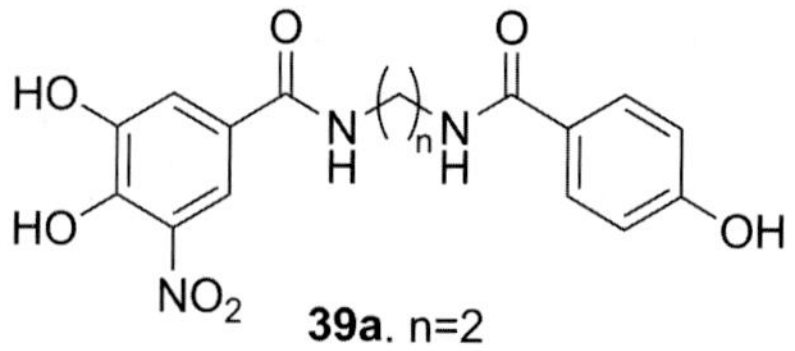

39a. n=2
39b. n=3
39c. n=5

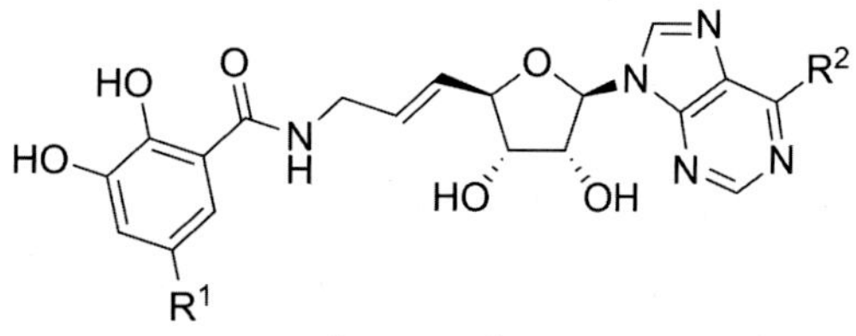

40a. R^1=NO_2, R^2=NH_2
40b. R^1=4-F-Ph, R^2=NH_2
40c. R^1=4-F-Ph, R^2=CH_3

FIG. 13 Bifunctional and bisubstrate COMT inhibitors.

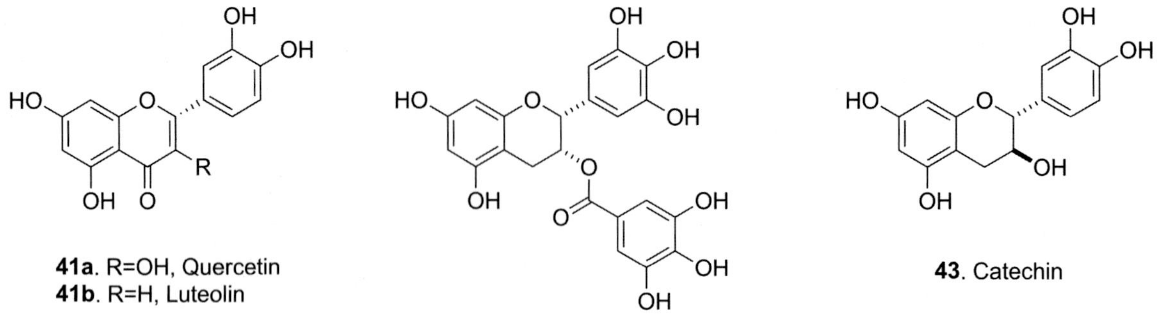

41a. R=OH, Quercetin
41b. R=H, Luteolin

42. Epigallocatechin-3-gallate

43. Catechin

FIG. 14 Naturally occurring polyphenols as COMT inhibitors.

endowed with an IC_{50} value of 9 nM of rat liver homogenates. This compound inhibited the enzyme with a competitive and more complex mechanism of action, and its dual interaction with both catechol and SAM binding site was confirmed through the X-ray analysis of the crystal structure of **40a** in complex with rat COMT [127]. By substituting the nitro group of **40a** with the 4-fluorophenyl residue, other inhibitors of COMT were selected (**40b–c**, Fig. 13), endowed with optimal in vitro potency and promising bisubstrate features [128]. In particular, compound **40c** was also modified on the adenine moiety, replacing the amino functionality in position 1 with a methyl group and achieving the most potent analogue of the series. Nonetheless, no experimental characterization in vivo has been performed.

– **Natural products**: nature has also represented a source of inspiration to find compounds potentially endowed with inhibitory activity toward COMT. A series of naturally derived compounds, bearing di- and trihydroxybenzene moieties, have been tested against this enzyme and a number of flavonoids and polyphenols showed interesting properties in vitro. The COMT-mediated *O*-methylation reaction of quercetin and luteolin (**41a** and **41b**, respectively, Fig. 14) was investigated with computational studies, highlighting the potential for these compounds to be a substrate for the enzyme [129]. Quercetin was found to effectively inhibit COMT, both in vitro and in vivo, and to protect against oxidative hippocampal injury. Also, catechins such as epigallocatechin-3-gallate and catechin (**42** and **43**, Fig. 14), common polyphenols present in tea, inhibited COMT in vitro and showed a protective effect against oxidative insult. However, discordant information has been reported on the in vivo application. At first, they were described as useful food supplements in L-DOPA treatments [130], but they have been recently reevaluated and resulted ineffective in modulating L-DOPA *O*-methylation process [131]. More in general, these polyphenols exerted beneficial effects on oxidative stress and neurodegeneration.

4 Clinically used agents or compounds in clinical development

As described previously, COMT inhibitors are mainly prescribed to limit "wearing off" motor fluctuations, in PD patients treated with levodopa, interfering with daily activities. This class of drugs has no direct impact on symptoms, but it can extend the effectiveness of primary medications and lower the administered doses by hampering systemic metabolism. They can be used in association with therapies involving direct or indirect dopaminergic activity in the brain: carbidopa-levodopa, highly selective MAO-B inhibitors (safinamide, selegiline, rasagiline), DOPA agonists, anticholinergics, and surgery for deep brain stimulation. Carbidopa-levodopa association is the most efficacious approach to counteract the motor symptoms in PD, despite patients experience less effect over time. Temporal fluctuations in this main therapy are recognized as "on" (minimal symptoms) and "off" (worsened symptoms) episodes [132]. Examples of first-line COMT inhibitors currently in the clinic are nitrocatechol derivatives (Table 1): Comtan (entacapone) approved in the 1990s, Tasmar (tolcapone) approved in the 1990s, and Ongentys (opicapone) approved recently in the United States and some European countries [105,133]. A commercial combination (carbidopa-L-DOPA-entacapone branded as Stalevo) is also available. A phase III noninferiority trial revealed a similar efficacy on "off" episodes and a promising reduction on the Unified Parkinson's Disease Rating Scale motor section for all three drugs when compared with the placebo. Data on patients' quality of life suggested positive results [134–136]. As evidenced by clinical trials and real-life observational studies, the most common side effects comprehended dyskinesia (as an exaggeration of levodopa side effects), abdominal pain, confusion and hallucinations, back pain, constipation, and nausea. Conversely, diarrhea and urine discoloration (reddish brown or rust-colored) were only registered with Entacapone and Tolcapone [137,138]. Moreover, Tolcapone treatment may induce potentially fatal liver failure, and after a first withdrawal, it was recommercialized under some restrictions (regular blood tests of liver function) [139].

TABLE 1 Comparison of the main characteristics of COMT inhibitors.

COMT inhibitor	cLogP	Doses (mg)	BBB permeability	Mayor side effects	Pharmacokinetic data
Entacapone	2.16	200	No		Low bioavailability and short-acting inhibitory profile
Tolcapone	3.17	100–200	Yes	Risk of drug-induced liver injury	Highly protein bound
Opicapone	2.95	50–100	No		Safe and long-acting inhibitory profile

These three compounds shared the same nitrocatechol pharmacophore, which is required for their tight-binding inhibition, but which may pose potential toxicity. The NO_2 EWG improved their potency reducing the pK_a of the phenol. In addition, they were not eligible as substrates by the enzyme itself. Toxicity was also related to the nitrocatechol CYP450-mediated metabolism to reactive and electrophilic quinone-imines in the liver, but differences registered among the COMT inhibitors suggested that the presence of this toxicophore was not inherently harmful. Indeed, nitrocatechols were also similar to nitrophenols known to act as protonophores able to transfer H^+ across lipidic membrane bilayers. Similarly, in mitochondria, protonophores uncoupled the proton gradient across the inner membrane. This damage was proportional to lipophilicity (cLogP reported in Table 1), which could guarantee better diffusion through these membranes. An exhaustive analysis of the clinical trials with published results from 2004 to nowadays, available at https://clinicaltrials.gov/ct2/results?term=COMT+inhibitor&age_v=&gndr=&type=&rslt=With&Search=Apply (accessed on 14.01.2023), revealed 47 documents. All the studies have evaluated the efficacity and safety of known COMT inhibitors toward the following dopamine-based pathologies: impulsive behavior, pathological gambling, alcohol (ab)use disorder, obsessive-compulsive disorder, schizophrenia, tobacco use disorder and dependence, methamphetamine dependence, cocaine dependence, multiple system atrophy, epilepsy, and Autonomic Nervous System diseases in comparison with placebo, L-DOPA, L-DOPA/carbidopa, and other yet approved combinations. Opicapone (named under the name BIA 9-1067) was also used as a micronized powder improving bioavailability.

5 Conclusions

COMT emerged in the late 1970s as a promising target in peripheral and central nervous system disorders. Several efforts have been dedicated to the discovery of novel, selective, and efficacious COMT inhibitors, mainly to be coadministered in anti-PD therapies. In fact, COMT was found to be involved in the metabolism of L-DOPA, the first-line treatment for PD, and its inhibition could modulate plasma and brain levels of this drug. Different classes of inhibitors have been proposed, behaving as centrally or peripherally acting agents, and those based on a nitrocatechol scaffold led to the discovery of a few candidates with favorable in vitro and in vivo characteristics, which successfully passed clinical trials. Entacapone and Tolcapone were the first approved drugs in PD therapies, in combination with L-DOPA and other AADC inhibitors. The last approved compound of this class was Opicapone, which reached the market in 2016. Despite the pressing demand for new and safer COMT inhibitors to improve the pipeline, the efforts made to expand the therapeutic efficacy of this class of compounds are mainly devoted to the field of dual or combinations of drugs.

References

[1] Petrossian TC, Clarke SG. Uncovering the human methyltransferasome. Mol Cell Proteomics 2011 Jan;10(1). M110.000976.

[2] Schubert HL, Blumenthal RM, Cheng X. Many paths to methyltransfer: a chronicle of convergence. Trends Biochem Sci 2003 Jun;28(6):329–35.

[3] Axelrod J, Tomchick R. Enzymatic O-methylation of epinephrine and other catechols. J Biol Chem 1958 Sep;233(3):702–5.

[4] Männistö PT, Ulmanen I, Lundström K, Taskinen J, Tenhunen J, Tilgmann C, et al. Characteristics of catechol O-methyltransferase (COMT) and properties of selective COMT inhibitors. In: Progress in drug research/Fortschritte der Arzneimittelforschung/Progrès des recherches pharmaceutiques. Basel: Birkhäuser Basel; 1992. p. 291–350.

[5] Masjost B, Ballmer P, Borroni E, Zürcher G, Winkler FK, Jakob-Roetne R, et al. Structure-based design, synthesis, and in vitro evaluation of bisubstrate inhibitors for catechol O-methyltransferase (COMT). Chem Eur J 2000 Mar 17;6(6):971–82.

[6] Zheng YJ, Bruice TC. A theoretical examination of the factors controlling the catalytic efficiency of a transmethylation enzyme: catechol *O*-methyltransferase. J Am Chem Soc 1997 Sep 1;119(35):8137–45.

[7] Lundström K, Tenhunen J, Tilgmann C, Karhunen T, Panula P, Ulmanen I. Cloning, expression and structure of catechol-O-methyltransferase. Biochim Biophys Acta 1995 Aug;1251(1):1–10.
[8] Männistö PT, Kaakkola S. Catechol-O-methyltransferase (COMT): biochemistry, molecular biology, pharmacology, and clinical efficacy of the new selective COMT inhibitors. Pharmacol Rev 1999 Dec;51(4):593–628.
[9] Law BJC, Bennett MR, Thompson ML, Levy C, Shepherd SA, Leys D, et al. Effects of active-site modification and quaternary structure on the regioselectivity of catechol- *O*-methyltransferase. Angew Chem Int Ed 2016 Feb 18;55(8):2683–7.
[10] Lotta T, Vidgren J, Tilgmann C, Ulmanen I, Melen K, Julkunen I, et al. Kinetics of human soluble and membrane-bound catechol O-methyltransferase: a revised mechanism and description of the thermolabile variant of the enzyme. Biochemistry 1995 Apr 4;34 (13):4202–10.
[11] Jeffery DR, Roth JA. Kinetic reaction mechanism for magnesium binding to membrane-bound and soluble catechol O-methyltransferase. Biochemistry 1987 May 19;26(10):2955–8.
[12] Tsao D, Diatchenko L, Dokholyan NV. Structural mechanism of S-adenosyl methionine binding to catechol O-methyltransferase. PLoS One 2011 Aug 31;6(8), e24287.
[13] Guldberg HC, Marsden CA. Catechol-O-methyl transferase: pharmacological aspects and physiological role. Pharmacol Rev 1975 Jun;27(2):135–206.
[14] Ball P, Knuppen R, Haupt M, Breuer H. Interactions between estrogens and catechol amines III. Studies on the methylation of catechol estrogens, catechol amines and other catechols by the Catechol- *O*-Methyltransferase1 of Human Liver. J Clin Endocrinol Metab 1972 Apr;34(4):736–46.
[15] Bonifati V, Meco G. New, selective catechol-O-methyltransferase inhibitors as therapeutic agents in Parkinson's disease. Pharmacol Ther 1999 Jan;81(1):1–36.
[16] Tenhunen J, Salminen M, Lundstrom K, Kiviluoto T, Savolainen R, Ulmanen I. Genomic organization of the human catechol O-methyltransferase gene and its expression from two distinct promoters. Eur J Biochem 1994 Aug;223(3):1049–59.
[17] Huh MM, Friedhoff AJ. Multiple molecular forms of catechol-O-methyltransferase. Evidence for two distinct forms, and their purification and physical characterization. J Biol Chem 1979 Jan 25;254 (2):299–308.
[18] Tenhunen J, Salminen M, Jalanko A, Ukkonen S, Ulmanen I. Structure of the rat catechol- *O*-methyltransferase gene: separate promoters are used to produce mRNAs for soluble and membrane-bound forms of the enzyme. DNA Cell Biol 1993 Apr;12(3):253–63.
[19] Salminen M, Lundström K, Tilgmann C, Savolainen R, Kalkkinen N, Ulmanen I. Molecular cloning and characterization of rat liver catechol-O-methyltransferase. Gene 1990 Jan;93(2):241–7.
[20] Ehler A, Benz J, Schlatter D, Rudolph MG. Mapping the conformational space accessible to catechol-*O*-methyltransferase. Acta Crystallogr D Biol Crystallogr 2014 Aug 1;70(8):2163–74.
[21] Martin J. SAM (dependent) I AM: the S-adenosylmethionine-dependent methyltransferase fold. Curr Opin Struct Biol 2002 Dec 1;12(6):783–93.
[22] Vidgren J, Svensson LA, Liljas A. Crystal structure of catechol O-methyltransferase. Nature 1994 Mar;368(6469):354–8.
[23] Delprato A, Xiao E, Manoj D. Connecting DCX, COMT and FMR1 in social behavior and cognitive impairment. Behav Brain Funct 2022 Dec 19;18(1):7.
[24] Fung MM, Viveros OH, O'Connor DT. Diseases of the adrenal medulla. Acta Physiol 2007 Nov 16;192(2):325–35.
[25] Pigoni A, Lazzaretti M, Mandolini GM, Delvecchio G, Altamura AC, Soares JC, et al. The impact of COMT polymorphisms on cognition in bipolar disorder: a review. J Affect Disord 2019 Jan;243:545–51.
[26] Tiihonen J, Hallikainen T, Lachman H, Saito T, Volavka J, Kauhanen J, et al. Association between the functional variant of the catechol-O-methyltransferase (COMT) gene and type 1 alcoholism. Mol Psychiatry 1999 May 1;4(3):286–9.
[27] Chaudhary A, Kumar P, Rai V. Catechol-O-methyltransferase (COMT) Val158Met polymorphism and susceptibility to alcohol dependence. Indian J Clin Biochem 2021 Jul 2;36(3):257–65.
[28] Abou Al Hassan S, Cutinha D, Mattar L. The impact of COMT, BDNF and 5-HTT brain-genes on the development of anorexia nervosa: a systematic review. Eat Weight Disord 2021 Jun 11;26 (5):1323–44.
[29] Männistö PT, Kaakkola S. New selective COMT inhibitors: useful adjuncts for Parkinson's disease? Trends Pharmacol Sci 1989 Feb;10(2):54–6.
[30] Jiménez-Jiménez FJ, Alonso-Navarro H, García-Martín E, Agúndez JAG. COMT gene and risk for Parkinson's disease. Pharmacogenet Genomics 2014 Jul;24(7):331–9.
[31] Ira E, Zanoni M, Ruggeri M, Dazzan P, Tosato S. COMT, neuropsychological function and brain structure in schizophrenia: a systematic review and neurobiological interpretation. J Psychiatry Neurosci 2013 Nov 1;38(6):366–80.
[32] Misir E, Ozbek MM, Halac E, Turan S, Alkas GE, Ciray RO, et al. The effects of *catechol-O-methyltransferase* single nucleotide polymorphisms on positive and negative symptoms of schizophrenia: a systematic review and meta-analysis. Psych J 2022 Dec 31;11 (6):779–91.
[33] Williams HJ, Owen MJ, O'Donovan MC. Is COMT a susceptibility gene for schizophrenia? Schizophr Bull 2007 Mar 19;33(3):635–41.
[34] Antypa N, Drago A, Serretti A. The role of COMT gene variants in depression: bridging neuropsychological, behavioral and clinical phenotypes. Neurosci Biobehav Rev 2013 Sep;37 (8):1597–610.
[35] Perkovic MN, Strac DS, Tudor L, Konjevod M, Erjavec GN, Pivac N. Catechol-O-methyltransferase, cognition and Alzheimer's disease. Curr Alzheimer Res 2018 Mar 14;15(5):408–19.
[36] Sharif AA. Entacapone in restless legs syndrome. Mov Disord 2002 Mar;17(2):421.
[37] Fava M, Rosenbaum JF, Kolsky AR, Alpert JE, Nierenberg AA, Spillmann M, et al. Open study of the catechol-O-methyltransferase inhibitor tolcapone in major depressive disorder. J Clin Psychopharmacol 1999 Aug;19(4):329–35.
[38] Fiocco AJ, Lindquist K, Ferrell R, Li R, Simonsick EM, Nalls M, et al. COMT genotype and cognitive function: an 8-year longitudinal study in white and black elders. Neurology 2010 Apr 20;74 (16):1296–302.
[39] Mullins N, Bigdeli TB, Børglum AD, Coleman JRI, Demontis D, Mehta D, et al. GWAS of suicide attempt in psychiatric disorders and association with major depression polygenic risk scores. Am J Psychiatr 2019 Aug;176(8):651–60.
[40] Hettema JM, An SS, Bukszar J, van den Oord EJCG, Neale MC, Kendler KS, et al. Catechol-O-methyltransferase contributes to genetic susceptibility shared among anxiety spectrum phenotypes. Biol Psychiatry 2008 Aug;64(4):302–10.

[41] Rothe C, Koszycki D, Bradwejn J, King N, Deluca V, Tharmalingam S, et al. Association of the Val158Met catechol O-methyltransferase genetic polymorphism with panic disorder. Neuropsychopharmacology 2006 Oct 8;31(10):2237–42.

[42] Sampaio AS, Hounie AG, Petribú K, Cappi C, Morais I, Vallada H, et al. COMT and MAO-A polymorphisms and obsessive-compulsive disorder: A Family-Based Association Study. PLoS One 2015 Mar 20;10(3), e0119592.

[43] Sun H, Yuan F, Shen X, Xiong G, Wu J. Role of COMT in ADHD: a systematic meta-analysis. Mol Neurobiol 2014 Feb 2;49(1):251–61.

[44] Koster R, Panagiotou OA, Wheeler WA, Karlins E, Gastier-Foster JM, Caminada de Toledo SR, et al. Genome-wide association study identifies the *GLDC/IL33* locus associated with survival of osteosarcoma patients. Int J Cancer 2018 Apr 15;142(8):1594–601.

[45] Liu J, Zhou Y, Liu S, Song X, Yang XZ, Fan Y, et al. The coexistence of copy number variations (CNVs) and single nucleotide polymorphisms (SNPs) at a locus can result in distorted calculations of the significance in associating SNPs to disease. Hum Genet 2018 Jul 17;137(6–7):553–67.

[46] Schlosser P, Li Y, Sekula P, Raffler J, Grundner-Culemann F, Pietzner M, et al. Genetic studies of urinary metabolites illuminate mechanisms of detoxification and excretion in humans. Nat Genet 2020 Feb 20;52(2):167–76.

[47] Almas A, Forsell Y, Millischer V, Möller J, Lavebratt C. Association of catechol-O-methyltransferase (COMT Val158Met) with future risk of cardiovascular disease in depressed individuals - a Swedish population-based cohort study. BMC Med Genet 2018 Dec 25;19 (1):126.

[48] Karling P, Danielsson Å, Wikgren M, Söderström I, Del-Favero J, Adolfsson R, et al. The relationship between the Val158Met catechol-O-methyltransferase (COMT) polymorphism and irritable bowel syndrome. PLoS One 2011 Mar 18;6(3), e18035.

[49] Gothelf D, Law AJ, Frisch A, Chen J, Zarchi O, Michaelovsky E, et al. Biological effects of COMT haplotypes and psychosis risk in 22q11.2 deletion syndrome. Biol Psychiatry 2014 Mar;75(5):406–13.

[50] Phoswa WN. Dopamine in the pathophysiology of preeclampsia and gestational hypertension: monoamine oxidase (MAO) and catechol-O-methyl transferase (COMT) as possible mechanisms. Oxid Med Cell Longev 2019 Nov 28;2019:1–8.

[51] Bastos P, Gomes T, Ribeiro L. Catechol-O-methyltransferase (COMT): an update on its role in cancer, neurological and cardiovascular diseases. Rev Physiol Biochem Pharmacol 2017;1–39. https://doi.org/10.1007/112_2017_2.

[52] Karhunen T, Tilgmann C, Ulmanen I, Panula P. Neuronal and non-neuronal catechol-*O*-methyltransferase in primary cultures of rat brain cells. Int J Dev Neurosci 1995 Dec;13(8):825–34.

[53] Redell JB, Dash PK. Traumatic brain injury stimulates hippocampal catechol-O-methyl transferase expression in microglia. Neurosci Lett 2007 Feb;413(1):36–41.

[54] Matsumoto M, Weickert CS, Akil M, Lipska BK, Hyde TM, Herman MM, et al. Catechol O-methyltransferase mRNA expression in human and rat brain: evidence for a role in cortical neuronal function. Neuroscience 2003 Jan;116(1):127–37.

[55] Guimarães S, Moura D. Vascular adrenoceptors: an update. Pharmacol Rev 2001 Jun;53(2):319–56.

[56] Adameova A, Abdellatif Y, Dhalla NS. Role of the excessive amounts of circulating catecholamines and glucocorticoids in stress-induced heart disease. Can J Physiol Pharmacol 2009 Jul;87 (7):493–514.

[57] Akhtar MJ, Yar MS, Grover G, Nath R. Neurological and psychiatric management using COMT inhibitors: a review. Bioorg Chem 2020 Jan;94, 103418.

[58] Marino BLB, de Souza LR, Sousa KPA, Ferreira JV, Padilha EC, da Silva CHTP, et al. Parkinson's disease: a review from pathophysiology to treatment. Mini-reviews. Med Chem 2020 May 27;20 (9):754–67.

[59] Wood AJJ, Calne DB. Treatment of Parkinson's disease. N Engl J Med 1993 Sep 30;329(14):1021–7.

[60] Goetz CG. Influence of COMT inhibition on levodopa pharmacology and therapy. Neurology 1998 May 1;50(Issue 5, Supplement 5):S26–30.

[61] Masri MS, Booth AN, DeEds F. O-methylation in vitro of dihydroxy- and trihydroxy-phenolic compounds by liver slices. Biochim Biophys Acta 1962 Dec;65(3):495–500.

[62] Angel A, Rogers KJ. Convulsant action of polyphenols. Nature 1968 Jan;217(5123):84–5.

[63] Giles RE, Miller JW. The catechol-O-methyl transferase activity and endogenous catecholamine content of various tissues in the rat and the effect of administration of U-0521 (3',4'-dihydroxy-2-methyl propiophenone). J Pharmacol Exp Ther 1967 Nov;158(2):189–94.

[64] Bacq ZM, Gosselin L, Dresse A, Renson J. Inhibition of O-methyltransferase by catechol and sensitization to epinephrine. Science 1979;130(3373):453–4. 1959 Aug 21.

[65] Zhu BT, Wang P, Nagai M, Wen Y, Bai HW. Inhibition of human catechol-O-methyltransferase (COMT)-mediated O-methylation of catechol estrogens by major polyphenolic components present in coffee. J Steroid Biochem Mol Biol 2009 Jan;113(1–2):65–74.

[66] Wang LH, Hsu KY, Uang YS, Hsu FL, Yang LM, Lin SJ. Caffeic acid improves the bioavailability of l-dopa in rabbit plasma. Phytother Res 2010 Jun;24(6):852–8.

[67] Archer S, Arnold A, Kullnig RK, Wylie DW. The enzymic methylation of pyrogallol. Arch Biochem Biophys 1960 Mar;87(1):153–4.

[68] Axelrod J, Laroche MJ. Inhibitor of O-methylation of epinephrine and norepinephrine in vitro and in vivo. Science 1979;130 (3378):800. 1959 Sep 25.

[69] Booth AN, Masri MS, Robbins DJ, Emerson OH, Jones FT, DeEds F. The metabolic fate of gallic acid and related compounds. J Biol Chem 1959 Nov;234(11):3014–6.

[70] Ericsson AD. Potentiation of the l-Dopa effect in man by the use of catechol-O-methyltransferase inhibitors. J Neurol Sci 1971 Oct;14 (2):193–7.

[71] Simpson GM, Varga V. An investigation of the clinical effect of GPA-1714, a catechol-O-methyl transferase inhibitor. J Clin Pharmacol New Drugs 1972 Oct 11;12(10):417–21.

[72] Backstrom R, Honkanen E, Pippuri A, Kairisalo P, Pystynen J, Heinola K, et al. Synthesis of some novel potent and selective catechol O-methyltransferase inhibitors. J Med Chem 1989 Apr 1;32(4):841–6.

[73] Borgulya J, Bruderer H, Bernauer K, Zürcher G, Prada M. Catechol-*O*-methyltransferase-inhibiting pyrocatechol derivatives: synthesis and structure-activity studies. Helv Chim Acta 1989 Aug 9;72 (5):952–68.

[74] Baeckstoem RJ, Heinola KF, Honkanen EJ, Kaakkola SK, Kairisalo PJ, Linden IBY, et al. Pharmacologically active compounds, methods for the preparation thereof and compositions containing the same. US4963590; 1990.

[75] Nissinen E, Lindén IB, Schultz E, Kaakkola S, Männistö PT, Pohto P. Inhibition of catechol-O-methyltransferase activity by two novel disubstituted catechols in the rat. Eur J Pharmacol 1988 Aug;153(2–3):263–9.

[76] Schultz E, Nissinen E. Inhibition of rat liver and duodenum soluble catechol-O-methyltransferase by a tight-binding inhibitor OR-462. Biochem Pharmacol 1989 Nov;38(22):3953–6.
[77] Männistö PT, Tuomainen P, Tuominen RK. Different in vivo properties of three new inhibitors of catechol O-methyltransferase in the rat. Br J Pharmacol 1992 Mar;105(3):569–74.
[78] Cedarbaum JM, Léger G, Reches A, Guttman M. Effect of nitecapone (OR-462) on the pharmacokinetics of levodopa and 3-O-methyldopa formation in cynomolgus monkeys. Clin Neuropharmacol 1990 Dec;13(6):544–52.
[79] Nissinen E, Linden IB, Schultz E, Pohto P. Biochemical and pharmacological properties of a peripherally acting catechol-O-methyltransferase inhibitor entacapone. Naunyn Schmiedeberg's Arch Pharmacol 1992 Sep;346(3).
[80] Keränen T, Gordin A, Karlsson M, Korpela K, Pentikäinen PJ, Rita H, et al. Inhibition of soluble catechol-O-methyltransferase and single-dose pharmacokinetics after oral and intravenous administration of entacapone. Eur J Clin Pharmacol 1994 Mar;46(2).
[81] Savolainen J, Forsberg M, Taipale H, Männistö PT, Järvinen K, Gynther J, et al. Effects of aqueous solubility and dissolution characteristics on oral bioavailability of entacapone. Drug Dev Res 2000 Apr;49(4):238–44.
[82] Novaroli L, Bouchard Doulakas G, Reist M, Rolando B, Fruttero R, Gasco A, et al. The lipophilicity behavior of three catechol-O-methyltransferase (COMT) inhibitors and simple analogues. Helv Chim Acta 2006 Jan;89(1):144–52.
[83] Wikberg T, Vuorela A, Ottoila P, Taskinen J. Identification of major metabolites of the catechol-O-methyltransferase inhibitor entacapone in rats and humans. Drug Metab Dispos 1993;21(1):81–92.
[84] Aho P, Pohto P, Linden IB, Backstrom R, Honkanen EJ, Nissinen E. New use of catechol-O-methyl transferase (COMT) inhibitors and their physiologically acceptable salts and esters. EP0323162; 1989.
[85] Linden IB, Ranta S, Hinonen E, Kaakkola S, Nissinen E, Pohto P. Use of catechol-O-methyl transferase (COMT) inhibitors as anti-cancer agents. GB2220569; 1990.
[86] Aperia A, Linden IB. Use of COMT inhibitors for the manifacture of a medicament for the prevention of diabetic vascular dysfunction. WO1998027973A1; 1998.
[87] Nissinen H, Vahteristo M, Kuoppamaeki M, Ellmen J, Leinonen M. Treatment of symptoms of Parkinson's disease. WO2007034024; 2007.
[88] Ellmen J, Karvinen J, Vahteristo M. Treatment of restless syndrome. WO2006051154; 2006.
[89] Bernauer K, Borgulya J, Bruderer H, Daprada M, Zuercher G. Catechol derivatives. US5236952; 1993.
[90] Männistö PT. Clinical potential of catechol-O-methyltransferase (COMT) inhibitors as adjuvants in Parkinson's disease. CNS Drugs 1994 Mar;1(3):172–9.
[91] Zürcher G, Dingemanse J, da Prada M. Potent COMT inhibition by Ro 40-7592 in the periphery and in the brain. Preclinical and clinical findings. Adv Neurol 1993;60:641–7.
[92] Zürcher G, Keller HH, Kettler R, Borgulya J, Bonetti EP, Eigenmann R, et al. Ro 40-7592, a novel, very potent, and orally active inhibitor of catechol-O-methyltransferase: a pharmacological study in rats. Adv Neurol 1990;53:497–503.
[93] Artusi CA, Sarro L, Imbalzano G, Fabbri M, Lopiano L. Safety and efficacy of tolcapone in Parkinson's disease: systematic review. Eur J Clin Pharmacol 2021 Jun 7;77(6):817–29.
[94] Assal F, Spahr L, Hadengue A, Rubbici-Brandt L, Burkhard PR. Tolcapone and fulminant hepatitis. Lancet 1998 Sep;352(9132):958.
[95] Olanow CW. Tolcapone and hepatotoxic effects. Arch Neurol 2000 Feb 1;57(2):263.
[96] Jorga K, Fotteler B, Heizmann P, Gasser R. Metabolism and excretion of tolcapone, a novel inhibitor of catechol-*O*-methyltransferase. Br J Clin Pharmacol 1999 Oct;48(4):513–20.
[97] Smith KS, Smith PL, Heady TN, Trugman JM, Harman WD, Macdonald TL. In vitro metabolism of tolcapone to reactive intermediates: relevance to tolcapone liver toxicity. Chem Res Toxicol 2003 Feb 1;16(2):123–8.
[98] Benes J, Soares Da Silva M, Learmonth A. Substituted 2-phenyl-1-(3,4-dihydroxy-5-nitrophenyl)-1-ethanones, their use in the treatment of some central and peripheral nervous system disorders and pharmaceutical compositions containing them. WO2000037423A1; 2000.
[99] Learmonth DA, Vieira-Coelho MA, Benes J, Alves PC, Borges N, Freitas AP, et al. Synthesis of 1-(3,4-dihydroxy-5-nitrophenyl)-2-phenyl-ethanone and derivatives as potent and long-acting peripheral inhibitors of catechol-*O*-methyltransferase. J Med Chem 2002 Jan 1;45(3):685–95.
[100] Learmonth DA, Soares Da Silva M. Substituted nitrated catechols, their use in the treatment of some central and peripheral nervous system disorders and pharmaceutical compositions containing them. WO/2001/098250; 2001.
[101] Learmonth DA, Palma PN, Vieira-Coelho MA, Soares-da-Silva P. Synthesis, biological evaluation, and molecular modeling studies of a novel, peripherally selective inhibitor of catechol-*O*-methyltransferase. J Med Chem 2004 Dec 1;47(25):6207–17.
[102] Kiss LE, Ferreira HS, Torrão L, Bonifácio MJ, Palma PN, Soares-da-Silva P, et al. Discovery of a long-acting, peripherally selective inhibitor of catechol-*O*-methyltransferase. J Med Chem 2010 Apr 22;53(8):3396–411.
[103] Almeida L, Rocha JF, Falcão A, Nuno Palma P, Loureiro AI, Pinto R, et al. Pharmacokinetics, pharmacodynamics and tolerability of opicapone, a novel catechol-O-methyltransferase inhibitor, in healthy subjects. Clin Pharmacokinet 2013 Feb 18;52(2):139–51.
[104] Bonifácio MJ, Sutcliffe JS, Torrão L, Wright LC, Soares-da-Silva P. Brain and peripheral pharmacokinetics of levodopa in the cynomolgus monkey following administration of opicapone, a third generation nitrocatechol COMT inhibitor. Neuropharmacology 2014 Feb;77:334–41.
[105] Greenwood J, Pham H, Rey J. Opicapone: a third generation COMT inhibitor. Clin Park Relat Disord 2021;4, 100083. https://doi.org/10.1016/j.prdoa.2020.100083.
[106] Berger AA, Winnick A, Izygon J, Jacob BM, Kaye JS, Kaye RJ, et al. Opicapone, a novel catechol-O-methyl transferase inhibitor, for treatment of Parkinson's disease "Off" episodes. Health Psychol Res 2022 Jun 28;10(5).
[107] Marsh W. Tropolone. In: xPharm: the comprehensive pharmacology reference. Elsevier; 2007. p. 1–3.
[108] Broch OJ. The in vivo effect of tropolone on noradrenaline metabolism and catechol-O-methyl transferase activity in tissues of the rat. Acta Pharmacol Toxicol (Copenh) 2009 Mar 13;33(5–6):417–28.
[109] Borchardt RT. Catechol O-methyltransferase. 4. In vitro inhibition by 3-hydroxy-4-pyrones, 3-hydroxy-2-pyridones, and 3-hydroxy-4-pyridones. J Med Chem 1973 May 1;16(5):581–3.
[110] Buchler I, Akuma D, Au V, Carr G, de León P, DePasquale M, et al. Optimization of 8-hydroxyquinolines as inhibitors of catechol *O*-methyltransferase. J Med Chem 2018 Nov 8;61(21):9647–65.
[111] Bieck PR, Antonin KH, Farger G, Nilsson EB, Schmidt EK, Dostert P, et al. Clinical pharmacology of the new COMT inhibitor CGP 28 014. Neurochem Res 1993 Nov;18(11):1163–7.

[112] Bieck P, Nilsson E, Antonin KH. Effect of the new selective COMT inhibitor CGP 28014 A on the formation of 3-O-methyldopa (3OMD) in plasma of healthy subjects. In: Amine oxidases and their impact on neurobiology. Vienna: Springer Vienna; 1990. p. 387–91.

[113] Ishikawa T, Nishimura T, Inoue H, Tanaka N, Muranaka H. Novel catechol derivative, pharmaceutical composition containing same, and uses of those. WO2007063789; 2007.

[114] Pérez RA, Fernández-Alvarez E, Nieto O, Piedrafita FJ. Inhibition of catechol-O-methyltransferase by 1-vinyl derivatives of nitrocatechols and nitroguaiacols. Biochem Pharmacol 1993 Apr;45(10):1973–81.

[115] Perez RA, Fernandez-Alvarez E, Nieto O, Javier PF. Kinetics of the reversible tight-binding inhibition of pig liver catechol-O-methyltransferase by [2-(3,4-dihydroxy-2-nitrophenyl) vinyl] phenyl ketone. J Enzym Inhib 1994 Jan 27;8(2):123–31.

[116] Rivas E, de Ceballos ML, Nieto O, Fontenla JA. *In vivo* effects of new inhibitors of catechol-*O*-methyl transferase. Br J Pharmacol 1999 Apr;126(7):1667–73.

[117] Learmonth DA, Bonifácio MJ, Soares-da-Silva P. Synthesis and biological evaluation of a novel series of "ortho-nitrated" inhibitors of catechol-*O*-methyltransferase. J Med Chem 2005 Dec 1;48 (25):8070–8.

[118] Pystynen J, Luiro A, Lotta T, Ovaska M, Vidgren J. New catechol derivatives. WO9637456A1; 1996.

[119] Ishikawa T, Inoue H, Kobayashi S, Yoshida M, Shiohara H, Ueno Y, et al. Novel catechol derivative, pharmaceutical composition containing the same, use of the catechol derivative, and use of the pharmaceutical composition. WO2009081891; 2009.

[120] Pystynen J, Ovaska M, Vidgren J, Lotta T, Yliperttula-Ikonen M. Coumarin derivatives with COMT inhibiting activity. WO2002002548A1; 2002.

[121] Ahlmark M, Baeckstroem R, Luiro A, Pystynen J, Tiainen E. New pharmaceutical compounds. WO2007010085; 2007.

[122] Rautio J, Leppänen J, Lehtonen M, Laine K, Koskinen M, Pystynen J, et al. Design, synthesis and in vitro/in vivo evaluation of orally bioavailable prodrugs of a catechol-O-methyltransferase inhibitor. Bioorg Med Chem Lett 2010 Apr;20(8):2614–6.

[123] Brevitt SE, Tan EW. Synthesis and *in vitro* evaluation of two progressive series of bifunctional polyhydroxybenzamide catechol-*O*-methyltransferase inhibitors. J Med Chem 1997 Jun 1;40 (13):2035–9.

[124] Bailey K, Tan EW. Synthesis and evaluation of bifunctional nitrocatechol inhibitors of pig liver catechol-O-methyltransferase. Bioorg Med Chem 2005 Oct;13(20):5740–9.

[125] Lerner C, Ruf A, Gramlich V, Masjost B, Zürcher G, Jakob-Roetne R, et al. X-ray crystal structure of a bisubstrate inhibitor bound to the enzyme catechol-O-methyltransferase: A dramatic effect of inhibitor preorganization on binding affinity. Angew Chem Int Ed Eng 2001 Nov 5;40(21):4040–2.

[126] Lerner C, Masjost B, Ruf A, Gramlich V, Jakob-Roetne R, Zürcher G, et al. Bisubstrate inhibitors for the enzyme catechol-O-methyltransferase (COMT): influence of inhibitor preorganisation and linker length between the two substrate moieties on binding affinity. Org Biomol Chem 2003;1(1):42–9. https://doi.org/10.1039/B208690P.

[127] Rodrigues ML, Bonifácio MJ, Soares-da-Silva P, Carrondo MA, Archer M. Crystallization and preliminary X-ray diffraction studies of a catechol-*O*-methyltransferase/inhibitor complex. Acta Crystallogr Sect F Struct Biol Cryst Commun 2005 Jan 1;61(1):118–20.

[128] Ellermann M, Paulini R, Jakob-Roetne R, Lerner C, Borroni E, Roth D, et al. Molecular recognition at the active site of catechol-*O*-methyltransferase (COMT): Adenine replacements in bisubstrate inhibitors. Chemistry 2011 May 27;17(23):6369–81.

[129] Cao Y, Chen ZJ, Jiang HD, Chen JZ. Computational studies of the regioselectivities of COMT-catalyzed *meta-/para*-O methylations of luteolin and quercetin. J Phys Chem B 2014 Jan 16;118 (2):470–81.

[130] Kang KS, Yamabe N, Wen Y, Fukui M, Zhu BT. Beneficial effects of natural phenolics on levodopa methylation and oxidative neurodegeneration. Brain Res 2013 Feb;1497:1–14.

[131] Lorenz M, Paul F, Moobed M, Baumann G, Zimmermann BF, Stangl K, et al. The activity of catechol-O-methyltransferase (COMT) is not impaired by high doses of epigallocatechin-3-gallate (EGCG) in vivo. Eur J Pharmacol 2014 Oct;740:645–51.

[132] Fabbri M, Ferreira JJ, Rascol O. COMT inhibitors in the management of Parkinson's disease. CNS Drugs 2022 Mar 25;36 (3):261–82.

[133] Scott LJ. Opicapone: a review in Parkinson's disease. CNS Drugs 2021 Jan 11;35(1):121–31.

[134] Jenner P, Rocha JF, Ferreira JJ, Rascol O, Soares-da-Silva P. Redefining the strategy for the use of COMT inhibitors in Parkinson's disease: the role of opicapone. Expert Rev Neurother 2021 Sep 2;21(9):1019–33.

[135] Berger AA, Robinson C, Winnick A, Izygon J, Jacob BM, Noonan MJ, et al. Opicapone for the treatment of parkinson's disease "Off" episodes: pharmacology and clinical considerations. Clin Drug Investig 2022 Feb 21;42(2):127–35.

[136] Jost WH. Evaluating Opicapone as add-on treatment to levodopa/DDCI in patients with Parkinson's disease. Neuropsychiatr Dis Treat 2022 Aug;18:1603–18.

[137] Gonçalves D, Alves G, Fortuna A, Soares-da-Silva P, Falcão A. Pharmacokinetics of opicapone, a third-generation COMT inhibitor, after single and multiple oral administration: a comparative study in the rat. Toxicol Appl Pharmacol 2017 May;323:9–15.

[138] Azevedo Kauppila L, Pimenta Silva D, Ferreira JJ. Clinical utility of opicapone in the management of Parkinson's disease: a short review on emerging data and place in therapy. Degener Neurol Neuromuscul Dis 2021 May;11:29–40.

[139] Silva TB, Borges F, Serrão MP, Soares-da-Silva P. Liver says no: the ongoing search for safe catechol O-methyltransferase inhibitors to replace tolcapone. Drug Discov Today 2020 Oct;25(10):1846–54.

Chapter 2.6

D-Alanine-D-alanine ligase

Alessio Nocentini
NEUROFARBA Department, Pharmaceutical and Nutraceutical Section, University of Florence, Firenze, Italy

1 Introduction

The bacterial cell wall is a thick and robust coating representing the outer structure of bacterial cells that fixes the cell shape and protects the cell from damages [1]. The cell wall absence in mammalian cells made the enzymes involved in the biosynthetic pathway of this bacterial structure an attractive target for antibacterial drugs [2–4]. Peptidoglycan (PG) is a unique large macromolecule, a polysaccharide, consisting of sugars and amino acids forming a mesh layer outside the plasma membrane and plays a crucial role in the physiological functions of the cell wall and builds a rigid structure allowing bacteria to survive in hypotonic environments [5]. The cell walls of almost all bacteria contain PG in percentages ranging from 30% to 95% in Gram-positive bacteria and from 5% to 20% in Gram-negative bacteria [6,7]. The biosynthetic pathway of PG in bacteria was gradually unveiled, with every enzyme involved in this pathway becoming a potential target of antibacterial agents [8]. In detail, the biosynthesis of PG is divided into three consecutive stages: (i) the synthesis of the peptidoglycan precursors in the cytoplasm; (ii) the synthesis of lipid-linked intermediates on the inner side of the cell membrane; (iii) the polymerization reaction taking place on the outer side of the cell membrane [9]. In the latter step, the process requires the cross-linking of peptidyl moieties present on adjacent glycan strands. The enzyme D-alanine-D-alanine transpeptidase catalyzes this cross-linking and is the target of β-lactam antibiotics, such as penicillins and cephalosporins. Instead, glycopeptide antibiotics, such as vancomycin, bind to the D-alanine-D-alanine PG termini and prevent subsequent cross-linking by the transpeptidase [10].

To date, only two intracellular enzymes among those involved in PG biosynthesis were validated as targets of bacterial agents in clinical use: UDP-*N*-acetylglucosamine-enolpyruvyl transferase, target of the antibiotic fosfomycin, and D-alanine-D-alanine ligase (Ddl, EC 6.3.2.4), primary target of the second-line antituberculosis agent D-cycloserine (DCS). Ddl is an ATP-dependent bacterial enzyme that catalyzes the ligation of two D-alanine residues to a D-alanyl-D-alanine product, which in the next step is incorporated into the final intracellular peptidoglycan precursor UDP-*N*-acetylmuramoyl pentapeptide [8]. It was demonstrated that inhibition of Ddl directly produces a significant loss of strength of the bacterial cell wall and, as a result, a more likely disruption of the bacterial cell [11]. Hence, Ddl inhibitors have become attractive agents and hot spots for the development of new antibacterial drugs [12,13]. This chapter speaks on structure and function of Ddl and reports the various classes of Ddl inhibitors currently being used and studied as antibacterial agents.

2 Structure and function of Ddl

Ddl is expressed by the ddl gene and is widely distributed in Gram-positive and Gram-negative bacteria [14,15]. Two isoforms of Ddl were identified, which are DdlA and DdlB [16] and show similar kinetic features, substrate specificity, and sensitivity to inhibitors [17]. The isoform DdlB from *Escherichia coli* is the most studied one. Ddl is among the first characterized enzymes of the ATP-grasp superfamily, which share an atypical nucleotide-binding site known as the ATP-grasp fold [18]. Crystal structures of several species revealed that Ddl is a homodimer (Fig. 1) [19].

In the catalytically active conformation, Ddl shows an ATP-binding pocket located between the center and the C-terminal domain and a bimolecular D-Ala binding pocket between the N-terminal and C-terminal domains (Fig. 2) [20]. The binding sites form a large cavity spanning the centrally located boundaries of all three domains in the protein. Each Ddl monomer can be subdivided into three α/β domains: an N-terminal domain (NTD), a central domain (MD), and a C-terminal domain (CTD, Fig. 2). Two magnesium ions (Mg^{2+}) at the center of the cavity are crucial for the anchorage of the ATP phosphate group (Fig. 3A) [21,22]. A structure-based mutagenesis study with DdlB from *E. coli* from Shi et al. in 1995 demonstrated that the amino acid residues Lys144, Leu183, Lys215, Glu270, and Asp257 bind to ATP, while residues Glu15, Ser150, His63, Arg255 and Tyr216, Asp 257, Ser281, and Leu282 interact with the two D-alanine ligands (Fig. 3A) [22].

Metalloenzymes. https://doi.org/10.1016/B978-0-12-823974-2.00009-7

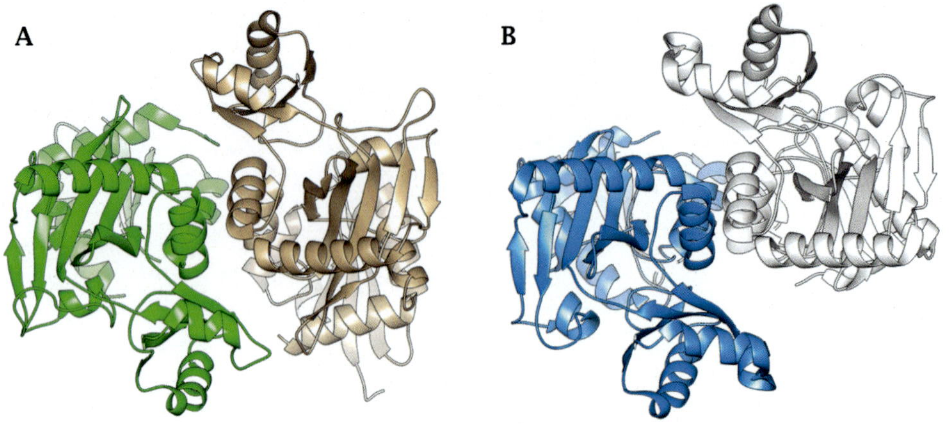

FIG. 1 Crystal structure of Ddl. Aligned ribbon view of the homodimer of (A) Ddl from *Mycobacterium tuberculosis* (PDB 3LWB) and (B) DdlB from *Escherichia coli* (PDB 4C5C).

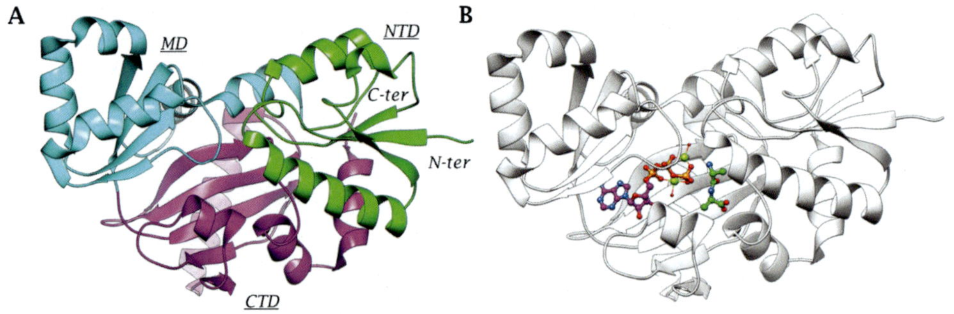

FIG. 2 Ribbon model of DdlB from *E. coli* (PDB 4C5C): (A) Monomer subdomain representation—i.e., the N-terminal, central, and C-terminal domains are colored *green*, *cyan*, and *magenta*, respectively; (B) overall monomer view in complex with ATP *(magenta)* and D-alanyl-D-alanine *(green)*. Mg^{2+} ions are represented as *light green spheres*. Metal-coordinating water molecules are shown as *small red spheres*.

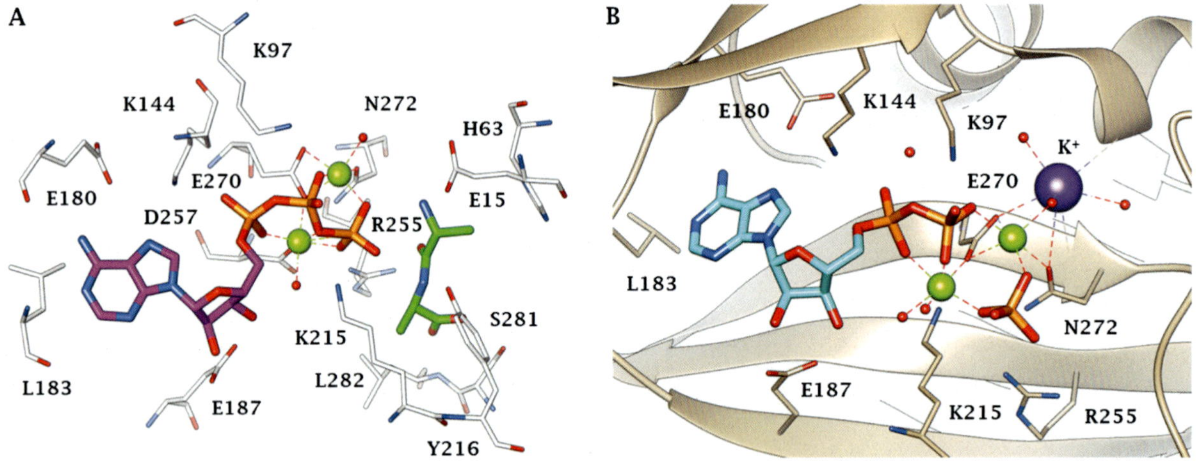

FIG. 3 (A) Active site stick view of DdlB from *E. coli* (PDB 4C5C) in complex with ATP *(magenta)* and D-alanyl-D-alanine *(green)*. (B) Active site ribbon view of Ddl enzyme from *Thermus thermophilus* (TtDdl) (PDB 6U1H) in complex with ADP *(cyan)*, phosphate and K^+ *(purple sphere)*. Coordination bonds are shown as *bicolored dashed lines*. Mg^{2+} ions are represented as *light green spheres*. Metal-coordinating water molecules are shown as *small red spheres*.

The catalytic mechanism of Ddl proceeds by an ordered *ter-ter* path including two half-reactions and involves formation of two reaction intermediates [23]: an acylphosphate intermediate and a tetrahedral intermediate (Fig. 4). In detail, in the first half-reaction, the carboxylate of D-Ala$_1$ attacks the γ-phosphate of ATP, leading to formation of the acylphosphate intermediate D-Ala-phosphate. In the second half-reaction, a second D-Ala$_2$ enters the active site, and its primary amine, in its deprotonated form (NH_2), reacts with D-Ala-phosphate. The deprotonated

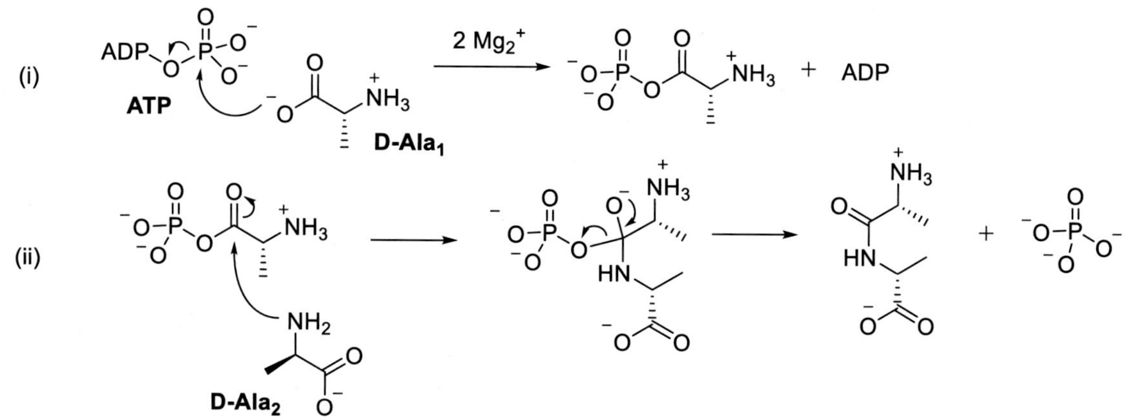

FIG. 4 Catalytic mechanism of Ddl, proceeding by an ordered *ter-ter* pathway (three substrates and three products) upon two half-reactions. The first half-reaction (i) involves the phosphorylation of D-Ala_1 to form the acylphosphate intermediate D-Ala-phosphate. In the second half-reaction (ii) D-Ala-phosphate is attacked by D-Ala_2 to form a tetrahedral intermediate, D-Ala-D-Ala-phosphate, which collapses within the active site to form D-Ala-D-Ala and Pi. The MVC K^+ is required for optimal activity.

D-Ala_2 attacks the phosphorylated carbonyl carbon of D-Ala_1, resulting in formation of a tetrahedral intermediate, D-Ala-D-Ala-phosphate. The tetrahedral intermediate then collapses to form the D-Ala-D-Ala dipeptide and phosphate [23]. It was shown that Ddl is activated by the monovalent cation (MVC) potassium (K^+) [18]. Ddl is a type II MVC activated enzyme, because the potassium ion binds at an allosteric site, causing an increase in substrate binding affinity and/or enzyme activity (Fig. 3B). Some kinetic studies revealed that the MVC K^+ was required for optimal activity of Ddl of bacteria such as *Enterococcus faecalis*, *Mycobacterium tuberculosis*, and *T. thermophilus*. In fact, these isoforms were characterized by increased k_{cat} and decreased K_m for the D-Ala substrate in the presence of KCl. NH_4^+ is a less potent activator than K^+, whereas Na^+ did not affect enzyme activity. In 2020, Pederick et al. captured crystal structures of Ddl from *T. thermophilus* at distinct stages of the catalytic mechanism and revealed minimal conformational change in the presence of K^+, suggesting that a conformational change is not responsible for the activating effect (Fig. 3B) [18]. The authors thus proposed that activation occurs with K^+ ions altering the charge distribution within the active site, increasing the binding affinity for the carboxylate substrate D-Ala_1 and promoting the formation of the acylphosphate intermediate.

3 Ddl inhibitors

(*R*)-4-Amino-3-isoxazolidinone, named D-cycloserine (DCS, **1**, Fig. 5), is a broad-spectrum antibiotic discovered in 1952 and approved by the United States Food and Drug Administration in 1956 [24]. DCS is a natural product of *Streptomyces garyphalus* and *Streptomyces lavendulae*. It is a second-line drug used for the treatment of multidrug-resistant tuberculosis (MDR-TB), when first-line drugs such as isoniazid, rifampicin, ethambutol, and streptomycin showed weak activity [25]. DCS mimics D-alanine, competitively targeting D-alanine racemase (Alr) and Ddl, and thus

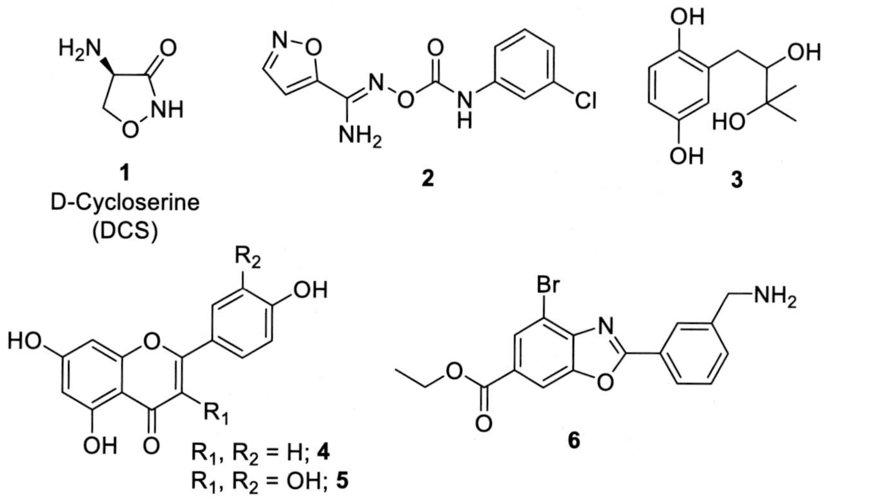

FIG. 5 Chemical structures of Ddl inhibitors **1–6**.

inhibiting the conversion of L-Ala to D-Ala and the subsequent synthesis of D-Ala-D-Ala, which weakens the bacterial cell wall and its resistance to acids [19]. In 2017, Batson et al. showed that the inhibition of D-alanine-D-alanine ligase by DCS proceeds via a distinct phosphorylated form of the drug (DSCP) [26]. In detail, the DCSP moiety mimics D-alanyl phosphate, which is a mandatory intermediate of the phosphoryl transfer during the first catalytic step (Fig. 6A). DCSP exploits most of the interactions within high-affinity site for the first D-Ala substrate (D-Ala$_1$). The amino group of DCSP forms a strong hydrogen bond with the carboxylate group of Glu15, mimicking the analog interaction made by the α-amino group of D-Ala$_1$. The ring oxygen in position 1 of DCSP is in H-bond distance with G276 backbone NH, while the adjacent nonprotonated ring nitrogen mimics the bifurcated interactions with G276 NH and R255 NH1 made by this atom in the product complex. The DCSP phosphate group is H-bonded to R255 (NH1 and NH2), K215, and the amide nitrogen of S150, replicating the interactions observed with the γ-phosphate of ATP, and it forms additional links with the adjacent ADP via two coordinated magnesium ions that bridge between the DCSP and ADP molecules (Fig. 6A) [26].

Notably, the rate of drug resistance of *M. tuberculosis* to DCS is only 7.4%, and the drug MIC value against the standard strain, *M. tuberculosis* H37Rof, is 25 μg/mL [27]. DCS exhibits a broad-spectrum antibacterial activity, with inhibitory effects against most Gram-negative bacteria, Gram-positive bacteria, and nontuberculous mycobacteria [28]. DCS shows a high bioavailability of 70%–90% and is widely distributed in human tissues and body fluids. DCS exhibits low protein-binding rate and is mostly excreted via glomerular filtration. However, DSC utility is limited because the drug is also a co-agonist of the *N*-methyl-D-aspartic acid (NMDA) receptor in the brain. In fact, psychiatric side effects such as anxiety, hallucination, depression, euphoria, and suicide are reported in 9.7%–50% of patients in DCS therapy [26].

In 2012, Hrast et al. identified isoxazole hit compounds, such as **2** (Fig. 5) by a virtual screening simulation [29]. These compounds were shown to bind to the bimolecular D-alanine binding pocket of *E. coli* Ddl as competitive inhibitors. The ligand IC_{50} values were in the range of 378–656 μM.

In 2009, Murakami et al. showed that the hydroquinone **3** (Fig. 5), extracted from the fungus *Acremonium murorum*, acts as inhibitor of the D-Ala-D-Ala pathway (Alr and Ddl), with an IC_{50} value of 20 μM [30].

Other natural compounds active as effective Ddl inhibitors were identified among flavonoid derivatives. Apigenin **4** (Fig. 5) was shown to be a potent Ddl inhibitor by Wu et al. in 2008, with a IC_{50} value against *Helicobacter pylori* Ddl of 132.7 μM, even lower compared to DCS (IC_{50} of 299.0 μM) [31]. Apigenin competes with ATP and displays MIC values of 25, 25, and 200 μg/mL against *H. pylori* SS1, 43504, and JM109 strains, respectively. As well, quercetin **5** (Fig. 5) is an ATP-competitive inhibitor of Ddl, with IC_{50} values of 48.5 and 19.9 μM against *H. pylori* and *E. coli* Ddl, respectively, but shows a worse antibacterial action than apigenin, likely because of a lower cell penetration resulting from an increased hydrophilicity, with MIC values of 200, 100, 300, and 300 μg/mL against *H. pylori* SS1, 43504, and JM109 strains, and *E. coli* 25922, respectively [31].

In 2009, Tytgat et al. developed an in silico method, which recognized that benzoxazole scaffolds can increase the interaction between inhibitors and Ddl [32]. Among the identified ligands, compound **6** (Fig. 5) was the most effective benzoxazole Ddl inhibitor, with an IC_{50} value of

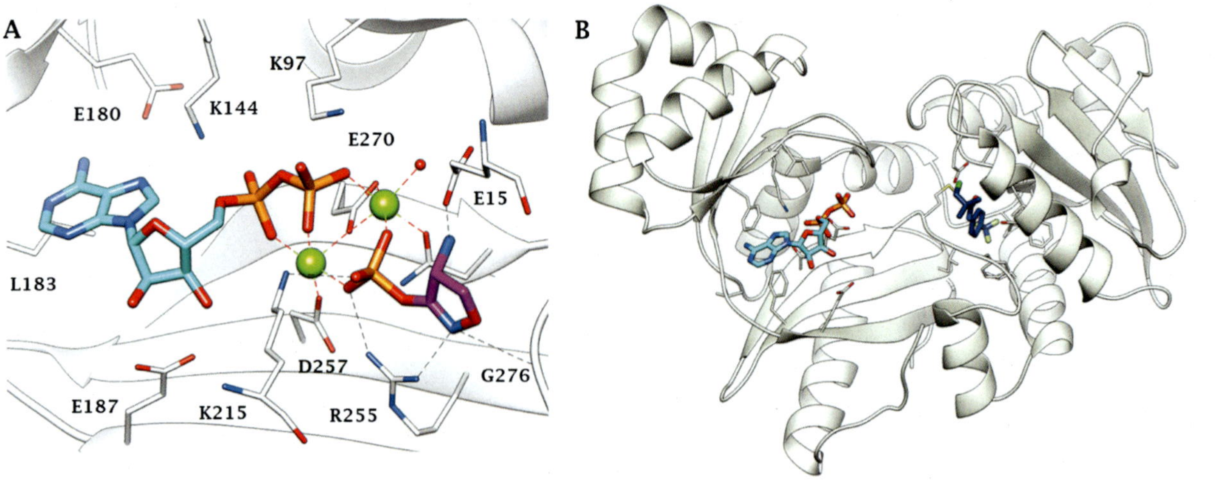

FIG. 6 (A) Active site stick view DdlB from *E. coli* (PDB 4C5A) in complex with ADP *(cyan)* and DCSP *(magenta)*. Coordination bonds are shown as *bicolored dashed lines*. Mg^{2+} ions are represented as *light green spheres*. Metal-coordinating water molecules are shown as *small red spheres*. (B) Overall ribbon view of Ddl enzyme from *Staphylococcus aureus* (PDB 2I80) in complex with ATP *(cyan)* and compound **32** *(dark blue)*.

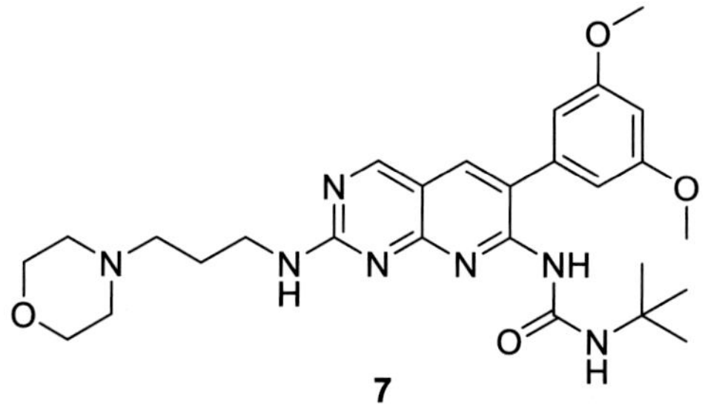

7

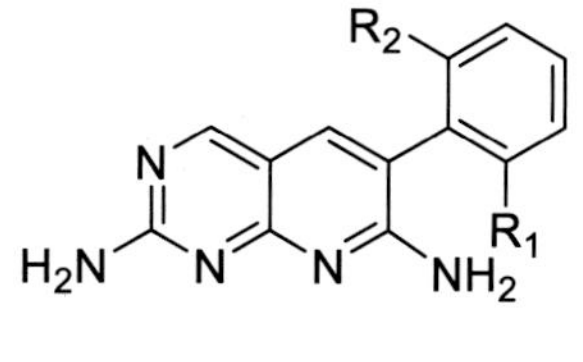

R_1, R_2 = F; **8**
R_1, R_2 = Br; **9**

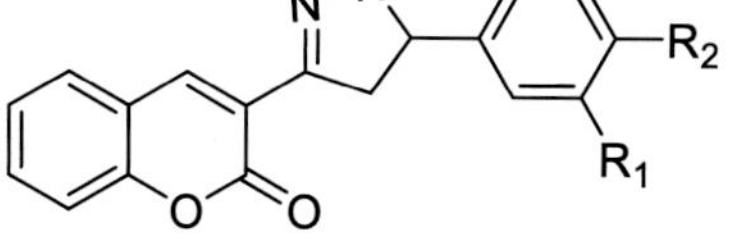

R_1 = OCH_3, R_2 = OH; **10**
R_1 = H, R_2 = CH_3; **11**

FIG. 7 Chemical structures of Ddl inhibitors **7–11**.

432 μM, and MIC values of 128, 32, and 32 μg/mL against *E. coli* ATCC 25922, *Staphylococcus aureus* ATCC 25923, and *E. faecalis* ATCC 29212.

In 2012, Škedelj et al. reported a series of 6-arylpyrido [2,3-*d*]pyrimidine as inhibitors of *E. coli* DdlB [33]. Among them, compound **7** (Fig. 7) showed the best action with an IC_{50} value of 133 μM against *E. coli* DdlB. Surface plasmon resonance experiments showed that the tested inhibitor quickly and tightly bound to DdlB. Steady-state kinetic studies revealed that these compounds are ATP-competitive inhibitors of DdlB. The evaluation of the antibacterial activity showed that compounds **8** and **9** (Fig. 7), and not **7**, had inhibitory effects against *E. coli* 1411 and *E. coli* AB734, with MIC values of 64 and 32 μg/mL, respectively.

Two biologically active pharmacophores were connected by Chate et al. in 2019 to provide a series of coumarin-pyrazolines acting as Ddl inhibitors (Fig. 7) [34]. Compound **10** exhibited the greater antibacterial effect, with MIC values of 14, 14, and 32 μg/mL against *E. coli* 1411, and SM1411 strains, and *S. aureus* NCIM-2901, respectively. Enzyme activity assay results demonstrated that compounds **10** and **11** had IC_{50} values of 106 and 111 μM, which are lower, as *E. coli* DdlB inhibitors, than DCS (IC_{50} of 276 μM).

A structure-based virtual screening set up by Kovac et al. in 2008 allowed the identification of pyridocarbazole *E. coli* DdlB inhibitors such as **12** and **13** (Fig. 8) [35]. The latter exhibited a IC_{50} value of 70 μM and effectively inhibited *E. coli* 1411, *E. coli* SM1411, and *S. aureus* SH1000, with MIC values of 32, 8, and 32 μg/mL, respectively. The indole moiety of the pyridocarbazole scaffold was predicted to locate in the D-alanine binding hydrophobic pocket. **12** showed instead a lower IC_{50}, that is, 43 μM, and greater activity against *E. coli* SM1411. In 2011, the same authors optimized such a series of compounds, with the *N*-methylation of pyridine and bromine substitution of carbazole contributing to achieve lower IC_{50} values, such as with **14** (Fig. 8) with a value of 23 μM, that reported also twofold increased action than **12** against *E. coli* 1411, *E. coli* SM1411, and *S. aureus* SH1000.

Compound **15** (Fig. 8) is an acridinylamine derivative acting as an ATP-noncompetitive inhibitor of *E. coli* DdlB, with an IC_{50} value of 162 μM [36]. It shows antibacterial activity against *E. coli* SM1411 and *S. aureus* 8325-4, with MIC values of 64 and 32 μg/mL, respectively. Modification of the lead compound by the authors was beneficial for increasing the inhibitory effect on the target enzyme, but most congeners did not show any antibacterial activity.

Compound **16** (Fig. 8) is instead an anthraquinone *E. coli* DdlB inhibitor, identified via virtual screening by Kovac et al. in 2008 [35]. **16** is a nogalamycin-derivative and a competitive inhibitor of ATP with a K_i value of 42 μM. It showed, however, no antibacterial activity against *E. coli* strains and poor action against *S. aureus* 8325-4, because of a low membrane permeability.

Phosphonic acid derivatives are among the first identified Ddl inhibitors. For instance, compound **17** (Fig. 9) was reported by Parsons et al. in 1988 to be a type of irreversible inhibitor binding, in the presence of ATP, to *Streptococcus faecalis* Ddl, with an IC_{50} value of 4 μM [37]. Therefore, Chakravarty et al. described **18** and **19** (Fig. 9) as effective inhibitors for *E. faecalis* Ddl with K_i values

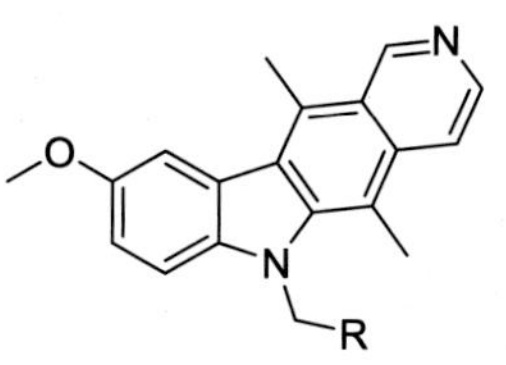

R = H; **12**
R = $(CH_2)_3NH_2$; **13**

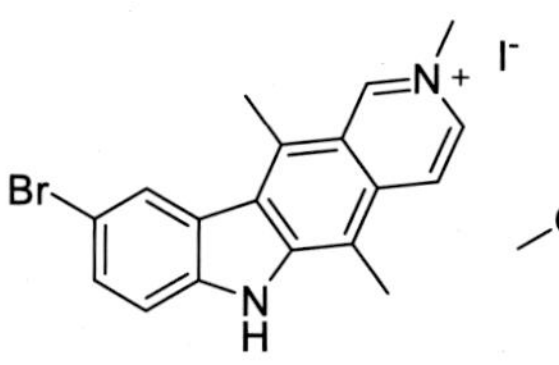

14

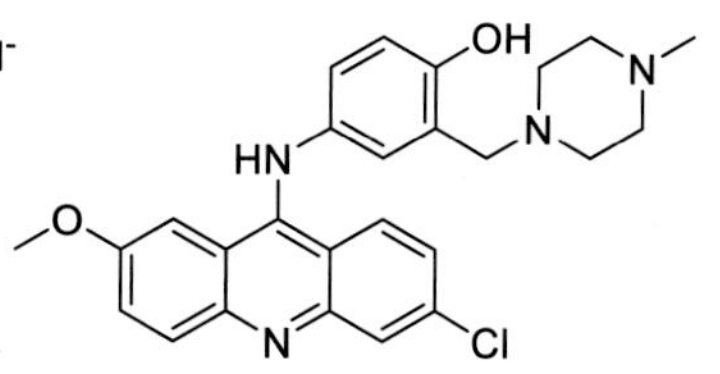

15

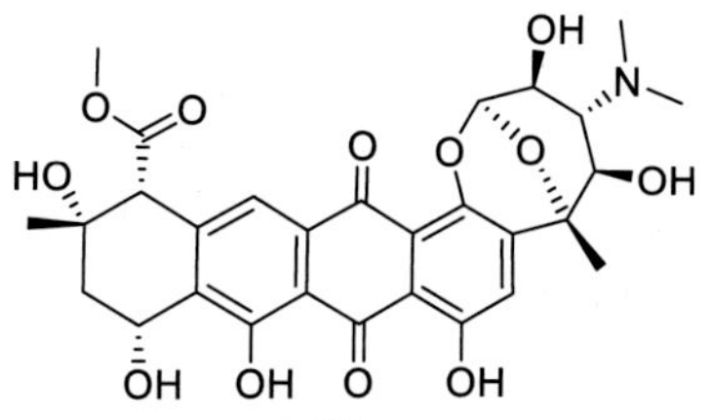

16

FIG. 8 Chemical structures of Ddl inhibitors **12–16**.

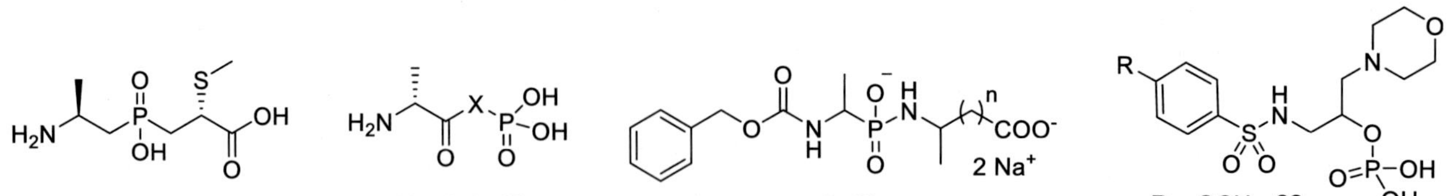

17; X = CH_2; 18; X = NH; 19; n = 0; 20; n = 1; 21; R = OCH_3; 22; R = F; 23

FIG. 9 Chemical structures of Ddl inhibitors **17–23**.

of 510 and 50 μM, respectively [38]. In 1991, Lacoste et al. developed time-dependent irreversible Ddl inhibitors, which are **20** and **21** (Fig. 9), with K_i values of 4 μM, respectively, against *E. faecalis* Ddl [39]. This class of compound was implemented in 2009 by Sova et al. reporting dual-target inhibitors of bacteria DdlB and D-alanyl-D-lactate ligase, such as **22** and **23** (Fig. 9), able to reverse bacterial resistance to vancomycin [40].

The potent thiosemicarbazide inhibitor **24** (Fig. 10) of *E. faecalis* Ddl was identified by a wide screening by Ameryckx et al. in 2018, with an IC_{50} value of 0.80 μM [41]. It exhibited antibacterial activity against *S. aureus* ATCC 25923 and *E. faecalis* ATCC 29212, with MIC values of 128 and 512 μg/mL, respectively. To implement the activity *in cellulo*, several benzoyl thiosemicarbazide Ddl inhibitors were developed, such as **25–29** (Fig. 10). Compound **25** was the most active among them with an IC_{50} value of 0.1 μM against *E. coli* Ddl, while compound **29** was inactive. Replacing the benzene ring with hydrophobic pentyl or naphthyl groups led to compounds **28** and **27**, effective as Ddl inhibitors with IC_{50} values of 0.93 and 0.52 μM, respectively. Instead, the replacement of the benzene ring with morpholinopropyl or pyridine groups markedly reduced the Ddl inhibitory action of similar such compounds. The switching of the thiourea linker between the two benzene rings with linkers produced the vanishing of the inhibitory effect. This class of thiourea Ddl inhibitors has no inhibitory effect on Gram-negative bacteria, such as *Klebsiella pneumoniae* ATCC 700603, *E. coli* ATCC 25922, and *Pseudomonas aeruginosa* PAO1. In contrast, analog such compounds showed effective action against a chorus of Gram-positive bacteria with MIC values ranging between 25 and 50 μg/mL. Derivative **26** was shown to act as a reversible and selective inhibitor of Ddl, with a mixed noncompetitive inhibition of D-Ala and ATP. In 2020, the same authors changed the benzoylthiosemicarbazide linker with other ones containing a thiourea moiety [42]. As a result, compounds such as **30** and **31** (Fig. 10) were attained, with IC_{50} values of 1.48 and 0.81 μM, respectively, against the *E. coli* Ddl. Compound **30** exhibits a relevant antibacterial activity, with MIC values of 1.06–2.10 μg/mL against vancomycin-resistant *Enterococcus* and 4.25–8.50 μg/mL against methicillin- and vancomycin-resistant *S. aureus*.

The propionamide derivative **32** (Fig. 11) was reported in 2006 as an inhibitor of *S. aureus* Ddl, with a K_i value of 4 μM [15]. It was demonstrated to produce the inhibition of the enzyme activity of *S. aureus* Ddl by forming an unproductive enzyme-substrate-inhibitor complex with no interference with the binding of D-alanine and ATP (Fig. 6B). The detailed binding of the inhibitor to the target was elucidated by X-ray crystallography (PDB 2I80).

In 2007, Kovac et al. described a class of diazenedicarboxamide as *E. coli* Ddl inhibitors (Fig. 11) [43]. The best derivatives, that is **33**, showed an IC_{50} value of 15 μM. However, compound **34** with an IC_{50} value of 106 μM

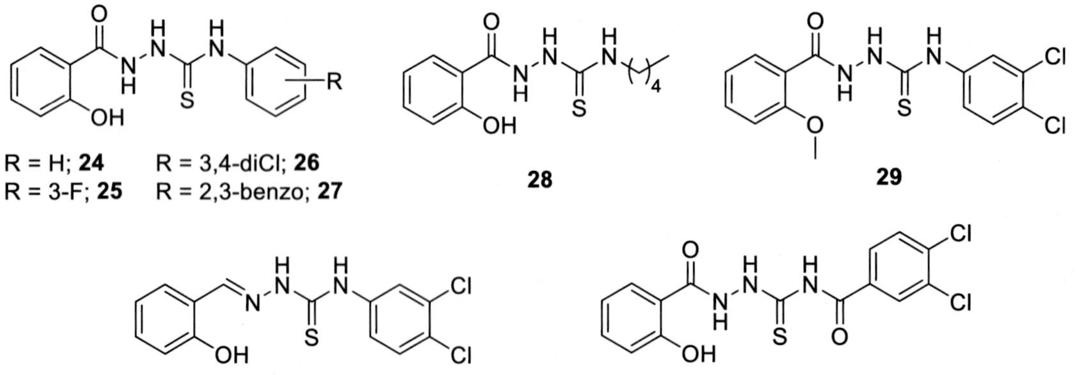

R = H; 24; R = 3-F; 25; R = 3,4-diCl; 26; R = 2,3-benzo; 27; 28; 29; 30; 31

FIG. 10 Chemical structures of Ddl inhibitors **24–31**.

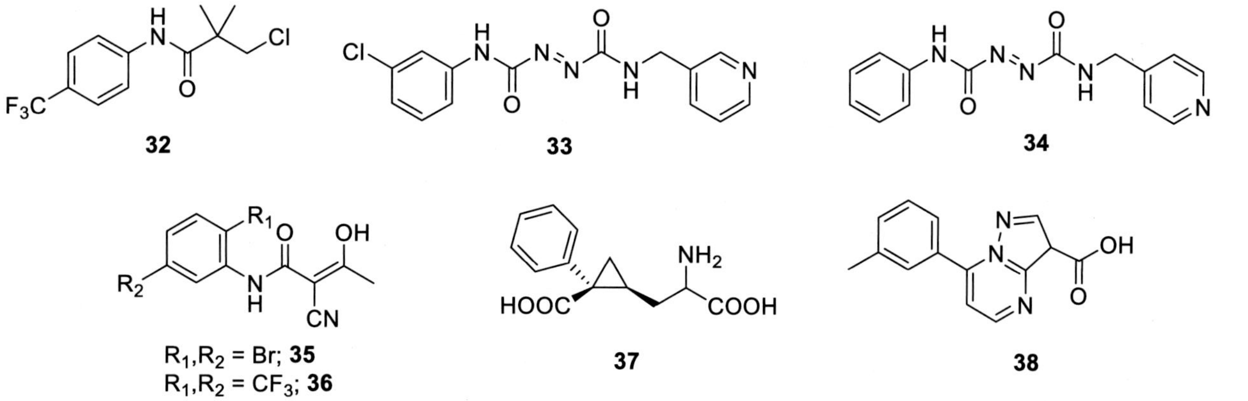

FIG. 11 Chemical structures of Ddl inhibitors **32–38**.

was the most active compound against *E. coli* 1411, *E. coli* 1411 AcrAB$^-$, and *S. aureus* 8325-4, with MIC values of 64, 32, and 128 µg/mL, respectively.

In 2009, Triola et al. pointed out that the tyrosine kinase inhibitor, α-cyano-β-hydroxy-β-methyl-*N*-(2,5-dibromophenyl)propenamide **35** (Fig. 11) was also an ATP-competitive *E. coli* DdlB inhibitor of bacterial with a K_i value of 185 µM [44]. Thus, the authors developed other butenamide Ddl inhibitors, among which **36** (Fig. 11) showed a K_i value of 60 µM.

Besong et al. reported a de novo design method based on the X-ray crystal structure of DdlB from *E. coli*, which identified the cyclopropyl-based amino acid **37** (Fig. 11) as a selective inhibitor of the enzyme with a K_i value of 12.5 µM [45]. This value was however measured for diastereomeric mixture of compound **37**, making the identified enantiomer a likely more effective DdlB inhibitor.

In 2023, Proj et al. used a fragment-based drug discovery approach, which recognized the fragment hit **38** (Fig. 11) showing an IC_{50} value of 168 µM against *E. coli* DdlB [46]. The binding of **38** to the target was confirmed with surface plasmon resonance analysis and steady-state kinetics showed **38** to be a competitive inhibitor with respect to ATP with a K_i value of 20.7 µM.

4 Conclusions

D-Alanine-D-alanine ligase is an essential enzyme in bacterial cell wall biosynthesis and an important target for developing new antibiotics. This chapter reviewed structure and function of Ddl and reported the recent advances over Ddl inhibitors. DCS remains the most recommended drug for the treatment of MDR-TB, ever since its approval by the FDA, and is the unique Ddl inhibitor used in clinics. Beside DCS, virtual screening and de novo design were the most adopted methods for discovering new Ddl inhibitors, which belong to a *plethora* of chemotypes. Among these, thiourea derivatives demonstrate the most relevant in cell antibacterial activity being able to inhibit even the growth of erythromycin, linezolid, methicillin, ciprofloxacin, and moxifloxacin-resistant bacteria. However, some issues pose limitations to widespread use of Ddl inhibitors, such as a limited correlation between enzyme inhibitory activity and antibacterial activity, mostly resulting from the inhibitor low bacterial membrane permeability. Therefore, optimization of lead compounds is required for implementing the in-cell efficacy of Ddl inhibitors as well as investigations in vivo will be needed to pose more solid bases for the safe clinical use of this class of antibiotic drugs.

References

[1] Green DW. The bacterial cell wall as a source of antibacterial targets. Expert Opin Ther Targets 2002;6:1–19.

[2] Bugg TD, Braddick D, Dowson CG, Roper DI. Bacterial cell wall assembly: still an attractive antibacterial target. Trends Biotechnol 2011;29:167–73.

[3] Schneider T, Sahl HG. An oldie but a goodie—cell wall biosynthesis as antibiotic target pathway. Int J Med Microbiol 2010;300(2–3):161–9.

[4] Kohanski MA, Dwyer DJ, Collins JJ. How antibiotics kill bacteria: from targets to networks. Nat Rev Microbiol 2010;8:423–35.

[5] Schumann P, Rainey F, Oren A. Peptidoglycan structure. In: Methods in microbiology. Taxonomy of prokaryotes, vol. 38. Elsevier Academic Press Inc; 2011. p. 101–29.

[6] Egan AJF, Errington J, Vollmer W. Regulation of peptidoglycan synthesis and remodelling. Nat Rev Microbiol 2020;18:446–60.

[7] Maitra A, Munshi T, Healy J, Martin LT, Vollmer W, Keep NH, Bhakta S. Cell wall peptidoglycan in *Mycobacterium tuberculosis*: an Achilles' heel for the TB-causing pathogen. FEMS Microbiol Rev 2019;43:548–75.

[8] Gautam A, Vyas R, Tewari R. Peptidoglycan biosynthesis machinery: a rich source of drug targets. Crit Rev Biotechnol 2011;31:295–336.

[9] Barreteau H, Kovac A, Boniface A, Sova M, Gobec S, Blanot D. Cytoplasmic steps of peptidoglycan biosynthesis. FEMS Microbiol Rev 2008;32:168–207.

[10] Kuzin AP, Sun T, Jorczak-Baillass J, Healy VL, Walsh CT, Knox JR. Enzymes of vancomycin resistance: the structure of D-alanine-D-lactate ligase of naturally resistant *Leuconostoc mesenteroides*. Structure 2000;8:463–70.

[11] Chen Y, Xu Y, Yang S, Li S, Ding W, Zhang W. Deficiency of D-alanyl-D-alanine ligase A attenuated cell division and greatly altered the proteome of *Mycobacterium smegmatis*. MicrobiologyOpen 2019;8:e00819.

[12] Neuhaus FC, Hammes WP. Inhibition of cell wall biosynthesis by analogues and alanine. Pharmacol Ther 1981;14:265–319.

[13] Qin Y, Xu L, Teng Y, Wang Y, Ma P. Discovery of novel antibacterial agents: recent developments in D-alanyl-D-alanine ligase inhibitors. Chem Biol Drug Des 2021;98:305–22.

[14] Albar OA, O'Connor CD, Giles IG, Akhtar M. D-alanine: D-alanine ligase of *Escherichia coli*. Expression, purification and inhibitory studies on the cloned enzyme. Biochem J 1992;282:747–52.

[15] Liu S, Chang JS, Herberg JT, Horng MM, Tomich PK, Lin AH, Marotti KR. Allosteric inhibition of *Staphylococcus aureus* D-alanine:D-alanine ligase revealed by crystallographic studies. Proc Natl Acad Sci U S A 2006;103:15178–83.

[16] Batson S, Rea D, Fülöp V, Roper DI. Crystallization and preliminary X-ray analysis of a D-alanyl-D-alanine ligase (EcDdlB) from *Escherichia coli*. Acta Crystallogr Sect F Struct Biol Cryst Commun 2010;66:405–8.

[17] Zawadzke LE, Bugg TD, Walsh CT. Existence of two D-alanine:D-alanine ligases in *Escherichia coli*: cloning and sequencing of the ddlA gene and purification and characterization of the DdlA and DdlB enzymes. Biochemistry 1991;30:1673–82.

[18] Pederick JL, Thompson AP, Bell SG, Bruning JB. d-Alanine-d-alanine ligase as a model for the activation of ATP-grasp enzymes by monovalent cations. J Biol Chem 2020;295:7894–904.

[19] Bruning JB, Murillo AC, Chacon O, Barletta RG, Sacchettini JC. Structure of the *Mycobacterium tuberculosis* D-alanine:D-alanine ligase, a target of the antituberculosis drug D-cycloserine. Antimicrob Agents Chemother 2011;55:291–301.

[20] Fan C, Park IS, Walsh CT, Knox JR. D-alanine:D-alanine ligase: phosphonate and phosphinate intermediates with wild type and the Y216F mutant. Biochemistry 1997;36:2531–8.

[21] Galperin MY, Koonin EV. A diverse superfamily of enzymes with ATP-dependent carboxylate-amine/thiol ligase activity. Protein Sci 1997;6:2639–43.

[22] Shi Y, Walsh CT. Active site mapping of *Escherichia coli* D-Ala-D-Ala ligase by structure-based mutagenesis. Biochemistry 1995;34:2768–76.

[23] Healy VL, Mullins LS, Li X, Hall SE, Raushel FM, Walsh CT. D-Ala-D-X ligases: evaluation of D-alanyl phosphate intermediate by MIX, PIX and rapid quench studies. Chem Biol 2000;7:505–14.

[24] Kumari S, Ram VJ. Advances in molecular targets and chemotherapy of tuberculosis. Drugs Today 2004;40:487–500.

[25] Lee SH, Seo KA, Lee YM, Lee HK, Kim JH, Shin C, Ghim JR, Shin JG, Kim DH. Low serum concentrations of moxifloxacin, prothionamide, and cycloserine on sputum conversion in multi-drug resistant TB. Yonsei Med J 2015;56:961–7.

[26] Batson S, de Chiara C, Majce V, Lloyd AJ, Gobec S, Rea D, Fülöp V, Thoroughgood CW, Simmons KJ, Dowson CG, Fishwick CWG, de Carvalho LPS, Roper DI. Inhibition of D-Ala:D-Ala ligase through a phosphorylated form of the antibiotic D-cycloserine. Nat Commun 2017;8:1939.

[27] Nakatani Y, Opel-Reading HK, Merker M, Machado D, Andres S, Kumar SS, Moradigaravand D, Coll F, Perdigão J, Portugal I, Schön T, Nair D, Devi KRU, Kohl TA, Beckert P, Clark TG, Maphalala G, Khumalo D, Diel R, Klaos K, Aung HL, Cook GM, Parkhill J, Peacock SJ, Swaminathan S, Viveiros M, Niemann S, Krause KL, Köser CU. Role of alanine racemase mutations in *Mycobacterium tuberculosis* d-cycloserine resistance. Antimicrob Agents Chemother 2017;61:e01575-17.

[28] Khosravi AD, Mirsaeidi M, Farahani A, Tabandeh MR, Mohajeri P, Shoja S, Hoseini Lar KhosroShahi SR. Prevalence of nontuberculous mycobacteria and high efficacy of d-cycloserine and its synergistic effect with clarithromycin against *Mycobacterium fortuitum* and *Mycobacterium abscessus*. Infect Drug Resist 2018;11:2521–32.

[29] Hrast M, Vehar B, Turk S, Konc J, Gobec S, Janežič D. Function of the D-alanine:D-alanine ligase lid loop: a molecular modeling and bioactivity study. J Med Chem 2012;55:6849–56.

[30] Murakami R, Muramatsu Y, Minami E, Masuda K, Sakaida Y, Endo S, Suzuki T, Ishida O, Takatsu T, Miyakoshi S, Inukai M, Isono F. A novel assay of bacterial peptidoglycan synthesis for natural product screening. J Antibiot (Tokyo) 2009;62:153–8.

[31] Wu D, Kong Y, Han C, Chen J, Hu L, Jiang H, Shen X. D-Alanine:D-alanine ligase as a new target for the flavonoids quercetin and apigenin. Int J Antimicrob Agents 2008;32:421–6.

[32] Tytgat I, Vandevuer S, Ortmans I, Sirockin F, Colacino E, Van Bambeke F, Duez C, Poupaert JH, Tulkens PM, Dejaegere A, Prévost M. Structure-based design of benzoxazoles as new inhibitors for D-alanyl-D-alanine ligase. QSAR Comb Sci 2009;28:1394–404.

[33] Škedelj V, Arsovska E, Tomašić T, Kroflič A, Hodnik V, Hrast M, Bešter-Rogač M, Anderluh G, Gobec S, Bostock J, Chopra I, O'Neill AJ, Randall C, Zega A. 6-Arylpyrido[2,3-d]pyrimidines as novel ATP-competitive inhibitors of bacterial D-alanine:D-alanine ligase. PLoS One 2012;7:e39922.

[34] Chate AV, Redlawar AA, Bondle GM, Sarkate AP, Tiwari SV, Lokwani DK. A new efficient domino approach for the synthesis of coumarin-pyrazolines as antimicrobial agents targeting bacterial D-alanine-D-alanine ligase. New J Chem 2009;43:9002–11.

[35] Kovac A, Konc J, Vehar B, Bostock JM, Chopra I, Janezic D, Gobec S. Discovery of new inhibitors of D-alanine:D-alanine ligase by structure-based virtual screening. J Med Chem 2008;51:7442–8.

[36] Vehar B, Hrast M, Kovač A, Konc J, Mariner K, Chopra I, O'Neill A, Janežič D, Gobec S. Ellipticines and 9-acridinylamines as inhibitors of D-alanine:D-alanine ligase. Bioorg Med Chem 2011;19:5137–46.

[37] Parsons WH, Patchett AA, Bull HG, Schoen WR, Taub D, Davidson J, Combs PL, Springer JP, Gadebusch H, Weissberger B, et al. Phosphinic acid inhibitors of D-alanyl-D-alanine ligase. J Med Chem 1988;31:1772–8.

[38] Chakravarty PK, Greenlee WJ, Parsons WH, Patchett AA, Combs P, Roth A, Busch RD, Mellin TN. (3-Amino-2-oxoalkyl)phosphonic acids and their analogues as novel inhibitors of D-alanine:D-alanine ligase. J Med Chem 1989;32:1886–90.

[39] Lacoste AM, Cholletgravey AM, Quang LV, Quang YV, Legoffic F. -Time-dependent inhibition of *Streptococcus faecalis* d-alanine: D-alanine ligase by α-aminophosphonamidic acids. Eur J Med Chem 1991;26:255–60.

[40] Sova M, Cadez G, Turk S, Majce V, Polanc S, Batson S, Lloyd AJ, Roper DI, Fishwick CW, Gobec S. Design and synthesis of new hydroxyethylamines as inhibitors of D-alanyl-D-lactate ligase (VanA) and D-alanyl-D-alanine ligase (DdlB). Bioorg Med Chem Lett 2009;19:1376–9.

[41] Ameryckx A, Thabault L, Pochet L, Leimanis S, Poupaert JH, Wouters J, Joris B, Van Bambeke F, Frédérick R. 1-(2-Hydroxybenzoyl)-thiosemicarbazides are promising antimicrobial agents

targeting d-alanine-d-alanine ligase in bacterio. Eur J Med Chem 2018;159:324–38.

[42] Ameryckx A, Pochet L, Wang G, Yildiz E, Saadi BE, Wouters J, Van Bambeke F, Frédérick R. Pharmacomodulations of the benzoyl-thiosemicarbazide scaffold reveal antimicrobial agents targeting d-alanyl-d-alanine ligase in bacterio. Eur J Med Chem 2020;200:112444.

[43] Kovac A, Majce V, Lenarsic R, Bombek S, Bostock JM, Chopra I, Polanc S, Gobec S. Diazenedicarboxamides as inhibitors of D-alanine-D-alanine ligase (Ddl). Bioorg Med Chem Lett 2007;17:2047–54.

[44] Triola G, Wetzel S, Ellinger B, Koch MA, Hübel K, Rauh D, Waldmann H. ATP competitive inhibitors of D-alanine-D-alanine ligase based on protein kinase inhibitor scaffolds. Bioorg Med Chem 2009;17:1079–87.

[45] Besong GE, Bostock JM, Stubbings W, Chopra I, Roper DI, Lloyd AJ, Fishwick CW, Johnson AP. A de novo designed inhibitor of D-Ala-D-Ala ligase from *E. coli*. Angew Chem Int Ed Engl 2005;44:6403–6.

[46] Proj M, Hrast M, Bajc G, Frlan R, Meden A, Butala M, Gobec S. Discovery of a fragment hit compound targeting D-Ala:D-Ala ligase of bacterial peptidoglycan biosynthesis. J Enzyme Inhib Med Chem 2023;38:387–97.

Chapter 2.7

Paraoxonases

Claudiu T. Supuran
NEUROFARBA Department, Section of Pharmaceutical and Nutraceutical Sciences, University of Florence, Florence, Italy

1 Introduction

Paraoxonase (PON, EC 3.1.8.1) activity has already been reported in the 1940s [1] as an enzyme being able to hydrolyze organophosphorus pesticides, which in that period started to be massively used in agriculture [2]. Most such compounds are phosphotriesters, thiophosphates, phosphorothiolates, or fluorophosphates. One of the most used (in the past) is parathion **1**, which is hydrolyzed to the active metabolite paraoxon **2** (Fig. 1), both compounds, and the entire class, which includes hundreds of representatives, acting as acetylcholinesterase irreversible inhibitors, and being extremely toxic to most organisms including vertebrates [2,3]. An enzyme hydrolyzing **2**, isolated by Mazur [1], was thus originally considered as a detoxifying mechanism against organophosphates, which explains the name PON.

Later, Mackness et al. [4] observed a distinction between the "A-esterase" (substrates paraoxon and other organophosphates) and arylesterase (substrate phenyl acetate) activity present in the plasma of birds and mammals (including humans), proposing an arylesterase activity for PON, as that enzyme (presumably PON1, see later in the text) could hydrolyze both organophosphates and phenyl acetate. As it will be shown later, PONs, of which three different isoforms are known today (PON1–PON3) [5–12], in fact possess a very large number of substrates and thus a promiscuous enzymatic activity [2,5]. Indeed, among the many other substrates of PON are also lactones such as dihydrocoumarin **3** and *N*-acyl-homoserine lactones **4** (R = long aliphatic chains), AHLs, which in fact seem to be the best substrates of the enzyme, which has been classified as a lactonases/lactonizing enzyme [5,10–12].

Human PON1, the best studied member of the family, is a circulating Ca^{2+}-dependent enzyme with a molecular mass of 43 kDa and incorporates a 354 amino acids long polypeptide chain [2,5–7]. It is synthesized in the liver, and in lower amounts in the kidneys and colon, and thereafter secreted into the bloodstream, where it is tightly bound to high-density lipoprotein (HDL) particles, one of the five major groups of lipoproteins [7,10]. PON2, a 354 amino acid residues protein with the molecular weight of 40–43 kDa, seems to be the most ancestral member of the family from which PON1 and PON3 evolved, is ubiquitously expressed in human tissues, being a membrane-bound enzyme [9,13]. PON3 was isolated in 1999 from rabbit liver microsomes, and it has around 60% homology to PON1. It is a glycoprotein of molecular weight of 40–45 kDa and incorporates 354 amino acid residues, similar to the other two isoforms mentioned above [8,14]. As it will be shown shortly, the catalytic activity of the three enzymes is quite diverse.

2 Structure and functions of CAs

The X-ray crystal structure of a PON1 variant as well as of some of its active-site mutants (bearing mutations mainly at amino acid 115) has been reported by Tawfik's group [6,11,15]. The overall fold of the enzyme is that of a six-bladed β-propeller (Fig. 2A) with each propeller subunit being made of four β-pleated sheets, and the two calcium ions, essential for the catalytic activity, being located in a central water filled core of the protein (Fig. 2B). One of the Ca(II) ions (in green), which is located deeper within the active site, is the so-called structural calcium and is not involved in catalysis, but stabilizes the protein fold. It is coordinated by a backbone oxygen from a carbonyl moiety (of residue I117) and by the carboxylates from residues Asp169 and Asp54 from PON1 as well as three water molecules, in an octahedral geometry of the metal ion (Fig. 2B).

The second calcium ion (in pink, Fig. 2), the catalytic one, which is more solvent exposed than the structural calcium, is coordinated, again in octahedral geometry by five amino acid residues (Asp269, Glu53, Asn168, Asn224, and Asn270) and a water molecule (Fig. 2B). Actually, the structure of the enzyme was solved with a phosphate bound to the second calcium ion (not shown in Fig. 2) [6], which substituted the water molecule involved in the catalytic cycle (see later in the text). The substrate-binding pocket of the enzyme is also shown in Fig. 2C (considering the organophosphatase activity of PON [2]), with the small group pocket shown in cyan, the leaving group

Metalloenzymes. https://doi.org/10.1016/B978-0-12-823974-2.00027-9

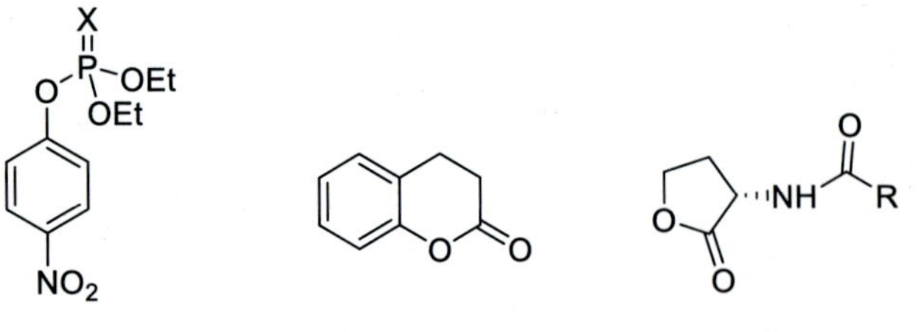

FIG. 1 Structure of organophosphorus pesticides parathion **1**, paraoxon **2**, dihydrocoumarin **3**, and *N*-acyl-homoserine lactones **4**, AHLs (R = long aliphatic chains), all acting as PON substrates.

pocket in green, and large group pocket residues colored in orange [2,6]. The two α-helixes shown in the upper part of Fig. 2A are thought to be involved in vivo in the anchoring of the enzyme to HDL [15–18]. Two histidine residues (His115 and His134, Fig. 2C) from the active site constitute the catalytic histidine dyad, which is involved in deprotonation of the calcium-bound water molecule, initially mediated by His115, thereafter the proton being transferred to His134 [19]. This is similar with the situation seen in another metalloenzyme, the carbonic anhydrase, CA (which contains Zn(II) at its active site [20]), case in which the proton transfer from a zinc-coordinated water molecule to the environment (rate-determining step in the catalytic cycle [20]) is achieved by a His residue positioned in the middle of the active site (His64 in human isoform CA II) and thereafter by a His cluster situated on the rim/surface of the active site cavity [21]. In both cases, for the calcium- and zinc-enzymes mentioned here, the nucleophilicity of the water molecule coordinated to the metal ion is thus highly enhanced for an efficient nucleophilic attack on the substrate(s) properly orientated in the active site [22]. During the catalysis (see next section), the PON catalytic calcium cation aligns/coordinates the substrate molecule via its carbonyl oxygen (C=O, in the case of lactones such as **3** or **4**) or phosphorus oxygen (P=O, in the case of organophosphates such as **2**, Fig. 1) and stabilizes the oxyanion tetrahedral/trigonal bipyramidal intermediate that is formed upon the nucleophilic attack.

3 Catalyzed reactions

A detailed study on the reactions catalyzed by PONs has been performed by Draganov et al. [5], the group that in fact also discovered the lactonase activity of these enzymes [12]. There are three main substrate types for these enzymes: (i) organophosphates (such as the pesticides mentioned above, of which paraoxon **2** is the most relevant one, but also other compounds, such as chlorpyrifos, oxon, diazoxon, and even sarin or other nerve gas acetylcholinesterase inhibitors [2,5]); (ii) aryl esters of C_1-C_4 aliphatic acids with phenol, 4-nitrophenol and related derivatives [5]; and (iii) lactones with a variety of structural variations, but mainly γ- and δ-lactones [5,12,23].

Some of the lactones that act as good substrates for various PON isoforms are shown in Fig. 3 and include simple compounds such as 2-coumarone **5**, homogentisic acid lactone **6**, lactones incorporating long aliphatic chains such as δ-undecanoic acid lactone **7**, angelica lactone **8**, and also

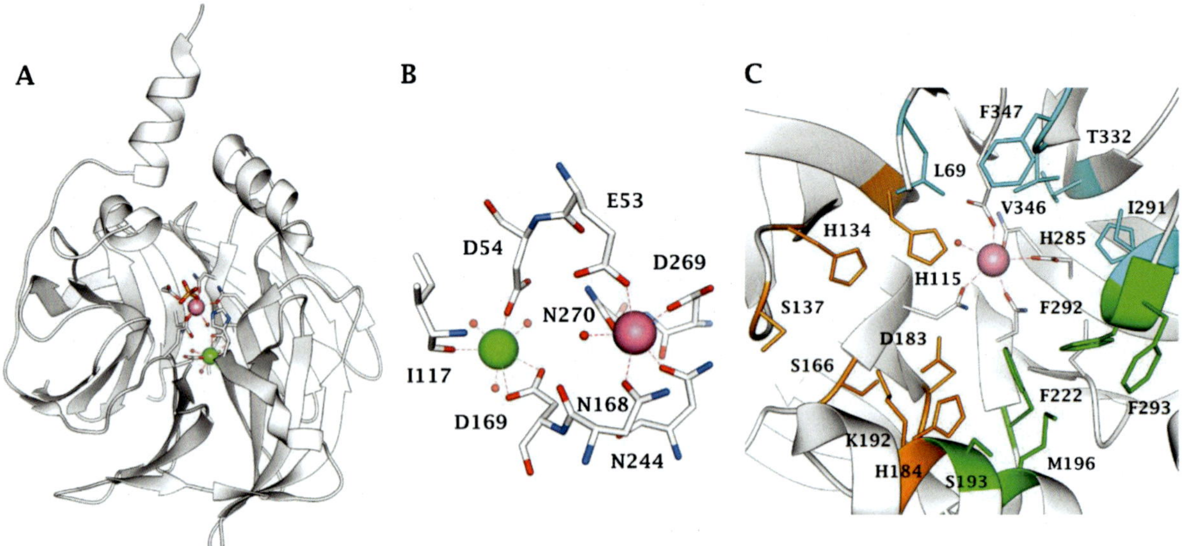

FIG. 2 X-ray solved crystal structure of PON1 (PDB 1V04). (A) Side view of β-propeller fold with HDL anchoring helices extending above. Catalytic and structural calcium ions are represented as a *pink and green sphere*, respectively. Water molecules are shown as *red spheres*. (B) Metal centers and coordinating residues of PON1. (C) Substrate-binding pockets of PON1. Large and small pocket residues are colored *orange* and *cyan*, respectively. The leaving group pocket is colored *green*. Residues coordinating the catalytic calcium ion are shown as *white sticks*.

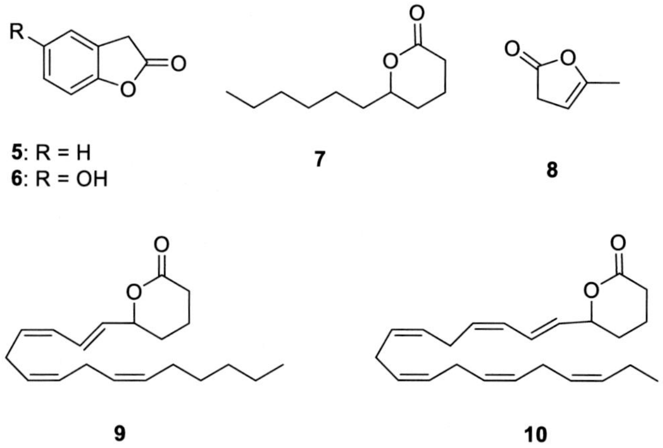

FIG. 3 Lactone substrates of PON of types **5**–**10**.

more complex, physiologically relevant compounds such as 5-hydroxy-6*E*,8*Z*,11*Z*,14*Z*-eicosatetraenoic acid1,5-lactone, 5-HETEL (**9**), or the corresponding lactone derived from 4-hydroxy-5*E*,7*Z*,10*Z*,13*Z*,16*Z*,19*Z*-docosahexaenoic acid.

The enzymatic activities of the three PON isoforms are very diverse for the three types of substrates mentioned above [5]. Thus, PON1 has all three activities, with organophosphates, aryl esters, and lactones as substrates, but the measured specific activities are very diverse. As shown by Draganov et al. [5], the best substrate is phenyl acetate (specific activity of 1120 U/mg) [5], followed by some lactones of the types mentioned here (130 U/mg for dihydrocoumarin **3**, 136 U/mg for **5**, 329.5 U/mg for **6**, 183 U/mg for **8**, and 75 U/mg for **9**). PON1 is also able to catalyze the reversible reaction, i.e., lactonization of 4-hydroxy-5*E*,7*Z*,10*Z*,13*Z*,16*Z*,19*Z*-docosahexaenoic acid to the lactone **10** (Fig. 3), with a low catalytic efficacy (1.5 U/mg) [5]. It should be mentioned that the efficacy of organophosphates hydrolysis catalyzed by PON1 is quite low, with a specific activity of only 1.9 U/mg for paraoxon **2**, but other phosphates such as chlopyrifos or diazoxozon are better substrates (specific activities of 41–113 U/mg) [5]. PON2 has a much lower enzyme activity compared to PON1 and also a more limited range of substrates [5]. Thus, PON2 does not show organophosphatase activity with any phosphate pesticide investigated so far [5], it has a very low arylesterase activity (specific activity of 0.086–1.4 U/mg with phenyl or 4-nitrophenyl acetates, propyonates, or butyrates) and also poor or no activity with most γ- or δ-lactones except for **3** and **5** (specific activity of 3.1–10.9 U/mg, with the best substrate being 2-coumaronone **5**) [5]. However, PON2 has lactonase activity with most investigated AHLs **4**, among which the hexanoyl, heptanoiyl, dodecanoyl, and tetradecanoyl derivatives **4**, for which specific activities of 0.26–0.51 U/mg were reported [5]. PON3 has a poor organophosphatase activity (specific activity of 0.205 U/mg with paraoxon **2** as substrate), acts as a rather effective arylesterase (with the substrates mentioned above), with specific activity of 4–39 U/mg [5], and also shows a broad lactonase activity with most simple and complex lactones mentioned here. Among the best PON3 substrates are γ-heptalactone, δ-undecanoic acid lactone **7**, and δ-decanolactone (specific activity of 27.2–84.4 U/mg [5]). 5-HETEL is also efficiently hydrolyzed (specific activity of 27.5 U/mg), whereas AHLs are much poorer substrates, with specific activities <0.09 EU/mg [5]. It is interesting to note that clinically used drugs incorporating five- or six-membered lactone moieties, such as the anticholesterol agent Lovastatin **11**, the diuretics Spironolactone **12** and Canrenone **13** (Fig. 4) are also substrates of PON3 but not of PON1 and PON2, with specific activities of 0.011–0.266 U/mg [5].

The proposed catalytic mechanism of paraoxonase, for the two main reactions that they catalyze, lactonase and organophosphatase, is shown in Fig. 5. The substrate (a chiral AHL, Fig. 5A) or paraoxon 2 (Fig. 5B) is aligned/coordinated to the catalytic calcium ion, which also interacts with the water molecule, which will nucleophilically attack the carbonyl or phosphorus atom from the substrate (first step). In this way, the tetrahedral/trigonal bipyramidal intermediates are formed, with the transfer of a proton (from the water molecule) to His115, which as imidazolium ion stabilizes the transition state (Step 2). In the last step, the tetrahedral/trigonal bipyramidal intermediates collapse, with generation of the γ-hydroxy acid for the lactone and diethylphosphate plus 4-nitrophenol, for paraoxon (Fig. 5). The arylesterase activity of the enzyme obviously works according to a similar mechanism.

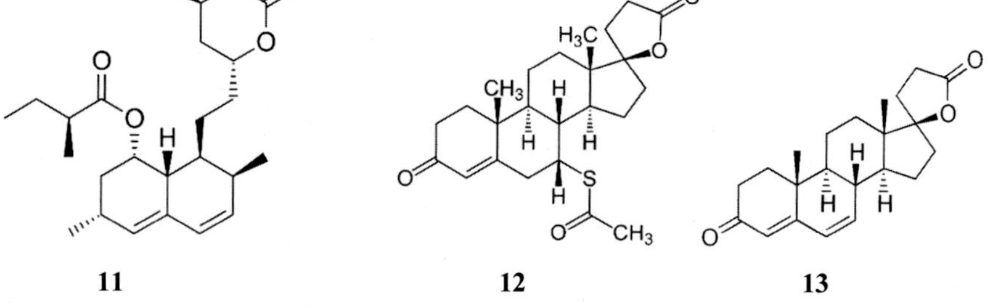

FIG. 4 Clinically used drugs incorporating lactone moieties: Lovastatin **11**, Spironolactone **12**, and Canrenone **13**.

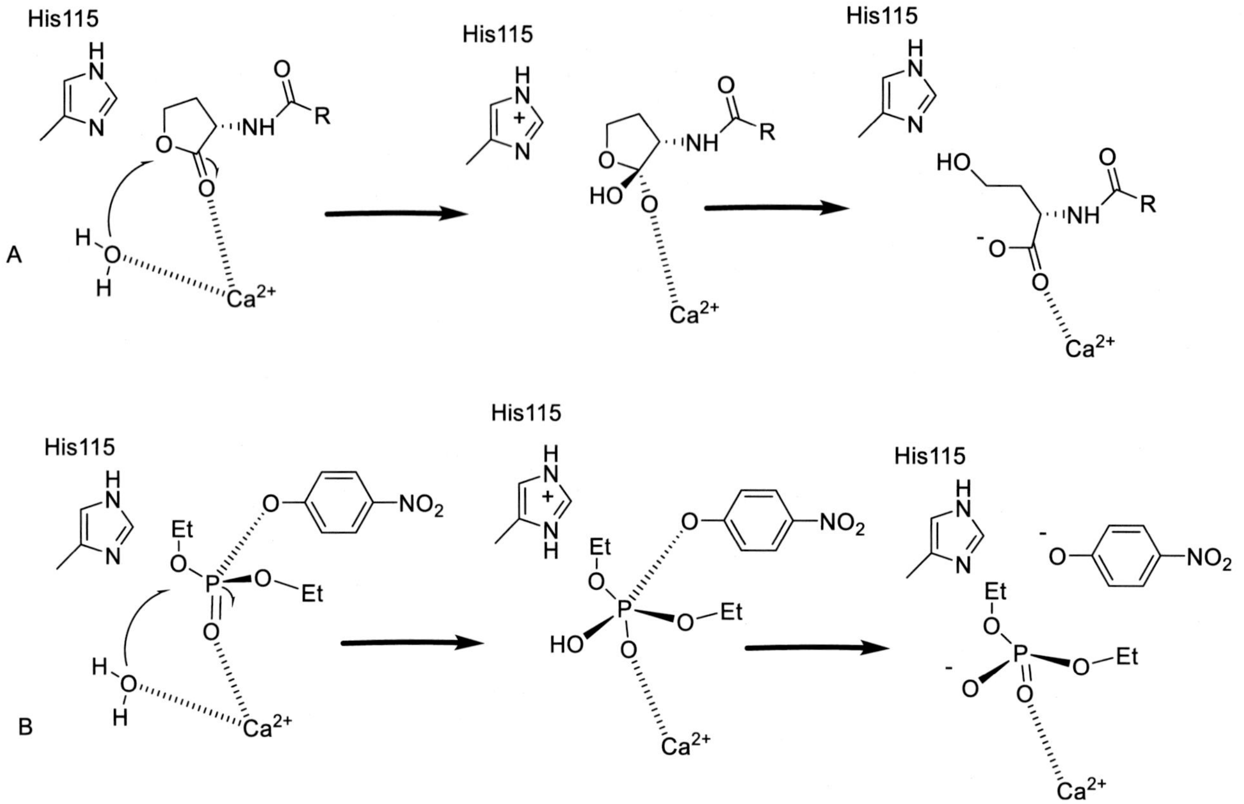

FIG. 5 PON catalytic mechanism as lactonase (A) and organophosphatase (B) with an AHL and paraoxon as substrates, respectively.

Detailed kinetic (k_{cat}) and thermodynamic (K_M) parameters for PON1–PON3 and diverse substrates were recently published [24].

4 PON inhibition

There are several studies in which the effects of various clinically used drugs were investigated on the inhibition of paraoxon hydrolysis catalyzed by PON1 and more rarely other isoforms [22,25–28]. Such drugs include sulfonamides (sulfacetamide, sulfasalazine, acetazolamide, furosemide, etc., which showed IC_{50}-s in the millimolar range) [22]; anabolic compounds (17β-estradiol, diethylstilbestrol, oxytocin, and trenbolone; again with IC_{50}-s in the millimolar range) [25]; antibiotics (gentamycin sulfate and cefazolin sodium salt, both millimolar inhibitors) [26], antidepressants and analgesics (haloperidol, fluoxetine hydrochloride, diazepam, and acepromazine) [27], nonsteroidal antiinflammatory drugs (diclofenac sodium, showed an IC_{50} of 0.845 mM after 6–12 h of incubation with the enzyme, whereas tenoxicam had an IC_{50} of 0.74 mM after 6 h of incubation with PON2) [28]. It may be observed that most of these structurally very different compounds show rather similar IC_{50} values, and by considering the rather low-organophosphatase activity of the enzyme, it is rather probable that these studies bring little novelty to the knowledge of PON enzymes pharmacology. However, apart the lactone-containing drugs mentioned above (Fig. 4), polymorphism in the PON genes is involved in the metabolism of clopidogrel, an antiplatelet drug metabolized by CYP450 enzymes and PON1 [29].

5 Physiological/pathological roles of PONs

There is much debate regarding the physiological and/or pathological roles of these enzymes in various conditions [5,7–9,16]. The reported and rather highly investigated antioxidant and antiinflammatory activities of PON (mainly PON1 and PON3) have been connected to the ability of these enzymes to protect LDL from oxidative stress [18,24,30]. However, recombinant human PONs (all isoforms) were shown not to protect LDL against copper ion–induced oxidation [5]. Correlated with the fact that the physiological substrates of these enzymes are still not precisely known [24], much doubt arose regarding the real role of these enzymes in many pathologies on which they were investigated extensively, such as cardiovascular diseases [31], cancer [32,33], and inflammation [34]. A relevant rescue was however furnished by studies by Tawfik's group [16,18], which showed that the stability and enzymatic activity of PON1 are dramatically stimulated when associated with HDL particles. The HDL-binding residue from PON active site was detected to be Asn168, which is a key amino acid residing at around 15 Å from the HDL-contacting interface. Asn168 coordinates to the catalytic calcium ion and aligns the lactone substrate for catalysis. HDL binding restrains the overall motion of the active site and particularly of Asn168, thus reducing the catalytic activation energy barrier [16]. Thus, disturbing this network, even at its most periphery, undermines PON1 catalytic activity. Membrane binding thus immobilizes long-range interactions via second- and third-shell residues that reduce the active site's floppiness and preorganize the catalytic residues [16]. ApoA-I, the major protein in HDL, was also shown to stabilize PON1, binding it with very high (nanomolar) affinity, and thus selectively stimulating its lactonase activity by up to 20-fold relative to the delipidated form of the enzyme [18]. Wild-type PON1 bound to recombinant HDL inhibited macrophage-mediated LDL oxidation and stimulated cholesterol efflux from the cells to 2.3- and 3.2-fold greater extents, respectively, compared with PON1 bound to phosphatidylcholine/free cholesterol particles without apoA-I [18]. Whether this association of PON with HDL is indeed the phenomenon explaining biological effects of the enzyme is still to be investigated in deeper detail. However, as already mentioned here, PON1 and PON3 seem to have evolved to hydrolyze 5-hydroxy-6*E*,8*Z*,11*Z*,14*Z*-eicosatetraenoic acid1,5-lactone, 5-HETEL (**9**), or the corresponding lactone derived from 4-hydroxy-5*E*,7*Z*,10*Z*,13*Z*,16*Z*,19*Z*-docosahexaenoic acid, thus regulating the lipid mediators derived from polyunsaturated fatty acids [23].

The physiological function of PON2 seems on the other hand to be better understood compared to those of PON1 and PON3. As shown in this chapter, PON2 has no catalytic activity as organophosphatase, has a rather weak activity as arylesterase, whereas its lactonase activity is absent or very low with most lactones (of types **5**–**10**) except *N*-acyl-homoserine lactones **4**, AHLs [5]. Starting with 2005, it became apparent that 3-oxohexanoyl homoserine lactone and structurally related AHLs [35–38] act as quorum sensing molecules in both gram-positive and gram-negative bacteria. Quorum sensing is a momentous communication system for the microbial community, which connects the bacteria to their surrounding environment [38]. Quorum sensing relies on the production, release, and detection of extracellular signaling molecules, which are called autoinducers, with the AHLs being relevant such compounds [35–38]. The ability of PON2 (and in a lower degree also of the other two PON isoforms) to hydrolyze AHLs makes these enzymes of relevance as antiinfective modulators, since quorum sensing is, for example, involved in biofilm formation, which is more difficult to treat with antiinfective agents compared with planktonic-type infections [39–44]. Indeed, AHLs contribute to regulate the transcription of specific genes and therefore expression of specific phenotypes, including growth, virulence, biofilm formation, bioluminescence, production of exopolysaccharide (EPS) [40–43]. For example, a recent study showed that dimetridazole and ribavirin, two drugs showing poor or no antibacterial activities on *Pseudomonas aeruginosa* isolates, may interfere with AHLs signaling and induce a potentiated antimicrobial activity of antibiotics such as polymyxin B, meropenem, and kanamycin, to which *P. aeruginosa* is resistant [44]. Thus, interference with PON2 and AHLs may lead to the development of strategies for fighting antibiotic drug resistance.

6 Conclusions

Relevant advances have been registered over the past two decades in understanding the biochemistry, substrate specificity, and catalytic mechanism of the calcium enzymes PON1–PON3. These phenomena are now well understood at molecular level, but still the precise physiological substrates of these enzymes are not know. It seems that PON2 is a specific lactonase of *N*-acyl-homoserine lactones, signaling molecules for quorum sensing in many bacterial species, and as thus might be a relevant drug target for designing novel antibacterial strategies. As the *N*-acyl-homoserine lactones are substrates for this enzyme, compounds that may enhance the catalytic efficiency of the enzyme, i.e., PON2 activators, would be of great interest. However, at the moment no activators of these enzymes were reported, and even the inhibition studies are rather scarce.

The involvement of the other isoforms PON1 and PON3 in various pathologies, such as cardiovascular diseases, cancer, and inflammation, was extensively studied in the past three decades, but without clear-cut results, probably also because, as mentioned above, the precise physiological substrates of these enzymes are not known. Relevant biochemical studies showed that PON enzymes alone do not possess antioxidant activity, whereas when they are associated with HDL, this activity is present. Although some molecular details were furnished for explaining these contradictory findings, much research should be performed for understanding all details and for a translation of such findings to clinics. Until then, these interesting enzymes remain "orphan," in the search of their precise physiologic and pathologic roles.

References

[1] Mazur A. An enzyme in the animal organism capable of hydrolysing the phosphorus–fluorine bond of alkyl fluorophosphates. J Biol Chem 1946;164:271–89.

[2] Bigley AN, Raushel FM. Catalytic mechanisms for phosphotriesterases. Biochim Biophys Acta 2013;1834(1):443–53.

[3] Wang H, Leeming MG, Cochran BJ, Hook JM, Ho J, Nguyen GTH, Zhong L, Supuran CT, Donald WA. Nontargeted identification of plasma proteins O-, N-, and S-transmethylated by O-methyl organophosphates. Anal Chem 2020;92(23):15420–8.

[4] Mackness MI, Thompson HM, Hardy AR, Walker CH. Distinction between 'A'-esterases and arylesterases. Implications for esterase classification. Biochem J 1987;245(1):293–6.

[5] Draganov DI, Teiber JF, Speelman A, Osawa Y, Sunahara R, La Du BN. Human paraoxonases (PON1, PON2, and PON3) are lactonases with overlapping and distinct substrate specificities. J Lipid Res 2005;46(6):1239–47.

[6] Harel M, Aharoni A, Gaidukov L, Brumshtein B, Khersonsky O, Meged R, Dvir H, Ravelli RB, McCarthy A, Toker L, Silman I, Sussman JL, Tawfik DS. Structure and evolution of the serum paraoxonase family of detoxifying and anti-atherosclerotic enzymes. Nat Struct Mol Biol 2004;11(5):412–9.

[7] Mackness M, Mackness B. Human paraoxonase-1 (PON1): gene structure and expression, promiscuous activities and multiple physiological roles. Gene 2015;567(1):12–21.

[8] Taler-Verčič A, Goličnik M, Bavec A. The structure and function of paraoxonase-1 and its comparison to paraoxonase-2 and -3. Molecules 2020;25(24):5980.

[9] Furlong CE, Marsillach J, Jarvik GP, Costa LG. Paraoxonases-1, -2 and -3: what are their functions? Chem Biol Interact 2016;259(Pt B):51–62.

[10] Khersonsky O, Tawfik DS. Structure–reactivity studies of serum paraoxonase PON1 suggest that its native activity is lactonase. Biochemistry 2005;44(16):6371–82.

[11] Khersonsky O, Tawfik DS. The histidine 115-histidine 134 dyad mediates the lactonase activity of mammalian serum paraoxonases. J Biol Chem 2006;281(11):7649–56.

[12] Billecke S, Draganov D, Counsell R, Stetson P, Watson C, Hsu C, La Du BN. Human serum paraoxonase (PON1) isozymes Q and R hydrolyze lactones and cyclic carbonate esters. Drug Metab Dispos 2000;28(11):1335–42.

[13] Mochizuki H, Scherer SW, Xi T, Nickle DC, Majer M, Huizenga JJ, Tsui LC, Prochazka M. Human PON2 gene at 7q21.3: cloning, multiple mRNA forms, and missense polymorphisms in the coding sequence. Gene 1998;213:149–57.

[14] Ozols J. Isolation and complete covalent structure of liver microsomal paraoxonase. Biochem J 1999;338:265–72.

[15] Ben-David M, Wieczorek G, Elias M, Silman I, Sussman JL, Tawfik DS. Catalytic metal ion rearrangements underline promiscuity and evolvability of a metalloenzyme. J Mol Biol 2013;425(6):1028–38.

[16] Rosenblat M, Gaidukov L, Khersonsky O, Vaya J, Oren R, Tawfik DS, Aviram M. The catalytic histidine dyad of high density lipoprotein-associated serum paraoxonase-1 (PON1) is essential for PON1-mediated inhibition of low density lipoprotein oxidation and stimulation of macrophage cholesterol efflux. J Biol Chem 2006;281(11):7657–65.

[17] Bar-Rogovsky H, Hugenmatter A, Tawfik DS. The evolutionary origins of detoxifying enzymes: the mammalian serum paraoxonases (PONs) relate to bacterial homoserine lactonases. J Biol Chem 2013;288(33):23914–27.

[18] Ben-David M, Sussman JL, Maxwell CI, Szeler K, Kamerlin SCL, Tawfik DS. Catalytic stimulation by restrained active-site floppiness—the case of high density lipoprotein-bound serum paraoxonase-1. J Mol Biol 2015;427(6 Pt B):1359–74.

[19] Elias M, Tawfik DS. Divergence and convergence in enzyme evolution: parallel evolution of paraoxonases from quorum-quenching lactonases. J Biol Chem 2012;287(1):11–20.

[20] Supuran CT. Carbonic anhydrases: novel therapeutic applications for inhibitors and activators. Nat Rev Drug Discov 2008;7(2):168–81.

[21] Briganti F, Mangani S, Orioli P, Scozzafava A, Vernaglione G, Supuran CT. Carbonic anhydrase activators: X-ray crystallographic and spectroscopic investigations for the interaction of isozymes I and II with histamine. Biochemistry 1997;36(34):10384–92.

[22] Ekinci D, Sentürk M, Beydemir S, Küfrevioğlu OI, Supuran CT. An alternative purification method for human serum paraoxonase 1 and its interactions with sulfonamides. Chem Biol Drug Des 2010;76(6):552–8.

[23] Teiber JF, Xiao J, Kramer GL, Ogawa S, Ebner C, Wolleb H, Carreira EM, Shih DM, Haley RW. Identification of biologically active δ-lactone eicosanoids as paraoxonase substrates. Biochem Biophys Res Commun 2018;505(1):87–92.

[24] Mohammed CJ, Lamichhane S, Connolly JA, Soehnlen SM, Khalaf FK, Malhotra D, Haller ST, Isailovic D, Kennedy DJ. A PON for all seasons: comparing paraoxonase enzyme substrates, activity and action including the role of PON3 in health and disease. Antioxidants (Basel) 2022;11(3):590.

[25] Demir D, Gencer N, Arslan O. An alternative purification method for human serum paraoxonase 1 and its interactions with anabolic compounds. J Enzyme Inhib Med Chem 2016;31(2):247–52.

[26] Sinan S, Kockar F, Arslan O. Novel purification strategy for human PON1 and inhibition of the activity by cephalosporin and aminoglikozide derived antibiotics. Biochimie 2006;88(5):565–74.

[27] Avcikurt AS, Sinan S, Kockar F. Antidepressant and antipsychotic drugs differentially affect PON1 enzyme activity. J Enzyme Inhib Med Chem 2015;30(2):245–9.

[28] Solmaz Avcıkurt A, Korkut O. Effect of certain non-steroidal anti-inflammatory drugs on the paraoxonase 2 (PON2) in human monocytic cell line U937. Arch Physiol Biochem 2018;124(4):378–82.

[29] Saiz-Rodríguez M, Belmonte C, Caniego JL, Koller D, Zubiaur P, Bárcena E, Romero-Palacián D, Eugene AR, Ochoa D, Abad-Santos F. Influence of CYP450 enzymes, CES1, PON1, ABCB1, and P2RY12 polymorphisms on clopidogrel response in patients subjected to a percutaneous neurointervention. Clin Ther 2019;41(6):1199–212.

[30] Manco G, Porzio E, Carusone TM. Human paraoxonase-2 (PON2): protein functions and modulation. Antioxidants (Basel) 2021;10(2):256.

[31] Kowalska K, Socha E, Milnerowicz H. Review: The role of paraoxonase in cardiovascular diseases. Ann Clin Lab Sci 2015;45(2):226–33.

[32] Medina-Díaz IM, Ponce-Ruíz N, Rojas-García AE, Zambrano-Zargoza JF, Bernal-Hernández YY, González-Arias CA, Barrón-Vivanco BS, Herrera-Moreno JF. The relationship between cancer and paraoxonase 1. Antioxidants (Basel) 2022;11(4):697.

[33] Arenas M, Rodríguez E, Sahebkar A, Sabater S, Rizo D, Pallisé O, Hernández M, Riu F, Camps J, Joven J. Paraoxonase-1 activity in patients with cancer: a systematic review and meta-analysis. Crit Rev Oncol Hematol 2018;127:6–14.

[34] Erre GL, Bassu S, Giordo R, Mangoni AA, Carru C, Pintus G, Zinellu A. Association between paraoxonase/arylesterase activity of serum PON-1 enzyme and rheumatoid arthritis: a systematic review and meta-analysis. Antioxidants (Basel) 2022;11(12):2317.

[35] Pearson JP, Passador L, Iglewski BH, Greenberg EP. A second N-acylhomoserine lactone signal produced by *Pseudomonas aeruginosa*. Proc Natl Acad Sci U S A 1995;92(5):1490–4.

[36] Winson MK, Camara M, Latifi A, Foglino M, Chhabra SR, Daykin M, Bally M, Chapon V, Salmond GP, Bycroft BW. Multiple N-acyl-L-homoserine lactone signal molecules regulate production of virulence determinants and secondary metabolites in *Pseudomonas aeruginosa*. Proc Natl Acad Sci U S A 1995;92 (20):9427–31.

[37] Smith D, Wang J-H, Swatton JE, et al. Variations on a theme: diverse N-acyl homoserine lactone-mediated quorum sensing mechanisms in Gram-negative bacteria. Sci Prog 2006;89(3–4):167–211.

[38] Yi L, Dong X, Grenier D, Wang K, Wang Y. Research progress of bacterial quorum sensing receptors: classification, structure, function and characteristics. Sci Total Environ 2021;763:143031.

[39] Camps J, Iftimie S, García-Heredia A, Castro A, Joven J. Paraoxonases and infectious diseases. Clin Biochem 2017;50(13–14):804–11.

[40] Papenfort K, Bassler BL. Quorum sensing signal–response systems in Gram-negative bacteria. Nat Rev Microbiol 2016;14:576–88.

[41] Miranda SW, Asfahl KL, Dandekar AA, Greenberg EP. *Pseudomonas aeruginosa* quorum sensing. Adv Exp Med Biol 2022;1386:95–115.

[42] Armes AC, Walton JL, Buchan A. Quorum sensing and antimicrobial production orchestrate biofilm dynamics in multispecies bacterial communities. Microbiol Spectr 2022;10(6):e0261522.

[43] Wang Y, Bian Z, Wang Y. Biofilm formation and inhibition mediated by bacterial quorum sensing. Appl Microbiol Biotechnol 2022;106 (19–20):6365–81.

[44] Yuan Y, Yang X, Zeng Q, Li H, Fu R, Du L, Liu W, Zhang Y, Zhou X, Chu Y, Zhang X, Zhao K. Repurposing Dimetridazole and Ribavirin to disarm *Pseudomonas aeruginosa* virulence by targeting the quorum sensing system. Front Microbiol 2022;13:978502.

Chapter 2.8

Phospholipases A2

Shibbir Ahmed Khan and Marc A. Ilies

Department of Pharmaceutical Sciences and Moulder Center for Drug Discovery Research, Temple University School of Pharmacy, Philadelphia, PA, United States

1 Structure and function of the PLA2 superfamily of enzymes

Phospholipases A2 (PLA2s) are lipolytic enzymes that hydrolyze the sn-2 ester bond of the phospholipids, releasing lysophospholipids (LPLs) and fatty acids (FAs) (Fig. 1). All these species are amphiphilic and form supramolecular structures such as micelles, vesicles, and liposomes in water and aqueous solutions. As a consequence, PLA2s are amphiphilic too and differ significantly from the classical hydrolases that act on water-soluble substrates [1,2]. They are able to dock at the water-lipid interface of the phospholipid assemblies via interfacial activation [3] and perform the hydrolysis of the individual phospholipids within these assemblies. Calcium ions are essential cofactors and play structural and/or catalytic roles (vide infra).

Focusing on the hydrolytic products, one can note that fatty acids released by PLA2 can be transported into mitochondria and degraded through β-oxidation to generate ATP [4]. Alternatively, fatty acids such as arachidonic acid (AA) can be further processed by cyclooxygenases (COXs), lipoxygenases (LOXs), and cytochrome p450 (CYP450) enzymes into prostaglandins (PGs), thromboxanes (TXs), leukotrienes (LTs), and other eicosanoids, which can act as potent inflammatory mediators, cell signaling molecules, etc. Lysophospholipids (LPLs) are important intermediates in the process of phospholipid remodeling, when fatty acids are removed from the sn-2 position and are replaced with different ones. Since the type (saturated/unsaturated) and chain length of FA strongly affect the physicochemical and self-assembling properties of the corresponding GPL, this process play an important role in membrane perturbation, and cell signaling. Alternatively, LPLs can be further hydrolyzed by lysophospholipase D (LPLD) to lysophosphatidic acid (LPA), which plays an important role in cell proliferation, survival, and migration [1,5,6].

The PLA2 superfamily currently contains 16 groups (Group I–XVI), following the chronology of their discovery [7]. The members of these PLA2 groups can be also divided into six subfamilies based on their substrate specificity, subcellular localization and differences in physiologic function, namely into secreted PLA2s (sPLA2s) (Groups I, II, III, V, IX, X, XI, XII, XIII, XIV), cytosolic PLA2s (cPLA2s) (Group IV), Ca^{2+}-independent PLA2s (iPLA2s) (Group VI), platelet-activating factors acetylhydrolase PLA2s (PAF PLA2s) (Groups VII, VIII), lysosomal PLA2s (LPLA2s) (Group XV), and adipose tissue-specific PLA2s (AdPLA2s) (Group XVI) [3,7]. sPLA2, cPLA2, and iPLA2 play critical roles in inflammation and cancer-related diseases, while PAF-AH PLA2, LPLA2, and AdPLA2 subfamilies are involved in the development and progression of obesity and atherosclerosis [3,8,9].

1.1 Secreted PLA2 (sPLA2)

The secreted PLA2s were isolated and purified from the venom of different snakes, scorpions and other venomous reptiles, as early as 1890, being studied in great detail ever since [7]. Thus, sPLA2s isolated from the venom of old world snakes (cobras) were classified as group I sPLA2s and those isolated from the venom of new world snakes (rattlesnakes) were classified as group II sPLA2s [10]. sPLA2s discovered in mammalian tissues or secreted pancreatic juices were found similar to that of the old-world snakes PLA2 (Group I, specifically Group IB). The sPLA2s discovered in the human synovial fluid was structurally similar to the new world snakes sPLA2s and belongs to group II SPLA2s [10]. Currently, mammalian sPLA2 family contains ten catalytically active isoforms namely IB, IIA, IIC, IID, IIE, IIF, III, V, X, and XII [1,3].

1.1.1 sPLA2 structure

Structurally, sPLA2s are low-molecular-weight enzymes, with a molecular weight ranging from 13 to 19 kDa and containing seven disulfide bonds (Fig. 2) [1]. Both groups I and II sPLA2s contain six common disulfide bonds in their native fold, placed in similar locations. Group I PLA2s have a seventh disulfide bond formed between residues 11 and

Metalloenzymes. https://doi.org/10.1016/B978-0-12-823974-2.00018-8

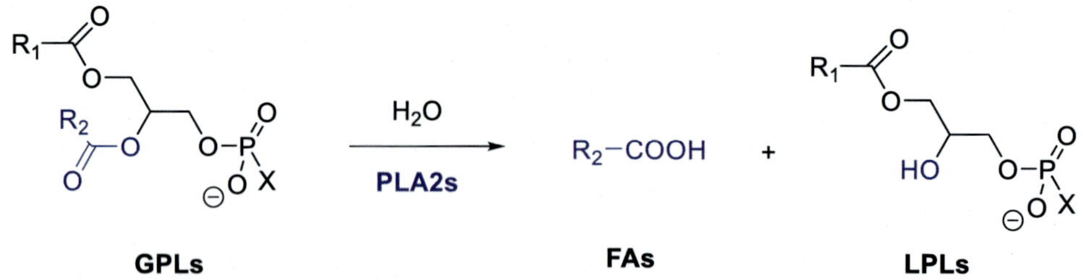

FIG. 1 PLA2 superfamily of enzymes catalyze the hydrolysis of the glycerophospholipids (GPLs) at the sn-2 position to yield fatty acids (FAs) and lysophospholipids (LPLs). Both FAs and LPLs are precursors for inflammatory mediators and intermediates involves in cell signaling, proliferation, survival.

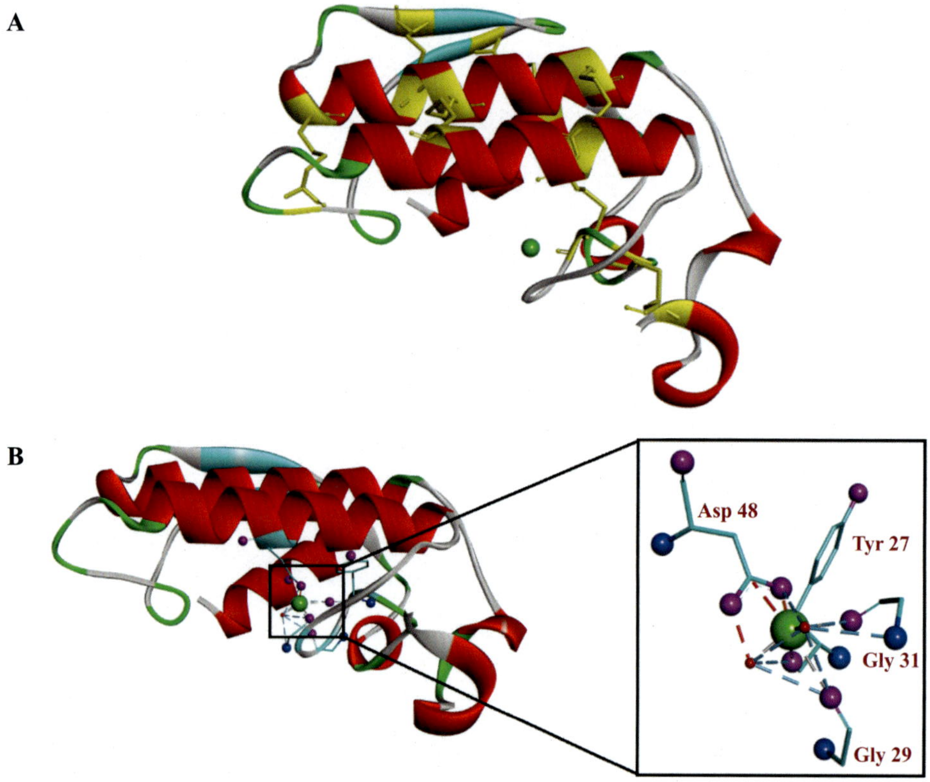

FIG. 2 (A) Crystal structure of Group IA sPLA2 (from *Naja naja*, PDB: 1PSH). The α-helices and β-sheets are shown in red and light blue. The cysteines forming the seven disulfide bonds are shown in yellow on the protein backbone and the calcium ion cofactor is shown as a green sphere; (B) Side/tilted view of the same sPLA2 structure, revealing the Ca^{2+} coordination by the carboxy group from Asp48 of one α-helix and by the C=O backbone groups of Tyr27, Gly29, and Gly31 from the opposite loop (magnified in the insert). The oxygen atoms and the nitrogen atoms of the amino acids coordinated with Ca^{2+} are colored pink and blue, respectively [1]. *(Image generated using BIOVIA Discovery studio.)*

77, while group II PLA2 has the seventh disulfide bond formed between residues 50 and 138 [7]. Another difference between group I and II sPLA2s is represented by an additional group of 6-7 negatively charged amino acid residues at the C terminal of group II, absent in group I, which makes the representatives of group II enzymes more negatively charged than Group I sPLA2s [11]. All the isoforms of sPLA2 have three long α-helices (red), two-stranded β-sheets (light blue) and a highly conserved Ca^{+2} binding loop (green sphere) as common structural features (Fig. 3A). A His/Asp catalytic dyad present in their active site polarizes a bound H_2O, which will subsequently attack the carbonyl group at the sn-2 position of phospholipids [3]. The Ca^{+2} ion cofactor, essential for the catalytic act, stabilizes the transition state by coordinating the carbonyl group and the negative charge from the phosphate group (Fig. 2) [3].

1.1.2 sPLA2 catalytic mechanism

The Ca^{2+} ion is hepta-coordinated with a pentagonal bipyramidal coordination geometry (Fig. 2B) and is required for both binding of the substrate and for catalysis [13]. Five of the ligands are amino acid residues from the protein backbone - one bidentate Asp48 carboxylic group side chain

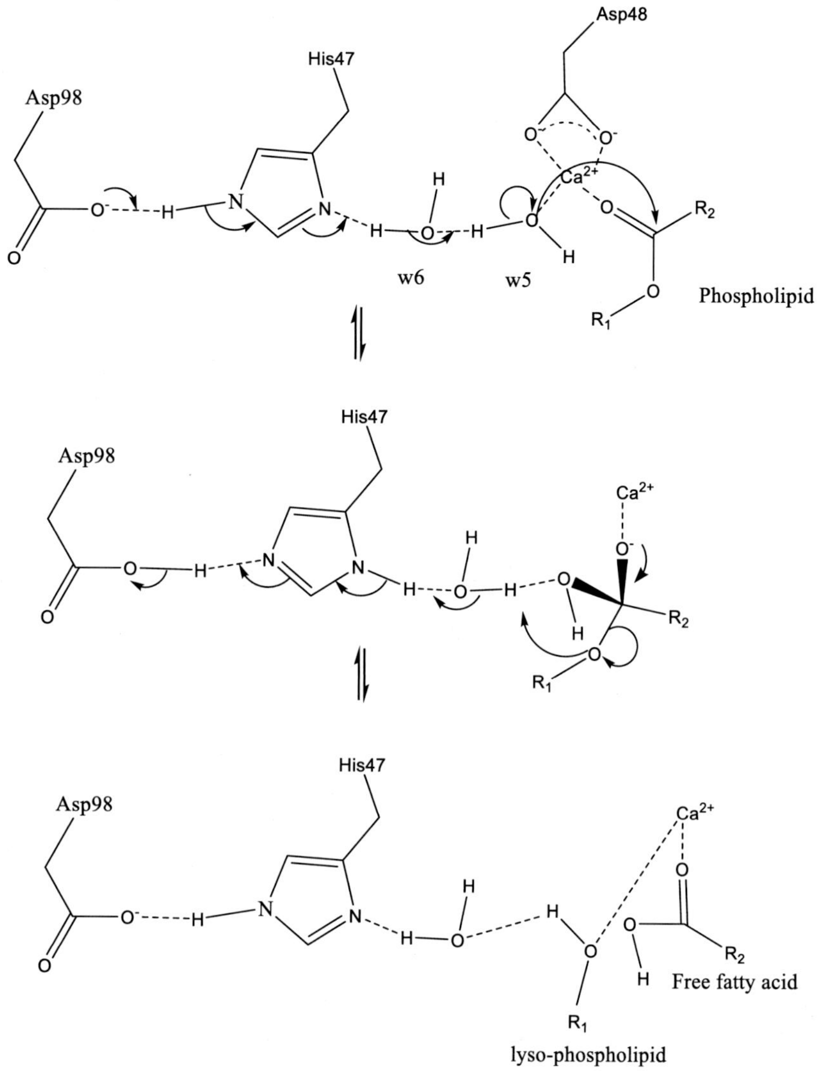

FIG. 3 Catalytic mechanism for sPLA2, with the catalytic dyad Asp98 and His47 central to catalysis H-bonded and with His47 further connected with the Ca^{2+} ion cofactor via two water molecules (w5 and w6). His47 facilitates the deprotonation of w5 water molecule polarized by Ca^{2+}, forming a hydroxide ion, which subsequently attacks the carbonyl group of the sn-2 ester of the substrate and generates a tetrahedral intermediate. The rate-limiting step is the decomposition of the tetrahedral intermediate with the formation of the products [1,12].

from the α-helix loop and three C=O backbone groups of Tyr27, Gly29, and Gly31 from the calcium-binding loop. The other ligands are two water molecules, one in the axial region and the other one in the equatorial region (Fig. 2B) [13,14].

Upon binding of glycerophospholipid substrate to the enzyme the axial water is replaced by the sn-3 phosphate group of the substrate. The equatorial water molecule coordinated to the Ca^{2+} ion (w5) is polarized by the metal ion and deprotonated by His47 (properly oriented by Asp98), generating a hydroxyl ion, which subsequently attacks the sn-2 ester carbonyl group and forms a tetrahedral intermediate (Fig. 3) [12]. The decomposition of the tetrahedral intermediate is the rate-limiting step of the enzymatic reaction, with the formation of the two products, which move out of the active site, being replaced by a new equatorial water molecule w5, thus closing the catalytic cycle (Fig. 3) [12].

1.1.3 sPLA2 substrate binding and interfacial kinetics

sPLA2s are interfacial enzymes that act at the water/oil interface at the surface of phospholipid self-assembled structures in water and aqueous solutions (micelles, mixed micelles, liposomes etc.) [1]. The active site of these enzymes has a common hydrophobic pocket about 15 Å deep that can accommodate a single phospholipid that is properly positioned next to the catalytic dyad and the Ca^{2+} ion for achieving the hydrolysis of the ester bond. The area surrounding the active site pocket of the protein forms the interfacial surface of the enzyme, known as the i-face. When the enzyme docks onto the lipid bilayer it displaces several water molecules from the region where it docks and creates a microenvironment that allows for the diffusion of the phospholipid substrate into the active site. The i-face has a planar surface composed of approximately 20 amino acids and a total surface area of about 1500 Å that remains in direct contact (docked) onto the membrane phospholipids (Fig. 8) [15,16]. The strong interfacial binding interactions required by a stable docking with the lipid bilayer are ensured by hydrophobic and amphiphilic amino acids within the i-face of the protein, together with charged amino acids that can interact electrostatically and through hydrogen bonds with the charged moieties of the phospholipids polar heads (Fig. 4) [17].

When the enzyme binds to the bilayer with the help of its i-face interface, it undergoes an allosteric structural modification, represented by E* in Fig. 5. The ability of the enzyme to dock onto the lipid bilayer ($E \rightarrow E^*$) is characterized by the equilibrium dissociation constant, K_d. The interfacial binding is essential for catalysis and can vary based on the composition and overall physicochemical properties of the bilayer vesicles. For example, the dissociation constant (K_d) of porcine pancreatic PLA2 on anionic vesicles is < 0.1 pM, whereas on zwitterionic vesicles it is more than 10 mM [18], explaining why some sPLA2 shows preference for anionic vesicles over the zwitterionic ones. Once bound, E* will accommodate one phospholipid molecule into the active site, hydrolyze it, releasing the products, and continuing the catalytic cycle as long as the enzyme remains docked onto the lipid bilayer (E*) (scooting mode), see below [12]. The rate of hydrolysis of different substrates by E* is controlled by the relative interfacial specificity constant (k^*_{cat}/K^*_S) (Fig. 5) [12,15,16].

1.1.4 Mode of action of sPLA2: scooting vs. hopping

The interaction of sPLA2 with the lipid bilayer/vesicles can be depicted considering two extreme modes of action, the scooting mode and the hopping mode (Fig. 6) [1]. In the scooting mode, the enzyme remains docked on the surface of the lipid bilayer/vesicle after each catalytic turnover. All the enzyme-containing vesicles will behave similarly if the number of enzymes per vesicle is similar. Alternatively, in the hopping mode, the enzyme dissociates from one vesicle after each catalytic turnover and moves to another bilayer/vesicle. As a consequence, all vesicles will hydrolyze uniformly. In practice it is hard to evaluate which particular mode the enzyme might prefer on a given vesicle. The residence time of the enzyme at the interface is not constant over the time course as the reaction progresses (Fig. 6) [12].

1.1.5 Substrate preference

sPLA2s show substrate specificity, which is similar to representatives of the different groups they belong to [1]. Group II subfamily of sPLA2 hydrolyze preferentially phosphatidylethanolamine, much more efficient than phosphatidylcholine, whereas groups III, V and X have a higher tendency of hydrolyzing phosphatidylcholine. While sPLA2- IB, IIA, and IIE do not show discrimination for

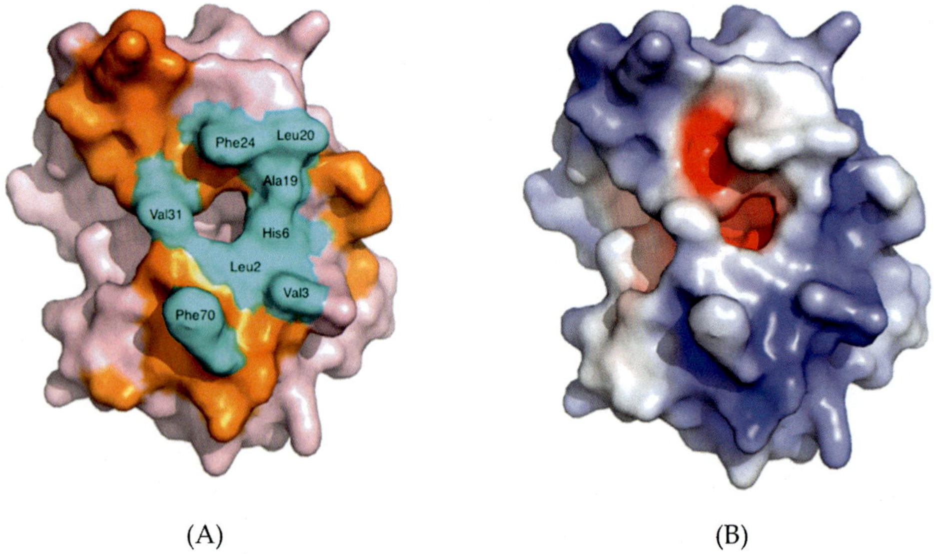

FIG. 4 (A) Space-filling model of human Group IIA sPLA2 (PDB 3U8B) depicting the amino acids, which form the i-face (colored orange) and the ones that flank the entrance into the active site (colored blue). (B) The distribution of electrostatic charge in the same sPLA2 with positively charged amino acid depicted in blue and with the negatively charged ones depicted in red. The white areas depict nonpolar, neutral amino acids [1,17].

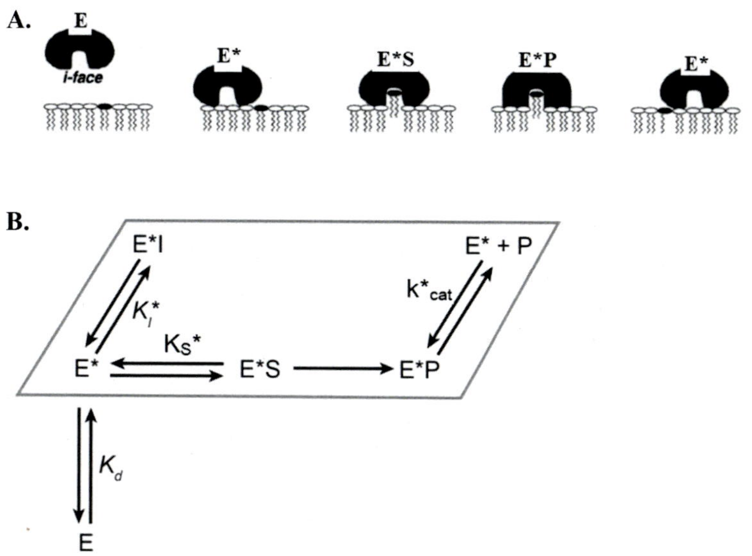

FIG. 5 Interfacial binding and catalytic action of sPLA2. (A) Cartoon depicting the binding of the sPLA2 at the lipid/water interface through its i-face, docking onto the bilayer and accommodating a single phospholipid molecule into its active site. Following hydrolysis, the LPL and FA products are released from the active site and the enzyme can bring another phospholipid into its active site [12]. (B) The kinetic scheme for the catalytic cycle of sPLA2, with the lipid bilayer represented by a parallelogram box. The free enzyme in the aqueous phase (E) must dock onto the lipid bilayer (E^*), an interaction characterized by the interfacial dissociation constant for the enzyme at the interface (K_d). E*S and E*P denote the enzyme-bound substrate and enzyme-bound product, while K_S, K_{cat}, and K_I are the dissociation constants for substrate, product, and inhibitor (when present) [1,15,16].

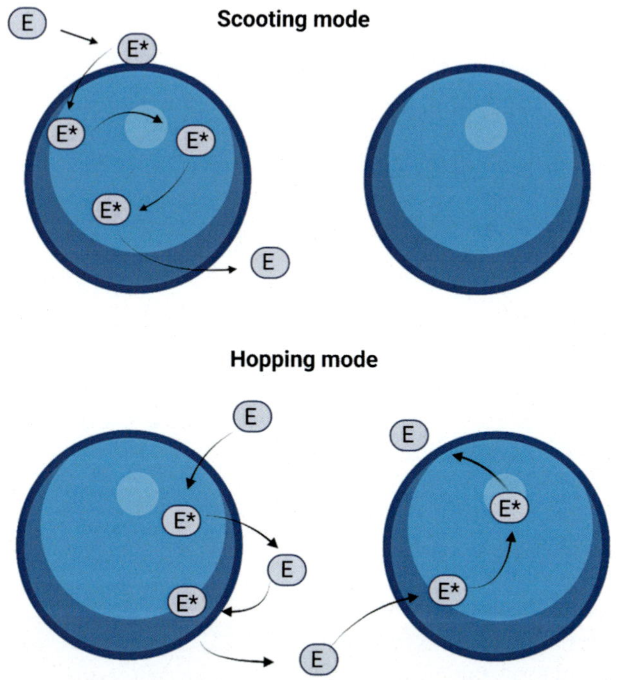

FIG. 6 Cartoon depicting the scooting mode (top) vs. the hopping mode (bottom) through which sPLA2 enzyme interacts with the lipid vesicles. In the scooting mode the enzyme remains docked onto the vesicles after each catalytic cycle until running out of substrate, while in hopping mode the enzyme leaves the bilayer/vesicle after each catalytic cycle and moves to another bilayer/vesicle [1,12]. *(Created with BioRender.com.)*

the sn-2 fatty acid of the GPLs, sPLA2-V hydrolyzed preferentially lipids with unsaturated fatty acids in the sn-2 position that contain a low degree of unsaturation, such as oleic acid (C18:1), and linoleic acid (C18:2). sPLA2-IID, IIF, and X have a preference for processing lipids with polyunsaturated fatty acids (PUFAs) in the sn-2 position, such as ω6 arachidonic acid (C20:4) and ω3 docosahexaenoic acid (22:6) [19].

1.2 Cytosolic PLA2 (cPLA2)

The cytosolic PLA2s (cPLA2s) are intercellular enzymes that belong to group IV of PLA2 superfamily and consists of six subgroups, denoted as cPLA2α, cPLA2β, cPLA2γ, cPLA2δ, cPLA2ε, and cPLA2ζ respectively, with molecular weighs ranging from 85 to 114kDa [1,20]. Representatives of this class are found in platelets, neutrophils (from where were first discovered in 1986 [21,22], in macrophages, and in organ tissues such as the brain, heart, liver, and pancreas [3]. They hydrolyzed phosphatidylcholines, phosphatidylethanolamines, and phosphatidylinositols, having a high specificity for phospholipids containing arachidonic acid (AA) at the sn-2 position [3].

1.2.1 cPLA2 general structure

The enzyme consists of a Ca^{2+} dependent lipid-binding domain (CaLB), also known as the C2 domain, attached

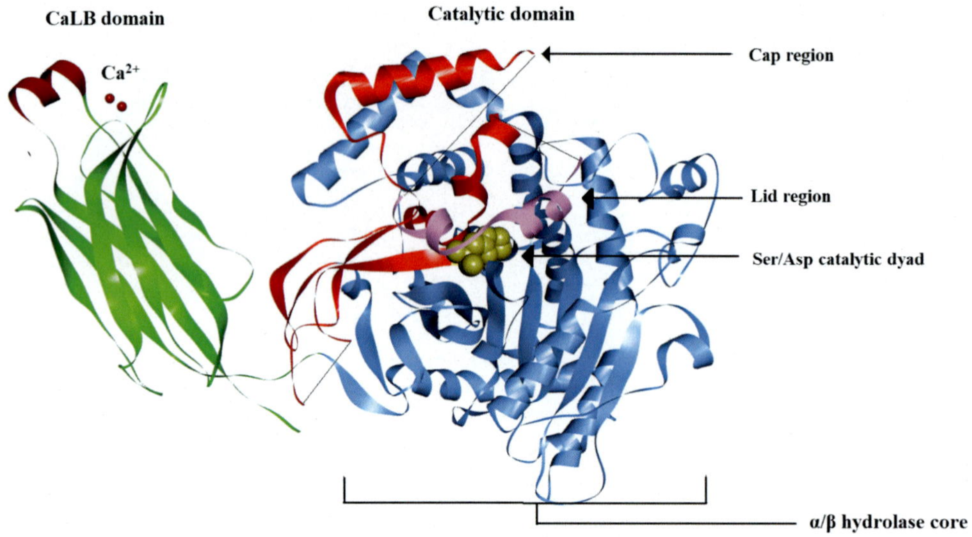

FIG. 7 Ribbon diagram of cPLA2 (PDB: 1CJY), revealing the Ca^{2+}-dependent lipid-binding (CaLB) domain (green), with the two bound Ca^{2+} ions (required for protein translocation to the membrane and activation), depicted as red spheres, the catalytic domain with the α/β hydrolase core (blue) and the catalytic dyad Ser/Asp (yellow). The cPLA2 cap is shown in red and the lid region covering the active site is shown in pink [1,22]. *(Image generated using BIOVIA Discovery studio.)*

via a flexible hinge to a catalytic domain (Fig. 7) [22]. The CaLB domain contains eight antiparallel β-sheets folded into a sandwich-like structure. The catalytic domain contains the active site with the Ser/Asp catalytic dyad. It consists of 14 β strands and 13 α helices, out of which 10 β-sheets and 9 α-helices form the central core of the catalytic domain (α/β hydrolase core). The remaining 4 β strands and the 4 α-helices (residues 370-548) form a "cap" region, which has a string of amino acids (residues 413-457) that blocks the lipid substrate from entering the active site of the inactive form of the enzyme, acting as a closed "lid." The lid is organized into a small loop, followed by a small helix (Fc) and a short turn. Residues 408-412 and the residues 434-457 do not have a traceable electron density in the X-ray crystal structure of the cPLA2, suggesting that these flexible regions form the "lid hinges," which can change conformation when enzyme comes in contact with the lipid substrate [22,23]. Importantly, both the CaLB domain and the catalytic domain are required for the full activity of the enzyme. Recruitment of two Ca^{2+} ions, which bind to the CaLB domain, facilitates the anchoring of the enzyme to the lipid bilayer [22,24]. Unlike the sPLA2s, the Ca^{2+} ion does not contribute directly to the catalytic activity of cPLA2. The catalytic domain remains in close contact with the membrane surface through specific electrostatic interaction with the membrane phospholipids and does not penetrate deeply into the lipid bilayer [22]. The cap region contains positively charged amino acid residues, which make electrostatic contacts with the negatively charged moieties of membrane phospholipids. The lid covers the active site when enzyme is inactive and prevents any lipid substrate to reach the active site. Binding of Ca^{2+} to the CaLB domain triggers the relocation of the enzyme to the lipid bilayer and its activation, when the lid opens up, allowing the binding of lipid substrates with the active site [23]. The active site is about 7 Å in diameter and 8 Å in depth, lined with hydrophobic residues, which can help accommodate the fatty acyl chain of the membrane phospholipids, which are believed to be "lifted" from the membrane until the first cis double bond of the unsaturated chains (frequently found attached to the sn-2 position of GPLs). This "lifting" process may explain the cPLA2s preference for phospholipids containing arachidonic acid at the sn-2 position [22]. At the bottom of the active site is located Ser228, which plays a key role in catalysis (Fig. 7).

1.2.2 cPLA2 activation and catalytic mechanism

cPLA2s are activated via Ca^{2+} ion binding, phosphorylation of specific cPLA2 residues and interaction with secondary messengers such as phosphatidyl inositol bisphosphate (PIP2) and ceramide-1 phosphate (C1P) (Fig. 8) [3]. Intracellular Ca^{2+} ion concentration $[Ca^{2+}]$ is tightly regulated inside the cell and consequently the cPLA2 is normally inactive. When $[Ca^{2+}]$ increases, two Ca^{2+} ions will bind to the CaLB domain and facilitate translocation of the enzyme from the cytosol to the membrane region (plasmalemma, perinuclear membrane etc.) [22]. The two Ca^{2+} ions bound to an anionic hole of CaLB domain, rigidifying it and optimizing the local charge and backbone conformation of the cPLA2 to facilitate the interaction with the phospholipid membrane [25].

The phosphorylation of amino acid residues (especially Ser505) located in the flexible linker between the CaLB

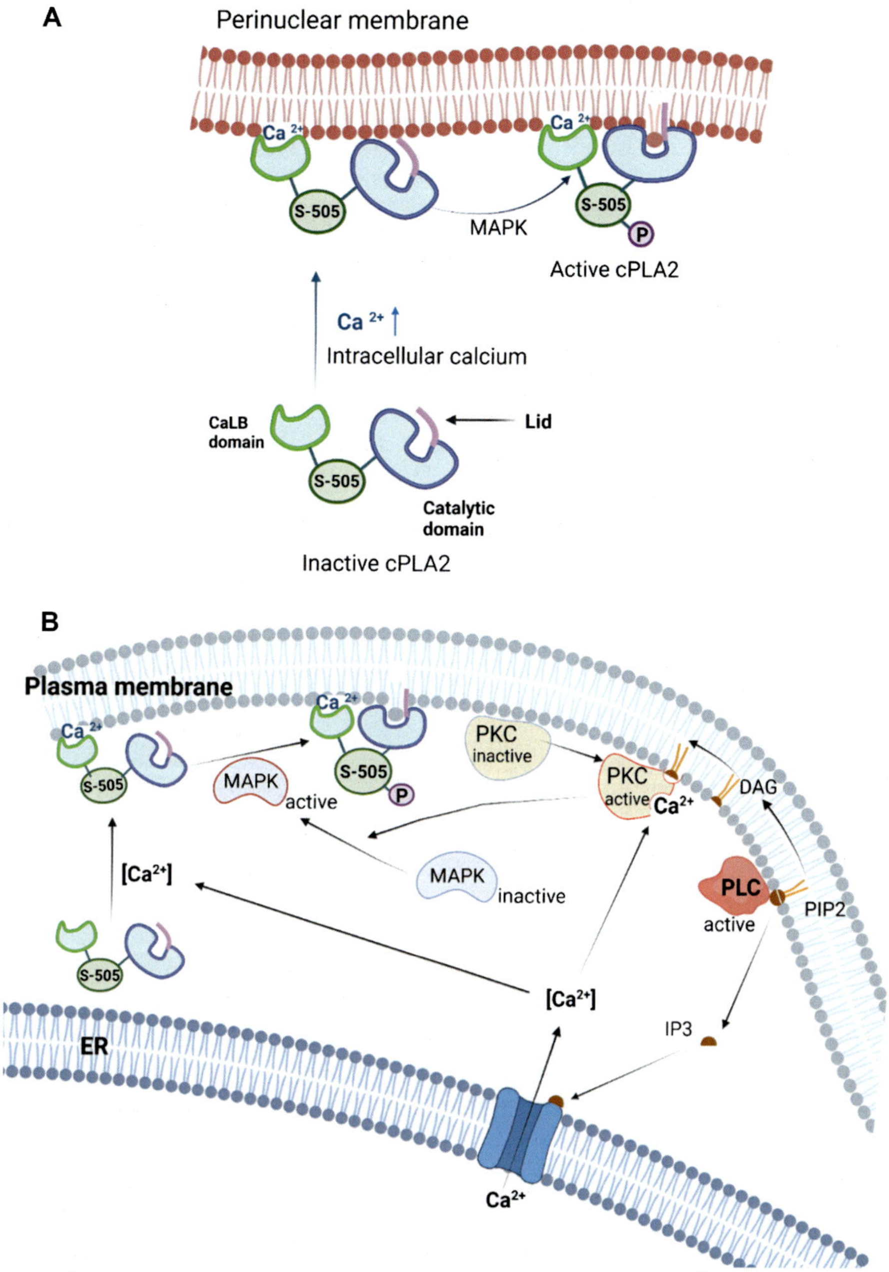

FIG. 8 Activation pathways for cPLA2 via calcium, phosphorylation, and secondary messengers. (A) Ca^{2+} binding in the CaLB domain of the cPLA2 triggers relocation to the membrane. Subsequent phosphorylation of Ser505 in the flexible linker of the cPLA2 via MAPK changes the conformation of the catalytic domain, which penetrates the membrane and starts the hydrolysis process. (B) Regulation of cPLA2 via hydrolysis of PIP2 by PLC, which generates secondary messengers IP3 and DAG. IP3 diffuses through the cytoplasm and opens the ER Ca^{2+} ion channels, releasing Ca^{2+} stored in ER and increasing the intracellular Ca^{2+} concentration. The DAG and the elevated cytoplasmic [Ca^{2+}] simultaneously activate PKC, which subsequently activates MAPK. The active MAPK will phosphorylate the Ser505 in cPLA2, which further enhances the cPLA2 activity [1]. *(Created with BioRender.com.)*

domain and catalytic domain, triggered by the same elevated [Ca^{2+}] concentration and intracellular signaling (Fig. 8), causes a conformational change of the interdomain region of the enzyme, moving the catalytic domain closer to the membrane surface and increase further the binding affinity and the catalytic efficiency of the enzyme (Fig. 8A) [26,27]. Phosphorylation of cPLA2 is done primarily by mitogen-activated protein kinase (MAPK), which phosphorylates cPLA2 at Ser505 residue in the flexible linker (Fig. 8) [28]. If Ser505 residue is mutated, one can observe a delayed translocation of the mutant enzyme to the membrane, even at elevated levels of Ca^{2+} and PIP2. This suggests that phosphorylation of Ser505 works in tandem and probably potentiates Ca^{2+} -mediated activation [29], probably through a change of the catalytic domain conformation [26]. In smooth muscle, MAPK- and Erk1/2-dependent cPLA2 activation was shown to occur when stimulated by muscarinic receptors m1 and m2 [28], while in macrophages and neutrophils, colony-stimulating factor (CSF1) stimulates MAPK to phosphorylate cPLA2 [30]. Importantly, accumulation of oligomeric amyloid-beta in patients with Alzheimer's disease was shown to induces the MAPK-mediated activation of brain cPLA2 in a spatial-specific manner [31].

Protein Kinase C (PKC) is another important kinase that regulates cPLA2 via phosphorylation (Fig. 8B), either through ERK-dependent and/or ERK-independent pathways [31]. One isoforms of PKC, namely PKCα, can phosphorylate directly cPLA2 and release AA independent of MAPK, while other isoforms of PKC (PKCβ1, PKCβ2, PKCγ) phosphorylate the MAP kinase, activating it and triggering subsequently phosphorylation of cPLA2. Studies in MDCK-D1 cells have revealed the central role played by P_{2U} receptor, a G-protein-coupled receptor (GPCR), which activates PKC that subsequently activates MAPK to phosphorylates cPLA2 [32].

Another activator of cPLA2, acting in tandem with Ca^{2+}, is ceramide-1-phosphate (C1P), an bioactive sphingolipid synthesized by ceramide kinase via phosphorylation of ceramide [33]. C1P action triggers the translocation of the C2 domain of the cPLA2 enzyme into the perinuclear membrane from the cytosol, where it starts releasing AA acid [27,34]. Inflammatory cytokines such as interleukin-1β activate sphingomyelinase (a specific form of PLC) to hydrolyze sphingomyelin to form ceramide. Ceramide produced mixes poorly with existing cholesterol in the membrane and segregates into microdomains, reducing the local rigidity of the membrane and increasing the penetration of cPLA2 into the local membrane environment via association of C2 domain with the C1P [35].

Phosphatidylinositol 4,5-bisphosphate (PIP2) hydrolysis pathway is an important pathway that can also trigger activation of cPLA2, both directly and indirectly [36,37]. The indirect pathway involves hydrolysis of plasma membrane PIP2 by phospholipase C (PLC) to produce diacyl glycerol (DAG) and inositol triphosphate (IP3)—secondary messengers acting inside the cell. IP3 is water-soluble and moves through the cytoplasm to trigger the release of Ca^{2+} from ER. DAG is a lipid that migrates laterally in the membrane, where it activates membrane-associated protein kinase C (PKC) in tandem with elevated cytosolic [Ca^{2+}]. PKC subsequently activates MAPK, which phosphorylates cPLA2 that is already translocated to the membrane in response to elevated cytosolic [Ca^{2+}] (Fig. 8B) [1,36,37].

The direct pathway of cPLA2 regulation by PIP2 involves binding of PIP2 to the catalytic domain of the cPLA2 after its translocation to the membrane, which increases its catalytic activity dramatically (up to 20-fold). The association of PIP2 with cPLA2 is mediated by a cluster of cationic amino acid residues (Lys-541, Lys-543, Lys-544, and Arg-488) present in the catalytic domain of cPLA2 that interact with anionic PIP2. Replacing these amino acids with neutral ones was shown to inhibit direct activation of cPLA2 via PIP2 binding [38–40].

1.3 Ca^{2+} independent PLA2 (iPLA2)

Ca^{2+} independent PLA2 (iPLA2) enzymes are members of the PLA2 superfamily belonging to the Group VI PLA2s [1]. The iPLA2 enzymes hydrolyze GPLs at the sn-2 position, but they also display hydrolytic activities normally associated with PLA1, lysophospholipase, transacylase, and PAF acetylhydrolase, being therefore more promiscuous as compared with their PLA2 congeners [41–43]. There iPLA2s comprise are six members (A, B, C, D, E, F), denominated iPLA2β, iPLA2γ, iPLA2 δ, iPLA2ε, iPLA2ζ, and iPLA2, with molecular weight ranging from 27 kDa (iPLA2) to 85 kDa (iPLA2 β), and as high as 146 kDa (iPLA2δ) [3]. Their cellular localization varies. The iPLA2β was found intracellularly in the cytosol (resting cells) of the B lymphocytes, pancreas islets, brain, and testis, while iPLA2γ is membrane-associated and found in the heart and skeletal muscles [41–44]. The iPLA2δ is localized in the ER and Golgi complex of the neurons, while iPLA2ε and iPLA2ζ were found in the liver adipocytes, white and brown fat adipocytes. iPLA2s are ubiquitously spread in all cells of the human body [45].

1.3.1 *iPLA2s structure, regulation*

The iPLA2α is a patatin and patatin-like homolog found in the potato tubers. It is not found in human body and bears a structural similarity to cPLA2α [41–43]. The iPLA2β (Group VI A) gene found in the human genome has multiple splice variants, with at least two isoforms (A-1 and A-2) active in the human body [3]. Korolev et al. solved the crystal structure of iPLA2β (Fig. 9A) [46] and showed that its structure can be divided into the N terminal domain, nine ankyrin repeats, and a catalytic domain (CAT) (Fig. 9A). The ankyrin repeat motif (ANK) contains 33 amino acid residues, organized into a helix-turn-helix structure followed by a hairpin loop [46]. These ankyrin repeats are found only the two variants of iPLA2β (VI-A1, VI-A2), while the other isoforms do not have any ankyrin repeats [45]. The ankyrin repeats are believed to mediate the interaction with integral membrane proteins such as ion channels and cell adhesion/signaling molecules. Unlike cPLA2, the iPLA2 enzymes do not have any CaLB domain, so their translocation from the cytosol to the membrane is done through the interaction of the ankyrin repeats with the membrane-bound proteins [41–43]. The ANK domain also has an ATP binding site, which helps in regulating the activity of the iPLA2. The ANK

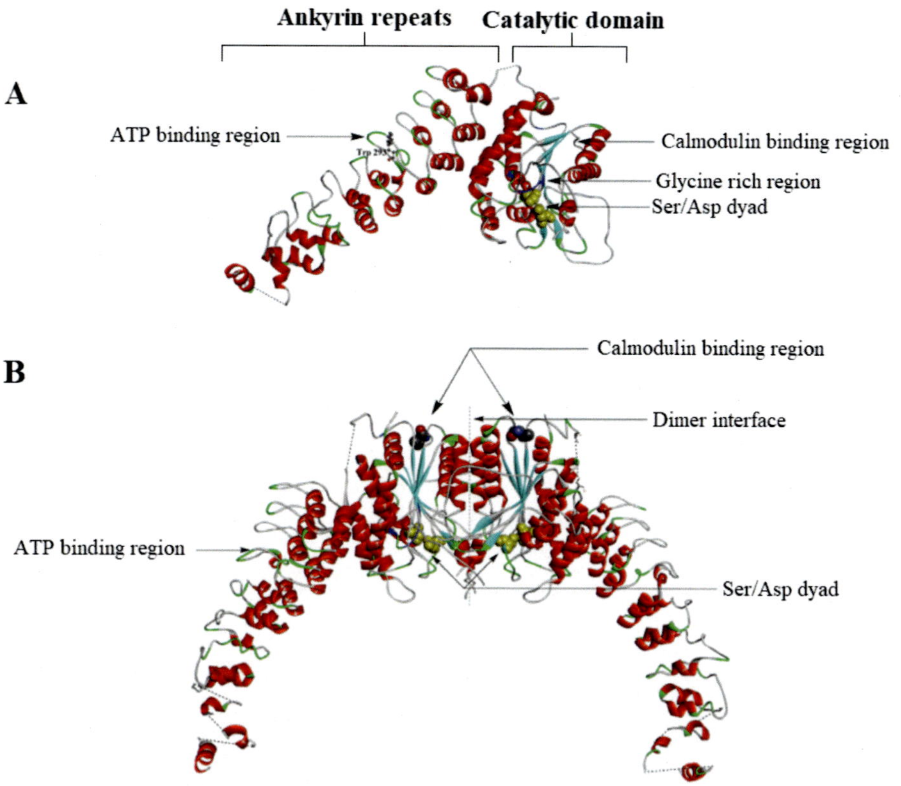

FIG. 9 (A) The crystal structure of iPLA2β in monomeric form (PDB- 6AUN), with the ankyrin repeats shown on the left side and the catalytic domain on the right side, connected by a flexible linker (not resolved in the X-ray crystal structure). The ATP binding site is located near the 6th ankyrin repeat, at Trp293 residue. The glycine-rich region, which forms a rigid handle, is shown in blue and the catalytic dyad (Ser465 and Asp598) is shown in purple, within the catalytic domain. (B) The dimeric structure of the iPLA2, generated when one CaM molecule binds two iPLA2 molecules. Binding, facilitated by Ca^{2+}, stabilizes a closed confirmation of the active sites of the enzyme, inhibiting its activity. Note the positioning of the two active sites very close to the dimeric interface [1]. *(Image generated using BIOVIA Discovery studio.)*

domain is attached to the CAT domain at the opposite side of the membrane-binding surface. The linker between the ANK and CAT is quite flexible and because of that is it unresolved in the crystal structure. The 7th and 8th ankyrin repeats were found to form hydrophobic interaction with the CAT domain [46]. (Fig. 9B). The catalytic domain has a large hydrophobic interface (~2800 Å), which triggers the dimerization of protein and the formation of a stable iPLA2 dimer. The catalytic domain active site contains a Ser465/Asp598 dyad. The active sites of the dimeric form of the enzyme are located very closely to the dimer interface and in proximity to each other (Fig. 9B) [46].

These iPLA2 enzymes are regulated by ATP binding, calmodulin (CaM) binding, and by caspase cleavage of the ankyrin repeats [45]. Thus, ATP-binding stabilizes the iPLA2β isozyme structure and activates it. Thus, the activity of iPLA2β expressed in Chinese Hamster Ovary (CHO) cells increases 2–4-fold upon binding of ATP [41–43]. It is the only isozyme that has been reported to be regulated by ATP binding [3]. The ATP-binding site was initially placed in the catalytic domain of the iPLA2, which has a highly conserved glycine-rich motif (GXGXXG) (Fig. 9A) [41–43], but a recent crystallography study revealed the ATP-binding site close to the 6th ankyrin repeat (AR6) near Trp293 residue, with this ANK repeat having a different conformation than the other ones [46,47].

Calmodulin (CaM) binding is another way of regulating iPLA2, as this translates into an inhibition of the activity of iPLA2 in the presence of Ca^{2+}. CaM is essential for dimerization of iPLA2β monomers (Fig. 13B) [49]. It binds at the interface of the iPLA2 dimer and changes the conformation of this interface, together with both active sites (Fig. 9B). The calmodulin binding domain of iPLA2β contains multiple contact points in the C-terminal region near residue 630 of the enzyme. Binding of CaM to iPLA2, in the presence of Ca^{2+}, forms a catalytically inactive ternary complex CaM/Ca^{2+}/iPLA2, which prevents proteolysis. The active site of iPLA2 is closed in the ternary complex, preventing the lipid substrate to access the catalytic dyad Dissociation of the CaM from the iPLA2 opens up the active site and allows lipid substrates to enter the active site and to be hydrolyzed [46].

The activity of iPLA2s can be also modulated through caspase cleavage of the ankyrin repeats. Caspase-3 is a protease enzyme of the caspase superfamily, which is activated during cell apoptosis by caspases 8, 9, and 10. The activated caspase-3 cleaves the ankyrin repeats of the iPLA2 to yield a truncated iPLA2 (MW ~ 70 kDa) with increased catalytic properties. The increased activity of iPLA2 leads to generation of lyso-PL in large amounts, which will subsequently act as a signaling molecule on the surface of the apoptotic cells, leading to their recognition and phagocytosis by macrophages. Interestingly, caspase-3 can also cleave cPLA2, but in that case, the event leads to the inactivation of the cPLA2 [50].

1.3.2 Mechanism for catalysis for cPLA2 and iPLA2

The iPLA2s contain a Ser/Asp catalytic dyad, similar with cPLA2 and different from the His/Asp dyad of sPLA2. Therefore, iPLA2s share a common catalytic mechanism for hydrolysis with cPLA2s (Fig. 10) [51]. Once the phospholipid is inserted into the active site pocket of the enzyme, the phosphate group of the phospholipid is stabilized by an arginine (cPLA2) or by a lysine (iPLA2). The hydrolysis mechanism involves the deprotonation of the Ser by the Asp, followed by a nucleophilic attack of the OH^- to the ester bond to generate a tetrahedral intermediate B (Fig. 15). The tetrahedral intermediate is stabilized by the oxyanion hole, formed by two glycines. The tetrahedral intermediate eliminates the lyso-PC, protonating it from the Asp. The Asp again removes a proton from a water molecule forming a hydroxyl group, which attacks the ester bond between the Ser and fatty acid (FA), generating a second tetrahedral group (D). The second tetrahedral group eliminates the FA, which leaves the active site and the Ser takes an H^+ from the Asp, regenerating the initial active form of the enzyme [51].

1.4 Other PLA2s

Lp-PLA2s (Group VII and VIII PLA2s) catalyze the hydrolysis of the acetyl group from the sn-2 position of PAF, generating lyso-PAF and acetate. Group VII-A Lp-PLA2 hydrolyzes substrates with shorter chain residues at sn-2 more efficiently than ones having longer chains attached at sn-2 position [3]. Lp-PLA2s have a molecular weight of 45 kDa and are primarily secreted by the macrophages, which circulate in the blood and form complexes with LDL and HDL. Structurally, these enzymes and have a characteristic GXSXG motif, which is also found in lipases and

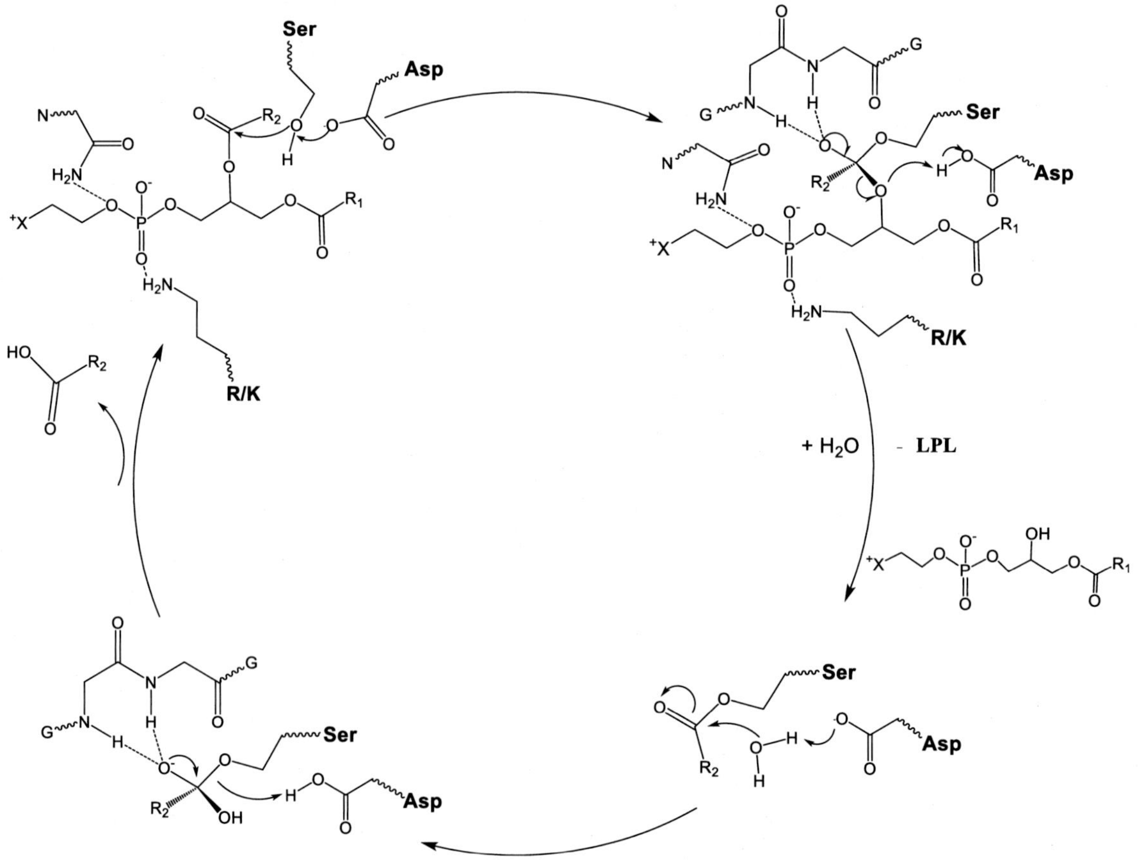

FIG. 10 The ping-pong catalytic mechanism of iPLA2s (common with cPLA2s) for hydrolysis of phospholipids, emphasizing the central role played by the Ser/Asp dyad [1].

esterases. They active site has a catalytic triad of Ser/Asp/His, lacks a lid region and has an open conformation of the active site, which can accommodate various acyl chain sizes at the sn-2 position of the phospholipid substrate [52,53]. In particular, Lp-PLA2 VIIA is secreted by the macrophage, monocytes, mast cells, and T-lymphocytes and its expression and secretion increase significantly when the macrophages are activated by exogenous stimuli such as cytokines, TNFα, and IFNɣ [52,54]. Group VII-B Lp-PLA2s are intracellular enzymes that are mainly expressed in the epithelial cells from proximal and distal tubules of the kidney, intestinal tract, and hepatocytes [55].

Lysosomal PLA2s (LPLA2s) are members of group XV PLA2s localized on intracellular vesicles such as lysosomes and endosomes [56]. Lysosomal PLA2 has a molecular weight of 45 kDa and is ubiquitously expressed throughout the body, being found in large amounts in the alveolar macrophages responsible for lung surfactant catabolism [57]. In some cases, LPLA2s have shown 50 times higher activity in alveolar macrophages as compared with peritoneal/peripheral macrophages [57]. It is believed that LPLA2s possess both PLA2 and transacylase activity, being able to transfer the acyl group from the sn-2 position of the phospholipid to a water molecule (phospholipaseA2 activity) or to a ceramide molecule, at C1 position (transacylase activity) [56].

Human adipose tissue PLA2s (AdPLA2s) are expressed ubiquitously throughout the body with a very high expression in the adipose tissue [9]. AdPLA2s have a molecular weight of 18 kDa and, interestingly, a Cys-His-His catalytic triad. They exhibit calcium-independent phospholipase activity for PCs and PEs [58]. A recent study has also revealed the expression of AdPLA2 in the skeletal muscles [59]. They are involved in the lipid metabolism at the level of adipose tissue and skeletal muscles (vide infra).

2 Physiologic and pathologic roles of PLA2 enzymes

The PLA2 superfamily of enzymes are ubiquitously expressed throughout the human body, as presented above, and play complex roles both in physiology and various pathologic conditions [1]. Research is in progress to fully elucidate these roles. In what follows we are providing a brief presentation of the known physiologic and pathologic processes involving PLA2s.

2.1 sPLA2 physiologic and pathologic roles

As presented above, sPLA2s are enzyme secreted outside the cells that require Ca^{2+} as catalytic cofactor. These metalloenzymes target extracellular substrates, many times located far away from the point where they were secreted [1,19]. The sPLA-IB, synthesized by the pancreatic acinar cells, is secreted as inactive zymogen into pancreatic juice (Fig. 11). It is activated by trypsin/plasmin and in the active form digests both dietary phospholipids and phosphatidylcholine (PC) from the bile, producing lysophosphatidylcholine (LPC) and free FA [60]. The free fatty acids can permeate the membranes of the intestinal mucosa, are absorbed, and processed. The LPC can be further hydrolyzed by lysophospholipase D into choline and lysophosphatidic acid. Alternatively, LPC can be passively absorbed into enterocytes, where it is involved in intracellular lipid trafficking, chylomicron assembly and secretion [61,62]. Other phospholipases expressed in the GI tract are sPLA2-X and phospholipase B (PLB). sPLA2-X is expressed in the enterocytes of the proximal small intestine and, in tandem with the pancreatic sPLA2-IB, hydrolyzes PC to produce LPC and FAs. Phospholipase B (PLB) is expressed in the distal intestine (ileum), where it hydrolyzes

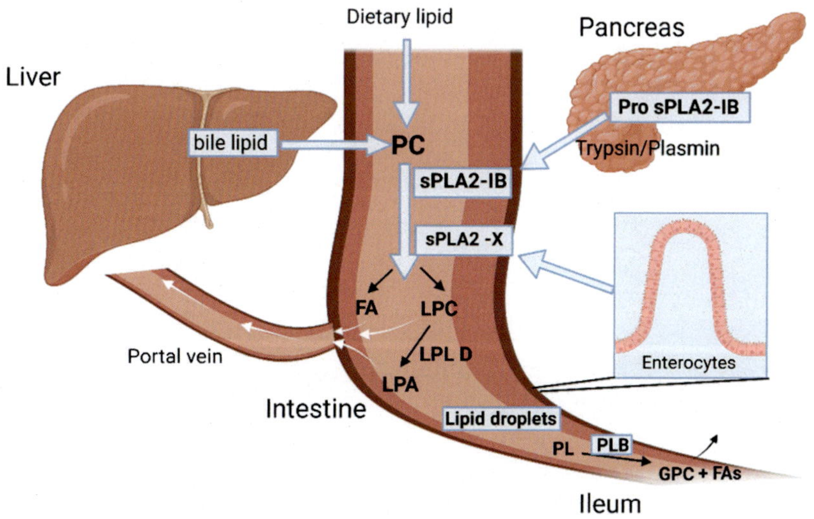

FIG. 11 Physiologic role of phospholipases in GI tract, emphasizing the main representatives sPLA2-IB, sPLA2-X, and phospholipase B, together with their site of generation and activation, substrates processed and the flux of products [1]. *(Created with BioRender.com.)*

the remaining phospholipids at both sn-1 and sn-2 fatty acyl chain positions and generates glycerophosphorylcholine (GPC) and FAs (Fig. 11) [1,63].

The sPLA-IIA plays an important role in inflammation and proinflammatory responses in a variety of cells and tissues, being often referred to as "inflammatory sPLA2" [1]. It is secreted by monocyte, macrophages, mast cells, neutrophils, T-cells, and in the synovial fluids [3,19,60]. Many times, they act far away from the point of secretion. Thus, activated platelets or leukocytes from inflammation sites can release exosomes and other micro-vesicles, known as extracellular vesicles (EV) (Fig. 12) [1]. These EVs were found in blood plasma, tears, synovial fluid, and in bronchoalveolar lavage fluid [64] and contain a large variety of phospholipid substrates, which are hydrolyzed by sPLA2s. In this context, it was demonstrated that the phospholipids of the EVs present in the synovial fluid of rheumatoid arthritis patients can be hydrolyzed by sPLA2, releasing arachidonic acid (AA), subsequently processed by COX enzyme, to generate inflammatory eicosanoids that potentiate local inflammation. Arachidonic acid (AA) can be also processed by platelet-type 12-lipoxygenase, which leads to 12-hydoxyecosatetraenoic acid (HETE) and inflammation [1,19]. Group IIA sPLA2s act on mitochondrial membranes of extracellular mitochondria, releasing AA, lyso-cardiolipin and the encapsulated mitochondrial DNA, and promoting inflammation. These extracellular mitochondria are generated (released) from activated platelets, neutrophils, lymphocytes, mast cells, but also from hepatocytes, cells from damaged tissues or organs. Mitochondrial damage-associated molecular patterns (DAMP) were identified in pathological conditions such as rheumatoid arthritis, systemic lupus erythematosus, and burn injuries (Fig. 12) [1,64].

sPLA2-IIA plays a central role in antimicrobial host defense, where it degrades bacterial membrane by hydrolyzing phosphatidylethanolamine and phosphatidylglycerol—both very abundant in the bacterial membrane (Fig. 12) [19]. This is the reason for which sPLA2-IIA is highly expressed in human tears [65]. In sepsis, peritonitis, and bacterial infection, the blood plasma concentration of sPLA2-IIA can increase up to 500-fold as compared to healthy persons [66].

The release of AA by sPLA2 and subsequent inflammation can occur via heparan sulfate proteoglycans (HSPG)-dependent and independent pathways. HSPG are glycoproteins found on the cell surface, involved in cell adhesion, signaling, and endocytosis [17]. In HSPG-dependent pathways, sPLA2-IIA binds the cell surface anionic HSPG and gets internalized into an intracellular vesicular compartment of the activated cells through a caveolae-dependent endocytic pathway [67]. Especially glycosylphosphatidylinositol (GPI) anchored HSPG were shown to act as an adaptor for sPLA2-IIA and facilitate

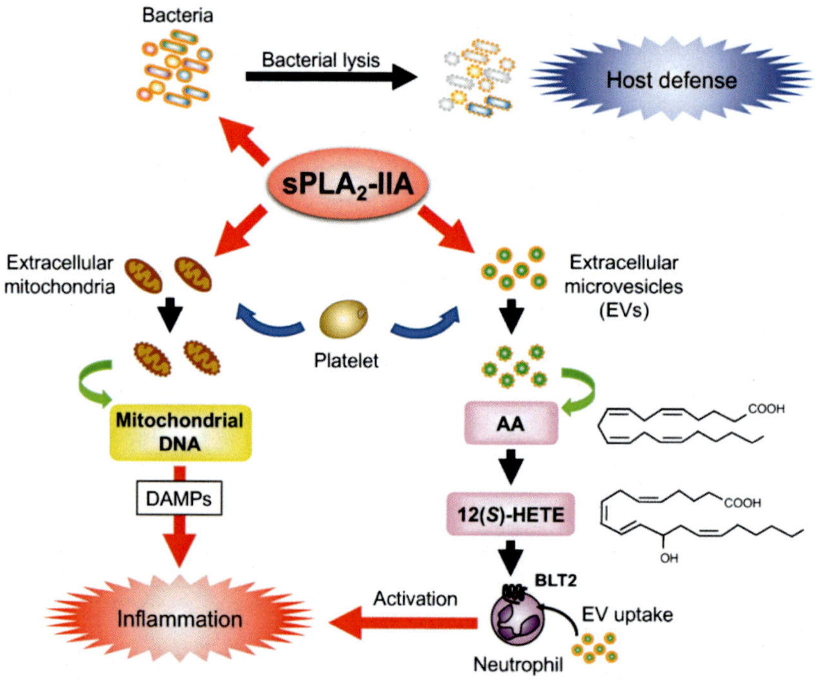

FIG. 12 Physiologic roles of sPLA2-IIA against foreign self-assembled lipid structures (bacterial membranes) in the blood, and in inflammation, where it hydrolyzes arachidonic acid from extracellular microvesicles, subsequently processed in several inflammatory pathways. It can also act on extracellular mitochondria (released from activated neutrophils, platelets, mast cells, lymphocytes etc.), cleaving cardiolipin and liberating mitochondrial DNA that potentiates inflammation [1,19].

the trafficking of sPLA2-IIA into the subcellular compartments of the cells, where they can release AA [67]. Other sPLA2 isoforms such as IB, IIC, and IIE bind weakly the heparanoids and are not trafficked and cannot release AA from the cellular compartments. Furthermore, in HSPG-independent pathways, sPLA2-X, can liberate AA from the outer leaflet of plasmalemma, rich in PCs [68].

Secreted PLA2s IIA, V, and X play also an important role in atherosclerosis. In a study performed on young adults (age 24–39 years) with cardiovascular disease, an increased level of sPLA2-IIA was detected in blood plasma, which may contribute to atherogenicity [69]. Transgenic mice expressing sPLA2 IIA displayed increased atherosclerotic lesions compared to the nontransgenic littermate's mice, irrespective of their feeding habit. In the same study, it was shown that sPLA2-IIA transgenic mice exhibited a lower level of high-density lipoprotein (HDL) and a higher level of low-density lipoprotein (LDL), pointing toward interaction of sPLA2 with those lipoproteins as a possible mechanisms for atherosclerosis [70].

sPLA2s are also overexpressed in various tumors and cancer cells [8], playing a pro-tumorigenic role in the colon [70], breast [71,72], lung [72], ovaries [73,74], and prostate cancer [75]. Interestingly, they have an antitumor effect in gastric cancer [76]. In terms of representatives with high expression and enzymatic activity related to cancer progression, the sPLA2 IIA and sPLA2 X were the most frequently identified and associated with these malignancies. The arachidonic acid production by sPLA2 IIA can contribute to cellular proliferation and tumor progression by feeding the prostaglandins and leukotrienes biosynthetic pathways. Lysophosphatidic acid (LPA) can stimulate cell proliferation, tumor invasion, and differentiation [76]. In this context, sPLA2-IIA was found to be overexpressed in human breast cancer tissues and was identified in nipple aspirate fluid (NAF) [77], and in blood serum from breast cancer patients [78]. It was shown that sPLA2 IIA concentration in NAF sample from breast cancer patients was positively correlated with the tumor stage. Higher expression of sPLA2 IIA were detected in stages III and IV, as compared to stage I and II cancers [77]. The sPLA2s exhibited a differential expression in various breast cancer cell lines, suggesting a different involvement in various breast cancer subtypes [78]. The serum level of PLA2 IIA in healthy individuals within the age group 18 to 63 was determined to be 2.2 ± 0.1 ng/mL, with a range of concentration of 1.4–4.2 ng/mL. This level was significantly increased in the serum of various cancer patients in a similar age group: lung (9.5 ± 4.0 ng/mL), breast (9.1 ± 3.8 ng/mL), esophagus (8.0 ± 2.8 ng/mL), colon (13.1 ± 4.4 ng/mL) liver ((31.4 ± 12.5), pancreas (9.7 ± 3.2 ng/mL), and bile duct (9.2 ± 3.3 ng/mL) [1,78].

2.2 cPLA2 physiologic and pathologic roles

As presented above, cPLA2s are expressed ubiquitously in human body, where they hydrolyze phospholipids and release arachidonic acid and lysophospholipids and play a major role in inflammation. The arachidonic acid released inside the cells is processed further via either cyclooxygenase (COX) or lipoxygenase (LOX) pathways to generate prostaglandins, thromboxanes, and leukotrienes as presented above [1]. Lysophospholipids produced by cPLA2-mediated phospholipid hydrolysis can act as second messengers for GPCR signaling [3].

In the epithelial cells, cPLA2s are localized on the surface of endoplasmic reticulum and Golgi complex, where they regulate the formation of tubules, Golgi structures and mediate intra-Golgi transport. Thus, upon binding on the Golgi structure, activated cPLA2 (vide supra) releases FA and lysophospholipids. LPLs affect the positive curvature of the Golgi membrane, leading to the formation of tubules, which cause Golgi stacking and promote intra-Golgi transport. cPLA2s were also shown to regulates the transport of proteins involved in formation of tight junctions and adherent junctions from the Golgi complex to cell-cell contacts, in confluent endothelial cells [1,19]. In liver cells and adipose tissue cells cPLA2s were shown to regulate lipid storage.

cPLA2s are also expressed in CNS cells, where they are involved in neuronal homeostasis and can play a critical role in spinal cord injury, postischemic brain injuries and neurodegenerative diseases [79]. In Alzheimer's disease (AD) an increased accumulation of β-amyloid (Aβ) oligomers, plaques and neurofibrillary tangles in the brain interferes with the synaptic transmission in the neurons and induces inflammatory pathways [80,81]. It was shown that Aβ peptides open up calcium ion channels, increase [Ca^{2+}] and induce MAPK pathways, activating cPLA2 as presented above (Fig. 8) [31,82]. The active cPLA2 enzyme contributes to the progression of the AD pathway mainly through the release of arachidonic acid, transformed by COX and LOX enzymes into leukotrienes, prostaglandins, lipoxins, and hydroperoxyacids, which lead to widespread inflammation [82]. Aβ also activates NFκB, a transcriptional factor present in neurons, astrocytes, and microglia through an oxidative pathway involving reactive oxygen species (ROS) in AD patients [83]. NFκB upregulates cPLA2 expression in microglia and increases NADPH oxidase activity, production of superoxide, PGE_2, and Nitric Oxide (NO) in microglia. Concomitantly, cPLA2 also reduces microglia's ability to phagocyte Aβ in the AD brain, which further potentiates the AD pathological evolution [83]. It was shown that silencing cPLA2 in a mouse model of Alzheimer's disease was able to reduce Aβ-induced neurotoxicity and improve the cognitive function of the animals [84].

Another key regulator of the brain architecture is Cdk5, a cyclin-dependent kinase involved in brain development, learning, and memory. Cdk5 was found to be deregulated in neurodegenerative diseases such as AD, Parkinson's and Huntington's diseases [85]. Cdk5/p25 (p25 an activator of Cdk5 protein) was shown to increase neurotoxicity in the AD brain. Cdk5/p25 upregulates cPLA2 expression in microglial cells and facilitates the release of lysophosphatidylcholine, which causes astrogliosis, increased production of cytokines and subsequent generation of neuroinflammation and neurodegeneration [86].

cPLA2 plays a key role in inflammation and cancer-related diseases through the release of arachidonic acids, which contributes to carcinogenesis, as mentioned above. cPLA2 is regulated by mitogen-activated protein kinase (MAPK) pathway, a tyrosine kinase receptor. Transforming growth factor-β (TGF-β) initiates cPLA2 phosphorylation through MAPK pathway via p38 and ERK, which leads to increased production of AA-derived eicosanoids in cancer cells, where they activate the GPCR-mediated signaling pathways and promote transcription, translation, and release of cytokines such as TNF-α and IL6s. These cytokines induce NF-κB-dependent gene expression, which further upregulates the expression of cPLA2s and sPLA2s. Thus, the AA release cycle continues, and tumor cells continue to proliferate [87–89]. Therefore, it is not surprising that cPLA2 has a pro-tumorigenic role in most cancers. In colon cancer, downregulating cPLA2 was shown to significantly suppress the formation of tumors in the intestine [8]. Cytosolic phospholipase A2α (cPLA2α) was found to be overexpressed in human breast cancer (BC) tissues, where it plays a central role in cancer metastasis. cPLA2 significantly increased the migration and invasion of breast cancer cell lines MDA-MB-231 and T47D. Silencing the cPLA2 gene reduced tumorigenesis and cancer metastasis [90].

cPLA2 has been proposed as a potential biomarker in breast cancer. In a study using fluorescent-tagged arachidonic acid substrates, the cPLA2 activity was measured by fluorescent microscopy in different breast cancer cells expressing these enzymes. The highest cPLA2 level was found in the basal-like triple-negative cells, followed by the HER2 cells and the luminal-like cells [90]. cPLA2 expression is strongly correlated with mTOR (mammalian target of rapamycin) signaling pathway, a pathway that is upregulated in many cancers and plays a crucial role in the development of tumors, including breast cancers. The tumorigenesis mechanism is not fully elucidated. It was proposed that the increased release of arachidonic acid activates mTOR, which subsequently leads to angiogenesis and breast cancer tumorigenesis [42].

2.3 iPLA2 physiologic and pathologic roles

iPLA2s play a key role in membrane homeostasis and phospholipid remodeling during cell cycle progression, with representatives involved in cell proliferation, apoptosis, bone formation, sperm development, cardiolipin acetylation, glucose-induced insulin secretion, and monocyte recruitment [1]. Importantly, in addition to phospholipase A2 activity, iPLA2s also exhibit lysophospholipase and transacylase activity, as presented above [41–43]. iPLA2β and iPLAγ are membrane-associated, localized on the mitochondria and peroxisome, where they help maintaining the integrity of the membrane of these organelles, preventing them from rupturing. iPLA2β are involved in diabetes, being involved in beta-cell apoptosis, superoxide production, and glucose-induced RhoA/Rho-kinase activation. iPLA2δ (Group VI C), a transmembrane protein that is expressed in neurons, is the largest protein among the iPLA2s. It is localized on the endoplasmic reticulum and Golgi apparatus of the neurons, displays esterase activity and important regulatory properties. Group VI D, VI E, and VI F iPLA2s are smaller iPLA2s localized in adipocytes, which exhibit triacylglycerol hydrolase, and transacylase activity [1,3].

iPLA2β (PLA2 VIA) is known to regulate pathophysiology during myocardial ischemic conditions through an unknown mechanism [91]. These ischemic conditions (myocardial ischemia, MI) occur when coronary arteries supplying the heart muscle are partially or completely obstructed due to atherosclerosis, thrombosis, and other causes. They generate oxidative stress, increased intracellular [Ca^{2+}], and inflammation, disturbing the normal functioning of the ER, especially its protein folding ability, and causing accumulation of misfolded proteins and ER stress [91]. In this context, it was shown that iPLA2β is overexpressed during myocardial ischemia, and that the resulting ER stress causes the translocation of iPLA2β to the ER, triggering apoptosis. Consequently, it was found that inhibition of iPLA2β improves ER stress and decreases cell death [91]. On the other hand, knocking down iPLA2γ—another iPLA2 associated with membranes of mitochondria and peroxisomes—was shown to decrease production of oxidized fatty acid in the ischemic cells and reduce the infarct size in a mouse model of MI [91]. Furthermore, a study performed in a rabbit animal model has shown that catalytically active iPLA2 was localized on the outer face of the inner mitochondrial membrane, and its increased activity during MI cold trigger apoptosis. The pro-apoptotic mechanism involved the accelerated catabolism of mitochondrial phospholipids by activated iPLA2, causing loss of mitochondrial Ca^{2+} homeostasis and permeabilization of mitochondrial membranes with release of cytochrome c. Inhibition of the iPLA2 with a specific iPLA2- inhibitor (bromoenol lactone (BEL) vide infra), proved to be cardioprotective and to significantly decrease the infarct size [92].

iPLA2s were found to be overexpressed in many cancers. Thus, ovarian cell lines SKOV-3 and Dov-13 overexpresses iPLA2s and the activity of these PLA2 was associated with enhanced cellular activity and growth [93]. It is believed that activation of iPLA2s occur via caspase-3 via a proteolytic process, which results in ankyrin-repeat domain truncated iPLA2 with an enhanced activity that will generate arachidonic acid and lysophosphatidylcholine, as presented above. Both products were shown to potentiate migratory activity of cancer cells and inhibit apoptosis [94]. Knocking down iPLA2β expression via RNA interference halted the cancer cell proliferation and tumorigenicity in a nude mice model of ovarian cancer [94]. Along the same lines, several studies examined the influence of iPLA2s in breast cancer and attempted to correlate their expression with cigarette smoking. Cigarette smoking was shown to upregulate the expression of COX-2 and PGE2 synthase, which facilitate the metabolic conversion of arachidonic acid into prostaglandin E2. Smoking was also proved to increase the accumulation of PAF in breast cancer tissues. The PGE2 and PAF, when present simultaneously, were found to potentiate cancer progression. Inhibition of iPLA2 in the breast cancer tissues via an iPLA2 specific inhibitor rescinded the metastasis of the cancer cells [95]. Another study showed that nicotine induces the expression of MMP-9 and iPLA2β in breast cancer, which can facilitate cell proliferation and lung metastasis. Inhibition of iPLA2 activity by specific inhibitor bromoenol lactone (BEL) reduced nicotine-induced tumor growth [96]. iPLA2 knock down in mice was shown to inhibit the metastasis of breast cancer to the lung [97].

2.4 Lipoprotein-associated PLA2 (Lp-PLA2) physiologic and pathologic roles

Lp-PLA2s were used as biomarkers in cardiovascular diseases (CVDs) since their expression increases the risk of CVD and strokes two folds [98]. It was found that Lp-PLA2s mediates vascular inflammation through the control of lipid metabolism in the blood, especially relevant in atherosclerosis. Thus, Lp-PLA2s hydrolyze the oxidized phospholipids on the LDL particles within the arterial intima and produce inflammatory mediators such as LPC and oxidized FA, which can initiate a cascade of events leading to atherosclerotic plaque formation. This includes upregulation of adhesion molecules and cytokines, differentiation of monocytes into macrophages, and recruitment of macrophages into the arterial intimal space, which engulf sthe oxidized LDL [98,99]. Rabbit animal models are particularly valuable for elucidation of these biological effects of Lp-PLA2s since the enzymes were found in the macrophages of both human and rabbit atherosclerotic lesions. Thus, a high Lp-PLA2 activity was detected in the atherosclerotic aortas of hyperlipidemic rabbits when compared to normal aortas of control (normal) animals. Inhibiting Lp-PLA2 with specific inhibitor darapladib (vide infra) was found to significantly reduce Lp-PLA2 activity in the lesion and to attenuate the progression of atherosclerotic plaque in the coronary arteries of experimental animals [99].

On the other hand, Lp-PLA2 VII-B, overexpressed in CHO-K1 cells, were shown to be able to suppress oxidative stress-induced cell death by hydrolyzing oxidized phospholipids and acting as antioxidant phospholipases [100]. They are also overexpressed in neurons after an ischemic event in the CNS. Group VIII Lp-PLA2s are expressed intracellularly (in the Golgi complex) in brain cells and are believed to regulate the functional organization of the Golgi apparatus with significant impact in brain development [3].

2.5 Lysosomal PLA2 (LPLA2) physiologic and pathologic roles

As mentioned above, LPLA2s are lysosomal and endosomal enzymes ubiquitously expressed throughout the body [56]. In alveolar macrophages they control the lung surfactant catabolism [57]. LPLA2-deficient mice developed phospholipidosis (a lysosomal storage disorder characterized by increased accumulation of intracellular phospholipids) in the lung [101]. They are responsible for lipid turnover in the ocular region. Thus, a higher activity of LPLA2 was evidenced in glaucoma patient's eyes as compared with normal subjects. It is believed that LPLA2 metabolizes the undigested substances filtered on the trabecular meshwork (TM) and maintains the balance of intraocular pressure (IOP) between the aqueous humor and TM, although the precise mechanism has not been clarified [102]. LPLA2 also displayed increased activity in the anterior chamber of the eye in an eye inflammation animal model in rats (endotoxin-induced uveitis). However, LPLA2 activity was not significantly increased in the rat serum or in the cerebrospinal fluid [102]. Cataract formation is a common complication in patients with uveitis, which evolves from the initial chronic intraocular inflammation and from the use of corticosteroids to treat this inflammation [103]. In his context, it was shown that samples collected from patients with a history of uveitis showed significantly higher activity of LPLA2 than from senile patients with cataracts [104].

2.6 Adipose specific PLA2 (Ad-PLA2) physiologic and pathologic roles

Human AdPLA2s are ubiquitously expressed throughout the body, being predominantly found in the adipose tissue and skeletal muscles. In the adipose tissue, they are believed to play an important role in obesity. Thus, AdPLA2 knockout mice exhibit resistance to obesity induced by a high-fat diet, with a high lipolytic activity evidenced in the white adipose tissue as compared to the wild-type animals [9]. AdPLA2 has been also categorized as a class II tumor suppressor, evidencing that the AdPLA2 gene (also known as

H-REV-107) is downregulated in ovarian carcinoma cell lines. It was shown that the expression of this gene in rats resulted in growth inhibition of RAS transformed cells, both in vitro and in vivo [58]. At the level of skeletal muscles, a recent study has shown that AdPLA2 is downregulated in a peripheral artery disease condition, which may impeded normal walking due to its abnormal effects on skeletal muscle lipid metabolism [59].

3 Classes of PLA2 inhibitors and their design

3.1 Inhibitors of secreted PLA2 (sPLA2)

3.1.1 Phospholipid analogs

One of the earliest attempts to develop PLA2 inhibitors was through synthetic phospholipid analogs. Two distinct features characterized these phospholipid analogs: the sn-2 ester moiety was replaced by a hydrolysis-resistant amide group and the acyloxy group at sn-1 position was replaced with a more hydrophobic, hydrolytically stable, alkyloxy group [105]. Thus, 1-stearyl-2-stearoylaminodeoxy phosphatidylcholine **1** was reported as one of the first PLA2 inhibitors in the 1980s (Chart 1). This modified phospholipid analog was shown to bind more tightly than DPPC in the active site of the protein and inhibited sPLA2 IA (purified from Cobra venom) with a K_I of approximately 40 μM [105]. Following this initial study, a series of fluorinated phospholipid analogs with tight binding with the sPLA2 were reported. These substrate analogs contained fluoroketone-, 1,2 diketone-, and difluoromethyleneketone moieties, which were designed to form tetrahedral intermediates upon complexation with the enzyme (transition-state analog inhibitors) [106]. Within this category, compound **2**, a short-chain phospholipid with a difluoroketone at sn-2 position, proved to be a simple competitive inhibitor (K_I= 50 μM) of sPLA2 IA that binds 300-fold tighter than the analog substrate 1,2-dibutyryl-glycero-3-phosphatidylethanolamine (K_m = 14 mM) [106]. A subsequent study from the same group [107] showed that single-chain fluoroketone analogs such as **3** inhibit better sPLA2 than double-chained fluoroketone compounds **4**.

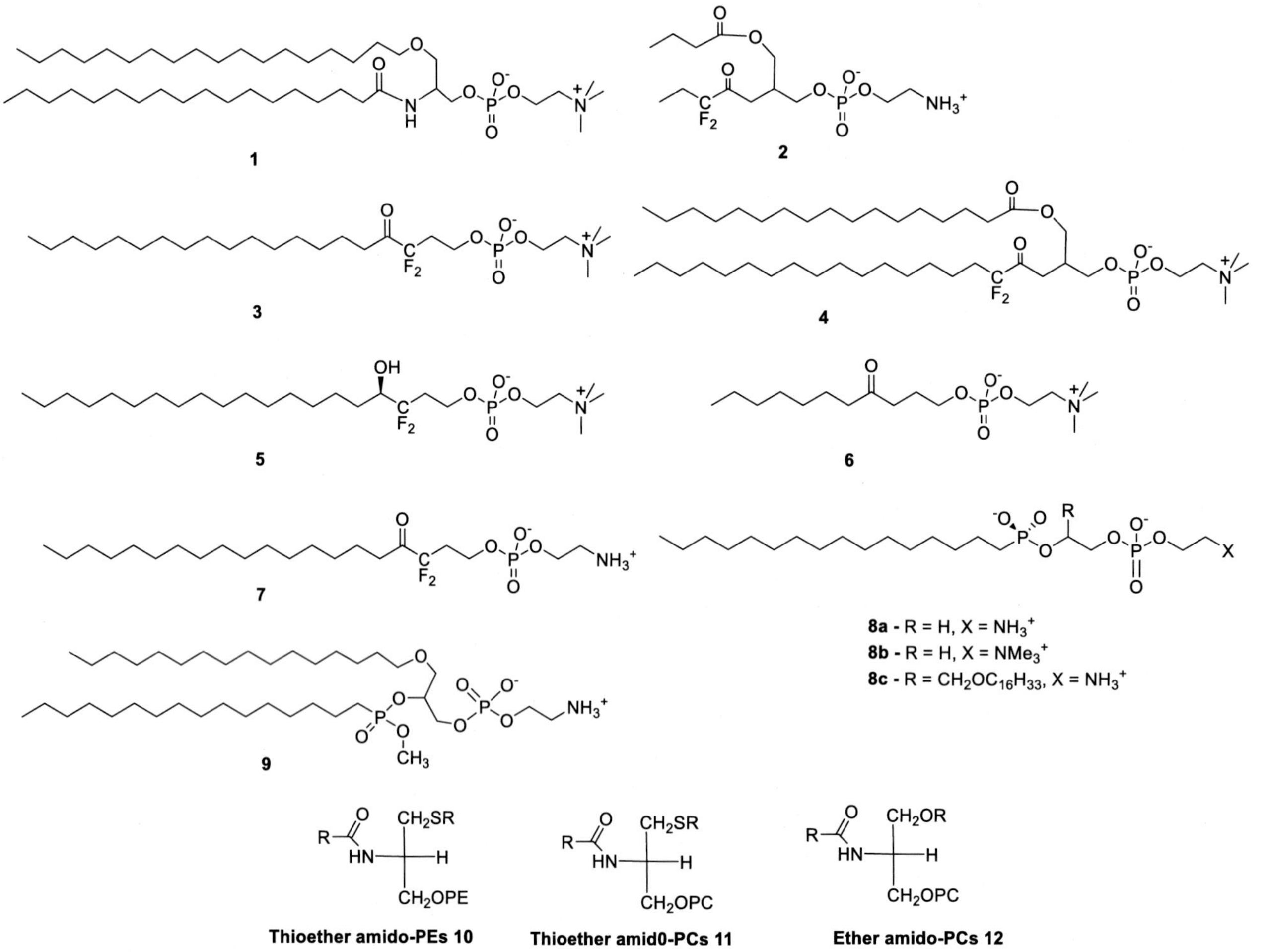

CHART 1 Phospholipid analogs proved efficient as sPLA2 inhibitors.

With the help of ^{19}F NMR analysis of the analogs, the study found that single-chain analogs formed partially hydrated fluoroketone, whereas double chain phospholipid analogs had less than 0.1% hydrated fluoroketone. Thus, it was suggested that hydration of the fluoroketone was essential for inhibition. The study also demonstrated the importance of fluorinated ketone groups in the inhibition of sPLA2. Reducing the difluorinated ketone group of **3** to difluorinated alcohol **5** dropped the potency of the inhibitor. Furthermore, replacing the fluorinated ketone group with a regular, nonfluorinated, ketone group in **6** also resulted in a much less potent sPLA2 inhibitor (Chart 1) [107]. Another interesting finding of the same study was that replacing the choline group with ethanolamine **7** increased the potency of the inhibitor 10 folds. It was concluded that the choline moiety has an activation effect on the enzyme and thus compounds incorporating choline can act as both activators and inhibitors, whereas the ethanolamine does not activate the enzyme, so compounds incorporating this moiety only have the inhibition properties [107].

Following these initial studies, a series of phosphonate containing phospholipid analogs **8** were synthesized [108]. The working hypothesis was that these phosphorus-containing species would mimic high-energy tetrahedral intermediates formed in the active site of PLA2s, which would not be hydrolyzed by the enzymes. Compound **8a–c** was synthesized and demonstrated effective inhibition of PLA2 in mixed micelle phospholipid hydrolysis assays. Compound **8c** ($IC_{50} = 5\,\mu M$) proved to be more effective than the amide and fluorinated analogs synthesized previously. Compound **8a** and **8b** displayed an IC_{50} of 0.75 mM and 2.3 mM, which is significantly higher than the IC_{50} for **8c**, suggesting that the upper alkyl chain in **8c** has an important role in binding with the active site of the enzyme. A congener of **8c**, the methylated phosphonate **9** (Chart 1), had an IC_{50} of 1.25 mM, thus showing the importance of phosphonate anion specific interaction with the enzyme's active site, where it coordinates the calcium ion present in the catalytic site [108].

One of the most efficient representatives from the phospholipid analogs class were the thioether amido-phosphatidylethanolamines **10**, introduced in the early 1990s, which displayed an IC_{50} as low as $0.45\,\mu M$ [109]. These were potent reversible inhibitors of PLA2. The rationale behind the design of these inhibitors was that the sn-2 amide analogs had a higher binding energy than natural phospholipids in the enzyme's active site, since the amide group forms hydrogen bonds with the water molecule also present in the active site of the enzyme. These inhibitors contained polar head group phosphoethanolamine (PE), which was even more potent than its corresponding phosphocholine-containing analogs (PC). The thioether amido-phosphatidylcholines **11** were also potent inhibitors of sPLA2, with IC_{50} as low as $2\,\mu M$. Notably, the congeners ether amido PCs **12** were less potent, having the lowest IC_{50} of $38\,\mu M$ (Chart 1). These inhibition data suggest that replacing the sn-1 function group with a more hydrophobic group increases the affinity of the analogs for the enzyme [109]. In a subsequent study from the same group, congeners of **11** having different numbers of methylene groups in the sn-2 acyl chain analogs were synthesized and evaluated as sPLA2 inhibitors. The inhibition study demonstrated that the sn-2 acyl chain is essential for binding the inhibitors with the enzyme and that the catalytic site of the enzyme could interact with the first 10 methylene groups of the chain. Moreover, each methylene group added to the sn-2 acyl chain of inhibitors, from C5 to C10, increased the binding energy by about 665 cal/mol. Further elongation of the alkyl chain did not increase the binding energy of the inhibitors [110].

3.1.2 *Dicarboxylic acids*

One of the first specific inhibitors of human sPLA2 was a dicarboxylic acid analog reported by Bristol-Myers Squibb (BMS). Thus, BMS-181162 **13** was designed as a novel compound for the treatment of inflammation. This compound specifically inhibits the 14 kDa human platelet PLA2 IIA (HP-sPLA2) and demonstrates efficacy in blocking the phorbol ester-induced skin inflammation in mice [111].

BMS-181162 **13** was shown to inhibit HP-sPLA2 in a dose-dependent manner with an IC_{50} of $40\,\mu M$, while the IC_{50} against other phospholipases such as PLA1, PLC, and sphingomyelinase was higher than $100\,\mu M$. Interestingly, phospholipase D on the other hand was stimulated by BMS-181162. This compound effectively blocked the edema in the ear of the mice with an ED_{50} of 160 μg/ear, whereas the 5-lipoxygenase inhibitor lonapalene displayed an ED_{50} of 800 μg/ear in the same study [111]. Furthermore, biological studies done with BMS-181162 **13** showed inhibition efficacy against human synovial fluid PLA2 (Group IIA) with an IC_{50} of $8\,\mu M$. It was shown that **13** can completely block the release of arachidonic acid from isolated human polymorphonuclear lymphocytes (PMN) at a concentration of $100\,\mu M$, displaying an IC_{50} value of $\sim 10\,\mu M$. BMS-181162 **13** also inhibited the biosynthesis of PAF from human PMNs with IC_{50} of $5\,\mu M$. These data suggested a potential antiinflammatory activity of the compound [112]. The mechanism of inhibition of BMS-181162 **13** was studied in another study, where they showed that the compound inhibits the enzyme reversibly in a dose-dependent manner. Importantly, its activity against the enzyme peaked when the enzyme was located at the lipid/water interface, partially associated with the lipid bilayer. The compound did not interact strongly with the lipid bilayers as did other phospholipid-based inhibitors; thus, it was easier for the enzyme to "extract" the inhibitor from the bilayer. BM—181162 **13** represents an important step forward in designing potent inhibitors of human sPLA2 pointing toward the advantages conferred by a compound inhibiting the enzyme at the interface that has a weak interaction with the phospholipid bilayer [113]. Several other derivatives of diacids were synthesized and characterized as potent inhibitors of human sPLA2 (nonpancreatic, class IIA) with strong activity in phorbol-ester-induced mice ear edema assay, summarized in Table 1 [114,115]. Compound **19** was selected for further development based on its good HP-sPLA2 inhibition, as well as excellent antiinflammatory activity in vivo [114].

3.1.3 Amides

As resented above, inhibitors of both pancreatic sPLA2 and HSF PLA2 were generated from congeners of natural phospholipids containing primary amides at the sn-2 position (vide supra) [3]. Nonphospholipid amide inhibitors were developed based on these first-generation phospholipid analogs. The synthetic effort was driven by the fact that phospholipid analogs were potent inhibitors but had poor metabolic stability. Thus, a series of compounds were

TABLE 1 Diacid-type PLA2 inhibitors and their biochemical and biological activity.

General structure	Compound	% Inhibition of HP-sPLA2 at 100/10 μM	IC_{50} (μM)	ED_{50} (μg/ear per dose)
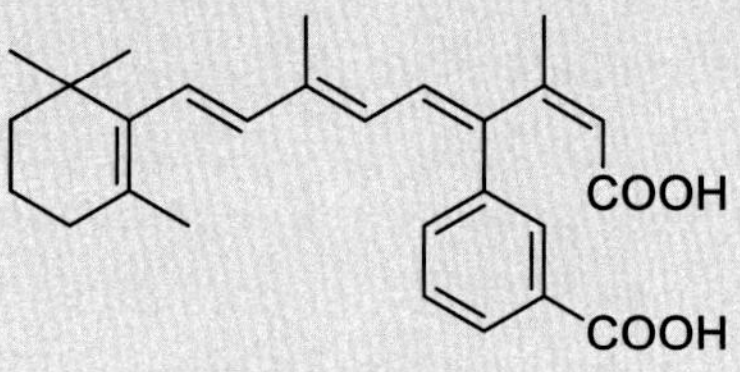 **BMS-181162**	**13**	77/12	45	180
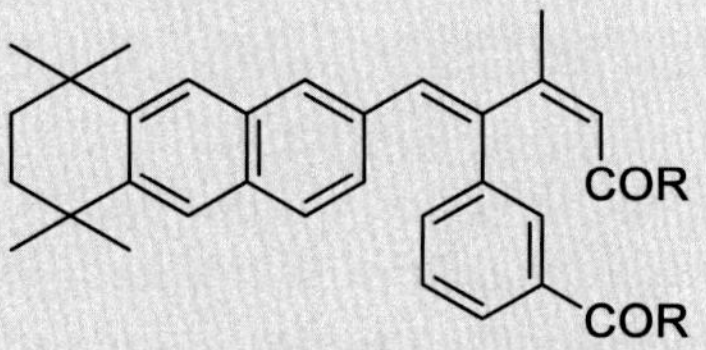	**14** R=OH	98/48	17	9, 37
	15 $R=NH_2$	98/68	4	186
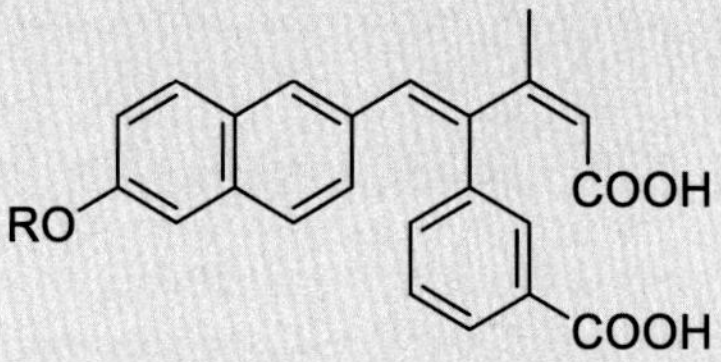	**16** $R=n\text{-}C_{10}H_{21}\text{-}$	100/45	14	31
	17 $R=Ph(CH_2)_3\text{-}$	100/44	17	30, 90
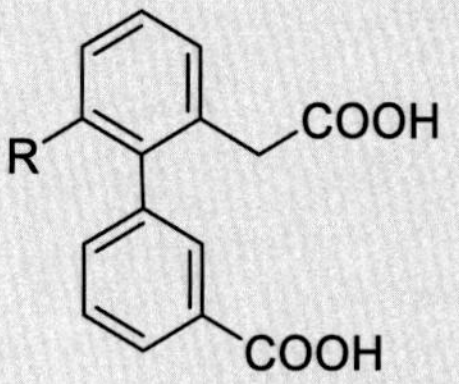	**18** R= 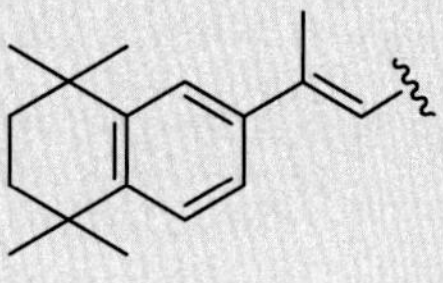	96/46	10	116
	19 R= 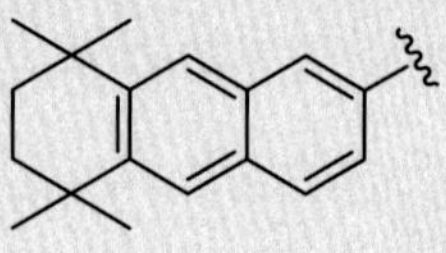	93/38	18	79
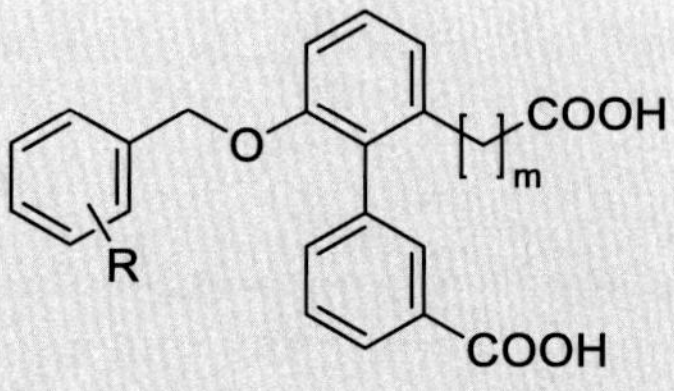	**20** m=0, R=3,4-C $(CH_3)_2CH_2CH_2(CH_3)_2C\text{-}$	54/22	>100	147
	21 m=1, R=3,4- bis($nC_5H_{11}O$-)	96/35	8	32
	22 m=1, R=4-$nC_{10}H_{21}O$-	99/77	4	73

synthesized replacing the phosphocholine moiety with a carboxylate group. The lead compound of the series was **23a**, which showed similar potency as the earlier analogs. By reducing the carbon chain length, and further optimizing the backbone, the potency increased 10 folds (**23b**, **23c**). The most potent of this series was compound **24** also known as FPL67047XX, which showed activity against both porcine pancreatic and human synovial fluid PLA2, with an IC_{50} of 21 ± 4 nM (Chart 2 and Table 2) [116].

A high-resolution X-ray crystallographic study was conducted to elucidate the binding of the inhibitor **FPL67047XX 24** in the active site of the human (nonpancreatic) sPLA2 (Fig. 13) [48].

The phenyl rings of the inhibitor make extensive interactions with the nonpolar amino acids in the hydrophobic channel of the enzyme. Both sn-1 and sn-2 chains of the compound were accommodated within the hydrophobic channel of the enzyme without any significant conformational change of the enzyme. The carboxylate group forms a coordinating bond with the calcium ion, similar to the phosphate group of natural phospholipid substrate. The chelating oxygen atoms of the Asp-49, Gly-30, Gly-32 and the amide carbonyl groups are in the same equatorial plane. The amide oxygen atom is also coordinating the calcium ion and makes a hydrogen bond with Gly-30. All three water molecules in the hydrophobic channel of the uninhibited enzyme were excluded when **FPL67047XX** bound the enzyme [48].

Following the success of these synthetic analogs, a series of chiral D-tyrosine derivatives **25** containing an amide group was synthesized and evaluated as for inhibition of human sPLA2 [117]. The compounds proved very potent sPLA2 inhibitors, with an IC50 as low as 29 nM (**25a**) and 19 nM (**25d**), as shown in Table 3. The crystal structure of inhibitor **25a** in the active site of the enzyme showed chelation with the Ca^{2+} ion through the carboxylate and amide oxygen atoms. The amide group of the inhibitor also formed a hydrogen bond with His48 of the enzyme. The phenyl groups stabilized the binding of the inhibitor in the active site of sPLA2 by extensive hydrophobic interactions with the hydrophobic channel of the enzyme [117].

The above-mentioned sPLA2 inhibitors **25** were evaluated for their antiinflammatory activity in rats (arthritis, ischemia-reperfusion injury models), and a strong correlation was found between their sPLA2 inhibition potency and antiinflammatory effects. Inhibitor **25b** (R= 2-picolyl-) when administered orally in rats showed promising antiinflammatory results, reducing their rear paw swelling to 0.18 ± 0.03 cm compared to 0.80 ± 0.06 cm in the control group [117].

Another potent nonphospholipid amide analogs introduced as PLA2 inhibitors were the 2-oxoamides. The first representatives of this class of inhibitors showed potency against all three major PLA2s (sPLA2, cPLA2, and iPLA2). Among them, the two 2-oxoamides esters **26** and **27** demonstrated significant inhibition against G-V sPLA2, G-IV

TABLE 2 Second-generation amide sPLA2 inhibitors and their biochemical and biological activity.

	IC_{50} (μM)	
Compound	**Porcine sPLA2 (IB)**	**Human sPLA2 (IIA)**
23a	0.19	Not tested
23b	0.023	Not tested
23c	0.016	1.85
FPL67047XX **24**	0.015	0.21

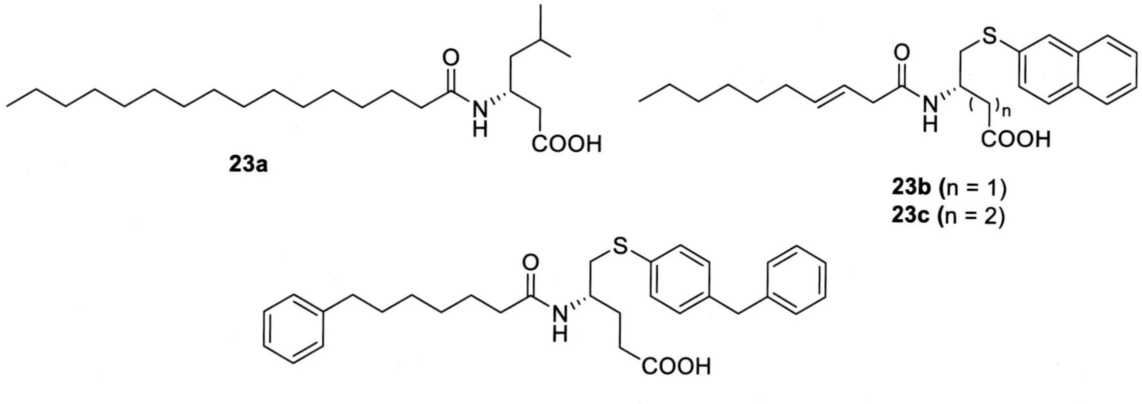

FPL67047XX **20**

CHART 2 Second-generation amide sPLA2 inhibitors.

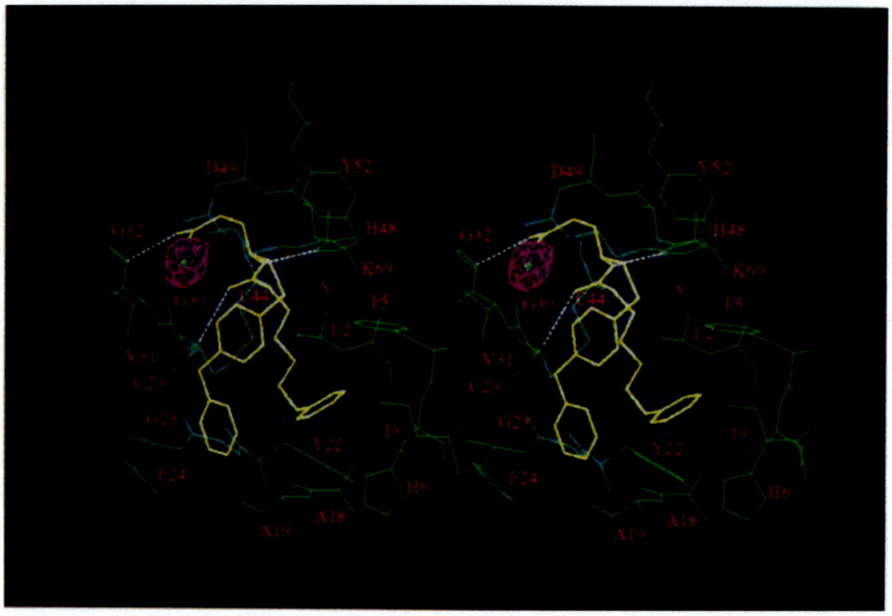

FIG. 13 X-ray crystal structure of inhibitor **FPL67047XX 20** bound into the human sPLA2 IIA active site (PDB 1kvo) [48]. The electron density map of the inhibitor is shown in blue, and the catalytic calcium ion is shown in purple. The inhibitor is highlighted in yellow, and the protein atoms are shown in green. The labeled amino acid residues are in contact with the inhibitor ($d < 4.3$ Å). The sulfur atom of the inhibitor is labeled as "S". The hydrogen bonds are shown as dotted lines. The resolution of the electron density map of the $sPLA_2$/inhibitor complex was 2.0 Å. *(Reprinted (adapted) with permission from Cha S-S, Lee D, Adams J, Kurdyla JT, Jones CS, Marshall LA, et al. High-resolution X-ray crystallography reveals precise binding interactions between human nonpancreatic secreted phospholipase A2 and a highly potent inhibitor (FPL67047XX). J Med Chem 1996;39(20):3878–81. Copyright 1996 American Chemical Society.)*

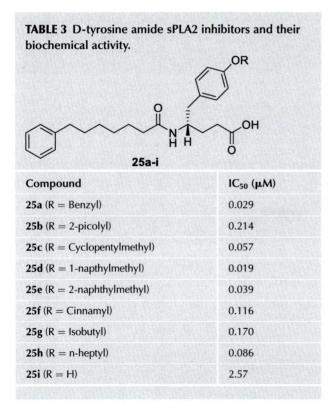

TABLE 3 D-tyrosine amide sPLA2 inhibitors and their biochemical activity.

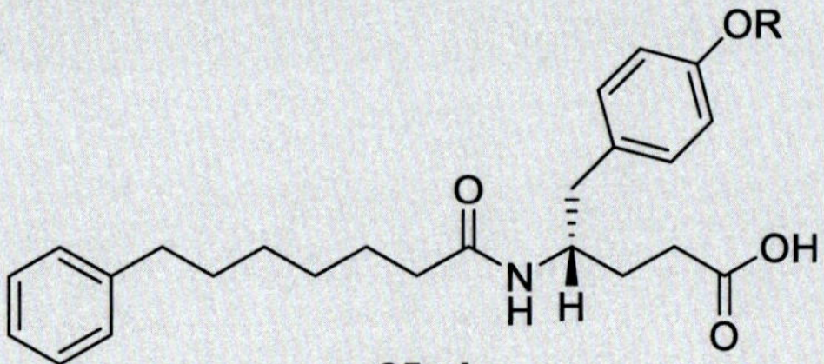

25a-i

Compound	IC_{50} (μM)
25a (R = Benzyl)	0.029
25b (R = 2-picolyl)	0.214
25c (R = Cyclopentylmethyl)	0.057
25d (R = 1-napthylmethyl)	0.019
25e (R = 2-naphthylmethyl)	0.039
25f (R = Cinnamyl)	0.116
25g (R = Isobutyl)	0.170
25h (R = n-heptyl)	0.086
25i (R = H)	2.57

cPLA2, and G-VI iPLA2 (Table 4) [118]. When the long aliphatic chain of those compounds was replaced with a shorter chain, the activity dropped significantly against cPLA2 and iPLA2. For example, compound **28** demonstrated that an amide-based analog with a short linear side chain (7-phenylheptanoyl) could be a potent and selective inhibitor of Group-V sPLA2 without affecting cPLA2 and iPLA2 activity significantly [118].

Computer-aided drug design is an important tool for the rational designing of PLA2 inhibitors. In a molecular docking study, improved GIIA sPLA2 inhibitors were developed by retaining the 2-oxoamide functional group while changing other structural features [119]. The design of the new 2-oxoamides derivatives was based on the lead compound **29** (Fig. 14). The docking results showed that derivatives containing α-amino acids with an R absolute configuration did not have any favorable interaction with the enzyme, whereas derivatives containing nonpolar α-amino acids with and S absolute configuration had many favorable binding modes, with good binding scores (Fig. 14) [119]. Based on these conclusions yielded by the docking studies, an S-valine derivative of 2-oxoamides inhibitors **GK241 30** was synthesized. Biological evaluation of **GK241 30** showed high potency and selectivity toward GIIA sPLA2. The IC_{50} of the compound **GK241 30** against GIIA sPLA2 was 143 nM, being thus twice more

TABLE 4 2-Oxoamide amide sPLA2 inhibitors and their biochemical activity.

Compound	Structure	% Inhibition (at 0.091 mole fraction, in mixed micelle assay)		
		sPLA2 (V)	cPLA2 (IVA)	iPLA2 (VI A)
26	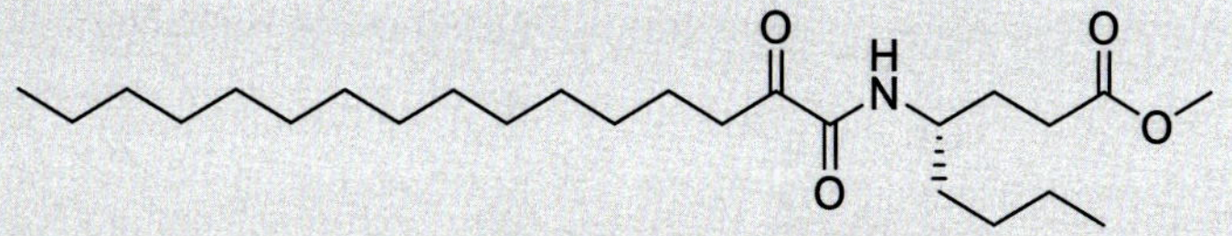	86±6	87±13	78±9
27	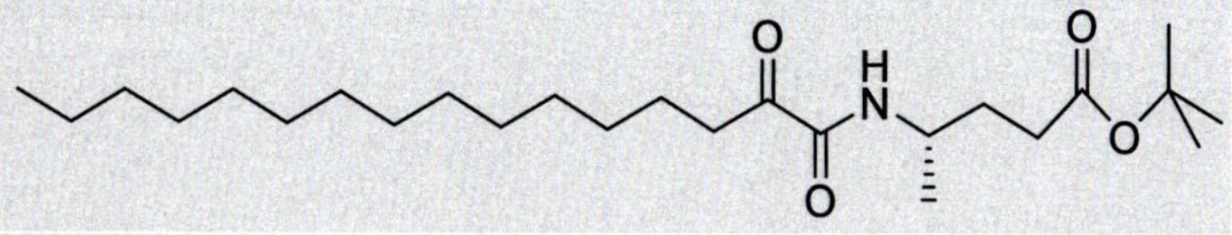	87±4	85±5	72±2
28	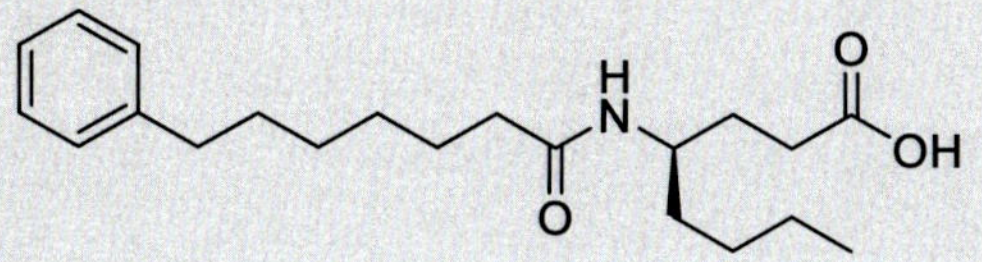	95	ND	ND

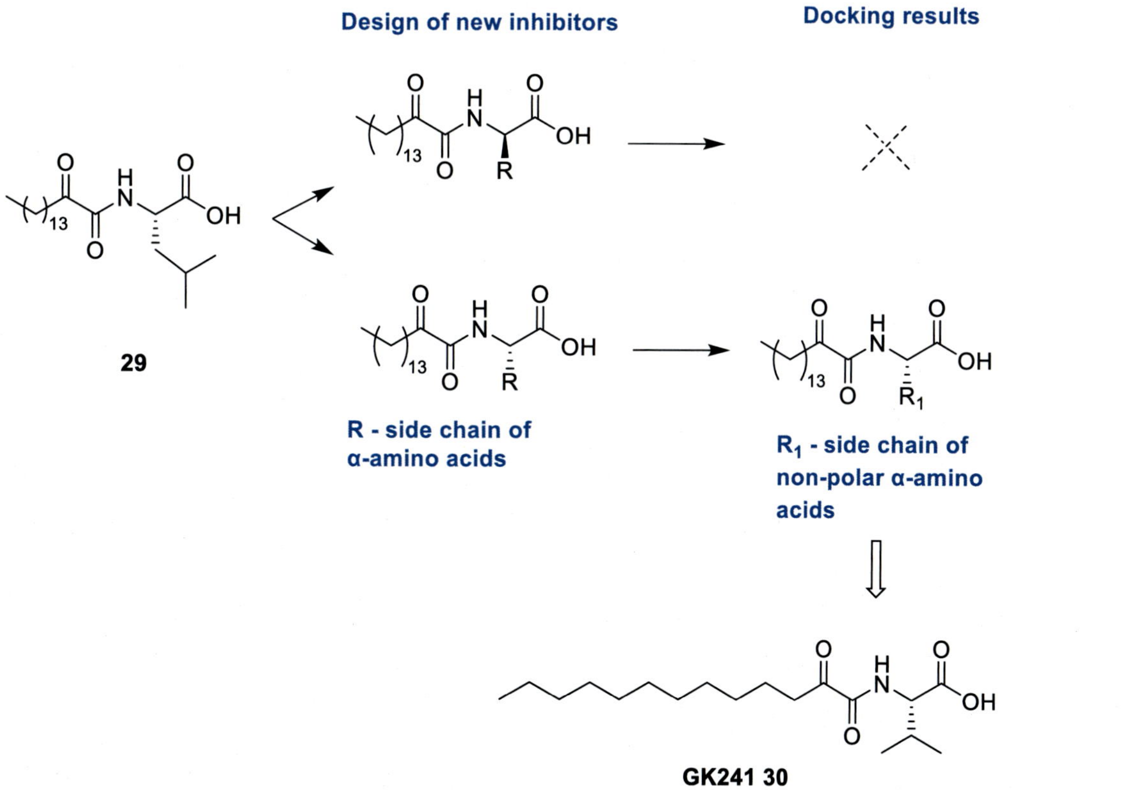

FIG. 14 The workflow for the rational development of sPLA2 inhibitors starting from the lead compound **29** with the help of molecular, to yield potent inhibitor GK241 **30** [119].

potent than the lead compound **29** and about 8 times more selective for GIIA sPLA2 vs. GV sPLA2 ($IC_{50}=1200\,nM$) [119].

3.1.4 Indoles

This new class of sPLA2inhibitors was first introduced an in 1995 by a team of researchers from Lilly Research Laboratories [120], which used the crystal structure of a complex between human nonpancreatic secretory phospholipase A2 (hnps-PLA2) with a lead compound derived from screening to develop a series of potent indole inhibitor derivatives. Additional crystal structures of hnps-PLA2 complexed with increasingly potent synthetic indole inhibitors were used to understand the binding of this class of inhibitors in the active site of the enzyme and to provide insights for further development. The application of structure-based drug design has generated inhibitor **LY311727 31**, three orders of magnitude more potent than the initial lead compound and also displaying 1,500-fold selectivity against hnps-PLA2 as than against porcine pancreatic s-PLA2 (Chart 3) [120].

The following year scientists from Eli Lilly synthesized a large library of indole sPLA2 inhibitors that yielded Varespladib **32** (**LY315920**), with the highest potency against Group-IIA sPLA2 known at that time ($IC_{50}=9\,nM$) (Chart 3) [121–123]. Furthermore, a pharmacologic study done with **LY315920 32** showed that the compound was 40 times less active against pancreatic sPLA2 IB and

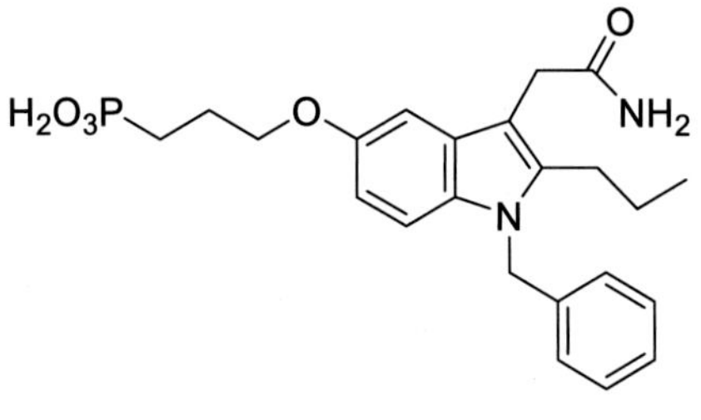

LY311727 31

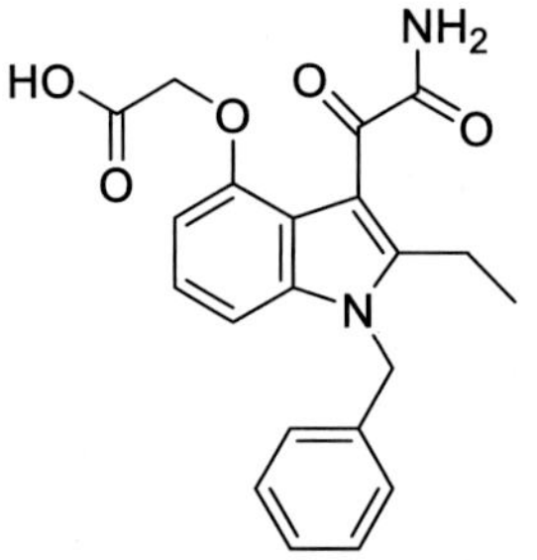

LY315920 32
(Varespladib)

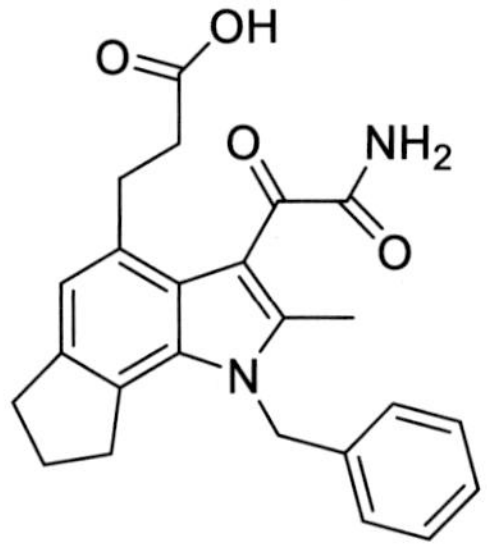

33

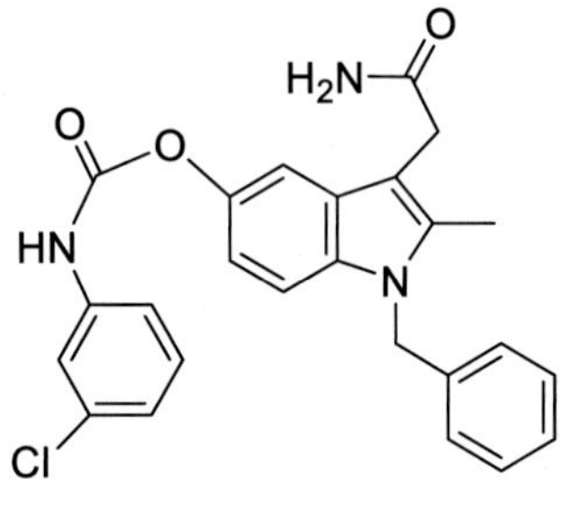

34

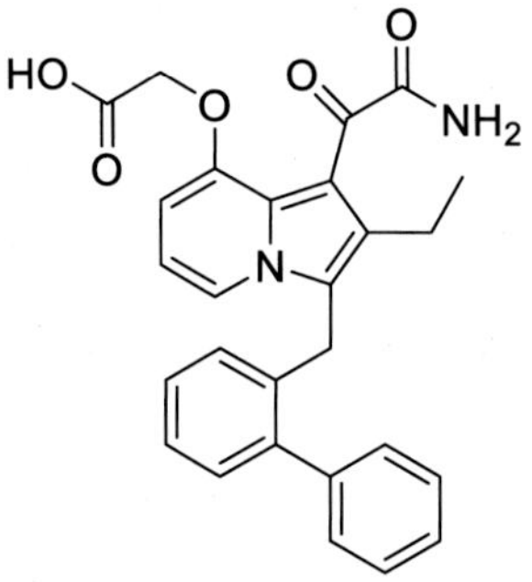

Indoxam 35

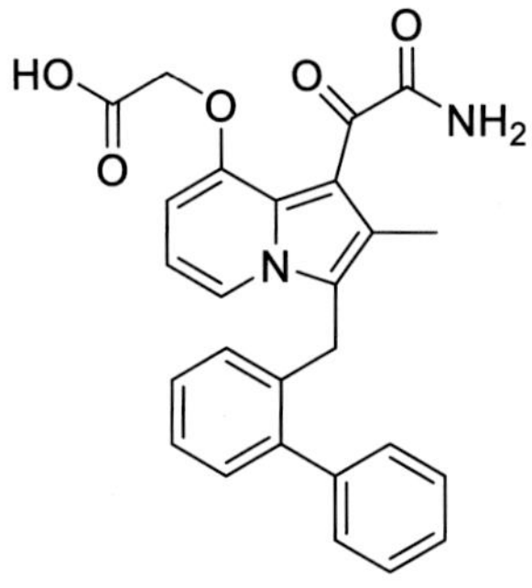

Methyl-Indoxam 36

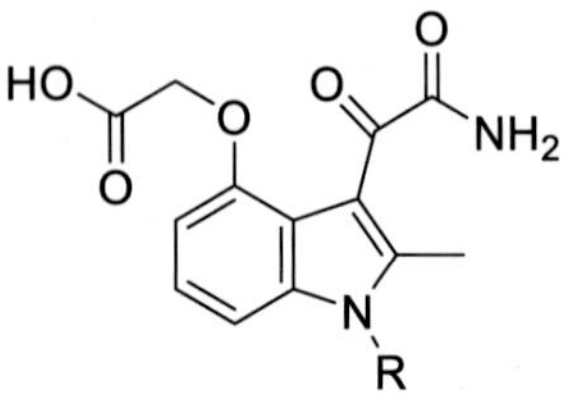

37 R = $C_6H_5COCH_2$
38 R = (S)-(+)-$CH_3CH_2CH(CH_3)CH_2$

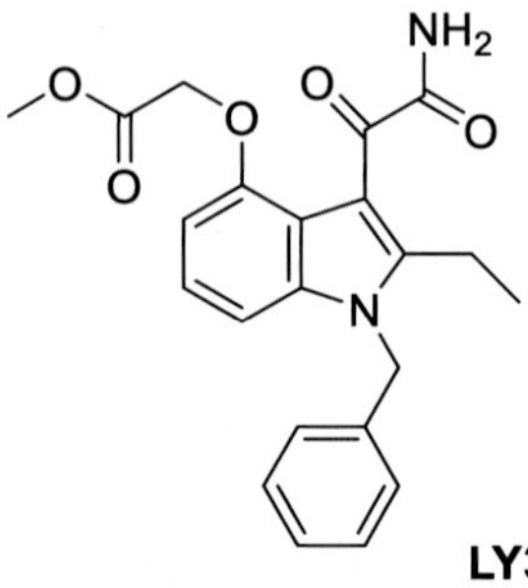

LY333013 39
(Methyl varespladib)

CHART 3 Representatives of the indole class of sPLA2 inhibitors.

inactive against cPLA2. **LY315920 32** also demonstrated promising antiinflammatory activity in two separate studies [124]. First, the IV administration of **LY315920 32** inhibited sPLA2-induced thromboxane A2 production in mice, with an ED_{50} of 16.1 mg/kg. Secondly, oral administration of **LY315920 32** in transgenic mice expressing human sPLA2 inhibited serum sPLA2 activity in a dose-dependent manner [124].

A few other indole derivatives were synthesized and evaluated as sPLA2 inhibitors including carbocyclic[g]-indole derivatives and indole-5-phenylcarbamate derivatives. These compounds were designed using molecular modeling and docking studies, which yielded cyclopent[g] indole **33** with an IC_{50} of 10 nM when assayed in DOC/PC assay system. The phenylcarbamate inhibitor **34** displayed an IC_{50} of 1.81 μM in a fluorescence-based assay of synovial-fluid PLA2 (Chart 3) [125,126].

Indoxam **35** and methyl-indoxam **36** are selective inhibitors of sPLA2 belonging to the class of 1-oxamoylindolizine, which were synthesized and characterized in the 1990s [127]. Indoxam **35** showed high potency in inhibiting Group IB and IIA sPLA2s. An in-vivo study performed in mice showed that indoxam **35** was able to suppress endotoxic shock and systemic inflammatory pathways involving proinflammatory mediators such as TNF-α, IL-1β, IL-6 in experimental animals. Indoxam blocked the production of the proinflammatory cytokines during the shock by blocking the PLA2 binding with the PLA2-receptor protein (K_I = 30 nM for sPLA2) [127]. Me-indoxam **36** is a derivative of the indoxam **35**. The X-ray structure of human-Group X PLA2 complexed with methyl-indoxam **36** was used in combination with molecular modeling to design a library of potent methyl-indoxam analog inhibitors [128]. A methyl-indoxam analog containing 4-(2-oxy-ethanoic acid) was used to synthesize a series of 48 compounds, which were evaluated for inhibition potency against several sPLA2. Several members of this group of compounds showed potent inhibition (IC_{50} < 50 nM) against human Group IIA, Group IIE, and Group V, intermediate inhibition against human Group IB, Group X (IC_{50} 0.5–5 μM) and poor inhibition against Group IID, Group IIF, and Group XIIA (IC_{50} > 5 μM). It was also observed that the indole analogs containing non-benzyl substituents at N_1-position generally demonstrated poor inhibition against sPLA2s (both human and murine). Interestingly, the combined use of methyl-indoxam **36** with its analogs showed more selectivity toward particular subtypes of PLA2 than a single compound. Thus, the combo methyl-indoxam **36**/analog **37** was fivefold more active against Group IIA than Group-V sPLA2. Also, the combination of Me-Indoxam **36** with analog **38** was 10-fold more selective toward Group-V as compared with Group IIA PLA2 (Chart 3) [128]. Furthermore, it was found that methyl-indoxam represses diet-induced obesity and glucose intolerance in mice. The sPLA2 IB secreted in the intestine plays a role in the phospholipid digestion from the diet and enhances the absorption of the lipid into the body. As Me-Indoxam **36** is a potent inhibitor of the sPLA2 IB it decreased lipid digestion in the intestine and thus helped in suppressing diet-induced obesity in the mice [129].

Varespladib **32** is one of the most studied indole-type inhibitors of sPLA2, with an IC_{50} of 9 nM (vide supra). Lilly scientists also synthesized the prodrug methyl varespladib **LY333013 39** (Chart 3), which is rapidly converted to varespladib **32** after being injected into the body [123,130]. Subsequently, a randomized double-blinded clinical study was conducted in 251 human patients with **LY333013 39** (administered orally) in patients with rheumatoid arthritis (RA), as an adjuvant drug for disease-modifying antirheumatoid drugs (DMARD), for 12 weeks long. The study concluded that **LY333013 39** was well tolerated but was not effective as an adjunct to DMARDs in the treatment of active RA [131]. The comparatively lower distribution of varespladib **LY315920 32** in the synovial fluid of RA inadequately inhibited sPLA2, considering that the concentration of sPLA2 in the synovial fluid is much higher than in the plasma [131].

Another study showed that varespladib **32** inhibits sPLA2 activity in the bronchoalveolar lavage sample collected after neonatal lung injury. sPLA2 was found to be overexpressed in acute respiratory distress syndrome and associated with surfactant catabolism and lung inflammation. In an ex-vivo study, varespladib **32** was able to lower the sPLA2 activity in the bronchoalveolar lavage fluid, displaying an IC_{50} of 87 μM [132]. A Phase-II clinical trial was conducted with methyl varespladib **39** to treat vascular inflammation in patients with Acute Coronary Syndrome (ACS). The study concluded that varespladib and its prodrug could not reduce the risk of recurrent cardiovascular events in patients with ACS. Importantly, it increased the risk of myocardial infarction significantly [133].

Varespladib **32** has also received a lot of attention as a potent broad-spectrum inhibitor of snake venom PLA2. Both varespladib **32** and its methyl prodrug **39** demonstrated inhibition against snake venom PLA2 at nanomolar and picomolar concentrations against 28 medically important snake venoms [134]. Varespladib **32** was able to suppress venom-induced sPLA2 activity in the rats when injected with 100% lethal doses of *Micrurus fulvius* venom. Even though snake venom comprises of different family of toxins with independent effects on morbidity and mortality on the host, varespladib **32** was able to protect against harmful effects of hemolysis, hemorrhage, and other tissue destruction. The study suggested that varespladib **32** can be used as a first-line treatment for different snakebites, either alone, or in combination with metalloprotease, serine protease, and other inhibitors [134]. In a more recent murine study, it was showed that both an IV bolus dose of varespladib **32** and an oral dose of methyl varespladib **39** were effective in reducing toxic effects when administered immediately and also at various time intervals after envenoming the mice subcutaneously with lethal doses of snake venoms. Varespladib **32** successfully suppressed the severe paralytic response in mice when injected with venoms of *O. scutellatus*, *B. multicinctus*, and *C.d. terrificus*. Both varespladib **32** and methyl varespladib **39** were effective in either delaying or stopping the neurotoxic events induced by the snake venoms [135].

3.1.5 Biphenyl derivatives

A series of biphenyl derivatives were synthesized and evaluated as potent sPLA2 inhibitors by Astrazeneca scientists [136]. The lead compound **40**, which inhibited sPLA2-IIA with IC_{50} 24 μM in an enzymatic assay and human plasma sPLA2 activity with an $IC_{u,50}=0.9$ μM, was optimized into inhibitor **41** (Fig. 15). The R-enantiomer of **41** proved to be most active against sPLA2-IIA, -V, X with IC_{50} of 10, 40, and 400 nM, respectively. Compound **(R)-41** (AZD2716) was also active against plasma sPLA2 with $IC_{u,50}=0.1$ nM. Additional in-vitro studies demonstrated that **(R)-41** suppressed sPLA2 activity in HepG2 cells ($IC_{50}<14$ nM), and in atherosclerotic plaque homogenates obtained from carotid endarterectomy of coronary artery patients (IC_{50} 56 ± 10 nM). The compound was also administered in mice, rats, dogs, and cynomolgus monkeys to evaluate its PK profile. When an oral dose of 30 mg of the compound was given to the cynomolgus monkeys, it showed a concentration-dependent inhibition of plasma sPLA2 activity, with $IC_{u,80}=13\pm3$ nM. Based on the suitable PK properties, in-vivo efficacy, and minimal drug–drug interactions, the (R)-**41** (AZD2716) was selected as the candidate drug for further clinical studies in the treatment of coronary artery disease [136].

3.2 Inhibitors of cytosolic PLA2 (cPLA2)

3.2.1 Fatty acid trifluoromethyl ketones

In the quest for efficient cPLA2 inhibitors, an arachidonic acid (AA) analog was synthesized by replacing the terminal -COOH group of AA with a trifluoromethyl ketone group ($COCF_3$), thus forming $AACOCF_3$ **42** (Chart 4). The trifluoroketone analog showed tight binding and delayed inhibition of cPLA2, in a phospholipid/TritonX100 mixed micelles systems [137]. A few other analogs of AA generated by replacing the COOH of arachidonic acid with $COCH_3$, $CH(OH)CF_3$, CHO, and $CONH_2$ were also tested for inhibition study using the same assay, but did not generate a significant inhibition of cPLA2. ^{19}F NMR analysis showed that analogs such as $AACOCF_2CF_3$ and $AACOCF_2Cl$ were poorly hydrated in the mixed micelle system compared to $AACOCF_3$ [137]. Furthermore, the ^{19}F-NMR and ^{13}C-NMR study on the cPLA2/$AACOCF_3$ **42** complex provided some insights into the inhibitor-bound active site of the enzyme. Thus, one molecule of inhibitor binds tightly to the active site, whereas other (excess) molecules remain loosely associated and nonspecifically bind to the hydrophobic region of the protein [138].

The cPLA2 specific inhibitors such as $AACOCF_3$ **42** were used to study the role of cPLA2 in the pathogenesis of inflammatory diseases such as multiple sclerosis (MS) and experimental autoimmune encephalomyelitis (EAE). When the experimental animals were treated with $AACOCF_3$ **42**, one could observe reduced inflammation, a significantly reduced axonal damage and prevention of further remission of the disease in the animals, indicating an important role of cPLA2 in the pathogenesis of EAE and MS [139]. $AACOCF_3$ **42** was also used in determining the role of cPLA2 in the progression of Alzheimer's disease in a mouse model. Using lipidomic analysis, the study found increased activity of cPLA2 in the brain tissue of mice expressing human amyloid precursor protein (hAPP). Amyloid-β (Aβ) peptide caused a dose-dependent increase in the activation of cPLA2 in the neural cells. Inhibiting the cPLA2 with specific inhibitors such as $AACOCF_3$ 42, was shown to reduce Aβ induced neurotoxicity in the mice brain. Protecting the hAPP mice by inhibiting cPLA2 or by genetically knocking out the cPLA2 improved learning capacity, behavioral alterations and premature mortality in the experimental animals [85].

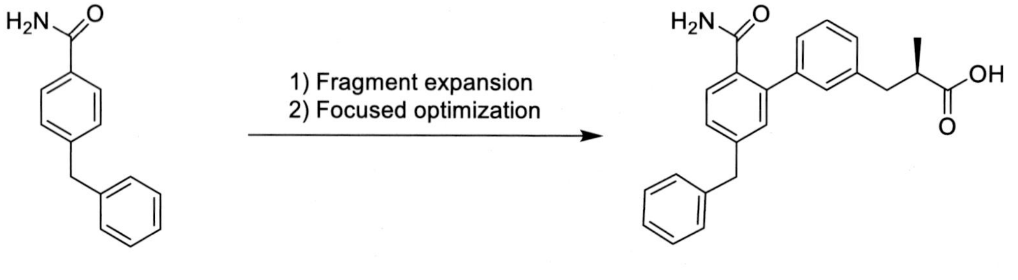

Lead compound 40 **(R)-41** (AZD2716)

FIG. 15 Optimization of lead compound **40** by Astrazeneca scientists has yielded AZD2716 **41**, the main representative of the biphenyl class of sPLA2 inhibitors.

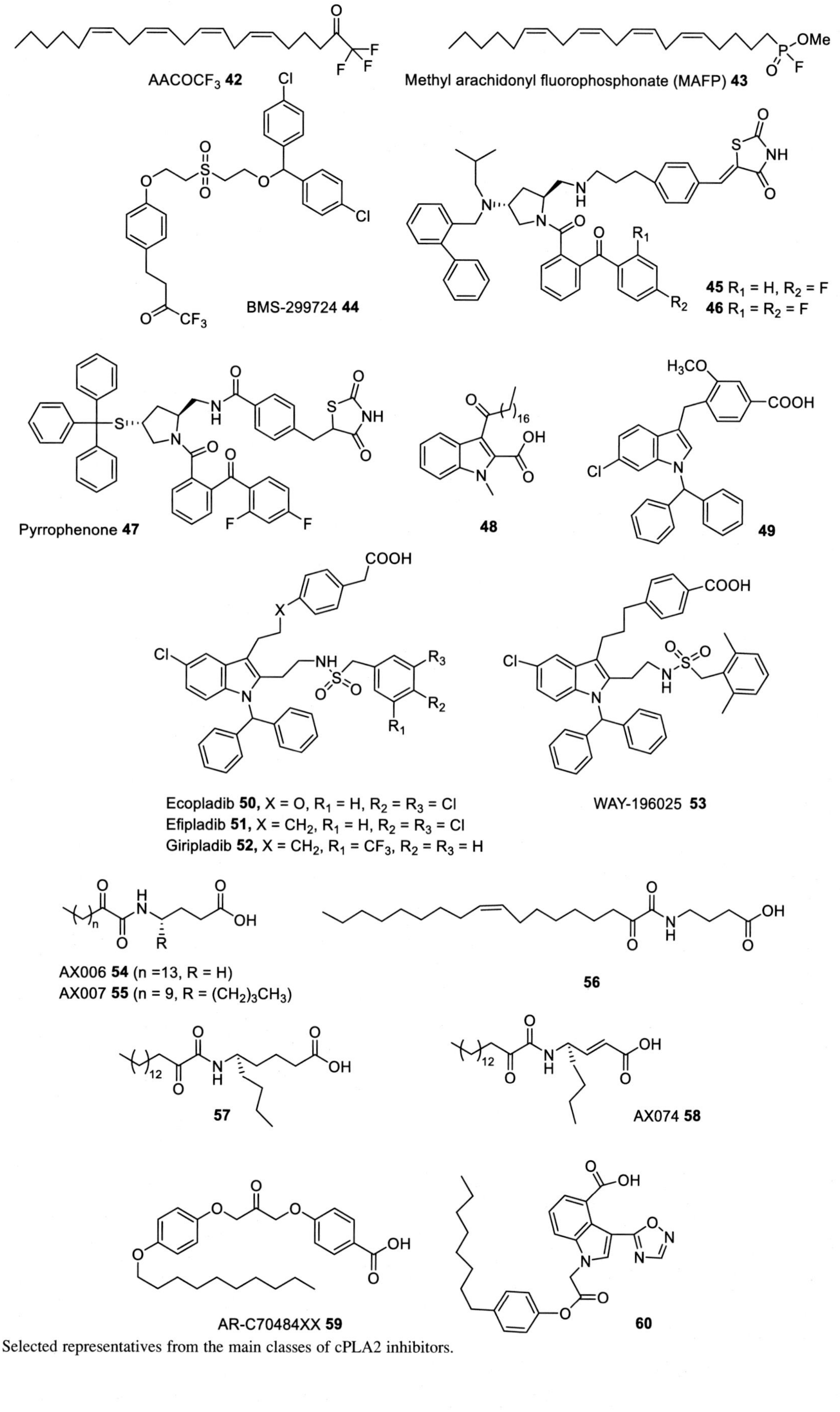

CHART 4 Selected representatives from the main classes of cPLA2 inhibitors.

3.2.2 Methyl Arachidonyl Fluorophosphonate

A series of methyl fluorophosphonates and alkyl trifluoromethyl ketones were synthesized and tested as inhibitors against cPLA2 and iPLA2 in three different assay systems – phospholipid vesicle assay, mixed micelles assay, and natural membrane-based assay. Out of all compounds tested, the analog methyl arachidonyl fluorophosphonate (MAFP) **43** (Chart 4) showed the highest potency against cPLA2, in a time-dependent irreversible inhibition of cPLA2 [140]. In another study, MAFP **43** was used to suppress the effect of cPLA2 and iPLA2 in spinal cord hyperalgesia and determine the sPLA2 extent of the effect on spinal cord inflammation [141].

3.2.3 Trifluoromethyl ketones

Bristol-Myers Squibb (BMS) reported a series of trifluoromethyl ketones as potent inhibitors of cPLA2 for treating inflammatory diseases. The most potent inhibitor was BMS-229724 **44** (Chart 4), which showed tight binding against cPLA2 with an IC_{50} of 2.8 μM. It showed selective inhibition against cPLA2 when assessed against sPLA2, PLC, and PLD in cells. This potent inhibitor demonstrated antiinflammatory activity against phorbol ester-induced chronic inflammation in mice ear models when applied topically (5% w/v). The compound successfully reduced prostaglandin and leukotriene levels in the skin of the mice [142].

3.2.4 Pyrrolidines

Scientists from Shionogi laboratories found two lead compounds belonging to the pyrrolidines class as potent inhibitors of cPLA2. Using the crystal structure of cPLA2 with these leads and subsequent SAR studies, they synthesized a library of compounds, which were tested toward both cPLA2 inhibition (enzymatic assay) and toward arachidonic acid release from A23187-stimulated THP-1 cell lines. Two of the most potent compounds **45** and **46** (Chart 4) showed excellent inhibitory potency against cPLA2 with $IC_{50} \sim 2$ nM in the enzymatic assay and inhibited the release of arachidonic acid from the THP-1 cell lines with $IC_{50} < 30$ nM, which was almost 4000 times more potent than $AACOCF_3$ **42** [143]. Furthermore, compound **45** showed selective potency against cPLA2α, being much less potent against cPLA2γ, iPLA2, sPLA2- Group-IIA, X, and V [144]. Pyrrophenone **47** was one of the more recently synthesized cPLA2 inhibitors from the pyrrolidine group, with IC_{50} 4.2 nM and also showing strong inhibition for arachidonic acid release, prostaglandin E2, leukotriene B4, and thromboxane B2 production in the human blood [145]. Unlike $AACOCF_3$ **42**, pyrrophenone **47** demonstrated reversible inhibition of cPLA2 without any slow binding effect. It also inhibited the esterase and lysophospholipase activity of cPLA2α [146]. In more recent studies these selective cPLA2 inhibitors were used to investigate the role of cPLA2α in various physiological pathways. For instance, pyrrophenone **47** was used to investigate the role of cPLA2α in relation to cell death in lung fibroblasts. The study revealed the off-target effect of pyrrophenone toward blocking the calcium release from the ER in a cPLA2α knockout mouse fibroblast cells [147]. Another study showed that inhibiting the cPLA2α increases chemosensitivity in cervical carcinoma. Interestingly, the efficacy of a chemotherapeutic drug such as paclitaxel increased significantly in resistant cell lines when given in combination with pyrrophenone **47** [148].

3.2.5 Indoles

A series of indole compounds were synthesized and evaluated as inhibitors of cPLA2 by Lehr in 1996 [149]. Out of this series of compounds, the most potent one was 1-methyl-3-octadecanoylindole-2-carboxylic acid **48** (Chart 4) with IC_{50} of 8 μM [150]. Another indole series was developed based on the earlier hit compounds and evaluated in a coumarin substrate assay and MC-9 cell-based assay. Compound **49** (Chart 4) is one of the examples from that series, which has an inhibition potency of IC_{50} of 0.8 μM in both the assay. The molecule was docked into the crystal structure of cPLA2α and the study indicated that the benzoate group was immersed into the active site pocket, in the area where the phosphate group normally binds. The benzhydryl group was found to interact with the α-helical lid of the cPLA2, which covers the active site of the enzyme [149].

The three most studied inhibitors from the indole group are ecopladib **50**, efipladib **51**, and giripladib **52** (Chart 4) [151–153]. Ecopladib **50** showed sub-micromolar inhibition potency against cPLA2α in the GLU micelle assay and also in the whole blood assay in rats. It also showed efficacy in rat carrageenan air pouch and carrageenan-induced paw edema animal models when given orally [151]. Giripladib **52** was found effective in treating inflammation for osteoarthritis. It significantly reduced clinical and histological symptoms of joint inflammation in two different RA mouse models [152]. Giripladib **52** was advanced up to phase II clinical trial for osteoarthritis but halted due to gastroenterological side effects. A recent study showed that co-administering giripladib with chemotherapeutics such as doxorubicin, cisplatin or tamoxifen to breast cancer stem cells effectively reduced tumor progression [154]. A detailed SAR study was done on efipladib **51** and its derivative **WAY-190625 53** (Chart 4), with the two compounds showing selective potency against cPLA2α in multiple studies, including isolated enzymatic assays, cell-based assays, and whole blood assays in humans and rats. The compounds also showed efficacy in reducing inflammation in RA rat models [153].

3.2.6 2-Oxoamides

This class of cPLA2 inhibitors were introduced in 2002 by Kokotos, Dennis, and collaborators [155], who designed them so that the 2-oxoamide functional group would replace the ester bond of the phospholipid substrate. The additional free carboxylic acid present in their structure mimicked the negative phosphate group of the natural phospholipid substrate. When these inhibitors were evaluated for their potency against cPLA2, it was found that attaching a long alkyl chain onto the 2-oxoacyl moiety will further increase the potency of the compounds. Inhibitors such as AX006 **54** and AX007 **55** acted in a fast and reversible manner in the in-vitro studies and were able to block the production of arachidonic acid and prostaglandin E2 in the cells. Both compounds were able to inhibit cPLA2 without the presence of PIP_2 in a mixed micelle assay study, with a very low mole fraction of inhibitor [156]. In another study aimed at developing selective inhibitors that would discern between cPLA2 and iPLA2, compounds AX006 **54** and AX007 **55** were found to be selective inhibitors of cPLA2. The selectivity for cPLA2 was conferred by the negative charged moiety of the carboxylate present in their structure [157]. Subsequent SAR studies on the 2-oxoamide series of inhibitors provided insights into the development of cPLA2 selective inhibitors. The most potent inhibitors were derivatives of δ- and γ- amino acid-2-oxoamide analogs with a short nonpolar aliphatic side chain [158]. Importantly, as discussed earlier, the α-amino acid derivatives (S-configuration) of 2-oxoamides with short aliphatic chains were selective toward sPLA2 [118,119]. From the new series of 2-oxoamide, two compounds **56** and **57** have shown selective potency toward cPLA2 and have displayed potent therapeutic effects in in rat models of pain and inflammation (Chart 4). Thus, inhibitors **56** and **57** demonstrated an antiinflammatory activity in rats with an ED_{50} of 0.0001 and 0.00005 mmol/kg, two orders of magnitude lower than AX006 **54**. They also showed significant reduction in edema models at 0.0001 mmol/kg dose, 100-folds lower than control drug indomethacin [158].

AX007 **55** was used to dock in the active site of the cPLA2 in a molecular dynamic (MD) simulation, in order to gain insights into the binding conformation of this 2-oxoamide inhibitor [159]. In addition, a series of 2-oxoamide inhibitors were also docked in the active site of the enzyme and the binding score was calculated. A combination of MD simulation with deuterium exchange mass spectrometry (DXMS) applied on the AX007 **55**-cPLA2 complex (Fig. 16) revealed a hydrogen bond between carboxylic acid moiety of the inhibitor and enzyme residue Arg200. The long aliphatic chain was filling to the hydrophobic region of the active site and interacted with several residues such as Phe199, Pro263, Leu264, and Phe 683. The compound AX074 **58** (Chart 4) from the 2-oxoamide analog class showed highest binding affinity and simulated

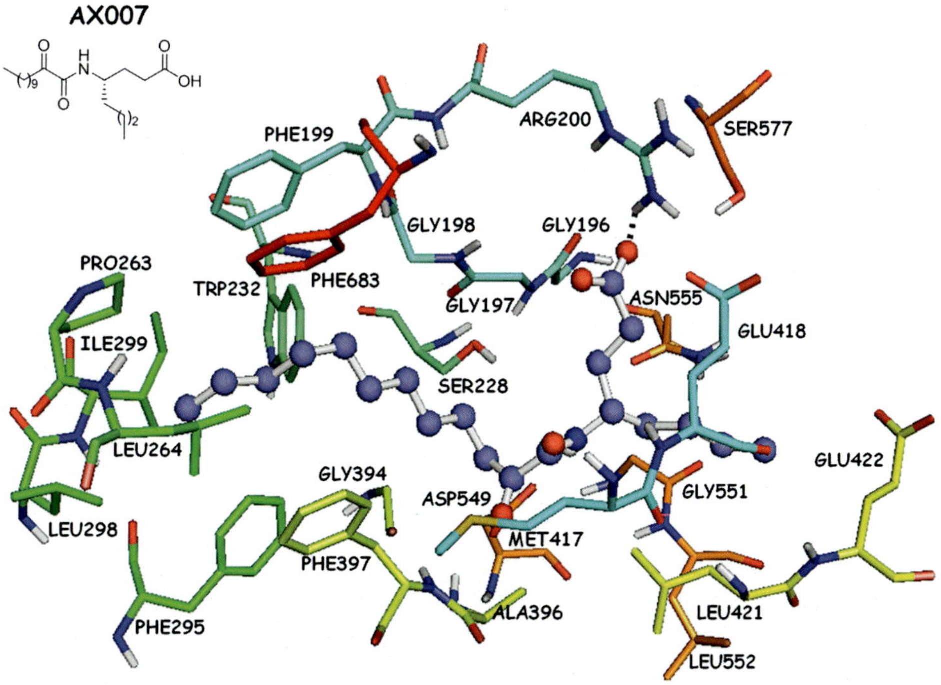

FIG. 16 AX007 **55**-GIVA cPLA2 complex, generated by a combination of molecular dynamics simulations and DXMS. *(Reprinted with permission from Mouchlis VD, Michopoulou V, Constantinou-Kokotou V, Mavromoustakos T, Dennis EA, Kokotos G. Binding conformation of 2-oxoamide inhibitors to group IVA cytosolic phospholipase A2 determined by molecular docking combined with molecular dynamics. J Chem Inf Model 2012;52(1):243–54. Copyright 2012 American Chemical Society.)*

inhibitory activity when docked in the cPLA2 crystal structure. The hydrogen bond distance of AX074 **58**/cPLA2 was smaller than the corresponding one observed in AX007 **55**/cPLA2. The aliphatic chain of AX074 **58** is four carbon atoms longer than in AX007 **55**, thus enhancing the hydrophobic interaction of AX074 **58** as compared with AX007 **55** [159].

3.2.7 1,3-Disubstituted propane-2-ones

A series of potent cPLA2 inhibitors were designed based on 1,3-disubstituted propan-2-one, with compound **59** (AR-C70484XX, Chart 4), which contain an aliphatic chain on one side and a benzoic acid group on the other side, inhibiting the enzyme with an IC_{50} of 8 nM in a bilayer assay, 30 nM in a soluble assay and 2.8 μM in a whole-cell assay. The most potent compound from this class of inhibitor was **60**, which contains a 3-methyl-1,2,4-oxadiazol-5-yl-moiety (Chart 4) with an IC_{50} of 2.1 nM and excellent metabolic stability [3].

3.3 Inhibitors of Ca^{2+} independent PLA2 (iPLA2)

3.3.1 Trifluoromethyl ketones and methyl fluorophosphonates

Both cPLA2 and iPLA2 have a similar active site, with a serine residue playing a central role in catalysis (vide supra). Thus, these two enzymes are inhibited by the same classes of inhibitors such as trifluoromethyl ketones (TFMK). Thus, both palmitoyl TFMK **61** and arachidonyl TFMK $AACOCF_3$ **42** were reported as potent iPLA2 inhibitors, with palmitoylTFMK **61** being 4-fold more potent than the $AACOCF_3$ **42**, with IC50 values of 3.8 μM (0.0075 mol fraction) and 15 μM (0.028 mol fraction), respectively [160]. While both enzymes were inhibited by the same inhibitor, the mechanism of action is slightly different for each enzyme. In case of cPLA2, $AACOCF_3$ **42** might form a stable hemiketal bond with the active site serine residue, whereas in the case of iPLA2, the enzyme is inhibited by a tetrahedral hydrated bond or with hemiketal bond with a different amino acid residue [161]. Methyl arachidonyl fluorophosphonates MAFP **43** was reported to be an irreversible inhibitor of both cPLA2 and iPLA2. However, MAFP **43** did not show any inhibition of sPLA2, or any other acyltransferase, transacylase and synthetase [162]. Among the MAFPs, analogs such as $Ph(CH_2)_4COCF_3$ and $Ph(CH_2)_4PO(OMe)F$ showed strong inhibition against iPLA2 but failed to show efficacy against cPLA2 [140].

3.3.2 Bromoenol Lactones

Bromoenol lactone **62** was the earliest selective inhibitor discovered against iPLA2. BEL **62** was used to identify and separate iPLA2 from other isoforms of PLA2 in murine macrophage-like cell line P388D [45,160]. BEL **62** was also used to inhibit another enzyme called magnesium-dependent phosphatidate phosphohydrolase-1 (PAP-1). A detailed cellular study showed that BEL **62** was able to induce apoptosis through a mechanism independent of iPLA2, via inhibition of PAP-1 [163]. Bromoenol lactone **62** has been also used in studying the physiological role of iPLA2. One of the studies characterized the role of iPLA2 in the development of cerebral cortex of a fetal rat brain. Upon treatment with inhibitor **62**, the brain showed significant damage in the cerebral cortex, along with loss of neurites and impaired cell body. The effect could not be reversed by adding additional arachidonic acid. This study suggested that the iPLA2 plays a housekeeping role in the development of the cerebral cortex via phospholipid membrane remodeling rather than hydrolyzing them and generating free fatty acid and lysophospholipids [164]. The role of iPLA2 in acute inflammatory condition hyperalgesia was also studied in a rat model with the help of the either cPLA2 specific inhibitor AACOCF3 **42** or iPLA2 specific inhibitor BEL **62**. The drug was administered locally in the hind limb of the rat in a carrageenan-induced inflammation model. BEL **62** significantly reduced prostaglandin levels and improved hyperalgesic response within 1-3 hours of carrageenan injection. On the other hand, AACOCF3 **42** did neither suppressed prostaglandin production nor improved hyperalgesic response, thus proving the central role of iPLA2 in hyperalgesic inflammation [165]. A nicotine-induced breast cancer cell proliferation and migration study demonstrated that BEL **62** significantly reduced both basal and nicotine-induced 4T1 breast cancer cell proliferation and migration. It also attenuates nicotine induced lung metastasis [97]. Along the same lines, a recent study was done to investigate the role of cPLA2 and iPLA2 in neuropathic pain in the spinal cord of a rat model with the help of cPLA2 and iPLA2 specific inhibitors. The study found that intrathecal injection of BEL **62** significantly improved the ipsilateral hindpaw withdrawal response when the rats were subjected to a behavioral study. It reduced neuronal firing signals in the dorsal neurons of the spinal cord, suggesting an analgesic effect on neuropathic pain. The study concluded that iPLA2 specific inhibitors can decrease sensitization in the dorsal region of the spinal cord and produce an analgesic effect in neuropathic pain for rats [166]. However, despite these promising results, the toxicity displayed by BEL **62** was quite significant. BEL inhibits iPLA2 through acylation of a critical cysteine [167] and can react with cysteines in other proteins generating unwanted side effects.

3.3.3 Polyfluoroketones

Polyfluoroketones were developed as one of the newer series of inhibitors of iPLA2. However, the initial compounds synthesized did not discern between iPLA2 and cPLA2 because of the similarity of their internal active site. An SAR study found that introducing a naphthyl group in the structure of the compound (FKGK18 **63**, Chart 5) makes it more potent against iPLA2. Another strategy tested was the introduction of more fluorine atoms adjacent to a carbonyl group to increase the reactivity of the carbonyl group and increase the binding affinity of the compound within the enzyme active site. However, this strategy does not enhance selectivity for iPLA2 [168]. In a more recent study, another series of substituted polyfluoroalkyl ketone compounds was synthesized and evaluated as iPLA2 inhibitors. Several polyfluoroalkyl ketone lead compounds were identified in this study, which showed more potency than FKGK18 **63**. Importantly, these compounds also showed much more selectivity toward iPLA2, especially pentafluoroethyl ketone GK187 **64** (Chart 5), which showed 25% less inhibition against cPLA2 and 32.8% less inhibition against sPLA2 when compared with other compounds in the series [169].

The mechanism of inhibition of iPLA2 by the fluoroketones was studied using molecular dynamic (MD) simulations and deuterium exchange mass spectrometry (DXMS). The iPLA2 model was built based on a homology with the known structure of patatin. The empty pockets of the enzyme were identified and the ability of the inhibitors to fit into the pocket was also determined. The molecular docking technique showed that the compound 1,1,1,3-tetrafluoro-7-phenylheptan-2-one (PHFK) **65** (Chart 5) showed excellent fitting into the active-site pocket of the enzyme and blocked the entrance of the phospholipid substrates. The polar fluoroketone headgroup formed stable hydrogen bonds with Gly486, Gly487, and Ser519 residues, while the nonpolar aliphatic chain and the phenyl moiety interacted with hydrophobic amino acid residues Met544, Val548, Phe549, Leu560, and Ala640 (Fig. 17) [170].

More recently, a new set of fluoroketone inhibitors was synthesized and studied for their binding mode with group IVA cPLA2 and group VIA iPLA2 [171]. The conclusions of this study led to the development of new potent and selective thioether fluoroketone inhibitors, as well as a potent thioether keto-1,2,4-oxadiazole inhibitor **66** (Chart 5) for GVIA iPLA2 with selectivity against group IVA cPLA2 and GV sPLA2. MD simulations showed that the sulfur atom increases the inhibitory potency of **66** versus its alkyl congener **67** (Chart 5) via interactions with Tyr643, Phe722, Leu770 as well as the "oxyanion hole" (Gly486/Gly487) (Fig. 6). It appears that the ability of the sulfur atom to interact via $\pi-\pi$ stacking interactions and other interaction with aromatic or nonaromatic residues of the enzyme helped the compounds to form a tighter bond with the enzyme during the simulation, as shown in Fig. 18 [171].

4 Conclusions

The superfamily of phospholipases A2 comprises a very diverse collection of members, evolved to perform the same hydrolytic reaction in different specific environments. We reviewed the diversity of its members, presenting comparatively their structure, mechanism of activation and of action, substrate specificity, and relating these properties with subcellular localization and tissue distribution of the isozymes. We have also highlighted the role played by different PLA2s in key physiologic processes in the human body and in different pathologies, with the aim of providing

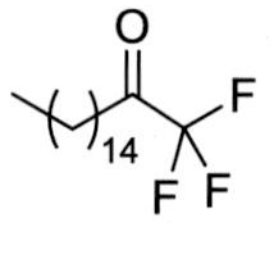

PalmitoylTFMK **61**

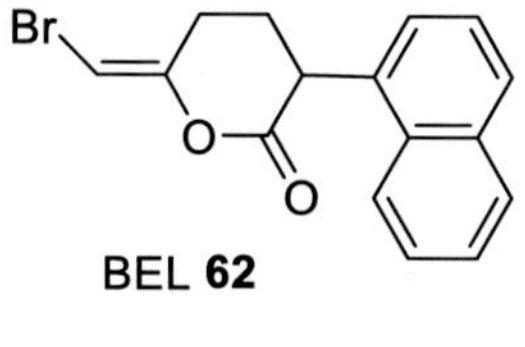

BEL **62**

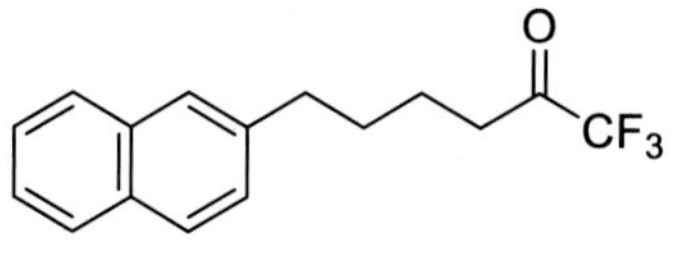

FKGK18 **63**

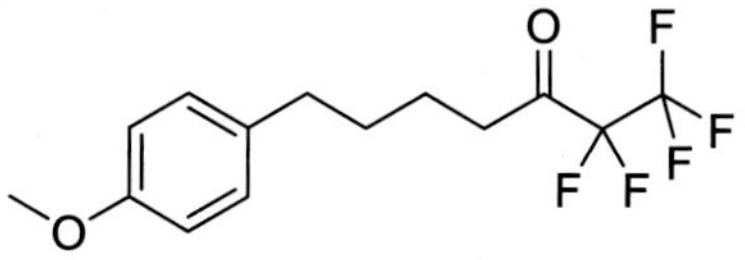

GK187 **64**

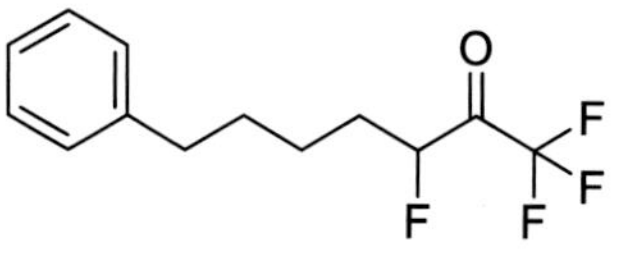

PHFK **65**

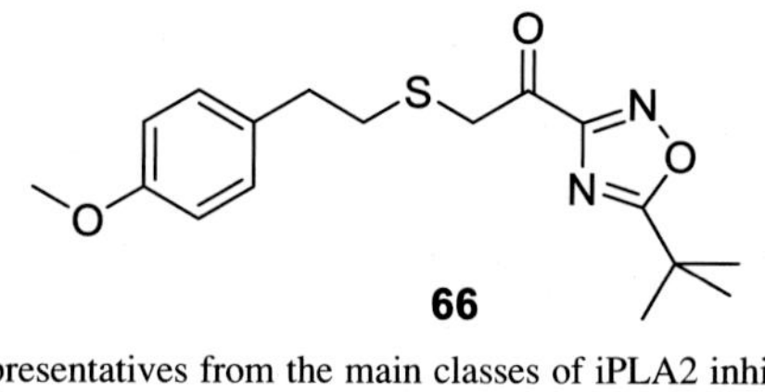

66

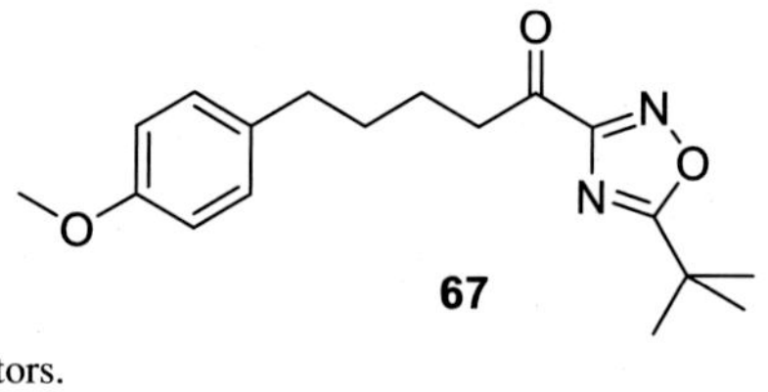

67

CHART 5 Selected representatives from the main classes of iPLA2 inhibitors.

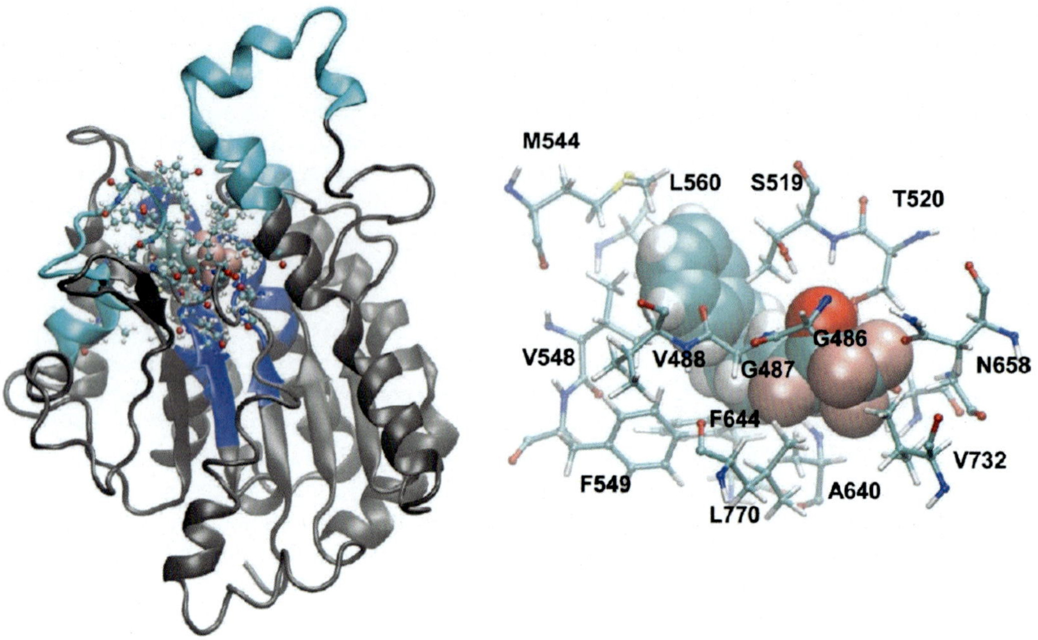

FIG. 17 Characterization of the binding of inhibitor PHFK 65 in the catalytic domain of iPLA2 [170].

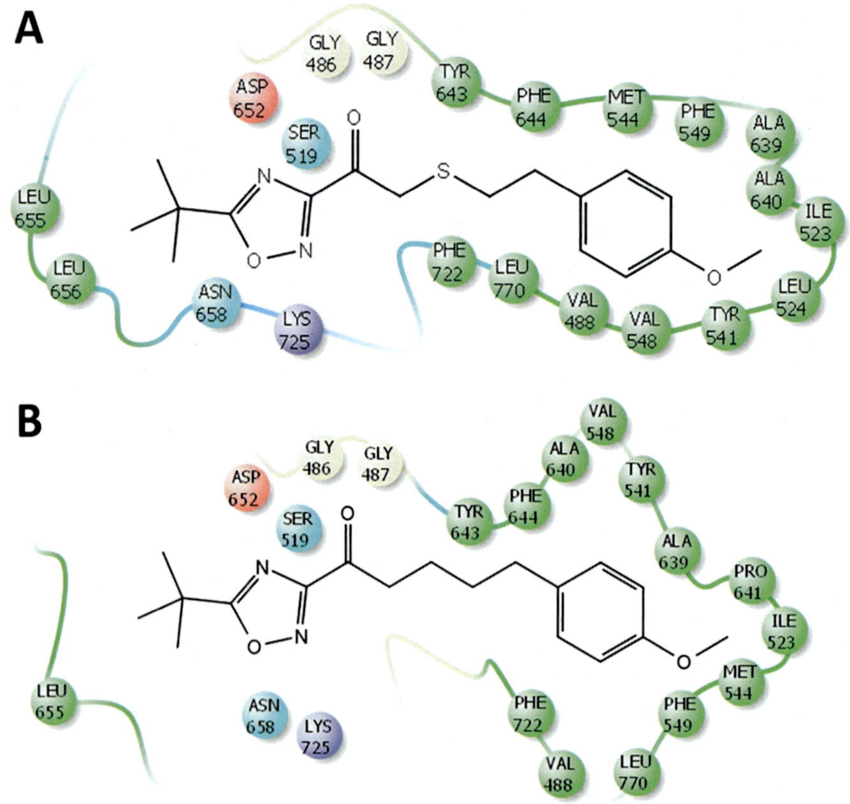

FIG. 18 Binding mode occurring during the MD simulations for (A) compound **66** and for (B) compound **67** in the GVIA iPLA2 binding site. *(Reprinted with permission from Mouchlis V.D., Limnios D., Kokotou M.G., Barbayianni E., Kokotos G., McCammon J.A., et al. Development of potent and selective inhibitors for group VIA calcium-independent phospholipase A2 guided by molecular dynamics and structure–activity relationships. J Med Chem 2016;59 (9):4403–4414. Copyright 2016 American Chemical Society.)*

an up-to-date overview of the impact of this class of enzymes in human health and diseases. Despite tremendous synthetic efforts during the last 5 decades, and thousands of inhibitors synthesized and evaluated in vitro and in vivo, also reviewed in detail in this work, currently there is no inhibitor in clinical use. However, progresses toward understanding of the structure, function, biodistribution, physiologic role of PLA2, coupled with advances in medicinal chemistry and selective PLA2 inhibitor design laid a solid foundation toward achieving this goal in the near future.

Acknowledgments

This research was funded by NIH (grant R03EB026189) and by Temple University School of Pharmacy-Dean's Office. S. A. K. acknowledges a scholarship from TU Graduate School.

References

[1] Khan SA, Ilies MA. The phospholipase A2 superfamily: structure, isozymes, catalysis, physiologic and pathologic roles. IJMS 2023;24(1353).

[2] Lam L, Ilies MA. Evaluation of the impact of esterases and lipases from the circulatory system against substrates of different lipophilicity. IJMS 2022;23(3):1262.

[3] Dennis EA, Cao J, Hsu Y-H, Magrioti V, Kokotos G. Phospholipase a enzymes: physical structure, biological function, disease implication, chemical inhibition, and therapeutic intervention. Chem Rev 2011;111(10):6130–85.

[4] Murakami M, Kudo I. Phospholipase A2. J Biochem 2002;131 (3):285–92.

[5] Moolenaar WH, van Meeteren LA, Giepmans BNG. The ins and outs of lysophosphatidic acid signaling. Bioessays 2004;26(8):870–81.

[6] Cathcart M-C, Lysaght J, Pidgeon GP. Eicosanoid signalling pathways in the development and progression of colorectal cancer: novel approaches for prevention/intervention. Cancer Metastasis Rev 2011;30(3–4):363–85.

[7] Six DA, Dennis EA. The expanding superfamily of phospholipase A2 enzymes: classification and characterization. Biochim Biophys Acta (BBA)-Mol Cell Biol Lipids 2000;1488(1–2):1–19.

[8] Peng Z, Chang Y, Fan J, Ji W, Su C. Phospholipase A2 superfamily in cancer. Cancer Lett 2021;497:165–77.

[9] Abbott MJ, Tang T, Sul HS. The role of phospholipase A2-derived mediators in obesity. Drug Discov Today: Dis Mech 2010;7(3–4), e213-e8.

[10] Burke JE, Dennis EA. Phospholipase A2 structure/function, mechanism, and signaling. J Lipid Res 2009;50, S237-S42.

[11] Heinrikson RL, Krueger ET, Keim PS. Amino acid sequence of phospholipase A2-alpha from the venom of *Crotalus adamanteus*. A new classification of phospholipases A2 based upon structural determinants. J Biol Chem 1977;252(14):4913–21.

[12] Berg OG, Gelb MH, Tsai M-D, Jain MK. Interfacial enzymology: the secreted phospholipase A—paradigm. Chem Rev 2001;101 (9):2613–54.

[13] Pang X-Y, Cao J, Addington L, Lovell S, Battaile KP, Zhang N, et al. Structure/function relationships of adipose phospholipase A2 containing a Cys-His-His catalytic triad. J Biol Chem 2012;287 (42):35260–74.

[14] Scott DL, White SP, Otwinowski Z, Yuan W, Gelb MH, Sigler PB. Interfacial catalysis: the mechanism of phospholipase A2. Science 1990;250(4987):1541–6.

[15] Lambeau G, Gelb MH. Biochemistry and physiology of mammalian secreted phospholipases A_2. Annu Rev Biochem 2008;77(1): 495–520.

[16] Winget JM, Pan YH, Bahnson BJ. The interfacial binding surface of phospholipase A2s. Biochim Biophys Acta (BBA) - Mol Cell Biol Lipids 2006;1761(11):1260–9.

[17] Jain MK, Berg OG. Coupling of the i-face and the active site of phospholipase A2 for interfacial activation. Curr Opin Chem Biol 2006;10(5):473–9.

[18] Kim RR, Chen Z, Mann TJ, Bastard K, Scott KF, Church WB. Structural and functional aspects of targeting the secreted human group IIA phospholipase A2. Molecules 2020;25(19):4459.

[19] Cao J, Hsu Y-H, Li S, Woods VL, Dennis EA. Lipoprotein-associated phospholipase A2 interacts with phospholipid vesicles via a surface-disposed hydrophobic α-helix. Biochemistry 2011;50 (23):5314–21.

[20] Ramirez F, Jain MK. Phospholipase A2 at the bilayer interface. Proteins: Struct Funct Bioinform 1991;9(4):229–39.

[21] Kramer RM, Checani GC, Deykin A, Pritzker CR, Deykin D. Solubilization and properties of Ca2+-dependent human platelet phospholipase A2. Biochim Biophys Acta (BBA) - Lipids Lipid Metab 1986;878(3):394–403.

[22] Alonso F, Henson PM, Leslie CC. A cytosolic phospholipase in human neutrophils that hydrolyzes arachidonoyl-containing phosphatidylcholine. Biochim Biophys Acta (BBA) - Lipids Lipid Metab 1986;878(2):273–80.

[23] Dessen A, Tang J, Schmidt H, Stahl M, Clark JD, Seehra J, et al. Crystal structure of human cytosolic phospholipase A2 reveals a novel topology and catalytic mechanism. Cell 1999;97(3):349–60.

[24] Burke JE, Hsu Y-H, Deems RA, Li S, Woods VL, Dennis EA. A phospholipid substrate molecule residing in the membrane surface mediates opening of the lid region in Group IVA cytosolic phospholipase A2. J Biol Chem 2008;283(45):31227–36.

[25] Dessen A, Somers WS, Stahl ML, Seehra JS, inventors; Genetics Institute LLC, assignee. Crystal structure of cPLA2 and methods of identifying agonists and antagonists using same patent US6801860B1; 2004. 2004/10/05.

[26] Hsu Y-H, Burke JE, Stephens DL, Deems RA, Li S, Asmus KM, et al. Calcium binding rigidifies the C2 domain and the intradomain interaction of GIVA phospholipase A2 as revealed by hydrogen/deuterium exchange mass spectrometry. J Biol Chem 2008;283 (15):9820–7.

[27] Pettus BJ, Bielawska A, Subramanian P, Wijesinghe DS, Maceyka M, Leslie CC, et al. Ceramide 1-phosphate is a direct activator of cytosolic phospholipase A2. J Biol Chem 2004;279(12):11320–6.

[28] Lin L-L, Wartmann M, Lin AY, Knopf JL, Seth A, Davis RJ. cPLA2 is phosphorylated and activated by MAP kinase. Cell 1993;72 (2):269–78.

[29] Das S, Rafter JD, Kim KP, Gygi SP, Cho W. Mechanism of Group IVA cytosolic phospholipase A2 activation by phosphorylation*. J Biol Chem 2003;278(42):41431–42.

[30] Zhou H, Das S, Murthy KS. Erk1/2- and p38 MAP kinase-dependent phosphorylation and activation of cPLA by m3 and m2 receptors. Am J Physiol-Gastrointest Liver Physiol 2003;284(3), G472-G80.

[31] Palavicini JP, Wang C, Chen L, Hosang K, Wang J, Tomiyama T, et al. Oligomeric amyloid-beta induces MAPK-mediated activation

of brain cytosolic and calcium-independent phospholipase A2 in a spatial-specific manner. Acta Neuropathol Commun 2017;5(1):56.
[32] Xu J, Weng Y-I, Simonyi A, Krugh BW, Liao Z, Weisman GA, et al. Role of PKC and MAPK in cytosolic PLA2 phosphorylation and arachadonic acid release in primary murine astrocytes. J Neurochem 2002;83(2):259–70.
[33] Xing M, Firestein BL, Shen GH, Insel PA. Dual role of protein kinase C in the regulation of cPLA2-mediated arachidonic acid release by P2U receptors in MDCK-D1 cells: involvement of MAP kinase-dependent and -independent pathways. J Clin Invest 1997;99(4):805–14.
[34] Hoeferlin LA, Wijesinghe DS, Chalfant CE. The role of ceramide-1-phosphate in biological functions. Handb Exp Pharmacol 2013;215:153–66.
[35] Nakamura H, Hirabayashi T, Shimizu M, Murayama T. Ceramide-1-phosphate activates cytosolic phospholipase A2alpha directly and by PKC pathway. Biochem Pharmacol 2006;71(6):850–7.
[36] Mencarelli C, Martinez-Martinez P. Ceramide function in the brain: when a slight tilt is enough. Cell Mol Life Sci 2013;70(2):181–203.
[37] Berridge MJ. The inositol trisphosphate/calcium signaling pathway in health and disease. Physiol Rev 2016;96(4):1261–96.
[38] Katan M, Cockcroft S. Phosphatidylinositol(4,5)bisphosphate: diverse functions at the plasma membrane. Essays Biochem 2020;64(3):513–31.
[39] Das S, Cho W. Roles of catalytic domain residues in interfacial binding and activation of Group IV cytosolic phospholipase A2*. J Biol Chem 2002;277(26):23838–46.
[40] Mosior M, Six DA, Dennis EA. Group IV cytosolic phospholipase A2 binds with high affinity and specificity to phosphatidylinositol 4,5-bisphosphate resulting in dramatic increases in activity*. J Biol Chem 1998;273(4):2184–91.
[41] Balsinde J, Winstead MV, Dennis EA. Phospholipase A2 regulation of arachidonic acid mobilization. FEBS Lett 2002;531(1):2–6.
[42] Wen ZH, Su YC, Lai PL, Zhang Y, Xu YF, Zhao A, et al. Critical role of arachidonic acid-activated mTOR signaling in breast carcinogenesis and angiogenesis. Oncogene 2013;32(2):160–70.
[43] Balsinde J, Balboa MA. Cellular regulation and proposed biological functions of group VIA calcium-independent phospholipase A2 in activated cells. Cell Signal 2005;17(9):1052–62.
[44] Casas J, Gijón MA, Vigo AG, Crespo MS, Balsinde J, Balboa MA. Phosphatidylinositol 4,5-bisphosphate anchors cytosolic group IVA phospholipase A2 to perinuclear membranes and decreases its calcium requirement for translocation in live cells. MBoC 2006;17(1):155–62.
[45] Ackermann EJ, Kempner ES, Dennis EA. Ca(2+)-independent cytosolic phospholipase A2 from macrophage-like P388D1 cells. Isolation and characterization. J Biol Chem 1994;269(12):9227–33.
[46] Ramanadham S, Ali T, Ashley JW, Bone RN, Hancock WD, Lei X. Calcium-independent phospholipases A2 and their roles in biological processes and diseases. J Lipid Res 2015;56(9):1643–68.
[47] Malley KR, Koroleva O, Miller I, Sanishvili R, Jenkins CM, Gross RW, et al. The structure of iPLA2β reveals dimeric active sites and suggests mechanisms of regulation and localization. Nat Commun 2018;9(1):765.
[48] Cha S-S, Lee D, Adams J, Kurdyla JT, Jones CS, Marshall LA, et al. High-resolution X-ray crystallography reveals precise binding interactions between human nonpancreatic secreted phospholipase A2 and a highly potent inhibitor (FPL67047XX). J Med Chem 1996;39(20):3878–81.
[49] Balsinde J, Pérez R, Balboa MA. Calcium-independent phospholipase A2 and apoptosis. Biochim Biophys Acta 2006;1761(11):1344–50.
[50] Hsu Y-H, Burke JE, Li S, Woods VL, Dennis EA. Localizing the membrane binding region of group VIA Ca2+-independent phospholipase A2 using peptide amide hydrogen/deuterium exchange mass spectrometry. J Biol Chem 2009;284(35):23652–61.
[51] Jenkins CM, Wolf MJ, Mancuso DJ, Gross RW. Identification of the calmodulin-binding domain of recombinant calcium-independent phospholipase A2β. J Biol Chem 2001;276(10):7129–35.
[52] McHowat J, Gullickson G, Hoover RG, Sharma J, Turk J, Kornbluth J. Platelet-activating factor and metastasis: calcium-independent phospholipase $A_2\beta$ deficiency protects against breast cancer metastasis to the lung. Am J Physiol-Cell Physiol 2011;300(4), C825-C32.
[53] Huang F, Wang K, Shen J. Lipoprotein-associated phospholipase A2: the story continues. Med Res Rev 2020;40(1):79–134.
[54] Ma Z, Turk J. The molecular biology of the group VIA Ca2+-independent phospholipase A2. Prog Nucleic Acid Res Mol Biol 2001;67:1–33. Elsevier.
[55] Mallat Z, Lambeau G, Tedgui A. Lipoprotein-associated and secreted phospholipases A2 in cardiovascular disease: roles as biological effectors and biomarkers. Circulation 2010;122(21):2183–200.
[56] Talmud PJ, Holmes MV. Deciphering the causal role of sPLA2s and Lp-PLA2 in coronary heart disease. Arterioscler Thromb Vasc Biol 2015;35(11):2281–9.
[57] Matsuzawa A, Hattori K, Aoki J, Arai H, Inoue K. Protection against oxidative stress-induced cell death by intracellular platelet-activating factor-acetylhydrolase II. J Biol Chem 1997;272(51):32315–20.
[58] Baheti U, Siddique SS, Foster CS. Cataract surgery in patients with history of uveitis. Saudi J Ophthalmol 2012;26(1):55–60.
[59] Sers C, Husmann K, Nazarenko I, Zhumabayeva B, Adhikari P. The class II tumour suppressor gene H-REV107-1 is a target of interferon-regulatory factor-1 and is involved in IFNg-induced cell death in human ovarian carcinoma cells. Oncogene 2002;21(18):2829–39.
[60] Murakami M, Sato H, Taketomi Y. Updating phospholipase A2 biology. Biomolecules 2020;10(10):1457.
[61] Murakami M, Taketomi Y, Girard C, Yamamoto K, Lambeau G. Emerging roles of secreted phospholipase A2 enzymes: Lessons from transgenic and knockout mice. Biochimie 2010;92(6):561–82.
[62] Hui DY. Intestinal phospholipid and lysophospholipid metabolism in cardiometabolic disease. Curr Opin Lipidol 2016;27(5):507–12.
[63] Ko C-W, Qu J, Black DD, Tso P. Regulation of intestinal lipid metabolism: current concepts and relevance to disease. Nat Rev Gastroenterol Hepatol 2020;17(3):169–83.
[64] Hui DY. Group 1B phospholipase A2 in metabolic and inflammatory disease modulation. Biochim Biophys Acta Mol Cell Biol Lipids 2019;1864(6):784–8.
[65] Dore E, Boilard E. Roles of secreted phospholipase A2 group IIA in inflammation and host defense. Biochim Biophys Acta (BBA) - Mol Cell Biol Lipids 2019;1864(6):789–802.
[66] Nevalainen TJ, Aho HJ, Peuravuori H. Secretion of Group 2 phospholipase A2 by lacrimal glands. Invest Ophthalmol Vis Sci 1994;35(2):417–21.
[67] Grönroos Juha O, Laine Veli JO, Nevalainen Timo J. Bactericidal Group IIA phospholipase A2 in serum of patients with bacterial infections. J Infect Dis 2002;185(12):1767–72.

[68] Murakami M, Kambe T, Shimbara S, Yamamoto S, Kuwata H, Kudo I. Functional association of type IIA secretory phospholipase A2 with the glycosylphosphatidylinositol-anchored heparan sulfate proteoglycan in the cyclooxygenase-2-mediated delayed prostanoid-biosynthetic pathway. J Biol Chem 1999;274(42):29927–36.

[69] Murakami M, Koduri RS, Enomoto A, Shimbara S, Seki M, Yoshihara K, et al. Distinct arachidonate-releasing functions of mammalian secreted phospholipase A2s in human embryonic kidney 293 and rat mastocytoma RBL-2H3 cells through heparan sulfate shuttling and external plasma membrane mechanisms. J Biol Chem 2001;276(13):10083–96.

[70] Ivandic B, Castellani LW, Wang X-P, Qiao J-H, Mehrabian M, Navab M, et al. Role of group II secretory phospholipase A2 in atherosclerosis. Arterioscler Thromb Vasc Biol 1999;19(5):1284–90.

[71] Yamashita SI, Yamashita JI, Ogawa M. Overexpression of group II phospholipase A2 in human breast cancer tissues is closely associated with their malignant potency. Br J Cancer 1994;69 (6):1166–70.

[72] Qu J, Zhao X, Wang J, Liu C, Sun Y, Cai H, et al. Plasma phospholipase A2 activity may serve as a novel diagnostic biomarker for the diagnosis of breast cancer. Oncol Lett 2018;15(4):5236–42.

[73] Yu JA, Li H, Meng X, Fullerton DA, Nemenoff RA, Mitchell JD, et al. Group IIa secretory phospholipase expression correlates with group IIa secretory phospholipase inhibition-mediated cell death in K-ras mutant lung cancer cells. J Thorac Cardiovasc Surg 2012;144(6). https://doi.org/10.1016/j.jtcvs.2012.08.064.

[74] Cai Q, Zhao Z, Antalis C, Yan L, Del Priore G, Hamed AH, et al. Elevated and secreted phospholipase A2 activities as new potential therapeutic targets in human epithelial ovarian cancer. FASEB J 2012;26(8):3306–20.

[75] Graff JR, Konicek BW, Deddens JA, Chedid M, Hurst BM, Colligan B, et al. Expression of group IIa secretory phospholipase A2 increases with prostate tumor grade. Clin Cancer Res 2001;7 (12):3857–61.

[76] Xing X-F, Li H, Zhong X-Y, Zhang L-H, Wang X-H, Liu Y-Q, et al. Phospholipase A2 group IIA expression correlates with prolonged survival in gastric cancer. Histopathology 2011;59(2):198–206.

[77] Brglez V, Lambeau G, Petan T. Secreted phospholipases A2 in cancer: diverse mechanisms of action. Biochimie 2014;107:114–23.

[78] Yamashita S-i, Ogawa M, Sakamoto K, Abe T, Arakawa H, Yamashita J-i. Elevation of serum group II phospholipase A2 levels in patients with advanced cancer. Clin Chim Acta 1994;228(2):91–9.

[79] Brglez V, Pucer A, Pungerčar J, Lambeau G, Petan T. Secreted phospholipases A2 are differentially expressed and epigenetically silenced in human breast cancer cells. Biochem Biophys Res Commun 2014;445(1):230–5.

[80] Leslie CC. Cytosolic phospholipase A2: physiological function and role in disease. J Lipid Res 2015;56(8):1386–402.

[81] Wenk GL. Neuropathol Changes Alzheimer's Dis 2003;4.

[82] Anfuso CD, Assero G, Lupo G, Nicotra A, Cannavò G, Strosznajder RP, et al. Amyloid beta(1–42) and its beta(25–35) fragment induce activation and membrane translocation of cytosolic phospholipase A2 in bovine retina capillary pericytes. Biochim Biophys Acta 2004;1686(1–2):125–38.

[83] Kaltschmidt B, Uherek M, Volk B, Baeuerle PA, Kaltschmidt C. Transcription factor NF-κB is activated in primary neurons by amyloid β peptides and in neurons surrounding early plaques from patients with Alzheimer disease. Proc Natl Acad Sci U S A 1997;94(6):2642–7.

[84] Teng T, Dong L, Ridgley DM, Ghura S, Tobin MK, Sun GY, et al. Cytosolic phospholipase A2 facilitates oligomeric amyloid-β peptide association with microglia via regulation of membrane-cytoskeleton connectivity. Mol Neurobiol 2019;56(5):3222–34.

[85] Sanchez-Mejia RO, Newman JW, Toh S, Yu G-Q, Zhou Y, Halabisky B, et al. Phospholipase A2 reduction ameliorates cognitive deficits in a mouse model of Alzheimer's disease. Nat Neurosci 2008;11(11):1311–8.

[86] Shah K, Lahiri DK. Cdk5 activity in the brain—multiple paths of regulation. J Cell Sci 2014;127(11):2391–400.

[87] Sundaram JR, Chan ES, Poore CP, Pareek TK, Cheong WF, Shui G, et al. Cdk5/p25-induced cytosolic PLA2-mediated lysophosphatidylcholine production regulates neuroinflammation and triggers neurodegeneration. J Neurosci 2012;32(3):1020–34.

[88] Han C, Bowen WC, Li G, Demetris AJ, Michalopoulos GK, Wu T. Cytosolic phospholipase A2α and peroxisome proliferator-activated receptor γ signaling pathway counteracts transforming growth factor β–mediated inhibition of primary and transformed hepatocyte growth. Hepatology 2010;52(2):644–55.

[89] Ns Y, Bishayee A, Vadlakonda L, Chintala R, Duddukuri G, Pallu R, et al. Phospholipase A2 isoforms as novel targets for prevention and treatment of inflammatory and oncologic diseases. Curr Drug Targets 2016;17(16):1940–62.

[90] Chen L, Fu H, Luo Y, Chen L, Cheng R, Zhang N, et al. cPLA2α mediates TGF-β-induced epithelial–mesenchymal transition in breast cancer through PI3k/Akt signaling. Cell Death Dis 2017;8(4), e2728-e.

[91] Jin T, Lin J, Gong Y, Bi X, Hu S, Lv Q, et al. iPLA2β contributes to ER stress-induced apoptosis during myocardial ischemia/reperfusion injury. Cells 2021;10(6):1446.

[92] Moon SH, Mancuso DJ, Sims HF, Liu X, Nguyen AL, Yang K, et al. Cardiac myocyte-specific knock-out of calcium-independent phospholipase A2γ (iPLA2γ) decreases oxidized fatty acids during ischemia/reperfusion and reduces infarct size. J Biol Chem 2016;291(37):19687–700.

[93] Williams SD, Gottlieb RA. Inhibition of mitochondrial calcium-independent phospholipase A2 (iPLA2) attenuates mitochondrial phospholipid loss and is cardioprotective. Biochem J 2002;362(Pt 1):23–32.

[94] Zhao X, Wang D, Zhao Z, Xiao Y, Sengupta S, Xiao Y, et al. Caspase-3-dependent activation of calcium-independent phospholipase A2 enhances cell migration in non-apoptotic ovarian cancer cells. J Biol Chem 2006;281(39):29357–68.

[95] Song Y, Wilkins P, Hu W, Murthy Karnam S, Chen J, Lee Z, et al. Inhibition of calcium-independent phospholipase A2 suppresses proliferation and tumorigenicity of ovarian carcinoma cells. Biochem J 2007;406(Pt 3):427–36.

[96] Kispert S, Schwartz T, McHowat J. Cigarette smoke regulates calcium-independent phospholipase A2 metabolic pathways in breast cancer. Am J Pathol 2017;187(8):1855–66.

[97] Calderon LE, Liu S, Arnold N, Breakall B, Rollins J, Ndinguri M. Bromoenol lactone attenuates nicotine-induced breast cancer cell proliferation and migration. PloS One 2015;10(11), e0143277.

[98] Rosenson RS, Stafforini DM. Modulation of oxidative stress, inflammation, and atherosclerosis by lipoprotein-associated phospholipase A2. J Lipid Res 2012;53(9):1767–82.

[99] Braun LT, Davidson MH. Lp-PLA2: a new target for statin therapy. Curr Atheroscler Rep 2010;12(1):29–33.

[100] Kono N, Inoue T, Yoshida Y, Sato H, Matsusue T, Itabe H, et al. Protection against oxidative stress-induced hepatic injury by

intracellular type II platelet-activating factor acetylhydrolase by metabolism of oxidized phospholipids in vivo. J Biol Chem 2008;283(3):1628–36.
[101] Abe A, Hiraoka M, Wild S, Wilcoxen SE, Paine R, Shayman JA. Lysosomal phospholipase A2 is selectively expressed in alveolar macrophages. J Biol Chem 2004;279(41):42605–11.
[102] Hiraoka M, Abe A, Lu Y, Yang K, Han X, Gross RW, et al. Lysosomal phospholipase A2 and phospholipidosis. Mol Cell Biol 2006;26(16):6139–48.
[103] Hiraoka M, Abe A, Lennikov A, Kitaichi N, Ishida S, Ohguro H. Increase of lysosomal phospholipase A2 in aqueous humor by uveitis. Exp Eye Res 2014;118:13–9.
[104] Hiraoka M, Abe A, Inatomi S, Sawada K, Ohguro H. Augmentation of lysosomal phospholipase A2 activity in the anterior chamber in glaucoma. Curr Eye Res 2016;41(5):683–8.
[105] Davidson FF, Hajdu J, Dennis EA. 1-Stearyl, 2-stearoylaminodeoxy phosphatidylcholine, a potent reversible inhibitor of phospholipase A2. Biochem Biophys Res Commun 1986;137(2):587–92.
[106] Gelb MH. Fluoro ketone phospholipid analogs: new inhibitors of phospholipase A2. J Am Chem Soc 1986;108(11):3146–7.
[107] Yuan W, Berman RJ, Gelb MH. Synthesis and evaluation of phospholipid analogs as inhibitors of cobra venom phospholipase A2. J Am Chem Soc 1987;109(26):8071–81.
[108] Yuan W, Gelb MH. Phosphonate-containing phospholipid analogs as tight-binding inhibitors of phospholipase-A2. J Am Chem Soc 1988;110(8):2665–6.
[109] Yu L, Deems R, Hajdu J, Dennis E. The interaction of phospholipase A2 with phospholipid analogues and inhibitors. J Biol Chem 1990;265(5):2657–64.
[110] Yu L, Dennis EA. Defining the dimensions of the catalytic site of phospholipase A2 using amide substrate analogs. J Am Chem Soc 1992;114(23):8757–63.
[111] Tramposch KM, Steiner SA, Stanley PL, Nettleton DO, Franson R, Lewin AH, et al. Novel inhibitor of phospholipase A2 with topical anti-inflammatory activity. Biochem Biophys Res Commun 1992;189(1):272–9.
[112] Tramposch K, Chilton F, Stanley P, Franson R, Havens M, Nettleton D, et al. Inhibitor of phospholipase A2 blocks eicosanoid and platelet activating factor biosynthesis and has topical anti-inflammatory activity. J Pharmacol Exp Therap 1994;271(2):852–9.
[113] Burke JR, Gregor KR, Tramposch KM. Mechanism of inhibition of human nonpancreatic secreted phospholipase A2 by the anti-inflammatory agent BMS-181162*(*). J Biol Chem 1995;270 (1):274–80.
[114] Springer DM, Luh B-Y, D'Andrea SV, Bronson JJ, Mansuri MM, Burke JR, et al. Dicarboxylic acid inhibitors of phospholipase A2. Bioorg Med Chem Lett 1997;7(7):793–8.
[115] Springer DM, Luh B-Y, Bronson JJ, McElhone KE, Mansuri MM, Gregor KR, et al. Biaryl diacid inhibitors of humans-PLA2 with anti-inflammatory activity. Bioorg Med Chem 2000;8(5):1087–109.
[116] Beaton HG, Bennion C, Connolly S, Cook AR, Gensmantel NP, Hallam C, et al. Discovery of new non-phospholipid inhibitors of the secretory phospholipases A2. J Med Chem 1994;37(5):557–9.
[117] Hansford KA, Reid RC, Clark CI, Tyndall JD, Whitehouse MW, Guthrie T, et al. D-Tyrosine as a chiral precusor to potent inhibitors of human nonpancreatic secretory phospholipase A2 (IIa) with anti-inflammatory activity. ChemBioChem 2003;4(2–3):181–5.
[118] Antonopoulou G, Barbayianni E, Magrioti V, Cotton N, Stephens D, Constantinou-Kokotou V, et al. Structure–activity relationships of natural and non-natural amino acid-based amide and 2-oxoamide inhibitors of human phospholipase A2 enzymes. Bioorg Med Chem 2008;16(24):10257–69.
[119] Vasilakaki S, Barbayianni E, Leonis G, Papadopoulos MG, Mavromoustakos T, Gelb MH, et al. Development of a potent 2-oxoamide inhibitor of secreted phospholipase A2 guided by molecular docking calculations and molecular dynamics simulations. Bioorg Med Chem 2016;24(8):1683–95.
[120] Schevitz R, Bach N, Carlson D, Chirgadze N, Clawson D, Dillard R, et al. Structure-based design of the first potent and selective inhibitor of human non-pancreatic secretory phospholipase A2. Nat Struct Biol 1995;2(6):458–65.
[121] Dillard RD, Bach NJ, Draheim SE, Berry DR, Carlson DG, Chirgadze NY, et al. Indole inhibitors of human nonpancreatic secretory phospholipase A2. 1. Indole-3-acetamides. J Med Chem 1996;39 (26):5119–36.
[122] Dillard RD, Bach NJ, Draheim SE, Berry DR, Carlson DG, Chirgadze NY, et al. Indole inhibitors of human nonpancreatic secretory phospholipase A2. 2. Indole-3-acetamides with additional functionality. J Med Chem 1996;39(26):5137–58.
[123] Draheim SE, Bach NJ, Dillard RD, Berry DR, Carlson DG, Chirgadze NY, et al. Indole inhibitors of human nonpancreatic secretory phospholipase A. 3. Indole-3-glyoxamides. J Med Chem 1996;39 (26):5159–75.
[124] Snyder DW, Bach NJ, Dillard RD, Draheim SE, Carlson DG, Fox N, et al. Pharmacology of LY315920/S-5920,[[3-(Aminooxoacetyl)-2-ethyl-1-(phenylmethyl)-1H-indol-4-yl] oxy] acetate, a potent and selective secretory phospholipase A2 inhibitor: a new class of anti-inflammatory drugs, SPI. J Pharmacol Exp Therap 1999;288 (3):1117–24.
[125] Liu Y, Han X-f, Huang C-k, Hao X, Lai L-H. Indole-5-phenylcarbamate derivatives as human non-pancreatic secretory phospholipase A2 inhibitor. Bioorg Med Chem Lett 2005;15 (20):4540–2.
[126] Sawyer JS, Beight DW, Smith EC, Snyder DW, Chastain MK, Tielking RL, et al. Carbocyclic [g] indole inhibitors of human non-pancreatic s-PLA2. J Med Chem 2005;48(3):893–6.
[127] Yokota Y, Hanasaki K, Ono T, Nakazato H, Kobayashi T, Arita H. Suppression of murine endotoxic shock by sPLA2 inhibitor, indoxam, through group IIA sPLA2-independent mechanisms. Biochim Biophys Acta (BBA)-Mol Cell Biol Lipids 1999;1438 (2):213–22.
[128] Smart BP, Pan YH, Weeks AK, Bollinger JG, Bahnson BJ, Gelb MH. Inhibition of the complete set of mammalian secreted phospholipases A2 by indole analogues: a structure-guided study. Bioorg Med Chem 2004;12(7):1737–49.
[129] Hui D, Cope M, Labonte E, Chang HT, Shao J, Goka E, et al. The phospholipase A2 inhibitor methyl indoxam suppresses diet-induced obesity and glucose intolerance in mice. Br J Pharmacol 2009;157 (7):1263–9.
[130] Tomita Y, Jyoyama H, Kobayashi M, Kuwabara K, Furue S, Ueno M, et al. Role of group IIA phospholipase A2 in rat colitis induced by dextran sulfate sodium. Eur J Pharmacol 2003;472 (1–2):147–58.
[131] Bradley JD, Dmitrienko AA, Kivitz AJ, Gluck OS, Weaver AL, Wiesenhutter C, et al. A randomized, double-blinded, placebo-controlled clinical trial of LY333013, a selective inhibitor of group II secretory phospholipase A2, in the treatment of rheumatoid arthritis. J Rheumatol 2005;32(3):417–23.
[132] De Luca D, Minucci A, Trias J, Tripodi D, Conti G, Zuppi C, et al. Varespladib inhibits secretory phospholipase A2 in bronchoalveolar

lavage of different types of neonatal lung injury. J Clin Pharmacol 2012;52(5):729–37.
[133] Nicholls SJ, Kastelein JJ, Schwartz GG, Bash D, Rosenson RS, Cavender MA, et al. Varespladib and cardiovascular events in patients with an acute coronary syndrome: the VISTA-16 randomized clinical trial. JAMA 2014;311(3):252–62.
[134] Lewin M, Samuel S, Merkel J, Bickler P. Varespladib (LY315920) appears to be a potent, broad-spectrum, inhibitor of snake venom phospholipase A2 and a possible pre-referral treatment for envenomation. Toxins 2016;8(9):248.
[135] Gutiérrez JM, Lewin MR, Williams DJ, Lomonte B. Varespladib (LY315920) and methyl varespladib (LY333013) abrogate or delay lethality induced by presynaptically acting neurotoxic snake venoms. Toxins 2020;12(2):131.
[136] Giordanetto F, Pettersen D, Starke I, Nordberg P, Dahlström M, Knerr L, et al. Discovery of AZD2716: a novel secreted phospholipase A2 (sPLA2) inhibitor for the treatment of coronary artery disease. ACS Med Chem Lett 2016;7(10):884–9.
[137] Street IP, Lin HK, Laliberte F, Ghomashchi F, Wang Z, Perrier H, et al. Slow- and tight-binding inhibitors of the 85-kDa human phospholipase A2. Biochemistry 1993;32(23):5935–40.
[138] Trimble LA, Street IP, Perrier H, Tremblay NM, Weech PK, Bernstein MA. NMR structural studies of the tight complex between a trifluoromethyl ketone inhibitor and the 85-kDa human phospholipase A2. Biochemistry 1993;32(47):12560–5.
[139] Kalyvas A, David S. Cytosolic phospholipase A2 plays a key role in the pathogenesis of multiple sclerosis-like disease. Neuron 2004;41 (3):323–35.
[140] Ghomashchi F, Loo R, Balsinde J, Bartoli F, Apitz-Castro R, Clark JD, et al. Trifluoromethyl ketones and methyl fluorophosphonates as inhibitors of group IV and VI phospholipases A2: structure-function studies with vesicle, micelle, and membrane assays. Biochim Biophys Acta (BBA)-Biomembr 1999;1420(1–2):45–56.
[141] Svensson C, Lucas K, Hua X-Y, Powell H, Dennis E, Yaksh T. Spinal phospholipase A2 in inflammatory hyperalgesia: role of the small, secretory phospholipase A2. Neuroscience 2005;133 (2):543–53.
[142] Burke JR, Davern LB, Stanley PL, Gregor KR, Banville J, Remillard R, et al. BMS-229724 is a tight-binding inhibitor of cytosolic phospholipase A2 that acts at the lipid/water interface and possesses anti-inflammatory activity in skin inflammation models. J Pharmacol Exp Therap 2001;298(1):376–85.
[143] Seno K, Okuno T, Nishi K, Murakami Y, Watanabe F, Matsuura T, et al. Pyrrolidine inhibitors of human cytosolic phospholipase A (2). J Med Chem 2000;43(6):1041–4.
[144] Ghomashchi F, Stewart A, Hefner Y, Ramanadham S, Turk J, Leslie CC, et al. A pyrrolidine-based specific inhibitor of cytosolic phospholipase A2α blocks arachidonic acid release in a variety of mammalian cells. Biochim Biophys Acta (BBA)-Biomembr 2001;1513 (2):160–6.
[145] Seno K, Okuno T, Nishi K, Murakami Y, Yamada K, Nakamoto S, et al. Pyrrolidine inhibitors of human cytosolic phospholipase A2. Part 2: synthesis of potent and crystallized 4-triphenylmethylthio derivative 'pyrrophenone'. Bioorg Med Chem Lett 2001;11 (4):587–90.
[146] Ono T, Yamada K, Chikazawa Y, Ueno M, Nakamoto S, Okuno T, et al. Characterization of a novel inhibitor of cytosolic phospholipase A2α, pyrrophenone. Biochem J 2002;363(3):727–35.
[147] Yun B, Lee H, Ewing H, Gelb MH, Leslie CC. Off-target effect of the cPLA2α inhibitor pyrrophenone: inhibition of calcium release from the endoplasmic reticulum. Biochem Biophys Res Commun 2016;479(1):61–6.
[148] Xu H, Sun Y, Zeng L, Li Y, Hu S, He S, et al. Inhibition of cytosolic phospholipase A2 alpha increases chemosensitivity in cervical carcinoma through suppressing β-catenin signaling. Cancer Biol Ther 2019;20(6):912–21.
[149] McKew JC, Foley MA, Thakker P, Behnke ML, Lovering FE, Sum F-W, et al. Inhibition of cytosolic phospholipase A2α: hit to lead optimization. J Med Chem 2006;49(1):135–58.
[150] Lehr M. 3-(Octadecanoylaminomethyl) indole-2-carboxylic acid derivatives and 1-methyl-3-octadecanoylindole-2-carboxylic acid as inhibitors of cytosolic phospholipase A2. Arch Pharm 1996;329 (8–9):386–92.
[151] Lee KL, Foley MA, Chen L, Behnke ML, Lovering FE, Kirincich SJ, et al. Discovery of ecopladib, an indole inhibitor of cytosolic phospholipase A2α. J Med Chem 2007;50(6):1380–400.
[152] Lamothe J, Lee K, Schelling S, Stedman N, Leach M, McKew J, et al, Sa. 29. Efficacy of giripladib, a novel inhibitor of cytosolic phospholipase A2α, in two different mouse models of rheumatoid arthritis. Clin Immunol 2008;127:S89–90.
[153] McKew JC, Lee KL, Shen MW, Thakker P, Foley MA, Behnke ML, et al. Indole cytosolic phospholipase A2 α inhibitors: discovery and in vitro and in vivo characterization of 4-{3-[5-chloro-2-(2-{[(3, 4-dichlorobenzyl) sulfonyl] amino} ethyl)-1-(diphenylmethyl)-1 H-indol-3-yl] propyl} benzoic acid. Efipladib J Med Chem 2008;51(12):3388–413.
[154] Liu S, Sun Y, Hou Y, Yang L, Wan X, Qin Y, et al. A novel lncRNA ROPM-mediated lipid metabolism governs breast cancer stem cell properties. J Hematol Oncol 2021;14(1):1–23.
[155] Kokotos G, Kotsovolou S, Six DA, Constantinou-Kokotou V, Beltzner CC, Dennis EA. Novel 2-oxoamide inhibitors of human group IVA phospholipase A2. J Med Chem 2002;45(14):2891–3.
[156] Kokotos G, Six DA, Loukas V, Smith T, Constantinou-Kokotou V, Hadjipavlou-Litina D, et al. Inhibition of group IVA cytosolic phospholipase A2 by novel 2-oxoamides in vitro, in cells, and in vivo. J Med Chem 2004;47(14):3615–28.
[157] Stephens D, Barbayianni E, Constantinou-Kokotou V, Peristeraki A, Six DA, Cooper J, et al. Differential inhibition of group IVA and group VIA phospholipases A2 by 2-oxoamides. J Med Chem 2006;49(9):2821–8.
[158] Six DA, Barbayianni E, Loukas V, Constantinou-Kokotou V, Hadjipavlou-Litina D, Stephens D, et al. Structure– activity relationship of 2-oxoamide inhibition of group IVA cytosolic phospholipase A2 and group V secreted phospholipase A2. J Med Chem 2007;50(17):4222–35.
[159] Mouchlis VD, Michopoulou V, Constantinou-Kokotou V, Mavromoustakos T, Dennis EA, Kokotos G. Binding conformation of 2-oxoamide inhibitors to group IVA cytosolic phospholipase A2 determined by molecular docking combined with molecular dynamics. J Chem Inf Model 2012;52(1):243–54.
[160] Ackermann EJ, Conde-Frieboes K, Dennis EA. Inhibition of macrophage Ca^{2+}-independent phospholipase A2 by bromoenol lactone and trifluoromethyl ketones (*). J Biol Chem 1995;270 (1):445–50.
[161] Conde-Frieboes K, Reynolds LJ, Lio Y-C, Hale MR, Wasserman HH, Dennis EA. Activated ketones as inhibitors of intracellular

Ca2+-dependent and Ca2+-independent phospholipase A2. J Am Chem Soc 1996;118(24):5519–25.

[162] Lio Y-C, Reynolds LJ, Balsinde J, Dennis EA. Irreversible inhibition of Ca2+-independent phospholipase A2 by methyl arachidonyl fluorophosphonate. Biochim Biophys Acta (BBA)-Lipids Lipid Metabol 1996;1302(1):55–60.

[163] Fuentes L, Pérez R, Nieto ML, Balsinde J, MaA B. Bromoenol lactone promotes cell death by a mechanism involving phosphatidate phosphohydrolase-1 rather than calcium-independent phospholipase A2. J Biol Chem 2003;278(45):44683–90.

[164] Kurusu S, Matsui K, Watanabe T, Tsunou T, Kawaminami M. The cytotoxic effect of bromoenol lactone, a calcium-independent phospholipase A2 inhibitor, on rat cortical neurons in culture. Cell Mol Neurobiol 2008;28(8):1109–18.

[165] Tsuchida K, Ibuki T, Matsumura K. Bromoenol lactone, an inhibitor of calcium-independent phospholipase A2, suppresses carrageenan-induced prostaglandin production and hyperalgesia in rat hind paw. Mediators Inflamm 2015;2015.

[166] Gwak YS, Chen G, Abdi S, Kim HK. Calcium-independent phospholipase A2 inhibitor produces an analgesic effect in a rat model of neuropathic pain by reducing central sensitization in the dorsal horn. Neurol Res 2021;43(8):683–92.

[167] Song H, Ramanadham S, Bao S, Hsu FF, Turk J. A bromoenol lactone suicide substrate inactivates group VIA phospholipase A2 by generating a diffusible bromomethyl keto acid that alkylates cysteine thiols. Biochemistry 2006;45(3):1061–73.

[168] Kokotos G, Hsu Y-H, Burke JE, Baskakis C, Kokotos CG, Magrioti V, et al. Potent and selective fluoroketone inhibitors of group VIA calcium-independent phospholipase A2. J Med Chem 2010;53(9):3602–10.

[169] Magrioti V, Nikolaou A, Smyrniotou A, Shah I, Constantinou-Kokotou V, Dennis EA, et al. New potent and selective polyfluoroalkyl ketone inhibitors of GVIA calcium-independent phospholipase A2. Bioorg Med Chem 2013;21(18):5823–9.

[170] Hsu Y-H, Bucher D, Cao J, Li S, Yang S-W, Kokotos G, et al. Fluoroketone inhibition of Ca2+-independent phospholipase A2 through binding pocket association defined by hydrogen/deuterium exchange and molecular dynamics. J Am Chem Soc 2013;135(4):1330–7.

[171] Mouchlis VD, Limnios D, Kokotou MG, Barbayianni E, Kokotos G, McCammon JA, et al. Development of potent and selective inhibitors for group VIA calcium-independent phospholipase A2 guided by molecular dynamics and structure–activity relationships. J Med Chem 2016;59(9):4403–14.

Section C

Zinc enzymes

Chapter 3.1

Carbonic anhydrases

Claudiu T. Supuran
NEUROFARBA Department, Section of Pharmaceutical and Nutraceutical Sciences, University of Florence, Florence, Italy

1 Introduction

Carbonic anhydrases (CAs, EC 4.2.1.1) catalyze a rather simple physiological reaction, the hydration of CO_2 to bicarbonate and protons [1,2]. As this process is fundamental in most living systems, these enzymes are widespread in both prokaryotes and eukaryotes, with eight genetic families encoding them reported to date, the α-, β-, γ-, δ-, ζ-, η-, θ-, and ι-classes [3–17]. The α-CAs are present in vertebrates, protozoa, algae, cytoplasm of green plants, and many Gram-negative *Bacteria* [1–3]; the β-CAs are found in both Gram-negative and -positive *Bacteria*, algae, and chloroplasts of mono- as well as dicotyledons, and also in many fungi and some *Archaea* [3–6]. The γ-CAs were found in *Archaea*, cyanobacteria, and most types of *Bacteria* [3,4], the δ- and ζ-CAs seem to be present only in marine diatoms [12,14], whereas the η-CAs in protozoa [13]. The θ- and ι-classes are also widely distributed in diatoms [15,16], but the last class has also been found in many bacteria [17]. In these organisms, CAs are involved in crucial physiological processes connected with pH and CO_2 homeostasis/sensing; biosynthetic reactions, such as gluconeogenesis, lipogenesis, and ureagenesis, respiration and transport of CO_2/bicarbonate, electrolyte secretion in a variety of tissues/organs, bone resorption, calcification, tumorigenicity, and many other physiological or pathological processes (thoroughly studied in vertebrates and some pathogens) [1–3,7]. In algae, plants, and some bacteria, CAs play an important role in photosynthesis and biosynthetic reactions connected to it [3–5]. In diatoms, δ-, ζ-, θ-, and ι-CAs play a crucial role in carbon dioxide fixation (from CO_2, COS, or CS_2 [18], see latter in the text), whereas in protozoans, the role of the η-CAs is poorly understood for the moment, but they may be involved in de novo purine/pyrimidine biosynthetic pathways [13].

2 Catalyzed reactions, structure and functions of CAs

Apart the physiological reaction mentioned previously (reaction 1, Chart 1), some classes of CAs possess a certain catalytic versatility, with the possibility to hydrate small molecules similar to CO_2 such as COS (reaction *2*) [18], CS_2 (reaction 3) [19], and cyanamide (reaction 4) [20,21] leading to H_2S (and CO_2) for the first reactions and urea for the last one. Aldehydes were also shown to be hydrated to *gem* diols (reaction 5) [22], whereas the esterase activity with carboxylic acid esters (reaction 6) [23–25], sulfonic acid esters (reaction 7) [26,27], phosphate esters (reaction 8) [24,25], and thioesters (reaction 9) [28,29] was also reported. Some other less investigated hydrolytic processes (reactions 10 and 11 in Chart 1) were also found for at least the α-CAs [30]. It is presently not known whether other reactions than the CO_2 hydration/bicarbonate dehydration may have physiological relevance, although the recently reported thioesterase activity [29] may interfere with the generation/hydrolysis of acyl-coenzyme A derivatives, and thus possess an important physiological role. The latest catalytic activity documented so far is the selenolesterase activity of α-CAs [31], reaction 12 in Chart 1.

$O{=}C{=}O + H_2O \Leftrightarrow HCO_3^- + H^+$	*(1)*
$O{=}C{=}S + H_2O \Leftrightarrow H_2S + CO_2$	*(2)*
$S{=}C{=}S + 2H_2O \Leftrightarrow 2\,H_2S + CO_2$	*(3)*
$HN{=}C{=}NH + H_2O \Leftrightarrow H_2NCONH_2$	*(4)*
$RCHO + H_2O \Leftrightarrow RCH(OH)_2$	*(5)*
$RCOOAr + H_2O \Leftrightarrow RCOOH + ArOH$	*(6)*
$RSO_3Ar + H_2O \Leftrightarrow RSO_3H + ArOH$	*(7)*
$ArOPO_3H_2 + H_2O \Leftrightarrow ArOH + H_3PO_4$	*(8)*
$R_2NCSSR' + H_2O \Leftrightarrow R_2NH + R'SH + COS$	*(9)*
$PhCH_2OCOCl + H_2O \Leftrightarrow PhCH_2OH + CO_2 + HCl$	*(10)*
$RSO_2Cl + H_2O \Leftrightarrow RSO_3H + HCl$	*(11)*
$ArCOSeR + H_2O \Leftrightarrow ArCOOH + RSeH$	*(12)*

CHART 1 Reactions catalyzed by CAs. The physiological one (reaction 1) is the CO_2 hydration to bicarbonate and protons, whereas reactions 2–12 were mainly documented for α-CAs and are probably without a physiological role in mammals. In some archaea, bacteria and diatoms reactions 2 and 3 may also have physiological relevance.

Metalloenzymes. https://doi.org/10.1016/B978-0-12-823974-2.00014-0

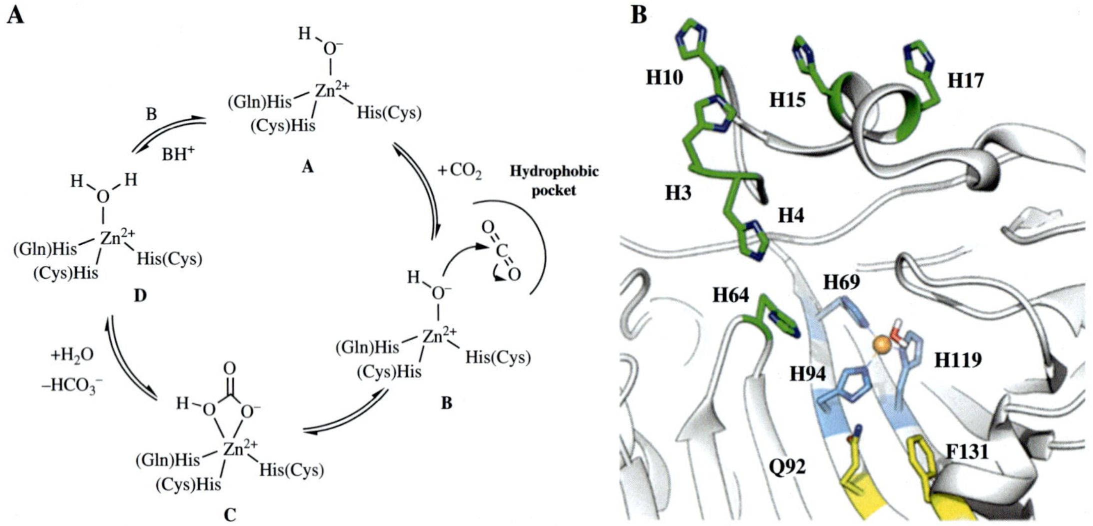

FIG. 1 (A) Catalytic mechanism of CAs for the CO_2 hydration reaction. A general zinc coordination pattern is shown. (B) Active site details for human (h) isoform hCA II (α-class enzyme), with the zinc ion *(gold sphere)*, its ligands and proton shuttling residues evidenced.

CAs are catalytically effective only with one metal ion bound within the active site cavity (at least for the first seven genetic families mentioned previously), as the apoenzymes are totally devoid of any catalytic action [1–3,6,10]. The active center normally comprises M(II) ions in a tetrahedral geometry, with three amino acid residues as ligands, in addition to a water molecule/hydroxide ion coordinating the metal (Fig. 1). Zn(II) is the metal ion that may be present in at least seven CA genetic families, but Cd(II) is interchangeable with Zn(II) in the ζ-CAs [1–3,14,18], Fe(II) seems to be present in γ-CAs, at least in anaerobic conditions [32], whereas Co(II) may substitute the zinc ion in many α-CAs without a significant loss of the catalytic activity [1–3,6]. These metal ions were rarely observed also in trigonal, bipyramidal, or octahedral coordination geometries within the CA active site [1–3,6], but the catalytically effective species are probably the tetrahedral ones. Indeed, in the α-, γ-, and δ-CAs, the Zn(II) is coordinated by three His residues and a water molecule/hydroxide ion. In type I β-CAs (i.e., enzymes with an open active site), the Zn(II) is coordinated by two Cys and one His residue, with water/hydroxide as the fourth ligand, whereas in the type II β-CAs (i.e., enzymes with a closed active site), an aspartate residue is the fourth zinc ligand. Such enzymes are devoid of catalytic activity due to the lack of a nucleophilic water molecule/hydroxide ion coordinated to the zinc. In the ζ-CAs, the metal ion coordination is similar with the one present in the type I β-CAs, except that the metal ion may be Cd(II) or Zn(II) [14,18]. In the η-CAs, the pattern of the metal ion coordination, (which presumably is Zn(II)) seems to be completely different from those present in the other five classes, with two His and one Gln residue in addition to the water molecule/hydroxide ion binding the Zn(II), as suggested by a computational approach [33]. Examples of structures for diverse CAs belonging to all classes for which the crystallographic structure was reported will be shown in the next sections of the chapter.

In all CA classes, a metal hydroxide derivative (L_3-M^{2+}-OH^-, where L is the coordinated amino acid residue, see Fig. 1A) of the enzyme is the catalytically active species, acting as a strong nucleophile (at neutral pH) on the CO_2 molecule bound in a hydrophobic pocket nearby [1,2,6]. This metal hydroxide species is generated from a water coordinated to the metal ion, which by itself is not nucleophilic enough to act as a catalyst (Fig. 1A). This nucleophilic, zinc hydroxide species of the enzyme (Fig. 1A, step A) attacks the CO_2 molecule bound within a hydrophobic pocket nearby (step B), generating bicarbonate bound to the zinc ion (Step C in Fig. 1A). An incoming water molecule replaces then the coordinated bicarbonate, which is liberated into solution with the generation of the acidic form of the enzyme (step D in Fig. 1A). This is catalytically inefficient and must lose one proton, for generating the zinc hydroxide species shown in Fig. 1A, step A. This is the rate-determining step of the catalytic cycle, and it is assisted by a His residue placed in the middle of the active site (His64) and presumably by a His cluster found in some catalytically very efficient isoforms, such as hCA II (illustrated in Fig. 1B) [34,35].

As observed in Fig. 2, the binding of the two substrates (CO_2 and bicarbonate) has been revealed in detail by using high-resolution X-ray crystallography, for the corresponding two compounds bound to hCA II [36,37]. Indeed, by using cryo-crystallography at high pressure, CO_2 was observed not to be coordinated to the metal ion, whereas the hydrophobic pocket, which accommodates it being formed by residues Val121, Val143, Trp209, and Leu198—Fig. 2 [37]. Bicarbonate was on the other hand observed coordinated to the zinc ion [38] and participating to a hydrogen bond with the OH of Thr199, which in turns also hydrogen bonds with the carboxylate of Glu106. These two residues (the Thr199-Glu106 dyad) are called the gate keepers and orientate the substrate(s) and inhibitors for the

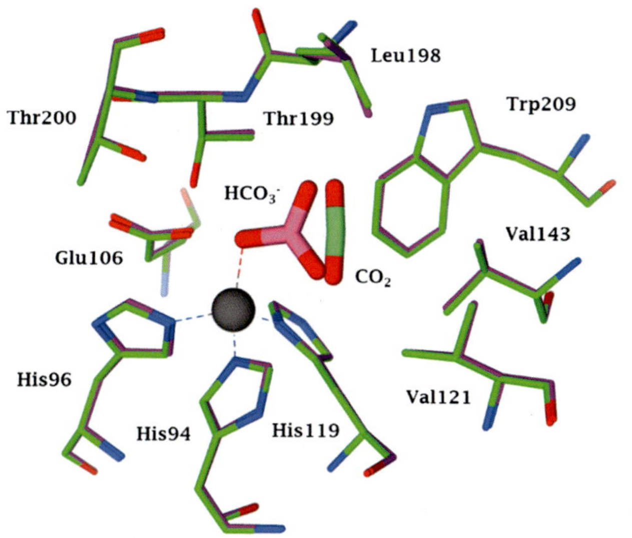

FIG. 2 Binding of CO_2 (*green*, pdb 2VVA) and bicarbonate (*pink*, pdb 2VVB) superimposed within the hCA II active site [36,37]. The zinc ion (shown as a *gray sphere*) with its three coordinated His ligands *(green/red)*, His94, 96, and 119, as well as residues involved in the binding of these substrates/inhibitors are also shown.

proper binding within the enzyme active site [1–3,37,38]. Presumably, in all other CA classes, this type of catalytic mechanism is also valid.

Detailed structural features of a typical α-CA, hCA II, are shown in Fig. 3. Most α-CAs are monomers, but homodimers were also reported for some human and bacterial enzymes [36–40]. As shown Fig. 3A for the human (h) isoform hCA II, the protein has an egg-like shape with dimensions of $50 \times 40 \times 40\,\text{Å}^3$ and a typical fold characterized by a central 10-stranded antiparallel β-sheet surrounded by several helices and additional β-strands [38–40]. Based on the high sequence identity of the cytosolic hCAs, their three-dimensional structures are rather similar, with all secondary structure elements strictly conserved [1,2,38–40]. However, an accurate structural comparison of all hCAs crystallized so far revealed a number of small local structural differences, which were mainly localized around residues 127–136 (hCA II amino acid numbering system), thus occurring both on the surface of the protein (residues 125–130) and in the middle of the active site (residue 131). The active site is located in a large, cone-shaped cavity that reaches the center of the molecule (Fig. 3B), where the Zn(II) ion coordinated by three His residues and a water molecule is placed [38–40]. An extended network of ordered water molecules is also present nearby the zinc [36–40].

The β-CAs are catalytically active mostly as dimers (Fig. 4A) [3,41–46], or tetramers (dimers of dimers) or as multiples of such homodimers, especially for the higher plant enzymes [41]. This is also true for the COS hydrolase from *Thiobacillus thioparus* [47], and the CS_2 hydrolase from *Acidianus* A1-3 [19] also belonging to the β-CA class, which are tetramers and octadecamers, respectively, with an occluded active site entrance, which induces the discrimination for the binding of COS and CS_2 over the more hydrophilic CO_2 to these enzymes [19,47]. In most β-CAs (e.g., from *Pisum sativum*, *Methanobacterium thermoautotrophicum*, *Mycobacterium tuberculosis* Rv1284, *C. neoformans*, etc.) [41–44], the active site zinc ion is coordinated by one histidine and two cysteine residues, with a fourth coordination site occupied by water molecule (the so-called open conformation, as shown in Fig. 4B). In contrast, the other subclass

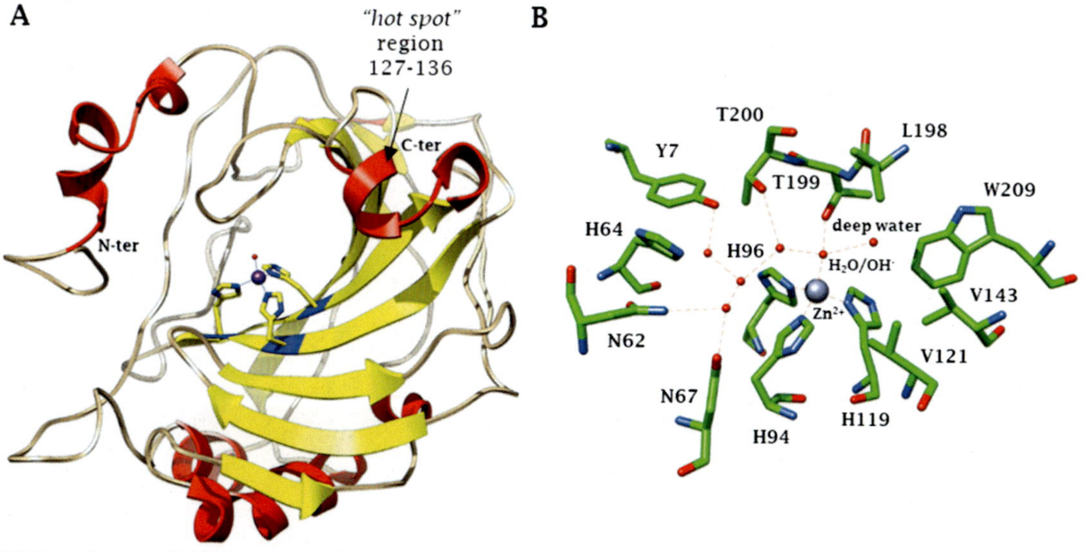

FIG. 3 (A) Ribbon diagram of hCA II structure (pdb code 1CA2), which has been chosen as representative α-CA. The active site Zn^{2+} coordination is also shown. Helix and β-strand regions are colored in *red* and *yellow*, respectively. (B) View of hCA II active site. The Zn^{2+} *(gray sphere)* is coordinated by three histidines residues and a water molecule/hydroxide ion *(red sphere)*, which is engaged in an extended network of hydrogen bonds with other water molecules, shown as *red spheres*.

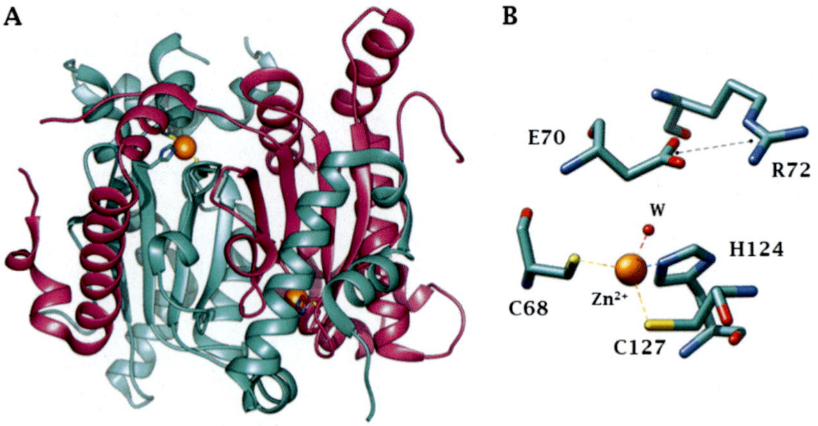

FIG. 4 (A) X-ray crystal structure of a representative dimeric β-CA from the pathogenic fungus *Cryptococcus neoformans* (Can2) dimer. One monomer is colored *magenta*, while the other one is shown is *turquoise*. The Zn^{2+} ions are evidenced as *orange spheres*. (B) Detail of Can2 active site. The Zn(II) ion *(gold sphere)* is coordinated by C68, H124, C127, and a water molecule *(red sphere)*, whereas the D70-R72 dyad (forming a salt bridge) is also evidenced.

of β-CAs (e.g., the enzymes from *Haemophilus influenzae*, *Escherichia coli*, *Porphyridium purpureum*, *Mycobacterium tuberculosis* Rv3588c, or *Vibrio cholerae*) [45,46] has a diverse zinc coordination geometry, where the water molecule being replaced by an aspartate side chain, forming a noncanonical CA active site (the closed conformation). However, the protein fold of the two types β-CAs is quite similar. Each monomer of the catalytically active dimer (or tetramer) adopts an α/β fold, being composed of an N-terminal arm, which extends away from the rest of the molecule and makes significant contacts with an adjacent monomer, a conserved zinc-binding core and a C-terminal domain dominated by α-helices (see Fig. 4A, in which the *C. neoformans* dimeric enzyme Can2 is shown) [42]). The active site is located at one end of the central β-sheet at the interface between dimers [42]. In the Can2 active site, the zinc ion is coordinated to residues, Cys68, His124, Cys127, in an almost tetrahedral geometry (Fig. 4B). The N-terminal subdomain of the Can2 core is formed by four antiparallel α-helices (α1-4), whereas the C-terminal subdomain containing a five-stranded β-sheet, with four parallel β-strands (β1-4) and β5, formed by the C-terminus, attached in an antiparallel orientation [42]. The C-terminal subdomain is packed between the N- and C-terminal domains of a partner monomer, resulting in a tight dimer interaction [42], Fig. 4A.

The prototype γ-CA has been initially characterized from the methanogenic archaeon *Methanosarcina thermophila*, by Ferry's group [48–50]. This enzyme, Cam, adopts a left-handed parallel β-helix fold, as shown by X-ray crystallography (Fig. 5A). This fold is of particular interest since it contains only left-handed crossover connections between the parallel β-strands, which have been infrequently observed so far in proteins. The active form of the enzyme is a trimer with three zinc-containing active sites (Fig. 5A), each located at the interface between two monomers, shown in Fig. 5B. There are structural similarities in the zinc coordination environment of the γ- and α-CAs, suggestive of convergent evolution dictated by the chemical requirements for catalysis of the same reaction [48–50]. A subsequent work from the same group, presenting a structure at higher resolution, showed the side chains of Glu62 and Glu84 to

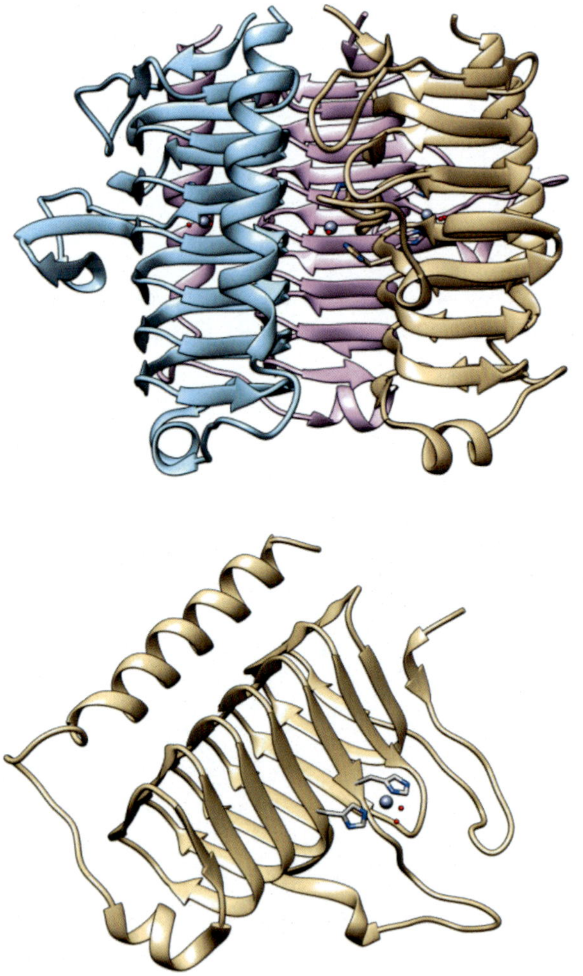

A

B

FIG. 5 (A) X-ray crystal structure of a homotrimeric γ-CA, Cam, from the archaeon *Methanosarcina thermophila*. (B) Protein fold of the Cam monomer, pdb file 1QRG.

share a proton. Furthermore, Glu84 exhibited multiple conformations, suggesting that this residue may act as a proton shuttle (similar to His64 in α-CAs), which seems to be an important aspect of the reaction mechanism of all CA classes. A hydrophobic pocket on the surface of the enzyme was also detected, which probably participates in the trapping of CO_2 at the active site [50]. Few other γ-CA structures except Cam were reported to date, one of which is the enzyme from the pathogenic bacterium *Burkholderia pseudomallei* [51].

The δ- and η-CAs have not been crystallized so far, whereas the cadmium-containing ζ-CA class is discussed in another chapter of the book, and thus, we will not deal with the structures of these enzymes. The θ-CAs seem to have a crystallographic structure very similar with the β-CAs [52], whereas the ι-CAs, as already mentioned, are devoid of metal ions within the active site.

3 CA inhibition mechanisms, classes of inhibitors

There are several CA inhibition mechanisms [1,6,8–10], which are schematically presented in Fig. 6. They are valid probably for all CA classes and are exemplified here (Fig. 6) for the α-CAs. These are the four main CA inhibition mechanisms reported to date:

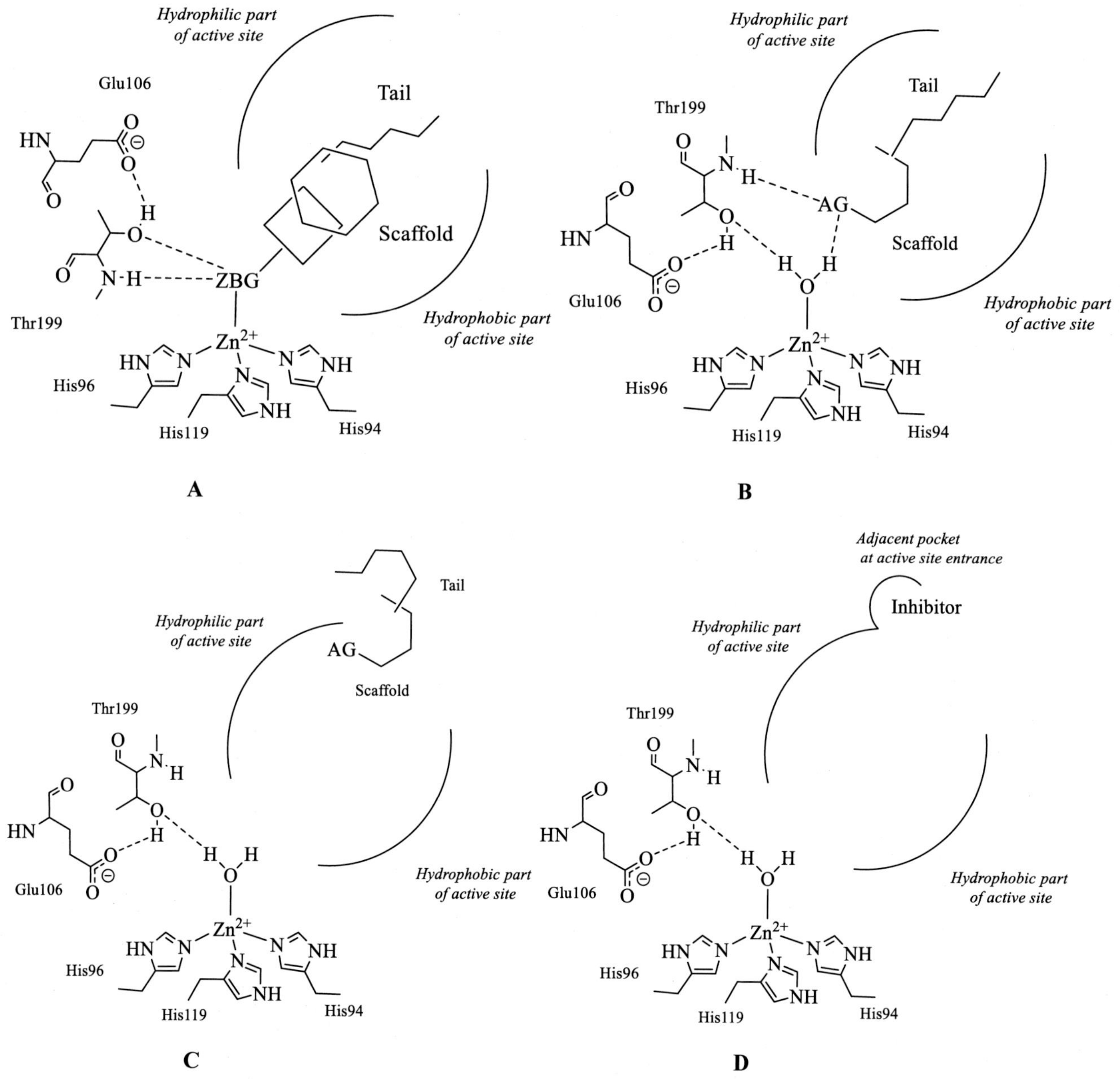

FIG. 6 CA inhibition mechanisms. (A) Zinc binding. (B) Anchoring to the zinc-coordinated water. (C) Occlusion of the active site entrance. (D) Out of the active site binding. The examples refer to an α-class enzyme but are probably valid for all other CA families.

(i) *Zinc-binding inhibitors* (Fig. 6A). These CA inhibitors (CAIs) incorporate a zinc-binding group (ZBG), or more generally, a metal-binding group (MBG), which is coordinated to the active site metal ion, which usually remains in its tetrahedral geometry, as shown in Fig. 6A, although in few cases, the presence of a water molecule coordinated to the metal ion was also evidenced, leading to a trigonal and bipyramidal geometry of the metal center [6,8–10,53]. This is the classical CA inhibition mechanism, and many classes of inhibitors bind in this way, among which are most inorganic anions [6,53], sulfonamides and their isosteres [1,6,8–10], phosphonamidates [54], dithiocarbamates [55] and their derivatives (monothiocarbamates [56], xanthates and trithiocarbonates [57]), selenols [58,59], carboxylates [60], hydroxamates [61], benzoxaboroles [62–64], carbamates [65], and ninhydrins [66].

(ii) *Inhibitors anchoring to the zinc-coordinated water* (Fig. 6B). These CAIs possess an anchoring group (AG) by which they hydrogen bond to the metal ion coordinated water molecule/hydroxide ion [1,6,8–10]. Originally discovered for phenols [67–73], this inhibition mechanism has been thereafter documented (by X-ray crystallography and other biophysical measurements) for catechols [74], polyamines [75], sulfocoumarins (which bind in hydrolyzed form as sulfonates) [76–79], and thioxocoumarins [80].

(iii) *Compounds that occlude the entrance of the active site* (Fig. 6C). These CAIs bind at the entrance of the active site cavity, in the region that for α-CAs is the most variable one from the viewpoint of amino acid sequence and also possessing AGs by which they interact with amino acid residues or water molecules in that region of the active site [1–3]. The first type of such an inhibitor was a natural product coumarin [81], but subsequent studies showed that all coumarins and structurally related mono-/bicyclis lactones possess this type of mechanism, which in fact leads to the hydrolysis of the lactone ring and formation of hydroxy-cinnamic acids (in the case of the coumarins) or aliphatic hydroxy-acids for the 5-/6-membered lactones [82–90]. The mechanism by which these inhibitors are generated is shown schematically in Chart 2.

(iv) *Out of the active site binding* (Fig. 6D). These inhibitors bind in an adjacent pocket at the entrance but outside the active site cavity and, by hydrogen bonding with His64 (the proton shuttle residue in α-CAs), lead to the collapse of the catalytic cycle, as demonstrated by using crystallography and kinetic measurements for a benzoic acid derivative possessing an *ortho*-benzylsulfoxide moiety [91].

Fig. 7 shows the detailed binding for some of these compounds (which inhibit CAs by mechanism (ii) mentioned previously), whereas Fig. 8 presents two classes of CAIs that bind by coordinating to the metal ion: a dithiocarbamate and a benzoxaborole derivative, again as determined by crystallographic techniques [55,62].

In Fig. 9, the X-ray crystal structures of two coumarin derivatives bound to hCA II are also shown [82,83] together

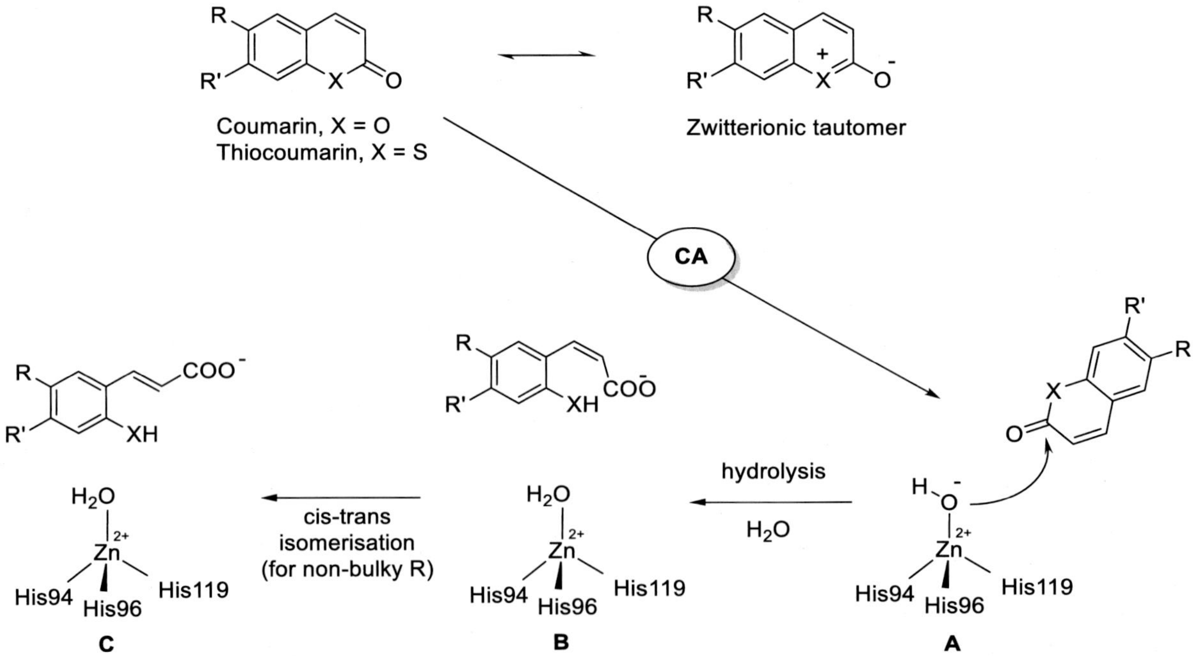

CHART 2 Proposed mechanism [80] for hydrolysis of (thio)coumarins with generation of the inhibitor (steps A and B), which thereafter occludes the active site 8 as schematically shown in C.

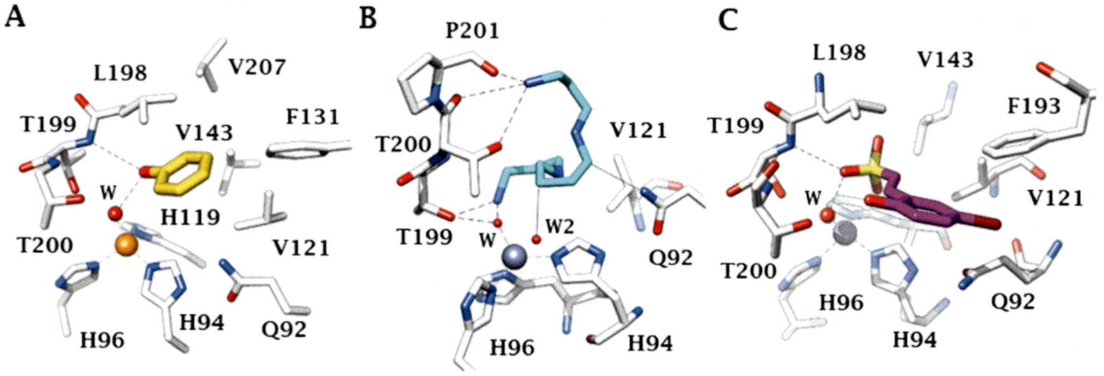

FIG. 7 X-ray crystal structures of compounds binding by mechanism (ii): (A) phenol. (B) spermine (pdb 3KWA). (C) hydroxyphenyl-ω-ethenylsulfonic acid derived from sulfocoumarin hydrolysis (pdb 4BCW).

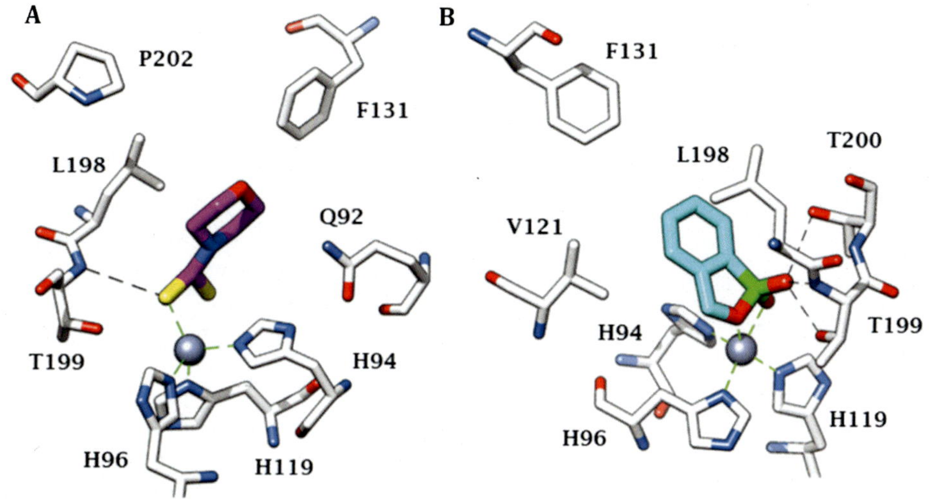

FIG. 8 Active site view of the hCA II adduct with (A) the dithiocarbamate derived from morpholine (pdb 3P5A) and (B) simple benzoxaborole (pdb 5JQT).

with the formation of the real CAI by active site-mediated hydrolysis of the lactone ring and generation of the 2-hydroxy-cinnamic acids, which bind at the entrance of the cavity either as *cis*- or *trans* isomers.

4 CA inhibitors in clinical use

Due to their ubiquitous presence in various tissues and organs in most vertebrates, including humans, hCAs have a variety of physiological functions (Fig. 10) and their inhibition or activation leads to pharmacological responses, which have been exploited therapeutically for more than 70 years [1,6–11]. As shown in Fig. 10, various isoforms are involved in diverse physiological/pathological processes, and their selective inhibition is highly desirable for obtaining drugs with few side effects [1].

CAIs were used as diuretics [1,11,92–94] and antiglaucoma agents [95–97] for more than seven decades, and they still have a firm place for the management of these conditions [11]. However, in the last decade, several other highly relevant applications emerged, probably due to the availability of much more isoform-selective inhibitors. The clinically used inhibitors are shown in Fig. 11. They include pan-inhibitors such as acetazolamide **AAZ**, methazolamide **MZA**, ethoxzolamide **EZA**, dichlorophenamide **DCP**; topically acting second generation antiglaucoma sulfonamides (dorzolamide **DZA**, brinzolamide **BRZ**); the orphan drug benzolamode **BZA**; antiepileptics such as topiramate **TPM** and zonisamide **ZNS**, as well as drugs that have been approved for other targets, among which are sulpiride **SLP**, indisulam **IND**, valdecoxib **VLX**, celecoxib **CLX**, sulthiame **SLT**, hydrochlorothoazide **HCT**, famotidine **FAM,** and epacadostat **EPA**, Fig. 11 [1,11]. Most of them act as nonselective CAIs for all human isoforms [1,11]. Their inhibition constants against 12 hCAs are reported in [1].

Although the CAIs shown in Fig. 1 are still in clinical use as diuretics [11,92–94], antiglaucoma [95–97], antiepileptics

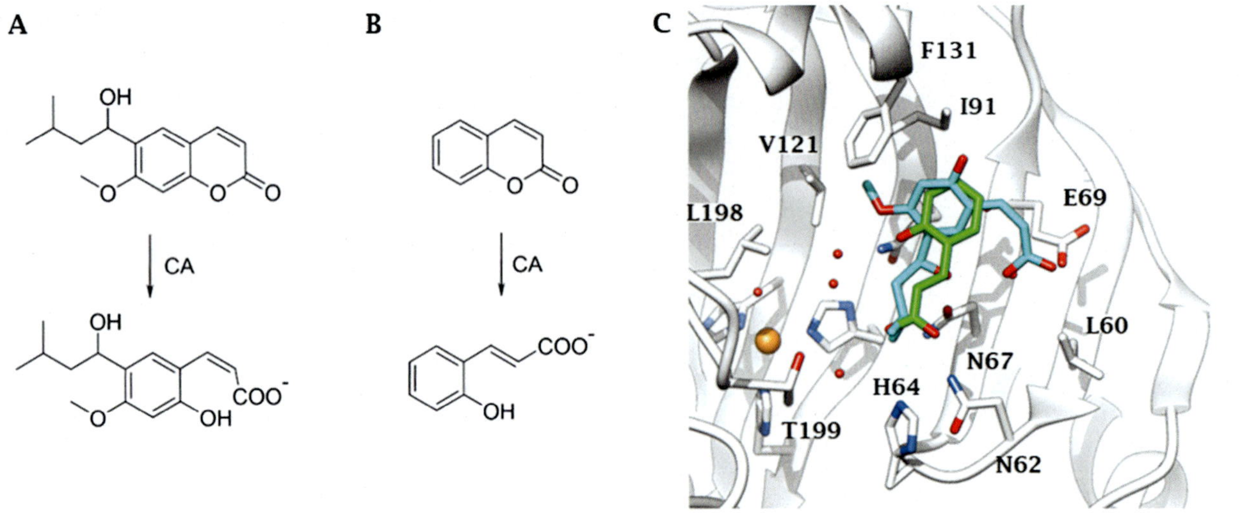

FIG. 9 CA-mediated hydrolysis of (A) the natural product coumarin isolated from *Leionema ellipticum* [82] and (B) unsubstituted coumarin. (C) Active site view of the superimposed hCA II adducts with coumarins hydrolysis products shown in panel A—*cis* isomer (*blue*, pdb 3F8E) and B—*trans* isomer (*green* pdb 5BNL).

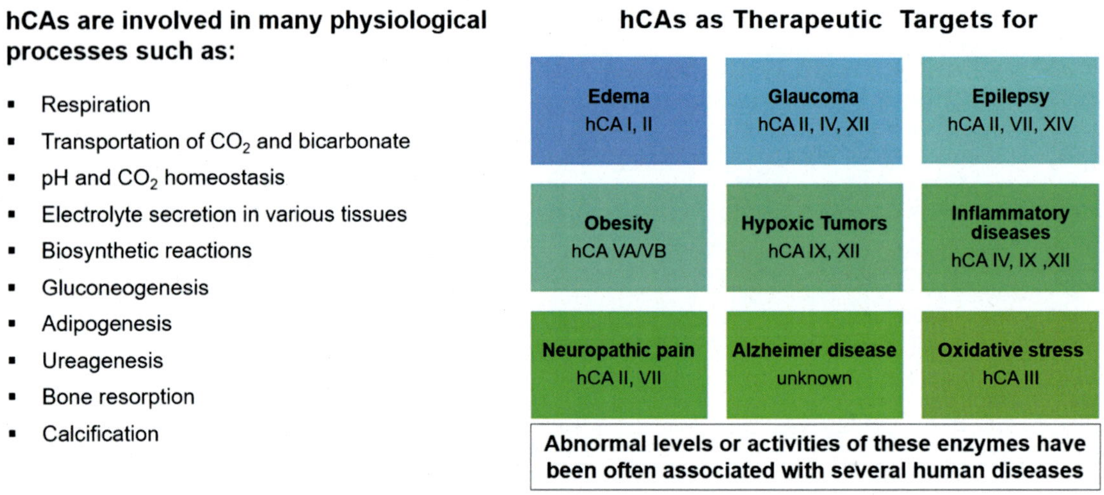

FIG. 10 Physiologic processes in which hCAs are involved and their therapeutic uses.

[98–100], and antiobesity agents [101,102], the need to have isoform-selective compounds that target a limited number of CA isoforms involved in specific diseases led to the discovery of the tail approach [103], which indeed afforded compounds able to achieve this goal [104].

Designing isoform-selective CAIs thus became the main issue in the revival of the CA field in the mid to late 1990s for obtaining compounds with an improved safety profile and efficacy compared to the inhibitors available at that time, as shown in Fig. 11 [11,103]. The tail approach has thus emerged in that period as an innovative modality for developing CA-selective inhibitors [103]. The idea is very simple: the attachment of moieties that may induce the desired physicochemical properties (e.g., enhanced hydro- or liposolubility) to scaffolds of simple aromatic/heterocyclic sulfonamides of types **1–24** shown in Fig. 12 (e.g., orthanilamide metanilamide, sulfanilamide **1–3** and their derivatives **4–12**; 5-amino-1,3,4-thiadiazole-2-sulfonamide **13**, 4-methyl-5-imino-1,3,4-thiadiazoline-2-sulfonamide **14,** and many other simple sulfonamides such as compounds **15–24** in Fig. 12) might lead to interactions with the external part of the CA active site and not only with residues near the zinc catalytic center that are the most highly conserved in all hCA isoforms. Indeed, the region at the entrance of the active site cavity is the most variable between the different human isoforms [1–3]. Detailed kinetic studies were thereafter performed on over 10,000 different sulfonamide derivatives [1,6,105–115], which demonstrated that the hypothesis was correct and that it is possible to design sulfonamide CAIs that are selective for

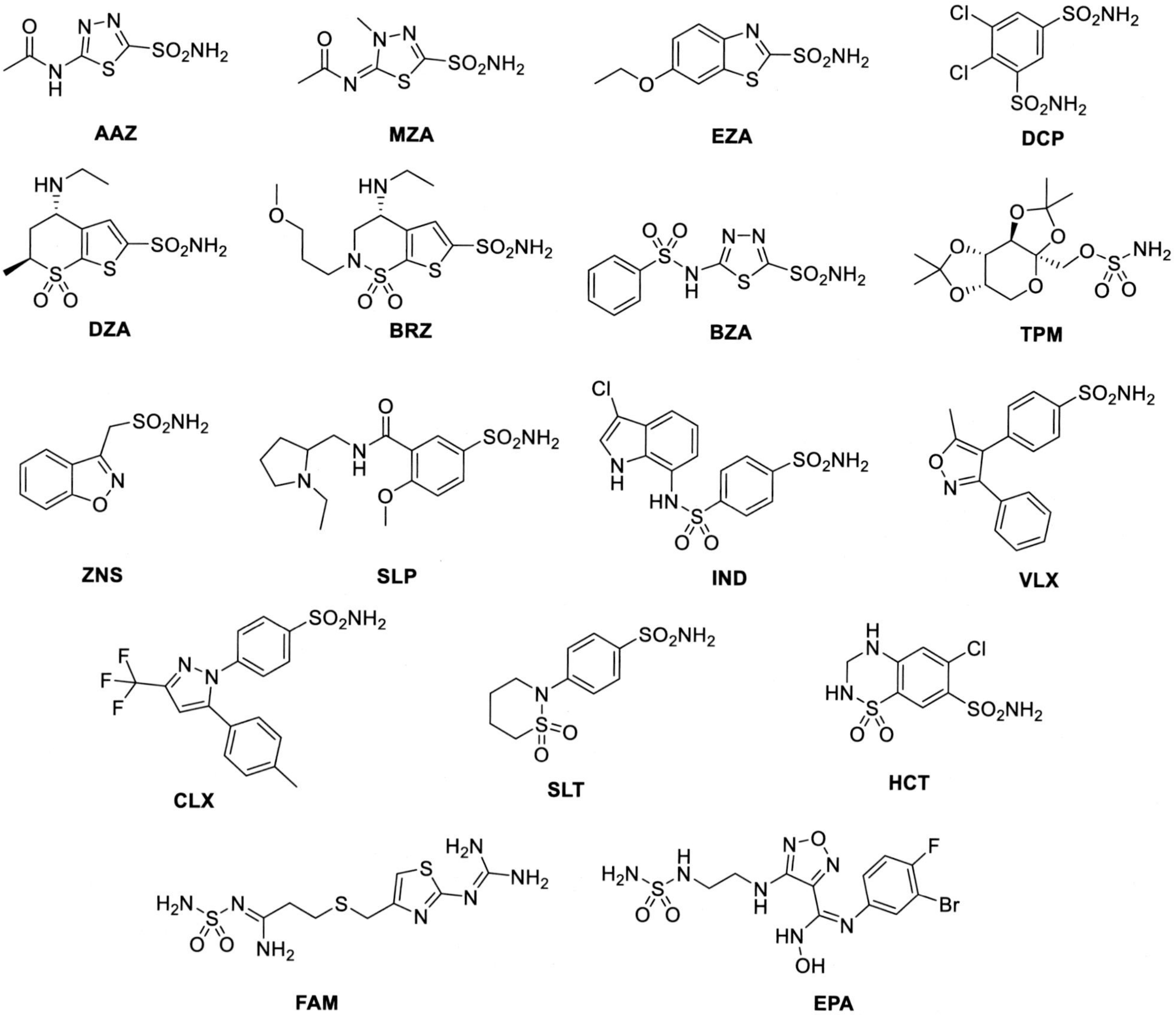

FIG. 11 Clinically used sulfonamide, sulfamate, or sulfamide CAIs **AAZ-EPA**.

each isoform of interest by tailoring the dimensions (length, bulkiness, etc.) as well as the chemical nature of the various tails [103–115]. This research led to the discovery of many new sulfonamide, sulfamate, and sulfamide classes, including the ureido-benzene-sulfonamides [113,115] as well as derivatives possessing more than one tail attached to the scaffold, such as dual-tailed or tri-tailed CAIs [113–115]. Of the many compounds discovered in this way, SLC-0111, compound **25**, was shown to be a CA IX/XII-selective inhibitor, now in clinical trials as an antitumor/antimetastatic agent (see discussion later in the text). Another notable example is the three-tailed sulfonamide **26** (Fig. 13) [75]. Both SLC-0111 and sulfonamide **26** strongly inhibit and bind to hCA IX (Fig. 13) [113–116].

The fact that isoforms hCA IX and XII are overexpressed in hypoxic tumors led to drug design campaigns and relevant proof-of-concept biological studies, which validated these two isoforms as new drug targets for the management of metastatic hypoxic tumors [117–120]. SLC-0111 **25** binds selectively to hCA IX by coordination of its SO_2NH^- moiety to the positively charged Zn(II) ion in the CA active site (Fig. 13A). In addition, a second strong contact with the Zn(II) ion involves an oxygen of the sulfonamide. In contrast, these interactions are either weak or absent in the case of the hCA II isoform [114]. Indeed the extensive preclinical models carried out by several independent groups with the lead CA IX/XII inhibitors, including SLC-0111, have demonstrated that the use of such inhibitors in combination with chemotherapy agents, immunotherapy, and radiotherapy [121–125] is highly important and desirable for sustained therapeutic antitumor response. The extensive studies reported in these and other papers, utilizing multiple in depth in vivo models, provided solid positive preclinical data to warrant the initiation of Phase 1

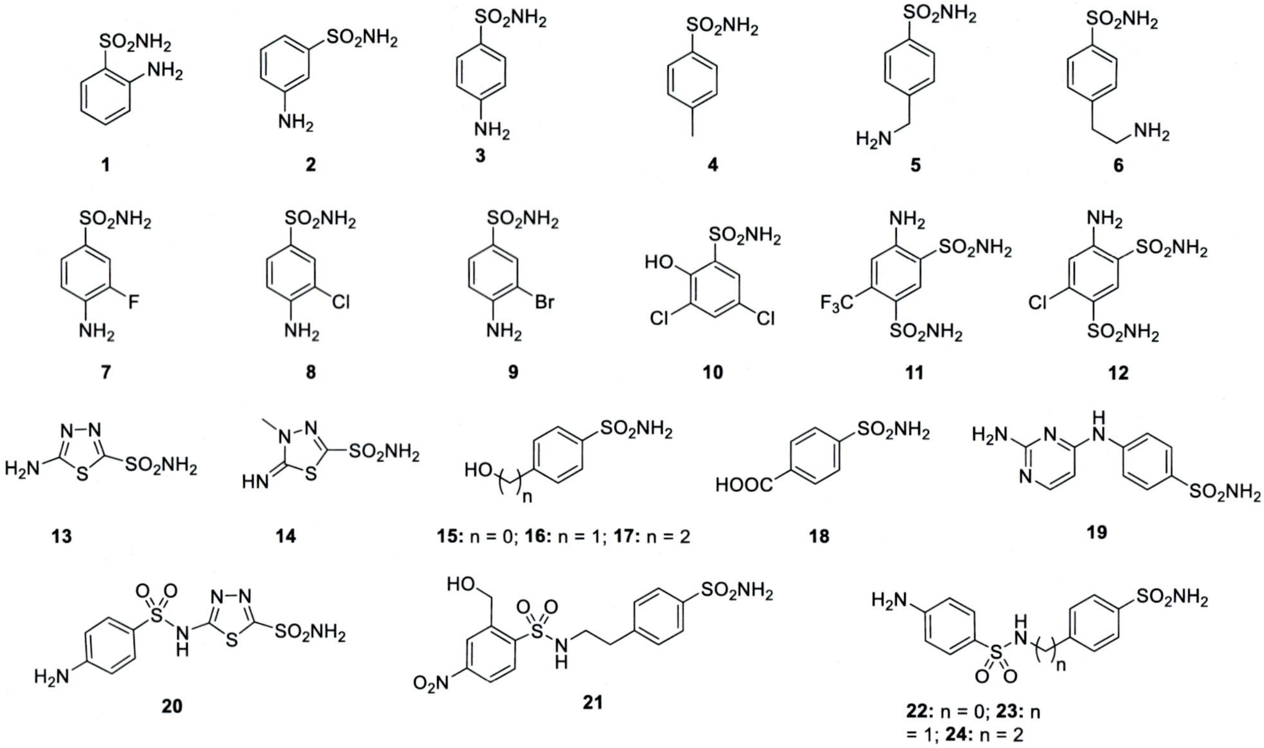

FIG. 12 Simple sulfonamides **1–24** used for derivatization reactions in the tail approach.

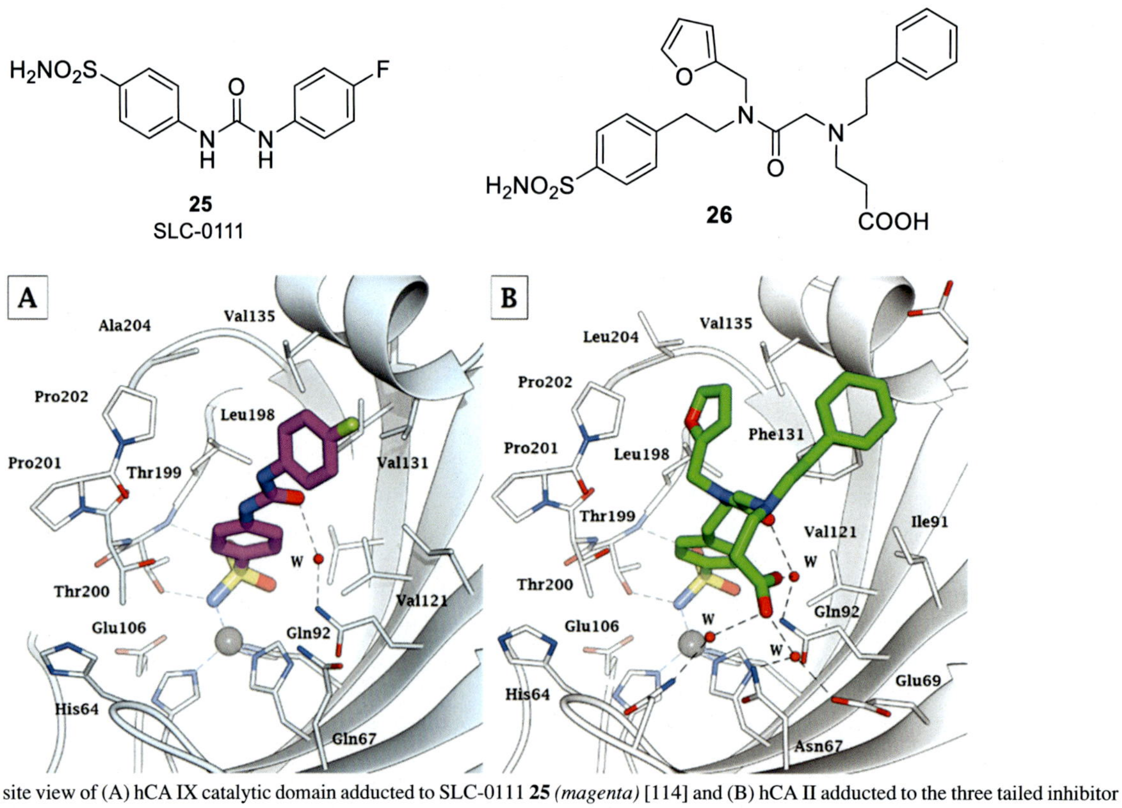

FIG. 13 Active site view of (A) hCA IX catalytic domain adducted to SLC-0111 **25** *(magenta)* [114] and (B) hCA II adducted to the three tailed inhibitor **26** (in *green*) [116]. H-bonds are represented as *black dashed lines*. The Zn(II) ion *(gray sphere)* and some important amino acid residues involved in the binding of inhibitors are shown.

clinical trials in 2014, of which a Phase 1 safety trial with SLC-0111 (as a monotherapeutic agent) has been completed [121], and a Phase 1b trial is currently underway to evaluate SLC-0111 in combination with gemcitabine in metastatic pancreatic cancer patients whose tumors are CA IX positive (ClinicalTrials.gov Identifier: NCT03450018). As a frontrunner selective inhibitor for the tumor-associated isoform CA IX, SLC-0111 has been utilized as a lead CAI for the development of novel promising small molecules with selective inhibitory activity toward CA IX and with good druggability and lead-likeness characters. Several drug design approaches based on the tail approach have been utilized to develop a range of new SLC-0111 analogues, which have been reviewed elsewhere [126]. Recently, it has also been demonstrated that SLC-0111 and some of its congeners possess an additional antitumor mechanism, by interfering with ferroptosis [127]. CAIX inhibition with these sulfonamides led to acidification of the intracellular pH, increased cellular reactive oxygen species accumulation, and induced susceptibility to alterations in iron homeostasis. This is due to the inhibition of bicarbonate production by CA IX or sodium-driven bicarbonate transporters, which also interferes with the cysteine/cystine transporter xCT, decreased adenosine 5′-monophosphate-activated protein kinase activation, and increased acetyl-coenzyme A carboxylase 1 activation [127]. Thus, an alkaline intracellular pH plays a critical role in suppressing ferroptosis, a finding that may lead to the development of innovative therapeutic strategies for solid tumors to overcome hypoxia- and acidosis-mediated tumor progression and therapeutic resistance [127].

Apart all these developments highlighted previously, nowadays the CAIs were proposed to be used or are already used for the management of the following pathologies, in addition to the classical ones mentioned previously: (i) treatment of idiopathic intracranial hypertension, by targeting hCA I and II but presumably also brain/cerebrospinal fluid isoforms [128]; (ii) cerebral ischemia/diabetic cerebrovascular pathologies (the targeted isoforms seem to be hCA IX and XII, possibly also hCA VA) [129]; (iii) neuropathic pain, by targeting the bran/peripheral nervous system isoforms hCA II and VII [130,131]; (iv) inflammation/arthritis, by targeting hCA IX/XII and possibly hCA IV [132–137]; and (v) hypoxic tumors, as already mentioned previously [11,117–120]. This is in fact the field in which the highest number of studies have been published in the past decades [117–126].

5 CA activators

The carbonic anhydrase activators (CAAs) have been the focus of controversy for a long time up until 1997 when the first X-ray crystal structure of a complex of an activator (histamine) bound to hCA II was reported, and the activation mechanism was revealed in molecular level detail [35]. The activator, which was observed to be bound within the enzyme active site, where it promotes the formation of an enzyme-activator complex, in which the proton shuttling moieties present in the activator participate in the rate-determining step of the catalytic cycle, that is, the transfer of a proton from the zinc-coordinated water to the external reaction medium, similar to the natural proton shuttle, which is residue His64 (in many CA isoforms), as mentioned previously [34,35]. In such enzyme-activator complexes, the proton transfer is intramolecular, being more efficient compared to the intermolecular transfer to buffer molecules, which are not bound within the enzyme cavity [35]. Many X-ray crystal structures with amines and amino acid activators were reported thereafter, in addition to histamine [138], including L- and D-His bound to hCA II and hCA I, L- and D-Phe, D-Trp, L-adrenaline and pyridinium derivatives of histamine bound to hCA II, which confirmed this general CA activation mechanism outlined previously [139–145]. The 13 catalytically active mammalian CAs (e.g., CA I–VA, VB, VI, VII, IX, XII–XV) were also investigated for their interaction with a rather large library of amino acids and amines [144], whereas several drug design studies have also been performed [146–152].

Why are these studies relevant? In the last few years, it has been demonstrated that CAAs of the amino acid type (e.g., D-Phe) can enhance memory and learning, which is antagonized by the simultaneous administration of a CAI of the sulfonamide type (e.g., acetazolamide) [153]. A recent study showed that administration of the D-Phe to rats rapidly activated the extracellular signal-regulated kinase (ERK) pathways, which are involved in critical steps of memory formation, both in the cortex and the hippocampus, two brain areas crucially involved in memory processing [153,154]. Even more recently, the same groups demonstrated that CAAs play a crucial role in extinction of contextual fear memory [155], opening the way toward the validation of CAs as new targets for improving cognition, but also in therapeutic areas, such as phobias, obsessive-compulsive disorder, generalized anxiety, and posttraumatic stress disorders, for which few efficient therapies are available to date [153–158].

6 Conclusions and future prospects

CAs are thoroughly investigated enzymes, with many new representatives and new enzyme families being constantly discovered. Almost each 2–3 years, a new CA class has been discovered in the last decade, and probably this trend will continue also in the future. The human (mammalian) enzymes are well characterized, although not all of them have been crystallized in detail (the structures of CA VB, X, and XI are still not known). The catalytic, inhibition and activation mechanisms of these enzymes are also well understood [1,6–11]. Starting from the classical sulfonamide inhibitors known for decades, in the last 20 years, new compounds have been discovered, together with a strategy to design isoform-selective inhibitors, the tail approach [1,6,103]. By using this drug design strategy, but also by discovering new CA inhibitory chemotypes [6], nowadays a large number of highly isoform-selective CAIs are available, which led to relevant progress in targeting various isoforms and diverse diseases in which they are involved [11]. The field of mammalian (human) CAs and their modulators are in constant development and progress, with new generation inhibitors/activators affording tools and possibilities for novel applications, as described here for antitumor agents, antiobesity drugs as well as totally novel applications (neuropathic pain, inflammation/arthritis, degenerative and other neurological disorders [11,155–159]). However, the prokaryotic/pathogenic CAs were less investigated till recent year. However, also in this field notable progress has been achieved, and probably this is the field that in future years the most notable achievements will be done.

It has been known for many years that pathogenic (but also nonpathogenic) bacteria, yeasts, protozoans, and worms encode for CAs belonging to various classes [3,4,13–17]. It is nowadays well established that some of these CAs may be important drug targets for designing anti-infectives with a novel mechanism of action, devoid of the drug resistance problems encountered with clinically used antiinfective drugs [160–173]. Proof-of-concept studies have been so far provided for the antibacterial, antifungal, and antiprotozoan action of some CAIs, but presumably in the nearby future more detailed such data will become available[174–178].

Overall, the CA field is a very dynamic one with a huge number of exciting discoveries being reported constantly, which will surely lead to relevant discoveries translatable to new drugs with more efficacy and less side effects. The future may also see the validation of CAs as targets for neurodegenerative disorders and the understanding of what are the isoforms involved in these processes. Combining the drug design studies with biochemical and structural investigations, as already done over the last decades, may lead to relevant developments in understanding better these enzymes and providing pharmacological tools/drugs with an innovative mechanism of action.

References

[1] Supuran CT. Carbonic anhydrases: novel therapeutic applications for inhibitors and activators. Nat Rev Drug Discov 2008;7(2):168–81.

[2] Aspatwar A, Tolvanen MEE, Barker H, Syrjänen L, Valanne S, Purmonen S, Waheed A, Sly WS, Parkkila S. Carbonic anhydrases in metazoan model organisms: molecules, mechanisms, and physiology. Physiol Rev 2022;102:1327–83.

[3] Capasso C, Supuran CT. An overview of the alpha-, beta- and gamma-carbonic anhydrases from bacteria: can bacterial carbonic anhydrases shed new light on evolution of bacteria? J Enzyme Inhib Med Chem 2015;30:325–32.

[4] Capasso C, Supuran CT. Bacterial, fungal and protozoan carbonic anhydrases as drug targets. Expert Opin Ther Targets 2015;19:1689–704.

[5] Supuran CT, Capasso C. Biomedical applications of prokaryotic carbonic anhydrases. Expert Opin Ther Pat 2018;28:745–54.

[6] Nocentini A, Angeli A, Carta F, Winum JY, Zalubovskis R, Carradori S, Capasso C, Donald WA, Supuran CT. Reconsidering anion inhibitors in the general context of drug design studies of modulators of activity of the classical enzyme carbonic anhydrase. J Enzyme Inhib Med Chem 2021;36:561–80.

[7] Supuran CT. Carbonic anhydrases and metabolism. Metabolites 2018;8(2):25.

[8] Supuran CT. How many carbonic anhydrase inhibition mechanisms exist? J Enzyme Inhib Med Chem 2016;31:345–60.

[9] Mishra CB, Tiwari M, Supuran CT. Progress in the development of human carbonic anhydrase inhibitors and their pharmacological applications: where are we today? Med Res Rev 2020;40:2485–565.

[10] Supuran CT. Structure and function of carbonic anhydrases. Biochem J 2016;473(14):2023–32.

[11] Supuran CT. Emerging role of carbonic anhydrase inhibitors. Clin Sci (Lond) 2021;135(10):1233–49.

[12] Alterio V, Langella E, Viparelli F, Vullo D, Ascione G, Dathan NA, Morel FM, Supuran CT, De Simone G, Monti SM. Structural and inhibition insights into carbonic anhydrase CDCA1 from the marine diatom *Thalassiosira weissflogii*. Biochimie 2012;94(5):1232–41.

[13] Supuran CT, Capasso C. The eta-class carbonic anhydrases as drug targets for antimalarial agents. Expert Opin Ther Targets 2015;19:551–63.

[14] Del Prete S, Vullo D, De Luca V, Supuran CT, Capasso C. Biochemical characterization of the δ-carbonic anhydrase from the marine diatom *Thalassiosira weissflogii*, TweCA. J Enzyme Inhib Med Chem 2014;29(6):906–11.

[15] Jensen EL, Clement R, Kosta A, Maberly SC, Gontero B. A new widespread subclass of carbonic anhydrase in marine phytoplankton. ISME J 2019;13(8):2094–106.

[16] Hirakawa Y, Senda M, Fukuda K, Yu HY, Ishida M, Taira M, Kinbara K, Senda T. Characterization of a novel type of carbonic anhydrase that acts without metal cofactors. BMC Biol 2021;19(1):105.

[17] Nocentini A, Supuran CT, Capasso C. An overview on the recently discovered iota-carbonic anhydrases. J Enzyme Inhib Med Chem 2021;36(1):1988–95.

[18] Alterio V, Langella E, Buonanno M, Esposito D, Nocentini A, Berrino E, Bua S, Polentarutti M, Supuran CT, Monti SM, De Simone G. Zeta-carbonic anhydrases show CS_2 hydrolase activity: a new metabolic carbon acquisition pathway in diatoms? Comput Struct Biotechnol J 2021;19:3427–36.

[19] Smeulders MJ, Barends TR, Pol A, Scherer A, Zandvoort MH, Udvarhelyi A, Khadem AF, Menzel A, Hermans J, Shoeman RL, Wessels HJ, van den Heuvel LP, Russ L, Schlichting I, Jetten MS, Op den Camp HJ. Evolution of a new enzyme for carbon disulphide conversion by an acidothermophilic archaeon. Nature 2011;478 (7369):412–6.

[20] Briganti F, Mangani S, Scozzafava A, Vernaglione G, Supuran CT. Carbonic anhydrase catalyzes cyanamide hydration to urea: is it mimicking the physiological reaction? J Biol Inorg Chem 1999;4 (5):528–36.

[21] Guerri A, Briganti F, Scozzafava A, Supuran CT, Mangani S. Mechanism of cyanamide hydration catalyzed by carbonic anhydrase II suggested by cryogenic X-ray diffraction. Biochemistry 2000;39(40):12391–7.

[22] Pocker Y, Meany JE. The catalytic versatility of carbonic anhydrase from erythrocytes. The enzymatic catalyzed hydration of acetaldehyde. J Am Chem Soc 1965;87:1809–11.

[23] Pocker Y, Stone JT. The catalytic versatility of erythrocyte carbonic anhydrase. The enzyme-catalyzed hydrolysis of *Para*-nitrophenyl acetate. J Am Chem Soc 1965;87:5497–8.

[24] Innocenti A, Supuran CT. Paraoxon, 4-nitrophenyl phosphate and acetate are substrates of α- but not of β-, γ- and ζ-carbonic anhydrases. Bioorg Med Chem Lett 2010;20:6208–12.

[25] Innocenti A, Scozzafava A, Parkkila S, Puccetti L, De Simone G, Supuran CT. Investigations of the esterase, phosphatase, and sulfatase activities of the cytosolic mammalian carbonic anhydrase isoforms I, II, and XIII with 4-nitrophenyl esters as substrates. Bioorg Med Chem Lett 2008;18(7):2267–71.

[26] Kazancioğlu EA, Güney M, Şentürk M, Supuran CT. Simple methanesulfonates are hydrolyzed by the sulfatase carbonic anhydrase activity. J Enzyme Inhib Med Chem 2012;27(6):880–5.

[27] Cavdar H, Ekinci D, Talaz O, Saraçoğlu N, Sentürk M, Supuran CT. α-Carbonic anhydrases are sulfatases with cyclic diol monosulfate esters. J Enzyme Inhib Med Chem 2012;27(1):148–54.

[28] Carta F, Maresca A, Scozzafava A, Supuran CT. 5- and 6-Membered (thio)lactones are prodrug type carbonic anhydrase inhibitors. Bioorg Med Chem Lett 2012;22(1):267–70.

[29] Tanc M, Carta F, Scozzafava A, Supuran CT. α-Carbonic anhydrases possess Thioesterase activity. ACS Med Chem Lett 2015;6(3):292–5.

[30] Supuran CT, Conroy CW, Maren TH. Is cyanate a carbonic anhydrase substrate? Proteins 1997;27:272–8.

[31] Angeli A, Carta F, Donnini S, Capperucci A, Ferraroni M, Tanini D, Supuran CT. Selenolesterase enzyme activity of carbonic anhydrases. Chem Commun (Camb) 2020;56(32):4444–7.

[32] Tripp BC, Bell 3rd CB, Cruz F, Krebs C, Ferry JG. A role for iron in an ancient carbonic anhydrase. J Biol Chem 2004;279(8):6683–7.

[33] De Simone G, Di Fiore A, Capasso C, Supuran CT. The zinc coordination pattern in the η-carbonic anhydrase from *plasmodium falciparum* is different from all other carbonic anhydrase genetic families. Bioorg Med Chem Lett 2015;25(7):1385–9.

[34] Tu CK, Silverman DN, Forsman C, Jonsson BH, Lindskog S. Role of histidine 64 in the catalytic mechanism of human carbonic anhydrase II studied with a site-specific mutant. Biochemistry 1989;28(19):7913–8.

[35] Briganti F, Mangani S, Orioli P, Scozzafava A, Vernaglione G, Supuran CT. Carbonic anhydrase activators: X-ray crystallographic and spectroscopic investigations for the interaction of isozymes I and II with histamine. Biochemistry 1997;36(34):10384–92.

[36] Domsic JF, Avvaru BS, Kim CU, Gruner SM, Agbandje-McKenna M, Silverman DN, McKenna R. Entrapment of carbon dioxide in the active site of carbonic anhydrase II. J Biol Chem 2008;283(45):30766–71.

[37] Sjöblom B, Polentarutti M, Djinovic-Carugo K. Structural study of X-ray induced activation of carbonic anhydrase. Proc Natl Acad Sci U S A 2009;106:10609–13.

[38] Jönsson BM, Håkansson K, Liljas A. The structure of human carbonic anhydrase II in complex with bromide and azide. FEBS Lett 1993;322:186–90.

[39] Alterio V, Hilvo M, Di Fiore A, Supuran CT, Pan P, Parkkila S, Scaloni A, Pastorek J, Pastorekova S, Pedone C, Scozzafava A, Monti SM, De Simone G. Crystal structure of the catalytic domain of the tumor-associated human carbonic anhydrase IX. Proc Natl Acad Sci U S A 2009;106(38):16233–8.

[40] Whittington DA, Waheed A, Ulmasov B, Shah GN, Grubb JH, Sly WS, Christianson DW. Crystal structure of the dimeric extracellular domain of human carbonic anhydrase XII, a bitopic membrane protein overexpressed in certain cancer tumor cells. Proc Natl Acad Sci U S A 2001;98(17):9545–50.

[41] Graham D, Reed ML, Patterson BD, Hockley DG, Dwyer MR. Chemical properties, distribution, and physiology of plant and algal carbonic anhydrases. Ann N Y Acad Sci 1984;429:222–37.

[42] Schlicker C, Hall RA, Vullo D, Middelhaufe S, Gertz M, Supuran CT, Mühlschlegel FA, Steegborn C. Structure and inhibition of the CO_2-sensing carbonic anhydrase Can2 from the pathogenic fungus *Cryptococcus neoformans*. J Mol Biol 2009;385(4):1207–20.

[43] Angeli A, Ferraroni M, Pinteala M, Maier SS, Simionescu BC, Carta F, Del Prete S, Capasso C, Supuran CT. Crystal structure of a tetrameric type II β-carbonic anhydrase from the pathogenic bacterium *Burkholderia pseudomallei*. Molecules 2020;25(10):2269.

[44] Lehneck R, Neumann P, Vullo D, Elleuche S, Supuran CT, Ficner R, Pöggeler S. Crystal structures of two tetrameric β-carbonic anhydrases from the filamentous ascomycete Sordaria macrospora. FEBS J 2014;281(7):1759–72.

[45] Mitsuhashi S, Mizushima T, Yamashita E, Yamamoto M, Kumasaka T, Moriyama H, Ueki T, Miyachi S, Tsukihara T. X-ray structure of beta-carbonic anhydrase from the red alga, Porphyridium purpureum, reveals a novel catalytic site for CO(2) hydration. J Biol Chem 2000;275(8):5521–6.

[46] Ferraroni M, Del Prete S, Vullo D, Capasso C, Supuran CT. Crystal structure and kinetic studies of a tetrameric type II β-carbonic anhydrase from the pathogenic bacterium vibrio cholerae. Acta Crystallogr D Biol Crystallogr 2015;71(Pt 12):2449–56.

[47] Ogawa T, Noguchi K, Saito M, Nagahata Y, Kato H, Ohtaki A, Nakayama H, Dohmae N, Matsushita Y, Odaka M, Yohda M, Nyunoya H, Katayama Y. Carbonyl sulfide hydrolase from *Thiobacillus thioparus* strain THI115 is one of the β-carbonic anhydrase family enzymes. J Am Chem Soc 2013;135(10):3818–25.

[48] Alber BE, Ferry JG. A carbonic anhydrase from the archaeon *Methanosarcina thermophila*. Proc Natl Acad Sci U S A 1994;91:6909–13.

[49] Kisker C, Schindelin H, Alber BE, Ferry JG, Rees DC. A left-hand beta-helix revealed by the crystal structure of a carbonic anhydrase from the archaeon *Methanosarcina thermophila*. EMBO J 1996;15(10):2323–30.

[50] Iverson TM, Alber BE, Kisker C, Ferry JG, Rees DC. A closer look at the active site of gamma-class carbonic anhydrases: high-resolution crystallographic studies of the carbonic anhydrase from *Methanosarcina thermophila*. Biochemistry 2000;39(31):9222–31.

[51] Di Fiore A, De Luca V, Langella E, Nocentini A, Buonanno M, Monti SM, Supuran CT, Capasso C, De Simone G. Biochemical, structural, and computational studies of a γ-carbonic anhydrase from the pathogenic bacterium *Burkholderia pseudomallei*. Comput Struct Biotechnol J 2022;20:4185–94.

[52] Jin S, Vullo D, Bua S, Nocentini A, Supuran CT, Gao YG. Structural and biochemical characterization of novel carbonic anhydrases from *Phaeodactylum tricornutum*. Acta Crystallogr D Struct Biol 2020;76(Pt 7):676–86.

[53] De Simone G, Supuran CT. (In)organic anions as carbonic anhydrase inhibitors. J Inorg Biochem 2012;111:117–29.

[54] Nocentini A, Gratteri P, Supuran CT. Phosphorus versus sulfur: discovery of benzenephosphonamidates as versatile sulfonamide-mimic

chemotypes acting as carbonic anhydrase inhibitors. Chemistry 2019;25:1188–92.

[55] Carta F, Aggarwal M, Maresca A, Scozzafava A, McKenna R, Supuran CT. Dithiocarbamates: a new class of carbonic anhydrase inhibitors. Crystallographic and kinetic investigations. Chem Commun 2012;48:1868–70.

[56] Vullo D, Durante M, Di Leva FS, Cosconati S, Masini E, Scozzafava A, Novellino E, Supuran CT, Carta F. Monothiocarbamates strongly inhibit carbonic anhydrases in vitro and possess intraocular pressure lowering activity in an animal model of glaucoma. J Med Chem 2016;59:5857–67.

[57] Carta F, Akdemir A, Scozzafava A, Masini E, Supuran CT. Xanthates and trithiocarbonates strongly inhibit carbonic anhydrases and show antiglaucoma effects in vivo. J Med Chem 2013;56:4691–700.

[58] Angeli A, Tanini D, Nocentini A, Capperucci A, Ferraroni M, Gratteri P, Supuran CT. Selenols: a new class of carbonic anhydrase inhibitors. Chem Commun (Camb) 2019;5(5):648–51.

[59] Tanini D, Capperucci A, Ferraroni M, Carta F, Angeli A, Supuran CT. Direct and straightforward access to substituted alkyl selenols as novel carbonic anhydrase inhibitors. Eur J Med Chem 2020;185, 111811.

[60] Langella E, D'Ambrosio K, D'Ascenzio M, Carradori S, Monti SM, Supuran CT, De Simone G. A combined crystallographic and theoretical study explains the capability of carboxylic acids to adopt multiple binding modes in the active site of carbonic anhydrases. Chemistry 2016;22:97–100.

[61] Di Fiore A, Maresca A, Supuran CT, De Simone G. Hydroxamate represents a versatile zinc binding group for the development of new carbonic anhydrase inhibitors. Chem Commun (Camb) 2012;48:8838–40.

[62] Alterio V, Cadoni R, Esposito D, Vullo D, Fiore AD, Monti SM, Caporale A, Ruvo M, Sechi M, Dumy P, Supuran CT, De Simone G, Winum JY. Benzoxaborole as a new chemotype for carbonic anhydrase inhibition. Chem Commun (Camb) 2016;52:11983–6.

[63] Langella E, Alterio V, D'Ambrosio K, Cadoni R, Winum JY, Supuran CT, Monti SM, De Simone G, Di Fiore A. Exploring benzoxaborole derivatives as carbonic anhydrase inhibitors: a structural and computational analysis reveals their conformational variability as a tool to increase enzyme selectivity. J Enzyme Inhib Med Chem 2019;34:1498–505.

[64] Nocentini A, Supuran CT, Winum JY. Benzoxaborole compounds for therapeutic uses: a patent review (2010-2018). Expert Opin Ther Pat 2018;28:493–504.

[65] De Simone G, Angeli A, Bozdag M, Supuran CT, Winum JY, Monti SM, Alterio V. Inhibition of carbonic anhydrases by a substrate analog: benzyl carbamate directly coordinates the catalytic zinc ion mimicking bicarbonate binding. Chem Commun (Camb) 2018;54:10312–5.

[66] Bouzina A, Berredjem M, Nocentini A, Bua S, Bouaziz Z, Jose J, Le Borgne M, Marminon C, Gratteri P, Supuran CT. Ninhydrins inhibit carbonic anhydrases directly binding to the metal ion. Eur J Med Chem 2021;209, 112875.

[67] Nair SK, Ludwig PA, Christianson DW. Two-site binding of phenol in the active site of human carbonic anhydrase II: structural implications for substrate association. J Am Chem Soc 1994;116:3659–60.

[68] Innocenti A, Vullo D, Scozzafava A, Supuran CT. Carbonic anhydrase inhibitors. Interactions of phenols with the 12 catalytically active mammalian isoforms (CA I–XIV). Bioorg Med Chem Lett 2008;18:1583–7.

[69] Innocenti A, Vullo D, Scozzafava A, Supuran CT. Carbonic anhydrase inhibitors. Inhibition of mammalian isoforms I–XIV with a series of substituted phenols including paracetamol and salicylic acid. Bioorg Med Chem 2008;16:7424–8.

[70] Bayram E, Senturk M, Kufrevioglu OI, Supuran CT. In vitro effects of salicylic acid derivatives on human cytosolic carbonic anhydrase isozymesI and II. Bioorg Med Chem 2008;16:9101–5.

[71] Innocenti A, Gülçin I, Scozzafava A, Supuran CT. Carbonic anhydrase inhibitors. Antioxidant polyphenols effectively inhibit mammalian isoforms I-XV. Bioorg Med Chem Lett 2010;20:5050–3.

[72] Nocentini A, Bonardi A, Gratteri P, Cerra B, Gioiello A, Supuran CT. Steroids interfere with human carbonic anhydrase activity by using alternative binding mechanisms. J Enzyme Inhib Med Chem 2018;33:1453–9.

[73] Karioti A, Carta F, Supuran CT. Phenols and polyphenols as carbonic anhydrase inhibitors. Molecules 2016;21:E1649.

[74] D'Ambrosio K, Carradori S, Cesa S, Angeli A, Monti SM, Supuran CT, De Simone G. Catechols: a new class of carbonic anhydrase inhibitors. Chem Commun (Camb) 2020;56:13033–6.

[75] Carta F, Temperini C, Innocenti A, Scozzafava A, Kaila K, Supuran CT. Polyamines inhibit carbonic anhydrases by anchoring to the zinc-coordinated water molecule. J Med Chem 2010;53:5511–22.

[76] Tars K, Vullo D, Kazaks A, Leitans J, Lends A, Grandane A, Zalubovskis R, Scozzafava A, Supuran CT. Sulfocoumarins (1,2-benzoxathiine-2,2-dioxides): a class of potent and isoform-selective inhibitors of tumor-associated carbonic anhydrases. J Med Chem 2013;56:293–300.

[77] Grandane A, Tanc M, Žalubovskis R, Supuran CT. Synthesis of 6-aryl-substituted sulfocoumarins and investigation of their carbonic anhydrase inhibitory action. Bioorg Med Chem 2015;23:1430–6.

[78] Pustenko A, Stepanovs D, Žalubovskis R, Vullo D, Kazaks A, Leitans J, Tars K, Supuran CT. 3H-1,2-benzoxathiepine 2,2-dioxides: a new class of isoform-selective carbonic anhydrase inhibitors. J Enzyme Inhib Med Chem 2017;32(1):767–75.

[79] Grandane A, Nocentini A, Werner T, Zalubovskis R, Supuran CT. Benzoxepinones: a new isoform-selective class of tumor associated carbonic anhydrase inhibitors. Bioorg Med Chem 2020;28(11), 115496.

[80] Ferraroni M, Carta F, Scozzafava A, Supuran CT. Thioxocoumarins show an alternative carbonic anhydrase inhibition mechanism compared to coumarins. J Med Chem 2016;59:462–73.

[81] Maresca A, Temperini C, Vu H, Pham NB, Poulsen SA, Scozzafava A, Quinn RJ, Supuran CT. Non-zinc mediated inhibition of carbonic anhydrases: coumarins are a new class of suicide inhibitors. J Am Chem Soc 2009;131:3057–62.

[82] Maresca A, Temperini C, Pochet L, Masereel B, Scozzafava A, Supuran CT. Deciphering the mechanism of carbonic anhydrase inhibition with coumarins and thiocoumarins. J Med Chem 2010;53:335–44.

[83] Temperini C, Innocenti A, Scozzafava A, Parkkila S, Supuran CT. The coumarin-binding site in carbonic anhydrase accommodates structurally diverse inhibitors: the antiepileptic lacosamide as an example. J Med Chem 2010;53:850–4.

[84] Supuran CT. Coumarin carbonic anhydrase inhibitors from natural sources. J Enzyme Inhib Med Chem 2020;35:1462–70.

[85] Touisni N, Maresca A, McDonald PC, Lou Y, Scozzafava A, Dedhar S, Winum JY, Supuran CT. Glycosyl coumarin carbonic anhydrase IX and XII inhibitors strongly attenuate the growth of primary breast tumors. J Med Chem 2011;54:8271–7.

[86] Davis RA, Vullo D, Maresca A, Supuran CT, Poulsen SA. Natural product coumarins that inhibit human carbonic anhydrases. Bioorg Med Chem 2013;21:1539–43.

[87] De Luca L, Mancuso F, Ferro S, Buemi MR, Angeli A, Del Prete S, Capasso C, Supuran CT, Gitto R. Inhibitory effects and structural insights for a novel series of coumarin-based compounds that selectively target human CA IX and CA XII carbonic anhydrases. Eur J Med Chem 2018;143:276–82.

[88] Maresca A, Scozzafava A, Supuran CT. 7,8-Disubstituted- but not 6,7-disubstituted coumarins selectively inhibit the transmembrane, tumor-associated carbonic anhydrase isoforms IX and XII over the cytosolic ones I and II in the low nanomolar/subnanomolar range. Bioorg Med Chem Lett 2010;20:7255–8.

[89] Maresca A, Supuran CT. Coumarins incorporating hydroxy- and chloro-moieties selectively inhibit the transmembrane, tumor-associated carbonic anhydrase isoforms IX and XII over the cytosolic ones I and II. Bioorg Med Chem Lett 2010;20:4511–4.

[90] Bonneau A, Maresca A, Winum JY, Supuran CT. Metronidazole-coumarin conjugates and 3-cyano-7-hydroxy-coumarin act as isoform-selective carbonic anhydrase inhibitors. J Enzyme Inhib Med Chem 2013;28:397–401.

[91] D'Ambrosio K, Carradori S, Monti SM, Buonanno M, Secci D, Vullo D, Supuran CT, De Simone G. Out of the active site binding pocket for carbonic anhydrase inhibitors. Chem Commun (Camb) 2015;51:302–5.

[92] Carta F, Supuran CT. Diuretics with carbonic anhydrase inhibitory action: a patent and literature review (2005-2013). Expert Opin Ther Pat 2013;23:681–91.

[93] Supuran CT. Carbonic anhydrase inhibitors and their potential in a range of therapeutic areas. Expert Opin Ther Pat 2018;28:709–12.

[94] Supuran CT. Applications of carbonic anhydrases inhibitors in renal and central nervous system diseases. Expert Opin Ther Pat 2018;28:713–21.

[95] Supuran CT, Altamimi ASA, Carta F. Carbonic anhydrase inhibition and the management of glaucoma: a literature and patent review 2013-2019. Expert Opin Ther Pat 2019;29:781–92.

[96] Supuran CT. The management of glaucoma and macular degeneration. Expert Opin Ther Pat 2019;29:745–7.

[97] Mincione F, Nocentini A, Supuran CT. Advances in the discovery of novel agents for the treatment of glaucoma. Expert Opin Drug Discov 2021;16(10):1209–25.

[98] Aggarwal M, Kondeti B, McKenna R. Anticonvulsant/antiepileptic carbonic anhydrase inhibitors: a patent review. Expert Opin Ther Pat 2013;23:717–24.

[99] Thiry A, Dogné JM, Supuran CT, Masereel B. Anticonvulsant sulfonamides/sulfamates/sulfamides with carbonic anhydrase inhibitory activity: drug design and mechanism of action. Curr Pharm Des 2008;14:661–71.

[100] Thiry A, Dogné JM, Supuran CT, Masereel B. Carbonic anhydrase inhibitors as anticonvulsant agents. Curr Top Med Chem 2007;7:855–64.

[101] Scozzafava A, Supuran CT, Carta F. Antiobesity carbonic anhydrase inhibitors: a literature and patent review. Expert Opin Ther Pat 2013;23:725–35.

[102] Supuran CT. Anti-obesity carbonic anhydrase inhibitors: challenges and opportunities. J Enzyme Inhib Med Chem 2022;37(1):2478–88.

[103] Scozzafava A, Menabuoni L, Mincione F, Briganti F, Mincione G, Supuran CT. Carbonic anhydrase inhibitors. Synthesis of water-soluble, topically effective, intraocular pressure-lowering aromatic/heterocyclic sulfonamides containing cationic or anionic moieties: is the tail more important than the ring? J Med Chem 1999;42:2641–50.

[104] Kumar A, Siwach K, Supuran CT, Sharma PK. A decade of tail-approach based design of selective as well as potent tumor associated carbonic anhydrase inhibitors. Bioorg Chem 2022;126, 105920.

[105] D'Ascenzio M, Secci D, Carradori S, Zara S, Guglielmi P, Cirilli R, Pierini M, Poli G, Tuccinardi T, Angeli A, Supuran CT. 1,3-Dipolar cycloaddition, HPLC enantioseparation, and docking studies of saccharin/isoxazole and saccharin/isoxazoline derivatives as selective carbonic anhydrase IX and XII inhibitors. J Med Chem 2020;63:2470–88.

[106] Güzel O, Temperini C, Innocenti A, Scozzafava A, Salman A, Supuran CT. Carbonic anhydrase inhibitors. Interaction of 2-(hydrazinocarbonyl)-3-phenyl-1H-indole-5-sulfonamide with 12 mammalian isoforms: kinetic and X-ray crystallographic studies. Bioorg Med Chem Lett 2008;18:152–8.

[107] Biswas S, Aggarwal M, Güzel Ö, Scozzafava A, McKenna R, Supuran CT. Conformational variability of different sulfonamide inhibitors with thienyl-acetamido moieties attributes to differential binding in the active site of cytosolic human carbonic anhydrase isoforms. Bioorg Med Chem 2011;19:3732–8.

[108] Wagner J, Avvaru BS, Robbins AH, Scozzafava A, Supuran CT, McKenna R. Coumarinyl-substituted sulfonamides strongly inhibit several human carbonic anhydrase isoforms: solution and crystallographic investigations. Bioorg Med Chem 2010;18:4873–8.

[109] Avvaru BS, Wagner JM, Maresca A, Scozzafava A, Robbins AH, Supuran CT, McKenna R. Carbonic anhydrase inhibitors. The X-ray crystal structure of human isoform II in adduct with an adamantyl analogue of acetazolamide resides in a less utilized binding pocket than most hydrophobic inhibitors. Bioorg Med Chem Lett 2010;20:4376–81.

[110] Carta F, Garaj V, Maresca A, Wagner J, Avvaru BS, Robbins AH, Scozzafava A, McKenna R, Supuran CT. Sulfonamides incorporating 1,3,5-triazine moieties selectively and potently inhibit carbonic anhydrase transmembrane isoforms IX, XII and XIV over cytosolic isoforms I and II: solution and X-ray crystallographic studies. Bioorg Med Chem 2011;19:3105–19.

[111] Carta F, Birkmann A, Pfaff T, Buschmann H, Schwab W, Zimmermann H, Maresca A, Supuran CT. Lead development of thiazolylsulfonamides with carbonic anhydrase inhibitory action. J Med Chem 2017;60:3154–64.

[112] Alterio V, Vitale RM, Monti SM, Pedone C, Scozzafava A, Cecchi A, De Simone G, Supuran CT. Carbonic anhydrase inhibitors: X-ray and molecular modeling study for the interaction of a fluorescent antitumor sulfonamide with isozyme II and IX. J Am Chem Soc 2006;128:8329–35.

[113] Pacchiano F, Aggarwal M, Avvaru BS, Robbins AH, Scozzafava A, McKenna R, Supuran CT. Selective hydrophobic pocket binding observed within the carbonic anhydrase II active site accommodate different 4-substituted-ureido-benzenesulfonamides and correlate to inhibitor potency. Chem Commun (Camb) 2010;46:8371–3.

[114] Mboge MY, Mahon BP, Lamas N, Socorro L, Carta F, Supuran CT, Frost SC, McKenna R. Structure activity study of carbonic anhydrase IX: selective inhibition with ureido-substituted benzenesulfonamides. Eur J Med Chem 2017;132:184–91.

[115] Pacchiano F, Carta F, McDonald PC, Lou Y, Vullo D, Scozzafava A, Dedhar S, Supuran CT. Ureido-substituted benzenesulfonamides potently inhibit carbonic anhydrase IX and show antimetastatic activity in a model of breast cancer metastasis. J Med Chem 2011;54:1896–902.

[116] Bonardi A, Nocentini A, Bua S, Combs J, Lomelino C, Andring J, Lucarini L, Sgambellone S, Masini E, McKenna R, Gratteri P, Supuran CT. Sulfonamide inhibitors of human carbonic anhydrases designed through a three-tails approach: improving ligand/isoform matching and selectivity of action. J Med Chem 2020;63:7422–44.

[117] McDonald PC, Chafe SC, Supuran CT, Dedhar S. Cancer therapeutic targeting of hypoxia induced carbonic anhydrase IX: from bench to bedside. Cancers (Basel) 2022;14(14):3297.

[118] Supuran CT. Carbonic anhydrase inhibitors as emerging agents for the treatment and imaging of hypoxic tumors. Expert Opin Investig Drugs 2018;27(12):963–70.

[119] Supuran CT. Carbonic anhydrase inhibitors: an update on experimental agents for the treatment and imaging of hypoxic tumors. Expert Opin Investig Drugs 2021;30(12):1197–208.

[120] Supuran CT. Experimental carbonic anhydrase inhibitors for the treatment of hypoxic tumors. J Exp Pharmacol 2020;12:603–17.

[121] McDonald PC, Chia S, Bedard PL, Chu Q, Lyle M, Tang L, Singh M, Zhang Z, Supuran CT, Renouf DJ, Dedhar S. A phase 1 study of SLC-0111, a novel inhibitor of carbonic anhydrase IX, in patients with advanced solid tumors. Am J Clin Oncol 2020;43:484–90.

[122] Lock FE, McDonald PC, Lou Y, Serrano I, Chafe SC, Ostlund C, Aparicio S, Winum JY, Supuran CT, Dedhar S. Targeting carbonic anhydrase IX depletes breast cancer stem cells within the hypoxic niche. Oncogene 2013;32:5210–9.

[123] Swayampakula M, McDonald PC, Vallejo M, Coyaud E, Chafe SC, Westerback A, Venkateswaran G, Shankar J, Gao G, Laurent EMN, Lou Y, Bennewith KL, Supuran CT, Nabi IR, Raught B, Dedhar S. The interactome of metabolic enzyme carbonic anhydrase IX reveals novel roles in tumor cell migration and invadopodia/MMP14-mediated invasion. Oncogene 2017;36:6244–61.

[124] Ciccone V, Filippelli A, Angeli A, Supuran CT, Morbidelli L. Pharmacological inhibition of CA-IX impairs tumor cell proliferation, migration and invasiveness. Int J Mol Sci 2020;21:2983.

[125] McDonald PC, Chafe SC, Brown WS, Saberi S, Swayampakula M, Venkateswaran G, Nemirovsky O, Gillespie JA, Karasinska JM, Kalloger SE, Supuran CT, Schaeffer DF, Bashashati A, Shah SP, Topham JT, Yapp DT, Li J, Renouf DJ, Stanger BZ, Dedhar S. Regulation of pH by carbonic anhydrase 9 mediates survival of pancreatic cancer cells with activated KRAS in response to hypoxia. Gastroenterology 2019;157:823–37.

[126] Angeli A, Carta F, Nocentini A, Winum JY, Zalubovskis R, Akdemir A, Onnis V, Eldehna WM, Capasso C, Simone G, Monti SM, Carradori S, Donald WA, Dedhar S, Supuran CT. Carbonic anhydrase inhibitors targeting metabolism and tumor microenvironment. Metabolites 2020;10(10):412.

[127] Chafe SC, Vizeacoumar FS, Venkateswaran G, Nemirovsky O, Awrey S, Brown WS, McDonald PC, Carta F, Metcalfe A, Karasinska JM, Huang L, Muthuswamy SK, Schaeffer DF, Renouf DJ, Supuran CT, Vizeacoumar FJ, Dedhar S. Genome-wide synthetic lethal screen unveils novel CAIX-NFS1/xCT axis as a targetable vulnerability in hypoxic solid tumors. Sci Adv 2021;7(35):eabj0364.

[128] Supuran CT. Acetazolamide for the treatment of idiopathic intracranial hypertension. Expert Rev Neurother 2015;15:851–6.

[129] Di Cesare ML, Micheli L, Carta F, Cozzi A, Ghelardini C, Supuran CT. Carbonic anhydrase inhibition for the management of cerebral ischemia: in vivo evaluation of sulfonamide and coumarin inhibitors. J Enzyme Inhib Med Chem 2016;31:894–9.

[130] Carta F, Di Cesare ML, Pinard M, Ghelardini C, Scozzafava A, McKenna R, Supuran CT. A class of sulfonamide carbonic anhydrase inhibitors with neuropathic pain modulating effects. Bioorg Med Chem 2015;23:1828–40.

[131] Supuran CT. Carbonic anhydrase inhibition and the management of neuropathic pain. Expert Rev Neurother 2016;16:961–8.

[132] Margheri F, Ceruso M, Carta F, Laurenzana A, Maggi L, Lazzeri S, Simonini G, Annunziato F, Del Rosso M, Supuran CT, Cimaz R. Overexpression of the transmembrane carbonic anhydrase isoforms IX and XII in the inflamed synovium. J Enzyme Inhib Med Chem 2016;31(sup4):60–3.

[133] Bua S, Di Cesare ML, Vullo D, Ghelardini C, Bartolucci G, Scozzafava A, Supuran CT, Carta F. Design and synthesis of novel nonsteroidal anti-inflammatory drugs and carbonic anhydrase inhibitors hybrids (NSAIDs-CAIs) for the treatment of rheumatoid arthritis. J Med Chem 2017;60:1159–70.

[134] Bua S, Lucarini L, Micheli L, Menicatti M, Bartolucci G, Selleri S, Di Cesare ML, Ghelardini C, Masini E, Carta F, Gratteri P, Nocentini A, Supuran CT. Bioisosteric development of multitarget nonsteroidal anti-inflammatory drug-carbonic anhydrases inhibitor hybrids for the management of rheumatoid arthritis. J Med Chem 2020;63:2325–42.

[135] Akgul O, Di Cesare ML, Vullo D, Angeli A, Ghelardini C, Bartolucci G, Alfawaz Altamimi AS, Scozzafava A, Supuran CT, Carta F. Discovery of novel nonsteroidal anti-inflammatory drugs and carbonic anhydrase inhibitors hybrids (NSAIDs-CAIs) for the management of rheumatoid arthritis. J Med Chem 2018;61:4961–77.

[136] Lucarini L, Durante M, Sgambellone S, Lanzi C, Bigagli E, Akgul O, Masini E, Supuran CT, Carta F. Effects of new NSAID-CAI hybrid compounds in inflammation and lung fibrosis. Biomolecules 2020;10(9):1307.

[137] Berrino E, Milazzo L, Micheli L, Vullo D, Angeli A, Bozdag M, Nocentini A, Menicatti M, Bartolucci G, di Cesare ML, Ghelardini C, Supuran CT, Carta F. Synthesis and evaluation of carbonic anhydrase inhibitors with carbon monoxide releasing properties for the management of rheumatoid arthritis. J Med Chem 2019;62:7233–49.

[138] Supuran CT. Carbonic anhydrase activators. Future Med Chem 2018;10:561–73.

[139] Temperini C, Scozzafava A, Vullo D, Supuran CT. Carbonic anhydrase activators. Activation of isozymes I, II, IV, VA, VII, and XIV with l- and d-histidine and crystallographic analysis of their adducts with isoform II: engineering proton-transfer processes within the active site of an enzyme. Chemistry 2006;12:7057–66.

[140] Temperini C, Scozzafava A, Supuran CT. Carbonic anhydrase activation and the drug design. Curr Pharm Des 2008;14:708–15.

[141] Temperini C, Innocenti A, Scozzafava A, Mastrolorenzo A, Supuran CT. Carbonic anhydrase activators: L-adrenaline plugs the active site entrance of isozyme II, activating better isoforms I, IV, VA, VII, and XIV. Bioorg Med Chem Lett 2007;17:628–35.

[142] Temperini C, Scozzafava A, Vullo D, Supuran CT. Carbonic anhydrase activators. Activation of isoforms I, II, IV, VA, VII, and XIV with L- and D-phenylalanine and crystallographic analysis of their adducts with isozyme II: stereospecific recognition within the active site of an enzyme and its consequences for the drug design. J Med Chem 2006;49:3019–27.

[143] Temperini C, Innocenti A, Scozzafava A, Supuran CT. Carbonic anhydrase activators: kinetic and x-ray crystallographic study for the interaction of D- and L-tryptophan with the mammalian isoforms I-XIV. Bioorg Med Chem 2008;16:8373–837.

[144] Dave K, Ilies MA, Scozzafava A, Temperini C, Vullo D, Supuran CT. An inhibitor-like binding mode of a carbonic anhydrase activator within the active site of isoform II. Bioorg Med Chem Lett 2011;21:2764–8.

[145] Bhatt A, Mondal UK, Supuran CT, lies MA, McKenna R. Crystal structure of carbonic anhydrase II in complex with an activating ligand: implications in neuronal function. Mol Neurobiol 2018;55:7431–7.

[146] Saada MC, Vullo D, Montero JL, Scozzafava A, Winum JY, Supuran CT. Carbonic anhydrase I and II activation with mono- and

dihalogenated histamine derivatives. Bioorg Med Chem Lett 2001;21:4884–7.

[147] Saada MC, Vullo D, Montero JL, Scozzafava A, Supuran CT, Winum JY. Mono- and di-halogenated histamine, histidine and carnosine derivatives are potent carbonic anhydrase I, II, VII, XII and XIV activators. Bioorg Med Chem 2014;22:4752–8.

[148] Vistoli G, Aldini G, Fumagalli L, Dallanoce C, Angeli A, Supuran CT. Activation effects of carnosine- and histidine-containing dipeptides on human carbonic anhydrases: a comprehensive study. Int J Mol Sci 2020;21:1761.

[149] Supuran CT, Dinculescu A, Balaban AT. Carbonic anhydrase activators. Part 5. CA II activation by 2,4,6-trisubstituted pyridinium cations with 1-(ω-aminoalkyl) side chains. Rev Roum Chim 1993;38:343–9.

[150] Draghici B, Vullo D, Akocak S, Walker EA, Supuran CT, Ilies MA. Ethylene bis-imidazoles are highly potent and selective activators for isozymes VA and VII of carbonic anhydrase, with a potential nootropic effect. Chem Commun (Camb) 2014;50:5980–3.

[151] Akocak S, Lolak N, Bua S, Nocentini A, Supuran CT. Activation of human α-carbonic anhydrase isoforms I, II, IV and VII with bis-histamine schiff bases and bis-spinaceamine substituted derivatives. J Enzyme Inhib Med Chem 2019;34:1193–8.

[152] Akocak S, Lolak N, Bua S, Nocentini A, Karakoc G, Supuran CT. α-Carbonic anhydrases are strongly activated by spinaceamine derivatives. Bioorg Med Chem 2019;27:800–4.

[153] Canto de Souza L, Provensi G, Vullo D, Carta F, Scozzafava A, Costa A, Schmidt SD, Passani MB, Supuran CT, Blandina P. Carbonic anhydrase activation enhances object recognition memory in mice through phosphorylation of the extracellular signal-regulated kinase in the cortex and the hippocampus. Neuropharmacology 2017;118:148–56.

[154] Blandina P, Provensi G, Passani MB, Supuran CT. Carbonic anhydrase modulation of emotional memory. Implications for the treatment of cognitive disorders. J Enzyme Inhib Med Chem 2020;35:1206–14.

[155] Schmidt SD, Costa A, Rani B, Godfried Nachtigall E, Passani MB, Carta F, Nocentini A, de Carvalho MJ, Furini CRG, Supuran CT, Izquierdo I, Blandina P, Provensi G. The role of carbonic anhydrases in extinction of contextual fear memory. Proc Natl Acad Sci U S A 2020;117:16000–8.

[156] Provensi G, Nocentini A, Passani MB, Blandina P, Supuran CT. Activation of carbonic anhydrase isoforms involved in modulation of emotional memory and cognitive disorders with histamine agonists, antagonists and derivatives. J Enzyme Inhib Med Chem 2021;36(1):719–26.

[157] Schmidt SD, Nachtigall EG, Marcondes LA, Zanluchi A, Furini CRG, Passani MB, Supuran CT, Blandina P, Izquierdo I, Provensi G, de Carvalho MJ. Modulation of Carbonic anhydrases activity in the Hippocampus or prefrontal cortex differentially affects social recognition memory in rats. Neuroscience 2022;497:184–95.

[158] Provensi G, Costa A, Rani B, Becagli MV, Vaiano F, Passani MB, Tanini D, Capperucci A, Carradori S, Petzer JP, Petzer A, Vullo D, Costantino G, Blandina P, Angeli A, Supuran CT. New β-arylchalcogeno amines with procognitive properties targeting Carbonic anhydrases and monoamine oxidases. Eur J Med Chem 2022;244, 114828.

[159] Provensi G, Carta F, Nocentini A, Supuran CT, Casamenti F, Passani MB, Fossati S. A new kid on the block? Carbonic anhydrases as possible new targets in Alzheimer's disease. Int J Mol Sci 2019;20 (19):4724.

[160] Abutaleb NS, Elhassanny AEM, Nocentini A, Hewitt CS, Elkashif A, Cooper BR, Supuran CT, Seleem MN, Flaherty DP. Repurposing FDA-approved sulphonamide carbonic anhydrase inhibitors for treatment of Neisseria gonorrhoeae. J Enzyme Inhib Med Chem 2022;37(1):51–61.

[161] Flaherty DP, Seleem MN, Supuran CT. Bacterial carbonic anhydrases: underexploited antibacterial therapeutic targets. Future Med Chem 2021;13(19):1619–22.

[162] Nocentini A, Hewitt CS, Mastrolorenzo MD, Flaherty DP, Supuran CT. Anion inhibition studies of the α-carbonic anhydrases from Neisseria gonorrhoeae. J Enzyme Inhib Med Chem 2021;36 (1):1061–6.

[163] Abdoli M, De Luca V, Capasso C, Supuran CT, Žalubovskis R. Benzenesulfonamides incorporating hydantoin moieties effectively inhibit Eukaryoticand human carbonic anhydrases. Int J Mol Sci 2022;23(22):14115.

[164] Angeli A, Velluzzi A, Selleri S, Capasso C, Spadini C, Iannarelli M, Cabassi CS, Carta F, Supuran CT. Seleno containing compounds as potent and selective antifungal agents. ACS Infect Dis 2022;8 (9):1905–19.

[165] De Luca V, Angeli A, Mazzone V, Adelfio C, Carginale V, Scaloni A, Carta F, Selleri S, Supuran CT, Capasso C. Heterologous expression and biochemical characterisation of the recombinant β-carbonic anhydrase (MpaCA) from the warm-blooded vertebrate pathogen malassezia pachydermatis. J Enzyme Inhib Med Chem 2022;37(1):62–8.

[166] Del Prete S, Angeli A, Ghobril C, Hitce J, Clavaud C, Marat X, Supuran CT, Capasso C. Sulfonamide inhibition profile of the β-carbonic anhydrase from Malassezia restricta, An opportunistic pathogen triggering scalp conditions. Metabolites 2020;10(1):39.

[167] Beatriz Vermelho A, Rodrigues GC, Nocentini A, Mansoldo FRP, Supuran CT. Discovery of novel drugs for chagas disease: is carbonic anhydrase a target for antiprotozoal drugs? Expert Opin Drug Discov 2022;17(10):1147–58.

[168] Mansoldo FRP, Carta F, Angeli A, Cardoso VDS, Supuran CT, Vermelho AB. Chagas disease: perspectives on the past and present and challenges in drug discovery. Molecules 2020;25(22):5483.

[169] Nocentini A, Osman SM, Almeida IA, Cardoso V, Alasmary FAS, AlOthman Z, Vermelho AB, Gratteri P, Supuran CT. Appraisal of anti-protozoan activity of nitroaromatic benzenesulfonamides inhibiting carbonic anhydrases from *Trypanosoma cruzi* and *Leishmania donovani*. J Enzyme Inhib Med Chem 2019;34(1):1164–71.

[170] da Silva CV, Vermelho AB, Ricci Junior E, Almeida Rodrigues I, Mazotto AM, Supuran CT. Antileishmanial activity of sulphonamide nanoemulsions targeting the β-carbonic anhydrase from Leish*m*ania species. J Enzyme Inhib Med Chem 2018;33(1):850–7.

[171] Da'dara AA, Angeli A, Ferraroni M, Supuran CT, Skelly PJ. Crystal structure and chemical inhibition of essential schistosome host-interactive virulence factor carbonic anhydrase SmCA. Commun Biol 2019;2:333.

[172] Angeli A, Ferraroni M, Da'dara AA, Selleri S, Pinteala M, Carta F, Skelly PJ, Supuran CT. Structural insights into schistosoma mansoni carbonic anhydrase (SmCA) inhibition by selenoureido-substituted benzenesulfonamides. J Med Chem 2021;64(14):10418–28.

[173] Aspatwar A, Barker H, Aisala H, Zueva K, Kuuslahti M, Tolvanen M, Primmer CR, Lumme J, Bonardi A, Tripathi A, Parkkila S, Supuran CT. Cloning, purification, kinetic and anion inhibition studies of a recombinant β-carbonic anhydrase from the Atlantic salmon parasite platyhelminth *Gyrodactylus salaris*. J Enzyme Inhib Med Chem 2022;37(1):1577–86.

[174] Supuran CT. Bacterial carbonic anhydrases as drug targets: toward novel antibiotics? Front Pharmacol 2011;2:34.

[175] Del Prete S, Nocentini A, Supuran CT, Capasso C. Bacterial ι-carbonic anhydrase: a new active class of carbonic anhydrase identified in the genome of the gram-negative bacterium Burkholderia territorii. J Enzyme Inhib Med Chem 2020;35:1060–8.

[176] An W, Holly KJ, Nocentini A, Imhoff RD, Hewitt CS, Abutaleb NS, Cao X, Seleem MN, Supuran CT, Flaherty DP. Structure-activity relationship studies for inhibitors for vancomycin-resistant Enterococcus and human carbonic anhydrases. J Enzyme Inhib Med Chem 2022;37(1):1838–44.

[177] Giovannuzzi S, Hewitt CS, Nocentini A, Capasso C, Costantino G, Flaherty DP, Supuran CT. Inhibition studies of bacterial α-carbonic anhydrases with phenols. J Enzyme Inhib Med Chem 2022;37(1):666–71.

[178] Giovannuzzi S, Hewitt CS, Nocentini A, Capasso C, Flaherty DP, Supuran CT. Coumarins effectively inhibit bacterial α-carbonic anhydrases. J Enzyme Inhib Med Chem 2022;37(1):333–8.

Chapter 3.2

Metallo-β-lactamases*

Elsa Denakpo[a], Guillaume Arlet[b], Alain Philippon[c], and Bogdan I. Iorga[a]
[a]Université Paris-Saclay, CNRS, Institut de Chimie des Substances Naturelles, Gif-sur-Yvette, France, [b]Sorbonne Université, U1135, CIMI-Paris, Paris, France, [c]Faculté de Médecine, Bactériologie, Université de Paris-Cité, Paris, France

1 Introduction

Metallo-β-lactamases (MBLs) are a major public health concern as they are able to hydrolyze most β-lactam antibiotics, including penicillins, cephalosporins, and carbapenems, leading to treatment failure and increased morbidity and mortality. These enzymes have been identified in a wide range of Gram-negative and Gram-positive bacteria and have been reported in many parts of the world. The emergence and spread of MBL-producing bacteria have led to a rise in treatment-resistant infections and a decrease in the effectiveness of antibiotics.

β-Lactamases can be classified into two different ways, according to their sequence similarity (molecular classification) or to the activity profile (functional classification). The former is known as Ambler classification [1,2] and organizes the β-lactamases in classes A, C, and D, which are serine-β-lactamases (SBLs), and class B, which are MBLs. The MBLs are further organized in three subclasses, B1, B2, and B3 (Fig. 1). The latter is a functional classification [4,5], based on the specific hydrolysis profile against different β-lactam substrates.

High-quality and comprehensive reviews focused on clinically relevant β-lactamases were published over the years, with a special focus on carbapenemases [5–12] or on MBLs [13–28].

2 Structure and function of metallo-β-lactamases

2.1 Phylogenetic comparison and evolution

The MBLs are organized phylogenically in three distinct clusters, which correspond to subclasses B1, B2, and B3 (Fig. 1). Subclasses B1 and B2 form a distinct phylogenic group [29], separated from the cluster B3 by an intermediary region with enzymes possessing variable active sites and mixed substrate hydrolysis profiles (Fig. 1). A phylogenetic comparison of all members of the MBL superfamily suggested that the β-lactamase activity appeared and evolved separately in the B1+B2 and B3 groups [28,30].

A recent study evidenced the evolutionary traits acquired by several clinically relevant MBLs in order to adapt to their native environment, depending on if they are soluble or membrane-bound proteins. An important factor influencing this evolution is the amount of metal available in the cell under zinc starvation conditions [31].

An in silico analysis based on protein sequence similarity networks (SSNs) applied to the superfamily of MBLs, combined with traditional cladistic and phenetic phylogenetic analysis, evidenced the zinc-coordinating motifs that are specific for each enzymatic reaction. Additionally, the local network interconnectivity was measured using the neighborhood connectivity coloring to detect new protein families within the SSN clusters. This approach evidenced the important number of unexplored protein families in this superfamily [32].

2.2 Primary structure/sequence analysis

The sequences of MBLs are not well conserved, with sequence identities as low as 10% for the most distant members. The most conserved residues are those coordinating the zinc ions (Fig. 2). The B1 subclass is characterized by the motifs ^{116}HxHxDx, ^{196}H, ^{221}C, and ^{263}H. The motifs present in the subclass B2 are very similar, differing only at the residue in position 116: ^{116}NxHxDx, ^{196}H, ^{221}C, and ^{263}H. The differences are more significant in the subclass B3. First, the Cys residue in position 221 is absent, and replaced by a His in position 121. Second, most sequences have the motif ^{116}HxHxDH, but there are some notable exceptions, with some different motifs: ^{116}QxHxDH (GOB, CPS, ESP, PLN, SIQ, LRA12, LRA17, LRA19, and PEDO-1/2 families), ^{116}ExHxDH (SIE and SSE families), and ^{116}HxRxDQ (SPR, CSR, and SER families). For the B3 subclass, it was proposed to use the active site amino

* Dedicated to the memory of Dr. Roger Labia (1941–2022), whose dedication and contributions to bacteriology were longstanding and far-reaching.

Metalloenzymes. https://doi.org/10.1016/B978-0-12-823974-2.00001-2

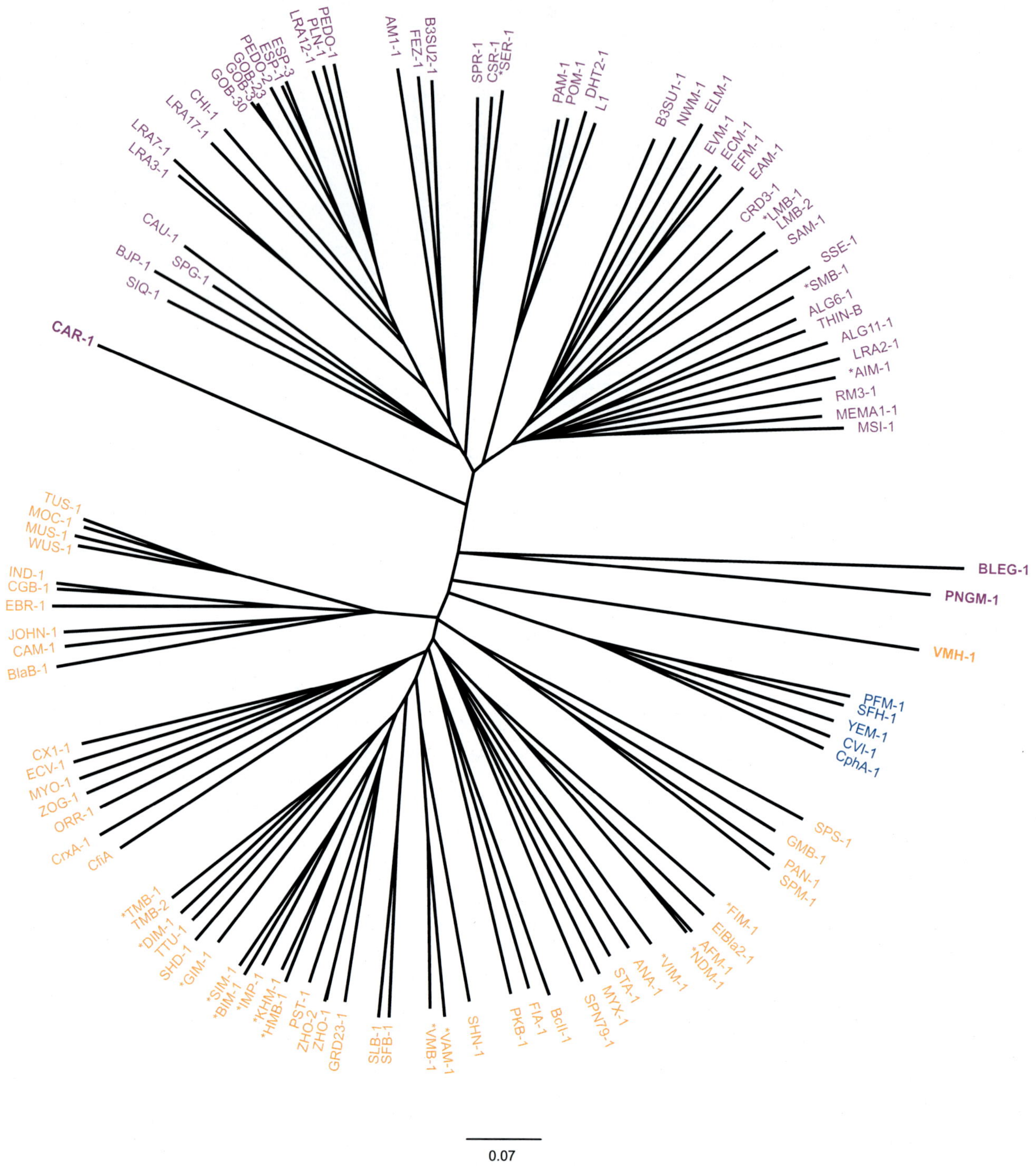

FIG. 1 Phylogenetic tree of MBLs. The enzymes belonging to subclasses B1, B2, and B3 are colored in *orange*, *blue*, and *magenta*, respectively. The plasmidic MBLs are labeled with a *, and those situated in the intermediary region are highlighted in *bold*. The tree was generated using SeaView [3] and the phylogram generated using FigTree (version 1.4.3). *(Credit: The authors (Elsa Denakpo, Guillaume Arlet, Alain Philippon, Bogdan I. Iorga).)*

acid changes to distinguish the variants (i.e., B3-RQK, B3-Q, B3-E) [34].

A standard numbering scheme for MBLs (named BBL) was developed in 2001 and updated in 2004 [35,36]. This scheme is based on the structural alignment of representative members of each subclass and allows the comparison of residues from equivalent positions, which otherwise would have different numbers, specific to each family. The representative MBLs from each subclass (B1, B2, and B3) are presented in Table 1.

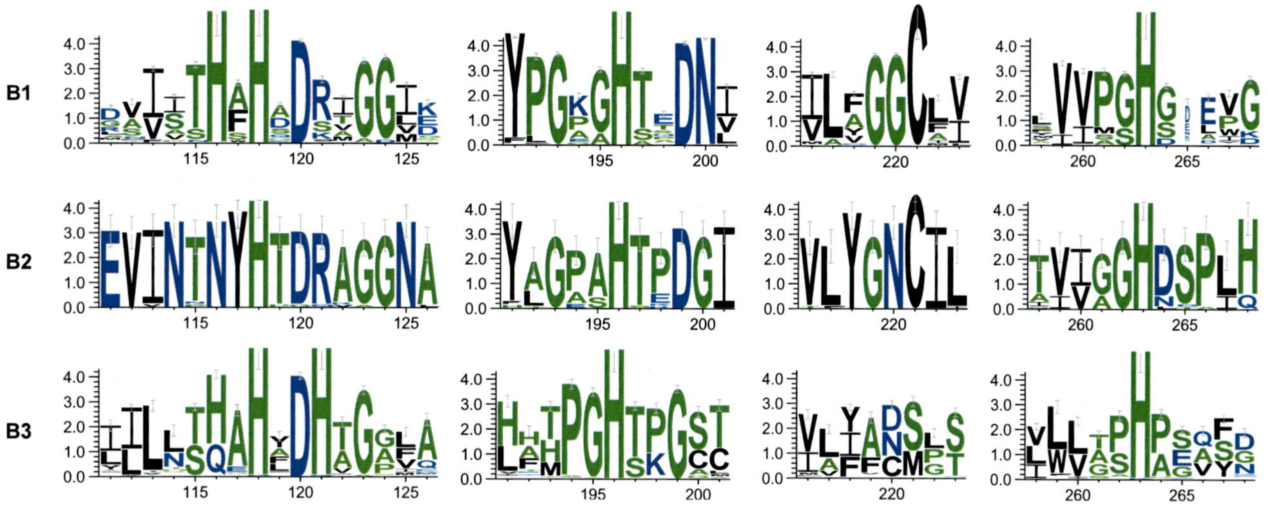

FIG. 2 Residue conservation in the sequences of MBLs in the regions neighboring the zinc-coordinating residues (positions 116, 118, 120, 196, 221, and 263 for subclasses B1 and B2, and positions 116, 118, 120, 121, 196, and 263 for subclass B3). The image was generated with WebLogo version 3 [33]. *(Credit: The authors (Elsa Denakpo, Guillaume Arlet, Alain Philippon, Bogdan I. Iorga).)*

2.3 Mutations and spectrum of activity

Historically, the class A molecular β-lactamases, particularly TEM enzymes, constituted an excellent model of molecular evolution when the first third-generation cephalosporins (ceftazidime and cefotaxime) were introduced into clinical practice in the 1980s. The substitution of certain amino acids in ESBLs (Extended-Spectrum Beta-Lactamases) such as Glu104Lys, Arg164Ser/His, Ala237Thr, Gly238Ser, Glu240Lys/Arg led to the extension of the spectrum due to an improved affinity [109,110]. On the other hand, other substitutions (Met69Leu/Val/Ile, Trp165Arg, Met182Thr, Arg244Cys/Ser/Thr, Val261Ile, Arg275Leu/Gln, Asn276Asp) in the same type of enzyme had an opposite effect, namely a decrease in affinity and therefore resistance to inhibitors such as clavulanic acid and sulbactam (defined as IRT, Inhibitor Resistant TEM) [111].

The evolutionary potential of MBLs appears to be less significant if judged by the extended inactivation spectrum of these enzymes with respect to β-lactams (except for monobactams such as aztreonam), including carbapenems and cefepime, and on the other hand, by a smaller number of variants, especially in transferable MBLs such as IMP, NDM, and VIM enzymes, which are mainly involved in clinical medicine. Thus, 249 variants have been identified for TEM type enzymes, 48 for NDM, 83 for VIM, and 98 for IMP [112]. Protein sequence heterogeneity is the lowest in the NDM family, ranging from 74% to 100%, and several substitution locations (one to three amino acids) have been identified without notable differences in β-lactam resistance levels and kinetic constants (P28A, V88L, D95N, D130G, E152K, M154L, G222D, R264H). Several substitution locations have also been evaluated by directed mutagenesis as indicated in Table 2 for several NDM-type enzymes, allowing for a better understanding of this enzyme family, including the importance of some amino acids away from the active site [113–118]. It is important to emphasize the absence of mutations that generate a higher level of resistance, especially with respect to carbapenems or aztreonam or cefepime.

Regarding the IMP family, protein heterogeneity ranges from 80% to 100%, and several substitutions (in positions 59, 67, 115, 119, 120, 121, 218, 233, 262) have been identified without notable differences in resistance to β-lactams, including carbapenems [119–124].

Several sites tested by directed mutagenesis were also evaluated (Table 3). The potential for expanding the spectrum remains limited again in the absence of mutants generating a higher level of resistance, particularly against carbapenems or aztreonam [119,125–128].

The VIM family consists of 80 variants, with a sequence identity ranging from 75% to 100% and which can be grouped in at least three clusters, including two dominant ones. The evolutionary potential of this family toward increased resistance to some β-lactams is again limited. This is true for the naturally occurring mutants, which are observed in clinic, but also for synthetic mutants obtained through directed mutagenesis [129–138].

The hyperproduction of the enzyme induced by a strong promoter, obtained by mutation, is a well-known mechanism reported in molecular classes A and C β-lactamases. However, this mechanism remains little described in transferable MBLs (e.g., IMP, NDM, or VIM) [139–145].

TABLE 1 Representative class B β-lactamases.

Subclass	Metallo-β-lactamase	Origin of name[a]	Accession	Gene[b]	Organism	Number of residues	References
B1	AFM-1	*Alcaligenes faecalis* Metallo-β-lactamase	AYV97588	P	*Alcaligenes faecalis*	267	[37]
B1	ANA-1	*Anaeromyxobacter* sp.	WP_041449074	CHR	*Anaeromyxobacter* sp.	255	[38]
B1	BcII-1	*Bacillus cereus* type II	AAA22276	CHR	*Bacillus cereus*	257	[39]
B1	BIM-1	Belem imipenemase	ANY85569	In	*Pseudomonas putida*	245	–
B1	BlaB-1	Beta-lactamase class B	AAF89154	CHR	*Elizabethkingia meningoseptica*	249	[40]
B1	CAM-1	Central Alberta Metallo-β-lactamase	AVX51087	CHR	*Pseudomonas aeruginosa*	241	[41]
B1	CEMC19-1	Name of clone	UVT38118	CHR	Uncultured bacteria	252	[42]
B1	CfiA	Cefoxitin-imipenem-hydrolyzing enzyme	AAA22907	CHR	*Bacteroides fragilis*	249	[43,44]
B1	CGB-1	*Chryseobacterium gleum* class B	AAL55263	CHR	*Chryseobacterium gleum*	242	[45]
B1	CrxA-1	Carbapenem resistant *B. xylanisolvens*	MBS5055441		*Bacteroides xylanisolvens*	250	[46]
B1	CX1-1	Name of clone	AGU01687	CHR	Uncultured bacteria	246	[47]
B1	DIM-1	Dutch imipenemase	AGC92784	P, In	*Pseudomonas stutzeri*	251	[48]
B1	EBR-1	*Empedobacter brevis* resistant	AAN32638	CHR	*Empedobacter brevis*	235	[49]
B1	ECV-1	*Echinicola vietnamensis*	AGA78874	CHR	*Echinicola vietnamensis*	257	[38]
B1	ElBla2	*Erythrobacter litoralis* Beta-lactamase 2	ABC63608	CHR	*Erythrobacter litoralis*	261	[50]
B1	FIA-1	*Fibrella aestuarina*	WP_041258349	CHR	*Fibrella aestuarina*	249	[38]
B1	FIM-1	Florence imipenemase	AFV91534	ISCR	*Pseudomonas aeruginosa*	262	[51]
B1	GIM-1	German imipenemase	ALO69078	P, In	*Pseudomonas aeruginosa*	250	[52]
B1	GMB-1	German Metallo-Beta-lactamase	QKF95688	CHR	*Citrobacter freundii*	270	[53]
B1	GRD23-1	Name of clone	APR64493	CHR	Uncultured bacteria	241	[54]
B1	HMB-1	Hamburg Metallo-Beta-lactamase	AMY61250	Tn	*Pseudomonas aeruginosa*	241	[55]
B1	IMP-1	Imipenem-hydrolyzing β-lactamase	ABK27309	Tn, IS	*Serratia marcescens*	246	[56]
B1	IND-1	*Chryseobacterium indologenes*	AAD20273	CHR	*Chryseobacterium indologenes*	239	[57]

B1	JOHN-1	*Flavobacterium johnsoniae*	AAK38324	CHR	*Flavobacterium johnsoniae*	248	[58]
B1	KHM-1	Kyorin Health Science MBL1	BAF91108	P	*Citrobacter freundii*	241	[59]
B1	MOC-1	*Myroides odoratus* Carbapenemase	ANJ59787	CHR	*Myroides odoratus*	248	–
B1	MUS-1	*Myroides odoratimimus*	AAN63647	CHR	*Myroides odoratimimus*	246	[60]
B1	MYO-1	*Myroides odoratimimus*	WP_081048762	CHR	*Myroides odoratimimus*	266	[38]
B1	MYX-1	*Myxococcus xanthus*	ABF86854	CHR	*Myxococcus xanthus*	261	[38]
B1	NDM-1	New Delhi Metallo-β-lactamase	AHM26723	P, In	*Klebsiella pneumoniae*	270	[61]
B1	ORR-1	*Ornithobacterium rhinotracheale*	WP_109545042	CHR	*Ornithobacterium rhinotracheale*	255	[38]
B1	PAN-1	*Pseudobacteriovorax antillogorgiicola*	SMF47935	CHR	*Pseudobacteriovorax antillogorgiicola*	270	[62]
B1	PEDO-3	*Pedobacter roseus*	AJP77076	CHR	*Pedobacter kyungheensis*	237	[63]
B1	PST-1	*Pseudomonas stutzeri*	WP_043942497	CHR	*Stutzerimonas stutzeri*	242	[38]
B1	SFB-1	*Shewanella frigidimarina* Beta-lactamase	AAT90847	CHR	*Shewanella frigidimarina*	244	[64]
B1	SHD-1	*Shewanella denitrificans*	ABE54111	CHR	*Shewanella denitrificans*	269	[38]
B1	SHN-1	*Shewanella denitrificans*	ABE56430	CHR	*Shewanella denitrificans*	237	[38]
B1	SIM-1	Seoul imipenemase	AER61546	In	*Acinetobacter baumannii*	246	[65]
B1	SLB-1	*Shewanella livingstonensis* Beta-lactamase	AAT90846	CHR	*Shewanella livingstonensis*	249	[64]
B1	SPM-1	Sao Paulo Metallo-β-lactamase	AAR15341	Tn	*Pseudomonas aeruginosa*	276	[66]
B1	SPN79-1	Name of clone	APR64486	CHR	*Phenylobacterium*/ Proteobacteria	278	[54]
B1	SPS-1	*Sediminispirochaeta smaragdinae*	ADK81930	CHR	*Sediminispirochaeta smaragdinae*	276	[38]
B1	STA-1	*Stigmatella aurantiaca*	WP_109545039	CHR	*Stigmatella aurantiaca*	261	[38]
B1	SZM-1	Shenzhen Metallo-β-lactamase	JAPDUI000000000[c]	CHR	*Arenimonas* metagenome	231	[67]
B1	TTU-1	*Teredinibacter turnerae*	ACR12883	CHR	*Teredinibacter turnerae*	243	[38]
B1	TMB-1	Tripoli Metallo-Beta-lactamase	CBY88906	In	*Achromobacter xylosoxidans*	245	[68]
B1	TMB-2	Tripoli Metallo-Beta-lactamase	BAM73613	CHR	*Acinetobacter courvalinii*	245	[69]
B1	TUS-1	*Myroides odoratus*	EKB08120	CHR	*Myroides odoratus*	248	[60]
B1	VAM-1	*Vibrio alginolyticus* Metallo-β-lactamase	QTJ60982	P, In	*Vibrio alginolyticus*	246	[70,71]

Continued

TABLE 1 Representative class B β-lactamases—cont'd

Subclass	Metallo-β-lactamase	Origin of name	Accession	Gene	Organism	Number of residues	References
B1	VIM-1	Verona Imipenemase	CAC35170	In	*Pseudomonas aeruginosa/ A. denitificans*	266	[72]
B1	VMB-1	*Vibrio* Metallo-Beta-lactamase	QGQ32905	In	*Vibrio alginolyticus*	247	[73]
B1	VMH-1	*Vibrio* MetalloHydrolase	ASM98011	CHR	*Vibrio vulnificus*	281	[74]
B1	WUS-1	*Wenzhou Medical University*	UPP01678	CHR	*Myroides albus*	246	[75]
B1	ZHO-1	*Zhongshania aliphaticivorans*	WP_082793656	CHR	*Zhongshania aliphaticivorans*	246	[76]
B1	ZOG-1	*Zobellia galactanivorans*	CAZ94871	CHR	*Zobellia galactanivorans*	247	[38]
B2	CphA	Carbapenem-hydrolyzing Metallo-β-lactamase	CAA40386	CHR	*Aeromonas hydrophila*	254	[77]
B2	ImiS (CphA-11)	Imipenemase *Aeromonas veronii* bv. sobria	CAA71441	CHR	*Aeromonas veronii* bv. sobria	255	[78]
B2	ImiH (CphA-13)	Imipenemase *Aeromonas hydrophila*	CAD69003	CHR	*Aeromonas hydrophila*	254	[79]
B2	CVI-1	*Chromobacterium violaceum*	WP_164461288	CHR	*Chromobacterium violaceum*	255	–
B2	PFM-1	*Pseudomonas fluorescens* Metallo-β-lactamase	QDC33502	CHR	*Pseudomonas fluorescens* complex	253	[80]
B2	SFH-1	*Serratia fonticola* carbapenem hydrolase	WP_024531368	CHR	*Serratia fonticola*	255	[81]
B2	YEM-1	*Yersinia mollaretii*	CQH06348	CHR	*Yersinia mollaretii*	246	[82]
B3	AIM-1	Adelaide imipenemase	CAQ53840	IS	*Pseudomonas aeruginosa*	303	[83]
B3	ALG6-1	Name of clone	APR64488	CHR	Uncultured bacterium	318	[54]
B3	ALG11-1	Name of clone	APR64489	CHR	Uncultured bacterium	295	[54]
B3	AM1-1	Name of clone	AGU01669	CHR	Uncultured bacterium	311	[47]
B3	BJP-1	*Bradyrhizobium japonicum*	BAC51495	CHR	*Bradyrhizobium diazoefficiens*	294	[84]
B3	BLEG-1	*Bacillus lehensis* strain G1	AIC95013	CHR	*Alkalihalobacillus lehensis*	428	[85]
B3	CAR-1	*Erwinia carotovora*	AIA71664	CHR	*Pectobacterium carotovorum*	342	[86]
B3	CAU-1	*Caulobacter crescentus*	CAC87665	CHR	*Caulobacter crescentus*	289	[87]
B3	CPS-1	*Chryseobacterium piscium*	AJP77054	CHR	*Chryseobacterium piscium*	290	[63]

B3	CRD3-1	Name of clone	APR64487	CHR	Uncultured bacterium	303	[54]
B3	CSR-1	*Cronobacter sakazakii* resistant	AKE96626	CHR	*Cronobacter sakazakii*	318	[34]
B3	DHT2-1	Name of clone	APR64485	CHR	Uncultured bacterium	302	[54]
B3	EAM-1	*Erythrobacter aquimaris* Metallo-β-lactamase	AFN85388	CHR	*Erythrobacter aquimaris*	318	[88]
B3	ECM-1	*Erythrobacter citreus* Metallo-β-lactamase	WP_122630824	CHR	*Erythrobacter citreus*	303	[88]
B3	EFM-1	*Erythrobacter flavus* Metallo-β-lactamase	AFN85384	CHR	*Erythrobacter flavus*	302	[88]
B3	ELM-1	*Erythrobacter longus* Metallo-β-lactamase	AFN85386	CHR	*Erythrobacter longus*	313	[88]
B3	ESP-1	*Epilithonimonas* sp.	AJP77085	CHR	*Epilithonimonas tenax*	290	[63]
B3	EVM-1	*Erythrobacter vulgaris* Metallo-β-lactamase	AFN85385	CHR	*Qipengyuania vulgaris*	305	[88]
B3	FEZ-1	*Fluoribacter gormanii* endogenous Zn β-lactamase	CAB96921	CHR	*Fluoribacter gormanii*	282	[89]
B3	GOB-1	*Elizabethkingia meningoseptica* class B	AAF04458	CHR	*Elizabethkingia meningoseptica*	290	[90]
B3	L1	Labile β-lactamase from *S. maltophilia*	CAA52968	CHR	*Stenotrophomonas maltophilia*	290	[91]
B3	LMB-1	Linz Metallo-Beta-lactamase	AMK49163	P	*Enterobacter cloacae*	304	[92,93]
B3	LMB-2	Linz Metallo-Beta-lactamase	SEI10464	CHR	*Rheinheimera pacifica*	304	–
B3	LRA2-1	Beta-lactam resistance from Alaska	ACH58985	CHR	Uncultured bacterium	302	[94]
B3	LRA3-1	Beta-lactam resistance from Alaska	ACH58987	CHR	Uncultured bacterium	302	[94]
B3	LRA7-1	Beta-lactam resistance from Alaska	ACH58998	CHR	Uncultured bacterium	302	[94]
B3	LRA8-1	Beta-lactam resistance from Alaska	ACH58988	CHR	Uncultured bacterium	305	[94,95]
B3	LRA12-1	Beta-lactam resistance from Alaska	ACH58990	CHR	Uncultured bacterium	293	[94]
B3	LRA17-1	Beta-lactam resistance from Alaska	ACH58994	CHR	Uncultured bacterium	294	[94]
B3	LRA19-1	Beta-lactam resistance from Alaska	ACH59005	CHR	Uncultured bacterium	300	[94]
B3	MEMA1-1	Meropenem resistance A1	KY705336[c]	CHR	Uncultured bacterium	285	[96]
B3	MIM-1	Maynooth Imipenemase-1	AIT78529	CHR	*Novosphingobium pentaromativorans*	299	[97]
B3	MSI-1	*Massilia* sp.	AJP77057	CHR	*Massilia oculi*	341	[63]
B3	NWM-1	North Rhine-Westphalia Metallo-β-lactamase	QXM27671	CHR	*Pseudomonas aeruginosa*	310	–

Continued

TABLE 1 Representative class B β-lactamases—cont'd

Subclass	Metallo-β-lactamase	Origin of name	Accession	Gene	Organism	Number of residues	References
B3	PAM-1	*Pseudomonas alcaligenes* Metallo-β-lactamase	BAO01151	CHR	*Pseudomonas alcaligenes*	287	[98]
B3	PAM-2	*Pseudomonas alcaligenes* Metallo-β-lactamase	BCG24254	CHR	*Pseudomonas tohonis*	287	[99]
B3	PEDO-1	*Pedobacter* sp.	AJP77059	CHR	*Pedobacter roseus*	286	[63]
B3	PEDO-2	*Pedobacter* sp.	AJP77071	CHR	*Pedobacter borealis*	285	[63]
B3	PJM-1	*Pseudoxanthomonas japonensis* Metallo-β-lactamase	WP_213603971	CHR	*Pseudoxanthomonas japonensis*	301	[100]
B3	PLN-1	*Pedobacter lusitanus*	KIO75746	CHR	*Pedobacter lusitanus*	289	[101]
B3	POM-1	*Pseudomonas otitidis* Metallo-β-lactamase	ABY56045	CHR	*Pseudomonas otitidis*	286	[102]
B3	PNGM-1	Papua New Guinea Metallo-β-lactamase	AWN09461	CHR	Uncultured bacterium	386	[103]
B3	RM3-1	Name of clone	AGU01679	CHR	Uncultured bacterium	302	[47,104]
B3	SAM-1	*Simiduia agarivorans* Metallo-β-lactamase	AFV00127	CHR	*Simiduia agarivorans*	295	[97]
B3	SER-1	*Salmonella enterica*	APY61104	?	*Salmonella enterica* ser Hillingdon	294	[34]
B3	SIE-1	*Sphingobium indicum* B3-E	KEY99315	CHR	*Sphingobium indicum*	307	[105]
B3	SIQ-1	*Sphingobium indicum* B3-Q	SMF77924	CHR	*Allosphingosinicella indica*	290	[34]
B3	SMB-1	*Serratia* Metallo-Beta-lactamase	BAL14456	IS	*Serratia marcescens*	280	[106]
B3	SPG-1	*Sphingomonas* sp.	AJP77080	CHR	*Sphingomonas* sp.	285	[63]
B3	SPR-1	*Serratia proteamaculans*	ABV42357	CHR	*Serratia proteamaculans*	298	[107]
B3	SSE-1		SBV32245	CHR	Uncultured *Sphingopyxis* sp.	288	[34]
B3	B3SU1-1	Subclass B3 and author	QWJ89341	CHR	Uncultured bacterium	303	–
B3	B3SU2-1	Subclass B3 and author	QWJ89342	CHR	Uncultured bacterium	293	–
B3	THIN-B	*Janthinobacterium lividum* class B	CAC33832	CHR	*Janthinobacterium lividum*	316	[108]

[a] *The underlined letters form the name of the β-lactamase.*
[b] *Chr, chromosome; In, integron; IS, insertion sequence; P, plasmid; Tn, transposon; –, unknown.*
[c] *No GenBank accession available.*

TABLE 2 Synthetic mutants of NDM-1: Bacteriological (MIC) and enzymatic (kinetic constants) changes.

	MIC[a] ($\mu g\,mL^{-1}$)						K_m (μM); k_{cat} (s^{-1}); k_{cat}/K_m ($\mu M^{-1}\,s^{-1}$)						
Mutation	AMP[b]	FOT[b]	CAZ[b]	FOX[b]	IMP[b]	MER[b]	AMP[b]	FOT[b]	CAZ[b]	FOX[b]	IMP[b]	MER[b]	Reference
WT	4096	256			16		143	6			65		[113]
							550	85			238		
							3.84	13.79			3.71		
F64H	4096→	256→			16→		188	6			164		[113]
							793	83			409		
							4.27	14.80			2.50		
D84E	256↘	16↘			1↘		82	16			70		[113]
							597	113			153		
							7.29	6.77			2.20		
W87P	2048↘	32↘			2↘		518	6			295		[113]
							1823	46			351		
							3.53	8.50			1.20		
K121R	512↘	64↘			4↘		61	4			31		[113]
							26	13			36		
							0.43	3.69			1.20		
K121T	512↘	64↘			2↘		231	10			117		[113]
							246	50			50		
							1.06	5.13			0.43		
G195A	4096→	256→			8↘		57	4			52		[113]
							248	32			94		
							4.43	7.74			1.84		
D199E	1024↘	64↘			2↘		11	6			6		[113]
							23	48			17		
							2.10	7.65			2.81		
K224H	256↘	128↘			0.5↘		906	26			ND		[113]
							330	111			ND		
							0.36	4.24			0.086		

Continued

TABLE 2 Synthetic mutants of NDM-1: Bacteriological (MIC) and enzymatic (kinetic constants) changes—cont'd

Mutation	MIC ($\mu g\,mL^{-1}$)						K_m (μM); k_{cat} (s^{-1}); k_{cat}/K_m ($\mu M^{-1}\,s^{-1}$)						Reference
	AMP	FOT	CAZ	FOX	IMP	MER	AMP	FOT	CAZ	FOX	IMP	MER	
K224R	2048↘	256→			4↘		179	3			69		[113]
							294	39			62		
							1.65	12.81			0.90		
D225E	4096→	128↘			16→		41	4			50		[113]
							102	38			117		
							2.54	9.78			2.37		
G232A	2048↘	64↘			1↘		178	24			ND		[113]
							417	90			ND		
							2.39	3.80			0.63		
N233H	4096→	256→			8↘		56	3			134		[113]
							591	55			286		
							10.58	20.42			2.22		
N233Q	2048↘	256→			2↘		160	1.1			565		[113]
							279	25			288		
							1.74	22.95			0.51		
WT	>256	>256	64	32	8	16	94.5	53	99.5	41.3	73.5	53.5	[114]
							471	332	114	23	567	237	
							4.98	6.26	1.15	0.56	7.71	4.43	
W93A	32↘	32↘	16↘	4↘	2↘	2↘	128.9	81.9	121.6	67	98.8	87.9	[114]
							394	286	92	17	439	197	
							3.06	3.49	0.76	0.25	4.44	2.24	
WT	>256					16	193					59	[115]
							139					35	
							0.72					0.59	
Y229W	128↘					8↘	297					268	[115]
							253					435	
							0.85					1.62	

WT		128	>128	8	>64	>64					35	80	[116,117]
											64	75	
											1.83	0.94	
L209F		1↘	16↘	8→	0.25↘	<0.0625↘					33	50	[116,117]
											0.55	0.86	
											0.017	0.017	
Y229W		>128→	>128→	8→	>64→	>64→					81	259	[117]
											38	820	
											0.47	3.17	
L209F/Y229W		32↘	32↘	4↘	64↘	>64→					120	88	[117]
											55	208	
											0.46	2.36	
V88L/M154L (NDM-5)	>500				>50	>50	117.5				6.32	11.8	[118]
							1912				44	190.7	
							16				7	16.1	
V88L/E152A/M154L	8↘				0.8↘	0.4↘	207.3				39.3	33.2	[118]
							239				94	81.1	
							1.15				2.3	2.4	

[a]*The arrows represent the variation of CMI compared to WT.*
[b]AMP, *ampicillin;* CAZ, *ceftazidime;* FOT, *cefotaxime;* FOX, *cefoxitin;* IMP, *imipenem;* MER, *meropenem.*

TABLE 3 Synthetic mutants of IMP-1: Bacteriological (MIC) and enzymatic (kinetic constants) changes.

	MIC[a] ($\mu g\,mL^{-1}$)						K_m (μM); k_{cat} (s^{-1}); k_{cat}/K_m ($\mu M^{-1}\,s^{-1}$)						
Mutation	AMP[b]	FOT[b]	CAZ[b]	FOX[b]	IMP[b]	MER[b]	AMP[b]	FOT[b]	CAZ[b]	FOX[b]	IMP[b]	MER[b]	Reference
WT	128		256		2	2	300		16.7		27	9	[125]
							46		1.1		3.9	3.7	
							0.15		0.066		0.144	0.41	
S262G/E126G (IMP-3)	8↘		64↘		0.25↘	8↗							[125]
S262G (IMP-6)	8↘		64↘		0.25↘	4↗	300		91		170	6.5	[125]
							4.9		5.9		11.6	10	
							0.016		0.065		0.07	1.54	
V67F (IMP-10)	16↘		>512↗		4↘	8↗	ND		36		92	17	[125]
							ND		7.2		15.2	17.9	
							0.006		0.2		0.17	1.1	
S262G/G235S (IMP-25)	8↘		128↘		0.25↘	8↗							[125]
E59K (IMP-30)	64↘		256→		1↘	1↘							[125]
E126G (IMP-34)	128→		256→		2→	2→							[125]
V67F/F87S (IMP-40)	4↘		256→		4↗	8↗							[125]
G63R (IMP-42)	64↘		512↗		1↘	2→							[125]
L82I/V96G (IMP-52)	128→		256→		1↘	4↗							[125]
E207K (IMP-60)	64↘		128↘		0.5↘	0.5↘							[125]
N129I (IMP-61)	128→		256→		2→	4↗							[125]
V66F (IMP-66)	32↘		128↘		1↘	8↗							[125]
L304I (IMP-70)	128→		256→		2→	1↘							[125]
V67A (IMP-76)	8↘		128↘		0.5↘	8↗							[125]
V67F/L250S (IMP-77)	16↘		>512↗		2→	8↗							[125]
V67F/S262G (IMP-78)	4↘		128↘		0.25↘	32↗	600		132		270	20.2	[125]
							2.3		20.5		20.3	24.8	
							0.004		0.16		0.075	1.23	
K249R (IMP-79)	128→		256→		2→	4↗							[125]
V67F/S262A (IMP-80)	8↘		512↗		1↘	8↗							[125]
L82I	128→		256→		2→	2→							[125]

WT										1.6	19		[126]
										12.7	41		
										8	2.2		
K161R										480	>1000		[126]
										28	>1		
										0.57	0.0013		
K161A										800	>1000		[126]
										9	>7.1		
										0.011	0.0077		
K161E										1100	>1000		[126]
										1.3	>2.8		
										0.003	0.0045		
N167A										33	>1000		[126]
										90	>140		
										2.7	0.34		
N167D										77	>1000		[126]
										38	>100		
										0.5	0.16		
WT	125	250			3		97	0.98			59		[119]
							62	16			98		
							0.64	17			1.7		
N233A	31↘	63↘			0.25↘		110	30			>1000		[119]
							230	495			>970		
							2.1	16			0.97		
N233C	<2↘	<2↘			0.25↘		1098	76			>2000		[119]
							9.1	51			>5.4		
							0.0083	0.67			0.0027		
N233D	250↗	500↗			3→								[119]
N233E	125→	250→			3→		92	8			178		[119]
							416	34			111		
							4.5	4.3			0.62		

Continued

TABLE 3 Synthetic mutants of IMP-1: Bacteriological (MIC) and enzymatic (kinetic constants) changes—cont'd

Mutation	MIC ($\mu g\,mL^{-1}$)						K_m (μM); k_{cat} (s^{-1}); k_{cat}/K_m ($\mu M^{-1}\,s^{-1}$)						Reference
	AMP	FOT	CAZ	FOX	IMP	MER	AMP	FOT	CAZ	FOX	IMP	MER	
N233F	125→	250→			3→								[119]
N233G	125→	500↗			0.38↘								[119]
N233H	125→	125↘			0.5↘								[119]
N233I	16↘	125↘			0.38↘								[119]
N233K	16↘	63↘			0.38↘		1445	98			>2000		[119]
							132	595			>336		
							0.091	6.1			0.17		
N233L	16↘	250→			0.5↘								[119]
N233M	16↘	500↗			0.75↘								[119]
N233P	125→	125↘			0.38↘								[119]
N233Q	125→	500↗			0.2↘								[119]
N233R	31↘	250→			0.38↘		501	73			>2000		[119]
							83	391			>134		
							0.17	5.3			0.067		
N233S	125→	500↗			1.5↘								[119]
N233T	16↘	500↗			0.38↘								[119]
N233V	31↘	500↗			0.38↘								[119]
N233W	250↗	63↘			1.5↘								[119]
N233Y	125→	125↘			3→								[119]

[a]*The arrows represent the variation of CMI compared to WT.* [b]AMP, *ampicillin;* CAZ, *ceftazidime;* FOT, *cefotaxime;* FOX, *cefoxitin;* IMP, *imipenem;* MER, *meropenem.*

2.4 Secondary and tertiary structure

The typical three-dimensional structure of class B β-lactamase contains a characteristic αβ/βα motif known as the MBL fold. Structurally, class B β-lactamases belong to the *Lactamase_B* PFAM family (PF00753) and share the same overall fold with many other enzymes that catalyze very different biochemical reactions [28]. Indeed, MBLs are part of a larger group of binuclear metallohydrolases that require two closely spaced transition metal ions to carry out a great variety of hydrolytic reactions [146]. It was proposed that MBLs present a modular structure in their active sites that can be dissected into two halves: one providing the attacking nucleophile, and the second one stabilizing a negatively charged reaction intermediate [147].

The X-ray structure of the MBL from *Bacillus cereus* BcII (strain 569/H/9) was published in 1995 [148], revealing a new fold. This represents the first structure of a protein from this family (subclass B1). In 1998 was published the first structure of a B3 subclass MBL (L1 from the opportunistic pathogen *Stenotrophomonas maltophilia*) [149], and in 2005 the first structure of a B2 subclass MBL (CphA from *Aeromonas hydrophila*) [150].

The active site is organized around one (subclass B2) or two (subclasses B1 and B3) zinc ions that coordinate six very conserved residues specific for each subclass (Fig. 3). These zinc ions are essential for the enzymatic activity and play key roles in the catalytic mechanism, including facilitating nucleophilic attack on the amide carbonyl by the zinc-bound hydroxide ion, stabilizing the anionic tetrahedral intermediate, and coordinating the departing amine nitrogen [151]. In some cases, steric constraints imposed by the N-terminus may limit the affinity of the active site for β-lactams [34,104].

Quantum mechanical and molecular dynamics simulations provided insight into the conformational flexibility of these enzymes, the mechanism for carbapenem hydrolysis, and the binding mode of different MBL substrates and inhibitors [152–154].

The number of MBL structures deposited in the Protein Data Bank (PDB) has increased significantly during the recent years, especially for the subclass B1 (Fig. 4). This increase is likely due to the importance of structural studies to guide and accompany the synthetic efforts focused on the development of efficient inhibitors targeting clinically relevant MBLs (e.g., NDM-1, VIM-1, IMP-1, all belonging to subclass B1).

3 Physiologic and pathologic role of metallo-β-lactamases

3.1 Metallo-β-lactamases in clinical microbiology

The emergence of these enzymes in human infections is the consequence of the use of carbapenems in therapy. Indeed, MBLs are able to hydrolyze all β-lactams except monobactams such as aztreonam. The transferable carbapenemases that currently predominate are KPCs (class A), OXA-48-like (class D), and MBLs (class B). Among these MBLs, the most important ones belong to subclass B1, IMP (Imipenemase), VIM (Verona integron-encoded metallo-β-lactamase), and NDM (New Delhi metallo-β-lactamase) [155]. IMP-1 was described in the 1990s in Japan in *Pseudomonas aeruginosa* and *Serratia marcescens* and currently has more than 90 variants (http://bldb.eu/BLDB.php?prot=B1#IMP) [27,112]. The first VIM-like enzyme was described in Italy in 1997 in *P. aeruginosa*, and more than 80 variants are currently reported (http://bldb.eu/BLDB.php?prot=B1#VIM) [27,112]. NDM-1 was identified in Sweden in 2008 (patient returning from India) in *Klebsiella pneumoniae*, and more than 45 variants were characterized to date (http://bldb.eu/BLDB.php?prot=B1#NDM) [27,112].

The IMP and VIM enzymes are encoded by gene cassettes in integrons, which are usually integrated into transposons [156]. The bla_{NDM} genes are carried by composite

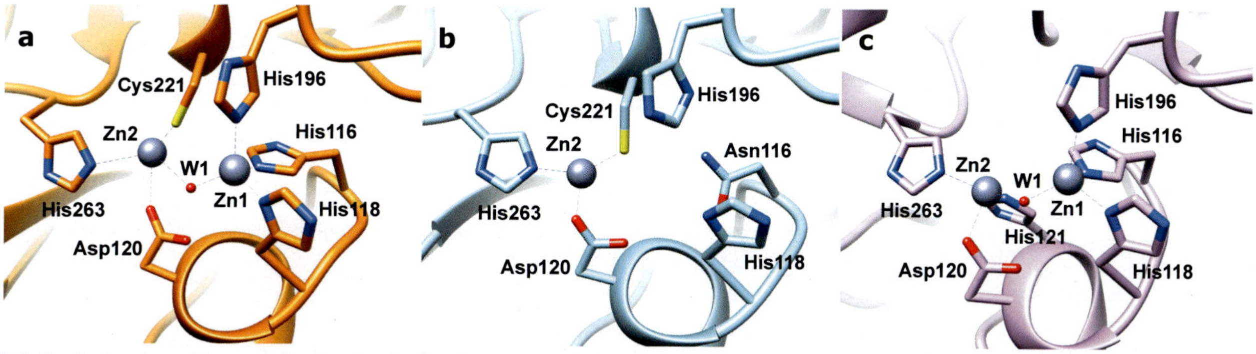

FIG. 3 Active site residues coordinating the zinc ions in representative members of subclass B1 (A, BcII, *orange*), B2 (B, CphA, *cyan*), and B3 (C, L1, *pink*) MBLs. The zinc ions and the bridged water molecules are represented as *gray* and *red* spheres, respectively. The residues are numbered according to the BBL standard numbering scheme [35,36]. *(Credit: The authors (Elsa Denakpo, Guillaume Arlet, Alain Philippon, Bogdan I. Iorga).)*

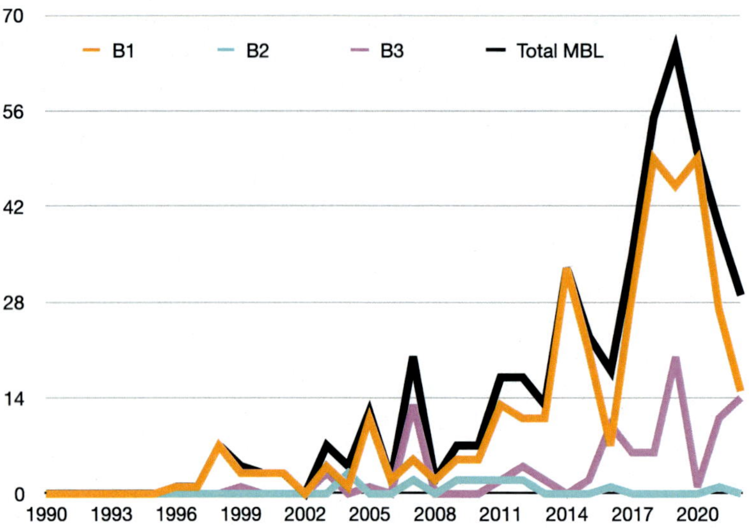

FIG. 4 Evolution of the number of MBL structures deposited in the Protein Data Bank (PDB) over the years. Data were extracted from the Beta-Lactamase DataBase (http://bldb.eu/BLDB/S-BLDB.php). *(Credit: The authors (Elsa Denakpo, Guillaume Arlet, Alain Philippon, Bogdan I. Iorga).)*

transposons such as *Tn*125 bordered by insertion sequences, in particular *IS*Aba125, but multiple recombinations were subsequently observed [26].

While these enzymes have been described all over the world, it seems that IMP- and VIM-type enzymes are predominant in some specific countries: IMP in Japan, Australia, and Thailand, and VIM in Greece, Italy, and neighboring countries [27,155].

NDM-like enzymes are endemic in India and Pakistan and neighboring countries. Since 2010, we assist to a worldwide dissemination of NDM-1 [25–27]. These genes are carried by a wide variety of plasmids (both narrow and broad spectrum) and in particular from the IncX3 group [25,26]. On these plasmids, NDM enzymes are often associated with resistances to other families of antibiotics (aminoglycosides, fluoroquinolones) [25]. The hosts harboring these plasmids are both Enterobacterales and particularly *Escherichia coli* and *K. pneumoniae* as well as *P. aeruginosa* and *Acinetobacter baumannii*. In addition, NDM enzymes have been also reported in animals and in the environment [157].

3.2 Metallo-β-lactamases in veterinary medicine and environment

Considering the animal ecosystem and the enzymatic carbapenem resistance (ECR) mediated by MBLs, numerous publications showed the implication of several transferable types such as IMP-1, NDM-5, VIM-1, and even SIM-1 produced by Enterobacterales (mainly *E. coli*, *K. pneumoniae*, and *Salmonella enterica*) and also by *P. aeruginosa* and *Acinetobacter* spp. [158–165]. Various animal species were implicated such as chickens, pigs, calves, boars, flies, birds, and particularly companion animals (dogs, cats). Such enzymatic resistance is widespread in countries such as Algeria, Australia, Brazil, China, Egypt, India, Japan, United States, and Europe. Considering that carbapenems are not authorized in veterinary medicine, such results indicate the importance of a rational use of antibiotics in this field and their prohibition as growth promoters. Nevertheless, the impact of animal reservoirs on human health still remains unclear and debatable; there are several examples of direct links suggesting a potential transmission way of the ECR plasmids between humans and companion animals [158,166].

A wide range of Gram-negative bacteria issued from nonclinical samples, distributed in various ecosystems such as soil, sediments, sewage, tap water, river, sea water, or costal mud, are able to naturally produce various MBLs (see Table 1 for representative examples). The phylum Bacteroidota (formerly Bacteroidetes) is composed of several classes of bacteria that can be opportunistic pathogens, including the following genera: *Bacteroides*, *Chitinophaga*, *Chryseobacterium*, *Echinicola*, *Elizabethkingia*, *Empedobacter*, *Epilithonimonas*, *Fibrella*, *Flavobacterium*, *Myroides*, *Myxococcus*, *Ornithobacterium*, *Pedobacter*, *Pontibacter*, *Sphingobacterium*, and *Zobellia*. Nevertheless, other MBL-producing Proteobacteria are also mentioned in Table 1, such as *Chromobacterium*, *Erythrobacter*, *Hirschia*, *Pseudobacteriovora*, *Shewanella*, *Sediminispirochaeta*, *Stigmatella*, and *Teredinibacter*. Well-known transferable carbapenem resistance genes (IMP-, NDM-, and VIM-types) have been previously reported to spread rapidly to humans via pathogen bacteria such as Enterobacterales, *P. aeruginosa*, and *Acinetobacter* spp. [167–171]. Such resistance genes encoding MBLs were mainly located on mobile genetic elements (plasmids, transposons, integrons, insertion sequences), which likely contribute to their rapid spread to different bacteria. Such global dissemination of carbapenem-resistant potential opportunistic bacteria is now recognized as a worldwide threat to public health, particularly from a One Health perspective, and a continuous monitoring of some environments is required [172,173].

3.3 Treatment options for infections mediated by metallo-β-lactamases

The treatment options for MBLs-producers and in particular for bacteria producing NDM are very limited. For serious infections, combination therapy including a polymyxin is the preferred option. However, resistance to polymyxins is emerging [174]. Although the introduction of new antibiotics allows the treatment of certain strains of carbapenem-resistant Enterobacterales (CRE) and carbapenem-resistant *P. aeruginosa* (CRPA) with polymyxin-sparing regimens, the use of polymyxins is currently still necessary in carbapenem-resistant *A. baumannii* (CRAB) and in CRE and CRPA harboring MBLs [175]. The activity of β-lactam antibiotics against MBL-producing Enterobacterales in animal infection models was recently reviewed [176].

Alternatively, cefiderocol is a siderophore cephalosporin antibiotic with potent activity against carbapenem-resistant Gram-negative bacteria, including both serine- and metallo-carbapenemase-positive strains [177,178]. Cefiderocol is the only of the four new agents with efficacy against both MBLs and resistant *Acinetobacter* species, but comparator studies using best available therapy for carbapenem-resistant Gram-negative bacterial infections show higher mortality rates with this new drug, making its role in clinical therapy still to be determined [179].

With the exception of taniborbactam (VNRX-5133) and xeruborbactam (QPX7728), none of the novel β-lactam-β-lactamase inhibitor combinations (avibactam, nacubactam, taniborbactam, zidebactam) present in vitro activity against Enterobacterales or *P. aeruginosa* producing MBLs or against carbapenemase-producing *A. baumannii* [180]. However, aztreonam can be used in combination with these inhibitors against multidrug-resistant (MDR) Enterobacterales and *P. aeruginosa*-producing MBLs [181]. The aztreonam/avibactam combination is efficient against MBL-producing enteric bacteria owing to the stability of the monobactam to these enzymes, but resistance is still an issue for MBL-producing nonfermentative bacteria [182].

4 Classes of metallo-β-lactamase inhibitors and their design

For many years, the design of clinically useful MBL inhibitors was an elusive task [183]. Many excellent reviews included a comprehensive description of these studies, on β-lactamase inhibitors in general [184–188] or more specifically on MBL inhibitors [189–206]. A detailed description of these studies is outside the scope of this chapter, and only a few representative examples will be detailed below.

4.1 Metallo-β-lactamase specific inhibitors

In an early study, Kurosaki et al. have exploited the common features of substrate and inhibitor binding to MBLs to generate a new class of irreversible thiol inhibitor (e.g., 3-(3-mercaptopropionylsulfanyl)propionic acid pentafluorophenyl ester) [207,208].

Bisthiazolidines (BTZs) were designed as substrate-mimicking scaffolds for the inhibition of the NDM-1 carbapenemase. Inspired by known interactions of MBLs with β-lactams, these compounds share some features with the β-lactam substrates, which can be modified with metal-binding groups to target the MBL active site [209–212]. Further studies showed that 2-mercaptomethyl thiazolidines (MMTZs) inhibit all MBL subclasses (B1, B2, B3) by maintaining a conserved binding mode [213,214].

Another MBL inhibitor, ANT431, was shown to potentiate the activity of meropenem against a broad range of MBL-producing CRE and restore its efficacy against an *E. coli* NDM-1-producing strain in a murine thigh infection model [215,216]. An optimized derivative, ANT2681, is a specific, competitive inhibitor of MBLs with potent activity against NDM enzymes. Susceptibility studies evidenced the efficacy of ANT2681 in combination with meropenem against MBL-positive Enterobacterales (including NDM-CRE) [217,218].

Some of the novel inhibitory approaches involve the use of chelating agents or metal-based drugs that displace the native metal ion [219]. For example, a fungal natural product, aspergillomarasmine A (AMA), was identified as a rapid and potent inhibitor of two clinically relevant MBLs, NDM-1 and VIM-2. AMA also fully restored the activity of meropenem against Enterobacterales, *Acinetobacter* spp., and *Pseudomonas* spp. possessing either VIM or NDM-type alleles [220–222]. Mechanistic studies showed that AMA inhibits the MBLs by removal of the active site metal ions required for β-lactam hydrolysis [223,224]. Significant improvements were obtained through the total synthesis [225–229], solid-phase synthesis [230], chemoenzymatic synthesis [231], and fragment-based drug discovery (FBDD) [232] of AMA and its derivatives.

A series of 1,2,4-triazole-3-thione compounds was designed and synthesized as inhibitors of dizinc MBLs [233–239]. Computational studies revealed the intermolecular interactions of these compounds with the extended recognition site of VIM-2 [240]. Some compounds showed submicromolar activities against VIM-type enzymes and strong NDM-1 inhibition and the mechanism of this inhibition involved at least partially the stripping of the catalytic zinc ions. A strong synergistic activity with meropenem was observed (16–1000-fold minimum inhibitory concentration (MIC) reduction) against VIM-type- and NDM-1-producing ultraresistant clinical isolates, including Enterobacterales and *P. aeruginosa*, indicating an efficient penetration of the bacterial cells. Furthermore, selected compounds exhibited no or moderate toxicity toward HeLa cells, favorable absorption, distribution, metabolism, excretion (ADME) properties, and no or modest inhibition of several mammalian metalloenzymes [241].

4.2 Covalent metallo-β-lactamase inhibitors

A recent review summarizes the current knowledge on the covalent inhibitors of MBLs and their inhibition modes, highlighting the importance of the rational design of covalent MBL inhibitors and the development of dual-action covalent MBL/SBL inhibitors based on lysine residue of MBLs and serine residue of SBLs [242].

Compounds from the ebsulfur family were reported to inhibit NDM-1 with IC_{50} values ranging from 0.16 to 9 μM. Labeling of NDM-1 using a constructed fluorescent sulfur derivative (Ebs-R) suggested that the inhibitor is covalently bound to the target. Moreover, labeling NDM-1 in living *E. coli* cells with Ebs-R by confocal microscopic imaging showed the real-time distribution of the intracellular recombinant protein [243,244].

The *O*-aryloxycarbonyl hydroxamates are known inactivators of *Enterobacter cloacae* P99 class C SBL with an unusual covalent mechanism targeting both active-site Ser and Lys residues. Unexpectedly, a recent study evidenced that these compounds can also serve as a classical affinity label for NDM-1. Mass spectrometry and ultraviolet photodissociation for extensive fragmentation show a stoichiometric covalent labeling that occurs specifically at Lys211. Moreover, the X-ray crystal structure of the inactivated enzyme reveals that the covalent adduct is bound at the substrate-binding site but not directly coordinated to the active-site zinc cluster [244,245].

Additional studies also reported 3-bromopyruvate [246], 1,2-isoselenazol-3(2*H*)-one derivatives [247], and disulfiram [248] as potent covalent NDM-1 inhibitors.

4.3 Conjugate metallo-β-lactamase inhibitors

Cephalosporin prodrugs containing several MBL inhibitors (thiomandelic acid, dipicolinic acid, 8-thioquinoline) were designed and synthesized for the delivery of the active compounds in a spatiotemporally controlled fashion. These conjugates demonstrated potent inhibition of NDM, VIM, and IMP families of MBLs and displayed potent synergy with meropenem against MBL-expressing clinical isolates of *K. pneumoniae* and *E. coli* [249,250]. Kinetic experiments showed that thiomandelic acid acts as a slowly turned-over substrate [249].

Similarly, a cephalosporin-tripodalamine conjugate (DPASC) was shown to inhibit different MBLs with high efficacy and low toxicity. The cephalosporin tag blocks the ligand binding site to reduce toxicity and is cleaved by MBLs to release active ligands to inhibit MBLs in situ [251].

4.4 Dual serine/metallo-β-lactamase inhibitors

A recent review describes the current knowledge about the dual SBL/MBL inhibitors, including their modes of inhibition and the crystal structures when available [252].

A series of azetidinimines, imino-analogs of β-lactams, was reported as efficient noncovalent inhibitors of all four classes (A, B, C, D) of β-lactamases. Despite the structural and mechanistic differences between the clinically relevant carbapenemases KPC-2 and OXA-48 SBLs and NDM-1 MBL, all three enzymes were inhibited with K_i values below 0.3 μM, while the class C cephalosporinase CMY-2 showed 86% inhibition at 10 μM. Selected azetidinimines were also able repotentiate imipenem against a resistant strain of *E. coli* expressing NDM-1 [253].

Similarly, compounds derived from the benzo-[*b*]-thiophene-2-boronic acid (BZB) and 4-amino-1,2,4-triazole-3-thione scaffolds showed cross-class micromolar inhibition against class A SBL KPC-2 and class B1 MBLs VIM-1 and IMP-1 [254,255].

The bicyclic boronate VNRX-5133 (taniborbactam) is a new type of β-lactamase inhibitor in clinical development combined with cefepime, for the treatment of infections caused by β-lactamase-producing CRE and CRPA. VNRX-5133 inhibits SBLs and some clinically important MBLs, including NDM-1, VIM-1, and VIM-2. However, VNRX-5133's activity against IMP-1, and against other B2 and B3 MBLs tested, was lower or not observed. Inhibition is achieved by mimicking the transition state structure and exploiting interactions with highly conserved active-site residues, while employing distinct mechanisms to inhibit both SBLs and MBLs. It is a reversible covalent inhibitor of SBLs with slow dissociation and a prolonged active-site residence time, while in MBLs it behaves as a competitive inhibitor [256–258].

Xeruborbactam (QPX7728) is an ultra-broad-spectrum β-lactamase inhibitor that is currently in clinical development. In addition to potent inhibition of clinically important SBLs, it also inhibits many MBLs. This compound displays a remarkably broad spectrum of inhibition, including class B and class D enzymes, and is little affected by porin modifications and efflux. QPX7728 combinations with several β-lactam antibiotics show broad coverage of Enterobacterales, *A. baumannii*, and *P. aeruginosa*. An important aspect is that QPX7728 can also be delivered orally [259–268].

5 Metallo-β-lactamase inhibitors in clinical development

Currently, the most advanced clinical trial targeting MBL-producing bacteria involves the combination cefepime/VNRX-5133 (taniborbactam), which has completed the phase 3 with the sponsorship of Venatorx Pharmaceuticals (code: NCT03840148). The combination aztreonam/avibactam is tested in phase 3 since December 2020 with the sponsorship of Pfizer (code: NCT03580044). Finally, the combination QPX2014/QPX7728 (xeruborbactam) has completed the phase 1 with the sponsorship of Qpex Biopharma (codes: NCT04380207 and NCT05072444). This information was last updated in January 2023.

6 Conclusion

The emergence of MBL-producing organisms has become a significant public health concern due to their ability to cause multidrug-resistant infections. While there are compounds that have been identified as MBL inhibitors, none are currently clinically approved. However, several compounds are being evaluated in clinical trials and may be approved in the near future.

Acknowledgments

This work was supported, in part, by grants from the Laboratory of Excellence in Research on Medication and Innovative Therapeutics (LERMIT, ANR-10-LABX-33), the Joint Programming Initiative on Antimicrobial Resistance (JPIAMR, ANR-14-JAMR-0002), and the PPR Antibioresistance (ANR-20-PAMR-0010).

References

[1] Ambler RP. The structure of β-lactamases. Philos Trans R Soc Lond B Biol Sci 1980;289(1036):321–31.

[2] Bush K. The ABCD's of β-lactamase nomenclature. J Infect Chemother 2013;19(4):549–59.

[3] Galtier N, Gouy M, Gautier C. SEAVIEW and PHYLO_WIN: two graphic tools for sequence alignment and molecular phylogeny. Comput Appl Biosci 1996;12(6):543–8.

[4] Bush K, Jacoby GA, Medeiros AA. A functional classification scheme for β-lactamases and its correlation with molecular structure. Antimicrob Agents Chemother 1995;39(6):1211–33.

[5] Bush K. Past and present perspectives on β-lactamases. Antimicrob Agents Chemother 2018;62(10):e01076-18.

[6] Philippon A, Dusart J, Joris B, Frère JM. The diversity, structure and regulation of β-lactamases. Cell Mol Life Sci 1998;54(4):341–6.

[7] Helfand MS, Bonomo RA. β-Lactamases: a survey of protein diversity. Curr Drug Targets Infect Disord 2003;3(1):9–23.

[8] Queenan AM, Bush K. Carbapenemases: the versatile β-lactamases. Clin Microbiol Rev 2007;20(3):440–58.

[9] Bush K. Bench-to-bedside review: the role of β-lactamases in antibiotic-resistant Gram-negative infections. Crit Care 2010;14(3):224.

[10] Bush K, Fisher JF. Epidemiological expansion, structural studies, and clinical challenges of new β-lactamases from Gram-negative bacteria. Annu Rev Microbiol 2011;65:455–78.

[11] Bush K. Proliferation and significance of clinically relevant β-lactamases. Ann N Y Acad Sci 2013;1277:84–90.

[12] Frère JM, Sauvage E, Kerff F. From "an enzyme able to destroy penicillin" to carbapenemases: 70 years of β-lactamase misbehaviour. Curr Drug Targets 2016;17(9):974–82.

[13] Bush K. Metallo-β-lactamases: a class apart. Clin Infect Dis 1998;27 (Suppl. 1):S48–53.

[14] Cricco JA, Vila AJ. Class B β-lactamases: the importance of being metallic. Curr Pharm Des 1999;5(11):915–27.

[15] Walsh TR, Toleman MA, Poirel L, Nordmann P. Metallo-β-lactamases: the quiet before the storm? Clin Microbiol Rev 2005;18 (2):306–25.

[16] Walsh TR. The emergence and implications of metallo-β-lactamases in Gram-negative bacteria. Clin Microbiol Infect 2005;11 (Suppl. 6):2–9.

[17] Crowder MW, Spencer J, Vila AJ. Metallo-β-lactamases: novel weaponry for antibiotic resistance in bacteria. Acc Chem Res 2006;39(10):721–8.

[18] Bebrone C. Metallo-β-lactamases (classification, activity, genetic organization, structure, zinc coordination) and their superfamily. Biochem Pharmacol 2007;74(12):1686–701.

[19] Cornaglia G, Akova M, Amicosante G, Cantón R, Cauda R, Docquier JD, Edelstein M, Frère JM, Fuzi M, Galleni M, Giamarellou H, Gniadkowski M, Koncan R, Libisch B, Luzzaro F, Miriagou V, Navarro F, Nordmann P, Pagani L, Peixe L, Poirel L, Souli M, Tacconelli E, Vatopoulos A, Rossolini GM. Metallo-β-lactamases as emerging resistance determinants in Gram-negative pathogens: open issues. Int J Antimicrob Agents 2007;29(4):380–8.

[20] Cornaglia G, Giamarellou H, Rossolini GM. Metallo-β-lactamases: a last frontier for β-lactams? Lancet Infect Dis 2011;11 (5):381–93.

[21] Nordmann P, Poirel L, Walsh TR, Livermore DM. The emerging NDM carbapenemases. Trends Microbiol 2011;19(12):588–95.

[22] Palzkill T. Metallo-β-lactamase structure and function. Ann N Y Acad Sci 2013;1277:91–104.

[23] Mojica MF, Bonomo RA, Fast W. B1-metallo-β-lactamases: where do we stand? Curr Drug Targets 2016;17(9):1029–50.

[24] Groundwater PW, Xu S, Lai F, Váradi L, Tan J, Perry JD, Hibbs DE. New Delhi metallo-β-lactamase-1: structure, inhibitors and detection of producers. Future Med Chem 2016;8(9):993–1012.

[25] Khan AU, Maryam L, Zarrilli R. Structure, genetics and worldwide spread of New Delhi Metallo-β-lactamase (NDM): a threat to public health. BMC Microbiol 2017;17(1):101.

[26] Wu W, Feng Y, Tang G, Qiao F, McNally A, Zong Z. NDM metallo-β-lactamases and their bacterial producers in health care settings. Clin Microbiol Rev 2019;32(2):e00115-18.

[27] Boyd SE, Livermore DM, Hooper DC, Hope WW. Metallo-β-lactamases: structure, function, epidemiology, treatment options, and the development pipeline. Antimicrob Agents Chemother 2020;64(10): e00397-20.

[28] Bahr G, González LJ, Vila AJ. Metallo-β-lactamases in the age of multidrug resistance: from structure and mechanism to evolution, dissemination, and inhibitor design. Chem Rev 2021;121 (13):7957–8094.

[29] Hall BG, Salipante SJ, Barlow M. The metallo-β-lactamases fall into two distinct phylogenetic groups. J Mol Evol 2003;57(3):249–54.

[30] Hall BG, Salipante SJ, Barlow M. Independent origins of subgroup B1 + B2 and subgroup B3 metallo-β-lactamases. J Mol Evol 2004;59 (1):133–41.

[31] López C, Delmonti J, Bonomo RA, Vila AJ. Deciphering the evolution of metallo-β-lactamases: a journey from the test tube to the bacterial periplasm. J Biol Chem 2022;298(3):101665.

[32] González JM. Visualizing the superfamily of metallo-β-lactamases through sequence similarity network neighborhood connectivity analysis. Heliyon 2021;7(1):e05867.

[33] Crooks GE, Hon G, Chandonia JM, Brenner SE. WebLogo: a sequence logo generator. Genome Res 2004;14(6):1188–90.

[34] Pedroso MM, Waite DW, Melse O, Wilson L, Mitić N, McGeary RP, Antes I, Guddat LW, Hugenholtz P, Schenk G. Broad spectrum antibiotic-degrading metallo-β-lactamases are phylogenetically diverse. Protein Cell 2020;11(8):613–7.

[35] Galleni M, Lamotte-Brasseur J, Rossolini GM, Spencer J, Dideberg O, Frère JM. Standard numbering scheme for class B β-lactamases. Antimicrob Agents Chemother 2001;45(3):660–3.

[36] Garau G, García-Sáez I, Bebrone C, Anne C, Mercuri P, Galleni M, Frère JM, Dideberg O. Update of the standard numbering scheme for class B β-lactamases. Antimicrob Agents Chemother 2004;48 (7):2347–9.

[37] Lin X, Lu J, Qian C, Lin H, Li Q, Zhang X, Liu H, Sun Z, Zhou D, Lu W, Zhu M, Zhang H, Xu T, Li K, Bao Q, Lin L. Molecular and functional characterization of a novel plasmid-borne bla_{NDM}-like gene, bla_{AFM-1}, in a clinical strain of *Aeromonas hydrophila*. Infect Drug Resist 2021;14:1613–22.

[38] Berglund F, Marathe NP, Österlund T, Bengtsson-Palme J, Kotsakis S, Flach CF, Larsson DGJ, Kristiansson E. Identification of 76 novel B1 metallo-β-lactamases through large-scale screening of genomic and metagenomic data. Microbiome 2017;5(1):134.

[39] Hussain M, Carlino A, Madonna MJ, Lampen JO. Cloning and sequencing of the metallothioprotein β-lactamase II gene of *Bacillus cereus* 569/H in *Escherichia coli*. J Bacteriol 1985;164(1):223–9.

[40] Rossolini GM, Franceschini N, Riccio ML, Mercuri PS, Perilli M, Galleni M, Frère JM, Amicosante G. Characterization and sequence of the *Chryseobacterium* (*Flavobacterium*) *meningosepticum* carbapenemase: a new molecular class B β-lactamase showing a broad substrate profile. Biochem J 1998;332(Pt 1):145–52.

[41] Boyd DA, Lisboa LF, Rennie R, Zhanel GG, Dingle TC, Mulvey MR. Identification of a novel metallo-β-lactamase, CAM-1, in clinical *Pseudomonas aeruginosa* isolates from Canada. J Antimicrob Chemother 2019;74(6):1563–7.

[42] Álvarez-Marín MT, Zarzuela L, Camacho EM, Santero E, Flores A. Detection by metagenomic functional analysis and improvement by experimental evolution of β-lactams resistance genes present in oil contaminated soils. Sci Rep 2022;12(1):10059.

[43] Thompson JS, Malamy MH. Sequencing the gene for an imipenem-cefoxitin-hydrolyzing enzyme (CfiA) from *Bacteroides fragilis* TAL2480 reveals strong similarity between CfiA and *Bacillus cereus* β-lactamase II. J Bacteriol 1990;172(5):2584–93.

[44] Rasmussen BA, Gluzman Y, Tally FP. Cloning and sequencing of the class B β-lactamase gene (ccrA) from *Bacteroides fragilis* TAL3636. Antimicrob Agents Chemother 1990;34(8):1590–2.

[45] Bellais S, Naas T, Nordmann P. Genetic and biochemical characterization of CGB-1, an Ambler class B carbapenem-hydrolyzing β-lactamase from *Chryseobacterium gleum*. Antimicrob Agents Chemother 2002;46(9):2791–6.

[46] Sóki J, Lang U, Schumacher U, Nagy I, Berényi Á, Fehér T, Burián K, Nagy E. A novel *Bacteroides* metallo-β-lactamase (MBL) and its gene (crxA) in *Bacteroides xylanisolvens* revealed by genomic sequencing and functional analysis. J Antimicrob Chemother 2022;77(6):1553–6.

[47] Zhang L, Calvo-Bado L, Murray AK, Amos GCA, Hawkey PM, Wellington EM, Gaze WH. Novel clinically relevant antibiotic resistance genes associated with sewage sludge and industrial waste streams revealed by functional metagenomic screening. Environ Int 2019;132:105120.

[48] Poirel L, Rodríguez-Martínez JM, Al Naiemi N, Debets-Ossenkopp YJ, Nordmann P. Characterization of DIM-1, an integron-encoded metallo-β-lactamase from a *Pseudomonas stutzeri* clinical isolate in the Netherlands. Antimicrob Agents Chemother 2010;54 (6):2420–4.

[49] Bellais S, Girlich D, Karim A, Nordmann P. EBR-1, a novel Ambler subclass B1 β-lactamase from *Empedobacter brevis*. Antimicrob Agents Chemother 2002;46(10):3223–7.

[50] Jiang XW, Cheng H, Huo YY, Xu L, Wu YH, Liu WH, Tao FF, Cui XJ, Zheng BW. Biochemical and genetic characterization of a novel metallo-β-lactamase from marine bacterium *Erythrobacter litoralis* HTCC 2594. Sci Rep 2018;8(1):803.

[51] Pollini S, Maradei S, Pecile P, Olivo G, Luzzaro F, Docquier JD, Rossolini GM. FIM-1, a new acquired metallo-β-lactamase from a *Pseudomonas aeruginosa* clinical isolate from Italy. Antimicrob Agents Chemother 2013;57(1):410–6.

[52] Castanheira M, Toleman MA, Jones RN, Schmidt FJ, Walsh TR. Molecular characterization of a β-lactamase gene, bla_{GIM-1}, encoding a new subclass of metallo-β-lactamase. Antimicrob Agents Chemother 2004;48(12):4654–61.

[53] Schauer J, Gatermann SG, Eisfeld J, Hans JB, Ziesing S, Schlüter D, Pfennigwerth N. Characterization of GMB-1, a novel metallo-β-lactamase (MBL) found in three different Enterobacterales species. J Antimicrob Chemother 2022;77(5):1247–53.

[54] Gudeta DD, Bortolaia V, Pollini S, Docquier JD, Rossolini GM, Amos GC, Wellington EM, Guardabassi L. Expanding the repertoire of carbapenem-hydrolyzing metallo-β-lactamases by functional metagenomic analysis of soil microbiota. Front Microbiol 2016;7:1985.

[55] Pfennigwerth N, Lange F, Belmar Campos C, Hentschke M, Gatermann SG, Kaase M. Genetic and biochemical characterization of HMB-1, a novel subclass B1 metallo-β-lactamase found in a *Pseudomonas aeruginosa* clinical isolate. J Antimicrob Chemother 2017;72(4):1068–73.

[56] Osano E, Arakawa Y, Wacharotayankun R, Ohta M, Horii T, Ito H, Yoshimura F, Kato N. Molecular characterization of an enterobacterial metallo β-lactamase found in a clinical isolate of *Serratia marcescens* that shows imipenem resistance. Antimicrob Agents Chemother 1994;38(1):71–8.

[57] Bellais S, Léotard S, Poirel L, Naas T, Nordmann P. Molecular characterization of a carbapenem-hydrolyzing β-lactamase from *Chryseobacterium* (*Flavobacterium*) *indologenes*. FEMS Microbiol Lett 1999;171(2):127–32.

[58] Naas T, Bellais S, Nordmann P. Molecular and biochemical characterization of a carbapenem-hydrolysing β-lactamase from *Flavobacterium johnsoniae*. J Antimicrob Chemother 2003;51(2):267–73.

[59] Sekiguchi J, Morita K, Kitao T, Watanabe N, Okazaki M, Miyoshi-Akiyama T, Kanamori M, Kirikae T. KHM-1, a novel

plasmid-mediated metallo-β-lactamase from a *Citrobacter freundii* clinical isolate. Antimicrob Agents Chemother 2008;52(11):4194–7.
[60] Mammeri H, Bellais S, Nordmann P. Chromosome-encoded β-lactamases TUS-1 and MUS-1 from *Myroides odoratus* and *Myroides odoratimimus* (formerly *Flavobacterium odoratum*), new members of the lineage of molecular subclass B1 metalloenzymes. Antimicrob Agents Chemother 2002;46(11):3561–7.
[61] Yong D, Toleman MA, Giske CG, Cho HS, Sundman K, Lee K, Walsh TR. Characterization of a new metallo-β-lactamase gene, bla_{NDM-1}, and a novel erythromycin esterase gene carried on a unique genetic structure in *Klebsiella pneumoniae* sequence type 14 from India. Antimicrob Agents Chemother 2009;53(12):5046–54.
[62] Kieffer N, Poirel L, Fournier C, Haltli B, Kerr R, Nordmann P. Characterization of PAN-1, a carbapenem-hydrolyzing class B β-lactamase from the environmental Gram-negative *Pseudobacteriovorax antillogorgiicola*. Front Microbiol 2019;10:1673.
[63] Gudeta DD, Bortolaia V, Amos G, Wellington EM, Brandt KK, Poirel L, Nielsen JB, Westh H, Guardabassi L. The soil microbiota harbors a diversity of carbapenem-hydrolyzing β-lactamases of potential clinical relevance. Antimicrob Agents Chemother 2016;60(1):151–60.
[64] Poirel L, Héritier C, Nordmann P. Genetic and biochemical characterization of the chromosome-encoded class B β-lactamases from *Shewanella livingstonensis* (SLB-1) and *Shewanella frigidimarina* (SFB-1). J Antimicrob Chemother 2005;55(5):680–5.
[65] Lee K, Yum JH, Yong D, Lee HM, Kim HD, Docquier JD, Rossolini GM, Chong Y. Novel acquired metallo-β-lactamase gene, bla_{SIM-1}, in a class 1 integron from *Acinetobacter baumannii* clinical isolates from Korea. Antimicrob Agents Chemother 2005;49(11):4485–91.
[66] Poirel L, Magalhaes M, Lopes M, Nordmann P. Molecular analysis of metallo-β-lactamase gene bla_{SPM-1}-surrounding sequences from disseminated *Pseudomonas aeruginosa* isolates in Recife, Brazil. Antimicrob Agents Chemother 2004;48(4):1406–9.
[67] Fang L, Liu Z, Lu Z, Huang R, Xiang R. Identification and characterization of a novel metallo β-lactamase, SZM-1, in Shenzhen Bay, South China. Front Microbiol 2022;13:996834.
[68] El Salabi A, Borra PS, Toleman MA, Samuelsen Ø, Walsh TR. Genetic and biochemical characterization of a novel metallo-β-lactamase, TMB-1, from an *Achromobacter xylosoxidans* strain isolated in Tripoli, Libya. Antimicrob Agents Chemother 2012;56(5):2241–5.
[69] Suzuki S, Matsui M, Suzuki M, Sugita A, Kosuge Y, Kodama N, Ichise Y, Shibayama K. Detection of Tripoli metallo-β-lactamase 2 (TMB-2), a variant of bla_{TMB-1}, in clinical isolates of *Acinetobacter* spp. in Japan. J Antimicrob Chemother 2013;68(6):1441–2.
[70] Cheng Q, Zheng Z, Ye L, Chen S. Identification of a novel metallo-β-lactamase, VAM-1, in a foodborne *Vibrio alginolyticus* isolate from China. Antimicrob Agents Chemother 2021;65(11):e0112921.
[71] Liu M, Zhang W, Peng K, Wang Z, Li R. Identification of a novel plasmid-mediated carbapenemase-encoding gene, bla_{VMB-2}, in *Vibrio diabolicus*. Antimicrob Agents Chemother 2021;65(8):e0020621.
[72] Lauretti L, Riccio ML, Mazzariol A, Cornaglia G, Amicosante G, Fontana R, Rossolini GM. Cloning and characterization of bla_{VIM}, a new integron-borne metallo-β-lactamase gene from a *Pseudomonas aeruginosa* clinical isolate. Antimicrob Agents Chemother 1999;43(7):1584–90.
[73] Zheng Z, Cheng Q, Chan EW, Chen S. Genetic and biochemical characterization of VMB-1, a novel metallo-β-lactamase encoded by a conjugative, broad-host range IncC plasmid from *Vibrio* spp. Adv Biosyst 2020;4(3):e1900221.
[74] Lu WJ, Hsu PH, Lin HV. A novel cooperative metallo-β-lactamase fold metallohydrolase from pathogen *Vibrio vulnificus* exhibits β-lactam antibiotic-degrading activities. Antimicrob Agents Chemother 2021;65(9):e0032621.
[75] Liu S, Zhang L, Feng C, Zhu J, Li A, Zhao J, Zhang Y, Gao M, Shi W, Li Q, Zhang X, Zhang H, Xu T, Lu J, Bao Q. Characterization and identification of a novel chromosome-encoded metallo-β-lactamase WUS-1 in *Myroides albus* P34. Front Microbiol 2022;13:1059997.
[76] Kieffer N, Guzmán-Puche J, Poirel L, Kang HJ, Jeon CO, Nordmann P. ZHO-1, an intrinsic MBL from the environmental Gram-negative species *Zhongshania aliphaticivorans*. J Antimicrob Chemother 2019;74(6):1568–71.
[77] Massidda O, Rossolini GM, Satta G. The *Aeromonas hydrophila* cphA gene: molecular heterogeneity among class B metallo-β-lactamases. J Bacteriol 1991;173(15):4611–7.
[78] Walsh TR, Neville WA, Haran MH, Tolson D, Payne DJ, Bateson JH, MacGowan AP, Bennett PM. Nucleotide and amino acid sequences of the metallo-β-lactamase, ImiS, from *Aeromonas veronii* bv. sobria. Antimicrob Agents Chemother 1998;42(2):436–9.
[79] Niumsup P, Simm AM, Nurmahomed K, Walsh TR, Bennett PM, Avison MB. Genetic linkage of the penicillinase gene, *amp*, and *blr*AB, encoding the regulator of β-lactamase expression in *Aeromonas* spp. J Antimicrob Chemother 2003;51(6):1351–8.
[80] Poirel L, Palmieri M, Brilhante M, Masseron A, Perreten V, Nordmann P. PFM-like enzymes are a novel family of subclass B2 metallo-β-lactamases from *Pseudomonas synxantha* belonging to the *Pseudomonas fluorescens* complex. Antimicrob Agents Chemother 2020;64(2):e01700-19.
[81] Saavedra MJ, Peixe L, Sousa JC, Henriques I, Alves A, Correia A. Sfh-I, a subclass B2 metallo-β-lactamase from a *Serratia fonticola* environmental isolate. Antimicrob Agents Chemother 2003;47(7):2330–3.
[82] Mercuri PS, Esposito R, Blétard S, Di Costanzo S, Perilli M, Kerff F, Galleni M. Mutational effects on carbapenem hydrolysis of YEM-1, a new subclass B2 metallo-β-lactamase from *Yersinia mollaretii*. Antimicrob Agents Chemother 2020;64(9):e00105-20.
[83] Yong D, Toleman MA, Bell J, Ritchie B, Pratt R, Ryley H, Walsh TR. Genetic and biochemical characterization of an acquired subgroup B3 metallo-β-lactamase gene, bla_{AIM-1}, and its unique genetic context in *Pseudomonas aeruginosa* from Australia. Antimicrob Agents Chemother 2012;56(12):6154–9.
[84] Stoczko M, Frère JM, Rossolini GM, Docquier JD. Postgenomic scan of metallo-β-lactamase homologues in rhizobacteria: identification and characterization of BJP-1, a subclass B3 ortholog from *Bradyrhizobium japonicum*. Antimicrob Agents Chemother 2006;50(6):1973–81.
[85] Au SX, Dzulkifly NS, Muhd Noor ND, Matsumura H, Raja Abdul Rahman RNZ, Normi YM. Dual activity BLEG-1 from *Bacillus lehensis* G1 revealed structural resemblance to B3 metallo-β-lactamase and glyoxalase II: an insight into its enzyme promiscuity and evolutionary divergence. Int J Mol Sci 2021;22(17):9377.
[86] Stoczko M, Frère JM, Rossolini GM, Docquier JD. Functional diversity among metallo-β-lactamases: characterization of the

CAR-1 enzyme of *Erwinia carotovora*. Antimicrob Agents Chemother 2008;52(7):2473–9.

[87] Docquier JD, Pantanella F, Giuliani F, Thaller MC, Amicosante G, Galleni M, Frère JM, Bush K, Rossolini GM. CAU-1, a subclass B3 metallo-β-lactamase of low substrate affinity encoded by an ortholog present in the *Caulobacter crescentus* chromosome. Antimicrob Agents Chemother 2002;46(6):1823–30.

[88] Girlich D, Poirel L, Nordmann P. Diversity of naturally occurring Ambler class B metallo-β-lactamases in *Erythrobacter* spp. J Antimicrob Chemother 2012;67(11):2661–4.

[89] Boschi L, Mercuri PS, Riccio ML, Amicosante G, Galleni M, Frère JM, Rossolini GM. The *Legionella* (*Fluoribacter*) *gormanii* metallo-β-lactamase: a new member of the highly divergent lineage of molecular-subclass B3 β-lactamases. Antimicrob Agents Chemother 2000;44(6):1538–43.

[90] Bellais S, Aubert D, Naas T, Nordmann P. Molecular and biochemical heterogeneity of class B carbapenem-hydrolyzing β-lactamases in *Chryseobacterium meningosepticum*. Antimicrob Agents Chemother 2000;44(7):1878–86.

[91] Walsh TR, Hall L, Assinder SJ, Nichols WW, Cartwright SJ, MacGowan AP, Bennett PM. Sequence analysis of the L1 metallo-β-lactamase from *Xanthomonas maltophilia*. Biochim Biophys Acta 1994;1218(2):199–201.

[92] Lange F, Pfennigwerth N, Hartl R, Kerschner H, Achleitner D, Gatermann SG, Kaase M. LMB-1, a novel family of class B3 MBLs from an isolate of *Enterobacter cloacae*. J Antimicrob Chemother 2018;73(9):2331–5.

[93] Dabos L, Rodriguez CH, Nastro M, Dortet L, Bonnin RA, Famiglietti A, Iorga BI, Vay C, Naas T. LMB-1 producing *Citrobacter freundii* from Argentina, a novel player in the field of MBLs. Int J Antimicrob Agents 2020;55(2):105857.

[94] Allen HK, Moe LA, Rodbumrer J, Gaarder A, Handelsman J. Functional metagenomics reveals diverse β-lactamases in a remote Alaskan soil. ISME J 2009;3(2):243–51.

[95] Pedroso MM, Selleck C, Enculescu C, Harmer JR, Mitić N, Craig WR, Helweh W, Hugenholtz P, Tyson GW, Tierney DL, Larrabee JA, Schenk G. Characterization of a highly efficient antibiotic-degrading metallo-β-lactamase obtained from an uncultured member of a permafrost community. Metallomics 2017;9 (8):1157–68.

[96] Lau CH, van Engelen K, Gordon S, Renaud J, Topp E. Novel antibiotic resistance determinants from agricultural soil exposed to antibiotics widely used in human medicine and animal farming. Appl Environ Microbiol 2017;83(16):e00989-17.

[97] Miraula M, Schenk G, Mitić N. Promiscuous metallo-β-lactamases: MIM-1 and MIM-2 may play an essential role in quorum sensing networks. J Inorg Biochem 2016;162:366–75.

[98] Suzuki M, Suzuki S, Matsui M, Hiraki Y, Kawano F, Shibayama K. A subclass B3 metallo-β-lactamase found in *Pseudomonas alcaligenes*. J Antimicrob Chemother 2014;69(5):1430–2.

[99] Yamada K, Yoshizumi A, Nagasawa T, Aoki K, Sasaki M, Murakami H, Morita T, Ishii Y, Tateda K. Molecular and biochemical characterization of novel PAM-like MBL variants, PAM-2 and PAM-3, from clinical isolates of *Pseudomonas tohonis*. J Antimicrob Chemother 2022;77(9):2414–8.

[100] Yamada K, Ishii Y, Tateda K. Biochemical characterization of the subclass B3 metallo-β-lactamase PJM-1 from *Pseudoxanthomonas japonensis*. Antimicrob Agents Chemother 2022;66(9):e0069122.

[101] Viana AT, Caetano T, Covas C, Santos T, Mendo S. Environmental superbugs: the case study of *Pedobacter* spp. Environ Pollut 2018;241:1048–55.

[102] Thaller MC, Borgianni L, Di Lallo G, Chong Y, Lee K, Dajcs J, Stroman D, Rossolini GM. Metallo-β-lactamase production by *Pseudomonas otitidis*: a species-related trait. Antimicrob Agents Chemother 2011;55(1):118–23.

[103] Park KS, Kim TY, Kim JH, Lee JH, Jeon JH, Karim AM, Malik SK, Lee SH. PNGM-1, a novel subclass B3 metallo-β-lactamase from a deep-sea sediment metagenome. J Glob Antimicrob Resist 2018;14:302–5.

[104] Salimraj R, Zhang L, Hinchliffe P, Wellington EM, Brem J, Schofield CJ, Gaze WH, Spencer J. Structural and biochemical characterization of Rm3, a subclass B3 metallo-β-lactamase identified from a functional metagenomic study. Antimicrob Agents Chemother 2016;60(10):5828–40.

[105] Wilson LA, Knaven EG, Morris MT, Monteiro Pedroso M, Schofield CJ, Brück TB, Boden M, Waite DW, Hugenholtz P, Guddat L, Schenk G. Kinetic and structural characterization of the first B3 metallo-β-lactamase with an active-site glutamic acid. Antimicrob Agents Chemother 2021;65(10):e0093621.

[106] Wachino J, Yoshida H, Yamane K, Suzuki S, Matsui M, Yamagishi T, Tsutsui A, Konda T, Shibayama K, Arakawa Y. SMB-1, a novel subclass B3 metallo-β-lactamase, associated with ISCR1 and a class 1 integron, from a carbapenem-resistant *Serratia marcescens* clinical isolate. Antimicrob Agents Chemother 2011;55(11):5143–9.

[107] Vella P, Miraula M, Phelan E, Leung EW, Ely F, Ollis DL, McGeary RP, Schenk G, Mitić N. Identification and characterization of an unusual metallo-β-lactamase from *Serratia proteamaculans*. J Biol Inorg Chem 2013;18(7):855–63.

[108] Rossolini GM, Condemi MA, Pantanella F, Docquier JD, Amicosante G, Thaller MC. Metallo-β-lactamase producers in environmental microbiota: new molecular class B enzyme in *Janthinobacterium lividum*. Antimicrob Agents Chemother 2001;45(3):837–44.

[109] Philippon A, Labia R, Jacoby G. Extended-spectrum β-lactamases. Antimicrob Agents Chemother 1989;33(8):1131–6.

[110] Matagne A, Lamotte-Brasseur J, Frère JM. Catalytic properties of class A β-lactamases: efficiency and diversity. Biochem J 1998;330(Pt 2):581–98.

[111] Chaïbi EB, Sirot D, Paul G, Labia R. Inhibitor-resistant TEM β-lactamases: phenotypic, genetic and biochemical characteristics. J Antimicrob Chemother 1999;43(4):447–58.

[112] Naas T, Oueslati S, Bonnin RA, Dabos ML, Zavala A, Dortet L, Retailleau P, Iorga BI. Beta-lactamase database (BLDB)—structure and function. J Enzyme Inhib Med Chem 2017;32(1):917–9.

[113] Sun Z, Hu L, Sankaran B, Prasad BVV, Palzkill T. Differential active site requirements for NDM-1 β-lactamase hydrolysis of carbapenem versus penicillin and cephalosporin antibiotics. Nat Commun 2018;9 (1):4524.

[114] Khan AU, Rehman MT. Role of non-active-site residue Trp-93 in the function and stability of New Delhi Metallo-β-Lactamase 1. Antimicrob Agents Chemother 2016;60(1):356–60.

[115] Chen J, Chen H, Shi Y, Hu F, Lao X, Gao X, Zheng H, Yao W. Probing the effect of the non-active-site mutation Y229W in New Delhi metallo-β-lactamase-1 by site-directed mutagenesis, kinetic studies, and molecular dynamics simulations. PLoS One 2013;8(12):e82080.

[116] Marcoccia F, Leiros HS, Aschi M, Amicosante G, Perilli M. Exploring the role of L209 residue in the active site of NDM-1 metallo-β-lactamase. PLoS One 2018;13(1):e0189686.

[117] Piccirilli A, Brisdelli F, Aschi M, Celenza G, Amicosante G, Perilli M. Kinetic profile and molecular dynamic studies show that Y229W substitution in an NDM-1/L209F variant restores the hydrolytic activity of the enzyme toward penicillins, cephalosporins, and carbapenems. Antimicrob Agents Chemother 2019;63(4):e02270-18.

[118] Kumar G, Issa B, Kar D, Biswal S, Ghosh AS. E152A substitution drastically affects NDM-5 activity. FEMS Microbiol Lett 2017;364 (3):fnx008.

[119] Brown NG, Horton LB, Huang W, Vongpunsawad S, Palzkill T. Analysis of the functional contributions of Asn233 in metallo-β-lactamase IMP-1. Antimicrob Agents Chemother 2011;55 (12):5696–702.

[120] Hall BG. *In vitro* evolution predicts that the IMP-1 metallo-β-lactamase does not have the potential to evolve increased activity against imipenem. Antimicrob Agents Chemother 2004;48 (3):1032–3.

[121] Liu EM, Pegg KM, Oelschlaeger P. The sequence-activity relationship between metallo-β-lactamases IMP-1, IMP-6, and IMP-25 suggests an evolutionary adaptation to meropenem exposure. Antimicrob Agents Chemother 2012;56(12):6403–6.

[122] Oelschlaeger P, Mayo SL, Pleiss J. Impact of remote mutations on metallo-β-lactamase substrate specificity: implications for the evolution of antibiotic resistance. Protein Sci 2005;14(3):765–74.

[123] Pegg KM, Liu EM, George AC, LaCuran AE, Bethel CR, Bonomo RA, Oelschlaeger P. Understanding the determinants of substrate specificity in IMP family metallo-β-lactamases: the importance of residue 262. Protein Sci 2014;23(10):1451–60.

[124] Zhang CJ, Faheem M, Dang P, Morris MN, Kumar P, Oelschlaeger P. Mutation S115T in IMP-type metallo-β-lactamases compensates for decreased expression levels caused by mutation S119G. Biomolecules 2019;9(11):724.

[125] Cheng Z, Bethel CR, Thomas PW, Shurina BA, Alao JP, Thomas CA, Yang K, Marshall SH, Zhang H, Sturgill AM, Kravats AN, Page RC, Fast W, Bonomo RA, Crowder MW. Carbapenem use is driving the evolution of imipenemase 1 variants. Antimicrob Agents Chemother 2021;65(4):e01714-20.

[126] Haruta S, Yamamoto ET, Eriguchi Y, Sawai T. Characterization of the active-site residues asparagine 167 and lysine 161 of the IMP-1 metallo β-lactamase. FEMS Microbiol Lett 2001;197(1):85–9.

[127] Yamaguchi Y, Kuroki T, Yasuzawa H, Higashi T, Jin W, Kawanami A, Yamagata Y, Arakawa Y, Goto M, Kurosaki H. Probing the role of Asp-120(81) of metallo-β-lactamase (IMP-1) by site-directed mutagenesis, kinetic studies, and X-ray crystallography. J Biol Chem 2005;280(21):20824–32.

[128] Haruta S, Yamaguchi H, Yamamoto ET, Eriguchi Y, Nukaga M, O'Hara K, Sawai T. Functional analysis of the active site of a metallo-β-lactamase proliferating in Japan. Antimicrob Agents Chemother 2000;44(9):2304–9.

[129] Martínez-García L, González-Alba JM, Baquero F, Cantón R, Galán JC. Ceftazidime is the key diversification and selection driver of VIM-type carbapenemases. MBio 2018;9(3):e02109-17.

[130] Cheng Z, Shurina BA, Bethel CR, Thomas PW, Marshall SH, Thomas CA, Yang K, Kimble RL, Montgomery JS, Orischak MG, Miller CM, Tennenbaum JL, Nix JC, Tierney DL, Fast W, Bonomo RA, Page RC, Crowder MW. A single salt bridge in VIM-20 increases protein stability and antibiotic resistance under low-zinc conditions. MBio 2019;10(6):e02412-19.

[131] Borgianni L, Vandenameele J, Matagne A, Bini L, Bonomo RA, Frère JM, Rossolini GM, Docquier JD. Mutational analysis of VIM-2 reveals an essential determinant for metallo-β-lactamase stability and folding. Antimicrob Agents Chemother 2010;54(8):3197–204.

[132] Chen JZ, Fowler DM, Tokuriki N. Comprehensive exploration of the translocation, stability and substrate recognition requirements in VIM-2 lactamase. Elife 2020;9:e56707.

[133] Samuelsen Ø, Castanheira M, Walsh TR, Spencer J. Kinetic characterization of VIM-7, a divergent member of the VIM metallo-β-lactamase family. Antimicrob Agents Chemother 2008;52 (8):2905–8.

[134] Marchiaro P, Tomatis PE, Mussi MA, Pasteran F, Viale AM, Limansky AS, Vila AJ. Biochemical characterization of metallo-β-lactamase VIM-11 from a *Pseudomonas aeruginosa* clinical strain. Antimicrob Agents Chemother 2008;52(6):2250–2.

[135] Liu Z, Zhang R, Li W, Yang L, Liu D, Wang S, Shen J, Wang Y. Amino acid changes at the VIM-48 C-terminus result in increased carbapenem resistance, enzyme activity and protein stability. J Antimicrob Chemother 2019;74(4):885–93.

[136] Merino M, Pérez-Llarena FJ, Kerff F, Poza M, Mallo S, Rumbo-Feal S, Beceiro A, Juan C, Oliver A, Bou G. Role of changes in the L3 loop of the active site in the evolution of enzymatic activity of VIM-type metallo-β-lactamases. J Antimicrob Chemother 2010;65 (9):1950–4.

[137] Lassaux P, Traoré DA, Loisel E, Favier A, Docquier JD, Sohier JS, Laurent C, Bebrone C, Frère JM, Ferrer JL, Galleni M. Biochemical and structural characterization of the subclass B1 metallo-β-lactamase VIM-4. Antimicrob Agents Chemother 2011;55 (3):1248–55.

[138] Rodriguez-Martinez JM, Nordmann P, Fortineau N, Poirel L. VIM-19, a metallo-β-lactamase with increased carbapenemase activity from *Escherichia coli* and *Klebsiella pneumoniae*. Antimicrob Agents Chemother 2010;54(1):471–6.

[139] Ikonomidis A, Ntokou E, Maniatis AN, Tsakris A, Pournaras S. Hidden VIM-1 metallo-β-lactamase phenotypes among *Acinetobacter baumannii* clinical isolates. J Clin Microbiol 2008;46(1):346–9.

[140] Falcone M, Mezzatesta ML, Perilli M, Forcella C, Giordano A, Cafiso V, Amicosante G, Stefani S, Venditti M. Infections with VIM-1 metallo-β-lactamase-producing *Enterobacter cloacae* and their correlation with clinical outcome. J Clin Microbiol 2009;47 (11):3514–9.

[141] Zhao WH, Chen G, Ito R, Hu ZQ. Relevance of resistance levels to carbapenems and integron-borne $bla_{\text{IMP-1}}$, $bla_{\text{IMP-7}}$, $bla_{\text{IMP-10}}$ and $bla_{\text{VIM-2}}$ in clinical isolates of *Pseudomonas aeruginosa*. J Med Microbiol 2009;58(Pt 8):1080–5.

[142] Hu Z, Zhao WH. Identification of plasmid- and integron-borne $bla_{\text{IMP-1}}$ and $bla_{\text{IMP-10}}$ in clinical isolates of *Serratia marcescens*. J Med Microbiol 2009;58(Pt 2):217–21.

[143] Singh T, Singh PK, Das S, Wani S, Jawed A, Dar SA. Transcriptome analysis of β-lactamase genes in diarrheagenic *Escherichia coli*. Sci Rep 2019;9(1):3626.

[144] Cheung CHP, Alorabi M, Hamilton F, Takebayashi Y, Mounsey O, Heesom KJ, Williams PB, Williams OM, Albur M, MacGowan AP, Avison MB. Trade-offs between antibacterial resistance and fitness

cost in the production of metallo-β-lactamases by enteric bacteria manifest as sporadic emergence of carbapenem resistance in a clinical setting. Antimicrob Agents Chemother 2021;65(8): e0241220.

[145] Li Y, Zhang R, Wang S. A natural novel mutation in the *bla*$_{NDM-5}$ promoter reducing carbapenems resistance in a clinical *Escherichia coli* strain. Microbiol Spectr 2022;10(1):e0118321.

[146] Schenk G, Mitić N, Gahan LR, Ollis DL, McGeary RP, Guddat LW. Binuclear metallohydrolases: complex mechanistic strategies for a simple chemical reaction. Acc Chem Res 2012;45(9):1593–603.

[147] Meini MR, Llarrull LI, Vila AJ. Overcoming differences: the catalytic mechanism of metallo-β-lactamases. FEBS Lett 2015;589 (22):3419–32.

[148] Carfi A, Pares S, Duée E, Galleni M, Duez C, Frère JM, Dideberg O. The 3-D structure of a zinc metallo-β-lactamase from *Bacillus cereus* reveals a new type of protein fold. EMBO J 1995;14 (20):4914–21.

[149] Ullah JH, Walsh TR, Taylor IA, Emery DC, Verma CS, Gamblin SJ, Spencer J. The crystal structure of the L1 metallo-β-lactamase from *Stenotrophomonas maltophilia* at 1.7 A resolution. J Mol Biol 1998;284(1):125–36.

[150] Garau G, Bebrone C, Anne C, Galleni M, Frère JM, Dideberg O. A metallo-β-lactamase enzyme in action: crystal structures of the monozinc carbapenemase CphA and its complex with biapenem. J Mol Biol 2005;345(4):785–95.

[151] Karsisiotis AI, Damblon CF, Roberts GC. A variety of roles for versatile zinc in metallo-β-lactamases. Metallomics 2014;6(7):1181–97.

[152] Estiu G, Suárez D, Merz KM. Quantum mechanical and molecular dynamics simulations of ureases and Zn β-lactamases. J Comput Chem 2006;27(12):1240–62.

[153] Lisa MN, Palacios AR, Aitha M, González MM, Moreno DM, Crowder MW, Bonomo RA, Spencer J, Tierney DL, Llarrull LI, Vila AJ. A general reaction mechanism for carbapenem hydrolysis by mononuclear and binuclear metallo-β-lactamases. Nat Commun 2017;8(1):538.

[154] Gervasoni S, Spencer J, Hinchliffe P, Pedretti A, Vairoletti F, Mahler G, Mulholland AJ. A multiscale approach to predict the binding mode of metallo β-lactamase inhibitors. Proteins 2022;90(2):372–84.

[155] Bush K, Bradford PA. Epidemiology of β-lactamase-producing pathogens. Clin Microbiol Rev 2020;33(2):e00047-19.

[156] Zhao WH, Hu ZQ. Acquired metallo-β-lactamases and their genetic association with class 1 integrons and ISCR elements in Gram-negative bacteria. Future Microbiol 2015;10(5):873–87.

[157] Das S. The crisis of carbapenemase-mediated carbapenem resistance across the human-animal-environmental interface in India. Infect Dis Now 2023;53(1):104628.

[158] Madec JY, Haenni M, Nordmann P, Poirel L. Extended-spectrum β-lactamase/AmpC- and carbapenemase-producing Enterobacteriaceae in animals: a threat for humans? Clin Microbiol Infect 2017;23(11):826–33.

[159] Roschanski N, Hadziabdic S, Borowiak M, Malorny B, Tenhagen BA, Projahn M, Kaesbohrer A, Guenther S, Szabo I, Roesler U, Fischer J. Detection of VIM-1-producing *Enterobacter cloacae* and *Salmonella enterica* serovars Infantis and Goldcoast at a breeding pig farm in Germany in 2017 and their molecular relationship to former VIM-1-producing *S. infantis* isolates in German livestock production. mSphere 2019;4(3):e00089-19.

[160] Cole SD, Peak L, Tyson GH, Reimschuessel R, Ceric O, Rankin SC. New Delhi metallo-β-lactamase-5-producing *Escherichia coli* in companion animals, United States. Emerg Infect Dis 2020;26 (2):381–3.

[161] Ouchar Mahamat O, Kempf M, Lounnas M, Tidjani A, Hide M, Benavides JA, et al. Epidemiology and prevalence of extended-spectrum β-lactamase- and carbapenemase-producing *Enterobacteriaceae* in humans, animals and the environment in West and Central Africa. Int J Antimicrob Agents 2021;57(1):106203.

[162] Kuang X, Zhang Y, Liu J, Yang RS, Qiu ZY, Sun J, Liao XP, Liu YH, Yu Y. Molecular epidemiology of New Delhi metallo-β-lactamase-producing *Escherichia coli* in food-producing animals in China. Front Microbiol 2022;13:912260.

[163] Kyung SM, Choi SW, Lim J, Shim S, Kim S, Im YB, Lee NE, Hwang CY, Kim D, Yoo HS. Comparative genomic analysis of plasmids encoding metallo-β-lactamase NDM-5 in Enterobacterales Korean isolates from companion dogs. Sci Rep 2022;12(1):1569.

[164] Selmi R, Tayh G, Srairi S, Mamlouk A, Ben Chehida F, Lahmar S, Bouslama M, Daaloul-Jedidi M, Messadi L. Prevalence, risk factors and emergence of extended-spectrum β-lactamase producing-, carbapenem- and colistin-resistant Enterobacterales isolated from wild boar (*Sus scrofa*) in Tunisia. Microb Pathog 2022;163:105385.

[165] Ahlstrom CA, Woksepp H, Sandegren L, Mohsin M, Hasan B, Muzyka D, et al. Genomically diverse carbapenem resistant *Enterobacteriaceae* from wild birds provide insight into global patterns of spatiotemporal dissemination. Sci Total Environ 2022;824:153632.

[166] Hong JS, Song W, Park HM, Oh JY, Chae JC, Han JI, Jeong SH. First detection of New Delhi metallo-β-lactamase-5-producing *Escherichia coli* from companion animals in Korea. Microb Drug Resist 2019;25(3):344–9.

[167] Poirel L, Pitout JD, Nordmann P. Carbapenemases: molecular diversity and clinical consequences. Future Microbiol 2007;2 (5):501–12.

[168] Nordmann P, Dortet L, Poirel L. Carbapenem resistance in *Enterobacteriaceae*: here is the storm! Trends Mol Med 2012;18(5):263–72.

[169] Mills MC, Lee J. The threat of carbapenem-resistant bacteria in the environment: evidence of widespread contamination of reservoirs at a global scale. Environ Pollut 2019;255(Pt 1):113143.

[170] Dewi D, Götz B, Thomas T. Diversity and genetic basis for carbapenem resistance in a coastal marine environment. Appl Environ Microbiol 2020;86(10):e02939-19.

[171] Ranjan R, Thatikonda S. β-Lactam resistance gene NDM-1 in the aquatic environment: a review. Curr Microbiol 2021;78(10):3634–43.

[172] Suzuki Y, Ida M, Kubota H, Ariyoshi T, Murakami K, Kobayashi M, Kato R, Hirai A, Suzuki J, Sadamasu K. Multiple β-lactam resistance gene-carrying plasmid harbored by *Klebsiella quasipneumoniae* isolated from urban sewage in Japan. mSphere 2019;4(5):e00391-19.

[173] Sung GH, Kim SH, Park EH, Hwang SN, Kim JD, Kim GR, Kim EY, Jeong J, Kim S, Shin JH. Association of carbapenemase-producing Enterobacterales detected in stream and clinical samples. Front Microbiol 2022;13:923979.

[174] Wailan AM, Paterson DL. The spread and acquisition of NDM-1: a multifactorial problem. Expert Rev Anti Infect Ther 2014;12 (1):91–115.

[175] Perez F, El Chakhtoura NG, Yasmin M, Bonomo RA. Polymyxins: to combine or not to combine? Antibiotics (Basel) 2019;8(2):38.

[176] Asempa TE, Abdelraouf K, Nicolau DP. Activity of β-lactam antibiotics against metallo-β-lactamase-producing Enterobacterales in animal infection models: a current state of affairs. Antimicrob Agents Chemother 2021;65(6):e02271-20.

[177] Sato T, Yamawaki K. Cefiderocol: discovery, chemistry, and *in vivo* profiles of a novel siderophore cephalosporin. Clin Infect Dis 2019;69(Suppl. 7):S538–43.

[178] McCreary EK, Heil EL, Tamma PD. New perspectives on antimicrobial agents: Cefiderocol. Antimicrob Agents Chemother 2021;65(8):e0217120.

[179] Noval M, Banoub M, Claeys KC, Heil E. The battle is on: new β-lactams for the treatment of multidrug-resistant Gram-negative organisms. Curr Infect Dis Rep 2020;22(1):1.

[180] Yahav D, Giske CG, Grāmatniece A, Abodakpi H, Tam VH, Leibovici L. New β-lactam-β-lactamase inhibitor combinations. Clin Microbiol Rev 2020;34(1):e00115-20.

[181] Le Terrier C, Nordmann P, Poirel L. *In vitro* activity of aztreonam in combination with newly developed β-lactamase inhibitors against MDR Enterobacterales and *Pseudomonas aeruginosa* producing metallo-β-lactamases. J Antimicrob Chemother 2022;78(1):101–7.

[182] Bush K. A resurgence of β-lactamase inhibitor combinations effective against multidrug-resistant Gram-negative pathogens. Int J Antimicrob Agents 2015;46(5):483–93.

[183] González MM, Vila AJ. An elusive task: a clinically useful inhibitor of metallo-β-lactamases. In: Supuran CT, Capasso C, editors. Zinc enzyme inhibitors: Enzymes from microorganisms. Cham: Springer International Publishing; 2017. p. 1–34.

[184] Buynak JD. β-Lactamase inhibitors: a review of the patent literature (2010–2013). Expert Opin Ther Pat 2013;23(11):1469–81.

[185] Chen J, Shang X, Hu F, Lao X, Gao X, Zheng H, Yao W. β-Lactamase inhibitors: an update. Mini Rev Med Chem 2013;13 (13):1846–61.

[186] Tooke CL, Hinchliffe P, Bragginton EC, Colenso CK, Hirvonen VHA, Takebayashi Y, Spencer J. β-Lactamases and β-lactamase inhibitors in the 21st century. J Mol Biol 2019;431(18):3472–500.

[187] Davies DT, Everett M. Designing inhibitors of β-lactamase enzymes to overcome carbapenem resistance in Gram-negative bacteria. Acc Chem Res 2021;54(9):2055–64.

[188] Li R, Chen X, Zhou C, Dai QQ, Yang L. Recent advances in β-lactamase inhibitor chemotypes and inhibition modes. Eur J Med Chem 2022;242:114677.

[189] Toney JH, Moloughney JG. Metallo-β-lactamase inhibitors: promise for the future? Curr Opin Investig Drugs 2004;5(8):823–6.

[190] Fast W, Sutton LD. Metallo-β-lactamase: inhibitors and reporter substrates. Biochim Biophys Acta 2013;1834(8):1648–59.

[191] King DT, Strynadka NC. Targeting metallo-β-lactamase enzymes in antibiotic resistance. Future Med Chem 2013;5(11):1243–63.

[192] Guo Z, Ma S. Recent advances in the discovery of metallo-β-lactamase inhibitors for β-lactam antibiotic-resistant reversing agents. Curr Drug Targets 2014;15(7):689–702.

[193] Somboro AM, Osei Sekyere J, Amoako DG, Essack SY, Bester LA. Diversity and proliferation of metallo-β-lactamases: a clarion call for clinically effective metallo-β-lactamase inhibitors. Appl Environ Microbiol 2018;84:e00698-18–18.

[194] McGeary RP, Tan DT, Schenk G. Progress toward inhibitors of metallo-β-lactamases. Future Med Chem 2017;9(7):673–91.

[195] Rotondo CM, Wright GD. Inhibitors of metallo-β-lactamases. Curr Opin Microbiol 2017;39:96–105.

[196] Wang P, Cheng J, Liu CC, Tang K, Xu F, Yu Z, Yu B, Chang J. The development of new small-molecule inhibitors targeting bacterial metallo-β-lactamases. Curr Top Med Chem 2018;18(10):834–43.

[197] Ju LC, Cheng Z, Fast W, Bonomo RA, Crowder MW. The continuing challenge of metallo-β-lactamase inhibition: mechanism matters. Trends Pharmacol Sci 2018;39(7):635–47.

[198] Linciano P, Cendron L, Gianquinto E, Spyrakis F, Tondi D. Ten years with New Delhi metallo-β-lactamase-1 (NDM-1): from structural insights to inhibitor design. ACS Infect Dis 2019;5(1):9–34.

[199] Shi C, Chen J, Kang X, Shen X, Lao X, Zheng H. Approaches for the discovery of metallo-β-lactamase inhibitors: a review. Chem Biol Drug Des 2019;94(2):1427–40.

[200] Kaushik A, Kaushik M, Lather V, Dua JS. Recent review on subclass B1 metallo-β-lactamases inhibitors: sword for antimicrobial resistance. Curr Drug Targets 2019;20(7):756–62.

[201] Reddy N, Shungube M, Arvidsson PI, Baijnath S, Kruger HG, Govender T, Naicker T. A 2018-2019 patent review of metallo β-lactamase inhibitors. Expert Opin Ther Pat 2020;30(7):541–55.

[202] Nagulapalli Venkata KC, Ellebrecht M, Tripathi SK. Efforts towards the inhibitor design for New Delhi metallo-β-lactamase (NDM-1). Eur J Med Chem 2021;225:113747.

[203] Wang T, Xu K, Zhao L, Tong R, Xiong L, Shi J. Recent research and development of NDM-1 inhibitors. Eur J Med Chem 2021;223:113667.

[204] Li X, Zhao J, Zhang B, Duan X, Jiao J, Wu W, Zhou Y, Wang H. Drug development concerning metallo-β-lactamases in Gram-negative bacteria. Front Microbiol 2022;13:959107.

[205] Mojica MF, Rossi MA, Vila AJ, Bonomo RA. The urgent need for metallo-β-lactamase inhibitors: an unattended global threat. Lancet Infect Dis 2022;22(1):e28–34.

[206] Gu X, Zheng M, Chen L, Li H. The development of New Delhi metallo-β-lactamase-1 inhibitors since 2018. Microbiol Res 2022;261:127079.

[207] Kurosaki H, Yamaguchi Y, Higashi T, Soga K, Matsueda S, Yumoto H, Misumi S, Yamagata Y, Arakawa Y, Goto M. Irreversible inhibition of metallo-β-lactamase (IMP-1) by 3-(3-mercaptopropionylsulfanyl)propionic acid pentafluorophenyl ester. Angew Chem Int Ed Engl 2005;44(25):3861–4.

[208] Spencer J, Walsh TR. A new approach to the inhibition of metallo-β-lactamases. Angew Chem Int Ed Engl 2006;45(7):1022–6.

[209] González MM, Kosmopoulou M, Mojica MF, Castillo V, Hinchliffe P, Pettinati I, Brem J, Schofield CJ, Mahler G, Bonomo RA, Llarrull LI, Spencer J, Vila AJ. Bisthiazolidines: a substrate-mimicking scaffold as an inhibitor of the NDM-1 carbapenemase. ACS Infect Dis 2015;1(11):544–54.

[210] Hinchliffe P, González MM, Mojica MF, González JM, Castillo V, Saiz C, et al. Cross-class metallo-β-lactamase inhibition by bisthiazolidines reveals multiple binding modes. Proc Natl Acad Sci USA 2016;113(26):E3745–54.

[211] Mojica MF, Mahler SG, Bethel CR, Taracila MA, Kosmopoulou M, Papp-Wallace KM, Llarrull LI, Wilson BM, Marshall SH, Wallace CJ, Villegas MV, Harris ME, Vila AJ, Spencer J, Bonomo RA. Exploring the role of residue 228 in substrate and inhibitor recognition by VIM metallo-β-lactamases. Biochemistry 2015;54(20):3183–96.

[212] Saiz C, Villamil V, González MM, Rossi MA, Martínez L, Suescun L, Vila AJ, Mahler G. Enantioselective synthesis of new oxazolidinylthiazolidines as enzyme inhibitors. Tetrahedron Asymmetry 2017;28(1):110–7.

[213] Hinchliffe P, Moreno DM, Rossi MA, Mojica MF, Martinez V, Villamil V, Spellberg B, Drusano GL, Banchio C, Mahler G, Bonomo RA, Vila AJ, Spencer J. 2-Mercaptomethyl thiazolidines (MMTZs) inhibit all metallo-β-lactamase classes by maintaining a conserved binding mode. ACS Infect Dis 2021;7(9):2697–706.

[214] Rossi MA, Martinez V, Hinchliffe P, Mojica MF, Castillo V, Moreno DM, Smith R, Spellberg B, Drusano GL, Banchio C, Bonomo RA, Spencer J, Vila AJ, Mahler G. 2-Mercaptomethyl-thiazolidines use conserved aromatic-S interactions to achieve broad-range inhibition of metallo-β-lactamases. Chem Sci 2021;12(8):2898–908.

[215] Everett M, Sprynski N, Coelho A, Castandet J, Bayet M, Bougnon J, et al. Discovery of a novel metallo-β-lactamase inhibitor that potentiates meropenem activity against carbapenem-resistant *Enterobacteriaceae*. Antimicrob Agents Chemother 2018;62(5), e00074-18.

[216] Leiris S, Coelho A, Castandet J, Bayet M, Lozano C, Bougnon J, et al. SAR studies leading to the identification of a novel series of metallo-β-lactamase inhibitors for the treatment of carbapenem-resistant *Enterobacteriaceae* infections that display efficacy in an animal infection model. ACS Infect Dis 2019;5(1):131–40.

[217] Davies DT, Leiris S, Sprynski N, Castandet J, Lozano C, Bousquet J, et al. ANT2681: SAR studies leading to the identification of a metallo-β-lactamase inhibitor with potential for clinical use in combination with meropenem for the treatment of infections caused by NDM-producing *Enterobacteriaceae*. ACS Infect Dis 2020;6 (9):2419–30.

[218] Zalacain M, Lozano C, Llanos A, Sprynski N, Valmont T, De Piano C, Davies D, Leiris S, Sable C, Ledoux A, Morrissey I, Lemonnier M, Everett M. Novel specific metallo-β-lactamase inhibitor ANT2681 restores meropenem activity to clinically effective levels against NDM-positive Enterobacterales. Antimicrob Agents Chemother 2021;65(6):e00203-21.

[219] Bahr G, González LJ, Vila AJ. Metallo-β-lactamases and a tug-of-war for the available zinc at the host-pathogen interface. Curr Opin Chem Biol 2022;66:102103.

[220] King AM, Reid-Yu SA, Wang W, King DT, De Pascale G, Strynadka NC, Walsh TR, Coombes BK, Wright GD. Aspergillomarasmine A overcomes metallo-β-lactamase antibiotic resistance. Nature 2014;510(7506):503–6.

[221] von Nussbaum F, Schiffer G. Aspergillomarasmine A, an inhibitor of bacterial metallo-β-lactamases conferring bla_{NDM} and bla_{VIM} resistance. Angew Chem Int Ed Engl 2014;53(44):11696–8.

[222] Rotondo CM, Sychantha D, Koteva K, Wright GD. Suppression of β-lactam resistance by aspergillomarasmine A is influenced by both the metallo-β-lactamase target and the antibiotic partner. Antimicrob Agents Chemother 2020;64(4):e01386-19.

[223] Bergstrom A, Katko A, Adkins Z, Hill J, Cheng Z, Burnett M, Yang H, Aitha M, Mehaffey MR, Brodbelt JS, Tehrani K, Martin NI, Bonomo RA, Page RC, Tierney DL, Fast W, Wright GD, Crowder MW. Probing the interaction of aspergillomarasmine A with metallo-β-lactamases NDM-1, VIM-2, and IMP-7. ACS Infect Dis 2018;4(2):135–45.

[224] Sychantha D, Rotondo CM, Tehrani K, Martin NI, Wright GD. Aspergillomarasmine A inhibits metallo-β-lactamases by selectively sequestering Zn^{2+}. J Biol Chem 2021;297(2):100918.

[225] Albu SA, Koteva K, King AM, Al-Karmi S, Wright GD, Capretta A. Total synthesis of aspergillomarasmine A and related compounds: a sulfamidate approach enables exploration of structure-activity relationships. Angew Chem Int Ed Engl 2016;55(42):13259–62.

[226] Koteva K, King AM, Capretta A, Wright GD. Total synthesis and activity of the metallo-β-lactamase inhibitor aspergillomarasmine A. Angew Chem Int Ed Engl 2016;55(6):2210–2.

[227] Liao D, Yang S, Wang J, Zhang J, Hong B, Wu F, Lei X. Total synthesis and structural reassignment of aspergillomarasmine A. Angew Chem Int Ed Engl 2016;55(13):4291–5.

[228] Zhang J, Wang S, Bai Y, Guo Q, Zhou J, Lei X. Total syntheses of natural metallophores staphylopine and aspergillomarasmine A. J Org Chem 2017;82(24):13643–8.

[229] Zhang J, Wang S, Wei Q, Guo Q, Bai Y, Yang S, Song F, Zhang L, Lei X. Synthesis and biological evaluation of aspergillomarasmine A derivatives as novel NDM-1 inhibitor to overcome antibiotics resistance. Bioorg Med Chem 2017;25(19):5133–41.

[230] Koteva K, Sychantha D, Rotondo CM, Hobson C, Britten JF, Wright GD. Three-dimensional structure and optimization of the metallo-β-lactamase inhibitor aspergillomarasmine A. ACS Omega 2022;7 (5):4170–84.

[231] Tehrani K, Fu H, Brüchle NC, Mashayekhi V, Prats Luján A, van Haren MJ, Poelarends GJ, Martin NI. Aminocarboxylic acids related to aspergillomarasmine A (AMA) and ethylenediamine-*N,N'*-disuccinic acid (EDDS) are strong zinc-binders and inhibitors of the metallo-β-lactamase NDM-1. Chem Commun (Camb) 2020;56 (20):3047–9.

[232] Chen AY, Thomas CA, Thomas PW, Yang K, Cheng Z, Fast W, Crowder MW, Cohen SM. Iminodiacetic acid as a novel metal-binding pharmacophore for New Delhi metallo-β-lactamase inhibitor development. ChemMedChem 2020;15(14):1272–82.

[233] Sevaille L, Gavara L, Bebrone C, De Luca F, Nauton L, Achard M, Mercuri P, Tanfoni S, Borgianni L, Guyon C, Lonjon P, Turan-Zitouni G, Dzieciolowski J, Becker K, Bénard L, Condon C, Maillard L, Martinez J, Frère JM, Dideberg O, Galleni M, Docquier JD, Hernandez JF. 1,2,4-Triazole-3-thione compounds as inhibitors of dizinc metallo-β-lactamases. ChemMedChem 2017;12(12):972–85.

[234] Gavara L, Sevaille L, De Luca F, Mercuri P, Bebrone C, Feller G, Legru A, Cerboni G, Tanfoni S, Baud D, Cutolo G, Bestgen B, Chelini G, Verdirosa F, Sannio F, Pozzi C, Benvenuti M, Kwapien K, Fischer M, Becker K, Frère JM, Mangani S, Gresh N, Berthomieu D, Galleni M, Docquier JD, Hernandez JF. 4-Amino-1,2,4-triazole-3-thione-derived Schiff bases as metallo-β-lactamase inhibitors. Eur J Med Chem 2020;208:112720.

[235] Gavara L, Verdirosa F, Legru A, Mercuri PS, Nauton L, Sevaille L, Feller G, Berthomieu D, Sannio F, Marcoccia F, Tanfoni S, De Luca F, Gresh N, Galleni M, Docquier JD, Hernandez JF. 4-(*N*-Alkyl- and -acyl-amino)-1,2,4-triazole-3-thione analogs as metallo-β-lactamase inhibitors: impact of 4-linker on potency and spectrum of inhibition. Biomolecules 2020;10(8):1094.

[236] Gavara L, Legru A, Verdirosa F, Sevaille L, Nauton L, Corsica G, Mercuri PS, Sannio F, Feller G, Coulon R, De Luca F, Cerboni G, Tanfoni S, Chelini G, Galleni M, Docquier JD, Hernandez JF. 4-Alkyl-1,2,4-triazole-3-thione analogues as metallo-β-lactamase inhibitors. Bioorg Chem 2021;113:105024.

[237] Legru A, Verdirosa F, Hernandez JF, Tassone G, Sannio F, Benvenuti M, Conde PA, Bossis G, Thomas CA, Crowder MW, Dillenberger M, Becker K, Pozzi C, Mangani S, Docquier JD, Gavara L. 1,2,4-Triazole-3-thione compounds with a 4-ethyl alkyl/aryl sulfide substituent are broad-spectrum metallo-β-lactamase inhibitors with re-sensitization activity. Eur J Med Chem 2021;226:113873.

[238] Gavara L, Verdirosa F, Sevaille L, Legru A, Corsica G, Nauton L, Sandra Mercuri P, Sannio F, De Luca F, Hadjadj M, Cerboni G, Vo Hoang Y, Licznar-Fajardo P, Galleni M, Docquier JD, Hernandez JF. 1,2,4-Triazole-3-thione analogues with an arylakyl group at position 4 as metallo-β-lactamase inhibitors. Bioorg Med Chem 2022;72:116964.
[239] Verdirosa F, Gavara L, Sevaille L, Tassone G, Corsica G, Legru A, Feller G, Chelini G, Mercuri PS, Tanfoni S, Sannio F, Benvenuti M, Cerboni G, De Luca F, Bouajila E, Vo Hoang Y, Licznar-Fajardo P, Galleni M, Pozzi C, Mangani S, Docquier JD, Hernandez JF. 1,2,4-Triazole-3-thione analogues with a 2-ethylbenzoic acid at position 4 as VIM-type metallo-β-lactamase inhibitors. ChemMedChem 2022;17(7):e202100699.
[240] Kwapien K, Gavara L, Docquier JD, Berthomieu D, Hernandez JF, Gresh N. Intermolecular interactions of the extended recognition site of VIM-2 metallo-β-lactamase with 1,2,4-triazole-3-thione inhibitors. Validations of a polarizable molecular mechanics potential by *ab initio* QC. J Comput Chem 2021;42(2):86–106.
[241] Legru A, Verdirosa F, Vo-Hoang Y, Tassone G, Vascon F, Thomas CA, Sannio F, Corsica G, Benvenuti M, Feller G, Coulon R, Marcoccia F, Devente SR, Bouajila E, Piveteau C, Leroux F, Deprez-Poulain R, Deprez B, Licznar-Fajardo P, Crowder MW, Cendron L, Pozzi C, Mangani S, Docquier JD, Hernandez JF, Gavara L. Optimization of 1,2,4-triazole-3-thiones toward broad-spectrum metallo-β-lactamase inhibitors showing potent synergistic activity on VIM- and NDM-1-producing clinical isolates. J Med Chem 2022;65(24):16392–419.
[242] Chen C, Yang KW. Structure-based design of covalent inhibitors targeting metallo-β-lactamases. Eur J Med Chem 2020;203:112573.
[243] Su J, Liu J, Chen C, Zhang Y, Yang K. Ebsulfur as a potent scaffold for inhibition and labelling of New Delhi metallo-β-lactamase-1 *in vitro* and *in vivo*. Bioorg Chem 2019;84:192–201.
[244] Mehaffey MR, Ahn YC, Rivera DD, Thomas PW, Cheng Z, Crowder MW, Pratt RF, Fast W, Brodbelt JS. Elusive structural changes of New Delhi metallo-β-lactamase revealed by ultraviolet photodissociation mass spectrometry. Chem Sci 2020;11(33):8999–9010.
[245] Thomas PW, Cammarata M, Brodbelt JS, Monzingo AF, Pratt RF, Fast W. A lysine-targeted affinity label for serine-β-lactamase also covalently modifies New Delhi metallo-β-lactamase-1 (NDM-1). Biochemistry 2019;58(25):2834–43.
[246] Kang PW, Su JP, Sun LY, Gao H, Yang KW. 3-Bromopyruvate as a potent covalently reversible inhibitor of New Delhi metallo-β-lactamase-1 (NDM-1). Eur J Pharm Sci 2020;142:105161.
[247] Yue K, Xu C, Wang Z, Liu W, Liu C, Xu X, Xing Y, Chen S, Li X, Wan S. 1,2-Isoselenazol-3(2*H*)-one derivatives as NDM-1 inhibitors displaying synergistic antimicrobial effects with meropenem on NDM-1 producing clinical isolates. Bioorg Chem 2022;129:106153.
[248] Chen C, Yang KW, Wu LY, Li JQ, Sun LY. Disulfiram as a potent metallo-β-lactamase inhibitor with dual functional mechanisms. Chem Commun (Camb) 2020;56(18):2755–8.
[249] Tehrani K, Wade N, Mashayekhi V, Brüchle NC, Jespers W, Voskuil K, Pesce D, van Haren MJ, van Westen GJP, Martin NI. Novel cephalosporin conjugates display potent and selective inhibition of imipenemase-type metallo-β-lactamases. J Med Chem 2021;64 (13):9141–51.
[250] van Haren MJ, Tehrani K, Kotsogianni I, Wade N, Brüchle NC, Mashayekhi V, Martin NI. Cephalosporin prodrug inhibitors overcome metallo-β-lactamase driven antibiotic resistance. Chemistry 2021;27(11):3806–11.
[251] Tian H, Wang Y, Dai Y, Mao A, Zhou W, Cao X, Deng H, Lu H, Ding L, Shen H, Wang X. A cephalosporin-tripodalamine conjugate inhibits metallo-β-lactamase with high efficacy and low toxicity. Antimicrob Agents Chemother 2022;e0035222.
[252] Chen C, Oelschlaeger P, Wang D, Xu H, Wang Q, Wang C, Zhao A, Yang KW. Structure and mechanism-guided design of dual serine/metallo-carbapenemase inhibitors. J Med Chem 2022;65(8):5954–74.
[253] Romero E, Oueslati S, Benchekroun M, D'Hollander ACA, Ventre S, Vijayakumar K, Minard C, Exilie C, Tlili L, Retailleau P, Zavala A, Elisée E, Selwa E, Nguyen LA, Pruvost A, Naas T, Iorga BI, Dodd RH, Cariou K. Azetidinimines as a novel series of non-covalent broad-spectrum inhibitors of β-lactamases with submicromolar activities against carbapenemases KPC-2 (class A), NDM-1 (class B) and OXA-48 (class D). Eur J Med Chem 2021;219:113418.
[254] Santucci M, Spyrakis F, Cross S, Quotadamo A, Farina D, Tondi D, De Luca F, Docquier JD, Prieto AI, Ibacache C, Blázquez J, Venturelli A, Cruciani G, Costi MP. Computational and biological profile of boronic acids for the detection of bacterial serine- and metallo-β-lactamases. Sci Rep 2017;7(1):17716.
[255] Linciano P, Gianquinto E, Montanari M, Maso L, Bellio P, Cebrián-Sastre E, Celenza G, Blázquez J, Cendron L, Spyrakis F, Tondi D. 4-Amino-1,2,4-triazole-3-thione as a promising scaffold for the inhibition of serine and metallo-β-lactamases. Pharmaceuticals (Basel) 2020;13(3):52.
[256] Krajnc A, Brem J, Hinchliffe P, Calvopiña K, Panduwawala TD, Lang PA, Kamps J, Tyrrell JM, Widlake E, Saward BG, Walsh TR, Spencer J, Schofield CJ. Bicyclic boronate VNRX-5133 inhibits metallo- and serine-β-lactamases. J Med Chem 2019;62(18):8544–56.
[257] Hamrick JC, Docquier JD, Uehara T, Myers CL, Six DA, Chatwin CL, John KJ, Vernacchio SF, Cusick SM, Trout REL, Pozzi C, De Luca F, Benvenuti M, Mangani S, Liu B, Jackson RW, Moeck G, Xerri L, Burns CJ, Pevear DC, Daigle DM. VNRX-5133 (taniborbactam), a broad-spectrum inhibitor of serine- and metallo-β-lactamases, restores activity of cefepime in Enterobacterales and *Pseudomonas aeruginosa*. Antimicrob Agents Chemother 2020;64 (3):e01963-19.
[258] Liu B, Trout REL, Chu GH, McGarry D, Jackson RW, Hamrick JC, Daigle DM, Cusick SM, Pozzi C, De Luca F, Benvenuti M, Mangani S, Docquier JD, Weiss WJ, Pevear DC, Xerri L, Burns CJ. Discovery of taniborbactam (VNRX-5133): a broad-spectrum serine- and metallo-β-lactamase inhibitor for carbapenem-resistant bacterial infections. J Med Chem 2020;63(6):2789–801.
[259] Hecker SJ, Reddy KR, Lomovskaya O, Griffith DC, Rubio-Aparicio D, Nelson K, Tsivkovski R, Sun D, Sabet M, Tarazi Z, Parkinson J, Totrov M, Boyer SH, Glinka TW, Pemberton OA, Chen Y, Dudley MN. Discovery of cyclic boronic acid QPX7728, an ultrabroad-spectrum inhibitor of serine and metallo-β-lactamases. J Med Chem 2020;63(14):7491–507.
[260] Lomovskaya O, Nelson K, Rubio-Aparicio D, Tsivkovski R, Sun D, Dudley MN. Impact of intrinsic resistance mechanisms on potency of QPX7728, a new ultrabroad-spectrum β-lactamase inhibitor of serine and metallo-β-lactamases in Enterobacteriaceae, *Pseudomonas aeruginosa*, and *Acinetobacter baumannii*. Antimicrob Agents Chemother 2020;64(6):e00552-20.
[261] Lomovskaya O, Rubio-Aparicio D, Nelson K, Sun D, Tsivkovski R, Castanheira M, Lindley J, Loutit J, Dudley M. *In vitro* activity of the

ultrabroad-spectrum β-lactamase inhibitor QPX7728 in combination with multiple β-lactam antibiotics against *Pseudomonas aeruginosa*. Antimicrob Agents Chemother 2021;65(6):e00210-21.

[262] Lomovskaya O, Tsivkovski R, Nelson K, Rubio-Aparicio D, Sun D, Totrov M, Dudley MN. Spectrum of β-lactamase inhibition by the cyclic boronate QPX7728, an ultrabroad-spectrum β-lactamase inhibitor of serine and metallo-β-lactamases: enhancement of activity of multiple antibiotics against isogenic strains expressing single β-lactamases. Antimicrob Agents Chemother 2020;64(6):e00212-20.

[263] Nelson K, Rubio-Aparicio D, Sun D, Dudley M, Lomovskaya O. *In vitro* activity of the ultrabroad-spectrum-β-lactamase inhibitor QPX7728 against carbapenem-resistant Enterobacterales with varying intrinsic and acquired resistance mechanisms. Antimicrob Agents Chemother 2020;64(8):e00757-20.

[264] Nelson K, Rubio-Aparicio D, Tsivkovski R, Sun D, Totrov M, Dudley M, Lomovskaya O. *In vitro* activity of the ultra-broad-spectrum β-lactamase inhibitor QPX7728 in combination with meropenem against clinical isolates of carbapenem-resistant *Acinetobacter baumannii*. Antimicrob Agents Chemother 2020;64(11):e01406-20.

[265] Sabet M, Tarazi Z, Griffith DC. *In vivo* activity of QPX7728, an ultrabroad-spectrum β-lactamase inhibitor, in combination with β-lactams against carbapenem-resistant *Klebsiella pneumoniae*. Antimicrob Agents Chemother 2020;64(11):e01267-20.

[266] Tsivkovski R, Totrov M, Lomovskaya O. Biochemical characterization of QPX7728, a new ultrabroad-spectrum β-lactamase inhibitor of serine and metallo-β-lactamases. Antimicrob Agents Chemother 2020;64(6):e00130-20.

[267] Lomovskaya O, Tsivkovski R, Sun D, Reddy R, Totrov M, Hecker S, Griffith D, Loutit J, Dudley M. QPX7728, an ultra-broad-spectrum β-lactamase inhibitor for intravenous and oral therapy: overview of biochemical and microbiological characteristics. Front Microbiol 2021;12:697180.

[268] Lomovskaya O, Rubio-Aparicio D, Tsivkovski R, Loutit J, Dudley M. The ultrabroad-spectrum β-lactamase inhibitor QPX7728 restores the potency of multiple oral β-lactam antibiotics against β-lactamase-producing strains of resistant Enterobacterales. Antimicrob Agents Chemother 2022;66(2):e0216821.

Chapter 3.3

Bacterial zinc proteases

Clemente Capasso[a] and Claudiu T. Supuran[b]
[a]*Department of Biology, Agriculture and Food Sciences, CNR, Institute of Biosciences and Bioresources, Napoli, Italy,* [b]*NEUROFARBA Department, Section of Pharmaceutical and Nutraceutical Sciences, University of Florence, Florence, Italy*

1 Introduction

Infectious diseases continue to be the leading cause of death among humans all over the world [1]. As a result of the introduction and spread of pathogenic bacterial strains resistant to most antibiotic classes used in clinical practice, there is a pressing need for the development of innovative therapeutic medicines. Moreover, infectious diseases that are resistant to antibiotics also substantially impact the provision of public health services [2]. There are numerous techniques for dealing with antibiotic resistance, such as community- and healthcare-based infection control and prevention, vaccine production, decrease in the administration of antibiotics in livestock farms, proper antibiotic use, maintaining funding to keep second-line antibiotics available, and finally, developing of innovative antibiotics and novel antiinfective drugs [3–6]. The last point is the primary objective of biochemical research, whose investigation in the twenty-first century is firmly obligated to find ways to create new antiinfective medications with a novel mode of action and without cross-resistance to the existing drugs [7]. In this context, bacterial genome research has helped to define new critical bacterial genes and revealed several structural information about proteins in bacteria that are crucial to disease, such as identifying zinc proteases necessary for bacterial development but not for mammalian cells [8]. Metalloproteases (MPRs) are promising targets for antimicrobial medicines and the foundation for future therapeutics [9–16]. This chapter will cover the bacterial MPRs, such as the clostridial collagenases, pseudolysin (Elastase E), anthrax lethal factors (LFs), botulinum, and tetanus toxins (see Table 1).

1.1 Bacterial proteases

Proteases, also known as proteinases or peptidases, are biomolecules able to hydrolyze the peptide bond joining two successive alpha-amino acids throughout a peptide or protein chain [10–16]. These biocatalysts are ubiquitous in bacteria and, because of their hydrolase activity, play crucial roles in a wide variety of bacterial functions, including host colonization, evasion of host immune responses, nutrition acquisition for bacterial growth, proliferation, facilitation of dispersion, and tissue damage during infection [17–19]. The importance of developing bacterial protease inhibitors to block microbial host invasion and the ongoing emergence of antibiotic resistance is readily apparent. In bacteria, five types of proteases have been identified, which follow Barrett and coworkers' classification: serine proteinases, threonine proteinases, cysteine proteinases, aspartic proteinases, and MPRs [20,21]. The five types are divided into two categories. One group includes serine, threonine, and cysteine proteinases characterized by a part of an amino acid as the nucleophile of the catalytic site, producing an acyl-enzyme intermediate and making these enzymes operate as transferases [20,21]. In contrast, the second group is formed by cysteine proteinases and MPRs, with an activated water molecule as a nucleophile [20,21]. Many researchers have embraced the protease specificity proposed by Berger and Schechter, in which the active site residues in the protease are composed of contiguous pockets termed subsites, which bind to a corresponding residue in the substrate sequence [22]. According to this definition, amino acid residues in the substrate sequence are consecutively numbered outward from the cleavage sites as P_3-P_2-P_1-P_1'-P_2'- ... P_n' (P_1 and P_1' positions represent the scissile bound), while the enzyme subsites are correspondingly labeled as S_3-S_2-S_1-S_1' -S_2' ... S_n' (S_1-S_1' subsites represent the catalytic site) toward the C-terminus (Fig. 1). The same logic applies to enzyme inhibitors that attach to the catalytic site [10–16].

2 Bacterial metalloproteases

As aforementioned, bacterial MPRs are crucial virulence factors and have many harmful effects during a bacterial infection [23–25]. Here are the following examples. MPRs cause necrotic or hemorrhagic tissue damage by digesting structural tissue components in local bacterial infections such as keratitis, dermatitis, and pneumonia [26]. MPRs operate as a synergistic virulence factor in systemic

Metalloenzymes. https://doi.org/10.1016/B978-0-12-823974-2.00019-X

TABLE 1 List of bacterial metalloproteases considered in the present chapter. Enzyme nomenclature, scissile peptide bond (P1-P1′), and microorganism source are reported.

EC and family	Microorganism source	Metalloprotease name	P1-P1′ Scissile bond
EC 3.4.24.3 (Family M9)	*Clostridium histolyticum*	Collagenase	Xaa-Gly
EC 3.4.24.26 (Family M4)	*Pseudomonas aeruginosa*	Pseudolysin (Elastase B)	Phe-Xaa; Gly-Leu
EC 3.4.24.68 (Family M27)	*Clostridium tetani*	Tentoxilysin (Tetanus neurotoxin)	Gln-Phe
EC 3.4.24.69 (Family M27)	*Clostridium botulinum* *Clostridium barati* *Clostridium butyricum*	Bontoxilysin (Botulinum neurotoxin)	Gln-Phe Gln-Arg Lys-Ala Arg-Ile
EC 3.4.24.83 (Family M34)	*Bacillus anthracis*	Anthrax toxin lethal factor	Pro-Xaa

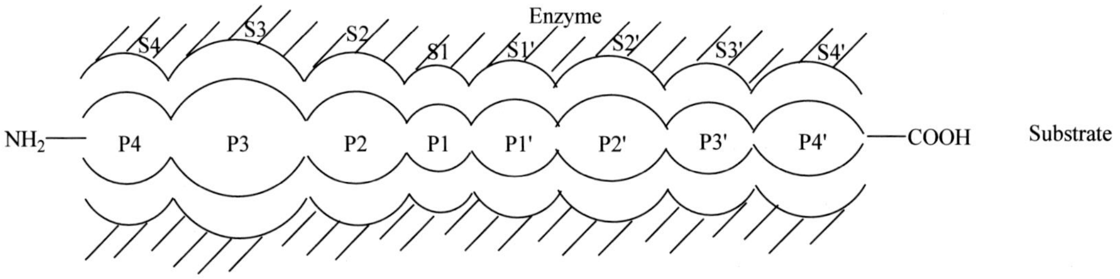

FIG. 1 Standard nomenclature of the protease amino acid residues (P_n-$P_{(n-1)}$- - - -P_1-P_1- - -$P_{n'}$) and the corresponding pockets (S_n- - - -$S_{n'}$) where they are accommodated within the enzyme active site.

infections such as septicemia, inducing aberrant proteolysis of plasma proteins and proteinase–proteinase inhibitor imbalances, disrupting physiological homeostasis, and causing immunocompromised host states [26]. Enterotoxins, such as cholera toxin, are triggered by MPRs [27]. Again, Botulinum (BoNT), and tetanus neurotoxins (TeNT) are metal proteases [27]. All these illustrated cases explain why MPRs should be considered crucial targets for developing new classes of antibacterial acting with a novel mechanism of action with respect to the commune FDA-approved antibiotics [10–16].

2.1 Types of bacterial MPRs

In the MPRS, the nucleophilic activated water molecule, which attacks the scissile peptide, is found to be coordinated to a divalent metal ion (Zn^{2+}) or two Zn^{2+} and, sometimes, one Zn and one Co^{2+}/Mn^{2+} ions. The other three metal-coordinated ligands are generally the amino acid moieties of His, Glu, and Asp, which are present within the active site [28]. Most MPRs require one metal ion for catalysis, but certain families are characterized by a binuclear metal center, where the two metal ions operate cocatalytically [28]. In this last case, five amino acid residues function as ligands (mostly carboxylate moieties), with one acting as a bridge because both metal ions coordinate it. MPRs reported so far as having two catalytic metal ions are exopeptidases; however, MPRs that have just one catalytic metal ion could either be exo- or endopeptidases [28]. Zinc-containing metalloproteases characterized by the HEXXH zinc-binding motif in their amino acid sequence represent the "zincins superfamily." This superfamily is further divided into several other families based on the position of the third zinc ligand, considering mainly five distinct groups: thermolysin (Family M4), collagenases (Family M9), serralysin (Family M10), neurotoxin (Family M27), and the anthrax toxin lethal factor (Family M34) with prototype enzymes from *Bacillus thermoproteolytics*, *Clostridium histolyticum*, *Serratia marcescens*, *Clostridium botulinum* or *Clostridium tetani*, and *Bacillus anthracis*, respectively [28].

The glutamic acid acts as the third zinc ligand in the thermolysin family, which includes the pseudolysin (elastase B) from *Pseudomonas aeruginosa* and MPRs from *Legionella pneumophila*, *Vibrio cholerae*, and *Vibrio vulnificus*. This family hydrolyzes the peptide bond at the amino group side of the P1′ amino acid residue, usually a hydrophobic residue [26].

Metalloprotease from *S. marcescens* and alkaline protease from *P. aeruginosa*, belong to the family of serralysin. This family has the HEXXHXXGXXH pattern for binding zinc. The third histidine in the motif serves as the third zinc ligand, while the water molecule is the fourth ligand. There is the existence of a putative fifth ligand represented by tyrosine. They hydrolyze the peptide bond at the carboxy group side of the P1 amino acid residue [26].

The HEXXH zinc-binding characterizes the neurotoxin family as the thermolysin family and might probably prefer proline-containing substrates [26,29].

3 Bacterial collagenase

3.1 Functional properties and prototype enzyme

About the bacterial collagenases, the enzyme (ChC) from *C. histolyticum* is the best studied in this family. All the *Clostridium* strains produce a collagenase (ChC), which may be used to enter hosts and digest host protein used for nourishment [30]. Purified collagenase isozymes range from 68 to 130 kDa. Based on collagen and synthetic substrate activity, they are classified as class I (α, β, γ, and η) and class II (δ, ε, and ξ) enzymes [31]. ChC is synthesized as precursors and is a secreted enzyme. Two zinc ligands are in the HEXXH motif, while the third is the Glu amino acid residue [31]. Like the vertebrate matrix metalloproteinases, ChC is a multiunit protein consisting of four segments, S1, S2a, S2b, and S3, with S1 incorporating the catalytic domain [32].

Gas gangrene and bacterial corneal keratitis are just two of the many illnesses caused by this anaerobe bacterium [30], and the ChC isoforms play an essential role in the pathogenicity of this and similar clostridia, which includes, among others, *Clostridium perfringens*, which is one of the most common causes of food poisoning [33]. ChC crude homogenate of *C. histolyticum* represents one of the best systems for destroying connective tissue. In contrast to the vertebrate collagenases, it breaks all three collagens' types (I, II, and III) into tiny peptides. Moreover, the collagen degradation carried out by ChC is hyperreactive due to the collagen fold structural characteristics [34].

3.2 Collagenase inhibitors

3.2.1 Hydroxamates inhibitors

In 2000, Supuran and coworkers started a research program to obtain ChC inhibitors based on hydroxamates as zinc binding functions, due to their well-established binding mode to the metal ion found in numerous zinc enzymes, such as matrix metalloproteases (MMPs) and ChC [35] (Fig. 2). As reported in the literature, a vast series of sulfonylated amino acid hydroxamate ChC inhibitors were reported (Fig. 3), using MMP inhibitors (MMPIs) as lead molecules for high-affinity ligands for this bacterial protease [13,14,35]. The structurally comparable arylsulfonylureas, arylureas, or sulfenamido-4-nitrobenzyl-Gly derivatives all block ChC (k_I in the range of nM), with the sulfonylated amino acid hydroxamates being the most effective. The most excellent ChC inhibitory characteristics were shown to be connected to the presence of an S1′ anchoring group, such as perfluoroalkylsulfonyl, perfluorophenylsulfonyl, 3-trifluoromethylphenylsulfonyl, 3-chloro-4-nitro-phenylsulfonyl, 3-/4-protected-aminophenylsulfonyl, or 3-/4-carboxy. These compounds showed k_I against ChC in the 5–10 nM range [13,14,35].

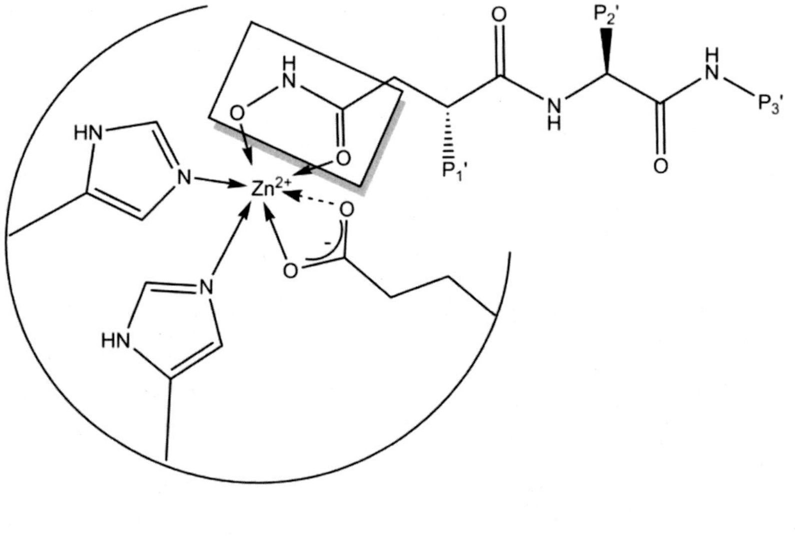

FIG. 2 Schematic binding of peptidomimetic hydroxamates to the zinc ion within ChC active site. The same binding also occurs in other zinc proteases, such as the matrix metalloproteases (MMPs) [13,14].

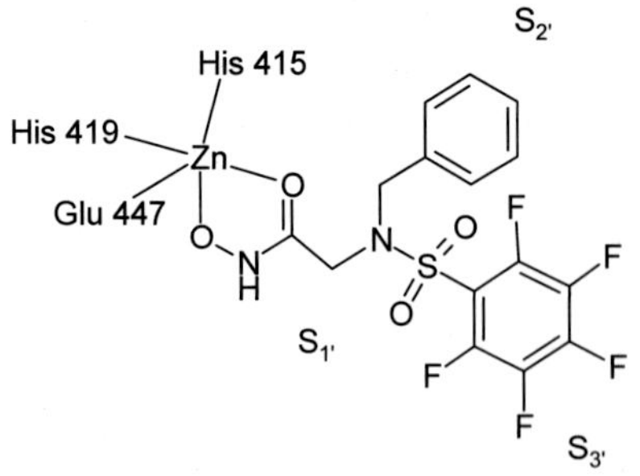

FIG. 3 Proposed binding of a sulfonylated amino acid hydroxamate incorporating a pentaafluorophenylsulfonamide moiety to the ChC active site [13].

3.2.2 1,3,4-Thiadiazole-2-thiones inhibitors

Inhibitors of ChC that incorporate zinc-binding capabilities based on 5-amino-2-mercapto-1,3,4-thiadiazole of type **1** have also been identified [11]. By reacting arylsulfonyl isocyanates or arylsulfonyl halides with phenylalanylalanine, followed by coupling with 5-amino-2-mercapto-1,3,4-thiadiazole in the presence of carbodiimides, a series of compounds were produced (Table 2). These novel chemicals inhibited human matrix metalloproteinase MMP-1, MMP-2, MMP-8, and MMP-9 and the collagenase from *C. histolyticum*. Depending on the substitution pattern at the arylsulfonyl(ureido) moieties, the novel derivatives were shown to be potent inhibitors of these metalloproteases, with activity in the low micromolar range for several of the enzymes mentioned previously.

3.2.3 Fatty acids inhibitors

Research into the ability of specific fatty acids to suppress ChC activity has also been conducted due to the similarities between the carboxylate zinc-binding group and the hydroxamate one [36]. At doses between 50 and 500 mM, it was shown that various fatty acids inhibited collagenase. Unbranched fatty acids with a carbon chain length of 16–19 were shown to be the most effective inhibitors of ChC [36]. Unfortunately, aside from this groundbreaking initial finding, research into ChC inhibitors has been limited.

3.2.4 New inhibitors of ChC

Recently, it has been demonstrated that N-aryl-3-mercaptosuccinimide inhibitors can inhibit ChC in vitro and have strong selectivity for bacterial metalloproteases over human enzymes [37]. Interestingly, in a zebrafish embryo toxicity model and certain human cell lines, several of these inhibitors are noncytotoxic. Furthermore, in an ex vivo pig-skin model, the most potent N-aryl-3-mercaptosuccinimide ChC inhibitor reduced collagen breakdown [37].

Drug repurposing involves employing currently approved medications for purposes other than those for which they were initially approved. In this context, in 2022, Nitulescu and his coworkers worked on drug repurposing that might directly inactivate bacterial collagenase A from *C. perfringens* [38]. The authors identified three already-approved drugs, benzthiazide, entacapone, and lodoxamide, as repurposed antivirulence treatments due to their high likelihood of blocking bacterial collagenase [38]. Of course, additional research into the chosen compounds is required to verify their expected biological action and evaluate their effectiveness in treating *C. perfringens* infections.

3.3 Potential uses of the ChC inhibitors

The ChC inhibitors might be highly beneficial in the fight against bacterial keratitis [39]. It has been discovered that collagen shields put on the corneas of individuals suffering from bacterial keratitis break down very quickly, typically within a few hours. Collagen shields compete for collagenase on the ocular surface to prevent corneal collagen breakdown in viral ulcers and melting diseases [39]. Combining this inhibitory action with an endogenous collagenase inhibitor would accelerate healing in this devastating eye disease. Another application for ChC inhibitors could be in treating burn wounds, which are interested in an imbalance of proteinase inhibitors and proteinase activity that regulates the destruction and regeneration of extracellular matrix proteins [40]. In this case, the ChC inhibitors can restore the right collagenase activity balance of the burn wounds.

4 Pseudolysin

4.1 Pseudolysin production and its action on the host

P. aeruginosa is an opportunistic pathogen that can cause fatal infections in immunocompromised patients with preexisting respiratory conditions such as bronchiectasis, cystic fibrosis, and diffuse panbronchiolitis [41–43]. During the infectious stage in the host, *P. aeruginosa* produces the endopeptidase called pseudolysin or *P. aeruginosa* elastase B, which has as wide broad substrate specificities. The tissue-damaging proteolytic activity of *P. aeruginosa* elastase can degrade plasma proteins such as immunoglobulins, complement factors, and cytokines [41,42]. The first indication that *P. aeruginosa* secretes an elastinolytic protease was arterial elastic laminae destruction in human systemic infections. Furthermore, pseudolysin is the primary extracellular virulence agent produced by *P. aeruginosa* [43]. This protease may cause tissue death, cell damage, or pathogenicity by interfering with host defensive mechanisms [44,45].

TABLE 2 ChC/MMP inhibitors incorporating 2-mercapto-1,3,4-thiadiazole moieties as zinc-binding groups of types 1a–1u [11].

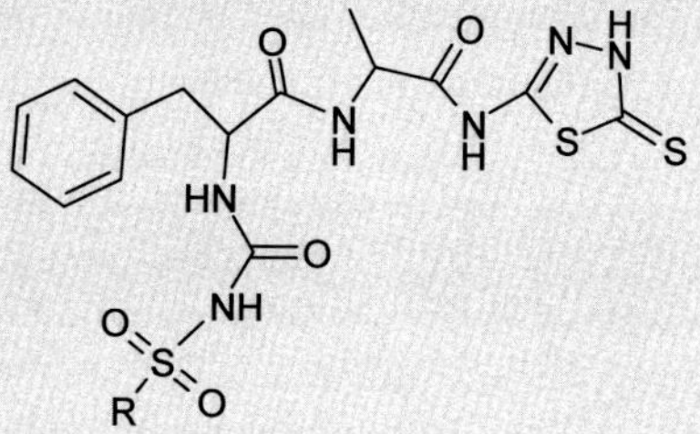

1a–1u

No.	R	K_I [a] (μM)				
		MMP-1[b]	MMP-2[b]	MMP-8[b]	MMP-9[b]	ChC[c]
1a	C_6H_5-	22	18	24	15	15
1b	$PhCH_2$-	24	16	25	16	19
1c	4-F-C_6H_4-	19	17	18	14	21
1d	4-Cl-C_6H_4-	20	16	18	15	20
1e	4-Br-C_6H_4-	24	15	20	21	16
1f	4-I-C_6H_4-	21	17	19	18	14
1g	4-CH_3-C_6H_4-	30	20	23	20	15
1h	4-O_2N-C_6H_4-	15	12	10	18	13
1i	3-O_2N-C_6H_4-	18	12	13	18	12
1j	2-O_2N-C_6H_4-	29	15	17	14	15
1k	4-AcNH-C_6H_4-	14	8	9	10	16
1m	C_6F_5-	9	6	1	2	5
1n	3-CF_3-C_6H_4	13	11	3	5	9
1o	2,5-$Cl_2C_6H_3$	19	10	9	17	15
1p	4-MeO-C_6H_4-	28	18	20	21	14
1q	2,4,6-$Me_3C_6H_2$-	33	24	25	29	17
1r	1-Naphthyl	80	36	40	44	35
1s	2-Naphthyl	72	33	45	53	40
1t	2-Thienyl	14	13	12	10	9
1u	Quinoline-8-yl	74	41	39	45	36

4.2 Pseudolysin inhibitors

Pseudolysin activity is inhibited by metal chelators such as EDTA, EGTA, 1,10-phenanthroline, and tetraethylene pentamine. Phosphoramidon, phosphoryl-dipeptides such as phosphoryl-Leu-Phe and phosphoryl-Leu-Trp, and peptides containing thiol or hydroxamate groups such as $HSCH_2CONH$-Phe-Leu, $HSCH_2CH$-($CH_2C_6H_5$)CO-Ala-Gly-NH_2, or HONHCOCH($CH_2C_6H_5$)-CO-Ala- Gly-NH_2, are effective reversible inhibitors [46]. Because these chemicals inhibit other PRs of the M4 family, they are not specific to pseudolysin [46].

Four small molecules, such as H_2N-Gly-Ala-$COCH_2$-CONHOH, benzyloxycarbonyl-L-Leu-OH, benzylo xycarbonyl-L-PheOH, and phosphoramidon, were investigated for their pseudolysin inhibition and the affinity analyzed by MD simulations and theoretical affinity predictions. These inhibitors showed a k_I in the range of 0.2–340 μM [47]. The differences in inhibitor affinity are reflected in the number and contact surface area of stabilizing hydrophobic, aromatic, and hydrogen bonding contacts. As a result, the free energy of association between inhibitors and pseudolysin is greatly influenced by interactions that occur both within and outside the rigid active site loop [47].

5 The neurotoxins produced by tetanus and botulinum

5.1 Clostridium-producing neurotoxins and their mechanism of action

Both tetanus and botulism are caused by strictly anaerobic microorganisms, with tetanus caused by *C. tetani* and botulism by *C. botulinum* (but also *Clostridium barati* and *Clostridium butirycum*) [48]. Tetanus is a deadly spastic paralysis disease of vertebrates characterized by skeletal muscular contractions antagonistic to one another [48]. The dysfunction of peripheral cholinergic synapses, which is highly variable, going from little undetectable consequences to generalized flaccid paralysis, is a hallmark of botulism. Both disorders are caused solely by clostridial neurotoxins (metalloproteases): tetanus neurotoxin (TeNT) and botulinum neurotoxin (BoNT). The last has seven serotypes indicated with the letter A–G [48]. Spores of toxigenic strains of *C. tetani* occur everywhere, but they are primarily concentrated in the feces of many animals, including domesticated ones. Spores germinate without oxygen, producing tetanus neurotoxin, which is then released during autolysis [48,49]. Clostridial spores with the genes encoding botulinum neurotoxins germinate in anaerobic foods, producing and releasing BoNTs. Botulism is caused by ingesting tainted food, leading to the transcytosis of botulinum neurotoxin across the intestinal epithelial layer and the subsequent spread of the toxin throughout the body [48,49]. TeNT attaches to the presynaptic membrane of the neuromuscular junction, which is then absorbed and delivered to the spinal cord via the retroaxonal pathway. The spastic paralysis that the poison caused was caused by a blockage of neurotransmitter release from spinal inhibitory interneurons [48,49]. The seven different forms of BoNTs exert their effects at the body's periphery by inhibiting the release of acetylcholine at the neuromuscular junction. This results in flaccid paralysis [48,49]. Tetanus and botulism are among the most "bulky" metalloproteases examined due to the depth of their protease active site, which may accommodate at least 16 amino acid residues, with far-reaching implications for developing effective inhibitors of BoNT and TeNT [50]. A 120-residue protein, called VAMP (vesicle-associated membrane protein) or synaptobrevin, or cellubrevin, is the sole known proteolytic substrate of TeNT. Similarly, BoNTs cleave specifically VAMP at different single peptide bonds [49]. A complete description of the active site and a model for interactions between the toxin and its cell surface receptor have been provided as a result of the determination of the intact neurotoxin BoNT/B's crystal structures and its complex with sialyllactose [51]. Moreover, in 2005, the first three-dimensional structure of the catalytic domain of tetanus neurotoxin was presented, elucidating the significance of nucleophilic water and the function of the zinc ion in the active site of TeNT proteolytic activity [52].

5.2 Inhibitors of BoNT and TeNT

Since there is no medication therapy to prevent the progression of tetanus or botulism after intoxication or infection, membrane-permeant protease inhibitors would be helpful as therapeutic agents. Since chelating drugs such as EDTA, 1,10-phenanthroline, or captopril inhibit BoNT and TeNT in millimolar concentrations, these substances are not beneficial for treating illnesses brought on by these neurotoxins. It has been demonstrated that the protease activity of the TeNT light chain is inhibited by a variety of β-aminothiols **2** and **3** at micromolar concentrations ($k_I = 35–250\,\mu M$) [53] (Fig. 4). The most effective inhibitors were those containing the free sulfonamide group and those integrating the aromatic ring. It is interesting to note that the most active derivatives in the series reported by a French group were those having the -SO_2NH_2 moiety (Fig. 5), indicating that, like CA, these sulfonamides bind to the Zn(II) ion in the metalloprotease active site [53,54].

Similarly, excellent inhibitory activities against both TeNT and BoNT/B are displayed by the dipeptidyl-based aminothiols including aliphatic/aromatic sulfonamide moieties [54]. This discovery demonstrates the feasibility of using synthetically created, potent inhibitors of these metalloproteases in treating botulism botulism and tetanus, the first such examples of their kind. Burnett et al. used a high-throughput assay to identify the small molecules that inhibit the metalloprotease activity of BoNT/A. Among them, the active structures (in the micromolar range) were silver sulfadiazine, michellamine B, NSC 357756, NSC 119889, and one dinitroderivative [55]. Additional inhibitors of BoNT serotype A (BoNT/A) light chain (LC) were discovered lately by the same team. Some of these

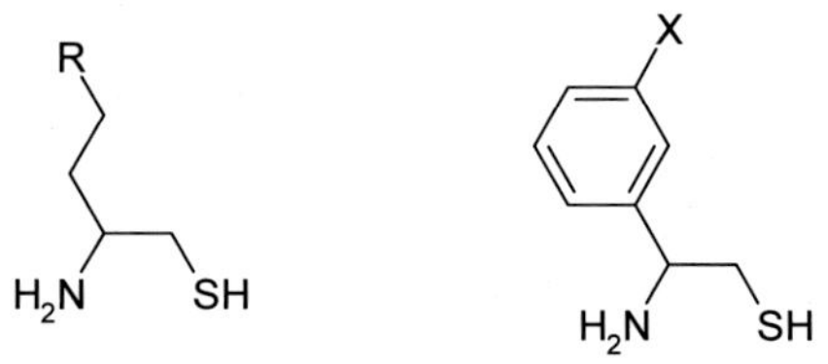

2 **3**

FIG. 4 Aminothiol TeNT of types **2** and **3** inhibitors reported in Ref. [53].

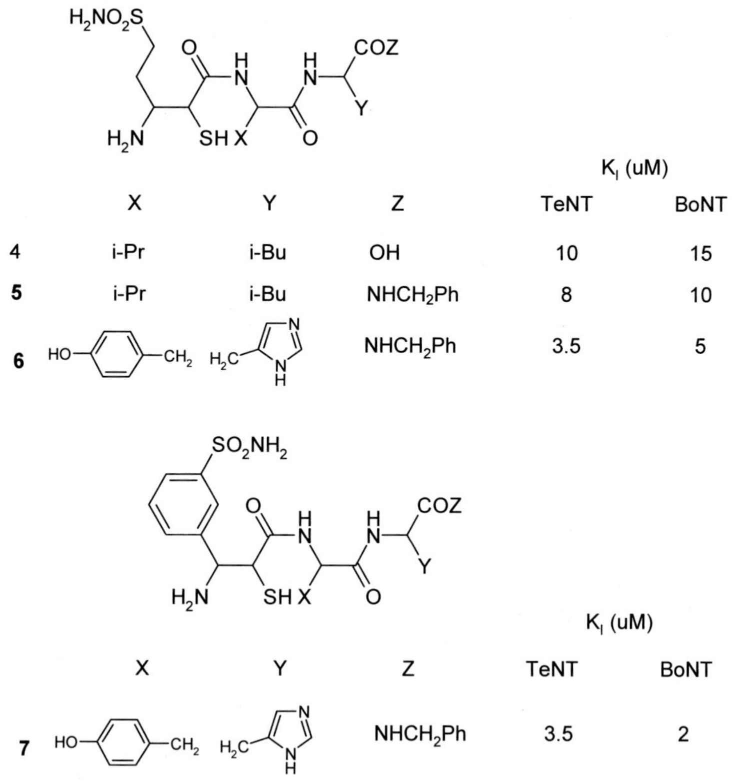

FIG. 5 Sulfonamides acting as micromolar NoNT and TeNT inhibitors.

compounds (such as antimalarial medicines) had a 4-amino-7-chloroquinoline substructure and an additional positively charged ionizable amine. In addition to blocking malaria parasites, these antimalarial drugs have been shown to inhibit BoNT/A translocation into neurons by increasing pH in the toxin-mediated endosome [56]. It follows that the structural class of small compounds studied (IC_{50} values ranging from 3.2 to 17 mM) may be helpful as dual-function BoNT/A inhibitors [56].

Small-molecule, nonpeptidic inhibitors offer the best opportunity to develop post-exposure therapeutics. For example, monoclonal antibodies, soluble versions of the BoNT receptors (with antitoxin function), and peptide-based inhibitors suffer from their large molecular size and metabolic instability, which limit their ability to reach the BoNT endopeptidase within neurons [57]. Unfortunately, the described drugs resulted in inadequate efficacy, poor cell membrane permeability, cytotoxicity, or pharmacokinetic features. However, several may have prolonged the mean time to death. Thus, new drug-like small molecule botulinum inhibitors with low toxicity are needed.

6 Anthrax toxin lethal factor

6.1 Bacillus anthracis and its lethal toxin

B. anthracis—a gram-positive, spore-forming rod-like bacterium—causes anthrax. Anthrax is predominantly a disease of grazing herbivores and rarely affects humans, generally through animal products [58]. Besides, in 2001, *B. anthracis* spores were mailed, demonstrating its power as a biological weapon [58–60]. The poly-D-glutamic acid capsule and the anthrax toxins encoded on the enormous plasmids pXO1 and pXO2 contribute significantly to *B. anthracis* pathogenicity [61]. The anthrax toxins are composed of three polypeptides: protective antigen (PA), lethal factor (LF), and edema factor (EF). A cell-associated, furin-like protease proteolytically activates PA in two proteins, PA_{83} (83 kDa) and PA_{63} (63 kDa) [61,62]. PA63 undergoes polymerization, which results in the formation of oligomeric structures. These structures bind EF or LF, speeding up their entry into the cell [61,62]. The edema toxin is a calmodulin-dependent adenylate cyclase that converts intracellular ATP to cAMP [63]. This results in a significant increase in cAMP levels, ultimately leading to edema development. The lethal toxin, also known as LeTx, is a zinc metalloprotease that cleaves cellular mitogen-activated protein kinase kinases (cleaves specifically isoforms 1, 2, and 3 of MAPK kinases within their N-terminal tail) [64]. This results in the cell's transcriptional machinery becoming deregulated, ultimately leading to the cell's death [65]. LF and its complex with MAPKK-2's N terminus were crystallized [66]. LF has four domains: domain I, which binds the membrane-translocating component of anthrax toxin, the protective antigen (PA); domains II, III, and IV form a long deep groove that traps the 16-residue N-terminal tail of

MAPKK-2 before the cleavage, domain II improves the substrate identification, domain III is inserted into the domain I and seems to be derived from it, and, finally, domain IV contains the catalytic core. The Zn(II) ion is coordinated by His686, His690, and Glu735 (placed on a different alpha helix), whereas the fourth ligand is a water molecule/hydroxide ion that acts as a nucleophile on the scissile peptide link [66]. Tyr728 is a crucial catalytic residue and is located near the zinc ion in the LF active site [66,67]. Tyr728 may help to stabilize the negative charge on the carbonyl oxygen of the scissile bond in the transition state [67].

6.2 Anthrax lethal factor metalloproteinase inhibitors

Because antibiotic therapy for anthrax infection has had limited success, researchers are looking into supplementary therapeutic techniques that target the bacteria's released toxin [68]. The zinc-dependent LF is a crucial component of anthrax toxin. In this context, several efforts have been attempted in recent years to develop LF inhibitors, ranging from the generation of traditional metal-chelating substrate analogs (e.g., hydroxamates) to the random screening of different chemical libraries [69]. The initial compounds utilized in designing LF inhibitors were peptidomimetics that were quite similar to the protease substrate [69]. These compounds also incorporated a hydroxamate as a zinc-binding group. Using this hydroxamate as a lead, it has been synthesized and tested with a pervasive series of hydroxamates with a reduced peptidomimetic character as in vitro LF inhibitors. These tests led to the discovery of the low nanomolar inhibitor (2R)-2-[(4-fluoro-3-methylphenyl)sulfonylamino]-N-hydroxy-2-(tetrahydro-2H-pyran-4-yl) acetamide (Fig. 6), which has a K_I of 24 nM and is not a potent inhibitor of host (MMP) proteinases either [70,71]. Because hydroxamates are toxic and cannot discriminate among most of metalloenzymes, numerous efforts have been made to identify nonhydroxamate competitive LF inhibitors [68,72]. Some of these compounds, which are also active as BoNT inhibitors, and other derivatives, belong to quite different chemical families and exhibit action in the nanomolar–micromolar range (Fig. 6) [68,72]. Thus, numerous distinct classes of specific nanomolar range inhibitors have been produced, some of which are shown in Fig. 6. Among them, some compounds have proven to be effective in anthrax toxemia and infection animal models [68]. It is envisaged these bacterial protease inhibitors should be therapeutically available soon since they have the potential to serve as the foundation for future medicines.

7 Conclusions

There is a wealth of information regarding zinc proteases in various bacterial species, some of which are responsible for developing severe illnesses. In addition, many of these diseases have developed resistance to the antibiotics that are now employed in clinical practice and mainly target the manufacture of bacterial cell walls or influence protein, DNA, and RNA synthesis. In these species, MPRs play critical roles in processes linked to colonization and evasion of host immune systems, acquiring resources for growth and proliferation, the encouragement of dispersion, or the damage to tissue that occurs during infection. Much fascinating new information has surfaced regarding the cloning, purification, and biochemical characterization of numerous zinc proteases from both gram-positive and gram-negative pathogens. However, there have been reports of a few potent inhibitors unique to bacterial proteases, except for *Clostridium* collagenase, pseudolysin, botulinum, tetanus neurotoxin, and deadly anthrax factor. These inhibitors are currently being researched.

The sulfonamide moiety substitution pattern made many hydroxamates efficient bacterial collagenase inhibitors. Pentafluorophenylsulfonyl, 3- and 4-protected-aminoph enylsulfonyl P1 groups, among others, were the best ChC inhibitors with low nanomolar affinities. Other low micromolar ChC inhibitors have 1,3,4-thiadiazole-2-thiol zinc binding groups. Fatty acids with zinc-binding groups were also tested for ChC inhibition.

Four small molecules, such as H_2N-Gly-Ala-$COCH_2$-CONHOH, benzyloxycarbonyl-L-Leu-OH, benzyloxy carbonyl-L-PheOH, and phosphoramidon, were investigated for their pseudolysin inhibition and the affinity. These inhibitors showed a k_I in the range of 0.2–340 μM.

Membrane-permeant protease inhibitors would be helpful therapeutic agents because there is no pharmacological therapy to stop the progression of tetanus or botulism following intoxication or infection. It has been shown that a variety of inhibitors with the free sulfonamide group and those integrating the aromatic ring block the protease activity of the TeNT light chain, showing that, like CA, these sulfonamides bind to the Zn(II) ion in the metalloprotease active site. The botulin toxin, particularly the isoform BoNT/A, was inhibited by small molecules such as silver sulfadiazine, michellamine B, NSC 357756, NSC 119889, and one dinitro derivative. In addition, some compounds having the 4-amino-7-chloroquinoline substructure and an additional positively charged, ionizable amine resulted in suitable inhibitors of BoNT/A translocation into neurons by increasing pH in the toxin-mediated endosome.

The anthrax toxin's zinc-dependent LF is an essential component, and thus, several attempts to create LF

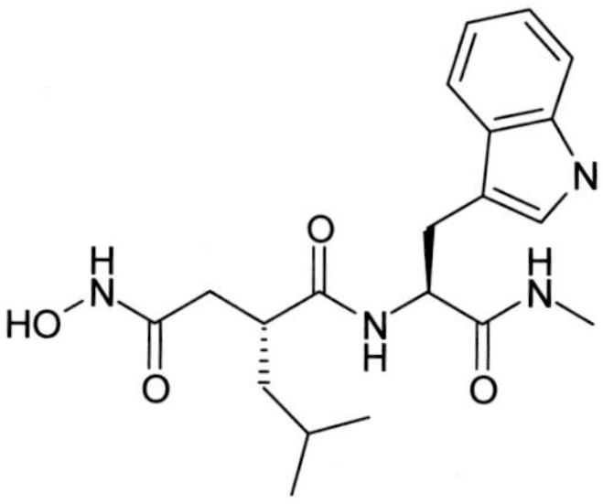

8 (Ki = 2.1 uM)

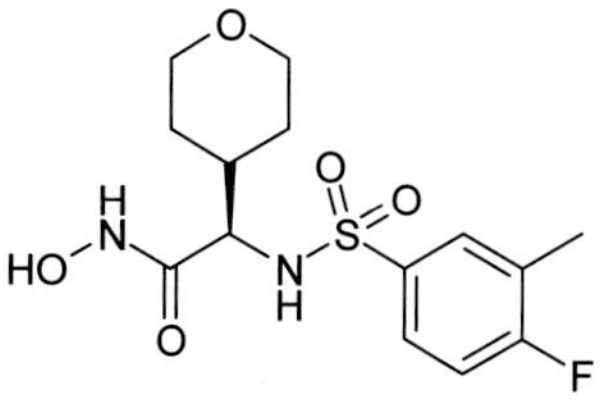

9 (Ki = 24 nM)

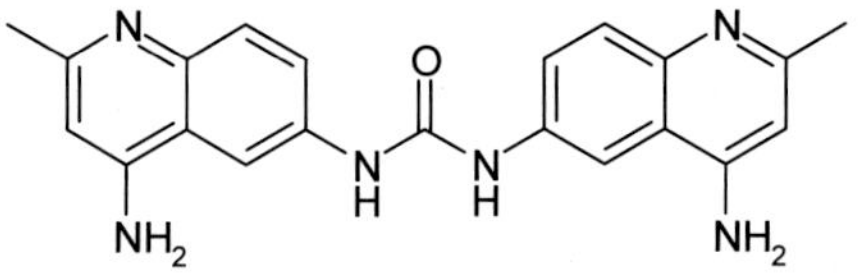

10 (Ki = 0.5 uM)

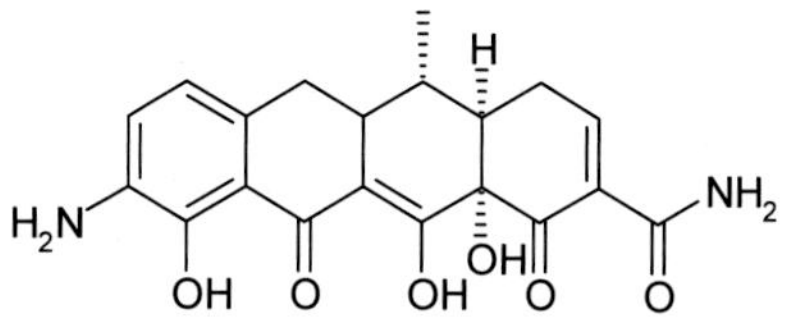

11 (Ki = 5 uM)

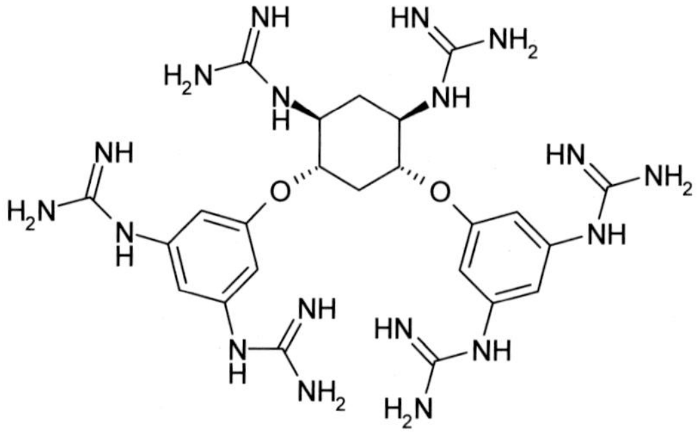

12 (Ki = 65 nM)

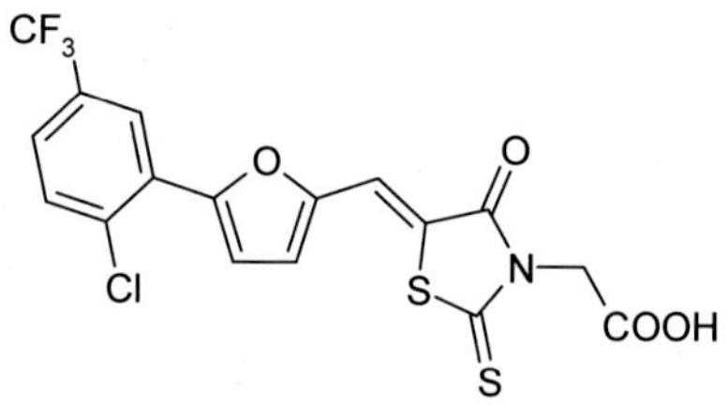

13 (Ki = 32 nM)

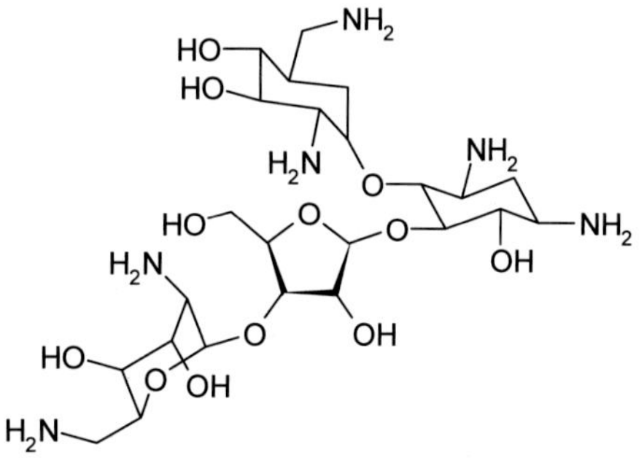

14 (Ki = 37 nM)

FIG. 6 Reported LF inhibitors and their inhibition constants (Ki-s) against the bacterial protease.

inhibitors have been made. In this context, hydroxamates and nonhydroxamates have been proposed.

Therefore, several inhibitor classes have been designed. Since they could form the basis of future drugs, these bacterial protease inhibitors should be seriously considered by the pharmaceutical industry as possible therapeutic drugs.

References

[1] Doolan JA, Williams GT, Hilton KLF, Chaudhari R, Fossey JS, Goult BT, Hiscock JR. Advancements in antimicrobial nanoscale materials and self-assembling systems. Chem Soc Rev 2022;51(20):8696–755.

[2] Reardon S. Resistance to last-ditch antibiotic has spread farther than anticipated. Nature 2017. https://doi.org/10.1038/nature.2017.22140.

[3] da Silva TH, Hachigian TZ, Lee J, King MD. Using computers to ESKAPE the antibiotic resistance crisis. Drug Discov Today 2022;27(2):456–70.

[4] Nataraj BH, Mallappa RH. Antibiotic resistance crisis: an update on antagonistic interactions between probiotics and methicillin-resistant *Staphylococcus aureus* (MRSA). Curr Microbiol 2021;78 (6):2194–211.

[5] Aslam B, Wang W, Arshad MI, Khurshid M, Muzammil S, Rasool MH, Nisar MA, Alvi RF, Aslam MA, Qamar MU, Salamat MKF, Baloch Z. Antibiotic resistance: a rundown of a global crisis. Infect Drug Resist 2018;11:1645–58.

[6] Hansen MP, Hoffmann TC, McCullough AR, van Driel ML, Del Mar CB. Antibiotic resistance: What are the opportunities for primary care in alleviating the crisis? Front Public Health 2015;3:35.

[7] Sun DX, Gao W, Hu HX, Zhou SM. Why 90% of clinical drug development fails and how to improve it? Acta Pharm Sin B 2022;12 (7):3049–62.

[8] Overington JP, Al-Lazikani B, Hopkins AL. Opinion—How many drug targets are there? Nat Rev Drug Discov 2006;5(12):993–6.

[9] Frees D, Brondsted L, Ingmer H. Bacterial proteases and virulence. Subcell Biochem 2013;66:161–92.

[10] Clare BW, Scozzafava A, Supuran CT. Protease inhibitors: synthesis of a series of bacterial collagenase inhibitors of the sulfonyl amino acyl hydroxamate type. J Med Chem 2001;44(13):2253–8.

[11] Ilies M, Banciu MD, Scozzafava A, Ilies MA, Caproiu MT, Supuran CT. Protease inhibitors: synthesis of bacterial collagenase and matrix metalloproteinase inhibitors incorporating arylsulfonylureido and 5-dibenzo-suberenyl/suberyl moieties. Bioorg Med Chem 2003;11 (10):2227–39.

[12] Santos MA, Marques S, Gil M, Tegoni M, Scozzafava A, Supuran CT. Protease inhibitors: synthesis of bacterial collagenase and matrix metalloproteinase inhibitors incorporating succinyl hydroxamate and iminodiacetic acid hydroxamate moieties. J Enzyme Inhib Med Chem 2003;18(3):233–42.

[13] Scozzafava A, Ilies MA, Manole G, Supuran CT. Protease inhibitors. Part 12. Synthesis of potent matrix metalloproteinase and bacterial collagenase inhibitors incorporating sulfonylated N-4-nitrobenzyl-beta-alanine hydroxamate moieties. Eur J Pharm Sci 2000;11 (1):69–79.

[14] Scozzafava A, Supuran CT. Protease inhibitors: synthesis of potent bacterial collagenase and matrix metalloproteinase inhibitors incorporating N-4-nitrobenzylsulfonylglycine hydroxamate moieties. J Med Chem 2000;43(9):1858–65.

[15] Scozzafava A, Supuran CT. Protease inhibitors: synthesis of matrix metalloproteinase and bacterial collagenase inhibitors incorporating 5-amino-2-mercapto-1,3,4-thiadiazole zinc binding functions. Bioorg Med Chem Lett 2002;12(19):2667–72.

[16] Supuran CT, Scozzafava A, Clare BW. Bacterial protease inhibitors. Med Res Rev 2002;22(4):329–72.

[17] Corre MH, Bachmann V, Kohn T. Bacterial matrix metalloproteases and serine proteases contribute to the extra-host inactivation of enteroviruses in lake water. ISME J 2022;16(8):1970–9.

[18] Rahman F, Wushur I, Malla N, Astrand OAH, Rongved P, Winberg JO, Sylte I. Zinc-chelating compounds as inhibitors of human and bacterial zinc metalloproteases. Molecules 2022;27(1).

[19] Shanks RMQ, Al Higaylan M, Stella N, Brothers KM, Thibodeau PH. Bacterial metalloproteases inhibit epithelial cell migration and wound healing. Invest Ophthalmol Vis Sci 2021;62(8).

[20] Barrett AJ. Bioinformatics of proteases in the MEROPS database. Curr Opin Drug Disc 2004;7(3):334–41.

[21] Barrett AJ, Mcdonald JK. Nomenclature—Protease, proteinase and peptidase. Biochem J 1986;237(3):935.

[22] Berger A, Schechter I. Mapping the active site of papain with the aid of peptide substrates and inhibitors. Philos Trans R Soc Lond B Biol Sci 1970;257(813):249–64.

[23] Lebrun I, Marques-Porto R, Pereira AS, Pereira A, Perpetuo EA. Bacterial toxins: an overview on bacterial proteases and their action as virulence factors. Mini Rev Med Chem 2009;9(7):820–8.

[24] Bozhokina ES, Tsaplina OA, Efremova TN, Kever LV, Demidyuk IV, Kostrov SV, Adam T, Komissarchik YY, Khaitlina SY. Bacterial invasion of eukaryotic cells can be mediated by actin-hydrolysing metalloproteases grimelysin and protealysin. Cell Biol Int 2011;35 (2):111–8.

[25] Okabe A, Matsushita O, Minami J. Bacterial metalloproteases. Nihon Saikingaku Zasshi 1995;50(4):971–89.

[26] Miyoshi S, Shinoda S. Microbial metalloproteases and pathogenesis. Microbes Infect 2000;2(1):91–8.

[27] Popoff MR. Bacterial toxins, current perspectives. Toxins (Basel) 2020;12(9).

[28] Coleman JE. Zinc enzymes. Curr Opin Chem Biol 1998;2(2):222–34.

[29] Hammond SE, Hanna PC. Lethal factor active-site mutations affect catalytic activity in vitro. Infect Immun 1998;66(5):2374–8.

[30] Graivier M, Hill D, Katz B, Boehm KA, Fisher J, Battista C. Collagenase clostridium histolyticum for the treatment of cellulite in the buttocks and thigh: early insights from clinical practice. Aesthet Surg J Open Forum 2022;4:ojac057.

[31] Bond MD, Vanwart HE. Purification and separation of individual collagenases of clostridium-histolyticum using red-dye ligand chromatography. Biochemistry-US 1984;23(13):3077–85.

[32] Bauer R, Wilson JJ, Philominathan STL, Davis D, Matsushita O, Sakon J. Structural comparison of ColH and ColG collagen-binding domains from *Clostridium histolyticum*. J Bacteriol 2013;195 (2):318–27.

[33] Ghoneim NH, Hamza DA. Epidemiological studies on Clostridium perfringens food poisoning in retail foods. Rev Sci Tech 2017;36 (3):1025–32.

[34] Eckhard U, Schonauer E, Ducka P, Briza P, Nuss D, Brandstetter H. Biochemical characterization of the catalytic domains of three different clostridial collagenases. Biol Chem 2009;390(1):11–8.

[35] Scozzafava A, Supuran CT. Protease inhibitors: synthesis of clostridium histolyticum collagenase inhibitors incorporating sulfonyl-L-alanine hydroxamate moieties. Bioorg Med Chem Lett 2000;10 (5):499–502.

[36] Rennert B, Melzig MF. Free fatty acids inhibit the activity of Clostridium histolyticum collagenase and human neutrophil elastase. Planta Med 2002;68(9):767–9.

[37] Konstantinovic J, Yahiaoui S, Alhayek A, Haupenthal J, Schonauer E, Andreas A, Kany AM, Muller R, Koehnke J, Berger FK, Bischoff M, Hartmann RW, Brandstetter H, Hirsch AKH. N-Aryl-3-mercaptosuccinimides as antivirulence agents targeting *Pseudomonas aeruginosa* elastase and clostridium collagenases. J Med Chem 2020;63(15):8359–68.

[38] Nitulescu G, Nitulescu GM, Zanfirescu A, Mihai DP, Gradinaru D. Candidates for repurposing as anti-virulence agents based on the structural profile analysis of microbial collagenase inhibitors. Pharmaceutics 2022;14:(1).

[39] Kuwano M, Horibe Y, Kawashima Y. Effect of collagen cross-linking in collagen corneal shields on ocular drug delivery. J Ocul Pharmacol Ther 1997;13(1):31–40.

[40] Metzmacher I, Ruth P, Abel M, Friess W. In vitro binding of matrix metalloproteinase-2 (MMP-2), MMP-9, and bacterial collagenase on collagenous wound dressings. Wound Repair Regen 2007;15(4):549–55.

[41] Li Y, Wang Y, Li C, Zhao D, Hu Q, Zhou M, Du M, Li J, Wan P. The role of elastase in corneal epithelial barrier dysfunction caused by *Pseudomonas aeruginosa* exoproteins. Invest Ophthalmol Vis Sci 2021;62(9):7.

[42] Yanagihara K, Tomono K, Kaneko Y, Miyazaki Y, Tsukamoto K, Hirakata Y, Mukae H, Kadota JI, Murata I, Kohno S. Role of elastase in a mouse model of chronic respiratory Pseudomonas aeruginosa infection that mimics diffuse panbronchiolitis. J Med Microbiol 2003;52(Pt 6):531–5.

[43] Azghani AO. Pseudomonas aeruginosa and epithelial permeability: role of virulence factors elastase and exotoxin A. Am J Respir Cell Mol Biol 1996;15(1):132–40.

[44] Kon Y, Tsukada H, Hasegawa T, Igarashi K, Wada K, Suzuki E, Arakawa M, Gejyo F. The role of Pseudomonas aeruginosa elastase as a potent inflammatory factor in a rat air pouch inflammation model. FEMS Immunol Med Microbiol 1999;25(3):313–21.

[45] Galdino ACM, de Oliveira MP, Ramalho TC, de Castro AA, Branquinha MH, Santos ALS. Anti-virulence strategy against the multidrug-resistant bacterial pathogen *Pseudomonas aeruginosa*: pseudolysin (Elastase B) as a potential druggable target. Curr Protein Pept Sci 2019;20(5):471–87.

[46] Kessler E, Ohman DE. Pseudolysin. In: Handbook of Proteolytic Enzymes. 3rd ed, vols. 1 and 2; 2013. p. 582–92.

[47] Adekoya OA, Willassen NP, Sylte I. Molecular insight into pseudolysin inhibition using the MM-PBSA and LIE methods. J Struct Biol 2006;153(2):129–44.

[48] Schiavo G, Matteoli M, Montecucco C. Neurotoxins affecting neuroexocytosis. Physiol Rev 2000;80(2):717–66.

[49] Humeau Y, Doussau F, Grant NJ, Poulain B. How botulinum and tetanus neurotoxins block neurotransmitter release. Biochimie 2000;82(5):427–46.

[50] Lacy DB, Tepp W, Cohen AC, DasGupta BR, Stevens RC. Crystal structure of botulinum neuro-toxin type A and implications for toxicity. Nat Struct Biol 1998;5(10):898–902.

[51] Swaminathan S, Eswaramoorthy S. Structural analysis of the catalytic and binding sites of Clostridium botulinum neurotoxin B. Nat Struct Biol 2000;7(8):693–9.

[52] Rao KN, Kumaran D, Binz T, Swaminathan S. Structural analysis of the catalytic domain of tetanus neurotoxin. Toxicon 2005;45(7):929–39.

[53] Martin L, Cornille F, Coric P, Roques BP, Fournie-Zaluski MC. beta-amino-thiols inhibit the zinc metallopeptidase activity of tetanus toxin light chain. J Med Chem 1998;41(18):3450–60.

[54] Martin L, Cornille F, Turcaud S, Meudal H, Roques BP, Fournie-Zaluski MC. Metallopeptidase inhibitors of tetanus toxin: a combinatorial approach. J Med Chem 1999;42(3):515–25.

[55] Burnett JC, Schmidt JJ, Stafford RG, Panchal RG, Nguyen TL, Hermone AR, Vennerstrom JL, McGrath CF, Lane DJ, Sausville EA, Zaharevitz DW, Gussio R, Bavari S. Novel small molecule inhibitors of botulinum neurotoxin A metalloprotease activity. Biochem Bioph Res Co 2003;310(1):84–93.

[56] Burnett JC, Opsenica D, Sriraghavan K, Panchal RG, Ruthel G, Hermone AR, Nguyen TL, Kenny TA, Lane DJ, McGrath CF, Schmidt JJ, Vennerstrom JL, Gussio R, Solaja BA, Bavari S. A refined pharmacophore identifies potent 4-amino-7-chloroquinoline-based inhibitors of the botulinum neurotoxin serotype a metalloprotease. J Med Chem 2007;50(9):2127–36.

[57] Li B, Peet NP, Butler MM, Burnett JC, Moir DT, Bowlin TL. Small molecule inhibitors as countermeasures for botulinum neurotoxin intoxication. Molecules 2011;16(1):202–20.

[58] Pita R, Gunaratna R. Anthrax as a biological weapon: from World War I to the Amerithrax Investigation. Int J Intell Counter 2010;23 (1):61–103.

[59] Inglesby TV. Anthrax as a biological weapon: Medical and public health management (vol. 281, p. 1735, 1999). Jama—J Am Med Assoc 2000;283(15):1963.

[60] Inglesby TV, Henderson DA, Bartlett JG, Ascher MS, Eitzen E, Friedlander AM, Hauer J, McDade J, Osterholm MT, O'Toole T, Parker G, Perl TM, Russell PK, Tonat K, Biodefense WGC. Anthrax as a biological weapon - Medical and public health management. Jama—J Am Med Assoc 1999;281(18):1735–45.

[61] Mock M, Fouet A. Anthrax. Annu Rev Microbiol 2001;55:647–71.

[62] Mock M, Mignot T. Anthrax toxins and the host: a story of intimacy. Cell Microbiol 2003;5(1):15–23.

[63] Leppla SH. Anthrax toxin edema factor—A bacterial adenylate-cyclase that increases cyclic-amp concentrations in eukaryotic cells. P Natl Acad Sci-Biol 1982;79(10):3162–6.

[64] Pellizzari R, Guidi-Rontani C, Vitale G, Mock M, Montecucco C. Anthrax lethal factor cleaves MKK3 in macrophages and inhibits the LPS/IFNgamma-induced release of NO and TNFalpha. FEBS Lett 1999;462(1–2):199–204.

[65] Leppla SH. Bacillus anthracis calmodulin-dependent adenylate cyclase: chemical and enzymatic properties and interactions with eucaryotic cells. Adv Cyclic Nucleotide Protein Phosphorylation Res 1984;17:189–98.

[66] Pannifer AD, Wong TY, Schwarzenbacher R, Renatus M, Petosa C, Bienkowska J, Lacy DB, Collier RJ, Park S, Leppla SH, Hanna P, Liddington RC. Crystal structure of the anthrax lethal factor. Nature 2001;414(6860):229–33.

[67] Tonello F, Naletto L, Romanello V, Dal Molin F, Montecucco C. - Tyrosine-728 and glutamic acid-735 are essential for the metalloproteolytic activity of the lethal factor of *Bacillus anthracis*. Biochem Biophys Res Commun 2004;313(3):496–502.

[68] Turk BE. Discovery and development of anthrax lethal factor metalloproteinase inhibitors. Curr Pharm Biotechnol 2008;9(1):24–33.

[69] Gowravaram MR, Tomczuk BE, Johnson JS, Delecki D, Cook ER, Ghose AK, Mathiowetz AM, Spurlino JC, Rubin B, Smith DL, Pulvino T, Wahl RC. Inhibition of matrix metalloproteinases by hydroxamates containing heteroatom-based modifications of the P-1' Group. J Med Chem 1995;38(14):2570–81.

[70] Shoop WL, Xiong Y, Wiltsie J, Woods A, Guo J, Pivnichny JV, Felcetto T, Michael BF, Bansal A, Cummings RT, Cunningham BR, Friedlander AM, Douglas CM, Patel SB, Wisniewski D, Scapin G, Salowe SP, Zaller DM, Chapman KT, Scolnick EM, Schmatz DM, Bartizal K, MacCoss M, Hermes JD. Anthrax lethal factor inhibition. P Natl Acad Sci USA 2005;102(22):7958–63.

[71] Xiong YS, Wiltsie J, Woods A, Guo J, Pivnichny JV, Tang W, Bansal A, Cummings RT, Cunningham BR, Friedlander AM, Douglas CM, Salowe SP, Zaller DM, Scolnick EM, Schmatz DM, Bartizal K, Hermes JD, MacCoss M, Chapman KT. The discovery of a potent and selective lethal factor inhibitor for adjunct therapy of anthrax infection. Bioorg Med Chem Lett 2006;16(4):964–8.

[72] Johnson SL, Jung D, Forino M, Chen Y, Satterthwait A, Rozanov DV, Strongin AY, Pellecchia M. Anthrax lethal factor protease inhibitors: synthesis, SAR, and structure-based 3D QSAR studies. J Med Chem 2006;49(1):27–30.

Chapter 3.4

Matrix metalloproteases

Andrea Trabocchi and Elena Lenci
Department of Chemistry "Ugo Schiff", University of Florence, Sesto Fiorentino, Italy

1 Structure and function of the enzyme(s)

Matrix metalloproteases (MMPs) are enzymes involved in the degradation of the extracellular matrix and display a critical role in a variety of physiological and pathological processes [1]. The MMP family is composed of 23 different proteins: each MMP has a different substrate specificity and a different role in physiological and pathological processes. These 23 enzymes have been subdivided into six classes, depending on their substrate specificity, their primary structures, and their cellular localization, as follows [2]:

- Collagenases (MMP1, 8, 13, and 18).
- Gelatinases (MMP2 and 9).
- Stromelysins (MMP3, 10, and 11).
- Matrilysins (MMP7 and 26).
- Membrane-type MMPs (MT-MMPs) (MMP14, 15, 16, 17, 24, and 25).
- Others (MMP12, 19, 20, 21, 22, 23, 28, and 29).

Despite that, they share a common structure, based on three domains: a pro-peptide domain, a catalytic domain that contain a Zn^{2+} ion coordinated by a tris(histidine) motif, and a hemopexin-like C-terminal domain, which is linked to the catalytic domain by a flexible hinge region (Fig. 1). The catalytic domains of MMPs share a sequential similarity ranging between 30% and 90%; nevertheless, the main differences in the structure of the diverse MMPs can be found in the subsites around the catalytic cleft, which differs for the depth, length, and amino acid composition. S1′ pocket is the most important as a key factor for substrate specificity, being characterized by high hydrophobicity, variable depth, and by the presence of a Ω-loop.

The main biological function of MMPs consists of degrading proteins and glycoproteins, such as extracellular matrix (ECM) components, membrane receptors, cytokines, and growth factors [3]. Also, MMPs are involved and regulate several processes, such as tissue repair, cell proliferation, differentiation and migration, embryo- and morphogenesis, angiogenesis, wound healing, apoptosis, and reproductive events.

2 Physiologic/pathologic role

The class of MMP enzymes have been studied for decades for their key role in cancer, particularly regulating pathways such as apoptosis, immunity, cellular migration, and angiogenesis. Nevertheless, they are nowadays experiencing a renewed interest in the scientific and industrial communities [4], due to the recent discoveries of their involvement in pathologies different from cancer, including inflammatory diseases, neurological, and immunological disorders [5]. Also, for many years it has been thought that MMPs can act only in the extracellular matrix or on the membrane surface, but recent reports about their role in the intracellular environment have opened the way to new potential applications in the field of inflammation, antiviral, and antibiotics research [6].

2.1 Collagenases

Collagenase-1 (MMP1), collagenase-2 (MMP8), and collagenase-3 (MMP13) belong to the subfamily of MMPs that cleave preferentially fibrillar collagens [7], with a selective preference for type III collagen (in the case of MMP1), type I collagen (in the case of MMP8), and type II collagen (in the case of MMP13) [8]. Also, these enzymes can cleave and activate various nonmatrix proteins, including cytokines, chemokines, and growth factors, thus regulating many pathophysiological processes involved in tissue regeneration and skeletal development [9], including osteoarthritic cartilage, chronic cutaneous ulcers, periodontitis, atherosclerosis, intestinal ulcerations, and aortic aneurysms [10].

2.2 Gelatinases

The subfamily of gelatinases (MMP2 and MMP9) is the most studied subgroup of MMPs, mainly due to their well-established role in oncology, as showed by the extensive number of gelatinase inhibitors that have been reported both in papers and patents [11,12]. In fact, high levels of these two enzymes have been found in several human tumors, including those of breast, brain, pancreas,

Metalloenzymes. https://doi.org/10.1016/B978-0-12-823974-2.00033-4

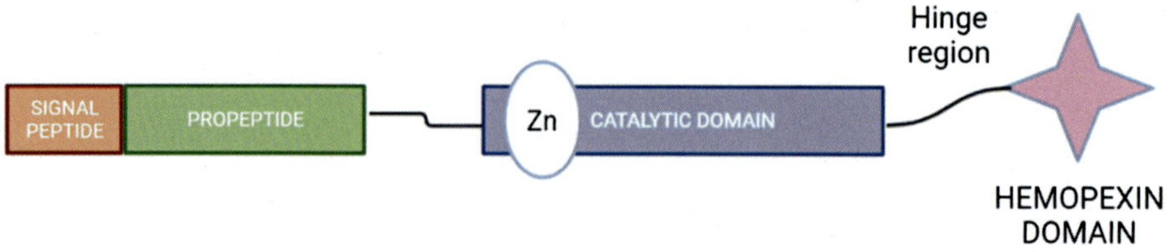

FIG. 1 General structure of MMP enzymes, highlighting the different protein domains.

colon-rectum, lung, bladder, skin, and prostate [13]. As they play a primary role in the angiogenic switch and cell invasiveness [14], the overexpression of gelatinases is often correlated with high aggressiveness and poor prognosis of the relative tumor. The preferential substrate of the two gelatinases, as indicated by their name, is gelatin; however, it is now widely accepted that they are able to degrade many other cytokines and extracellular matrix components, thus addressing several pro-angiogenic factors, including the transforming growth factor beta (TGF-β) or the hypoxia-induced transcription factor-1α (HIF-1α) [15,16]. Despite their high structural similarity, MMP2 and MMP9 are known to activate different cell signaling cascades, so that their inhibition can show beneficial or detrimental effects depending on the stage of tumor progression [17]. For example, MMP2 interacts with integrin $\alpha_v\beta_3$ that activates the expression of the vascular endothelial growth factor (VEGF), promoting angiogenesis (Fig. 2A) [18]. Differently, MMP9 interacts with receptor CD44 (Fig. 2B), inducing the interaction between the phosphatidyl-inositol 3-kinase (PI3K) and the serine/threonine kinase AKT, favoring cancer cell migration and invasion [19].

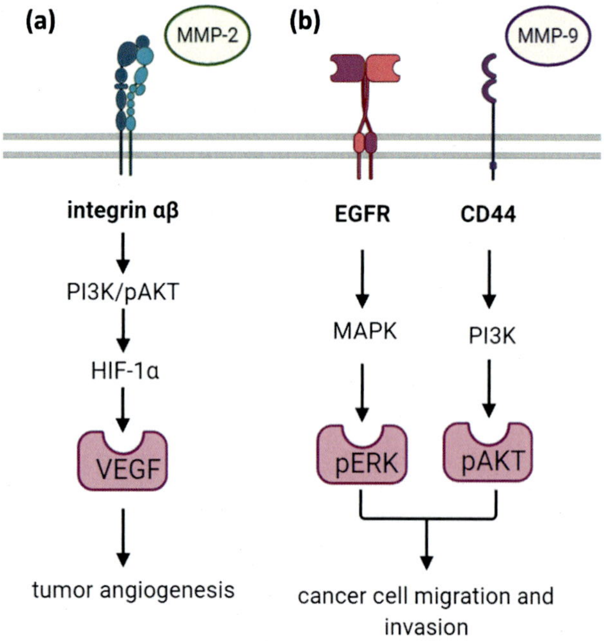

FIG. 2 Schematic representations of the diverse cell signal cascades activated by (A) MMP2 and (B) MMP9.

2.3 Membrane-type MMPs

The subfamily of membrane-type MMPs (MT-MMPs) is relatively less studied as compared to the other ones. It includes four transmembrane MMPs: MMP14 (MT1-MMP), MMP15 (MT2-MMP), MMP16 (MT3-MMP), and MMP24 (MT5-MMP), as well as two glycosyl-phosphatidylinositol-anchored MMPs: MMP17 (MT4-MMP) and MMP25 (MT6-MMP). Among this family, MMP14 (MT1-MMP) is the most important one, because its dysregulation has been found in many dysfunctional connective tissue metabolisms and in various type of tumors [20]. This enzyme has wide substrate specificity, digesting different types of collagens and other ECM components, including dermatan, fibronectin, laminin, several cytokines, as well the propeptide of other MMPs, in particular gelatinases. The activation of proMMP2 by MMP14 is one of the most relevant cellular processes involved in the local degradation of extracellular matrix during cell invasion and metastasis, especially in the case of CNS disorders and tumors [21], and the MMP14/MMP2/integrin $\alpha_v\beta_3$ axis represents an attractive target for therapeutic applications for CNS tumors [22].

Also, MMP14 influences both intercellular and cell-matrix communication by regulating the activity of many plasma membrane-anchored and extracellular proteins [23].

2.4 Human macrophage elastase (MMP12)

Among the other MMPs that do not belong to any particular subfamily, macrophage elastase (MMP12) is of particular interest, as it exhibits all the characteristics of other MMPs, but it is preferentially produced by macrophages infiltrating into tissues where injury or remodeling is occurring [24]. Its preferential substrate is elastin, but it can also digest type IV collagen, fibronectin, vitronectin, and other extracellular matrix components. Although it seems to be mainly involved in chronic obstructive pulmonary disorder (COPD) [25], MMP12 has a role in the malignant progression of several type of cancer [26,27], as well as in allergic asthma [28], emphysema [29], acute lung injury, and others.

3 Classes of inhibitors/activators and their design

Developing MMP inhibitors (MMPi) has attracted the interest of the scientific community since the beginning of the 1980s and first clinical trials of broad-spectrum MMPi were reported early in the 1990s, mainly for cancer therapeutic applications [30].

Main reasons for the failure of first-generation MMPi (such as Ilomastat, Marimastat, and CGS-27023A) include the low bioavailability and poor metabolic profile of these compounds, and their lack of specificity within the MMP family, versus other metalloenzymes, such as ADAMs, that resulted in multiple side effects, including arthralgia, myalgia, tendinitis, musculoskeletal syndrome (MSS), and gastrointestinal disorders [31].

Although MMPi research development experienced a setback at the beginning of this century, due to the failure of several high-profile clinical trials [32], these enzymes are today experiencing a renewed interest due to their involvement in other pathologies, including inflammatory diseases, neurological, and immunological disorders [5].

It is now widely accepted that to develop efficient and therapeutically worthy compounds, MMPi should be highly selective against the specific MMP believed to be involved in a peculiar pathology [33,34]. As shown in Fig. 3, effective inhibitors should contain:

- A functional group able to chelate the catalytic Zn^{2+} atom. Most of first-generation inhibitors contained the hydroxamic element as zinc-binding group (ZBG). This functional group is highly susceptible to human metabolic reactions and is easily transformed into carboxylic group with the release of toxic hydroxylamine. Thus, in second-generation MMP inhibitors, several research groups developed compounds that contain different ZBG, such as thiolates (S-), carboxylates (COO-), and phosphonyls (PO_2-) groups or even nitrogen-based and heterocyclic chelators, compensating the lower zinc-binding efficacy by introducing more hydrophobic interacting elements. The nonhydroxamate-based inhibitors possess different spectrum of affinities for the different MMPs and are poorly recognised by other metalloenzymes, thus showing a significant reduction of severe side effects, such as those of musculoskeletal syndrome (MSS) [35].
- One or more side chains experiencing van der Waals interactions and hydrophobic bridges with enzyme subsites. One subsite (S1′ pocket) is particularly relevant for achieving selective inhibitors, as it has been widely studied how this pocket differ for the size, shape, and flexibility in the diverse MMPs, especially in the case of the so-called medium-size S1′ pockets (namely MMP2, MMP8, MMP9, MMP12, and MMP14) [36]. In particular, it has been demonstrated that the incorporation of long hydrophobic chains in R1 position can promote the selectivity of the inhibitors, even in the case of two different enzymes of the same subfamily (such as for the two gelatinases MMP2 and MMP9), by promoting conformational changes of the S1′ pocket that open additional channels at the bottom of this subsite [37].
- One or more functional groups that can undergo hydrogen bonding with the protein backbone.
- A linear or cyclic skeleton that can orient the pharmacophoric elements in the correct positions. The first-generation MMPi were designed starting from the amino acid sequence around the cleavage site of the natural substrate and were based on a linear peptide scaffold. With the advancement on knowledge about peptidomimetic chemistry [38], different strategies for increasing the rigidity and the conformation constraints of the scaffold were applied to maximise both the affinity within the catalytic cleft and the in vivo oral availability. Next-

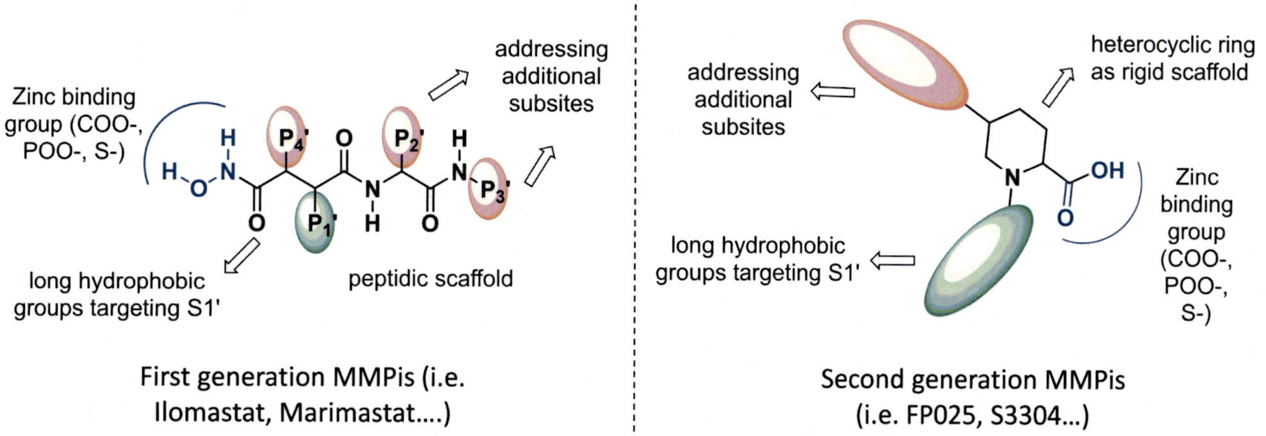

FIG. 3 Comparison between the general structure of first (left) and second generation (right) MMP inhibitors, highlighting the pharmacophoric groups relevant for the interaction within the catalytic cleft of MMPs.

generation MMPi are based on heterocyclic scaffolds that act as type III functional-structural peptidomimetics, such as proline, piperidine, and heteroaromatic skeletons.

Despite much research being devoted to the development of collagenase inhibitors and several candidates being reported in the literature so far, especially for MMP13 [39,40], no collagenase inhibitors have received FDA approval yet, mainly because most of them are not selective or show poor bioavailability. However, two compounds, Cipemastat and CL-82198 (Fig. 4), entered clinical trials in the last decades, although these studies were terminated prematurely for lack of efficacy and/or adverse tolerability. Cipemastat (also named Trocade) is a potent selective inhibitor of all three collagenases (MMP1, 8, and 13), displaying the higher activity versus MMP1 (Ki values of 3.0, and of 4.4 and 3.4 nM against MMP8 and 13, respectively). This compound is a first-generation MMP inhibitor, characterized by the presence of the hydroxamate group as ZBG and three heterocyclic elements (piperidine, cyclopropane, and substituted imidazolidine-2,4-dione) as interacting appendages. It was developed by Roche (RO-323555) and entered clinical trials for the treatment of patients with rheumatoid arthritis and osteoarthritis, as it was found to inhibit interleukin-1α (IL-1α)-induced collagen degradation with an $IC_{50} = 60\,nM$ [41]. However, clinical trials were terminated prematurely because it did not prevent progression of joint damage in patients with rheumatoid arthritis [42]. On the contrary, CL-82198 is a very selective inhibitor, being able of addressing only the S1′ pocket of MMP13, without showing any activity against all the other MMPs and also against the other two collagenases. It is a relative weak binder, with $IC_{50} = 3.2\,\mu M$, but its high selectivity made it a good candidate for the pharmacologic treatment of osteoarthritis (OA) progression. This compound is a second-generation MMP inhibitor that does not possess any zinc-binding element but possesses the right size and shape to fit the S1′ pocket of MMP13, which has a peculiar extended nature near the surface of the protein [43].

Thus, developing inhibitors that display selectivity for only one of the two gelatinases is of high relevance for a successful clinical trial [44]. In this context, Trabocchi's research group has developed highly selective MMP2 and MMP9 inhibitors based on the D-proline scaffold. In particular, D-proline-derived hydroxamic acids containing long appendages at the amino group, composed by three aromatic rings, were found to be sub-nanomolar inhibitors of MMP2 with selectivity ratio up to 730 against MMP9 (Fig. 5, left) [45]. Also, the synthesis and evaluation of D-proline derivatives containing amino appendages at C-4 enabled the identification of a compound containing the lysine fragment (Fig. 5, right) that showed >200 selectivity ratio against

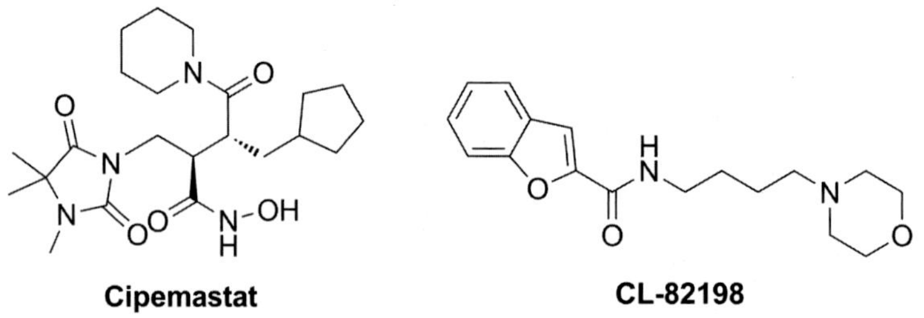

FIG. 4 Collagenase inhibitors that entered clinical trials in the last few years.

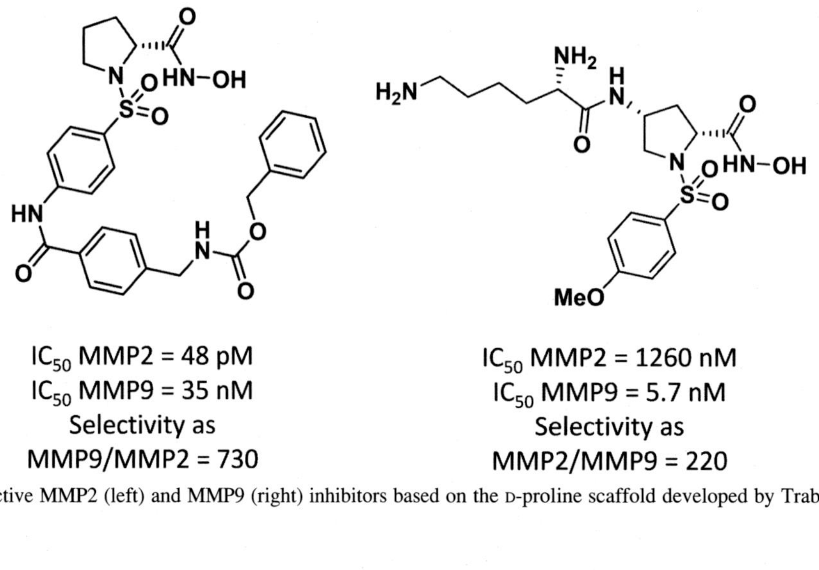

FIG. 5 Structure of selective MMP2 (left) and MMP9 (right) inhibitors based on the D-proline scaffold developed by Trabocchi's research group.

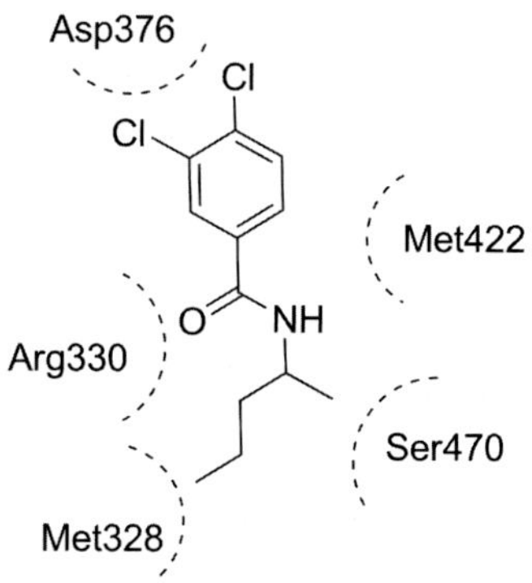

FIG. 6 Structure of the MMP14 allosteric inhibitor NSC405020 along with key interacting amino acids in the binding site of MMP14 hemopexin domain.

MMP2 [46]. These compounds paved the way for the development of dual inhibitors, acting against gelatinases and other molecular targets involved in the tumor microenvironment, such as carbonic anhydrases (CAs) and integrins. In particular, Trabocchi's research group developed a D-proline derivative containing a biphenyl sulfonamido moiety able to inhibit either MMP9 and CAII, with potential application in gastric and colorectal cancer [47], and dual $\alpha_v\beta_3$-MMP2 peptidomimetic inhibitors based on the tyrosine scaffold useful in melanoma cancer [48].

Several inhibitors have appeared in the literature against MT1-MMP [49,50]. Nevertheless, only one compound has been investigated in advanced preclinical studies, pentanylbenzamide compound NSC405020 (Fig. 6). This compound does not act as a canonical MMP inhibitor, as it binds selectively to the hemopexin-like domain of MMP14, locking its catalytic cleft into a less favorable conformation for binding the collagen substrate [51]. NSC405020 interacts with the PEX domain of MMP14 through the key residues Met328, Arg330, Asp376, Met422, and Ser470 (Fig. 6) [52]. This kind of allosteric/exosite inhibitor shows a noncompetitive inhibition mode and has the advantage of avoiding off-target inhibition with limited side effects. In particular, NSC405020, by blocking the collagenolytic activity of MMP14, is able to inhibit MCF7-β3/MT tumor growth in xenografted mice and can be used as a chemical probe to investigate vestibular Schwannoma, hemostasis, and thrombosis [53].

Concerning elastases, several inhibitors have been developed over the years against MMP12 [54] for the treatment of allergic asthma, where the accumulation of macrophages associated with inflammation can be treated by inhibiting this enzyme.

4 Clinically used agents or compounds in clinical development

Main drug candidates developed over last years that are currently under clinical trials or have been already approved by FDA for the treatment of cancer [55–57] and/or other disorders, including arthritis [58], neurological disorders [59], and periodontal disease are herein reported. A representative chart of these MMP inhibitors is reported in Table 1. To date, the only two compounds of this series that have been approved by the FDA are Periostat (doxycycline) for the treatment of periodontal diseases [61], and Abatemapir, (commercialized under the name of Xeglyze) for the treatment of head lice infestation [68]. Abetamapir is a 5,5′-dimethyl-2,2′-bipyridyl compound that inhibits all the different MMPs of the family by chelating different metal ions, including the zinc cation necessary for the function of these enzymes. As MMPs are critical for all stages of head lice ova development [69], Abametapir has a direct ovicidal activity and can be used for the treatment of head lice infestation, with no adverse effects. As for Periostat and Abametapir, most of the compounds reported in Table 1 are natural products or antibiotics that have been modified and applied in the screening against MMPs, that resulted to be potent and selective in inhibiting the activity or modifying the expression of these enzymes. Finally, novel inhibitors that appeared in the literature in recent years and are in advanced stage of preclinical studies are discussed through the chapter and classified depending which subfamily of MMPs they are targeting.

A class of compounds that have shown a relevant role as antitumoral MMPi are tetracyclines antibiotics. These compounds, when chemically modified by removing the dimethylamino group from C-4 position, do not show antimicrobial activity, but can chelate the zinc ion, bind to active MMPs and inhibit their activity by disrupting their conformation [70]. This class of compounds (which include Periostat, Metastat and minocylcline reported in Table 1) can inhibit different MMPs, including gelatinases, stromelysins, collagenases, and membrane-type MMPs, but show a high selectivity against gelatinases, thus representing good candidates for the treatment of different type of tumors. For example, Metastat (COL-3) is a selective inhibitor of MMP2 and 9 that is now being evaluated in Phase I clinical trials for the treatment of sarcoma patients [65], and in Phase II clinical trials for the management of recurrent high-grade glioma and AIDS-related Kaposi's sarcoma [66]. Interestingly, tetracycline compounds, as well as other MMPi, have been clinically evaluated only for the treatment of solid tumors, and no examples are reported in clinical trials, by far, about the use of MMPi to treat haematological cancer, such as leukemias. For example, the other gelatinase inhibitor that is now under clinical trial as antitumoral compound, S3304, is being evaluated for the treatment of patients with locally advanced nonsmall cell lung cancer [67]. S3304 is a second-generation MMPi, based on the D-tryptophan amino acid scaffold, that contains the carboxylic acid group as zinc binding motif and possess the long *p*-tolylethynyl-thiophenyl chain as long hydrophobic group for the S1′ pocket [71]. S3304 is highly selective for MMP9 and does not show activity for MMPs involved in the appearance of musculoskeletal side effects, such as MMP3 and MMP7.

TABLE 1 Second-generation MMP inhibitors that are currently under clinical trials (at the end of December 2022) or have been approved by the FDA.

Compound	Target	Application	Status	Ref.
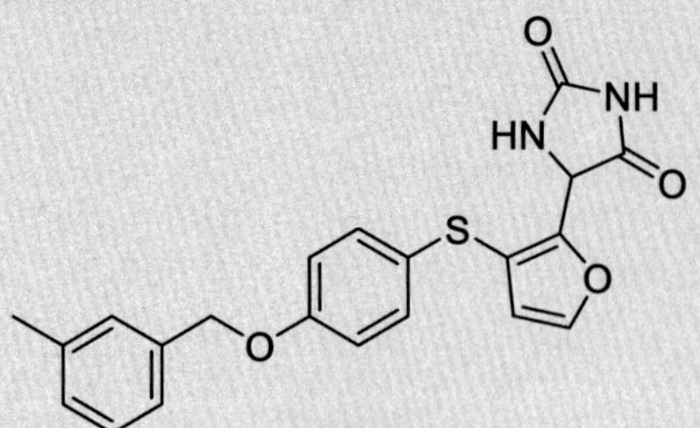 **FP025**	MMP12	Treatment of asthma and COPD	Under phase II clinical trials	[60]
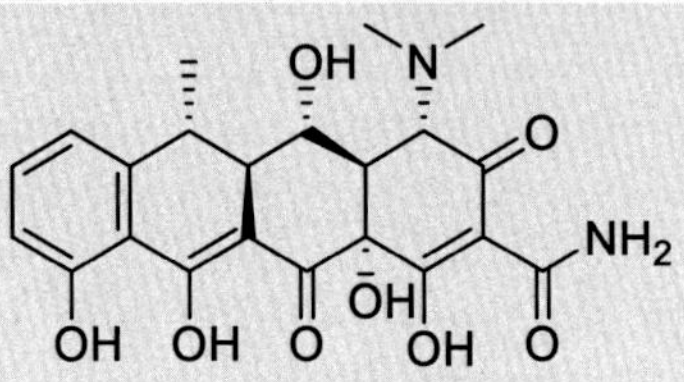 **Periostat (doxycycline)**	MMP9	Treatment of periodontal diseases	Approved	[61]
	MMP8	Prevention of myocardial infarction risk	Under clinical trials	[62]
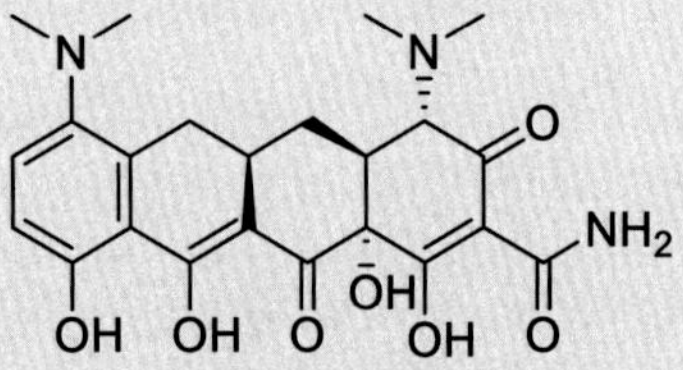 **Minocycline**	MMP9	Treatment of fragile X syndrome	Under clinical trials	[63]
	MMP9	Treatment of acute intracerebral hemorrhage	Under clinical trials	[64]
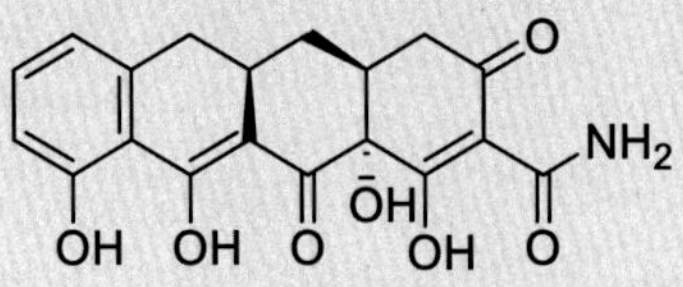 **Metastat (COL-3)**	MMP2 and MMP9	Treatment of sarcoma	Under phase I clinical trials	[65]
		Treatment of high grade glioma	Under phase II clinical trials	[66]
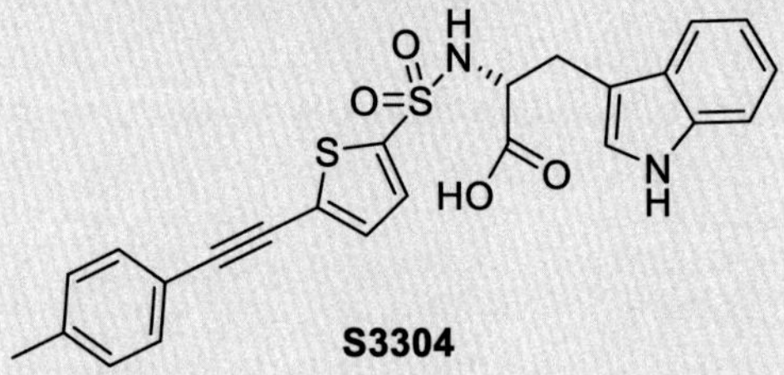 **S3304**	MMP2 and MMP9	Treatment of lung cancers/nonsmall cell lung cancer	Terminated phase I clinical trials	[67]
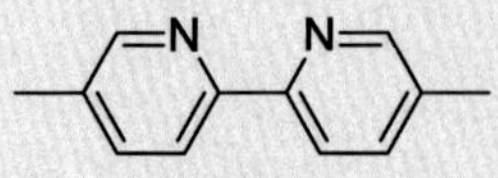 **Abatemapir**	Broad spectrum	Treatment of head lice infestation	Approved	[68]

Gelatinase inhibitors have a variety of different therapeutic applications that go beyond the development of antitumor drugs. For example, MMP2 inhibitors can be used in the treatment of heart injury, strokes and ischemic events [72], whereas MMP9 inhibitors can find application in treating inflammatory and neurodegenerative disorders [73]. In the case of tetracycline compounds, the above mentioned Periostat (Doxycycline, Table 1) is the only MMP inhibitor approved by the FDA for the management of periodontal diseases, thanks to its ability to improve the periodontal wound repair [61]. Also, it has showed to be effective against rheumatoid arthritis and atherosclerosis and it is now under clinical trials for evaluating its usefulness in preventing the risk of myocardial infarction, thanks to its ability of reducing the systemic inflammation by downregulating MMP8 level [62]. Also, minocycline is being evaluated in patients with fragile X syndrome (FXS) [63], and it is now under phase II clinical trials for the treatment of human spinal cord injury [64].

Finally, in the context of gelatinase inhibitors used for treating pathologies other than cancer, a relevant compound is (*R*)-ND-336 (Fig. 7), a gelatinase inhibitor that has been recently developed for the treatment of Diabetic Foot Ulcers (DFUs), one of the most common complications found in patients with chronic diabetes [74]. In fact, by far, the only available treatment of DFUs is the surgical amputation of the damaged tissue or of the entire foot, even though more that 30% of individuals who undergo DFU surgical amputation show 5-year mortality after surgery. Very recently, Chang and coworkers reported that high MMP9 levels are found in severe infected DFUs and that the use of the MMP9 inhibitor (*R*)-ND-336, alone or in combination with the antibiotic linezolid, was able to accelerate wound healing in animal models [75]. This compound successfully underwent toxicological studies [76] and it is now under assessment by the FDA for entering Phase I clinical trials to become the first topical gel treatment for DFUs, reducing the need for amputations and decreasing the mortality rates in diabetic patients.

With reference to elastase inhibitors as clinical candidates, FP-025 (Table 1) is a potent second-generation inhibitor of MMP12 which shows high selectivity (over three orders of magnitude) versus other MMP family members. This compound can reduce peribronchial and periarterial cellular infiltrates in lungs of HDM-sensitized mice [77], and it is now under Phase II clinical trials for the treatment of asthma and COPD [60].

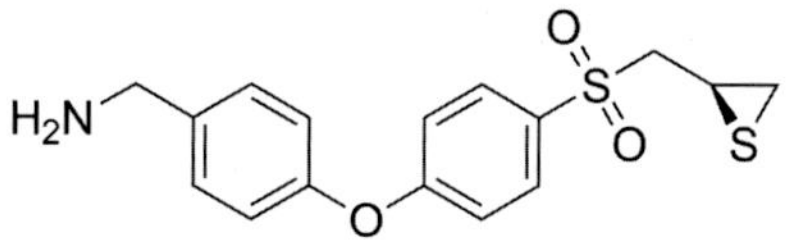

FIG. 7 Structure of the MMP9 inhibitor (*R*)-ND-336 for the treatment of Diabetic Foot Ulcers (DFU).

5 Conclusion and outlook

The renewed interest in MMPs has caused a boost in the number of papers reporting novel MMPs inhibitors with improved selectivity. However, as observed also for patents describing new MMPi between 2014 and 2020 [78], only collagenases, gelatinases, membrane-type MMP and elastase are being studied for the development of selective inhibitors, whereas no compounds have been recently reported against stromelysins and matrilisins. The targets of these compounds are few MMPs if compared with the whole class, as they specifically refer to MMP1, 2, 3, 8, 9, 12, 13, 14. These compounds have been developed not only for the potential applications as chemotherapeutic agents but also for the treatment of inflammatory and neurological disorders, chronic obstructive pulmonary diseases, asthma and more. Despite that, most of the newly developed compounds, as well as the first-generation MMPi, still fail to succeed in phase III clinical trials, mainly because of the metabolic instability of the compounds or due to the huge differences that are found between the animal models used in preclinical studies and the human models of clinical trials. Also, despite recent advances, the complex mechanisms of all MMP functions and the different role of each enzyme in different pathologies is still not well explored, causing the occurrence of side effects due to cross-inhibition of other MMPs or other metalloenzymes. By far, the only two compounds that have been approved by the FDA are Periostat (doxycycline) and Abatemapir, for the treatment of periodontal disease and head lice infestation, respectively. However, we expect that in the coming years, more MMPi will enter clinical trials, due to the increasing interest toward these enzymes as therapeutic targets in a wide range of pathologies beyond cancer.

References

[1] Fingleton B. Matrix metalloproteinases as valid clinical targets. Curr Pharm Des 2007;13:333–46.

[2] Cui N, Hu M, Khalil RA. Biochemical and biological attributes of matrix metalloproteinases. Prog Mol Biol Transl Sci 2017;147:1–73.

[3] Laronha H, Caldeira J. Structure and function of human matrix metalloproteinases. Cells 2020;9(5):1076.

[4] Vandenbroucke RC, Libert C. Is there new hope for therapeutic matrix metalloproteinase inhibition? Nat Rev Drug Discov 2014;13:904–27.

[5] Khokha R, Murthy A, Weiss A. Metalloproteinases and their natural inhibitors in inflammation and immunity. Nat Rev Immunol 2013;13:649–65.

[6] Bassiouni W, Ali MAM, Schulz R. Multifunctional intracellular matrix metalloproteinases: implications in disease. FEBS J 2021;288:7162–82.

[7] Ala-aho R, Kähäri V-M. Collagenases in cancer. Biochimie 2005;87 (3):273–86.

[8] Egeblad M, Werb Z. New functions for the matrix metalloproteinases in cancer progression. Nat Rev Cancer 2002;2:161–74.

[9] Inada M, Wang Y, Byrne MH, et al. Critical roles for collagenase-3 (MMP13) in development of growth plate cartilage and in endochondral ossification. Proc Natl Acad Sci U S A 2004;101:17192–7.
[10] Alipour H, Raz A, Zakeri S, Djadid ND. Therapeutic applications of collagenase (metalloproteases): a review. Asian Pac J Trop Biomed 2016;6(11):975–81.
[11] Xun L. Gelatinase inhibitors: a patent review (2011–2017). Expert Opin Ther Pat 2018;28:31–46.
[12] Kumar Baidya S, Amin SA, Jha T. Outline of gelatinase inhibitors as anti-cancer agents: a patent mini-review for 2010-present. Eur J Med Chem 2021;213, 113044.
[13] Bauvois B. New facets of matrix metalloproteinases MMP-2 and MMP-9 as cell surface transducers: Outside-in signaling and relationship to tumor progression. BBA Rev Cancer 2012;1825:29–36.
[14] Vihinen P, Ala-aho R, Kahari VM. Matrix metalloproteinases as therapeutic targets in cancer. Curr Cancer Drug Targets 2005;5: 203–20.
[15] Mu D, Cambier S, Fjellbirkeland L, Baron JL, Munger JS, Kawakatsu H, Sheppard D, Broaddus VC, Nishimura SL. The integrin alpha(v) beta8 mediates epithelial homeostasis through MT1-MMP-dependent activation of TGF-beta1. J Cell Biol 2002;157:493–507.
[16] Bergers G, Brekken R, Mcmahon G, Vu TH, Itoh T, Tamaki K, Tanzawa K, Thorpe P, Itohara S, Werb Z, Hanahan D. Matrix metalloproteinase-9 triggers the angiogenic switch during carcinogenesis. Nat Cell Biol 2000;2:737–44.
[17] Higashi S, Miyazaki K. Identification of amino acid residues of the matrix metalloproteinase-2 essential for its selective inhibition by beta-amyloid precursor protein-derived inhibitor. Biol Chem 2008; 283:10068–78.
[18] Chetty C, Lakka SS, Bhoopathi P, Rao JS. MMP-2 alters VEGF expression via αVβ3 integrin-mediated PI3K/AKT signaling in A549 lung cancer cells. Int J Cancer 2010;127:1081–95.
[19] Dufour A, Zucker S, Sampson NS, Kuscu C, Cao J. Role of matrix metalloproteinase-9 dimers in cell migration. J Biol Chem 2010;285:35944–56.
[20] Bartolome RA, Ferreiro S, Miquilena-Colina ME, Martinez-Prats L, Soto-Montenegro ML, García-Bernal D, Vaquero JJ, Agami R, Delgado R, Desco M, Sánchez-Mateos P, Teixidó J. The chemokine receptor CXCR4 and the metalloproteinase MT1-MMP are mutually required during melanoma metastasis to lungs. Am J Pathol 2009;174:602–12.
[21] Lakhan SE, Kirchgessner A, Tepper D, Leonard A. Matrix metalloproteinases and blood-brain barrier disruption in acute ischemic stroke. Front Neurol 2013;4:32.
[22] Yosef G, Arkadash V, Papo N. Targeting the MMP-14/MMP-2/integrin αvβ3 axis with multispecific N-TIMP2-based antagonists for cancer therapy. J Biol Chem 2018;293(34):13310–26.
[23] Niland S, Riscanevo AX, Eble JA. Matrix metalloproteinases shape the tumor microenvironment in cancer progression. Int J Mol Sci 2021;23(1):146.
[24] Lagente V, Le Quement C, Boichot E. Macrophage metalloelastase (MMP-12) as a target for inflammatory respiratory diseases. Exp Opin Ther Targets 2009;13:287–95.
[25] Demkow U, van Overveld FJ. Role of elastases in the pathogenesis of chronic obstructive pulmonary disease: Implications for treatment. Eur J Med Res 2010;15:27–35.
[26] Ng KT, Qi X, Kong KL, Cheung BY-Y, Lo C-M, Poon RT-P, Fan S-T, Man K. Overexpression of matrix metalloproteinase-12 (MMP-12) correlates with poor prognosis of hepatocellular carcinoma. Eur J Cancer 2011;47(15):2299–305.
[27] He MK, Le Y, Zhang YF, Jian P-E, Yu Z-S, Wang L-J, Shi M. Matrix metalloproteinase 12 expression is associated with tumor FOXP3+ regulatory T cell infiltration and poor prognosis in hepatocellular carcinoma. Oncol Lett 2018;16:475–82.
[28] Warner RL, Lukacs NW, Shapiro SD, Bhagarvathula N, Nerusu KC, Varani J, Johnson KJ. Role of metalloelastase in a model of allergic lung responses induced by cockroach allergen. Am J Pathol 2004;165:1921–30.
[29] Churg A, Wang R, Wang X, Onnervik P-O, Thim K, Wright JL. Effect of an MMP-9/MMP-12 inhibitor on smoke-induced emphysema and airway remodelling in guinea pigs. Thorax 2007;62(8):706–11.
[30] Dove A. MMP inhibitors: glimmers of hope amidst clinical failures. Nat Med 2002;8:95.
[31] Rudek MA, Venitz J, Figg WD. Matrix metalloproteinase inhibitors: Do they have a place in anticancer therapy? Pharmacotherapy 2002;22:705–20.
[32] Fields GB. The rebirth of matrix metalloproteinase inhibitors: moving beyond the dogma. Cells 2019;8:984.
[33] Murphy G, Nagase H. Progress in matrix metalloproteinase research. Mol Aspects Med 2008;29:290–308.
[34] Laronha H, Carpinteiro I, Portugal J, Azul A, Polido M, Petrova KT, Salema-Oom M, Caldeira J. Challenges in matrix metalloproteinases inhibition. Biomolecules 2020;10(5):717.
[35] Li K, Tay FR, Yiu CKY. The past, present and future perspectives of matrix metalloproteinase inhibitors. Pharmacol Ther 2020;207, 107465.
[36] Baidya SK, Banerjee S, Adhikari N, Jha T. Selective inhibitors of medium-size S1′ pocket matrix metalloproteinases: a stepping stone of future drug discovery. J Med Chem 2022;65 (16):10709–54.
[37] Fabre B, Ramos A, de Pascual-Teresa B. Targeting matrix metalloproteinases: exploring the dynamics of the S1′ pocket in the design of selective, small molecule inhibitors. J Med Chem 2014; 57:10205–19.
[38] Lenci E, Trabocchi A. Peptidomimetic toolbox for drug discovery. Chem Soc Rev 2020;49(11):3262.
[39] Nara H, Kori M. Discovery of novel, highly potent and selective matrix metalloproteinase (MMP-13) inhibitors with a 1,2,4-triazol-3-yl moiety as a zinc binding group using a structure-based design approach. J Med Chem 2017;60:608–26.
[40] Nara H, Sato K, Kaieda A, et al. Design, synthesis, and biological activity of novel, potent, and highly selective fused pyrimidine-2-carboxamide-4-one-based matrix metalloproteinase (MMP)-13 zinc-binding inhibitors. Bioorg Med Chem 2016;24:6149–65.
[41] Lewis EJ, Bishop J, Bottomley KM, Bradshaw D, Brewster M, Broadhurst MJ, Brown PA, Budd JM, Elliott L, Greenham AK, Johnson WH, Nixon JS, Rose F, Sutton B, Wilson K. Ro 32-3555, an orally active collagenase inhibitor, prevents cartilage breakdown in vitro and in vivo. Br J Pharmacol 1997;121(3):540–6.
[42] Hemmings FJ, Farhan M, Rowland J, Banken L, Jain R. Tolerability and pharmacokinetics of the collagenase-selective inhibitor Trocade™ in patients with rheumatoid arthritis. Rheumatology 2001;40(5):537–43.
[43] Chen JM, Nelson FC, Levin JI, Mobilio D, Moy FJ, Nilakantan R, Zask A, Powers R. Structure-based design of a novel, potent, and selective inhibitor for MMP-13 utilizing NMR spectroscopy

and computer-aided molecular design. J Am Chem Soc 2000;122: 9648–54.
[44] Deryugina EI, Quigley JP. Matrix metalloproteinases and tumor metastasis. Cancer Metastasis Rev 2006;25:9–34.
[45] Lenci E, Innocenti R, Di Francescantonio T, Menchi G, Bianchini F, Contini A, Trabocchi A. Identification of highly potent and selective MMP2 inhibitors addressing the S1′ subsite with d-proline-based compounds. Bioorg Med Chem 2019;27:1891–902.
[46] Lenci E, Contini A, Trabocchi A. Discovery of a d-pro-lys peptidomimetic inhibitor of MMP9: addressing the gelatinase selectivity beyond S1' subsite. Bioorg Med Chem Lett 2020;30, 127467.
[47] Lenci E, Angeli A, Calugi L, Innocenti R, Carta F, Supuran CT, Trabocchi A. Multitargeting application of proline-derived peptidomimetics addressing cancer-related human matrix metalloproteinase 9 and carbonic anhydrase II. Eur J Med Chem 2021;214: 13260.
[48] Baldini L, Lenci E, Bianchini F, Trabocchi A. Identification of a common pharmacophore for binding to MMP2 and RGD integrin: towards a multitarget approach to inhibit cancer angiogenesis and metastasis. Molecules 2022;27(4):124.
[49] Cepeda MA, Pelling JJH, Evered CL, Williams KC, Freedman Z, Stan I, Willson JA, Leong HS, Damjanovski S. Less is more: low expression of MT1-MMP is optimal to promote migration and tumourigenesis of breast cancer cells. Mol Cancer 2016;15:65.
[50] Pahwa S, Stawikowski MJ, Fields GB. Monitoring and inhibiting MT1-MMP during cancer initiation and progression. Cancers 2014;6:416–35.
[51] Juban G, Saclier M, Yacoub-Youssef H, Kernou A, Arnold L, Boisson C, Ben Larbi S, Magnan M, Cuvellier S, Théret BJ, Petrof BJ, Desguerre I, Gondin J, Mounier R, Chazaud B. AMPK activation regulates LTBP4-dependent TGF-β1 secretion by pro-inflammatory macrophages and controls fibrosis in Duchenne muscular dystrophy. Cell Rep 2018;25(8):2163–76.
[52] Remacle AG, Golubkov VS, Shiryaev SA, Dahl R, Stebbins JL, Chernov AV, Cheltsov AV, Pellecchia M, Strongin AY. Novel MT1-MMP small-molecule inhibitors based on insights into hemopexin domain function in tumor growth. Cancer Res 2012;72 (9):2339–49.
[53] Zarrabi K, Dufour A, Li J, Kuscu C, Pulkoski-Gross A, Zhi J, Hu Y, Sampson NS, Zucker S, Cao J. Inhibition of matrix metalloproteinase 14 (MMP-14)-mediated cancer cell migration. J Biol Chem 2011;286 (38):33167–77.
[54] Norman P. Selective MMP-12 inhibitors: WO-2008057254. Exp Opin Ther Pat 2009;19(7):1029–34.
[55] Fields GB. Mechanisms of action of novel drugs targeting angiogenesis-promoting matrix metalloproteinases. Front Immunol 2019;10:1278.
[56] Lia N-G, Shib Z-H, Tang Y-P, Duan J-A. Selective matrix metallo proteinase inhibitors for cancer. Curr Med Chem 2009;16:3805–27.
[57] Tauro M, McGuire J, Lynch CC. New approaches to selectively target cancer-associated matrix metalloproteinase activity. Cancer Metastasis Rev 2014;33:1043–57.
[58] Martel-Pelletier J, Welsch DJ, Pelletier JP. Metalloproteases and inhibitors in arthritic diseases. Best Pract Res Clin Rheumatol 2001;15:805–29.
[59] Chopra S, Overall CM, Dufour A. Matrix metalloproteinases in the CNS: Interferons get nervous. Cell Mol Life Sci 2019;76:3083–95.
[60] Amar S, Fields GB. Potential clinical implications of recent matrix metalloproteinase inhibitor design strategies. Expert Rev Proteomics 2015;12(5):445–7.
[61] Boelen GJ, Boute L, d'Hoop J, et al. Matrix metalloproteinases and inhibitors in dentistry. Clin Oral Investig 2019;23:2823–35.
[62] Kormi I, Alfakry H, Tervahartiala T, Pussinen PJ, Sinisalo J, Sorsa T. The effect of prolonged systemic doxycycline therapy on serum tissue degrading proteinases in coronary bypass patients: a randomized, double-masked, placebo-controlled clinical trial. Inflamm Res 2014;63:329–34.
[63] Dziembowska M, Pretto DI, Janusz A, Kaczmarek L, Leigh MJ, Gabriel N, Durbin-Johnson B, Hagerman RJ, Tassone F. High MMP-9 activity levels in fragile X syndrome are lowered by minocycline. Am J Med Genet A 2013;161:1897–903.
[64] Chang JJ, Kim-Tenser M, Emanuel BA, Jones GM, Chapple K, Alikhani A, Sanossian N, Mack WJ, Tsivgoulis G, Alexandrov AV, Pourmotabbed T. Minocycline and matrix metalloproteinase inhibition in acute intracerebral hemorrhage: a pilot study. Eur J Neurol 2017;24:1384–91.
[65] Syed S, Takimoto C, Hidalgo M, Rizzo J, Kuhn JG, Hammond LA, Schwartz G, Tolcher A, Patnaik A, Eckhardt SG, Rowinsky EK. A phase I and pharmacokinetic study of Col-3 (Metastat), an oral tetracycline derivative with potent matrix metalloproteinase and antitumor properties. Clin Cancer Res 2004;10(19):6512–21.
[66] Rudek MA, New P, Mikkelsen T, Phuphanich S, Alavi JB, Nabors LB, Piantadosi S, Fisher JD, Grossman SA. Phase I and pharmacokinetic study of COL-3 in patients with recurrent high-grade gliomas. J Neurooncol 2011;105:375–81.
[67] Chiappori AA, Eckhardt SG, Bukowski R, Sullivan DM, Ikeda M, Yano Y, Yamada-Sawada T, Kambayashi Y, Tanaka K, Javle MM, Mekhail T, O'bryant CL, Creaven PJ. A phase I pharmacokinetic and pharmacodynamic study of s-3304, a novel matrix metalloproteinase inhibitor, in patients with advanced and refractory solid tumors. Clin Cancer Res 2007;13(7):2091–9.
[68] Woods AD, Porter CL, Feldman SR. Abametapir for the treatment of head lice: a drug review. Ann Pharmacother 2022;56(3):352–7.
[69] Bowles VM, Young AR, Barker SC. Metalloproteases and egg-hatching in *Pediculus humanus*, the body (clothes) louse of humans (Phthiraptera: Insecta). Parasitology 2008;135(Pt 1):125–30.
[70] Nuti E, Tuccinardi T, Rossello A. Matrix metalloproteinase inhibitors: New challenges in the era of post broad-spectrum inhibitors. Curr Pharm Des 2007;13:2087–100.
[71] Mant TG, Bradford D, Amin DM, Pisupati J, Kambayashi Y, Yano Y, Tanaka K, Yamada-Sawada T. Pharmacokinetics and safety assessments of high-dose and 4-week treatment with S-3304, a novel matrix metalloproteinase inhibitor, in healthy volunteers. Br J Clin Pharmacol 2007;63(5):512–26.
[72] Rempe RG, Hartz AM, Bauer B. Matrix metalloproteinases in the brain and blood-brain barrier: versatile breakers and makers. J Cereb Blood Flow Metab 2016;36(9):1481–507.
[73] Brkic M, Balusu S, Libert C, Vandenbroucke RE. Friends or foes: matrix metalloproteinases and their multifaceted roles in neurodegenerative diseases. Mediators Inflamm 2015;2015, 620581.
[74] Peng Z, Nguyen TT, Song W, Anderson B, Wolter WR, Schroeder VA, Hesek D, Lee M, Mobashery S, Chang M. The selective MMP-9 inhibitor (*R*)-ND-336 alone or in combination with linezolid accelerates wound healing in infected diabetic mice. ACS Pharmacol Transl Sci 2020;4(1):107–17.

[75] Nguyen TT, Ding D, Wolter WR, Pérez RL, Champion MM, Mahasenan KV, Hesek D, Lee M, Schroeder VA, Jones JI, Lastochkin E, Rose MK, Peterson CE, Suckow MA, Mobashery S, Chang M. Validation of matrix metalloproteinase-9 (MMP-9) as a novel target for treatment of diabetic foot ulcers in humans and discovery of a potent and selective small-molecule MMP-9 inhibitor that accelerates healing. J Med Chem 2018;61:8825–37.

[76] Gabriel CE, Nguyen TT, Gargano EM, Fisher JF, Chang M, Mobashery S. Metabolism of the selective matrix metalloproteinase-9 inhibitor (*R*)-ND-336. ACS Pharmacol Transl Sci 2021;4(3):1204–13.

[77] Ravanetti L, Dekker T, Guo L, Dijkhuis A, Dierdorp BS, Diamant Z, Florquin S, Lutter R. Efficacy of FP-025: A novel matrix metalloproteinase-12 (MMP-12) inhibitor in murine allergic asthma. Allergy 2022. https://doi.org/10.1111/all.15513.

[78] Lenci E, Cosottini L, Trabocchi A. Novel matrix metalloproteinase inhibitors: an updated patent review (2014–2020). Expert Opin Ther Pat 2021;31(6):509–23.

Chapter 3.5

A disintegrin and metalloproteinases (ADAMs) and tumor necrosis factor-alpha-converting enzyme (TACE)

Doretta Cuffaro[a], Simone D. Scilabra[b], Donatella P. Spanò[b,c], Matteo Calligaris[a,b], Elisa Nuti[a], and Armando Rossello[a]

[a]*Department of Pharmacy, University of Pisa, Pisa, Italy,* [b]*Proteomics Group of Fondazione Ri.MED, Research Department IRCCS ISMETT (Istituto Mediterraneo per i Trapianti e Terapie ad Alta Specializzazione), Palermo, Italy,* [c]*STEBICEF (Dipartimento di Scienze e Tecnologie Biologiche Chimiche e Farmaceutiche), University of Palermo, Palermo, Italy*

1 General features of ADAMs

A disintegrin and metalloproteinases (ADAMs) are membrane-tethered proteases specialized in the proteolytic cleavage of transmembrane proteins, the so-called ectodomain shedding. As a consequence of this cleavage, the extracellular portion of the protein (ectodomain) is released into the extracellular milieu, while a small peptide fragment remains anchored to the membrane [1].

The human genome encodes for 30 ADAMs, of which 12 are predicted to be proteolytically active (ADAM8, 9, 10, 12, 15, 17, 19, 20, 21, 28, 30, 33) [2]. ADAMs are expressed in all mammalian tissues, where they regulate a variety of biological processes, including development, cell signaling, and inflammation. Similar to the first ADAMs ever discovered, ADAM1 and ADAM2, also known as fertilin α and β, almost half of the ADAM members are specifically or predominantly expressed in the reproductive system, indicating an important function for these proteins in mammalian reproduction and fertilization [2,3].

All the proteolytically active members of the ADAM family share a highly conserved catalytic domain in which a Zn^{2+} ion is used for the nucleophilic attack of the substrate peptide bond. Within the catalytic domain, the Zn^{2+} ion is coordinated with three histidines and one glutamic acid of the consensus sequence HEXXHXXGXXH [2,3]. ADAMs are synthesized as inactive pro-protein precursors in which a pro-domain constrains the protease activity through a cysteine-switch mechanism. This mechanism is not only common among ADAMs but also shared by the majority of metzincins, such as matrix metalloproteinases (MMPs) and disintegrin metalloproteases with thrombospondin domains (ADAMTSs) [4]. The cysteine-switch is based on a pivotal cysteine contained within the conserved PRCGXPD motif of the pro-domain that coordinates the catalytic Zn^{2+} ion, thereby blocking the enzyme activity [2,3]. Pro-domain removal and activation of ADAMs, including ADAM10 and ADAM17, involve the action of furin and occur in the Golgi apparatus. This ensures ADAMs to be inactive in the ER and at early stages of the secretory pathway and prevents them from cleaving proteins in an unspecific and deregulated manner, which would have detrimental effects for the cell. In addition to the catalytic domain and pro-domain, ADAMs share a similar multidomain molecular structure that comprises a disintegrin domain, a cysteine-rich domain, a stalk region, a transmembrane domain, and a small cytoplasmic domain (Fig. 1) [2,3]. Specific functions of these domains, when known, will be discussed for separate ADAMs in the following sections.

2 ADAM17

2.1 ADAM17 structure

Since its discovery as the tumor necrosis α (TNFα) convertase, ADAM17 has been one of the most investigated ADAMs, not only in the context of inflammation but also in other diseases, such as cancer. Its multidomain structure is similar to that of other ADAMs [5]. In addition to the disintegrin domain, which is involved in molecular interactions with integrins, and the cysteine-rich domain, before the transmembrane domain, ADAM17 possesses a membrane proximal domain (MPD) and a short stalk domain called CANDIS (Conserved Adam seventeeN Dynamic Interaction Sequence), which play an important role in regulating the activity of the protease [6,7]. Indeed, MDP and the CANDIS can regulate ADAM17 conformation, and

Metalloenzymes. https://doi.org/10.1016/B978-0-12-823974-2.00016-4

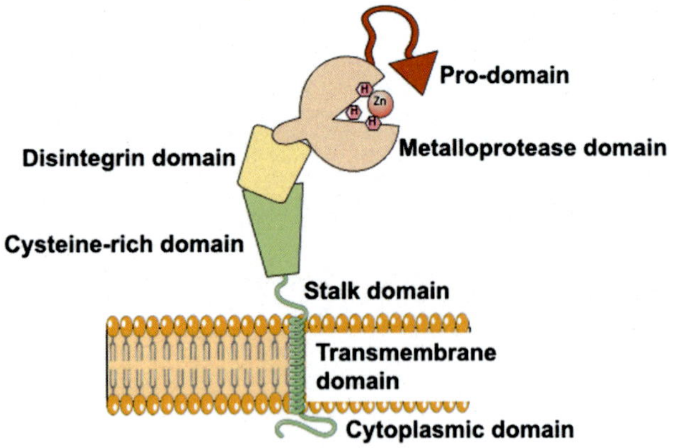

FIG. 1 Schematic representation of ADAM multidomain structure.

therefore its activity, by forming electrostatic interactions with the plasma membrane. ADAM17 interacts through the transmembrane domain with its essential cofactors iRhom1 and iRhom2 [8,9]. These seven-membrane spanning proteins emerged to control ADAM17 maturation through the secretory pathway, activation in response to specific stimuli and address its activity toward some, but not all ADAM17 substrates [10–14] Finally, it has recently emerged that the intracellular domain of ADAM17 has an important role in maintaining the stability of the protease.

2.2 ADAM17 functions

The function of a protease is strictly connected to the collection of substrates that it cleaves. Thus, given that over 80 different substrates have been identified for ADAM17 to date, it is clear how this enzyme could play a role in a large number of biological processes, spanning from development to immunity (Table 1). Herein, we will mostly encompass functions that have been proven in vivo and in disease.

2.2.1 EGFR-signaling

ADAM17 is expressed in all mammalian tissues [72]. Its genetic ablation leads to in utero or perinatal death (few days after birth), indicating for the protease a clear function in development [72]. In addition, ADAM17-deficient mice show open eye lids at birth. This phenotype resembles that shown by epidermal growth factor receptor (EGFR)-null mice, indicating that ADAM17 has an essential function in regulating the EGFR signaling pathway. Indeed, ADAM17 controls shedding of EGFR ligands, such as EGF, amphiregulin, transforming growth factor α (TGFα), and heparin binding (HB)-EGF [3,5,72]. Moreover, in line with the ADAM17-deficient mouse, a rare loss of function mutation in ADAM17 identified in human caused hair and epithelial defects associated with absence of EGFR-signaling, in addition to the inability of mounting efficient inflammatory responses, which is another distinctive feature of ADAM17-deficient mice and described in the following section [92]. Due to the early mortality of ADAM17-deficient mice, most of the in vivo functions of ADAM17 have been elucidated by using conditional knockouts, or a hypomorphic ADAM17 mouse (ADAM17$^{ex/ex}$), which is viable, although it develops a number of defects in the eye, heart, and skin, as a consequence of impaired EGFR signaling [93]. Interestingly, the hypomorphic ADAM17 mouse expresses reduced levels of ADAM17 transcripts and protein and behaves like a full ADAM17-null mouse when specifically challenged with cues leading to the activation of the protease. ADAM17 knockout mice displayed defects in epidermal barrier integrity and in endochondral ossification, when the protease was ablated in keratinocytes and chondrocytes, respectively [94,95]. Both phenotypes were phenocopied by ablation of EGFR and rescued by TGFα, indicating for the ADAM17/EGFR axis a pivotal role in these processes.

2.2.2 ADAM17 in immune responses

ADAM17 was first identified as the TNFα converting enzyme (and therefore also known as TACE), which means being the enzyme responsible for cleavage of TNF, a cytokine that is synthesized as a transmembrane precursor and requires to be released from the cell surface in order to elicit its proinflammatory potential [54,55]. Thus, since its discovery, a role of ADAM17 in inflammation has been clear, and corroborated over time by a number of experiments in vivo. As expected from its role in releasing TNF, specific ablation of ADAM17 in myeloid cells led to strong protection from endotoxin shock lethality [96]. Interestingly, ADAM17 function in inflammation is not limited to its ability to release TNF. Indeed, several key players in inflammation are regulated by the protease, including the TNF receptors TNFR1 and 2, and L-selectin [97,98]. While TNFR1 transduces the proinflammatory cascade initiated by binding soluble TNF (sTNF), and its ADAM17-mediated cleavage antagonizes its activation and leads to inflammation resolution, TNFR2 triggers an antiinflammatory cascade by binding the transmembrane pro-TNF, and its ADAM17-mediated cleavage has opposite consequences to that of TNFR1 cleavage [99]. In addition, ADAM17 loss in leukocytes prevents shedding of L-selectin and this was shown to increase neutrophil recruitment and higher pathogens clearance in a murine model of *Escherichia coli*-mediated sepsis [98]. Moreover, ADAM17 sheds the ectodomain of IL-6R, thus being a main regulator of IL-6 classic and trans-signaling pathways that promote resolution of inflammation or activation of immune responses, respectively [100]. Today, it is clear that the function of ADAM17 in inflammation goes much

TABLE 1 List of validated ADAM17 substrates.

Cytokine	Cell-to-cell communication	Signaling-receptor	Cell adhesion	Cellular transport	Enzyme	Others
CSF-1 [15]	Amphiregulin [16,17]	Axl [18]	ALCAM [19]	IGF-2R [20]	ACE-2 [21,22]	APLP-2 [23]
CX3CL1 [24]	DLL-1 [25]	CD16 [26]	CD44 [19]	LDL-R [27]	Carbonic hydrolase 9 [28]	APP [29,30]
IL-8 [31]	Epigen [32]	CD163 [33]	Collagen XVII [34]	LRP-1 [35]	Klotho [36]	Preadypocyte factor [37]
KL-1 [38]	Epiregulin [17]	CD30 (TNFRSF8) [39]	Desmoglein-2 [19]	SORCS-1 [40]	NPR-1 [41]	PMEL-17 [42]
KL-2 [38]	HB-EGF [17,43]	CD40 (TNFRSF5) [44]	EpCam [19]	SORCS-3 [40]		PrP [45]
Lymphotoxin-α [46]	ICOS-L [47]	CD89 [48]	GP-1ba [49]	SORL-A [40]		Syndecan-1 [50]
RANKL [51]	IL-15R [52]	EPCR [53]	GP-5 [49]	SORT-1 [40]		Syndecan-4 [50]
TNFα [54,55]	IL-1R2 [56]	ErbB4 [57,58]	GP-6 [49]	TREM-2 [59]		Vasorin [60]
	IL-6R [61]	GHRH receptor [62,63]	ICAM-1 [64]			
	Jagged [65]	M-CSFR [66]	JAM-A [67]			
	LAG-3 [68]	Notch-1 [65]	L1-CAM [69]			
	MIC-A [70]	NRP-1 [71]	L-selectin [72]			
	MIC-B [70]	NTRK1 [73]	LYPD3 [74]			
	NRG-1 [75]	PTK7 [76]	MUC-1 [77]			
	PD-L1 [78]	PTPRF [79]	NCAM [80]			
	TGFα [17,72,81–83]	PTPRZ [79]	Nectin-4 [84]			
	TIM-1 [85]	SEMA-4D [86]	SynCAM-1 [87]			
	TIM-4 [85]	TNF-R1 [88]	Thrombospondin-1 [89]			
		TNF-R2 [88]	VACM-1 [90,91]			
		VEGF-R2 [71]				

beyond the initially identified activation of TNF and pro-inflammatory responses. ADAM17 can trigger immune responses, as well as their resolution, and it is evident that ADAM17 activity must be timely and spatially regulated in order to avoid aberrant inflammation and culminate in pathological conditions.

2.3 ADAM17 in diseases

2.3.1 Rheumatoid arthritis and osteoarthritis

Aberrant TNF release is associated with a number of inflammatory conditions, such as rheumatoid arthritis, highlighting the pivotal role that ADAM17 plays in the development of such diseases [101]. Ablation of ADAM17 in the myeloid cells of a murine model of rheumatoid arthritis (RA) protected against development of the disease to a similar extent as TNF loss [102]. Moreover, the RA pathology of a knock-in mouse carrying an uncleavable form of TNF was further improved compared to that of ADAM17-deficient or TNF-deficient mice [103]. Altogether, these studies clearly displayed the ADAM17 capability of switching on and off inflammation and contribute to the progression of rheumatoid arthritis when its activity is not accurately balanced. For this reason, ADAM17 has been historically considered an attractive target for the development of therapeutics to cure arthritis.

Although it is not considered a canonical inflammatory disease, osteoarthritis is also characterized by aberrant TNF release, which, in turn, enhances the production of cartilage-degrading proteases in the joint, including collagenases and aggrecanases. In agreement, loss of ADAM17 in both ex vivo and in vivo models of osteoarthritis prevented cartilage breakdown, a hallmark of osteoarthritis, thus indicating a key role for ADAM17 in the development of the disease [104].

2.3.2 Lung and kidney pathology

Similar to rheumatoid arthritis, acute lung inflammation is characterized by high levels of TNF [105]. ADAM17 ablation in leukocytes reduced levels of alveolar TNF and had protective effects in a model of lung injury induced by LPS inhalation, including decreased inflammation and neutrophil infiltration [106]. Moreover, ADAM17 has proven to contribute to the progression of a number of pathologies of the lung, such as asthma, chronic obstructive pulmonary disease (COPD), and cystic fibrosis (CF). Differently from lung inflammation these diseases are driven by aberrant ADAM17-mediated activation of the EGFR pathway, which, in turn, leads to airway epithelial cell wound healing, abnormal airway proliferation, and progressive lung tissue scarring [107]. Similarly, as a consequence of sustained EGFR activation, fibrosis is developed in kidneys after injury, and both ADAM17 and its substrate amphiregulin (AREG) positively correlate with fibrosis in acute kidney injury and chronic kidney disease [108]. Moreover, ADAM17 was shown to play a detrimental role in the development of fibrosis in murine models of acute or chronic kidney injury, further indicating ADAM17 as a therapeutic target for kidney disease.

2.3.3 Atherosclerosis

The role of ADAM17 in atherosclerosis is quite controversial. While genetic ablation of TNF reduces atherosclerosis in experimental mouse models, either pharmacological TNF blockage or ADAM17 ablation were not protective against development of the disease [109]. In fact, when the ADAM17 hypomorphic mouse ($ADAM17^{ex/ex}$) was crossed with a mouse model of atherosclerosis (the Ldlr−/− mouse), it displayed augmented atherosclerotic lesions than its wild type control [110].

2.3.4 Ulcerative colitis and pancreatitis

Similar to atherosclerosis, ulcerative colitis (UC) is characterized by high levels of TNF and anti-TNF therapy is effectively used in its treatment [111,112]. However, the role of ADAM17 in the progression of UC seems to be more complex than solely mediating TNF release. For example, ADAM17 was found to be protective, rather than detrimental, in a model of dextran sulfate sodium-induced colitis, and this was due to its ability to activate EGFR signaling, which is pivotal for epithelial proliferation and to restore functions of the intestinal barrier [113]. Conversely, evidence supports a negative role for ADAM17 in pancreatitis. Expression of ADAM17 and activation of the IL-6 trans-signaling pathway positively correlate with pancreatitis in patients. Moreover, loss of ADAM17 reduced necrosis and fibrosis of the pancreas, together with other features of the pancreatitis' pathology in a murine model of the disease [114].

2.3.5 Alzheimer's and central nervous system

Growing evidence indicates that neuroinflammation is strongly associated with neuron loss in Alzheimer's and other neurodegenerative diseases [115]. In the brain neuroinflammation is characterized by activation of glial cells, including resident microglia and astrocytes. These cells begin to produce TNF, which has damaging effects in neurodegeneration. In support of this evidence, anti-TNF therapy leads to clinical improvement in patients with Alzheimer's and iRhom2, the essential ADAM17 cofactor that is required for its maturation and activity, has been identified as a genetic risk factor for the disease [116,117]. However, despite this evidence, the role of ADAM17 in Alzheimer's is not yet clearly understood, and comprehensive studies to assess its contribution to neurodegeneration are still missing. ADAM17 was reported to be the "stimulated α-secretase" of the amyloid precursor protein (APP), i.e., the protein responsible for the initiation of the amyloid cascade in Alzheimer's [118]. As such, ADAM17 triggers the protective "nonamyloidogenic pathway" by cleaving APP. This makes evident how complex the function of ADAM17 in Alzheimer's is, as the protease is able to initiate both neurotoxic and neuroprotective molecular pathways.

TNF release and neuroinflammation are known to play a negative role in spinal cord injury (SCI). By subjecting the $ADAM17^{ex/ex}$ mouse to SCI, it was found that ADAM17 deficiency significantly increased nerve recovery by promoting scavenging of apoptotic cells and axon pruning

[119]. Ablation of ADAM17 in microglia, but not in other cells, led to similar results as those seen in the hypomorphic ADAM17$^{ex/ex}$ mouse. This further confirmed the detrimental role of microglial ADAM17 in neuroinflammation, mainly elicited by sustaining TNF release.

2.3.6 Cancer

ADAM17 has historically been associated with cancer progression for its capability to trigger the epidermal growth factor receptor (EGFR) pathway by shedding EGFR ligands [82]. Nowadays, it is quite clear that its role in cancer is much more complex, and the outcome of ADAM17 activity can depend on different tumors and stages of the disease. High levels of soluble EGFR ligands and elevated expression of ADAM17 correlate with poor cancer prognosis [120]. Ablation of ADAM17, and subsequent reduced shedding of the EGFR ligands amphiregulin and TGFα, diminished progression of colon and breast cancer, respectively [121,122]. Additional to its role in promoting cancer cell proliferation by activation of EGFR signaling, ADAM17 on endothelial cells has been recently shown to have a pivotal role in tumor cell extravasation and metastasis. Endothelial cell necroptosis is a crucial driver of this process, which is activated by the ADAM17-mediated shedding of TNFR1 [123]. In agreement, genetic loss of ADAM17 or its pharmacological inhibition was shown to prevent metastases formation and cancer progression [120]. More recently, it has emerged a role of ADAM17 in modulating immune evasion of cancer cells by shedding the programmed death-ligand 1 (PD-L1). PD-L1 is highly expressed in several cancers where its primary role is to suppress adaptive immune responses and favor cancer immune evasion. For this reason, PD-L1 is a main target of immunotherapy, where monoclonal antibodies against PD-L1 are used to block its inhibitory activity and boost the immune response against cancer cells. Interestingly, ADAM17-mediated shedding of PD-L1 is detrimental in immunotherapy as increased soluble PD-L1 promotes resistance to anti-PD-L1 inhibitors, thereby lowering the efficacy of the treatment [124]. Additionally, ADAM17 promotes immune evasion by releasing CD16A, a human IgG Fc receptor expressed on NK cells that activate their cytotoxic potential [125]. Thus, CD16A shedding reduces its levels on the NK cell membrane and, subsequently, CD16A signaling, and cytokine production.

3 ADAM8

3.1 ADAM8 structure

ADAM8 (also referred to as MS2 or CD156a) is one of the proteolytically active membrane-anchored proteases belonging to the family of ADAMs [2,3]. Its gene is located in chromosome 10q26.3 and encodes an 824 amino acid protein [126]. Similar to the other active ADAMs, ADAM8 structure consists of a pro-domain, a metalloproteinase domain, a disintegrin domain, a cysteine-rich domain, an epidermal growth factor (EGF)-like repeat-containing domain, a transmembrane region, and a cytoplasmic tail. ADAM8 is synthetized as an inactive precursor (proADAM8), in which the pro-domain inhibits its catalytic activity. Differently from other ADAMs that required furin to get the pro-domain cleaved off for the activation, the pro-domain of ADAM8 is removed through an autocatalytic mechanism [127]. Moreover, the catalytic domain is insensitive to tissue inhibitors of metalloproteinases (TIMPs), the endogenous inhibitor proteins that regulate most ADAMs by forming 1:1 complex with the active enzymes [128]. In addition to its proteolytic function played by the metalloproteinase domain, ADAM8 mediates cell-cell and cell-ECM interactions via the disintegrin/cysteine-rich domains, the latter also important for substrate recognition in concert with the EGF-like repeat-containing domain [129]. Although it has not yet been confirmed by experimental data, in silico modeling identified SH3 binding sites in the cytoplasmic tail of ADAM8 that could mediate the interaction with intracellular proteins involved in signal transduction, trafficking, and intracellular structure [126,130].

3.2 ADAM8 functions

Compared to ADAM17, the substrate repertoire of ADAM8 is much more restricted, with less than 10 substrates being so far validated for the protease (Table 2). Nevertheless, a number of pathophysiological processes regulated by ADAM8 have been described, especially in the context of inflammation and cancer.

3.2.1 Inflammation and immune responses

ADAM8 was originally identified in murine macrophages and macrophage-like cell lines by screening a cDNA library that enriched genes upregulated in response to LPS [130]. During mice development, ADAM8 is localized in the

TABLE 2 List of validated ADAM8 substrates.

Cytokine	Signaling-receptor	Cell adhesion	Others
CD30-L [131,132]	CD23 [131,132]	CHL-1 [133]	ADAM8 (prodomain) [127]
	TNF-R1 [134]	L-selectin [135]	APP [136]
		PSGL-1 [137]	PrP [138]

trophoblast, then, at later stages, it is predominantly expressed in gonads, thymus, cartilage and bone, brain and spinal cord, lymphatic, and venous vessels [139]. ADAM8 was found to play a role in bone and cartilage development, as its interaction with integrin $\alpha9\beta1$ regulated osteoclast formation and function [140]. In humans, ADAM8 expression is high in immune cells (excluding T cells), indicating its essential role in tuning inflammation and immune responses [126]. Being a "sheddase," ADAM8 mediates ectodomain shedding of several proteins, including itself, to contribute in a wide range of processes both physiological and pathological [134]. Adam8$^{-/-}$ mice showed no evident abnormal phenotype neither in developmental stages nor in adulthood [139]. As mentioned above, ADAM8 is preferentially expressed on hematopoietic cell lines and many of the identified substrates are leukocytic surface proteins participating in the inflammatory response. For this reason, ADAM8 levels positively correlate with the burden of autoimmune disorders. In knee joints of patients with RA were found both membrane-anchored and soluble ADAM8. Here the sheddase was shown to cleave L-selectin, thus allowing neutrophil infiltration and intense local inflammatory response [135]. The low-affinity IgE receptor CD23 was also found to be shed by ADAM8, thus the metalloprotease has been linked with allergic airway inflammation and lung inflammation, as asthma [131]. By inducing asthma through different allergens, peribronchial and perivascular inflammatory cells and bronchiolar epithelial cells showed an upregulation of ADAM8 levels, causing lung inflammation [141]. Surprisingly, an ADAM8 inhibitor peptide attenuated airway hyperresponsiveness to allergens stimulation in a murine model of asthma [142]. In acute lung inflammation, ADAM8 is required on leukocytes to promote chemotaxis and transepithelial migration [143]. Moreover, ADAM8 enhanced invasiveness of neutrophils through cleavage of P-selectin glycoprotein ligand-1 (PSGL-1), facilitating infiltration, and regenerative myogenesis [137].

Finally, it was uncovered a role for ADAM8 in osteoarthritis pathology and chronic low-back pain. This was due to the protease capability to cleave fibronectin and induce cartilage catabolism [144,145].

3.2.2 Neurodegeneration

In line with its function in immune responses, by using a murine model of neurodegeneration it was demonstrated that ADAM8 levels increased in response to TNF in areas of brain and spinal cord involved in neurodegenerative processes [2,146]. In these models, ADAM8 reduced TNF signaling through TNFR-1 shedding [134]. ADAM8 was proven to have an α-secretase activity, and therefore able to cleave the amyloid precursor protein (APP), as well as the prion protein (PrP), another protein that is closely associated with neurodegeneration [138,147]. In addition, in vitro studies demonstrated ADAM8 shedding of myelin basic protein (MBP) and the neural cell adhesion molecule "close homologue of L1" (CHL-1), suggesting a role for the sheddase in myelination, and regulation of neurite outgrowth and neuronal cell death [133,147,148]. In conclusion, all these evidences suggest that ADAM8 plays an important role in the central nervous system neuroinflammation.

3.2.3 Cancer

The catalytic activity of ADAM8 is associated with different tumor types, mostly in promoting proliferation and migration. Increased ADAM8 levels were found in both human primary brain tumors and tumor cells, positively correlating with invasiveness and poor prognosis [149]. Similarly, ADAM8 higher expression in human triple-negative breast cancer (TNBC) was linked to an aggressive phenotype developing brain metastases [149]. ADAM8 knockdown on TNBC repressed tumor expansion. More recent studies evaluated ADAM8 in gastrointestinal cancers (GI), such as gastric cancer [150], pancreatic cancer [151], hepatocellular carcinoma [152], and colorectal carcinoma [153]. These showed that an upregulation of ADAM8 transcript and protein enhanced their migration and invasive activity, further suggesting ADAM8 as a potential prognostic biomarker in some of GI cancers [154,155]. Schlomann et al. identified a specific ADAM8 inhibitor peptide that reduced invasiveness of pancreatic carcinoma in vivo [155]. An ADAM8 role in vascular remodeling and angiogenesis was observed not only in cancer but also in spinal cord injury and recovery from myocardial infarction [137,156].

4 ADAM10

4.1 ADAM10 structure

ADAM10 is ubiquitously expressed in mammals. Similar to other ADAMs, ADAM10 is synthesized as an inactive protease that gets activated by removal of its pro-domain, which occurs in the Golgi apparatus by action of furin, and shows a multidomain structure comprising a conserved metalloprotease catalytic domain, a disintegrin domain, a cysteine-rich domain, a transmembrane domain, and an intracellular domain [157]. ADAM10 was reported to be the receptor for the *Staphylococcus aureus* alpha-hemolysin (Hla), a cytotoxin that drives staphylococcal diseases. Although it is not clear if the disintegrin domain of ADAM10 could bind integrins, Wilke and colleagues reported the association between Hla and $\beta1$-integrin, and raised the hypothesis that the Hla-ADAM10 complex may interact with integrins and alter integrin-mediated signaling [158]. The cytoplasmic domain of ADAM10 is involved in trafficking and endocytosis of the protease [159,160]. TspanC8 tetraspanins, a group of 6 four-membrane-spanning proteins, have recently emerged as essential regulators of ADAM10 maturation through the secretory pathway and activation [161–163]. Differently from the ADAM17/iRhoms interaction, which occurs through the transmembrane domain of the protease, TspanC8 tetraspanins interact with the extracellular region of ADAM10 [161].

4.2 ADAM10 functions

ADAM10 can cleave over 50 different transmembrane proteins, spanning from adhesion molecules to signaling receptors, thereby regulating a vast variety of biological processes (Table 3) [158]. ADAM10 cleavage is a prerequisite for regulate intramembrane proteolysis (RIP), a molecular process needed to maintain the turnover of transmembrane proteins and transmit information to the nucleus [200]. RIP comprises two sequential proteolytic cleavages of transmembrane proteins. The first cleavage, mediated by ADAM10, releases the extracellular domain of the protein and maintains a smaller peptide fragment anchored to the membrane. The second intramembrane cleavage, which is mediated by the γ-secretase, liberates a small peptide fragment, known as the intracellular domain (ICD), that is ultimately translocated into the nucleus where it functions as a transcription factor and regulates gene expression. A well-investigated example of ADAM10-mediated RIPping is that of Notch [188]. Notch has a fundamental role in development. Upon binding of its ligands, including Jagged-1 (JAG1), Jagged-2 (JAG2), and Delta-1 (DLL1), Notch is cleaved by ADAM10 and undergoes RIPping, and ultimately activates genes for differentiation, proliferation, and cell-fate determination. Notch ablation in mouse is embryonically lethal, and ADAM10 loss phenocopies Notch ablation in that ADAM10-null mice die embryonically (E9.5) and show major developmental

TABLE 3 List of validated ADAM10 substrates.

Cytokine	Cell-to-cell communication	Signaling-receptor	Cell adhesion	Cellular transport	Enzyme	Others
CX3CL1 [24,164]	Betacellulin [17]	CD23 [132,165,166]	CD44 [167]	LRP-1 [35]	Klotho [36,168]	APP [169–171]
CX3CL16 [24,172]	C-Met [173]	CD30 (TNFRS8) [174]	CDH-1 [175]		MEPRIN β [176]	PrP [177,178]
FAS-L [179,180]	DLL-1 [181,182]	ErbB2 [183]	CDH-2 [184]			APLP-2 [23]
RANKL [185]	EGF [17]	Notch-1 [186–188]	CDH-5 [189]			ITM2b [190]
	LAG-3 [68]	TSHR [191]	Collagen XVII [34]			
	NG-2 [192]		Desmoglein-2 [19]			
	NRG-1 [75]		Ephrin-A2 [193]			
			Ephrin-A5 [194]			
			L1-CAM [69,195]			
			NCAM [196]			
			Nectin-1 [197]			
			NGL-1 [198]			
			PCDHγC3 [199]			

defects [187,200]. Recently, it has been demonstrated that a loss of ADAM10 proteolytic activity by either inhibition or loss of function mutations induces removal of the protease from the cell surface and the whole cell, and functionally means a loss of ADAM10 substrate shedding [201].

4.3 ADAM10 in diseases

ADAM10 activity has been widely investigated in Alzheimer's disease, as it emerged as the physiologically relevant α-secretase that triggers the protective "nonamyloidogenic" processing of APP [171]. In addition, due to its ubiquitous expression, ADAM10 has been associated with a number of different pathological conditions, including inflammatory diseases and cancer [202].

4.3.1 ADAM10 in Alzheimer's and other brain disorders

ADAM10 has been extensively investigated in the brain, not only for its link to Alzheimer's but also for its function in neurogenesis and neuronal migration, elicited by cleavage of its substrate Notch. Moreover, ADAM10 controls axon sprouting and synapsis formation through the cleavage of a vast number of key players involved in these processes, such as ephrins and ephrin receptors, the neuronal and the neural glial-related cell adhesion molecule (NCAM and NrCAM), neuroligin (NLGN) and neurexin (NRXN), and many others [157].

One of most investigated ADAM10 substrates in the brain is the amyloid precursor protein (APP), whose processing triggers the amyloid cascade leading to Alzheimer's. Based on the amyloid hypothesis, Alzheimer's pathology is initiated by the formation of Aβ oligomers, which trigger neuroinflammation, deposition of neurofibrillary tangles, neuronal cell death and ultimately dementia [203]. BACE1 was identified as the protease responsible for the cleavage of APP leading to Aβ formation and initiation of the amyloid cascade [204]. BACE1 cleavage of APP releases its ectodomain and generates a membrane anchored fragment, known as C99. Such a cleavage is followed by γ-secreatse's, that liberates the APP intracellular domain (AICD) and a small 40–42 amino acid long peptide in the extracellular milieu, the Aβ peptide. ADAM10 competes with BACE1 for APP cleavage [118]. Differently from BACE1, though, its cleavage generates a shorter membrane anchored fragment (the C83), which does not give rise to a toxic Aβ peptide when cleaved by the γ-secretase. Thus, ADAM10 has been historically considered protective in Alzheimer's disease and a valuable therapeutic target for its ability to trigger the protective non amyloidogenic processing of APP. In agreement, mice overexpressing ADAM10 in neurons of a mouse model of Alzheimer's showed reduced Aβ plaques and improved the pathology, and acitretin, a drug for the therapy of psoriasis that promotes overexpression of the protease, increased the release of soluble APPα in the cerebrospinal fluid of Alzheimer's patients [205,206].

In addition to APP, ADAM10 was shown to be the physiological sheddase of the cellular prion protein (PrPc) [177,207]. PrPc can undergo a conformational change into a pathological and misfolded form (PrPsc). PrPsc can spread, producing neuroinflammation and eventually neurodegeneration, two major features of prion disease. ADAM10 was shown to play a dual role in prion disease. On the one hand, ADAM10 releases PrPc, preventing its conversion into the toxic PrPsc form. Conversely, by shedding PrPsc, ADAM10 contributes to its spreading and progression of the disease [178].

ADAM10 was found to play a role in the development of another neurodegenerative disease, Huntington's disease. Levels of ADAM10 augmented in a mouse model of Huntington's disease, especially in the postsynaptic densities, a cytoskeletal specialization of the postsynaptic membrane where glutamate-receptor channels are clustered [208]. Here, ADAM10 provoked excess cleavage of N-cadherin and synapses loss, and its pharmacological inhibition or genetic ablation rescued the synaptic function, thus ameliorating the cognitive deficit in these mice [209,210].

4.3.2 Immune disorders

ADAM10 is highly expressed in the bone marrow and lymphoid tissues, where it cleaves a number of immune proteins, including cytokines and chemoattractant, such as TNF, IL-6R, CXCL16, and CX3CL1 [202]. As a consequence, its role in immunity has been largely investigated as a potential drug target for a number of immune disorders.

Rheumatoid arthritis. ADAM10 was found to play a major role in development of rheumatoid arthritis by mediating the shedding of the low-affinity immunoglobulin E (IgE) receptor CD23 [165,211]. CD23 is indeed a key mediator of the allergic response and antigen presentation of IgE antigen complexes, and its cleavage from B cells by ADAM10 generates a soluble form of CD23 (sCD23) that activates immune cells to produce TNF and other pro-inflammatory cytokines, thereby contributing to the progression of the disease [212]. As well as in immune cells, ADAM10 expression is high in endothelial cells in RA [211]. In these cells, ADAM10 can cleave a number of adhesion molecules (e.g., JAM-A and JAM-C), cytokines and chemokines (CXCL16 and CX3CL1), which might favor the infiltration of immune cells and inflammatory process. Altogether, this evidence clearly suggests that ADAM10 inhibition could have a favorable outcome in RA.

Psoriasis. ADAM10, which is upregulated in psoriatic lesions, slowed down progression of the disease in a murine model [213]. For this reason, Acitretin, a drug that stimulates the expression of ADAM10, has been approved for treatment of psoriasis [214]. Mechanistically, this is due to a negative regulation of Notch signaling in keratinocyte proliferation, which is a hallmark of the disease. Both ADAM10 and Notch ablation in the keratinocytes of a mouse model of psoriasis enhanced their hyperproliferation and subsequent lesions [215,216].

Systemic lupus erythematous (SLE). Differently from psoriasis, which is driven by both inflammatory and proliferative cues, SLE is a fully autoimmune disease, in which an antibody-induced malfunction of the immune system causes widespread inflammation and tissue damage. ADAM10 expression positively correlated with SLE, indicating for the protease a role in the pathology that is majorly driven by its ability to cleave Axl, a tyrosin-kinase receptor that stimulates cell proliferation and survival [124]. Axl is highly expressed in macrophages, and, upon binding of its ligand growth arrest-specific 6 (GAS6), suppresses the NF-kB pathway and therefore inflammation. ADAM10 mediated shedding of Axl worsened SLE, since its soluble form sAXL could function as a decoy receptor for GAS6, thus enhancing inflammation. Given its role, ADAM10 inhibition was proven beneficial in SLE [217].

Atherosclerosis. ADAM10 is commonly found highly expressed in atherosclerotic plaques, and Van der Vorst and colleagues found a dual role for macrophage ADAM10 in the development of the disease [218]. On the one hand, they showed that ablation of ADAM10 in the macrophages of a murine model of atherosclerosis led to an increased content of collagen in the atherosclerotic plaques, due to decreased activity of matrix metalloproteinases, including MMP-2, -9, and -13. This, in turn, inhibited the migration of macrophages toward the plaques. On the other hand, ADAM10 loss inhibited the production of pro-inflammatory cytokines, including TNF and IL-12. Thus, these data suggested a negative role for ADAM10 in the development of atherosclerosis and indicated it as a potential therapeutic target for the disease.

4.3.3 Cancer

Glioblastoma. Glioblastoma (GBM) is one of the most aggressive cancers that results from malignant astrocytes in the brain. The expression of ADAM10 is enhanced in GBM, and it negatively correlated with patient prognosis [219]. There are several mechanisms that are regulated by ADAM10 in glioblastoma that can favor the tumor aggressiveness. First, ADAM10 increased glioblastoma cell migration and invasion by shedding adhesion molecules, such as L1 and N-cadherin, and, in line, an ADAM10 inhibitor decreased metastasis [220,221]. The second mechanism involves neuroligin-3 (NLGN3) a protein commonly involved in synaptic transmission that has been linked with high grade GBM. NLGN3 is released from neurons by ADAM10, and it promoted glioma proliferation through the PI3K-mTOR pathway [222]. Again, pharmacological ADAM10 inhibition prevented the release of NLGN3 into the tumor microenvironment, thereby impairing growth of glioblastoma. Next, ADAM10 regulates NK activation and glioblastoma cell lysis. NK cells recognize glioblastoma through its activating receptor NKG2D that is capable of binding the stress receptors of MHC class I-like superfamily, such as MICA, MICB, and ULBP1. These molecules can be cleaved by ADAM10 in glioblastoma, thereby favoring immune evasion of the tumor [223]. Finally, ADAM10 expression correlated with that of Notch in patient specimens of glioblastoma, suggesting that ADAM10 could contribute to glioblastoma proliferation and progression by activating the Notch signaling pathway [224].

Lymphoma and leukemia. Hodgkin's lymphoma (HL) is an aggressive type of cancer that affects the lymphatic system and is characterized by Reed-Sternberg cells, i.e., large, abnormal lymphocytes that may contain more than one nucleus. Similar to glioblastoma, HL can evade the immunosurveillance of NK cells through cleavage of the stress receptors MICB and the ULBP2. This process is mediated by ADAM10, and its inhibition increased sensitivity of HL to NK cytotoxicity [225]. T-cell acute lymphoblastic leukemia (T-ALL) is a type of acute leukemia with aggressive malignant neoplasm of the bone marrow. One of the causes of T-ALL is the aberrant activation of Notch1, which induces the malignant transformation of T-cells [226]. Through Notch1 shedding, ADAM10 regulates T-cells proliferation and contributes to the progression of T-ALL [227,228]. Thus, the protease is considered as a valuable drug target also in this type of cancer.

Breast Carcinoma. The human epidermal growth factor receptor 2 (HER2) is a key player in the development of breast carcinoma. Elevated expression of HER2 occurs in 10%–20% of all invasive breast carcinomas, and its expression negatively correlates with prognosis [229]. Mechanistically, HER2, upon dimerization triggers through its tyrosin kinase domain proliferative and antiapoptotic signaling, and therefore must be tightly regulated to prevent cancerous cell growth. HER2 is the target of the monoclonal antibody trastuzumab, which inhibits dimerization of the receptor and therefore its signaling function [230]. ADAM10 contributes to breast cancer progression by a dual mechanism. First, its capability to shed HER2 ligands, such as betacellulin, contributes to its malignant activation [231]. Secondly, ADAM10 can shed HER2 itself, thereby producing a truncated form of HER2 that is constitutively active [183]. In addition, ADAM10 can also dampen effects of the trastuzumab therapy and increase trastuzumab resistance. Indeed, ADAM10 was found upregulated by trastuzumab treatment, and the ectodomains of HER2 generated by ADAM10 cleavage can act as decoy receptors for the drug, thereby decreasing its effect on the membrane tethered active receptor [232]. Altogether, this evidence show that ADAM10 inhibition could be beneficial in the treatment of breast cancer, especially in combination with the trastuzumab therapy.

5 ADAM inhibitors

The high structural homology among the ADAM and MMP catalytic domains, but in general among metzincin catalytic domains, is one of the factors that have hindered the development of selective inhibitors. The similarity in the binding sites, in sequence and overall shape, makes selective inhibitors a really challenging achievement. Nonetheless, some specific structural features such as the shape of S1′ specificity pocket, could be exploited to design selective ADAM inhibitors.

The X-ray crystal structure studies of ADAM8 [95], ADAM10 [233], and ADAM17 [234] have revealed a specific ellipsoid shape of catalytic domains. The catalytic zinc ion is arranged in an active site cleft (cleavage site), a specific region which separates two different subdomains: the "primed" *C*-terminal side, from the "unprimed" *N*-terminal side. In the unprimed side are positioned the classical S1, S2, and S3 pockets of the enzyme, corresponding to the P1, P2, and P3 substrate subsites, while on the primed side of the zinc the S1′, S2′, and S3′ pockets correspond to the subsites P1′, P2′, and P3′.

The majority of the small molecules developed so far as ADAM inhibitors, are constituted by a peptide-like backbone allocated on the "primed" side, a lipophilic substituent fitting the S1′ specificity pocket and a zinc-binding group (ZBG). Most reported inhibitors contain sulfonamide or amide groups representing hydrogen bond acceptors, aromatic or polyaromatic groups able to interact with the S1′ pocket and hydroxamic or carboxylic acids as ZBGs. The selectivity is obtained targeting definite structural elements, typical of a specific ADAM. For example, a selective ADAM17 inhibitor able to spare the other ADAMs and MMPs, requires functional interaction with the typical "L-shaped" S1′ pocket of ADAM17 binding site [235].

The most classical approach to design a metzincin inhibitor, relies on zinc metal chelation and an aromatic backbone, able to interact with the S1′ and/or S2′ pockets of the enzyme [236–239]. As a result, inhibitors containing Zn-chelator groups such as a hydroxamic acid present high activity in the nanomolar range, but often these molecules are nonselective, inhibiting either the nontarget ADAMs or MMPs, responsible for off-target toxicity.

Several reviews analyzed the ADAM inhibitors developed so far [240]. In this paragraph, we classify the ADAM inhibitors reported in literature during the last decade on the basis of their target enzyme, in: ADAM8, ADAM10, and ADAM17 inhibitors. In particular, synthetic small-molecules, natural compounds, and peptide-based inhibitors will be taken in consideration.

5.1 ADAM8 inhibitors

The nonessential function in physiological pathways, confers to ADAM8 the opportunity to become a promising novel target for cancer therapy due to a limited toxicity caused by ADAM8 inhibition. To date, the high homology of metzincin binding sites have hampered the development of selective ADAM8 inhibitors, and the majority of reported ADAM8 inhibitors present a broad-spectrum activity [241].

The catalytic domain of human ADAM8 co-crystalized in complex with the hydroxamic acid inhibitor batimastat gave important information about the structure of ADAM8 binding site [95].

ADAM8 binding site possesses a central five-stranded β-sheet and a catalytic Zn^{2+} ion, similarly to other ADAM family members, whereas the S1′ binding pocket and the disintegrin domain present differences which could be exploited for the design of ADAM8 selective inhibitors.

In contrast with other MMPs and ADAMs, ADAM8 is activated by autocatalysis and in order to show in vivo activity requires homophilic multimerization of at least two ADAM8 disintegrin domain monomers on cell surface [127]. The prevention of ADAM8 multimeric binding and the consequent autoactivation process is a novel approach to selectively target ADAM8.

Schlomann et al. reported the critical role of ADAM8 disintegrin domain in multimerization and through homology modeling identified the interactions implicated in ADAM8 dimerization [155]. The ADAM8 selective inhibition was achieved using an exosite inhibitor obtained by disintegrin domain structural modeling targeting the "KDX" motive in the exposed loop of ADAM8. The exosite inhibitor consisted of a short cyclic peptide **1**, called BK-1361 (Table 4), able to interact with the DIS domain and thus blocking ADAM8 activation.

BK-1361 exhibited a significant selectivity over other metzincins and nanomolar potency for ADAM8 measured by CD23 cell-based shedding assays. Several studies reported significant results of BK-131 in pancreatic ductal adenocarcinoma (PDAC). In PDAC cells, BK-1361 importantly reduced migration and invasiveness, and decreased the activation of MMPs and ERK1/2 pathway. Furthermore, in a in vivo model of xenograft pancreatic tumor, BK-1361 reduced tumor burden and metastasis of implanted pancreatic tumor cells, improving clinical symptoms and survival rates.

BK-1361 was evaluated also as therapeutic agent in asthma reporting promising results [242]. Chen and coworkers reported that treatment with BK-1361, in a murine asthma model attenuated airway responsiveness to methacholine stimulation, airway wall tissue remodeling

TABLE 4 Chemical structure and inhibitory activity (IC_{50}, nM) of ADAM8, ADAM10, and ADAM17 inhibitors.

Compd	Structure	ADAM8	ADAM10	ADAM17	Ref.
ADAM8 inhibitors					
1	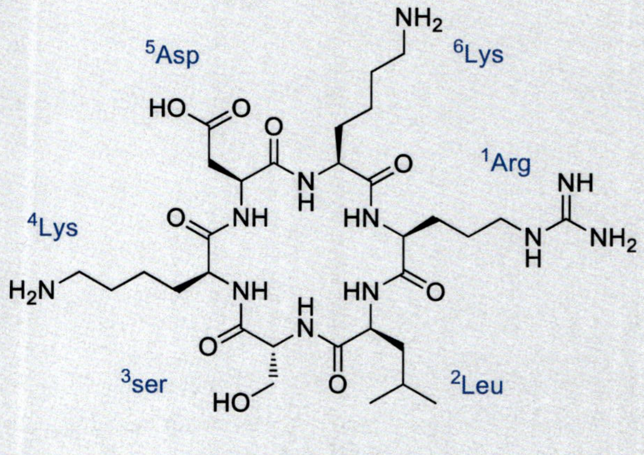 cyclo(RLsKDK)	182	–	–	[142,242–244]
2	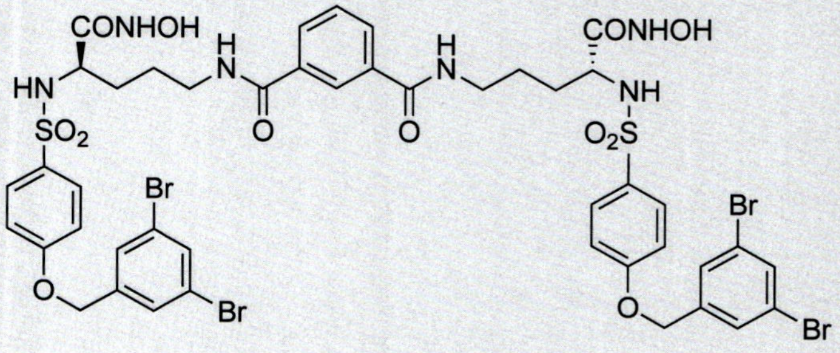	73	10,000	24	[245]
ADAM10 inhibitors					
3	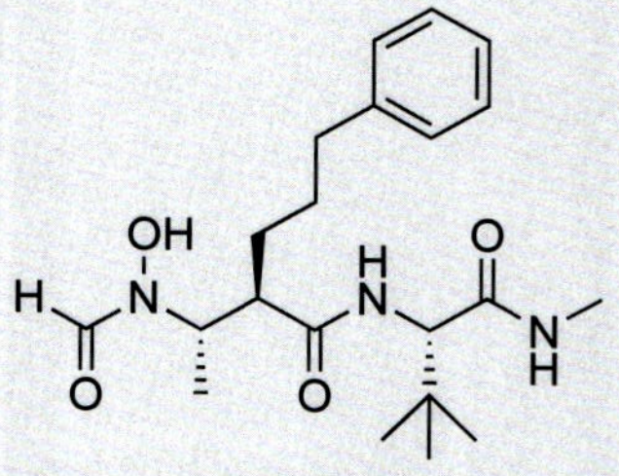 **GI254023X**	–	5.3	541	[246–248]

Continued

TABLE 4 Chemical structure and inhibitory activity (IC_{50}, nM) of ADAM8, ADAM10, and ADAM17 inhibitors—cont'd

Compd	Structure	ADAM8	ADAM10	ADAM17	Ref.
4	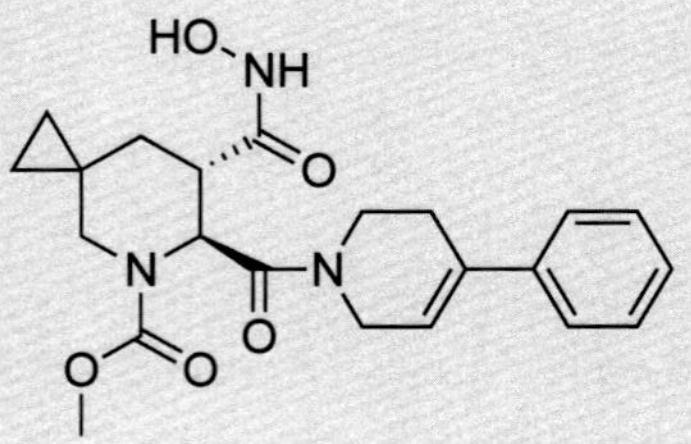 **INCB3619**	–	22	14	[249–251]
5	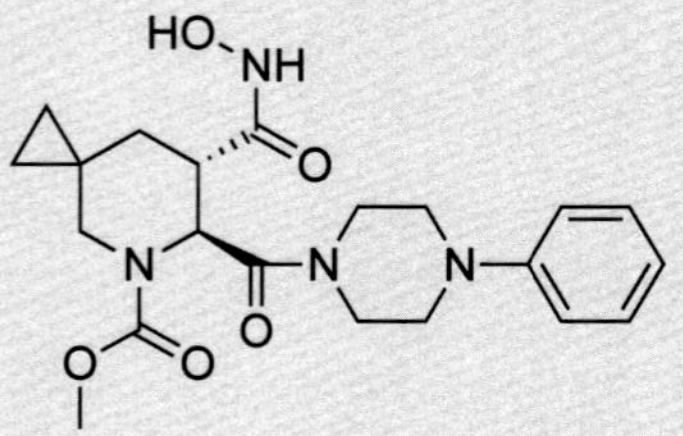 **INCB7839**	–	–	–	[249–253]
6	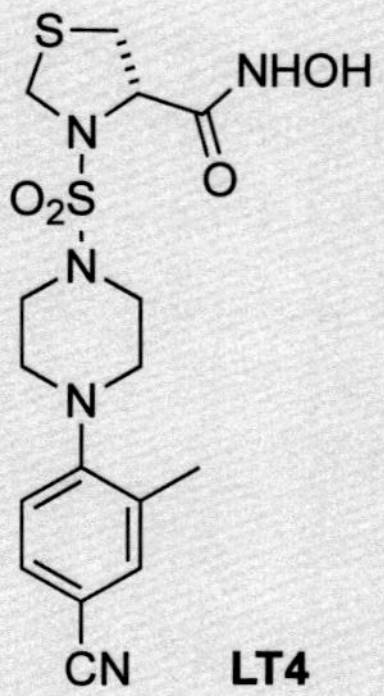 **LT4**	–	40	1500	[254,255]

7	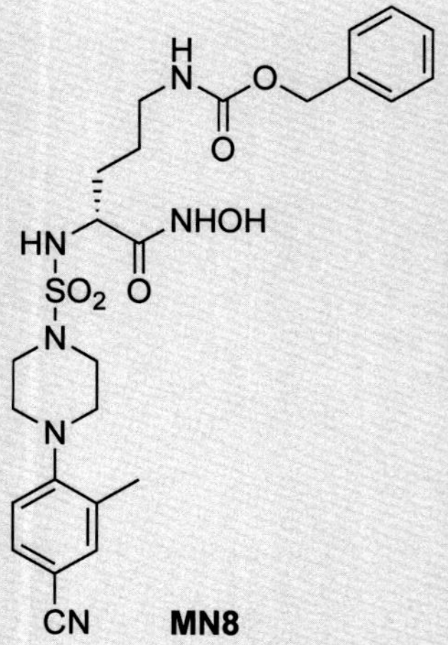 **MN8**	–	9.2	90	[254,255]
8	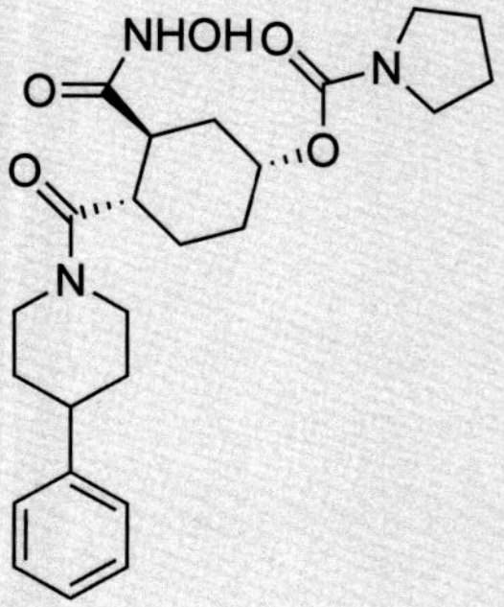	–	73 (Ki)	900 (Ki)	[256]
ADAM17 inhibitors					
11	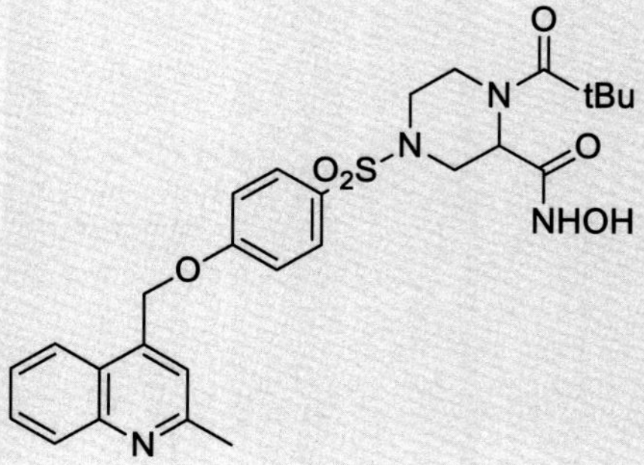	–	>10,000	12	[257]

Continued

TABLE 4 Chemical structure and inhibitory activity (IC_{50}, nM) of ADAM8, ADAM10, and ADAM17 inhibitors—cont'd

Compd	Structure	ADAM8	ADAM10	ADAM17	Ref.
12	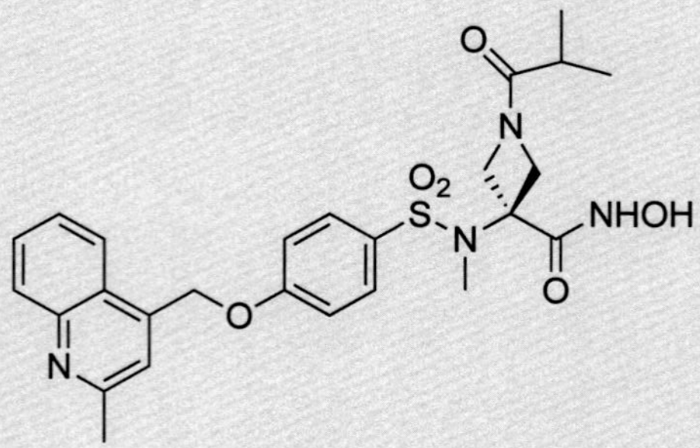	–	950	4	[258]
13	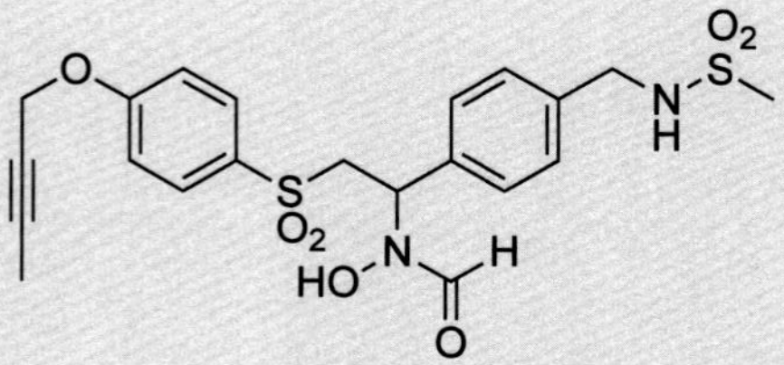 KP-475	–	748	11	[259]
14	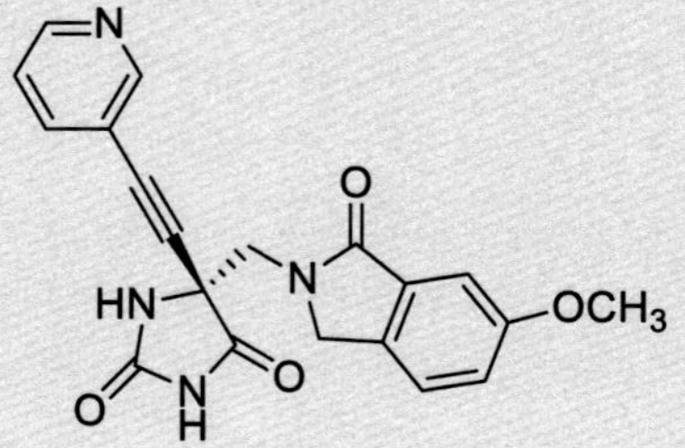	–	22 (*Ki*)	0.62	[260]
15	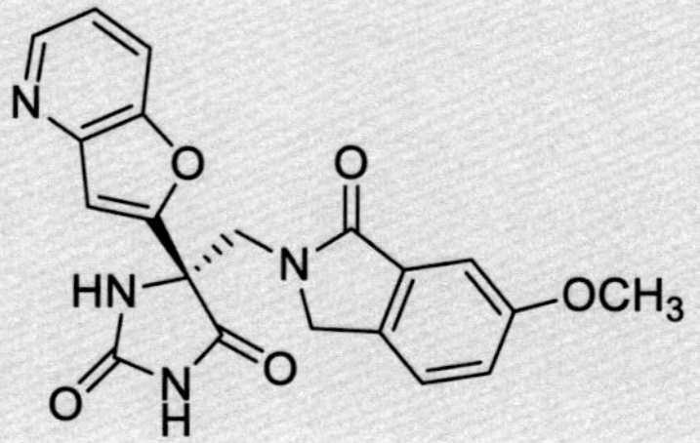	–	–	0.5 (*Ki*)	[261]

16	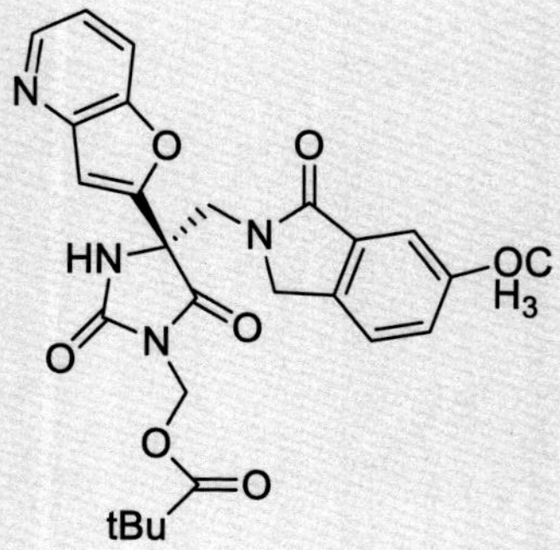	–	–	–	[262]
17	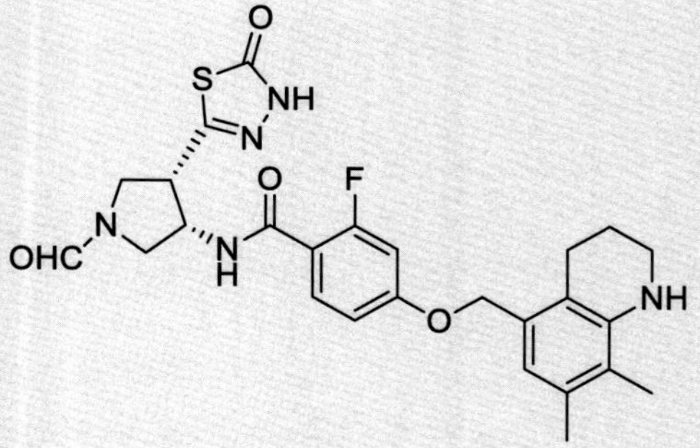	–	10,000	5.4	[263]
18	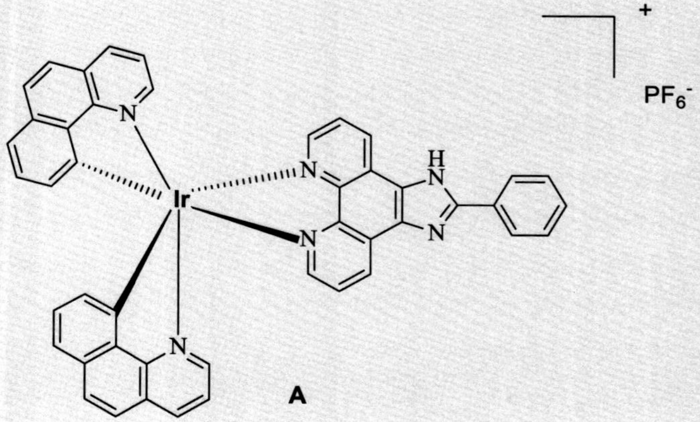	–	–	28,000	[264]

Continued

TABLE 4 Chemical structure and inhibitory activity (IC_{50}, nM) of ADAM8, ADAM10, and ADAM17 inhibitors—cont'd

Compd	Structure	ADAM8	ADAM10	ADAM17	Ref.
19	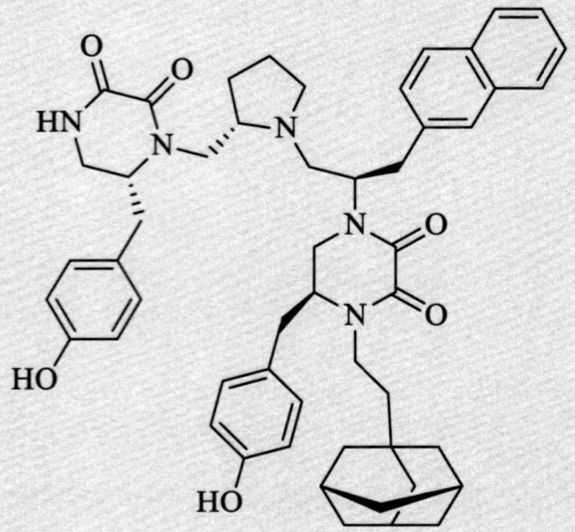	–	>100,000	4200	[265]
20	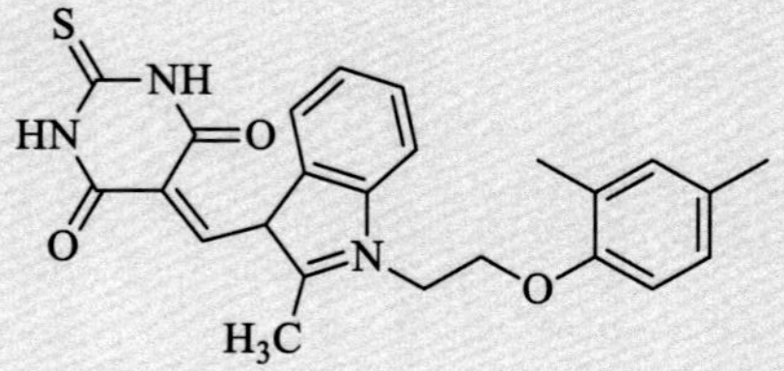 ZLDI-8	–	–	72,000	[266–268]
21	Hxm-Phe-Arg-Gln	–	240 (Ki)	47 (Ki)	[269]
22	Hxm-Phe-Ser-Asn	–	670 (Ki)	92 (Ki)	[269]
23	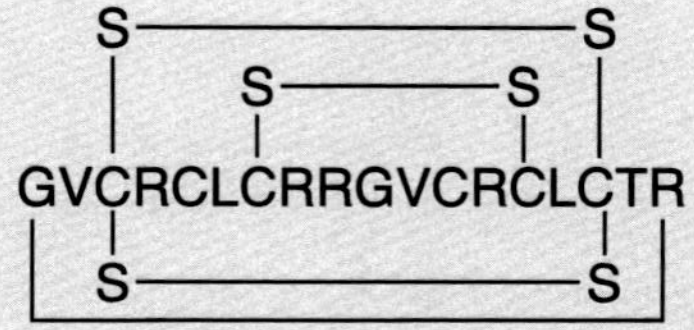 RTD-2	–	–	55	[270]
24	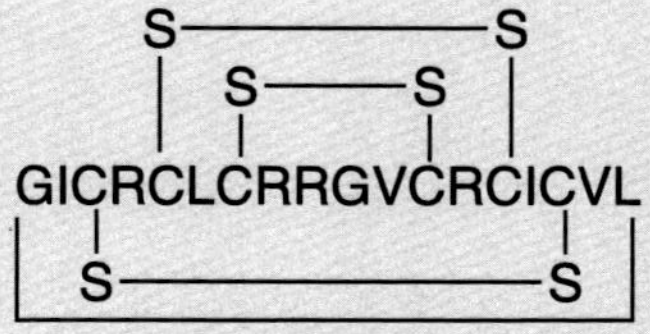 RTD-5	–	–	52	[270]

and bronchial inflammation, suppressing the expression of inflammatory cells and factors (i.e., sCD23), important T_H2 cytokines (i.e., IL-5), and ADAM8 positive eosinophiles in the lung [142].

Yim et al. further explored the BK-1361 structure [243], also called cyclo(RLsKDK), reporting two different libraries of cyclic RLsKDK-peptidomimetic analogs. The optimization of cyclo peptide series, containing six amino acids, has been carried out to evaluate the effect of the insertion of diverse in length α-chains and of altering the spatial conformation of the side chain. Moreover, a fundamental point was to understand how the amino acid different conformation could affect the binding affinity. The new peptidomimetic ADAM8 inhibitors showed inhibitory activity comparable to RLsKDK, thus representing new potential lead structures for further development.

The inhibition of dimerization was used as a strategy to inhibit activation of several MMPs, that necessarily need to dimerize to promote pathological effects [271–273]. Using the same approach, a bifunctional ADAM8 inhibitor, binding two ADAM8 homodimer catalytic sites, might efficiently reduce ADAM8 activation and impair ADAM8-promoted tumor invasiveness and growth. This strategy was adopted by Nuti et al. [245] in 2021 in order to develop dimeric arylsulfonamides as potent inhibitors of ADAM8. Considering the already reported high homology between ADAM17 and 8 binding site, [244] bifunctional ligands were designed starting from a nanomolar inhibitor of ADAM17 with a good selectivity over MMPs and ADAM10 and with good binding with ADAM8, as indicated by docking studies. The new bifunctional ADAM8 inhibitors presented nanomolar activity for ADAM8 and ADAM17, sparing MMPs. The most promising compound was compound **2** (Table 4), which showed good results not only on ADAM8 inhibition but also on CD23 shedding in HEK cells and inhibited the invasiveness of MDA-MB-231 breast cancer cells.

5.2 ADAM10 modulators

The emerging role of ADAM10 in various diseases revealed how could be significant to address the research versus ADAM10 as a potential drug target. The increasing or the decreasing of ADAM10 activity become interesting depending on the pathological context, as already explained in Section 4.3.

5.2.1 ADAM10 activators

Although protease overexpression is frequently linked to the development of pathological effects, in the last years some mechanism regarding protease activation are under investigation and consequently protease activators have been developed [274–276]. An improvement in ADAM10 mediated proteolysis role is significant in the modulation of the Aβ neurotoxin excessive production, one of the major responsible of Alzheimer's disease (AD). In fact, the use of ADAM10 activators is strictly connected with the alfa-secretase activation pathway that increases the generation of sAßPPα, a neurotrophic, neuroprotective agent, involved in the maintenance of dendritic integrity in the hippocampus, attenuating synaptic deficits [277].

Therefore, the strategy of an ADAM10 activator focused mainly on increasing ADAM10 expression, maturation, and activation specifically in the brain, attempting to tackle a brain and disease-specific mechanism to regulate ADAM10 activity.

In the recent years, the use of molecules of natural origin able to modulate ADAM10 activity has been reported. The strategy is to promote the long term intake of natural molecules through the consumption by diet [278]. ADAM10 activators resulted as disease-modifying agents since they could increase the cognitive process and lead to slow neuronal loss and functional memory decline.

Moreover, ADAM10 is a suitable therapeutic target for other cognitive loss diseases such as Fragile X Syndrome (FXS) [157] and autism spectrum disorder [279]. Nevertheless, further research is required to understand the fundamentals of these molecular mechanisms involved in ADAM10 activation with the consequent beneficial or adverse effects in AD and other neurological diseases.

Among natural molecules, the study of retinoids as ADAM10 activators revealed beneficial effects in brain functions but also in the treatment of psoriasis. Clinically, acitretin (Fig. 2) is an oral retinoid approved for psoriasis treatment. Tippman and colleagues demonstrated the effect of acitretin in the stimulation of ADAM10 by increasing mature ADAM10 levels and also α-secretase activity [280]. In psoriasis the upregulation of ADAM10 promoted by acitretin is supposed to restore normal keratocyte differentiation and epidermal architecture. Likewise in AD patients acitretin treatment significantly increased sAßPPα levels, confirming with preclinical data of an increase of ADAM10 activity [206].

Concerning ADAM10 maturation, Obregon et al. [281] reported that (−) epigallocatechin-3-gallate (EGCG) (Fig. 2), a natural compound derived from green tea, is able to promote ADAM10 maturation in cell-based models. At present, the mechanism of action for EGCG remains unidentified, but it is basically demonstrated to be secondary to a tyrosine kinase pathway.

Some of the just described ADAM10 activation mechanisms are also modulated by synthetic activators. Similarly to acitretin, Tamibarotene, and Cilostazol (Fig. 2) are able to positively regulate ADAM10 expression via retinoid acid receptor activation [282,283].

Etazolate (Fig. 2), an anxiolytic derivative, activates ADAM10 through a mediating receptor. This molecule firstly stimulates GABA-A receptor and secondly positively regulates ADAM10 displaying significant tropism to cells of central nervous system [284].

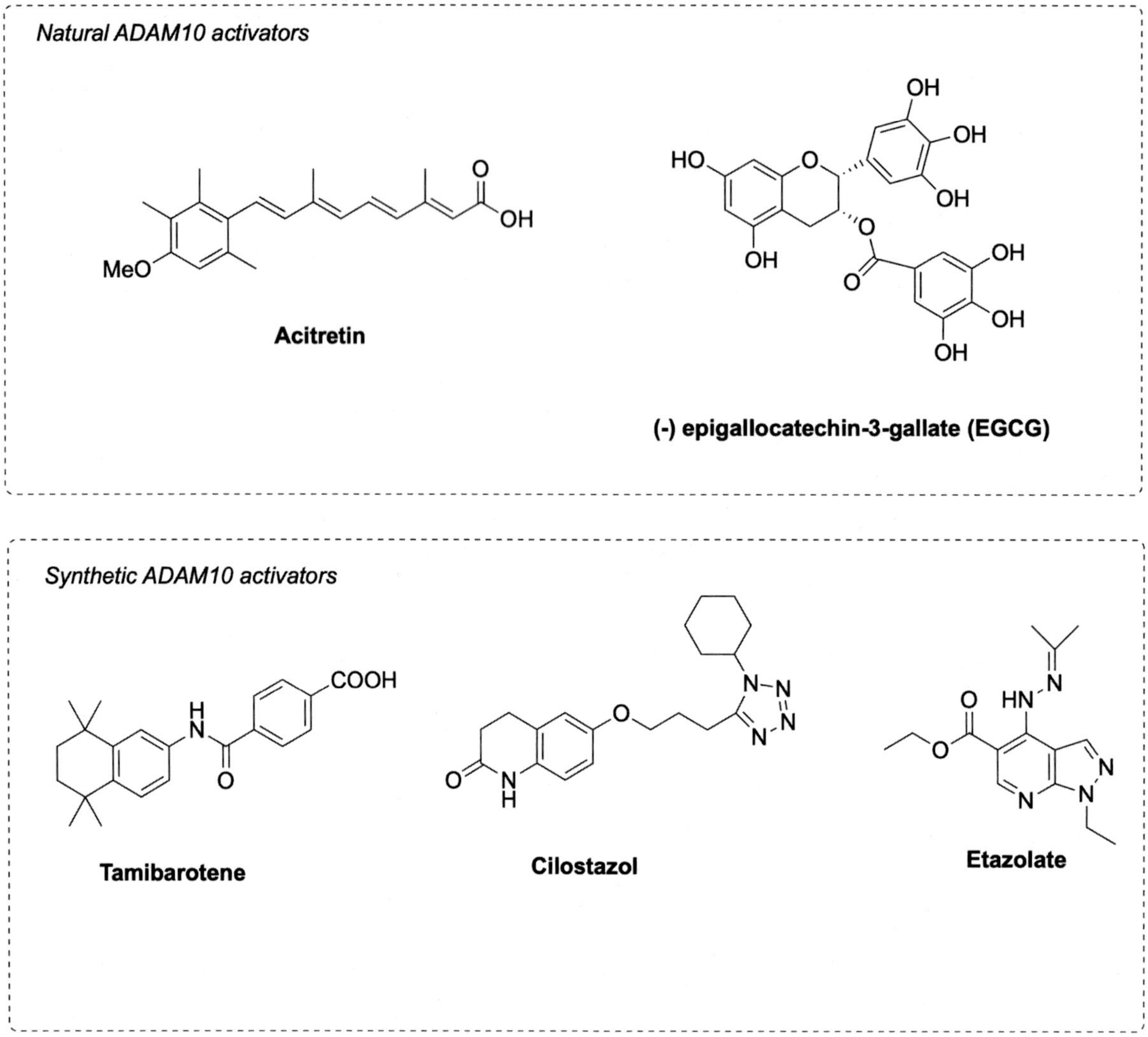

FIG. 2 Chemical structure of molecules able to enhance ADAM10 activity.

Latest findings about ADAM10 activation involved the development of ADAM10 endocytosis inhibitors for the treatment of AD. Musardo et al. [285] developed a cell-permeable peptide (PEP3) able to interfere with ADAM10 endocytosis, upregulating the postsynaptic localization and activity of ADAM10, with a strong safety profile. PEP3 is constituted by a peptide (QPPRQRPRG) linked to the TAT sequence (YGRKKRRQRR) by a short linker composed by three glycines and one serine (GGSG).

5.2.2 ADAM10 inhibitors

A selective targeting of ADAM10 is highly desirable in order to avoid off-target effects due to the simultaneous inhibition of other ADAMs or MMPs.

- *ADAM10 inhibition promoted by natural molecules*

Among natural compounds, two molecules have been recently reported as ADAM10 inhibitors: Rapamycin and Triptolide (Fig. 3). Rapamycin is a macrocyclic antibiotic produced by the bacterium *Streptomyces hygroscopicus* and used as an immunosuppressant drug to avoid rejection in organ transplantation. Zhang et al. [286] demonstrated a decreased ADAM10 activation promoted by Rapamycin, which stimulated the amyloidogenic amyloid precursor protein (APP) pathway in murine neuron-like cells (N2a) transfected with the human "Swedish" mutant APP, as further confirmed by in vivo studies. Another natural compound is the diterpenoid epoxide Triptolide [287] from *Tripterygium wilfordii*, reported to downregulate ADAM10 expression in different cell lines.

- *ADAM10 inhibition promoted by synthetic compounds*

The first class of reported ADAM10 inhibitors is represented by the hydroxamate-based zinc-chelating inhibitors, able to block ADAM10 activity thanks to the coordination of the catalytic zinc ion in the protease active site. Commonly, ADAM10 synthetic inhibitor structure, is designed focusing on the binding with the prime subsites S_1'–S_3' and the chelation of the zinc through a suitable zinc-binding group (ZBG).

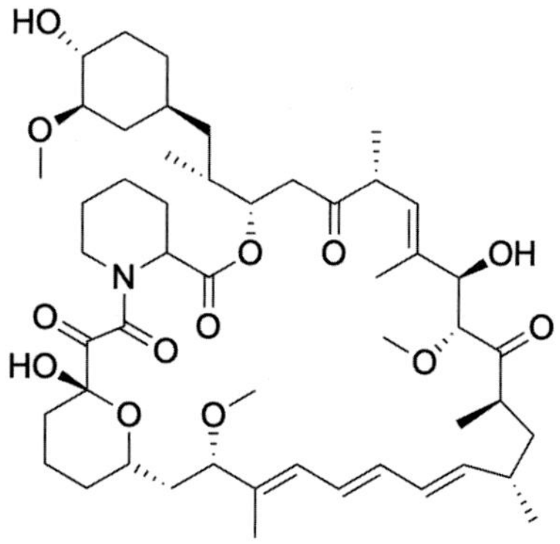

Rapamycin

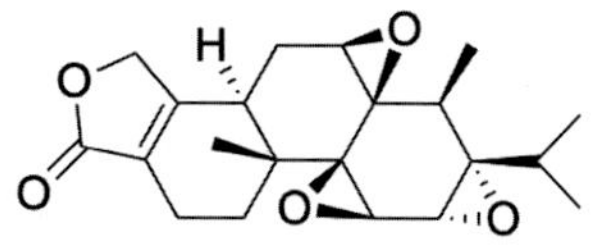

Triptolide

FIG. 3 Chemical structure of natural molecules able to inhibit ADAM10 activity.

Among hydroxamate-based ADAM10 inhibitors, one of the most studied is GI254023X (**3**, Table 3) developed by GlaxoSmithKline. Nowadays, GI254023X is the ADAM10 inhibitor most used in different cell culture experiments.

In preclinical studies, GI254023X demonstrated a potent activity in different types of cancer such as glioblastoma [246] and breast cancer [247] and recently in the neurological and histopathological outcome of traumatic brain injury (TBI) [248]. Unfortunately, during clinical studies the insufficient selectivity of GI254023X caused different adverse effects following systemic administration. In fact, the reported hepatotoxicity, caused the failure of preclinical development [288].

Several studies documented the ADAM10 inhibitory activity of compounds INCB3619 (**4**, Table 4) and INCB7839 (**5**, Tables 4 and 5) developed by Incyte Corporation. INCB3619 and INCB7839 (aderbasib), are not selective ADAM10/ADAM17 inhibitors, as demonstrated by reduced shedding of both ADAM10 and ADAM17 substrates in different cell-based experiments [249–251]. INCB3619 is an orally active compound, reporting an IC_{50} values of 14 (ADAM10) and 22 (ADAM17) nM, respectively sparing other MMPs and ADAMs. INCB7839 resulted particularly active on HER2+ breast cancer in association with trastuzumab, with consistent results even in clinical studies (Table 5). These molecules have shown promising effects in reducing tumor growth as single agents or in combination with specific chemotherapeutic drugs when tested in preclinical models of breast cancer and NSCLC (Table 5. Clinical trials: NCT01254136) [252,253,291]. Nevertheless, their efficacy in clinical studies has been stunted due to safety and toxicity concerns caused by the induction of deep vein thrombosis in some patients [292]. The inhibitor INCB7839 is currently under investigation in the clinical trial NCT02141451 (Table 5), in combination with Rituximab (a monoclonal antibody targeting CD20), for the treatment of Diffuse Large B Cell Non-Hodgkin Lymphoma. Moreover, preliminary studies on INCB7839 revealed that it is a well-tolerated drug by systemic administration and further clinical studies are ongoing on glioblastoma (NCT04295759, Table 5). Generally, both the Incyte molecules are used as reference nonselective ADAM10/ADAM17 inhibitors in different field of investigation regarding ADAM enzyme involvement in various pathologies [293].

In 2018 Rossello et al. [254] reported the discovery of two promising ADAM10 selective inhibitors, hydroxamates LT4 and MN8 (**6** and **7**, Table 4). Biological assays have been conducted in vitro on human recombinant enzymes, showing nanomolar activity against ADAM10 and an excellent selectivity profile sparing MMPs and the other tested ADAMs. LT4 and MN8 were tested on Hodgkin Lymphoma (HL) cell lines demonstrating to strongly reduce the NKG2D-L shedding mediated by ADAM10, thus restoring the sensitivity of HL cell lines to NKG2D-dependent cell killing, exerted by natural killer and γδ T cells. Interestingly, Tosetti et al. reported [255] that the treatment with LT4 and MN8 resulted in the increase of membrane CD30 levels, which restored sensitivity to anti-CD30 monoclonal therapies used in HL, such as Iratumumab. In fact, LT4 and MN8 not only interfered with HL cell growth but also enhanced the antilymphoma effect of the anti-CD30 ADC Brentuximab-Vedotin in 3D culture systems very similar to those approved as preclinical models. This was evident at low and ineffective doses of the ADC as well, indicating a possible combined scheme to potentiate ADC-based lymphoma therapy [294].

In 2018 Mahasenan et al. [256] reported their studies on ADAM10 selectivity by small molecules bearing a hydroxamate ZBG. Selective inhibition of ADAM10 was achieved by binding with S1 subsite whereas the S1′ subsite did not provide opportunity to exploit selectivity. For these reasons, after a structure-based computational analysis they identified compound **8** (Table 4) as a nanomolar selective ADAM10 inhibitor, endowed with good pharmacokinetic properties.

TABLE 5 Chemical structure of ADAM10 and ADAM17 inhibitors entered in clinical phase so far.

Compd	Structure	ID	Status	Phase	Disease
5	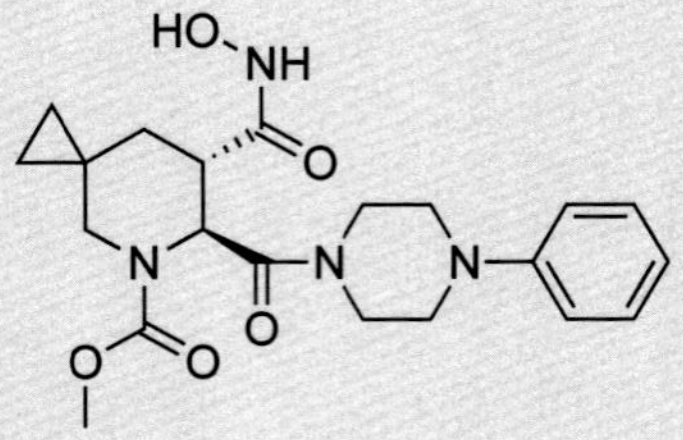 INCB7839	NCT01254136	Terminated (Incyte has suspended development of the compound.)	I/II	INCB007839 with trastuzumab and vinorelbine in patients with metastatic HER2+ breast cancer
		NCT02141451	Completed	I/II	Hematopoietic cell transplant (HCT) for patients with diffuse large B cell lymphoma (DLBCL).
		NCT04295759	Recruiting	I	Treating children with recurrent/progressive high-grade gliomas
9	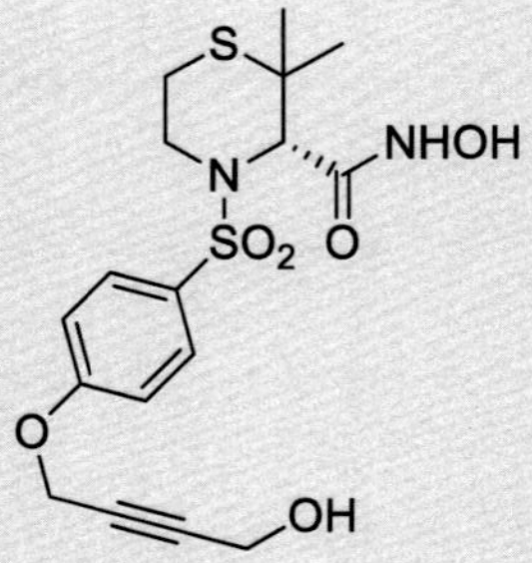 Apratastat	NCT00095342	Completed	II	Active rheumatoid arthritis
10	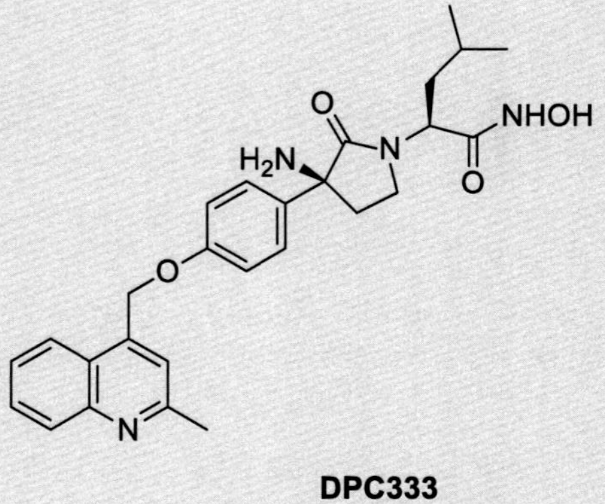 DPC333	[289,290]	Completed	I/II	Pharmacokinetics and pharmacodynamics of DPC333 in human, and rheumatoid arthritis

5.3 ADAM17 inhibitors

As mentioned in the previous paragraphs, ADAM17 regulates different pathological and physiological cellular processes because is involved in the shedding of many substrates. Therefore, considering its multiple activities, it is highly desirable to obtain selective ADAM17 inhibition.

Despite many efforts to develop selective ADAM17 small molecule inhibitors, no clinical candidate has still reached the market.

Only few ADAM17 inhibitors developed by multinational pharmaceutical companies entered in clinical trials, even if they have been withdrawn later on owing to their toxicity. Apratastat [295] (**9**, Table 5, Wyeth pharmaceuticals, NCT00095342), DPC 333 [289] (**10**, Table 5, Bristol-Myers Squibb Company), and INCB7839 [252] (**5**, Tables 4 and 5, Incyte corporation, *see* ADAM10 paragraphs) are the most famous and studied ADAM17 inhibitors, failed in phase-I/II of clinical trials.

The vast majority of small molecule inhibitors of ADAM17 reported in the last 10 years, have been deeply reviewed in several papers [120,240,290,296]. Here, we discuss the recent advances in ADAM17 drug discovery published starting from 2015, classifying them on the basis of mechanism of interaction with the catalytic domain in: hydroxamate-based and nonhydroxamate-based inhibitors.

5.3.1 *Hydroxamate-based ADAM17 inhibitors*

Ouvry et al. [257] deeply investigated the fundamental groups of the two potent ADAM17 inhibitors Apratastat and DPC-333 in order to understand how to improve ADAM17 activity and selectivity for topical application. The new class of inhibitors derived from Apratastat and DPC-333, and comprised the insertion of novel cyclic linkers attached to the hydroxamate function. Compound **11** (Table 4) was identified as the first ADAM17 inhibitor reporting nanomolar activity for the target as antipsoriasis agent. Unfortunately, results in cell model assays revealed a weak potency that precluded a further investigation.

To directly evaluate the inhibitory activity of a new series of sulfonamido-based hydroxamate derivatives, avoiding the cell drop off of compound **11**, a new TNFα inhibition cell model in human peripheral blood mononuclear cells was adopted [258]. In this series the quinoline-based **12** (Table 4) with *N*-acetylated azetidinhydroxyacetamide linker reported the most promising in vitro inhibitory activity ($IC_{50} = 4\,nM$) and selectivity profile, confirmed also by an oxazolone-induced chronic skin inflammation model. The excellent results supported the selection of compound **12** as a clinical candidate for further biological investigations.

Hirata et al. [259] in 2017 reported an innovative and alternative zinc-binding group: the reverse hydroxamate. The reverse hydroxamate KP-457 (**13**, Table 4) demonstrated nanomolar activity for ADAM17 and a good selectivity over ADAM10 and other MMPs. Furthermore, KP457 was able to reduce the glycoprotein GPIbα expression in vivo, in a human induced pluripotent stem cell-derived (iPSCs) platelets model.

5.3.2 *Nonhydroxamate-based ADAM17 inhibitors*

A strategy to avoid the classical loss of selectivity and the consequent off-target effects due to inhibition of other metzincins, [297] consisted in introducing mild ZBGs in the inhibitor scaffold. Nonhydroxamate-based ADAM17 inhibitors were developed, where the hydroxamate moiety was replaced with other ZBGs such as hydantoin.

Girijavallabhan et al. [260] actively worked in developing the first examples of hydantoin-based ADAM17 inhibitors. The first series of inhibitors presented an hydantoin moiety conjugated with a pendant acetylene group (compound **14**, Table 4). Compound **14** was the hit compound used to explore hydantoin as ZBG, developing inhibitors with improved pharmacokinetic properties and bioavailability. A few years later, Tong et al. [261] published a new hydantoin-based series where the pendant acetylene was replaced by an azabenzofuran group (compound **15**, Table 4). The azabenzofuran group affected the potency in human whole blood assay (hWBA) and improved the pharmacokinetic profile of the new compounds.

Furthermore, the insertion of polar functionalities, such as basic groups and H-bonding donor groups in the benzofuran ring, further improved the potency. The azabenzofuran hydantoin **15** was identified as the lead compound and further investigated in term of pharmacokinetic properties in order to improve the oral absorption and the membrane permeation [262]. Moreover, the ketoamide on hydantoin ring was used to design a prodrug approach. In fact, **15** was further modified by inserting in the ketoamide NH different removable substituents. The pivalate prodrug **16** (Table 4), presented the best results with a rat AUC rate of 13.1 μM h and appreciable DMPK properties evaluated through oral administration in fasted rats, dogs, and monkeys.

In 2017 a thiadiazolone-derivative JTP-96193 [263] (**17**, Table 4) was reported as nonhydroxamate ADAM17 inhibitor. JTP-96193 revealed an excellent activity and selectivity profile, with a nanomolar activity for ADAM17 and a 180-fold selectivity over ADAM10.

Furthermore, this molecule reported interesting results on a mouse model of obesity, on type 2 diabetes, and diabetic peripheral neuropathy (DPN). In this context, JTP-96193 significantly reduced the TNFα release from fat tissue, increasing the insulin resistance and consequently preventing diabetes development. Moreover, the administration of JTP-96193 prevented the development of DPN in streptozotocin (STZ)-induced diabetic mice without effect on glucose blood level [263].

Another approach in the development of nonhydroxamate ADAM17 inhibitors was described by Leung et al. [264] which reported the first metal-based inhibitor of ADAM17, iridium(III)-based cyclometalated complex (**18**, Table 4). The authors screened an in-house library of diverse complexes, identifying how the charge localization in the metal complex affected the target inhibition. The best compound was compound **18**, which showed an ADAM17 inhibitory activity in the μM range and ability to inhibit TNFα secretion and p-38 phosphorylation MAP kinase on a monocyte THP-1 cell model. Unfortunately, no data regarding the selectivity profile were reported.

The targeting of a secondary exosite of the enzyme was the strategy adopted by Minond et al. [265] in order to achieve selectivity, pursuing development of ADAM17 inhibitors acting via no zinc-binding mechanism. The purpose was to target a specific area of ADAM protease, usually interacting with various substrates, commonly carbohydrates. A new series of ADAM17 exosite-targeting inhibitors was reported, with an accurate structure-activity relationship study. The most promising exosite inhibitors were further studied by biochemical and cell-based assays. Data indicated compound **19** as the best exosite-binding ADAM17 inhibitor of this series. Compound **19** selectively inhibited ADAM17 in cell-based assay, demonstrating also an unusual substrate selectivity by sparing ADAM17-mediated cleavage of TNFα.

In 2018 the thioxodihydro pyrimidindione ZLDI-8 (**20**, Table 4), was reported as a nonzinc binder ADAM17 inhibitor by computational study. Compound **20** was identified by virtual screening and demonstrated good ADAM17 inhibitory activity in the treatment hepatocellular carcinoma (HCC) [266–268,298].

In fact, ZLDI-8 interrupted the Notch pathway in HCC cells, eluding the NICD (Intracellular domain of Notch) accumulation in the nucleus, and inhibiting the epithelial-mesenchymal transition (EMT) process of HCC cells. Importantly, The Notch path inhibition promoted by ZLDI-8 reduced the expression of specific antiapoptosis, pro-survival, and EMT-related genes. Furthermore, after ZLDI-8 administration a major sensibility of HCC cells to other drugs such as Sorafenib, Etoposide, and paclitaxel was observed.

Further studies [267] demonstrated the anticancer activity of ZLDI-8, which was able to block in vitro the migration and invasion in highly aggressive type of HCC cells (MHCC97-H and LM3), and to inhibit in vivo lung metastasis. Moreover, ZLDI-8 was demonstrated [268] to affect chemo-resistant nonsmall cell lung cancer (A549 and A549-Taxol cells), inducing apoptosis and attenuating migration and invasion in an in vivo xenograft of multidrug resistant lung cancer model.

An emerging class of molecules which deserve a mention are peptide-based inhibitors. To date, peptide-based inhibitors were successfully developed only for ADAM8 (see previous paragraph) and ADAM17. Different libraries were designed to mimic the sequence of ADAM17 cleavage site in collagen type II, as exhaustively reviewed by Pluda et al. [299] The most recent peptide-based inhibitors were reported by Wang et al. in 2016 [269] and by Schaal et al. in 2017–18 [270,300]. Wang and colleagues identified two linear peptides, Hxm-Phe-Arg-Gln and Hxm-Phe-Ser-Asn (**21** and **22**, Table 4), exhibiting high activity for ADAM17 with Ki in the nM range (Ki: 47 nM and 92 nM, respectively) and a five to sevenfold selectivity over ADAM10 [269]. Schaal and colleagues screened θ-defensins, a family of 18 amino acid macrocyclic peptides, expressed in granulocytes. The octapeptide rhesus θ defensin 2 (RTD-2, **23** Table 4) and 5 (RTD-5, **24** Table 4) [300] were identified as nanomolar inhibitors of ADAM17 (IC_{50}: 52 nM and 55 nM, respectively) and they were able to inhibit the TNFα shedding in THP1 and HT-29 cell lines, typical of blood leukocytes.

6 Conclusions

ADAM proteases are membrane-bound proteases that mediate ectodomain shedding of transmembrane proteins. To date, the ectodomain shedding is considered an important dysregulation process involved in many pathological mechanisms such as inflammation, cancer, and immune disorders. For these reasons, ADAM proteases are considered an important therapeutic target in drug discovery.

In the last years, many research efforts have been focused on ADAM inhibitors facing several obstacles. In particular, the lack of selectivity is responsible for different adverse effects and an unexpected toxicity was the undefeatable limit in the clinical phase development.

In this chapter are described the structure, biological functions, and pathological involvement of the most studied ADAMs: ADAM8, ADAM10, and ADAM17. Furthermore, ADAM8, 10, and 17 selective inhibitors have been reported, examining the results of the most promising and innovative molecules. Although no drug is still available on the market, some recent papers identified promising candidates which deserve further investigation in clinical studies.

References

[1] Lichtenthaler SF, Lemberg MK, Fluhrer R. Proteolytic ectodomain shedding of membrane proteins in mammals-hardware, concepts, and recent developments. EMBO J 2018;37:e99456.

[2] Hsia H-E, Tüshaus J, Brummer T, Zheng Y, Scilabra SD, Lichtenthaler SF. Functions of "A disintegrin and metalloproteases

(ADAMs)" in the mammalian nervous system. Cell Mol Life Sci 2019;76:3055–81.

[3] Blobel CP. ADAMs: key components in EGFR signalling and development. Nat Rev Mol Cell Biol 2005;6:32–43.

[4] Cuffaro D, Ciccone L, Rossello A, Nuti E, Santamaria S. Targeting aggrecanases for osteoarthritis therapy: from zinc chelation to exosite inhibition. J Med Chem 2022;65:13505–32.

[5] Calligaris M, Cuffaro D, Bonelli S, Spanò DP, Rossello A, Nuti E, Scilabra SD. Strategies to target ADAM17 in disease: from its discovery to the iRhom revolution. Molecules 2021;26:944.

[6] Düsterhöft S, Michalek M, Kordowski F, Oldefest M, Sommer A, Röseler J, Reiss K, Grötzinger J, Lorenzen I. Extracellular juxtamembrane segment of ADAM17 interacts with membranes and is essential for its shedding activity. Biochemistry 2015;54:5791–801.

[7] Düsterhöft S, Höbel K, Oldefest M, et al. A disintegrin and metalloprotease 17 dynamic interaction sequence, the sweet tooth for the human interleukin 6 receptor. J Biol Chem 2014;289:16336–48.

[8] Christova Y, Adrain C, Bambrough P, Ibrahim A, Freeman M. Mammalian iRhoms have distinct physiological functions including an essential role in TACE regulation. EMBO Rep 2013;14:884–90.

[9] Li X, Maretzky T, Weskamp G, et al. iRhoms 1 and 2 are essential upstream regulators of ADAM17-dependent EGFR signaling. Proc Natl Acad Sci U S A 2015;112:6080–5.

[10] Lora J, Weskamp G, Li TM, Maretzky T, Shola DTN, Monette S, Lichtenthaler SF, Lu TT, Yang C, Blobel CP. Targeted truncation of the ADAM17 cytoplasmic domain in mice results in protein destabilization and a hypomorphic phenotype. J Biol Chem 2021;296:100733.

[11] Adrain C, Zettl M, Christova Y, Taylor N, Freeman M. Tumor necrosis factor signaling requires iRhom2 to promote trafficking and activation of TACE. Science 2012;335:225–8.

[12] Dulloo I, Muliyil S, Freeman M. The molecular, cellular and pathophysiological roles of iRhom pseudoproteases. Open Biol 2019;9:190003.

[13] Maretzky T, McIlwain DR, Issuree PDA, Li X, Malapeira J, Amin S, Lang PA, Mak TW, Blobel CP. iRhom2 controls the substrate selectivity of stimulated ADAM17-dependent ectodomain shedding. Proc Natl Acad Sci U S A 2013;110:11433–8.

[14] McIlwain DR, Lang PA, Maretzky T, et al. iRhom2 regulation of TACE controls TNF-mediated protection against *Listeria* and responses to LPS. Science 2012;335:229–32.

[15] Horiuchi K, Le Gall S, Schulte M, Yamaguchi T, Reiss K, Murphy G, Toyama Y, Hartmann D, Saftig P, Blobel CP. Substrate selectivity of epidermal growth factor-receptor ligand sheddases and their regulation by phorbol esters and calcium influx. Mol Biol Cell 2007;18:176–88.

[16] Gschwind A, Hart S, Fischer OM, Ullrich A. TACE cleavage of proamphiregulin regulates GPCR-induced proliferation and motility of cancer cells. EMBO J 2003;22:2411–21.

[17] Sahin U, Weskamp G, Kelly K, Zhou H-M, Higashiyama S, Peschon J, Hartmann D, Saftig P, Blobel CP. Distinct roles for ADAM10 and ADAM17 in ectodomain shedding of six EGFR ligands. J Cell Biol 2004;164:769–79.

[18] Weinger JG, Omari KM, Marsden K, Raine CS, Shafit-Zagardo B. Up-regulation of soluble Axl and Mer receptor tyrosine kinases negatively correlates with Gas6 in established multiple sclerosis lesions. Am J Pathol 2009. https://doi.org/10.2353/ajpath.2009.080807.

[19] Bech-Serra JJ, Santiago-Josefat B, Esselens C, Saftig P, Baselga J, Arribas J, Canals F. Proteomic identification of desmoglein 2 and activated leukocyte cell adhesion molecule as substrates of ADAM17 and ADAM10 by difference gel electrophoresis. Mol Cell Biol 2006;26:5086–95.

[20] Leksa V, Loewe R, Binder B, et al. Soluble M6P/IGF2R released by TACE controls angiogenesis via blocking plasminogen activation. Circ Res 2011;108:676–85.

[21] Lambert DW, Yarski M, Warner FJ, Thornhill P, Parkin ET, Smith AI, Hooper NM, Turner AJ. Tumor necrosis factor-alpha convertase (ADAM17) mediates regulated ectodomain shedding of the severe-acute respiratory syndrome-coronavirus (SARS-CoV) receptor, angiotensin-converting enzyme-2 (ACE2). J Biol Chem 2005;280:30113–9.

[22] Haga S, Yamamoto N, Nakai-Murakami C, Osawa Y, Tokunaga K, Sata T, Yamamoto N, Sasazuki T, Ishizaka Y. Modulation of TNF-alpha-converting enzyme by the spike protein of SARS-CoV and ACE2 induces TNF-alpha production and facilitates viral entry. Proc Natl Acad Sci U S A 2008;105:7809–14.

[23] Endres K, Postina R, Schroeder A, Mueller U, Fahrenholz F. Shedding of the amyloid precursor protein-like protein APLP2 by disintegrin-metalloproteinases. FEBS J 2005;272:5808–20.

[24] Schulte A, Schulz B, Andrzejewski MG, et al. Sequential processing of the transmembrane chemokines CX3CL1 and CXCL16 by alpha- and gamma-secretases. Biochem Biophys Res Commun 2007;358:233–40.

[25] Dyczynska E, Sun D, Yi H, Sehara-Fujisawa A, Blobel CP, Zolkiewska A. Proteolytic processing of delta-like 1 by ADAM proteases. J Biol Chem 2007;282:436–44.

[26] Wang Y, Wu J, Newton R, Bahaie NS, Long C, Walcheck B. ADAM17 cleaves CD16b (FcγRIIIb) in human neutrophils. Biochim Biophys Acta 2013;1833:680–5.

[27] Scilabra SD, Troeberg L, Yamamoto K, Emonard H, Thøgersen I, Enghild JJ, Strickland DK, Nagase H. Differential regulation of extracellular tissue inhibitor of metalloproteinases-3 levels by cell membrane-bound and shed low density lipoprotein receptor-related protein 1. J Biol Chem 2013;288:332–42.

[28] Zatovicova M, Sedlakova O, Svastova E, Ohradanova A, Ciampor F, Arribas J, Pastorek J, Pastorekova S. Ectodomain shedding of the hypoxia-induced carbonic anhydrase IX is a metalloprotease-dependent process regulated by TACE/ADAM17. Br J Cancer 2005;93:1267–76.

[29] Buxbaum JD, Liu KN, Luo Y, Slack JL, Stocking KL, Peschon JJ, Johnson RS, Castner BJ, Cerretti DP, Black RA. Evidence that tumor necrosis factor alpha converting enzyme is involved in regulated alpha-secretase cleavage of the Alzheimer amyloid protein precursor. J Biol Chem 1998;273:27765–7.

[30] Slack BE, Ma LK, Seah CC. Constitutive shedding of the amyloid precursor protein ectodomain is up-regulated by tumour necrosis factor-alpha converting enzyme. Biochem J 2001;357:787–94.

[31] Patel IR, Attur MG, Patel RN, Stuchin SA, Abagyan RA, Abramson SB, Amin AR. TNF-alpha convertase enzyme from human arthritis-affected cartilage: isolation of cDNA by differential display, expression of the active enzyme, and regulation of TNF-alpha. J Immunol 1998;160:4570–9.

[32] Sahin U, Blobel CP. Ectodomain shedding of the EGF-receptor ligand epigen is mediated by ADAM17. FEBS Lett 2007;581:41–4.

[33] Etzerodt A, Maniecki MB, Møller K, Møller HJ, Moestrup SK. Tumor necrosis factor α-converting enzyme (TACE/ADAM17)

mediates ectodomain shedding of the scavenger receptor CD163. J Leukoc Biol 2010;88:1201–5.

[34] Franzke C-W, Tasanen K, Borradori L, Huotari V, Bruckner-Tuderman L. Shedding of collagen XVII/BP180: structural motifs influence cleavage from cell surface. J Biol Chem 2004;279:24521–9.

[35] Liu Q, Zhang J, Tran H, Verbeek MM, Reiss K, Estus S, Bu G. LRP1 shedding in human brain: roles of ADAM10 and ADAM17. Mol Neurodegener 2009;4:17.

[36] Chen C-D, Podvin S, Gillespie E, Leeman SE, Abraham CR. Insulin stimulates the cleavage and release of the extracellular domain of Klotho by ADAM10 and ADAM17. Proc Natl Acad Sci U S A 2007;104:19796–801.

[37] Wang Y, Sul HS. Ectodomain shedding of preadipocyte factor 1 (Pref-1) by tumor necrosis factor alpha converting enzyme (TACE) and inhibition of adipocyte differentiation. Mol Cell Biol 2006;26:5421–35.

[38] Kawaguchi N, Horiuchi K, Becherer JD, Toyama Y, Besmer P, Blobel CP. Different ADAMs have distinct influences on Kit ligand processing: phorbol-ester-stimulated ectodomain shedding of Kitl1 by ADAM17 is reduced by ADAM19. J Cell Sci 2007;120:943–52.

[39] Hansen HP, Recke A, Reineke U, Von Tresckow B, Borchmann P, Von Strandmann EP, Lange H, Lemke H, Engert A. The ectodomain shedding of CD30 is specifically regulated by peptide motifs in its cysteine-rich domains 2 and 5. FASEB J 2004;18:893–5.

[40] Nyborg AC, Ladd TB, Zwizinski CW, Lah JJ, Golde TE. Sortilin, SorCS1b, and SorLA Vps10p sorting receptors, are novel gamma-secretase substrates. Mol Neurodegener 2006;1:3.

[41] Cho RW, Park JM, Wolff SBE, et al. mGluR1/5-dependent long-term depression requires the regulated ectodomain cleavage of neuronal pentraxin NPR by TACE. Neuron 2008;57:858–71.

[42] Kawaguchi M, Hozumi Y, Suzuki T. ADAM protease inhibitors reduce melanogenesis by regulating PMEL17 processing in human melanocytes. J Dermatol Sci 2015;78:133–42.

[43] Schäfer B, Gschwind A, Ullrich A. Multiple G-protein-coupled receptor signals converge on the epidermal growth factor receptor to promote migration and invasion. Oncogene 2004;23:991–9.

[44] Contin C, Pitard V, Itai T, Nagata S, Moreau J-F, Déchanet-Merville J. Membrane-anchored CD40 is processed by the tumor necrosis factor-alpha-converting enzyme. Implications for CD40 signaling. J Biol Chem 2003;278:32801–9.

[45] Alfa Cissé M, Sunyach C, Slack BE, Fisher A, Vincent B, Checler F. M1 and M3 muscarinic receptors control physiological processing of cellular prion by modulating ADAM17 phosphorylation and activity. J Neurosci 2007;27:4083–92.

[46] Young J, Yu X, Wolslegel K, et al. Lymphotoxin-alphabeta heterotrimers are cleaved by metalloproteinases and contribute to synovitis in rheumatoid arthritis. Cytokine 2010;51:78–86.

[47] Marczynska J, Ozga A, Wlodarczyk A, et al. The role of metalloproteinase ADAM17 in regulating ICOS ligand-mediated humoral immune responses. J Immunol 2014;193:2753–63.

[48] Peng M, Guo S, Yin N, Xue J, Shen L, Zhao Q, Zhang W. Ectodomain shedding of Fcalpha receptor is mediated by ADAM10 and ADAM17. Immunology 2010;130:83–91.

[49] Fox JE. Shedding of adhesion receptors from the surface of activated platelets. Blood Coagul Fibrinolysis 1994;5:291–304.

[50] Pruessmeyer J, Martin C, Hess FM, et al. A disintegrin and metalloproteinase 17 (ADAM17) mediates inflammation-induced shedding of syndecan-1 and -4 by lung epithelial cells. J Biol Chem 2010;285:555–64.

[51] Lum L, Wong BR, Josien R, Becherer JD, Erdjument-Bromage H, Schlöndorff J, Tempst P, Choi Y, Blobel CP. Evidence for a role of a tumor necrosis factor-alpha (TNF-alpha)-converting enzyme-like protease in shedding of TRANCE, a TNF family member involved in osteoclastogenesis and dendritic cell survival. J Biol Chem 1999;274:13613–8.

[52] Badoual C, Bouchaud G, Agueznay NEH, et al. The soluble alpha chain of interleukin-15 receptor: a proinflammatory molecule associated with tumor progression in head and neck cancer. Cancer Res 2008;68:3907–14.

[53] Qu D, Wang Y, Esmon NL, Esmon CT. Regulated endothelial protein C receptor shedding is mediated by tumor necrosis factor-alpha converting enzyme/ADAM17. J Thromb Haemost 2007;5:395–402.

[54] Black RA, Rauch CT, Kozlosky CJ, et al. A metalloproteinase disintegrin that releases tumour-necrosis factor-alpha from cells. Nature 1997;385:729–33.

[55] Moss ML, Jin SL, Milla ME, et al. Cloning of a disintegrin metalloproteinase that processes precursor tumour-necrosis factor-alpha. Nature 1997;385:733–6.

[56] Reddy P, Slack JL, Davis R, Cerretti DP, Kozlosky CJ, Blanton RA, Shows D, Peschon JJ, Black RA. Functional analysis of the domain structure of tumor necrosis factor-alpha converting enzyme. J Biol Chem 2000;275:14608–14.

[57] Määttä JA, Sundvall M, Junttila TT, Peri L, Laine VJO, Isola J, Egeblad M, Elenius K. Proteolytic cleavage and phosphorylation of a tumor-associated ErbB4 isoform promote ligand-independent survival and cancer cell growth. Mol Biol Cell 2006;17:67–79.

[58] Rio C, Buxbaum JD, Peschon JJ, Corfas G. Tumor necrosis factor-alpha-converting enzyme is required for cleavage of erbB4/HER4. J Biol Chem 2000;275:10379–87.

[59] Wunderlich P, Glebov K, Kemmerling N, Tien NT, Neumann H, Walter J. Sequential proteolytic processing of the triggering receptor expressed on myeloid cells-2 (TREM2) protein by ectodomain shedding and γ-secretase-dependent intramembranous cleavage. J Biol Chem 2013;288:33027–36.

[60] Malapeira J, Esselens C, Bech-Serra JJ, Canals F, Arribas J. ADAM17 (TACE) regulates TGFβ signaling through the cleavage of vasorin. Oncogene 2011. https://doi.org/10.1038/onc.2010.565.

[61] Chalaris A, Rabe B, Paliga K, Lange H, Laskay T, Fielding CA, Jones SA, Rose-John S, Scheller J. Apoptosis is a natural stimulus of IL6R shedding and contributes to the proinflammatory trans-signaling function of neutrophils. Blood 2007;110:1748–55.

[62] Schantl JA, Roza M, Van Kerkhof P, Strous GJ. The growth hormone receptor interacts with its sheddase, the tumour necrosis factor-alpha-converting enzyme (TACE). Biochem J 2004;377:379–84.

[63] Zhang Q, Thomas SM, Xi S, Smithgall TE, Siegfried JM, Kamens J, Gooding WE, Grandis JR. SRC family kinases mediate epidermal growth factor receptor ligand cleavage, proliferation, and invasion of head and neck cancer cells. Cancer Res 2004;64:6166–73.

[64] Tsakadze NL, Sithu SD, Sen U, English WR, Murphy G, D'Souza SE. Tumor necrosis factor-alpha-converting enzyme (TACE/ADAM-17) mediates the ectodomain cleavage of intercellular adhesion molecule-1 (ICAM-1). J Biol Chem 2006;281:3157–64.

[65] Brou C, Logeat F, Gupta N, Bessia C, LeBail O, Doedens JR, Cumano A, Roux P, Black RA, Israël A. A novel proteolytic cleavage involved in Notch signaling: the role of the disintegrin-metalloprotease TACE. Mol Cell 2000;5:207–16.

[66] Rovida E, Paccagnini A, Del Rosso M, Peschon J, Dello Sbarba P. TNF-alpha-converting enzyme cleaves the macrophage colony-stimulating factor receptor in macrophages undergoing activation. J Immunol 2001;166:1583–9.

[67] Koenen RR, Pruessmeyer J, Soehnlein O, et al. Regulated release and functional modulation of junctional adhesion molecule A by disintegrin metalloproteinases. Blood 2009;113:4799–809.

[68] Li N, Wang Y, Forbes K, et al. Metalloproteases regulate T-cell proliferation and effector function via LAG-3. EMBO J 2007;26:494–504.

[69] Maretzky T, Schulte M, Ludwig A, Rose-John S, Blobel C, Hartmann D, Altevogt P, Saftig P, Reiss K. L1 is sequentially processed by two differently activated metalloproteases and presenilin/gamma-secretase and regulates neural cell adhesion, cell migration, and neurite outgrowth. Mol Cell Biol 2005;25:9040–53.

[70] Chitadze G, Lettau M, Bhat J, et al. Shedding of endogenous MHC class I-related chain molecules A and B from different human tumor entities: heterogeneous involvement of the "a disintegrin and metalloproteases" 10 and 17. Int J Cancer 2013;133:1557–66.

[71] Swendeman S, Mendelson K, Weskamp G, Horiuchi K, Deutsch U, Scherle P, Hooper A, Rafii S, Blobel CP. VEGF-A stimulates ADAM17-dependent shedding of VEGFR2 and crosstalk between VEGFR2 and ERK signaling. Circ Res 2008;103:916–8.

[72] Peschon JJ, Slack JL, Reddy P, et al. An essential role for ectodomain shedding in mammalian development. Science 1998;282:1281–4.

[73] Díaz-Rodríguez E, Montero JC, Esparís-Ogando A, Yuste L, Pandiella A. Extracellular signal-regulated kinase phosphorylates tumor necrosis factor alpha-converting enzyme at threonine 735: a potential role in regulated shedding. Mol Biol Cell 2002;13:2031–44.

[74] Esselens CW, Malapeira J, Colomé N, Moss M, Canals F, Arribas J. Metastasis-associated C4.4A, a GPI-anchored protein cleaved by ADAM10 and ADAM17. Biol Chem 2008;389:1075–84.

[75] Fleck D, van Bebber F, Colombo A, et al. Dual cleavage of neuregulin 1 type III by BACE1 and ADAM17 liberates its EGF-like domain and allows paracrine signaling. J Neurosci 2013;33:7856–69.

[76] Na H-W, Shin W-S, Ludwig A, Lee S-T. The cytosolic domain of protein-tyrosine kinase 7 (PTK7), generated from sequential cleavage by a disintegrin and metalloprotease 17 (ADAM17) and γ-secretase, enhances cell proliferation and migration in colon cancer cells. J Biol Chem 2012;287:25001–9.

[77] Julian J, Dharmaraj N, Carson DD. MUC1 is a substrate for gamma-secretase. J Cell Biochem 2009;108:802–15.

[78] Romero Y, Wise R, Zolkiewska A. Proteolytic processing of PD-L1 by ADAM proteases in breast cancer cells. Cancer Immunol Immunother 2020;69:43–55.

[79] Ruhe JE, Streit S, Hart S, Ullrich A. EGFR signaling leads to downregulation of PTP-LAR via TACE-mediated proteolytic processing. Cell Signal 2006;18:1515–27.

[80] Kalus I, Bormann U, Mzoughi M, Schachner M, Kleene R. Proteolytic cleavage of the neural cell adhesion molecule by ADAM17/TACE is involved in neurite outgrowth. J Neurochem 2006;98:78–88.

[81] Althoff K, Müllberg J, Aasland D, Voltz N, Kallen K, Grötzinger J, Rose-John S. Recognition sequences and structural elements contribute to shedding susceptibility of membrane proteins. Biochem J 2001;353:663–72.

[82] Borrell-Pagès M, Rojo F, Albanell J, Baselga J, Arribas J. TACE is required for the activation of the EGFR by TGF-alpha in tumors. EMBO J 2003;22:1114–24.

[83] Kenny PA, Bissell MJ. Targeting TACE-dependent EGFR ligand shedding in breast cancer. J Clin Invest 2007;117:337–45.

[84] Fabre-Lafay S, Garrido-Urbani S, Reymond N, Gonçalves A, Dubreuil P, Lopez M. Nectin-4, a new serological breast cancer marker, is a substrate for tumor necrosis factor-alpha-converting enzyme (TACE)/ADAM-17. J Biol Chem 2005;280:19543–50.

[85] Schweigert O, Dewitz C, Möller-Hackbarth K, Trad A, Garbers C, Rose-John S, Scheller J. Soluble T cell immunoglobulin and mucin domain (TIM)-1 and -4 generated by A Disintegrin And Metalloprotease (ADAM)-10 and -17 bind to phosphatidylserine. Biochim Biophys Acta 2014;1843:275–87.

[86] Zhu L, Bergmeier W, Wu J, et al. Regulated surface expression and shedding support a dual role for semaphorin 4D in platelet responses to vascular injury. Proc Natl Acad Sci U S A 2007;104:1621–6.

[87] Tanabe Y, Kasahara T, Momoi T, Fujita E. Neuronal RA175/SynCAM1 isoforms are processed by tumor necrosis factor-alpha-converting enzyme (TACE)/ADAM17-like proteases. Neurosci Lett 2008;444:16–21.

[88] Ermert M, Pantazis C, Duncker H-R, Grimminger F, Seeger W, Ermert L. In situ localization of TNFalpha/beta, TACE and TNF receptors TNF-R1 and TNF-R2 in control and LPS-treated lung tissue. Cytokine 2003;22:89–100.

[89] Caolo V, Swennen G, Chalaris A, Wagenaar A, Verbruggen S, Rose-John S, Molin DGM, Vooijs M, Post MJ. ADAM10 and ADAM17 have opposite roles during sprouting angiogenesis. Angiogenesis 2015;18:13–22.

[90] Garton KJ, Gough PJ, Raines EW. Emerging roles for ectodomain shedding in the regulation of inflammatory responses. J Leukoc Biol 2006;79:1105–16.

[91] Singh RJR, Mason JC, Lidington EA, Edwards DR, Nuttall RK, Khokha R, Knauper V, Murphy G, Gavrilovic J. Cytokine stimulated vascular cell adhesion molecule-1 (VCAM-1) ectodomain release is regulated by TIMP-3. Cardiovasc Res 2005;67:39–49.

[92] Bandsma RHJ, van Goor H, Yourshaw M, et al. Loss of ADAM17 is associated with severe multiorgan dysfunction. Hum Pathol 2015;46:923–8.

[93] Srour N, Lebel A, McMahon S, Fournier I, Fugère M, Day R, Dubois CM. TACE/ADAM-17 maturation and activation of sheddase activity require proprotein convertase activity. FEBS Lett 2003;554:275–83.

[94] Franzke C-W, Cobzaru C, Triantafyllopoulou A, Löffek S, Horiuchi K, Threadgill DW, Kurz T, van Rooijen N, Bruckner-Tuderman L, Blobel CP. Epidermal ADAM17 maintains the skin barrier by regulating EGFR ligand-dependent terminal keratinocyte differentiation. J Exp Med 2012;209:1105–19.

[95] Hall T, Shieh HS, Day JE, et al. Structure of human ADAM-8 catalytic domain complexed with batimastat. Acta Crystallogr Sect F Struct Biol Cryst Commun 2012;68:616–21.

[96] Horiuchi K, Kimura T, Miyamoto T, Takaishi H, Okada Y, Toyama Y, Blobel CP. Cutting edge: TNF-alpha-converting enzyme (TACE/ADAM17) inactivation in mouse myeloid cells prevents lethality from endotoxin shock. J Immunol 2007;179:2686–9.

[97] Bell JH, Herrera AH, Li Y, Walcheck B. Role of ADAM17 in the ectodomain shedding of TNF-alpha and its receptors by neutrophils and macrophages. J Leukoc Biol 2007;82:173–6.

[98] Li Y, Brazzell J, Herrera A, Walcheck B. ADAM17 deficiency by mature neutrophils has differential effects on L-selectin shedding. Blood 2006;108:2275–9.

[99] MacEwan DJ. TNF ligands and receptors—a matter of life and death. Br J Pharmacol 2002;135:855–75.

[100] Schumertl T, Lokau J, Rose-John S, Garbers C. Function and proteolytic generation of the soluble interleukin-6 receptor in health and disease. Biochim Biophys Acta Mol Cell Res 2022;1869:119143.

[101] Ishii S, Isozaki T, Furuya H, Takeuchi H, Tsubokura Y, Inagaki K, Kasama T. ADAM-17 is expressed on rheumatoid arthritis fibroblast-like synoviocytes and regulates proinflammatory mediator expression and monocyte adhesion. Arthritis Res Ther 2018;20:159.

[102] Issuree PDA, Maretzky T, McIlwain DR, et al. iRHOM2 is a critical pathogenic mediator of inflammatory arthritis. J Clin Invest 2013;123:928–32.

[103] Alexopoulou L, Kranidioti K, Xanthoulea S, Denis M, Kotanidou A, Douni E, Blackshear PJ, Kontoyiannis DL, Kollias G. Transmembrane TNF protects mutant mice against intracellular bacterial infections, chronic inflammation and autoimmunity. Eur J Immunol 2006;36:2768–80.

[104] Kaneko T, Horiuchi K, Chijimatsu R, et al. Regulation of osteoarthritis development by ADAM17/Tace in articular cartilage. J Bone Miner Metab 2022;40:196–207.

[105] Mukhopadhyay S, Hoidal JR, Mukherjee TK. Role of TNFalpha in pulmonary pathophysiology. Respir Res 2006;7:125.

[106] Dreymueller D, Martin C, Kogel T, Pruessmeyer J, Hess FM, Horiuchi K, Uhlig S, Ludwig A. Lung endothelial ADAM17 regulates the acute inflammatory response to lipopolysaccharide. EMBO Mol Med 2012;4:412–23.

[107] Shiomi T, Tschumperlin DJ, Park J-A, Sunnarborg SW, Horiuchi K, Blobel CP, Drazen JM. TNF-α-converting enzyme/a disintegrin and metalloprotease-17 mediates mechanotransduction in murine tracheal epithelial cells. Am J Respir Cell Mol Biol 2011;45:376–85.

[108] Kefaloyianni E, Muthu ML, Kaeppler J, et al. ADAM17 substrate release in proximal tubule drives kidney fibrosis. JCI Insight 2016;1:e87023.

[109] McKellar GE, McCarey DW, Sattar N, McInnes IB. Role for TNF in atherosclerosis? Lessons from autoimmune disease. Nat Rev Cardiol 2009;6:410–7.

[110] Nicolaou A, Zhao Z, Northoff BH, et al. Adam17 deficiency promotes atherosclerosis by enhanced TNFR2 signaling in mice. Arterioscler Thromb Vasc Biol 2017;37:247–57.

[111] Berns M, Hommes DW. Anti-TNF-α therapies for the treatment of Crohn's disease: the past, present and future. Expert Opin Investig Drugs 2016;25:129–43.

[112] Reinecker HC, Steffen M, Witthoeft T, Pflueger I, Schreiber S, MacDermott RP, Raedler A. Enhanced secretion of tumour necrosis factor-alpha, IL-6, and IL-1 beta by isolated lamina propria mononuclear cells from patients with ulcerative colitis and Crohn's disease. Clin Exp Immunol 1993;94:174–81.

[113] Shimoda M, Horiuchi K, Sasaki A, Tsukamoto T, Okabayashi K, Hasegawa H, Kitagawa Y, Okada Y. Epithelial cell-derived a disintegrin and metalloproteinase-17 confers resistance to colonic inflammation through EGFR activation. EBioMedicine 2016;5:114–24.

[114] Saad MI, Weng T, Lundy J, et al. Blockade of the protease ADAM17 ameliorates experimental pancreatitis. Proc Natl Acad Sci U S A 2022;119. e2213744119.

[115] Prokop S, Miller KR, Heppner FL. Microglia actions in Alzheimer's disease. Acta Neuropathol 2013;126:461–77.

[116] De Jager PL, Srivastava G, Lunnon K, et al. Alzheimer's disease: early alterations in brain DNA methylation at ANK1, BIN1, RHBDF2 and other loci. Nat Neurosci 2014;17:1156–63.

[117] Tobinick E. Perispinal etanercept for treatment of Alzheimer's disease. Curr Alzheimer Res 2007;4:550–2.

[118] Lammich S, Kojro E, Postina R, Gilbert S, Pfeiffer R, Jasionowski M, Haass C, Fahrenholz F. Constitutive and regulated alpha-secretase cleavage of Alzheimer's amyloid precursor protein by a disintegrin metalloprotease. Proc Natl Acad Sci U S A 1999;96:3922–7.

[119] Sommer D, Corstjens I, Sanchez S, et al. ADAM17-deficiency on microglia but not on macrophages promotes phagocytosis and functional recovery after spinal cord injury. Brain Behav Immun 2019;80:129–45.

[120] Rossello A, Nuti E, Ferrini S, Fabbi M. Targeting ADAM17 sheddase activity in cancer. Curr Drug Targets 2016;17:1908–27.

[121] Schmidt S, Schumacher N, Schwarz J, et al. ADAM17 is required for EGF-R-induced intestinal tumors via IL-6 trans-signaling. J Exp Med 2018;215:1205–25.

[122] Gao M-Q, Kim BG, Kang S, Choi YP, Yoon J-H, Cho NH. Human breast cancer-associated fibroblasts enhance cancer cell proliferation through increased TGF-α cleavage by ADAM17. Cancer Lett 2013;336:240–6.

[123] Bolik J, Krause F, Stevanovic M, et al. Inhibition of ADAM17 impairs endothelial cell necroptosis and blocks metastasis. J Exp Med 2022;219:e20201039.

[124] Orme JJ, Jazieh KA, Xie T, et al. ADAM10 and ADAM17 cleave PD-L1 to mediate PD-(L)1 inhibitor resistance. Oncoimmunology 2020;9:1744980.

[125] Romee R, Foley B, Lenvik T, et al. NK cell CD16 surface expression and function is regulated by a disintegrin and metalloprotease-17 (ADAM17). Blood 2013;121:3599–608.

[126] Yoshiyama K, Higuchi Y, Kataoka M, Matsuura K, Yamamoto S. CD156 (human ADAM8): expression, primary amino acid sequence, and gene location. Genomics 1997;41:56–62.

[127] Schlomann U, Wildeboer D, Webster A, et al. The metalloprotease disintegrin ADAM8. Processing by autocatalysis is required for proteolytic activity and cell adhesion. J Biol Chem 2002;277:48210–9.

[128] Amour A, Knight CG, English WR, Webster A, Slocombe PM, Knäuper V, Docherty AJP, Becherer JD, Blobel CP, Murphy G. The enzymatic activity of ADAM8 and ADAM9 is not regulated by TIMPs. FEBS Lett 2002;524:154–8.

[129] Rao H, Lu G, Kajiya H, et al. Alpha9beta1: a novel osteoclast integrin that regulates osteoclast formation and function. J Bone Miner Res 2006;21:1657–65.

[130] Yoshida S, Setoguchi M, Higuchi Y, Akizuki S, Yamamoto S. Molecular cloning of cDNA encoding MS2 antigen, a novel cell surface antigen strongly expressed in murine monocytic lineage. Int Immunol 1990;2:585–91.

[131] Fourie AM, Coles F, Moreno V, Karlsson L. Catalytic activity of ADAM8, ADAM15, and MDC-L (ADAM28) on synthetic peptide substrates and in ectodomain cleavage of CD23. J Biol Chem 2003;278:30469–77.

[132] Moss ML, Rasmussen FH. Fluorescent substrates for the proteinases ADAM17, ADAM10, ADAM8, and ADAM12 useful for high-throughput inhibitor screening. Anal Biochem 2007;366:144–8.

[133] Naus S, Richter M, Wildeboer D, Moss M, Schachner M, Bartsch JW. Ectodomain shedding of the neural recognition molecule CHL1 by the metalloprotease-disintegrin ADAM8 promotes neurite outgrowth and suppresses neuronal cell death *. J Biol Chem 2004;279:16083–90.

[134] Bartsch JW, Wildeboer D, Koller G, Naus S, Rittger A, Moss ML, Minai Y, Jockusch H. Tumor necrosis factor-α (TNF-α) regulates shedding of TNF-α receptor 1 by the metalloprotease-disintegrin ADAM8: evidence for a protease-regulated feedback loop in neuroprotection. J Neurosci 2010;30:12210–8.

[135] Gómez-Gaviro M, Domínguez-Luis M, Canchado J, et al. Expression and regulation of the metalloproteinase ADAM-8 during human neutrophil pathophysiological activation and its catalytic activity on L-selectin shedding. J Immunol 2007;178:8053–63.

[136] Naus S, Reipschläger S, Wildeboer D, Lichtenthaler SF, Mitterreiter S. Identification of candidate substrates for ectodomain shedding by the metalloprotease-disintegrin ADAM8. Biol Chem 2006. https://doi.org/10.1515/BC.2006.045.

[137] Nishimura D, Sakai H, Sato T, Sato F, Nishimura S, Toyama-Sorimachi N, Bartsch JW, Sehara-Fujisawa A. Roles of ADAM8 in elimination of injured muscle fibers prior to skeletal muscle regeneration. Mech Dev 2015;135:58–67.

[138] Liang J, Wang W, Sorensen D, Medina S, Ilchenko S, Kiselar J, Surewicz WK, Booth SA, Kong Q. Cellular prion protein regulates its own α-cleavage through ADAM8 in skeletal muscle *. J Biol Chem 2012;287:16510–20.

[139] Kelly K, Hutchinson G, Nebenius-Oosthuizen D, Smith AJH, Bartsch JW, Horiuchi K, Rittger A, Manova K, Docherty AJP, Blobel CP. Metalloprotease-disintegrin ADAM8: expression analysis and targeted deletion in mice. Dev Dyn 2005;232:221–31.

[140] Choi SJ, Han JH, Roodman GD. ADAM8: a novel osteoclast stimulating factor. J Bone Miner Res 2001;16:814–22.

[141] King NE, Zimmermann N, Pope SM, Fulkerson PC, Nikolaidis NM, Mishra A, Witte DP, Rothenberg ME. Expression and regulation of a disintegrin and metalloproteinase (ADAM) 8 in experimental asthma. Am J Respir Cell Mol Biol 2004;31:257–65.

[142] Chen J, Deng L, Dreymüller D, et al. A novel peptide ADAM8 inhibitor attenuates bronchial hyperresponsiveness and Th2 cytokine mediated inflammation of murine asthmatic models. Sci Rep 2016;6:30451.

[143] Dreymueller D, Pruessmeyer J, Schumacher J, Fellendorf S, Hess FM, Seifert A, Babendreyer A, Bartsch JW, Ludwig A. The metalloproteinase ADAM8 promotes leukocyte recruitment in vitro and in acute lung inflammation. Am J Physiol Lung Cell Mol Physiol 2017;313:L602–14.

[144] Zack MD, Malfait A-M, Skepner AP, et al. ADAM-8 isolated from human osteoarthritic chondrocytes cleaves fibronectin at Ala(271). Arthritis Rheum 2009;60:2704–13.

[145] Zhang Y, Tian Z, Gerard D, Yao L, Shofer FS, Cs-Szabo G, Qin L, Pacifici M, Enomoto-Iwamoto M. Elevated inflammatory gene expression in intervertebral disc tissues in mice with ADAM8 inactivated. Sci Rep 2021;11:1804.

[146] Schlomann U, Rathke-Hartlieb S, Yamamoto S, Jockusch H, Bartsch JW. Tumor necrosis factor alpha induces a metalloprotease-disintegrin, ADAM8 (CD 156): implications for neuron-glia interactions during neurodegeneration. J Neurosci 2000;20:7964–71.

[147] Naus S, Reipschläger S, Wildeboer D, Lichtenthaler SF, Mitterreiter S, Guan Z, Moss ML, Bartsch JW. Identification of candidate substrates for ectodomain shedding by the metalloprotease-disintegrin ADAM8. Biol Chem 2006;387:337–46.

[148] Chantry A, Gregson NA, Glynn P. A novel metalloproteinase associated with brain myelin membranes. Isolation and characterization. J Biol Chem 1989;264:21603–7.

[149] Wildeboer D, Naus S, Amy Sang Q-X, Bartsch JW, Pagenstecher A. Metalloproteinase disintegrins ADAM8 and ADAM19 are highly regulated in human primary brain tumors and their expression levels and activities are associated with invasiveness. J Neuropathol Exp Neurol 2006;65:516–27.

[150] Huang J, Bai Y, Huo L, Xiao J, Fan X, Yang Z, Chen H, Yang Z. Upregulation of a disintegrin and metalloprotease 8 is associated with progression and prognosis of patients with gastric cancer. Transl Res 2015;166:602–13.

[151] Valkovskaya N, Kayed H, Felix K, Hartmann D, Giese NA, Osinsky SP, Friess H, Kleeff J. ADAM8 expression is associated with increased invasiveness and reduced patient survival in pancreatic cancer. J Cell Mol Med 2007;11:1162–74.

[152] Zhang Y, Tan Y-F, Jiang C, Zhang K, Zha T-Z, Zhang M. High ADAM8 expression is associated with poor prognosis in patients with hepatocellular carcinoma. Pathol Oncol Res 2013;19:79–88.

[153] Yang Z, Bai Y, Huo L, Chen H, Huang J, Li J, Fan X, Yang Z, Wang L, Wang J. Expression of A disintegrin and metalloprotease 8 is associated with cell growth and poor survival in colorectal cancer. BMC Cancer 2014;14:568.

[154] Errico A. Gastrointestinal cancer: ADAM8 provides new hope in pancreatic cancer. Nat Rev Clin Oncol 2015;12:126.

[155] Schlomann U, Koller G, Conrad C, et al. ADAM8 as a drug target in pancreatic cancer. Nat Commun 2015;6:6175.

[156] Mahoney ET, Benton RL, Maddie MA, Whittemore SR, Hagg T. ADAM8 is selectively up-regulated in endothelial cells and is associated with angiogenesis after spinal cord injury in adult mice. J Comp Neurol 2009;512:243–55.

[157] Saftig P, Lichtenthaler SF. The alpha secretase ADAM10: a metalloprotease with multiple functions in the brain. Prog Neurobiol 2015;135:1–20.

[158] Wilke GA, Bubeck Wardenburg J. Role of a disintegrin and metalloprotease 10 in *Staphylococcus aureus* alpha-hemolysin-mediated cellular injury. Proc Natl Acad Sci U S A 2010;107:13473–8.

[159] Marcello E, Saraceno C, Musardo S, et al. Endocytosis of synaptic ADAM10 in neuronal plasticity and Alzheimer's disease. J Clin Invest 2013;123:2523–38.

[160] Wild-Bode C, Fellerer K, Kugler J, Haass C, Capell A. A basolateral sorting signal directs ADAM10 to adherens junctions and is required for its function in cell migration. J Biol Chem 2006;281:23824–9.

[161] Matthews AL, Szyroka J, Collier R, Noy PJ, Tomlinson MG. Scissor sisters: regulation of ADAM10 by the TspanC8 tetraspanins. Biochem Soc Trans 2017;45:719–30.

[162] Jouannet S, Saint-Pol J, Fernandez L, Nguyen V, Charrin S, Boucheix C, Brou C, Milhiet P-E, Rubinstein E. TspanC8 tetraspanins differentially regulate the cleavage of ADAM10 substrates, Notch activation and ADAM10 membrane compartmentalization. Cell Mol Life Sci 2016;73:1895–915.

[163] Haining EJ, Yang J, Bailey RL, Khan K, Collier R, Tsai S, Watson SP, Frampton J, Garcia P, Tomlinson MG. The TspanC8 subgroup of tetraspanins interacts with A disintegrin and metalloprotease 10 (ADAM10) and regulates its maturation and cell surface expression. J Biol Chem 2012;287:39753–65.

[164] Hundhausen C, Schulte A, Schulz B, et al. Regulated shedding of transmembrane chemokines by the disintegrin and metalloproteinase 10 facilitates detachment of adherent leukocytes. J Immunol 2007;178:8064–72.

[165] Weskamp G, Ford JW, Sturgill J, et al. ADAM10 is a principal "sheddase" of the low-affinity immunoglobulin E receptor CD23. Nat Immunol 2006;7:1293–8.

[166] Lemieux GA, Blumenkron F, Yeung N, Zhou P, Williams J, Grammer AC, Petrovich R, Lipsky PE, Moss ML, Werb Z. The low affinity IgE receptor (CD23) is cleaved by the metalloproteinase ADAM10. J Biol Chem 2007;282:14836–44.

[167] Nagano O, Murakami D, Hartmann D, De Strooper B, Saftig P, Iwatsubo T, Nakajima M, Shinohara M, Saya H. Cell-matrix interaction via CD44 is independently regulated by different metalloproteinases activated in response to extracellular Ca(2+) influx and PKC activation. J Cell Biol 2004;165:893–902.

[168] Bloch L, Sineshchekova O, Reichenbach D, Reiss K, Saftig P, Kuro-o M, Kaether C. Klotho is a substrate for alpha-, beta- and gamma-secretase. FEBS Lett 2009;583:3221–4.

[169] Colombo A, Wang H, Kuhn P-H, Page R, Kremmer E, Dempsey PJ, Crawford HC, Lichtenthaler SF. Constitutive α- and β-secretase cleavages of the amyloid precursor protein are partially coupled in neurons, but not in frequently used cell lines. Neurobiol Dis 2013;49:137–47.

[170] Jorissen E, Prox J, Bernreuther C, et al. The disintegrin/metalloproteinase ADAM10 is essential for the establishment of the brain cortex. J Neurosci 2010;30:4833–44.

[171] Kuhn P-H, Wang H, Dislich B, Colombo A, Zeitschel U, Ellwart JW, Kremmer E, Rossner S, Lichtenthaler SF. ADAM10 is the physiologically relevant, constitutive alpha-secretase of the amyloid precursor protein in primary neurons. EMBO J 2010;29:3020–32.

[172] Scholz F, Schulte A, Adamski F, Hundhausen C, Mittag J, Schwarz A, Kruse M-L, Proksch E, Ludwig A. Constitutive expression and regulated release of the transmembrane chemokine CXCL16 in human and murine skin. J Invest Dermatol 2007;127:1444–55.

[173] Kopitz C, Gerg M, Bandapalli OR, et al. Tissue inhibitor of metalloproteinases-1 promotes liver metastasis by induction of hepatocyte growth factor signaling. Cancer Res 2007;67:8615–23.

[174] Eichenauer DA, Simhadri VL, von Strandmann EP, et al. ADAM10 inhibition of human CD30 shedding increases specificity of targeted immunotherapy in vitro. Cancer Res 2007;67:332–8.

[175] Maretzky T, Reiss K, Ludwig A, Buchholz J, Scholz F, Proksch E, de Strooper B, Hartmann D, Saftig P. ADAM10 mediates E-cadherin shedding and regulates epithelial cell-cell adhesion, migration, and beta-catenin translocation. Proc Natl Acad Sci U S A 2005;102:9182–7.

[176] Herzog C, Haun RS, Ludwig A, Shah SV, Kaushal GP. ADAM10 is the major sheddase responsible for the release of membrane-associated meprin A. J Biol Chem 2014. https://doi.org/10.1074/jbc.M114.559088.

[177] Altmeppen HC, Prox J, Puig B, et al. Lack of a-disintegrin-and-metalloproteinase ADAM10 leads to intracellular accumulation and loss of shedding of the cellular prion protein in vivo. Mol Neurodegener 2011;6:36.

[178] Altmeppen HC, Prox J, Krasemann S, et al. The sheddase ADAM10 is a potent modulator of prion disease. Elife 2015;4:e04260.

[179] Kirkin V, Cahuzac N, Guardiola-Serrano F, et al. The Fas ligand intracellular domain is released by ADAM10 and SPPL2a cleavage in T-cells. Cell Death Differ 2007;14:1678–87.

[180] Schulte M, Reiss K, Lettau M, Maretzky T, Ludwig A, Hartmann D, de Strooper B, Janssen O, Saftig P. ADAM10 regulates FasL cell surface expression and modulates FasL-induced cytotoxicity and activation-induced cell death. Cell Death Differ 2007;14:1040–9.

[181] Qi H, Rand MD, Wu X, Sestan N, Wang W, Rakic P, Xu T, Artavanis-Tsakonas S. Processing of the notch ligand delta by the metalloprotease Kuzbanian. Science 1999;283:91–4.

[182] Six E, Ndiaye D, Laabi Y, Brou C, Gupta-Rossi N, Israel A, Logeat F. The Notch ligand Delta1 is sequentially cleaved by an ADAM protease and gamma-secretase. Proc Natl Acad Sci U S A 2003;100:7638–43.

[183] Liu PCC, Liu X, Li Y, et al. Identification of ADAM10 as a major source of HER2 ectodomain sheddase activity in HER2 overexpressing breast cancer cells. Cancer Biol Ther 2006;5:657–64.

[184] Reiss K, Maretzky T, Ludwig A, Tousseyn T, de Strooper B, Hartmann D, Saftig P. ADAM10 cleavage of N-cadherin and regulation of cell-cell adhesion and beta-catenin nuclear signalling. EMBO J 2005;24:742–52.

[185] Hikita A, Yana I, Wakeyama H, Nakamura M, Kadono Y, Oshima Y, Nakamura K, Seiki M, Tanaka S. Negative regulation of osteoclastogenesis by ectodomain shedding of receptor activator of NF-kappaB ligand. J Biol Chem 2006;281:36846–55.

[186] Bozkulak EC, Weinmaster G. Selective use of ADAM10 and ADAM17 in activation of Notch1 signaling. Mol Cell Biol 2009;29:5679–95.

[187] Hartmann D, de Strooper B, Serneels L, et al. The disintegrin/metalloprotease ADAM 10 is essential for Notch signalling but not for alpha-secretase activity in fibroblasts. Hum Mol Genet 2002;11:2615–24.

[188] van Tetering G, van Diest P, Verlaan I, van der Wall E, Kopan R, Vooijs M. Metalloprotease ADAM10 is required for Notch1 site 2 cleavage. J Biol Chem 2009;284:31018–27.

[189] Schulz B, Pruessmeyer J, Maretzky T, Ludwig A, Blobel CP, Saftig P, Reiss K. ADAM10 regulates endothelial permeability and T-cell transmigration by proteolysis of vascular endothelial cadherin. Circ Res 2008;102:1192–201.

[190] Martin L, Fluhrer R, Reiss K, Kremmer E, Saftig P, Haass C. Regulated intramembrane proteolysis of Bri2 (Itm2b) by ADAM10 and SPPL2a/SPPL2b. J Biol Chem 2008;283:1644–52.

[191] Kaczur V, Puskas LG, Nagy ZU, Miled N, Rebai A, Juhasz F, Kupihar Z, Zvara A, Hackler L, Farid NR. Cleavage of the human thyrotropin receptor by ADAM10 is regulated by thyrotropin. J Mol Recognit 2007;20:392–404.

[192] Sakry D, Neitz A, Singh J, et al. Oligodendrocyte precursor cells modulate the neuronal network by activity-dependent ectodomain cleavage of glial NG2. PLoS Biol 2014. https://doi.org/10.1371/journal.pbio.1001993.

[193] Hattori M, Osterfield M, Flanagan JG. Regulated cleavage of a contact-mediated axon repellent. Science 2000;289:1360–5.

[194] Janes PW, Wimmer-Kleikamp SH, Frangakis AS, et al. Cytoplasmic relaxation of active Eph controls ephrin shedding by ADAM10. PLoS Biol 2009;7:e1000215.

[195] Mechtersheimer S, Gutwein P, Agmon-Levin N, et al. Ectodomain shedding of L1 adhesion molecule promotes cell migration by autocrine binding to integrins. J Cell Biol 2001;155:661–73.

[196] Brennaman LH, Moss ML, Maness PF. EphrinA/EphA-induced ectodomain shedding of neural cell adhesion molecule regulates growth cone repulsion through ADAM10 metalloprotease. J Neurochem 2014;128:267–79.

[197] Kim J, Lilliehook C, Dudak A, Prox J, Saftig P, Federoff HJ, Lim ST. Activity-dependent alpha-cleavage of nectin-1 is mediated by a disintegrin and metalloprotease 10 (ADAM10). J Biol Chem 2010;285:22919–26.

[198] Suzuki K, Hayashi Y, Nakahara S, et al. Activity-dependent proteolytic cleavage of neuroligin-1. Neuron 2012;76:410–22.

[199] Reiss K, Maretzky T, Haas IG, Schulte M, Ludwig A, Frank M, Saftig P. Regulated ADAM10-dependent ectodomain shedding of gamma-protocadherin C3 modulates cell-cell adhesion. J Biol Chem 2006;281:21735–44.

[200] Saftig P, Hartmann D. ADAM10: a major membrane protein ectodomain sheddase involved in regulated intramembrane proteolysis. In: Hooper NM, Lendeckel U, editors. The ADAM family of proteases. Springer Science & Business Media; 2006. p. 85–121. ISBN: s978-0-387-25151-6, 978-0-387-25149-3 https://doi.org/10.1007/b106833.

[201] Seifert A, Düsterhöft S, Wozniak J, Koo CZ, Tomlinson MG, Nuti E, Rossello A, Cuffaro D, Yildiz D, Ludwig A. The metalloproteinase ADAM10 requires its activity to sustain surface expression. Cell Mol Life Sci 2021;78:715–32.

[202] Wetzel S, Seipold L, Saftig P. The metalloproteinase ADAM10: a useful therapeutic target? Biochim Biophys Acta Mol Cell Res 2017;1864:2071–81.

[203] Golde TE, Dickson D, Hutton M. Filling the gaps in the abeta cascade hypothesis of Alzheimer's disease. Curr Alzheimer Res 2006;3:421–30.

[204] Vassar R. BACE1: the beta-secretase enzyme in Alzheimer's disease. J Mol Neurosci 2004;23:105–14.

[205] Postina R, Schroeder A, Dewachter I, et al. A disintegrin-metalloproteinase prevents amyloid plaque formation and hippocampal defects in an Alzheimer disease mouse model. J Clin Invest 2004;113:1456–64.

[206] Endres K, Fahrenholz F, Lotz J, Hiemke C, Teipel S, Lieb K, Tüscher O, Fellgiebel A. Increased CSF APPs-α levels in patients with Alzheimer disease treated with acitretin. Neurology 2014;83:1930–5.

[207] Linsenmeier L, Mohammadi B, Wetzel S, et al. Structural and mechanistic aspects influencing the ADAM10-mediated shedding of the prion protein. Mol Neurodegener 2018;13:18.

[208] Cozzolino F, Vezzoli E, Cheroni C, et al. ADAM10 hyperactivation acts on piccolo to deplete synaptic vesicle stores in Huntington's disease. Hum Mol Genet 2021;30:1175–87.

[209] Vezzoli E, Caron I, Talpo F, et al. Inhibiting pathologically active ADAM10 rescues synaptic and cognitive decline in Huntington's disease. J Clin Invest 2019;129:2390–403.

[210] Lo Sardo V, Zuccato C, Gaudenzi G, et al. An evolutionary recent neuroepithelial cell adhesion function of huntingtin implicates ADAM10-Ncadherin. Nat Neurosci 2012;15:713–21.

[211] Isozaki T, Rabquer BJ, Ruth JH, Haines GK, Koch AE. ADAM-10 is overexpressed in rheumatoid arthritis synovial tissue and mediates angiogenesis. Arthritis Rheum 2013;65:98–108.

[212] Lecoanet-Henchoz S, Gauchat JF, Aubry JP, Graber P, Life P, Paul-Eugene N, Ferrua B, Corbi AL, Dugas B, Plater-Zyberk C. CD23 regulates monocyte activation through a novel interaction with the adhesion molecules CD11b-CD18 and CD11c-CD18. Immunity 1995;3:119–25.

[213] Oh ST, Schramme A, Stark A, Tilgen W, Gutwein P, Reichrath J. Overexpression of ADAM 10 and ADAM 12 in lesional psoriatic skin. Br J Dermatol 2008;158:1371–3.

[214] Dogra S, Yadav S. Acitretin in psoriasis: an evolving scenario. Int J Dermatol 2014;53:525–38.

[215] Thélu J, Rossio P, Favier B. Notch signalling is linked to epidermal cell differentiation level in basal cell carcinoma, psoriasis and wound healing. BMC Dermatol 2002;2:7.

[216] Weber S, Niessen MT, Prox J, et al. The disintegrin/metalloproteinase Adam10 is essential for epidermal integrity and Notch-mediated signaling. Development 2011;138:495–505.

[217] Ekman C, Jönsen A, Sturfelt G, Bengtsson AA, Dahlbäck B. Plasma concentrations of Gas6 and sAxl correlate with disease activity in systemic lupus erythematosus. Rheumatology (Oxford) 2011;50:1064–9.

[218] van der Vorst EPC, Jeurissen M, Wolfs IMJ, et al. Myeloid A disintegrin and metalloproteinase domain 10 deficiency modulates atherosclerotic plaque composition by shifting the balance from inflammation toward fibrosis. Am J Pathol 2015;185:1145–55.

[219] Qu M, Qiu BO, Xiong W, Chen D, Wu A. Expression of a-disintegrin and metalloproteinase 10 correlates with grade of malignancy in human glioma. Oncol Lett 2015;9:2157–62.

[220] Yang M, Li Y, Chilukuri K, Brady OA, Boulos MI, Kappes JC, Galileo DS. L1 stimulation of human glioma cell motility correlates with FAK activation. J Neurooncol 2011;105:27–44.

[221] Kohutek ZA, diPierro CG, Redpath GT, Hussaini IM. ADAM-10-mediated N-cadherin cleavage is protein kinase C-alpha dependent and promotes glioblastoma cell migration. J Neurosci 2009;29:4605–15.

[222] Venkatesh HS, Tam LT, Woo PJ, et al. Targeting neuronal activity-regulated neuroligin-3 dependency in high-grade glioma. Nature 2017;549:533–7.

[223] Waldhauer I, Goehlsdorf D, Gieseke F, Weinschenk T, Wittenbrink M, Ludwig A, Stevanovic S, Rammensee H-G, Steinle A. -Tumor-associated MICA is shed by ADAM proteases. Cancer Res 2008;68:6368–76.

[224] Kanaya K, Sakai K, Hongo K, Fukushima M, Kawakubo M, Nakayama J. High expression of ADAM10 predicts a poor prognosis for patients with glioblastoma. Int J Clin Exp Pathol 2017;10:618–24.

[225] Zingoni A, Vulpis E, Loconte L, Santoni A. NKG2D ligand shedding in response to stress: role of ADAM10. Front Immunol 2020;11:447.

[226] Tosello V, Ferrando AA. The NOTCH signaling pathway: role in the pathogenesis of T-cell acute lymphoblastic leukemia and implication for therapy. Ther Adv Hematol 2013;4:199–210.

[227] Tian L, Wu X, Chi C, Han M, Xu T, Zhuang Y. ADAM10 is essential for proteolytic activation of Notch during thymocyte development. Int Immunol 2008;20:1181–7.

[228] Sulis ML, Saftig P, Ferrando AA. Redundancy and specificity of the metalloprotease system mediating oncogenic NOTCH1 activation in T-ALL. Leukemia 2011;25:1564–9.

[229] Ménard S, Tagliabue E, Campiglio M, Pupa SM. Role of HER2 gene overexpression in breast carcinoma. J Cell Physiol 2000; 182:150–62.

[230] Valabrega G, Montemurro F, Aglietta M. Trastuzumab: mechanism of action, resistance and future perspectives in HER2-overexpressing breast cancer. Ann Oncol 2007;18:977–84.

[231] Sanderson MP, Erickson SN, Gough PJ, Garton KJ, Wille PT, Raines EW, Dunbar AJ, Dempsey PJ. ADAM10 mediates ectodomain shedding of the betacellulin precursor activated by p-aminophenylmercuric acetate and extracellular calcium influx. J Biol Chem 2005;280:1826–37.

[232] Feldinger K, Generali D, Kramer-Marek G, et al. ADAM10 mediates trastuzumab resistance and is correlated with survival in HER2 positive breast cancer. Oncotarget 2014;5:6633–46.

[233] Seegar TCM, Killingsworth LB, Saha N, et al. Structural basis for regulated proteolysis by the α-secretase ADAM10. Cell 2017;171:1638–1648.e7.

[234] Maskos K, Fernandez-Catalan C, Huber R, et al. Crystal structure of the catalytic domain of human tumor necrosis factor-alpha-converting enzyme. Proc Natl Acad Sci U S A 1998;95:3408–12.

[235] Nuti E, Casalini F, Santamaria S, et al. Selective arylsulfonamide inhibitors of ADAM-17: hit optimization and activity in ovarian cancer cell models. J Med Chem 2013;56:8089–103.

[236] Fischer T, Senn N, Riedl R. Design and structural evolution of matrix metalloproteinase inhibitors. Chemistry 2019;25:7960–80.

[237] Cuffaro D, Camodeca C, D'Andrea F, Piragine E, Testai L, Calderone V, Orlandini E, Nuti E, Rossello A. Matrix metalloproteinase-12 inhibitors: synthesis, structure-activity relationships and intestinal absorption of novel sugar-based biphenylsulfonamide carboxylates. Bioorg Med Chem 2018;26:5804–15.

[238] Nuti E, Cuffaro D, Bernardini E, et al. Development of thioaryl-based matrix metalloproteinase-12 inhibitors with alternative zinc-binding groups: synthesis, potentiometric, NMR, and crystallographic studies. J Med Chem 2018;61:4421–35.

[239] Cuffaro D, Nuti E, D'Andrea F, Rossello A. Developments in carbohydrate-based metzincin inhibitors. Pharmaceuticals (Basel) 2020;13:376.

[240] Camodeca C, Cuffaro D, Nuti E, Rossello A. ADAM metalloproteinases as potential drug targets. Curr Med Chem 2019;26:2661–89.

[241] Dormán G, Cseh S, Hajdú I, Barna L, Kónya D, Kupai K, Kovács L, Ferdinandy P. Matrix metalloproteinase inhibitors: a critical appraisal of design principles and proposed therapeutic utility. Drugs 2010;70:949–64.

[242] Knolle MD, Owen CA. ADAM8: a new therapeutic target for asthma. Expert Opin Ther Targets 2009;13:523–40.

[243] Yim V, Noisier AFM, Hung K-Y, Bartsch JW, Schlomann U, Brimble MA. Synthesis and biological evaluation of analogues of the potent ADAM8 inhibitor cyclo(RLsKDK) for the treatment of inflammatory diseases and cancer metastasis. Bioorg Med Chem 2016;24:4032–7.

[244] Schlomann U, Dorzweiler K, Nuti E, Tuccinardi T, Rossello A, Bartsch JW. Metalloprotease inhibitor profiles of human ADAM8 in vitro and in cell-based assays. Biol Chem 2019;400:801–10.

[245] Cuffaro D, Camodeca C, Tuccinardi T, Ciccone L, Bartsch JW, Kellermann T, Cook L, Nuti E, Rossello A. Discovery of dimeric arylsulfonamides as potent ADAM8 inhibitors. ACS Med Chem Lett 2021;12:1787–93.

[246] Wolpert F, Tritschler I, Steinle A, Weller M, Eisele G. A disintegrin and metalloproteinases 10 and 17 modulate the immunogenicity of glioblastoma-initiating cells. Neuro Oncol 2014;16:382–91.

[247] Mullooly M, McGowan PM, Kennedy SA, Madden SF, Crown J, O'Donovan N, Duffy MJ. ADAM10: a new player in breast cancer progression? Br J Cancer 2015;113:945–51.

[248] Appel D, Hummel R, Weidemeier M, Endres K, Gölz C, Schäfer MKE. Pharmacologic inhibition of ADAM10 attenuates brain tissue loss, axonal injury and pro-inflammatory gene expression following traumatic brain injury in mice. Front Cell Dev Biol 2021;9:661462.

[249] Wege AK, Dreyer TF, Teoman A, Ortmann O, Brockhoff G, Bronger H. CX3CL1 overexpression prevents the formation of lung metastases in trastuzumab-treated MDA-MB-453-based humanized tumor mice (HTM). Cancers (Basel) 2021;13:2459.

[250] Smith TM, Tharakan A, Martin RK. Targeting ADAM10 in cancer and autoimmunity. Front Immunol 2020;11:499.

[251] Cai C, Zhang M, Liu L, Zhang H, Guo Y, Lan T, Xu Y, Ma P, Li S. ADAM10-cleaved ephrin-A5 contributes to prostate cancer metastasis. Cell Death Dis 2022;13:453.

[252] Newton RC, Bradley EC, Levy RS, Doval D, Bondarde S, Sahoo TP, Lokanatha D, Julka PK, Nagarkar R, Friedman SM. Clinical benefit of INCB7839, a potent and selective ADAM inhibitor, in combination with trastuzumab in patients with metastatic HER2+ breast cancer. J Clin Oncol 2010;28:3025.

[253] Friedman S, Levy R, Garrett W, et al. Clinical benefit of INCB7839, a potent and selective inhibitor of ADAM10 and ADAM17, in combination with trastuzumab in metastatic HER2 positive breast cancer patients. Cancer Res 2009;69:5056.

[254] Camodeca C, Nuti E, Tepshi L, Boero S, Tuccinardi T, Stura EA, Poggi A, Zocchi MR, Rossello A. Discovery of a new selective inhibitor of A Disintegrin And Metalloprotease 10 (ADAM-10) able to reduce the shedding of NKG2D ligands in Hodgkin's lymphoma cell models. Eur J Med Chem 2016;111:193–201.

[255] Tosetti F, Venè R, Camodeca C, Nuti E, Rossello A, D'Arrigo C, Galante D, Ferrari N, Poggi A, Zocchi MR. Specific ADAM10 inhibitors localize in exosome-like vesicles released by Hodgkin lymphoma and stromal cells and prevent sheddase activity carried to bystander cells. Oncoimmunology 2018;7:e1421889.

[256] Mahasenan KV, Ding D, Gao M, Nguyen TT, Suckow MA, Schroeder VA, Wolter WR, Chang M, Mobashery S. In search of selectivity in inhibition of ADAM10. ACS Med Chem Lett 2018;9:708–13.

[257] Ouvry G, Berton Y, Bhurruth-Alcor Y, et al. Identification of novel TACE inhibitors compatible with topical application. Bioorg Med Chem Lett 2017;27:1848–53.

[258] Boiteau J-G, Ouvry G, Arlabosse J-M, et al. Discovery and process development of a novel TACE inhibitor for the topical treatment of psoriasis. Bioorg Med Chem 2018;26:945–56.

[259] Hirata S, Murata T, Suzuki D, Nakamura S, Jono-Ohnishi R, Hirose H, Sawaguchi A, Nishimura S, Sugimoto N, Eto K. Selective inhibition of ADAM17 efficiently mediates glycoprotein Ibα retention during ex vivo generation of human induced pluripotent stem cell-derived platelets. Stem Cells Transl Med 2017;6:720–30.

[260] Girijavallabhan VM, Chen L, Dai C, et al. Novel TNF-α converting enzyme (TACE) inhibitors as potential treatment for inflammatory diseases. Bioorg Med Chem Lett 2010;20:7283–7.

[261] Tong L, Kim SH, Rosner K, et al. Fused bi-heteroaryl substituted hydantoin compounds as TACE inhibitors. Bioorg Med Chem Lett 2017;27:3037–42.

[262] Tong L, Kim SH, Chen L, et al. Development of a prodrug of hydantoin based TACE inhibitor. Bioorg Med Chem Lett 2017;27:3704–8.

[263] Maekawa M, Tadaki H, Tomimoto D, et al. A novel TNF-α converting enzyme (TACE) selective inhibitor JTP-96193 prevents insulin resistance in KK-ay type 2 diabetic mice and diabetic peripheral neuropathy in type 1 diabetic mice. Biol Pharm Bull 2019;42:1906–12.

[264] Leung C-H, Liu L-J, Lu L, He B, Kwong DWJ, Wong C-Y, Ma D-L. A metal-based tumour necrosis factor-alpha converting enzyme inhibitor. Chem Commun (Camb) 2015;51:3973–6.

[265] Knapinska AM, Dreymuller D, Ludwig A, et al. SAR studies of exosite-binding substrate-selective inhibitors of a disintegrin and metalloprotease 17 (ADAM17) and application as selective in vitro probes. J Med Chem 2015;58:5808–24.

[266] Zhang Y, Li D, Jiang Q, et al. Novel ADAM-17 inhibitor ZLDI-8 enhances the in vitro and in vivo chemotherapeutic effects of Sorafenib on hepatocellular carcinoma cells. Cell Death Dis 2018;9:743.

[267] Lu H-Y, Chu H-X, Tan Y-X, Qin X-C, Liu M-Y, Li J-D, Ren T-S, Zhang Y-S, Zhao Q-C. Novel ADAM-17 inhibitor ZLDI-8 inhibits the metastasis of hepatocellular carcinoma by reversing epithelial-mesenchymal transition in vitro and in vivo. Life Sci 2020;244:117343.

[268] Lu H-Y, Zu Y-X, Jiang X-W, Sun X-T, Liu T-Y, Li R-L, Wu Q, Zhang Y-S, Zhao Q-C. Novel ADAM-17 inhibitor ZLDI-8 inhibits the proliferation and metastasis of chemo-resistant non-small-cell lung cancer by reversing Notch and epithelial mesenchymal transition in vitro and in vivo. Pharmacol Res 2019;148:104406.

[269] Wang Z, Wang L, Fan R, Zhou J, Zhong J. Molecular design and structural optimization of potent peptide hydroxamate inhibitors to selectively target human ADAM metallopeptidase domain 17. Comput Biol Chem 2016;61:15–22.

[270] Schaal JB, Tran DQ, Subramanian A, et al. Suppression and resolution of autoimmune arthritis by rhesus θ-defensin-1, an immunomodulatory macrocyclic peptide. PLoS One 2017;12:e0187868.

[271] Nuti E, Rosalia L, Cuffaro D, et al. Bifunctional inhibitors as a new tool to reduce cancer cell invasion by impairing MMP-9 homodimerization. ACS Med Chem Lett 2017;8:293–8.

[272] Nuti E, Rossello A, Cuffaro D, et al. Bivalent inhibitor with selectivity for trimeric MMP-9 amplifies neutrophil chemotaxis and enables functional studies on MMP-9 proteoforms. Cells 2020;9:E1634.

[273] Cuffaro D, Nuti E, Gifford V, Ito N, Camodeca C, Tuccinardi T, Nencetti S, Orlandini E, Itoh Y, Rossello A. Design, synthesis and biological evaluation of bifunctional inhibitors of membrane type 1 matrix metalloproteinase (MT1-MMP). Bioorg Med Chem 2019;27:196–207.

[274] Kim H-R, Tagirasa R, Yoo E. Covalent small molecule immunomodulators targeting the protease active site. J Med Chem 2021;64:5291–322.

[275] Nocentini A, Cuffaro D, Ciccone L, Orlandini E, Nencetti S, Nuti E, Rossello A, Supuran CT. Activation of carbonic anhydrases from human brain by amino alcohol oxime ethers: towards human carbonic anhydrase VII selective activators. J Enzyme Inhib Med Chem 2021;36:48–57.

[276] Cuffaro D, Di Leo R, Ciccone L, Nocentini A, Supuran CT, Nuti E, Rossello A. New isoxazolidinyl-based N-alkylethanolamines as new activators of human brain carbonic anhydrases. J Enzyme Inhib Med Chem 2023;38:2164574.

[277] Yuan X-Z, Sun S, Tan C-C, Yu J-T, Tan L. The role of ADAM10 in Alzheimer's disease. J Alzheimers Dis 2017;58:303–22.

[278] Manzine PR, Ettcheto M, Cano A, et al. ADAM10 in Alzheimer's disease: pharmacological modulation by natural compounds and its role as a peripheral marker. Biomed Pharmacother 2019;113:108661.

[279] Chen J, Yu S, Fu Y, Li X. Synaptic proteins and receptors defects in autism spectrum disorders. Front Cell Neurosci 2014;8:276.

[280] Tippmann F, Hundt J, Schneider A, Endres K, Fahrenholz F. Up-regulation of the alpha-secretase ADAM10 by retinoic acid receptors and acitretin. FASEB J 2009;23:1643–54.

[281] Obregon DF, Rezai-Zadeh K, Bai Y, et al. ADAM10 activation is required for green tea (−)-epigallocatechin-3-gallate-induced alpha-secretase cleavage of amyloid precursor protein. J Biol Chem 2006;281:16419–27.

[282] Lee HR, Shin HK, Park SY, Kim HY, Lee WS, Rhim BY, Hong KW, Kim CD. Cilostazol suppresses β-amyloid production by activating a disintegrin and metalloproteinase 10 via the upregulation of SIRT1-coupled retinoic acid receptor-β. J Neurosci Res 2014;92:1581–90.

[283] Fukasawa H, Nakagomi M, Yamagata N, Katsuki H, Kawahara K, Kitaoka K, Miki T, Shudo K. Tamibarotene: a candidate retinoid drug for Alzheimer's disease. Biol Pharm Bull 2012;35:1206–12.

[284] Marcade M, Bourdin J, Loiseau N, Peillon H, Rayer A, Drouin D, Schweighoffer F, Désiré L. Etazolate, a neuroprotective drug linking GABA(A) receptor pharmacology to amyloid precursor protein processing. J Neurochem 2008;106:392–404.

[285] Musardo S, Therin S, Pelucchi S, et al. The development of ADAM10 endocytosis inhibitors for the treatment of Alzheimer's disease. Mol Ther 2022;30:2474–90.

[286] Zhang S, Salemi J, Hou H, Zhu Y, Mori T, Giunta B, Obregon D, Tan J. Rapamycin promotes beta-amyloid production via ADAM-10 inhibition. Biochem Biophys Res Commun 2010;398:337–41.

[287] Soundararajan R, Sayat R, Robertson GS, Marignani PA. Triptolide: an inhibitor of a disintegrin and metalloproteinase 10 (ADAM10) in cancer cells. Cancer Biol Ther 2009;8:2054–62.

[288] Duffy MJ, Mullooly M, O'Donovan N, Sukor S, Crown J, Pierce A, McGowan PM. The ADAMs family of proteases: new biomarkers and therapeutic targets for cancer? Clin Proteomics 2011;8:9.

[289] Qian M, Bai SA, Brogdon B, et al. Pharmacokinetics and pharmacodynamics of DPC 333 ((2R)-2-((3R)-3-amino-3{4-[2-methyl-4-quinolinyl) methoxy] phenyl}-2-oxopyrrolidinyl)-N-hydroxy-4-methylpentanamide)), a potent and selective inhibitor of tumor necrosis factor alpha-converting enzyme in rodents, dogs, chimpanzees, and humans. Drug Metab Dispos 2007;35:1916–25.

[290] Moss ML, Sklair-Tavron L, Nudelman R. Drug insight: tumor necrosis factor-converting enzyme as a pharmaceutical target for rheumatoid arthritis. Nat Clin Pract Rheumatol 2008;4:300–9.

[291] Zhou B-BS, Peyton M, He B, et al. Targeting ADAM-mediated ligand cleavage to inhibit HER3 and EGFR pathways in non-small cell lung cancer. Cancer Cell 2006;10:39–50.

[292] Infante J, Burris HA, Lewis N. A multicenter phase Ib study of the safety, pharmacokinetics, biological activity and clinical efficacy of INCB7839, a potent and selective inhibitor of ADAM10 and ADAM17. Breast Cancer Res Treat 2007;106:S269.

[293] Wang J-N, Cao X-J. Targeting ADAM10 in renal diseases. Curr Mol Med 2023.

[294] Pece R, Tavella S, Costa D, et al. Inhibitors of ADAM10 reduce Hodgkin lymphoma cell growth in 3D microenvironments and enhance brentuximab-vedotin effect. Haematologica 2022;107:909–20.

[295] Thabet MM, Huizinga TWJ. Drug evaluation: apratastat, a novel TACE/MMP inhibitor for rheumatoid arthritis. Curr Opin Investig Drugs 2006;7:1014–9.

[296] Murumkar PR, Ghuge RB, Chauhan M, Barot RR, Sorathiya S, Choudhary KM, Joshi KD, Yadav MR. Recent developments and strategies for the discovery of TACE inhibitors. Expert Opin Drug Discov 2020;15:779–801.

[297] Laronha H, Carpinteiro I, Portugal J, Azul A, Polido M, Petrova KT, Salema-Oom M, Caldeira J. Challenges in matrix metalloproteinases inhibition. Biomolecules 2020;10:717.

[298] Li D-D, Zhao C-H, Ding H-W, Wu Q, Ren T-S, Wang J, Chen C-Q, Zhao Q-C. A novel inhibitor of ADAM17 sensitizes colorectal cancer cells to 5-Fluorouracil by reversing Notch and epithelial-mesenchymal transition in vitro and in vivo. Cell Prolif 2018;51:e12480.

[299] Pluda S, Mazzocato Y, Angelini A. Peptide-based inhibitors of ADAM and ADAMTS metalloproteinases. Front Mol Biosci 2021;8:703715.

[300] Schaal JB, Maretzky T, Tran DQ, Tran PA, Tongaonkar P, Blobel CP, Ouellette AJ, Selsted ME. Macrocyclic θ-defensins suppress tumor necrosis factor-α (TNF-α) shedding by inhibition of TNF-α–converting enzyme. J Biol Chem 2018;293:2725–34.

Chapter 3.6

Angiotensin-converting enzyme

Francesca Arrighi, Emanuela Berrino, and Daniela Secci
Department of Drug Chemistry and Technologies, Sapienza University of Rome, Rome, Italy

1 Structure and function of angiotensin-converting enzyme (ACE)

ACE, also known as peptidyl-dipeptidase A or dipeptidyl carboxypeptidase (EC 3.4.15.1), is a zinc-containing non-specific metallopeptidase that can cleave two amino acids from its substrates at the C-terminal end [1,2]. It is best known for its role in the blood pressure (BP) regulation through the renin-angiotensin system (RAS), where it converts the angiotensin I (Ang I) in the eight amino acid peptide angiotensin II (Ang II) hydrolyzing the Phe8-His9 peptide bond [2]. This potent vasopressor and main actor of the classical pathway of the RAS system exerts its action through binding of different AT receptors [3]. Moreover, ACE is also involved in the regulation of the Kinin system through the inactivation of bradykinin, peptide with a potent vasodilator activity [4]. Due to its wide substrate specificity, ACE can metabolize many other biologically active peptides of various lengths. It is therefore involved in many other physiological processes not related to the regulation of blood pressure, such as hematopoiesis, immunity, reproduction, and neuropeptide regulation [1]. It was reported for the first time by Skeggs Jr. in 1956, more than 50 years after the identification of renin [2]. In 1966, ACE bradykinin-degrading activity was identified, initially reconducted to another enzyme named kininases II [5]. The structure of ACE was identified in 1988, when Corvol and colleagues cloned the enzyme from human endothelial cells while Bernstein and colleagues in 1989 did the same starting from mouse kidney [6,7].

In human cells, two different forms of ACE have been identified, known as somatic ACE (sACE) and germinal ACE (gACE). sACE is composed of two homologous domains, named as N and C domains, both containing the catalytic zinc ion [6] (Fig. 1). As in many other metalloproteases, the metal-binding site is characterized by the HEXXH motif, which in ACE is composed of His-Glu-Met-Gly-His [8]. High homology was found between the two domains, with 60% of sequence identity. However, differences have been found as for chloride dependency and, most importantly, for catalytic efficiency and substrate specificity [6,9]. C-terminal domain was found to be a better site of conversion of Ang I in Ang II [10], whereas N-terminal domain was mostly responsible for Ang (1-7) cleavage and for the conversion of β-amyloid (1-42) in its less neurotoxic form β-amyloid (1-40) [11,12]. N-terminal domain is also responsible for the cleavage of the immuno-suppressive and antifibrotic peptide Ac-Ser-Asp-Lys-Pro (AcSDKP) [13]. gACE, also known as testis ACE (tACE), is identical to the C domain of sACE, only differing from it for 36 amino acids, which forms the amino terminus [14] (Fig. 1). Both ACE isoforms can be classified as type I membrane protein, being bound to the cellular membrane through a hydrophobic anchor located at the C terminal end of the enzyme and possessing an extracellular orientation [15]. A soluble ACE was also identified in blood and other biological fluids [16]. As for the expression, sACE is mostly expressed in the endothelial surface of lung and vessels, but also in other endothelial cell types as smooth muscle cells, monocyte, T lymphocytes, whereas gACE is located exclusively in testis [1]. Crystal structures of gACE (homologous to the C domain of sACE) and N domain of sACE, with and without ACE inhibitors, have been obtained in 2003 and 2006, respectively, by the same research group [17,18].

Besides the classical RAS system, composed by the ACE/Ang II/AT1R axes, a so-called "nonclassic RAS system" was identified in the 2000s, represented by the ACE2/Ang (1-7)/MAS axis [19,20]. Although others ACE homologous proteins were found in other species (i.e., *Drosophila melanogaster* genome contains six genes that encode ACE-like proteins), ACE2 was the first ACE homolog to be found in human [21]. It acts as zinc metalloprotease, possessing a single HEXXH zinc-binding domain, homologous to the C terminal domain of ACE, with an overall identity of 40% found between the two enzymes. However, ACE2 acts as a single carboxypeptidase, removing a single amino acidic residue from the C terminal of its substrate [20]. ACE2 can be considered as a critical enzyme in counter-regulatory response to RAS, being able to remove the terminal Phe residue from Ang II, producing Ang (1-7), a peptide possessing vasodilatory properties.

Metalloenzymes. https://doi.org/10.1016/B978-0-12-823974-2.00017-6

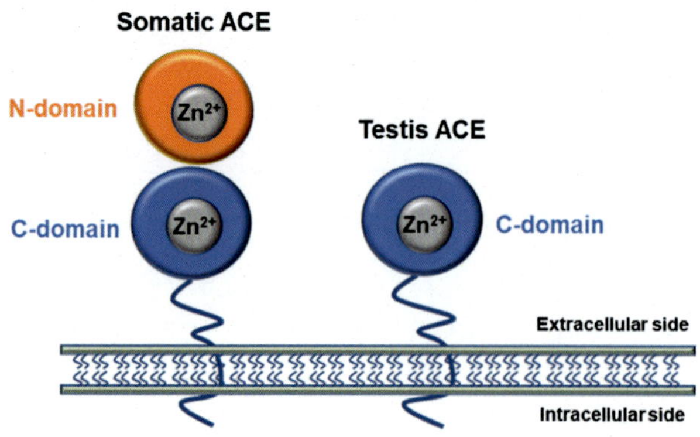

FIG. 1 Model of somatic and testis ACE.

Moreover, ACE2 can also cleave the terminal Leu from Ang I, producing Ang (1-9), which can be further metabolized by ACE or Neutral Endopeptidase (NEP), yielding Ang (1-7) [20,22,23].

When compared to ACE, which has a near ubiquitous expression, ACE2 can be mostly found in the heart, kidney, and testes. Different types of cells are able to produce this enzyme, such as macrophages, smooth muscle cells, myocytes, and endothelial cells. Interestingly, ACE2 is not inhibited by ACE inhibitors [20,24].

2 Physiologic/pathologic roles

The first implication discovered for ACE was related to its involvement in the **regulation of cardiovascular system and blood pressure**, through generation of Ang II. Binding of Ang II to receptor AT1 produces hypertensive responses (i.e., vasoconstriction, cell growth, sodium and fluid retention, and activation of sympathetic system) [2] (Fig. 2). AT1 receptors are therefore implicated in many pathological conditions, which include hypertension, cardiac arrhythmias, atherosclerosis, and stroke [25,26]. As for AT2 receptors, they seem to be involved in the control of cell differentiation, development, and angiogenesis [25,27]. The expression of these two receptors is strictly controlled [25]. Potentiation of ACE hypertensive effect is also mediated by its bradykinin-degrading activity [4]. Bradykinin is actually degraded more rapidly by ACE when compared to Ang I ((k_{cat}/K_m) of 3900–5000 $mM^{-1}\,s^{-1}$ vs 147–189 $mM^{-1}\,s^{-1}$, respectively) [28]. NEP and ACE are indeed the two major enzymes involved in bradykinin degradation, suppressing its vasodilatory activity explicated through PGI_2 and NO production, tissue plasminogen activator release, suppression of apoptosis in endothelial cells, and improved survival of endothelial progenitor cells [29,30].

On the other hand, ACE2 exerts opposite actions when compared to its homologous enzyme ACE. ACE2/Ang (1-7)/Mas axis induces indeed a vasodilatory answer [22,23,31]. ACE2 can be considered as a key protective regulator of cardiovascular function, removing Ang II from circulation and producing Ang (1-7), peptide with opposite effects when compared to Ang II (Fig. 2). Moreover, soluble circulating ACE2 can be a biomarker in hypertension and heart failure [31–33]. ACE2 expression increased also in the patients with type 1 or type 2 diabetes [34].

Other members of the RAS were also recently identified, represented by the peptides Angiotensine A (Ang A) and Alamandine [35] (Fig. 2). Ang A is an octapeptide highly similar to Ang II, only differing in the N-terminal in which the Asp is decarboxylated into Ala. It has higher affinity for AT2R when compared to Ang II and a smaller

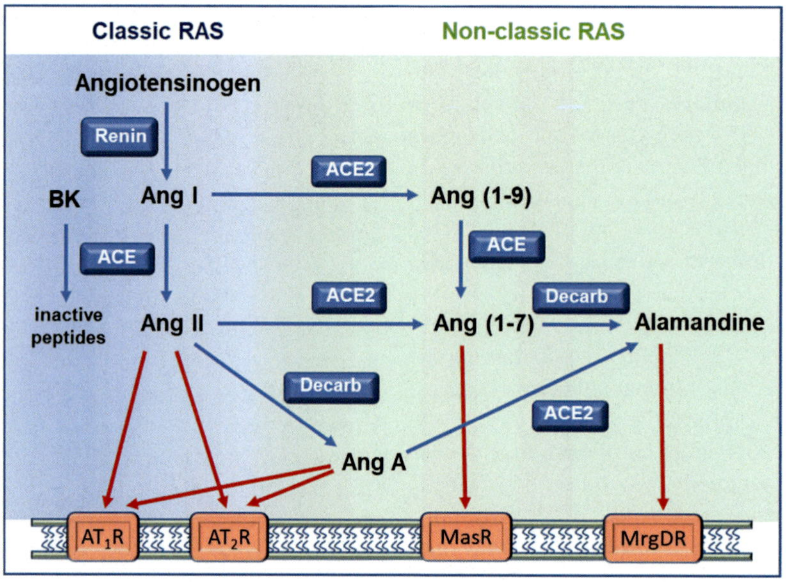

FIG. 2 Classic and nonclassic RAS. *Blue arrows* indicate enzymatic reactions; *red arrows* indicate the binding of the peptides to the receptors.

vasoconstricting activity [36]. Alamandine is an heptapeptide that can be formed from Ang A through hydrolysis by ACE2 or directly from Angiotensin-(1-7), through decarboxylation [37]. Alamandine produces vasodilatory activities and long lasting antihypertensive effect in Spontaneously Hypertensive Rat model, and these effects are independent of Mas and can be mediated by a Mas-related receptor, MrgD [37].

Even if the great part of the studies of the past years on ACE (or RAS in general) were focused on its role in the blood pressure and cardiovascular system, nowadays many efforts are spent toward the understanding of ACE involvement in other physiological and pathological pathways [38,39]. This phenomenon was also pushed forward by the discovery that RAS components can be found in other organs not related to blood pressure regulation (i.e., brain, immune cells), and that their cells are able to independently regulate their own Ang II levels [40–44].

Studies conducted on mice and humans revealed that ACE possess an important role in **kidney development**. In humans, mutations in ACE, or any functional interruption of the RAS, can result indeed in renal tubular dysgenesis (RTD) [45–47]. ACE and RAS in general were also found to play an important role in **hematopoietic cell development** [48]. As discussed before, testis ACE is fundamental in the regulation and maintenance of **stable male fertility.** In this prospective, the role of the C-domain is crucial in order to preserve testis ACE activity [47,49]. gACE likely affects male reproduction acting on a substrate other than angiotensin I. However, the exact substrate of testis ACE is currently unknown and this area deserves further investigation [49,50].

Another field where ACE is known to play a crucial role is in **immune responses and inflammation** [44]. Connection between ACE and immune function dates back in the 1970s, when increased serum ACE levels were detected in patients with active sarcoidosis, when compared to the levels observed in patients where the disease was treated or resolved [51]. Since then, many studies further supported the idea that ACE plays a role in both innate and adaptive immunity, modulating macrophage and neutrophil functions [42,44]. Macrophages overexpressing ACE showed an enhanced immune response in both innate and adaptive immunity models [52]. Overexpression on ACE in neutrophiles from mouse models showed an increased resistance to bacterial infections and an enhanced bacteria killing activity, most likely induced by an increased reactive oxygen species (ROS) production mediated by ACE [53]. ACE also affects the presentation of major histocompatibility complex (MHC) class I and MHC class II peptides [54,55]. The improved immune response induced by ACE overexpression relies on the catalytic activity of the enzyme. However, only few of these activities can be ascribed to Ang II (i.e., leukocyte recruitment, increased production of ROS, activation of TLR4, and regulation of TNF expression) [52,56,57]. Undefined substrates and products of this enzyme are likely to mediate these responses but the biochemical pathway is still unknown. More insights into the role of ACE expression and activity in myeloid cells could be crucial for the treatment of several diseases, such as infections and tumors [44].

All these evidences highlight a protective role played by the classical ACE/Ang II/AT1R pathway in the early stages of **inflammatory diseases and infections** [44,58]. However, exaggerate inflammation can determine tissue alteration and fibrotic remodeling, with critical consequences in many diseases (i.e., lung diseases) [59]. ACE2 enzyme seems to be critical in this step, attenuating the immune response and promoting tissue repair [59–62]. ACE2 is particularly important in lungs and widely expressed on the barrier sides of bronchial or skin cells [63]. Several pathogens have shown to decrease the expression and/or activity of ACE2, increasing Ang II, [des-Arg 9]-BK (DABK) or Ang A bioavailability as well as decreasing levels of counterregulatory molecules, such as Ang 1-7 or alamandine, inducing a worsening of the infection [64]. Moreover, some pathogens, such as Coronavirus family viruses (i.e., SARS-CoV-2) enter the host cells exploiting the membrane bound ACE2, for which they have high affinity [65,66]. After the interaction and virus internalization, ACE2 expression was found to be downregulated, to prevent co-infections by other intracellular pathogens, but this has the effect of inducing an enhanced activation of the pro-inflammatory classic RAS pathway [67–69]. Interestingly, soluble ACE2 retain the ability to bind the virus, preventing its internalization into the cells [70,71]. Another interesting phenomenon studied in depth very recently is the perturbation of the crosstalk of ACE and ACE2 when the cells are attacked by the coronavirus, which contributes to the characteristic symptoms [72]. In particular, the spike protein of the SARS-CoV-2 Delta variant revealed to have a much greater impact on the crosstalk than the wild type [72].

The physio/pathological role of ACE in **malignancies** is quite debated, due to the diverse functions played by this enzyme in different cell types within the tumor microenvironment [73,74]. On one side, increasing evidences suggest that Ang II mediated angiogenesis through activation of the vascular endothelial growth factor (VEGF) is the major mechanism behind ACE contribution to carcinogenesis and tumor growth, acting on tumor microvasculature [75,76]. However, in vivo and clinical studies using different tumor types often give controversial results [52,77,78]. Therefore, the ability of ACE to modulate tumor immunology, acting on myelomonocytic and T cell functions, should be taken into account, considering the complex immune network within the tumor microenvironment [44,73]. In this way, ACE overexpression in myelocytes is reported to attenuate tumor growth, in a Ang II independent manner [52].

The role of Ang II in promoting vascular remodeling, inflammation and oxidative stress can be also applied to the Central Nervous System (CNS), where dysregulation of cerebral blood flow, can be often related to pathological conditions including **anxiety, learning, memory, physiological responses to stress and ischemic stroke** [79–82]. The discovery of RAS components expression in the CNS, with a specific regulation distinct from the peripheral one, ensured also by the blood-brain barrier, opened the way for many studies in this field [38,79,83]. On the other hand, the nonclassical RAS axis, also found to be largely expressed in the brain, was reported to produce neuroprotective effects, mediating the release of NO and promoting antiinflammatory, antifibrotic, and vasodilatation effects, and acting as antagonistic pathway of the classical RAS [84]. In particular, it was found that classic RAS pathway is involved in depression promoting inflammation, oxidative stress, and stress responses and reducing brain-derived neurotrophic factor (BDNF) levels [85]. These effects can be also related to the progress of neurodegenerative disorders, such as **Alzheimer's disease**, foreseeing a pathological role of ACE in worsening neurodegeneration [86,87]. However, studies conducted using mouse models possessing increased ACE levels in macrophages showed how improved adaptive immune response mediated by ACE overexpression led to the reduction of plaque size in mice aged 8 months, without affecting cognitive functions [88]. Accumulation of amyloid plaques within the central nervous system is considered a major pathological feature of the disease, although its pathogenesis is very complex and still unclear [89]. Moreover, it must be considered that both ACE and ACE2 metabolize β-amyloid peptide to its less neurotoxic form, possibly reducing the risk of dementia [12,23,90].

3 Classes of modulators and their design

ACE inhibitors (ACEI) can be considered as the first example in Medicinal Chemistry of rationally designed enzyme inhibitors, even though their discovery was more the result of a "rational intuition" rather than of a rational design as we would intend it today [91]. When Captopril was designed, there was indeed any knowledge of the sequence or three-dimensional structure of the enzyme [92]. However, some crucial events and important observations and intuitions led to the design and synthesis of the first ACE inhibitor [93,94]. First of all, the discovery made by Ferreira that a substance (subsequently called bradykinin potentiating factor, BPF) found in the venom of the South American pit viper *Bothrops jararaca* had a potentiating effect on the activity of bradykinin, and that BPF was able to inhibit ACE [95,96]. Similar results were obtained by Cushman and Ondetti working on the same venom, looking for peptides with ACE inhibitory properties, and led to the isolation, characterization and synthesis of the most potent of them, **SQ 20881** or teprotide, the first peptide-based ACE inhibitor [97]. Starting from its structure, Cushman and Ondetti performed a deep structure-activity analysis, to define the shape of the ACE catalytic site and, eventually, design a low molecular weight ACE inhibitor that could have been orally administered, overcoming the limitations associated with peptide-based inhibitors. The presence of a zinc ion within the active site was just an hypothesis back then, suggested by the ability of EDTA to suppress the enzyme activity [2]. However, starting from this consideration, ACE active site was modeled on the catalytic site of carboxypeptidase A, a zinc metalloenzyme for which the crystal structure was known [92]. All these studies led to the definition of some structural determinants for the design of efficient ACEI: the optimal carboxyl-terminal amino acid sequence was **Phe-Ala-Pro**; a free C-terminal carboxyl group was important in binding to the enzyme; the positioning of aromatic amino acids in the antepenultimate position enhanced inhibitory activity; the N terminus of any inhibitor had to provide a critical interaction with the enzyme.

However, a crucial insight for the design of ACEIs came from the paper from Byers and Wolfenden where they described the development of carboxypeptidase A inhibitors, based on the structure of benzylsuccinic acid [98]. This led to the design of D-2-methylsuccinyl-Pro (**SQ 13,297**), in which substitution of the N-terminal carboxyl with a sulfhydryl group, endowed of better zinc-chelating ability, produced **captopril** (**SQ 14225**), the first example of mercaptoacyl amino acid inhibitor and the first clinically used ACE inhibitor, approved by FDA in 1981 for essential hypertension [92,99].

Typically, ACE inhibitors are composed of four parts: P1 group, P1′ group, P2 group, and P2′ group, which are bound to the ACE binding pockets S1, S1′, S2, and S2′, respectively [91] (Fig. 3). Different zinc chelating groups were explored over time, which allow us to classify the first generation of ACEIs in three main classes: **Sulfhydryl-Containing**, **Carboxyl-Containing**, and **Phosphoryl-Containing ACE inhibitors** [91,100,101]. Subsequently, more zinc-binding moieties were explored, including **phosphonates** [102], **hydroxamates** [103], **ketones** [104], and **silanediols** [105]. In addition, ACE inhibitors can be divided not only according to the zinc ligand group but, also, to the strength of the binding and the inhibitory activity according on the number of the binding sites occupied and by the duration of the inhibitory activity, which depends on the dissociation rate [106].

The first ACEI reported, captopril, did not occupy all the binding sites available. Therefore, structure-activity studies were performed on a series of *N*-carboxyalkyl peptides in order to extend the interactions to other binding sites, increasing inhibitory potency [107] (Fig. 3). The best

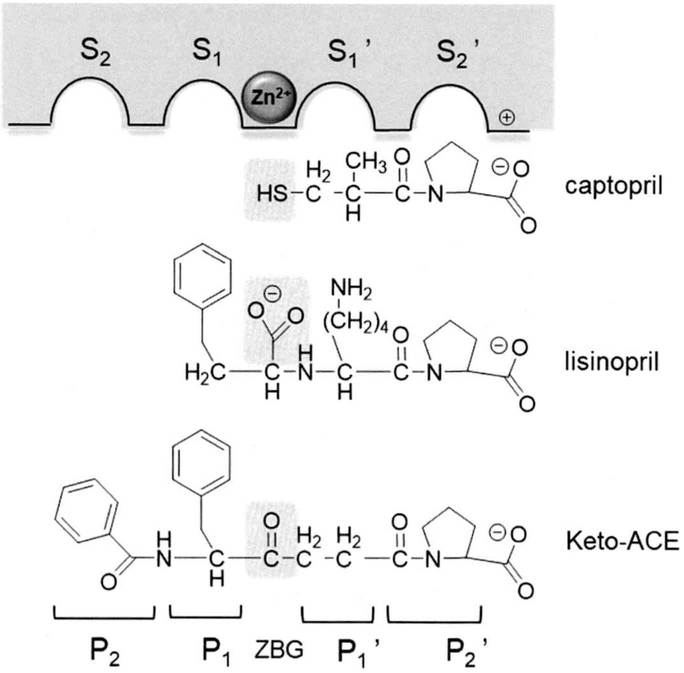

FIG. 3 Models of interactions between inhibitors and the active sites of ACE.

substituent to occupy S1 pocket was found to be the benzylmethylene moiety. P1′ and P2′ were also widely changed, with **enalaprilat** and **lisinopril** containing the dipeptide Ala-Pro and Lys-Pro, respectively [107,108]. In general, it can be said that charged amino acids such as lysine and arginine are preferred as P1′ groups. Aromatic amino acids and proline and their derivatives, with large side chains, are preferred as P2′ groups. For the P1 group, amino acid residues containing many hydrophobic groups in the side chains are more favorable [91,107].

More information on ACE-binding site was obtained when both N and C ACE domains were crystalized, with and without ACEIs [17,18]. From that moment, a more precise analysis of the interactions established by the compounds within the binding site became possible, allowing to design more potent inhibitors, which could also possess selectivity toward the different ACE domains [91,109] (Fig. 4).

First generation of ACEIs are indeed essentially mixed N- and C-domain inhibitors [91], although in some cases a slight selectivity was reported (i.e., captopril being modestly N-domain selective and lisinopril more C-selective) [28,110,111]. However, these differences are not clinically significant. Since each domain differ for substrate selectivity and have in some way different biological functions, developing domain-selective inhibitors could present several advantages, both in terms of activity (N-selective inhibitors could open to the treatment of new therapeutic areas) and safety profile of the compounds (C-selective compounds could control blood pressure with less impact on the BK and nonclassical RAS pathway metabolism) [102,112–115]. Many reviews discussed the approaches used to design selective inhibitors, and new compounds were also recently reported [91,116–118]. In general, it can be said that bulky P1′ and P2′ groups and a large, neutral or basic P2 group confer C-selectivity, whereas N-selectivity is conferred by an amidated C-terminal carboxyl and an acidic P2 group [91,116]. The use of weakly zinc-coordinating group, such as **phosphinic acid moiety** (**—PO_2-CH_2—**) inserted in place of a peptide bond, can also represent an advantage in the seeking of domain selective inhibitors [119,120] (Fig. 4). Weaker coordination allows other chemical determinants within ACE inhibitor to have greater contribution in the relative binding with the different domains [121]. Moreover, phosphinic peptides seems to mimic the transition state formed during substrate hydrolysis mediated by the enzyme [119,121].

As for **ACE2**, its cardioprotective properties limited the development of inhibitors for lack of potential therapeutic applications [19,23,31,122]. However, some derivatives, such as **MLN-4760 (GL1001)**, **DX-600**, and **416F2** were developed, based on the knowledge of the structure, mechanism and specificity of the enzyme [31,123–125] (Fig. 5). With the recent spread of the new SARS-CoV-2, the possibility to inhibit ACE2 in order to limit the virus internalization became appealing [23,126,127]. Many studies applied a repurposing approach to identify compounds that could effectively bind ACE2. Among them, vincristine, vinblastine and bisoctrizole showed to bind to ACE2 [128]. Moreover, rationally designed ACE2-derived peptides

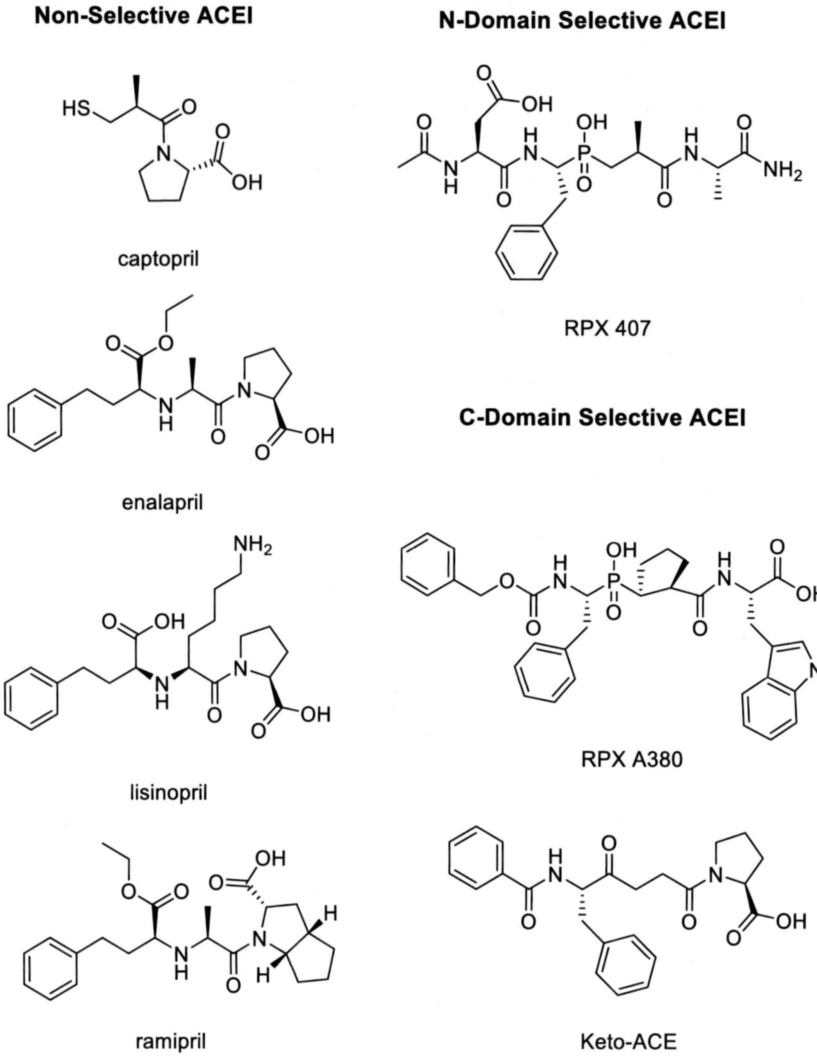

FIG. 4 Selected chemical structures of selective and nonselective ACEIs.

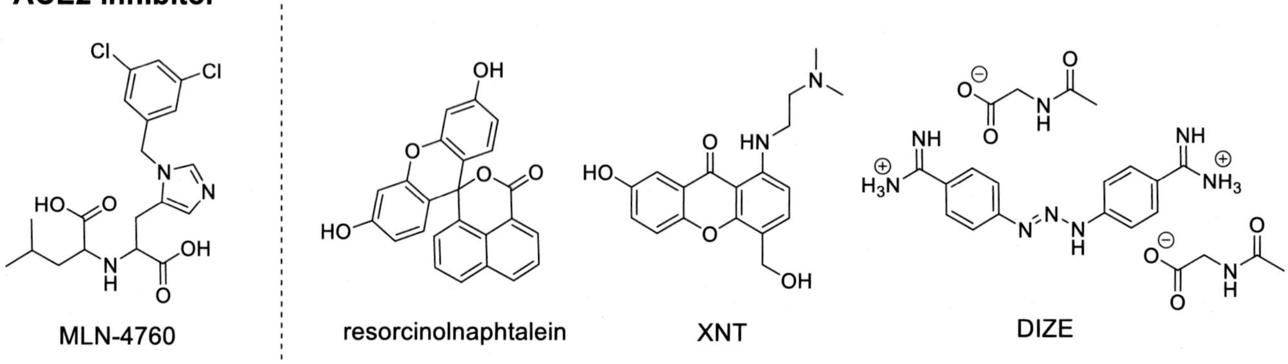

FIG. 5 Chemical structures of selected ACE2 inhibitors and activators.

based on the Receptor Binding Domain (RBD)-ACE2 binding interfaces of SARS-CoV-2 and SARS-CoV have been reported. Two peptides proved to be particularly effective. Among them, **SAP6**, containing the minimal conserved EDLFYQ sequence, showed how the N-terminal $\alpha 1$ helix of ACE2 is important for interaction with Spike protein [129].

Another approach to the quest for potent and safer ACEIs is represented by the exploitation of **natural sources** [130,131]. This method, which is definitely not new (also considering that the first peptide-based ACEI was isolated from snake venom), led to the identification of many bioactive peptides. One of the richest sources is represented by milk [132]. However, in some cases, poor correlation between the in vitro ACE inhibitory activity of milk-derived peptides and the in vivo antihypertensive activity has been observed. This can be partly due to digestion, which renders less active peptide sequences and/or due to their low bioavailability. The activity of these peptides depends on their inherent amino acid composition and sequence [130,131]. Other studies reported that natural ACE inhibitors from plant sources are able to inhibit ACE via competitive, noncompetitive, and uncompetitive binding [131,133,134]. ACE inhibitory plant metabolites mostly belong to the groups of peptides, protein hydrolysates, phenolics, flavonoids, terpenoids, and alkaloids. This topic is extensively reviewed here [133].

Other natural products studied as ACE inhibitors are represented by some natural glucosides, including: calceolarioside B, junipediol A, and eugenol [135]. The saccharide unit seems to be the main responsible for interaction with the zinc in the ACE active site. While the mechanisms of action of these bioactive phytocompounds appear to resemble those of synthetic ACE inhibitors, the details of the interactions between phytochemicals and the enzyme are poorly understood. Moreover, studies on other aspects, such as dosage, mode of delivery, toxicity, and cost-effectiveness should be conducted when developing widely acceptable antihypertensive herbal drugs [135].

Plants and food products were also explored as natural sources of ACE2 inhibitors [136]. In particular, they ability to influence viral tropism was evaluated. Flavonoids, steroids and coumarins were found to possess antiviral activities, potentially mitigating effects against viral entry by directly or indirectly modulating the ACE2 activity at the Spike-RBD/TMPRSS2/ACE2 axis [136].

A quite recent field with very interesting therapeutic perspectives is the use of ACE2 activators. Considering the general cytoprotective activities mediated by the RAS nonconventional pathway, the possibility to positively modulate ACE2 activity seems intriguing [31,137,138]. Both small molecules and peptidic activators have been developed and studied in several disease models, with important beneficial roles observed in cardiovascular diseases or cancer [31,137,138]. Structure-based drug design was applied for the identification of ACE2 activators. Among them, we can list resorcinolnaphthalein, XNT, and DIZE (Fig. 5) [137–141].

The use of exogenous ACE2 (recombinant human ACE2, rhACE2) was also reported and is currently considered for treating Acute Respiratory Distress Syndrome (ARDS), pulmonary arterial hypertension, and heart failure [142–144]. Recently it was also considered as a potential treatment for SARS-CoV-2 infection, not only limiting the virus spread blocking its interaction with ACE2 on the host cell, but also protecting the lung from developing ARDS by modulating RAS and therefore SARS-CoV-2-related underlying comorbidities [142,145].

4 Clinically used agents or compounds in clinical development

ACEIs were initially developed for the treatment of hypertension [106,146]. However, over the years many other therapeutic applications have been studied, eventually leading to their approval also for the treatment of heart failure, myocardial infarction, and renal dysfunction [147–151]. Renoprotective activities were observed especially in diabetic patients with or without hypertension [150]. The prophylactic use in cardiovascular diseases as component of the polypill (fixed dose combination of aspirin, statin, and an antihypertensive agent) also gave very positive outcomes, although being still under investigation [150,152–154]. Currently, 10 ACE inhibitors are used in clinics, with different therapeutic indications [101,155]. The differences between ACE inhibitors according to their chemical structure reflect differences in compounds potency, bioavailability, plasma half-life, and route of elimination. With the only exception represented by captopril and lisinopril, all the other ACEIs in clinical use are prodrugs esters, which undergo hepatic hydrolysis to give the active metabolites [106,156].

A clinical study currently active is evaluating the possible beneficial effects of ACEI (ramipril) in the clinical evolution and ventricular remodeling after transcatheter aortic valve implantation (RASTAVI, NCT03201185) [157]. Another study started in 2018 and expected to be completed in 2026 is comparing the effects of beta blockers, ACEIs and Angiotensin Receptor Blockers (ARBs) in myocardial infarction with nonobstructive coronary arteries (MINOCA-BAT, NCT03686696) [158,159]. The beneficial effects of ACEI in glomerulopathy when compared to renin inhibitors is also under evaluation. Aliskiren and Enalapril are the drugs selected for this study (NCT04183101) [160]. The use of ACEIs for heart failure (HF) is clinically approved and well established (Table 1). More recently, the possibility to stimulate the nonclassic RAS pathway

TABLE 1 Clinically approved ACE inhibitors and therapeutic uses.

Indications	Benazepril hydrochloride	Captopril	Enalapril maleate	Fosinopril sodium	Lisinopril	Moexipril hydrochloride	Perindopril erbumine	Quinapril hydrochloride	Ramipril	Trandolapril
Renal impairment dosing adjustment	Yes	Yes	Yes		Yes	Yes	Yes	Yes	Yes	Yes
Hepatic impairment dosing adjustment				Yes		Yes		Yes		Yes
Diabetic nephropathy	Yes*	Yes	Yes*		Yes*				Yes*	
Heart failure		Yes	Yes	Yes	Yes		Yes*	Yes	Yes	Yes
Hypertension	Yes	Yes	Yes	Yes	Yes	Yes	Yes	Yes	Yes	Yes
Hypertensive emergency		Yes*	Yes*							
Hypertensive urgency		Yes*	Yes*							
Myocardial infarction prophylaxis							Yes		Yes	
Proteinuria		Yes	Yes	Yes*	Yes*					
Reduction of cardiovascular mortality							Yes		Yes	
Stroke prophylaxis									Yes	

Yes: Labeled; Yes*: off-label, recommended [101,155].

using hrACE2 (GSK2586881) appeared to be a promising strategy for HF, including phenotypes with preserved and mildly reduced ejection fraction [144,161]. Preclinical studies reported the ability of hrACE2 to reduce Ang II-mediated cardiac remodeling and myocardial fibrosis in wild-type mice [162], and to have a protective effect in murine models of Ang II-induced HF [144]. A good safety profile was also shown in two completed phase I (NCT00886353) and phase II (NCT01597635) clinical trials, where no cardiovascular effects were evident [163–165].

An interesting aspect to be considered when studying the therapeutic activities of ACEIs is the extent of BK potentiation contribution to the beneficial effects observed during the treatment [166]. Usually, the bradykinase activity of ACE and its inhibition during treatment with ACEIs is seen as a disadvantage, being the main responsible of cough, the most common side effect in patients using ACEIs [167]. However, considering the activities mediated by BK (i.e., release of NO and PGI2, tissue plasminogen activator release, suppression of apoptosis in endothelial cells, among others), it is likely that inhibition of its degradation mediated by ACE could be beneficial for cardiovascular diseases [168,169]. However, few studies measured the levels of BK during ACEIs treatment, so that this aspect remains to be clarified [166,168,169].

The most common side effect in patients using ACE inhibitors is cough and, as stated above, it is thought to be induced by high levels of bradykinins due to inhibition of its degradation to inactive peptides. The reported incidence of cough varies widely in different studies but is generally accepted to be around 5%–20% [167,170]. Angioedema is another complication of RAAS blockers. Although it occurs with much less frequency, it is a much more serious complication. In a metaanalysis by Makani et al., the incidence of angioedema with ACE inhibitors was 0.3%, while the incidence with ARBs was 0.13%. The risk of angioedema was twice as high in ACE inhibitors as compared to ARBs, and there was no statistically significant difference between ARBs and placebo [171]. A clinical study currently recruiting (**KIN-ACE, NCT04763577**) aims to study the contribution of ACE BK degrading activity to angioedema caused by ACEIs [172].

New clinical areas where classic and nonclassic ACE pathways are involved have been reported in Section 2. Here, we will focus on the new therapeutic applications for which ACE modulators are currently in clinical development.

As discussed in the previous sections, the physio/pathological role of ACE in tumors is quite controversial [73]. The involvement or RAS in tumor growth, proliferation and metastasis justified the exploration of ACEIs as possible candidates to enhance the effect of chemo-radiotherapy and targeted therapy efficacy [173]. However, there are conflicting reports as to the use of ARB/ACEI to affect tumor growth. Most of the clinical studies conducted so far, also in association with other antitumor agents, seems to support the role of ARB/ACEI drugs in tumor treatment considering RAS role in angiogenesis, extracellular matrix (ECM), tumor microenvironment (TME), and hypoxia [173]. ACEIs represent also a possible therapeutic option for radioprotection, but additional studies are needed to evaluate effects on bladder radiotoxicity and in relation to hypertension [174]. A recent study evaluated how ACEIs can inhibit fibrogenesis in animal models. The aim of the study was to evaluate the impact of these drugs on the risk of liver cancer and cirrhosis in patients with nonalcoholic fatty liver disease (NAFLD). The study underlined that ACEI treatment, rather than ARB, was associated with lower risk of liver-related events (LREs) and NAFLD, especially among those with chronic kidney disease (CKD) [175].

Another field where the therapeutic use of compounds acting on the RAS, including ACEI, is under evaluation is represented by the treatment of **cognitive disorders** [87,176]. Antihypertensive drug therapies have been documented to be associated with a lower risk of developing dementia and early cognitive impairment [87,176,177]. However, many studies, including some randomized clinical trials, produced mixed results about the impact of hypertension treatment on the risk of Alzheimer's Disease Related Dementias (ADRD) [87]. Recently, a metaanalysis regarding the prevention of different types of dementia mediated by the use of RAS inhibitors as a whole and of ACEIs and ARBs was conducted [87]. The results showed that RAS antagonists are associated with reduced risk of dementia when compared to other classes of antihypertensives. ARBs, but not of ACEIs, revealed to produce a protective role against both AD and dementia in general, probably because of the distinct actions toward independent receptor pathways or to differential influences on amyloid metabolism. This protective effect could be ascribed to the influences of the RAS on structure and function of the blood-brain barrier, neuroinflammation, and oxidative stress [87].

RAS role in the pathogenesis and development of pulmonary diseases has been also object of intense interest among the scientific community, with many clinical studies reported so far analyzing the effects of RAS inhibitors in modulating evolution of lung injury [59].

In this context, acting on the nonclassical RAS pathway proved to be another promising strategy [61,62,164,178]. As said above, the use of rhACE2 could be also beneficial in the context of the ongoing SARS-CoV-2 pandemic, targeting the RAS and reducing the pathogen's cell entry [142]. Several clinical studies supported the relevance of ACE2 physiology in COVID-19, both as receptor for SARS-CoV-2 (ClinicalTrials.gov identifiers: NCT04335136) and as generator of

Ang (1-7) (ClinicalTrials.gov identifiers: NCT04311177, NCT04312009, NCT04338009, NCT04338009, NCT04394117, and NCT04394117) [127]. Therefore, compounds mimicking sACE2 or blocking ACE2-spike protein interaction could be beneficial for the infection treatment. Some candidates are currently in clinical evaluation [127].

An important aspect to be considered is the role of the classic ACE pathway during SARS-CoV-2. In consideration of the complex interplay between ACE and ACE2 in physio/pathological conditions, during the pandemic it was hypnotized that patients with COVID-19 treated with ACEIs might have a poorer prognosis [179]. However, large observational retrospective studies in humans and metaanalyses have revealed that there is no harmful effect of RAS blockers on COVID-19 susceptibility, severity, or mortality [126]. Instead, ACEIs and ARBs are expected to improve COVID-19 clinical outcomes and reduce complication [126,179]. Moreover, the very recent studies exploring the interplay between ACE and ACE2 during SARS-CoV-2 showed how the binding of the spike protein and ACE2 could weaken the crosstalk effect of ACE and ACE2, leading to accumulation of Ang II, further supporting the beneficial use of ACEIs during the infection [72].

In conclusion, this chapter aimed to report the last advances in the classical and nonclassical RAS pathways, with particular meaning for the ACE and ACE2 physio/pathological roles and pharmacological modulation.

Since the isolation of ACE in the 1950s, and the development of the first ACE Inhibitors in the 1970s, many new discoveries in the field of RAS occurred over the years. New proteins and peptidic mediators have been discovered. A deeper understanding of ACE structure has allowed to develop new inhibitors, some of them being domain-selective. Increased understanding of the RAS, especially the nonclassic system, and its involvement in a plethora of biological pathways not strictly related to the regulation of cardiovascular system has opened the way to the possibility to target ACE and ACE2 for the treatment of complex and harmful diseases. Although being known since a long time, there is still so much to discover about these enzymes. We are sure that other aspects of the RAS will continue to be discovered in the future.

Conflicts of interest

The authors declare no conflict of interest.

References

[1] Sturrock ED, Anthony CS, Danilov SM. Peptidyl-dipeptidase a/angiotensin I-converting enzyme. In: Handbook of proteolytic enzymes. Academic Press; 2013. p. 480–94.

[2] Skeggs LT, Kahn JR, Shumwa NP. The preparation and function of the hypertensin-converting enzyme. J Exp Med 1956;103(3):295–9.

[3] Carey RM, Siragy HM. Newly recognized components of the renin-angiotensin system: potential roles in cardiovascular and renal regulation. Endocr Rev 2003;24(3):261–71.

[4] Yang HYT, Erdös EG, Levin Y. A dipeptidyl carboxypeptidase that converts angiotensin I and inactivates bradykinin. Biochim Biophys Acta Protein Struct 1970;214(2):374–6.

[5] Erdős EG, Yang HYT. An enzyme in microsomal fraction of kidney that inactivates bradykinin. Life Sci 1967;6(6):569–74.

[6] Soubrier F, Alhenc-Gelas F, Hubert C, Allegrini J, John M, Tregear G, et al. Two putative active centers in human angiotensin I-converting enzyme revealed by molecular cloning. Proc Natl Acad Sci 1988;85(24):9386–90.

[7] Bernstein KE, Martin BM, Edwards AS, Bernstein EA. Mouse angiotensin-converting enzyme is a protein composed of two homologous domains. J Biol Chem 1989;264(20):11945–51.

[8] Vallee BL, Auld DS. Zinc coordination, function, and structure of zinc enzymes and other proteins. Biochemistry 1990;29(24):5647–59.

[9] Masuyer G, Yates CJ, Sturrock ED, Acharya KR. Angiotensin-I converting enzyme (ACE): structure, biological roles, and molecular basis for chloride ion dependence. Biol Chem 2014;395(10):1135–49.

[10] Fuchs S, Xiao HD, Hubert C, Michaud A, Campbell DJ, Adams JW, et al. Angiotensin-converting enzyme C-terminal catalytic domain is the main site of angiotensin I cleavage in vivo. Hypertension 2008;51(2):267–74.

[11] Deddish PA, Marcic B, Jackman HL, Wang HZ, Skidgel RA, Erdös EG. N-domain–specific substrate and C-domain inhibitors of angiotensin-converting enzyme. Hypertension 1998;31(4):912–7.

[12] Zou K, Maeda T, Watanabe A, Liu J, Liu S, Oba R, et al. Aβ42-to-Aβ40- and angiotensin-converting activities in different domains of angiotensin-converting enzyme. J Biol Chem 2009;284(46):31914–20.

[13] Rousseau A, Michaud A, Chauvet MT, Lenfant M, Corvol P. The hemoregulatory peptide N-acetyl-Ser-Asp-Lys-Pro is a natural and specific substrate of the N-terminal active site of human angiotensin- converting enzyme. J Biol Chem 1995;270(8):3656–61.

[14] Ehlers MRW, Fox EA, Strydom DJ, Riordan JF. Molecular cloning of human testicular angiotensin-converting enzyme: the testis isozyme is identical to the C-terminal half of endothelial angiotensin-converting enzyme. Proc Natl Acad Sci U S A 1989;86(20):7741–5.

[15] Chubb AJ, Schwager SLU, Woodman ZL, Ehlers MRW, Sturrock ED. Defining the boundaries of the testis angiotensin I-converting enzyme ectodomain. Biochem Biophys Res Commun 2002;297 (5):1225–30.

[16] Oppong SY, Hooper NM. Characterization of a secretase activity which releases angiotensin-converting enzyme from the membrane. Biochem J 1993;292(2):597–603.

[17] Natesh R, Schwager SLU, Sturrock ED, Acharya KR. Crystal structure of the human angiotensin-converting enzyme–lisinopril complex. Nature 2003;421(6922):551–4.

[18] Corradi HR, Schwager SLU, Nchinda AT, Sturrock ED, Acharya KR. Crystal structure of the N domain of human somatic angiotensin I-converting enzyme provides a structural basis for domain-specific inhibitor design. J Mol Biol 2006;357(3):964–74.

[19] Clarke NE, Hooper NM, Turner AJ. Angiotensin-converting enzyme-2. In: Handbook of proteolytic enzymes. Academic Press; 2013. p. 499–504.

[20] Tipnis SR, Hooper NM, Hyde R, Karran E, Christie G, Turner AJ. A human homolog of angiotensin-converting enzyme. J Biol Chem 2000;275(43):33238–43.

[21] Siviter RJ, Nachman RJ, Dani MP, Keen JN, Shirras AD, Isaac RE. Peptidyl dipeptidases (Ance and Acer) of *Drosophila melanogaster*:

major differences in the substrate specificity of two homologs of human angiotensin I-converting enzyme. Peptides (NY) 2002;23 (11):2025–34.
[22] Santos RAS, Sampaio WO, Alzamora AC, Motta-Santos D, Alenina N, Bader M, et al. The ACE2/Angiotensin-(1-7)/Mas axis of the renin-angiotensin system: focus on Angiotensin-(1-7). Physiol Rev 2018;98(1):505–53.
[23] Turner AJ, Nalivaeva NN. Angiotensin-converting enzyme 2 (ACE2): two decades of revelations and re-evaluation. Peptides (NY) 2022;151:170766.
[24] Towler P, Staker B, Prasad SG, Menon S, Tang J, Parsons T, et al. ACE2 X-ray structures reveal a large hinge-bending motion important for inhibitor binding and catalysis. J Biol Chem 2004;279(17):17996–8007.
[25] Kaschina E, Unger T. Angiotensin AT1/AT2 receptors: regulation, signalling and function. Blood Press 2009;12(2):70–88. https://doi.org/10.1080/08037050310001057.
[26] Murphy TJ, Alexander RW, Griendling KK, Runge MS, Bernstein KE. Isolation of a cDNA encoding the vascular type-1 angiotensin II receptor. Nature 1991;351(6323):233–6.
[27] Nouet S, Nahmias C. Signal transduction from the angiotensin II AT2 receptor. Trends Endocrinol Metab 2000;11(1):1–6.
[28] Ehlers MRW, Riordan JF. Angiotensin-converting enzyme: zinc- and inhibitor-binding stoichiometries of the somatic and testis isozymes. Biochemistry 1991;30(29):7118–26.
[29] Moreau ME, Garbacki N, Molinaro G, Brown NJ, Marceau F, Adam A. The kallikrein-kinin system: current and future pharmacological targets. J Pharmacol Sci 2005;99(1):6–38.
[30] Erdös EG, Skidgel RA. Metabolism of bradykinin by peptidases in health and disease. In: The kinin system. Academic Press; 1997. p. 111–41.
[31] Turner AJ. ACE2 cell biology, regulation, and physiological functions. In: The protective arm of the renin angiotensin system (RAS): Functional aspects and therapeutic implications. Academic Press; 2015. p. 185–9.
[32] Ramchand J, Burrell LM. Circulating ACE2: a novel biomarker of cardiovascular risk. Lancet 2020;396(10256):937–9.
[33] Úri K, Fagyas M, Siket IM, Kertész A, Csanádi Z, Sándorfi G, et al. New perspectives in the renin-angiotensin-aldosterone system (RAAS) IV: circulating ACE2 as a biomarker of systolic dysfunction in human hypertension and heart failure. PloS One 2014;9(4):e87845.
[34] Bindom SM, Hans CP, Xia H, Boulares AH, Lazartigues E. Angiotensin I–converting enzyme type 2 (ACE2) gene therapy improves glycemic control in diabetic mice. Diabetes 2010;59 (10):2540–8.
[35] Santos RAS, Oudit GY, Verano-Braga T, Canta G, Steckelings UM, Bader M. The renin-angiotensin system: going beyond the classical paradigms. Am J Physiol Heart Circ Physiol 2019;316(5):H958–70.
[36] Jankowski V, Vanholder R, van der Giet M, Tölle M, Karadogan S, Gobom J, et al. Mass-spectrometric identification of a novel angiotensin peptide in human plasma. Arterioscler Thromb Vasc Biol 2007;27(2):297–302.
[37] Villela DC, Passos-Silva DG, Santos RAS. Alamandine: a new member of the angiotensin family. Curr Opin Nephrol Hypertens 2014;23(2):130–4.
[38] Giani JF, Veiras LC, Shen JZY, Bernstein EA, Cao DY, Okwan-Duodu D, et al. Novel roles of the renal angiotensin-converting enzyme. Mol Cell Endocrinol 2021;529.
[39] Bernstein KE, Ong FS, Blackwell WLB, Shah KH, Giani JF, Gonzalez-Villalobos RA, et al. A modern understanding of the traditional and nontraditional biological functions of angiotensin-converting enzyme. Pharmacol Rev 2013;65(1):1–46.
[40] Ganten D, Minnich JL, Granger P, Hayduk K, Brecht HM, Barblau A, et al. Angiotensin-forming enzyme in brain tissue. Science (1979) 1971;173(3991):64–5.
[41] McKinley MJ, Albiston AL, Allen AM, Mathai ML, May CN, McAllen RM, et al. The brain renin–angiotensin system: location and physiological roles. Int J Biochem Cell Biol 2003;35 (6):901–18.
[42] Cao DY, Saito S, Veiras LC, Okwan-Duodu D, Bernstein EA, Giani JF, et al. Role of angiotensin-converting enzyme in myeloid cell immune responses. Cell Mol Biol Lett 2020;25(1):1–12.
[43] Veiras LC, Cao DY, Saito S, Peng Z, Bernstein EA, Shen JZY, et al. Overexpression of ACE in myeloid cells increases immune effectiveness and leads to a new way of considering inflammation in acute and chronic diseases. Curr Hypertens Rep 2020;22(1):1–7.
[44] Bernstein KE, Khan Z, Giani JF, Cao DY, Bernstein EA, Shen XZ. Angiotensin-converting enzyme in innate and adaptive immunity. Nat Rev Nephrol 2018;14(5):325–36.
[45] Kim HS, Krege JH, Kluckman KD, Hagaman JR, Hodgin JB, Best CF, et al. Genetic control of blood pressure and the angiotensinogen locus. Proc Natl Acad Sci 1995;92(7):2735–9.
[46] Niimura F, Labosky PA, Kakuchi J, Okubo S, Yoshida H, Oikawa T, et al. Gene targeting in mice reveals a requirement for angiotensin in the development and maintenance of kidney morphology and growth factor regulation. J Clin Invest 1995;96(6):2947–54.
[47] Krege JH, John SWM, Langenbach LL, Hodgin JB, Hagaman JR, Bachman ES, et al. Male–female differences in fertility and blood pressure in ACE-deficient mice. Nature 1995;375(6527):146–8.
[48] Hubert C, Savary K, Gasc JM, Corvol P. The hematopoietic system: a new niche for the renin–angiotensin system. Nat Clin Pract Cardiovasc Med 2006;3(2):80–5.
[49] Fuchs S, Frenzel K, Hubert C, Lyng R, Muller L, Michaud A, et al. Male fertility is dependent on dipeptidase activity of testis ACE [5]. Nat Med 2005;11(11):1140–2.
[50] Hagaman JR, Moyer JS, Bachman ES, Sibony M, Magyar PL, Welch JE, et al. Angiotensin-converting enzyme and male fertility. Proc Natl Acad Sci U S A 1998;95(5):2552–7.
[51] Lieberman J. Elevation of serum angiotension-converting-enzyme (ACE) level in sarcoidosis. Am J Med 1975;59(3):365–72.
[52] Shen XZ, Li P, Weiss D, Fuchs S, Xiao HD, Adams JA, et al. Mice with enhanced macrophage angiotensin-converting enzyme are resistant to melanoma. Am J Pathol 2007;170(6):2122–34.
[53] Khan Z, Shen XZ, Bernstein EA, Giani JF, Eriguchi M, Zhao TV, et al. Angiotensin-converting enzyme enhances the oxidative response and bactericidal activity of neutrophils. Blood 2017;130(3):328–39.
[54] Shen XZ, Lukacher AE, Billet S, Williams IR, Bernstein KE. Expression of angiotensin-converting enzyme changes major histocompatibility complex class I peptide presentation by modifying C termini of peptide precursors. J Biol Chem 2008;283(15):9957–65.
[55] Zhao T, Bernstein KE, Fang J, Shen XZ. Angiotensin-converting enzyme affects the presentation of MHC class II antigens. Lab Invest 2017;97(7):764–71.
[56] Benigni A, Cassis P, Remuzzi G. Angiotensin II revisited: new roles in inflammation, immunology and aging. EMBO Mol Med 2010;2 (7):247–57.
[57] Montezano AC, Nguyen Dinh Cat A, Rios FJ, Touyz RM. Angiotensin II and vascular injury. Curr Hypertens Rep 2014;16(6):1–11.
[58] Sodhi CP, Nguyen J, Yamaguchi Y, Werts AD, Lu P, Ladd MR, et al. A dynamic variation of pulmonary ACE2 is required to modulate

neutrophilic inflammation in response to *Pseudomonas aeruginosa* lung infection in mice. J Immunol 2019;203(11):3000–12.
[59] Hrenak J, Simko F. Renin–angiotensin system: an important player in the pathogenesis of acute respiratory distress syndrome. Int J Mol Sci 2020;21(21):8038.
[60] Gaddam RR, Chambers S, Bhatia M. ACE and ACE2 in inflammation: a tale of two enzymes. In: Inflammation and allergy—Drug targets. Bentham Science Publishers B.V.; 2014. p. 224–34.
[61] Ye R, Liu Z. ACE2 exhibits protective effects against LPS-induced acute lung injury in mice by inhibiting the LPS-TLR4 pathway. Exp Mol Pathol 2020;113:104350.
[62] Zambelli V, Bellani G, Borsa R, Pozzi F, Grassi A, Scanziani M, et al. Angiotensin-(1-7) improves oxygenation, while reducing cellular infiltrate and fibrosis in experimental Acute Respiratory Distress Syndrome. Intensive Care Med Exp 2015; 3(1):1–17.
[63] Radzikowska U, Ding M, Tan G, Zhakparov D, Peng Y, Wawrzyniak P, et al. Distribution of ACE2, CD147, CD26, and other SARS-CoV-2 associated molecules in tissues and immune cells in health and in asthma, COPD, obesity, hypertension, and COVID-19 risk factors. Allergy 2020;75(11):2829–45.
[64] Sodhi CP, Wohlford-Lenane C, Yamaguchi Y, Prindle T, Fulton WB, Wang S, et al. Attenuation of pulmonary ACE2 activity impairs inactivation of des-arg9 bradykinin/BKB1R axis and facilitates LPS-induced neutrophil infiltration. Am J Physiol Lung Cell Mol Physiol 2018;314(1):L17–31.
[65] Li W, Moore MJ, Vasllieva N, Sui J, Wong SK, Berne MA, et al. Angiotensin-converting enzyme 2 is a functional receptor for the SARS coronavirus. Nature 2003;426(6965):450–4.
[66] Xu H, Zhong L, Deng J, Peng J, Dan H, Zeng X, et al. High expression of ACE2 receptor of 2019-nCoV on the epithelial cells of oral mucosa. Int J Oral Sci 2020;12(1):1–5.
[67] Glowacka I, Bertram S, Herzog P, Pfefferle S, Steffen I, Muench MO, et al. Differential downregulation of ACE2 by the spike proteins of severe acute respiratory syndrome coronavirus and human coronavirus NL63. J Virol 2010;84(2):1198–205.
[68] Haga S, Yamamoto N, Nakai-Murakami C, Osawa Y, Tokunaga K, Sata T, et al. Modulation of TNF-α-converting enzyme by the spike protein of SARS-CoV and ACE2 induces TNF-α production and facilitates viral entry. Proc Natl Acad Sci U S A 2008;105 (22):7809–14.
[69] Kuba K, Imai Y, Rao S, Gao H, Guo F, Guan B, et al. A crucial role of angiotensin converting enzyme 2 (ACE2) in SARS coronavirus–induced lung injury. Nat Med 2005;11(8):875–9.
[70] Hong PJ, Look DC, Tan P, Shi L, Hickey M, Gakhar L, et al. Ectodomain shedding of angiotensin converting enzyme 2 in human airway epithelia. Am J Physiol Lung Cell Mol Physiol 2009;297 (1):84–96.
[71] Hofmann H, Geier M, Marzi A, Krumbiegel M, Peipp M, Fey GH, et al. Susceptibility to SARS coronavirus S protein-driven infection correlates with expression of angiotensin converting enzyme 2 and infection can be blocked by soluble receptor. Biochem Biophys Res Commun 2004;319(4):1216–21.
[72] Jiang J, Li MY, Wu XY, Ying YL, Han HX, Long YT. Protein nanopore reveals the renin–angiotensin system crosstalk with single-amino-acid resolution. Nat Chem 2023;1–9.
[73] Okwan-Duodu D, Landry J, Shen XZ, Diaz R. Angiotensin-converting enzyme and the tumor microenvironment: mechanisms beyond angiogenesis. Am J Physiol Regul Integr Comp Physiol 2013;305(3):205–15.
[74] de Alvarenga EC, de Castro FM, Carvalho CC, Florentino RM, França A, Matias E, et al. Angiotensin converting enzyme regulates cell proliferation and migration. PloS One 2016;11(12).
[75] Egami K, Murohara T, Shimada T, Sasaki KI, Shintani S, Sugaya T, et al. Role of host angiotensin II type 1 receptor in tumor angiogenesis and growth. J Clin Invest 2003;112(1):67–75.
[76] Fujita M, Hayashi I, Yamashina S, Fukamizu A, Itoman M, Majima M. Angiotensin type 1a receptor signaling-dependent induction of vascular endothelial growth factor in stroma is relevant to tumor-associated angiogenesis and tumor growth. Carcinogenesis 2005;26(2):271–9.
[77] Sipahi I, Debanne SM, Rowland DY, Simon DI, Fang JC. Angiotensin-receptor blockade and risk of cancer: meta-analysis of randomised controlled trials. Lancet Oncol 2010;11(7):627–36.
[78] Yoon C, Yang HS, Jeon I, Chang Y, Park SM. Use of angiotensin-converting-enzyme inhibitors or angiotensin-receptor blockers and cancer risk: a meta-analysis of observational studies. Can Med Assoc J 2011;183(14):E1073–84.
[79] von Bohlen Und Halbach O, Albrecht D. The CNS renin-angiotensin system. Cell Tissue Res 2006;326(2):599–616.
[80] Wright JW, Harding JW. Brain renin-angiotensin—a new look at an old system. Prog Neurobiol 2011;95(1):49–67.
[81] Welcome MO, Mastorakis NE. Stress-induced blood brain barrier disruption: molecular mechanisms and signaling pathways. Pharmacol Res 2020;157:104769.
[82] Gong S, Deng F. Renin-angiotensin system: the underlying mechanisms and promising therapeutical target for depression and anxiety. Front Immunol 2023;13:7691.
[83] Rocha NP, Simoes e Silva AC, Prestes TRR, Feracin V, Machado CA, Ferreira RN, et al. RAS in the central nervous system: potential role in neuropsychiatric disorders. Curr Med Chem 2018;25 (28):3333–52.
[84] Wang L, de Kloet AD, Pati D, Hiller H, Smith JA, Pioquinto DJ, et al. Increasing brain angiotensin converting enzyme 2 activity decreases anxiety-like behavior in male mice by activating central Mas receptors. Neuropharmacology 2016;(105):114–23.
[85] Park HS, You MJ, Yang B, Jang KB, Yoo J, Choi HJ, et al. Chronically infused angiotensin II induces depressive-like behavior via microglia activation. Sci Rep 2020;10(1):1–17.
[86] Kehoe PG. The coming of age of the angiotensin hypothesis in Alzheimer's disease: progress toward disease prevention and treatment? J Alzheimers Dis 2018;62(3):1443–66.
[87] Scotti L, Bassi L, Soranna D, Verde F, Silani V, Torsello A, et al. Association between renin-angiotensin-aldosterone system inhibitors and risk of dementia: a meta-analysis. Pharmacol Res 2021;166:105515.
[88] Bernstein KE, Koronyo Y, Salumbides BC, Sheyn J, Pelissier L, Lopes DHJ, et al. Angiotensin-converting enzyme overexpression in myelomonocytes prevents Alzheimer's-like cognitive decline. J Clin Invest 2014;124(3):1000–12.
[89] Selkoe DJ, Hardy J. The amyloid hypothesis of Alzheimer's disease at 25 years. EMBO Mol Med 2016;8(6):595–608.
[90] Kehoe PG, Wong S, al Mulhim N, Palmer LE, Miners JS. Angiotensin-converting enzyme 2 is reduced in Alzheimer's disease in association with increasing amyloid-β and tau pathology. Alzheimers Res Ther 2016;8(1):1–10.

[91] Acharya KR, Sturrock ED, Riordan JF, Ehlers MRW. Ace revisited: a new target for structure-based drug design. Nat Rev Drug Discov 2003;2(11):891–902.

[92] Ondetti MA, Rubin B, Cushman DW. Design of specific inhibitors of angiotensin-converting enzyme: new class of orally active antihypertensive agents. Science (1979) 1977;4288(196):441–4.

[93] Cushman DW, Ondetti MA. Design of angiotensin converting enzyme inhibitors. Nat Med 1999;5(10):1110–2.

[94] Bakhle YS. How ACE inhibitors transformed the renin–angiotensin system. Br J Pharmacol 2020;177(12):2657–65.

[95] Ferreira SH, Prito SPR, Paulo ES. A bradykinin-potentiating factor (BPF) present in the venom of *Bothrops jararaca*. Br J Pharmacol Chemother 1965;24(1):163–9.

[96] Ferreira SH, Greene LJ, Alabaster VA, Bakhle YS, Vane JR. Activity of various fractions of bradykinin potentiating factor against angiotensin I converting enzyme. Nature 1970;225(5230):379–80.

[97] Ondetti MA, Williams NJ, Sabo EF, Pluscec J, Weaver ER, Kocy O. Angiotensin-converting enzyme inhibitors from the venom of *Bothrops jararaca*. Isolation, elucidation of structure, and synthesis. Biochemistry 1971;10(22):4033–9.

[98] Byers LD, Wolfenden R. Binding of the by-product analog benzylsuccinic acid by carboxypeptidase A. Biochemistry 1973;12(11):2070–8.

[99] Zweifler AJ, Julius S, Nicholls MG. Efficacy of an oral angiotensin-converting enzyme inhibitor (captopril) in severe hypertension. Arch Intern Med 1981;141(7):907–10. [Internet]. [cited 2023 Feb 25]. Available from: https://jamanetwork.com/journals/jamainternalmedicine/fullarticle/601168.

[100] Holmquist B, Vallee BL. Metal-coordinating substrate analogs as inhibitors of metalloenzymes. Proc Natl Acad Sci 1979;76 (12):6216–20.

[101] Zaman MA, Oparil S, Calhoun DA. Drugs targeting the renin–angiotensin–aldosterone system. Nat Rev Drug Discov 2002;1 (8):621–36.

[102] Dive V, Cotton J, Yiotakis A, Michaud A, Vassiliou S, Jiracek J, et al. RXP 407, a phosphinic peptide, is a potent inhibitor of angiotensin I converting enzyme able to differentiate between its two active sites. Proc Natl Acad Sci U S A 1999;96(8):4330–5.

[103] Harris RB, Strong PDM, Wilson IB. Dipeptide-hydroxamates are good inhibitors of the angiotensin I-converting enzyme. Biochem Biophys Res Commun 1983;116(2):394–9.

[104] Almquist RG, Chao WR, Mitoma C, Rossi DJ, Panasevich RE, Matthews RJ. Synthesis and biological activity of ketomethylene-containing nonapeptide analogues of Snake venom angiotensin converting enzyme inhibitors. Pharmacol Rev 1988;31(5):60.

[105] Mutahi MW, Nittoli T, Guo L, Sieburth SMN. Silicon-based metalloprotease inhibitors: synthesis and evaluation of silanol and silanediol peptide analogues as inhibitors of angiotensin-converting enzyme. J Am Chem Soc 2002;124(25):7363–75.

[106] Piepho RW. Overview of the angiotensin-converting-enzyme inhibitors. Am J Health Syst Pharm 2000;57(Suppl. 1).

[107] Patchett AA, Harris E, Tristram EW, Wyvratt MJ, Wu MT, Taub D, et al. A new class of angiotensin-converting enzyme inhibitors. Nature 1980;288(5788):280–3.

[108] Bull HG, Thornberry NA, Cordes MHJ, Patchett AA. Inhibition of rabbit lung angiotensin-converting enzyme by N(α)-[(S)-1-carboxy-3-phenylpropyl]L-alanyl-L-proline and N(α)-[(S)-1-carboxy-3-phenylpropyl]L-lysyl-L-proline. J Biol Chem 1985;260 (5):2952–62.

[109] S. Anthony C, Masuyer G, D. Sturrock E, R. Acharya K. Structure based drug design of angiotensin-I converting enzyme inhibitors. Curr Med Chem 2012;19(6):845–55.

[110] Michaud A, Williams TA, Chauvet MT, Corvol P. Substrate dependence of angiotensin I-converting enzyme inhibition: captopril displays a partial selectivity for inhibition of N-acetyl-seryl-aspartyl-lysyl-proline hydrolysis compared with that of angiotensin I. Mol Pharmacol 1997;51(6):1070–6.

[111] Wei L, Clauser E, Alhenc-Gelas F, Corvol P. The two homologous domains of human angiotensin I-converting enzyme interact differently with competitive inhibitors. J Biol Chem 1992;267 (19):13398–405.

[112] Georgiadis D, Beau F, Czarny B, Cotton J, Yiotakis A, Dive V. Roles of the two active sites of somatic angiotensin-converting enzyme in the cleavage of angiotensin I and bradykinin. Circ Res 2003;93 (2):148–54.

[113] Deddish PA, Marcic B, Jackman HL, Wang HZ, Skidgel RA, Erdös EG. N-domain specific substrate and C-domain inhibitors of angiotensin-converting enzyme: angiotensin-(1 7) and keto-ACE. Hypertension 1998;31(4):912–7.

[114] Azizi M, Junot C, Ezan E, Ménard J. Angiotensin I-converting enzyme and metabolism of the haematological peptide N-acetyl-seryl-aspartyl-lysyl-proline. Clin Exp Pharmacol Physiol 2001;28 (12):1066–9.

[115] Rhaleb NE, Peng H, Harding P, Tayeh M, LaPointe MC, Carretero OA. Effect of N-acetyl-seryl-aspartyl-lysyl-proline on DNA and collagen synthesis in rat cardiac fibroblasts. Hypertension 2001;37 (3):827–32.

[116] Polakovičová M, Jampílek J. Advances in structural biology of ACE and development of domain selective ACE-inhibitors. Med Chem (Los Angeles) 2019;15(6):574–87.

[117] Zheng W, Tian E, Liu Z, Zhou C, Yang P, Tian K, et al. Small molecule angiotensin converting enzyme inhibitors: a medicinal chemistry perspective. Front Pharmacol 2022;13:968104.

[118] Redelinghuys P, Nchinda AT, Sturrock ED. Development of domain-selective angiotensin I-converting enzyme inhibitors. Ann N Y Acad Sci 2005;1056(1):160–75.

[119] Dive V, Georgiadis D, Matziari M, Makaritis A, Beau F, Cuniasse P, et al. Phosphinic peptides as zinc metalloproteinase inhibitors. Cell Mol Life Sci 2004;61(16):2010–9.

[120] Krapcho J, Turk C, Cushman DW, Powell JR, DeForrest JM, Spitzmiller ER, et al. Angiotensin-converting enzyme inhibitors. Mercaptan, carboxyalkyl dipeptide, and phosphinic acid inhibitors incorporating 4-substituted prolines. J Med Chem 1988;31(6):1148–60.

[121] Corradi HR, Chitapi I, Sewell BT, Georgiadis D, Dive V, Sturrock ED, et al. The structure of testis angiotensin-converting enzyme in complex with the C domain-specific inhibitor RXPA380. Biochemistry 2007;46(18):5473–8.

[122] Turner AJ. Angiotensin-converting enzyme 2: cardioprotective player in the renin-angiotensin system? Hypertension 2008;52 (5):816–7.

[123] Dales NA, Gould AE, Brown JA, Calderwood EF, Guan B, Minor CA, et al. Substrate-based design of the first class of angiotensin-converting enzyme-related carboxypeptidase (ACE2) inhibitors. J Am Chem Soc 2002;124(40):11852–3.

[124] Huang L, Sexton DJ, Skogerson K, Devlin M, Smith R, Sanyal I, et al. Novel peptide inhibitors of angiotensin-converting enzyme 2. J Biol Chem 2003;278(18):15532–40.

[125] Mores A, Matziari M, Beau F, Cuniasse P, Yiotakis A, Dive V. Development of potent and selective phosphinic peptide inhibitors of angiotensin-converting enzyme 2. J Med Chem 2008;51 (7):2216–26.
[126] Matsuzawa Y, Kimura K, Ogawa H, Tamura K. Impact of renin–angiotensin–aldosterone system inhibitors on COVID-19. Hypertens Res 2022;45(7):1147–53.
[127] Lim SP. Targeting SARS-CoV-2 and host cell receptor interactions. Antiviral Res 2023;210:105514.
[128] Vaseghi G, Golestaneh A, Jafari L, Ghasemi F. Drug repurposing against angiotensin-converting enzyme-related carboxypeptidase (ACE2) through computational approach. J Med Signals Sens 2022;12(4):341.
[129] Larue RC, Xing E, Kenney AD, Zhang Y, Tuazon JA, Li J, et al. Rationally designed ACE2-derived peptides inhibit SARS-CoV-2. Bioconjug Chem 2021;32(1):215–23.
[130] Martínez-Maqueda D, Miralles B, Recio I, Hernández-Ledesma B. Antihypertensive peptides from food proteins: a review. Food Funct 2012;3(4):350–61.
[131] Margalef M, Bravo FI, Arola-Arnal A, Muguerza B. Natural angiotensin converting enzyme (ACE) inhibitors with antihypertensive properties. In: Natural products targeting clinically relevant enzymes. John Wiley & Sons, Ltd; 2017. p. 45–67.
[132] Fitzgerald RJ, Meisel H. Milk protein-derived peptide inhibitors of angiotensin-I-converting enzyme. Br J Nutr 2000;84(S1):33–7.
[133] Chakraborty R, Roy S. Angiotensin-converting enzyme inhibitors from plants: a review of their diversity, modes of action, prospects, and concerns in the management of diabetes-centric complications. J Integr Med 2021;19(6):478–92.
[134] Jao CL, Huang SL, Hsu KC. Angiotensin I-converting enzyme inhibitory peptides: inhibition mode, bioavailability, and antihypertensive effects. Biomedicine (Taipei) 2012;2(4):130–6.
[135] Alvarenga DJ, Matias LMF, Cordeiro CF, Souza TBd, Lavorato SN, Pereira MGAG, et al. Synthesis of eugenol-derived glucosides and evaluation of their ability in inhibiting the angiotensin converting enzyme. Nat Prod Res 2022;36(9):2246–53.
[136] Abubakar MB, Usman D, El-Saber Batiha G, Cruz-Martins N, Malami I, Ibrahim KG, et al. Natural products modulating angiotensin converting enzyme 2 (ACE2) as potential COVID-19 therapies. Front Pharmacol 2021;12:627.
[137] Hernández Prada JA, Ferreira AJ, Katovich MJ, Shenoy V, Qi Y, Santos RAS, et al. Structure-based identification of small-molecule angiotensin-converting enzyme 2 activators as novel antihypertensive agents. Hypertension 2008;51(5):1312–7.
[138] Marquez A, Wysocki J, Pandit J, Batlle D. An update on ACE2 amplification and its therapeutic potential. Acta Physiol 2021;231 (1):e13513.
[139] Shenoy V, Gjymishka A, Jarajapu YP, Qi Y, Afzal A, Rigatto K, et al. Diminazene attenuates pulmonary hypertension and improves angiogenic progenitor cell functions in experimental models. Am J Respir Crit Care Med 2013;187(6):648–57.
[140] Qaradakhi T, Gadanec LK, McSweeney KR, Tacey A, Apostolopoulos V, Levinger I, et al. The potential actions of angiotensin-converting enzyme II (ACE2) activator diminazene aceturate (DIZE) in various diseases. Clin Exp Pharmacol Physiol 2020;47 (5):751–8.
[141] Chen P, Liu C, Zhang Z, Li Z, Chen S, Lu Y. Protocol for high-throughput screening of ACE2 enzymatic activators to treat COVID-19-induced metabolic complications. STAR Protoc 2022;3(3):101641.
[142] Pang X, Cui Y, Zhu Y. Recombinant human ACE2: potential therapeutics of SARS-CoV-2 infection and its complication. Acta Pharmacol Sin 2020;41(9):1255–7.
[143] Zhang H, Baker A. Recombinant human ACE2: acing out angiotensin II in ARDS therapy. Crit Care 2017;21(1):1–3.
[144] Patel VB, Lezutekong JN, Chen X, Oudit GY. Recombinant human ACE2 and the angiotensin 1-7 axis as potential new therapies for heart failure. Can J Cardiol 2017;33(7):943–6.
[145] Zoufaly A, Poglitsch M, Aberle JH, Hoepler W, Seitz T, Traugott M, et al. Human recombinant soluble ACE2 in severe COVID-19. Lancet Respir Med 2020;8(11):1154–8.
[146] Oates JA, Wood AJJ, Williams GH. Converting-enzyme inhibitors in the treatment of hypertension. N Engl J Med 1988;319 (23):1517–25.
[147] Schmieder RE, Martus P, Klingbeil A. Reversal of left ventricular hypertrophy in essential hypertension: a meta-analysis of randomized double-blind studies. JAMA 1996;275(19):1507–13.
[148] Pfeffer MA, Lamas GA, Vaughan DE, Parisi AF, Braunwald E. Effect of captopril on progressive ventricular dilatation after anterior myocardial infarction. N Engl J Med 1988;319(2):80–6.
[149] Jafar TH, Schmid CH, Landa M, Giatras I, Toto R, Remuzzi G, et al. Angiotensin-converting enzyme inhibitors and progression of nondiabetic renal disease. A meta-analysis of patient-level data. Ann Intern Med 2001;135(2):73–87.
[150] Lewis EJ, Hunsicker LG, Bain RP, Rohde RD. The effect of angiotensin-converting-enzyme inhibition on diabetic nephropathy. N Engl J Med 1993;15(5):69.
[151] Garg R, Yusuf S, Bussmann WD, Sleight P, Uprichard A, Massie B, et al. Overview of randomized trials of angiotensin-converting enzyme inhibitors on mortality and morbidity in patients with heart failure. JAMA 1995;273(18):1450–6.
[152] Selak V, Webster R, Stepien S, Bullen C, Patel A, Thom S, et al. Reaching cardiovascular prevention guideline targets with a polypill-based approach: a meta-analysis of randomised clinical trials. Heart 2019;105(1):42–8.
[153] Muñoz D, Uzoije P, Reynolds C, Miller R, Walkley D, Pappalardo S, et al. Polypill for cardiovascular disease prevention in an underserved population. N Engl J Med 2019;381(12):1114–23.
[154] Roshandel G, Khoshnia M, Poustchi H, Hemming K, Kamangar F, Gharavi A, et al. Effectiveness of polypill for primary and secondary prevention of cardiovascular diseases (PolyIran): a pragmatic, cluster-randomised trial. Lancet 2019;394(10199):672–83.
[155] Angiotensin-Converting Enzyme Inhibitor (ACE inhibitor) Drugs | FDA [Internet]. [cited 2023 Feb 27]. Available from: https://www.fda.gov/drugs/postmarket-drug-safety-information-patients-and-providers/angiotensin-converting-enzyme-inhibitor-ace-inhibitor-drugs.
[156] Leonetti G, Cuspidi C. Choosing the right ACE inhibitor: a guide to selection. Drugs 1995;49(4):516–35.
[157] Renin-angiotensin System Blockade Benefits in Clinical Evolution and Ventricular Remodeling After Transcatheter Aortic Valve Implantation (RASTAVI)—Full Text View—ClinicalTrials.gov [Internet]. [cited 2023 Feb 25]. Available from: https://clinicaltrials.gov/ct2/show/NCT03201185.
[158] Nordenskjöld AM, Agewall S, Atar D, Baron T, Beltrame J, Bergström O, et al. Randomized evaluation of beta blocker and ACE-inhibitor/

angiotensin receptor blocker treatment in patients with myocardial infarction with non-obstructive coronary arteries (MINOCA-BAT): rationale and design. Am Heart J 2021;231:96–104.
[159] Randomized Evaluation of Beta Blocker and ACEI/ARB Treatment in MINOCA Patients—MINOCA-BAT—Full Text View—ClinicalTrials.gov [Internet]. [cited 2023 Feb 25]. Available from: https://clinicaltrials.gov/ct2/show/NCT03686696.
[160] Evaluation of a Renin Inhibitor, Aliskiren, Compared to Enalapril, in C3 Glomerulopathy—Full Text View—ClinicalTrials.gov [Internet]. [cited 2023 Feb 25]. Available from: https://clinicaltrials.gov/ct2/show/NCT04183101.
[161] Basu R, Poglitsch M, Yogasundaram H, Thomas J, Rowe BH, Oudit GY. Roles of angiotensin peptides and recombinant human ACE2 in heart failure. J Am Coll Cardiol 2017;69(7):805–19.
[162] Zhong J, Basu R, Guo D, Chow FL, Byrns S, Schuster M, et al. Angiotensin-converting enzyme 2 suppresses pathological hypertrophy, myocardial fibrosis, and cardiac dysfunction. Circulation 2010;122(7):717–28.
[163] Safety and Tolerability Study of APN01 (Recombinant Human Angiotensin Converting Enzyme 2)—Full Text View—ClinicalTrials.gov [Internet]. [cited 2023 Feb 25]. Available from: https://clinicaltrials.gov/ct2/show/NCT00886353.
[164] The Safety, Tolerability, PK and PD of GSK2586881 in Patients With Acute Lung Injury—Full Text View—ClinicalTrials.gov [Internet]. [cited 2023 Feb 25]. Available from: https://www.clinicaltrials.gov/ct2/show/NCT01597635.
[165] Haschke M, Schuster M, Poglitsch M, Loibner H, Salzberg M, Bruggisser M, et al. Pharmacokinetics and pharmacodynamics of recombinant human angiotensin-converting enzyme 2 in healthy human subjects. Clin Pharmacokinet 2013;52(9):783–92.
[166] Taddei S, Bortolotto L. Unraveling the pivotal role of bradykinin in ACE inhibitor activity. Am J Cardiovasc Drugs 2016;16 (5):309–21.
[167] Luque CA, Vazquez OM. Treatment of ACE inhibitor-induced cough. Pharmacotherapy J Hum Pharmacol Drug Ther 1999;19(7):804–10.
[168] Ames J, Ainer VG, Ason J, Orrow DM, Ngela A, Oveland L, et al. Effect of bradykinin-receptor blockade on the response to angiotensin-converting–enzyme inhibitor in normotensive and hypertensive subjects. N Engl J Med 1998;339(18):1285–92.
[169] Murphey L, Vaughan D, Brown N. Contribution of bradykinin to the cardioprotective effects of ACE inhibitors. Eur Heart J Suppl 2003;5 (A):A37–41.
[170] Borghi C, Veronesi M. Cough and ACE inhibitors: the truth beyond placebo. Clin Pharmacol Ther 2019;105(3):550–2.
[171] Makani H, Messerli FH, Romero J, Wever-Pinzon O, Korniyenko A, Berrios RS, et al. Meta-analysis of randomized trials of angioedema as an adverse event of renin-angiotensin system inhibitors. Am J Cardiol 2012;110(3):383–91.
[172] Bradykinin-degradating Enzymes Activities in Angiotensin-Converting Enzyme Inhibitors-associated Angioedema—Full Text View—ClinicalTrials.gov [Internet]. [cited 2023 Feb 25]. Available from: https://clinicaltrials.gov/ct2/show/NCT04763577.
[173] Yang J, Yang X, Gao L, Zhang J, Yi C, Huang Y. The role of the renin-angiotensin system inhibitors in malignancy: a review. Am J Cancer Res 2021;11(3):884.
[174] Kerns SL, Amidon Morlang A, Lee SM, Peterson DR, Marples B, Zhang H, et al. Use of angiotensin converting enzyme inhibitors is associated with reduced risk of late bladder toxicity following radiotherapy for prostate cancer. Radiother Oncol 2022;168:75–82.
[175] Zhang X, Wong GLH, Yip TCF, Tse YK, Liang LY, Hui VWK, et al. Angiotensin-converting enzyme inhibitors prevent liver-related events in nonalcoholic fatty liver disease. Hepatology 2022;76(2):469–82.
[176] Ding J, Davis Plourde KL, Sedaghat S, Tully PJ, Wang W, Phillips C, et al. Antihypertensive medications and risk for incident dementia and Alzheimer's disease: a meta-analysis of individual participant data from prospective cohort studies. Lancet Neurol 2020;19(1):61–70.
[177] Pajewski NM, Elahi FM, Tamura MK, Hinman JD, Nasrallah IM, Ix JH, et al. Plasma amyloid beta, neurofilament light chain, and total tau in the Systolic Blood Pressure Intervention Trial (SPRINT). Alzheimers Dement 2022;18(8):1472–83. [Internet]. [cited 2023 Feb 25]. Available from: https://onlinelibrary.wiley.com/doi/full/10.1002/alz.12496.
[178] Khan A, Benthin C, Zeno B, Albertson TE, Boyd J, Christie JD, et al. A pilot clinical trial of recombinant human angiotensin-converting enzyme 2 in acute respiratory distress syndrome. Crit Care 2017;21(1).
[179] Sriram K, Insel PA. Risks of ACE inhibitor and ARB usage in COVID-19: evaluating the evidence. Clin Pharmacol Ther 2020;108(2):236–41.

Chapter 3.7

Histidinol dehydrogenase

Jean-Yves Winum
IBMM, Univ Montpellier, CNRS, ENSCM, Montpellier, France

1 Introduction

Antibiotic resistance and multi-resistance (AMR) among bacterial infections are currently a serious health concern on a global scale. AMR has been recognized as one of the top 10 worldwide public health hazards to humanity by the World Health Organization (WHO), which has made combating drug resistance a top priority [1]. The current era, commonly referred to as the "post-antibiotic era," is characterized by a surge in the hunt for new antibacterial drugs. Compounds having novel molecular targets may be particularly helpful. Therefore, to overcome bacterial resistance and to create potent new classes of antibacterial drugs that could go around the established resistance mechanism, it is becoming increasingly important to find and validate rational therapeutic targets in bacteria [2–4].

For many years, it has been thought that an effective therapeutic intervention method would be to inhibit the enzymes that catalyze certain phases of bacterial amino-acid production [5,6].

In this context, among the 21 proteinogenic amino acids, L-histidine is an essential nutrient for mammals but is produced de novo by plants and microorganisms. Due to its side-pK_a chain's value of 6.0, histidine is the only amino acid whose side chain may change from an unprotonated to a protonated form under neutral pH conditions. This property allows histidine residues to function in numerous cellular enzymatic activities as both a proton acceptor and a proton donor [7].

Highly conserved in lower eukaryotic and prokaryotic organisms, histidine biosynthesis pathway consists in the conversion of 5-phosphoribosyl-1-pyrophosphate (PRPP) to L-histidine in ten enzymatic steps and 10 different enzymes including HisG, HisE, HisI, HisA, HisH, HisF, HisC, HisB, and HisD (Fig. 1) [8]. Due to the reaction catalyzed by the products of *hisH* and *hisF* genes, this pathway is connected to nitrogen metabolism and the de novo production of purines [8–11].

This ancestral pathway has attracted attention of researchers as it is present in bacteria, as well as in archaebacteria, fungi, and plants, but not in mammals. In bacteria, it has been extensively studied mainly in *Escherichia coli* and *Salmonella typhimurium* under biochemical and genetic aspects showing similar features with eight biosynthetic genes clustered in a compact operon (hisGDC[NB]HAF[IE]) [12]. Three of the his genes code for bifunctional enzymes (i.e., *hisNB*, *hisD*, and *hisIE*), while another enzyme is heterodimeric, being composed of the *hisF* and *hisH* gene products [9]. Therefore, this biosynthetic pathway could provide a plethora of highly favorable prospective protein targets that can be investigated for the design of selective antibacterial drugs. [13].

Indeed, the genes coding for the enzymes of the His pathway are a subset of the virulence genes of the intramacrophagic pathogen and have been called the intramacrophagic virulome. They have a growing significance in modern medicine as they are required for the development, e.g., growth and/or virulence, of several facultative intramacrophagic pathogen species such as for example *S. typhimurium*, [14] *Burkholderia pseudomallei*, [15] *Brucella suis* [16], and *Mycobacterium tuberculosis* [17]. Virulence is correlated with the pathogen's ability to multiply inside the macrophage host cell and evade the host immune system.

Additionally, it is worth to mention that the essential character of histidine biosynthesis for virulence has also been demonstrated in intracellular pathogenic fungi such as *Aspergillus fumigatus* [18]

Therefore, the ability to specifically block one of the bacterial enzymes from histidine path within the host cell has provided an innovative way for the development of new potential antibacterial with a lower risk of adverse secondary effects on the host. Indeed, limiting the selective pressure to the intracellular niche and decreasing the likelihood of damaging the commensal flora are two benefits of defining the targets for antibacterial through investigation of the intracellular virulome. This confined approach of targeting proteins from specific pathways of pathogenic bacteria is a major contributor to the development of antibiotics, which differ from most of the currently available antibiotics that target peptidoglycan and its biosynthesis pathway [19–21].

L-histidinol dehydrogenase (HDH, EC 1.1.1.23), encoded by the *hisD* gene, has been shown to be crucial

Metalloenzymes. https://doi.org/10.1016/B978-0-12-823974-2.00020-6

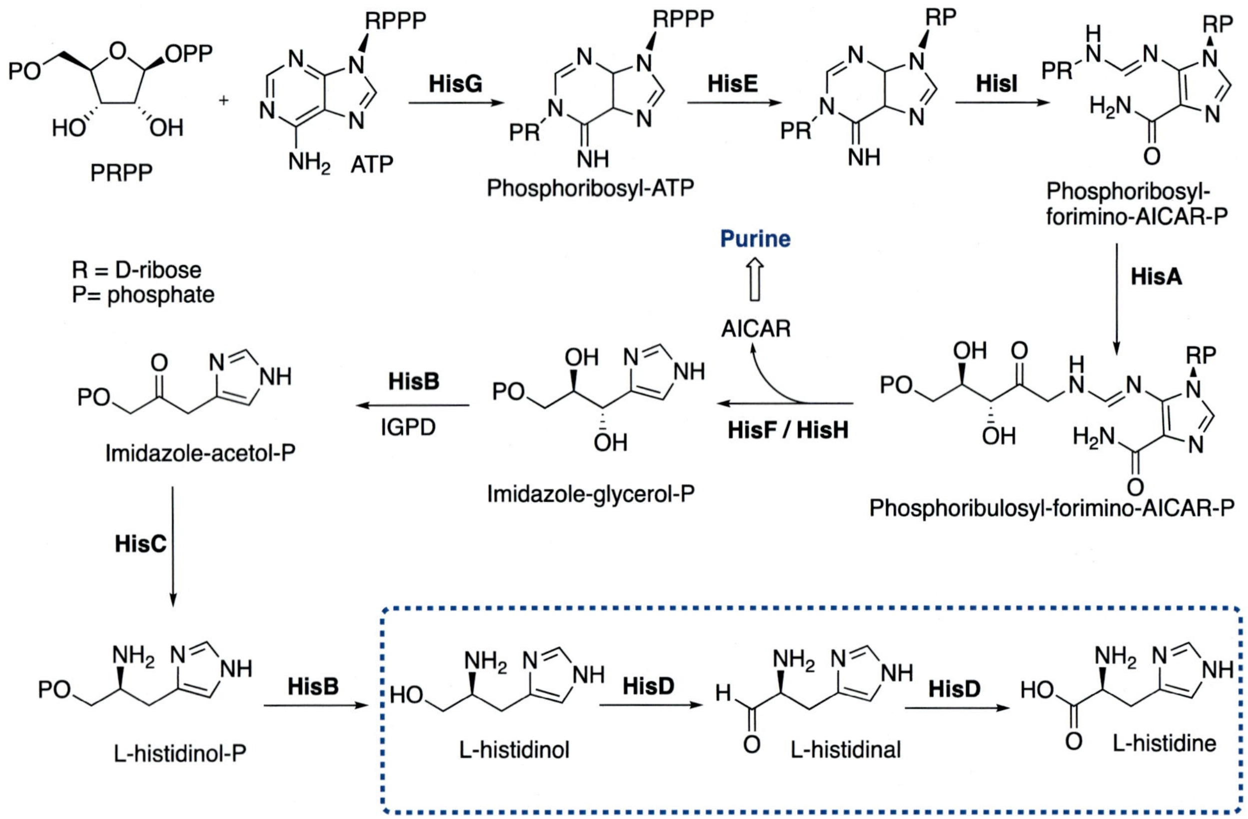

FIG. 1 Histidine biosynthetic pathway: *PRPP*, phosphoribosyl pyrophosphate; *ATP*, adenosine triphosphate; *HisG*, ATP phosphoribosyltransferase; *HisE*, phosphoribosyl-ATP pyrophosphatase; *HisI*, phosphoribosyl-AMP cyclohydrolase; *HisA*, 5′ProFAR isomerase; *HisF*, synthase subunit of IGP synthase; *HisH*, glutaminase subunit of IGP synthase; *HisB*, imidazoleglycerol-phosphate dehydratase; *HisC*, histidinol-phosphate aminotransferase; *HisN*, histidinol-phosphate phosphatase; *HisD*, histidinol dehydrogenase; *AICAR*, aminoimidazole carboxamide ribotide.

for the pathogenicity, growth, and survival inside the macrophage of *B. suis* [17] and *M. tuberculosis* [22], the etiological agents of brucellosis and tuberculosis.

The purpose of this chapter is to highlight the relevance of this enzyme for the development of new antibacterial agents against both *M. tuberculosis* and *B. suis*.

2 Structure and function of the enzyme HDH

In plants, archae, and bacteria, L-histidinol dehydrogenase (HDH, EC 1.1.1.23) is a well-conserved protein. It is a metalloprotein with a homodimeric structure with a weight of around 50 kDa/monomer and one Zn^{2+} cation per subunit. The presence of a divalent metal ion is necessary for the activity of an enzyme; hence, replacing the Zn^{2+} cation with Mg^{2+}, Ni^{2+}, Co^{2+}, or Cu^{2+} either decreases or increases the activity of an enzyme. Using a Bi-Uni-Uni-Bi PingPong kinetic mechanism, this metalloprotein catalyzes the consecutive NAD-dependent four electron oxidation reaction of L-histidinol to L-histidinal and from there to L-histidine in the final two steps of L-histidine biosynthesis [23–26] (Fig. 2)

In 1999, Grubmeyer proposed a mechanism based on *S. typhimurium* HDH kinetic study. L-Histidinol binds first, followed by NAD^+ [27,28]. While L-histidinal remains enzyme-bound, NADH and H^+ are released. The second NAD^+ then binds and is reduced, releasing $NADH+H^+$ once more before L-histidine is produced. The first half of the mechanism is a fast and reversible reaction and the second part is slower and irreversible, both partial reactions participate in the overall rate limitation. Experiments on *St*HDH suggested that His-261 and His-326 are engaged in proton transfers during catalysis while His-261 and His-418 are potential possibilities for zinc ion ligands [27,28].

Grubmeyer's study of the enzyme's kinetics [24] provided insights into the mechanism and was followed by protein mutations [29] that soon produced a comprehensive site map of the active site [30]. In fact, the function of Zn^{2+} and the presence of NAD+ in the active site, as well as the number of active sites in the dimeric, active enzyme, were clarified [23]. It was determined that there were two identical active sites, one in each component of the dimer, which improved our knowledge of the enzyme. Two histidine residues that are essential for the action were also discovered

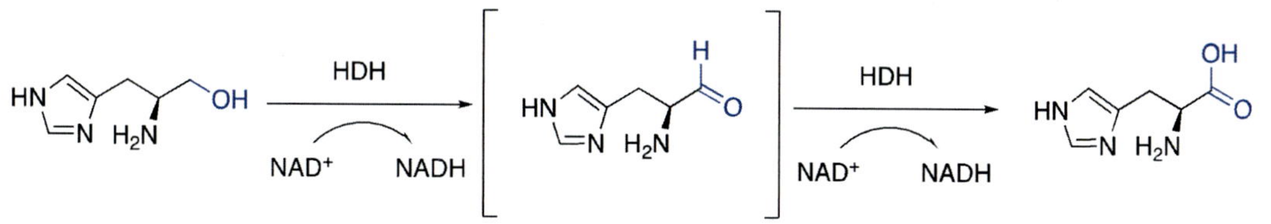

FIG. 2 Sequential 2-step oxidation reaction catalyzed by L-HDH.

as a result of this investigation. Additionally, substantial research has been done on the stereochemistry of the catalytic process using exchange reactions, NMR studies, and isotope effects. These experiments revealed that both reductions occur via R stereochemistry at NAD^+, but during oxidations the pro-*S* and pro-*R* hydrogen of L-histidinol were first eliminated [30].

Apart from *S. typhimurium*, only few HDH from pathogenic bacteria have been biochemically characterized including those from *Escherichia coli* (*Ec*HDH) [31], *M. tuberculosis* (*Mt*HDH) [32], and *B. suis* (*Bs*HDH) [33,34]. Monomer molecular weight and kinetic constants of the enzymatic mechanism have been determined for these four HDH and are reported in the Table 1 [34].

On a structural study point of view, the number of deposited structures is still limited since only seven entries can be found in the Protein Data Bank. *E. coli* HDH (*Ec*HDH) and *B. suis* (*Bs*HDH) are the only enzyme of this family for which a structural study has been published. Among these, four were obtained with different or no ligands from *E. coli* [35] and two were from the *B. suis* protein [33,34] with and without an "in house" inhibitor.

The first structural details on this family of enzymes were revealed by the crystal structure of recombinant *Ec*HDH [35], which was solved both in its apo form and in association with Zn^{2+}, NAD^+, and L-histidine or L-histamine. The dimeric nature of the enzyme and the binding of one Zn^{2+} ion per monomer were both verified by analysis of these structures. As the two monomers engage to create the dimer, there are several polar interactions that take place.

The active site, which is situated at the dimer interface between domains 1, 2, and 4 of each monomer, was clearly identified thanks to the structure of *Ec*HDH in association with Zn^{2+}, NAD^+, and L-histidinol. Two areas constitute the active site: a deep cavity that functions as the substrate binding site and a groove on the protein surface that recognizes NAD^+.

Four protein residues (Gln259, His262, Asp360, and His419*) and two atoms of L-histidinol are octahedrally coordinated to the Zn^{2+} ion, which is located at the bottom of the deep cavity.

In the case of *B. suis*, the crystal structure of a mutated form of the enzyme, where Cys366 was substituted by a serine residue, was solved and reported in 2014 (Fig. 3) [33]. Interestingly, Zn^{2+} was located in the substrate binding pocket, tetrahedrally coordinated to a water molecule and residues belonging to both subunits, e.g., Glu357 and Asp361 of one monomer and His420 of the other one.

According to the high sequence identity (39.9%) between *Bs*HDH and *Ec*HDH, these enzymes exhibit a high degree of three-dimensional similarity both in the monomeric and dimeric form, with the majority of the residues involved in the substrate and NAD^+ binding being conserved (Fig. 4). Except for residues 133–137 and 164–166, which were situated on the edge of the NAD^+ binding region, the majority of the structural differences were found on the protein surface. This suggests that the two enzymes differ in how they recognize cofactors.

Altogether, these discoveries expanded our understanding of two key aspects of HDHs' catalytic mechanism: first, NAD+ is only loosely bound to the enzyme, allowing for a quick exchange at the active site, which enables the reaction to proceed; second, the Zn^{2+} ion is crucial for the correct positioning of the substrate and the other reaction intermediates.

TABLE 1 kinetic parameters for the HDH of *S. typhimurium* (*St*HDH), *E. coli* (*Ec*HDH), *B. suis* (*Bs*HDH), and *M. tuberculosis* (*Mt*HDH).

HDH	Mw/monomer	K_{cat} (s^{-1})	K_M histidinol	K_M NAD^+
*St*HDH	47 kDa	16.7	11 μM	0.6 μM
*Ec*HDH	52 kDa	–	14 μM	0.57 mM
*Bs*HDH	49 kDa	12	12 μM	4.72 μM
*Mt*HDH	45 kDa	1.45	49 μM	14 mM

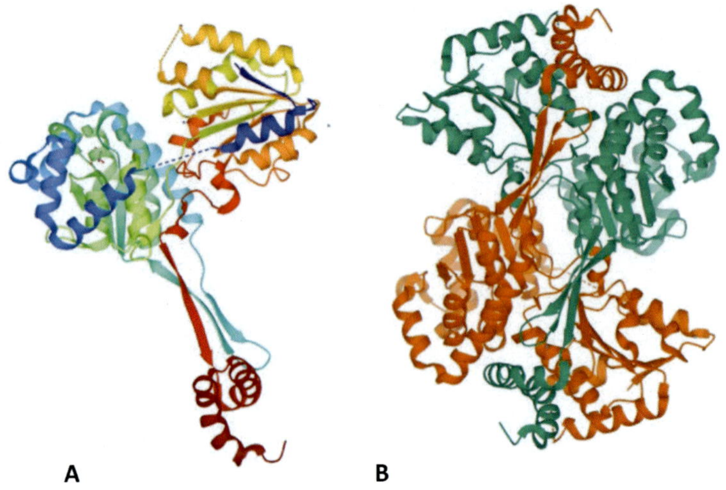

FIG. 3 (A) Overall fold of the mutated form of *Bs*HDH monomer. (B) Dimeric structure of the mutated form of *Bs*HDH with one monomer in *red* and the other in *green*.

```
S.typhimurium      MSFNTLIDWNSCSPEQQRALLTRPAISASDSITRTVSDILDNVKTRGDDALREYSAKFDKTEV--TALRVTPEEIAAAGARLSDELKQAMTAAVKNIETFH
E.coli             MSFNTIIDWNSCTAKQQRQLLMRPAISASESITRTVNDILDSVKARGDDALREYSAKFDKTTV--TALKVSAEEIAAASERLSDELKQAMAVAVKNIETFH
M.tuberculosis     MLTRIDLRGAELTAAELRAALPRGGADV-EAVLPTVRPIVAAVAERGAEAALDFGASFDGVRP--HAIRVPDAALDAALAGLDCDVCEALQVMVERTRAVH
B.suis             --MVTTLRQTDPDFEQKFAAFLSGKREVSEDVDRAVREIVDRVRREGDSALLDYSRRFDRIDLEKTGIAVTEAEIDAAFDAAPASTVEALKLARDRIEKHH
                          :   .    :    :      .. : :  :*  *:  *   * .*  ::.  **       .: *.   : **      .  :*:     ..   *

SAQTLPPVDVETQPGVRCQQVTRPVSSVGLYIPGGSAPLFSTVLMLATPARIAGCQKVVLCSPPP------IADEILYAAQLCGVQEIFNVGGAQAIAALAF-GSES----VPKVDKI
TAQKLPPVDVETQPGVRCQQVTRPVASVGLYIPGGSAPLFSTVLMLATPARIAGCKKVVLCSPPP------IADEILYAAQLCGVQDVFNVGGAQAIAALAF-GTES----VPKVDKI
SGQRRTDVTTTLGPGATVTERWVPVERVGLYVPGGNAVYPSSVVMNVVPAQAAGVDSLVVASPPQAQWDGMPHPTILAAARLLGVDEVWAVGGAQAVALLAYGGTDTDGAALTPVDMI
ARQLPKDDRYTDALGVELGSRWTAIEAVGLYVPGGTASYPSSVLMNAMPAKVAGVDRIVMVVPAP---DGNLNPLVLVAARLAGVSEIYRVGGAQAIAALAY-GTET----IRPVAKI
: *            *.   .   .:  ****:***.*   *:*:* . **. ** . :*:  *.              :* **.* **.::: ******:* **: *:::     :  *  *

FGPGNAFVTEAKRQVSQRLDGAAIDMPAGPSEVLVIADSGATPDFVASDLLSQAEHGPDSQVILLTPDADIARKVAEAVERQLAELPRADTARQALSA--SRLIVTKDLAQCVAISNQ
FGPGNAFVTEAKRQVSQRLDGAAIDMPAGPSEVLVIADSGATPDFVASDLLSQAEHGPDSQVILLTPDADMAHQVAEAVERQLAELPRAETARQALNA--SRLIVTKDLAQCVEISNQ
TGPGNIYVTAAKRLCRSRV---GIDAEAGPTEIAILADHTADPVHVAADLISQAEHDELAASVLVTPSEDLADATDAELAGQLQTTVHRERVTAALTGRQSAIVLVDDVDAAVLVVNA
VGPGNAYVAAAKRIVFGTV---GIDMIAGPSEVLIVADKDNNPDWIAADLLAQAEHDTAAQSILMTNDEAFAHAVEEAVERQLHTLARTETASASWRD-FGAVILVKDFEDAIPLANR
 **** :*: ***       :    .**  ***:*: ::**     *   :*:**::****.  :   :*:*  .  :*   .    :   **      . : .  :      . :::..*.   .: : *

YGPEHLIIQTRNARDLVDAITSAGSVFLGDWSPESAGDYASGTNHVLPTYGYTATCSSLGLADFQKRMTVQELSKAGFSALASTIETLAAAERLTAHKNAVTLRVNALKEQA
YGPEHLIIQTRNARELVDGITSAGSVFLGDWSPESAGDYASGTNHVLPTYGYTATCSSLGLADFQKRMTVQELSKVGFSALASTIETLAAAERLTAHKNAVTLRVNALKEQA
YAAEHLEIQTADAPQVASRIRSAGAIFVGPWSPVSLGDYCAGSNHVLPTAGCARHSSGLSVQTFLRGIHVVEYTEAALKDVSGHVITLATAEDLPAHGEAVRRRFER-----
IAAEHLEIAVADAEAFVPRIRNAGSIFIGGYTPEVIGDYVGGCNHVLPTARSARFSSGLSVLDYMKRTSLLKLGSEQLRALGPAAIEIARAEGLDAHAQSVAIRLNL-----
 ..*** *  .  :*   ..   *  .**::*:* ::*     ***  .*  ******    :   .*.*.:   :   .   : :   .   :  :.       :* ** * ** ::*   *.:
```

FIG. 4 Multialignment of the HDH amino acid sequences from the four bacterial species: *B. suis*, *M. tuberculosis*, *S. typhimurium*, and *E. coli*. Conserved histidines are in bold, and residues acting as bases in the catalytic mechanism are underlined. Amino acids involved in substrate binding are highlighted in *yellow* and those involved in cofactor recognition in *gray*. The first ones have been experimentally identified in *Ec*HDH, *Mt*HDH, and *Bs*HDH, while the second ones only in *Ec*HDH and *Mt*HDH. The asterisk (*) indicates identity at all aligned positions; the symbol (:) relates to conserved substitutions, while (.) means that semi-conserved substitutions are observed. The multialignment was performed with MUSCLE, version 3.8. Accession numbers: *S. typhimurium* (accession number: NP_461017.1), *E. coli* (accession number: NP_310848.1), *M. tuberculosis* (accession number: NP_216115.1), *B. suis* (accession number: WP_004687945.1).

These three-dimensional structures will allow for new perspectives in drug design, which could quickly result in the development of novel inhibitors.

3 Pathologic role of histidinol dehydrogenase

Brucellosis and tuberculosis are bacterial infectious diseases caused respectively by Brucella and Mycobacterium tuberculosis. Brucellosis is a highly prevalent zoonotic disease caused by *Brucella* spp., a gram-negative facultative intracellular bacterium, which is transmitted from animals to humans by contact or by ingestion of contaminated milk [36]. If left untreated, the disease can become chronic with serious consequences, in some cases leading to death. Although *Brucella* spp. infect humans as an incidental host, 500 000 new human infections occur annually, and no patient-friendly treatments or approved human vaccines are reported [37].

For tuberculosis, approximately 9 million new cases and 2 million deaths per year are reported [38].

Mycobacterium tuberculosis has the particularity to evade antimicrobial immunity and to persist within macrophages by interfering with multiple host cellular functions through its virulence factors, causing latent tuberculosis [39].

In humans, although the antibiotics currently used against these infections are still mostly effective, a growing number of clinical strains are (multi-)resistant to treatment and tuberculosis has become a real public health problem worldwide.

These intracellular pathogens have adapted to a life inside host cells, in which they use host nutrients to replicate and spread. As virulence is linked to the capacity of the bacteria to replicate inside the macrophage host cell and to escape from the host immune system, the analysis of virulence mechanisms of the intracellular pathogens *M. tuberculosis* and *Brucella* spp. has led to the identification of histidinol dehydrogenase (HDH) as an essential protein for the intramacrophage growth of these pathogens. Histidinol dehydrogenase (HDH; EC 1.1.1.23), encoded by the gene *hisD* (BR0252) in *B. suis* and *hisD* (Rv1599) in *M. tuberculosis* is essential for intramacrophagic replication of the pathogen inside the host [17,22].

L-HDH has no equivalent in mammals, so the specific inhibition of this zinc metalloprotein is likely to limit the selective pressure to the intracellular niche and also reduces the probability of affecting the commensal flora. In addition, L-HDH has been included by the TDR Targets Database as one of the top 50 targets. This ranking considers factors including the presence of a solved structure or structural model, essentiality in any species, the lack of a human ortholog, and expression during latency or dormancy [40].

In this way, inhibition of this metalloprotein represents an innovative and important antibacterial approach in the fight against brucellosis and tuberculosis with limited secondary adverse effects.

4 Classes of inhibitors and their design

Even if HDH was studied in details for some pathogens, few inhibitors were described in literature [41] (Fig. 5). Seminal work on the inhibitory study of bacterial HDH was published by Grubmeyer and coll. in 1989 [30]. This study examined the inhibitory effects of a library of histidinol analogs against *S. typhimurium* HDH, revealing that methyl ketone **1** was one of the most effective compounds, with an inhibitory activity in the low micromolar range.

Dancer et al. [42] reported in 1996 the synthesis of new inhibitors based on the structure analysis of **1** by lengthening the side chain to target a valuable neighboring lipophilic pocket near to the active site. Compounds **2a** and **2b** had the greatest inhibitory action against *E. coli* HDH (IC_{50} of **2a**=1 M) and cabbage HDH (IC_{50} of **2b**=0.04 M) among the limited inhibitor library presented.

In 2007 our group looked into inhibitor series against *B. suis* HDH in keeping with the structures previously disclosed by the Grubmeyer and Dancer research groups [43,44]. Eleven very potent *B. suis* L-HDH with IC_{50} below 30 nM were detected, with the best molecule **4** having an IC_{50} of 3 nM.

In vitro growth and infection experiments performed respectively in minimal medium devoid of sources of histidine and in human macrophage-like cells showed that **4** was the most effective on growth inhibition of *B. suis* in minimal medium. The effect of this inhibitor was specific of bacterial HDH, as addition of histidine to the medium

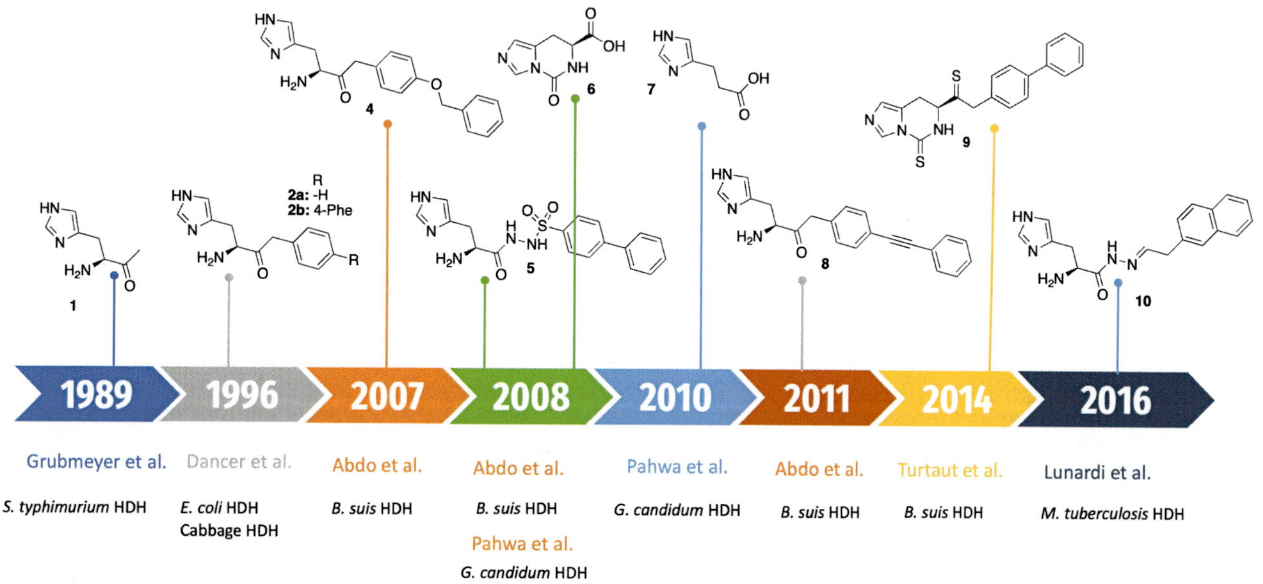

FIG. 5 Timeline of the discovery of HDH inhibitors.

relieved growth inhibition. Compound **4** also completely abolished the intracellular replication of Brucella in human THP-1 macrophage-like cells at concentrations between 10 and 25 μM indicating that the molecules had the capacity to cross the multiple cytoplasmic and vacuolar membrane barriers of the macrophage host cell as well as the membranes of the pathogen.

The same group has explored an additional library of inhibitors in 2011 [45], which expanded the structure by a second aromatic ring with length and geometry variations at the linker level, in order to increase the inhibitory effectiveness of the model inhibitor by exploiting neighboring lipophilic pocket in the active site. By taking advantage of molecular modeling, a series of inhibitors was designed and then synthesized using pallado-catalyzed chemistry. With an IC_{50} of less than 70 nM, good to very good inhibitory effects against *B. suis* HDH were detected, compound **8** being the most effective inhibitor with IC_{50} of 3 nM. This compound was able to inhibit growth of *B. suis* in minimal medium showing a 70%–100% inhibition of Brucella growth, as compared to an untreated control culture. Additionally, a strong inhibition of intramacrophagic replication of *B. suis* at a concentration of 5 μM with a reduction of bacteria number by a factor 103 over the infection period of 24 h. These findings unambiguously demonstrate that the HDH inhibitor diffuses through bacterial and eukaryotic membranes to the target enzyme within the bacterial cytoplasm, where it inhibits histidine production, which is crucial for *Brucella* development in an intracellular state [46].

Our group also published two additional series of inhibitors in 2008 [47] and 2014 [48] that, depending on the inhibitor series, either had an oxo- and thiooxo-imidazo [1,5-c]pyrimidine ring in place of the aminomethyl imidazole part or a sulfonyl hydrazide moiety as a linker of the side chains with aromatic rings. Despite the differences in structures, these changes failed to produce more potent inhibitors, with the most effective inhibitors, **5** and **9**, having IC_{50} values of 25 and 5 μM for *B. suis* HDH, respectively.

In the case of *M. tuberculosis* HDH inhibitors, only the paper of Lunardi and coll. reported in 2016 a small library of 11 hydrazone inhibitors whose structure was based on what was already described previously for *B. suis* HDH. This series of inhibitor exhibited micromolar IC_{50} values ranging from 1.1 to 29 μM, compound **10** with the 2-naphtyl substituent being the most active [49].

Besides research works on bacterial HDH, Pahwa et al. have disclosed other interesting series of inhibitors against fungal histidinol dehydrogenase in 2008 [50] and 2010 [51] *Geotrichum candidum* histidinol dehydrogenase was found to be inhibited by two novel compounds (**7** and **8**) that were developed based on docking results, with IC_{50} values about 3 μM.

5 Clinically used agents or compounds in clinical development

The emergence of multi-drug resistance pathogenic strains of intramacrophagic bacteria presents one of the difficulties for traditional treatments. Thus, it is of importance to investigate potential new therapeutic targets for the ground-breaking antivirulence therapy that makes use of the pathogen's vital proteins that operate as virulence factors. This is a crucial initial step in the development of novel drugs.

A variety of virulence factors are used by pathogens to establish their presence in their specific infectious environment. The ability of bacterial virulence mechanisms to regulate the infectious process is crucial for maximizing the effectiveness of virulence factors in promoting survival and undermining host defenses.

Anti-virulence strategies against bacterial infections have attracted significant interest from the scientific community [52–54]. The most appealing strategies include the suppression of quorum sensing and virulence control, virulence factor secretion and function, and bacterial adherence and colonization [55]. Nevertheless, the US Food and Drug Administration (FDA) has only approved three antivirulence treatments, all of which are antibody-based, others being currently in the clinical pipeline [56,57].

In the case of histidinol dehydrogenase, several unprecedent steps forward in terms of therapeutic design and optimization have been realized. The current available research works identified key structure-activity relationships between small-molecule substrates and the catalytic site zinc active site of HDH, focusing on extending interactions into the lipophilic subsite of the enzyme. However, none of these compounds has entered clinical trials as therapeutic. Anyhow the development of new potent inhibitors can now exploit structural and mechanistic analyses of interesting hit compounds.

Some of them showed efficient activities in vitro on the bacterial growth in minimal medium and on the replication of the pathogen in human macrophage-like THP-1 cells. However, these studies are still in their initial phase and further work should validate the model in order to detect inhibitors with high affinity and selectivity. Thus, bacterial HDH represents at this moment very promising zinc metalloenzyme target for obtaining anti-infective agent devoid of the resistance problems, but further studies are needed to validate this enzyme as novel drug target.

6 Conclusion

Given the infections' fast-growing resistance to current antibiotics, there is an urgent need for the development of new molecules having antibacterial action. More critically, finding emerging hits and therapeutic target are required in the ongoing struggle between drug development and new resistance [58].

The advancement of bacterial genomics and genome sequencing has led to the identification of several possible novel therapeutic targets. A promising strategy for the discovery of innovative anti-infectives is the targeting of bacterial metalloenzymes. Small molecules that can block metalloenzymes involved in the biosynthesis of amino-acids are of interest because they may be used alone or in combination with well-known antibiotics to prevent mutations and other types of drug resistance [59].

The contribution of amino acid biosynthetic genes to the virulence of intracellular pathogens has led considering amino-acid pathways, which is not found in humans but which essential for survival and infectivity of pathogenic bacteria within the host [60,61].

Using preliminary interdisciplinary strategy, different research groups in the fields of bacteriology, structural biology, and medicinal chemistry were able to find potent HDH inhibitors that had significant in vitro activity against the intramacrophagic bacteria *B. suis* and *M. tuberculosis*. The current available studies in literature showed the non human enzyme histidinol dehydrogenase as a potent viable and attractive antibacterial target. Despite the appealing results, this metalloenzyme has received very little attention in terms of biochemical and structural research, and only a small number of powerful inhibitors have been designed, synthesized, and tested [41,46]. Considering the features of the already produced inhibitors, some of which appear extremely promising with inhibition of intracellular bacterial growth in macrophage infection. They would need to be further investigated in vivo in order to move this issue forward. Moreover, a deeper understanding of the interactions between inhibitors and the active site of the pathogenic bacterial enzyme HDH will further reveal the rational design of new and potentially more active inhibitors. Additionally, studies are needed to develop HDH inhibitors for additional bacterial pathogens where HDH has been identified as a virulence component. Altogether, histidine biosynthetic pathway drives growth and virulence of intracellular pathogenic bacteria within the host and further investigations are required to gain a more comprehensive understanding of the infection-relevant processes and their potential use as antibacterial targets.

References

[1] WHO. Antibiotic resistance., 2023, https://www.who.int/news-room/fact-sheets/detail/antibiotic-resistance. [Accessed January 2023].

[2] Walesch S, Birkelbach J, Jézéquel G, Haeckl FPJ, Hegemann JD, Hesterkamp T, Hirsch AKH, Hammann P, Müller R. Fighting antibiotic resistance-strategies and (pre)clinical developments to find new antibacterials. EMBO Rep 2022; e56033.

[3] Silver LL. Appropriate targets for antibacterial drugs. Cold Spring Harb Perspect Med 2016;6(12), a030239.

[4] Payne DJ, Gwynn MN, Holmes DJ, Pompliano DL. Drugs for bad bugs: confronting the challenges of antibacterial discovery. Nat Rev Drug Discov 2007;6(1):29–40.

[5] McDevitt D, Payne DJ, Holmes DJ, Rosenberg M. Novel targets for the future development of antibacterial agents. J Appl Microbiol 2002;92(Suppl):28S–34S.

[6] Amorim Franco TM, Blanchard JS. Bacterial branched-chain amino acid biosynthesis: structures, mechanisms, and drugability. Biochemistry 2017;56(44):5849–65.

[7] Rebek J. On the structure of histidine and its role in enzyme active sites. Struct Chem 1990;1:129–31.

[8] Winkler ME, Ramos-Montañez S. Biosynthesis of histidine. EcoSal Plus 2009;3(2):10.

[9] Alifano P, Fani R, Liò P, Lazcano A, Bazzicalupo M, Carlomagno MS, et al. Histidine biosynthetic pathway and genes, structure, regulation, and evolution. Microbiol Rev 1996;60(1):44–69.

[10] Fani R, Brilli M, Fondi M, Lió P. The role of gene fusions in the evolution of metabolic pathways: the histidine biosynthesis case. BMC Evol Biol 2007;7(Suppl 2):S4.

[11] Del Duca S, Chioccioli S, Vassallo A, Castronovo LM, Fani R. The role of gene elongation in the evolution of histidine biosynthetic genes. Microorganisms 2020;8(5):732.

[12] Carlomagno MS, Chiariotti L, Alifano P, Nappo AG, Bruni CB. Structure and function of the *Salmonella typhimurium* and *Escherichia coli* K-12 histidine operons. J Mol Biol 1988;203(3): 585–606.

[13] Henriksen ST, Liu J, Estiu G, Oltvai ZN, Wiest O. Identification of novel bacterial histidine biosynthesis inhibitors using docking, ensemble rescoring, and whole-cell assays. Bioorg Med Chem 2010;18(14):5148–56.

[14] Anderson CJ, Kendall MM. *Salmonella enterica Serovar Typhimurium* STRATEGIES FOR HOST ADAPTATION. Front Microbiol 2017;8:1983.

[15] Pilatz S, Breitbach K, Hein N, Fehlhaber B, Schulze J, Brenneke B, Eberl L, Steinmetz I. Identification of *Burkholderia pseudomallei* genes required for the intracellular life cycle and *in vivo* virulence. Infect Immun 2006;74(6):3576–86.

[16] Lunardi J, Nunes JE, Bizarro CV, Basso LA, Santos DS, Machado P. Targeting the histidine pathway in *Mycobacterium tuberculosis*. Curr Top Med Chem 2013;13(22):2866–84.

[17] Kohler S, Foulongne V, Ouahrani-Bettache S, Bourg G, Teyssier J, Ramuz M, Liautard JP. The analysis of the intramacrophagic virulome of *Brucella suis* deciphers the environment encountered by the pathogen inside the macrophage host cell. Proc Natl Acad Sci U S A 2002;99:15711–6.

[18] Dietl AM, Amich J, Leal S, Beckmann N, Binder U, Beilhack A, Pearlman E, Haas H. Histidine biosynthesis plays a crucial role in metal homeostasis and virulence of *Aspergillus fumigatus*. Virulence 2016;7(4):465–76.

[19] Johnson BK, Abramovitch RB. Small molecules that sabotage bacterial virulence. Trends Pharmacol Sci 2017;38(4):339–62.

[20] Rasko DA, Sperandio V. Anti-virulence strategies to combat bacteria-mediated disease. Nat Rev Drug Discov 2010;9(2):117–28.

[21] Campos JC, Antunes LC, Ferreira RB. Global priority pathogens: virulence, antimicrobial resistance and prospective treatment options. Future Microbiol 2020;15:649–77.

[22] Parish T. Starvation survival response of *Mycobacterium tuberculosis*. J Bacteriol 2003;185(22):6702–6.
[23] Grubmeyer C, Skiadopoulos M, Senior AE. L-histidinol dehydrogenase, a Zn^{2+}-metalloenzyme. Arch Biochem Biophys 1989;272 (2):311–7.
[24] Grubmeyer CT, Chu KW, Insinga S. Kinetic mechanism of histidinol dehydrogenase: histidinol binding and exchange reactions. Biochemistry 1987;26(12):3369–73.
[25] Görisch H, Hölke W. Binding of histidinal to histidinol dehydrogenase. Eur J Biochem 1985;150(2):305–8.
[26] Görisch H. Steady-state investigations of the mechanism of histidinol dehydrogenase. Biochem J 1979;181(1):153–7.
[27] Grubmeyer C, Teng H. Mechanism of *Salmonella typhimurium* histidinol dehydrogenase: kinetic isotope effects and pH profiles. Biochemistry 1999;38(22):7355–62.
[28] Teng H, Grubmeyer C. Mutagenesis of histidinol dehydrogenase reveals roles for conserved histidine residues. Biochemistry 1999;38(22):7363–71.
[29] Lee SY, Grubmeyer CT. Purification and in vitro complementation of mutant histidinol dehydrogenases. J Bacteriol 1987;169(9):3938–44.
[30] Grubmeyer CT, Insinga S, Bhatia M, Moazami N. *Salmonella typhimurium* histidinol dehydrogenase: complete reaction stereochemistry and active site mapping. Biochemistry 1989;28(20):8174–80.
[31] Andorn N, Aronovitch J. Purification and properties of histidinol dehydrogenase from *Escherichia coli* B. J Gen Microbiol 1982;128 (3):579–84.
[32] Nunes JE, Ducati RG, Breda A, Rosado LA, de Souza BM, Palma MS, Santos DS, Basso LA. Molecular, kinetic, thermodynamic, and structural analyses of *Mycobacterium tuberculosis* hisD-encoded metal-dependent dimeric histidinol dehydrogenase (EC 1.1.1.23). Arch Biochem Biophys 2011;512(2):143–53.
[33] D'ambrosio K, Lopez M, Dathan NA, Ouahrani-Bettache S, Köhler S, Ascione G, Monti SM, Winum JY, De Simone G. Structural basis for the rational design of new anti-Brucella agents: the crystal structure of the C366S mutant of L-histidinol dehydrogenase from *Brucella suis*. Biochimie 2014;97:114–20.
[34] Monti SM, De Simone G, D'Ambrosio K. L-Histidinol dehydrogenase as a new target for old diseases. Curr Top Med Chem 2016;16(21):2369–78.
[35] Barbosa JA, Sivaraman J, Li Y, Larocque R, Matte A, Schrag JD, Cygler M. Mechanism of action and NAD+-binding mode revealed by the crystal structure of L-histidinol dehydrogenase. Proc Natl Acad Sci U S A 2002;99(4):1859–64.
[36] WHO. Brucellosis., 2023, https://www.who.int/news-room/fact-sheets/detail/brucellosis. [Accessed January 2023].
[37] Głowacka P, Żakowska D, Naylor K, Niemcewicz M, Bielawska-Drózd A. Brucella—virulence factors, pathogenesis and treatment. Pol J Microbiol 2018;67(2):151–61.
[38] WHO. Global tuberculosis report 2021., 2023, https://www.who.int/publications/i/item/9789240037021. [Accessed January 2023].
[39] Queval CJ, Brosch R, Simeone R. The macrophage: a disputed fortress in the battle against Mycobacterium tuberculosis. Front Microbiol 2017;8:2284.
[40] Agüero F, Al-Lazikani B, Aslett M, Berriman M, Buckner FS, Campbell RK, Carmona S, Carruthers IM, Chan AW, Chen F, Crowther GJ, Doyle MA, Hertz-Fowler C, Hopkins AL, McAllister G, Nwaka S, Overington JP, Pain A, Paolini GV, Pieper U, Ralph SA, Riechers A, Roos DS, Sali A, Shanmugam D, Suzuki T, Van Voorhis WC, Verlinde CL. Genomic-scale prioritization of drug targets: the TDR Targets database. Nat Rev Drug Discov 2008;7 (11):900–7.
[41] Köhler S, Dessolin J, Winum JY. Inhibitors of histidinol dehydrogenase. In: Supuran C, Capasso C, editors. Zinc enzyme inhibitors. Topics in medicinal chemistry, vol. 22. Cham: Springer; 2016.
[42] Dancer JE, Ford Mark J, Hamilton K, Kilkelly M, Lindell SD, O'Mahony MJ, Saville-Stones EA. Synthesis of potent inhibitors of histidinol dehydrogenase. Bioorg Med Chem Lett 1996;6:2131–6.
[43] Abdo MR, Joseph P, Boigegrain RA, Liautard JP, Montero JL, Köhler S, Winum JY. Brucella suis histidinol dehydrogenase: synthesis and inhibition studies of a series of substituted benzylic ketones derived from histidine. Bioorg Med Chem 2007;15(13):4427–33.
[44] Joseph P, Abdo MR, Boigegrain RA, Montero JL, Winum JY, Köhler S. Targeting of the *Brucella suis* virulence factor histidinol dehydrogenase by histidinol analogues results in inhibition of intramacrophagic multiplication of the pathogen. Antimicrob Agents Chemother 2007;51(10):3752–5.
[45] Abdo MR, Joseph P, Mortier J, Turtaut F, Montero JL, Masereel B, Köhler S, Winum JY. Anti-virulence strategy against Brucella suis: synthesis, biological evaluation and molecular modeling of selective histidinol dehydrogenase inhibitors. Org Biomol Chem 2011;9 (10):3681–90.
[46] Lopez M, Köhler S, Winum JY. Zinc metalloenzymes as new targets against the bacterial pathogen Brucella. J Inorg Biochem 2012;111:138–45.
[47] Abdo MR, Joseph P, Boigegrain RA, Montero JL, Köhler S, Winum JY. *Brucella suis* histidinol dehydrogenase: synthesis and inhibition studies of substituted N-L-histidinylphenylsulfonyl hydrazide. J Enzyme Inhib Med Chem 2008;23(3):357–61.
[48] Turtaut F, Lopez M, Ouahrani-Bettache S, Köhler S, Winum JY. Oxo- and thiooxo-imidazo[1,5-c]pyrimidine molecule library: beyond their interest in inhibition of *Brucella suis* histidinol dehydrogenase, a powerful protection tool in the synthesis of histidine analogues. Bioorg Med Chem Lett 2014;24(21):5008–10.
[49] Lunardi J, Kras Borges Martinelli L, Silva Raupp A, Sacconi Nunes JE, Rostirolla DC, Saraiva Macedo Timmers LF, Drumond Villela A, Pissinate K, Limberger J, de Souza ON, Basso LA, Santiago Santos D, Machado P. *Mycobacterium tuberculosis* histidinol dehydrogenase: biochemical characterization and inhibition studies. RSC Adv 2016;6:28406–18.
[50] Pahwa S, Chavan AG, Jain R, Roy N. Target-specific anti-fungal discovery by targeting *Geotrichum candidum* histidinol dehydrogenase: a hybrid approach. Chem Biol Drug Des 2008;72(3):229–34.
[51] Pahwa S, Kaur S, Jain R, Roy N. Structure based design of novel inhibitors for histidinol dehydrogenase from *Geotrichum candidum*. Bioorg Med Chem Lett 2010;20(13):3972–6.
[52] Clatworthy AE, Pierson E, Hung DT. Targeting virulence: a new paradigm for antimicrobial therapy. Nat Chem Biol 2007;3(9):541–8.
[53] Dickey SW, Cheung GYC, Otto M. Different drugs for bad bugs: antivirulence strategies in the age of antibiotic resistance. Nat Rev Drug Discov 2017;16(7):457–71.
[54] Neville N, Jia Z. Approaches to the structure-based design of antivirulence drugs: therapeutics for the post-antibiotic era. Molecules 2019;24(3):378.
[55] Dehbanipour R, Ghalavand Z. Anti-virulence therapeutic strategies against bacterial infections: recent advances. Germs 2022;12 (2):262–75.

[56] Butler MS, Gigante V, Sati H, Paulin S, Al-Sulaiman L, Rex JH, Fernandes P, Arias CA, Paul M, Thwaites GE, Czaplewski L, Alm RA, Lienhardt C, Spigelman M, Silver LL, Ohmagari N, Kozlov R, Harbarth S, Beyer P. Analysis of the clinical pipeline of treatments for drug-resistant bacterial infections: despite progress, more action is needed. Antimicrob Agents Chemother 2022;66(3), e0199121.

[57] Theuretzbacher U, Outterson K, Engel A, Karlén A. The global preclinical antibacterial pipeline. Nat Rev Microbiol 2020;18(5):275–85.

[58] Looper RE, Boger DL, Silver LL. Small molecular weapons against multi-drug resistance. Acc Chem Res 2021;54(13):2785–7.

[59] Plotniece A, Sobolev A, Supuran CT, Carta F, Björkling F, Franzyk H, Yli-Kauhaluoma J, Augustyns K, Cos P, De Vooght L, Govaerts M, Aizawa J, Tammela P, Žalubovskis R. Selected strategies to fight pathogenic bacteria. J Enzyme Inhib Med Chem 2023;38 (1):2155816.

[60] Singh SB, Young K, Silver LL. What is an "ideal" antibiotic? Discovery challenges and path forward. Biochem Pharmacol 2017;133:63–73.

[61] Murima P, McKinney JD, Pethe K. Targeting bacterial central metabolism for drug development. Chem Biol 2014;21(11):1423–32.

Chapter 3.8

Histone deacetylases and other epigenetic targets

Fabrizio Carta
NEUROFARBA Department, Section of Pharmaceutical and Nutraceutical Sciences, University of Florence, Florence, Italy

1 Introduction

The field of epigenetics relates to heritable alterations in gene expression, which however do not involve any modification to the DNA sequence [1–3]. Such changes are not necessarily irreversible, and they are usually induced by exposure to the most variegate environmental factors such as diet, stress, and toxins [3,4]. As result important diseases (i.e., cancers, neurological disorders, and autoimmune disorders) may take place [1,5,6]. Overall, research into epigenetics clearly established the magnitude of the impact from the surrounding biotic or abiotic environment on genetic expression [1,7]. It is therefore expected that understanding the mechanisms behind epigenetic regulation may help in conceiving new potential therapeutic approaches for the management of diseases, usually of the chronic type. From the biochemical viewpoint, epigenetic changes may occur through variegate mechanisms such as DNA methylation/alkylation, protein assembly modifications, and noncoding RNA regulations. In this context, currently investigated epigenetic targets in Medicinal Chemistry include:

- Histone deacetylases (HDACs; EC 3.5.1.98): are enzymes that remove acetyl groups from acetylated lysine residues in histone proteins, thus leading structural reorganization of the chromatin and reduction of gene expression. Since HDACs act on variegate substrates of epigenetic meaning bearing acetylated lysine residues, they are better refereed as KDACs [1,8].
- Histone acetyltransferases (HATs; EC 2.3.1.48): are enzymes that acetylate lysine amino acids on histone proteins from acetyl-CoA. DNA Overall, histone acetylation determines increases in gene expression. Also for HATs, the activity on variegate epigenetic substrates showing a lysine residue makes them properly fitted as KATs [9].
- DNA methyltransferases (DNMTs; EC 2.1.1.37): are enzymes that add methyl groups to the cytosine residues in DNA and this may usually lead to gene silencing [10].
- Histone methyltransferases (HMTs; EC 2.1.1.355; EC 2.1.1.367, EC 2.1.1.368, and EC 2.1.1.366): are enzymes that add methyl groups to histone proteins, which can lead to modulation of the genetic expression [11].
- Bromodomain and extra-terminal (BET) proteins: bind to acetylated histones and regulate gene transcription [12].
- RNA polymerase II (RNAPII): is a key enzyme involved in gene transcription [13].

Overall, targeting epigenetic regulators offers promising opportunities for the development of novel therapeutics. The aim of this chapter is to give an overview of the most significant research contributions in the field of KDAC modulators for biomedical purposes.

2 Histone deacetylase (KDACs/KDACs)

The histone deacetylase enzymes (KDACs/KDACs) are a class of enzymes that play an important role in regulating gene expression and chromatin structure. KDACs are primarily responsible for removing acetyl groups from lysine residues on histone proteins, thus leading to a more compact and condensed chromatin structure that is associated with gene repression. KDACs and histone acetyltransferases (HATs/KATs) activities are the main players in determining the pattern of histones acetylation and agreement to the transformation in Scheme 1 [1,9].

Although this chapter is focused on KDACs expressed in eukaryotic cells (i.e., specifically in mammals), it is worth considering that bacteria and yeast also do express such enzymes [14]. The main differences occurring between eukaryotic and prokaryotic KDACs are related to:

- *Substrate specificity*: Human KDACs are typically more specific on the substrates than bacterial KDACs. For instance, human KDAC isoforms tend to target specific histone aggregates, while the bacterial ones deacetylate a wider range of substrates.

Metalloenzymes. https://doi.org/10.1016/B978-0-12-823974-2.00037-1

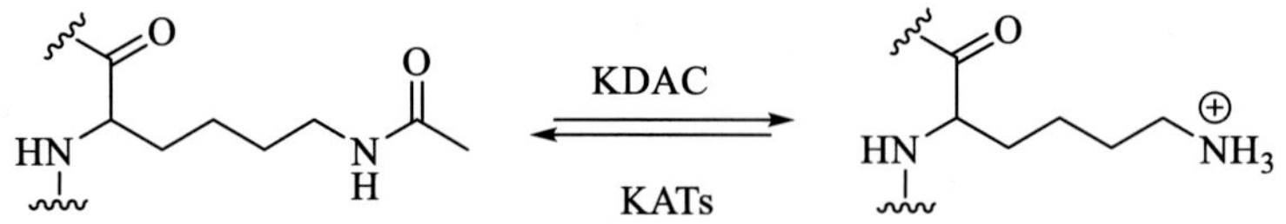

SCHEME 1 KDACs and KATs main reaction on the lysine amino acid substrate.

- *Regulation*: Human KDACs are often regulated by post-translational modifications, such as phosphorylation reactions. Changes in gene expression or protein-protein interactions are the typical regulatory pathways for the bacterial expressed KDACs.
- *Biological function*: Human KDACs have been reported implicated in a wide range of biological processes, including gene expression, cell differentiation and progression of chronic diseases. The knowledge on bacterial KDACs accounts for their role in virulence regulation or other aspects of bacterial physiology yet to be fully deciphered.
- *Structure*: Human KDACs are typically larger and more complex either for their monomeric or supramolecular structural organization when compared to the bacterial ones.

The level of structural complexity of eukaryotic KDACs is reflected into their classification, which accounts for four distinct classes based on the enzymatic structures, sequence homologies to their yeast counterparts as well as enzymatic activities [1–3,6]:

- Class I KDACs: are primarily expressed within the cellular nuclei and include KDAC1, KDAC2, KDAC3, and KDAC8. They are homologous to the yeast Rpd3 deacetylase and are directly involved in regulating gene expression, cell cycle progression, and cellular differentiation.
- Class II KDACs: include KDAC4, KDAC5, KDAC6, KDAC7, KDAC9, and KDAC10. They are all featured by larger sizes when compared to the other KDAC classes and do have the ability to shuttle between cellular cytoplasm and nuclei. They are homologous to the yeast Hda1 protein, and are involved in regulating gene expression, as well as protein stability and cytoskeleton dynamics. A subclassification into IIa and IIb resides on the presence of a single (i.e., KDAC4, KDAC5, KDAC7, and KDAC9) or double (i.e., KDAC6 and KDAC10) catalytic domain.
- Class III KDACs (also known as Sirtuins): They are structurally and enzymatically different from class I and II KDACs as do dependent from NAD^+ as cofactor. In humans the seven isoforms expressed (i.e., SIRT1–SIRT7) are involved in metabolism regulation, stress response modulation as well as aging.
- Class IV KDACs: This class accounts only for the KDAC11 enzyme, which merges class I and class II KDACs features. It is involved in gene expression regulation, immune response and cell cycle progression modulation.

Except for the class III, all the other KDACs are Zn (II) dependent enzymes and do share a common catalytic mechanism.

The structural heterogeneity and catalytic differentiation among classes and related subtypes endows KDACs of fine-tuning ability toward the catalyzed reaction, and that is clear confirmation of their importance in biology.

3 Class I and II KDAC structures

Among the Zn (II) dependent KDACs, the isoform 8 is the only one that does not require aggregation into functional multimeric complexes to exert its activity [15]. In addition, the elevated homology of the catalytic domains among such enzymes makes KDAC8 the ideal model to study essential structural organization and mechanism. Here we report the X-ray crystal structure of Class II KDAC8 in adduct with the experimental ligand **1** [16] (Fig. 1).

The head-to-head dimeric arrangement reported is likely not to occur in solution, as in agreement with size-exclusion chromatography and light-scattering experiments, but it is mediated by the tail moieties of the inhibitors (i.e., pyridine and thiophene groups), which interact each other by stacking contacts and with the Pro273 and Tyr306 of the opposite monomer. The monomeric structure of KDAC8 compared to its bacterial homologue HDLP [15] showed remarkable differences at level of the loops emerging from the core of the protein, the distal helices ($\alpha1$, $\alpha3$ and $\alpha4$) and the long extended loop (aminoacidic residues 325–356), which connects helices $\alpha10$ and $\alpha11$. As result KDAC8 presents a wider active site and thus far more accessible from the ligands when compared to its bacterial homologue. Since structural variabilities of the loops surrounding the highly conserved core in KDAC8 is found among the various isozymes within each class, it is expected that the substrate specificity pattern of each Zn-dependent KDAC relies at this level. As a matter of fact, such informations are used within the Medicinal Chemistry context to develop modulators endowed with isoform specific features. The structural organization of the active cleft is defined by a hydrophobic 12 Å long tunnel containing the catalytic site at the bottom. The walls of the tunnel are formed by Phe152, Phe208, His180, Gly151, Met274, and Tyr306

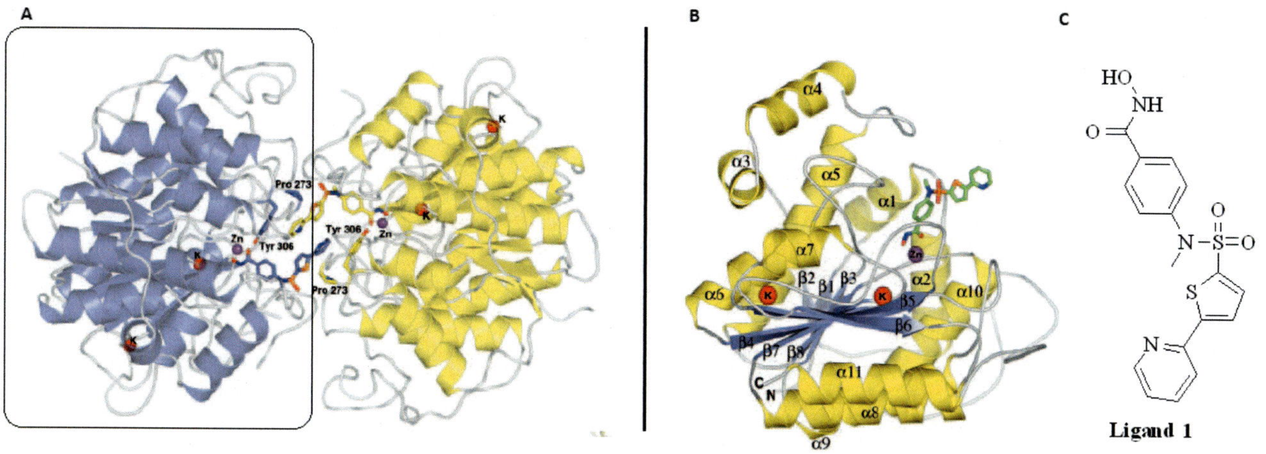

FIG. 1 (A) Ribbon diagram of human KDAC8-in adduct with compound **1**. The two monomers are in *indigo* and *yellow*, the two inhibitor molecules and the residues involved in packing are in stick representation. *Red*, oxygen; *blue*, nitrogen; *orange*, sulfur. Carbon atoms are in *indigo* and *yellow* for ligand A and B, respectively. *Violet* and *red* spheres are Zn (II) and K ions, respectively; (B) Ribbon diagram of human HDAC8 monomer; (C) Chemical structure of the hydroxamic acid ligand **1** [16].

residues, which are conserved across class I KDACs isozymes. The only exception is Met274, which is a Leu in the others [17]. At the bottom of the tunnel is a Zn (II) ion coordinated to Asp178, His180 and Asp267. Under natural substrate conditions two additional interactions are from oxygens of the acetyl moiety and a water molecule, whereas in the presence of the prototype inhibitors of the hydroxamic type the carbonyl and hydroxyl oxygens take place instead. Fig. 2.

The crystal structure of KDAC8 revealed two K ions bound at two sites named as **Site 1**, which is distant 7.0 Å from the zinc-binding site, and **Site 2** located 15 Å apart from **Site 1** (Fig. 3).

Since each site is organized according to a hexa-coordination geometry, which is typical for potassium binding proteins [18], and the crystallographic experimental conditions were performed at K^+ concentrations much lower than those present intracellularly (i.e., 75 mM) it is likely that the potassium-binding sites are biologically relevant. Close inspection at **Site 1** showed that the K ion is coordinated to six oxygen atoms from Asp176, Asp178, His180, Ser199, and Leu200 all arranged in a distorted octahedral geometry (Fig. 3). Since **Site 1** and the zinc-binding site share Asp178 and His180 residues, the presence of the K ion may be expected to have a role in the catalytic mechanism of the enzyme. It is reasonable to speculate that the ion increases the positive electrostatic potential within **Site 1** and thus participates to stabilize the oxyanion formed at the transition state during the deacetylation reaction and/or to give stabilization to the negatively charged

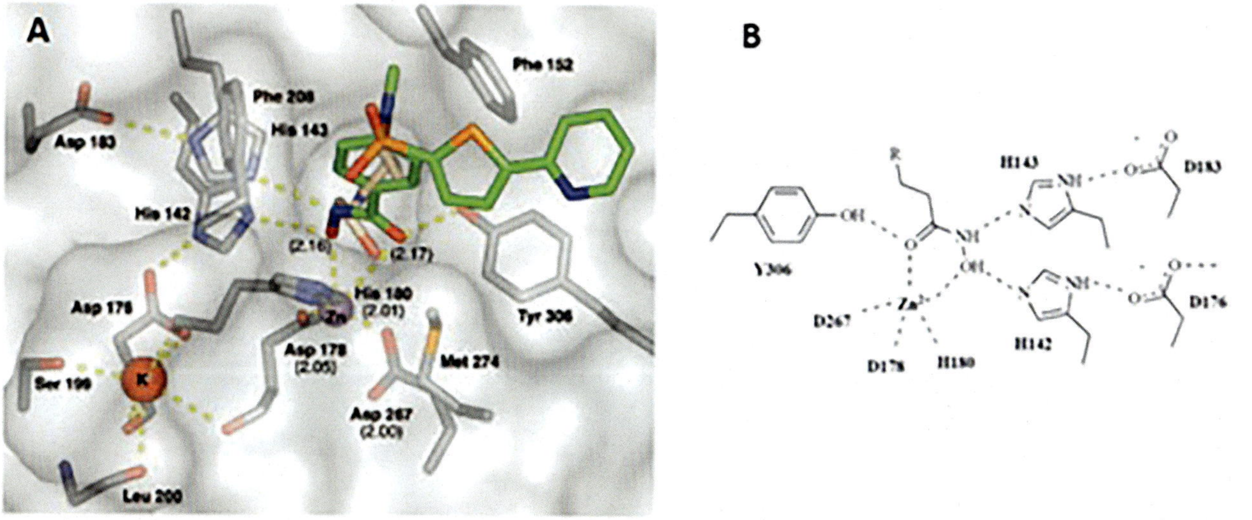

FIG. 2 (A) Representation of the KDAC8 active site in adduct with ligand **1**. Polar interactions are shown as dashed *yellow* lines. Numbers in parentheses represent the zinc-ligand distances [16]; (B) Schematic representation of the KDAC8 active bound to the hydroxamic acid moiety.

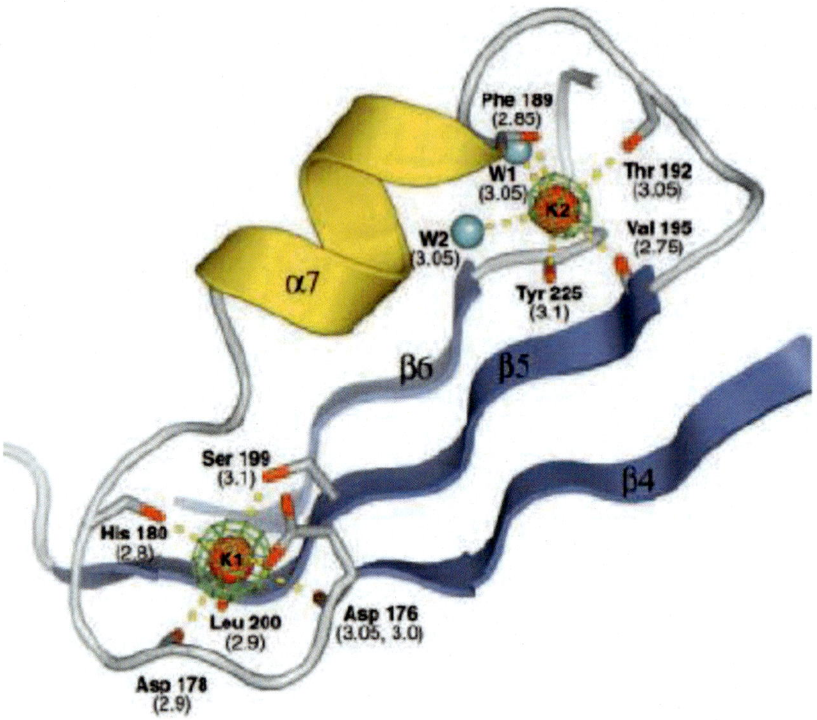

FIG. 3 HDAC8 structure of the two potassium-binding sites. Water molecules are W1 and W2, respectively. Coordination bonds are as dashed *yellow* lines, and corresponding distances are reported in brackets [16].

end-product (see the section "Class I and II KDACs mechanism on nucleosomal core histones"). Furthermore, coordination of the K ion to Asp176 induces its spatial repositioning with variation of its hydrogen bond coordination magnitudes with surrounding aminoacids [15]. Specifically, the modified interaction between Asp176 and His142 affects the pK_a of the latter with predictable interferences on the reaction mechanism [15]. **Site 2** is far more distant from the catalytic cavity and its implication into the enzyme activity is difficult to demonstrate. In this case the K ion is reasonably involved in maintaining the protein into a correct conformation. **Site 2** is defined by the main-chain carbonyl oxygen of Phe189, Thr192, Val195, Tyr225, and two water molecules. Quite interestingly **Site 1** is far more conserved than **Site 2** within KDAC I and II isozyme classes [16].

4 Structure of sirtuins (Class III KDACs)

Comparative structural analysis of sir2 from *Archaeoglobus fulgidus*, Hst2 from Yeast and the human expressed SIRT2 revealed high conserved organization structures (Fig. 4) [6,19].

All such proteins account for a catalytic tunnel, which is defined between a large domain with a Rossmann-type

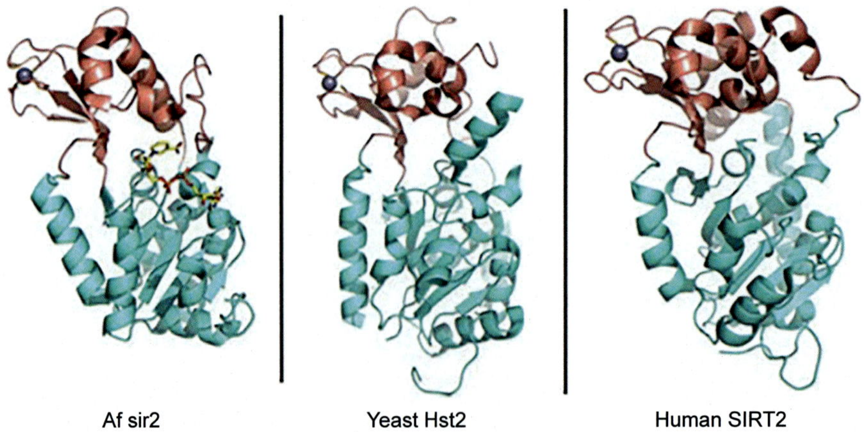

FIG. 4 Crystal structures of the Class III KDACs from *Archaeoglobus fulgidus* sir2 (PDB ID: 1ICI), Yeast Hst2 (PDB ID: 1Q14), and human SIRT2 (PDB ID: 1J8F). The cofactor NAD $^+$ is drawn in stick model and the Zn (II) ions are as gray spheres [6].

folding (i.e., typically found in NAD(H)/NADP(H)-binding enzymes) and a smaller zinc-binding domain [19]. At the bottom of the catalytic tunnel is the cofactor NAD^+, which interacts with the substrate terminal acetamido warhead. For sake of brevity the X-ray crystal structure of the firstly reported SIRT2 [19] will be discussed in this chapter section, thus no consideration on the disclosed SIRT free bound structures is given [20–22] or in adduct with the NAD cofactor/analogs and/or acetyl-lysine peptide substrate [23–31], also in consideration of the high structural homology among them [19,20].

The solved X-ray crystal structure of SIRT2 showed a large domain containing the tubular catalytic cleft and a large structural Rossmann arrangement [32]. A smaller domain bears a structural Zn (II) ion. A large groove is defined at the interfaces of the two domains and by three loops of the bigger one (Fig. 5).

The structural organization of the Rossmann-type accounts for six β-strands (β1–3 and β7–9) that form a parallel β-sheet and six α-helices (α1, α7, α8 and α10–α12). In addition, the typical NAD-binding features such as the highly conserved Gly-X-Gly sequence and a charged pocket necessary to allocate the ribose moieties are present. The NAD binding site is contained within the large groove, whose amino acidic composition results highly conserved among all Sir2 classes. Mutagenesis experiments on one or more of these residues resulted in severe disruption of the KDAC activity [19–31]. Closer inspection to the SIRT2 large domain revealed that it is similar to inverted Rossmann folded enzymes [33], and that is in agreement with the NAD cofactor bound according to an inverted orientation. The catalytic core is additionally defined by a 19 amino acidic residues part of the *N*-terminal extension, which however seems not to be essential for the catalytic activity to take place in vitro. Noteworthy the amphipathic features of such helix are better rationalized for protein-protein interaction with transcriptional regulators. As for the zinc-binding module it is enough to report in this section that there is a high degree of variability among isozymes although the four zinc-coordinating C residues are highly conserved among all KDAC classes [19–31].

5 Class I and II KDACs mechanism on nucleosomal core histones

The mechanism of deacetylation on ε-acetylamino residues catalyzed by Class I KDACs is reported in Fig. 6 for the KDAC8 as model, and it was experimentally proven by structural, biochemical and mutagenesis experiments.

This model accounts for the H143 residue acting as base to accept a proton from the zinc bound water molecule at the initial and rate-determining nucleophilic attack step. Then the same proton will be transferred to the amide nitrogen atom, which will facilitate the cleavage of the amide bond upon collapse of the tetrahedral intermediate [34]. As for the H142 residue, it acts as a general electrostatic catalyst by means of D176, which mediates the K ion effect located nearby at **Site 1** (see the section "Class I and II KDAC structures"). As matter of fact, KDAC8 with H143A mutation in inactive whereas the H142A still preserves residual catalytic activity [35]. This model is similar to the conventional metalloenzymes thermolysin [36] and carboxypeptidase A [37].

Class II KDACs show the Y306H residue mutation. The role of this tyrosine residue is particularly relevant for stabilizing the tetrahedral intermediate generated within the catalytic cleft as well as for polarizing the substrate carbonyl moiety for nucleophilic attack. The lower catalytic activity observed for all Class II KDACs when compared to the isozymes in Class I and IIb is attributable to such a mutation. A mutagenesis experiment on the KDAC4 Y306H allowed to restore enzymatic levels comparable Class I enzymes [38].

A slightly different reaction mechanism was reported for the KDAC8 analogue HDLP, being the H131 residue,

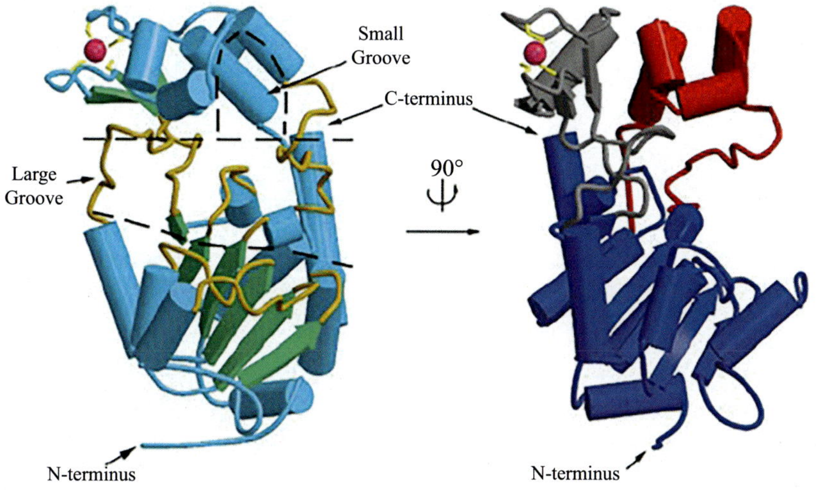

FIG. 5 Overall view of the SIRT2 structure. The NAD-binding site is constituted by a Rossmann-type folding (right, in *blue*), whereas the smaller domain one contains two modules. One module (*gray*) binds to a structural Zn (II) ion (*magenta*) and the other (*red*) contains a hydrophobic pocket [19].

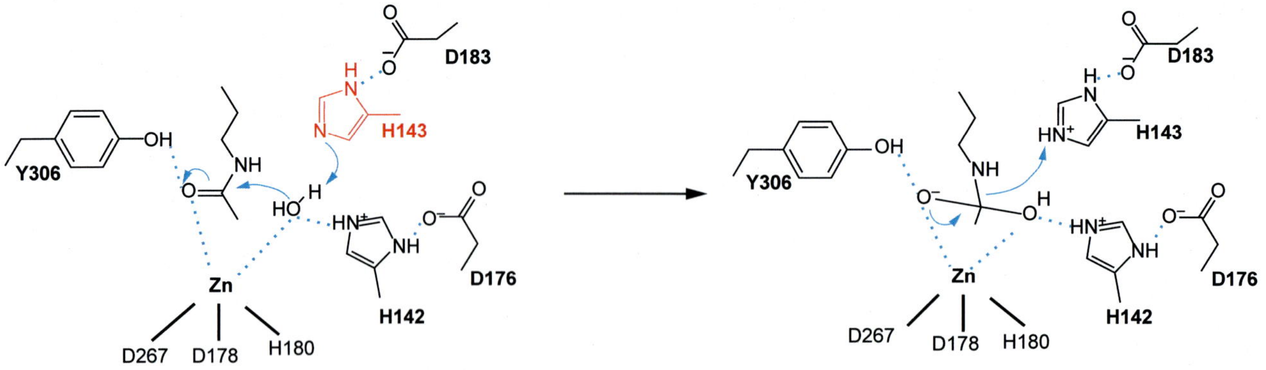

FIG. 6 Deacetylation mechanism of Zn (II)-dependent KDAC8 [6].

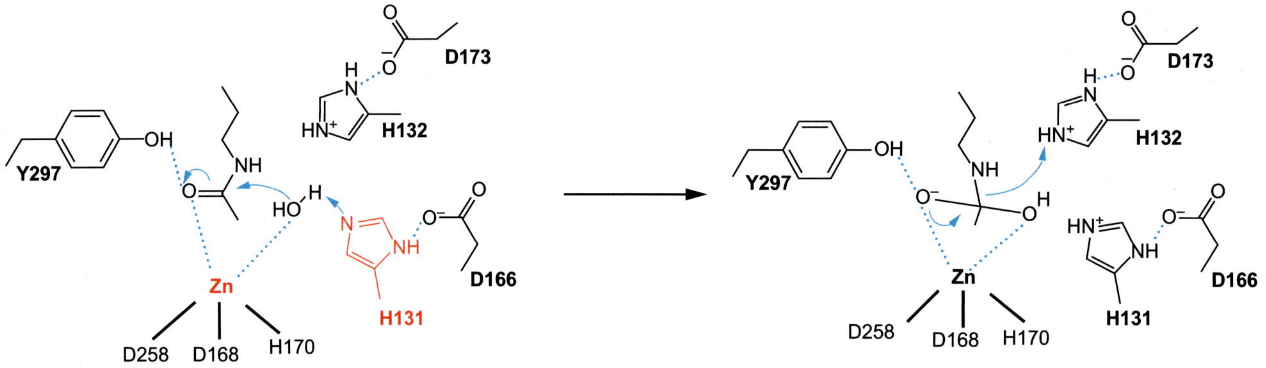

FIG. 7 Deacetylation mechanism of Zn (II) dependent HDLP [6].

coordinated to D166, the starter and rate limiting modulator of the entire deacetylation process (Fig. 7).

6 Catalytic mechanisms of sirtuins (Class III KDACs)

The mechanism for the deacetylation of lysine residues catalyzed from the sirtuins is reported in Fig. 8.

The first step of the reaction involves nucleophilic addition of the acetamide oxygen onto the electrophilic carbon of the NAD^+ cofactor to afford an *O*-alkylamidate intermediate and nicotinamide. The *O*-alkylamidate is then converted by an intramolecular rearrangement to a cyclic intermediate, which is cleaved to release the deacetylated lysine chain and the acetylated ribose. Overall, the reaction mechanism consists of a transfer of the acetyl moiety from the lysine substrate to the activated ribose acceptor [22].

Besides the specific catalytic mechanism of each class of KDACs, of relevance is also the KDAC recruiting mechanisms, which allows such enzymes to interact correctly with specific regions of the DNA genome in order to exert their catalytic activity. Overall, the recruitment is a very complex process, which accounts for interactions of KDACs with a large variety of ancillary proteins and chromatin-modifying enzymes. In addition, the specific mechanism of recruitment is strictly dependent from the cellular context and the regulatory pathways involved in gene expression. For the purposes of this chapter, it is enough to report briefly the main and known recruitment pathways:

- Co-repressor complex recruitment: KDACs are recruited toward the DNA regions of interest by co-repressor complexes, which among others contain the Sin3 [39], NCoR/SMRT [40] and CoREST [41,42] proteins. Such complexes in turn bind to specific transcription factors and then recruit KDACs.
- DNA binding proteins: KDACs may be recruited by means of interactions with DNA-binding proteins. The most common are the Mad/Mxi complex [43], YY1 [44], and the E2F [45].
- Chromatin-modifying enzymes: chromatin-modifying enzymes other than KDACs (i.e., HMTs and DNMTs) modify the chromatin structure and expose the genome for KDACs recruiting [10,11].
- Noncoding RNAs: noncoding RNAs, such as Xist [46] and Air [47], recruit KDACs to specific genomic regions.

The proteins exerting key roles in recruiting the KDACs are quite variegate and numerous [48–54]. In addition, the formation of such aggregates is known to influence the substrate specificity and/or enzymatic effectiveness for each

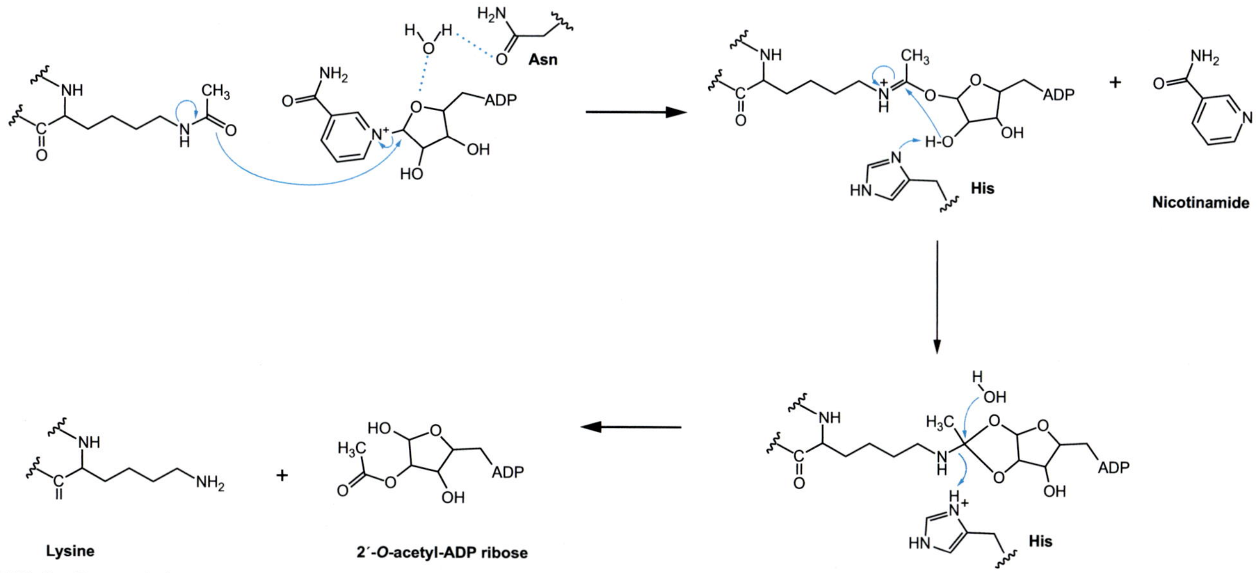

FIG. 8 Deacetylation mechanism of NAD-dependent KDACs. [6].

KDAC [55,56]. An additional intriguing feature of KDACs is the effect of post-translational modifications on the enzymatic activity. To date, acetylations, glycosylations, *S*-nitrosylations, phosphorylations, sumoylations, and ubiquitination are well-known post-translational events reported to take place on KDACs at cellular level [6,57–59]. Overall, the results are appreciable either on the enzymatic activities themselves (i.e., reduction/inhibition of KDAC deacetylation reactions) or disruption of the multimeric functional complexes. In the last case the observed effects at molecular level accounted for defects on the KDAC recruiting mechanisms, formation of multimeric complexed devoid of catalytic activity and/or loss of substrate specificity [48–54]. To date, the factors and the level of complexity, which regulate substrate specificity and enzymatic capability for the Zn (II) dependent KDACs are far to be rationalized as many key aspects are still missing or not adequately explored. As for the Sirtuins, the substrate specificity is better defined and schematically reported in Table 1.

TABLE 1 Histone specificity for the human expressed Sirtuins [6].

Sirtuin	Histone substrate
SIRT1	H3K9
	H3K14
	H3K56
	H4K16
	H1K26
SIRT2	H4K16
	H3K56
SIRT3	H4K16
SIRT4	None
SIRT5	None
SIRT6	H3K9
	H3K56
SIRT7	H3K18

7 KDACs on nonhistone proteins

KDAC catalyzed deacetylation reactions accounts for a vast range of substrates of the enzyme/structural type, which among others include transcription factors, cytoskeletal proteins, molecular chaperones as well as nuclear import factors [6]. Detailed considerations on the effects of KDACs on specific pathways/events at cellular level will not be discussed in this chapter for the sake of brevity. It is however important to highlight the key role of such enzymes in many physio/pathological events, and that makes KDACs privileged targets for the management of disease, usually of the chronic type (i.e., cancers and inflammations).

8 Zinc-dependent KDAC inhibitors (KDACis)

Since KDAC activity and substrate specificity may be tuned in multiple ways, potentially the management of KDAC-related diseases could be accomplished through the most

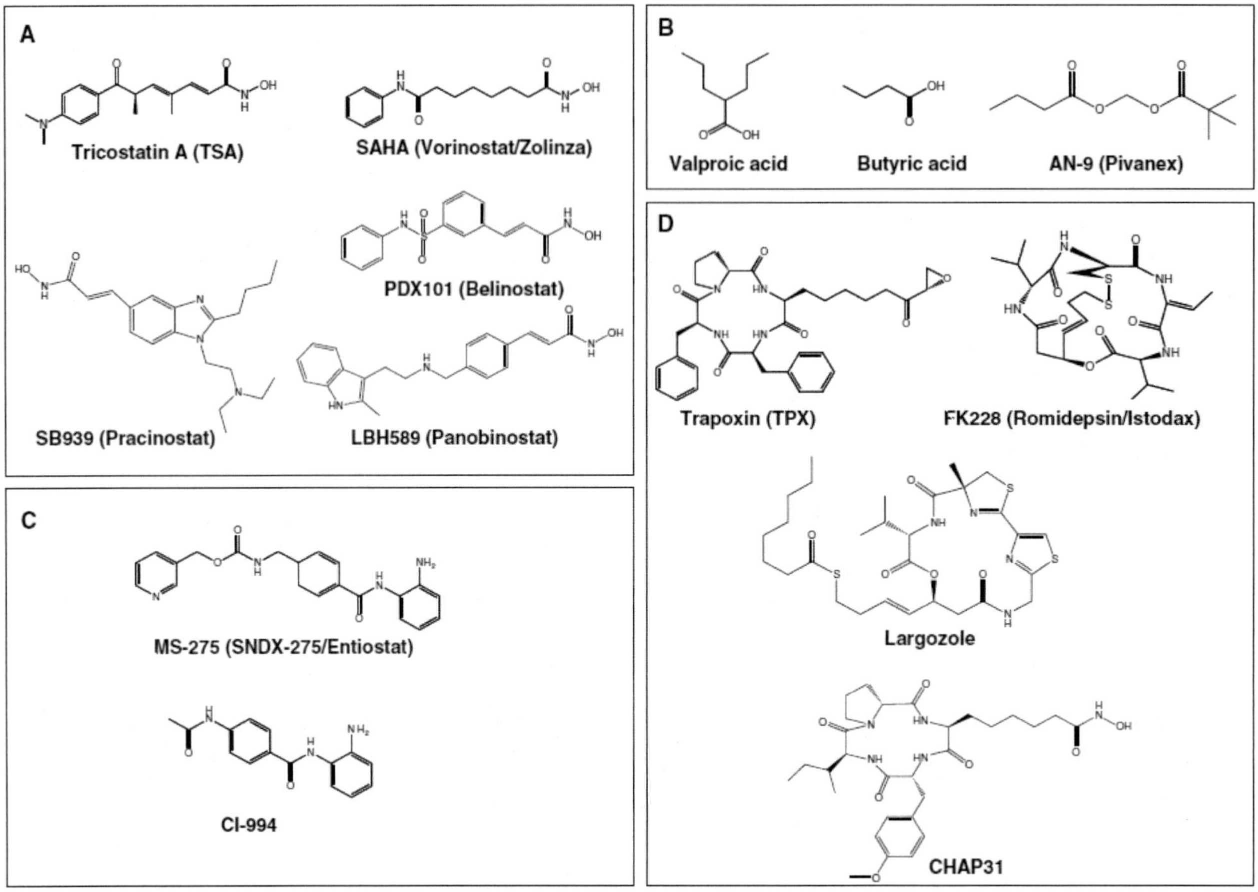

FIG. 9 KDAC inhibitors classified according to the chemical structures are (A) hydroxamic acids, (B) short-chain fatty acids, (C) benzamides, and (D) cyclic peptides [6].

variegate approaches at different biochemical levels, whose majority unfortunately still are out of control or scarcely known. The first and most reliable way to effectively modulate KDACs is represented by the use of molecular ligands of synthetic or natural origin able to inhibit the catalyzed reaction by interaction with the catalytic cycle (Fig. 9).

Classification of KDAC is based on pharmacophoric groups, accounts for four distinct classes, whose main representatives are reported in Fig. 9.

The most common KDACs are the hydroxamates, with the natural product **(*R*)-Trichostatin A (TSA)** being the first discovered from *Streptomyces hygroscopicus* isolates [60,61]. Immediate evidence of the effectiveness of the hydroxamic moiety in inhibiting the KDACs was obtained as the **TSA** corresponding *N*-glucopyranosyl derivative (i.e., **Trichostatin C**; **TSC**) did not affect the enzymatic activity in vitro [60,61] (Fig. 10).

TSA and **SAHA** represent the prototypical hydroxamate-type inhibitors for KDACs. They are widely considered as *pan*-inhibitors since do effectively interact with Class I, II, and IV enzymes, although recent contributions in the literature disagree with this assumption since unexpected isoform selectivity was retrieved even among compounds widely perceived as nonselective [62]. The crystal structures of **TSA** and **SAHA** in adduct with the class I and II KDAC homologue HDLP were reported in a seminal contribution by Finnin et al. [15], thus revealing the key structural features governing the binding modes of hydroxamic acids onto KDACs (Fig. 11).

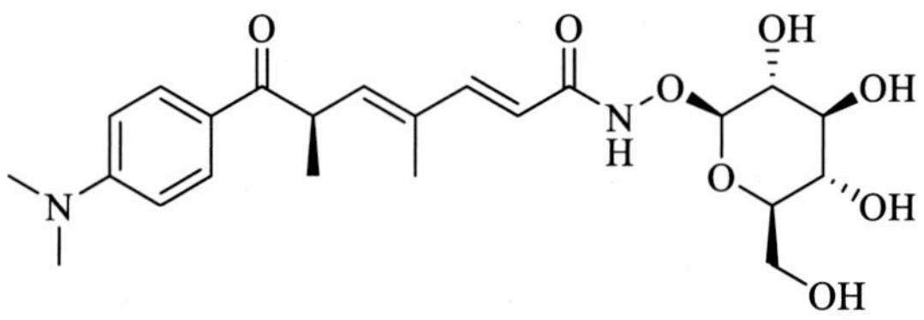

Trichostatin C (TSC)

FIG. 10 Chemical structure of the natural product **Trichostatin C** (**TSC**).

The adduct showed **TSA** bound deep inside the HDLP cavity with its aliphatic backbone engaged in multiple contacts to the tubular pocket and its hydroxamic moiety coordinated the Zn (II) ion in a bidentate fashion and further stabilized by additional interactions with H131, H132, and Y297. Further stabilization is ensured by the dimethylamino-phenyl moiety, which is engaged with amino acids at the pocket rim. As for **SAHA**, the binding

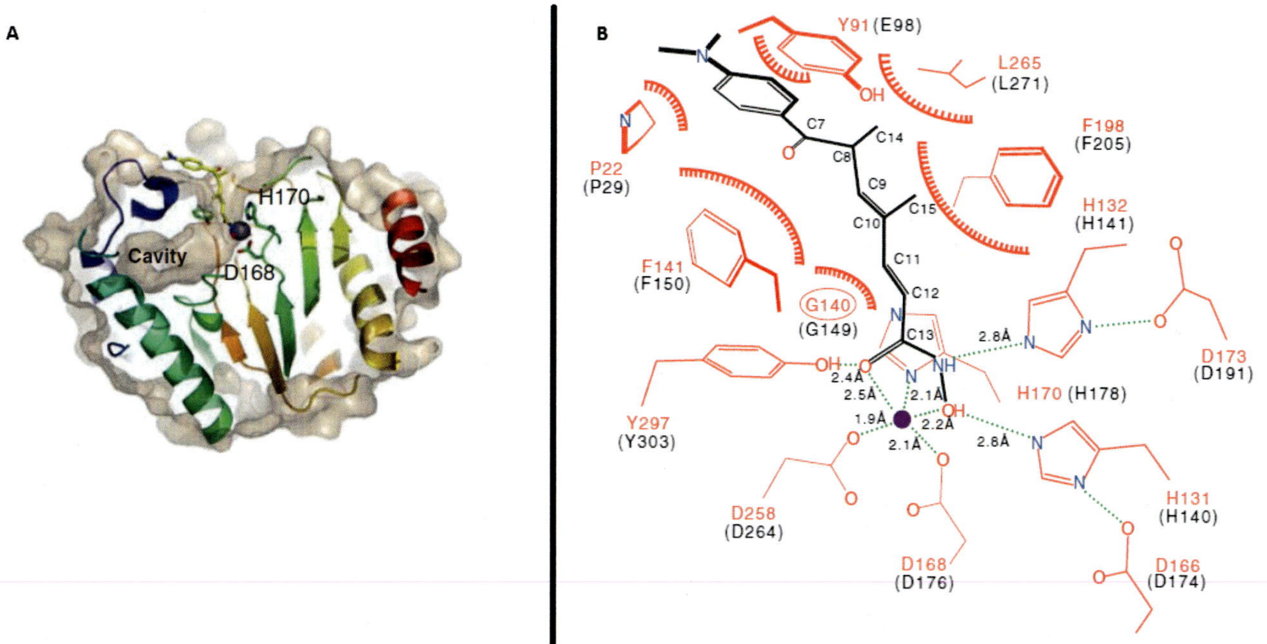

FIG. 11 (A) Molecular surfaces of **TSA** in adduct with HDLP (PDB ID: 1C3R); (B) Schematic representation of the ligand-protein interactions: residues are labeled in *red* and their counterparts in KDAC1 in *black*. Thatched semicircles indicate van der Waals contacts, *green* dashed lines signify hydrogen bonds [15].

mode into HDLP is quite similar to **TSA**. The main difference is related to the allocation of the six carbon-long aliphatic chain within the tubular hydrophobic cleft. The longer chain of **SAHA** when compared to TSA makes it less comfortably packed, thus with reduction of the van der Waals contacts with the cavity walls. The indeterminate electron density maps for the phenyl-amino ketone were indicative of a high degree of freedom of such moiety, which in turn suggest scarse interactions at this level between the ligand and the enzyme. All the above considerations properly justify **SAHA** as weaker binder to HDLP (i.e., about 30-fold) when compared to **TSA**. Nevertheless, **SAHA** had better pharmacological profile, which allowed its approval in 2006 by the US FDA for the treatment refractory cutaneous T-cell lymphoma (CTLC) [63,64]. The drug is admitted to the market only in USA but not in EU since the Committee for Medicinal Products for Human Use (CHMP) at the European Medicines Agency's (EMA) expressed doubts on the validation procedures of the drug on patients [65].

Belinostat (Beleodaq) was approved under accelerated circumstances in 2014 for the treatment of patients with relapsed or refractory peripheral T-cell lymphoma (PTCL).

Pracinostat is an investigational drug that both the US FDA and the EMA have granted the status of Orphan Drug to be used in combination with Azacitidine for the treatment of newly diagnosed AML patients unfit for standard intensive chemotherapy or older than 75 years. Patients recruitment was discontinued during phase III clinical trials as the primary results gathered did not meet the primary end point of overall survival (OS) [66,67].

Panobinostat in 2014 received accelerated approval from the US FDA for the treatment of patients with multiple myeloma (MM) in association with Bortezomib and Dexamethasone. [68,69].

Short aliphatic carboxylic acids, such as **valproic** and **butyric acid**, showed KDAC inhibition effectiveness at micromolar concentration ranges [70]. The binding mode of the carboxylic moiety with Zn (II) dependent KDACs is speculated to involve hydrogen bonding with amino acid residues present at the upper/middle section of the cavity, thus no direct interactions with the catalytic metal take place. Appreciable results in vivo were obtained by using pro-drugs of the esteric type. A typical example is represented by the experimental compound **Pivanex**. The high lipophilicity of such compound allows the true drug (i.e., *n*-butyric acid) to permeate through cellular membranes easier and thus to be more effective [70]. **Pivanex** was approved in 2009 by the US FDA for the treatment of relapsed or refractory aggressive non-Hodgkin lymphoma (NHL) in patients who have already received at least two prior therapies. However, in 2017, the results on clinical trials brought to its discontinuation.

The Benzamides (i.e., *o*-aminoanilides) became of interest as KDACi with the identification of the compound **MS-275** as a low micromolar inhibitor on the KDACs extracted from human leukemia K562 cells [71] and as it showed significant oral anticancer activity in animal models, clearly ascribed to inhibition of such enzymes without severe side effects [72]. Large series of benzamide containing compounds were reported in the literature with the aim to inhibit KDAC enzymes and possibly to discriminate among the various isoforms. A contribution of relevance in this field was from Schreiber et al., which made use of the combinatorial approach [73]. Among the 2400 benzamides synthesized **Histacin** and **PAOA** resulted selective inhibitors for KDAC6 in cells assays [74] (Fig. 12).

The binding mode of the experimental compound of the *o*-aminoanilide type *N*-(4-aminobiphenyl-3-yl)benzamide **2** with KDAC2 was revealed by means of crystal structure of the corresponding adduct [75]. Specifically, the authors did consider **2** as the as it contains the minimal structural requirements necessary to interact with KDAC enzymes (Fig. 13).

As shown, **2** is buried deep inside the catalytic cleft and makes use of an additional lipophilic pocket not yet reported for the other ligand classes. Specifically, the phenyl moiety in **2** interacted with P114, L144, and P155, which in turn was engaged with the benzamidic ring by means of a parallel π-stacking. The typical penta-coordination pattern of the metal ion in KDACs is ensured by the bidentate binding of the *o*-aminoanilide moiety, which is assisted from a twisted conformation of the two aromatic rings through the amide connection Fig. 13B.

The cyclic peptide class accounts for a large series variegate compounds, usually of natural or semisynthetic origin, containing within them functional groups able to interact with the Zn (II) ion in different ways thus including some previously discussed as for CHAP31 in Fig. 7. A seminal example is given by the marine natural cyclic depsipeptide **Largazole** in adduct with KDAC8 [76] (Fig. 14).

The electron density maps clearly reported that the rigid macrocyclic structure of the **Largazole** is placed at the entry of the active site with very minimal conformational changes upon binding to KDAC8 [77]. Then the linear alkyl side chain extends toward the bottom of the cavity with the free thiol group coordinated to the Zn (II) ion. In this case, the overall metal coordination geometry is of the tetrahedral

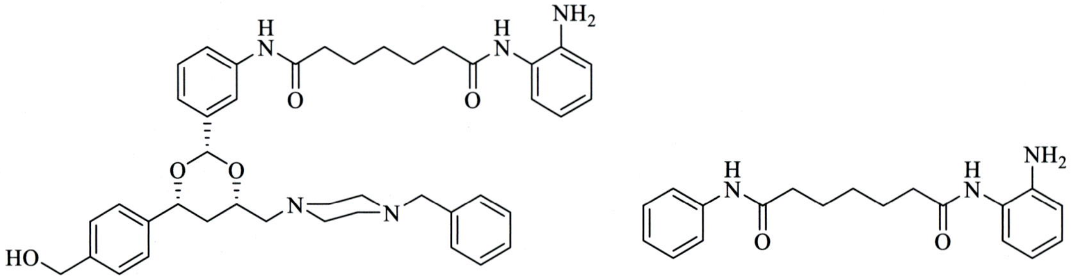

Histacin **PAOA**

FIG. 12 Chemical structures of KDAC6 selective inhibitors **Histacin** and **PAOA**.

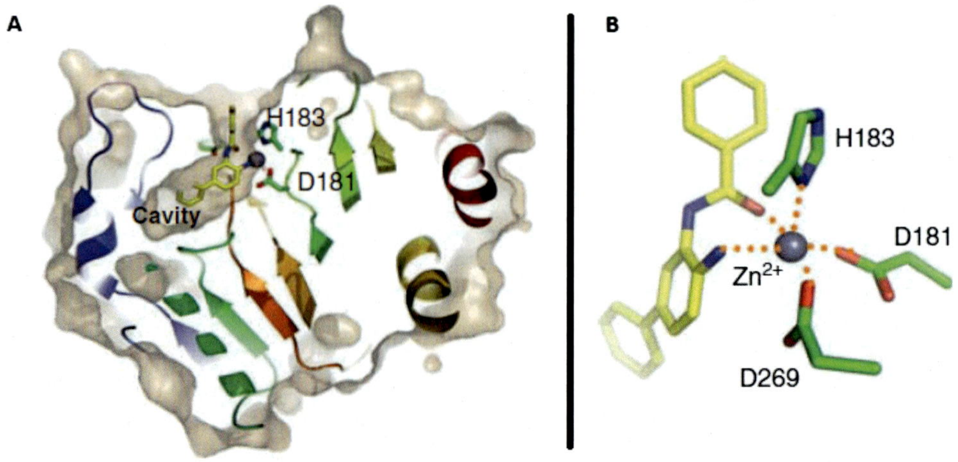

FIG. 13 (A) *N*-(4-Aminobiphenyl-3-yl)benzamide **2** in adduct with KDAC2 (PDB ID: 3MAX); (B) close up of **2** bound to the catalytic Zn (II) ion.

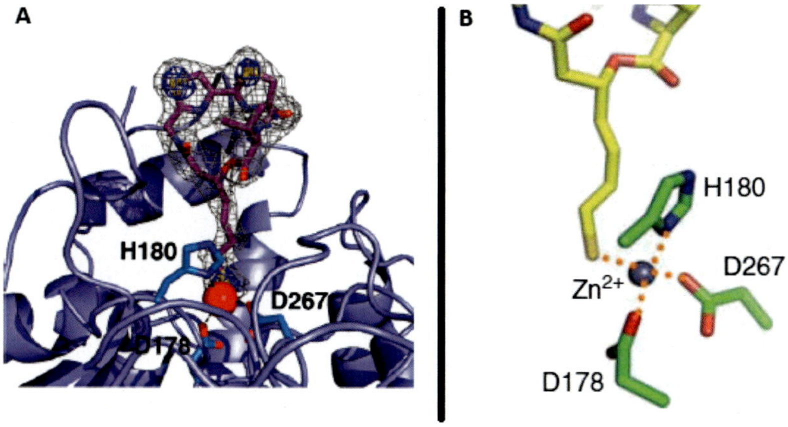

FIG. 14 (A) KDAC8-largazole thiol adduct. The catalytic Zn (II) ion (*red* sphere) is coordinated by D178, H180, and D267. Structural K^+ ions as *green* spheres (PDB ID: 3EWF); (B) Close up of the ligand-KDAC8 adduct (PDB ID: 3RQD) [76].

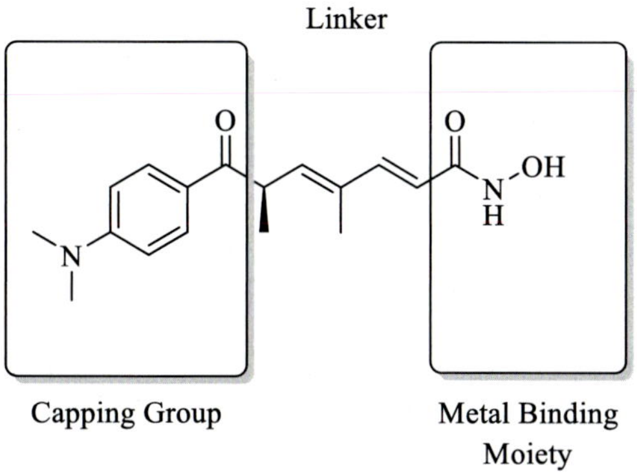

FIG. 15 Structural determinants for KDACis on the prototypic Trichostatin.

type [76]. Such binding mode is likely to be for the macrocyclic depsipeptide **Romidepsin** in Fig. 7, recently approved for the treatment of cutaneous T-cell lymphoma [78], upon disulfide bond cleavage in vivo [79]. To date, no crystal structure in adduct with KDACs is currently available.

The cyclic tetrapeptides **Trapoxin B** in Fig. 7 and its related structural analogues such as the **HC-toxin I** bear the epoxyketone moiety as KDAC inhibitor key feature [80,81]. In vitro experiments on murine KDACs clearly showed that **Trapoxin B** was able to inhibit KDACs irreversibly [80]. Since crystallographic evidences are yet to be produced, it was reasonably speculated that the epoxide moiety is attacked by an active site nucleophilic amino acid such as a histidine, to form a covalent adduct. It is still uncertain if the covalent adduct formed is effectively responsible for the inhibitory activity [80,81].

Overall, the above compounds considered as representative of the four main classes of KDACis, are all structurally related to a common blueprint, which is reported in Fig. 15 for the prototypic inhibitor **Trichostatin A**.

Three main sections may be considered: (i) the capping group is placed at the top entry of the catalytic cleft and it engages interactions with the amino acids defining the rim; (ii) a linker interconnects the capping group; and the (iii) metal binding moiety. All three parts contribute to give stabilization to the ligand-enzyme adduct. Specifically, the interactions occurring between the capping and the linker with the corresponding enzyme counterparts define the isoform specificity. Plenty of literature and reviews on design and development of selective KDACis are present and continuously updated, thus this Chapter will not consider such an aspect as it will bring the reader out of focus. Instead, it is worth considering that the molecular design based on the modular approach of the type in Fig. 15 gives space to a large variety of functional groups to be inserted as potential KDACis. Among others are electrophilic ketones [82–87], *N*-formyl hydroxylamines [88,89], mercapto amides [90–94], sulfones [90], and phosphones [95]. To the best of our knowledge, kinetic values and in silico prediction binding models of such moieties on KDACs are reported.

The binding mode of electrophilic ketones on the Zn (II) coordination sphere requires activation by means of the hydration reaction to form the gem-diol, which is the real inhibiting species (Fig. 16A), whereas in the case of the mercapto amides, sulfones, and phosphones, the bidentate interaction occurs upon displacement of the Zinc (II)-coordinated water (Fig. 16C–E).

9 Sirtuin inhibitors

Inhibitors of the KDACIII may be divided into ligands to the NAD^+ cofactor site and those that interact with the natural substrate acetyl-lysine binding pocket. Among the vast series of compounds acting as Sirtuins inhibitors, we report a selection of the most important in the field (Fig. 17).

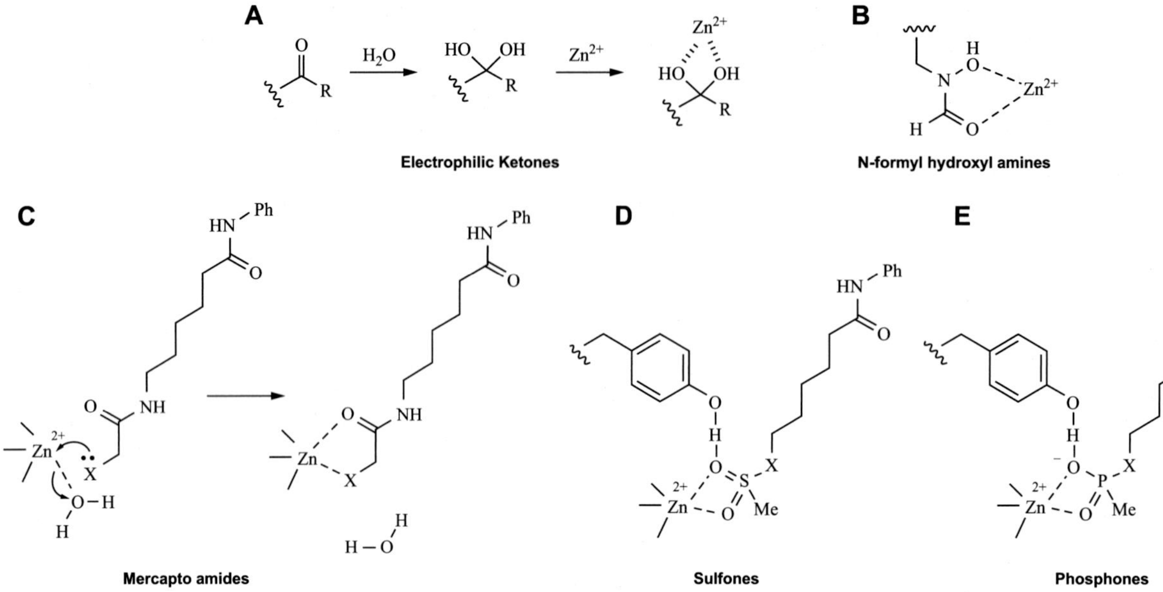

FIG. 16 Chemical structures of electrophilic ketones, *N*-formyl hydroxylamines, mercapto amides, sulfones, and phosphones.

Nicotinamide is an unspecific SIRT1 and 2 inhibitor, which binds to a conserved pocket adjacent to the NAD^+-binding pocket, thus blocking the hydrolysis of the cofactor [96,97]. More interestingly is the nicotinamide analogue **2-Anilinobenzamide,** which resulted in a SIRT1-specific inhibitor in vitro with noncompetitive profile for the cofactor NAD^+ but rather for the acetyl lysine peptide substrate [98].

The small-molecule **Splitomycin** was identified in a cell-based assay screening in *Saccharomyces cerevisiae* as Sir2p inhibitor [99]. Such a discovery paved the way for large drug design projects among which is the contribution of Posakony et al. [100]. Such a contribution is of particular relevance as a series of splitomycin derivatives were synthesized and identified as acetylating agents forming a covalent adduct with a nucleophilic amino acid of the enzyme. As follow-up Neugebauer reported for the first time splitomycin based compounds, which resulted selective SIRT2 inhibitors over SIRT1 [101].

Sirtinol was firstly identified as SIRT inhibitor by Grozinger et al. [102] and was better characterized for its biological effects by Heltweg B et al. [103]. Mai et al. gave important contributions in the field by means of the synthesis of series of sirtinol analogues, which were profiled in vitro on yeast Sir2, human expressed SIRT1 and 2 and on in vivo assay. The sirtinol derivatives **2** and **3** in Fig. 17 resulted up to 10-fold more potent inhibitors of sirtinol for SIRT1 and 2 isoforms in vitro, whereas in yeast in vivo assay resulted as much potent as potent as sirtinol [104]. Another interesting and related compound is the **Salermide,** which in vitro resulted a strong inhibitor for SIRT1 and SIRT2. **Salermide** revealed to be a potent apoptotic inducer at cellular level and therefore responsible of the antitumoral properties [105] (Fig. 17).

Compounds bearing the indole moiety resulted highly potent and selective SIRT1 inhibitors [106] and were identified as noncompetitive ligands for both the substrates. Among the large series of compounds reported in the literature, **EX-527** is one of the most potent SIRT1 inhibitors [106] (Fig. 17).

Since inhibitors of the kinase enzymes act as ATP mimetics, they are also able to inhibit SIRT enzymes by interacting with the adenosine-binding site of NAD^+. The compound bis(indolyl)maleinimide named **Ro31-8220** is the main representative of such class of compounds [107] (Fig. 17).

Various **Suramin** analogues were tested for inhibition of SIRT1/2 [108]. Among the compounds considered the amino-anthranilic acid derivative **NF675** resulted as very potent inhibitor of SIRT1 with a high selectivity over SIRT2 (Fig. 18).

An important class of compounds is represented by the **Tenovins**, which were reported as effective SIRT1 and SIRT2 in vitro inhibitors at concentrations in the low micromolar range. Besides the in vitro effectiveness Tenovins resulted active on mammalian cells at micromolar concentrations and were able decrease tumor growth in in vivo experiments. Specifically, **Tenovin-1** and **Tenovin-6** in Fig. 18 resulted very effective on melanoma cell lines without induction of significant toxic effects [109]. More recently **Tenovin-6** was reported to induce p53 acetylation, and to significantly reduce chronic myelogenous leukemia (CML) growth as well as to eliminate cancer stem cells when administered in combination with Imatinib [110].

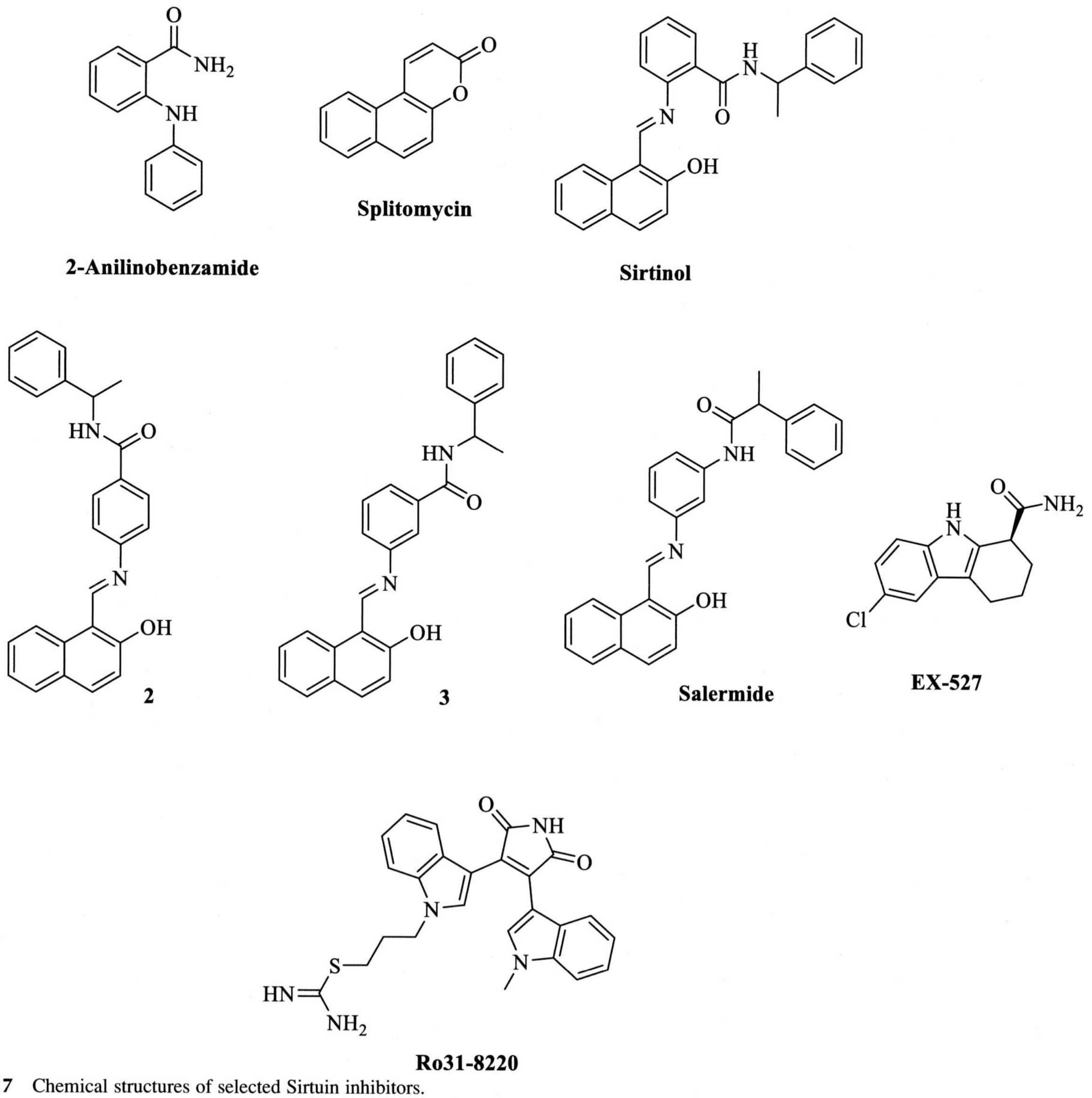

FIG. 17 Chemical structures of selected Sirtuin inhibitors.

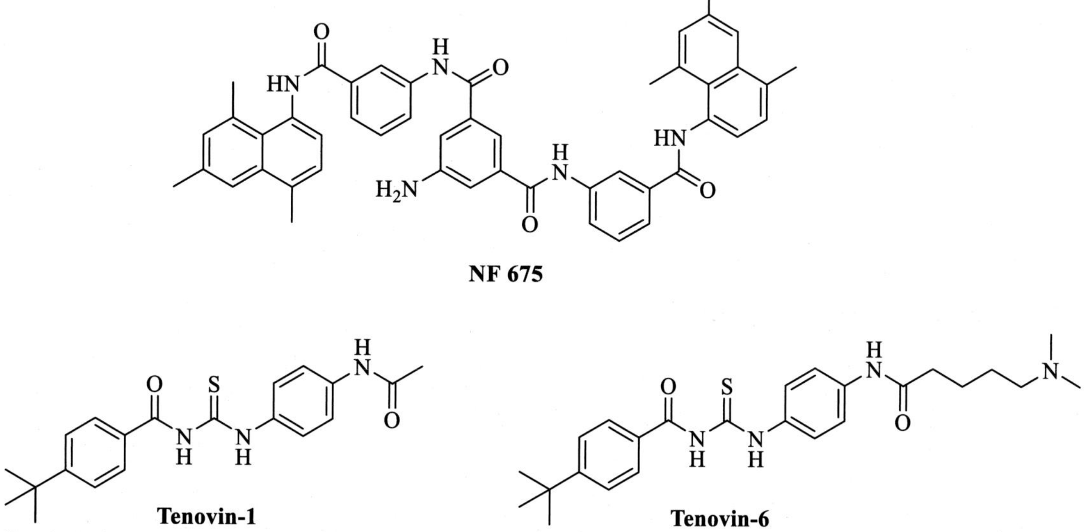

FIG. 18 Chemical structures of **NF 675**, **Tenovin-1** and **-6**.

10 Conclusions

There is increasing and striking evidence that epigenetic modulation is a valid approach for the management of important diseases such as cancers. The potential of this field is continuously renewed, although not exclusively, by means of inhibition of KDAC enzymes. Since the discovery of hydroxamic acids as prototypic KDACis, variegate chemical moieties were identified as effective modulators potentially useful for real biomedical applications. More importantly, deeper knowledge on the structural features as well as on the biology of such enzymes was produced. The actual playground is dominated by the search and development of isozyme selective KDAC modulators with the aim to obtain druggable molecules useful in Medicine. In such a context Medicinal Chemistry makes use of various scientific disciplines altogether to push toward advancements in drug discovery. As result quite interesting compounds are facing advanced clinical trials. Despite such progresses, new frontiers in modulation of KDACs is represented by targeting of KDAC associated effectors, recruiting proteins/complexes or post-translational modification processes. Such approaches, although at their infancy, may be exploited to produce new drugs in relatively short timeframes. New perspectives in the field are the use of KDAC modulators for the management of apparently epigenetically unrelated diseases such as the numerous evidences on host epigenetic modulation by microbiota [111] or KDACis repurposing as anti-infectives [112].

References

[1] Park S-Y, Kim J-S. A short guide to histone deacetylases including recent progress on class II enzymes. Exp Mol Med 2020;52:204–12.

[2] Di Gennaro E, Bruzzese F, Caraglia M, Abruzzese A, Budillon A. Acetylation of proteins as novel target for antitumor therapy: review article. Amino Acids 2004;26:435–41.

[3] Minucci S, Pelicci PG. Histone deacetylase inhibitors and the promise of epigenetic (and more) treatments for cancer. Nat Rev Cancer 2006;6:38–51.

[4] Vahid F, Zand H, Nosrat-Mirshekarlou E, Najafi R, Hekmatdoost A. The role dietary of bioactive compounds on the regulation of histone acetylases and deacetylases: a review. Gene 2015;562:8–15.

[5] Mihaylova MM, Shaw RJ. Metabolic reprogramming by class I and II histone deacetylases. Trends Endocrinol Metab 2013;24:48–57.

[6] Seto E, Yoshida M. Erasers of histone acetylation: the histone deacetylase enzymes. Cold Spring Harb Perspect Biol 2014;6, a018713.

[7] Bieliauskas AV, Pflum MKH. Isoform-selective histone deacetylase inhibitors. Chem Soc Rev 2008;37:1402–13.

[8] Van Dyke MW. Lysine deacetylase (KDAC) regulatory pathways: an alternative approach to selective modulation. Chem Med Chem 2014;3:511–22.

[9] Roth SJ, Denu JM, Allis CD. Histone acetyltransferases. Annu Rev Biochem 2001;70:81–120.

[10] Lyko F. The DNA methyltransferase family: a versatile toolkit for epigenetic regulation. Nat Rev Genet 2018;19:81–92.

[11] Huang S. Histone methyltransferases, diet nutrients and tumour suppressors. Nat Rev Cancer 2002;2:469–76.

[12] Ali HA, Li Y, Bilal AHM, Qin T, Yuan Z, Zhao W. A comprehensive review of BET protein biochemistry, physiology, and pathological roles. Front Pharmacol 2022;13, 818891.

[13] Osman S, Cramer P. Structural biology of RNA Polymerase II transcription: 20 years on. Annu Rev Cell Dev Biol 2020;36:1–34.

[14] Yang XJ, Seto E. The Rpd3/Hda1 family of lysine deacetylases: from bacteria and yeast to mice and men. Nat Rev Mol Cell Biol 2008;9:206–18.

[15] Finnin MS, Donigian JR, Cohen A, Richon VM, Rifkind RA, Marks PA, Breslow R, Pavletich NP. Structures of a histone deacetylase homologue bound to the TSA and SAHA inhibitors. Nature 1999;401:188–93.

[16] Vannini A, Volpari C, Filocamo G, Caroli Casavola E, Brunetti M, Renzoni D, Chakravarty P, Paolini C, De Francesco R, Gallinari P, Steinkuhler C, Di Marco S. Crystal structure of a eukaryotic zinc-dependent histone deacetylase, human HDAC8, complexed with a hydroxamic acid inhibitor. Proc Natl Acad Sci U S A 2004;101:15064–9.

[17] Lee H, Rezai-Zadeh N, Seto E. Negative regulation of histone deacetylase 8 activity by cyclic AMP-dependent protein kinase A. Mol Cell Biol 2004;2:765–73.

[18] Harding MM. Metal-ligand geometry relevant to proteins and in proteins: sodium and potassium. Acta Crystallogr D 2002;58:872–4.

[19] Finnin MS, Donigian JR, Pavletich NP. Structure of the histone deacetylase SIRT2. Nat Struct Biol 2001;8:621–5.

[20] Schuetz A, Min J, Antoshenko T, Wang C-L, Allali-Hassani A, Dong A, Loppnau P, Vedadi M, Bochkarev A, Sternglanz R, Plotnikov AN. Structural basis of inhibition of the human NAD+-dependent deacetylase SIRT5 by Suramin. Structure 2007;15:377–89.

[21] Zhao K, Chai X, Clements A, Marmorstein R. Structure and autoregulation of the yeast Hst2 homolog of Sir2. Nat Struct Biol 2003;10:864–71.

[22] Avalos JL, Boeke JD, Wolberger C. Structural basis for the mechanism and regulation of Sir2 enzymes. Mol Cell 2004;13:639–48.

[23] Min J, Landry J, Sternglanz R, Xu RM. Crystal structure of a SIR2 homolog-NAD complex. Cell 2001;105:269–79.

[24] Chang JH, Kim HC, Hwang KY, Lee JW, Jackson SP, Bell SD, Cho Y. Structural basis for the NAD-dependent deacetylase mechanism of Sir2. J Biol Chem 2002;277:34489–98.

[25] Zhao K, Chai X, Marmorstein R. Structure of the yeast Hst2 protein deacetylase in ternary complex with 20-O-acetyl ADP ribose and histone peptide. Structure 2003;11:1403–11.

[26] Zhao K, Chai X, Marmorstein R. Structure and substrate binding properties of cobB, a Sir2 homolog protein deacetylase from Escherichia coli. J Mol Biol 2004;337:731–41.

[27] Zhao K, Harshaw R, Chai X, Marmorstein R. Structural basis for nicotinamide cleavage and ADP-ribose transfer by NAD(+)-dependent Sir2 histone/protein deacetylases. Proc Natl Acad Sci U S A 2004;101:8563–8.

[28] Avalos JL, Celic I, Muhammad S, Cosgrove MS, Boeke JD, Wolberger C. Structure of a Sir2 enzyme bound to an acetylated p53 peptide. Mol Cell 2002;10:523–35.

[29] Avalos JL, Bever KM, Wolberger C. Mechanism of sirtuin inhibition by nicotinamide: altering the NAD(+) cosubstrate specificity of a Sir2 enzyme. Mol Cell 2005;17:855–68.

[30] Cosgrove MS, Bever K, Avalos JL, Muhammad S, Zhang X, Wolberger C. The structural basis of sirtuin substrate affinity. Biochemistry 2006;45:7511–21.

[31] Hoff KG, Avalos JL, Sens K, Wolberger C. Insights into the sirtuin mechanism from ternary complexes containing NAD+ and acetylated peptide. Structure 2006;14:1231–40.

[32] Bellamacina CR. The nicotinamide dinucleotide binding motif: a comparison of nucleotide binding proteins. FASEB J 1996;10:1257–69.

[33] Prasad GS, Sridhar V, Yamaguchi M, Hatefi Y, Stout CD. Crystal structure of transhydrogenase domain III at 1.2 A resolution. Nat Struct Biol 1999;12:1126–31.

[34] Wu R, Lu Z, Cao Z, Zhang Y. Zinc chelation with hydroxamate in histone deacetylases modulated by water access to the linker binding channel. J Am Chem Soc 2011;133:6110–3.

[35] Gantt SL, Joseph CG, Fierke CA. Activation and inhibition of histone deacetylase 8 by monovalent cations. J Biol Chem 2010;285:6036–43.

[36] Matthews BW. Structural basis of the action of thermolysin and related zinc peptidases. Acc Chem Res 1988;9:333–40.

[37] Wu S, Zhang C, Xu D, Guo H. Catalysis of carboxypeptidase A: promoted-water vs nucleophilic pathways. J Phys Chem B 2010;28:9259–67.

[38] Lahm A, Paolini C, Pallaoro M, Nardi MC, Jones P, Neddermann P, Sambucini S, Bottomley MJ, Lo Surdo P, Carfi A, Koch U, De Francesco R, Steinkühler C, Gallinari P. Unraveling the hidden catalytic activity of vertebrate class IIa histone deacetylases. Proc Natl Acad Sci 2007;104:17335–40.

[39] Grzenda A, Lomberk G, Zhang J-S, Urrutia R. Sin3: master scaffold and transcriptional corepressor. Biochim Biophys Acta 2009;0:443–50.

[40] Mottis A, Mouchiroud L, Auwerx J. Emerging roles of the corepressors NCoR1 and SMRT in homeostasis. Genes Dev 2013;8:819–35.

[41] Andrés ME, Burger C, Peral-Rubio MJ, Battaglioli E, Anderson ME, Grimes J, Dallman J, Ballas N, Mandel G. CoREST: a functional corepressor required for regulation of neural-specific gene expression. Proc Natl Acad Sci U S A 1999;17:9873–8.

[42] Garcia-Martinez L, Adams AM, Chan HL, Nakata Y, Weich N, Stransky S, Zhang Z, Alshalalfa M, Sarria L, Mahal BA, Kesmodel SB, Celià-Terrassa T, Liu Z, Minucci S, Bilbao D, Sidoli S, Verdun RE, Morey L. Endocrine resistance and breast cancer plasticity are controlled by CoREST. Nat Struct Mol Biol 2022;11:1122–35.

[43] Spronk CA, Tessari M, Kaan AM, Jansen JF, Vermeulen M, Stunnenberg HG, Vuister GW. The Mad1-Sin3B interaction involves a novel helical fold. Nat Struct Biol 2000;12:1100–4.

[44] Pazhani J, Veeraraghavan VP, Jayaraman S. Transcription factors: a potential therapeutic target in head and neck squamous cell carcinoma. Epigenomics 2023;2:57–60.

[45] Rayman JB, Takahashi Y, Indjeian VB, Dannenberg J-H, Catchpole S, Watson RJ, te Riele H, Dynlacht BD. E2F mediates cell cycle-dependent transcriptional repression in vivo by recruitment of an HDAC1/mSin3B corepressor complex. Genes Dev 2002;8:933–47.

[46] Chow JC, Hall LL, Baldry SEL, Thorogood NP, Lawrence JB, Brown CJ. Inducible XIST-dependent X-chromosome inactivation in human somatic cells is reversible. Proc Natl Acad Sci U S A 2007;24:10104–9.

[47] Sleutels F, Zwart R, Barlow DP. The non-coding Air RNA is required for silencing autosomal imprinted genes. Nature 2002;6873:810–3.

[48] Ng HH, Bird A. Histone deacetylases: silencers for hire. Trends Biochem Sci 2000;3:121–6.

[49] Jones PL, Veenstra GJ, Wade PA, Vermaak D, Kass SU, Landsberger N, Strouboulis J, Wolffe AP. Methylated DNA and MeCP2 recruit histone deacetylase to repress transcription. Nat Genet 1998;2:187–91.

[50] Nan X, Ng HH, Johnson CA, Laherty CD, Turner BM, Eisenman RN, Bird A. Transcriptional repression by the methyl-CpG-binding protein MeCP2 involves a histone deacetylase complex. Nature 1998;6683:386–9.

[51] Robertson KD, Ait-Si-Ali S, Yokochi T, Wade PA, Jones PL, Wolffe AP. DNMT1 forms a complex with Rb, E2F1 and HDAC1 and represses transcription from E2F-responsive promoters. Nat Genet 2000;3:338–42.

[52] Brehm A, Miska EA, McCance DJ, Reid JL, Bannister AJ, Kouzarides T. Retinoblastoma protein recruits histone deacetylase to repress transcription. Nature 1998;6667:597–601.

[53] Ferreira R, Magnaghi-Jaulin L, Robin P, Harel-Bellan A, Trouche D. The three members of the pocket proteins family share the ability to repress E2F activity through recruitment of a histone deacetylase. Proc Natl Acad Sci U S A 1998;18:10493–8.

[54] Kawai H, Li H, Avraham S, Jiang S, Karsenty Avraham H. Overexpression of histone deacetylase HDAC1 modulates breast cancer progression by negative regulation of estrogen receptor alpha. Int J Cancer 2003;3:353–8.

[55] Zhang X, Wharton W, Yuan Z, Tsai SC, Olashaw N, Seto E. Activation of the growth-differentiation factor 11 gene by the histone deacetylase (HDAC) inhibitor trichostatin A and repression by HDAC3. Mol Cell Biol 2004;24:5106–18.

[56] Vermeulen M, Carrozza MJ, Lasonder E, Workman JL, Logie C, Stunnenberg HG. In vitro targeting reveals intrinsic histone tail specificity of the Sin3/histone deacetylase and N-CoR/SMRT corepressor complexes. Mol Cell Biol 2004;24:2364–72.

[57] Pflum MK, Tong JK, Lane WS, Schreiber SL. Histone deacetylase 1 phosphorylation promotes enzymatic activity and complex formation. J Biol Chem 2001;276:47733–41.

[58] Ford J, Ahmed S, Allison S, Jiang M, Milner J. JNK2-dependent regulation of SIRT1 protein stability. Cell Cycle 2008;7:3091–7.

[59] Kang H, Jung JW, Kim MK, Chung JH. CK2 is the regulator of SIRT1 substrate-binding affinity, deacetylase activity and cellular response to DNA-damage. PLoS One 2009;4, e6611.

[60] Yoshida M, Kijima M, Akita M, Beppu T. Potent and specific inhibition of mammalian histone deacetylase both in vivo and in vitro by trichostatin A. J Biol Chem 1990;265:17174–9.

[61] Tsuji N, Kobayashi K, Nagashima Y, Wakisaka K, Koizumi J. A new antifungal antibiotic, trichostatin. J Antibiot 1976;29:1–6.

[62] Bradner JE, West N, Grachan ML, Greenberg EF, Haggarty SJ, Warnow T, Mazitschek R. Chemical phylogenetics of histone deacetylases. Nat Chem Biol 2010;3:238–43.

[63] Duvic M, Talpur R, Ni X, Zhang C, Hazarika P, Kelly C, Chiao JH, Reilly JF, Ricker JL, Richon VM, Frankel SR. Phase 2 trial of oral vorinostat (suberoylanilide hydroxamic acid, SAHA) for refractory cutaneous T-cell lymphoma (CTCL). Blood 2007;109:31–9.

[64] https://www.accessdata.fda.gov/drugsatfda_docs/label/2011/021991s002lbl.pdf (last access 01/04/2023).

[65] https://www.ema.europa.eu/en/documents/medicine-qa/questions-answers-withdrawal-marketing-application-vorinostat-msd_it.pdf (last access 01/04/2023).

[66] Helsinn Group and MEI Pharma Discontinue the Phase 3 Study with Pracinostat in AML after Completing Interim Analysis; Lugano, Switzerland, and San Diego, CA. Helsinn (last access 01/04/2023).

[67] Garcia-Manero G, Abaza Y, Takahashi K, Medeiros BC, Arellano M, Khaled SK, Patnaik M, Odenike O, Sayar H, Tummala M, Patel P, Maness-Harris L, Stuart R, Traer E, Karamlou K, Yacoub A, Ghalie R, Giorgino R, Atallah E. Pracinostat plus azacitidine in older patients with newly diagnosed acute myeloid leukemia: results of a phase 2 study. Blood Adv 2019;4:508–19.

[68] https://www.cancer.gov/news-events/cancer-currents-blog/2015/fda-approves-panobinostat (last access 01/04/2023).

[69] San-Miguel JF, Hungria VTM, Yoon S-S, Beksac M, Dimopoulos MA, Elghandour A, Jedrzejczak WW, Günther A, Nakorn TN, Siritanaratkul N, Corradini P, Chuncharunee S, Lee J-J, Schlossman R-L, Shelekhova T, Yong K, Tan D, Numbenjapon T, Cavenagh DJ, Hou J, LeBlanc R, Nahi H, Qiu L, Salwender H, Pulini S, Moreau P, Warzocha K, White D, Bladé J, Chen WM, de la Rubia J, Gimsing P, Lonial S, Kaufman JL, Ocio EM, Veskovski L, Sohn SK, Wang MC, Lee JH, Einsele H, Sopala M, Corrado C, Bengoudifa B-R, Binlich F, Richardson PG. Panobinostat plus bortezomib and dexamethasone versus placebo plus bortezomib and dexamethasone in patients with relapsed or relapsed and refractory multiple myeloma: a multicentre, randomised, double-blind phase 3 trial. Lancet Oncol 2014;11:1195–206.

[70] Rephaeli A, Zhuk R, Nudelman A. Prodrugs of butyric acid from bench to bedside: synthetic design, mechanisms of action, and clinical applications. Drug Dev Res 2000;50:379–91.

[71] Suzuki T, Ando T, Tsuchiya K, Fukazawa N, Saito A, Mariko Y, Yamashita T, Nakanishi O. Synthesis and histone deacetylase inhibitory activity of new benzamide derivatives. J Med Chem 1999;42:3001–3.

[72] Saito A, Yamashita T, Mariko Y, Nosaka Y, Tsuchiya K, Ando T, Suzuki T, Tsuruno T, Nakanishi O. A synthetic inhibitor of histone deacetylase, MS-27-275, with marked in vivo antitumor activity against human tumors. Proc Natl Acad Sci U S A 1999;96:4592–7.

[73] Sternson SM, Wong JC, Grozinger CM, Schreiber SL. Synthesis of 7200 small molecules based on a substructural analysis of the histone deacetylase inhibitors trichostatin and trapoxin. Org Lett 2001;3:4239–42.

[74] Wong JC, Hong R, Schreiber SL. Structural biasing elements for in-cell histone deacetylase paralog selectivity. J Am Chem Soc 2003;125:5586–7.

[75] Bressi JC, Jennings AJ, Skene R, Wu Y, Melkus R, De Jong R, O'Connell S, Grimshaw CE, Navre M, Gangloff AR. Exploration of the HDAC2 foot pocket: synthesis and SAR of substituted N-(2-aminophenyl) benzamides. Bioorg Med Chem Lett 2010;20:3142–5.

[76] Cole KE, Dowling DP, Boone MA, Phillips AJ, Christianson DW. Structural basis of the antiproliferative activity of largazole, a depsipeptide inhibitor of the histone deacetylases. J Am Chem Soc 2011;133:12474–7.

[77] Seiser T, Kamena F, Cramer N. Synthesis and biological activity of largazole and derivatives. Angew Chem 2008;47:6483–5.

[78] Guan P, Fang H. Clinical development of histone deacetylase inhibitor romidepsin. Drug Discovery Ther 2010;4:388–91.

[79] Furumai R, Matsuyama A, Kobashi N, Lee KH, Nishiyama M, Nakajima H, Tanaka A, Komatsu Y, Nishino N, Yoshida M, Horinouchi S. FK228 (depsipeptide) as a natural prodrug that inhibits class I histone deacetylases. Cancer Res 2002;62:4916–21.

[80] Kijima M, Yoshida M, Sugita K, Horinouchi S, Beppu T. Trapoxin, an antitumor cyclic tetrapeptide, is an irreversible inhibitor of mammalian histone deacetylase. J Biol Chem 1993;268:22429–35.

[81] Mai A, Massa S, Rango R, Cerbara I, Jesacher F, Loidl P, Brosch G. 3-(4-Aroyl-1-methyl-1H-2-pyrrolyl)-N-hydroxy-2-propenamides as a new class of synthetic histone deacetylase inhibitors. 2. Effect of pyrrole-C2 and/or -C4 substitutions on biological activity. J Med Chem 2003;46:512–24.

[82] Curtin M, Glaser K. Histone deacetylase inhibitors: the Abbott experience. Curr Med Chem 2003;10:2373–92.

[83] Frey RR, Wada CK, Garland RB, Curtin ML, Michaelides MR, Li J, Pease LJ, Glaser KB, Marcotte PA, Bouska JJ, Murphy SS, Davidsen SK. Trifluoromethyl ketones as inhibitors of histone deacetylase. Bioorg Med Chem Lett 2002;12:3443–7.

[84] Jose B, Oniki Y, Kato T, Nishino N, Sumida Y, Yoshida M. Novel histone deacetylase inhibitors: cyclic tetrapeptide with trifluoromethyl and pentafluoroethyl ketones. Bioorg Med Chem Lett 2004;14:5343–6.

[85] Wada CK, Frey RR, Ji Z, Curtin ML, Garland RB, Holms JH, Li J, Pease LJ, Guo J, Glaser KB, Marcotte PA, Richardson PL, Murphy SS, Bouska JJ, Tapang P, Magoc TJ, Albert DH, Davidsen SK, Michaelides MR. Alpha-keto amides as inhibitors of histone deacetylase. Bioorg Med Chem Lett 2003;13:3331–5.

[86] Vasudevan A, Ji Z, Frey RR, Wada CK, Steinman D, Heyman HR, Guo Y, Curtin ML, Guo J, Li J, Pease L, Glaser KB, Marcotte PA, Bouska JJ, Davidsen SK, Michaelides MR. Heterocyclic ketones as inhibitors of histone deacetylase. Bioorg Med Chem Lett 2003;13:3909–13.

[87] Glaser KB, Li J, Pease LJ, Staver MJ, Marcotte PA, Guo J, Frey RR, Garland RB, Heyman HR, Wada CK, Vasudevan A, Michaelides MR, Davidsen SK, Curtin ML. Differential protein acetylation induced by novel histone deacetylase inhibitors. Biochem Biophys Res Commun 2004;325:683–90.

[88] Wu TYH, Hassig C, Wu Y, Ding S, Schultz PG. Design, synthesis, and activity of HDAC inhibitors with a N-formyl hydroxylamine head group. Bioorg Med Chem Lett 2004;14:449–53.

[89] Nishino N, Yoshikawa D, Watanabe LA, Kato T, Jose B, Komatsu Y, Sumida Y, Yoshida M. Synthesis and histone deacetylase inhibitory activity of cyclic tetrapeptides containing a retrohydroxamate as zinc ligand. Bioorg Med Chem Lett 2004;14:2427–31.

[90] Suzuki T, Matsuura A, Kouketsu A, Nakagawa H, Miyata N. Identification of a potent non-hydroxamate histone deacetylase inhibitor by mechanism-based drug design. Bioorg Med Chem Lett 2005;15:331–5.

[91] Chen B, Petukhov PA, Jung M, Velena A, Eliseeva E, Dritschilo A, Kozikowski AP. Chemistry and biology of mercaptoacetamides as novel histone deacetylase inhibitors. Bioorg Med Chem Lett 2005;15:1389–92.

[92] Anandan S-K, Ward JS, Brokx RD, Bray MR, Patel DV, Xiao X-X. Mercaptoamide-based non-hydroxamic acid type histone deacetylase inhibitors. Bioorg Med Chem Lett 2005;15:1969–72.

[93] Rizvi NA, Humphrey JS, Ness EA, Johnson MD, Gupta E, Williams K, Daly DJ, Sonnichsen D, Conway D, Marshall J, Hurwitz H. A phase I study of oral BMS-275291, a novel nonhydroxamate sheddase-sparing matrix metalloproteinase inhibitor, in patients with advanced or metastatic cancer. Clin Cancer Res 2004;10:1963–70.

[94] Baxter AD, Bird J, Bhogal R, Massil T, Minton KJ, Montana J, Owen DA. A novel series of matrix metalloproteinase inhibitors for the treatment of inflammatory disorders. Bioorg Med Chem Lett 1997;7:897–902.

[95] Kapustin GV, Féjer G, Gronlund JL, McCafferty DG, Seto E, Etzkorn FA. Phosphorus-based SAHA analogues as histone deacetylase inhibitors. Org Lett 2003;5:3053–6.

[96] Sanders BD, Jackson B, Brent M, Taylor AM, Dang W, Berger SL, Schreiber SL, Howitz K, Marmorstein R. Identification and characterization of novel sirtuin inhibitor scaffolds. Bioorg Med Chem 2009;17:7031–41.
[97] Tervo AJ, Kyrylenko S, Niskanen P, Salminen A, Leppanen J, Nyronen TH, Jarvinen T, Poso A. An in silico approach to discovering novel inhibitors of human sirtuin type 2. J Med Chem 2004;47:6292–8.
[98] Suzuki T, Imai K, Nakagawa H, Miyata N. 2-Anilinobenzamides as SIRT inhibitors. Chem Med Chem 2006;1:1059–62.
[99] Bedalov A, Gatbonton T, Irvine WP, Gottschling DE, Simon JA. Identification of a small molecule inhibitor of Sir2p. Proc Natl Acad Sci U S A 2001;26:15113–8.
[100] Posakony J, Hirao M, Stevens S, Simon JA, Bedalov A. Inhibitors of Sir2: evaluation of splitomicin analogs. J Med Chem 2004;47:2635–44.
[101] Neugebauer RC, Uchiechowska U, Meier R, Hruby H, Valkov V, Verdin E, Sippl W, Jung M. Structure-activity studies on splitomicin derivatives as sirtuin inhibitors and computational prediction of binding mode. J Med Chem 2008;51:1203–13.
[102] Grozinger CM, Chao ED, Blackwell HE, Moazed D, Schreiber SL. Identification of a class of small molecule inhibitors of the sirtuin family of NAD-dependent deacetylases by phenotypic screening. J Biol Chem 2001;276:38837–43.
[103] Heltweg B, Gatbonton T, Schuler AD, Posakony J, Li H, Goehle S, Kollipara R, Depinho RA, Gu Y, Simon JA, Bedalov A. Antitumor activity of a small-molecule inhibitor of human silent information regulator 2 enzymes. Cancer Res 2006;66:4368–77.
[104] Mai A, Massa S, Lavu S, Pezzi R, Simeoni S, Ragno R, Mariotti FR, Chiani F, Camilloni G, Sinclair DA. Design, synthesis, and biological evaluation of sirtinol analogs as class III histone/protein deacetylase (sirtuin) inhibitors. J Med Chem 2005;48:7789–95.
[105] Lara E, Mai A, Calvanese V, Altucci L, Lopez-Nieva P, Martinez-Chantar ML, Varela-Rey M, Rotili D, Nebbioso A, Ropero S, Montoya G, Oyarzabal J, Velasco S, Serrano M, Witt M, Villar-Garea A, Inhof A, Mato JM, Esteller M, Fraga MF. Salermide, a sirtuin inhibitor with a strong cancer-specific proapoptotic effect. Oncogene 2009;28:781–91.
[106] Napper AD, Hixon J, McDonagh T, Keavey K, Pons JF, Barker J, Yau WT, Amouzegh P, Flegg A, Hamelin E, Thomas RJ, Kates M, Jones S, Navia MA, Saunders JO, DiStefano PS, Curtis R. Discovery of indoles as potent and selective inhibitors of the deacetylase SIRT1. J Med Chem 2005;48:8045–54.
[107] Trapp J, Jochum A, Meier R, Saunders L, Marshall B, Kunick C, Verdin E, Goekjian PG, Sippl W, Jung M. Adenosine mimetics as inhibitors of NAD1-dependent histone deacetylases, from kinase to sirtuin inhibition. J Med Chem 2006;49:7307–16.
[108] Howitz KT, Bitterman KJ, Cohen HY, Lamming DW, Lavu S, Wood JG, Zipkin RE, Chung P, Kisielewski A, Zhang LL, Scherer B, Sinclair DA. Small molecule activators of sirtuins extend Saccharomyces cerevisiae lifespan. Nature 2003;425:191–6.
[109] Lain S, Hollick JJ, Campbell J, Staples OD, Higgins M, Aoubala M, McCarthy A, Appleyard V, Murray KE, Baker L, Thompson A, Mathers J, Holland SJ, Stark MJ, Pass G, Woods J, Lane DP, Westwood NJ. Discovery, in vivo activity, and mechanism of action of a small-molecule p53 activator. Cancer Cell 2008;13:454–63.
[110] Li L, Wang L, Li L, Wang Z, Ho Y, McDonald T, Holyoake TL, Chen W, Bhatia R. Activation of p53 by SIRT1 inhibition enhances elimination of CML leukemia stem cells in combination with imatinib. Cancer Cell 2012;21:266–81.
[111] El-Sayed A, Aleya A, Kamel M. Microbiota and epigenetics: promising therapeutic approaches? Environ Sci Pollut Res 2021;28:49343–61.
[112] Schwartz L, Bochter MS, Simoni A, Bender K, de Dios Ruiz Rosado J, Cotzomi-Ortega I, Sanchez-Zamora YI, Becknell B, Linn S, Li B, Santoro N, Eichler T, Spencer JD. Repurposing HDAC inhibitors to enhance ribonuclease 4 and 7 expression and reduce urinary tract infection. Proc Natl Acad Sci U S A 2023;4, e2213363120.

Chapter 3.9

CD73 (5′-Ectonucleotidase)

Claudiu T. Supuran[a] and Clemente Capasso[b]
[a]*NEUROFARBA Department, Section of Pharmaceutical and Nutraceutical Sciences, University of Florence, Florence, Italy,* [b]*Department of Biology, Agriculture and Food Sciences, CNR, Institute of Biosciences and Bioresources, Napoli, Italy*

1 Introduction

The production and degradation of nucleic acids, as well as the maintenance of the nucleotide and nucleoside pool, which are each made up of a nucleobase and a pentacyclic sugar moiety, are all processes that involve a number of enzymes [1,2]. Among them, the ecto-5′-nucleotidases (e5NTs) hydrolyze 5′-nucleoside-monophosphates (NMPs) to generate nucleosides and inorganic phosphate [2] (Fig. 1).

There are two nucleosidases expressed in mammals: the ectonucleoside triphosphate diphoshohydrolase 1, known as CD39 (EC 3.6.1.5), and the ecto-5′-nucleotidase indicated with the acronym CD73 (EC 3.1.3.5) [1,2]. CD39 generates adenosine monophosphate (AMP) from the extracellular adenosine triphosphate (ATP) and/or adenosine diphosphate (ADP) by hydrolyzing the terminal phosphate moiety. CD 73 produces adenosine (A) from the extracellular AMP, also generating inorganic phosphate during the hydrolytic process [3]. Purinergic signaling, which employs the released ATP, ADP, AMP, as well as other nucleotides into the extracellular space, is triggered by tissue injury, cellular stress, or apoptosis [1,4]. Many cellular responses are regulated through the changes in extracellular nucleotides concentrations [5,6], which play significant physiological roles in the mammalian central nervous system, including neurotransmission and neuromodulation [5,6]. Furthermore, purinergic signaling is also correlated to the homeostasis of the immune system [1,4]. Balanced concentrations of ATP, ADP, AMP, and A are essential for controlling inflammatory reactions and as thus tissue damage, and for this reason, the regulation of enzymes that produce and/or degrade these molecules is a tightly controlled process [7,8]. The two isotypes of P_2 purinergic receptors P_{2X}Rs and P_{2Y}rs mediate inflammatory responses via regulation of ATP levels. Many cellular responses are regulated through the changes in extracellular nucleotides concentrations, which in turn act on their specific receptors, most of which are G-protein-coupled receptors (GPCRs). In the case of adenosine, there are at least four types of such GPCR receptors, the A_1, A_{2A}, A_{2B}, and A_3 adenosine receptors [5,6].

In bacteria, CD73 has been linked to an enhanced microbial virulence and pathogenicity [9–12]. This has been observed for various pathogenic microorganisms, including *Staphylococcus aureus*, *Vibrio cholerae*, *Streptococcus agalactiae*, and *Pseudomonas aeruginosa* [9–12] (see Section 3).

CD73 seems to represent an important pharmacological target, due to its role in regulating adenosine concentrations outside the cell, by hydrolyzing AMP to A and Pi [6,13,14] as well as the bacterial virulence.

2 Mammalian 5′-nucleotidases

e5NTs are widely expressed in prokaryotes and eukaryotes, with a large number of representatives present in bacteria, plants, and animals. The metal ions present in their active sites are Mg^{2+} or Zn^{2+}, and they represent cofactors used in the catalytic process. Often, a dinuclear metal center is required for the hydrolysis of phosphate esters mediated by these enzymes (Fig. 2). The animals' e5NTs are classified into two classes: the cytosolic and extracellular membrane-bound forms, which are structurally not related [15] The cytosolic forms are crucial for the nucleotides metabolism [5,6], whereas the membrane-bound enzymes are primarily involved in the synthesis of adenosine through the extracellular hydrolysis of AMP, but they are also implicated in transmembrane signaling, cell-matrix, and cell-cell interactions [6].

The X-ray crystal structure of the soluble form of hCD73 has been reported, with a resolution 2.2 Å. The protein is a homodimer, being composed of two identical 70 kDa polypeptide chains. They are connected by a flexible stretch of oligopeptides, which is important for allowing the enzyme to assume two very distinct conformations: a close one and an open one, making the overall structure to look like a butterfly, which is crucial for catalysis, but also for the binding of substrates and inhibitors [16]. The monomers composing hCD73 are each formed by two distinct domains: the *N*-terminal domain, involved in the binding of the two divalent cations essential for catalysis, and the *C*-terminal

Metalloenzymes. https://doi.org/10.1016/B978-0-12-823974-2.00031-0

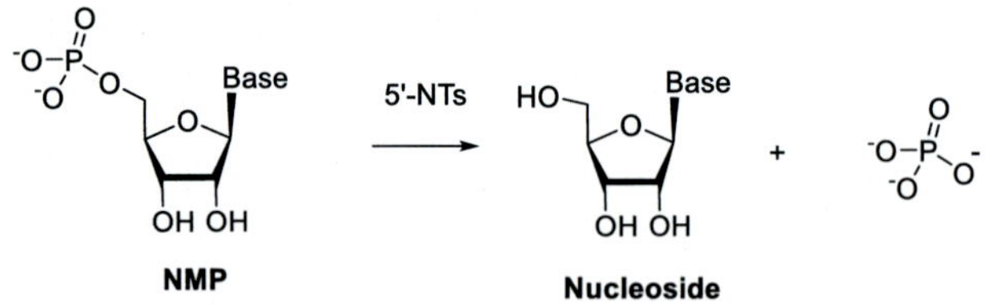

FIG. 1 Reactions catalyzed by 5′-NTs.

domain, responsible for the interaction with the nucleotide(s) substrate [16]. hCD73 seems to be a zinc-dependent enzyme, and the two cations are involved in the catalytic mechanism of this enzyme. It has been demonstrated that by adding Zn^{2+} salts to a solution of apo-hCD73 leads to reconstitution of the enzyme activity as ecto-5′-nucleotidase [16]. A disulfide bridge between two cysteine residues, Cys353 and Cys358, which are located in a loop area, is involved in the dimerization and formation of the quaternary structure of hCD73 [16]. Mutations of these two Cys residues leads to the loss of structural integrity and results in an enzyme without catalytic activity as ecto-5′-nucleotidase [16,17]. The enzyme hydrolytic activity only occurs when the protein adopts its close conformation, which places the nucleotide substrate within the middle of the *N*- and *C*-terminal domains constituting the active site. In this way, the phosphate moiety which will be hydrolyzed is placed nearby the catalytic zinc ions and the nucleophiles associated with them. Thus, the ecto-5′-nucleotidase active site is formed at the intersection of these two domains. One should mention that hCD73 is a highly glycosylated enzyme, which incorporates various oligosaccharide moieties on diverse amino acid residues of the protein [16].

3 Bacterial 5′-nucleotidases

The bacterial level of ribo- and deoxyribonucleoside monophosphates, as well as all the pool of ribo- and deoxyribonucleotides, is regulated by cytoplasmic 5′-NTs [18]. The 5′-NT activity has been identified in the cytoplasmic soluble fraction of many gram-positive and gram-negative bacteria. Only a few of these enzymes have been defined a functional role. Among the bacterial species, one well-studied microbe for its 5′-NTs is represented by the gram-negative opportunistic pathogen, *Escherichia coli*. The cytoplasm of *E. coli* contains four 5′-NTs (UmpH, SurE, YfbR, and YjjG) [19]. UmpH allows the degradation of UMP and is essential for bacterial pyrimidine homeostasis [20]; SurE is involved in the regulation of dNTP and NTP pools, playing a significant physiological role in the stress response, and is essential for the survival of cells in the stationary growth phase; YfbR plays a critical role in the deoxycytidine pathway and de novo synthesis of thymidylate; YjjG can dephosphorylate a wide range of noncanonical pyrimidine derivatives preventing the incorporation of these potentially mutagenic compounds into DNA and RNA.

Moreover, it has been hypothesized that. *E. coli* SuhB, a polypeptide chain that is homologous to mammalian inositol monophosphatase, could also function as 5′-nucleotidase since it can hydrolyze adenosine 2′-monophosphate, also considering that *E. coli* does not present the phosphatidylinositol [21], which is essential for eukaryotic membrane function, serving as organelle markers and signaling cascades [19]. Recently, it has been identified in the genome of *Bacillus subtilis* five gene encoding for 5′-NTs and heterologously expressed in *E. coli*. Surprisingly, the recombinant proteins displayed a 5′-nucleosidase activity with different intensity and substrate specificities [22]. For example, only one of them (YcsE) resulted in a powerful 5′-nucleotidase, having a broader specificity (IMP, inosine monophosphate; AMP, adenosine monophosphate; GMP, guanosine monophosphate; UMP, uridine monophosphate; CMP, cytidine monophosphate; G6P, glucose 6-phosphate; and β-glycerophosphate), while the others had a feeble 5′-nucleotidase activity [22]. Again, some of these enzymes have an essential role in the ability of the organism to withstand oxidative

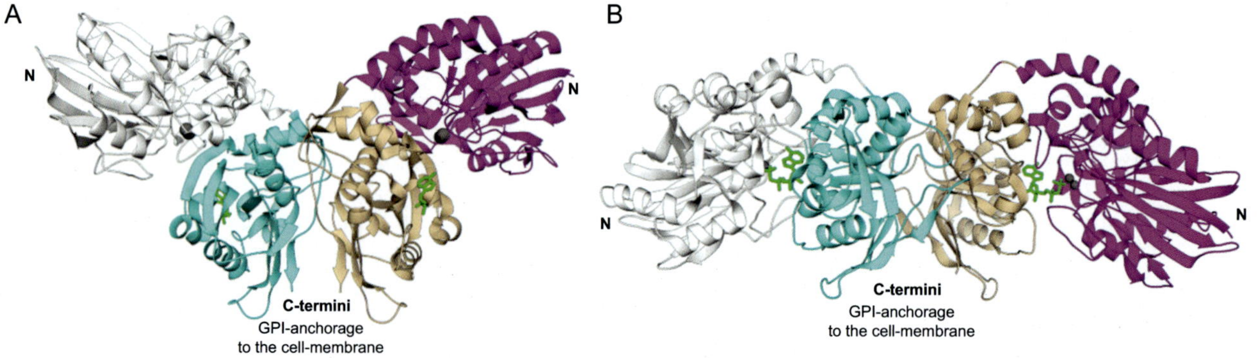

FIG. 2 Crystal structure of e5NT of human origin, in: (A) an open conformation (PDB 4H2G) butterfly-like and (B) closed conformation (PDB 4H2I) form. The *N*- and *C*-terminus domains of one subunit of the dimer are shown in gray and sky blue, respectively [15]. The *N*- and *C*-terminus domains of the other subunit are shown in magenta and brown, respectively. The ligands adenosine and adenosine (α,β)-methylene diphosphate (AMPCP, 1) in the open and closed forms respectively are shown as magenta sticks while the metal ions are shown as grey spheres.

stress, while others are crucial for development and proliferation of the microorganisms on solid media [22].

3.1 Membrane-bound and periplasmic 5′-NTs

Generally, the membrane-bound and periplasmic 5′-NTs have the physiological function of breaking down the extracellular nucleotides to meet the nutritional needs of the microorganism. DNA degradation provides phosphate, carbon, and nitrogen to the microbes. Thus, membrane-bound and periplasmic 5′-NT activities allow bacterial survival when phosphate is limited, but extracellular DNA is available [23]. As described in the literature, *E. coli* periplasmic 5′NTs are involved in the assimilation of exogenous nucleotides, CDP-alcohols, UDP-sugars, DNA, and other phosphorylated substances that cannot cross the cell membrane [24].

3.2 Structural features

Based on similarities in the primary sequence, it is readily apparent that the animal ecto-5′-NTs are related to the bacterial 5′-NTs, which are part of a group of proteins called metallophosphoesterases characterized by a dinuclear metal-binding center. The metal ion nature present in these enzymes is still rather controversial. The human CD73 enzyme contains two zinc ions, but the *E. coli* ortholog e5NT structure has been obtained with manganese(II) ions bound within the active site [25]. However, the enzyme comprises domains linked by a flexible stretch of amino acids, allowing it to take on a closed or open, butterfly-like shape [6,25]. The N-terminal domain (residues 25–342) binds the two metal ions and contains the catalytic Asp–His dyad. The C-terminal domain (residues 362–550) has a unique structure and binds the substrate [25]. In the closed conformation, the active site is located between the two domains, and the substrate, which is near the catalytically active metal ions (Zinc(II) or Manganese (II)), is hydrolyzed [2,6,25]. Interestingly, Xf50-Nt, which is the recombinant 5′-NT encoded by the *Xylella fastidiosa* genome, exhibited the highest affinity for Mg(II), followed by Mn(II) > Co (II) > Cu(II) > Ni(II) > Ca(II) > Zn(II) [26].

The 5′-nucleotidases encoded by the bacterial genome are polypeptides anchored to the bacterial outer membrane [27], located in the periplasmic space [28] or in the cellular cytoplasm [22]. Depending on the type of acceptor of the phosphoryl group, which can be an activated water molecule or a nucleophilic amino acid residue, the bacterial 5′NTs are catalytically categorized as a type I or type II, respectively [29] E. coli 5′-NT hydrolyzes AMP, ADP, ATP, and other 5′-ribo- and 5′-deoxyribonucleotides [6]. Indeed, the enzyme can hydrolyze the artificial substrate *p*-nitrophenyl phosphate, which is not a substrate for ecto-5′-NT [6].

4 CD73 catalytic mechanism

CD73 catalyzes the hydrolysis of monophosphate nucleoside substrates. The catalytic mechanism of this process has been proposed by Knöfel and Sträter [25]. The first step consists of a nucleophilic attack of a hydroxyl moiety bound to one of the two Zn^{2+} metal ions present in the active site, on the trigonal bipyramidal phosphorus of the substrate, which releases the nucleoside and leads to the formation of Pi [16] (Fig. 3). hCD73 contains two zinc ions, but the first crystallographic studies were performed on an ortholog e5NT enzyme from *Escherichia coli*. Those experiments were performed in the presence of Mn(II) salts, which in fact have been observed bound to the protein [25,30]. Thus, it is probable that the two metal ions (M1 and M2) present in the active site are coordinated in either a trigonal bipyramidal geometry (as for Zn^{2+}) or in an octahedral geometry (as for Mn^{2+}), as illustrated in Fig. 3. In the noninhibited enzyme, M1 is coordinated by residues Asp36, His38, Asp85 (which acts as bridging ligand between two metal ions), a hydroxide anion (W1, also acting as a bridging ligand), and two water molecules (W2 and W3) [30]. M2 was observed to be coordinated by Asn117, His220, His243, as well as the bridging ligands mentioned above (W1 and Asp85). In e5NT adducts with the nonhydrolyzable ADP analog adenosine (α,β)-methylene diphosphate (AMPCP), investigated by means of crystallographic experiments, the terminal phosphonate group of the ligand was seen bidentately bridged to the two metal ions, thus replacing W1 [25,30].

A catalytic dyad conserved in all e5NTs investigated so far, composed of residues His118 and Asp121, is probably involved in catalysis too, as it stabilizes the P(V) intermediate in its trigonal bipyramidal geometry, during the hydrolysis. What the attacking nucleophile is, it is also a matter of debate, but most probably it is the water molecule/hydroxide ion coordinated to one of the metal ions, M1, as it is placed in a favorable geometry for attacking the scissile phosphate bond [25,30]. Positively charged residues such as Arg354 and Arg395 seem to also take part in the catalytic process, and they clearly stabilize the transition state during the hydrolysis of the phosphate [25,30].

5 Inhibitors of the human CD73 (hCD73)

5.1 Nucleoside/nucleotide inhibitors

Muller's group reported most of the drug design studies of nucleotide- and nucleoside- hCD73 inhibitors [2,30,31]. ADP may serve as a substrate for other enzymes involved in purinergic signaling, and for this reason, a nonhydrolyzable derivative of it, designated as AMPCP (compound **1**, Fig. 3), constituted the starting point for designing hCD73 inhibitors, as this compound has an interesting

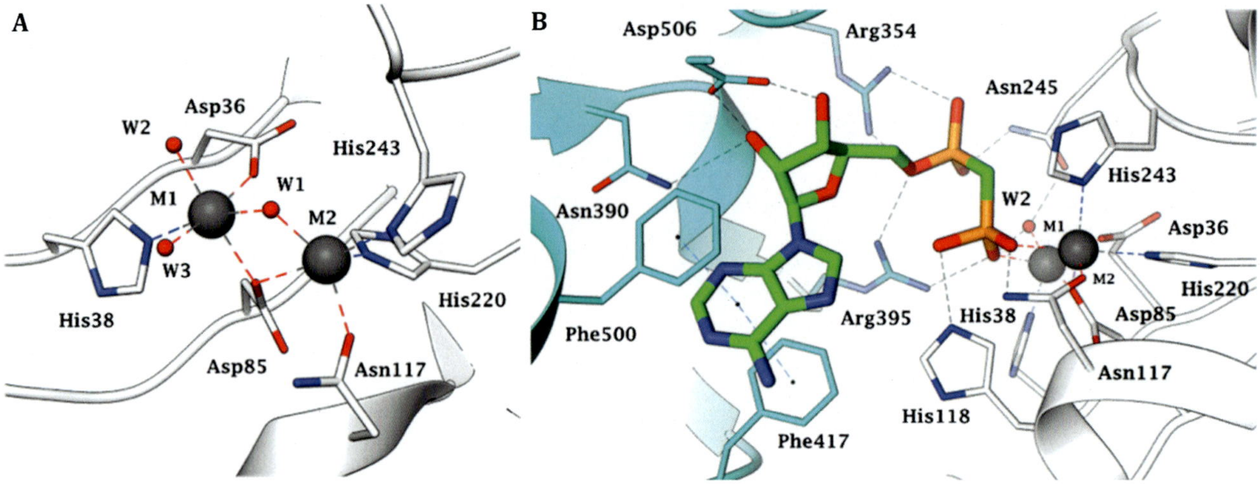

FIG. 3 (A) Active site view of apo-e5NT open conformation. (B) View of the ADP analog AMPCP (**1**) bound to the active site of e5NT in the closed conformation [15]. The two zinc ions are shown as grey spheres. The protein residues of the *C*-terminal domain forming the substrate binding site are shown in blue, whereas the amino acids of the *N*-terminal domain are depicted in gray. Water molecules (small red spheres), unless coordinating metal ions, are omitted for clarity. H-bonds and π-π interactions are depicted as black and blue dashed lines, respectively.

inhibition constant, $K_I = 0.87\,\mu M$ [2,30–34]. In this lead compound, the bridging oxygen present in the diphosphate group of ADP has been replaced by a methylene moiety, which represent a classical isosteric replacement on medicinal chemistry [15–19]. X-ray crystal structures of AMPCP bound to hCD73 afforded crucial structural information for the design of hCD73 inhibitors. [30] All fragments of this compound were observed to be important for inhibition of the enzyme: (i) the terminal phosphonate moiety is anchored to both metal ions from the active site. As thus, they act as zinc-binding groups (ZBGs), a case which is frequently observed in many other zinc-enzyme inhibitors; [35] (ii) the sugar moiety and the phosphate group near the sugar participate in hydrophobic interactions and hydrogen bonds with the following residues: Arg354, Arg395, and Asp506 (Fig. 2); and (iii) the purine ring of the inhibitor was observed to participate in π stacking interactions which involve two phenyl aromatic rings from the enzyme active site, of Phe417 and Phe500. They participate to a very effective π-stacking with the heterocyclic aromatic ring of the inhibitor. Thus, many drug design studies were focused on analogs of compound **1** (Fig. 3), in which all these three structural elements mentioned above were changed: the ZBG group, the sugar moiety, and the heterocyclic moiety [33]. The X-ray crystal structures of the close form of hCD73 with two different derivatives (**2** and **3**) bound within the active site were again reported by Muller's group. [33] Compound **2** incorporates a hydrazinyl functionality at the purine ring, and acts as a low nM inhibitor (K_I of 15.5 nM), whereas derivative **3**, with a bulkier substitution pattern (piperazine instead of the hydrazine group), was 10 times a less efficient inhibitor (K_I of 184 nM) compared to **2**, presumably due to steric clash involving the piperazine functionality **2**. Other interactions between different inhibitors and residues from hCD73 active site were observed to be similar to those mentioned for the AMPCP **1** adduct, providing thus an explanation for the favorable inhibitory profile reported for derivatives of this class.

In previous works, [32–34] Muller's group demonstrated that AMPCP structure may be used as lead and a large number of derivatives, among which compounds **4–12** (Fig. 4) show significant inhibitory activity against the enzyme. Many of these hCD73 inhibitors were highly efficient, with inhibition constants in the subnanomolar-low nanomolar range [18,19].

5.2 Nonnucleotide inhibitors

X-ray crystallography has been extremely useful also in the process of identifying nonnucleotide hCD73 inhibitors, with Muller's group accomplishing the most significant achievements. [30] As for the previously mentioned hCD73 inhibitors, the effectiveness of the binding is governed by the π-stacking interactions between aromatic ring(s) of the inhibitor and the phenyl moieties of Phe500 and Phe417 from the enzyme active site. Furthermore, the endocyclic oxygen and the OH of the ribose moiety were observed to participate in H-bonds with residues Arg395, Arg354, Asn390, and Asp506, as illustrated in Fig. 5. However, most of these inhibitors seem to lack a well-defined ZBG, which leads to weaker inhibition compared to the nucleoside/nucleotide derivatives, which possess the phosphate as a ZBG.

The natural product baicalin **13** [36] (Fig. 4), a flavone derivative acting as hCD73 inhibitor, was crystallized when bound to the open conformation of hCD73 (Fig. 6). [30] Baicalin was observed bound in the same active site region as

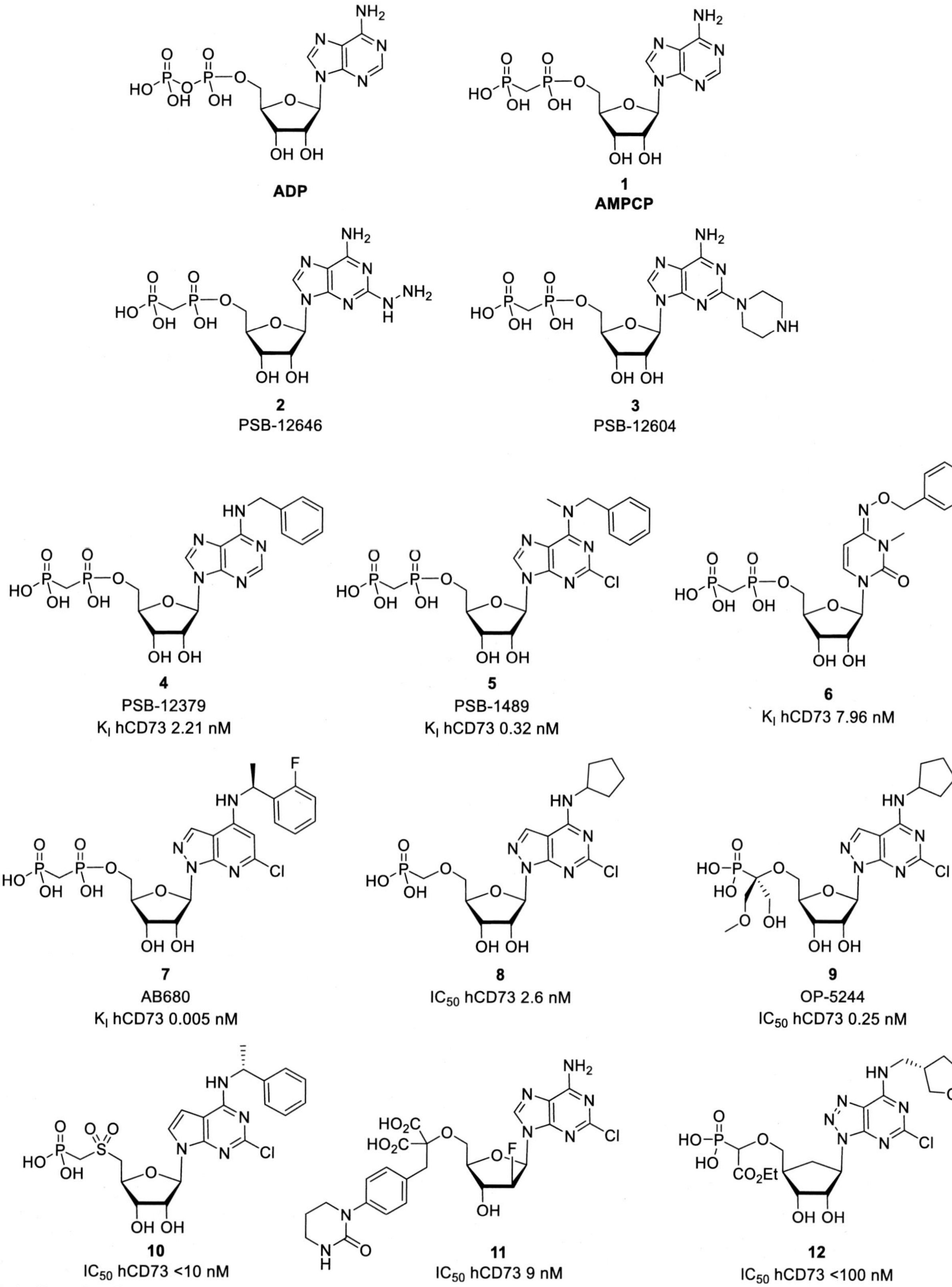

FIG. 4 Examples of compounds designed as hCD73 inhibitors considering ADP AMPCP, **1** and PSB-12646 **2**/PSB-12604 **3** as leads. The CD73 inhibitors **4–12** incorporate nucleoside/nucleotide scaffolds.

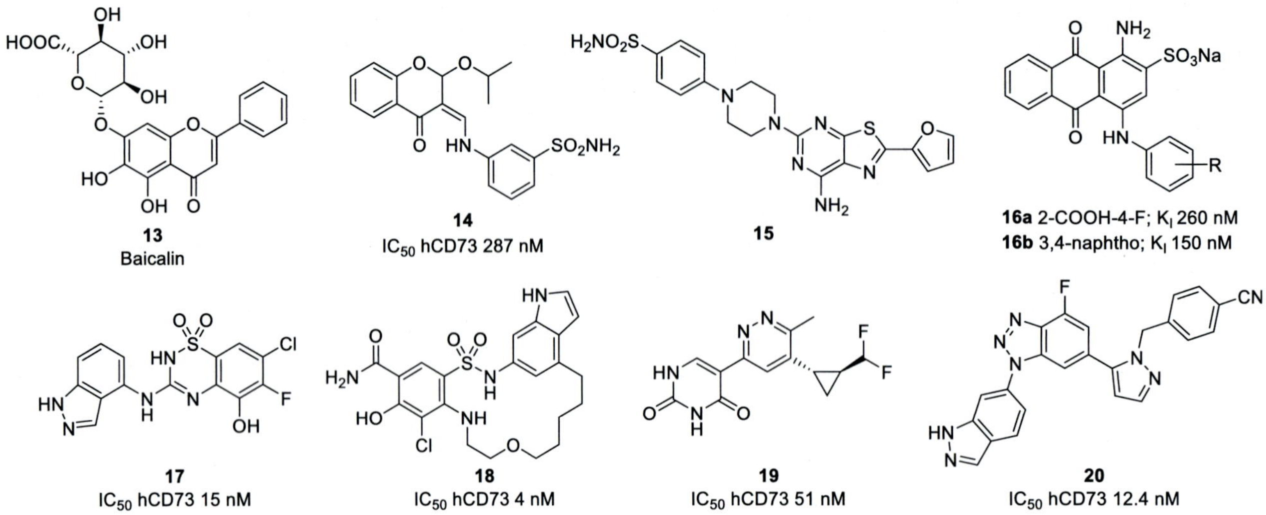

FIG. 5 Nonnucleoside hCD73 inhibitors **13–20**.

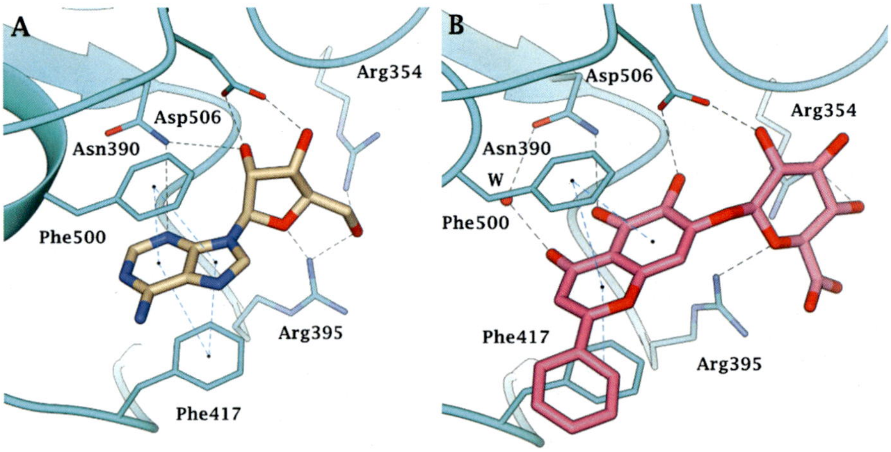

FIG. 6 The open conformation e5NT, its substrate binding site view in adduct with (A) adenosine (PDB 4H2G) and (B) baicalin (PDB 4H2B) [15]. The H-bonds and π-π interactions involving the inhibitor and active site amino acid residues are shown as black and blue dashed lines, respectively.

adenosine, with its benzopyrone ring interacting with Phe417 and Phe500 by means of π stacking, similar to the purine ring present in adenosine and discussed above. The glucuronic acid moiety from baicalin, is oriented toward the same part of the active site as the ribose of adenosine, although the two moieties are rather different. Indeed, in the case of baicalin the sugar moiety is bulkier than the ribose from adenosine, and thus, it is protruding toward the metal ions, without interacting with any of the two zinc ions from the enzyme active site.

Another interesting class of hCD73 inhibitors is constituted by the anthraquinone sulfonates.[31] Compounds such as 1-amino-4-[4-fluoro-2-carboxyphenylamino]-9,10-dioxo-9,10-dihydroanthracene-2-sulfonate (**16a**, PSB-0952, K_I of 260 nM) and 1-amino-4-[2-anthracenylamino]-9,10-dioxo-9,10-dihydroanthracene-2-sulfonate (**16b**, PSB-0963, K_I of 150 nM) show highly effective hCD73 inhibition among all nonnucleotide inhibitors described so far. [31]

5.3 Monoclonal antibodies (MAbs) as hCD73 inhibitors

Antibodies bind to antigens on a target cell triggering cell-membrane destruction, blocking cell growth, preventing blood vessel growth, blocking the immune response, or directly attacking abnormal, cancer cells. [37] Antibodies used to treat cancer recognize surface-bound antigens, including enzymes, such as CD73. They may also neutralize enzymes by inhibiting their catalytic activity or interfering with their signaling role in tumor cells. CD73 inhibition by MAbs was demonstrated to induce apoptosis of the tumor cells, which may constitute an anticancer treatment approach.

Various MAbs were developed against hCD73 [38] and, recently, several new such biological agents, e.g., Ab001/Ab002 and their humanized version Hu001/Hu002 were reported. [39] These MAbs showed high CD73 binding affinity, inhibited the enzyme activity, and in vivo, also inhibited the growth of RAS-mutant nonsmall-cell lung cancer (NSCLC) tumors, in an animal model of cancer. [39] Several antibody-drug conjugates (ADCs) of the two humanized MAbs mentioned above were also reported and showed enhanced efficacy compared to the agents alone, providing the proof of concept for using such therapeutics in clinical settings [39].

6 Inhibitors of the bacterial CD73

6.1 5′-NTs and bacterial virulence

Membrane-associated 5′-NTs are linked to virulence in several pathogenic bacteria since pathogenic microorganisms, and their hosts frequently interact via surface-exposed proteins. For example, the bacterial membrane-associated 5′-NT aided bacteria in their ability to infect the host by increasing extracellular quantities of potent immunosuppressive chemical compounds, i.e., decreasing the concentration of the pro-inflammatory ATP and/or increased concentrations of Ado. This observation has been reported for many microbes, such as *Staphylococcus aureus*, *Vibrio cholerae*, *Streptococcus agalactiae*, *Pseudomonas aeruginos*a, and others [9–12].

Periplasmic 5′-nucleotidases can alter the pathogenicity of these bacteria, enhancing the damage they cause when invading host cells [29]. For example, it has been reported that the periplasmic 5′-NT activity disrupts critical intermediates in the host's phospholipid biosynthetic pathways, such as CDP-ethanolamine and CDP-choline. As a result, the synthesis of cell wall components of the host is impaired [40]. It has been shown that *Burkholderia cepacia* and *P. aeruginosa* produce secreted ATP-hydrolyzing enzymes, which via activating purinergic nucleotide receptors, interfere with the ATP levels of macrophages and mast cells, causing their demise [41,42]. In 2013, the recombinant *Xylella fastidiosa* 5′-NT (Xf50 -Nt) was considered an essential enzyme in the biofilm formation of this microorganism. *X. fastidiosa* is a gram-negative plant microbe responsible for illnesses that affect significant crops all over the world [43]. Its biofilm formation is responsible for the occlusion of the plant xylem vessels, provoking devastating plant disease [26]. In 2015, a novel extracellular nucleotidase (S5nA) has been identified in *Streptococcus pyogenes*, an opportunistic pathogen that causes pharyngitis, impetigo, necrotizing fasciitis, and toxic shock syndrome [44]. S5nA hydrolyzes AMP and ADP, but not ATP, to generate Ado, demonstrating how the conversion of dAMP into dAdo resulted in a potent inhibitor of macrophages and monocytes. Thus, S5nA is being considered as a novel *S. pyogenes* virulence factor able to deregulate the host immune response [44]. However, in 2016, Fiedler's group demonstrated that S5nA was unessential for the growth of S. *pyogenes* in the human blood, as well as for the evasion of phagocytosis by neutrophils, bacterial biofilms formation, and virulence [45].

Intracellular 5′-NTs, as described above, are mainly involved in controlling intracellular nucleotide pools for DNA/RNA synthesis. It has been reported that the deletion of the nagD gene encoding for an intracellular 5′-NT in *Staphylococcus aureus* drastically reduces the virulence of this pathogen [46].

6.2 Bacterial 5′-NT inhibitors

Already in 1966 and 1967, the first 5′-NT inhibitors were discovered in the cytoplasm of several bacteria [47]. It was demonstrated that in *E. coli* the activity of the 5′-NT-UshA (a surface-located metallophosphoesterase/5′-nucleotidase) was blocked by proteinaceous substances which protect cytoplasmic nucleotides before 5′-NT-UshA secretion [47]. However, the complete characterization of such proteinaceous bacterial 5′-NT inhibitors is a poorly explored field and no new convincing results have been obtained in the past decades. On the other hand, it has been reported that polyphenolic natural compounds isolated from betel nuts (*Areca catechu*) or wine grapes also possess inhibitory effects against bacterial 5′-NTs [48–50].

Although these enzymes seem to show a relevant potential for the development of inhibitors with pharmacological activity as antibacterials, there are few drug design studies in the literature that considered bacterial 5′-NTs as drug targets. The most interesting one has been reported by Muller's group and targets a 5′-NT from *Legionella pneumophila*, the etiological agent of Legionnaire's disease [51,52]. The nucleoside triphosphate diphosphohydrolase (NTPDase) from *L. pneumophila*, denominated Lp1NTPDase, was found to be a structural and functional homolog of mammalian such enzymes and to catalyze the hydrolysis of ATP to ADP and ADP to AMP, and its activity contributes to the virulence of this pathogen [51,52]. Thus, the capillary electrophoresis (CE)-based enzyme assay for studying the Lp1NTPDase activity and inhibition was established by Muller's group, which led to the identification of several compounds belonging to 1-amino-4-ar(alk)ylamino-2-sulfoanthraquinones, possessing IC_{50} values in the low micromolar range.

7 Conclusions

hCD73 inhibitor research is focused on inflammation and cancer. Several clinical trials using anti-CD73 monoclonal antibodies, such as Oleclumab (MEDI9447), are currently being conducted for the treatment of solid tumors (pancreatic, breast, prostate, NSLCC, etc.), either alone or in combination with other anticancer agents [53].

However, mAbs have low penetrability in solid tumors and their pharmacokinetic are rather complex, which prompted research for the identification of small molecule compounds acting as hCD73 inhibitors. Such compounds might be more resilient for anticancer drug development for several reasons, including the ease of administration, lack of allergic reactions, etc. Indeed, the small molecule CD73 inhibitor AB680 [54] (compound **7**, $K_I = 0.005$ nM; Fig. 3) recently entered Phase I clinical trials as an antitumor agent. [55] This potent, competitive h CD73 inhibitor, is possibly the first of many derivatives that will be studied in clinical contexts in the next years. It should be noted that CD73 is a hypoxia-inducible factor (HIF) regulated enzyme and targeting it may be particularly relevant for a range of malignancies associated with hypoxia and extracellular acidification [56,57]. Thus, a combination of CD73 inhibitors with carbonic anhydrase (CA) IX/XII inhibitors may significantly improve cancer immunotherapy [58–60]. In general, the most recent breakthroughs in the CD73 inhibitors are particularly relevant for the development of cutting-edge cancer treatments and therapies.

The field of bacterial 5′-NTs is still wholly unknown compared to mammalian 5′-nucleotidases, which have been the subject of intensive research and have been characterized in detail. AMP, ADP, ATP, and other 5′-ribo- and 5′-deoxyribonucleotides are hydrolyzed by bacterial 5′-NTs. Intriguingly, numerous harmful bacteria have connected these enzymes to their infectiousness. For example, *Staphylococcus aureus*, *Vibrio cholerae*, *Streptococcus agalactiae*, and *Pseudomonas aeruginos*a membrane-associated 5′-NTs have been connected to microbial virulence since the surface-exposed proteins are typically the medium via which pathogenic microbes and hosts communicate. Again, it has been demonstrated that the host invasion by *Burkholderia cepacian*, *P. aeruginosa*, *Xylella fastidiosa*, and *Streptococcus pyogenes* is dependent on the activity of the periplasmic 5′-nucleotidases [46].

Unfortunately, very few inhibitors of the bacterial 5-NTs are reported. Among them, it is possible to list the following: protein-based material, natural polyphenolic compounds, and 1-amino-4-ar(alk)ylamino-2-sulfoanthraquinones showing low-micromolar IC50 values [51,52]. Thus, although these enzymes indicate potential as antibacterial therapeutic targets, still few drug design studies concerning the bacterial 5′-NTs have been available.

References

[1] Illes P, Xu GY, Tang Y. Purinergic signaling in the central nervous system in health and disease. Neurosci Bull 2020;36(11):1239–41.

[2] Nocentini A, Capasso C, Supuran CT. Small-molecule CD73 inhibitors for the immunotherapy of cancer: a patent and literature review (2017–present). Expert Opin Ther Pat 2021;31(10):867–76.

[3] Camici M, Garcia-Gil M, Tozzi MG. The inside story of adenosine. Int J Mol Sci 2018;19(3):784–97. https://doi.org/10.3390/ijms19030784.

[4] Cekic C, Linden J. Purinergic regulation of the immune system. Nat Rev Immunol 2016;16(3):177–92.

[5] Bogan KL, Brenner C. 5′-Nucleotidases and their new roles in NAD (+) and phosphate metabolism. New J Chem 2010;34(5):845–53.

[6] Strater N. Ecto-5′-nucleotidase: structure function relationships. Purinergic Signal 2006;2(2):343–50.

[7] Zimmermann H, Zebisch M, Strater N. Cellular function and molecular structure of ecto-nucleotidases. Purinergic Signal 2012;8 (3):437–502.

[8] Burnstock G. Purine and pyrimidine receptors. Cell Mol Life Sci 2007;64(12):1471–83.

[9] Thammavongsa V, Kern JW, Missiakas DM, Schneewind O. Staphylococcus aureus synthesizes adenosine to escape host immune responses. J Exp Med 2009;206(11):2417–27.

[10] Punj V, Zaborina O, Dhiman N, Falzari K, Bagdasarian M, Chakrabarty AM. Phagocytic cell killing mediated by secreted cytotoxic factors of Vibrio cholerae. Infect Immun 2000;68(9):4930–7.

[11] Firon A, Dinis M, Raynal B, Poyart C, Trieu-Cuot P, Kaminski PA. Extracellular nucleotide catabolism by the Group B Streptococcus ectonucleotidase NudP increases bacterial survival in blood. J Biol Chem 2014;289(9):5479–89.

[12] Zaborina O, Dhiman N, Ling Chen M, Kostal J, Holder IA, Chakrabarty AM. Secreted products of a nonmucoid Pseudomonas aeruginosa strain induce two modes of macrophage killing: external-ATP-dependent, P2Z-receptor-mediated necrosis and ATP-independent, caspase-mediated apoptosis. Microbiology (Reading) 2000;146(Pt 10):2521–30.

[13] Zimmermann H. 5′-Nucleotidase: molecular structure and functional aspects. Biochem J 1992;285(Pt 2):345–65.

[14] Zhang B. CD73: a novel target for cancer immunotherapy. Cancer Res 2010;70(16):6407–11.

[15] Burnstock G, Verkhratsky A. Long-term (trophic) purinergic signalling: purinoceptors control cell proliferation, differentiation and death. Cell Death Dis 2010;1, e9.

[16] Heuts DP, Weissenborn MJ, Olkhov RV, Shaw AM, Gummadova J, Levy C, Scrutton NS. Crystal structure of a soluble form of human CD73 with ecto-5′-nucleotidase activity. ChemBioChem 2012;13 (16):2384–91.

[17] St Hilaire C, Ziegler SG, Markello TC, Brusco A, Groden C, Gill F, Carlson-Donohoe H, Lederman RJ, Chen MY, Yang D, Siegenthaler MP, Arduino C, Mancini C, Freudenthal B, Stanescu HC, Zdebik AA, Chaganti RK, Nussbaum RL, Kleta R, Gahl WA, Boehm M. NT5E mutations and arterial calcifications. New Engl J Med 2011;364 (5):432–42.

[18] Proudfoot M, Kuznetsova E, Brown G, Rao NN, Kitagawa M, Mori H, Savchenko A, Yakunin AF. General enzymatic screens identify three new nucleotidases in *Escherichia coli*—Biochemical characterization of SurE, YfbR, and YjjG. J Biol Chem 2004;279(52):54687–94.

[19] Botero S, Chiaroni-Clarke R, Simon SM. *Escherichia coli* as a platform for the study of phosphoinositide biology. Sci Adv 2019;5 (3):1–16. https://doi.org/10.1126/sciadv.aat4872.

[20] Reaves ML, Young BD, Hosios AM, Xu YF, Rabinowitz JD. Pyrimidine homeostasis is accomplished by directed overflow metabolism. Nature 2013;500(7461):237.

[21] Kates M. Citation classic—bacterial lipids. Cc/Life Sci 1980;17:12.

[22] Terakawa A, Natsume A, Okada A, Nishihata S, Kuse J, Tanaka K, et al. Bacillus subtilis 5'-nucleotidases with various functions and substrate specificities. BMC Microbiol 2016;16:1–13. https://doi.org/10.1186/s12866-016-0866-5.

[23] Pinchuk GE, Ammons C, Culley DE, Li SMW, McLean JS, Romine MF, Nealson KH, Fredrickson JK, Beliaev AS. Utilization of DNA as a sole source of phosphorus, carbon, and energy by Shewanella spp.: Ecological and physiological implications for dissimilatory metal reduction. Appl Environ Microbiol 2008;74(4):1198–208.

[24] Kakehi M, Usuda Y, Tabira Y, Sugimoto S. Complete deficiency of 5′-nucleotidase activity in *Escherichia coli* leads to loss of growth on purine nucleotides but not of their excretion. J Mol Microbiol Biotechnol 2007;13(1–3):96–104.

[25] Knofel T, Strater N. Mechanism of hydrolysis of phosphate esters by the dimetal center of 5′-nucleotidase based on crystal structures. J Mol Biol 2001;309(1):239–54.

[26] Santos CA, Saraiva AM, Toledo MA, Beloti LL, Crucello A, Favaro MT, Horta MA, Santiago AS, Mendes JS, Souza AA, Souza AP. Initial biochemical and functional characterization of a 5′-nucleotidase from Xylella fastidiosa related to the human cytosolic 5′-nucleotidase I. Microb Pathog 2013;59–60:1–6.

[27] Tamao Y, Noguchi K, Sakai-Tomita Y, Hama H, Shimamoto T, Kanazawa H, Tsuda M, Tsuchiya T. Sequence analysis of nutA gene encoding membrane-bound Cl(-)-dependent 5′-nucleotidase of Vibrio parahaemolyticus. J Biochem 1991;109(1):24–9.

[28] Burns DM, Beacham IR. Nucleotide sequence and transcriptional analysis of the E. coli ushA gene, encoding periplasmic UDP-sugar hydrolase (5′-nucleotidase): regulation of the ushA gene, and the signal sequence of its encoded protein product. Nucleic Acids Res 1986;14(10):4325–42.

[29] Zakataeva NP. Microbial 5′-nucleotidases: their characteristics, roles in cellular metabolism, and possible practical applications. Appl Microbiol Biotechnol 2021;105(20):7661–81.

[30] Knapp K, Zebisch M, Pippel J, El-Tayeb A, Muller CE, Strater N. Crystal structure of the human ecto-5′-nucleotidase (CD73): insights into the regulation of purinergic signaling. Structure 2012;20(12):2161–73.

[31] Baqi Y, Lee SY, Iqbal J, Ripphausen P, Lehr A, Scheiff AB, Zimmermann H, Bajorath J, Muller CE. Development of potent and selective inhibitors of ecto-5′-nucleotidase based on an anthraquinone scaffold. J Med Chem 2010;53(5):2076–86.

[32] Bhattarai S, Freundlieb M, Pippel J, Meyer A, Abdelrahman A, Fiene A, Lee SY, Zimmermann H, Yegutkin GG, Strater N, El-Tayeb A, Muller CE. alpha,beta-Methylene-ADP (AOPCP) derivatives and analogues: development of potent and selective ecto-5′-nucleotidase (CD73) inhibitors. J Med Chem 2015;58(15):6248–63.

[33] Bhattarai S, Pippel J, Scaletti E, Idris R, Freundlieb M, Rolshoven G, Renn C, Lee SY, Abdelrahman A, Zimmermann H, El-Tayeb A, Muller CE, Strater N. 2-Substituted alpha,beta-methylene-ADP derivatives: potent competitive ecto-5′-nucleotidase (CD73) inhibitors with variable binding modes. J Med Chem 2020;63(6):2941–57.

[34] Junker A, Renn C, Dobelmann C, Namasivayam V, Jain S, Losenkova K, Irjala H, Duca S, Balasubramanian R, Chakraborty S, Borgel F, Zimmermann H, Yegutkin GG, Muller CE, Jacobson KA. Structure-activity relationship of purine and pyrimidine nucleotides as ecto-5′-nucleotidase (CD73) inhibitors. J Med Chem 2019;62(7):3677–95.

[35] Nocentini A, Angeli A, Carta F, Winum JY, Zalubovskis R, Carradori S, Capasso C, Donald WA, Supuran CT. Reconsidering anion inhibitors in the general context of drug design studies of modulators of activity of the classical enzyme carbonic anhydrase. J Enzym Inhib Med Ch 2021;36(1):561–80.

[36] Atanasov AG, Zotchev SB, Dirsch VM, Supuran CT, Taskforce INPS. Natural products in drug discovery: advances and opportunities. Nat Rev Drug Discov 2021;20(3):200–16.

[37] Zahavi D, Weiner L. Monoclonal antibodies in cancer therapy. Antibodies 2020;9(3):1–20. https://doi.org/10.3390/antib9030034.

[38] Al-Rashida M, Qazi SU, Batool N, Hameed A, Iqbal J. Ectonucleotidase inhibitors: a patent review (2011–2016). Expert Opin Ther Pat 2017;27(12):1291–304.

[39] Jin R, Liu L, Xing Y, Meng T, Ma LP, Pei JP, Cong Y, Zhang XS, Ren ZQ, Wang X, Shen JK, Yu K. Dual mechanisms of novel CD73-targeted antibody and antibody-drug conjugate in inhibiting lung tumor growth and promoting antitumor immune-effector function. Mol Cancer Ther 2020;19(11):2340–52.

[40] Alves-Pereira I, Canales J, Cabezas A, Cordero PM, Costas MJ, Cameselle JC. CDP-alcohol hydrolase, a very efficient activity of the 5′-nucleotidase/UDP-sugar hydrolase encoded by the ushA gene of yersinia intermedia and *Escherichia coli*. J Bacteriol 2008;190(18):6153–61.

[41] Melnikov A, Zaborina O, Dhiman N, Prabhakar BS, Chakrabarty AM, Hendrickson W. Clinical and environmental isolates of Burkholderia cepacia exhibit differential cytotoxicity towards macrophages and mast cells. Mol Microbiol 2000;36(6):1481–93.

[42] Zaborina O, Misra N, Kostal J, Kamath S, Kapatral V, El-Idrissi ME, Prabhakar BS, Chakrabarty AM. P2Z-independent and P2Z receptor-mediated macrophage killing by Pseudomonas aeruginosa isolated from cystic fibrosis patients. Infect Immun 1999;67(10):5231–42.

[43] Mansfield J, Genin S, Magori S, Citovsky V, Sriariyanum M, Ronald P, Dow M, Verdier V, Beer SV, Machado MA, Toth I, Salmond G, Foster GD. Top 10 plant pathogenic bacteria in molecular plant pathology. Mol Plant Pathol 2012;13(6):614–29.

[44] Zheng LS, Khemlani A, Lorenz N, Loh JMS, Langley RJ, Proft T. Streptococcal 5′-nucleotidase A (S5nA), a novel streptococcus pyogenes virulence factor that facilitates immune evasion. J Biol Chem 2015;290(52):31126–37.

[45] Dangel ML, Dettmann JC, Hasselbarth S, Krogull M, Schakat M, Kreikemeyer B, et al. The 5′-nucleotidase S5nA is dispensable for evasion of phagocytosis and biofilm formation in *Streptococcus pyogenes*. PloS One 2019;14(1). https://doi.org/10.1371/journal.pone.0211074.

[46] Begun J, Sifri CD, Goldman S, Calderwood SB, Ausubel FM. Staphylococcus aureus virulence factors identified by using a high-throughput Caenorhabditis elegans-killing model. Infect Immun 2005;73(2):872–7.

[47] Zimmermann H. 5′-Nucleotidase—molecular-structure and functional-aspects. Biochem J 1992;285:345–65.

[48] Iwamoto M, Matsuo T, Uchino K, Tonosaki Y, Fukuchi A. New 5′-nucleotidase inhibitors, Npf-86ia, Npf-86ib, Npf-86iia, and Npf-86iib from Areca-Catechu. 2. Anti-tumor effects. Planta Med 1988;54(5):422–5.

[49] Toukairin T, Uchino K, Iwamoto M, Murakami S, Tatebayashi T, Ogawara H, Tonosaki Y. New polyphenolic 5′-nucleotidase inhibitors isolated from the wine grape koshu and their biological effects. Chem Pharm Bull 1991;39(6):1480–3.

[50] Uchino K, Matsuo T, Iwamoto M, Tonosaki Y, Fukuchi A. New 5′-Nucleotidase Inhibitors, Npf-86ia, Npf-86ib, Npf-86iia, and Npf-86iib from Areca-Catechu. 1. Isolation and biological properties. Planta Med 1988;54(5):419–22.

[51] Sansom FM, Riedmaier P, Newton HJ, Dunstone MA, Muller CE, Stephan H, Byres E, Beddoe T, Rossjohn J, Cowan PJ, d'Apice AJF, Robson SC, Hartland EL. Enzymatic properties of an ecto-nucleoside triphosphate diphosphohydrolase from Legionella pneumophila—Substrate specificity and requirement for virulence. J Biol Chem 2008;283(19):12909–18.

[52] Fiene A, Baqi Y, Malik EM, Newton P, Li WJ, Lee SY, Hartland EL, Muller CE. Inhibitors for the bacterial ectonucleotidase Lp1NTPDase from Legionella pneumophila. Bioorg Med Chem 2016;24 (18):4363–71.

[53] Geoghegan JC, Diedrich G, Lu X, Rosenthal K, Sachsenmeier KF, Wu H, Dall'Acqua WF, Damschroder MM. Inhibition of CD73 AMP hydrolysis by a therapeutic antibody with a dual, non-competitive mechanism of action. MAbs 2016;8(3):454–67.

[54] Jeffrey JL, Lawson KV, Powers JP. Targeting metabolism of extracellular nucleotides via inhibition of ectonucleotidases CD73 and CD39. J Med Chem 2020;63(22):13444–65.

[55] Bowman CE, da Silva RG, Pham A, Young SW. An exceptionally potent inhibitor of human CD73. Biochemistry-Us 2019;58 (31):3331–4.

[56] Giatromanolaki A, Kouroupi M, Pouliliou S, Mitrakas A, Hasan F, Pappa A, et al. Ectonucleotidase CD73 and CD39 expression in non-small cell lung cancer relates to hypoxia and immunosuppressive pathways. Life Sci 2020;259:118389–97. https://doi.org/10.1016/j.lfs.2020.118389.

[57] Petruk N, Tuominen S, Akerfelt M, Mattsson J, Sandholm J, Nees M, et al. CD73 facilitates EMT progression and promotes lung metastases in triple-negative breast cancer. Sci Rep-Uk 2021;11(1):6035–47. https://doi.org/10.1038/s41598-021-85379-z.

[58] McDonald PC, Chia S, Bedard PL, Chu Q, Lyle M, Tang LR, Singh M, Zhang ZH, Supuran CT, Renouf DJ, Dedhar S. A phase 1 study of SLC-0111, a novel inhibitor of carbonic anhydrase IX, in patients with advanced solid tumors. Am J Clin Oncol-Canc 2020;43(7):484–90.

[59] Angeli A, Carta F, Nocentini A, Winum JY, Zalubovskis R, Akdemir A, et al. Carbonic anhydrase inhibitors targeting metabolism and tumor microenvironment. Metabolites 2020;10(10):412–32. https://doi.org/10.3390/metabo10100412.

[60] Supuran CT. Carbonic anhydrase inhibitors: an update on experimental agents for the treatment and imaging of hypoxic tumors. Expert Opin Investig Drugs 2021;30(12):1197–208.

Chapter 3.10

Glyoxalase II

Fabrizio Carta
NEUROFARBA Department, Section of Pharmaceutical and Nutraceutical Sciences, University of Florence, Florence, Italy

1 Introduction

Methylglyoxal (MG; IUPAC name 2-oxopropanal) shown in Fig. 1 is a highly reactive small-molecule compound largely produced in eukaryotic, prokaryotes, yeast as well as protoctista cells as result of carbohydrate, lipid, and protein metabolisms [1–3]. In humans, a roughly amount of 120 μM per day of MG is produced under physiological conditions [1–3], whereas when pathological circumstances occur, increased intracellular production and cellular outflow bring MG plasma concentration up to 400 μM [4]. Since MG content is highly variable among tissues and depending on physio/pathological status, its content is highly unreliable to detect [5].

As schematically shown in Fig. 1, the majority of MG is produced by the glycolytic pathway. Additional contributions are from glucose autoxidation, glycated protein degradation, lipid peroxidation, threonine catabolism, and acetone oxidation to ketone during diabetic ketoacidosis processing [4–6].

The generation of such considerable amounts of highly reactive MG is a threat up to systemic level as the majority of advanced glycation end-products (AGEs) and DNA adducts are from irreversible MG-modifications. From the chemical viewpoint, the most important MG-derived AGEs are mainly due to transformation of Arginine (Arg) side chains into hydroimidazolones (i.e., **MG-H1**, **MG-H2**, and **MG-H3**) or substituted pyrimidine derivatives (Scheme 1) [7].

A well-known example in humans relates to albumin proteins whose Arg410 residue is converted into the hydroimidazolones MG-H1, thus causing inhibition of the antioxidant features, esterase activity and alteration of affinity for drugs [8–10]. Overall, such modifications are reported to increase the outflow of pro-inflammatory markers such as interleukins-1β and -6, tumor necrosis factor α (TNF-α) and promote proteins degradation and endocytosis [4–6]. It has been demonstrated that MG accumulation causes even worse consequences which may result in mutagenesis and cytotoxicity [11,12].

Thus, in order to prevent abnormal levels of MG and thus to limit the content of MG-AGEs, multiple pathways are involved in its metabolism, and the Glyoxalase System (GS) results to be the most effective being able to detoxify about 99% of the total MG produced under physiological conditions in humans [13]. The remaining MG is converted to harmless products by the enzymes aldo-keto reductase (AKR) and aldehyde dehydrogenase (ALDH) to form hydroxy acetone and pyruvate respectively [11,14,15].

2 The Glyoxalase System (GS)

The GS encoding genes are virtually present among all living species on hearth, such as vertebrates, plants, yeasts, bacterial/prokaryotes, fungi as well as protoctista organisms [16]. Such a widespread distribution of the GS is a clear indication of its relevant functions in physiology. In eukaryotic cells the GS enzymatic pool is located at the cytosol, with the exception of some cancer cells, which do express some components within their nuclei [15]. It consists of the enzyme glyoxalase 1 (GLO1; EC 4.4.1.5), glyoxalase 2 (GLO2; EC 3.1.2.6,) and glutathione (GSH), which is required in catalytic amounts (Fig. 2).

The metabolic conversion of MG to D-Lactate through the GS metabolic pathway begins with the spontaneous addition of GSH to the formyl group of the reactive dicarbonyl species to afford the hemi thioacetal derivative. GLO1 catalyzes the isomerization (i.e., hydride shift) of the hemi thioacetal to *S*-D-lactoyl glutathione which in turn is hydrolyzed by GLO2 to D-lactate and GSH. In 2012, it was discovered in human patients affected of early onset Parkinson's disease a new member of the GS, named glyoxalase 3 (GLO3) [16], whose homologs were firstly identified in *Escherichia coli* and on superior organisms [17,18]. GLO3 catalyzes the conversion of MG into D-lactate in a single step and without any GSH cofactor [16–18].

It is generally accepted that the GS is way broader than two or three GLO enzymes implicated in converting MG into D-Lactate, since additional enzymes such as the glyoxalase domain-containing protein 4 (GLOD4) and glyoxalase domain-containing protein 5 (GLOD5) and the methylmalonyl-CoA epimerase (MCEE) have been identified. Although their features are largely unknown, it

Metalloenzymes. https://doi.org/10.1016/B978-0-12-823974-2.00015-2

FIG. 1 Metabolic sources of methylglyoxal (MG) in cells and its chemical structure [1].

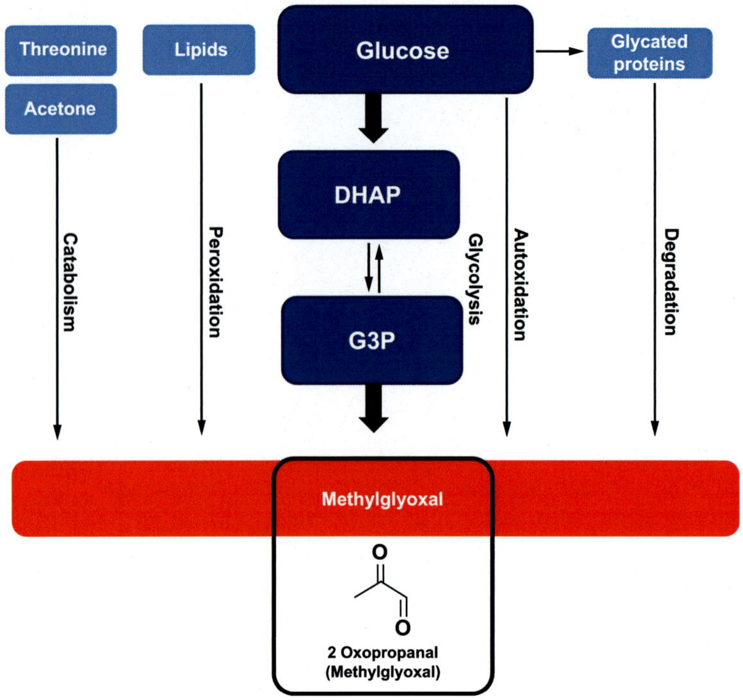

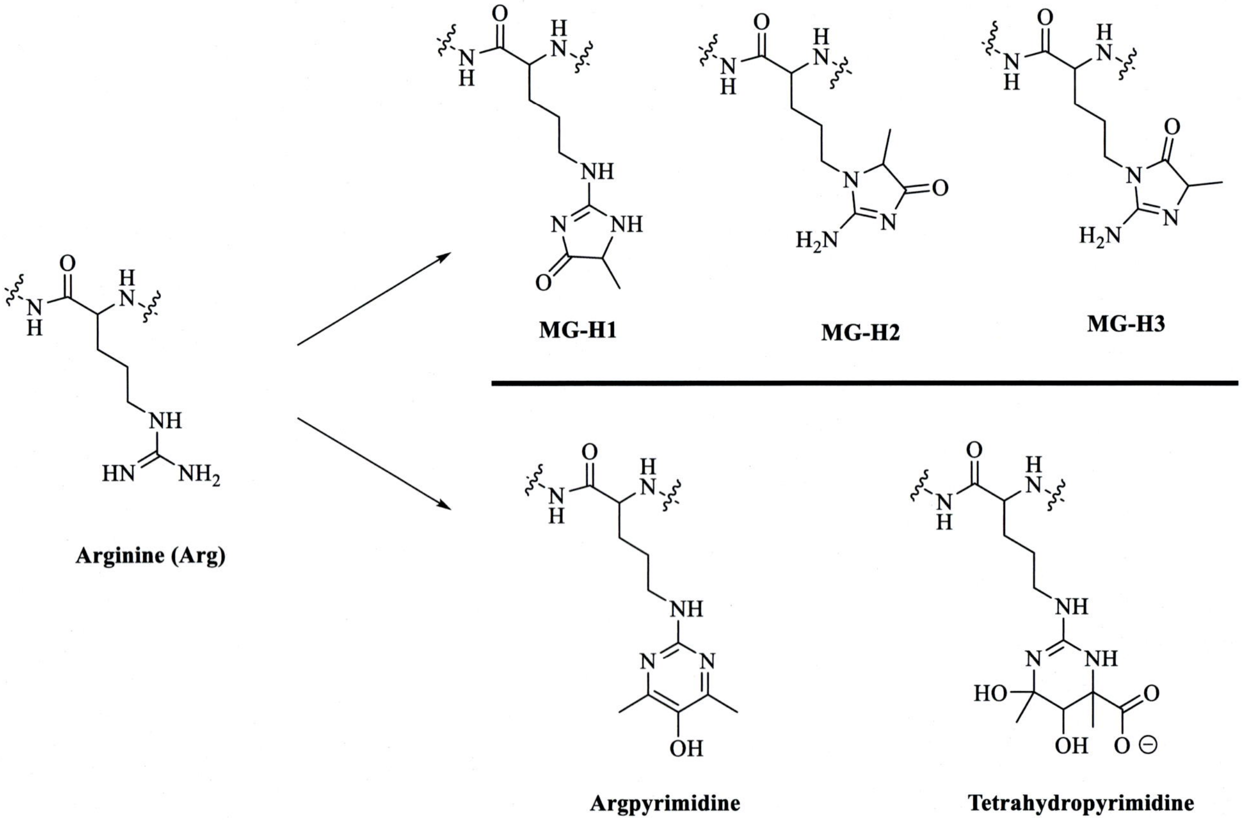

SCHEME 1 MG-derived AGEs on arginine [7].

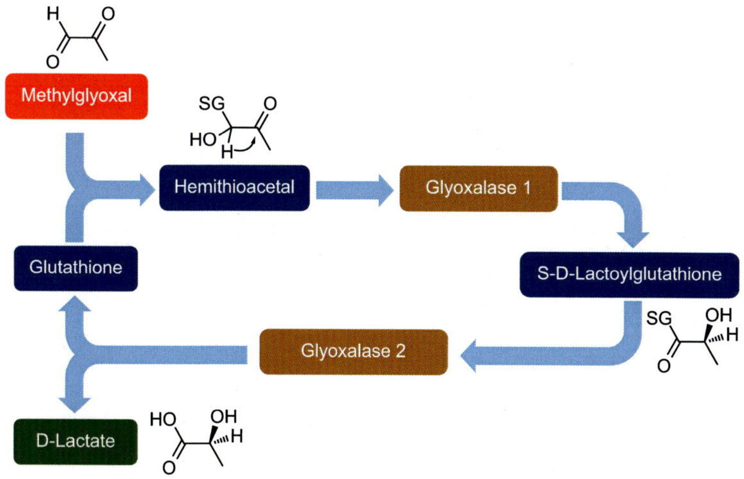

FIG. 2 The Glyoxalase System (GS) and its pathway for conversion of MG to D-Lactate [1].

is demonstrated that all such enzymes do possess broad functions in dicarbonyl stress metabolism, with a central focus on MG [19–21]. The elevated efficiency of the GS pathway depicted in Fig. 2 resides on both GLO1 and GLO2 acting cooperatively. However, disconnection of GS components activities was reported and revealed to be usually associated to important pathologies [15,21]. In this context, the majority of scientific contributions are on GLO1 dysfunctions and associated diseases [4,22–33]. This chapter gives deeper insight into the main features of the second the GS player (i.e., the GLO2) expressed in different organisms in order to give a knowledge on its main features herein considered relevant for understanding its mechanisms as well as its relation with important diseases.

3 Glyoxalase 2 enzymes (GLOs2)

Genomic metadata analyses on various organisms revealed extensive polymorphism in GS components gene encodings. For instance, in microbial and eukaryotic mammal cells, a single gene encodes for all GS-related enzymes [34–37], whereas in plants and yeasts, multiple genes were identified [38,39]. Specifically for the subject of this chapter, the GLOs2 cytosolic and mitochondrial isozymes in yeast and higher plants are encoded by separate genes, whereas in vertebrates, they are both encoded within a single gene [37]. Overall, plants offer the most variegate polymorphism for GLOs2, thus indicating the existence of tissue-specific GS-based MG detoxification systems. For instance, in rice, three Glo2 genes are present, in Glycine 12 [38]. To the best of our knowledge, the weed *Arabidopsis thaliana*, widely present in Eurasia and Africa, offers one of the clearest examples on GLOs2 heterogeneity and ongoing enzymatic functional diversification within an organism. *A. thaliana* genome encodes for five GLOs2 (i.e., GLX2-1, GLX2-4, and GLX2-5) [39–41] among which three isozymes are expressed at mitochondrial level [41]. Quite interestingly two out of the five *A. thaliana* expressed GLOs2 are devoid of the thiol-esterase activity (i.e., GLX2-1 and GLX2-3) [39,41,42]. The GLX2-5 is constitutively expressed under normal growing conditions, it possesses the thiol-esterase activity and shares high similarity (i.e., up to 88%) with the GLX2-1 isoform, which however is expressed only when abiotic stress sources are present. GLX2-1 lacks any glyoxalase activity, but it possesses a β-lactamase activity instead although plants do not produce any β-lactam containing moieties [43]. Therefore, GLX2-1 may be the result of a currently ongoing gene evolution leading to isozymes functionally diversified although originated from a common gene. Another example in *A. thaliana* is the GLX2-3 isoform also lacking glyoxalase activity, but it possesses sulfur dioxygenase activity, which is critical for seed development and sustainment of metabolic conditions involving elevated protein turnovers [44–50].

Significant differences between the GLOs2 enzymes expressed in parasites and hosts (i.e., humans) offer clever opportunities for drug development as in the case of *Plasmodium falciparum*. The first report of a GS pathway in apicomplexan parasites was on the *P. falciparum*-infected erythrocytes [51]. *P. falciparum* genome encodes for one GLO1 and two GLOs2, which were found in the cytosol (cGLO2) and in the apicoplast of the parasite (tGLO2), respectively [52]. The latter resulted expressed only in trophozoites and gametocytes [53]. The glucose uptake in infected red blood cells (RBCs) is significantly increased up to 75-fold when compared to healthy RBCs [54,55], and that is responsible in elevated production of MG in RBC affected patients [51]. In this context, the GS system by means of the tGLO2 isoform received a considerable attention as a selective and effective antimalarial target for drug development.

The human expressed GLO2 establishes the rate for the entire GS cycle in Fig. 2 and is the only known to date

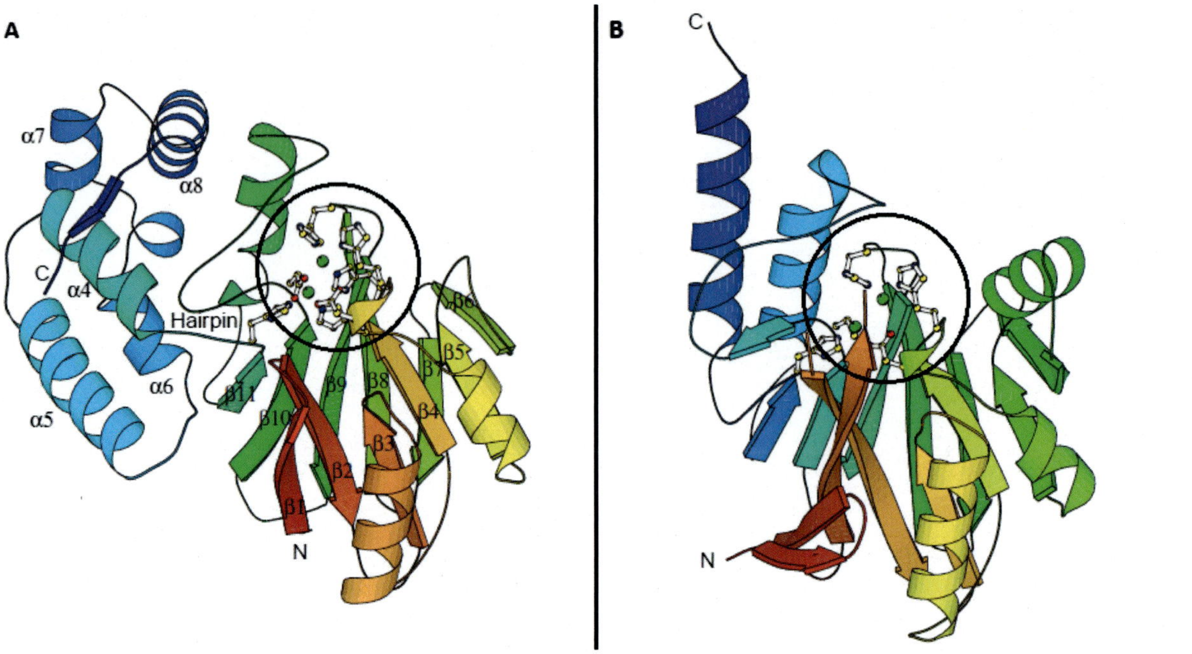

FIG. 3 (A) Overall schematic representation of the human expressed GLO2; (B) Similar view of the metallo-β-lactamase from *B. fragilis* [16] after scaling and superimposition with human GLO2. Color ramping according to residues starting with *red* at the *N*-terminus and finishing with *blue* at the *C*-terminus. In the circle is the binuclear metal binding site. Metal ions and the coordinating residues are represented by *balls and sticks* [62].

enzyme to able hydrolyze *S*-D-lactoyl-glutathione in humans [56]. From the genetic viewpoint, the human GLO2 does not belong to the glyoxalase gene family, instead it is evolutionarily related to the β-lactamases [57]. Although many aspects on the role of GLO2 in humans are still missing, overexpression of such an enzyme is reported in some forms of cancer [58,59] A specific example refers to prostate cancer cells lines, which do express GLO2 within their nuclei, whereas their normal counterparts are devoid [58,59].

4 Structural aspects of GLOs2

Alignment of primary structures of GLOs2 enzymes from various organisms revealed a highly conserved structural organization, which accounts for a β-lactamase-type and a hydroxy acyl glutathione hydrolase domain [35,60,61]. Various GLOs2 structures have been solved by means of X-ray crystallography experiments, and among them is the 29kDa human expressed GLO2 [35,61], which is herein discussed and compared to homolog GLOs2 expressed from other organisms. Cameron et al. reported two human GLO2 structures solved at 1.9 and 1.45Å, respectively: (i) GLO2 in adduct with acetate and cacodylate ions bound within the active site. Such ions are from the crystallization mixture used to perform the experiment; (ii) GLO2 in adduct with the glutathione thiolester substrate analog *S*-(*N*-hydroxy-*N*-bromophenylcarbamoyl)glutathione (HBPC-GSH) as substrate and obtained from classical soaking procedures [35].

An overall representation of the human expressed GLO2 enzyme is reported in Fig. 3A.

The high structural homology of the β-lactamase-type folding domain in human GLO2 with known metallo-β-lactamase enzymes is clear when superposition with the metallo-β-lactamase from *Bacteroides fragilis* [63] is operated (Fig. 3A and B) [35]. Such a result is not surprising as primary sequence alignments between the two enzymes accounted for 25% aminoacidic residues being identical [64,65]. Although important structural discrepancies between GLO2 and the metallo-β-lactamase from *B. fragilis* are present, the enzymatic cores structure and functionality are preserved [64,65]. Each human GLO2 monomer contains a binuclear metal binding site, which is located at a position topologically equivalent to that observed in the metallo-β-lactamase from *B. fragilis* and specifically at the top edge of the β sandwich cluster (Fig. 3A and B). One of the metal ions is less than 1Å from the rotation axis and the active site extends from this metal-binding site across the domain interface. Analysis of the metal content accounted for roughly 1.5 and 0.7mol per protein of Zn and Fe respectively [35]. Such a metal promiscuity is reasonably justified with metal exchange events, which occurred during any of the purification steps of the protein. Thus, the nature of Zn-dependent metalloenzyme for the human expressed GLO2 is confirmed also in consideration that the data on metal content are in line with those on metal bindings of its homolog from *A. thaliana* [40]. The coordination sphere of the two Zn(II) ions varies slightly

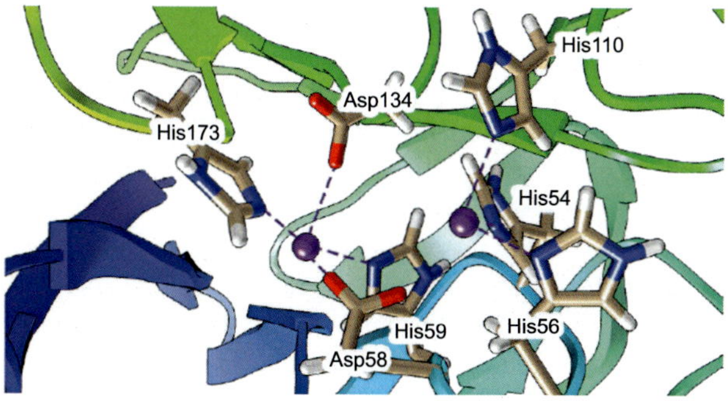

FIG. 4 Close-up of the Zn(II) binuclear center in human GLO2 active site (PDB ID 1QH5).

depending on the ligand bound within the active site. In all cases, seven amino acid residues and one water molecule interact directly with both Zn(II) ions, which are separated each other of roughly 3.5 Å and are bridged by a water molecule and an oxygen of Asp134. The assumed conformation of Asp134 is particularly favorable for the interaction with Zn(II)-1 rather than Zn(II)-2. The coordination sphere of Zn (II)-1 accounts for His54, His56, and His110, whereas for the Zn(II)-2 are the amino acids Asp58, His59, and His173 (Fig. 4).

High variabilities for the metal-binding sites between the human GLO2 and the metallo-β-lactamases were reported [41] and even among the latter significant differences are well-known and not discussed in this chapter for the sake of brevity [40,63,66–71]. Such a variability in structural organization for the metal binding sites in GLOs2 is reflected in the ion promiscuity as besides Zn(II), also Fe(II) and Mn(II) have been detected [35,36,40,41,57]. Relevant examples include GLOs2 from prokaryotes and eukaryotic cells from plants which contain a dual binding ion site for Fe(II) and Zn(II) [72]. In GLO2 from rice [73], *Leishmania infantum* [37], and *A. thaliana* [73], a Zn/Fe binuclear center was found. Specifically in *A. thaliana* cytoplasmic GLX2-2 Zn(II), Fe(II), or Mn(II) may be found [72,74–76].

The dimeric GLO2 in adduct HBPC-GSH, clearly revealed the ligand located in a groove on the surface of one monomer whereas on the associated monomeric counterpart only the thioesterase reaction product (i.e., GSH) was present. As for the second GLO2 dimeric crystal, a symmetric setting was reported as acetate and cacodylate were bound on both monomers [35].

Close inspection of the monomeric-GLO2/HBPC-GSH adduct shows the GSH tightly bound to the enzyme by means of its glycine and cysteine amino acid residues. Such a binding mode for GSH in human GLO2 differs from GLO1 as in the latter the additional interaction with the γ-glutamate residue is present [77] (Fig. 5).

The carboxylate of the glycine section interacts with Arg249, Lys252, and Lys143 residues. Arg249 is anchored in position by means of a hydrogen-bond with Asp253 and the carbonyl oxygen of Cys141. The hydrophobic part of the Lys252 side chain packs against the phenyl ring of Phe180. Lys143 is situated a little further from the carboxylate, and instead hydrogen bonds to the carbonyl oxygen of the γ-glutamate are established. The carbonyl oxygen of Cys141 can interact with GSH by means of a water molecule acting as bridge. Further stabilization of the GSH is ensured by interactions with Tyr175 and Tyr145 (Fig. 5). As for the HBPC ligand section, the main interaction with the enzyme is given by its phenyl moiety, which is involved in π-stacking with His56, and thus intercalates between the two metal ions. The occurrence of loose contacts between the HBPC section with the enzyme was revealed in crystallography from the less well-defined electron densities. The associated monomeric unit bound only to GSH showed the ligand assuming a very close conformation to its counterpart HBPC-GSH and thus suggesting that the looseness detected for the HBPC section may attributed to a hydrolysis reaction taking place associated to low affinity of the HBPC substrate for the human GLO2 [78].

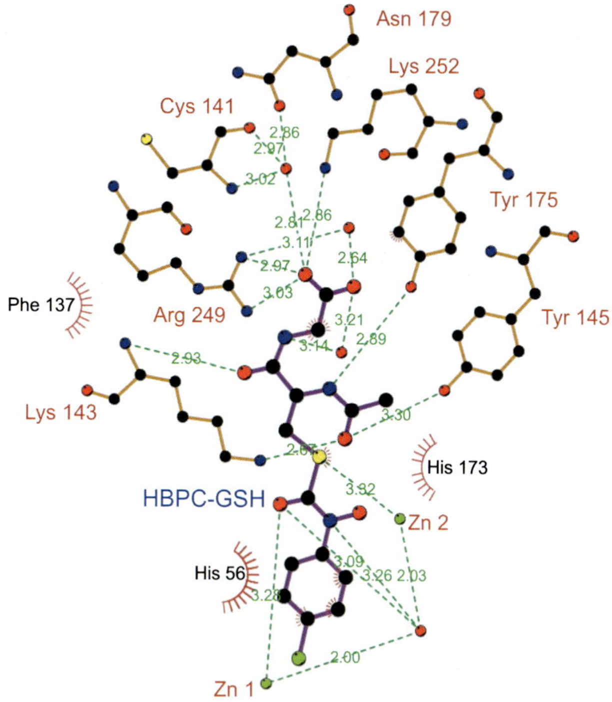

FIG. 5 Schematic diagram of the interactions in GLO2/HBPC-GSH adducts. Atoms and residues involved in hydrophobic contacts are shown as fanned by *red dashes* [35].

The second dimeric and symmetric GLO2/acetate-cacodylate adduct revealed that the ions placed within the enzyme do have a very similar conformation to the natural substrate GSH and its derivative with HBPC. Thus, the favorable electronic conformation explains the stability of such an adduct [35].

GLOs2 are a very efficient enzymes with k_{cat}/KM ratios for the *S*-D-lactoyl-glutathione as substrate close to the diffusion limit values and very limited differences among humans, plants and yeast [79–82] and large optimal pH values (i.e., 6.8–7.5) [83–85] and sensitive to variable concentrations of NaCl [82,83]. To date, GLOs2 mechanism was studied on human, *A. thaliana* and *P. falciparum* expressed enzymes and they all fit within a Theorell-Chance model, which accounts for an immediate release of the first product followed by the second which however dissociates slower. Such a mechanism clearly implies the substrate binding determines the rate-limiting step [61,76,83] In agreement with such a model, a theoretical study by Chen et al. gives support to the commonly accepted reaction mechanism for GLOs2 reported in Scheme 2 [86].

The proposed mechanism starts with the nucleophilic attach of the metal shared HO^- moiety toward the carbonyl of the ligand (i.e., *S*-D-lactoyl-glutathione) to form the tetrahedral intermediate, thus with the enzyme and the substrate covalently bound. Collapse of the adduct by means of the C—S bond disconnection results in the formation of D-lactic acid and GSH in the anionic form. Finally, the insertion of a water molecule allows: (i) displacement and thus release of D-lactic acid; (ii) release of GSH upon protonation; and (iii) regeneration of the enzyme active site with introduction of the HO^- moiety shared between the metal ions [86]. The suggested mechanism for GLOs2 based on homonuclear divalent Zn(II) ions, such as the human isoform, endows the two metals with additional distinct roles. Specifically, the Zn(II)-2 is involved in orienting the electrophilic carbonyl moiety of the substrate for HO^- attack and stabilizing the negatively charged tetrahedral intermediate, whereas the Zn(II)-1 drives the tetrahedral intermediate to collapse correctly by interaction with the electron rich sulfur atom in GSH. Experimental findings on GLOs2 from in vivo specimens containing promiscuous Fe(II)/Zn(II)/Ni(II) occupied binding sites, agree with the role of Zn(II)-1 [72,74,76,87,88]. Overall, the effect of different metal ions on the catalytic activity of GLOs2 accounts for enzymes able to maintain elevated enzymatic performances and that makes them extremely adaptive to the environmental conditions since the nature of the metals included in GLOs2 is exogenous [72,74,76,87,88]. In this context is the seminal contribution of Campos-Bermudez et al. on the metal-selective GLO2 products obtained when *S. thyphimurium* was exposed to Mn(II) sources [89], thus paving the way

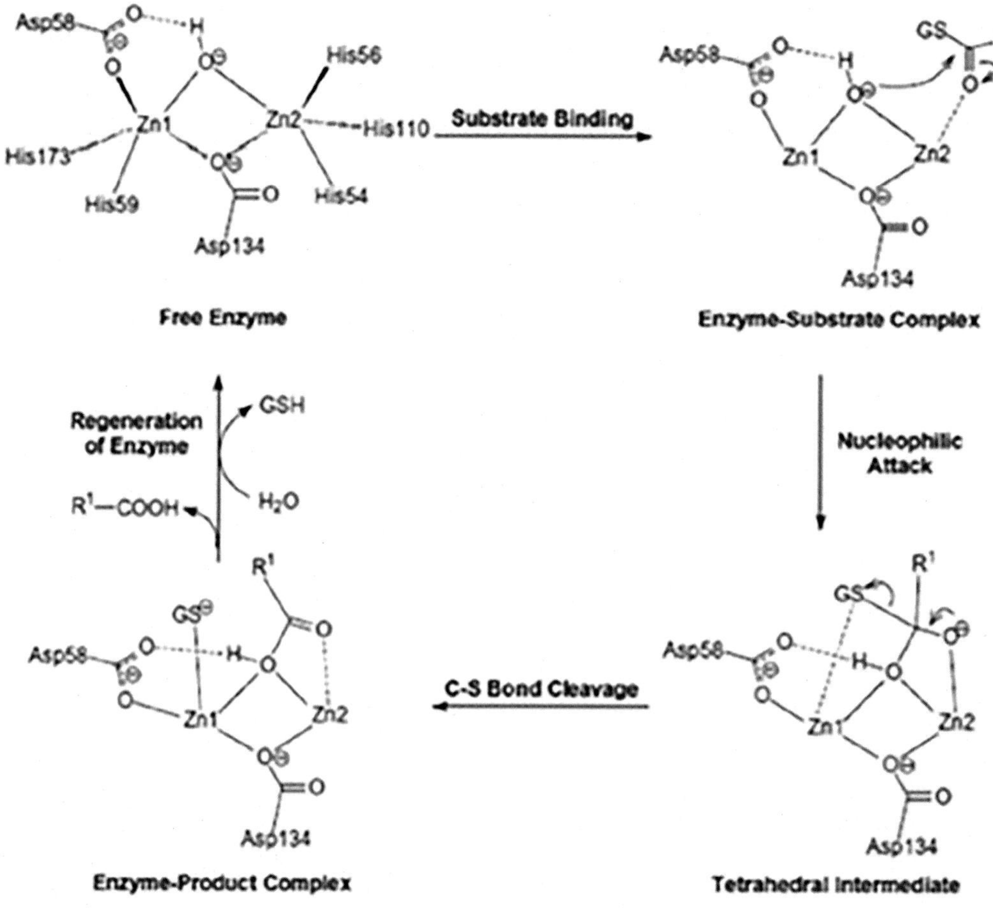

SCHEME 2 Proposed catalytic cycle for GLOs2 [86].

for a plausible way to modulate a key enzyme of an organism pathogenic for humans.

5 GLOs2 biological implications

Deep interests in understanding human GLOs2 signaling pathways deputed to regulate the expression of such enzymes as well as those triggered by them are currently ongoing. A contribution particularly relevant in this field was the evidence that in human breast cancer MCF7 cell lines, GLOs2 are up regulated by the genes p63 and p73 through a specific responsive element located in the intron 1 of *Glo2* gene [90]. In addition, the same Authors demonstrated in the same cell lines that the cytosolic expressed GLO2 and not the mitochondrial is responsible for inhibition of the apoptosis induced by overproduction of MG [90]. This is the first contribution that gives solid support to the p63/p73-Glo2 axis, which might have been a novel pathway in human carcinogenesis [23,91,92] and paved the way for estrogen-regulated tumors through the PTEN/PI3K/AKT/mTOR signaling [59,93,94]. The above contributions within the Medicinal Chemistry field triggered important efforts toward the identification of GLOs2 modulators of synthetic [95–97] or natural origin [98] for the control of cellular metabolism or structural protein assembly [99–105] in chronic diseases. A pivotal role of GLOs2 in cellular biology is represented by regulation of posttranslational modifications (PTMs) such as the *S*-glutathionylation. GLOs2 was proved to catalyze *S*-glutathionylation reactions on exposed Cys residues on specific substrates such as the actin and malate dehydrogenase (MDH) proteins. *S*-glutathionylation on Cys374 in actin was proved to reduce the ability of globular G-actin to polymerize into filamentous F-actin [106–109]. Modification of the dynamics of actin de-/polymerization is well known to affect cellular cytoskeletal rearrangements, which do regularly happen during physiological cellular cycle stages and/or when pathological conditions take place [106,110–114]. The effects of GLO2 catalyzed *S*-glutathionylation on MDH activity is quite evident as it resulted correlated with the correct induced supramolecular organization of the monomeric enzyme up to tetramers. On the contrary, exposure of MDH to the *S*-glutathionylating agent GSSG, thus non-GLO2 catalyzed, leads to nonspecific *S*-glutathionylation on all exposed Cys residues, which trigger the formation of nonfunctional quaternary complexes [115–117].

S-acetylation of cysteine thiol residues is another PTM, which deeply involves GLOs2. Specifically, the *S*-acetylated thiols do act as acetyl transfer agents for proximal Lys residues [118]. The *S*-acetylation activity of GLOs2 is of importance at mitochondrial level as it limits the AcCoA-dependent *S*-acetylation reactions with suppression of indiscriminate acetyl transfers and modulation of important metabolic transformations (i.e., Krebs cycle) or epigenetic targets [119–122].

GLOs2 in mitochondrial mammal cells is involved in modulation of the lactoylation of Lys residues by control of the concentration of the GLO1's produced product (i.e., the *S*-D-lactoyl glutathione). The role of GLOs2 in such a PTM event was firstly reported in cell lines lacking GLOs2 enzymes, and it was demonstrated that the accumulation of *S*-D-lactoyl glutathione determined indiscriminate and elevated levels of lactoyl-Lys derivatives with epigenetic target modifications [123]. GLOs2, being the only enzymes in mammalian cells able to hydrolyze *S*-D-lactoyl glutathione, under normal conditions prevent excessive accumulation of substrate and do maintain it within physiologically tolerable concentrations by conversion into D-lactate. The transient increase of *S*-D-lactoyl glutathione is however intense enough to promote lactoylations (i.e., deactivation) on key enzymes involved in glycolysis [124]. This is a clear example of GLOs2-catalyzed PTM with retroactive metabolic control with cytoprotective effects [125].

6 Conclusions

Dysfunctions of GSs are well known as cause of diseases, and in humans it is established that they contribute and sustain the development of chronic diseases of relevance, usually of the metabolic type. State of the art in the field of GLOs2 accounts for a growing interest on such enzyme with particular focus on the structural and kinetic features in relation to diseases of the chronic type and their implication on metabolisms. Many aspects on the functionality of such enzymes are yet to be understood and/or discovered. The majority of scientific efforts with the aim to figure out appropriate ways of manipulation and control for biomedical purposes are in this context.

References

[1] Phillips SA, Thornalley PJ. The formation of methylglyoxal from triose phosphates. Investigation using a specific assay for methylglyoxal. Eur J Biochem 1993;212:101–5.

[2] Sousa Silva M, Gomes RA, Ferreira AEN, Ponces Freire A, Cordeiro C. The glyoxalase pathway: the first hundred years... and beyond. Biochem J 2013;453:1–15.

[3] Thornalley PJ. Modification of the glyoxalase system in human red blood cells by glucose in vitro. Biochem J 1998;254:751–5.

[4] Maessen DEM, Stehouwer CDA, Schalkwijk CG. The role of methylglyoxal and the glyoxalase system in diabetes and other age-related diseases. Clin Sci 2015;128:839–61.

[5] Rabbani N, Thornalley PJ. Measurement of methylglyoxal by stable isotopic dilution analysis LC-MS/MS with corroborative prediction in physiological samples. Nat Protoc 2014;9:1969–79.

[6] Kalapos MP. Where does plasma methylglyoxal originate from? Diabetes Res Clin Pract 2013;99:260–71.

[7] Ahmed N, Thornalley PJ, Dawczynski J, Franke S, Strobel J, Stein G, Haik JM. Methylglyoxal-derived hydroimidazolone advanced glycation end-products of human lens proteins. Invest Ophthalmol Vis Sci 2003;12:5287–92.
[8] Ahmed N, Thornalley PJ. Peptide mapping of human serum albumin modified minimally by methylglyoxal in vitro and in vivo. Ann N Y Acad Sci 2005;1043:260–6.
[9] Watanabe H, Tanase S, Nakajou K, Maruyama T, Kragh-Hansen U, Otagiri M. Role of arg-410 and tyr-411 in human serum albumin for ligand binding and esterase-like activity. Biochem J 2000;349:813–9.
[10] Faure P, Troncy L, Lecomte M, Wiernsperger N, Lagarde M, Ruggiero D, Halimi S. Albumin antioxidant capacity is modified by methylglyoxal. Diabetes Metab 2005;31:169–77.
[11] Rabbani N, Thornalley PJ. Methylglyoxal, glyoxalase 1 and the dicarbonyl proteome. Amino Acids 2012;42:1133–42.
[12] Thornalley PJ, Waris S, Fleming T, Santarius T, Larkin SJ, Winklhofer-Roob BM, Stratton MR, Rabbani N. Imidazopurinones are markers of physiological genomic damage linked to DNA instability and glyoxalase 1-associated tumour multidrug resistance. Nucleic Acids Res 2010;38:5432–42.
[13] Rabbani N, Thornalley PJ. Dicarbonyl proteome and genome damage in metabolic and vascular disease. Biochem Soc Trans 2014;42:425–32.
[14] Izaguirre G, Kikonyogo A, Pietruszko R. Methylglyoxal as substrate and inhibitor of human aldehyde dehydrogenase: comparison of kinetic properties among the three isozymes. Comp Biochem Physiol B Biochem Mol Biol 1998;119:747–54.
[15] Baba SP, Barski OA, Ahmed Y, O'Toole TE, Conklin DJ, Bhatnagar A, Srivastava S. Reductive metabolism of AGE precursors: a metabolic route for preventing AGE accumulation in cardiovascular tissue. Diabetes 2009;58:2486–97.
[16] Lee JY, Song J, Kwon K, Jang S, Kim C, Baek K, Kim J, Park C. Human DJ-1 and its homologs are novel glyoxalases. Hum Mol Genet 2012;21:3215–25.
[17] Ghosh A, Kushwaha HR, Hasan MR, Pareek A, Sopory SK, Singla-Pareek SL. Presence of unique glyoxalase III proteins in plants indicates the existence of shorter route for methylglyoxal detoxification. Sci Rep 2016;6:18358.
[18] Kwon K, Choi D, Hyun JK, Jung HS, Baek K, Park C. Novel glyoxalases from *Arabidopsis thaliana*. FEBS J 2013;280:3328–39.
[19] Chaudhuri J, Bains Y, Guha S, Kahn A, Hall D, Bose N, Gugliucci A, Kapahi P. The role of advanced glycation end products in aging and metabolic diseases: bridging association and causality. Cell Metab 2018;28:337–52.
[20] Rabbani N, Xue M, Thornalley PJ. Dicarbonyl stress, protein glycation and the unfolded protein response. Glycoconj J 2021;38:331–40.
[21] Farrera DO, Galligan JJ. The human glyoxalase gene family in health and disease. Chem Res Toxicol 2022;10:1766–76.
[22] Thornalley PJ. Glyoxalase I—structure, function and a critical role in the enzymatic defence against glycation. Biochem Soc Trans 2003;31:1343–8.
[23] Antognelli C, Talesa VN. Glyoxalases in urological malignancies. Int J Mol Sci 2018;19:415.
[24] Arai M, Nihonmatsu-Kikuchi N, Itokawa M, Rabbani N, Thornalley PJ. Measurement of glyoxalase activities. Biochem Soc Trans 2014;42:491–4.
[25] Giacco F, Du X, D'Agati VD, Milne R, Sui G, Geoffrion M, Brownlee M. Knockdown of glyoxalase 1 mimics diabetic nephropathy in nondiabetic mice. Diabetes 2014;63291–9.
[26] Wu L, Juurlink BH. Increased methylglyoxal and oxidative stress in hypertensive rat vascular smooth muscle cells. Hypertension 2002;39:809–14.
[27] Wang X, Desai K, Clausen JT, Wu L. Increased methylglyoxal and advanced glycation end products in kidney from spontaneously hypertensive rats. Kidney Int 2004;66:2315–21.
[28] Wang X, Desai K, Chang T, Wu L. Vascular methylglyoxal metabolism and the development of hypertension. J Hypertens 2005;23:1565–73.
[29] Antognelli C, Trapani E, Monache SD, Perrelli A, Fornelli C, Retta F, Cassoni P, Talesa VN, Retta SF. Data in support of sustained upregulation of adaptive redox homeostasis mechanisms caused by KRIT1 loss-of-function. Data Brief 2018;16:929–38.
[30] Distler MG, Plant LD, Sokoloff G, Hawk AJ, Aneas I, Wuenschell GE, Termini J, Meredith SC, Nobrega MA, Palmer AA. Glyoxalase 1 increases anxiety by reducing GABAA receptor agonist methylglyoxal. J Clin Invest 2012;122:2306–15.
[31] Tatone C, Heizenrieder T, Emidio GD, Treffon P, Amicarelli F, Seidel T, Eichenlaub-Ritter U. Evidence that carbonyl stress by methylglyoxal exposure induces DNA damage and spindle aberrations, affects mitochondrial integrity in mammalian oocytes and contributes to oocyte ageing. Hum Reprod 2011;26:1843–59.
[32] Tatone C, Eichenlaub-Ritter U, Amicarelli F. Dicarbonyl stress and glyoxalases in ovarian function. Biochem Soc Trans 2014;42:433–8.
[33] Antognelli C, Mancuso F, Frosini R, Arato I, Calvitti M, Calafiore R, Talesa VN, Luca G. Testosterone and follicle stimulating hormone-dependent glyoxalase 1 up-regulation sustains the viability of porcine sertoli cells through the control of hydroimidazolone- and argpyrimidine-mediated NF-κB pathway. Am J Pathol 2018;188:2553–63.
[34] Cordell PA, Futers TS, Grant PJ, Pease RJ. The human hydroxyacylglutathione hydrolase (HAGH) gene encodes both cytosolic and mitochondrial forms of glyoxalase II. J Biol Chem 2004;279:28653–61.
[35] Cameron AD, Ridderstrom M, Olin B, Mannervik B. Crystal structure of human glyoxalase II and its complex with a glutathione thiolester substrate analogue. Structure 1999;7:1067–78.
[36] O'Young J, Sukdeo N, Honek JF. *Escherichia coli* glyoxalase II is a binuclear zinc-dependent metalloenzyme. Arch Biochem Biophys 2007;459:20–6.
[37] Silva MS, Barata L, Ferreira AE, Romao S, Tomas AM, Freire AP, Cordeiro C. Catalysis and structural properties of *Leishmania infantum* glyoxalase II: trypanothione specificity and phylogeny. Biochemistry 2008;47:195–204.
[38] Ghosh A, Islam T. Genome-wide analysis and expression profiling of glyoxalase gene families in soybean (*Glycine max*) indicate their development and abiotic stress specific response. BMC Plant Biol 2016;16:87.
[39] Mustafiz A, Singh AK, Pareek A, Sopory SK, Singla-Pareek SL. Genome-wide analysis of rice and *Arabidopsis* identifies two glyoxalase genes that are highly expressed in abiotic stresses. Funct Integr Genom 2011;11:293–305.
[40] Crowder MW, Maiti MK, Banovic L, Makaroff CA. Glyoxalase II from *A. thaliana* requires Zn(II) for catalytic activity. FEBS Lett 1997;418:351–4.

[41] Marasinghe GPK, Sander IM, Bennett B, Periyannan G, Yang KW, Makaroff CA, Crowder MW. Structural studies on a mitochondrial glyoxalase II. J Biol Chem 2005;280(49):40668–75.

[42] Yadav SK, Singla-Pareek SL, Kumar M, Pareek A, Saxena M, Sarin NB, Sopory SK. Characterization and functional validation of glyoxalase II from rice. Protein Expr Purif 2007;51:126–32.

[43] Devanathan S, Erban A, Perez-Torres Jr R, Kopka J, Makaroff CA. *Arabidopsis thaliana* glyoxalase 2-1 is required during abiotic stress but is not essential under normal plant growth. PLoS One 2014;9: e95971.

[44] Holdorf MM, Owen HA, Lieber SR, Yuan L, Adams N, Dabney-Smith C, Makaroff CA. *Arabidopsis* ETHE1 encodes a sulfur dioxygenase that is essential for embryo and endosperm development. Plant Physiol 2012;160:226–36.

[45] Bray EA, Bailey-Serres J, Weretilnyk E. Responses to abiotic stresses. In: Gruissem W, Buchannan B, Jones R, editors. Biochemistry and molecular biology of plants. Rockville, MD: American Society of Plant Physiologists; 2000. p. 1158–249.

[46] Apel K, Hirt H. Reactive oxygen species: metabolism, oxidative stress, and signal transduction. Annu Rev Plant Biol 2004;55:373–99.

[47] Hasanuzzaman M, Bhuyan M, Zulfiqar F, Raza A, Mohsin SM, Mahmud JA, Fujita M, Fotopoulos V. Reactive oxygen species and antioxidant defense in plants under abiotic stress: revisiting the crucial role of a universal defense regulator. Antioxidants 2020;9:681.

[48] Hoque TS, Hossain MA, Mostofa MG, Burritt DJ, Fujita M, Tran LS. Methylglyoxal: an emerging signaling molecule in plant abiotic stress responses and tolerance. Front Plant Sci 2016;7:1341.

[49] Hoque TS, Uraji M, Ye W, Hossain MA, Nakamura Y, Murata Y. Methylglyoxal-induced stomatal closure accompanied by peroxidase-mediated ROS production in *Arabidopsis*. J Plant Physiol 2012;169:979–86.

[50] Kaur C, Singla-Pareek SL, Sopory SK. Glyoxalase and methylglyoxal as biomarkers for plant stress tolerance. Crit Rev Plant Sci 2014;33:429–56.

[51] Vander Jagt DL, Hunsaker LA, Campos NM, Baack BR. D-lactate production in erythrocytes infected with *Plasmodium falciparum*. Mol Biochem Parasitol 1990;42:277–84.

[52] Urscher M, Przyborski JM, Imoto M, Deponte M. Distinct subcellular localization in the cytosol and apicoplast, unexpected dimerization and inhibition of *Plasmodium falciparum* glyoxalases. Mol Microbiol 2010;76:92–103.

[53] Le Roch KG, Johnson JR, Florens L, Zhou Y, Santrosyan A, Grainger M, Yan SF, Williamson KC, Holder AA, Carucci DJ, et al. Global analysis of transcript and protein levels across the *Plasmodium falciparum* life cycle. Genome Res 2004;14:2308–18.

[54] Sherman IW. Biochemistry of *Plasmodium* (malarial parasites). Microbiol Rev 1979;43:453–95.

[55] Kappe SH, Vaughan AM, Boddey JA, Cowman AF. That was then but this is now: malaria research in the time of an eradication agenda. Science 2010;328:862–6.

[56] Perez C, Barkley-Levenson AM, Dick BL, Glatt PF, Martinez Y, Siegel D, Momper JD, Palmer AA, Cohen SM. Metal-binding pharmacophore library yields the discovery of a glyoxalase 1 inhibitor. J Med Chem 2019;62:1609–25.

[57] Campos-Bermudez VA, Leite NR, Krog R, Costa-Filho AJ, Soncini FC, Oliva G, Vila AJ. Biochemical and structural characterization of *Salmonella typhimurium* glyoxalase II: new insights into metal ion selectivity. Biochemistry 2007;46:11069–79.

[58] Antognelli C, Ferri I, Bellezza G, Siccu P, Love HD, Talesa VN, Sidoni A. Glyoxalase 2 drives tumorigenesis in human prostate cells in a mechanism involving androgen receptor and p53–p21 axis. Mol Carcinog 2017;56:2112–26.

[59] Talesa VN, Ferri I, Bellezza G, Love HD, Sidoni A, Antognelli C. Glyoxalase 2 is involved in human prostate cancer progression as part of a mechanism driven by PTEN/PI3K/AKT/mTOR signaling with involvement of PKM2 and ERa. Prostate 2017;77:196–210.

[60] Neuwald AF, Liu JS, Lipman DJ, Lawrence CE. Extracting protein alignment models from the sequence database. Nucleic Acids Res 1997;25:1665–77.

[61] Uotila L. Purification and characterization of S-2-hydroxyacylglutathione hydrolase (glyoxalase II) from human liver. Biochemistry 1973;12:3944–51.

[62] Kraulis PJ. MolScript: a program to produce both detailed and schematic plots of protein structures. J Appl Cryst 1991;24:946–50.

[63] Concha NO, Rasmussen BA, Bush K, Herzberg O. Crystal structure of the wide-spectrum binuclear zinc beta-lactamase from *Bacteroides fragilis*. Structure 1996;4:823–36.

[64] Carfi A, Pares S, Duée E, Galleni M, Duez C, Frère JM, Dideberg O. The 3-D structure of a zinc metallo-beta-lactamase from *Bacillus cereus* reveals a new type of protein fold. EMBO J 1995;14:4914–21.

[65] Carfi A, Duée E, Paul-Soto R, Galleni M, Frère JM, Dideberg O. X-ray structure of the ZnII beta-lactamase from *Bacteroides fragilis* in an orthorhombic crystal form. Acta Crystallogr D 1998;54:45–57.

[66] Ullah JH, Walsh TR, Taylor IA, Emery DC, Verma CS, Gamblin SJ, Spencer J. The crystal structure of the L1 metallo-beta-lactamase from *Stenotrophomonas maltophilia* at 1.7 Å resolution. J Mol Biol 1998;284:125–36.

[67] Alberts IL, Nadassy K, Wodak SJ. Analysis of zinc binding sites in protein crystal structures. Prot Sci 1998;7:1700–16.

[68] Vallee BL, Auld DS. New perspectives on zinc biochemistry: cocatalytic sites in multi-zinc enzymes. Biochemistry 1993;32:6493–500.

[69] Sträter N, Lipscomb WN. Transition state analogue Lleucinephosphonic acid bound to bovine lens aminopeptidase: X-ray structure at 1.65 Å resolution in a new crystal form. Biochemistry 1995;34:9200–10.

[70] Sträter N, Klabunde T, Tucker P, Witzel H, Krebs B. Crystal structure of a purple acid-phosphatase containing a dinuclear Fe (III)-Zn(II) active-site. Science 1995;268:1489–92.

[71] Fabiane SM, Sohi MK, Wan T, Payne DJ, Bateson JH, Mitchell T, Sutton BJ. Crystal structure of the zinc-dependent betalactamase from *Bacillus cereus* at 1.9 Å resolution: binuclear active site with features of a mononuclear enzyme. Biochemistry 1998;37:12404–11.

[72] Schilling O, Wenzel N, Naylor M, Vogel A, Crowder M, Makaroff C, Meyer-Klaucke W. Flexible metal binding of the metallo-beta-lactamase domain: glyoxalase II incorporates iron, manganese, and zinc in vivo. Biochemistry 2003;42:11777–86.

[73] Ghosh A, Pareek A, Sopory SK, Singla-Pareek SL. A glutathione responsive rice glyoxalase II, OsGLYII-2, functions in salinity adaptation by maintaining better photosynthesis efficiency and anti-oxidant pool. Plant J 2014;80:93–105.

[74] Limphong P, McKinney RM, Adams NE, Makaroff CA, Bennett B, Crowder MW. The metal ion requirements of *Arabidopsis thaliana* Glx2-2 for catalytic activity. J Biol Inorg Chem 2010;15:249–58.

[75] Wenzel NF, Carenbauer AL, Pfiester MP, Schilling O, Meyer-Klaucke W, Makaroff CA, Crowder MW. The binding of iron and zinc to glyoxalase II occurs exclusively as di-metal centers and is unique within the metallo-beta-lactamase family. J Biol Inorg Chem 2004;9:429–38.

[76] Zang TM, Hollman DA, Crawford PA, Crowder MW, Makaroff CA. *Arabidopsis* glyoxalase II contains a zinc/iron binuclear metal center that is essential for substrate binding and catalysis. J Biol Chem 2001;276:4788–95.

[77] Cameron AD, Ridderström M, Olin B, Kavarana MJ, Creighton DJ, Mannervik B. The reaction mechanism of glyoxalase I explored by an X-ray crystallographic analysis of the human enzyme in complex with a transition state analogue. Biochemistry 1999;38:13480–90.

[78] Murthy NS, Bakeris T, Kavarana MJ, Hamilton DS, Lan Y, Creighton DJ. S-(N-aryl-N-hydroxycarbamoyl)glutathione derivatives are tight-binding inhibitors of glyoxalase I and slow substrates for glyoxalase II. J Med Chem 1994;37:2161–6.

[79] Guha MK, Vander Jagt DL, Creighton DJ. Diffusion-dependent rates for the hydrolysis reaction catalyzed by glyoxalase II from rat erythrocytes. Biochemistry 1988;27:8818–22.

[80] Bito A, Haider M, Hadler I, Breitenbach M. Identification and phenotypic analysis of two glyoxalase II encoding genes from *Saccharomyces cerevisiae*, GLO2 and GLO4, and intracellular localization of the corresponding proteins. J Biol Chem 1997;272:21509–19.

[81] Ridderstrom M, Mannervik B. Molecular cloning and characterization of the thiolesterase glyoxalase II from *Arabidopsis thaliana*. Biochem J 1997;322:449–54.

[82] Ridderstrom M, Saccucci F, Hellman U, Bergman T, Principato G, Mannervik B. Molecular cloning, heterologous expression, and characterization of human glyoxalase II. J Biol Chem 1996;271:319–23.

[83] Urscher M, Deponte M. *Plasmodium falciparum* glyoxalase II: Theorell-Chance product inhibition patterns, rate-limiting substrate binding via Arg(257)/Lys(260), and unmasking of acid-base catalysis. Biol Chem 2009;390:1171–83.

[84] Allen RE, Lo TW, Thornalley PJ. Purification and characterisation of glyoxalase II from human red blood cells. Eur J Biochem 1993;213:1261–7.

[85] Ball JC, Vander Jagt DL. S-2-hydroxyacylglutathione hydrolase (glyoxalase II): active-site mapping of a nonserine thiolesterase. Biochemistry 1981;20:899–905.

[86] Chen SL, Fang WH, Himo F. Reaction mechanism of the binuclear zinc enzyme glyoxalase II—A theoretical study. J Inorg Biochem 2009;103:274–81.

[87] Saxena M, Bisht R, Roy SD, Sopory SK, Bhalla-Sarin N. Cloning and characterization of a mitochondrial glyoxalase II from *Brassica juncea* that is upregulated by NaCl, Zn, and ABA. Biochem Biophys Res Commun 2005;336:813–9.

[88] Limphong P, Crowder MW, Bennett B, Makaroff CA. *Arabidopsis thaliana* GLX2-1 contains a dinuclear metal binding site, but is not a glyoxalase 2. Biochem J 2009;417:323–30.

[89] Campos-Bermudez VA, Moran-Barrio J, Costa-Filho AJ, Vila AJ. Metal-dependent inhibition of glyoxalase II: a possible mechanism to regulate the enzyme activity. J Inorg Biochem 2010;104:726–31.

[90] Xu Y, Chen X. Glyoxalase II, a detoxifying enzyme of glycolysis byproduct methylglyoxal and a target of p63 and p73, is a pro-survival factor of the p53 family. J Biol Chem 2006;281:26702–13.

[91] Alexandrova EM, Moll UM. Role of p53 family members p73 and p63 in human hematological malignancies. Leuk Lymphoma 2012;53:2116–29.

[92] Moll UM, Slade N. p63 and p73: roles in development and tumor formation. Mol Cancer Res 2004;2:371–86.

[93] Rulli A, Antognelli C, Prezzi E, Baldracchini F, Piva F, Giovannini E, Talesa V. A possible regulatory role of 17beta-estradiolband tamoxifen on glyoxalase I and glyoxalase II genes expression in MCF7 and BT20 human breast cancer cells. Breast Cancer Res Treat 2006;96:187–96.

[94] Antognelli C, Del Buono C, Baldracchini F, Talesa V, Cottini E, Brancadoro C, Zucchi A, Marini E. Alteration of glyoxalase genes expression in response to testosterone in LNCaP and PC3 human prostate cancer cells. Cancer Biol Ther 2007;6:1880–8.

[95] Bush PE, Norton SJ. S-(nitrocarbobenzoxy)glutathiones: potent competitive inhibitors of mammalian glyoxalase II. J Med Chem 1985;28:828–30.

[96] Chyan MK, Elia AC, Principato GB, Giovannini E, Rosi G, Norton SJ. S-fluorenylmethoxycarbonyl glutathione and diesters: inhibition of mammalian glyoxalase II. Enzym Protein 1994;48:164–73.

[97] Hsu YR, Norton SJ. S-carbobenzoxyglutathione: a competitive inhibitor of mammalian glyoxalase II. J Med Chem 1983;26:1784–5.

[98] Antognelli C, Frosini R, Santolla MF, Peirce MJ, Talesa VN. Oleuropein-induced apoptosis is mediated by mitochondrial glyoxalase 2 in NSCLC A549 cells: a mechanistic inside and a possible novel nonenzymatic role for an ancient enzyme. Oxid Med Cell Longev 2019;2019, 8576961.

[99] Gillespie E, Lichtenstein LM. Histamine release from human leukocytes: studies with deuterium oxide, colchicine, and cytochalasin B. J Clin Invest 1972;51:2941–7.

[100] Norton SJ, Elia AC, Chyan MK, Gillis G, Frenzel C, Principato GB. Inhibitors and inhibition studies of mammalian glyoxalase II activity. Biochem Soc Trans 1993;21:545–9.

[101] Clellan JD, Thornalley PJ. The potentiation of GTP-dependent assembly of microtubules by S-D-lactoylglutathione. Biochem Soc Trans 1993;21:160S.

[102] Di Simplicio P, Vignani R, Talesa V, Principato G. Evidence of glyoxalase II activity associated with microtubule polymerization in bovine brain. Pharmacol Res 1990;22:172.

[103] Chen W, Seefeldt T, Young A, Zhang X, Zhao Y, Ruffolo J, Kaushik RS, Guan X. Microtubule S-glutathionylation as a potential approach for antimitotic agents. BMC Cancer 2012;12:245.

[104] Carletti B, Passarelli C, Sparaco M, Tozzi G, Pastore A, Bertini E, Piemonte F. Effect of protein glutathionylation on neuronal cytoskeleton: a potential link to neurodegeneration. Neuroscience 2011;192:285–94.

[105] Gillespie E. Effects of S-lactoylglutathione and inhibitors of glyoxalase I on histamine release from human leukocytes. Nature 1979;277:135–7.

[106] Wang J, Boja ES, Tan W, Tekle E, Fales HM, English S, Mieyal JJ, Chock PB. Reversible glutathionylation regulates actin polymerization in A431 cells. J Biol Chem 2001;276:47763–6.

[107] Dalle-Donne I, Giustarini D, Rossi R, Colombo R, Milzani A. Reversible S-glutathionylation of Cys 374 regulates actin filament formation by inducing structural changes in the actin molecule. Free Radic Biol Med 2003;34:23–32.

[108] Dalle-Donne I, Rossi R, Giustarini D, Colombo R, Milzani A. Actin S-glutathionylation: evidence against a thiol-disulphide exchange mechanism. Free Radic Biol Med 2003;35:1185–93.

[109] Lassing I, Schmitzberger F, Bjornstedt M, Holmgren A, Nordlund P, Schutt CE, Lindberg U. Molecular and structural basis for redox regulation of beta-actin. J Mol Biol 2007;370:331–48.

[110] Xu Q, Huff LP, Fujii M, Griendling KK. Redox regulation of the actin cytoskeleton and its role in the vascular system. Free Radic Biol Med 2017;109:84–107.

[111] Cha SJ, Kim H, Choi HJ, Lee S, Kim K. Protein glutathionylation in the pathogenesis of neurodegenerative diseases. Oxid Med Cell Longev 2017;2017, 2818565.

[112] Kruyer A, Ball LE, Townsend DM, Kalivas PW, Uys JD. Post-translational S-glutathionylation of cofilin increases actin cycling during cocaine se. PLoS ONE 2019;14:e0223037.

[113] Pastore A, Tozzi G, Gaeta LM, Bertini E, Serafini V, Di Cesare S, Bonetto V, Casoni F, Carrozzo R, Federici G, et al. Actin glutathionylation increases in fibroblasts of patients with Friedreich's ataxia: a potential role in the pathogenesis of the disease. J Biol Chem 2003;278:42588–95.

[114] Varland S, Vandekerckhove J, Drazic A. Actin post-translational modifications: the cinderella of cytoskeletal control. Trends Biochem Sci 2019;44:502–16.

[115] Galeazzi R, Laudadio E, Falconi E, Massaccesi L, Ercolani L, Mobbili G, Minnelli C, Scire A, Cianfruglia L, Armeni T. Protein-protein interactions of human glyoxalase II: findings of a reliable docking protocol. Org Biomol Chem 2018;16:5167–77.

[116] Kojer K, Riemer J. Balancing oxidative protein folding: the influences of reducing pathways on disulfide bond formation. Biochim Biophys Acta 2014;1844:1383–90.

[117] Iglesias AA, Andreo CS. NADP-dependent malate dehydrogenase (decarboxylating) from sugar cane leaves. Kinetic properties of different oligomeric structures. Eur J Biochem 1990;192:729–33.

[118] James AM, Hoogewijs K, Logan A, Hall AR, Ding S, Fearnley IM, Murphy MP. Non-enzymatic N-acetylation of lysine residues by acetylCoA often occurs via a proximal S-acetylated thiol intermediate sensitive to glyoxalase II. Cell Rep 2017;18:2105–12.

[119] Wagner GR, Payne RM. Widespread and enzyme-independent Nepsilon-acetylation and Nepsilon-succinylation of proteins in the chemical conditions of the mitochondrial matrix. J Biol Chem 2013;288:29036–45.

[120] Wagner GR, Hirschey MD. Nonenzymatic protein acylation as a carbon stress regulated by sirtuin deacylases. Mol Cell 2014;54:5–16.

[121] Weinert BT, Moustafa T, Iesmantavicius V, Zechner R, Choudhary C. Analysis of acetylation stoichiometry suggests that SIRT3 repairs nonenzymatic acetylation lesions. EMBO J 2015;34:2620–32.

[122] Bracher PJ, Snyder PW, Bohall BR, Whitesides GM. The relative rates of thiol-thioester exchange and hydrolysis for alkyl and aryl thioalkanoates in water. Orig Life Evol Biosph 2011;41:399–412.

[123] Zhang D, Tang Z, Huang H, Zhou G, Cui C, Weng Y, Liu W, Kim S, Lee S, Perez-Neut M, Ding J, Czyz D, Hu R, Ye Z, He M, Zheng YG, Shuman HA, Dai L, Ren B, Roeder RG, Becker L, Zhao Y. Metabolic regulation of gene expression by histone lactylation. Nature 2019;574:575–80.

[124] Gaffney DO, Jennings EQ, Anderson CC, Marentette JO, Shi T, Schou Oxvig AM, Streeter MD, Johannsen M, Spiegel DA, Chapman E, Roede JR, Galligan JJ. Non-enzymatic lysine lactoylation of glycolytic enzymes. Cell Chem Biol 2020;27:206–13.

[125] Donnellan L, Young C, Simpson BS, Acland M, Dhillon VS, Costabile M, Fenech M, Hoffmann P, Deo P. Proteomic analysis of methylglyoxal modifications reveals susceptibility of glycolytic enzymes to dicarbonyl stress. Int J Mol Sci 2022;23:3689.

Chapter 3.11

Glutamate carboxypeptidase II

Giulia Barchielli, Antonella Capperucci, and Damiano Tanini
Department of Chemistry "Ugo Schiff", University of Florence, Florence, Italy

1 Introduction

The type-II membrane glycoprotein glutamate carboxypeptidase II (EC 3.4.17.21), known as GCP II, is a homodimeric dinuclear zinc carboxypeptidase with the highest expression levels found in the nervous and prostatic tissue [1,2].

GCP II catalyzes the hydrolysis of endogenous substrates bearing L-glutamate at the C-terminal P1' position. One of these substrates is the abundant brain peptide *N*-acetyl-L-aspartyl-L-glutamate (α-NAAG), whose cleavage provides *N*-acetyl aspartate and glutamate. GCP II also catalyzes the hydrolysis of C-terminal glutamate residues of folylpoly-γ-L-glutamate (also known as folate polyglutamate and as pteroylpoly-γ-L-glutamate). Owing to these catalytic activities, GCP II has also been formerly referred as *N*-acetylated-α-linked acidic dipeptidase (NAALADase) and as folate hydrolase (FolH1).

Glutamate carboxypeptidase II is also known as prostate-specific membrane antigen (PSMA). It is worth remembering that this membrane-bound exopeptidase (EC 3.4.17.21) is different from the prostate-specific antigen (PSA, EC 3.4.21.77), which refers to kallikrein-related peptidase 3 [1,3].

2 GCP II biological localization

Both in vitro immunoactivity and enzymatic essays showed that significant differences exist in the localization of GCP II between different species [3–11]. According the aforementioned studies, in healthy humans GCP II is mainly expressed in prostate (secretory acinar epithelium), central and peripheral nervous system (astrocytes and Schwann cells, respectively), kidney (proximal tubules), and small intestine (brush border membranes of jejunum) [3–9,12–18]. A lower expression was also detected in testis, ovaries, bladder, pancreas, heart, salivary glands, skin, liver, lung, and colon [4–9,16,19]. In vivo research studies, carried out by using GCP II small ligands or labeled antibodies, showed a restricted expression of the enzyme. This might be due to a limited accessibility of GCP II for test ligands in vivo or to the masking of the sites of interaction by other molecules or posttranslational modification [8,20–24]. The *N*-acetyl-aspartyl-glutamate (NAAG) distribution seems to influence the CGP II distribution in most regions of the human body [3].

2.1 Nervous system

NAAG is the majorly distributed peptide transmitter in the mammalian brain, and it is also the one with the main concentration (mM). It is synthetized in neurons and stored in presynaptic axon terminals in vesicles [25]. In the nervous tissue, NAAG exhibits a negative modulatory effect while glutamate has an excitatory role [3,26].

NAAG is released in the synaptic cleft after calcium-dependent depolarization, then it quickly diffuses in the extra synaptic space where it can take two different ways. In the first case, NAAG can act as an agonist at the metabotropic glutamate receptor 3 (mGluR3) on the presynaptic nerve terminal and astrocytes. By activating mGluR3, NAAG leads to a decrease in the cellular concentration of the second messengers cAMP and cGMP (via a G-protein-coupled pathway). In the presynaptic nerve, this causes a reduction in the amount of glutamate that will be released upon further nerve stimulation. In the astrocytes, the link of NAAG to mGluR3 supports the production and secretion of neuroprotective factors such as the transforming growth-factor β [27–31]. Alternatively, NAAG can undergo hydrolysis by GCP II into *N*-Ac-Asp and glutamate, which are transported into astrocytes and oligodendrocytes. In this case, NAAG is no longer able to perform the agonist activity on mGluR3 [8].

The GCP II enzymatic activity—and its possible regulation—may determine if the excitatory or inhibitory effects would prevail after the NAAG release in neural system [3].

"Glutamate excitotoxicity" is a phenomenon due to the nonphysiological increase in the glutamate concentration in synapsis. This causes hyperactivation of ionotropic glutamate channels NMDAR with consequent excessive calcium influx and triggering of several cellular responses including nitric oxide overproduction, ionic homeostasis imbalance, caspase activation, mitochondrial dysfunction

Metalloenzymes. https://doi.org/10.1016/B978-0-12-823974-2.00002-4

and associated free radical generation, which result in the necrotic or apoptotic death of the neuron. Consequently, neuronal dysfunction and degeneration both acute and chronic, including stroke—brain ischemia [8,32], amyotrophic lateral sclerosis, Alzheimer's and Parkinson's diseases, traumatic brain injury [8,33,34], inflammatory and neuropathic pain [35–37], peripheral neuropathy [38,39], and epilepsy [40] might occur.

A variety of studies on animal models showed that a lower stimulation of ionotropic glutamate channels might ameliorate serious pathological disorders both of peripherical and central nervous system [8]. This is also supported by the evidence of GCP II dysregulation in several neurologic and psychiatric disorders that involve glutaminergic neurotransmission [3,41–43].

The inhibition of GCP II increases the intact NAAG part and decreases the glutamate concentration. Furthermore, under GCP II inhibition the mGluR3 activation leads to additional reduction of released glutamate and to the secretion of protective peptides. Supporting this findings, GCP II knockout mice showed reduced sensitivity to central and peripheric nerve damage [44] and proved to be less sensitive to traumatic brain injury in terms of neuronal loss, oxidative stress and apoptosis [45,46] The involvement of the mGluR3 receptors was confirmed in animal models of memory, Alzheimer's, and ethanol intoxication; GCP II inhibition was found to improve cognition and to reduce the cognitive and motor deficits induced by ethanol [47].

The attempt to employ NMDAR (*N*-methyl-D-aspartate receptor) antagonist for therapeutical purposes has failed. In this context, the employ of small molecules as GCP II specific inhibitors might be effective, via the increase of extrasynaptic NAAG concentration and its agonist action on mGluR3. In several preliminary in vitro and animal model studies, the utility of GCP II inhibitors has been demonstrated with multiple potential clinical applications as discussed by different authors [8,48,49] for inflammatory and neuropathic pain [8,50–52], peripheral neuropathy [8], motoneuron disease [8], cognitive impairment in multiple sclerosis [53], brain ischemia [8], spinal cord and traumatic brain injury [8], perinatal injury [54], epilepsy [8], schizophrenia [55], ethanol intoxication [47], and drug abuse (reducing self-administration and priming induced drug seeking) [8,56–59]. The efficacy of GCP II inhibition has also been confirmed in preclinical models of neurological disorders, where glutamatergic excitotoxicity plays a role [60,61].

Very recently, the importance of GCP II inhibition as a potential strategy for the treatment of cognitive disorders associated with aging and/or neuroinflammation has been endorsed in studies with animals [62].

2.2 Kidney

In the kidney, GCP II is located in the brush border of the proximal convoluted tubules where it has a role (still not fully understood) in the selective circulating peptides reabsorption [3,5].

2.3 Small intestine

In the brush border of the small intestine cells, GCP II cleavages the dietary poly-γ-glutamyl folates to their mono-glutamyl form by removing the γ-linked glutamate. Mono-glutamyl folates can then be actively absorbed by specific carriers [3,8]. This activity is critical in human nutrition since humans require folates and folate metabolism has been reported to be possibly influenced by GCP II polymorphism. However, further epidemiologic studies are needed to clarify the actual role of GCP II in folate metabolism [8].

3 GCP II structure and reaction mechanism

GCP II is characterized by the presence of two Zn^{2+} ions in its active site within a (μ-aquo)(μ-carboxylato)dizinc(II) core. The two zinc ions are coordinated by two histidines (His377 and His 553), two aspartates (Asp387, bridging the two Zn^{2+} ions, and Asp453, which binds one Zn^{2+} ion in a bidentate mode), and one glutamate residue (Glu425). The zinc-zinc distance is of 3.3 A. In its substrate-free form, the two zinc ions of GCP II are bridged asymmetrically by a bidentate water or hydroxyl ligand. Thus, each of the two Zn^{2+} ions has the classical tetrahedral coordination sphere.

Although the mechanism of GCP II is not understood in detail, some information can be extrapolated from studies performed on other structurally related metallopeptidases. For example, experimental and computational data are available on aminopeptidase from *Aeromonas proteolytica* (AAP) and *Streptomyces griseus* (SGAP) [63,64].

Such studies suggested that in GCP II the bridging oxygen is donated by a hydroxide anion and that the Glu424 (equivalent of Glu 151 in AAP) is a proton shuttle between the bridging water molecule (deprotonated upon coordination to Zn^{2+}) and the scissile peptide bond. Additionally, docking studies performed on model dipeptides indicated that the substrate is bound to the dizinc cluster through the terminal amino functionality rather than the carbonyl moiety of the peptide bond. The bridging hydroxide can perform the nucleophilic attack at the carbonyl group without being shifted to the terminal position, which imply coordination at the Zn(1) only. DFT studies on the model of the active site of AAP also highlighted that the proton transfer from Glu151 to the nitrogen atom of the peptide bond is reasonably the rate-determining step. The following steps of the cycle, including the proton transfer from OH^- to Glu151 and the cleavage of the peptide bond, involve a sequence of intermediates and transition states with small reaction barriers. The catalytic Zn(1) ion stabilizes the anionic tetrahedral intermediate, whereas the cocatalytic Zn(2) is involved in the binding of the substrate and in the orientation of the peptide bond toward the nucleophile [65]. Thus, the main role of the Zn(1) ion is the binding and the activation of a water molecule for the nucleophilic attack onto the scissile peptide bond. On the other hand, the Zn(2) ion mainly polarizes the carbonyl functionality of the peptide bond, compensating for the absence of an oxyanion hole in GCP II. Both Zn ions act stabilizing the *gem*-diol transition state [66]. The distance between the two Zn^{2+} ions increases from 3.3 to 3.6–3.8 A in complexes with transition-state-analogue inhibitors [66,67]. A similar scenario may also occur in the *gem*-diol transition state of bound substrates.

Based on crystallographic, biochemical, and computational evidences, Lubkowski et al. proposed a detailed mechanism of substrate hydrolysis by human GCP II (Fig. 1) [68].

Because two of its lone pairs are occupied in the coordination with the two zinc ions of the active pocket of the enzyme, the hydroxide anion of GCP II is significantly less nucleophilic than a free hydroxide anion. The direct attack of the OH^- at the carbonyl carbon of the peptide bond is therefore disfavored, and the reaction reasonably proceeds concertedly with the proton activating the nitrogen atom of the peptide bond and the concomitant formation of a metastable tetrahedral intermediate. In this regard—as mentioned earlier—Glu424 behaves as a general base/acid, which shuttles the first proton from the activated water to the nitrogen atom of the peptide bond and then accepts the second proton from the bridging OH^-.

Several interactions take place between *N*-Ac-Asp-Glu and the substrate-binding cavity of GCP II. The β-carboxylate group of the Asp residue of the substrate is involved in ionic interactions with the guanidinium groups of Arg534 and Arg536. The presence of hydrogen-bonding interactions with Asn519 and with two water molecules is also suggested. The carbonyl oxygen of the acetyl group is engaged in interactions with the side chain of Arg536, Asp453, Asn519; additionally, the methyl group protrudes into a pocket delimitated by the side chains of Ile386, Asp387, Ser454, Glu457, and Tyr549 [68].

According DFT calculation-based studies, in transition state-1 (Fig. 1), the hydrogen of the hydroxyl group of Glu24 is in the near-attack conformation and the Zn—Zn distance is elongated (from 3.3 to 3.5 A). The carbonyl oxygen of the peptide bond is stabilized by Zn(1). Following the transition state-1, the system evolves toward the formation of the intermediate, which is structurally characterized by (i) the elongated Zn—Zn distance, (ii) a newly formed C—O bond between the oxygen atom of the bridging hydroxide and the carbon atom of the peptide bond

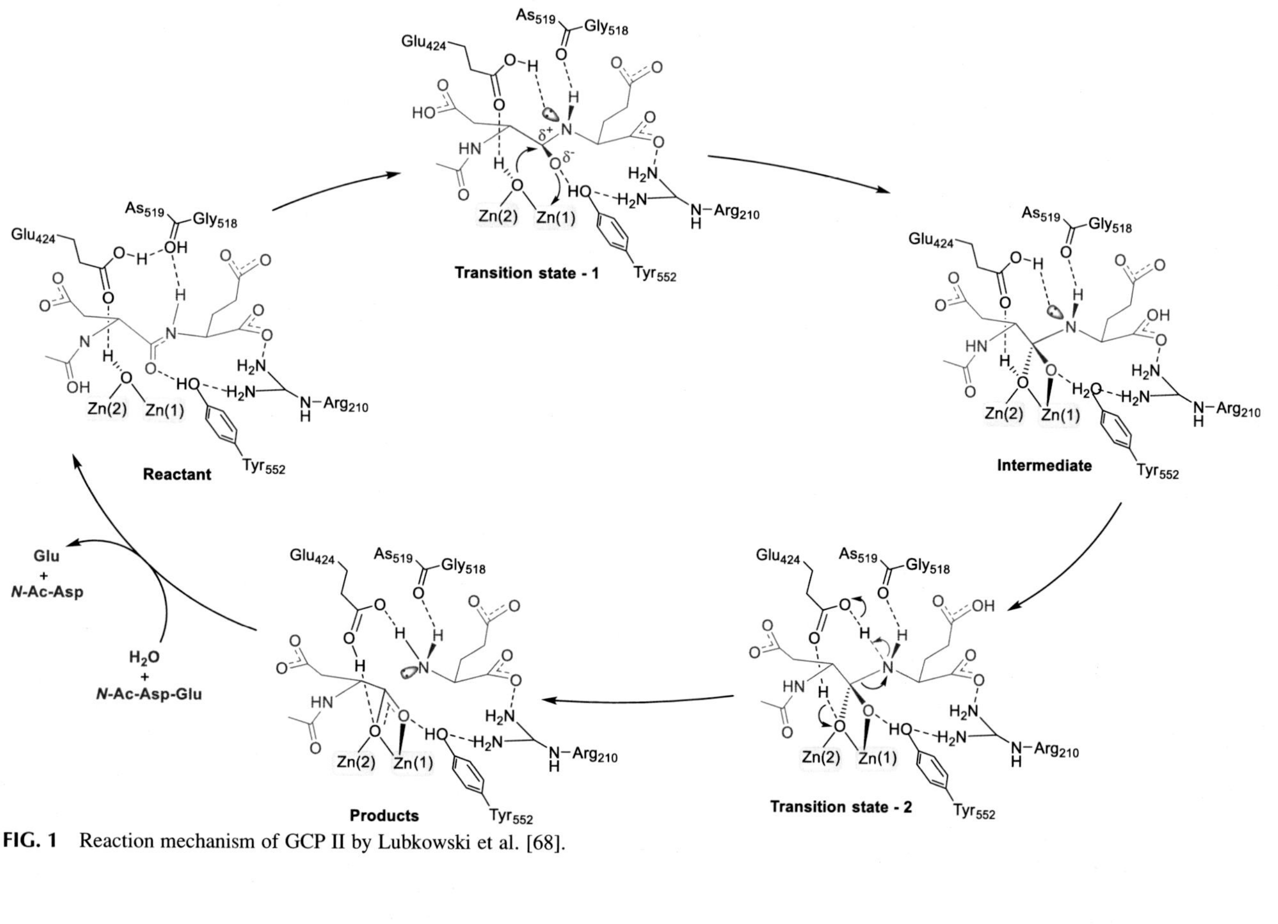

FIG. 1 Reaction mechanism of GCP II by Lubkowski et al. [68].

of the substrate, (iii) the stabilization of the carbonyl oxygen of the substrate by Zn(1) and Tyr552, and (iv) the repositioning of the protonated hydroxyl group of Glu424 (Fig. 1).

The transition state-2 (reasonably the rate-determining one) foresees the elongation of the hydrolyzed C—N bond. The reaction proceeds through a proton transfer from the hydroxyl group of Glu424 to the nitrogen atom of the peptide bond, a proton transfer from OH^- to Glu424 and, finally, the cleavage of the peptide bond. Following the transition state-2, the addition of a water molecule—which replaces the bridging hydroxide used in the hydrolysis—leads to the dissociation of *N*-Ac-Asp from the active site (Fig. 1). According to DFT calculations the release of *N*-Ac-Asp from the active site is promoted by the "pull effect" of several arginine residues (Arg463, Arg511, Arg534, Arg536) located in the exit channel of GCP II [66]. At this stage, the glutamate residue—which is the second product of the reaction—likely remains bound in the active pocket of the enzyme [68].

4 GCP II inhibition

Inhibitors of GCP II have been widely studied. Their physiological role and the possibility to use GCP II inhibitors for therapeutic purposes (i.e., neurological disorders, cancer) have been extensively reviewed [3,8,61,69–75].

Excess of glutamate is associated with various neurological disorders, as stroke, amyotrophic lateral sclerosis, spinal cord injury, epilepsy, chronic pain. The formation of glutamate by GPC II-catalyzed hydrolysis of *N*-acetylaspartylglutamate **1** (NAAG) is considered an important source of glutamate **3** (Fig. 2). Thus, besides conventional therapeutic approaches for the treatment of the above-mentioned diseases (use of small molecules to block postsynaptic glutamate receptors or upstream reduction of presynaptic glutamate), the use of potential GPC II inhibitors as therapeutic agents, to prevent glutamate release, was deeply investigated.

The first chemicals capable to inhibit GCP II were adaptable oxoanions such as phosphate and sulfate, as well as bidentate metal ion chelating agents such as EDTA and EGTA (CLAN MH 1).

Initially the attention was focused toward the synthesis of peptide analogues of NAAG **1** to evaluate their capability to inhibit the hydrolysis by NAALA dipeptidase (*N*-acetylated alpha-linked acidic dipeptidase). It was found that *N*-fumaryl-L-glutamic acid was the most active GCP II inhibitor (IC_{50} value of 0.8 μM) [76]. The kinetics of NAAG hydrolysis by the cloned human enzyme were also evaluated, showing a pharmacologic profile very similar to the endogenously expressed GCP II activity [77]. Glutamate and quisqualic acid derivatives were identified as GCP II inhibitors, active at μM concentration [22,23,78]. While peptide-based inhibitors were not used in evaluating the therapeutic usefulness of GPC II inhibition, their investigation was important to gain insight into the structural requirements for inhibition of this enzyme. It was showed that the spacing linker groups in the C-terminal glutamate residue are significant for binding to the active site of the enzyme. Later on, effective GCP II inhibitors mainly belonging to the classes of (i) phosphonate-based inhibitors (IC_{50} 300 pM), (ii) thiol-based inhibitors (IC_{50} 90 nM), (iii) hydroxamate-based inhibitors, (iv) urea-based inhibitors (IC_{50} 20 nM), and (v) sulfamide-based inhibitors have been described [8,70,73,79,80].

4.1 2-(Phosphonomethyl)pentanedioic acid (2-PMPA) inhibitors

Relevant results were reported by the Jackson's group, who first identified phosphorus-based inhibitors, such as **2-PMPA** (2-(phosphonomethyl)pentanedioic acid) (Fig. 3). **2-PMPA** proved to be a very active compound with a

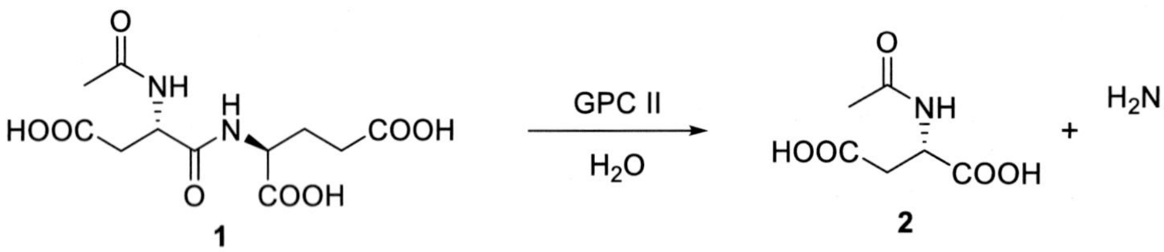

1 2 3

FIG. 2 GPC II-catalyzed hydrolysis of *N*-acetylaspartylglutamate (NAAG, **1**).

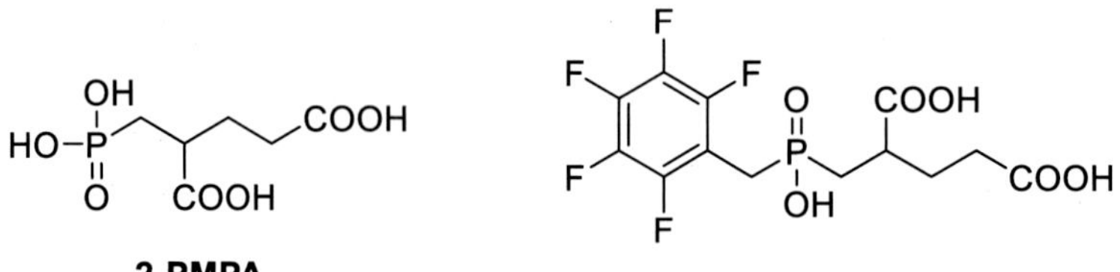

2-PMPA **GPI 5232**

FIG. 3 **2-PMPA** GCP II inhibitors.

$K_i = 0.275$ nM [81], with high selectivity for GCP II and high aqueous solubility/stability [32]. **2-PMPA** is a competitor inhibitor; its pentanedioic acid portion interacts with the glutarate recognition site of GCP II and the phosphonate group chelates the zinc ions at the active site.

Slusher and coworkers reported the therapeutic utility of GCP II inhibition by **2-PMPA**, able to protect against ischemic injury and to reduce the ischemia-induced rise in glutamate, thus contributing to neuroprotection in animal models [82–85]. However, the polar nature of **2-PMPA** caused a poor pharmacokinetic profile, thus limiting its use as a therapeutic drug. The fluorinated aryl-derivative **GPI 5232** (Fig. 3) also behaved as an effective inhibitor in vitro and in vivo, reducing brain injury in animal models [86]. Interestingly, the (*S*)-enantiomer of **GPI 5232** was more potent with respect to the (*R*)-enantiomer [87]. Crystal structure of the extracellular part of GCP II in complex with **2-PMPA** was described [67].

The activity of **2-PMPA** in several preclinical models of neurological disorders associated with glutamatergic excitotoxicity was recently reviewed [61]. Additionally, **2-PMPA** showed a broad and potent analgesic and neuroprotective effects in a variety of preclinical studies [8,52,88]. A novel gut-restricted GCP II inhibitor IBD3540 with high oral anticolitis efficacy in mouse models with a promising preclinical safety profile was also reported [89].

Besides the **2-PMPA** usefulness in several preclinical models, its clinical use was rather hampered for its high polarity and therefore for its low oral bioavailability. Indeed, **2-PMPA** and related structures were effective in treatment and useful to prove the role of GCP II in models of disease where an excess of glutamate is thought to be pathogenic, including amyotrophic lateral sclerosis, pain, ischemia, seizures, morphine tolerance, and aggression [8]. However, the low oral bioavailability and the limited brain penetration significantly limited the advance of the aforementioned compounds in clinical tests. Very recently, the therapeutic potential of **2-PMPA** in the treatment of Alzheimer's disease in animal models [47], glioblastoma—in vitro [90] and Amyotrophic Lateral Sclerosis—in vitro [91] was confirmed. New opportunities of **2-PMPA** as a possible nephron-protective strategy in PSMA-targeted prostate cancer radiotherapy also arose [92,93], giving new input to the research on these molecules.

Several attempts were performed in order to increase the oral bioavailability and the lipophilicity, thus improving the possibility for these molecules to cross the blood-brain barrier [8]. At first the new synthetized molecules showed a lower potency (in the nM range) without any significant improvement of the pharmacokinetic. The intranasal administration seemed to offer in some case a valid alternative to deliver therapeutic concentrations in the brain [94,95]. The synthesis of prodrugs as an alternative route has also been followed [61], with encouraging results in animal models for some **2-PMPA** prodrugs in terms of stability, pharmacokinetics and delivery to the plasma and brain [61,95,96].

For example, **2-PMPA** prodrugs were studied and synthesized through esterification of carboxylic or phosphonic groups with pivaloyloxymethyl (POM), isopropyloxycarbonyloxymethyl (isopropyl carbonate) (POC) and (5-methyl-2-oxo-1,3-dioxol-4-yl)methyl (ODOL) moieties, which can be functionalized by esterase enzymes [61]. Phosphonate esters with the hydrophobic POC (or POM) groups were prepared, leaving unsubstituted the α- and γ-carboxilic groups (Fig. 4, **4**). These derivatives were rather unstable, with low permeability. Esterification also of the carboxylic groups enabled to enhance their chemical stability. However, they revealed to be too stable *in vivo* and minimal release of **2-PMPA** was achieved. The introduction of POC and POM groups on both the phosphonic and the α-carboxylic groups provided tris-POC-2-PMPA, tris-POM-2-PMPA, and tetra-POM-2-PMPA (Fig. 4, **5–7**). Oral administration of **5** in mice for over 4h showed >20 fold enhancement of **2-PMPA** with respect to orally administered **2-PMPA** (molar equivalent dose), showing the first example of orally bioavailable prodrugs [97].

2-PMPA prodrugs bearing ODOL groups, an FDA-approved promoiety, were also prepared. Compounds containing two (**8**), three (**9**), or four (**10**) masked acidic groups (Fig. 5) were synthesized from suitably substituted benzyl esters, using the different reactivity of carboxylic and phosphonate esters [96]. Their stability in vitro, as well as their pharmacokinetics in vivo, was determined in mice and dog. It was found that prodrug **10** delivered the highest **2-PMPA** levels. These results suggest that ODOL-substituted compounds offer a promising strategy to increase the oral bioavailability, opening possible space for **2-PMPA** clinical translation.

4.2 Thiol-based GCP II inhibitors

Another approach to overcome the poor oral bioavailability of the phosphorous-containing GPC II inhibitors was the preparation of molecules with a reduced polarity, bearing a less polar thiol group instead of a phosphorylated moiety. A series of 2-(thioalkyl)pentanedioic acids **11a–f** (Fig. 6) were prepared and studied [98,99].

2-(Phosphonomethyl)pentanedioic acid (**2-MPPA**, Fig. 6) was the first compound identified in this class of inhibitors and it proved to be efficient after oral administration in animal models of neuropathic pain, familial amyotrophic lateral sclerosis, cocaine addiction, painful and sensory diabetic neuropathy [8]. When tested in next human studies, **2-MPPA** did not cause adverse effects on CNS and it was well tolerated at plasma concentration but it did not significantly advance in clinic due to its quite low potency and concerns about the potential immune reactivity (that is common for thiol-containing drugs) [8]. Very recently, **2-MPPA** was administrated in animal models, improving

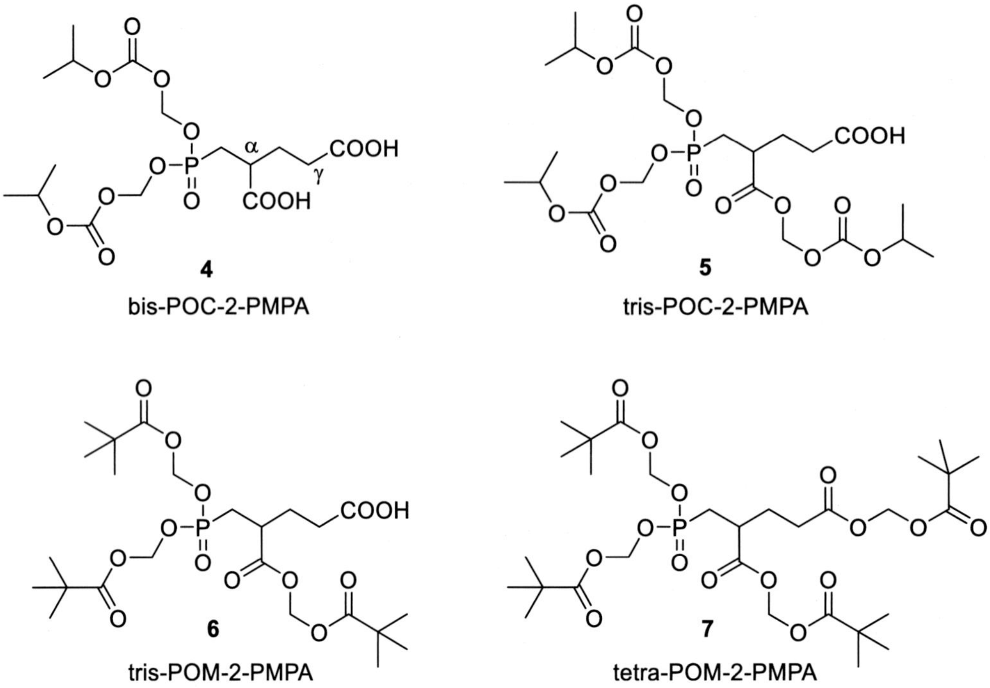

FIG. 4 **2-PMPA** prodrugs.

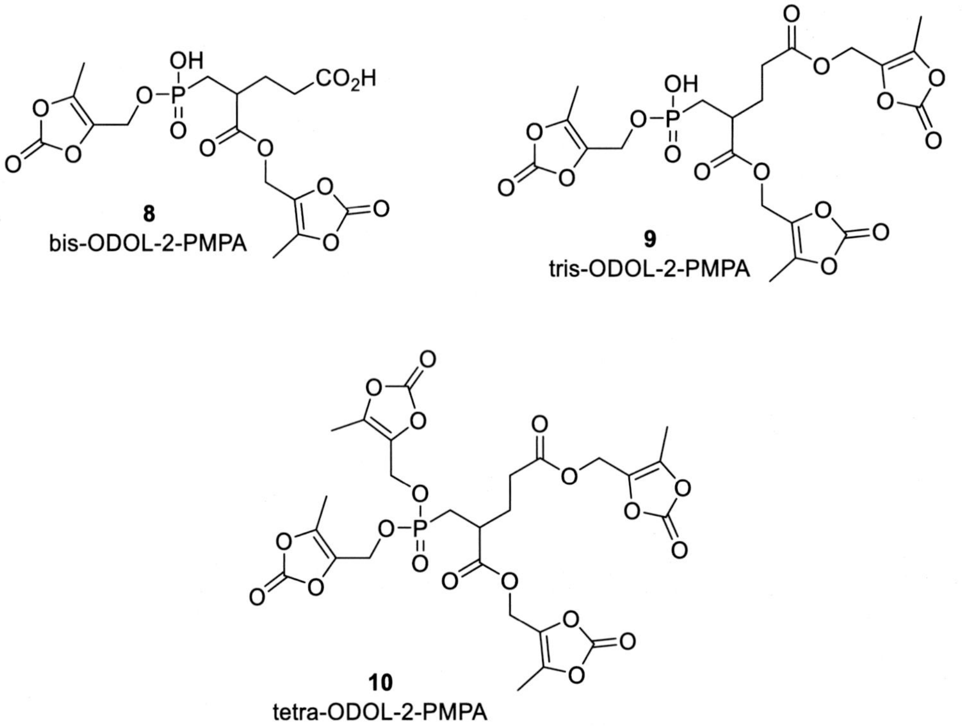

FIG. 5 ODOL derivatives of **2-PMPA**.

working memory performance in young and aged tested rats, and it also improved performance after local infusion into the medial prefrontal cortical [100]. **2-MPPA** showed also neuroprotective effects in superoxide dismutase transgenic (SOD) after induced hypoxia-ischemia [101]. Since this molecule proved to be well-tolerated, researchers suggested it could provide an important new direction for treatment of cognitive disorders associated with aging and/or inflammation and for the prevention of typical damage related to brain injury and inflammation.

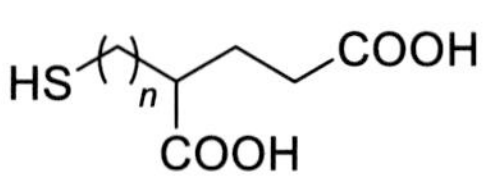

11a-f *(n = 0-5)*
Thiol-based inhibitors

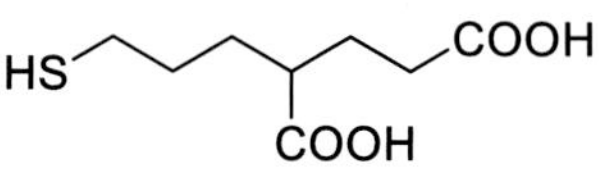

2-MPPA *(n = 3)*
(or **GPI 5693**)

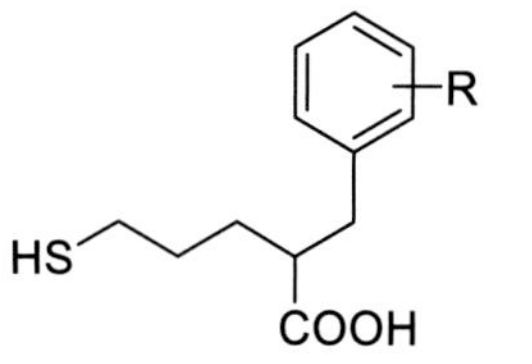

R = H, 2-CO_2H, 3-CO_2H, 4-CO_2H

12a-d
2-MPPA carboxybenzyl-analogues

FIG. 6 Thiol-based GCP II inhibitors.

Research efforts were directed toward the synthesis of more potent molecules [8], A series of derivatives of **2-MPPA** (Fig. 6, **11a–f**) with a different number of methylene groups between the thiol moiety and the pentanedioic acid residue were also synthesized through different approaches, aiming to establish the optimal position of the –SH group to achieve the best affinity to GCP II [98]. The biological evaluation of the inhibitory potency against the enzyme revealed that the efficacy was dependent on the number of CH_2 groups. Indeed, 2-(3-mercaptopropyl)pentanedioic acid (**2-MPPA**, also known as **GPI 5693**, Fig. 6, $n = 3$) was found to behave as the more active compound in an animal model of peripheral neuropathy by oral administration ($IC_{50} = 90\,nM$). However, none of the thiol-based compounds has been shown to be as potent as **2-PMPA**.

Both enantiomers of **2-MPPA** were also prepared and tested, showing comparable efficacy as GCP II inhibitors in the neuropathic pain model, by oral administration [87].

Garrido, Sanabria et al. investigated the functional role of **2-MPPA** with respect to its neuroprotective effects in hippocampal mossy fiber-CA3 pyramidal cell synapses [102]. Based on a mechanistic study, it is assumed that the action of **2-MPPA** in preclinical models of neurological disorders is due to a presynaptic regulation of glutamatergic neurotransmission.

In 2005, van Gerven and coworkers reported that **GPI 5693** was safe and fairly well tolerated at plasma exposures. Even if no important drug-related CNS adverse events were described, minor CNS effects were observed following the highest dose level [103].

Another series of thiol-based GCP II inhibitors containing a benzyl group, in place of the carboxyethyl group of **2-MPPA**, was also synthesized (Fig. 6, **12**) [104]. In vitro GCP II assay showed that the *meta*-substituted analogue 3-(2-carboxy-5-mercaptopentyl)benzoic acid (Fig. 6, **12c**, R = 3-CO_2H) was more potent than **2-MPPA** ($IC_{50} = 15\,nM$). On the contrary, the unsubstituted and the *ortho*-substituted derivatives (Fig. 6, **12a,b**, R = H, 2-CO_2H) were less active than **2-MPPA** in inhibiting GCP II. This result was not unexpected as the removal of one carboxylate group decreases the interaction with the glutamate recognition site of GCP II. However, these results represent the first example to enhance the activity of GCP II inhibitors by modifying the glutarate moiety.

A series of *N*-substituted 3-(2-mercaptoethyl)-1*H*-indole-2-carboxylic acids **13** (Fig. 7), having IC_{50} values in the range of 20–50 nM were also synthesized. Notably, at that time, derivatives **13** were the first achiral GCP II inhibitors structurally different from NAAG [105].

4.3 Hydroxamate-based inhibitors

Other different zinc-binding compounds, as hydroxamate-based derivatives, were also investigated (Fig. 8, **14a–c**). The succinyl hydroxamic acid derivative 2-(hydroxycarbamoylmethyl)-pentanedioic acid (hydroxamate-based inhibitor (**14a**) $n = 1$, Fig. 8) showed a high GCP II inhibitor activity with an IC_{50} value of 220 nM [79]. In 2016, Barinka and coworkers reported the synthesis and the structural characterization of novel hydroxamic acid-based inhibitors

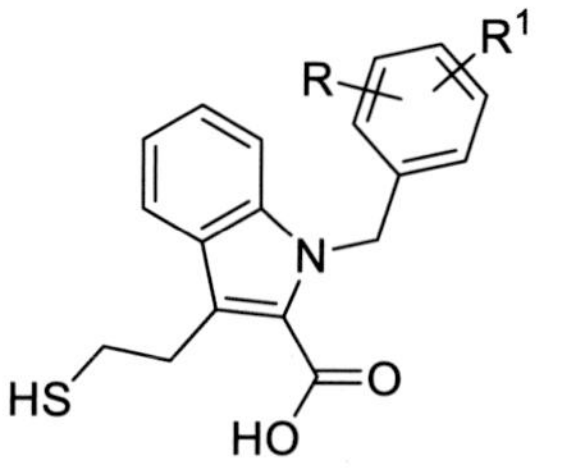

13a: R = R^1 = H; IC_{50} (nM): 12000 ± 5000
13b: R = 2-CO_2H, R^1 = H; IC_{50} (nM): 22 ± 21
13c: R = 3-CO_2H, R^1 = H; IC_{50} (nM): 22 ± 10
13d: R = 4-CO_2H, R^1 = H; IC_{50} (nM): 94 ± 12
13e: R = 2-Br, R^1 = 5-CO_2H; IC_{50} (nM): 54 ± 9
13f: R = 3-^{t}Bu, R^1 = 5-CO_2H; IC_{50} (nM): 34 ± 24
13g: R = 2-CO_2H, R^1 = 5-CO_2H; IC_{50} (nM): 93 ± 41

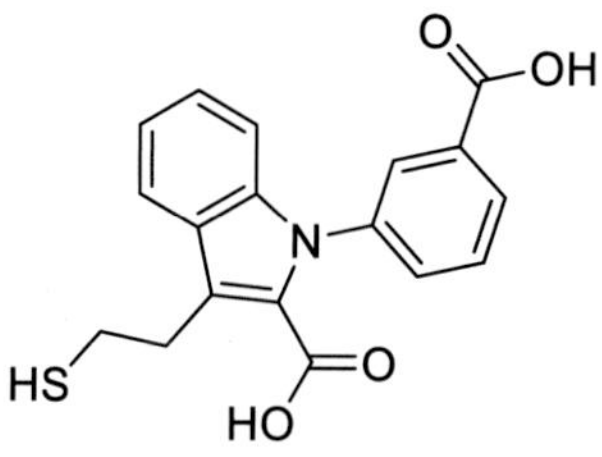

13h: IC_{50} (nM): 22 ± 4

FIG. 7 Structure of *N*-substituted 3-(2-mercaptoethyl)-1*H*-indole-2-carboxylic acids **13**.

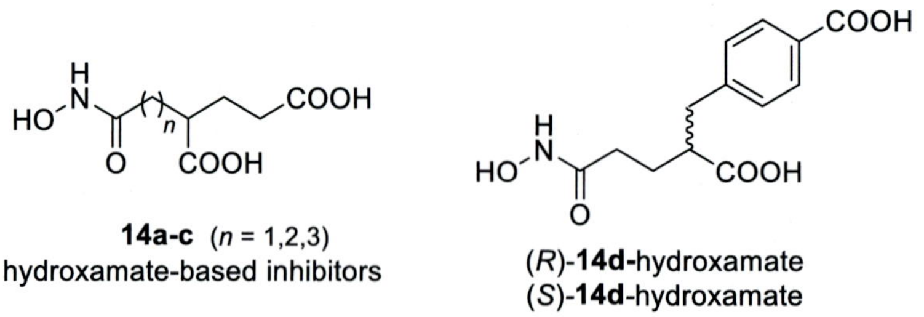

FIG. 8 Hydroxamate-based inhibitors.

(Fig. 8, **14d**) [106]. These compounds contain the Zn-binding group (hydroxamic acid) bonded to the benzoic acid moiety *via* a flexible linker (2-carboxybutyl group). The racemic mixture was synthesized in eight step (27% overall yield). Chiral separation of the enantiomers, by means of a chiral column, and cleavage of the benzyl groups by hydrogenation on Pd/C led to hydroxamates (*R*)-**14d** and (*S*)-**14d** (Fig. 8). Crystal structures evidenced a unique binding mode, which can explain the lack of enantiospecificity observed for the two isomers. The inhibitory properties were determined, confirming the use of such derivatives in the treatment of chronic neurological disorders, as neuropathic pain [106].

4.4 Urea-based GCP II inhibitors

Urea-based inhibitors are the third class of molecules, structurally similar to NAAG, where a urea linkage join 2 amino acids through their NH_2 groups. Urea represents here the zinc-binding group. In 2004, the urea-based NAAG analogue **ZJ-43** (Fig. 9) was synthesized, and its inhibitory activity on cloned human GCP II was determined using a fluorescent assay [107]. Notably, **ZJ-43** and other molecules of this group exhibited a low nM potency [8]. Remarkably, **ZJ-43** showed efficacy in various animal models of neuropathic pain, inflammatory peripheral pain, neurological disorders (i.e., schizophrenia), and traumatic brain injury. On the other hand, **ZJ-43** exhibited low oral bioavailability and minimal brain penetration [8].

Several research studies have been pursued in order to improve GCII inhibitors oral bioavailability and minimize the potential toxicity. In particular, the use of prodrugs has been sought after and alternative zinc-binding groups (i.e., sulfamides, sulfonamides, imidazoles, and other nitrogen-containing heterocycles) have been tested [8].

It is interesting to point out that while in case of neurological disease the employ of GCP II inhibitors increased NAAG and decreased glutamate in the brain, in the absence of disease, the inhibition had no effect of the basal glutamate transmission. This is very important from a therapeutical point of view because it means that limiting the excess of glutamate by GCP II inhibitors may provide neuroprotection without the undesired side effects typically observed with the use of glutamate receptors antagonists [8].

To overcome the limitations associated with poor oral bioavailability and negligible brain penetration of "traditional" GCP II inhibitors, recently, catechol-based inhibitors were developed. In particular D-DOPA offered a noncompetitive mode of inhibition with an excellent pharmacokinetic profile, which can be enhanced by coadministration with the DAAO inhibitor sodium benzoate, resulting in robust target engagement in the brain [108].

More recently, Šácha and coworkers developed potent urea-based GCP II inhibitors by structure-aided design [109]. The efficacy was enhanced by using a rigid linker, capable to increase the selectivity and to allow a suitable connection with different functional groups, such as a fluorophore or biotin. The replacement of the terminal PEG_{12}-biotin with the fluorescein (Fig. 9, **15**) enabled a further

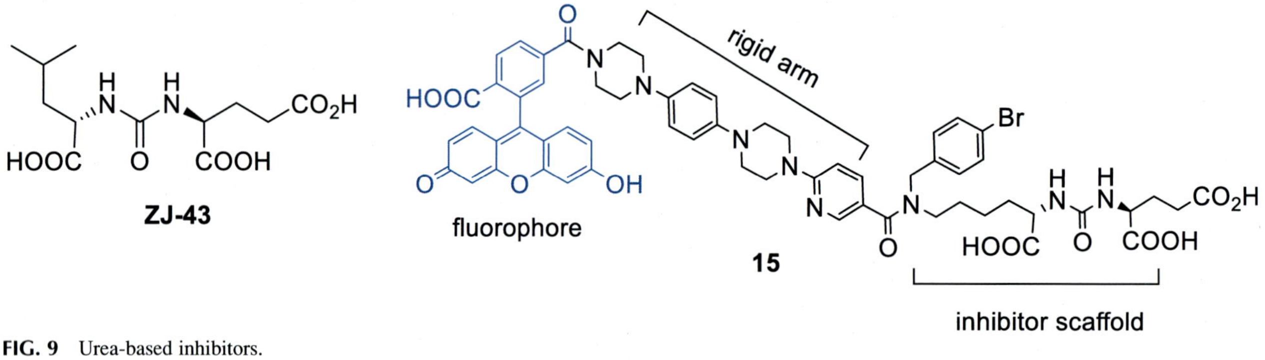

FIG. 9 Urea-based inhibitors.

improvement of inhibitory potency, with a $K_i = 8.6\,pM$, with respect to the biotin derivative ($K_i = 0.11\,nM$).

4.5 Sulfamide derivatives as GCP II inhibitors

In 2013 Berkman and coworkers reported a small library of sulfamide analogs of GCP II inhibitors scaffolds (Fig. 10) in order to evaluate the behavior of the sulfamide group as a zinc-binding moiety. Compounds **16** and **17** were the most efficient inhibitors, and only the aspartyl-glutamyl sulfamide **16** showed a submicromolar activity ($IC_{50} = 0.9\,\mu M$) [73].

5 GCP II and diseases

5.1 Cancer

Highest levels of GCP II/PSMA than normal were found in malignant tissues. In prostate cancer, the GCP II expression was found to be positively correlated with the Gleason score (cancer grade) and disease progression, increasing from the healthy tissue to androgen refractory malignancies through benign prostate hyperplasia and low grade to metastatic adenocarcinoma [3,8,48]. It is believed that, in prostate cancer, CGP II is negatively regulated by androgens and is promoted by other growth factors such as basic fibroblast growth factor, TGF (transforming growth factor), and EGF (Epidermal Growth Factor). The increased PSMA/GCP II expression in prostate cancer tissues is also related with an increased ability of cells to process folate [48].

An increased expression of GCP II was also detected in solid tumors derived from tissues that normally express the enzyme such as the Schwann cells, bladder, kidney, colon, and breast [3,8,69]. An increased expression of GCP II is observed in neovasculature of solid tumors, whereas it is absent in the vasculature of corresponding benign tissues. This suggests a possible role of GCP II in tumor angiogenesis [3,8,48,69]. However, some significant differences were observed between different tissues tumors and between non-small-cell cancers and small cell cancers [48].

GCP II/PSMA may reveal useful with theranostics purposes for these tumor types [69]. As a possible consequence of this, NAAG concentration in plasma could be a noninvasive measurement to monitor cancer progression [110].

GCP II higher expression increases folate uptake giving the cell a proliferative advantage [8,48]. Some studies highlighted that GCP II is required for carcinogenesis and cell invasion in prostate. Conversely, other studies found an inverse correlation between GCP II and prostate cancer invasiveness [8,48]. Additional investigations are required to clarify whether enzyme inhibition could block the invasiveness and growth of prostate cancer or not.

In this context, it is worth reminding the marked differences existing in the expression of GCP II between rodents and humans. The highest expression of GCP II in mice is in the kidney, brain, and salivary glands, while the enzyme is almost absent in rodent prostate and small intestine. Additionally, there are no reports about the presence of GCP II in neovasculature of solid tumors in rodents [8,111]. Mouse GCP II possesses lower catalytic efficiency but similar substrate specificity compared with the human enzyme. Differences between rodents and humans also exist in the link to plasma proteins of small-molecules GCP II ligands. For all these reason, mouse GCP II could approximate human GCP II in drug development but significant differences in GCP II tissue expression need be taken into account when developing novel GCP II-based anticancer and therapeutic approaches, including targeted anticancer drug delivery systems [111].

GCP II is a promising candidate both for tumors imaging/diagnosis and the delivery of toxic therapy, especially for prostatic cancer [3,8]. Monoclonal antibodies (mAb), including the ^{111}In-labeled ProstaScint—approved by FDA as an imaging for metastatic cancer—were designed for these purposes [3]. This antibody recognizes only an intracellular epitope of GCP II/PSMA and, thus, only dying or died necrotic cells can be detected. Furthermore, ProstaScint has to be administrated several days prior the test, with issues linked to background radioactivity [3,8]. To overcome the limitations of ProstaScint, antibodies that recognize the extracellular epitopes of GCP II/PSMA have been obtained. For example, J591 can be used in the imaging of both prostate and other solid tumors, owing to the presence of PSMA within the tumor vasculature [3,8]. Small radiolabeled molecules have also been used for imaging purposes. These molecules are characterized by high affinity to PSMA as well as by the possibility to be rapidly uptaken by the tumor tissue and washed-out to non-target sites [8,70].

Urea-based GCP II inhibitors have been used for imaging of prostate cancer [3,8]. ^{125}I, ^{123}I, ^{99m}Tc-labeled urea derivatives were successfully used for SPECT

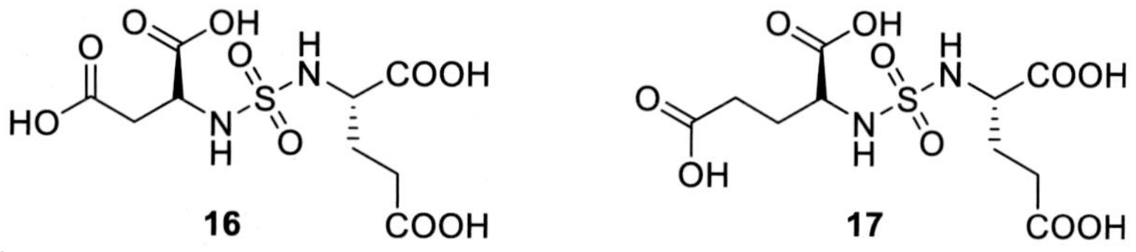

FIG. 10 Sulfamides derivatives.

techniques while ^{11}C, ^{68}Ga, and ^{18}F-labeled molecules were employed for PET imaging providing high target to non-target ratio [8].

Radiolabeled **^{18}F-DCFPyL** (Fig. 11) (PSMA) PET/CT was tested in women with advanced high-grade serous ovarian cancer (HGSOC). It showed higher specificity for the metastatic sites as compared to standard of care contrast-enhanced CT but detects fewer metastatic sites of disease, especially in the upper abdomen and along the gastrointestinal tract, limiting its clinical utility as a diagnostic tool in HGSOC [112].

GCP II antibodies with specific conjugated molecules can be used for cancer therapy, for example using radioactive elements as in the case of a derivative of ProstaScint conjugated with ^{90}Y instead of ^{111}In or ^{90}Y-J591 and ^{117}Lu-J591 [8].

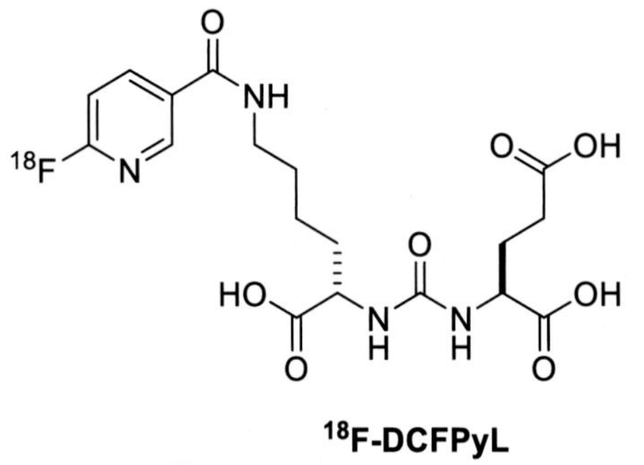

FIG. 11 Radiolabeled **^{18}F-DCFPyL.**

Immunotoxins can also be conjugated to antibodies, for example this is the case of **J591** conjugated with ricin A, melitin like peptide, monomethylauristatin E, and others [8]. These molecules showed promising results in vitro and in xenograft models, from the tumor growth inhibition to a full ablation of implanted tumors [8]. Recently PSMA-specific small-molecule carriers equipped by Doxorubicin (Dox) (Fig. 12) were synthesized and tested in animal models. Preliminary results showed that the novel compounds were able to release the active substance inside cancer cells thereby providing a relatively high Dox concentration in nuclei and a relevant cytotoxic effect [113].

The issues in clinical application include the possible immunogenicity of these conjugated antibodies and the difficulties faced in producing them in a large homogeneous scale. The use of recombinant DNA technologies might help to overcome, at least partially, the problem of immunogenicity, with the obtainment of smaller, single-chain antibody fragments [8].

Molecules belonging to the phosphoramidate peptidomimetic class has been radiolabeled with ^{18}F showing pseudoirreversible binding to GCP II [114]. Pseudoirreversibility is a feature of inhibitor binding that increase the internalization and that can thus be exploited for transporting therapies into PSMA-positive cells. This characteristic was not observed for urea based GCP II inhibitors [8].

All the PSMA-based imaging agents are unable to cross the blood-brain barrier because they are highly charged.

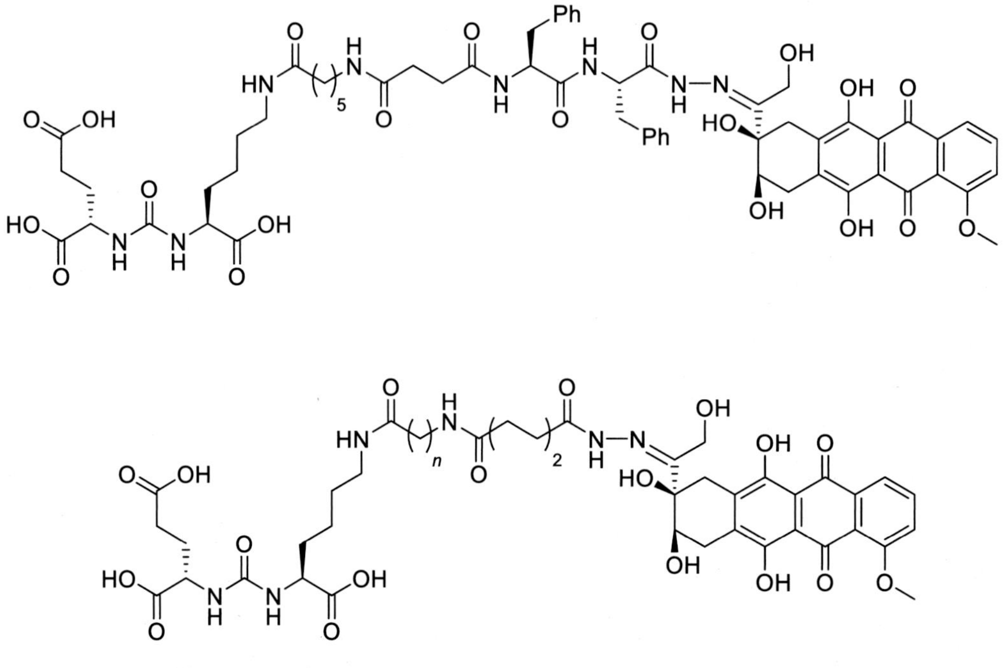

FIG. 12 Doxorubicin-derived compounds for cancer therapy.

Thus, although they are effective in the periphery, their use for imaging within the CNS is hampered. In order to use such derivatives in diagnosis and therapy in the brain, the removal of the tricarboxylic moiety is required [8,115].

Notably, PSMA targeting species labeled with infrared emitting fluorescent species can be used as guidance in surgeries [8,21].

5.2 Inflammatory bowel diseases

The role of GCP II in IBD (inflammatory bowel diseases) was first suggested by studies about the human disease, by genome-wide expression investigation [116]. Recently, the expression of the FOLH1 gene that codes for GCP II was confirmed to be strongly upregulated in biopsies of patients with IBD [69–118]. FOLH1 was described as a "hub" gene that has significant correlations with over a dozen of known IBD gene biomarkers [116].

GCP II enzymatic activity has been demonstrated to be significantly increased (by 300%–3000%) in both Crohn's disease and ulcerative colitis disease patient biopsies [69,119]. The pharmacological inhibition of this upregulated activity provided therapeutic benefit in preclinical IBD models [69,119,120]. In this regard, **2-PMPA** has been proposed as a novel treatment for inflammatory bowel diseases. Daily treatment with a hypotonic **2-PMPA** enema ameliorated macroscopic and microscopic symptoms of IBD in mouse model, thus highlighting the therapeutic potential of FOLH1/GCP II inhibitors for the local treatment of IBD [119,120].

5.3 Benign inflammatory states

The activity of GCP II in benign inflammatory states, including anal fistula, sarcoidosis, fasciitis, and cerebral infarction, has been observed. Some expression has also been found in in Paget disease, fractures, and synovitis. Although the role of the enzyme in these pathological states is poorly understood, it is interesting to target it for possible novel therapeutic strategies [69].

5.4 Male reproduction

Research on the urogenital system of aged mice highlighted that the PSMA/GCP II-deficient mouse model had increased propensity for enlarged seminal vesicles upon aging. Significant amounts of PSMA/GCP II within the mouse urogenital system were detected only in the epididymis. As the enzyme is also present in the human epididymis, these findings suggest a role of PSMA/GCP II on reproduction and provide a groundwork for further studies on humans [121].

6 Conclusions and perspectives

GCP II is currently a pharmacological and diagnostic target being pursued in the clinic for the regulation of glutamate and for imaging purposes. The inhibition of GCP II represents an interesting strategy and, although several GCP II inhibitors have been developed, some challenges remain ahead. The improvement of the oral bioavailability of GCP II inhibitors capable to penetrate the blood-brain barrier would enable a more effective treatment of neurological diseases. Indeed, although PSMA/GCP II represents a promising target for treating neurological disorders, issues in delivering drugs across the blood-brain barrier significantly hurdle its full exploration.

Additionally, the design and synthesis of specific small molecules that could discriminate between GCP II and its paralogs (e.g., GCP III, whose physiological role is almost unknown, but its substate specificity and pharmacologic profile are very similar to those of GCP II) would be highly desirable. In this context, more information about GCP II paralogs and orthologs in mammalian might help to better understand the role of the enzyme in healthy and diseases tissues.

Finally, the clarification of the specific function of GCP II in cancer cells would enable a more rational approach toward new potential therapeutic tools for tumor treatment.

References

[1] Mesters JR, Hilgenfeld R. Glutamate carboxypeptidase II. In: Handbook of metalloproteins. John Wiley & Sons, Ltd; 2006–2008.

[2] Carter RE, Feldman AR, Coyle JT. Prostate-specific membrane antigen is a hydrolase with substrate and pharmacologic characteristics of a neuropeptidase. Proc Natl Acad Sci USA 1996;93 (2):749–53.

[3] Slusher BS, Rojas C, Coyle JT. Glutamate carboxypeptidase II. In: Handbook of proteolytic enzymes. 3rd ed; 2013. Chapter 367 and reference cited therein.

[4] Robinson MB, Blakely RD, Couto R, Coyle JT. Hydrolysis of the brain dipeptide *N*-acetyl-L-aspartyl-L-glutamate. Identification and characterization of a novel *N*-acetylated alpha-linked acidic dipeptidase activity from rat brain. J Biol Chem 1987;262:14498–506.

[5] Slusher BS, Robinson MB, Tsai G, Simmons ML, Richards SS, Coyle JT. Rat brain *N*-acetylated alpha-linked acidic dipeptidase activity. Purification and immunologic characterization. J Biol Chem 1990;265(34):21297–301.

[6] Berger UV, Carter RE, McKee M, Coyle JT. *N*-acetylated alpha-linked acidic dipeptidase is expressed by non-myelinating Schwann cells in the peripheral nervous system. J Neurocytol 1995;24(2):99–109.

[7] Rovenská M, Hlouchová K, Sácha P, Mlcochová P, Horák V, Zámecník J, Barinka C, Konvalinka J. Tissue expression and enzymologic characterization of human prostate specific membrane antigen and its rat and pig orthologs. Prostate 2008;68(2):171–82.

[8] Bařinka C, Rojas C, Slusher B, Pomper M. Glutamate carboxypeptidase II in diagnosis and treatment of neurologic disorders and prostate cancer. Curr Med Chem 2012;19(6):856–70. and reference cited therein.

[9] Halsted CH, Ling E, Luthi-Carter R, Villanueva JA, Gardner J, Coyle JT. Folylpoly-gamma-glutamate carboxypeptidase from pig jejunum. Molecular characterization and relation to glutamate carboxypeptidase II. J Biol Chem 1988;273:20417–24.

[10] Aggarwal S, Ricklis RM, Williams SA, Denmeade SR. Comparative study of PSMA expression in the prostate of mouse, dog, monkey, and human. Prostate 2006;66(9):903–10.

[11] Shafizadeh TB, Halsted CH. Gamma-Glutamyl hydrolase, not glutamate carboxypeptidase II, hydrolyzes dietary folate in rat small intestine. J Nutr 2007;137(5):1149–53.

[12] Luthi-Carter R, Berger UV, Barczak AK, Enna M, Coyle JT. Isolation and expression of a rat brain cDNA encoding glutamate carboxypeptidase II. Proc Natl Acad Sci USA 1998;95(6):3215–20.

[13] Berger UV, Luthi-Carter R, Passani LA, Elkabes S, Black I, Konradi C, Coyle JT. Glutamate carboxypeptidase II is expressed by astrocytes in the adult rat nervous system. J Comp Neurol 1999;415 (1):52–64.

[14] Sácha P, Zámecník J, Barinka C, Hlouchová K, Vícha A, Mlcochová P, Hilgert I, Eckschlager T, Konvalinka J. Expression of glutamate carboxypeptidase II in human brain. Neuroscience 2007;144 (4):1361–72.

[15] Silver DA, Pellicer I, Fair WR, Heston WD, Cordon-Cardo C. Prostate-specific membrane antigen expression in normal and malignant human tissues. Clin Cancer Res 1997;3(1):81–5.

[16] Sokoloff RL, Norton KC, Gasior CL, Marker KM, Grauer LS. A dual-monoclonal sandwich assay for prostate-specific membrane antigen: levels in tissues, seminal fluid and urine. Prostate 2000;43(2):150–7.

[17] Bostwick DG, Pacelli A, Blute M, Roche P, Murphy GP. Prostate specific membrane antigen expression in prostatic intraepithelial neoplasia and adenocarcinoma: a study of 184 cases. Cancer 1998;82(11):2256–61.

[18] Chang SS, Reuter VE, Heston WD, Bander NH, Grauer LS, Gaudin PB. Five different anti-prostate-specific membrane antigen (PSMA) antibodies confirm PSMA expression in tumor-associated neovasculature. Cancer Res 1999;59(13):3192–8.

[19] Kinoshita Y, Kuratsukuri K, Landas S, Imaida K, Rovito Jr PM, Wang CY, Haas GP. Expression of prostate-specific membrane antigen in normal and malignant human tissues. World J Surg 2006;30(4):628–36.

[20] Bander NH. Technology insight: monoclonal antibody imaging of prostate cancer. Nat Clin Pract Urol 2006;3(4):216–25.

[21] Chen Y, Dhara S, Banerjee SR, Byun Y, Pullambhatla M, Mease RC, Pomper MG. A low molecular weight PSMA-based fluorescent imaging agent for cancer. Biochem Biophys Res Commun 2009;390(3):624–9.

[22] David KA, Milowsky MI, Kostakoglu L, Vallabhajosula S, Goldsmith SJ, Nanus DM, Bander NH. Clinical utility of radiolabeled monoclonal antibodies in prostate cancer. Clin Genitourin Cancer 2006;4(4):249–56.

[23] Foss CA, Mease RC, Fan H, Wang Y, Ravert HT, Dannals RF, Olszewski RT, Heston WD, Kozikowski AP, Pomper MG. Radiolabeled small-molecule ligands for prostate-specific membrane antigen: in vivo imaging in experimental models of prostate cancer. Clin Cancer Res 2005;11(11):4022–8.

[24] Mease RC, Dusich CL, Foss CA, Ravert HT, Dannals RF, Seidel J, Prideaux A, Fox JJ, Sgouros G, Kozikowski AP, Pomper MG. *N*-[*N*-[(*S*)-1,3-Dicarboxypropyl]carbamoyl]-4-[^{18}F]fluorobenzyl-L-cysteine, [^{18}F]DCFBC: a new imaging probe for prostate cancer. Clin Cancer Res 2008;14(10):3036–43.

[25] Becker I, Lodder J, Gieselmann V, Eckhardt M. Molecular characterization of *N*-acetylaspartylglutamate synthetase. J Biol Chem 2010;285(38):29156–64.

[26] Wroblewska B, Wroblewski JT, Saab OH, Neale JH. *N*-acetylaspartylglutamate inhibits forskolin-stimulated cyclic AMP levels via a metabotropic glutamate receptor in cultured cerebellar granule cells. J Neurochem 1993;61(3):943–8.

[27] Fricker AC, Mok MH, de la Flor R, Shah AJ, Woolley M, Dawson LA, Kew JN. Effects of *N*-acetylaspartylglutamate (NAAG) at group II mGluRs and NMDAR. Neuropharmacology 2009;56(6–7):1060–7.

[28] Wroblewska B, Wegorzewska IN, Bzdega T, Olszewski RT, Neale JH. Differential negative coupling of type 3 metabotropic glutamate receptor to cyclic GMP levels in neurons and astrocytes. J Neurochem 2006;96(4):1071–7.

[29] Bruno V, Battaglia G, Casabona G, Copani A, Caciagli F, Nicoletti F. Neuroprotection by glial metabotropic glutamate receptors is mediated by transforming growth factor-beta. J Neurosci 1998;18 (23):9594–600.

[30] Thomas AG, Liu W, Olkowski JL, Tang Z, Lin Q, Lu XC, Slusher BS. Neuroprotection mediated by glutamate carboxypeptidase II (NAALADase) inhibition requires TGF-beta. Eur J Pharmacol 2001;430(1):33–40.

[31] Niswender CM, Conn PJ. Metabotropic glutamate receptors: physiology, pharmacology, and disease. Annu Rev Pharmacol Toxicol 2010;50:295–322.

[32] Slusher BS, Vornov JJ, Thomas AG, Hurn PD, Harukuni I, Bhardwaj A, Traystman RJ, Robinson MB, Britton P, Lu XC, Tortella FC, Wozniak KM, Yudkoff M, Potter BM, Jackson PF. Selective inhibition of NAALADase, which converts NAAG to glutamate, reduces ischemic brain injury. Nat Med 1999;5(12):1396–402.

[33] Long JB, Yourick DL, Slusher BS, Robinson MB, Meyerhoff JL. Inhibition of glutamate carboxypeptidase II (NAALADase) protects against dynorphin A-induced ischemic spinal cord injury in rats. Eur J Pharmacol 2005;508(1–3):115–22.

[34] Zhong C, Zhao X, Sarva J, Kozikowski A, Neale JH, Lyeth BG. NAAG peptidase inhibitor reduces acute neuronal degeneration and astrocyte damage following lateral fluid percussion TBI in rats. J Neurotrauma 2005;22(2):266–76.

[35] Carpenter KJ, Sen S, Matthews EA, Flatters SL, Wozniak KM, Slusher BS, Dickenson AH. Effects of GCP-II inhibition on responses of dorsal horn neurones after inflammation and neuropathy: an electrophysiological study in the rat. Neuropeptides 2003;37(5):298–306.

[36] Chen SR, Wozniak KM, Slusher BS, Pan HL. Effect of 2-(phosphono-methyl)-pentanedioic acid on allodynia and afferent ectopic discharges in a rat model of neuropathic pain. J Pharmacol Exp Ther 2002;300(2):662–7.

[37] Yamamoto T, Hirasawa S, Wroblewska B, Grajkowska E, Zhou J, Kozikowski A, Wroblewski J, Neale JH. Antinociceptive effects of *N*-acetylaspartylglutamate (NAAG) peptidase inhibitors ZJ-11, ZJ-17 and ZJ-43 in the rat formalin test and in the rat neuropathic pain model. Eur J Neurosci 2004;20(2):483–94.

[38] Carozzi VA, Chiorazzi A, Canta A, Lapidus RG, Slusher BS, Wozniak KM, Cavaletti G. Glutamate carboxypeptidase inhibition

reduces the severity of chemotherapy-induced peripheral neurotoxicity in rat. Neurotox Res 2010;17(4):380–91.

[39] Zhang W, Murakawa Y, Wozniak KM, Slusher B, Sima AA. The preventive and therapeutic effects of GCPII (NAALADase) inhibition on painful and sensory diabetic neuropathy. J Neurol Sci 2006;247(2):217–23.

[40] Witkin JM, Gasior M, Schad C, Zapata A, Shippenberg T, Hartman T, Slusher BS. NAALADase (GCP II) inhibition prevents cocaine-kindled seizures. Neuropharmacology 2002;43(3):348–56.

[41] Tsai GC, Stauch-Slusher B, Sim L, Hedreen JC, Rothstein JD, Kuncl R, Coyle JT. Reductions in acidic amino acids and N-acetylaspartylglutamate in amyotrophic lateral sclerosis CNS. Brain Res 1991;556(1):151–6.

[42] Ghose S, Chin R, Gallegos A, Roberts R, Coyle J, Tamminga C. Localization of NAAG-related gene expression deficits to the anterior hippocampus in schizophrenia. Schizophr Res 2009;111 (1-3):131–7.

[43] Guilarte TR, Hammoud DA, McGlothan JL, Caffo BS, Foss CA, Kozikowski AP, Pomper MG. Dysregulation of glutamate carboxypeptidase II in psychiatric disease. Schizophr Res 2008;99(1-3):324–32.

[44] Bacich DJ, Wozniak KM, Lu XC, O'Keefe DS, Callizot N, Heston WD, Slusher BS. Mice lacking glutamate carboxypeptidase II are protected from peripheral neuropathy and ischemic brain injury. J Neurochem 2005;95(2):314–23.

[45] Gao Y, Xu S, Cui Z, Zhang M, Lin Y, Cai L, Wang Z, Luo X, Zheng Y, Wang Y, Luo Q, Jiang J, Neale JH, Zhong C. Mice lacking glutamate carboxypeptidase II develop normally, but are less susceptible to traumatic brain injury. J Neurochem 2015;134(2):340–53.

[46] Cao Y, Gao Y, Xu S, Bao J, Lin Y, Luo X, Wang Y, Luo Q, Jiang J, Neale JH, Zhong C. Glutamate carboxypeptidase II gene knockout attenuates oxidative stress and cortical apoptosis after traumatic brain injury. BMC Neurosci 2016;17:15.

[47] Olszewski RT, Janczura KJ, Bzdega T, Der EK, Venzor F, O'Rourke B, Hark TJ, Craddock KE, Balasubramanian S, Moussa C, Neale JH. NAAG Peptidase Inhibitors Act via mGluR3: animal models of memory, Alzheimer's, and ethanol intoxication. Neurochem Res 2017;42(9):2646–57.

[48] Evans JC, Malhotra M, Cryan JF, O'Driscoll CM. The therapeutic and diagnostic potential of the prostate specific membrane antigen/glutamate carboxypeptidase II (PSMA/GCPII) in cancer and neurological disease. Br J Pharmacol 2016;173(21):3041–79.

[49] Neale JH, Yamamoto T. *N*-acetylaspartylglutamate (NAAG) and glutamate carboxypeptidase II: An abundant peptide neurotransmitter-enzyme system with multiple clinical applications. Prog Neurobiol 2020;184, 101722.

[50] Sasson NJ, Turner-Brown LM, Holtzclaw TN, Lam KS, Bodfish JW. Children with autism demonstrate circumscribed attention during passive viewing of complex social and nonsocial picture arrays. Autism Res 2008;1(1):31–42.

[51] Yamamoto T, Kozikowski A, Zhou J, Neale JH. Intracerebroventricular administration of *N*-acetylaspartylglutamate (NAAG) peptidase inhibitors is analgesic in inflammatory pain. Mol Pain 2008;4:31.

[52] Nonaka T, Yamada T, Ishimura T, Zuo D, Moffett JR, Neale JH, Yamamoto T. A role for the locus coeruleus in the analgesic efficacy of *N*-acetylaspartylglutamate peptidase (GCPII) inhibitors ZJ43 and 2-PMPA. Mol Pain 2017;13. 1744806917697008.

[53] Hollinger KR, Alt J, Riehm AM, Slusher BS, Kaplin AI. Dose-dependent inhibition of GCPII to prevent and treat cognitive impairment in the EAE model of multiple sclerosis. Brain Res 2016;1635:105–12.

[54] Zhang Z, Bassam B, Thomas AG, Williams M, Liu J, Nance E, Rojas C, Slusher BS, Kannan S. Maternal inflammation leads to impaired glutamate homeostasis and up-regulation of glutamate carboxypeptidase II in activated microglia in the fetal/newborn rabbit brain. Neurobiol Dis 2016;94:116–28.

[55] Olszewski RT, Janczura KJ, Ball SR, Madore JC, Lavin KM, Lee JC, Lee MJ, Der EK, Hark TJ, Farago PR, Profaci CP, Bzdega T, Neale JH. NAAG peptidase inhibitors block cognitive deficit induced by MK-801 and motor activation induced by d-amphetamine in animal models of schizophrenia. Transl Psychiatry 2012;2(7), e145.

[56] Peng XQ, Li J, Gardner EL, Ashby Jr CR, Thomas A, Wozniak K, Slusher BS, Xi ZX. Oral administration of the NAALADase inhibitor GPI-5693 attenuates cocaine-induced reinstatement of drug-seeking behavior in rats. Eur J Pharmacol 2010;627(1–3):156–61.

[57] Xi ZX, Kiyatkin M, Li X, Peng XQ, Wiggins A, Spiller K, Li J, Gardner EL. *N*-acetylaspartylglutamate (NAAG) inhibits intravenous cocaine self-administration and cocaine-enhanced brain-stimulation reward in rats. Neuropharmacology 2010;58(1):304–13.

[58] Xi ZX, Li X, Peng XQ, Li J, Chun L, Gardner EL, Thomas AG, Slusher BS, Ashby Jr CR. Inhibition of NAALADase by 2-PMPA attenuates cocaine-induced relapse in rats: a NAAG-mGluR2/3-mediated mechanism. J Neurochem 2010;112(2):564–76.

[59] Zhu H, Lai M, Chen W, Mei D, Zhang F, Liu H, Zhou W. *N*-acetylaspartylglutamate inhibits heroin self-administration and heroin-seeking behaviors induced by cue or priming in rats. Neurosci Bull 2017;33(4):396–404.

[60] Morland C, Nordengen K. *N*-acetyl-aspartyl-glutamate in brain health and disease. Int J Mol Sci 2022;23(3):1268.

[61] Krečmerová M, Majer P, Rais R, Slusher BS. Phosphonates and phosphonate prodrugs in medicinal chemistry: past successes and future prospects. Front Chem 2022;10:88737.

[62] Yang S, Datta D, Woo E, Duque A, Morozov YM, Arellano J, Slusher BS, Wang M, Arnsten AFT. Inhibition of glutamate-carboxypeptidase-II in dorsolateral prefrontal cortex: potential therapeutic target for neuroinflammatory cognitive disorders. Mol Psychiatry 2022;27(10):4252–63.

[63] Desmarais W, Bienvenue DL, Bzymek KP, Petsko GA, Ringe D, Holz RC. The high-resolution structures of the neutral and the low pH crystals of aminopeptidase from Aeromonas proteolytica. J Biol Inorg Chem 2006;11:398–408.

[64] Schürer G, Horn AHC, Gedeck P, Clark T. The reaction mechanism of bovine lens leucine aminopeptidase. J Phys Chem B 2002;106:8815–30.

[65] Chen S, Marino T, Fang W, Russo N, Himo F. Peptide hydrolysis by the binuclear zinc enzyme aminopeptidase from Aeromonas proteolytica: a density functional theory study. J Phys Chem B 2008;112:2494–500.

[66] Mesters JR, Barinka C, Li W, Tsukamoto T, Majer P, Slusher BS, Konvalinka J, Hilgenfeld R. Structure of glutamate carboxypeptidase II, a drug target in neuronal damage and prostate cancer. EMBO J 2006;25(6):1375–84.

[67] Mesters JR, Henning K, Hilgenfeld R. Human glutamate carboxypeptidase II inhibition: structures of GCPII in complex with two potent inhibitors, quisqualate and 2-PMPA. Acta Crystallogr D Biol Crystallogr 2007;63:508–13.

[68] Klusák K, Bařinka C, Plechanovová A, Mlčochová P, Konvalinka J, Rulíšek L, Lubkowski J. Reaction mechanism of glutamate carboxypeptidase II revealed by mutagenesis, x-ray crystallography, and computational methods. Biochemistry 2009;48(19):4126–38.

[69] Vornov JJ, Peters D, Nedelcovych M, Hollinger K, Rais R, Slusher BS. Looking for drugs in all the wrong places: use of GCPII inhibitors outside the brain. Neurochem Res 2020;45(6):1256–67.

[70] Nikfarjam Z, Zargari F, Nowroozi A, Bavi O. Metamorphosis of prostate specific membrane antigen (PSMA) inhibitors. Biophys Rev 2022;14(1):303–15.

[71] Tsukamoto T, Wozniak KM, Slusher BS. Progress in the discovery and development of glutamate carboxypeptidase II inhibitors. Drug Discov Today 2007;12(17–18):767–76.

[72] Pavlícek J, Ptácek J, Barinka C. Glutamate carboxypeptidase II: an overview of structural studies and their importance for structure-based drug design and deciphering the reaction mechanism of the enzyme. Curr Med Chem 2012;19(9):1300–9.

[73] Choy CJ, Fulton MD, Davis AL, Hopkins M, Choi JK, Anderson MO, Berkman CE. Rationally designed sulfamides as glutamate carboxypeptidase II inhibitors. Chem Biol Drug Des 2013;82(5):612–9.

[74] Ferraris DV, Shukla K, Tsukamoto T. Structure-activity relationships of glutamate carboxypeptidase II (GCPII) inhibitors. Curr Med Chem 2012;19(9):1282–94.

[75] Zhou J, Neale JH, Pomper MG, Kozikowski AP. NAAG peptidase inhibitors and their potential for diagnosis and therapy. Nat Rev Drug Discov 2005;4(12):1015–26.

[76] Subasinghe N, Schulte M, Chan MY, Roon RJ, Koerner JF, Johnson RL. Synthesis of acyclic and dehydroaspartic acid analogues of Ac-Asp-Glu-OH and their inhibition of rat brain *N*-acetylated alpha-linked acidic dipeptidase (NAALA dipeptidase). J Med Chem 1990;33(10):2734–44.

[77] Luthi-Carter R, Barczak AK, Speno H, Coyle JT. Hydrolysis of the neuropeptide *N*-acetylaspartylglutamate (NAAG) by cloned human glutamate carboxypeptidase II. Brain Res 1998;795(1–2):341–8.

[78] Ghosh A, Heston WD. Effect of carbohydrate moieties on the folate hydrolysis activity of the prostate specific membrane antigen. Prostate 2003;57(2):140–51.

[79] Stoermer D, Liu Q, Hall MR, Flanary JM, Thomas AG, Rojas C, Slusher BS, Tsukamoto T. Synthesis and biological evaluation of hydroxamate-Based inhibitors of glutamate carboxypeptidase II. Bioorg Med Chem Lett 2003;13(13):2097–100.

[80] Park JD, Kim DH, Kim SJ, Woo JR, Ryu SE. Sulfamide-based inhibitors for carboxypeptidase A. Novel type transition state analogue inhibitors for zinc proteases. J Med Chem 2002;45(24):5295–302.

[81] Jackson PF, Cole DC, Slusher BS, Stetz SL, Ross LE, Donzanti BA, Trainor DA. Design, synthesis, and biological activity of a potent inhibitor of the neuropeptidase *N*-acetylated alpha-linked acidic dipeptidase. J Med Chem 1996;39(2):619–22.

[82] Lu XM, Tang Z, Liu W, Lin Q, Slusher BS. *N*-acetylaspartylglutamate protects against transient focal cerebral ischemia. Eur J Pharmacol 2000;408:233–9.

[83] Rojas C, Frazier ST, Flanary J, Slusher BS. Kinetics and inhibition of glutamate carboxypeptidase II using a microplate assay. Anal Biochem 2002;310(1):50–4.

[84] Tortella FC, Lin Y, Ved H, Slusher BS, Dave JR. Neuroprotection produced by the NAALADase inhibitor 2-PMPA in rat cerebellar neurons. Eur J Pharmacol 2000;402(1–2):31–7.

[85] Cai Z, Lin S, Rhodes PG. Neuroprotective effects of *N*-acetylaspartylglutamate in a neonatal rat model of hypoxia-ischemia. Eur J Pharmacol 2002;437(3):139–45.

[86] Williams AJ, Lu XM, Slusher B, Tortella FC. Electroencephalogram analysis and neuroprotective profile of the *N*-acetylated-alpha-linked acidic dipeptidase inhibitor, GPI5232, in normal and brain-injured rats. J Pharmacol Exp Ther 2001;299(1):48–57.

[87] Tsukamoto T, Majer P, Vitharana D, Ni C, Hin B, Lu XC, Thomas AG, Wozniak KM, Calvin DC, Wu Y, Slusher BS, Scarpetti D, Bonneville GW. Enantiospecificity of glutamate carboxypeptidase II inhibition. J Med Chem 2005;48(7):2319–24.

[88] Vornov JJ, Hollinger KR, Jackson PF, Wozniak KM, Farah MH, Majer P, Rais R, Slusher BS. Still NAAG'ing after all these years: the continuing pursuit of GCPII inhibitors. Adv Pharmacol 2016;76:215–55.

[89] Peters D, Norris L, Tenora L, Snajdr I, Zhu X, Sakamoto S, Thomas A, Majer P, Rais R, Slusher B. Inflamm Bowel Dis 2022;28(1):S4.

[90] Gao Y, Zheng H, Li L, Feng M, Chen X, Hao B, Lv Z, Zhou X, Cao Y. Prostate-specific membrane antigen (PSMA) promotes angiogenesis of glioblastoma through interacting with ITGB4 and regulating NF-κB signaling pathway. Front Cell Dev Biol 2021;9, 598377.

[91] Tallon C, Sharma A, Zhang Z, Thomas AG, Ng J, Zhu X, Donoghue A, Schulte M, Joe TR, Kambhampati SP, Sharma R, Liaw K, Kannan S, Kannan RM, Slusher BS. Dendrimer-2PMPA delays muscle function loss and denervation in a murine model of amyotrophic lateral sclerosis. Neurotherapeutics 2022;19(1):274–88.

[92] Kratochwil C, Giesel FL, Leotta K, Eder M, Hoppe-Tich T, Youssoufian H, Kopka K, Babich JW, Haberkorn U. PMPA for nephroprotection in PSMA-targeted radionuclide therapy of prostate cancer. J Nucl Med 2015;56(2):293–8.

[93] Chatalic KL, Heskamp S, Konijnenberg M, Molkenboer-Kuenen JD, Franssen GM, Clahsen-van Groningen MC, Schottelius M, Wester HJ, van Weerden WM, Boerman OC, de Jong M. Towards personalized treatment of prostate cancer: PSMA I&T, a promising prostate-specific membrane antigen-targeted theranostic agent. Theranostics 2016;6(6):849–61.

[94] Rais R, Wozniak K, Wu Y, Niwa M, Stathis M, Alt J, Giroux M, Sawa A, Rojas C, Slusher BS. Selective CNS uptake of the GCP-II inhibitor 2-PMPA following intranasal administration. PLoS One 2015;10(7), e0131861.

[95] Nedelcovych M, Dash RP, Tenora L, Zimmermann SC, Gadiano AJ, Garrett C, Alt J, Hollinger KR, Pommier E, Jančařík A, Rojas C, Thomas AG, Wu Y, Wozniak K, Majer P, Slusher BS, Rais R. Enhanced brain delivery of 2-(phosphonomethyl)pentanedioic acid following intranasal administration of its γ-substituted ester prodrugs. Mol Pharm 2017;14(10):3248–57.

[96] Dash RP, Tichý T, Veeravalli V, Lam J, Alt J, Wu Y, Tenora L, Majer P, Slusher BS, Rais R. Enhanced oral bioavailability of 2-(phosphonomethyl)-pentanedioic acid (2-PMPA) from its (5-methyl-2-oxo-1,3-dioxol-4-yl)methyl (ODOL)-based prodrugs. Mol Pharm 2019;16(10):4292–301.

[97] Majer P, Jančařík A, Krečmerová M, Tichý T, Tenora L, Wozniak K, Wu Y, Pommier E, Ferraris D, Rais R, Slusher BS. Discovery of orally available prodrugs of the glutamate carboxypeptidase II (GCPII) inhibitor 2-phosphonomethylpentanedioic acid (2-PMPA). J Med Chem 2016;59(6):2810–9.

[98] Majer P, Jackson PF, Delahanty G, Grella BS, Ko YS, Li W, Liu Q, Maclin KM, Poláková J, Shaffer KA, Stoermer D, Vitharana D, Wang EY, Zakrzewski A, Rojas C, Slusher BS, Wozniak KM, Burak E, Limsakun T, Tsukamoto T. Synthesis and biological evaluation of thiol-based inhibitors of glutamate carboxypeptidase II: discovery of an orally active GCP II inhibitor. J Med Chem 2003;46(10):1989–96.

[99] Garbiras BJ, Stephen M. Preparation of carboxythioalactones and their active derivatives. Synthesis 1999;270–4.

[100] Datta D, Leslie SN, Woo E, Amancharla N, Elmansy A, Lepe M, Mecca AP, Slusher BS, Nairn AC, Arnsten AFT. Glutamate carboxypeptidase II in aging rat prefrontal cortex impairs working memory performance. Front Aging Neurosci 2021;13, 760270.

[101] Arteaga Cabeza O, Zhang Z, Smith Khoury E, Sheldon RA, Sharma A, Zhang F, Slusher BS, Kannan RM, Kannan S, Ferriero DM. Neuroprotective effects of a dendrimer-based glutamate carboxypeptidase inhibitor on superoxide dismutase transgenic mice after neonatal hypoxic-ischemic brain injury. Neurobiol Dis 2021;148, 105201.

[102] Sanabria ER, Wozniak KM, Slusher BS, Keller A. GCP II (NAALADase) inhibition suppresses mossy fiber-CA3 synaptic neurotransmission by a presynaptic mechanism. J Neurophysiol 2004;91 (1):182–93.

[103] van der Post JP, de Visser SJ, de Kam ML, Woelfler M, Hilt DC, Vornov J, Burak ES, Bortey E, Slusher BS, Limsakun T, Cohen AF, van Gerven JM. The central nervous system effects, pharmacokinetics and safety of the NAALADase-inhibitor GPI 5693. Br J Clin Pharmacol 2005;60(2):128–36.

[104] Majer P, Hin B, Stoermer D, Adams J, Xu W, Duvall BR, Delahanty G, Liu Q, Stathis MJ, Wozniak KM, Slusher BS, Tsukamoto T. Structural optimization of thiol-based inhibitors of glutamate carboxypeptidase II by modification of the P1' side chain. J Med Chem 2006;49(10):2876–85.

[105] Grella B, Adams J, Berry JF, Delahanty G, Ferraris DV, Majer P, Ni C, Shukla K, Shuler SA, Slusher BS, Stathis M, Tsukamoto T. The discovery and structure–activity relationships of indole-based inhibitors of glutamate carboxypeptidase II. Bioorg Med Chem Lett 2010;20:7222–5.

[106] Novakova Z, Wozniak K, Jancarik A, Rais R, Wu Y, Pavlicek J, Ferraris D, Havlinova B, Ptacek J, Vavra J, Hin N, Rojas C, Majer P, Slusher BS, Tsukamoto T, Barinka C. Unprecedented binding mode of hydroxamate-based inhibitors of glutamate carboxypeptidase II: structural characterization and biological activity. J Med Chem 2016;59(10):4539–50.

[107] Olszewski RT, Bukhari N, Zhou J, Kozikowski AP, Wroblewski JT, Shamimi-Noori S, Wroblewska B, Bzdega T, Vicini S, Barton FB, Neale JH. NAAG peptidase inhibition reduces locomotor activity and some stereotypes in the PCP model of schizophrenia via group II mGluR. J Neurochem 2004;89(4):876–85.

[108] Gori SS, Thomas AG, Pal A, Wiseman R, Ferraris DV, Gao RD, Wu Y, Alt J, Tsukamoto T, Slusher BS, Rais R. D-DOPA is a potent, orally bioavailable, allosteric inhibitor of glutamate carboxypeptidase II. Pharmaceutics 2022;14(10):2018.

[109] Tykvart J, Schimer J, Jančařík A, Bařinková J, Navrátil V, Starková J, Šrámková K, Konvalinka J, Majer P, Šácha P. Design of highly potent urea-based, exosite-binding inhibitors selective for glutamate carboxypeptidase II. J Med Chem 2015;58(10):4357–63.

[110] Asaka R, Le A. Dual role of *N*-acetyl-aspartyl-glutamate metabolism in cancer monitor and therapy. Mol Cell Oncol 2019;6(5), e1627273.

[111] Knedlík T, Vorlová B, Navrátil V, Tykvart J, Sedlák F, Vaculín Š, Franěk M, Šácha P, Konvalinka J. Mouse glutamate carboxypeptidase II (GCPII) has a similar enzyme activity and inhibition profile but a different tissue distribution to human GCPII. FEBS Open Bio 2017;7(9):1362–78.

[112] Metser U, Kulanthaivelu R, Chawla T, Johnson S, Avery L, Hussey D, Veit-Haibach P, Bernardini M, Hogen L. ^{18}F-DCFPyL PET/CT in advanced high-grade epithelial ovarian cancer: a prospective pilot study. Front Oncol 2022;12:1025475.

[113] Ivanenkov YA, Machulkin AE, Garanina AS, Skvortsov DA, Uspenskaya AA, Deyneka EV, Trofimenko AV, Beloglazkina EK, Zyk NV, Koteliansky VE, Bezrukov DS, Aladinskaya AV, Vorobyeva NS, Puchinina MM, Riabykh GK, Sofronova AA, Malyshev AS, Majouga AG. Synthesis and biological evaluation of doxorubicin-containing conjugate targeting PSMA. Bioorg Med Chem Lett 2019;29(10):1246–55.

[114] Liu T, Toriyabe Y, Kazak M, Berkman CE. Pseudoirreversible inhibition of prostate-specific membrane antigen by phosphoramidate peptidomimetics. Biochemistry 2008;47(48):12658–60.

[115] Wang H, Byun Y, Barinka C, Pullambhatla M, Bhang HE, Fox JJ, Lubkowski J, Mease RC, Pomper MG. Bioisosterism of urea-based GCPII inhibitors: synthesis and structure-activity relationship studies. Bioorg Med Chem Lett 2010;20(1):392–7.

[116] Zhang T, Song B, Zhu W, Xu X, Gong QQ, Morando C, Dassopoulos T, Newberry RD, Hunt SR, Li E. An ileal Crohn's disease gene signature based on whole human genome expression profiles of disease unaffected ileal mucosal biopsies. PLoS One 2012;7(5), e37139.

[117] Noble CL, Abbas AR, Lees CW, Cornelius J, Toy K, Modrusan Z, Clark HF, Arnott ID, Penman ID, Satsangi J, Diehl L. Characterization of intestinal gene expression profiles in Crohn's disease by genome-wide microarray analysis. Inflamm Bowel Dis 2010;16(10):1717–28.

[118] Ben-Shachar S, Yanai H, Baram L, Elad H, Meirovithz E, Ofer A, Brazowski E, Tulchinsky H, Pasmanik-Chor M, Dotan I. Gene expression profiles of ileal inflammatory bowel disease correlate with disease phenotype and advance understanding of its immunopathogenesis. Inflamm Bowel Dis 2013;19(12):2509–21.

[119] Rais R, Jiang W, Zhai H, Wozniak KM, Stathis M, Hollinger KR, Thomas AG, Rojas C, Vornov JJ, Marohn M, Li X, Slusher BS. *FOLH1*/GCPII is elevated in IBD patients, and its inhibition ameliorates murine IBD abnormalities. JCI Insight 2016;1(12), e88634.

[120] Date AA, Rais R, Babu T, Ortiz J, Kanvinde P, Thomas AG, Zimmermann SC, Gadiano AJ, Halpert G, Slusher BS, Ensign LM. Local enema treatment to inhibit FOLH1/GCPII as a novel therapy for inflammatory bowel disease. J Control Release 2017;263:132–8.

[121] Vorlová B, Sedlák F, Kašpárek P, Šrámková K, Malý M, Zámečník J, Šácha P, Konvalinka J. A novel PSMA/GCPII-deficient mouse model shows enlarged seminal vesicles upon aging. Prostate 2019;79(2):126–39.

Chapter 3.12

Neutral endopeptidase (neprilysin)

Annamaria Mascolo[a,b], Liberata Sportiello[a,b], Maria Antonietta Riemma[a], Antonella De Angelis[a], Annalisa Capuano[a,b], and Liberato Berrino[a]

[a]*Department of Experimental Medicine—Section of Pharmacology "L. Donatelli", University of Campania "Luigi Vanvitelli", Naples, Italy,* [b]*Campania Regional Centre for Pharmacovigilance and Pharmacoepidemiology, Naples, Italy*

1 Structure and function of the enzyme

Neprilysin, also known as neutral endopeptidase (NEP), enkephalinase, CALLA, or CD10, is a mammalian type II integral membrane zinc endopeptidase, which inactivates different types of bioactive peptides. NEP belongs to the group of metallopeptidases including the endothelin-converting enzymes 1 and 2 (ECE-1 and -2), the erythrocyte surface antigen KELL, and the PEX gene product. All these metallopeptidases belong to the M13 subfamily of mammalian neutral endopeptidases and are constituted of a short cytoplasmic N-terminus, a single transmembrane helix, and a large extracellular C-terminus [1]. Therefore, the majority of NEP, including its active site, faces the extracellular space. The structural homology of NEP with ECE-1, ECE-2, KELL, and PEX is 35%, 30%, 31%, and 25%, respectively, but it can increase to 51%, 44%, 43%, and 39% if only the C-terminal end (about 250 amino acids) is considered. This level of similarity is enough to suppose a common origin and fold between these peptidases [1]. NEP is present as a noncovalently associated homodimer in some species [2], although a monomeric form has been found in rabbits [3]. The NEP structure consists of 750, 742, and 742 amino acids evidenced by cDNA cloning in rabbit [4], rat [5], and human [6], respectively. NEP structure and activity are maintained by the presence of intra-chain disulfide bridges [7]. Specifically, 12 cysteine residues are present in the NEP extracellular domain, and four of them are present in the first 32 amino acids immediately after the trans-membrane helix [8]. The active site of NEP, located in its C-terminus, can be divided into three interactive subsites: the S1 zing-binding moiety, and the S1′ and S2′ subsites [9]. Commonly to other peptidases, the zinc-binding sequence is HEXXH (histidine-glutamate-x-x-histidine). Specifically, the human histidines (His 583 and 587) altogether with a glutamate (Glu 646), in this sequence, are the three ligands for the zinc. Glu 584 is, instead, fundamental for its catalytic activity.

Studies have also described an alternative form of NEP, the soluble one (sNEP), which can be found in the plasma and urine [10,11] and has the same enzymatic activity of the transmembrane form [12].

A particular characteristic of NEP is the restricted active site that prevents the access of large compounds and explains its oligopeptidase activity. Among the active oligopeptide substrates for NEP, there are opioid peptides (such as enkephalins), natriuretic peptides, bradykinin, endothelin-1 (ET-1), adrenomedullin, substance P, glucagon, glucagon-like peptide, somatostatin, and angiotensin I (AI) and II (AII) [13,14]. All these substrates highlight the important role of NEP in the modulation of nociceptive and pressor responses, stimulating a growing interest in developing inhibitors of this enzyme to treat a variety of conditions varying from cardiovascular diseases to pain.

2 Physiologic/pathologic role

NEP expression is reported in different organ and tissues, such as the lungs, kidneys, liver, adipose tissue, brain, carotid body, vascular smooth muscle cells, endothelial cells, cardiac cells, fibroblasts, neutrophils, testes, flat bones of the skull, the mandibula, the vertebrae, the limb bones, articular cartilages, and synovia [12,15].

In the lung, NEP can participate to the pulmonary development and functions by hydrolyzing bombesin-like peptides [16]. NEP levels are, indeed, markers of lung dysfunctions [17]. Moreover, pulmonary NEP may modulate the inflammatory response to neuropeptides with a role in the constriction of airway smooth muscle cells, such as the substance P [18].

In the reproductive system, NEP was found in the testes, playing a role in the sperm formation and the processes related to fertility [19]; in the ovary, where it controls the maturation of follicles, the ovulation, and the blood flow

Metalloenzymes. https://doi.org/10.1016/B978-0-12-823974-2.00021-8

of ovaries [20]; and in the placenta, where by hydrolyzing the oxytocin [21], it regulates the peptide activity at the fetal-maternal interface [15]. Finally, high levels of NEP were found into the maternal circulation by placenta during particular conditions such as preeclampsia, which can predispose to several complications both in the mother and fetus, including the onset of hypertension and heart failure [22].

In the human brain, NEP is heterogeneously distributed with the highest amount found in the globus pallidus and pars reticulata of the substantia nigra [23]. NEP was also found on Schwann cells of peripheral nervous system [24], on axons, and synapses, suggesting that after its synthesis it is axonally translocated to synapses. In this regard, the NEP release by Schwann cells was found increased after axons damage, suggesting a role recovered in nerve growth and regeneration [25]. Moreover, NEP dysfunction was associated with the onset of Charcot-Marie-Tooth disease characterized by late-onset axonal neuropathy [26] with muscle weakness, atrophy, and sensory disturbance in the lower limbs. More recent data suggest a specific role of NEP in GABAergic and glutamatergic neurons [27], suggesting a potential in influencing these neurotransmissions. NEP can also metabolize different substrates in the brain including sensory and inflammatory neuropeptides such as tachykinins, encephalin, and neurokinins [28,29], recovering also a potential role in influencing pain perception [30]. Furthermore, by altering the effects of neuropeptides NEP can influence the movement regulation and the locomotor activity in the basal ganglia [31,32]. Finally, evidence showed a role of NEP inhibition in altering the cognitive function [15,32,33]. In fact, the ability of NEP in catabolizing the Amyloid β (Aβ) at multiple sites [34–36] suggests that an NEP deficiency can determine impairing of spatial working memory [37]. These findings made the increase in NEP activity a very attractive strategy for Alzheimer's disease.

In the cardiovascular system, NEP recovers a fundamental role in inactivating natriuretic peptides and in regulating the renin-angiotensin system (RAS). NEP cleaves all natriuretic peptides: the atrial natriuretic peptide (ANP), the brain natriuretic peptide (BNP), and the C-type natriuretic peptide (CNP) [38], but with a greater affinity the ANP and CNP [39]. By degrading natriuretic peptides, NEP limits their biological effects, which include natriuresis, diuresis, decrease in blood pressure, increase in endothelial permeability, and reduction of cardiac remodelling [40]. Moreover, NEP can influence the cardiovascular homeostasis by altering the RAS. In fact, NEP catabolizes AII and converts AI into angiotensin 1–7 (A1–7), which can induce vasodilation, natriuresis, and antiproliferative effects [40,41]. Based on this physiological role, the plasma sNEP is found to be a biomarker of heart failure, predicting the prognosis of cardiovascular death and hospitalization in patients with acute and chronic heart failure [12]. Furthermore, NEP can be a biomarker of chronic kidney disease at its early stage [42] and a study suggested that high NEP urinary levels are indicative of acute kidney injury [43].

In the adipose tissue, NEP can regulate adipocyte function and act as an adipokine regulator [44]. Moreover, NEP inhibition showed to improve whole-body insulin-mediated glucose disposal in obese insulin-resistant Zucker rats [45–47], proving that NEP is directly involved in insulin resistance. A study conducted in healthy subjects also showed that an increase in the plasma NEP activity is associated with insulin resistance and obesity [48]. Moreover, NEP showed to accelerate adipogenesis by enhancing the insulin-mediated PI3K-Akt activation [49]. Finally, the increased NEP activity in human microvascular endothelial cells was found related to hyperlipidaemia and hyperglycaemia associated with type 2 diabetes and delayed diabetic wound healing, supposing a role of NEP in impairing the normal tissue response to injuries [50].

The recent ongoing pandemic caused by the new coronavirus (SARS-CoV2) has also raised the question about a possible physiological role of NEP in the development of viral infection. The most relevant and severe complications of SARS-CoV2 infection are in the lungs and heart where the viral receptor, the angiotensin-converting enzyme-2 (ACE2), mediates the viral entry into the cells [51,52]. Considering the aforementioned role recovered by NEP into the lungs and heart in regulating the RAS, a protective effect may be hypothesized by this enzyme against the viral inflammation and fibrosis [53,54].

3 Classes of inhibitors and their design

Based on the important physiological role recovered by NEP and discovered over the years, the interest has been growing for designing compounds able to block this enzyme. The fundamental characteristics for a NEP inhibitor are the presence of: a zinc-binding group (P1), a hydrophobic group (P1′), and an amide group (or amide surrogate, P2′) available for establishing two hydrogen bonds. Generally, the zinc-coordinating group can be a carboxylic acid, or a thiol, or a phosphonic acid analog, which is divided from the chiral anchor atom by a single linker atom that can be nitrogen or carbon [1,55–57]. Potent NEP inhibitors can be either tri- and tetrapeptides or smaller molecules such as dipeptide mimics [58]. Over the years, different NEP inhibitors were identified, including thiorphan, candoxatril, ecadotril, and sacubitril [50]. However, they failed in showing efficacy probably due to the increased levels of AII and ET-1, which degradation is also mediated by NEP

[59,60]. For this reason, the NEP block alone appears unable to induce cardiovascular benefits but necessitates of a multi-target action on more vasoactive peptidases [61]. Therefore, attempts were made for designing compounds able to inhibit both NEP and Angiotensin-Converting Enzyme (ACE) that were also called vasopeptidase inhibitors [50]. These dual inhibitors were designed rationally based on the characteristics of single ACE and NEP inhibitors. Indeed, these molecules are characterized by a benzyl group in P1′, known to be important for NEP inhibition, and a proline group in P2′ present in ACE inhibitors such as captopril. From this chemical structure, a series of potent mercaptoacyl dipeptides with dual activity were developed, leading to the optimization of the final compound omapatrilat (7,6-fused bicyclic thiazepinone) [59]. However, this development was interrupted in clinical phases by findings showing that an NEP/ACE inhibition was not safe. Indeed, omapatrilat showed to increase the risk of serious angioedema when compared to enalapril in phase 3 clinical trials [62,63]. The main hypothesis for this unfavorable event was the excessive increase in bradykinin levels obtained by the simultaneous inhibition of these enzymes [50]. On this front, NEP and ECE inhibitors such as daglutril (previously known as SLV-306) and SLV338 were developed based on the discovery that phosphoramidon inhibited both ECE and NEP [59]. Among them, the most advanced developed was daglutril, an aromatic heteropolycyclic compound composed by a sequence of exactly two alpha-amino acids joined by a peptide bond (dipeptides). Despite daglutril showed preliminary efficacy data in increasing levels of the precursor of ET-1 (big ET-1) and natriuretic peptides, it was not further investigated due to the lack of efficacy in lowering the systemic blood pressure [64,65], suggesting that an effective strategy needs to incorporate the RAS blockade. In this regard, a triple inhibition of ACE, ECE, and NEP was also tested in experimental preclinical models. The most studied ACE/ECE/NEP inhibitor was CGS-35601, an α-mercaptodipeptide with a central cyclic non-natural amino acid and a tryptophan in P2′ that binds the S2′ subsite of all three enzymes [59]. This strategy despite showing initial positive results was later abandoned due to the safety concerns raised for vasopeptidase inhibitors [60].

Based on these findings of the lacking efficacy with NEP inhibitors alone and the necessity to also counteract the RAS activation mediated by the increase in AII, a new class of compounds was developed and introduced in clinical practice. This class is represented by a dual angiotensin receptor-neprilysin inhibitor (ARNi), which compound is LCZ696. This drug is an orally available single molecule, which releases after administration two compounds: the angiotensin receptor blocker (valsartan) and the pro-drug of the NEP inhibitor (AHU377) that is hydrolyzed by

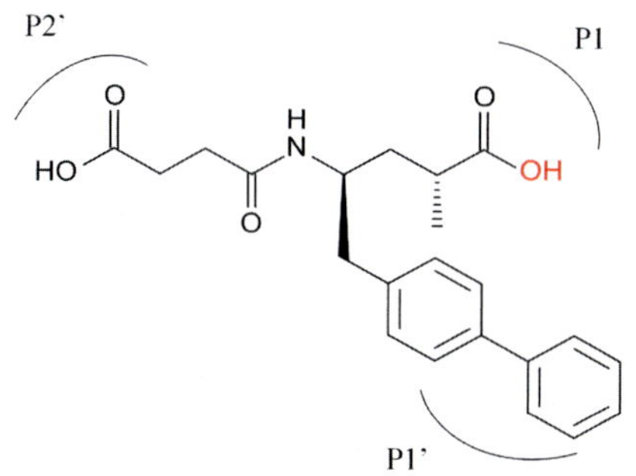

FIG. 1 Chemical structures of sacubitril. The portion P1 is the zinc coordinating group, P1′ and P2′ bind S1′ and S2′ moieties of NEP active site, respectively.

cleavage of the ethyl ester into the active form (LBQ657 or sacubitril), rapidly providing plasma exposure to both drugs [61]. Sacubitril (Fig. 1) binds noncovalently the active site of NEP through hydrogen and van der Waals interactions that involve all its functional groups, giving its high inhibitory potency (5 nM) [66]. A distinct feature of sacubitril compared to other NEP inhibitors is the presence of a longer linker between the zinc group (carboxylate, P1) and the chiral carbon atom binding the P1 and P2′ molecule moieties that anchor the S1′ and S2′ subsites of the active site, respectively. The succinic acid in P2′ is the best option when compared to shorter or longer chain. Indeed, it allows an optimal interaction of the carboxylic acid with Arg102 and Arg110, while the biphenyl in P1′ binds deeply the S1′ of NEP, allowing conformational modifications for perfect hydrophobic interactions [65]. A recent research investigated the chemistry optimization and validation of new compounds binding the NEP showed the synthesis of selective, orally bioavailable, and subnanomolar inhibitors that had a 17-fold high potency when a chlorine atom was added to P1′ [9]. However, this study encouraged to evaluate also other linkers and changes of S2′ binding moiety (P2′).

4 Clinically used agents or compounds in clinical development

Based on the biological activity of NEP, its inhibitors have been evaluated or are in evaluation for many therapeutic indications. The first successful developed strategy was the ARNi LCZ696 (sacubitril-valsartan) that has a dual cardioprotective action (NEP inhibition and angiotensin receptor blockade). LCZ696 showed to reduce heart failure hospitalizations and cardiac mortality when compared to the ACE inhibitor enalapril in the phase 3 PARADIGM-HF clinical trial [67]. Based on this result, this association sacubitril-valsartan was authorized for the treatment of heart failure with reduced ejection fraction. However, it failed in showing any efficacy for patients with a preserved ejection fraction [68]. Moreover, it was investigated that hypertension showed initial promising results in lowering the blood pressure [69,70]. The major safety concern with ARNi was the potential risk of Alzheimer's disease due to the role of NEP in the beta-amyloid degradation. Sacubitril could indeed determine an increase in beta-amyloid plaque deposition and potentially increase the risk of Alzheimer's disease. However, a reanalysis of the PARADIGM-HF trial found no increased dementia-related events with sacubitril-valsartan compared to enalapril, although longer follow-up and more sensitive tools may be needed to detect cognitive impairment [71]. It is ongoing a multicenter, randomized, double-blinded trial (the PERSPECTIVE trial; NCT02884206) to evaluate the safety and long-term neurocognitive effects of sacubitril/valsartan in patients with chronic heart failure and preserved ejection fraction.

In the setting of cardiovascular diseases, myocardial infarction has also raised interest for the development of NEP inhibitors. Natriuretic peptide levels rise during a myocardial infarction as response to the stimulus of the infarcted area and are associated with a reduced survival [72]. However, potentially, they should induce favorable effects by reducing the ischemia reperfusion injury, blunting sympathetic nerve activity, and inhibiting neutrophil degranulation [73]. Data on humans with anterior myocardial infarction showed that the infusion of ANP is associated with a reduction of cardiac sympathetic nerve activity and left ventricular remodeling [74], while the infusion of BNP with an improvement of left ventricular ejection fraction and less ventricular dilatation [75]. On this ground, ARNi may be considered for the treatment of post-myocardial infarction. The phase 3 PARADISE-MI clinical trial has indeed tested the hypothesis that sacubitril/valsartan was superior to the ACE inhibitor (ramipril) in patients with acute myocardial infarction complicated by a reduced left ventricular ejection fraction and pulmonary congestion, but it did not found a significant lower incidence of death from cardiovascular causes or incident heart failure in such patients [76]. Ongoing clinical trials of sacubitril/valsartan investigating new potential cardiac therapeutic indications are listed in Table 1.

NEP inhibitors also are in clinical development for their potential role in patients with type 2 diabetes mellitus. This use is supported by evidence showing that among NEP substrates there is the incretin glucagon-like peptide-1 (GLP-1), natriuretic peptides, and bradykinin, which all can modulate the glucose metabolism [77]. In this regard, three human studies supported the use of NEP inhibitors in the prevention and treatment of type 2 diabetes [78–80]. One study was a post hoc analysis of the PARADIGM-HF clinical trial that showed a greater reduction in HbA1c and fewer request of oral glucose-lowering medications or insulin therapy in patients with type 2 diabetes and heart failure treated with sacubitril-valsartan than enalapril [80]. Another study demonstrated an improvement in sensitivity and lipid mobilization in obese hypertensive patients treated with the ARNi than amlodipine (a calcium channel blocker) [78]. Finally, a study found that a change of treatment from an ACE inhibitor or angiotensin receptor blocker to ARNi for 3 months determined a reduction of plasma neprilysin activity and fructosamine levels, a marker of protein glycation [79]. However, considering that NEP substrates can be metabolized by other enzymes (such as DPP-4), resulting in a decreased efficacy for the use of a NEP inhibitor alone, combination strategy inhibiting more enzymes could be considered as a better option for treating type 2 diabetes mellitus [77].

NEP inhibitors are also potentially considered for the prevention of the renal dysfunction as an increase in natriuretic peptides, especially ANP, was found effective in increasing the glomerular filtration rate (GFR) by mediating a dilatation of the glomerular afferent arteriolar and a constriction of the efferent arteriolar [81]. Moreover, the administration of ANP and BNP in in healthy humans was associated with an improvement of GFR [81]. On this basis, clinical data showed that sacubitril-valsartan was associated with a less significant decline in eGFR and lower levels of serum creatinine compared to valsartan in the PARAMOUNT phase 2 trial of subjects with heart failure with preserved ejection fraction and hypertension [82]. Finally, sacubitril-valsartan was associated with less events of elevated creatinine or serum potassium in the PARADIGM trial [67]. These findings lay the foundations to clinically investigate NEP inhibitors in kidney diseases. The randomized UK HARP-III trial (United Kingdom Heart and Renal Protection-III) showed similar effects on kidney function and albuminuria with sacubitril-valsartan compared to irbesartan over 12 months, but with additional effect of sacubitril-valsartan in lowering blood pressure and cardiac biomarkers in subjects with chronic kidney disease [83].

Based on the role of NEP in nociception, NEP inhibitors are also hypothesized for the pain management. Specifically, considering that NEP substrates can be degraded from other peptidases, dual inhibitors such as PL37 and PL265 are in clinical development. PL37 is an oral dual enkephalinase inhibitor developed as a prodrug, rapidly converted into two active compounds able to inhibit NEP and aminopeptidase N (APN), thereby increasing the local concentrations of enkephalins [84]. PL37 is currently under investigation for postoperative pain after cataract surgery in phase I clinical trials and for neuropathic pain (diabetic neuropathy) and migraine in phase 2 clinical trials. PL265 is instead a single active compound able to block both NEP and APN that is in phase 2 clinical evaluation for the treatment of neuropathic pain and pain associated with dry eye [84].

TABLE 1 Characteristics of ongoing clinical trials on sacubitril-valsartan investigating cardiac outcomes.

Clinical trial identifier	Phase	Study design	Indication	Experimental arm	Comparator arm	Primary endpoint	Study completion date
NCT04912167	3	RCT, single blind	ST-elevation myocardial infarction	Sacubitril-valsartan	Enalapril	Change of the indexed left ventricular mass from baseline to 6-month follow-up on cardiovascular magnetic resonance (CMR)	June 2026
NCT04864145	–	RCT, single blind	Aortic valve disease	Transcatheter aortic valve implantation	Therapy with sacubitril-valsartan, diuretics, dihydropyridine calcium channel blocker, angiotensin-converting enzyme inhibitors/angiotensin receptor blockers, or beta blockers	A composite of all cause death, disabling stroke, or heart failure rehospitalization at 12 months	May 2031
NCT03832660	2	RCT, open-label	Hypertrophic Cardiomyopathy	Lifestyle and sacubitril-valsartan	No intervention	Change in exercise tolerance (peak oxygen consumption and anaerobic threshold) post intervention at 4 months	June 2022
NCT05212597	2	RCT, open-label	Aortic valve insufficiency	Sacubitril-valsartan	Amlodipine-losartan	Change of left ventricular end-diastolic volume index from baseline to 12 months follow-up	December 2024
NCT04800081	–	RCT, open-label	Hypertension in postmenopausal women	Sacubitril-valsartan	Valsartan	Blood pressure, 24 h automatic blood pressure monitoring, home blood pressure monitoring, office blood pressure at 12 weeks	June 2021
NCT05545059	3	RCT, double blind	Resistant hypertension	Sacubitril-valsartan	Valsartan	Change in 24 hours average ambulatory systolic pressure from baseline to 8 weeks after randomization	March 2023
NCT04149990	2	RCT, quadruple blind	Myocardial infarction, diastolic dysfunction	Sacubitril-valsartan	Placebo	The primary endpoint will be the ratio of mean pulmonary capillary wedge pressure (PCWP) at peak exercise divided by cardiac index at peak exercise at 26 weeks	May 2023
NCT04929600	4	RCT, quadruple blind	Hypertension, left ventricular hypertrophy	Sacubitril-valsartan	Amlodipine	Change in left ventricular global longitudinal strain (LVGLS) after 24 weeks	June 2023

Continued

TABLE 1 Characteristics of ongoing clinical trials on sacubitril-valsartan investigating cardiac outcomes—cont'd

Clinical trial identifier	Phase	Study design	Indication	Experimental arm	Comparator arm	Primary endpoint	Study completion date
NCT04853758	3	RCT, triple blind	Chagas cardiomyopathy	Sacubitril-valsartan	Enalapril	Change of left ventricular ejection fraction (LVEF) at 6 months	March 2023
NCT03553810	2	RCT, single blind	Hypertensive heart disease	Sacubitril-valsartan	Valsartan	Changes from baseline in fibrosis volume at 52 weeks	December 2023
NCT04971720	2/3	RCT, quadruple blind	Hypertension in a high-risk population (obesity, cardiovascular diseases)	Sacubitril-valsartan	Valsartan	Change in mean nocturnal systolic blood pressure	January 2027

References

[1] Oefner C, D'Arcy A, Hennig M, Winkler FK, Dale GE. Structure of human neutral endopeptidase (Neprilysin) complexed with phosphoramidon. J Mol Biol 2000;296(2):341–9. https://doi.org/10.1006/JMBI.1999.3492.

[2] Fulcher IS, Kenny AJ. Proteins of the kidney microvillar membrane. The amphipathic forms of endopeptidase purified from pig kidneys. Biochem J 1983;211(3):743–53. https://doi.org/10.1042/BJ2110743.

[3] Kerr MA, Kenny AJ. The molecular weight and properties of a neutral metallo-endopeptidase from rabbit kidney brush border. Biochem J 1974;137(3):489–95. https://doi.org/10.1042/BJ1370489.

[4] Devault A, Lazure C, Nault C, Le Moual H, Seidah NG, Chrétien M, Kahn P, Powell J, Mallet J, Beaumont A. Amino acid sequence of rabbit kidney neutral endopeptidase 24.11 (enkephalinase) deduced from a complementary DNA. EMBO J 1987;6(5):1317–22. https://doi.org/10.1002/J.1460-2075.1987.TB02370.X.

[5] Malfroy B, Schofield PR, Kuang WJ, Seeburg PH, Mason AJ, Henzel WJ. Molecular cloning and amino acid sequence of rat enkephalinase. Biochem Biophys Res Commun 1987;144(1):59–66. https://doi.org/10.1016/S0006-291X(87)80475-8.

[6] Malfroy B, Kuang WJ, Seeburg PH, Mason AJ, Schofield PR. Molecular cloning and amino acid sequence of human enkephalinase (neutral endopeptidase). FEBS Lett 1988;229(1):206–10. https://doi.org/10.1016/0014-5793(88)80828-7.

[7] Tam LT, Engelbrecht S, Talent JM, Gracy RW, Erdos EG. The importance of disulfide bridges in human endopeptidase (enkephalinase) after proteolytic cleavage. Biochem Biophys Res Commun 1985;133(3):1187–92. https://doi.org/10.1016/0006-291X(85)91262-8.

[8] Relton JM, Gee NS, Matsas R, Turner AJ, Kenny AJ. Purification of endopeptidase-24.11 ('enkephalinase') from pig brain by immunoadsorbent chromatography. Biochem J 1983;215(3):519–23. https://doi.org/10.1042/BJ2150519.

[9] Kawanami T, Karki RG, Cody E, Liu Q, Liang G, Ksander GM, Rigel DF, Schiering N, Gong Y, Coppola GM, Iwaki Y, Sun R, Neubert A, Fan L, Ingles S, D'Arcy A, Villard F, Ramage P, Jeng AY, Mogi M. - Structure-guided design of substituted biphenyl butanoic acid derivatives as neprilysin inhibitors. ACS Med Chem Lett 2020;11(2):188–94. https://doi.org/10.1021/ACSMEDCHEMLETT.9B00578.

[10] Aviv R, Gurbanov K, Hoffman A, Blumberg S, Winaver J. Urinary neutral endopeptidase 24.11 activity: modulation by chronic salt loading. Kidney Int 1995;47(3):855–60. https://doi.org/10.1038/KI.1995.128.

[11] Bayés-Genís A, Barallat J, Galán A, De Antonio M, Domingo M, Zamora E, Urrutia A, Lupón J. Soluble neprilysin is predictive of cardiovascular death and heart failure hospitalization in heart failure patients. J Am Coll Cardiol 2015;65(7):657–65. https://doi.org/10.1016/J.JACC.2014.11.048.

[12] Ramanathan K, Padmanabhan G. Soluble neprilysin: a versatile biomarker for heart failure, cardiovascular diseases and diabetic complications—a systematic review. Indian Heart J 2020;72(1):14–9. https://doi.org/10.1016/J.IHJ.2020.01.006.

[13] Cappetta, D., De Angelis, A., Flamini, S., Cozzolino, A., Bereshchenko, O., Ronchetti, S., Cianflone, E., Gagliardi, A., Ricci, E., Rafaniello, C., Rossi, F., Riccardi, C., Berrino, L., Bruscoli, S., & Urbanek, K. (2021). Deficit of glucocorticoid-induced leucine zipper amplifies angiotensin-induced cardiomyocyte hypertrophy and diastolic dysfunction. J Cell Mol Med, 25(1), 217–228. https://doi.org/https://doi.org/10.1111/JCMM.15913.

[14] Pavo N, Prausmüller S, Bartko PE, Goliasch G, Hülsmann M. Neprilysin as a biomarker: challenges and opportunities. Card Fail Rev 2020;6. https://doi.org/10.15420/CFR.2019.21.

[15] Nalivaeva NN, Zhuravin IA, Turner AJ. Neprilysin expression and functions in development, ageing and disease. Mech Ageing Dev 2020;192. https://doi.org/10.1016/J.MAD.2020.111363.

[16] Sunday ME, Hua J, Torday JS, Reyes B, Shipp MA. CD10/neutral endopeptidase 24.11 in developing human fetal lung. Patterns of expression and modulation of peptide-mediated proliferation. J Clin Invest 1992;90(6):2517–25. https://doi.org/10.1172/JCI116145.

[17] Wick MJ, Buesing EJ, Wehling CA, Loomis ZL, Cool CD, Zamora MR, Miller YE, Colgan SP, Hersh LB, Voelkel NF, Dempsey EC. Decreased neprilysin and pulmonary vascular remodeling in chronic obstructive pulmonary disease. Am J Respir Crit Care Med 2011;183(3):330. https://doi.org/10.1164/RCCM.201002-0154OC.

[18] Borson DB. Roles of neutral endopeptidase in airways. Am J Physiol 1991;260(4 Pt 1). https://doi.org/10.1152/AJPLUNG.1991.260.4.L212.

[19] Ghaddar G, Ruchon AF, Carpentier M, Marcinkiewicz M, Seidah NG, Crine P, Desgroseillers L, Boileau G. Molecular cloning and biochemical characterization of a new mouse testis soluble-zinc-metallopeptidase of the neprilysin family. Biochem J 2000;347(Pt 2):419. https://doi.org/10.1042/0264-6021:3470419.

[20] Zappulla JP, DesGroseillers L. Neutral endopeptidase is expressed on the follicular granulosa cells of rabbit ovaries. Comp Biochem Physiol B Biochem Mol Biol 2001;129(4):863–70. https://doi.org/10.1016/S1096-4959(01)00390-6.

[21] Johnson AR, Skidgel RA, Gafford JT, Erdös EG. Enzymes in placental microvilli: Angiotensin I converting enzyme, angiotensinase A, carboxypeptidase, and neutral endopeptidase ("enkephalinase"). Peptides 1984;5(4):789–96. https://doi.org/10.1016/0196-9781(84)90023-8.

[22] Gill M, Motta-Mejia C, Kandzija N, Cooke W, Zhang W, Cerdeira AS, Bastie C, Redman C, Vatish M. Placental syncytiotrophoblast-derived extracellular vesicles carry active NEP (neprilysin) and are increased in preeclampsia. Hypertension 2019;73(5):1112–9. https://doi.org/10.1161/HYPERTENSIONAHA.119.12707.

[23] Llorens C, Malfroy B, Schwartz JC, Gacel G, Roques BP, Roy J, Morgat JL, Javoy-Agid F, Agid Y. Enkephalin dipeptidyl carboxypeptidase (enkephalinase) activity: selective radioassay, properties, and regional distribution in human brain. J Neurochem 1982;39(4):1081–9. https://doi.org/10.1111/J.1471-4159.1982.TB11500.X.

[24] Barnes K, Doherty S, Turner AJ. Endopeptidase-24.11 is the integral membrane peptidase initiating degradation of somatostatin in the hippocampus. J Neurochem 1995;64(4):1826–32. https://doi.org/10.1046/J.1471-4159.1995.64041826.X.

[25] Kioussi C, Crine P, Matsas R. Endopeptidase-24.11 is suppressed in myelin-forming but not in non-myelin-forming Schwann cells during development of the rat sciatic nerve. Neuroscience 1992;50(1):69–83. https://doi.org/10.1016/0306-4522(92)90382-C.

[26] Higuchi Y, Hashiguchi A, Yuan J, Yoshimura A, Mitsui J, Ishiura H, Tanaka M, Ishihara S, Tanabe H, Nozuma S, Okamoto Y, Matsuura E, Ohkubo R, Inamizu S, Shiraishi W, Yamasaki R, Ohyagi Y, Kira JI, Oya Y, Takashima H. Mutations in MME cause an autosomal-recessive Charcot-Marie-Tooth disease type 2. Ann Neurol 2016;79(4):659–72. https://doi.org/10.1002/ANA.24612.

[27] Pacheco-Quinto J, Eckman CB, Eckman EA. Major amyloid-β-degrading enzymes, endothelin-converting enzyme-2 and neprilysin, are expressed by distinct populations of GABAergic interneurons in hippocampus and neocortex. Neurobiol Aging 2016;48:83–92. https://doi.org/10.1016/J.NEUROBIOLAGING.2016.08.011.

[28] Bayes-Genis A, Barallat J, Richards AM. A test in context: neprilysin: function, inhibition, and biomarker. J Am Coll Cardiol 2016;68 (6):639–53. https://doi.org/10.1016/J.JACC.2016.04.060.

[29] Turner AJ, Nalivaeva NN. New insights into the roles of metalloproteinases in neurodegeneration and neuroprotection. Int Rev Neurobiol 2007;82:113–35. https://doi.org/10.1016/S0074-7742(07)82006-X.

[30] Fischer HS, Zernig G, Hauser KF, Gerard C, Hersh LB, Saria A. Neutral endopeptidase knockout induces hyperalgesia in a model of visceral pain, an effect related to bradykinin and nitric oxide. J Mol Neurosci 2002;18(1–2):129–34. https://doi.org/10.1385/JMN:18:1-2:129.

[31] Chen XY, Xue Y, Chen H, Chen L. The globus pallidus as a target for neuropeptides and endocannabinoids participating in central activities. Peptides 2020;124. https://doi.org/10.1016/J.PEPTIDES.2019.170210.

[32] Fischer HS, Zernig G, Schuligoi R, Miczek KA, Hauser KF, Gerard C, Saria A. Alterations within the endogenous opioid system in mice with targeted deletion of the neutral endopeptidase ('enkephalinase') gene. Regul Pept 2000;96(1–2):53–8. https://doi.org/10.1016/S0167-0115 (00)00200-7.

[33] Madani R, Poirier R, Wolfer DP, Welzl H, Groscurth P, Lipp HP, Lu B, El Mouedden M, Mercken M, Nitsch RM, Mohajeri MH. Lack of neprilysin suffices to generate murine amyloid-like deposits in the brain and behavioral deficit in vivo. J Neurosci Res 2006;84 (8):1871–8. https://doi.org/10.1002/JNR.21074.

[34] Iwata N, Tsubuki S, Takaki Y, Shirotani K, Lu B, Gerard NP, Gerard C, Hama E, Lee HJ, Saido TC. Metabolic regulation of brain Abeta by neprilysin. Science 2001;292(5521):1550–2. https://doi.org/10.1126/ SCIENCE.1059946.

[35] Iwata N, Tsubuki S, Takaki Y, Watanabe K, Sekiguchi M, Hosoki E, Kawashima-Morishima M, Lee HJ, Hama E, Sekine-Aizawa Y, Saido TC. Identification of the major Abeta1-42-degrading catabolic pathway in brain parenchyma: suppression leads to biochemical and pathological deposition. Nat Med 2000;6(2):143–50. https:// doi.org/10.1038/72237.

[36] Zou LB, Mouri A, Iwata N, Saido TC, Wang D, Wang MW, Mizoguchi H, Noda Y, Nabeshima T. Inhibition of neprilysin by infusion of thiorphan into the hippocampus causes an accumulation of amyloid Beta and impairment of learning and memory. J Pharmacol Exp Ther 2006;317(1):334–40. https://doi.org/10.1124/JPET.105.095687.

[37] Hüttenrauch M, Baches S, Gerth J, Bayer TA, Weggen S, Wirths O. Neprilysin deficiency alters the neuropathological and behavioral phenotype in the 5XFAD mouse model of Alzheimer's disease. J Alzheimer's Dis 2015;44(4):1291–302. https://doi.org/10.3233/JAD-142463.

[38] Fossiez F, Lemay G, Labonte N, Parmentier-Lesage F, Boileau G, Crine P. Secretion of a functional soluble form of neutral endopeptidase-24.11 from a baculovirus-infected insect cell line. Biochem J 1992;284(Pt 1):53–9. https://doi.org/10.1042/BJ2840053.

[39] Almenoff J, Orlowski M. Membrane-bound kidney neutral metalloendopeptidase: interaction with synthetic substrates, natural peptides, and inhibitors. Biochemistry 1983;22(3):590–9. https://doi.org/ 10.1021/BI00272A011.

[40] Rossi F, Mascolo A, Mollace V. The pathophysiological role of natriuretic peptide-RAAS cross talk in heart failure. Int J Cardiol 2017;226:121–5. https://doi.org/10.1016/J.IJCARD.2016.03.080.

[41] Mascolo A, Urbanek K, De Angelis A, Sessa M, Scavone C, Berrino L, Rosano GMC, Capuano A, Rossi F. Angiotensin II and angiotensin 1-7: which is their role in atrial fibrillation? Heart Fail Rev 2020;25 (2):367–80. https://doi.org/10.1007/S10741-019-09837-7.

[42] Gutta S, Grobe N, Kumbaji M, Osman H, Saklayen M, Li G, Elased KM. Increased urinary angiotensin converting enzyme 2 and neprilysin in patients with type 2 diabetes. Am J Physiol Renal Physiol 2018;315 (2):F263–74. https://doi.org/10.1152/AJPRENAL.00565.2017.

[43] Pajenda S, Mechtler K, Wagner L. Urinary neprilysin in the critically ill patient. BMC Nephrol 2017;18(1). https://doi.org/10.1186/ S12882-017-0587-5.

[44] Schling P, Schäfer T. Human adipose tissue cells keep tight control on the angiotensin II levels in their vicinity. J Biol Chem 2002;277 (50):48066–75. https://doi.org/10.1074/JBC.M204058200.

[45] Arbin, V., Claperon, N., Fournié-Zaluski, M.C., Roques, B.P., & Peyroux, J. (2001). Acute effect of the dual angiotensin-converting enzyme and neutral endopeptidase 24-11 inhibitor mixanpril on insulin sensitivity in obese Zucker rat. Br J Pharmacol, 133(4), 495–502. https://doi.org/https://doi.org/10.1038/SJ.BJP.0704098.

[46] Arbin V, Claperon N, Fournié-Zaluski MC, Roques BP, Peyroux J. Effects of dual angiotensin-converting enzyme and neutral endopeptidase 24-11 chronic inhibition by mixanpril on insulin sensitivity in lean and obese Zucker rats. J Cardiovasc Pharmacol 2003;41 (2):254–64. https://doi.org/10.1097/00005344-200302000-00015.

[47] Wang CH, Leung N, Lapointe N, Szeto L, Uffelman KD, Giacca A, Rouleau JL, Lewis GF. Vasopeptidase inhibitor omapatrilat induces profound insulin sensitization and increases myocardial glucose uptake in Zucker fatty rats: studies comparing a vasopeptidase inhibitor, angiotensin-converting enzyme inhibitor, and angiotensin II type I receptor blocker. Circulation 2003;107(14):1923–9. https:// doi.org/10.1161/01.CIR.0000062646.09566.CC.

[48] Standeven KF, Hess K, Carter AM, Rice GI, Cordell PA, Balmforth AJ, Lu B, Scott DJ, Turner AJ, Hooper NM, Grant PJ. Neprilysin, obesity and the metabolic syndrome. Int J Obes 2010;35(8):1031–40. https://doi.org/10.1038/ijo.2010.227.

[49] Kim J, Han D, Byun SH, Kwon M, Cho SJ, Koh YH, Yoon K. Neprilysin facilitates adipogenesis through potentiation of the phosphatidylinositol 3-kinase (PI3K) signaling pathway. Mol Cell Biochem 2017;430(1–2):1–9. https://doi.org/10.1007/S11010-017-2948-6.

[50] Muangman P, Spenny ML, Tamura RN, Gibran NS. Fatty acids and glucose increase neutral endopeptidase activity in human microvascular endothelial cells. Shock (Augusta, Ga) 2003;19(6):508–12. https://doi.org/10.1097/01.SHK.0000055815.40894.16.

[51] Mascolo A, Scavone C, Rafaniello C, De Angelis A, Urbanek K, di Mauro G, Cappetta D, Berrino L, Rossi F, Capuano A. The role of renin-angiotensin-aldosterone system in the heart and lung: focus on COVID-19. Front Pharmacol 2021;12. https://doi.org/10.3389/ FPHAR.2021.667254.

[52] Mascolo A, Scavone C, Rafaniello C, Ferrajolo C, Racagni G, Berrino L, Paolisso G, Rossi F, Capuano A. Renin-angiotensin system and coronavirus disease 2019: a narrative review. Front Cardiovas Med 2020;7:143. https://doi.org/10.3389/FCVM.2020.00143/BIBTEX.

[53] Manolis AS, Manolis TA, Manolis AA, Melita H. The controversy of renin-angiotensin-system blocker facilitation versus countering COVID-19 infection. J Cardiovasc Pharmacol 2020;76(4):397–406. https://doi.org/10.1097/FJC.0000000000000894.

[54] Mohammed El Tabaa M, Mohammed El Tabaa M. Targeting neprilysin (NEP) pathways: a potential new hope to defeat COVID-19 ghost. Biochem Pharmacol 2020;178. https://doi.org/10.1016/J. BCP.2020.114057.

[55] Glossop MS, Bazin RJ, Dack KN, Fox DNA, MacDonald GA, Mills M, Owen DR, Phillips C, Reeves KA, Ringer TJ, Strang RS, Watson CAL. Synthesis and evaluation of heteroarylalanine diacids as potent and selective neutral endopeptidase inhibitors. Bioorg Med Chem Lett 2011;21(11):3404–6. https://doi.org/10.1016/J.BMCL.2011.03.109.

[56] Oefner C, Pierau S, Schulz H, Dale GE. Structural studies of a bifunctional inhibitor of neprilysin and DPP-IV. Acta Crystallogr D Biol Crystallogr 2007;63(Pt 9):975–81. https://doi.org/10.1107/S0907444907036281.

[57] Sahli S, Frank B, Schweizer WB, Diederich F, Blum-Kaelin D, Aebi JD, Böhm HJ, Oefner C, Dale GE. Second-generation inhibitors for the metalloprotease neprilysin based on bicyclic heteroaromatic scaffolds: synthesis, biological activity, and x-ray crystal-structure analysis. Helv Chim Acta 2005;88(4):731–50. https://doi.org/10.1002/HLCA.200590051.

[58] Bohacek R, De Lombaert S, McMartin C, Priestle J, Grütter M. Three-dimensional models of ACE and NEP inhibitors and their use in the design of potent dual ACE/NEP inhibitors. J Am Chem Soc 1996;118 (35):8231–49. https://doi.org/10.1021/JA950818Y/SUPPL_FILE/JA8231.PDF.

[59] Arendse LB, Jan Danser AH, Poglitsch M, Touyz RM, Burnett JC, Llorens-Cortes C, Ehlers MR, Sturrock ED. Novel therapeutic approaches targeting the renin-angiotensin system and associated peptides in hypertension and heart failure. Pharmacol Rev 2019;71 (4):539. https://doi.org/10.1124/PR.118.017129.

[60] Von Lueder TG, Atar D, Krum H. Current role of neprilysin inhibitors in hypertension and heart failure. Pharmacol Ther 2014;144(1):41–9. https://doi.org/10.1016/J.PHARMTHERA.2014.05.002.

[61] Campbell DJ. Vasopeptidase inhibition: a double-edged sword? Hypertension 2003;41(3 I):383–9. https://doi.org/10.1161/01.HYP.0000054215.71691.16.

[62] Kostis JB, Packer M, Black HR, Schmieder R, Henry D, Levy E. Omapatrilat and enalapril in patients with hypertension: the omapatrilat cardiovascular treatment vs. enalapril (OCTAVE) trial. Am J Hypertens 2004;17(2):103–11. https://doi.org/10.1016/J.AMJHYPER.2003.09.014.

[63] Packer M, Califf RM, Konstam MA, Krum H, McMurray JJ, Rouleau JL, Swedberg K. Comparison of omapatrilat and enalapril in patients with chronic heart failure: the Omapatrilat Versus Enalapril Randomized Trial of Utility in Reducing Events (OVERTURE). Circulation 2002;106(8):920–6. https://doi.org/10.1161/01.CIR.0000029801.86489.50.

[64] Dickstein K, De Voogd HJ, Miric MP, Willenbrock R, Mitrovic V, Pacher R, Koopman PA. Effect of single doses of SLV306, an inhibitor of both neutral endopeptidase and endothelin-converting enzyme, on pulmonary pressures in congestive heart failure. Am J Cardiol 2004;94(2):237–9. https://doi.org/10.1016/J.AMJCARD.2004.03.074.

[65] Schiering N, D'Arcy A, Villard F, Ramage P, Logel C, Cumin F, Ksander GM, Wiesmann C, Karki RG, Mogi M. Structure of neprilysin in complex with the active metabolite of sacubitril. Sci Rep 2016;6. https://doi.org/10.1038/SREP27909.

[66] Ksander GM, Ghai RD, deJesus R, Diefenbacher CG, Yuan A, Berry C, Sakane Y, Trapani A. Dicarboxylic acid dipeptide neutral endopeptidase inhibitors. J Med Chem 1995;38(10):1689–700. https://doi.org/10.1021/JM00010A014.

[67] McMurray JJ, Packer M, Desai AS, Gong J, Lefkowitz MP, Rizkala AR, Rouleau JL, Shi VC, Solomon SD, Swedberg K, Zile MR. Angiotensin-neprilysin inhibition versus enalapril in heart failure. N Engl J Med 2014;371(11):132–3. https://doi.org/10.1056/NEJMOA1409077.

[68] Mascolo A, di Mauro G, Cappetta D, De Angelis A, Torella D, Urbanek K, Berrino L, Nicoletti GF, Capuano A, Rossi F. Current and future therapeutic perspective in chronic heart failure. Pharmacol Res 2022;175. https://doi.org/10.1016/J.PHRS.2021.106035.

[69] Geng Q, Yan R, Wang Z, Hou F. Effects of LCZ696 (sacubitril/valsartan) on blood pressure in patients with hypertension: a meta-analysis of randomized controlled trials. Cardiology 2020;145 (9):589–98. https://doi.org/10.1159/000507327.

[70] Li Q, Li L, Wang F, Zhang W, Guo Y, Wang F, Liu Y, Jia J, Lin S. Effect and safety of LCZ696 in the treatment of hypertension: a meta-analysis of 9 RCT studies. Medicine 2019;98(28). https://doi.org/10.1097/MD.0000000000016093.

[71] Cannon JA, Shen L, Jhund PS, Kristensen SL, Køber L, Chen F, Gong J, Lefkowitz MP, Rouleau JL, Shi VC, Swedberg K, Zile MR, Solomon SD, Packer M, McMurray JJV. Dementia-related adverse events in PARADIGM-HF and other trials in heart failure with reduced ejection fraction. Eur J Heart Fail 2017;19(1):129–37. https://doi.org/10.1002/EJHF.687.

[72] De Lemos JA, Morrow DA. Brain natriuretic peptide measurement in acute coronary syndromes: ready for clinical application? Circulation 2002;106(23):2868–70. https://doi.org/10.1161/01.CIR.0000042763.07757.C0.

[73] Nishikimi T, Maeda N, Matsuoka H. The role of natriuretic peptides in cardioprotection. Cardiovasc Res 2006;69(2):318–28. https://doi.org/10.1016/J.CARDIORES.2005.10.001/2/69-2-318-FIG 2.GIF.

[74] Kasama S, Toyama T, Hatori T, Sumino H, Kumakura H, Takayama Y, Ichikawa S, Suzuki T, Kurabayashi M. Effects of intravenous atrial natriuretic peptide on cardiac sympathetic nerve activity and left ventricular remodeling in patients with first anterior acute myocardial infarction. J Am Coll Cardiol 2007;49(6):667–74. https://doi.org/10.1016/J.JACC.2006.09.048.

[75] Chen HH, Martin FL, Gibbons RJ, Schirger JA, Wright RS, Schears RM, Redfield MM, Simari RD, Lerman A, Cataliotti A, Burnett JC. Low-dose nesiritide in human anterior myocardial infarction suppresses aldosterone and preserves ventricular function and structure: a proof of concept study. Heart 2009;95(16):1315–9. https://doi.org/10.1136/HRT.2008.153916.

[76] Pfeffer MA, Claggett B, Lewis EF, Granger CB, Kober L, Maggiono AP, Mann DL, McMurray JJV, Rouleau JL, Solomon SD, Steg PG, Berwanger O, Cikes M, de Pasquale CG, East C, Fernandez A, Jering K, Landmesser U, Mehran R, Braunwald E. Angiotensin receptor-neprilysin inhibition in acute myocardial infarction. N Engl J Med 2021;385(20):11–2. https://doi.org/10.1056/NEJMOA2104508.

[77] Esser N, Zraika S. Neprilysin inhibition: a new therapeutic option for type 2 diabetes? Diabetologia 2019;62(7):1113. https://doi.org/10.1007/S00125-019-4889-Y.

[78] Jordan J, Stinkens R, Jax T, Engeli S, Blaak EE, May M, Havekes B, Schindler C, Albrecht D, Pal P, Heise T, Goossens GH, Langenickel TH. Improved insulin sensitivity with angiotensin receptor neprilysin inhibition in individuals with obesity and hypertension. Clin Pharmacol Ther 2017;101(2):254–63. https://doi.org/10.1002/CPT.455.

[79] Nougué H, Pezel T, Picard F, Sadoune M, Arrigo M, Beauvais F, Launay JM, Cohen-Solal A, Vodovar N, Logeart D. Effects of sacubitril/valsartan on neprilysin targets and the metabolism of natriuretic peptides in chronic heart failure: a mechanistic clinical study. Eur J Heart Fail 2019;21(5):598–605. https://doi.org/10.1002/EJHF.1342.

[80] Seferovic JP, Claggett B, Seidelmann SB, Seely EW, Packer M, Zile MR, Rouleau JL, Swedberg K, Lefkowitz M, Shi VC, Desai AS, McMurray JJV, Solomon SD. Effect of sacubitril/valsartan versus enalapril on glycaemic control in patients with heart failure and diabetes: a post-hoc analysis from the PARADIGM-HF trial. Lancet

Diab Endocrinol 2017;5(5):333–40. https://doi.org/10.1016/S2213-8587(17)30087-6.
[81] Riddell E, Vader JM. Potential expanded indications for neprilysin inhibitors. Curr Heart Fail Rep 2017;14(2):134. https://doi.org/10.1007/S11897-017-0327-Y.
[82] Voors AA, Gori M, Liu LCY, Claggett B, Zile MR, Pieske B, McMurray JJV, Packer M, Shi V, Lefkowitz MP, Solomon SD. Renal effects of the angiotensin receptor neprilysin inhibitor LCZ696 in patients with heart failure and preserved ejection fraction. Eur J Heart Fail 2015;17(5):510–7. https://doi.org/10.1002/EJHF.232.
[83] Haynes R, Judge PK, Staplin N, Herrington WG, Storey BC, Bethel A, Bowman L, Brunskill N, Cockwell P, Hill M, Kalra PA, McMurray JJV, Taal M, Wheeler DC, Landray MJ, Baigent C. Effects of sacubitril/valsartan versus irbesartan in patients with chronic kidney disease. Circulation 2018;138(15):1505–14. https://doi.org/10.1161/CIRCULATIONAHA.118.034818.
[84] Southerland WA, Gillis J, Kuppalli S, Fonseca A, Mendelson A, Horine SV, Bansal N, Gulati A. Dual enkephalinase inhibitors and their role in chronic pain management. Curr Pain Headache Rep 2021;25(5). https://doi.org/10.1007/S11916-021-00949-0.

Section D

Other metalloenzymes

Chapter 4.1

The role of arginase in human health and disease

Luigi F. Di Costanzo
Department of Agriculture - Department of Excellence - University of Naples Federico II - Palace of Portici - Piazza Carlo di Borbone, Portici (NA), Italy

1 Structure and function of the arginase isozymes

Arginase is manganese-containing enzyme performing the chemical change of L-arginine to urea and ornithine. The catalytic reaction performed by arginase consists of the activation of a water molecule bridging a manganese ion pair hosted in the enzyme active site to form a reactive hydroxide anion [1]. The transfer of this reactive group to the incoming planar guanidino group of arginine causes its hydrolytic cleavage and release of products. Arginase is classified as an arginine amidinohydrolase, EC 3.5.3.1, and represents the last step of the bilocated mitochondrion/cytoplasm **urea cycle**. This biochemical pathway utilizes six enzymatic reactions and two transporters to convert highly toxic ammonia to urea, a water-soluble compound, expelled by the kidney into the urine [2].

Urea and ornithine have two sustaining life functions. With its ~47% of nitrogen molecular content, **urea** detoxifies our body from nitrogenous waste [3]. For healthy individuals, most of the nitrogen removed in the form of urea represents most of the total intake of protein gained through the diet [3]. Of the two amino groups present in the urea molecule, one is formally derived from a glutamate residue, originated by the transamination reaction of most of the protein-degraded amino acids (except lysine), with alpha-ketoglutarate. The second urea amino group is derived from an aspartate residue in the glutamate biochemical pathway [4,5]. Finally, the bridging carbonyl group present in urea is derived from the Krebs' cycle. Once transferred from blood to the liver, the amino group of glutamate residue gets yanked and transferred.

The second arginase product, **ornithine**, represents a precursor of collagen, the most abundant structural protein in our body. As shown in Fig. 1, ornithine undergoes side chain cyclization resulting into proline, a main collagen-building block. In another biochemical process, decarboxylation of ornithine side chain results in putrescine, a precursor of linear aliphatic polyamines consisting of polycations, containing primary or secondary aminic nitrogen atom groups (Fig. 1). Owing to their positive charges at physiological pH, polyamines can bind nucleic acids and therefore, can contribute to cell cycle growth and proliferation, cell viability, and pathological responses (Section 2) [6–8].

Because of **arginine** recognition asarginase substrate, the enzyme has become a central ongoing research target [9]. Arginine is needed for metabolic and physiological functions and is considered conditionally essential in that is provided from the diet and protein turnover. The amino acid is also a precursor of creatine biosynthesis and regulation of collagen assembly [10]. Interestingly, arginine acts also as a substrate for another important enzyme: **nitric oxide synthase** (NOS, Fig. 2). This enzyme interconverts arginine into the radical molecule NO and citrulline, a precursor for the synthesis of arginine [11–13]. The NO molecule, the 1992 "Molecule of the Year," in is an important regulator of cardiovascular homeostasis and signaling molecule for vasodilation and other cellular processes.

Because of the shared substrate, arginase depletes NOS from its arginine pool. NO synthase is expressed in the form of three isozymes: neuronal NOS (abbreviated as nNOS or NOS1), inducible NOS (iNOS or NOS2), and endothelial NOS (eNOS or NOS3). The radical NO is also formed in the reduction of nitrate to nitrite and then to NO molecule [14]. In the conversion of arginine, NOS produces an intermediate, namely N^{ω}-hydroxy-L-arginine (NOHA), which is also an inhibitor of arginase (Fig. 2). Because of that shared small molecules recognition, arginase and NOS are mutually regulated [1]. In addition, from a biochemical point of view, arginase shows a K_M of ~1 mM, roughly 1000-fold smaller than NOS-3. On reverse, Vmax for NOS-3 is ~1000 times less than that of arginase, equalizing the two enzymes capability to metabolize L-arginine. Among the factors affecting arginase activity, the enzyme shows optimal activity at basic pH, with the best Vmax in the range of pH 9.0–9.5 [15].

Metalloenzymes. https://doi.org/10.1016/B978-0-12-823974-2.00008-5

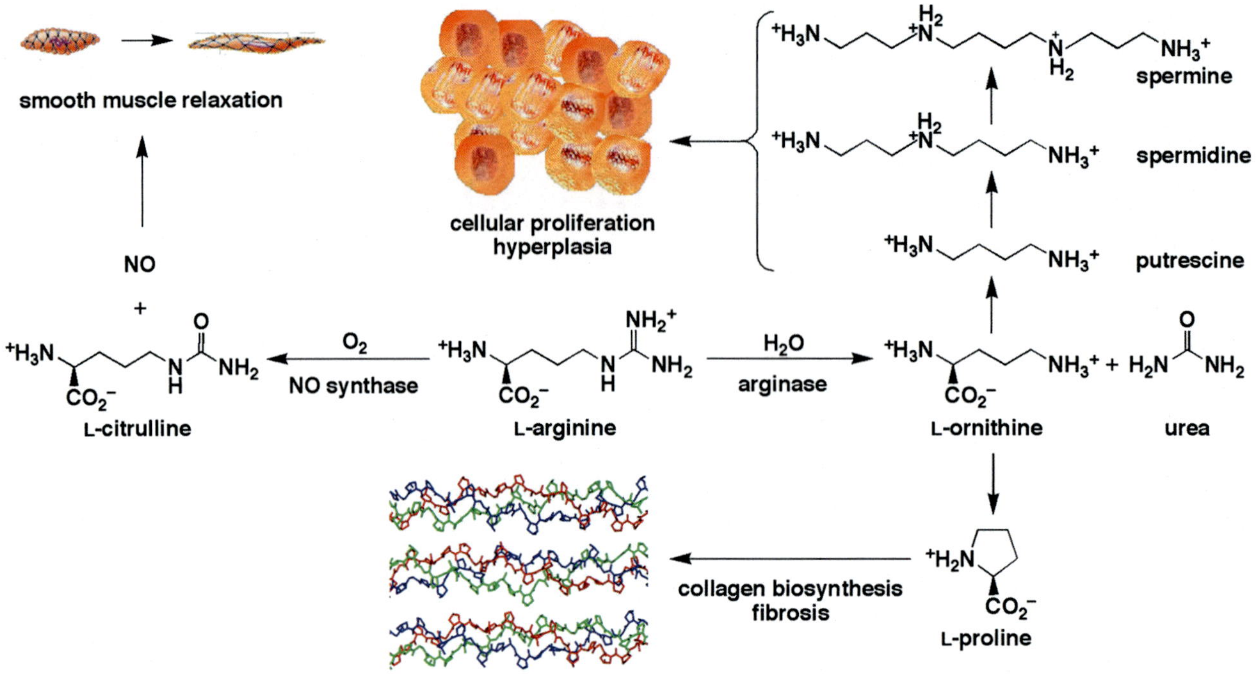

FIG. 1 Metabolic pathways of L-arginine. The central schematic reactions show the catabolism of arginine to L-ornithine and urea, catalyzed by arginase, or L-citrulline and nitric oxide (NO) catalyzed by nitric oxide synthase, NOS. The radical NO modulates smooth muscle relaxation (upper left). Also shown are the schematic pathways for the synthesis of polyamines (upper right) from the precursor L-ornithine. Polyamines facilitate cellular proliferation hyperplasia. The arginase product L-ornithine in turn can be anabolized to L-proline biosynthesis (down right) to support collagen production (colored sticks representation). *(Reprinted with permission from reference Ilies M, Di Costanzo L, North ML, Scott JA, Christianson DW. 2-Aminoimidazole amino acids as inhibitors of the binuclear manganese metalloenzyme human arginase i. J Med Chem 2010;53(10). Copyright 2010 American Chemical Society.)*

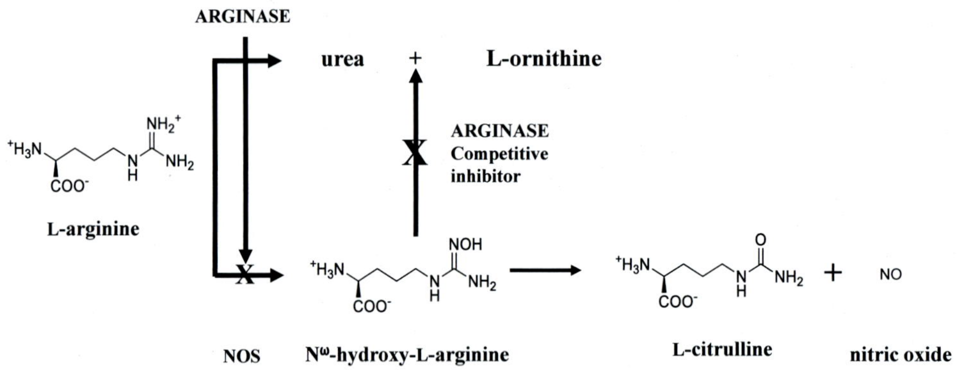

FIG. 2 Arginase and nitric oxide synthase (NOS) mutual regulation. The scheme illustrates how the two enzymes are mutually regulated. Arginase depletes NOS of its L-arginine substrate. In turn, N^{ω}-hydroxy-L-arginine (NOHA), a stable intermediate formed as first step N-hydroxylation of the L-arginine by NOS and represents a competitive inhibitor of arginase. The enzyme NOS produces L-citrulline and the radical molecule NO.

In humans, arginase is present as two isoforms known as **arginase I and arginase II**. Arginase I is mainly present in the cytoplasm of liver cells codified with a protein chain of 322 residues. Arginase II is a mitochondrial longer protein chain enzyme of 354 residues expressed in extra-hepatic tissues, especially the kidney, and other tissues, including the brain and retina. The two isoforms share a sequence identity of ~58%.

The structure of human arginase I consists of a trimeric quaternary structure (Fig. 3A) in which each monomer adopts an α/β fold consisting of a central eight-stranded parallel β-sheet flanked on both sides by several α-helices (Fig. 3A). Arginase fold partially resembles a Rossmann fold (α/β fold with strand order 321456, Fig. 3B) and shares similarity with HDAC, an enzyme that catalyzes the removal of acetyl groups from the lysine residues of histone and nonhistone proteins, suggesting an evolutionary divergence from a common metallo-protein hydrolase ancestor [16,17]. Each monomer has an approximate dimension of $40 \times 50 \times 50\,Å^3$. The assembly is

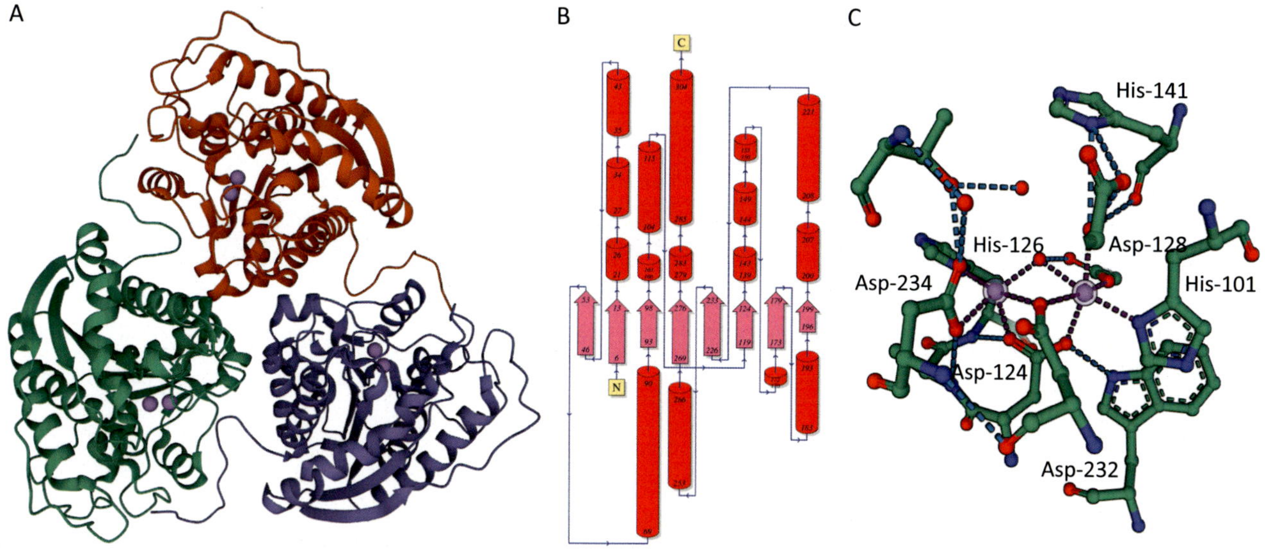

FIG. 3 (A) Ribbon plot representation of human arginase I (pdb entry 2aeb). The dinuclear center of Mn^{2+} ions is represented as purple color spheres. (B) Secondary structure of human arginase I. (C) Metal-binding site of human arginase I. The Mn^{2+} ions appear as purple spheres; metal-bound solvent molecules are shown as red spheres (pdb entry 2zav). The metal coordinative residues are shown in ball-and-stick representation.

characterized by a heterologous trimer with a C_3 assembly (Fig. 3A) where a characteristic S-shaped C-terminal tail is partially looping around a near symmetrical chain although is does not play a critical role in stabilizing the trimeric arrangement, nor to the activity of the enzyme [18–20].

The atomic details of arginase in complex with its substrate analogue inhibitor ABH (see Section 3, Fig. 4) have provided insights into the enzyme catalytic mechanism (Fig. 5). The first structure of rat arginase I (pdb entry 1rla) [21], human arginase I in complex with ABH (pdb entry 2aeb) [22], and II (pdb entry 1pq3) [23], together with a variety of arginase structures from several species can be explored from the Protein Data Bank (rcsb.org) [24]. These structural studies have led to the design of small molecules

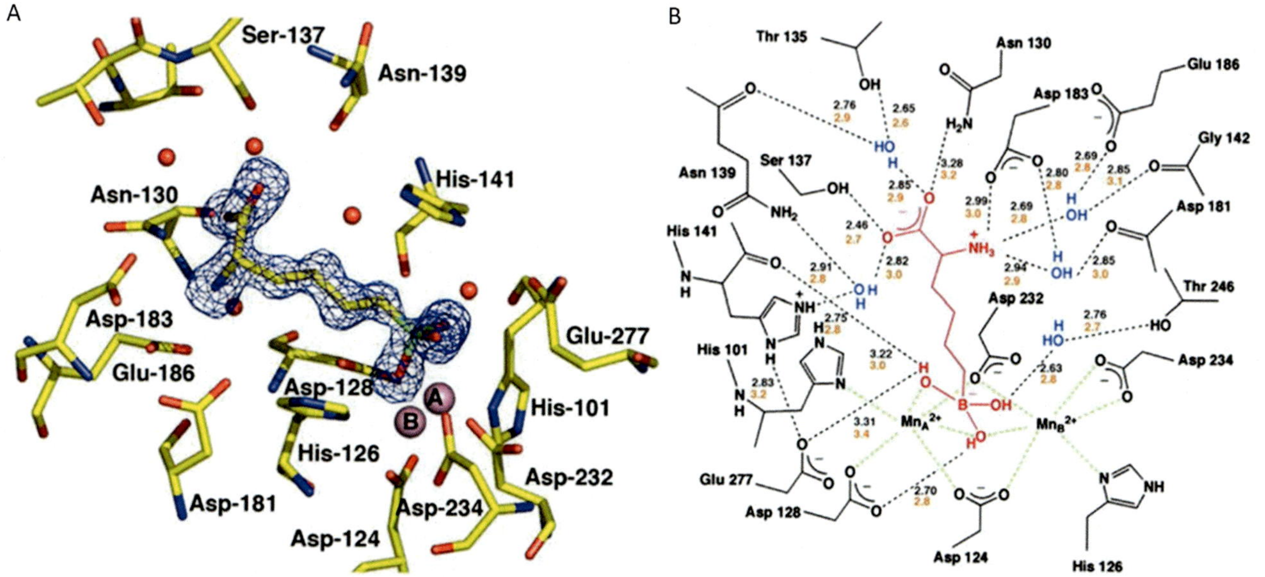

FIG. 4 (A) Electron density map of the human arginase I-ABH complex showing the resulting of metal-bridging hydroxide attack to the boronic acid group of ABH to yield the tetrahedral boronate anion. This adduct represents the intermediate as a postulated step of the catalytic mechanism outlined in this Fig. 5. (B) Arginase I-ABH complex. Summary of the average intermolecular hydrogen bond interactions in the human arginase I (*gray* numbers, pdb entry 2aeb). The Mn^{2+} ions coordination interactions are indicated by *green dashed lines*, and hydrogen bonds are indicated by *black dashed lines*. By comparison the average intermolecular interactions in the similar rat arginase I enzyme (*orange* numbers, pdb entry 1d3v). *(Reprinted with permission from Di Costanzo L., Sabio G., Mora A., Rodriguez P.C., Ochoa A.C., Centeno F., et al. Crystal structure of human arginase I at 1.29-Åresolution and exploration of inhibition in the immune response. Proc Natl Acad Sci U S A 2005;102(37).)*

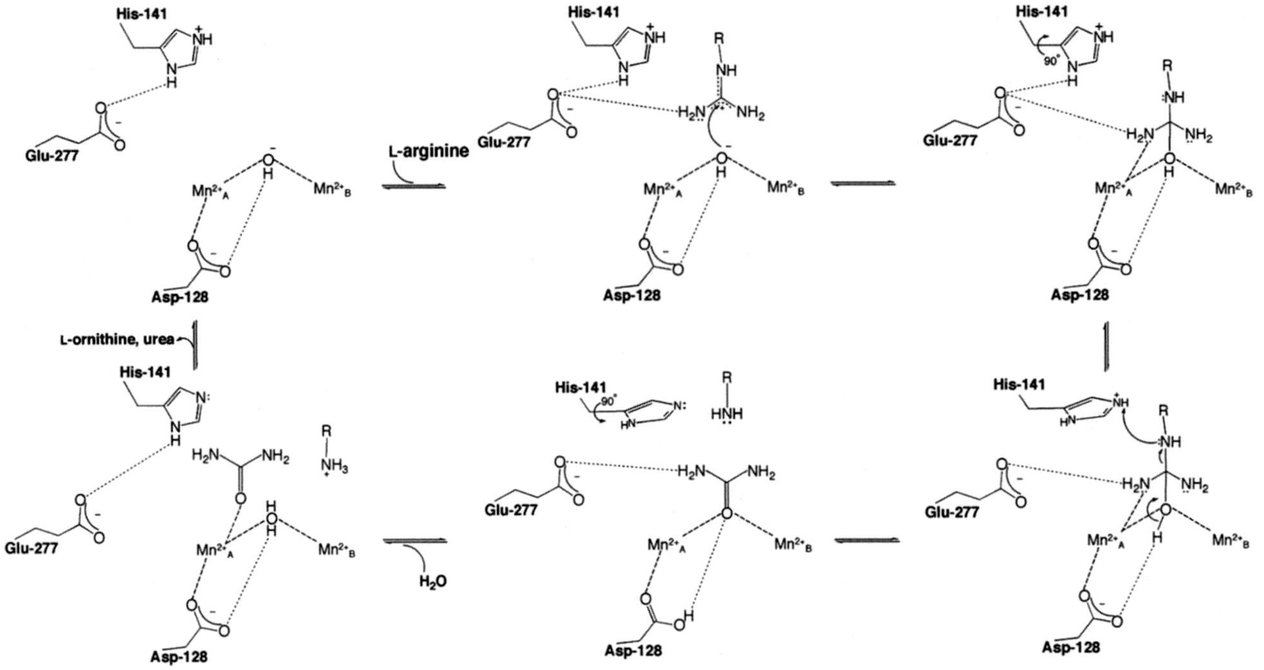

FIG. 5 Proposed mechanism of human arginase I. *(Reprinted with permission from Di Costanzo L., Sabio G., Mora A., Rodriguez P.C., Ochoa A. C., Centeno F., et al. Crystal structure of human arginase I at 1.29-Å resolution and exploration of inhibition in the immune response. Proc Natl Acad Sci U S A 2005;102(37).)*

and inspired the development of antibodies or shed light on the relation between pH and enzyme activity [15]. Considering the high sequence similarity between the arginase I and II isoforms, their structure is unsurprisingly similar.

Amino acid side chains that coordinate manganese ions in the arginase active site (Fig. 3C) are located on the edge of the central β-sheet in loop segments immediately adjacent to strands 8, 7, and 4 (Fig. 3B) [25]. The enzyme presents a binuclear cluster at the bottom of a ~15 Å deep cavity. This small pocket and structural similarity between the two arginase isozymes make it difficult to design specific inhibitors. The crystal structure of the unliganded human arginase I displays an octahedral coordination geometry of the two Mn^{2+} ions bridged by D124 and D232, and two Mn^{2+} ions present a separation of ~3.3 Å, as shown in Fig. 3C. The Mn^{2+} ions coordination spheres are completed by residues H101, H126, D234, D128, the metal-bridging solvent, and an additional solvent molecule is coordinated to one of the ions (Fig. 3C) [25]. The structure of unliganded human arginase I shows that the metal-bridging hydroxide ion and H141 interact through a hydrogen-bonded water molecule [22,25]. These interactions are consistent with the proposed role of H141 as a proton shuttle in the regeneration of the catalytic nucleophilic metal-bridging hydroxide ion [22]. Based on these observations, the proposed mechanism is summarized in Fig. 5.

The bioinorganic chemistry role of Mn^{2+} ion cluster has been investigated by arginase I **reconstitution with other metal ions** of the first transition series including Co^{2+}, Ni^{2+}, and Zn^{2+} ions. Biochemical and structural studies of the reconstituted arginase I with the different metals have demonstrated a lower K_M for arginine and an enhanced enzyme catalytic activity around physiological pH (~7.4) [26,27]. These investigations represent the foundation for the use of an engineered version of the cobalt-substituted arginase for therapeutic intervention [27]. Despite the essentially identical structures of arginase in complex with cobalt and manganese ions, the electrostatic interaction of the metal-bound solvent molecule could explain the functional difference observed for arginine affinity at different pH values [26].

2 Physiologic and pathologic role associated to arginase

Historically, the pathological role associated with arginase I is represented by the autosomal recessive disorder known as **argininemia**, caused by the enzyme deficiency. The disease leads to *hyperammonemia* characterized by accumulation of the arginine in the blood [28–30]. Argininemia develops in the neonatal period of life causing a progressive spasticity for the loss of psychomotor developmental and cognitive skills [31,32]. As a consequence of arginase deficiency, and any other enzyme of the urea cycle, toxic ammonia could accumulate in the brain leading to mental disorders and eventually death [33,34]. Unlike arginase I, the physiological role of arginase II is less understood, and it is thought to be linked to polyamines metabolism [35].

The current view of pathologies associated to human arginase goes beyond argininemia. Because of the central role of arginine metabolism, often arginase represents a checkpoint in many clinical studies of cancer, cardiovascular, immune system, and aging dysfunctions [9]. Most of these studies have identified the intracellular events leading to an overexpression of arginase isozymes including Th2 cytokines, especially interleukins IL-13, IL-4, IL-10, and tumor necrosis factor (TNF)-α [36,37]. A number of clinical studies have shown arginase co-expression with NOS and their mutual regulation [9]. However, expression of arginine transporters in different cell types may also play a role in dysfunctions associated with arginine metabolism [38].

Cancer cells are supported by a microenvironment containing an impaired differentiation of human T-cell biology cells known as myeloid-derived suppressor cells (MDSC) [39,40]. Cancers cells are characterized by strong immune suppressive activity, a hallmark for cancer growth. Several biochemical factors elicit MDSC differentiation including overexpression of arginase I and nitric oxide synthase [39]. A recent approach for potential cancer treatment is represented by the development of specific arginase I antibodies (see Section 3) [41].

The inner cellular layer of arteries, veins, and capillaries is represented by a monolayer of endothelial cells (EC) in direct contact with blood cells and other biochemical factors. Healthy vascular endothelial cells rely on proper blood flow and conversion of arginine into NO and citrulline. In healthy tissues, eNOS contributes to regulating production of NO molecule. In contrast, during inflammation, blood vessels express iNOS in addition to eNOS, and if its substrate and cofactors are not limited, iNOS may produce NO continuously leading to vascular dysfunction [42]. The interplay between arginase and eNOS is critical for healthy vascular tone and smooth muscle relaxation. In the aged animal model, iNOS causes arginase activation through its *S*-nitrosylation, which, in turn, contributes to endothelial dysfunction by impairing eNOS signaling [43,44]. Inhibition of arginase reverses endothelial dysfunction and vascular stiffness by production of glutamate, proline, and polyamines [45,46]. The effect of NO is not only limited to the EC, eNOS is also present in the circulating blood cells, and it contributes to the regulation of blood pressure [47]. Furthermore, ornithine produced by arginase undergoes biochemical change to polyamines and proline and therefore represents a modulator for the proliferation of smooth muscle cells [48]. As a consequence, arginase inhibition might have a role in the dysfunctions associated with hypertension and aging.

Several reports have supported the role of arginase in **erectile dysfunction** (ED). Within the corpora cavernosa (CC), the penis's internal sponge-like tissue, smooth muscle cells, regulated by NO, can undergo processes of relaxation or contraction (see Fig. 1). Despite the normal presence of both arginase isozymes in CC, an increase in the expression of the isoform II is observed in patients affected by ED [23,49]. The inhibition of arginase may play a role in restoring the undesired effect.

Myocardial infarction and heart attack are leading causes of death in the population. Animal model studies have shown an increase in the heart tissue arginase expression leading to an impaired endothelium-dependent vasorelaxation [50]. In clinical studies, use of an arginase inhibitor has reversed high blood pressure and decreased size of infarction [51,52].

Arginase is involved in **inflammation** as the result of cellular and vascular events causing asthma and allergic diseases [53,54]. In lungs affected by asthma, arginase I induced by IL-13, fuels the bioenergetic pathways that support pathogenic innate lymphoid cells (ILCs), a set of heterogeneous immune cells [55]. As consequence of the reduced pool of substrate for NOS, a compromised NO relaxation leads to bronchoconstriction. Arginase overexpression increases production of metabolites proline and polyamines, which in turn stimulate cellular proliferation and hyperplasia. Therefore, the use of inhibitors could block asthma symptoms [56].

Arginase activity is also upregulated in **COVID-19** patients, presenting a low arginine/ornithine ratio and a reduction of NO production as a result of lower NOS activity [57].

3 The development of arginase inhibitors and antibodies

Inhibitory studies of arginase I and II have been a research topic for many decades starting from the 1940s [58]. It was only later in the 1990s that macromolecular crystallography and cryo-microscopy provided details for the design and developments of small molecule inhibitors and specific antibodies (see Fig. 6). As discussed in Section 1, arginase is characterized by a relatively small active site pocket and the similarity between its two isozymes makes it a difficult task to identify selective inhibitors.

Typical arginase inhibitors are amino acids with low molecular weight and contain a boronic acid group. The chemical structures of the enzyme substrate (arginine), products (urea/ornithine) and the rapidly changing arginine adduct, undergoing nucleophilic attack by hydroxide anion (aka transition state analogue), have suggested possible inhibitors. The development of arginase inhibitors (Fig. 6) is characterized by the role of the transition state analogue inhibitor, 2(*S*)-amino-6-boronohexanoic acid (ABH), one the strongest arginase inhibitor. Much of the work in this

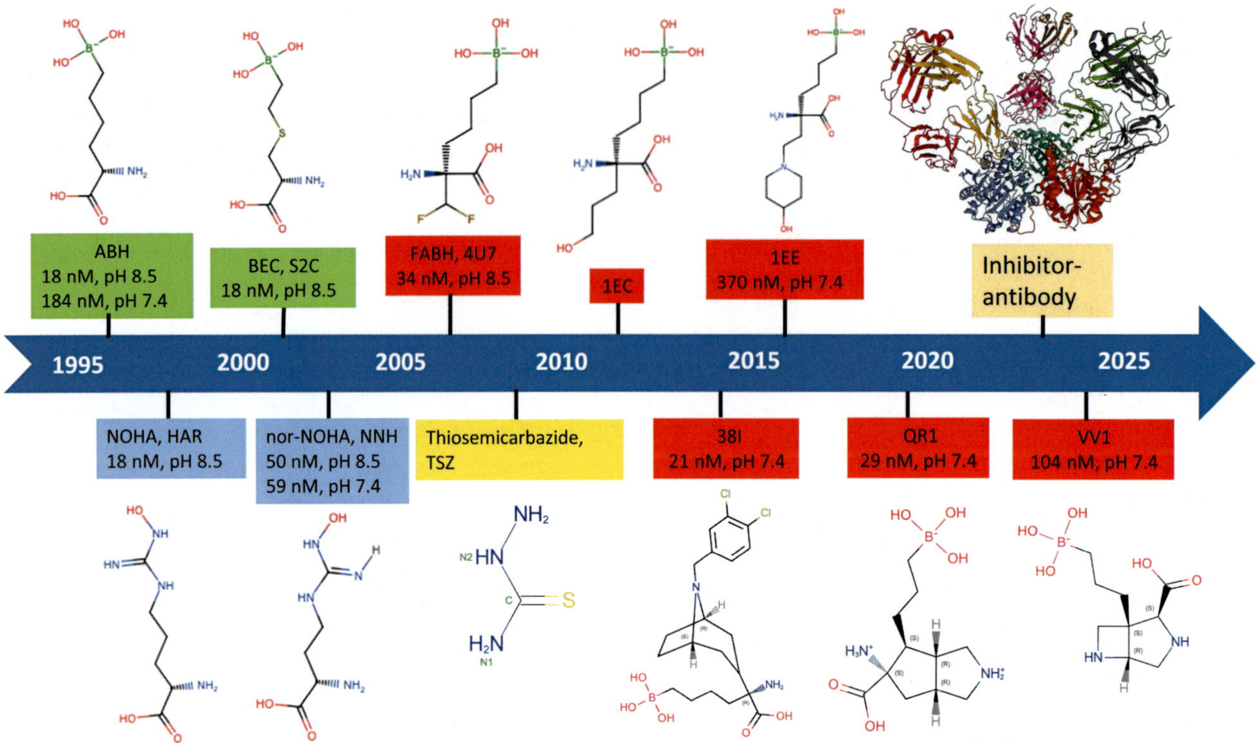

FIG. 6 Timeline of arginase inhibitors development. Typical arginase inhibitors are organized in the following groups: ABH and its related compound (*green color*); ABH derivatives (*red color*); substrate and product analogue inhibitors (*blue* and *yellow color*, respectively). The boronic acid group is represented as hydrated form, in agreement with the experimental structures of arginase in complex with these compounds. Affinities at different pHs for human arginase I are indicated. The arginase I-inhibitor antibody mAb5 complex (pdb entry 7ley), and inhibitors 3D-structure can be explored from the link: https://www.rcsb.org, using the accession codes indicated for each entity.

area has been pioneered by Dr. Christianson's lab at the University of Pennsylvania. Since 1997, it was demonstrated that a simple borate anion was able to display the manganese-bridging solvent molecule of the native enzyme [59]. The tetrahedral borate anion mimics the binding of the postulated tetrahedral transition state in the catalytic reaction (see Section 1, Fig. 4). This result, together with the atomic details of arginase in complex with the product ornithine, suggested the design of ABH [59]. This molecule is isosteric with arginase substrate arginine. The molecule ABH contains the trigonal planar boronic acid group that mimics the guanidino group of arginine. The classical boron Lewis acid in ABH makes the ideal compound undergoing hydroxide nucleophilic attack. The high affinity of ABH is the result of the structural similarity between its hydrated form (Fig. 6) and the proposed tetrahedral intermediate (flanking transition states). The enzyme active site represents provides the perfect geometrical framework and interactions for the binding of ABH molecule and its high affinity is a consequence of strong metal coordination and hydrogen bond interactions (Fig. 5) [22].

After ABH, one of its analogues was prepared, namely (*S*)-(2-boronoethyl)-L-cysteine (or BEC). BEC contains a —S— atom group on the gamma-position and presents a slightly lower affinity, as compared with ABH, for the effect of the covalent geometry around the sulfur atom group. The inhibitor ABH has inspired other reactive chemical groups in place of its boronic acid group. For instance, the inhibitor (*S*)-2-amino-7-oxoheptanoic acid (AOH; https://www.rcsb.org), contains a terminal aldehyde group able to react with a nucleophile. The crystal structure of arginase I in complex with AOH reveals the nucleophilic attack by the metal-bridging hydroxide ion forming a tetrahedral gem-diol coordinated to the binuclear manganese cluster. Although structural details are very noteworthy, AOH presents a modest binding affinity ($K_i = 60$ mM). Similar, inhibitors with a tetrahedral terminal group about to bind the metal arginase cluster are available [60].

The first generation of arginase I and II inhibitors also include those mimicking the substrate arginine and the product urea or ornithine [25,61,62]. Notably, the *N*-hydroxy-L-arginine (NOHA) and its analogue bearing one-methyl group shorter side chain, designated as nor-NOHA, are competitive inhibitors and contain an —OH terminal group. Crystal structures of arginase I in complex with these inhibitors reveal the presence of a bridging

-OH atom group and replacement of the solvent molecule present in the unliganded arginase structure [61]. Therefore, the design and synthesis of newer arginase inhibitors have been step-by-step progress, guided by a retrostructural design approach, which is still ongoing these days.

The initial structure of human arginase I and II in complex with ABH has inspired several analogous inhibitors containing the same boronic acid terminal group. The crystal structure of HAI in complex with Me-ABH and/or FABH (namely, the amino-6-borono-2-methylhexanoic acid and the 2-amino-6-borono-2-(difluoromethyl)hexanoic acid), where the hydrogen atom group at the Cα of ABH has been replaced by a methyl or $—CF_2$ group, respectively, opened the way for other laboratories to explore the interaction of these substituents with arginase surface residues or solvent molecules around the C-alpha atom group [56]. This strategy led to arginase binding of α,α-disubstituted amino acids bearing a boronic acid terminal group [56]. In the past decade, supported by these results, researchers have also envisioned substituents longer than the simple methyl group at the C-alpha atom group. For instance, they have enclosed the ABH α-amino group in a cyclic or bicyclic proline leading to structures included in Fig. 5 [51,63]. As hypothesized, structure analysis of the newer derivatives in complex with arginase I has revealed direct or water-mediated H-bonds with residues Asp183 and/or Asp181 (Fig. 4) [64] and similar interactions with nearby residues or solvent molecules. Of particular interest is the compound CB-1158 (accession code XC3, https://www.rcsb.org), an ABH derivative, that can be considered as a modified proline residue sharing the Cα-Cβ bond of ABH chemical structure. This compound is used in several clinical trials and structural studies have revealed the foundation for CB-1158's higher arginase affinity with respect ABH at physiological pH (~7.4). Inhibition of ABH and CB-1158 is reversed at pH=9.5 [15,65].

Although similarities between two isozymes represent an advantage to target with the same inhibitors both macromolecules, their different tissue distribution or function often requires a specific isoform's inhibitor for clinical purposes. It is a difficult task to address the specific enzyme inhibition. The recent discovery of human arginase monoclonal antibodies (mAbs) represents an easier route to address specific inhibition of a particular enzyme isoform [41]. By using a protein engineering technique, scientists at Merck were able to identify five inhibitor-antibodies of human arginase I through multiple rounds of selection in yeast starting from an available library [41]. They have structurally characterized the complexes of these inhibitor-antibodies with human arginase I, and the selected mAbs have shown the formation of large macromolecular complexes, as the one shown in Fig. 7. This approach could also allow selective inhibition of human arginase II.

4 Compounds and arginase formulation used in clinical development

Clinical intervention based on human arginase I and II includes all physiological and pathological aspects associated with the two isozymes and their specific tissue distribution. These therapeutic interventions cover a wide range of dysfunctions from arginase I deficiency in argininemia, to arginase overexpression/activity in a variety of dysfunctions [9]. Clinical trials are accumulating over the years and involve different treatments including arginase as drug, its substate arginine, and a selection of enzyme inhibitors. Some trials also include combination of drugs related to arginase and other interventions. A search of the term "arginase" within the web-based resource ClinicalTrials.gov reveals a total of 48 "interventional studies," as of this writing, most of which are in phase 1 or phase 2, and scientific literature is constantly reviewing the new findings [9]. Besides α-difluoromethylornithine (DFMO) an ornithine decarboxylase (ODC) inhibitor, and a weak inhibitor of arginase I, drugs approved by the FDA for other targets, have also been virtually screened to identify selective inhibitors of arginase I and II [66].

From these studies, arginase I is given as a drug in formulations known as pegzilarginase (aka AEB1102, or Co-ArgI-PEG) or other pegylated recombinant human arginase (aka 1PEG-BCT-100, arginase conjugated with Peg-5000). As an established technique, pegylated enzymes are adducts of an enzyme linked to a polymer, such as polyethylene glycol (e.g., Peg-5000), used for the therapeutic delivery of proteins with potential biotechnological applications [67,68].

The use of pegzilarginase has reached clinical phase 3 for arginase I deficiency. Other arginase formulations are also object of several ongoing studies. Pegzilarginase, is an engineered arginase 1 enzyme cobalt-substituted (see Section 1) linked to PEG that has an increased catalytic activity and a better stability with respect to the native manganese enzyme [27]. These investigational arginase I as therapeutic agents, tested in animal models, showed a significant lowering of the circulating arginine in patients affected by arginase 1 deficiency [69,70].

Besides arginase, several of the enzyme inhibitors, including ABH, BEC, NOHA, nor-NOHA (Fig. 6) and the more recent CB-1158, are commercially available and used in humans for clinical studies. The use of these inhibitors often shows promising results for patients being treated for coronary artery disease, heart failure, and hypertension [71–73]. Inhibition of arginase II with NOHA has led to a reduced cerebral blood flow as a result of traumatic brain injury in animal model [74]. ABH and BEC have also been used in studies to enhance smooth muscle relaxation in human penile corpus cavernosum in patients affected by

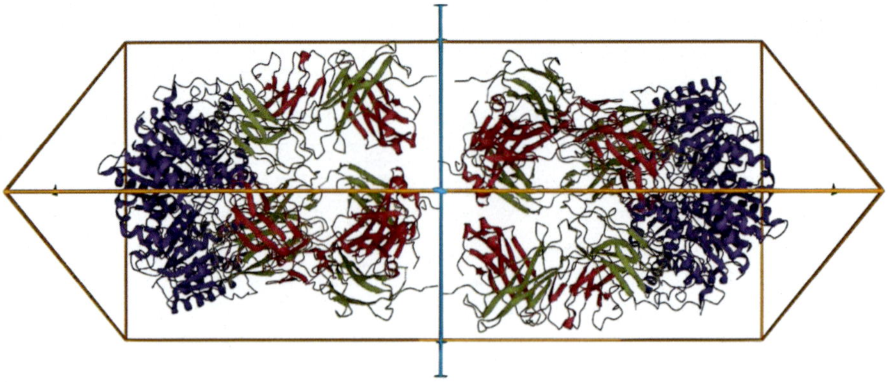

FIG. 7 Ribbon plot representation of human arginase I in complex with light and heavy chain of inhibitor antibody mAb1 (pdb entry 7lex). The structure is comprised of two symmetrically related complexes formed by the human arginase I trimer (*purple color*) interacting with three pairs of mAb1 light chain and heavy chain (*green* and *red colors*), one for each arginase monomer.

erectile dysfunction [75,76]. However, despite the increasing number of reports, the use of arginase inhibitors has been limited to a "proof-of-concept" application. Larger clinical trials need to test adverse effects on long-term treatment outcomes [9].

A significant segment of clinical studies using arginase inhibitors is aimed to the immunosuppression of myeloid cell-mediated tumor cells [77]. The inhibitor CB-1158 has been used as a single drug and/or in combination with other therapies to investigate its ability to block myeloid cell-mediated [65,78].

Arginine is also used in clinical studies. As a natural amino acid, arginine is commonly used as supplement among athletes and bodybuilders to enhance performance, respiratory ability, and muscle/weight control. Although, arginine efficacy is controversial and not fully understood, its use in clinical studies has shown improvement of cardiac functions in patients affected by myocardial ischemia [79,80].

In summary, arginase represents a fascinating metalloenzyme with multiple physiological and pathological roles, and therefore, a current target to identify small-molecule inhibitors and antibodies as therapeutic treatment in multiple areas of human health and disease.

References

[1] Ash DE. Structure and function of arginases. J Nutr 2004.

[2] Haskins N, Bhuvanendran S, Anselmi C, Gams A, Kanholm T, Kocher KM, et al. Mitochondrial enzymes of the urea cycle cluster at the inner mitochondrial membrane. Front Physiol 2021;11.

[3] Weiner ID, Mitch WE, Sands JM. Urea and ammonia metabolism and the control of renal nitrogen excretion. Clin J Am Soc Nephrol 2015;10(8).

[4] Brosnan JT. Glutamate, at the interface between amino acid and carbohydrate metabolism. J Nutr 2000.

[5] Brosnan ME, Brosnan JT. Hepatic glutamate metabolism: a tale of 2 hepatocytes. Am J Clin Nutr 2009.

[6] Igarashi K, Kashiwagi K. Effects of polyamines on protein synthesis and growth of Escherichia coli. J Biol Chem 2018;293.

[7] LeBrasseur N. Polyamines in inflammation. J Cell Biol 2003;162:2.

[8] Chia TY, Zolp A, Miska J. Polyamine immunometabolism: central regulators of inflammation, cancer and autoimmunity. Cells 2022;11.

[9] Caldwell RW, Rodriguez PC, Toque HA, Narayanan SP, Caldwell RB. Arginase: a multifaceted enzyme important in health and disease. Physiol Rev [Internet] 2018;98:641–65. Available from: www.prv.org.

[10] Albaugh VL, Arginine BA. Chapter 27—cellular and physiological effects of arginine in seniors. In: Nutrition and Functional Foods for Healthy Aging. 2017. p. 317–336 (Ref Modul Life Sci 4).

[11] Durante W. Regulation of L-arginine transport and metabolism in vascular smooth muscle cells. Cell Biochem Biophys 2001;35(1).

[12] Knowles RG, Moncada S. Nitric oxide synthases in mammals. Biochem J 1994;298.

[13] Gantner BN, LaFond KM, Bonini MG. Nitric oxide in cellular adaptation and disease. vol. 34. Redox Biology; 2020.

[14] DeMartino AW, Kim-Shapiro DB, Patel RP, Gladwin MT. Nitrite and nitrate chemical biology and signalling. Br J Pharmacol 2019;176.

[15] Grobben Y, Uitdehaag JCM, Willemsen-Seegers N, Tabak WWA, de Man J, Buijsman RC, et al. Structural insights into human Arginase-1 pH dependence and its inhibition by the small molecule inhibitor CB-1158. J Struct Biol X 2020;4.

[16] Dowling DP, Di Costanzo L, Gennadios HA, Christianson DW. Evolution of the arginase fold and functional diversity. Cell Mol Life Sci 2008;65.

[17] Seto E, Yoshida M. Erasers of histone acetylation: the histone deacetylase enzymes. Cold Spring Harb Perspect Biol 2014;6(4).

[18] Mora A, Del Ara RM, Fuentes JM, Soler G, Centeno F. Implications of the S-shaped domain in the quaternary structure of human arginase. Biochim Biophys Acta – Protein Struct Mol Enzymol 2000;1476(2).

[19] García D, Uribe E, Lobos M, Orellana MS, Carvajal N. Studies on the functional significance of a C-terminal S-shaped motif in human arginase type I: essentiality for cooperative effects. Arch Biochem Biophys 2009;481(1).

[20] Sabio G, Mora A, Rangel MA, Quesada A, Marcos CF, Alonso JC, et al. Glu-256 is a main structural determinant for oligomerisation of human arginase I. FEBS Lett 2001;501(2–3).

[21] Kanyo ZF, Scolnick LR, Ash DE, Christianson DW. Structure of a unique binuclear manganese cluster in arginase. Nature 1996;383 (6600).

[22] Di Costanzo L, Sabio G, Mora A, Rodriguez PC, Ochoa AC, Centeno F, et al. Crystal structure of human arginase I at 1.29-Å resolution and exploration of inhibition in the immune response. Proc Natl Acad Sci U S A 2005;102(37).

[23] Cama E, Colleluori DM, Emig FA, Shin H, Kim SW, Kim NN, et al. Human arginase II: crystal structure and physiological role in male and female sexual arousal. Biochemistry 2003;42(28).

[24] Goodsell DS, Zardecki C, Di Costanzo L, Duarte JM, Hudson BP, Persikova I, et al. RCSB protein data Bank: enabling biomedical research and drug discovery. Protein Sci 2020.

[25] Di Costanzo L, Pique ME, Christianson DW. Crystal structure of human arginase I complexed with thiosemicarbazide reveals an unusual thiocarbonyl μ-sulfide ligand in the binuclear manganese cluster. J Am Chem Soc 2007;129(20).

[26] L. D'Antonio E, Hai Y, W. Christianson D. Structure and function of non-native metal clusters in human arginase I. Biochemistry 2012;51 (42):8399–409.

[27] Stone EM, Glazer ES, Chantranupong L, Cherukuri P, Breece RM, Tierney DL, et al. Replacing Mn2+ with Co2+ in human arginase I enhances cytotoxicity toward l-arginine auxotrophic cancer cell lines. ACS Chem Biol 2010;5(3):333–42.

[28] Scaglia F, Brunetti-Pierri N, Kleppe S, Marini J, Carter S, Garlick P, et al. Clinical consequences of urea cycle enzyme deficiencies and potential links to arginine and nitric oxide metabolism. J Nutr 2004;134(10):2796S–7S.

[29] Ah Mew N, Simpson KL, Gropman AL, Lanpher BC, Chapman KA, Summar ML. Urea cycle disorders overview. In: GeneReviews® (Internet). Seattle, WA: University of Washington, Seattle; 2003-2017. p. 1993–2023. https://www.ncbi.nlm.nih.gov/books/NBK1217/.

[30] Mew NA, Pappa MB, Gropman AL. Urea cycle disorders. Rosenberg's Mol Genet Basis Neurol Psychiatr Dis Fifth Ed 2015;633–47.

[31] Rozenblatt-Rosen O, Stubbington MJT, Regev A, Teichmann SA. The human cell atlas: from vision to reality. Nature 2017;550:451–3.

[32] Uhlen M, Uhlén M, Fagerberg L, Hallström BM, Hallstrom BM, Lindskog C, et al. Proteomics. Tissue-based map of the human proteome. Science 2015;347(6220).

[33] Christopher R, Rajivnath V, Shetty KT. Arginase deficiency. Indian J Pediatr 1997;64:2.

[34] Soria LR, Ah Mew N, Brunetti-Pierri N. Progress and challenges in development of new therapies for urea cycle disorders. Hum Mol Genet 2019;28.

[35] Szeliga M, Albrecht J. Roles of nitric oxide and polyamines in brain tumor growth. Adv Med Sci 2021;66.

[36] Ochoa JB, Bernard AC, O'Brien WE, Griffen MM, Maley ME, Rockich AK, et al. Arginase I expression and activity in human mononuclear cells after injury. Ann Surg 2001;233(3).

[37] Pesce JT, Ramalingam TR, Mentink-Kane MM, Wilson MS, Kasmi KCE, Smith AM, et al. Arginase-1-expressing macrophages suppress Th2 cytokine-driven inflammation and fibrosis. PLoS Pathog 2009;5(4).

[38] Morris SM. Arginine metabolism revisited. J Nutr 2016;146:12.

[39] Kumar V, Patel S, Tcyganov E, Gabrilovich DI. The nature of myeloid-derived suppressor cells in the tumor microenvironment. Trends Immunol 2016;37.

[40] Colegio OR, Chu NQ, Szabo AL, Chu T, Rhebergen AM, Jairam V, et al. Functional polarization of tumour-associated macrophages by tumour-derived lactic acid. Nature 2014;513(7519).

[41] Palte RL, Juan V, Gomez-Llorente Y, Andre Bailly M, Chakravarthy K, Chen X, et al. Cryo-EM structures of inhibitory antibodies complexed with arginase 1 provide insight into mechanism of action. Available from:; 2021. https://doi.org/10.1038/s42003-021-02444-z.

[42] Gunnett CA, Lund DD, McDowell AK, Faraci FM, Heistad DD. Mechanisms of inducible nitric oxide synthase-mediated vascular dysfunction. Arterioscler Thromb Vasc Biol 2005;25(8).

[43] Jae HK, Bugaj LJ, Young JO, Bivalacqua TJ, Ryoo S, Soucy KG, et al. Arginase inhibition restores NOS coupling and reverses endothelial dysfunction and vascular stiffness in old rats. J Appl Physiol 2009;107(4).

[44] Santhanam L, Lim HK, Lim HK, Miriel V, Brown T, Patel M, et al. Inducible NO synthase-dependent S-nitrosylation and activation of arginase1 contribute to age-related endothelial dysfunction. Circ Res 2007;101(7).

[45] Li H, Meininger CJ, Hawker JR, Haynes TE, Kepka-Lenhart D, Mistry SK, et al. Regulatory role of arginase I and II in nitric oxide, polyamine, and proline syntheses in endothelial cells. Am J Physiol – Endocrinol Metab 2001;280(1). 43-1.

[46] Lucas R, Fulton D, Caldwell RW, Romero MJ. Arginase in the vascular endothelium: friend or foe? Front Immunol 2014;5.

[47] Wood KC, Cortese-Krott MM, Kovacic JC, Noguchi A, Liu VB, Wang X, et al. Circulating blood endothelial nitric oxide synthase contributes to the regulation of systemic blood pressure and nitrite homeostasis. Arterioscler Thromb Vasc Biol 2013;33(8).

[48] Durante W. Role of arginase in vessel wall remodeling. Front Immunol 2013;4.

[49] Castela Â, Costa C. Molecular mechanisms associated with diabetic endothelial-erectile dysfunction. Nat Rev Urol 2016;13.

[50] Mahdi A, Kövamees O, Pernow J. Improvement in endothelial function in cardiovascular disease—is arginase the target? Int J Cardiol 2020;301.

[51] Van Zandt MC, Whitehouse DL, Golebiowski A, Ji MK, Zhang M, Beckett RP, et al. Discovery of (R)-2-amino-6-borono-2-(2-(piperidin-1-yl)ethyl)hexanoic acid and congeners as highly potent inhibitors of human arginases i and II for treatment of myocardial reperfusion injury. J Med Chem 2013;56(6).

[52] Gonon AT, Jung C, Katz A, Westerblad H, Shemyakin A, Sjöquist PO, et al. Local arginase inhibition during early reperfusion mediates cardioprotection via increased nitric oxide production. PLoS ONE 2012;7(7).

[53] van den Berg MP, Meurs H, Gosens R. Targeting arginase and nitric oxide metabolism in chronic airway diseases and their co-morbidities. Curr Opin Pharmacol 2018;40.

[54] Monticelli LA, Buck MD, Flamar AL, Saenz SA, Wojno EDT, Yudanin NA, et al. Arginase 1 is an innate lymphoid-cell-intrinsic metabolic checkpoint controlling type 2 inflammation. Nat Immunol 2016;17(6).

[55] Zhou L, Lin Q, Sonnenberg GF. Metabolism nature metabolism metabolic control of innate lymphoid cells in health and disease. Nature 2022. https://doi.org/10.1038/s42255-022-00685-8.

[56] Ilies M, Di Costanzo L, Dowling DP, Thorn KJ, Christianson DW. Binding of α,α-disubstituted amino acids to arginase suggests new avenues for inhibitor design. J Med Chem 2011;54(15).

[57] Durante W. Targeting arginine in COVID-19-induced immunopathology and vasculopathy. Metabolites 2022;12.

[58] Pudlo M, Demougeot C, Girard-Thernier C. Arginase inhibitors: a rational approach over one century. Med Res Rev 2017;37.

[59] Baggio R, Elbaum D, Kanyo ZF, Carroll PJ, Cavalli RC, Ash DE, et al. Inhibition of Mn^{2+} 2-arginase by borate leads to the design of a transition state analogue inhibitor, 2(S)-amino-6-boronohexanoic acid. J Am Chem Soc 1997;119(34).
[60] Cama E, Shin H, Christianson DW. Design of amino acid sulfonamides as transition-state analogue inhibitors of arginase. J Am Chem Soc 2003;125(43).
[61] Di Costanzo L, Ilies M, Thorn KJ, Christianson DW. Inhibition of human arginase I by substrate and product analogues. Arch Biochem Biophys [Internet] 2010;496(2):101–8. Available from: http://www.rcsb.org/pdb.
[62] Ilies M, Di Costanzo L, North ML, Scott JA, Christianson DW. 2-Aminoimidazole amino acids as inhibitors of the binuclear manganese metalloenzyme human arginase i. J Med Chem 2010;53(10).
[63] Lu M, Zhang H, Li D, Childers M, Pu Q, Palte RL, et al. Structure-based discovery of proline-derived arginase inhibitors with improved oral bioavailability for immuno-oncology. ACS Med Chem Lett 2021;12(9).
[64] Li D, Zhang H, Lyons TW, Lu M, Achab A, Pu Q, et al. Comprehensive strategies to bicyclic prolines: applications in the synthesis of potent arginase inhibitors. ACS Med Chem Lett 2021;12 (11):1678–88.
[65] Steggerda SM, Bennett MK, Chen J, Emberley E, Huang T, Janes JR, et al. Inhibition of arginase by CB-1158 blocks myeloid cell-mediated immune suppression in the tumor microenvironment. J Immunother Cancer 2017;5(1).
[66] Detroja TS, Samson AO. Virtual screening for FDA-approved drugs that selectively inhibit arginase type 1 and 2. Molecules [Internet] 2022;27(16):5134. [cited 2022 December 11]. Available from: https://www.mdpi.com/1420-3049/27/16/5134.
[67] Tsui SM, Lam WM, Lam TL, Chong HC, So PK, Kwok SY, et al. Pegylated derivatives of recombinant human arginase (rhArg1) for sustained in vivo activity in cancer therapy: preparation, characterization and analysis of their pharmacodynamics in vivo and in vitro and action upon hepatocellular carcinoma cell (HCC). Cancer Cell Int 2009;9.
[68] Roberts MJ, Milton HJ. Attachment of degradable polyethylene glycol to proteins has the potential to increase therapeutic efficacy. J Pharm Sci 1998;87(11).
[69] Diaz GA, Schulze A, McNutt MC, Leão-Teles E, Merritt JL, Enns GM, et al. Clinical effect and safety profile of pegzilarginase in patients with arginase 1 deficiency. J Inherit Metab Dis 2021;44(4).
[70] Burrage LC, Sun Q, Elsea SH, Jiang MM, Nagamani SCS, Frankel AE, et al. Human recombinant arginase enzyme reduces plasma arginine in mouse models of arginase deficiency. Hum Mol Genet 2015;24(22).
[71] Tratsiakovich Y, Yang J, Gonon AT, Sjöquist PO, Pernow J. Arginase as a target for treatment of myocardial ischemia-reperfusion injury. Eur J Pharmacol 2013;720(1–3).
[72] Jung C, Gonon AT, Sjöquist PO, Lundberg JO, Pernow J. Arginase inhibition mediates cardioprotection during ischaemia-reperfusion. Cardiovasc Res 2010;85(1).
[73] Grönros J, Kiss A, Palmér M, Jung C, Berkowitz D, Pernow J. Arginase inhibition improves coronary microvascular function and reduces infarct size following ischaemia-reperfusion in a rat model. Acta Physiol 2013;208(2).
[74] Bitner BR, Brink DC, Mathew LC, Pautler RG, Robertson CS. Impact of arginase II on CBF in experimental cortical impact injury in mice using MRI. J Cereb Blood Flow Metab 2010;30(6).
[75] Kim NN, Cox JD, Baggio RF, Emig FA, Mistry SK, Harper SL, et al. Probing erectile function: S-(2-boronoethyl)-L-cysteine binds to arginase as a transition state analogue and enhances smooth muscle relaxation in human penile corpus cavernosum. Biochemistry 2001;40(9).
[76] Segal R, Hannan JL, Liu X, Kutlu O, Burnett AL, Champion HC, et al. Chronic oral administration of the arginase inhibitor 2(S)-amino-6-boronohexanoic acid (ABH) improves erectile function in aged rats. J Androl 2012;33(6).
[77] Ye C, Geng Z, Dominguez D, Chen S, Fan J, Qin L, et al. Targeting ornithine decarboxylase by α-difluoromethylornithine inhibits tumor growth by impairing myeloid-derived suppressor cells. J Immunol 2016;196(2).
[78] Papadopoulos KP, Tsai FY-C, Bauer TM, Muigai L, Liang Y, Bennett MK, et al. CX-1158-101: a first-in-human phase 1 study of CB-1158, a small molecule inhibitor of arginase, as monotherapy and in combination with an anti-PD-1 checkpoint inhibitor in patients (pts) with solid tumors. J Clin Oncol 2017;35(15_suppl).
[79] Salmani M, Alipoor E, Navid H, Farahbakhsh P, Yaseri M, Imani H. Effect of L-arginine on cardiac reverse remodeling and quality of life in patients with heart failure. Clin Nutr 2021;40(5).
[80] Tousoulis D, Antoniades C, Tentolouris C, Goumas G, Stefanadis C, Toutouzas P. L-Arginine in cardiovascular disease: dream or reality? Vasc Med 2002;7.

Chapter 4.2

Methionine aminopeptidases

Timo Heinrich, Frank T. Zenke, Jörg Bomke, Jakub Gunera, Ansgar Wegener, Manja Friese-Hamim, Philip Hewitt, Djordje Musil, and Felix Rohdich
Merck Healthcare KGaA, Darmstadt, Germany

1 Methionine aminopeptidases

1.1 Introduction

Ribosomal protein synthesis is universally initiated with a methionine or *N*-formyl methionine at the N-terminal of the nascent protein chain that is leaving the ribosomal exit tunnel. Thus N-terminal protein modifications (NMPs) are in principle the first possible co-translational alteration that can be found on the protein chains, and the removal of the initiator methionine is one of the most frequent modification events [1]. The N-terminal methionine excision (NME) is evolutionary conserved and affecting the majority of synthesized proteins in all kingdoms of life [2]. The nonprocessive methionine aminopeptidases (MAPs or MetAPs, EC 3.4.11.18) are the enzymes that specifically catalyze the hydrolytic cleavage in the NME for all organisms [3–6] MetAPs have been identified as ribosome-associated protein biogenesis factors in yeast [7] and bacteria [8], supporting the view they act co-translationally on proteins in situ. Independent of the NME processing, ribosome-associated *N*-acetyltransferases can acetylate the N-terminal amino group of the nascent polypeptide [9], and this acetylation can be found for up to 90% of total human proteins [10].

The NME has been the subject of extensive proteomic studies [11,12] demonstrating specific rules for substrate recognition by MetAPs. MetAPs only process terminal methionine residues and not *N*-formyl methionine or *N*-acetylated methionine. *N*-formyl methionine as initiation residue is usually obtained during protein expression in mitochondria or in prokaryotes and can be preprocessed by peptide deformylases (PDFs), turning the respective protein products into potential substrates of MetAPs. The extent of removal of methionyl from a protein is determined by its N-terminal peptide sequence. Several studies have led to the conclusion that both bacterial and eukaryotic MetAPs prefer small amino acids (e.g., Ala, Gly, Ser, Pro, Cys, Thr, and Val) as the penultimate residue [4,12–14]. Interestingly, an additional bias was found by the residue in the P2′ position and partly beyond. In the *Escherichia coli* MetAP2 protein, the amino acid Pro in position P2′ mostly prevents NME. In the case of human NME a P1′ position filled with Thr or Val biases toward processing through MetAP2 over MetAP1 and will lead to complete excision for any amino acid in the P2′ position despite Asp, Glu, or Pro [15].

For cells it is of fundamental importance to make and keep regulatory proteins available for the timepoint and timeframe they are required to execute their function, but also to maintain homeostasis of amino acid pools in general by recycling nonrequired proteins through targeted degradation. In particular, the NME has also been connected to methionine metabolism [16] and implicated with glutathione redox homeostasis [17]. The NMP was very early recognized as an important mechanism to bias the protein stability and turnover kinetics in a cellular context [18].

The variety of modified or unmodified destabilizing N-terminal residues constitutes the main determinants of N-degrons that are targeting the proteins for proteolytic degradation [19]. For eukaryotes this is mediated by N- or Ac/N-recognins, which are classes of E3 ligases binding specifically to N-degrons, facilitating ubiquitination and subsequent proteasomal degradation. In bacteria, the adaptor protein ClpS acts in a similar way to N-recognins for targeting proteins for degradation through ClpAP proteases [9].

MetAP substrates for NME are frequently acetylated at the neoNt residues and targeted for regulated degradation by the Ac/N-end rule pathway [1,19]. Natural Ac/N-degrons can be regulated through steric shielding in case the degron is structurally involved in intramolecular folding or complex interactions with other proteins. This conformational bias is an additional potential repressor of the otherwise efficient targeting for proteolytic degradation [20,21]. Here, the N-terminal methionine excision by MetAPs preceding the acetylation can directly influence the ability to engage in a shielding conformation and thus indirectly impacts the half-life of the target protein. MetAP activity on proteins with penultimate Cys usually leads to complete NME and, under cellular conditions that favor oxidation, the Nt Cys can be converted to Cys-sulfinate or Cys-sulfonate. These are in turn secondary destabilizing N-degrons (Cys-OX arginylation => R-Cys-Ox) of the

Metalloenzymes. https://doi.org/10.1016/B978-0-12-823974-2.00023-1

Arg/N-end rule pathway [22]. More recently, Nguyen et al. [23] reported that MetAP mediated N-terminal methionine excision can occur even at proteins bearing Met-Asn or Met-Gln at their N termini. The neo Nt residues after NME are tertiary destabilizing N-degrons (Asn/Gln deamination => Asp/Glu arginylation => R-Asp/R-Glu) of the Arg/N-end rule pathway.

Another important in vivo role for MetAPs is the priming activity required for protein myristoylation. The removal of the initial amino-terminal methionine exposes Gly as neoN-terminal residue for the covalent attachment of myristic acid by *N*-myristoyltransferases [24] Myristoylation has been shown to be required for proper functioning and localization of several signaling proteins including src tyrosine kinase family members [25,26], cyclic AMP-dependent kinase and the protein phosphatase calcineurin [27].

MetAPs were initially classified into type 1 and type 2 MetAPs designated MetAP1 and MetAP2 based on sequence alignments. Later studies based on solved crystal structures showed a distinct helical subdomain, approximately 60 residues long, in the C-terminal domain in MetAP2 that MetAP1 lacks [28,29]. In addition to type 1 and 2 classification, each class can be further divided into subclasses or isoforms based on the absence or presence of N-terminal domain extensions in both prokaryotic and eukaryotic MetAPs [3]. In general, although structurally similar, the corresponding sequence homology between MetAP1 and MetAP2 is low [30]. Full sequence alignment of human MetAP1 and MetAP2 shows only 10% sequence identity. Even in the substrate pocket, the sequence identity of amino acids is only around 20%. In contrast, a comparison between human MetAP1 and MetAP1 from *Escherichia coli* shows a 40% sequence identity overall and 73% sequence identity in the substrate pocket.

Type 1 MetAPs are divided into Type 1a, 1b and 1c. These subtypes not only differ structurally but are also expressed in different kingdoms of life: Type 1a and 1c MetAPs can only be found in prokaryotes, whereas type 1b is present only in eukaryotes [31]. Type 1b and 1c distinguish themselves from type 1a by 120 and 40 additional amino acids at their N-termini, respectively. Furthermore, expression of MetAP types 1 and 2 differs between eubacteria and archaebacteria: the former expresses only MetAP1 while the latter expresses only MetAP2. In contrast, eukaryotes were shown to possess cytosolic forms of both MetAP1 and MetAP2 [32,33]. Eukaryotic MetAPs (both type 1 and 2) differ from their prokaryotic counterparts by an additional N-terminal extension—two putative zinc finger motifs in MetAP1 and highly charged N-terminus with alternating polyacidic and polybasic regions in MetAP2, respectively (Fig. 1) [6]. Over the years, beginning with first structural studies in the early 1990s, more than 160 crystal structures of MetAPs have been deposited in the Protein Data Bank (PDB). Most of these structures belong to bacterial MetAPs (81 structures), only a few to Archaea (8 structures), with a total of 78 structures of Eukaryota MetAPs [34]. The strong increase in the number of published structures in the last 20 years, and the relevance of MetAPs as cancer drug targets, accelerated the gain of knowledge and our understanding of MetAPs on molecular mechanistical and physiological levels.

1.1.1 Cofactors

Initial characterization of the translated protein of the *map* gene from *Escherichia coli* revealed the necessity of metal ions for peptidase activity [4] and the substitution with cofactors in form of divalent metal ions turned out to be a common feature of all MetAPs.

MetAPs act as binuclear metallohydrolases. EPR studies with Mn^{2+} loaded *Ec*MetAP showed a protein species formed upon cooperative binding of two metal ions [35]. Correspondingly, MetAP crystal structures contain two metal ions in the active site [31,36–38]. However, a few examples of ligand co-crystal structures with just a single active site metal ion [39] or an auxiliary third metal ion are known as well [31,40–42]. For example, although *Ec*MetAP can be fully activated by only one equivalent of metal ions [39,43,44], the auxiliary third metal ion most probably comes piggybacked with metal-chelating ligands and is attributed to the crystallization conditions with high concentrations of metal ions in the medium [40–42].

The nature of native metal ions utilized by MetAPs is still under debate [40,42,45,46]. Co^{2+} was first assumed to be a general cofactor of MetAPs [4]. However, further metal ions like Zn^{2+}, Mn^{2+}, Fe^{2+}, Ni^{2+}, Mg^{2+}, Ca^{2+}, Cd^{2+} and Cu^{2+} were also considered [5,47–50]. In vitro EDTA can strip off the metal ions. Afterward MetAPs are rather promiscuous regarding their selectivity for metal ions. Some MetAPs accept a handful of the above-mentioned divalent ions enabling them to cleave peptide bonds of artificial substrates [41].

For human MetAP2, manganese is the relevant cofactor in vivo; in doing so two Mn^{2+} ions supplement the active site [51].

1.1.2 Structural features

With the advancements in X-ray crystallography methods, detailed structural information became more readily available to support drug discovery. The first crystal structure of *Escherichia coli* MetAP revealed a novel protein fold, a central antiparallel β-sheet covered by two pairs of α-helices (α1-α2, α3-α4) and by a C-terminal loop on one side, whereas the other face of the β-sheet forms the active site containing two cobalt ions about 2.9 Å apart [52]. Interestingly, the EcMetAP structure appeared to display a pseudo twofold symmetry structurally relating the first and

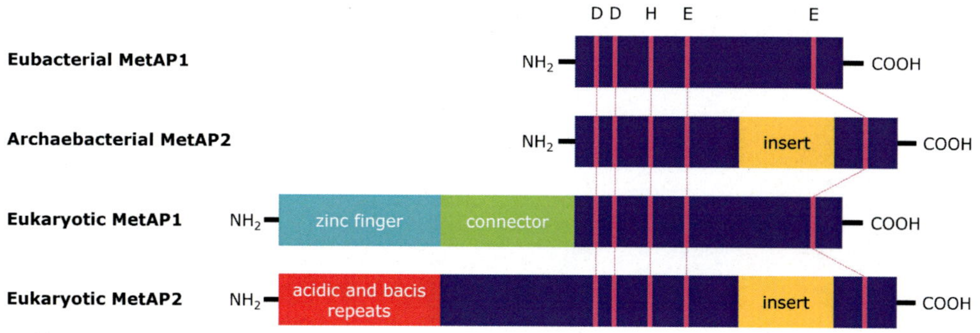

FIG. 1 Domain architecture of MetAPs. In *purple*: catalytic domain; in *yellow*: insert within the catalytic domain observed in MetAP2s; in *red*: N-terminal region with charged residues; in *blue*: N-terminal zinc finger domain; in *magenta* conserved metal chelating residues.

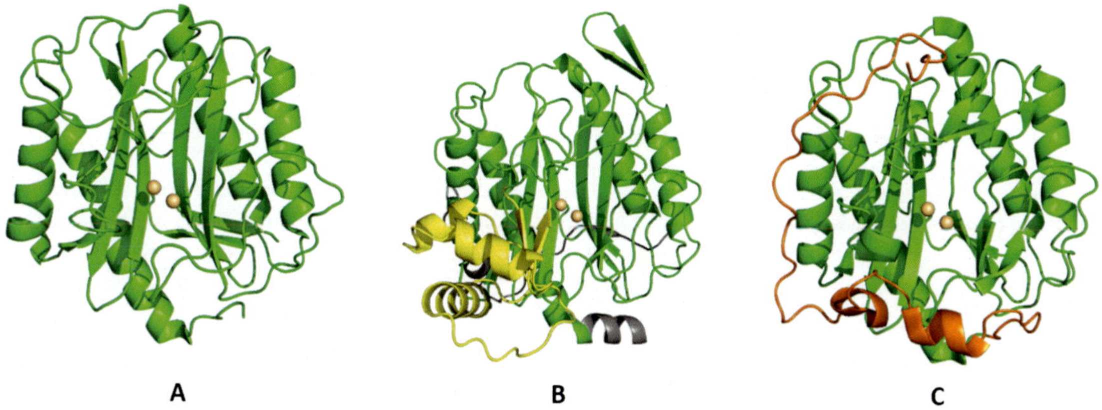

FIG. 2 (A) Overall structure of *Escherichia coli* MetAP (PDB ID: 2MAT) with protein represented by a ribbon diagram in *green* and the cobalt ions by *orange* spheres; (B) overall structure of human MetAP2 (PDB ID: 1B59) with protein shown as a ribbon diagram, polybasic N-terminal chain in *gray*, catalytic domain in *green*, insertion in the catalytic domain in *yellow*, and cobalt ions as *orange* spheres; (C) overall structure of human MetAP1 (PDB ID: 2B3K) with protein shown as a ribbon diagram, the connector domain in *orange*, the catalytic domain in *green*, and cobalt ions as *orange* spheres.

the second half of the polypeptide chain (Fig. 2A). This symmetry is believed to be evidence of an ancestral gene duplication and fusion. The same half barrel pita-bread-like fold [53] of the MetAP catalytic domains is also conserved in HsMetAP2 crystal structures. However, human MetAP2 has an around 160-residue-long N-terminal extension and an insertion (residues 381–444), compared to the EcMetAP structure [29]. This insertion and the N-terminal extension break the pseudo twofold symmetry observed in *Escherichia coli* MetAP (Fig. 2B). Similarly, the catalytic domain of human MetAP1 adopts the pita-bread-like fold characteristic for the MetAP enzymes [31]. So far, all published HsMetAP1 structures include, in addition to the catalytic domain, a connector region. The connector region is absent in HsMetAP2 and overlaps the concave side of the catalytic domain (Fig. 2C).

MetAPs are binuclear metallohydrolases with the catalytic site located near the middle of the central β-sheet on the concave side. They are formed by a pair of divalent cations coordinated by the protein side chains with a conserved D-D-H-E-E pattern, that is, Asp 97/251/229—Asp 108/262/240—His 171/331/303—Glu 204/364/336—Glu 235/459/367 for EcMetAP/HsMetAP2/HsMetAP1, respectively. In the apo MetAP structures there is an additional water molecule coordinating both metal ions simultaneously (Fig. 3).

Despite their differences in structure and sequence that allows for classification and subclassification of MetAPs, there are commonalities across all MetAPs. In the C-terminal catalytic domain five conserved amino acids are responsible for binding of up to two metal ions that serve as cofactors in the hydrolysis of the initiator methionine. In the prototypical MetAP1 of *Escherichia coli*, Asp97, Asp108, His171, Glu204, and Glu235 form a cluster of those metal chelating amino acids [52]. The overall fold of MetAPs—pita-bread-like fold—is shared between MetAP1 and MetAP2: N- and C-terminal domains consist of two α-helices and two antiparallel β-strands. The two domains are arranged in a pseudo twofold-symmetry fashion [52]. The aforementioned, conserved amino acids that form the metal centers are contributed by both domains. In contrast to the pseudo-symmetric arrangement of the two

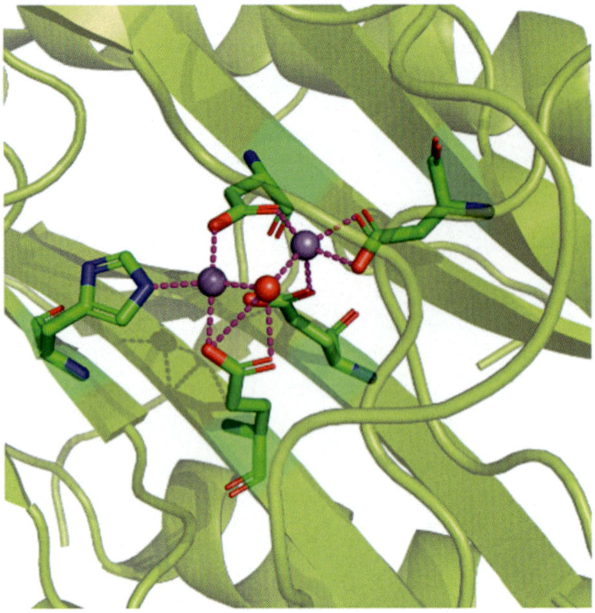

FIG. 3 Catalytic site of human MetAP2 with the protein structure shown as a ribbon diagram, metal-coordinating residues as sticks, Mn^{2+} ions as *magenta* spheres and coordinating water molecule as a *red* sphere.

domains and the bipartite contribution to metal chelating residues, the substrate pockets binding the methionine are built up by amino acids primarily from the N-terminal domain [54].

The first structure of human MetAP2 (with a small molecule ligand) was described in 1998 as a complex with fumagillin, a fungal derivative of the antiangiogenic compound TNP-470 [29]. MetAP2 had been discovered as the molecular target of fumagillin shortly before in a pulldown approach with biotinylated fumagillin [55]. Further structures of complexes with small molecule inhibitors will be detailed in the upcoming sections.

1.2 Classes of inhibitors and their design

A comprehensive literature search did not reveal any publications on activators of MetAPs. Accordingly, the review is focused on MetAP inhibitors.

1.2.1 MetAP1 inhibitors

An ample review from 2019 compiled the chemical matter systematically, and references to the original publications are provided [46]. In Fig. 4, the different classes are summarized: containing carboxylic and hydroxamic acids [56] and esters, including salicylic arrangement (**A**), peptide-based acids (**G**), chelating heteroaryl compounds (**B**), 1,2,4-triazol derivatives (**C**), quinolines (**D**) [57], thiazole-based inhibitors (**E**), prodrug-β-aminoketones (**H**), pyrocatechols (**F**), barbiturates, and more natural products containing motives, such as fumagillin, bengamides, and 2-hydroxy-3-aminoamide-bestatin-like analogs. The structures of the latter are depicted in the MetAP2 (see Section 1.2.2). Similar to MetAP2 inhibitors the contact with two metal ions in the active site of MetAP1 is crucial. For both peptidases it is still a matter of debate, which bivalent metal ions are present in the catalytic center, as discussed above (Section 1.1.1).

In addition to the review articles more recent publications have contributed to the understanding of these diverse inhibitors:

A sensitive, specific, and reproducible UPLC-MS/MS method was developed and validated to quantify the described MetAP1 inhibitor OJT007 in rat plasma after intravenous administration [58]. A polycyclic extension of the amino-thiazole class (**E**) was uncovered through a high-throughput screen, inhibiting MtMetAP1a and -1c with μM activity in the presence of manganese and cobalt ions [59,60]. 3D-quantitative structure activity relationship studies and molecular docking were used to investigate amino-thiazoles [61].

A novel series of 3,6-disubstituted 1,2,4-triazolo[3,4-*b*]-1,3,4-thiadiazoles was published, inhibiting *Escherichia coli* MetAP1. For the most active compounds, an EcMetAP1 inhibition of between 83% and 87% was found, but no information about screening concentration or assay conditions were given [62].

Distubstituted-1,3-thiazole derivatives were discussed in the context of bacterial MetAP inhibition. The antifungal activity against *Candida tropicalis* was twofold worse than fluconazole [63]. Substituted 3-amino-*N*-(thiazol-2-yl)pyrazine-2-carboxamides inhibited MtMetAP-1 in enzymatic assay efficiently, but probably because of missing solubility and permeability were weak inhibitors of whole-cell growth [64].

In addition to the above-mentioned quinoline scaffolds (**D**), quinoline-carbaldehyde derivatives were found to be novel inhibitors of *Leishmania donovani* MetAP1 (LdMetAP1). Structural studies highlighted key differences in the binding modes of the lead compounds to LdMetAP1 and HsMetAP1 providing structural basis for differences in inhibition (PDB: 4FUK, 2B3K) [65]. Quinoline-based thiazolidinone hybrid LdMetAP1 inhibitor QYT-4h, a (Z)-5-((Z)-benzylidine)-2-(quinolin-3-ylimino)thiazolidin-4-one derivative, was found with 20-fold less potency for human MetAP1 [57].

Besides the reviewed work on bengamides, further derivatives and properties were disclosed. Bioassay-guided fractionation identified bengamide B as the active component from *Tedania genus*, which displayed activity in the nanomolar range against both drug-sensitive and drug-resistant *Mycobacterium tuberculosis*. The active compound inhibited in vitro activity of *M. tuberculosis* MetAP1c protein, suggesting the potent inhibitory action may be due to interference with MetAP activity. *Tedania*-derived bengamide B was nontoxic against human

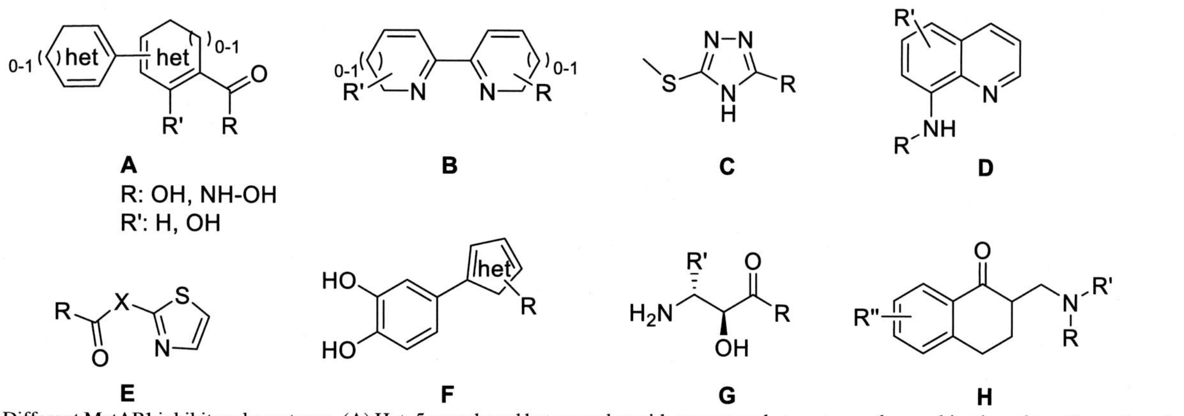

FIG. 4 Different MetAP1 inhibitor chemotypes. (A) Het: 5-membered heterocycles with one or two heteroatoms, the combination of two 5-membered or two 6-membered rings is not described, the carboxylate can be found on the 5-or the 6membered ring, salicylic acids can have a forth substituent like halogen; (B) second heteroatom can be located in one or both rings, identical and mixed ring sizes are described, both, R and R′ can also be another ring, or aliphatic, aromatic, halogen, alcohol or carboxylic; (C) mercapto-triazoles R: amino or alkyl-amino; (D) R can be alkyl or aryl sulfonyl or carboxyl, R′ can be H or halogen; (E) X is usually N but can be O, R is usually heteroaryl but also another amide; (F) het is thiophen or thiazole, R one or two substituents, aliphatic or carboxylic; (G) stereochemistry plays a role, depending on R′ (hetero-aliphatic), R: substituted -amide, −hydroxamic acid, amino acid, heteroaryl; (H) R and R′ can be identical or a rind, R″ is H or an electron rich substituent.

cell lines, synergized with rifampicin for in vitro inhibition of bacterial growth and reduced intracellular replication of *M. tuberculosis* [66].

The characterization of the biosynthetic pathway revealed that bacterial resistance to bengamides can be elicited by mutation of a cysteine or alanine to the more bulky Leu154 in the myxobacterial MetAP protein. A combination of semisynthesis of microbially derived bengamides and total synthesis resulted in an optimized derivative that combined high cellular potency in the nanomolar range with high metabolic stability. This translated into an improved half-life in mice and antitumor efficacy in a melanoma mouse model. Structural investigation supported the work (PDB: 3PKA, 1QZY) [67].

γ,δ-unsaturated α-amino-azepinones could be crystallized in EcMetAP and were superimposed with *Staphylococcus aureus* MAP, setting the base to design analogs, which also achieve *Staphylococcus aureus* inhibition. These unprecedented bengamide analogs were also crystallized (PDB: 4Z7M) [68].

In the context of fumagillin type MetAP1 inhibitors, ovalicin (Fig. 5) was found to inhibit two Type 1 wild-type MetAPs from *Streptococcus pneumoniae* and *Enterococcus faecalis* at low micromolar to nanomolar concentrations. F309 in the active site of Type 1 human MetAP (HsMetAP1b) seems to be the key to the activity, while ovalicin-sensitive Type 1 MetAPs have a methionine or isoleucine at this position. Type 2 human MetAP (HsMetAP2)

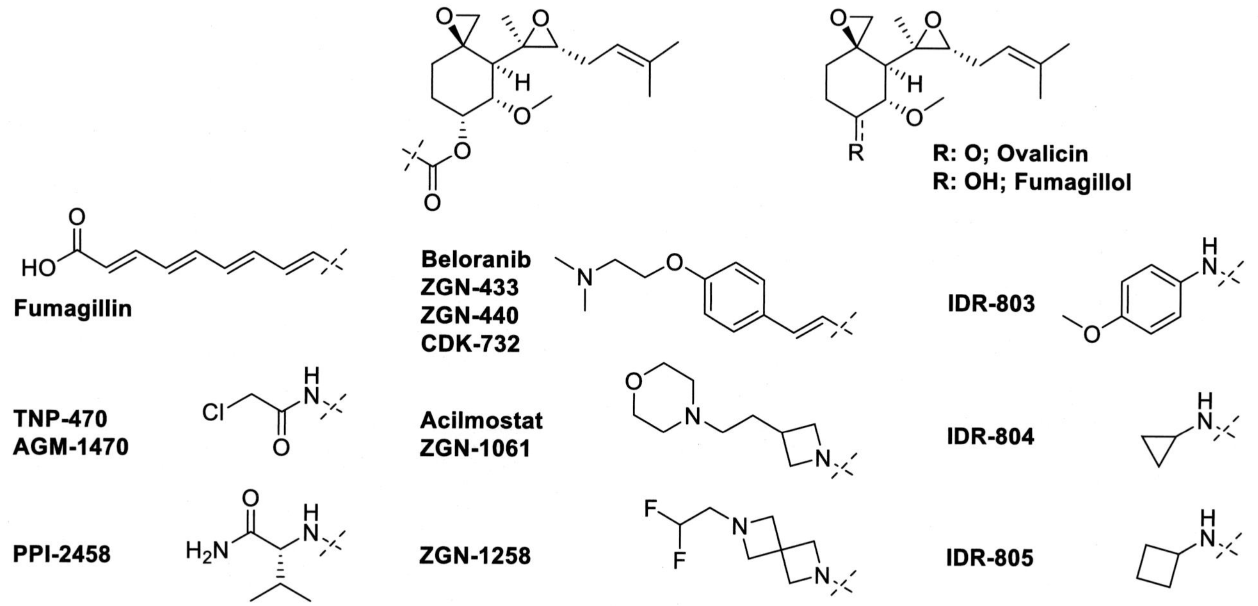

FIG. 5 Structure of fumagillin and closely related analogs.

also has isoleucine (I338) in the analogous position. Ovalicin inhibited F309M and F309I mutants of human MetAP1b at low micromolar concentration. Molecular dynamics simulations suggest that ovalicin is not stably bound in the active site of wild-type MetAP1b before covalent modification. In the case of F309M mutant and human Type 2 MetAP, the inhibitor remains longer in the active site, providing time for covalent modification. The structurally enabled work refers to X-ray data (PDB: 2GZ5, 4U6J, 5YKp, 5YR4, 5YR5, 5YR6) [69].

7-Bromo-5-chloroquinolin-8-ol (CLBQ14), a close relative of clioquinol (Fig. 7), was reported as a potent and selective inhibitor of two MetAPs from *M. tuberculosis*: MtMetAP1a and MtMetAP1c. CLBQ14 is potent against replicating and aged nongrowing *M. tuberculosis* at low micromolar concentrations. Furthermore, it was observed that the antibacterial activity of this pharmacophore correlates well with in vitro enzymatic inhibitory activity [70]. Other nitroxoline analogs showed excellent inhibition potency in *Burkholderia pseudomallei* MetAP1 enzyme activity assay with the lowest IC_{50} of 30 nM and inhibition of the growth of *B. pseudomallei* and *B. thailandensis* at concentrations $\geq$31 μM. Structural studies were explored to better understand the nature of inhibitor interactions within the active site (PDB: 3IU9, 3VH9) [71].

Diketoesters were used as temples for amino acid analogs, which were found to be active against different bacterial strains. Biochemical evaluation against MetAPs from *Streptococcus pneumonia* (SpMetAP), *M. tuberculosis* (MtMetAP), *Enterococcus faecalis* (EfMetAP) and human (HsMetAP) showed different behavior of the compounds against these four enzymes [72]. Diketo acids and their bioisosteres were optimized in another work as bacterial MetAP1a and 1c inhibitors. Optimized compounds exhibited antibacterial potential with minimal inhibitory concentration (MIC) values of 31 μg/mL (*Enterococcus faecalis*) and 63 μg/mL (*Streptococcus pneumoniae; Escherichia coli*) [69].

In a screen of 20 α-aminophosphonates, some derivatives were identified that selectively inhibit the *Streptococcus pneumonia* MetAP in low micromolar range but not the human enzyme. Structural information was used to rationalize the observations (PDB: 4KM3, 4UJ6, 4U75, 4U71, 4U70, 4U6C, 4U6E, 4U73, 4U69, 4U6Z, 4U1B, 4U6W) [73].

A virtual library of 5-arylidene barbituric acids was created and molecular docking was performed for identification of novel possible inhibitors of NDM-1 and MetAP-1. These nontoxic 5-arylidenepyrimidine-triones showed bacterioxtatic activity [74].

Besides high-throughput screening (HTS) runs, in silico work using the FlexX software was also published to identify bacterial MetAP inhibitors. From 200,000 considered compounds from the ZINC database nine were purchased and tested against *Staphylococcus aureus*, *Escherichia coli*, and *Pseudomonas aeruginosa* [75]. In an additional structure-based pharmacophore modeling approach, three diverse ZINC compounds with stable binding mode over time were identified [76].

MetAPs are essential enzymes in all living cells but due to the similarity in the active sites of the MetAP enzyme between pathogens and human, there is only limited success in discovering selective inhibitors as antimicrobial agents. New variants of MetAPs from *Vibrio* species, having two or three inserts within the catalytic domain and specific inhibitors were identified. Derived from scaffold **B**, the new pyrimidinylpyrimidine derivatives showed different biding affinity to *Vibrio* MetAPs compared with their human counterpart [77].

1.2.2 MetAP2 inhibitors

The history of MetAP2 inhibitors goes back to the middle of the 20th century when drug discovery usually started with a biologically active, natural compound, which could be further investigated and optimized in the absence of knowledge about the target, such as the structure and architecture of the active site of the target. Fumagillin (Fig. 5) was an intriguing example, as it was isolated and characterized as antiangiogenic. Accordingly, the design of analogs was mainly property driven. When the target of MetAP2 was identified, it was proven that the fumagillin-based compounds were all covalent MetAP2 inhibitors (His231 side chain) [55].

Fumagillin and prominent analogs like beloranib (CKD-732, ZGN-440; ZGN-433;), TNP-470 (AGM-1470), the orally active analogs PPI-2458 and aclimostat (ZGN-1061) have been described in the literature and reviewed extensively (Fig. 5). Additional not disclosed analoges, ZGN-1345 and ZGN-1136, were described as MetAP-2 inhibitors preventing liver fibrosis and hepatocellular carcinoma [78]. For PPI-2458 it could be explicitly shown that degradation of the parent compound by reaction of the spiroepoxide warhead was compensated by the formation of active equivalents leading to a pharmacologically useful level of MetAP2 inhibition [79]. The high activity of these derivatives in conjunction with their suboptimal property profile sparked the investigation of diverse formulation options, including lodamin (poly(ethylene-glycol)-poly(lactic)acid conjugate of TNP-470) [80], XMT-1107 (fumagillol conjugate) or SDX-7320 (fumagillin derivative conjugated to the polymer poly(*N*-hydroxypropylmethacrylamide) [81].

These compounds are based on the original fumagillin core structure with an epoxy-cyclohexyl scaffold and a second epoxide in one side chain. Attempts to optimize multicyclic fumagillin derivatives were not followed up further due to synthetic challenges of this approach [82].

A synthetically more attractive approach considered spiroepoxytriazoles, which showed improved stability in mouse plasma and microsomes relative to beloranib. The biochemical activity of these derivatives correlated with human umbilical vein endothelial cell (HUVEC) proliferation inhibition, and the most potent compound was found to have an EC_{50} in the low three-digit nanomolar range [83].

Compound library screening delivered bicyclic derivatives, which were optimized based on pharmacophore models and their activities against colon cancer and other tumor cells. The target of the most active compound 3-phenyl-4*H*-thieno[3,2-*b*]pyrrole-5-carboxylic acid (3-trifluoromethyl-phenyl)-amide was explored by target fishing strategy and validated by molecular docking and biological activity analysis. The results of apoptosis and flow cytometry demonstrated that the compound induced cell apoptosis probably by MetAP-2 inhibiton [84].

The organoselenium drug molecule ebselen has antiinflammatory, antioxidant, and cytoprotective activity (PZ51; Fig. 6). It has been shown to have antifungal activity against *Aspergillus fumigatus*. The compound has a pleiotropic mode of action; however, it could be shown that MetAP2 is one target, contributing to its activity. The optimized derivative 2-(2-chloro-4-methylphenyl)-1,2-benzoselenazol-3-one was characterized with low three-digit nanomolar activity in an enzymatic assay [85]. Another class of known drugs are actinomycin derivatives, which have been the subject of many studies without clear understanding of mode of action, and were investigated in enzymatic alanine and MetAP assays. Micromolar IC_{50} values were obtained for all derivatives against both enzymes. Molecular modeling and docking studies in the active sites confirmed on-target activity [86].

In addition to the potent and selective fumagillin class of inhibitors, structurally divergent MetAP2 inhibitors were also reviewed (Fig. 7).

2-Hydroxy-3-aminoamides (bestatin class of pseudosubstrate-like reversible inhibitors, like A-357300) were investigated with much effort. A-357300 is described with an IC_{50} of 0.12 μM in the MetAP2 assay and 57 μM in MetAP1 [87].

Fused caprolactam containing ketide-amino acid derivatives (bengamide class of modified marine natural product inhibitors, like LAF389) where characterized as MetAP2 inhibitors from a proteomics approach in 2003 but showed MetAP1 activity on a similar sub-μM level [88]. This structural class is still of interest in more recent structural analyses [89] and was reviewed recently [90].

Anthranilic acid sulfonamide analogs constitute a large class of MetAP2 inhibitors, including very potent derivatives, which show selectivity (like A-800141: MetAP2: 0.012 μM; MetAP1: 36 μM) [91]. Also, QSAR studies were performed to establish a potency predictive model. No examples were provided, which confirm if this approach supported the design of improved MetAP2 inhibitors [92]. Mono- or double substituted triazole derivatives (like SB-587094 or JNJ-4929821), were described as selective MetAP2 inhibitors, with some selectivity for the Co^{2+} metal cofactor in the enzyme (Fig. 8).

The pharmacophore of triazole-thioether was also considered in a new approach highlighted by the preferred core structure of seven-membered rings like those in diazepines. The extended structure of the resulting scaffold [1] is depicted in Fig. 8. All the functionalized derivatives were evaluated for in vitro antifungal activity using test panels of fungal strains compared with standard drug fluconazole. Inhibitory effects on *A. fumigatus*, with MIC value of 0.20 mg/mL for the most potent ones, were found. Docking studies against enzyme N-myristoyl transferase classified certain substitution patterns on the quinoline as favorable judged on C-score values. Moderate activity in a renal cell line proliferation assay motivated docking studies against the active site of MetAP2, indicating that a slightly different

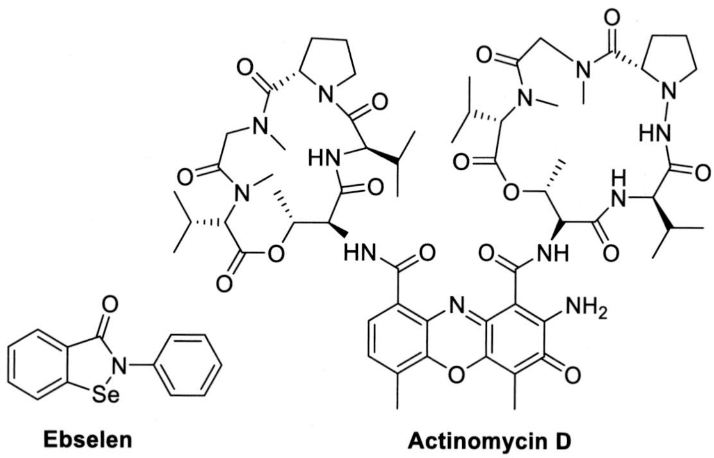

FIG. 6 Established drugs that were discovered to target MetAP2 much later.

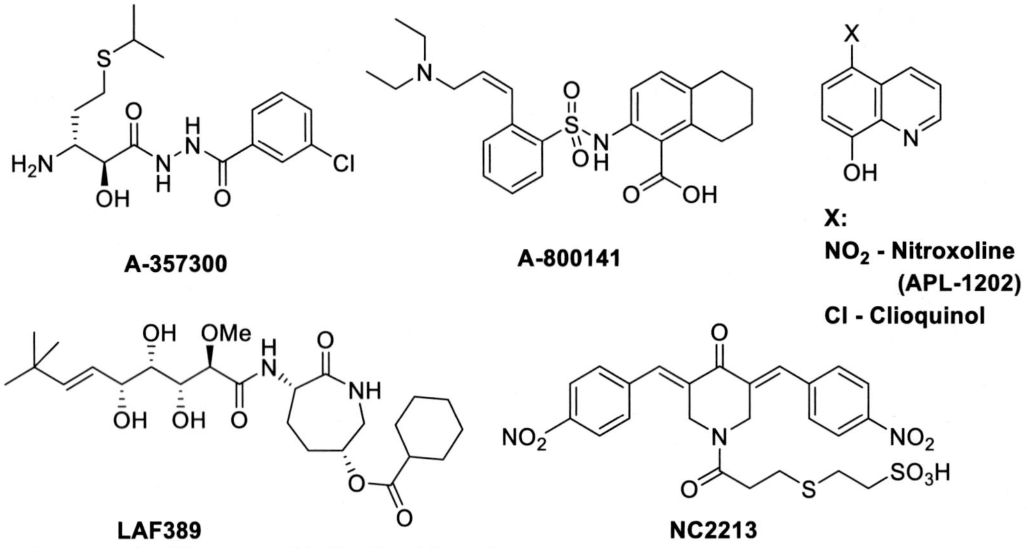

FIG. 7 Dedicated examples for different reversible MetAP2 inhibitor classes.

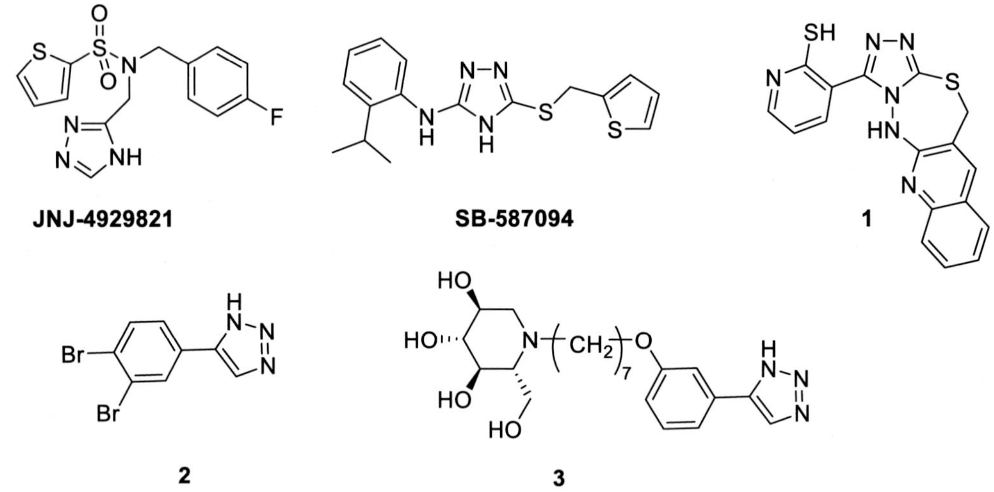

FIG. 8 Different triazole-scaffolds.

quinoline substitution pattern is beneficial for enzyme interaction [93]. The amino- and thioether-substituted triazoles of SB-587094 type were the basis for a QSAR study, which delivered findings, contradictory to the SB structure (besides confirmed NH on triazole for His231 and aniline for Asp264 interaction). The aniline should contain an electronegative group in the meta position and the aryl moiety of the benzylthioether should contain bulky groups in the ortho- and electronegative groups in the meta position. No examples were prepared to probe these CoMFA derived suggestions [94].

Triazole-based fragments like 4-(4-tolyl)-1,2,3-triazole had been found to inhibit MetAP2 (PDB: 2ADU) and **2** inhibited angiogenesis in vivo [95]. Hybridization of the aryl-triazole scaffold with the α-glucosidase inhibitor *N*-methyl-1-deoxynojirimycin (**3**) improved inhibition of proliferation of bovine aortic endothelial cells, but it was not proven that this improvement was still MetAP2 linked [96].

In alternative SB-587094-oriented approaches, the 4-phenyl-1,2,4-triazole core, containing a 1,4-benzodioxan substituent on one triazole carbon and a benzylthioether on the other triazole carbon was investigated. Micromolar activity in a MetAP2 inhibition assay as well as in cancer cell proliferation assays was reported. The most potent compound with ortho-F-benzylthioether inhibited the growth of HEPG2 cells with IC_{50} of 0.81 μM and inhibited the activity of MetAP2 with IC_{50} of 0.93 μM, which was comparable to the positive control TNP-470 [97]. Similarly, 1,4-benzodioxan substituted 1,3,4-oxadiazoles with styrene and no arylthioether moieties were investigated. Again, the most active compound contained an ortho-*F*-phenyl residue and inhibited the growth of HUVEC cells with IC_{50} of 1.16 μM and inhibited the activity of MetAP2 with IC_{50} of 2.08 μM, similar to TNP-470 [98]. The carboxylic acid function of COX inhibitor flurbiprofen was transferred into triazole thioethers of SB-587094 type and used in molecular modeling studies to investigate MetAP2 binding affinity. Nanomolar Ki values were calculated but no crystallization, nor enzymatic characterization was provided. Nevertheless, observed micromolar proliferation inhibition of cancer cell lines was attributed to MetAP2 inhibition [99]. An identical approach was pursued using etodolac as a triazole-building block with similar results for calculated MetAP2 affinity and cellular activity. Also, in this study, MetAP2 activity was based on in silico work [100]. In addition, the carboxylic acid function of naproxen was used to synthesize corresponding triazoles. In this approach the thioether moiety contained hydrazide-hydrazones. With these compounds not only cellular proliferation assays with micromolar activity were performed, and molecular modeling studies to insinuate MetAP2 binding, but also compound distribution of a labeled derivative in in vivo studies, demonstrating accumulation in prostate [101,102].

Hydroxy quinoline derivatives (clioquinol, nitroxoline, Fig. 7) [103,104] are structurally described and shown to bind to MetAP2 and inhibit enzymatic function [105]. In addition, these compounds also inhibit MetAP1. The described enzymatic activities are in the micromolar range, so it might be questioned whether observed effects in cellular assays or even in in vivo investigations are driven by MetAP2 inhibition. Recently, a report on a slightly modified hydroxy-quinoline derivative, CLBQ14 (7-bromo-5-chloroquinolin-8-ol), was published [106]. Limited solubility might be responsible for the difference in bioavailability of 39% and 90%, respectively, from oral and subcutaneous routes. In comprehensive SAR studies with oxines, it was proposed that other not yet identified targets might contribute significantly to the observed antiproliferative and antiangiogenic effects of this structural class [107].

Besides the above-mentioned compounds, further derivatives were published in the context of MetAP2 inhibition, but it is not fully clear whether the observed effects were MetAP2 related or even driven by its inhibition. The symmetric *N*-acyl piperidin-4-one NC2213 (Fig. 7) is observed to block expression of MetAP2 and inhibited proliferation of human colon cell lines in the single digit micromolar range [108].

More recently published MetAP2 inhibitors are summarized in Fig. 9. 5-Chloro-3-phenyl-indol-2-amide was identified from a screening campaign with a MetAP2 IC_{50} of about 300 nM. Structural investigations (PDB: 7A12) disclosed the amide oxygen interaction with the first metal ion and a water mediated interaction between the indole NH and the second metal in the active site of MetAP2. Investigation of alternative metal binding moieties revealed different carbonyl and carboxyl moieties, an oxazole and sulfone, but none was more potent than the free amide from the screening hit. Occupation of the buried substrate recognition P1 pocket with a fluorine in position 6 of the indole improved potency by an order of magnitude, and the introduction of an ortho isopropyl ether on the phenyl ring contributed equally. Property adjustment required the change of the 5-Cl into a 5-F and the addition of piperazine para to the indole at the 3-phenyl moiety to give the potent (MetAP2 IC_{50}: 2 nM) lead compound GSK-2229238, which was structurally confirmed (PDB: 6QEI), showed good bioavailability (89% at 10 mg/kg) and demonstrated reduction in the secondary immune response (anti-KLH specific IgG) in a mouse model [109].

Optimization of the fragment hit 8-Br-indazol-3-ol (MetAP2 IC_{50}: 2.5 μM), without immediate structural but in silico support, and SAR by catalog gave the 350-fold more potent lead compound **4**. The intermediate hit 4-(3-methylpyridin-4-yl)-6-(trifluoromethyl)-1H-indazole could be crystallized (PDB: 5JI6) and confirmed the hypothesized binding mode of the indazole nitrogen interaction with the manganese ions in the active site, the halogen occupation of the hydrophobic cleft between Tyr444 and His339 and the pyridine moiety pointing to the pocket entrance. Property optimization delivered potent (MetAP2 IC_{50}: 5 nM) lead compound **4** with good properties (AUC: 15,084 ng × h/mL; $t_{1/2}$: 2 h; F: 58%), which lowered body weight in DIO mice in a 28-day model by 14% at a dose of 30 mg/kg [110].

In a fragment-based drug discovery approach, a pyrazolo[4,3-*b*]indole core was designed. Optimized lead compound **5** could be crystallized and confirmed the expected binding mode (PDB: 5JFR). Activity (MetAP2 IC_{50}: 5 nM) and property profile (AUC: 2237 ng × h/mL; $t_{1/2}$: 6 h; F: 70%) triggered evaluation of the compound in a 7-day DIO-mouse model for obesity where it reduced body weight by 4% (30 mg/kg qd) [111].

The origin of compound **6** and derivatives, which differ mainly in the aryl substituent of the tricyclic core structure, was not disclosed. These compounds were recognized in a patent highlight including information about their μM enzymatic MetAP2 inhibition profiles [112].

An HTS approach had uncovered compound **7** with 1 μM MetAP2 inhibition in an enzymatic assay. Structural analysis (PDB: 5LYW) revealed the stereochemical requirements on the pyrrolidine moiety, because the 2S enantiomer was virtually inactive and probably would clash with the Ile338 side chain. The o-*Cl*-phenyl moiety was found to be sandwiched between Tyr444 and His339 side chains. This extended π-π interaction could be improved with isoquinolin-5-yl (fivefold more active). The purine nitrogens 3 and 9 are in contact with the manganese ions in the active site of MetAP2. Optimized nitrogen allocation in the bicyclic scaffold impacted electron distribution and

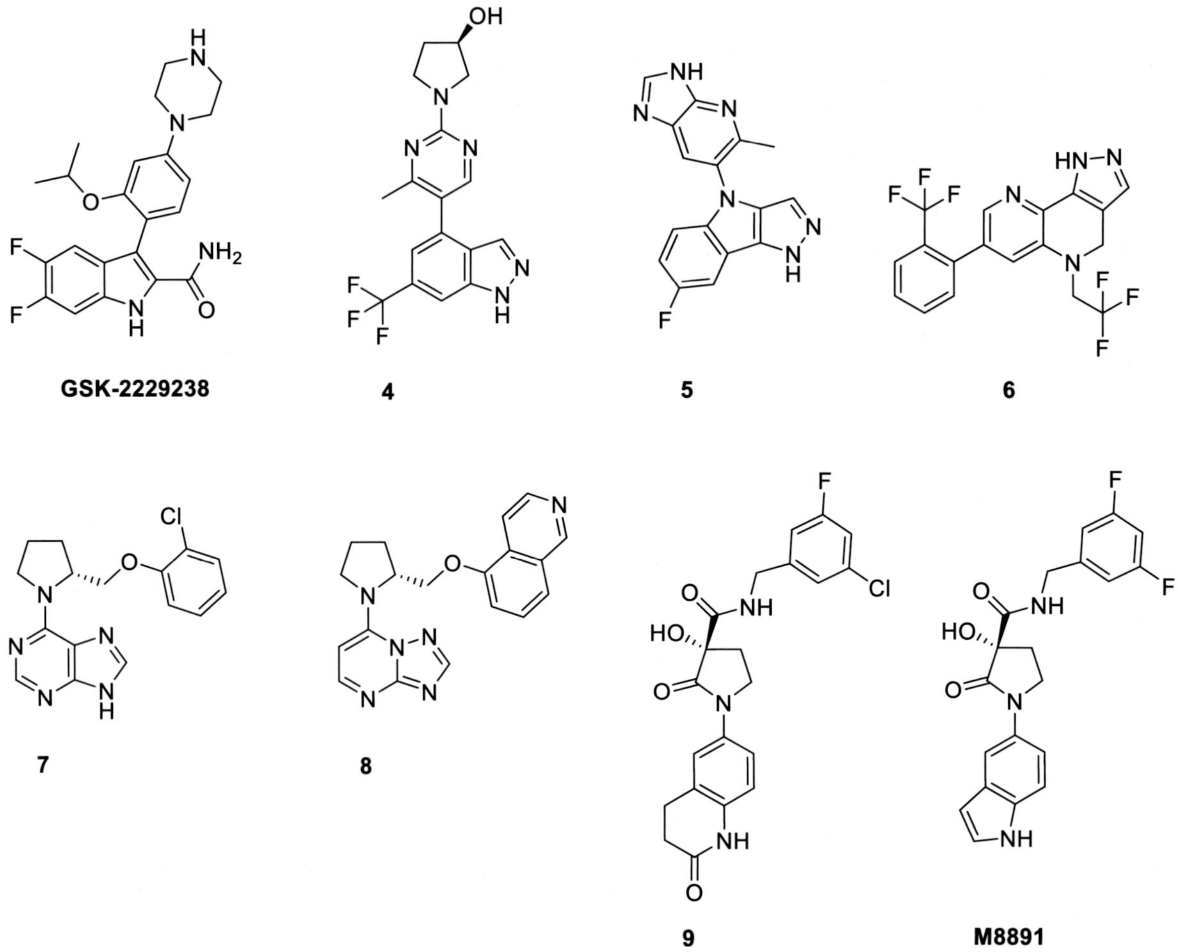

FIG. 9 New MetAP2 inhibitors.

prevented tautomerization needs, giving the two-digit nanomolar lead compound **8**. Obviously, it was not possible to translate this enzymatic activity into potent and consistent HUVEC proliferation inhibition. Surprisingly, structural investigations revealed that the bicyclic scaffold flipped by 180 degrees when substituting the purine with the triazolo-pyrimidine core (PDB: 5LYX) [113].

Serendipity played a central role during the identification of a completely new MetAP2 binding class. Commercial α-amidated lactam-based compounds oxidized, over years of storage, to cyclic tartronic diamides that were found to be active against MetAP2. Structural investigations revealed an unprecedented interaction pattern between the three oxygens of the scaffold and the two manganese ions in the active site and adjacent amino acid side chains of MetAP2. Further polar contacts at the entrance of the pocket and lipophilic interactions in the substrate recognition area were responsible for the high activity in enzymatic as well as cellular assays. Optimized lead compound **9** showed good solubility in biorelevant media (FaSSIF: 158 µg/mL; FeSSIF: 334 µg/mL). Favorable physicochemical properties contributed to good passive permeability and acceptable efflux in the Caco-2 assay ($a > b$: 16×10^{-6} cm/s; ER: 5) as well as favorable metabolic stability in liver microsomes from mouse and human (<10 µL/min/mg protein). After intravenous dosing in mice, compound **9** showed favorable pharmacokinetic (PK) characteristics (CL: 0.18 L/h/kg, V_{ss}: 0.78 L/kg, $t_{1/2}$: 3 h). The oral bioavailability was 35% at 0.5 mg/kg and increased to ~100% at 50 mg/kg, suggesting the saturation of a clearance mechanism. In an in vivo efficacy model (human glioblastoma, U87) a T/C value of 45% was achieved by daily dosing of 10 or 20 mg/kg [37]. Further PK investigations in nonrodent species suggested an enterohepatic circulation (EHC) of compound **9**, preventing further development [37]. Reduction of H-bond acceptor count optimized the EHC issue of the lead without compromise on activity and overall good properties, giving the development compound M8891 [NCT03138538]. As biophysical investigations seemed to be problematic by applying established techniques like surface plasmon resonance spectroscopy, innovative options were considered. Fluorescence cross correlation spectroscopy was found to be supportive and enabled the determination of affinity and residence time. Fluorescence correlation spectroscopy with intact A549 cells allowed for determination of target occupancy. In the U87 model M8891 achieved a T/C value of 19% after 20mg/kg once daily dosing [38].

Besides these hits and leads from practical approaches, virtual screening was also performed, to identify putative starting points for optimization of MetAP2 inhibitors. The Lipinski-guided filtering of a commercial library of 3 million compounds resulted in ~236,000 derivatives, which were virtually screened using PyRx, followed by docking with AutoDock4, and manual binding mode inspection with Biovia Discovery Studio 4.5 identified 10 best candidates. Molecular dynamics simulation finally delivered two potential new MetAP2 inhibitors, one pyrrolidine-sulfonamide and one piperazine-carboxamide, with calculated single and low two-digit nanomolar inhibition constants [114].

1.3 Physiology and pathophysiology

1.3.1 Physiological role of MetAPs

MetAP2 attracted significant attention after its seminal discovery as the pharmacological target of the natural compound fumagillin, which was isolated in 1949 from the fungus *A. fumigatus*. Several antibiotic activities were attributed to fumagillin in early studies [115,116] Fumagillin is an approved antibiotic for control of nosema disease, a microsporidian pathogen in honey bees [117]. Also, the semisynthetic fumagillin analog TNP-470 has shown in vivo activity against microsporidia [118]. Further antiparasitic activity of these derivates has been shown against malaria parasites, *L. donavani,* trypanosomes, and amoebas in murine models in vivo [119–122]. Giardiasis is a severe intestinal parasitic disease caused by *Giardia lamblia.* In preclinical in vivo mouse studies, fumagillin has shown anti-*Giardia* activity comparable to standard-of-care metronidazole [123].

However, the groundbreaking discovery of its antiangiogenic/antitumoral activity led to a boost of drug optimization culminating in the clinical investigation of fumagillin derivatives in several clinical trials in oncology [124].

Analyses of the RNA expression and protein abundance showed that all three human MetAP genes, MetAP1, MetAP2, and MetAP1D are ubiquitously expressed across all different human tissues (Human Protein Atlas: MetAP1; MetAP2; MetAP1D). MetAP2 gene expression can be modulated by mitogens or heat shock, and evidence for regulation of gene expression by c-myc exists [125–127]. The chromodomain helicase DNA binding protein CHD1L/ALC1, which is important for chromatin remodeling and overexpressed in a variety of different malignancies, has been implicated in transcriptional regulation of MetAP2 as well [128]. The MetAP2 promoter appears to be regulated by circadian rhythm through binding of clock proteins [129]. Overexpression of MetAP2 was reported in several cancers, such as colorectal cancer, mesothelioma, cholangiosarcoma, and glioma [130–134], while expression did not appear to be altered in relation to normal tissues in B cell lymphoma [135]. Notably, one study reported enhanced proliferation upon overexpression of MetAP2 in fibroblasts and epithelial cells (transformation of fibroblasts in soft agar) [136].

Evidence for posttranscriptional regulation of MetAP2 is sparse and not yet supported by solid scientific evidence. Data on autoproteolysis of purified rat MetAP2 into a p26 and p52 fragment have been reported; however, the physiological relevance of this autoproteolytic cleavage for MetAP2 function in cells or tissues is unclear [137]. In a rat neuronal ischemia/reperfusion model MetAP2 was shown to be proteolytically cleaved by calpain and inhibited by the specific inhibitor calpastatin, again, the relevance of calpain-mediated proteolysis remains to be further investigated [138]. MetAP2 was found to be glycosylated and, in reticulocyte lysates, deglycosylation correlated with increased eIF2a phosphorylation [139]. Subsequently, a glycosylation site located in the N-terminus (aa 60–63) could be identified [140]. The aminopeptidase activity of human MetAP2 can be reduced by nitration of tyrosine residue Tyr336, which is located adjacent to the catalytic binding pocket, however, evidence for this modification in situ has not been presented [141]. Furthermore, MetAP2 but not MetAP1, are subject to redox regulation of a cysteine disulfide bridge Cys(228)-Cys(448) [142]. Indeed, the reduced and oxidized forms show different proteolytic activities and substrate specificity, the first evidence for the existence of both forms was shown in glioblastoma cells. MetAP2 activity may also regulate the glutathione redox state of cells, as fumagillin-mediated inhibition of activity shifts the equilibrium toward the oxidized form [17].

It was estimated that, depending on the organism and cellular compartment, between 55% and 70% of proteins are subjected to removal of the initiator methionine [2]. Several proteomic studies were carried out using knockout cells, siRNA, or pharmacological inhibitors to capture the substrate space of MetAP1 and MetAP2 [17,88,143]. These studies provided evidence for overlapping substrate specificity but also notable differences, but overall could not provide an obvious explanation for the different phenotypes observed upon MetAP2 inhibition. Numerous substrates of MetAP2 have been identified in other studies and, in part, used to track pharmacodynamic MetAP2 inhibition in cellular systems and animal models; their relevance for the cellular phenotypes remain to be identified (Cyclophilin A, GAPDH [144]; 14-3-3γ [88]; SH3BGRL, eEF2, TRX-1 [145]; eNOS [146]; EF1α [147]). Notably, Rab37 represents an example of a MetAP2 substrate, which appears to be significantly stabilized upon pharmacological inhibition, and for which increased expression of Rab37 reduced endothelial network formation [148]. Another substrate functionally impacted by MetAP inhibition is the tyrosine kinase Src, which requires amino-terminal processing to attach a lipid group for proper membrane localization 3 [25]. Bengamides, which inhibit MetAP1 and MetAP2, indeed inhibit N-terminal myristoylation of Src, while MetAP2-specific inhibition is not sufficient [144].

On a separate academic path, studies exploring the regulation of translational activity in reticulocyte lysates had investigated p67, later identified as MetAP2, as an interacting protein of the translation initiation factor eIF2α [139,149]. Translational activity of eIF2α is regulated by phosphorylation by different protein kinases, and interaction with an N-terminal acidic region of MetAP2 may protect eIF2α from phosphorylation [137,150–152]. First data indicated a role of this interaction in cell cycle-dependent regulation of eIF2α, heat shock-mediated phosphorylation of eIF2α, or vaccinia virus PKR-mediated eIF2α phosphorylation [150,153,154]. The calcium binding protein S100A4 interacts with the amino acid 170–208 of MetAP2 [155,156]. Loading of cancer cells with a synthetic peptide corresponding to the S100A4-binding domain of MetAP2 (NBD peptide) reduced motility and invasion, and introduction of NBD into endothelial cells prevented capillary formation [157,158]. ERK1/2 have been shown to interact with MetAP2 and inhibition of MetAP2 by fumagillin led to a reduction in ERK activity [159].

To investigate the function and relevance of MetAP2 in different cell models, knockdown/knockout technologies, ectopic expression of wildtype protein or variants thereof, and different pharmacological inhibitors have been used. Downregulation of MetAP2 expression by different approaches has been shown to cause growth inhibition and/or apoptosis in different cell types. An increase in phosphorylation of eIF2a and a reduction in protein synthesis was observed in rat hepatoma cells KRC7 together with signs of G0/G1 cell cycle arrest and apoptosis [160]. A reduction in PMA-induced protein biosynthesis was also observed upon expression of antisense DNA in rat KRC7 cells [161]. Similarly, the viability of mesothelioma cell lines was reduced with signs of apoptotic cell death after transfection of antisense oligonucleotides [131]. Downregulation of MetAP2 by shRNA in SNB19 glioblastoma cells, reduced proliferation with concomitant G0/G1 cell cycle arrest, decreased protein synthesis, reduction in VEGF secretion [162]. The upregulation of the tumor suppressor VHL was suggested to be required for the reduction in VEGF expression/secretion. A catalytically inactive variant of MetAP2 was identified as acting as a dominant-negative variant on proliferation of HT1080 tumor cells [136]. Mixed results were reported after knockdown of MetAP2 in endothelial cells. One study reported that siRNA-mediated knockdown of MetAP2 did not alter growth of human endothelial cells and did not phenocopy the action of fumagillin or bengamides [163], while clear antiproliferative activity was observed in two other studies [164,165]. In one of the latter studies, the effect of single and double knockdown of MetAP1 and MetAP2 was investigated, with the double knockdown exhibiting a more severe antiproliferative effect versus the single knockdowns. This phenotype is suggestive of functional redundancy between the two cytoplasmic MetAPs, which was also observed in other eukaryotic organisms such as yeast [33].

With the advent of CRISPR and RNAi genomic data sets, made available by the Cancer Dependency Map initiative at the Broad Institute of MIT and Harvard, a broader look at the CRISPR-mediated knockout of the MetAP is possible (www.depmap.org). At the current stage of data release (Version 23Q2) both MetAP1 and MetAP2 scored as essential in 662/1095 and 666/1095 cancer cell lines, respectively. Interestingly, the top gene showing co-dependency with MetAP1 CRISPR is MetAP2 and vice versa, again demonstrating that an intimate functional redundancy between both peptidases exists.

It is important to note that pharmacological inhibitors, especially covalent inhibitors such as fumagillin or TNP-470, are to be used at appropriate concentrations, as it is well known that the biphasic behavior in proliferation assays is suggestive of activities probably driven by MetAP2 inhibition-independent effects at higher concentrations, above 30 μg/mL [166]. Based on multiple publications using MetAP2 inhibitors several inhibitor-sensitive cell lines from human or other species (bovine, rat, mouse) originating from different tissues (endothelial, fibroblast, smooth muscle, cancer cell lines) have been identified and characterized. However, a plethora of cell lines have been identified, which do not, or only weakly, respond to MetAP2 inhibitors, indicating a differential activity on proliferation [144]. HUVEC belongs to one of the most sensitive human endothelial cell lines and has been used frequently as a model to study the cellular consequences of MetAP2 inhibition. Cell growth and thymidine incorporation into DNA are potently inhibited by TNP-470 and cells accumulate in G0/G1 [166]. TNP-470 exposure led to induction of the tumor suppressor p53, and its transcriptionally regulated cell cycle inhibitor p21, which could explain the cytostatic effect of the inhibitor [167]. Conversely, knockout of p53 or p21 in mouse embryo fibroblasts triggered resistance to TNP-470. Growth-factor-induced DNA synthesis in bovine vascular smooth muscle cells was inhibited by TNP-470 and accompanied by a reduction in CDK2 mRNA and activity [168]. Studies with fumagillin showed similar antiproliferative activities in endothelial cells and weak to no activity in other cell types. A compensatory upregulation of MetAP2 expression, but not MetAP1, in sensitive endothelial cells was noted upon treatment with fumagillin [169]. Disruption of WNT planar cell polarity by inhibition of MetAP2 was accompanied by accumulation of Rab37 and was shown to contribute to endothelial growth arrest and dysfunction [148,170]. Mesothelioma cell lines were identified as exquisitely sensitive to fumagillin and showed inhibition of DNA synthesis as well as apoptosis induction [131]. An increase in production of nitric oxide was detected in hepatoma and bovine vascular endothelial cells [146]. Whether this is a consequence of the inhibition of N-terminal processing by MetAP2 remains to be further corroborated. A targeted fumagillin prodrug was shown to induce NO production in human endothelial cells, which in turn induced AMPK phosphorylation in cocultured macrophages [171]. Another derivative of fumagillin, PPI-2458, showed antiproliferative activity on several non-Hodgkin Lymphoma cell lines [172]. Antiangiogenic activity in more complex models, like vascular sprouting of rat aorta rings, was observed with both fumagillin as well as TNP-470 [173,174]. Several preclinical studies have shown cytostatic and cytotoxic effects against endothelial cells, leading to inhibition of new blood vessel formation and the reduction of tumor growth in a variety of preclinical cancer models [95,104,175–178]. In addition to the effect of MetAP2 inhibitors on angiogenesis, the role of MetAP2 in lymph-angiogenesis has been investigated [179]. Recent evidence suggests that pharmacological inhibition of MetAP2 may also affect vasculogenic mimicry contributing to blood supply of tumors [180].

Several potent noncovalent MetAP2 inhibitors have been developed in the past and, in general, the pharmacology of these compounds mirrored the effects seen by fumagillin-derived inhibitors. A-357300 (Fig. 7) inhibited the growth of HUVEC and human dermal microvascular endothelial cells (HMVEC), while other primary epithelial cells, fibroblasts, or astrocytes were not affected [87]. Sprouting and tube formation of HMVEC cells was potently inhibited by A-357300 as well. A differential activity was observed on the tested tumor cell lines; the activity pattern in cancer cell lines did not correlate with the p53 functional status. Overexpression of MetAP2 in the sensitive HT1080 fibrosarcoma cancer cells reduced the potency of MetAP2 inhibition. Growth inhibition was of cytostatic nature with cells accumulating in the G0/G1 cell cycle phase. The more potent and structurally distinct MetAP2 inhibitor A-800141 (Fig. 7) demonstrated a similar pharmacological profile to A-357300 [91]. A moderate upregulation of p53 and p21 after treatment with A-800141 was observed. M8891, the first potent and reversible MetAP2 inhibitor tested in a phase I clinical trial in cancer patients (Section 1.4), potently inhibited HUVEC cell proliferation as well as proliferation of more than 50% of patient-derived cancer cells tested in a 3D clonogenic assay. The sensitivity of the A549 lung adenocarcinoma cell line to M8891 was abolished upon CRISPR-mediated knockout of p53 and, conversely, knockout of MetAP1 sensitized the cells further to MetAP2 inhibition [147]. The mechanism of action of how inhibition of endothelial and cancer cell proliferation is mediated by MetAP2 inhibition is not fully understood. We attempted to capture the diverse findings related to MetAP2 inhibition that may or may not contribute to its antiproliferative activity (Fig. 10).

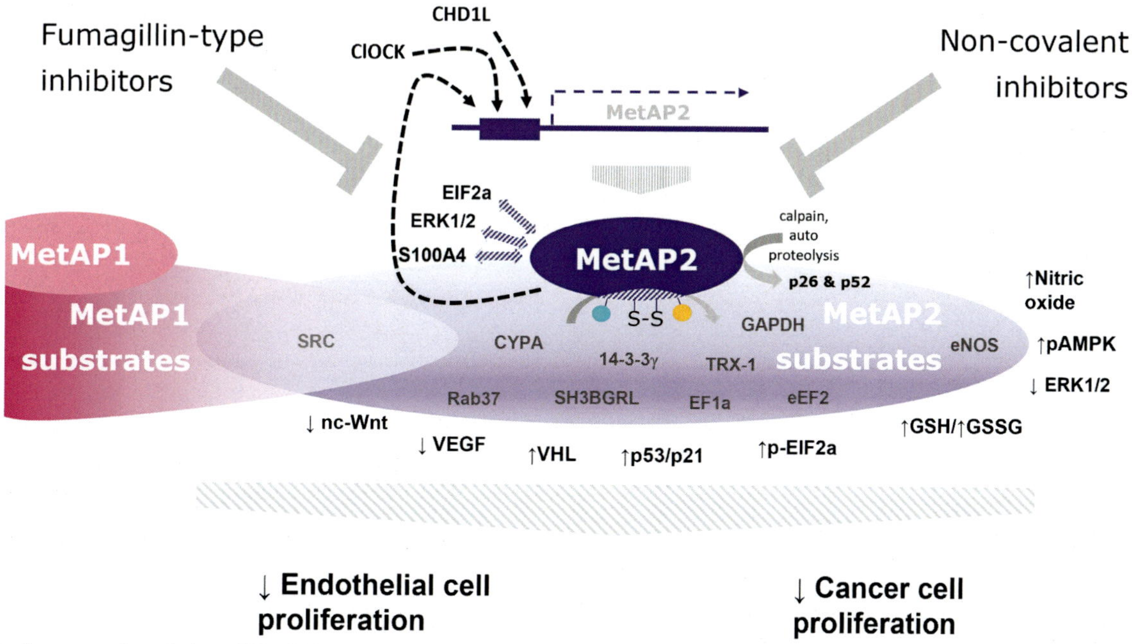

FIG. 10 Summary of regulation of MetAP2 expression and activity, substrates and biological consequences based on published findings. The main impact of MetAP2 inhibition on endothelial and cancer cell proliferation was highlighted. Posttranslational modifications of MetAP2 are labeled in *blue* (glycosylation) and *yellow* (nitration).

Covalent MetAP2 inhibitors gained further attention as potential antiobesity drugs [181]. The initially observed weight loss in animal studies and humans is mechanistically not well understood. It was hypothesized that a reduction in ERK activity results in decreased cholesterol and lipid biosynthesis and further hormonal and metabolic changes, contributing to the observed weight loss. All studies that initially investigated antiobesity activity have used fumagillin-based drugs. A recent study has compared covalent and noncovalent inhibitors with both classes demonstrating body weight loss and similar patterns of metabolic changes indicating lipolysis in brown adipocytes [182]. An excellent summary of the pharmacology of various MetAP2 inhibitors was published while this chapter was drafted [183].

1.3.2 In vivo pharmacological effects of MetAP inhibition

Different biological actions of MetAP2 were investigated in various disease-relevant preclinical animal models in vivo, with a strong focus on various cancer types.

Several preclinical studies conducted with fumagillin and derivatives have shown they exert cytostatic and cytotoxic effects on endothelial cells, leading to inhibition of new blood vessel formation using the Matrigel plug model in vivo and to reduction of tumor growth; this was investigated in a variety of preclinical cancer models [95,104,175–178].

Early anticancer drug discovery efforts have focused on analogs of fumagillin. Several fumagillin analogs, like CKD-732, TNP-470 and PPI-2458 (Fig. 5), were found to be potent selective inhibitors of MetAP2 leading to proliferation inhibition of endothelial cells and selected tumor cells [184], and were investigated in clinical trials for the treatment of different types of tumors. Subsequently, multiple reversible MetAP2 inhibitors, such as bengamides, 2-hydroxy-3-aminoamides, anthranilic acid sulfonamides (all Fig. 7), triazole analogs (Fig. 8) and M8891 (Fig. 9) have demonstrated their potential to inhibit angiogenesis and tumor growth in vivo [103,147].

Tumor vascularity is correlated with an aggressive disease phenotype in many cancers suggesting that angiogenesis inhibitors may be a useful addition to current therapeutic strategies [180]. It has been shown that, by irreversible MetAP2 inhibition using TNP-470, local and disseminated human neuroblastoma growth rates were suppressed in murine models. Similar antitumor effects have been shown by the reversible MetAP2 inhibitor, A-357300, in a CHP-134-derived neuroblastoma xenograft model. A-357300 also significantly inhibited establishment and growth rate of hematogenous metastatic deposits following tail vein inoculation of CHP-134 cells, and increased overall survival of mice. Lastly, A-357300 caused regression of established tumors in a genetically engineered murine model with progression-free survival of mice [185].

In an in vivo murine model xenografted with the SNU-398 hepatoma cell line, the effects on angiogenesis and tumor growth were assessed with the four highly potent MetAP2 inhibitors, IDR-803, IDR-804, IDR-805, and CKD-732 (all Fig. 5). It was shown that all compounds suppressed the growth of the engrafted tumor in vivo [178].

Preclinically, the MetAP2 inhibitor PPI-2458 (Fig. 5) has shown dose-dependent inhibition of B16F10 mouse melanoma cell proliferation in vitro and tumors in mice [186].

The orally active and highly specific MetAP2 inhibitor A-800141 (Fig. 7) exhibited an antiangiogenesis effect and a broad anticancer activity in a variety of tumor xenografts, including B cell lymphoma, neuroblastoma, and prostate and colon carcinomas, either as a single agent or in combination with cytotoxic agents. A-800141 blocked tumor growth and MetAP2 activity in a similar dose-response manner in mouse models, demonstrating the antitumor effects seen for A-800141, are causally connected to MetAP2 inhibition in vivo [91].

Identification of new biomarkers representing intrinsic features of malignant transformation and development of prognostic imaging technologies are critical for improving treatment decisions and patient survival. In a preclinical in vivo prostate cancer study, the first MetAP2-activated positron emission tomography (PET) imaging tracer for monitoring MetAP2 activity was analyzed. The nanoparticles assembled upon MetAP2 activation were imaged in single prostate cancer cells with postclick fluorescent labeling. The fluorine-18 labeled tracers successfully differentiated MetAP2 activity in both MetAP2 knockdown and MetAP2 inhibitor TNP-470 treated human PC3 prostate cancer xenografts by micro-PET/computerized tomography (CT) scanning. This highly sensitive imaging technology may provide a new tool for noninvasive early risk stratification of prostate cancer and monitoring the therapeutic effect of MetAP2 inhibitors as anticancer drugs [187].

However, targeting MetAP2 therapeutically has been reported across many different diseases. Therefore, different biological actions of MetAP2 were investigated in various disease-relevant preclinical animal models in vivo [184].

The ability of the MetAP2 inhibitor TNP-470 to increase the antidiabetic effect of sitagliptin (incretin-enhancer drug) was investigated in high fat fed (HFF) mice. TNP-470 and sitagliptin were administered alone or in combination. Individual therapy with TNP-470 or sitagliptin resulted in numerous metabolic benefits including reduced blood glucose, increased circulating and pancreatic insulin, and improved glucose tolerance, insulin sensitivity, pyruvate tolerance, overall pancreatic islet architecture, and reduced food intake and body weight [188].

MetAP2 inhibitors have been shown to reduce body weight in obese mice [189–192]. A study comparing covalent and noncovalent inhibitors with both classes demonstrated body weight loss and similar patterns of metabolic changes indicating lipolysis in brown adipocytes [182]. To understand the antiobesity mechanism of MetAP2 inhibition, three MetAP2 inhibitors with different binding modes (reversible versus irreversible) and chemical scaffolds were tested. Using chemically different MetAP2 compounds increases the chance that the antiobesity activities observed is related to MetAP2 inhibition and not compound-specific off-target effects. Covalent beloranib (Fig. 5), and two reversible MetAP2 inhibitors (A-357300 (Fig. 7) and compound **4** (Fig. 9)) showed that the effect on body weight and fat accumulation mediated by MetAP2 inhibition was apparent in obese animals but not in lean animals [182]. Pharmacological inhibition of MetAP2 causes rapid and sustained alterations to mitochondrial physiology with the potential to improve skeletal muscle function and metabolic fitness [193].

The ciliopathies Bardet-Biedl syndrome and Alström syndrome are genetically inherited pleiotropic disorders with hyperphagia and obesity as primary clinical features. To investigate the effects of MetAP2 inhibition, a mouse model of ciliopathy produced by conditional deletion of the Thm1 gene in adulthood has been used. These knockout mice, treated with ZGN-1258, a further fumagillin derivative (Fig. 5), showed decreased hypothalamic proopiomelanocortin expression as well as hyperphagia, obesity, metabolic disease, and hepatic steatosis. In addition, a reduction of daily food intake and body weight compared to control mice was observed. This was accompanied by decreased levels of blood glucose, insulin, and leptin. Further, MetAP2 inhibition reduced gonadal adipose depots and adipocyte size and improved liver morphology [194].

Idiopathic pulmonary fibrosis (IPF) is a progressive, scarring lung disease characterized by fibroblast accumulation and deposition of collagen. Factors that promote growth and/or survival of fibroblasts expressing MetAP2 in IPF lungs are potential therapeutic targets. In a preclinical bleomycin-induced pulmonary mouse fibrosis model, fumagillin attenuated collagen deposition in injured lungs of mice. Treatment with fumagillin caused a selective reduction in the numbers of myofibroblasts, but not type II alveolar epithelial cells, macrophages, or B- and T-lymphocytes in the lungs of bleomycin-treated mice. These data suggest that MetAP2 is a potential pharmacologic target in IPF [195].

Pulmonary Hypertension (pH) is a pathophysiologic condition characterized by hypoxemia and right ventricular strain. Proliferation of fibroblasts, smooth muscle cells and endothelial cells is central to the pathology of PH in animal models and in humans. To study the role of MetAP2 in PH, fumagillin was investigated in an experimental PH rat model. Early treatment after monocrotaline injury prevented PH and right ventricular remodeling by decreasing the thickness of the medial layer of the pulmonary arteries. Treatment with fumagillin beginning 2 weeks after monocrotaline injury did not prevent PH but was associated with decreased right ventricular mass and decreased cardiomyocyte hypertrophy, suggesting a direct effect of fumagillin on right ventricular remodeling [196].

Rheumatoid arthritis (RA) is a chronic inflammatory disease associated with increased synovial vascularity, and hence is a potential therapeutic target for angiogenesis inhibitors. PPI-2458 (Fig. 5) has been investigated in disease models of RA, namely acute and chronic collagen-induced arthritis (CIA) in mice and a peptidoglycan-polysaccharide-induced arthritis model in rats. PPI-2458 reduced clinical signs of arthritis in both acute and chronic CIA models. This reduction in arthritis was paralleled by decreased joint inflammation and destruction. Detailed mechanism-of-action studies demonstrated that PPI-2458 inhibited human endothelial cell proliferation and angiogenesis in vitro, without affecting production of inflammatory cytokines. These results highlight MetAP2 as a good candidate for therapeutic intervention in RA [197]. In another peptidoglycan-polysaccharide-induced arthritis rat model, PPI-2458 significantly attenuated paw swelling when therapeutically administered after the onset of chronic disease [198]. Pannus formation, in both RA and CIA, is angiogenesis-dependent. In addition to the previous cited in vivo studies, PPI-2458 was evaluated in Arthritic syngeneic rat models to investigate the potential to involute synovitis. Significant reduction in clinical severity scores compared with vehicle control was observed. Structural damage was virtually eliminated with PPI-2458. Continuous inhibition of MetAP2 was needed to maintain benefits, although pannus involution could be achieved. The study indicated that the MetAP2 inhibition by PPI-2458 can regress established CIA and that angiogenic mechanisms might be important targets in the treatment of other pannus-mediated diseases such as RA [199]. Additional evidence for angiogenesis contributing to rheumatoid arthritis RA has been generated in an inflammatory arthritis mouse model directly targeting endothelial cells. A lipase-labile fumagillin prodrug (Fum-PD) was developed and specifically delivered to angiogenic vessels using $\alpha v\beta 3$-integrin-targeted perfluorocarbon nanocarriers, which effectively suppressed clinical disease in this experimental RA model [171].

To evaluate a role for the immunomodulatory effects of MetAP2, the effect of the MetAP2 inhibitor PPI-2458 was investigated in in vitro and in vivo B cell mechanistic models. Immunohistochemical analysis of germinal centers from human, mouse and marmoset spleens showed a similar expression pattern of MetAP2 in the marmoset and man. In a marmoset T-cell-dependent immunization model, the MetAP2 inhibitor suppressed an antigen-specific antibody response. Furthermore, histological analysis showed loss of B cells in the spleen and disrupted germinal center formation. The results provide experimental evidence to support a role for MetAP2 in immunomodulation, which are mediated by disruption of the germinal center reaction and a block in the differentiation of B cells into plasma cells [200].

MetAP2 inhibition modifies Hemoglobin S (HbS) to delay polymerization and improve blood flow in sickle cell disease by retention of the initiator methionine to create a mixture of N-terminal modified HbS tetramers [201].

1.3.3 Side effects of MetAP inhibition

MetAP activity is essential for cellular growth and viability, any form of knockout of either MetAP1 or MetAP2 causes a decrease in growth rates while elimination of both genes is lethal, indicating that the two MetAPs play essential functions and are together essential for proliferation [184]. A conditional MetAP2 knockout mouse resulted in early gastrulation defects. Targeted deletion of MetAP2, specifically in the hemangioblast lineage, resulted in abnormal vascular development, and these embryos died at the midsomite stage. MetAP2-null mouse embryos have reduced cell proliferation and show significant growth delay, do not start gastrulation, indicating requirement for a functional MetAP2 for embryonic development and survival at the initiation of gastrulation. In addition, knockdown of MetAP2 using siRNA suppresses the proliferation of cultured endothelial cells, suggesting an essential role of MetAP2 in embryonic development and vasculogenesis/angiogenesis [165]. Further it was postulated that MetAP2 is not essential for embryonic cell proliferation itself, but it directly affects other essential processes during gastrulation, such as cell polarity and migration [202].

The majority of the published preclinical safety data associated with MetAP2 inhibitors are related to Fumagillin and several of its analogs (e.g., TNP-470). An in vivo study indicated genotoxic effects of fumagillin [203]. In addition, it was shown that fumagillin is clastogenic and cytotoxic to cultured human lymphocytes. It was concluded that this is probably not related to the direct effect on the target [204]. The clinical development of TNP-470, has been severely limited due to dose-limiting CNS toxicity [205–208]. Toxic doses of TNP-470, have been associated with local skin lesions, weight loss, and death. Clinical safety data for TNP-470 is summarized in Section 1.4.

In a neuroblastoma preclinical model, TNP-470 had significant associated toxicity at the clinically effective exposure [202]. In the Sprague Dawley rat, seizures in all animals at 60 mg/kg were reported after the second dose of TNP-470, which occurred within minutes after dosing and included ataxia, twitching, piloerection, salivation, muscle tetanus and exophthalmos. These animals were also noted to be easily excited and to have a heightened startle response. Seizures were initially mild with quick recovery, progressing to violent, with delayed recoveries observed. In contrast, no animals in the PPI-2458 treated groups (6 or 60 mg/kg) were observed to experience seizures after administration [209]. In contrast, the reversible MetAP2 inhibitor, A-357300, significantly inhibits CHP-134-derived neuroblastoma SC xenograft growth rate without inducing any signs of neurotoxicity [87].

Early stage bone induction via human bone morphogenic protein was shown to be reversibly inhibited by TNP-470 [210]. Further, it was suggested that acute exposure to TNP-470 impairs the development and functional capacity of the primate corpus luteum in a dose-dependent manner. The results are consistent with a critical role for angiogenesis in cyclic ovarian function in primates [211]. It was also hypothesized that TNP-470 induces microphthalmia, either indirectly by their known effects on placental morphology (and/or function) or directly via altering microvascular growth in the fetus [212].

PPI-2458 seems to have reduced toxicity and greater selectivity for angiogenesis inhibition versus inflammation as compared with TNP-470 [213].

Beloranib was generally well tolerated in preclinical testing. However, in clinical trials of beloranib in patients with obesity, Prader-Willi Syndrome, or type 2 diabetes, adverse events of venous thromboembolism occurred in treated patients (Section 1.4) [214]. The follower compound ZGN-1061 was developed to address these clinical adverse events observed for beloranib in preclinical animal safety studies. Both compounds were administered subcutaneously (SC) every 3 days for approximately 28 days in the rat. For beloranib, a no observed adverse effect level (NOAELs) of 0.6 mg/kg (males) and 2 mg/kg (females) was established. Mortality was observed at the highest dose tested. In contrast, ZGN-1061 doses ranged from 2 mg/kg to 25 mg/kg, no mortality was noted, and the highest dose was well tolerated. Toxicology studies of beloranib in dogs showed that animals developed impaired hemostasis (i.e., marked reductions in platelet count, bleeding gums, bloody stool) within 2 weeks at doses of 0.6 mg/kg SC every 3 days. This was accompanied by increased D-dimer concentrations and decreased thrombin time, antithrombin III, and platelet count. ZGN-1061 (2 mg/kg SC every 3 days) administered to dogs for 10 days showed no such effects, maybe because of shorter duration of systemic exposure compared to beloranib [215].

Several other related MetAP2 inhibitors have made it to the clinic, as described in Table 1. However, for these candidates (SDX-7320, XMT-1107, APL-1202, APL-1501, LAF389, and M8891) no preclinical safety data are publicly available.

TABLE 1 MetAP inhibitors in clinical development.

Drug	Indication	Clinical phase	Clinical outcome	Reference
Fumagillin	Microsporidiosis	Prospective efficacy study	Efficacy: 94% (n=112 out of 132 eligible patients) no spores; approved. Safety: Severe thrombocytopenia (<50 G/L) developed in 50 patients (29.6%), neutropenia (<1 G/L) in 20 patients (11.8%) and severe anemia (<8 g/dL) in 21 patients (12.4%)	[216]
TNP-470	AIDS-associated kaposi sarcoma	Phase I	Efficacy: 18% (n=7 out of 38) partial response, average duration of response 11 weeks. Safety: Neutropenia (n=2), hemorrhage (n=3), and urticaria (n=1)	[217]
	Cervical cancer	Phase I	Efficacy: 5% (n=1 out of 18) complete response; 17% stable disease (n=3). Safety: Grade 3 neurotoxicities consisting of weakness, nystagmus, diplopia, and ataxia in two patients	[205]
	Nonsmall-cell lung cancer	Phase I	Efficacy: Median survival duration 297 days; 24% partial response (n=4 out of 17), and 47% (n=8) stable disease. Safety: Hematological toxic effects similar to those expected with the chemotherapy doublet. All neurocognitive impairments graded as mild to moderate and reversed after discontinuation of TNP-470 administration. PK: No alterations in the pharmacokinetic disposition of carboplatin	[208,218]
	Colon adenocarcinoma	Phase I	Efficacy: 32% (n=2 out of 6) inhibition of disease progression for 13 months, 17% (n=1) stable disease for 27 weeks at dose level of 235 mg/m^2. Safety: Principal AEs were dizziness, lightheadedness, vertigo, ataxia, decrease in concentration and short-term memory, confusion, anxiety, and depression dose related; 2 patients treated at 235 mg/m^2 experienced DLTs in the form of grade III cerebellar neurotoxicity. PK: The mean plasma half-life $t_{1/2}$ of TNP-470 and its principal metabolite, AGM-1883, extremely short ($t_{1/2}$ of 2 min and 6 min, respectively)	[206]
	Renal cancer	Phase I	Efficacy: No definite antitumor activity (n=32); however, transient stimulation of the serum prostate-specific antigen concentration. Safety: The maximum tolerated dose was 70.88 mg/m^2 of body surface area. The dose-limiting toxic effect was a characteristic neuropsychiatric symptom complex (anesthesia, gait disturbance, and agitation) that resolved upon cessation of therapy	[207]
	Renal cancer	Phase II	Efficacy: No objective response (N=33); one partial response of short duration (response rate, 3%, 95% CI, 0% to 16%), 6 patients (18%) without disease progression for 6 or more months. Safety: Generally well tolerated, but asthenia, fatigue, vertigo, dizziness, sense of imbalance, and loss of concentration led to discontinuation	[219]
PPI-2458	Non-Hodgkin's lymphoma, unspecified solid tumor	Phase I	Efficacy: 22% (n=7 out of 32) stable disease. Safety: Well tolerated up to 15 mg. PK/PD: Elimination half-live 1.6 h; MetAP2 in whole blood 80% inactivated for up to 48 h	[220,221]

Continued

TABLE 1 MetAP inhibitors in clinical development—cont'd

Drug	Indication	Clinical phase	Clinical outcome	Reference
SDX-7320	Unspecified refractory or late-stage solid tumor	Phase I	Efficacy: 44% ($n=14$ out of 32) had stable disease for 2 cycles and three patients had stable disease for ≥6 cycles. Safety: Six patients had a combined total of 9 grade 3/4 TEAEs; thrombocytopenia (reversible); vasculitis. DLT was thrombocytopenia. Tumor-related and metabolic biomarkers: Average of each patient's maximum % change relative to baseline ± SD): bFGF (−53 ± 76%), VEGF-C (−35 ± 39%), and insulin (−55 ± 30%). PD: Inhibition of MetAP2 in whole blood was 100% for all doses observed	[222]
XMT-1107	Unspecified advanced solid tumor	Phase I	Efficacy: 50% ($n=26$ out of 52) stable disease with 12 patients stable for at least 4 cycles. Safety: well tolerated without CNS toxicity up to a dose of 770 mg/m^2; one DLT of grade 4 thrombocytopenia was seen at 245 mg/m^2; grade 3 ($n=4$) and grade 4 ($n=3$) thrombocytopenia at 105 mg/m^2 and above; grade 3 anemia ($n=3$); grade 3 transaminase elevation ($n=2$). PD: Complete MetAP2 inhibition at doses above 325 mg/m^2	[223]
Beloranib	Unspecified refractory solid tumor	Phase I	Efficacy: Reduction in soluble vascular endothelial growth factor receptor-3 (sVEGF-3) level correlated with a decrease in tumor size ($r=0.54$, $P=.045$). Safety: Confusion and insomnia were DLTs, and MTD was 15 mg/m^2. PK/PD: AUC and C_{max} increased dose dependently with increasing doses; the BED was 5 mg/m^2 according to ex vivo PD	[224]
	Metastatic colorectal cancer	Phase Ib	Efficacy:—. Safety: Grade 3 nausea, insomnia, and fatigue (2 DLTs); grade 3 insomnia MTD was 10 mg/m^2/day, and the clinically recommended dose was 5 mg/m^2/day in combination with XELOX. PK: AUC and C_{max} increased in a dose-dependent manner; no notable effects on the PK of XELOX-derived platinum	[225]
	Obesity	Phase I	Efficacy: −3.8 kg bodyweight (95% CI −5.1, −0.9; $n=8$) versus −0.6 kg with placebo (−4.5, −0.1; $n=6$) and improvements in lipids (−42% triglycerides and −18% LDL-cholesterol). Safety: Frequent AEs were headache, infusion site injury, nausea, and diarrhea; no clinically significant abnormal laboratory findings	[226]
	Obesity	Phase II	Efficacy: Dose-dependent progressive weight loss of −5.5 ± 0.5, −6.9 ± 0.6 and −10.9 ± 1.1 kg for the 0.6, 1.2, and 2.4 mg beloranib doses, respectively, compared with −0.4 ± 0.4 kg with placebo (all $P<.0001$ vs placebo); associated with corresponding reductions in waist circumference and body fat mass, as well as improvements in lipids, high-sensitivity C-reactive protein and blood pressure. Safety: Treatment appeared to be safe, 0.6 and 1.2 mg doses were generally well tolerated. The 2.4 mg dose was associated with increased sleep latency and mild to moderate gastrointestinal AEs	[227]
	Obesity and type II diabetes	Phase II	Efficacy: Least squares mean ± standard error weight change (baseline 111 kg) −3.1 ± 1.2% with placebo ($n=22$) versus −13.5 ± 1.1% and −12.7 ± 1.3% with 1.2 and 1.8 mg, respectively ($n=25$; $n=19$; $P<.0001$); change in HbA1c (baseline 67 mmol/mol [8.3%]) was −6.6 ± 2.2 mmol/mol (−0.6 ± 0.2%) with placebo versus −21.9 ± 2.2 mmol/mol (−2.0 ± 0.2%) or −21.9 ± 3.3 mmol/mol (−2.0 ± 0.3%) with 1.2 or 1.8 mg ($P<.0001$), respectively. Safety: Most common adverse events were sleep related; one nonfatal pulmonary embolism	[228]

TABLE 1 **MetAP inhibitors in clinical development—cont'd**

Drug	Indication	Clinical phase	Clinical outcome	Reference
	Obesity, Prader-Willi-syndrome	Phase III	Efficacy: Reduction in Hyperphagia Questionnaire for Clinical Trials (HQ-CT) total score (−7.0, 95% CI −10.5 to −3.6; $P=.0001$) versus placebo; weight loss (−9.5%, 95% CI −12.1 to −6.8; $P<.0001$) versus placebo. Safety: Two fatal events of pulmonary embolism and two events of deep vein thrombosis compared to placebo let to treatment stop and subsequent discontinuation of clinical development [229]	[230]
ZGN-1061	Obesity and type II diabetes	Phase I	Efficacy: Weight change −1.5 vs −0.2 kg placebo in the multiple ascending phase (ZGN-1061, $N=22$; placebo, $N=7$). Safety: ZGN-1061 was well tolerated across all doses, with the most frequent adverse events being mild headache and procedural-related irritation. There were no severe or serious AEs	[231]
	Obesity and type II diabetes	Phase II	Efficacy: Relative to placebo, the 0.9 and 1.8 mg doses induced clinically meaningful reductions in HbA1c of 0.6% (95% CI 0.2%–0.9%; $P=.0006$) and 1.0% (95% CI 0.6%–1.4%; $P<.0001$), respectively; the 1.8 mg dose also induced weight loss of 2.2% (95% CI 1.1%–3.3%; $P=.0002$). Safety: Most frequent TEAEs (≥10% in any group) were upper respiratory tract infection, contusion, injection site bruising, diarrhea, headache, nasopharyngitis, and pain in an extremity. Note: Given the high cardiovascular risk for beloranib clinical development of ZGN-1061 was discontinued	[232]
APL-1202	Recurrent nonmuscle invasive bladder cancer (NMIBC)	Phase I	Efficacy:—. Safety: Well tolerated. PK: comparable PK characteristics (plasma AUC_τ, C_{max} and $t_{1/2}$, and urinary excreted faction Ae%) in all five patients evaluated	[233]
	High-risk, relapsed NMIBC	Phase II	Efficacy: 1-year recurrence-free rate of 39% (33%–45%, 95% CI) with a median recurrence-free survival of 9 months. Safety: Safe and well tolerated at a daily dose up to 750 mg, 3 times daily. PK: Plasma exposure at steady state appears to increase proportionally to the increase of doses	[234]
	High-risk chemo-refractory NMIBC	Phase III	Recruiting	NCT04490993
M8891	Unspecified solid tumor	Phase I	Efficacy: Eight patients (30.8%) experienced stable disease for 30–93 days. Safety: Manageable safety profile; thrombocytopenia observed ($n=10$ out of 26 patients), but no bleeding events; two DLTs at 60 and 80 mg daily (both grade 4 thrombocytopenia). PK/PD: Low to moderate interpatient variability; exposure and target engagement increased dose-proportionally up to 35 mg daily, which was determined as RP2D.	[229,235]
LAF389	Advanced solid tumor	Phase I	Efficacy: No objective responses. Safety: Four cardiovascular DLTs at 30 mg ($n=2$ out of 2 patients) and 25 mg ($n=2$ out of 9), 8 additional patients at various dose levels with (cardio)vascular toxicity, probably drug related, and 1 patient died owing to pulmonary embolism at the 5 mg dose. PK: Pharmacokinetic parameters variable, although linear and without obvious accumulation. Note: Clinical development discontinued	[236]

Drug name, indication and clinical phase with outcome is summarized.

AE, adverse event; *AUC*, area under the curve; *BED*, biologically efficacious dose; *CI*, confidence interval; C_{max}, maximal plasma concentration; *CNS*, central nervous system; *DLT*, dose-limiting toxicity; *LDL*, low-density lipid; *MTD*, maximum tolerated dose; *PD*, pharmacodynamics; *PK*, pharmacokinetics; *SD*, standard deviation; $t_{1/2}$, half-life; *TEAE*, treatment-emergent AE; *XELOX*, capectibine and oxaliplatin.

In summary, irreversible inhibition of MetAP2 is often associated with adverse events (mainly neurotoxicity) in nonclinical species and in humans. However, studies in animal models with reversible inhibitors have shown good outcome without the associated toxicities, supporting the hypothesis that the toxicity of irreversible inhibitors (e.g., TNP-470) might not be related to MetAP2 inhibition.

1.4 MetAP2 inhibitors in clinical development

1.4.1 Fumagillin

Several clinical studies have been reported for fumagillin (Fig. 5); however, these studies investigated its application as an antibiotic, such as for the treatment of microsporidiosis [237,238]. In 2006, Fumagillin was registered for the treatment of intestinal microsporidiosis in immunocompromised patients by Sanofi-Aventis in France (Flisint) [239]; however, drug-induced hematological toxicity (thrombocytopenia, neutropenia) was frequently observed. Other reported adverse events were gastrointestinal disorders, elevation of hepatic and pancreatic enzymes, and drug-drug interaction with tacrolimus [240,241]. A prospective clinical study with 116 patients, performed from 2007 to 2018 in France, confirmed high efficacy (94% of patients with available stool examination ($n=132$) had no spores detected) with a low relapse rate (3 out of 99) but, again, severe hematological toxicity was observed (Table 1) [216].

Thus, a broader application and a prolonged oral administration seemed to be prohibited because of its severe toxicities, including those causing weight loss, and poor oral pharmacokinetics [240,242,243]. However, these data triggered the synthesis of new derivatives with improved antiangiogenic activities and reduced undesired effects (see below).

1.4.2 TNP-470 (AGM-1470)

Both fumagillin and its less toxic semisynthetic analog TNP-470 (or AGM-1470; Takeda; Fig. 5) possess potent antiangiogenic activity and block tumor growth and metastasis, with TNP-470 50-fold more effective in endothelial cell growth inhibition [244–246].

TNP-470 entered clinical development in 1998 and has been investigated in several clinical trials for treatment of different cancers since, including carcinoma of the cervix that metastasized to lung, high-grade sarcoma of the kidney, renal-cell carcinoma, androgen independent prostate cancer, colon adenocarcinoma, nonsmall-cell lung cancer and Kaposi's sarcoma (KS) [206,217,219,247,248].

In the initial phase I study, with patients diagnosed with AIDS-associated KS, complete and durable tumor regression was reported in some patients for which all conventional therapy failed. Given as a weekly infusion (10–70 mg/m^2), AGM-1470 was well tolerated, with 18% of patients (7 out of 89) obtaining a partial response, and an average duration of response at 11 weeks. In addition, adverse events such as neutropenia, hemorrhage, and urticaria were evaluated as moderate and manageable. However, it was recommended to further evaluate the drug in patients with AIDS-KS as a single agent or as a combination therapy with other drugs. PK was highly variable and the elimination half-life was very short, ranging from 0.01 to 0.61 h [217].

In an independent phase I study in patients with advanced cervical cancer, one complete response (1 out of 18, ~5%), and stable disease in three additional patients (~17%) was observed, respectively [205]. Neurotoxicity was the dose-limiting toxicity, which was classified as neurocognitive (anxiety, agitation) origin, but was reversible upon discontinuation of therapy. A recommended dose and regimen of 60 mg/m^2 3QW as IV infusion was reported.

In a further clinical phase I trial, with patients with nonsmall-cell lung cancer (NSCLC), in which TNP-470 was combined with paclitaxel and carboplatin, four patients (24%) had a partial response, and eight (47%) had stable disease [208,218].

Data from another clinical phase I study suggested that TNP-470 given as an IV infusion dose of 235 mg/m^2 once weekly inhibited progression of colon adenocarcinoma for 13 months in 2 (33%) and showed stable disease for 27 weeks in 1 out of 6 patients (17%) [206]. However, none of the later studies could confirm these results.

A further phase I study, in patients with metastatic and androgen-independent prostate cancer, showed characteristic neuropsychiatric symptoms but no definite antitumor activity [207].

A multicenter phase II study in patients with metastatic renal carcinoma revealed manageable toxicities but, again, did not show any significant objective responses [219].

Another phase II clinical trial, with the aim to evaluate the survival and patterns of failure in patients treated with Gemzar-based chemoradiation and TNP-470, started in 2002, but was terminated in 2004 due to a low recruitment rate [NCT00038701].

In summary, the benefit-risk ratio for treatment with TNP-470 seemed to be unfavorable because many patients experienced neurotoxic side effects, that is, cerebellar dysfunctions such as malaise, rare seizures, asthenia, dysphoria, dizziness, lightheadedness, vertigo, and ataxia at doses where antitumor activity was seen. Therefore, the clinical development of TNP-470 was terminated [207,208,249].

In the following years, a series of further analogs were designed through chemical modifications of the carboxylic side chain in order to overcome the pharmacokinetic liabilities of TNP-470, including both its rapid degradation by epoxide hydrolase and its uptake into the central nervous system (CNS), with the latter likely linked to the above-mentioned neurotoxicity profile (e.g., PPI-2458; see below).

1.4.3 PPI-2458

PPI-2458 (Fig. 5), is another fumagillin-derived irreversible MetAP2 inhibitor, developed by Praecis Pharmaceuticals (GlaxoSmithKline), which was shown to inhibit non-Hodgkin's lymphoma cell proliferation in vitro and in vivo [172], is orally available and potentially has a better safety profile by reducing potential CNS toxicities observed with TNP-470 [186,209].

The safety and tolerability of PPI-2458 in patients with non-Hodgkin's lymphoma and solid tumors was examined in a multicenter, open-label phase I clinical trial [NCT00100347]. PPI-2458 was administered to subjects orally every other day for two 28-day cycles. The drug seemed to be well tolerated up to 15mg, the elimination half-life was 1.6h and MetAP2 in white blood cells was substantially inhibited (80%) for up to 48h [220].

An interim analysis showed that 7 out of 32 subjects (22%) across the first 4 cohorts had stable disease at the end of the first 2 cycles of treatment [221].

Notably, exposure to the parent compound made up only ~10% of the overall drug-related circulating components, suggesting that active metabolites may contribute to the pharmacological effects [220]. However, this study was terminated and a phase II clinical trial [TrialTroveID-058242], which was planned in patients with unspecified cancer, was not initiated due to a business decision (details not reported).

It should be noted that PPI-2458 potently inhibited proliferation of human fibroblast-like synoviocytes derived from patients with RA and, thus, may have additional potential for the treatment of that condition [209].

1.4.4 SDX-7320/caplostatin

SDX-7320 is an TNP-470 derivative developed by SynDevRx in which the fumagillol moiety was coupled to a bio-compatible and stable, but cleavable, polymer (poly (*N*-hydroxypropylmethacrylamide, HPMA)) backbone, which was designed to limit CNS penetration and therefore reduce CNS toxicity. In addition, it was hypothesized to accumulate in tumors, due to enhanced permeability and retention. The cleaved, active moiety SDX-7539 would then be released inside the tumor to inhibit MetAP2 [250,251].

As well as treating solid tumors, metastatic breast cancer, metastatic colorectal cancer and metastatic prostate cancer, the company was also investigating the potential of SDX-7320 for the treatment of obesity, obesity-driven cancer and type 2 diabetes.

A phase Ia dose escalation study in patients with unspecified solid tumors, assessing safety and tolerability, was completed in December 2019 [NCT02743637]. SDX-7320 administered SC was well tolerated, with prolonged stable disease and some improvements in both tumor-related and metabolic biomarkers observed. Repeated administration resulted in dose-limiting thrombocytopenia, which was reversible upon cessation of dosing. The maximum tolerated dose (MTD) was determined to be 49 mg/m^2 on a Q14D schedule. Six patients had a combined total of nine Grade 3/4 treatment-emergent adverse events (TEAEs), which were considered to be SDX-7320 related. Inhibition of MetAP2 in whole blood was at 100% at all doses evaluated, while the time to reach 100% inhibition was inversely related to dose [222].

Additional clinical studies with SDX-7320 (in combination with other anticancer agents) in patients with solid tumors sensitive to metabolic hormones had been planned; however, up to now no further updates have been reported.

1.4.5 XMT-1107

Mersana Therapeutics and ex-Japanese licensee Teva Pharmaceuticals were developing a small-molecule MetAP2 and angiogenesis inhibitor, XMT-1107. This is composed of the small-molecule active agent and fumagillin analog XMT-1191 conjugated to a 70kDa biodegradable, hydrophilic polymer (poly(1-hydroxymethylethylene-hydroxymethylformal (PHF)) fleximer, for the treatment of cancer. XMT-1107 was assumed not to cross the blood-brain barrier, and thus decrease CNS exposure, and, in addition, to show prolonged PK and pharmacodynamic (PD) effects. Preclinically, XMT-1107, demonstrated improved anti-tumor activity compared to both unconjugated XMT-1191 or TNP-470 [252].

In April 2010, a phase I clinical trial [NCT01011972] was initiated in the US in patients with refractory advanced solid tumors, to assess the safety, PK, recommended phase II dose (RP2D), and MTD of XMT-1107 given as single agent IV every 3 weeks at escalating doses using a 3+3 design. XMT-1107 was well tolerated up to a dose of 770 mg/m^2 without significant CNS toxicity. Free and conjugated XMT-1191 concentrations increased linearly with XMT-1107 dose, and MetAP2 inhibition was detected in leukocytes at all dose levels, with complete inhibition throughout the 21-day cycle in six out of seven patients evaluated for PD at doses above 325 mg/m^2, supporting the intravenous dosing schedule. Twenty-six out of 52 patients had a best response of stable disease, with 12 patients stable for at least 4 cycles [223].

Although the overall safety profile and preliminary anti-tumor effect supported further development in combination with chemotherapy and/or other antiangiogenic agents, Mersana have been seeking to out-license rights for the drug since 2014. Since 2016, the drug was no longer listed on the company's pipeline.

1.4.6 Beloranib (ZGN-433, ZGN-440, CKD-732)

Beloranib, (Fig. 5) also called ZGN-433/440 or CKD-732, is another fumagillin-derived MetAP2 inhibitor, which was originally discovered by Chong Kun Dang (CKD)

Pharmaceuticals and its clinical development led by Zafgen Inc. (now Larimar Therapeutics Inc.). Preclinically, CKD-732 showed no effects on general behavior, spontaneous locomotor activity, motor coordination, analgesia, or convulsion, suggesting a better safety profile than other fumagillin derivatives, including TNP-470 [253].

The drug was originally designed as an angiogenesis inhibitor for the treatment of cancer [178] and, in 2008, a phase I dose escalation study was initiated in patients with refractory solid tumors to assess safety, PK and PD of CKD-732 [224]. The drug was administered as an SC infusion in a twice weekly regimen ranging from 1 to 15 mg/m^2 over 2 weeks.

Dose-limiting toxicities (DLTs) were of neurological nature (confusion, insomnia) and the MTD was determined to be 15 mg/m^2. The overall exposure (area under the curve [AUC]) and maximal plasma concentration (C_{max}) increased dose-dependently. The elimination half-life ($t_{1/2}$) was short, ranging from 0.7 to 3.4 h. The biologically efficacious dose (BED) was 5 mg/m^2 according to ex vivo PD measurements. A reduction in soluble vascular endothelial growth factor receptor-3 (sVEGF-3) level was correlated with a decrease in tumor size. A further clinical phase I study was performed with CKD-732 in combination with capectibine and oxaliplatin (XELOX) in metastatic colorectal cancer patients who progressed on irinotecan-based chemotherapy to evaluate the safety, tolerability, and PK [225]. The MTD was 10 mg/m^2/day, and the clinically recommended dose was 5 mg/m^2/day for CKD-732 in combination with XELOX. Frequently encountered nonhematological grade 3/4 adverse events included insomnia (22.2%), fatigue (11.1%), sensory neuropathy (11.1%), hyperbilirubinemia (11.1%), and dyspnea (11.1%). The AUC and C_{max} of CKD-732 increased in a dose-dependent manner. There were no notable effects of CKD-732 on the PK of XELOX-derived platinum.

However, once the potential antiobesity effects of CDK-732 became apparent, the clinical development began to focus on these effects and positive results in preliminary clinical trials for this indication were shown. CDK-732 was licensed by Zafgen, renamed to ZGN-433/440 and later to beloranib. In the subsequent clinical development, beloranib was evaluated in various parenteral formulations and administration forms, that is, as hemioxalate SC and intravenously (IV), respectively.

In a first phase I clinical trial for the indication obesity, published in 2013, beloranib was evaluated for safety, tolerability and efficacy in obese female volunteers [NCT01372761] [226]. In this study, beloranib was given IV at doses of 0.1, 0.3, or 0.9 mg/m^2 twice weekly for 4 weeks; treatment (0.9 mg/m^2) was well tolerated, associated with rapid weight loss as compared to placebo and improvements in lipids (−42% triglycerides and −18% LDL-cholesterol), C-reactive protein, and adiponectin.

Results from a phase II clinical trial for obesity were also promising, with clinically meaningful weight loss and improvements in cardiometabolic risk factors in the treatment group [227]. This double-blinded, randomized study investigated the effects of beloranib suspension (0.6, 1.2 and 2.4 mg) or placebo, administered SC, for 12 weeks in 147 participants. Beloranib appeared to be safe, and generally well tolerated at doses of 0.6 and 1.2 mg. At week 12, beloranib resulted in dose-dependent progressive weight loss as compared with placebo and was associated with corresponding reductions in waist circumference and body fat mass, as well as improvements in lipids, high-sensitivity C-reactive protein and blood pressure. The 2.4 mg dose was associated with increased sleep latency and mild-to-moderate gastrointestinal adverse events over the first month of treatment.

In a further randomized, placebo-controlled phase II clinical study, in patients with obesity and type II diabetes, the efficacy and safety of beloranib was evaluated [228]. Beloranib was given to 153 participants at 1.2 or 1.8 mg SC twice weekly for 26 weeks, and the treated group showed significant reduction in body weight and HbA1c levels as compared to placebo. The most common adverse events were sleep related.

Zafgen continued with a clinical phase III trial in patients with Prader-Willi syndrome (PWS) [214], a genetic disorder characterized by hyperphagia and excess body fat. Patients with PWS have reduced life expectancy due to increased risk of metabolic disease, cardiovascular events and complications from hyperphagia (e.g., gastric necrosis and choking) [227,230].

Beloranib treatment at 2.4 mg/m^2 twice weekly produced statistically significant and clinically meaningful improvements in hyperphagia-related behaviors and weight loss in participants with PWS versus placebo. However, dosing had to be stopped due to an imbalance in venous thrombotic events in beloranib-treated participants (two fatal events of pulmonary embolism and two events of deep vein thrombosis) compared to placebo. Discussions with the US Food and Drug Administration (FDA) indicated that the obstacles to gaining approval were insurmountable, and development of beloranib was ended in 2016 [254].

Zafgen developed two second-generation MetAP2 inhibitors; ZGN-1258, which was profiled only nonclinically, and ZGN-1061 (aclimostat), which was clinically evaluated up to phase II studies in patients with obesity and type II diabetes [215,231,232].

However, both programs were stopped in 2019 after receipt of a clinical hold by the FDA for the treatment of type 2 diabetes due to the high cardiovascular risk. Subsequently, the Investigational New Drug Application for aclimostat was withdrawn. No further development activities have been reported since 2019.

1.4.7 APL-1202 (APL, nitroxoline)

Asieris Pharmaceuticals, under license from Johns Hopkins University, is currently developing APL-1202, a reversible MetAP2 inhibitor (Fig. 7), as an oral tablet formulation for the potential treatment of nonmuscle invasive bladder cancer (NMIBC). This drug, originally named nitroxoline, an antibiotic used to treat urinary tract infections [255], was identified from a high-throughput screen of a library for MetAP2 inhibitors [105]. The drug was reported to inhibit MetAP2 activity in vitro and endothelial cell (HUVEC) proliferation, albeit with low potencies, that is, IC_{50} values of 55 nM and 1.9 μM, respectively.

APL-1202 entered clinical development with an open-label phase Ib trial in the US [NCT03672240], in patients with NMIBC who were resistant to intravesical Bacillus Calmette-Guerin (BCG) treatment, to evaluate the safety and PK characteristics [233]. APL-1202 was well tolerated at a dose of 750 mg daily (2 × 125 mg tablet, 3 times a day) for 12 weeks in patients with NMIBC who also received induction or maintenance BCG. PK results suggested similar PK characteristics (plasma AUC_{tau}, C_{max} and $t_{1/2}$, and urinary excreted faction $A_e\%$) in all five patients evaluated.

Asieris continued in 2014 with an open-label, single arm phase II clinical trial [NCT04498702] performed in China in patients with high-risk NMIBC who relapsed after intravesical therapies to evaluate PK, safety and efficacy of APL-1202 [234]. Patients were treated with APL-1202 tablets daily for 12 weeks, followed by 12 weeks of dosing, starting at months 6 and 12, at doses of 300 and 750 mg 3 times a day through a 3-plus-3 escalation scheme.

APL-1202 demonstrated good safety and was well tolerated at a daily dose up to 750 mg three times daily, and high-risk NMIBC patients treated with intravesical chemotherapies were shown to have a 1-year recurrence-free rate of 39% (33%–45%, 95% CI) with a median recurrence-free survival of 9 months.

A multicenter, randomized, placebo-controlled, pivotal phase III trial in China [NCT04490993] in patients with intermediate or high-risk NMIBC relapsed from other chemotherapies, to evaluate APL-1202 in combination with intravesical epirubicin chemotherapy is active, but, to date, no data have been published.

1.4.8 APL-1501

In early 2020, Asieris announced APL-1501 as a second-generation product of APL-1202, which is being developed as a prodrug. It has an oral sustained-release formulation, with improved PK characteristics compared to APL-1202, aiming to improve patient compliance by reducing dosing frequency (Asieris Website) [256]. This product is expected to treat not only bladder cancer, but also other diseases such as prostate cancer and urinary tract infection.

A clinical phase I study [NCT04601766] to evaluate the safety, tolerability, and PK characteristics of APL-1501 was approved in late 2020 in Australia, but recruitment has not started.

1.4.9 M8891

More recently, Merck KGaA, Darmstadt, Germany, developed a potent, selective, and reversible MetAP2 inhibitor, M8891 (Fig. 9), originated from a novel cyclic tartronic diamide scaffold [38]. The overall favorable properties together with its strong in vivo efficacy allowed it to enter clinical development in a phase I, open-label, dose-escalation study to determine the MTD, RP2D, safety, tolerability, PK, and PD profiles of M8891 monotherapy in patients with advanced solid tumors [NCT03138538] [229]. M8891 was administered orally in a once-daily regimen. Preliminary results showed that M8891 has a manageable safety profile (DLT: thrombocytopenia) and a favorable PK profile with low-to-moderate interpatient variability. Target engagement in tumors was observed at the starting dose of 7 mg once daily and increased with the dose and exposure in a dose-linear manner up to 35 mg once daily, which was also determined as RP2D based on overall safety, PK and PD profiles [235].

1.4.10 MetAP1 inhibitors

Clinical development of selective MetAP1 inhibitors has not been reported. LAF389 (Fig. 7), a specific bengamide derivative, which reversibly inhibits MetAP2, but is not selective against MetAP1, has entered clinical development. However, severe cardiovascular dose-limiting toxicities have been reported at 30 mg and 25 mg/day, and its phase I clinical trial was terminated [236].

In conclusion, to date no selective MetAP1 inhibitors, but various reversible and irreversible MetAP2 inhibitors have been clinically evaluated in different indications, such as oncology, obesity and diabetes, but none have reached market approval yet. However, MetAP2 inhibition is still considered as a valid mode of action that warrants further clinical evaluation due to its innovative first-in-class and broad application potential.

References

[1] Giglione C, Fieulaine S, Meinnel T. N-terminal protein modifications: bringing back into play the ribosome. Biochimie 2015;114:134–46.

[2] Giglione C, Boularot A, Meinnel T. Protein N-terminal methionine excision. Cell Mol Life Sci 2004;61(12):1455–74.

[3] Bradshaw RA, Brickey WW, Walker KW. N-terminal processing: the methionine aminopeptidase and N alpha-acetyl transferase families. Trends Biochem Sci 1998;23(7):263–7.

[4] Ben-Bassat A, Bauer K, Chang SY, Myambo K, Boosman A, Chang S. Processing of the initiation methionine from proteins: properties

of the *Escherichia coli* methionine aminopeptidase and its gene structure. J Bacteriol 1987;169(2):751–7.

[5] Kendall RL, Bradshaw RA. Isolation and characterization of the methionine aminopeptidase from porcine liver responsible for the co-translational processing of proteins. J Biol Chem 1992;267 (29):20667–73.

[6] Chang YH, Teichert U, Smith JA. Molecular cloning, sequencing, deletion, and overexpression of a methionine aminopeptidase gene from *Saccharomyces cerevisiae*. J Biol Chem 1992;267 (12):8007–11.

[7] Raue U, Oellerer S, Rospert S. Association of protein biogenesis factors at the yeast ribosomal tunnel exit is affected by the translational status and nascent polypeptide sequence. J Biol Chem 2007;282(11):7809–16.

[8] Bhakta S, Akbar S, Sengupta J. Cryo-EM structures reveal relocalization of MetAP in the presence of other protein biogenesis factors at the ribosomal tunnel exit. J Mol Biol 2019;431(7):1426–39.

[9] Kramer G, Boehringer D, Ban N, Bukau B. The ribosome as a platform for co-translational processing, folding and targeting of newly synthesized proteins. Nat Struct Mol Biol 2009;16 (6):589–97.

[10] Arnesen T, Van Damme P, Polevoda B, Helsens K, Evjenth R, Colaert N, et al. Proteomics analyses reveal the evolutionary conservation and divergence of N-terminal acetyltransferases from yeast and humans. Proc Natl Acad Sci U S A 2009;106(20):8157–62.

[11] Meinnel T, Giglione C. Tools for analyzing and predicting N-terminal protein modifications. Proteomics 2008;8(4):626–49.

[12] Frottin F, Martinez A, Peynot P, Mitra S, Holz RC, Giglione C, et al. The proteomics of N-terminal methionine cleavage. Mol Cell Proteomics 2006;5(12):2336–49.

[13] Hirel PH, Schmitter MJ, Dessen P, Fayat G, Blanquet S. Extent of N-terminal methionine excision from *Escherichia coli* proteins is governed by the side-chain length of the penultimate amino acid. Proc Natl Acad Sci U S A 1989;86(21):8247–51.

[14] Flinta C, Persson B, Jörnvall H, von Heijne G. Sequence determinants of cytosolic N-terminal protein processing. Eur J Biochem 1986;154(1):193–6.

[15] Xiao Q, Zhang F, Nacev BA, Liu JO, Pei D. Protein N-terminal processing: substrate specificity of *Escherichia coli* and human methionine aminopeptidases. Biochemistry 2010;49(26):5588–99.

[16] Dummitt B, Micka WS, Chang YH. N-terminal methionine removal and methionine metabolism in Saccharomyces cerevisiae. J Cell Biochem 2003;89(5):964–74.

[17] Frottin F, Bienvenut WV, Bignon J, Jacquet E, Vaca Jacome AS, Van Dorsselaer A, et al. MetAP1 and MetAP2 drive cell selectivity for a potent anti-cancer agent in synergy, by controlling glutathione redox state. Oncotarget 2016;7(39):63306–23.

[18] Arfin SM, Bradshaw RA. Cotranslational processing and protein turnover in eukaryotic cells. Biochemistry 1988;27(21):7979–84.

[19] Varshavsky A. The N-end rule pathway and regulation by proteolysis. Protein Sci 2011;20(8):1298–345.

[20] Kim HK, Kim RR, Oh JH, Cho H, Varshavsky A, Hwang CS. The N-terminal methionine of cellular proteins as a degradation signal. Cell 2014;156(1–2):158–69.

[21] Gibbs DJ, Bacardit J, Bachmair A, Holdsworth MJ. The eukaryotic N-end rule pathway: conserved mechanisms and diverse functions. Trends Cell Biol 2014;24(10):603–11.

[22] Hu RG, Sheng J, Qi X, Xu Z, Takahashi TT, Varshavsky A. The N-end rule pathway as a nitric oxide sensor controlling the levels of multiple regulators. Nature 2005;437(7061):981–6.

[23] Nguyen KT, Kim JM, Park SE, Hwang CS. N-terminal methionine excision of proteins creates tertiary destabilizing N-degrons of the Arg/N-end rule pathway. J Biol Chem 2019;294(12):4464–76.

[24] Meinnel T, Dian C, Giglione C. Myristoylation, an ancient protein modification mirroring Eukaryogenesis and evolution. Trends Biochem Sci 2020;45(7):619–32.

[25] Hu X, Dang Y, Tenney K, Crews P, Tsai CW, Sixt KM, et al. Regulation of c-Src nonreceptor tyrosine kinase activity by bengamide a through inhibition of methionine aminopeptidases. Chem Biol 2007;14(7):764–74.

[26] Peseckis SM, Deichaite I, Resh MD. Iodinated fatty acids as probes for myristate processing and function. Incorporation into pp60v-src. J Biol Chem 1993;268(7):5107–14.

[27] Aitken A, Cohen P, Santikarn S, Williams DH, Calder AG, Smith A, et al. Identification of the NH_2-terminal blocking group of calcineurin B as myristic acid. FEBS Lett 1982;150(2):314–8.

[28] Tahirov TH, Oki H, Tsukihara T, Ogasahara K, Yutani K, Ogata K, et al. Crystal structure of methionine aminopeptidase from hyperthermophile, *Pyrococcus furiosus*. J Mol Biol 1998;284(1):101–24.

[29] Liu S, Widom J, Kemp CW, Crews CM, Clardy J. Structure of human methionine aminopeptidase-2 complexed with fumagillin. Science 1998;282(5392):1324–7.

[30] Lowther WT, Brot N, Weissbach H, Matthews BW. Structure and mechanism of peptide methionine sulfoxide reductase, an "anti-oxidation" enzyme. Biochemistry 2000;39(44):13307–12.

[31] Addlagatta A, Hu X, Liu JO, Matthews BW. Structural basis for the functional differences between type I and type II human methionine aminopeptidases. Biochemistry 2005;44(45):14741–9.

[32] Arfin SM, Kendall RL, Hall L, Weaver LH, Stewart AE, Matthews BW, et al. Eukaryotic methionyl aminopeptidases: two classes of cobalt-dependent enzymes. Proc Natl Acad Sci U S A 1995;92 (17):7714–8.

[33] Li X, Chang YH. Amino-terminal protein processing in *Saccharomyces cerevisiae* is an essential function that requires two distinct methionine aminopeptidases. Proc Natl Acad Sci U S A 1995;92 (26):12357–61.

[34] Berman HM, Westbrook J, Feng Z, Gilliland G, Bhat TN, Weissig H, et al. The protein data bank. Nucleic Acids Res 2000;28(1):235–42.

[35] D'Souza VM, Brown RS, Bennett B, Holz RC. Characterization of the active site and insight into the binding mode of the anti-angiogenesis agent fumagillin to the manganese(II)-loaded methionyl aminopeptidase from *Escherichia coli*. J Biol Inorg Chem 2005;10(1):41–50.

[36] Lowther WT, Matthews BW. Structure and function of the methionine aminopeptidases. Biochim Biophys Acta 2000;1477(1–2):157–67.

[37] Heinrich T, Seenisamy J, Blume B, Bomke J, Calderini M, Eckert U, et al. Discovery and structure-based optimization of next-generation reversible methionine aminopeptidase-2 (MetAP-2) inhibitors. J Med Chem 2019;62(10):5025–39.

[38] Heinrich T, Seenisamy J, Becker F, Blume B, Bomke J, Dietz M, et al. Identification of methionine aminopeptidase-2 (MetAP-2) inhibitor M8891: a clinical compound for the treatment of Cancer. J Med Chem 2019;62(24):11119–34.

[39] Ye QZ, Xie SX, Ma ZQ, Huang M, Hanzlik RP. Structural basis of catalysis by monometalated methionine aminopeptidase. Proc Natl Acad Sci U S A 2006;103(25):9470–5.

[40] Helgren TR, Wangtrakuldee P, Staker BL, Hagen TJ. Advances in bacterial methionine aminopeptidase inhibition. Curr Top Med Chem 2016;16(4):397–414.

[41] Huang M, Xie SX, Ma ZQ, Hanzlik RP, Ye QZ. Metal mediated inhibition of methionine aminopeptidase by quinolinyl sulfonamides. Biochem Biophys Res Commun 2006;339(2):506–13.

[42] Schiffmann R, Heine A, Klebe G, Klein CD. Metal ions as cofactors for the binding of inhibitors to methionine aminopeptidase: a critical view of the relevance of in vitro metalloenzyme assays. Angew Chem Int Ed Engl 2005;44(23):3620–3.

[43] D'Souza VM, Bennett B, Copik AJ, Holz RC. Divalent metal binding properties of the methionyl aminopeptidase from *Escherichia coli*. Biochemistry 2000;39(13):3817–26.

[44] Chai SC, Ye QZ. Analysis of the stoichiometric metal activation of methionine aminopeptidase. BMC Biochem 2009;10:32.

[45] D'Souza VM, Holz RC. The methionyl aminopeptidase from *Escherichia coli* can function as an iron(II) enzyme. Biochemistry 1999;38(34):11079–85.

[46] Žalubovskis R, Winum JY. Inhibitors of selected bacterial metalloenzymes. Curr Med Chem 2019;26(15):2690–714.

[47] Marschner A, Klein CD. Metal promiscuity and metal-dependent substrate preferences of Trypanosoma brucei methionine aminopeptidase 1. Biochimie 2015;115:35–43.

[48] Walker KW, Bradshaw RA. Yeast methionine aminopeptidase I can utilize either Zn^{2+} or Co^{2+} as a cofactor: a case of mistaken identity? Protein Sci 1998;7(12):2684–7.

[49] Hu XV, Chen X, Han KC, Mildvan AS, Liu JO. Kinetic and mutational studies of the number of interacting divalent cations required by bacterial and human methionine aminopeptidases. Biochemistry 2007;46(44):12833–43.

[50] Sule N, Singh RK, Zhao P, Srivastava DK. Probing the metal ion selectivity in methionine aminopeptidase via changes in the luminescence properties of the enzyme bound europium ion. J Inorg Biochem 2012;106(1):84–9.

[51] Wang J, Sheppard GS, Lou P, Kawai M, Park C, Egan DA, et al. Physiologically relevant metal cofactor for methionine aminopeptidase-2 is manganese. Biochemistry 2003;42(17):5035–42.

[52] Roderick SL, Matthews BW. Structure of the cobalt-dependent methionine aminopeptidase from *Escherichia coli*: a new type of proteolytic enzyme. Biochemistry 1993;32(15):3907–12.

[53] Bazan JF, Weaver LH, Roderick SL, Huber R, Matthews BW. Sequence and structure comparison suggest that methionine aminopeptidase, prolidase, aminopeptidase P, and creatinase share a common fold. Proc Natl Acad Sci U S A 1994;91(7):2473–7.

[54] Lowther WT, Zhang Y, Sampson PB, Honek JF, Matthews BW. Insights into the mechanism of *Escherichia coli* methionine aminopeptidase from the structural analysis of reaction products and phosphorus-based transition-state analogues. Biochemistry 1999;38(45):14810–9.

[55] Sin N, Meng L, Wang MQ, Wen JJ, Bornmann WG, Crews CM. The anti-angiogenic agent fumagillin covalently binds and inhibits the methionine aminopeptidase, MetAP-2. Proc Natl Acad Sci U S A 1997;94(12):6099–103.

[56] Bala S, Yellamanda KV, Kadari A, Ravinuthala VSU, Kattula B, Singh OV, et al. Selective inhibition of helicobacter pylori methionine aminopeptidase by azaindole hydroxamic acid derivatives: design, synthesis, in vitro biochemical and structural studies. Bioorg Chem 2021;115, 105185.

[57] Bhat SY, Bhandari S, Thacker PS, Arifuddin M, Qureshi IA. Development of quinoline-based hybrid as inhibitor of methionine aminopeptidase 1 from *Leishmania donovani*. Chem Biol Drug Des 2021;97(2):315–24.

[58] Rincon Nigro M, Ma J, Awosemo OT, Xie H, Olaleye OA, Liang D. Development and validation of a sensitive, specific and reproducible UPLC-MS/MS method for the quantification of OJT007, a novel anti-leishmanial agent: application to a pharmacokinetic study. Int J Environ Res Public Health 2021;18(9).

[59] Krátký M, Vinšová J, Novotná E, Mandíková J, Wsól V, Trejtnar F, et al. Salicylanilide derivatives block *Mycobacterium tuberculosis* through inhibition of isocitrate lyase and methionine aminopeptidase. Tuberculosis (Edinb) 2012;92(5):434–9.

[60] Bhat S, Olaleye O, Meyer KJ, Shi W, Zhang Y, Liu JO. Analogs of N'-hydroxy-N-(4H,5H-naphtho[1,2-d]thiazol-2-yl)methanimidamide inhibit mycobacterium tuberculosis methionine aminopeptidases. Bioorg Med Chem 2012;20(14):4507–13.

[61] Meetei PA, Hauser AS, Raju PS, Rathore RS, Prabhu NP, Vindal V. Investigations and design of pyridine-2-carboxylic acid thiazol-2-ylamide analogs as methionine aminopeptidase inhibitors using 3D-QSAR and molecular docking. Med Chem Res 2014;23 (8):3861–75.

[62] Li YJ, Liu LJ, Jin K, Xu YT, Sun SQ. Synthesis and bioactivity of a novel series of 3,6-disubstituted 1,2,4-triazolo[3,4-b]-1,3,4-thiadiazoles. Chin Chem Lett 2010;21(3):293–6.

[63] Mir Mohammad Masood MI, Alam S, Hasan P, Queenb A, Shahidc S, Zahidc M, Azamb A, Abida M. Synthesis, antimicrobial evaluation and in silico studies of novel 2,4- disubstituted-1,3-thiazole derivatives. Lett Drug Des Discov 2019;16:160–73.

[64] Juhás M, Pallabothula VSK, Grabrijan K, Šimovičová M, Jandourek O, Konečná K, et al. Design, synthesis and biological evaluation of substituted 3-amino-N-(thiazol-2-yl)pyrazine-2-carboxamides as inhibitors of mycobacterial methionine aminopeptidase 1. Bioorg Chem 2022;118, 105489.

[65] Bhat SY, Jagruthi P, Srinivas A, Arifuddin M, Qureshi IA. Synthesis and characterization of quinoline-carbaldehyde derivatives as novel inhibitors for leishmanial methionine aminopeptidase 1. Eur J Med Chem 2020;186, 111860.

[66] Quan DH, Nagalingam G, Luck I, Proschogo N, Pillalamarri V, Addlagatta A, et al. Bengamides display potent activity against drug-resistant mycobacterium tuberculosis. Sci Rep 2019;9 (1):14396.

[67] Wenzel SC, Hoffmann H, Zhang J, Debussche L, Haag-Richter S, Kurz M, et al. Production of the bengamide class of marine natural products in myxobacteria: biosynthesis and structure-activity relationships. Angew Chem Int Ed Engl 2015;54(51):15560–4.

[68] Rose JA, Lahiri SD, McKinney DC, Albert R, Morningstar ML, Shapiro AB, et al. Novel broad-spectrum inhibitors of bacterial methionine aminopeptidase. Bioorg Med Chem Lett 2015;25 (16):3301–6.

[69] Hasan P, Pillalamarri VK, Aneja B, Irfan M, Azam M, Perwez A, et al. Synthesis and mechanistic studies of diketo acids and their bioisosteres as potential antibacterial agents. Eur J Med Chem 2019;163:67–82.

[70] Olaleye O, Raghunand TR, Bhat S, Chong C, Gu P, Zhou J, et al. Characterization of clioquinol and analogues as novel inhibitors of methionine aminopeptidases from Mycobacterium tuberculosis. Tuberculosis (Edinb) 2011;91(Suppl 1):S61–5.

[71] Wangtrakuldee P, Byrd MS, Campos CG, Henderson MW, Zhang Z, Clare M, et al. Discovery of inhibitors of *Burkholderia pseudomallei*

methionine aminopeptidase with antibacterial activity. ACS Med Chem Lett 2013;4(8):699–703.
[72] Masood MM, Pillalamarri VK, Irfan M, Aneja B, Jairajpuri MA, Zafaryab M, et al. Diketo acids and their amino acid/dipeptidic analogues as promising scaffolds for the development of bacterial methionine aminopeptidase inhibitors. RSC Adv 2015;5 (43):34173–83.
[73] Arya T, Reddi R, Kishor C, Ganji RJ, Bhukya S, Gumpena R, et al. Identification of the molecular basis of inhibitor selectivity between the human and streptococcal type I methionine aminopeptidases. J Med Chem 2015;58(5):2350–7.
[74] Tumskiy RS, Tumskaia AV, Pylaev TE, Avdeeva ES, Evstigneeva SS. Docking and antibacterial activity of novel nontoxic 5-arylidenepyrimidine-triones as inhibitors of NDM-1 and MetAP-1. Future Med Chem 2021;13(12):1041–55.
[75] Boucherit H, Chikhi A, Bensegueni A, Merzoug A, Bolla JM. The research of new inhibitors of bacterial methionine aminopeptidase by structure based virtual screening approach of zinc database and in vitro validation. Curr Comput Aided Drug Des 2020;16(4):389–401.
[76] Albayati S, Uba AI, Yelekçi K. Potential inhibitors of methionine aminopeptidase type II identified via structure-based pharmacophore modeling. Mol Divers 2022;26(2):1005–16.
[77] Pillalamarri V, Reddy CG, Bala SC, Jangam A, Kutty VV, Addlagatta A. Methionine aminopeptidases with short sequence inserts within the catalytic domain are differentially inhibited: structural and biochemical studies of three proteins from Vibrio spp. Eur J Med Chem 2021;209, 112883.
[78] Wang Y, Sojoodi M, Qiao G, Lin Z, Barrett SC, Zukerberg L, et al. Abstract 108: inhibiting methionine aminopeptidase 2 prevents liver fibrosis and hepatocellular carcinoma. Cancer Res 2021;81 (13_Supplement):108.
[79] Arico-Muendel CC, Blanchette H, Benjamin DR, Caiazzo TM, Centrella PA, DeLorey J, et al. Orally active fumagillin analogues: transformations of a reactive warhead in the gastric environment. ACS Med Chem Lett 2013;4(4):381–6.
[80] Yoshimura T, Benny O, Bazinet L, D'Amato RJ. Suppression of autoimmune retinal inflammation by an antiangiogenic drug. PloS One 2013;8(6), e66219.
[81] Sherbet GV. S100A4 has potenial benefits as a therapeutic target. In: Molecular approach to cancer management. Elsevier Inc; 2017. p. 211–21.
[82] Furness M, Robinson T, Ehlers T, Hubbard R, Arbiser J, Goldsmith D, et al. Antiangiogenic agents: studies on fumagillin and curcumin analogs. Curr Pharm Des 2005;11:357–73.
[83] Morgen M, Jöst C, Malz M, Janowski R, Niessing D, Klein CD, et al. Spiroepoxytriazoles are fumagillin-like irreversible inhibitors of MetAP2 with potent cellular activity. ACS Chem Biol 2016;11 (4):1001–11.
[84] Fang B, Hu C, Ding Y, Qin H, Luo Y, Xu Z, et al. Discovery of 4H-thieno[3,2-b]pyrrole derivatives as potential anticancer agents. J Heterocyclic Chem 2021;58(8):1610–27.
[85] Węglarz-Tomczak E, Burda-Grabowska M, Giurg M, Mucha A. Identification of methionine aminopeptidase 2 as a molecular target of the organoselenium drug ebselen and its derivatives/analogues: synthesis, inhibitory activity and molecular modeling study. Bioorg Med Chem Lett 2016;26(21):5254–9.
[86] Węglarz-Tomczak E, Talma M, Giurg M, Westerhoff HV, Janowski R, Mucha A. Neutral metalloaminopeptidases APN and MetAP2 as newly discovered anticancer molecular targets of actinomycin D and its simple analogs. Oncotarget 2018;9(50):29365–78.
[87] Wang J, Sheppard GS, Lou P, Kawai M, BaMaung N, Erickson SA, et al. Tumor suppression by a rationally designed reversible inhibitor of methionine aminopeptidase-2. Cancer Res 2003;63(22):7861–9.
[88] Towbin H, Bair KW, DeCaprio JA, Eck MJ, Kim S, Kinder FR, et al. Proteomics-based target identification: bengamides as a new class of methionine aminopeptidase inhibitors. J Biol Chem 2003;278 (52):52964–71.
[89] Xu W, Lu JP, Ye QZ. Structural analysis of bengamide derivatives as inhibitors of methionine aminopeptidases. J Med Chem 2012;55 (18):8021–7.
[90] White KN, Tenney K, Crews P. The bengamides: a mini-review of natural sources, analogues, biological properties, biosynthetic origins, and future prospects. J Nat Prod 2017;80(3):740–55.
[91] Wang J, Tucker LA, Stavropoulos J, Zhang Q, Wang YC, Bukofzer G, et al. Correlation of tumor growth suppression and methionine aminopetidase-2 activity blockade using an orally active inhibitor. Proc Natl Acad Sci U S A 2008;105(6):1838–43.
[92] Fassihi A, Shahlaei M, Moeinifard B, Sabet R. QSAR study of anthranilic acid sulfonamides as methionine aminopeptidase-2 inhibitors. Monatsh für Chem 2012;143(2):189–98.
[93] Shaikh SKJ, Kamble RR, Bayannavar PK, Somagond SM, Joshi SD. Triazolothiadizepinylquinolines as potential MetAP-2 and NMT inhibitors: microwave-assisted synthesis, pharmacological evaluation and molecular docking studies. J Mol Struct 2020;1203, 127445.
[94] Philip S, Inturi B, Bhavya K, Pujar GV, Purohit M. 3D-QSAR and molecular docking studies on 1, 2, 4 triazoles as METAP2 inhibitors. Int J Pharm Pharm Science 2014;6:205–12.
[95] Kallander LS, Lu Q, Chen W, Tomaszek T, Yang G, Tew D, et al. 4-Aryl-1,2,3-triazole: a novel template for a reversible methionine aminopeptidase 2 inhibitor, optimized to inhibit angiogenesis in vivo. J Med Chem 2005;48(18):5644–7.
[96] Zhou Y, Zhao Y, O'Boyle KM, Murphy PV. Hybrid angiogenesis inhibitors: synthesis and biological evaluation of bifunctional compounds based on 1-deoxynojirimycin and aryl-1,2,3-triazoles. Bioorg Med Chem Lett 2008;18(3):954–8.
[97] Hou YP, Sun J, Pang ZH, Lv PC, Li DD, Yan L, et al. Synthesis and antitumor activity of 1,2,4-triazoles having 1,4-benzodioxan fragment as a novel class of potent methionine aminopeptidase type II inhibitors. Bioorg Med Chem 2011;19(20):5948–54.
[98] Sun J, Li MH, Qian SS, Guo FJ, Dang XF, Wang XM, et al. Synthesis and antitumor activity of 1,3,4-oxadiazole possessing 1,4-benzodioxan moiety as a novel class of potent methionine aminopeptidase type II inhibitors. Bioorg Med Chem Lett 2013;23 (10):2876–9.
[99] Yılmaz Ö, Bayer B, Bekçi H, Uba AI, Cumaoğlu A, Yelekçi K, et al. Synthesis, Anticancer Activity on Prostate Cancer Cell Lines and Molecular Modeling Studies of Flurbiprofen-Thioether Derivatives as Potential Target of MetAP (Type II). Med Chem 2020;16(6):735–49.
[100] Çoruh I, Çevik Ö, Yelekçi K, Djikic T, Küçükgüzel ŞG. Synthesis, anticancer activity, and molecular modeling of etodolac-thioether derivatives as potent methionine aminopeptidase (type II) inhibitors. Arch Pharm (Weinheim) 2018;351(3–4), e1700195.
[101] Han MI, Bekçi H, Uba AI, Yıldırım Y, Karasulu E, Cumaoğlu A, et al. Synthesis, molecular modeling, in vivo study, and anticancer activity of 1,2,4-triazole containing hydrazide-hydrazones derived

from (S)-naproxen. Arch Pharm (Weinheim) 2019;352(6), e1800365.
[102] Birgül K, Yıldırım Y, Karasulu HY, Karasulu E, Uba AI, Yelekçi K, et al. Synthesis, molecular modeling, in vivo study and anticancer activity against prostate cancer of (+) (S)-naproxen derivatives. Eur J Med Chem 2020;208, 112841.
[103] Yin SQ, Wang JJ, Zhang CM, Liu ZP. The development of MetAP-2 inhibitors in cancer treatment. Curr Med Chem 2012;19(7):1021–35.
[104] Ehlers T, Furness S, Robinson TP, Zhong HA, Goldsmith D, Aribser J, et al. Methionine aminopeptidase type-2 inhibitors targeting angiogenesis. Curr Top Med Chem 2016;16(13):1478–88.
[105] Shim JS, Matsui Y, Bhat S, Nacev BA, Xu J, Bhang HE, et al. Effect of nitroxoline on angiogenesis and growth of human bladder cancer. J Natl Cancer Inst 2010;102(24):1855–73.
[106] Ekpenyong O, Gao X, Ma J, Cooper C, Nguyen L, Olaleye OA, et al. Pre-clinical pharmacokinetics, tissue distribution and physicochemical studies of CLBQ14, a novel methionine aminopeptidase inhibitor for the treatment of infectious diseases. Drug Des Devel Ther 2020;14:1263–77.
[107] Bhat S, Shim JS, Zhang F, Chong CR, Liu JO. Substituted oxines inhibit endothelial cell proliferation and angiogenesis. Org Biomol Chem 2012;10(15):2979–92.
[108] Selvakumar P, Lakshmikuttyamma A, Das U, Pati HN, Dimmock JR, Sharma RK. NC2213: a novel methionine aminopeptidase 2 inhibitor in human colon cancer HT29 cells. Mol Cancer 2009;8:65.
[109] Hirst DJ, Brandt M, Bruton G, Christodoulou E, Cutler L, Deeks N, et al. Structure-based optimisation of orally active & reversible MetAP-2 inhibitors maintaining a tight 'molecular budget'. Bioorg Med Chem Lett 2020;30(21), 127533.
[110] Cheruvallath Z, Tang M, McBride C, Komandla M, Miura J, Ton-Nu T, et al. Discovery of potent, reversible MetAP2 inhibitors via fragment based drug discovery and structure based drug design-part 1. Bioorg Med Chem Lett 2016;26(12):2774–8.
[111] McBride C, Cheruvallath Z, Komandla M, Tang M, Farrell P, Lawson JD, et al. Discovery of potent, reversible MetAP2 inhibitors via fragment based drug discovery and structure based drug design-part 2. Bioorg Med Chem Lett 2016;26(12):2779–83.
[112] Rosse G. Pyridinonaphthyridinone inhibitors of type 2 methionine aminopeptidase. ACS Med Chem Lett 2015;6(6):622–3.
[113] Heinrich T, Buchstaller HP, Cezanne B, Rohdich F, Bomke J, Friese-Hamim M, et al. Novel reversible methionine aminopeptidase-2 (MetAP-2) inhibitors based on purine and related bicyclic templates. Bioorg Med Chem Lett 2017;27(3):551–6.
[114] Weako J, Uba AI, Keskin Ö, Gürsoy A, Yelekçi K. Identification of potential inhibitors of human methionine aminopeptidase (type II) for cancer therapy: structure-based virtual screening, ADMET prediction and molecular dynamics studies. Comput Biol Chem 2020;86, 107244.
[115] Hanson FR, Eble TE. An antiphage agent isolated from aspergillus SP. J Bacteriol 1949;58(4):527–9.
[116] Killough JH, Magill GB, Smith RC. The treatment of amebiasis with fumagillin. Science 1952;115(2977):71–2.
[117] Huang WF, Solter LF, Yau PM, Imai BS. Nosema ceranae escapes fumagillin control in honey bees. PLoS Pathog 2013;9(3), e1003185.
[118] Zhang H, Huang H, Cali A, Takvorian PM, Feng X, Zhou G, et al. Investigations into microsporidian methionine aminopeptidase type 2: a therapeutic target for microsporidiosis. Folia Parasitol (Praha) 2005;52(1–2):182–92.
[119] Arico-Muendel C, Centrella PA, Contonio BD, Morgan BA, O'Donovan G, Paradise CL, et al. Antiparasitic activities of novel, orally available fumagillin analogs. Bioorg Med Chem Lett 2009;19(17):5128–31.
[120] Chen X, Xie S, Bhat S, Kumar N, Shapiro TA, Liu JO. Fumagillin and fumarranol interact with P. falciparum methionine aminopeptidase 2 and inhibit malaria parasite growth in vitro and in vivo. Chem Biol 2009;16(2):193–202.
[121] Zhang P, Nicholson DE, Bujnicki JM, Su X, Brendle JJ, Ferdig M, et al. Angiogenesis inhibitors specific for methionine aminopeptidase 2 as drugs for malaria and leishmaniasis. J Biomed Sci 2002;9(1):34–40.
[122] Lefkove B, Govindarajan B, Arbiser JL. Fumagillin: an anti-infective as a parent molecule for novel angiogenesis inhibitors. Expert Rev Anti Infect Ther 2007;5(4):573–9.
[123] Kulakova L, Galkin A, Chen CZ, Southall N, Marugan JJ, Zheng W, et al. Discovery of novel antigiardiasis drug candidates. Antimicrob Agents Chemother 2014;58(12):7303–11.
[124] Ingber D, Fujita T, Kishimoto S, Sudo K, Kanamaru T, Brem H, et al. Synthetic analogues of fumagillin that inhibit angiogenesis and suppress tumour growth. Nature 1990;348(6301):555–7.
[125] Menssen A, Hermeking H. Characterization of the c-MYC-regulated transcriptome by SAGE: identification and analysis of c-MYC target genes. Proc Natl Acad Sci U S A 2002;99(9):6274–9.
[126] Chatterjee M, Chatterjee N, Datta R, Datta B, Gupta NK. Expression and activity of p67 are induced during heat shock. Biochem Biophys Res Commun 1998;249(1):113–7.
[127] Gupta S, Bose A, Chatterjee N, Saha D, Wu S, Gupta NK. p67 transcription regulates translation in serum-starved and mitogen-activated KRC-7 cells. J Biol Chem 1997;272 (19):12699–704.
[128] He WP, Guo YY, Yang GP, Lai HL, Sun TT, Zhang ZW, et al. CHD1L promotes EOC cell invasiveness and metastasis via the regulation of METAP2. Int J Med Sci 2020;17(15):2387–95.
[129] Nakagawa H, Koyanagi S, Takiguchi T, Kuramoto Y, Soeda S, Shimeno H, et al. 24-hour oscillation of mouse methionine aminopeptidase2, a regulator of tumor progression, is regulated by clock gene proteins. Cancer Res 2004;64(22):8328–33.
[130] Selvakumar P, Lakshmikuttyamma A, Kanthan R, Kanthan SC, Dimmock JR, Sharma RK. High expression of methionine aminopeptidase 2 in human colorectal adenocarcinomas. Clin Cancer Res 2004;10(8):2771–5.
[131] Catalano A, Romano M, Robuffo I, Strizzi L, Procopio A. Methionine aminopeptidase-2 regulates human mesothelioma cell survival: role of Bcl-2 expression and telomerase activity. Am J Pathol 2001;159(2):721–31.
[132] Sawanyawisuth K, Wongkham C, Pairojkul C, Saeseow OT, Riggins GJ, Araki N, et al. Methionine aminopeptidase 2 over-expressed in cholangiocarcinoma: potential for drug target. Acta Oncol 2007;46 (3):378–85.
[133] Ho CY, Bar E, Giannini C, Marchionni L, Karajannis MA, Zagzag D, et al. MicroRNA profiling in pediatric pilocytic astrocytoma reveals biologically relevant targets, including PBX3, NFIB, and METAP2. Neuro Oncol 2013;15(1):69–82.
[134] Dasgupta B, Yi Y, Hegedus B, Weber JD, Gutmann DH. Cerebrospinal fluid proteomic analysis reveals dysregulation of methionine aminopeptidase-2 expression in human and mouse neurofibromatosis 1-associated glioma. Cancer Res 2005;65(21):9843–50.
[135] Kanno T, Endo H, Takeuchi K, Morishita Y, Fukayama M, Mori S. High expression of methionine aminopeptidase type 2 in germinal center B cells and their neoplastic counterparts. Lab Invest 2002;82 (7):893–901.

[136] Tucker LA, Zhang Q, Sheppard GS, Lou P, Jiang F, McKeegan E, et al. Ectopic expression of methionine aminopeptidase-2 causes cell transformation and stimulates proliferation. Oncogene 2008;27 (28):3967–76.

[137] Datta B, Ghosh A, Majumdar A, Datta R. Autoproteolysis of rat p67 generates several peptide fragments: the N-terminal fragment, p26, is required for the protection of eIF2alpha from phosphorylation. Biochemistry 2007;46(11):3465–75.

[138] Clinkinbeard T, Ghoshal S, Craddock S, Creed Pettigrew L, Guttmann RP. Calpain cleaves methionine aminopeptidase-2 in a rat model of ischemia/reperfusion. Brain Res 2013;1499:129–35.

[139] Chakraborty A, Saha D, Bose A, Chatterjee M, Gupta NK. Regulation of eIF-2 alpha-subunit phosphorylation in reticulocyte lysate. Biochemistry 1994;33(21):6700–6.

[140] Datta R, Choudhury P, Ghosh A, Datta B. A glycosylation site, 60SGTS63, of p67 is required for its ability to regulate the phosphorylation and activity of eukaryotic initiation factor 2alpha. Biochemistry 2003;42(18):5453–60.

[141] Ng JY, Chiu J, Hogg PJ, Wong JW. Tyrosine nitration moderates the peptidase activity of human methionyl aminopeptidase 2. Biochem Biophys Res Commun 2013;440(1):37–42.

[142] Chiu J, Wong JW, Hogg PJ. Redox regulation of methionine aminopeptidase 2 activity. J Biol Chem 2014;289(21):15035–43.

[143] Jonckheere V, Fijałkowska D, Van Damme P. Omics assisted N-terminal proteoform and protein expression profiling on methionine aminopeptidase 1 (MetAP1) deletion. Mol Cell Proteomics 2018;17 (4):694–708.

[144] Turk BE, Griffith EC, Wolf S, Biemann K, Chang YH, Liu JO. Selective inhibition of amino-terminal methionine processing by TNP-470 and ovalicin in endothelial cells. Chem Biol 1999;6 (11):823–33.

[145] Warder SE, Tucker LA, McLoughlin SM, Strelitzer TJ, Meuth JL, Zhang Q, et al. Discovery, identification, and characterization of candidate pharmacodynamic markers of methionine aminopeptidase-2 inhibition. J Proteome Res 2008;7(11):4807–20.

[146] Yoshida T, Kaneko Y, Tsukamoto A, Han K, Ichinose M, Kimura S. Suppression of hepatoma growth and angiogenesis by a fumagillin derivative TNP470: possible involvement of nitric oxide synthase. Cancer Res 1998;58(16):3751–6.

[147] Friese-Hamim M, Bogatyrova O, Ortiz MJ, Wienke D, Heinrich T, Rohdich F, et al. Antitumor activity of M8891, a potent and reversible inhibitor of methionine aminopeptidase 2. Cancer Res 2019;79(13 Supplement). Abstract 3075.

[148] Sundberg TB, Darricarrere N, Cirone P, Li X, McDonald L, Mei X, et al. Disruption of Wnt planar cell polarity signaling by aberrant accumulation of the MetAP-2 substrate Rab37. Chem Biol 2011;18(10):1300–11.

[149] Wu S, Rehemtulla A, Gupta NK, Kaufman RJ. A eukaryotic translation initiation factor 2-associated 67 kDa glycoprotein partially reverses protein synthesis inhibition by activated double-stranded RNA-dependent protein kinase in intact cells. Biochemistry 1996;35(25):8275–80.

[150] Datta B, Datta R. Mutation at the acidic residue-rich domain of eukaryotic initiation factor 2 (eIF2alpha)-associated glycoprotein p67 increases the protection of eIF2alpha phosphorylation during heat shock. Arch Biochem Biophys 2003;413(1):116–22.

[151] Datta R, Choudhury P, Bhattacharya M, Soto Leon F, Zhou Y, Datta B. Protection of translation initiation factor eIF2 phosphorylation correlates with eIF2-associated glycoprotein p67 levels and requires the lysine-rich domain I of p67. Biochimie 2001;83(10):919–31.

[152] Datta R, Tammali R, Datta B. Negative regulation of the protection of eIF2alpha phosphorylation activity by a unique acidic domain present at the N-terminus of p67. Exp Cell Res 2003;283(2):237–46.

[153] Datta B, Datta R, Mukherjee S, Zhang Z. Increased phosphorylation of eukaryotic initiation factor 2alpha at the G2/M boundary in human osteosarcoma cells correlates with deglycosylation of p67 and a decreased rate of protein synthesis. Exp Cell Res 1999;250 (1):223–30.

[154] Gil J, Esteban M, Roth D. In vivo regulation of the dsRNA-dependent protein kinase PKR by the cellular glycoprotein p67. Biochemistry 2000;39(51):16016–25.

[155] Endo H, Takenaga K, Kanno T, Satoh H, Mori S. Methionine aminopeptidase 2 is a new target for the metastasis-associated protein, S100A4. J Biol Chem 2002;277(29):26396–402.

[156] Katagiri N, Nagatoishi S, Tsumoto K, Endo H. Structural features of methionine aminopeptidase2-active core peptide essential for binding with S100A4. Biochem Biophys Res Commun 2019;516 (4):1123–9.

[157] Takenaga K, Ochiya T, Endo H. Inhibition of the invasion and metastasis of mammary carcinoma cells by NBD peptide targeting S100A4 via the suppression of the Sp1/MMP-14 axis. Int J Oncol 2021;58(3):397–408.

[158] Ochiya T, Takenaga K, Asagiri M, Nakano K, Satoh H, Watanabe T, et al. Efficient inhibition of tumor angiogenesis and growth by a synthetic peptide blocking S100A4-methionine aminopeptidase 2 interaction. Mol Ther Methods Clin Dev 2015;2:15008.

[159] Datta B, Majumdar A, Datta R, Balusu R. Treatment of cells with the angiogenic inhibitor fumagillin results in increased stability of eukaryotic initiation factor 2-associated glycoprotein, p67, and reduced phosphorylation of extracellular signal-regulated kinases. Biochemistry 2004;43(46):14821–31.

[160] Datta B, Datta R. Induction of apoptosis due to lowering the level of eukaryotic initiation factor 2-associated protein, p67, from mammalian cells by antisense approach. Exp Cell Res 1999;246 (2):376–83.

[161] Gupta S, Wu S, Chatterjee N, Ilan J, Ilan J, Osterman JC, et al. Regulation of an eukaryotic initiation factor-2 (eIF-2) associated 67 kDa glycoprotein (p67) and its requirement in protein synthesis. Gene Expr 1995;5(2):113–22.

[162] Lin M, Zhang X, Jia B, Guan S. Suppression of glioblastoma growth and angiogenesis through molecular targeting of methionine aminopeptidase-2. J Neurooncol 2018;136(2):243–54.

[163] Kim S, LaMontagne K, Sabio M, Sharma S, Versace RW, Yusuff N, et al. Depletion of methionine aminopeptidase 2 does not alter cell response to fumagillin or bengamides. Cancer Res 2004;64 (9):2984–7.

[164] Bernier SG, Taghizadeh N, Thompson CD, Westlin WF, Hannig G. Methionine aminopeptidases type I and type II are essential to control cell proliferation. J Cell Biochem 2005;95(6):1191–203.

[165] Yeh JJ, Ju R, Brdlik CM, Zhang W, Zhang Y, Matyskiela ME, et al. Targeted gene disruption of methionine aminopeptidase 2 results in an embryonic gastrulation defect and endothelial cell growth arrest. Proc Natl Acad Sci U S A 2006;103(27):10379–84.

[166] Kusaka M, Sudo K, Matsutani E, Kozai Y, Marui S, Fujita T, et al. Cytostatic inhibition of endothelial cell growth by the angiogenesis inhibitor TNP-470 (AGM-1470). Br J Cancer 1994;69(2):212–6.

[167] Zhang Y, Griffith EC, Sage J, Jacks T, Liu JO. Cell cycle inhibition by the anti-angiogenic agent TNP-470 is mediated by p53 and p21WAF1/CIP1. Proc Natl Acad Sci U S A 2000;97 (12):6427–32.
[168] Koyama H, Nishizawa Y, Hosoi M, Fukumoto S, Kogawa K, Shioi A, et al. The fumagillin analogue TNP-470 inhibits DNA synthesis of vascular smooth muscle cells stimulated by platelet-derived growth factor and insulin-like growth factor-I. Possible involvement of cyclin-dependent kinase 2. Circ Res 1996;79(4):757–64.
[169] Wang J, Lou P, Henkin J. Selective inhibition of endothelial cell proliferation by fumagillin is not due to differential expression of methionine aminopeptidases. J Cell Biochem 2000;77(3):465–73.
[170] Zhang Y, Yeh JR, Mara A, Ju R, Hines JF, Cirone P, et al. A chemical and genetic approach to the mode of action of fumagillin. Chem Biol 2006;13(9):1001–9.
[171] Zhou HF, Yan H, Hu Y, Springer LE, Yang X, Wickline SA, et al. Fumagillin prodrug nanotherapy suppresses macrophage inflammatory response via endothelial nitric oxide. ACS Nano 2014;8 (7):7305–17.
[172] Cooper AC, Karp RM, Clark EJ, Taghizadeh NR, Hoyt JG, Labenski MT, et al. A novel methionine aminopeptidase-2 inhibitor, PPI-2458, inhibits non-Hodgkin's lymphoma cell proliferation in vitro and in vivo. Clin Cancer Res 2006;12(8):2583–90.
[173] Garrabrant T, Tuman RW, Ludovici D, Tominovich R, Simoneaux RL, Galemmo Jr RA, et al. Small molecule inhibitors of methionine aminopeptidase type 2 (MetAP-2). Angiogenesis 2004;7(2):91–6.
[174] Laschke MW, Vorsterman van Oijen AE, Scheuer C, Menger MD. In vitro and in vivo evaluation of the anti-angiogenic actions of 4-hydroxybenzyl alcohol. Br J Pharmacol 2011;163(4):835–44.
[175] Winter PM, Schmieder AH, Caruthers SD, Keene JL, Zhang H, Wickline SA, et al. Minute dosages of alpha(nu)beta3-targeted fumagillin nanoparticles impair Vx-2 tumor angiogenesis and development in rabbits. FASEB J 2008;22(8):2758–67.
[176] Sheen IS, Jeng KS, Jeng WJ, Jeng CJ, Wang YC, Gu SL, et al. Fumagillin treatment of hepatocellular carcinoma in rats: an in vivo study of antiangiogenesis. World J Gastroenterol 2005;11(6):771–7.
[177] Lu J, Chong CR, Hu X, Liu JO. Fumarranol, a rearranged fumagillin analogue that inhibits angiogenesis in vivo. J Med Chem 2006;49 (19):5645–8.
[178] Chun E, Han CK, Yoon JH, Sim TB, Kim YK, Lee KY. Novel inhibitors targeted to methionine aminopeptidase 2 (MetAP2) strongly inhibit the growth of cancers in xenografted nude model. Int J Cancer 2005;114(1):124–30.
[179] Esa R, Steinberg E, Dror D, Schwob O, Khajavi M, Maoz M, et al. The role of methionine aminopeptidase 2 in lymphangiogenesis. Int J Mol Sci 2020;21(14).
[180] Shimizu S, Kawahara R, Simizu S. Methionine aminopeptidase-2 is a pivotal regulator of vasculogenic mimicry. Oncol Rep 2022;47(2).
[181] Joharapurkar AA, Dhanesha NA, Jain MR. Inhibition of the methionine aminopeptidase 2 enzyme for the treatment of obesity. Diabetes Metab Syndr Obes 2014;7:73–84.
[182] Huang HJ, Holub C, Rolzin P, Bilakovics J, Fanjul A, Satomi Y, et al. MetAP2 inhibition increases energy expenditure through direct action on brown adipocytes. J Biol Chem 2019;294(24):9567–75.
[183] Goya Grocin A, Kallemeijn WW, Tate EW. Targeting methionine aminopeptidase 2 in cancer, obesity, and autoimmunity. Trends Pharmacol Sci 2021;42(10):870–82.
[184] Selvakumar P, Lakshmikuttyamma A, Dimmock JR, Sharma RK. Methionine aminopeptidase 2 and cancer. Biochim Biophys Acta 2006;1765(2):148–54.
[185] Morowitz MJ, Barr R, Wang Q, King R, Rhodin N, Pawel B, et al. Methionine aminopeptidase 2 inhibition is an effective treatment strategy for neuroblastoma in preclinical models. Clin Cancer Res 2005;11(7):2680–5.
[186] Hannig G, Lazarus DD, Bernier SG, Karp RM, Lorusso J, Qiu D, et al. Inhibition of melanoma tumor growth by a pharmacological inhibitor of MetAP-2, PPI-2458. Int J Oncol 2006;28(4):955–63.
[187] Xie J, Rice MA, Chen Z, Cheng Y, Hsu EC, Chen M, et al. In vivo imaging of methionine aminopeptidase II for prostate cancer risk stratification. Cancer Res 2021;81(9):2510–21.
[188] Craig SL, Gault VA, Flatt PR, Irwin N. The methionine aminopeptidase 2 inhibitor, TNP-470, enhances the antidiabetic properties of sitagliptin in mice by upregulating xenin. Biochem Pharmacol 2021;183, 114355.
[189] Rupnick MA, Panigrahy D, Zhang CY, Dallabrida SM, Lowell BB, Langer R, et al. Adipose tissue mass can be regulated through the vasculature. Proc Natl Acad Sci U S A 2002;99(16):10730–5.
[190] Bråkenhielm E, Cao R, Gao B, Angelin B, Cannon B, Parini P, et al. Angiogenesis inhibitor, TNP-470, prevents diet-induced and genetic obesity in mice. Circ Res 2004;94(12):1579–88.
[191] Lijnen HR, Frederix L, Van Hoef B. Fumagillin reduces adipose tissue formation in murine models of nutritionally induced obesity. Obesity (Silver Spring) 2010;18(12):2241–6.
[192] White HM, Acton AJ, Considine RV. The angiogenic inhibitor TNP-470 decreases caloric intake and weight gain in high-fat fed mice. Obesity (Silver Spring) 2012;20(10):2003–9.
[193] Wei Jung J, MacArthur MR, Mitchell SJ. Pharmacological inhibition of MetAP2 causes rapid and sustained changes to mitochondria. J Am Coll Surg 2020;231(4), e115.
[194] Pottorf TS, Fagan MP, Burkey BF, Cho DJ, Vath JE, Tran PV. MetAP2 inhibition reduces food intake and body weight in a ciliopathy mouse model of obesity. JCI Insight 2020;5(2).
[195] Kass D, Bridges RS, Borczuk A, Greenberg S. Methionine aminopeptidase-2 as a selective target of myofibroblasts in pulmonary fibrosis. Am J Respir Cell Mol Biol 2007;37(2):193–201.
[196] Kass DJ, Rattigan E, Kahloon R, Loh K, Yu L, Savir A, et al. Early treatment with fumagillin, an inhibitor of methionine aminopeptidase-2, prevents pulmonary hypertension in monocrotaline-injured rats. PloS One 2012;7(4), e35388.
[197] Bainbridge J, Madden L, Essex D, Binks M, Malhotra R, Paleolog EM. Methionine aminopeptidase-2 blockade reduces chronic collagen-induced arthritis: potential role for angiogenesis inhibition. Arthritis Res Ther 2007;9(6):R127.
[198] Lazarus DD, Doyle EG, Bernier SG, Rogers AB, Labenski MT, Wakefield JD, et al. An inhibitor of methionine aminopeptidase type-2, PPI-2458, ameliorates the pathophysiological disease processes of rheumatoid arthritis. Inflamm Res 2008;57(1):18–27.
[199] Brahn E, Schoettler N, Lee S, Banquerigo ML. Involution of collagen-induced arthritis with an angiogenesis inhibitor, PPI-2458. J Pharmacol Exp Ther 2009;329(2):615–24.
[200] Priest RC, Spaull J, Buckton J, Grimley RL, Sims M, Binks M, et al. Immunomodulatory activity of a methionine aminopeptidase-2 inhibitor on B cell differentiation. Clin Exp Immunol 2009;155 (3):514–22.
[201] Demers M, Sturtevant S, Guertin KR, Gupta D, Desai K, Vieira BF, et al. MetAP2 inhibition modifies hemoglobin S to delay polymerization and improves blood flow in sickle cell disease. Blood Adv 2021;5(5):1388–402.
[202] Shusterman S, Grupp SA, Barr R, Carpentieri D, Zhao H, Maris JM. The angiogenesis inhibitor tnp-470 effectively inhibits human

neuroblastoma xenograft growth, especially in the setting of subclinical disease. Clin Cancer Res 2001;7(4):977–84.

[203] Stanimirovic Z, Stevanovic J, Bajic V, Radovic I. Evaluation of genotoxic effects of fumagillin by cytogenetic tests in vivo. Mutat Res 2007;628(1):1–10.

[204] Stevanovic J, Stanimirovic Z, Radakovic M, Stojic V. In vitro evaluation of the clastogenicity of fumagillin. Environ Mol Mutagen 2008;49(8):594–601.

[205] Kudelka AP, Levy T, Verschraegen CF, Edwards CL, Piamsomboon S, Termrungruanglert W, et al. A phase I study of TNP-470 administered to patients with advanced squamous cell cancer of the cervix. Clin Cancer Res 1997;3(9):1501–5.

[206] Bhargava P, Marshall JL, Rizvi N, Dahut W, Yoe J, Figuera M, et al. A phase I and pharmacokinetic study of TNP-470 administered weekly to patients with advanced cancer. Clin Cancer Res 1999;5 (8):1989–95.

[207] Logothetis CJ, Wu KK, Finn LD, Daliani D, Figg W, Ghaddar H, et al. Phase I trial of the angiogenesis inhibitor TNP-470 for progressive androgen-independent prostate cancer. Clin Cancer Res 2001;7(5):1198–203.

[208] Herbst RS, Madden TL, Tran HT, Blumenschein Jr GR, Meyers CA, Seabrooke LF, et al. Safety and pharmacokinetic effects of TNP-470, an angiogenesis inhibitor, combined with paclitaxel in patients with solid tumors: evidence for activity in non-small-cell lung cancer. J Clin Oncol Off J Am Soc Clin Oncol 2002;20(22):4440–7.

[209] Bernier SG, Lazarus DD, Clark E, Doyle B, Labenski MT, Thompson CD, et al. A methionine aminopeptidase-2 inhibitor, PPI-2458, for the treatment of rheumatoid arthritis. Proc Natl Acad Sci U S A 2004;101(29):10768–73.

[210] Mori S, Yoshikawa H, Hashimoto J, Ueda T, Funai H, Kato M, et al. Antiangiogenic agent (TNP-470) inhibition of ectopic bone formation induced by bone morphogenetic protein-2. Bone 1998;22 (2):99–105.

[211] Hazzard TM, Rohan RM, Molskness TA, Fanton JW, D'Amato RJ, Stouffer RL. Injection of antiangiogenic agents into the macaque preovulatory follicle: disruption of corpus luteum development and function. Endocrine 2002;17(3):199–206.

[212] Rutland CS, Jiang K, Soff GA, Mitchell CA. Maternal administration of anti-angiogenic agents, TNP-470 and Angiostatin4.5, induces fetal microphthalmia. Mol Vis 2009;15:1260–9.

[213] Ashraf S, Mapp PI, Walsh DA. Angiogenesis and the persistence of inflammation in a rat model of proliferative synovitis. Arthritis Rheum 2010;62(7):1890–8.

[214] McCandless SE, Yanovski JA, Miller J, Fu C, Bird LM, Salehi P, et al. Effects of MetAP2 inhibition on hyperphagia and body weight in Prader-Willi syndrome: a randomized, double-blind, placebo-controlled trial. Diabetes Obes Metab 2017;19(12):1751–61.

[215] Burkey BF, Hoglen NC, Inskeep P, Wyman M, Hughes TE, Vath JE. Preclinical efficacy and safety of the novel antidiabetic, antiobesity MetAP2 inhibitor ZGN-1061. J Pharmacol Exp Ther 2018;365 (2):301–13.

[216] Maillard A, Scemla A, Laffy B, Mahloul N, Molina JM. Safety and efficacy of fumagillin for the treatment of intestinal microsporidiosis. A French prospective cohort study. J Antimicrob Chemother 2021;76(2):487–94.

[217] Dezube BJ, Von Roenn JH, Holden-Wiltse J, Cheung TW, Remick SC, Cooley TP, et al. Fumagillin analog in the treatment of Kaposi's sarcoma: a phase I AIDS Clinical Trial Group Study. AIDS Clinical Trial Group no. 215 team. J Clin Oncol Off J Am Soc Clin Oncol 1998;16(4):1444–9.

[218] Tran HT, Blumenschein Jr GR, Lu C, Meyers CA, Papadimitrakopoulou V, Fossella FV, et al. Clinical and pharmacokinetic study of TNP-470, an angiogenesis inhibitor, in combination with paclitaxel and carboplatin in patients with solid tumors. Cancer Chemother Pharmacol 2004;54(4):308–14.

[219] Stadler WM, Kuzel T, Shapiro C, Sosman J, Clark J, Vogelzang NJ. Multi-institutional study of the angiogenesis inhibitor TNP-470 in metastatic renal carcinoma. J Clin Oncol Off J Am Soc Clin Oncol 1999;17(8):2541–5.

[220] Arico-Muendel CC, Belanger B, Benjamin D, Blanchette HS, Caiazzo TM, Centrella PA, et al. Metabolites of PPI-2458, a selective, irreversible inhibitor of methionine aminopeptidase-2: structure determination and in vivo activity. Drug Metab Dispos 2013;41(4):814–26.

[221] Stiede K, Eder J, Anthony S, Conkling P, Fayad L, Petrylak D, et al. Phase I dose escalation safety/tolerance study of PPl-2458 in subjects with non-Hodgkin's lymphoma or solid tumors (poster 139). Eur J Cancer Suppl 2006;4(12):45.

[222] Mita MM, Bendell J, Mita AC, Gordon M, Sachdev J, Carver BJ, et al. SDX-7320 elicits improvements in tumor-related and metabolic biomarkers: results of a phase 1 dose-escalation study in patients with advanced refractory or late-stage solid tumors. Cancer Res 2020;80(16 Supplement). Abstract CT153.

[223] Bendell JC, Shapiro G, Sausville EA, Jones SF, Hilton JF, Shkolny D, et al. A phase 1 first-in-human study of XMT-1107, a polymer-conjugated fumagillol derivative, in patients (pts) with advanced solid tumors. J Clin Oncol Off J Am Soc Clin Oncol 2014;32(15):2526.

[224] Shin SJ, Jeung HC, Ahn JB, Rha SY, Roh JK, Park KS, et al. A phase I pharmacokinetic and pharmacodynamic study of CKD-732, an antiangiogenic agent, in patients with refractory solid cancer. Invest New Drugs 2010;28(5):650–8.

[225] Shin SJ, Ahn JB, Park KS, Lee YJ, Hong YS, Kim TW, et al. A phase Ib pharmacokinetic study of the anti-angiogenic agent CKD-732 used in combination with capecitabine and oxaliplatin (XELOX) in metastatic colorectal cancer patients who progressed on irinotecan-based chemotherapy. Invest New Drugs 2012;30 (2):672–80.

[226] Hughes TE, Kim DD, Marjason J, Proietto J, Whitehead JP, Vath JE. Ascending dose-controlled trial of beloranib, a novel obesity treatment for safety, tolerability, and weight loss in obese women. Obesity (Silver Spring) 2013;21(9):1782–8.

[227] Kim DD, Krishnarajah J, Lillioja S, de Looze F, Marjason J, Proietto J, et al. Efficacy and safety of beloranib for weight loss in obese adults: a randomized controlled trial. Diabetes Obes Metab 2015;17(6):566–72.

[228] Proietto J, Malloy J, Zhuang D, Arya M, Cohen ND, de Looze FJ, et al. Efficacy and safety of methionine aminopeptidase 2 inhibition in type 2 diabetes: a randomised, placebo-controlled clinical trial. Diabetologia 2018;61(9):1918–22.

[229] Carducci M, Wang D, Habermehl C, Bödding M, Rohdich F, Stinchi S, et al. A multicenter, open-label, dose-escalation, first-in-man study of MetAP2 inhibitor M8891 in patients with advanced solid tumours. Ann Oncol 2020;31(abstract 566P):S486.

[230] Shoemaker A, Proietto J, Abuzzahab MJ, Markovic T, Malloy J, Kim DD. A randomized, placebo-controlled trial of beloranib for the treatment of hypothalamic injury-associated obesity. Diabetes Obes Metab 2017;19(8):1165–70.

[231] Malloy J, Zhuang D, Kim T, Inskeep P, Kim D, Taylor K. Single and multiple dose evaluation of a novel MetAP2 inhibitor: results of a

randomized, double-blind, placebo-controlled clinical trial. Diabetes Obes Metab 2018;20(8):1878–84.

[232] Wentworth JM, Colman PG. The methionine aminopeptidase 2 inhibitor ZGN-1061 improves glucose control and weight in overweight and obese individuals with type 2 diabetes: a randomized, placebo-controlled trial. Diabetes Obes Metab 2020;22(7):1215–9.

[233] Sfakianos J, Shore ND, Zhuang J. Phase Ib study: APL-1202 (APL) in combination with bacillus calmette-guerin (BCG) in recurrent non-muscle invasive bladder cancer (NMIBC). J Clin Oncol 2020;38(15) [abstract e17039].

[234] Ye D, Yao X, Wang G, Pu J, Yao X, Zhou F, et al. An oral methionine aminopeptidase II inhibitor for high-risk non-muscle invasive bladder cancer relapsed after intravesical therapies: update of a phase II trial. J Clin Oncol 2017;35(6).

[235] Carducci M, Ding W, Christina H, Matthias B, Felix R, Floriane L, et al. A first-in-human, dose escalation study of the methionine aminopeptidase 2 inhibitor M8891 in patients with advanced solid tumors. Cancer Res Commun 2023. In press.

[236] Dumez H, Gall H, Capdeville R, Dutreix C, van Oosterom AT, Giaccone G. A phase I and pharmacokinetic study of LAF389 administered to patients with advanced cancer. Anticancer Drugs 2007;18(2):219–25.

[237] Molina JM, Goguel J, Sarfati C, Chastang C, Desportes-Livage I, Michiels JF, et al. Potential efficacy of fumagillin in intestinal microsporidiosis due to *Enterocytozoon bieneusi* in patients with HIV infection: results of a drug screening study. AIDS (The French Microsporidiosis Study Group Aids) 1997;11(13):1603–10.

[238] Guruceaga X, Perez-Cuesta U, Abad-Diaz de Cerio A, Gonzalez O, Alonso RM, Hernando FL, et al. Fumagillin, a mycotoxin of *Aspergillus fumigatus*: biosynthesis, biological activities, detection, and applications. Toxins 2019;12(1):7.

[239] EMA. Fumagillin for the treatment of diarrhoea associated with intestinal microsporidial infection. EMA; 2013.

[240] Champion L, Durrbach A, Lang P, Delahousse M, Chauvet C, Sarfati C, et al. Fumagillin for treatment of intestinal microsporidiosis in renal transplant recipients. Am J Transplant 2010;10(8):1925–30.

[241] Arzouk N, Michelon H, Snanoudj R, Taburet AM, Durrbach A, Furlan V. Interaction between tacrolimus and fumagillin in two kidney transplant recipients. Transplantation 2006;81(1):136–7.

[242] An J, Wang L, Patnode ML, Ridaura VK, Haldeman JM, Stevens RD, et al. Physiological mechanisms of sustained fumagillin-induced weight loss. JCI Insight 2018;3(5).

[243] Conteas CN, Berlin OG, Ash LR, Pruthi JS. Therapy for human gastrointestinal microsporidiosis. Am J Trop Med Hyg 2000;63(3–4):121–7.

[244] Yamamoto T, Sudo K, Fujita T. Significant inhibition of endothelial cell growth in tumor vasculature by an angiogenesis inhibitor, TNP-470 (AGM-1470). Anticancer Res 1994;14(1a):1–3.

[245] Slichenmyer WJ, Elliott WL, Fry DW. CI-1033, a pan-erbB tyrosine kinase inhibitor. Semin Oncol 2001;28(5 Suppl 16):80–5.

[246] Kurebayashi J, Kurosumi M, Dickson RB, Sonoo H. Angiogenesis inhibitor O-(Chloroacetyl-carbamoyl) fumagillol (TNP-470) inhibits tumor angiogenesis, growth and spontaneous metastasis of MKL-4 human breast Cancer cells in female athymic nude mice. Breast Cancer 1994;1(2):109–15.

[247] Kudelka AP, Verschraegen CF, Loyer E. Complete remission of metastatic cervical cancer with the angiogenesis inhibitor TNP-470. N Engl J Med 1998;338(14):991–2.

[248] Zukiwski A, Gutterman J, Bui C, Sella A, Ellerhorst J, Tu S, et al. Phase I trial of the angiogenesis inhibitor TNP-470 (AGM-1470) in patients with androgen independent prostate cancer. Proc am soc. Clin Oncol 1994;13(51). Abstract 795.

[249] Kruger EA, Figg WD. TNP-470: an angiogenesis inhibitor in clinical development for cancer. Expert Opin Investig Drugs 2000;9(6):1383–96.

[250] Satchi-Fainaro R, Mamluk R, Wang L, Short SM, Nagy JA, Feng D, et al. Inhibition of vessel permeability by TNP-470 and its polymer conjugate, caplostatin. Cancer Cell 2005;7(3):251–61.

[251] Cornelius P, Petersen JS, Mayes B, Turnquist D, Sullivan K, Anderson-Villaluz A, et al. Preclinical activity of SDX-7320 in mouse models of obesity and obesity-driven cancer. Cancer Res 2018;78(13 Supplement). Abstract 4919.

[252] Akullian L, Stevenson C, Lowinger T, Fram R. Anti-angiogenic and antitumor activity of XMT-1107, a fumagillin-derived polymer conjugate, and its in vivo release product XMT-1191. Cancer Res 2009;69(9 Supplement). Abstract 670.

[253] Kim EJ, Shin WH. General pharmacology of CKD-732, a new anticancer agent: effects on central nervous, cardiovascular, and respiratory system. Biol Pharm Bull 2005;28(2):217–23.

[254] Nasdaq Press Release. Zafgen halts development of beloranib, to cut jobs by ~34%: Zacks, 2016. Available from: https://www.nasdaq.com/articles/zafgen-halts-development-of-beloranib-to-cut-jobs-by-34-2016-07-20.

[255] Mrhar A, Kopitar Z, Kozjek F, Presl V, Karba R. Clinical pharmacokinetics of nitroxoline. Int J Clin Pharmacol Biopharm 1979;17(12):476–81.

[256] Cision Press Release. Asieris' APL-1501 approved for phase I clinical trial in Australia, 2020. [updated 10/28/2020. Available from: https://www.prnewswire.com/news-releases/asieris-apl-1501-approved-for-phase-i-clinical-trial-in-australia-301162402.html.

Chapter 4.3

1-Deoxy-D-xylulose 5-phosphate reductoisomerase, the first committed enzyme in the MEP terpenoid biosynthetic pathway—Its chemical mechanism and inhibition

Wen-Yun Gao and Heng Li
College of Life Sciences, Northwest University, Xi'an, PR China

Terpenoid natural products are found essentially in all forms of life. More than 45,000 naturally occurring terpenoids comprise physiologically important compounds such as vitamins A and D, cholesterol, steroid hormones, the side chain of chlorophylls, and carotenoids and medically significant constituents like taxol, artemisinin, and ginkgolides, to mention just a few. The pioneering studies of Bloch, Cornforth, Lynen, and many others by using yeast and animal cells have led to the elucidation of the classic mevalonate (MVA) pathway of terpenoid biosynthesis [1–3]. The research carried out independently by Arigoni, Rohmer, Zenk, and coworkers more than 20 years ago made the discovery by studies on certain bacteria that there is a second biochemical route of terpenoid biosynthesis, namely the 2-methylerythritol-4-phosphate (MEP) pathway [4,5]. The succeeding investigations of terpenoid biosynthesis have indicated how widespread the MEP pathway is and are even suggesting the new pathway may be used in nature much more frequently than the MVA pathway. There is a large dispersion of the non-MVA pathway among bacteria and also a dominant distribution of it in higher plants. It has been shown explicitly the coexistence in two different cell compartments of both biosynthetic routes for terpenoid biosynthesis in higher plant cells with the MVA route being present in the cytosol and the alternative pathway being confined to the plastids. Further experiments confirmed that in higher plants, the MVA pathway affords sterols and the alternative pathway furnishes a wide variety of hemiterpenes, monoterpenes, diterpenes, and tetraterpenes [4,6]. Although the MEP pathway is dominating in terpenoid production in higher plants and bacteria including human pathogenic organisms, it is absent in animals and human beings. Moreover, the pathway is essential for the survival of the causative agents because disruption of it is lethal. This characteristic makes the MEP pathway a promising target for screening new antibiotics with unique mode of action [4,5,7].

The MEP terpenoid biosynthetic pathway starts from the condensation of one molecule each of pyruvate and D-glyceraldehyde 3-phosphate (D-GAP) to generate 1-deoxy-D-xylulose-5-phosphate (DXP, **1**), which is catalyzed by DXP synthase (DXS) in the presence of thiamine diphosphate (ThDP) and a bivalent metal ion (Mg^{2+}, Mn^{2+}, etc.). Subsequently, the straight-chain C5 compound DXP is converted into MEP (**2**) by DXP reductoisomerase (DXR) under aid of a bivalent metal ion (Mg^{2+}, Mn^{2+}, and Co^{2+}) using NADPH (β-nicotinamide adenine dinucleotide 2′-phosphate reduced form) as a hydrogen donor, creating a branch-chain C5 carbon skeleton, which remains unchanged till the final products of the pathway, namely the two isoprenyl diphosphates. In the ensuing steps, the intermediate MEP is transformed in five consecutive reactions mediated by the proteins ispD, E, F, G, and H, respectively into the universal building blocks of all the natural terpenoids, isopentenyl diphosphate (IPP) and its isomer dimethylallyl diphosphate (DMAPP) (Fig. 1) [4,6].

Among all the enzymes implicated in the MEP pathway, the first committed enzyme DXR has drawn the most attention because (i) it is a vital enzyme for natural terpenoid biosynthesis; (ii) it possesses unique druggable properties [4,8]. Therefore, DXR has been accepted as one of the most promising targets in the search for antibiotics and antimalarials and intensive investigations have been carried out to elucidate its catalytic mechanism and seeking for its efficacious inhibitors, which resulted in the setup of the

Metalloenzymes. https://doi.org/10.1016/B978-0-12-823974-2.00026-7

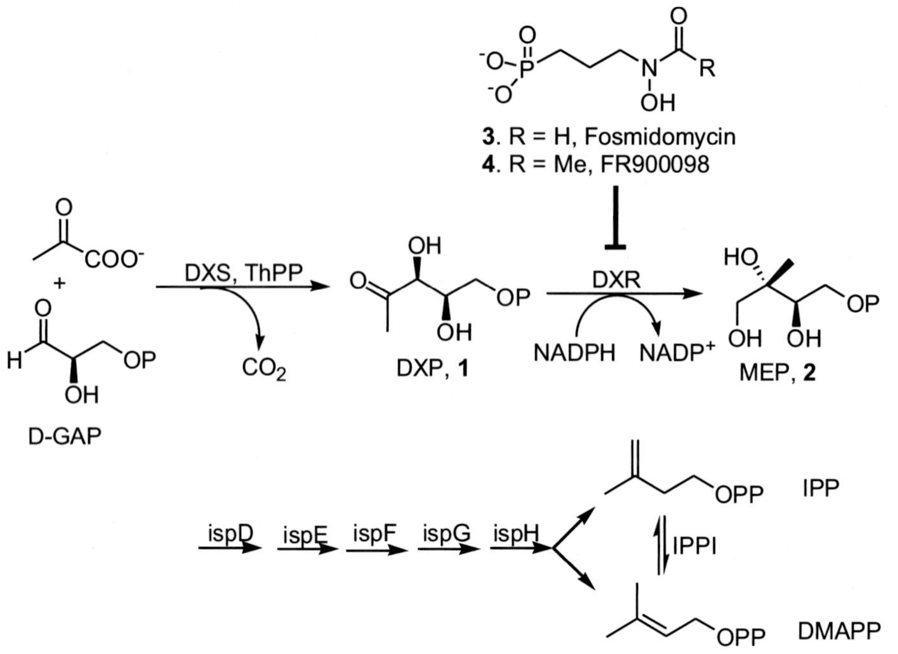

FIG. 1 The MEP terpenoid biosynthetic pathway and DXR inhibitors.

retroaldol-aldol sequence as its mode of action and the discovery of fosmidomycin (**3**) and its congoner FR900098 (**4**) as its potent, specific inhibitors [9–11]. Up to date, a suite of DXR proteins from various sources including microorganisms, algaes, and higher plants have been well-documented [4,12,13]. In a recent study, a DXR-like (DRL or type II DXR) enzyme was identified in *Brucella abortus*, which utilizes the MEP pathway to synthesize terpenoids but lacks the gene encoding DXR. Although in-depth studies have indicated that DRL possesses no significant sequence identity with DXR and its crystal structure shows only slightly structural similarity to DXR and uncovers a distinct architecture of the active site, it indeed holds identical functions with DXR: it catalyzes the same biochemical reaction as DXR, its activity is inhibited by fosmidomicin, and it can compensate a DXR-deficient *Escherichia coli* strain [14–16]. These results reveal how finite is still our understanding about the biosynthesis of terpenoids and also DXR protein. Extensive research should therefore be performed to expand our knowledge on the topic. This article is going to provide an overview on the DXR enzyme with emphasis on its chemical mechanism, substrate binding, catalytic cycle, and inhibition.

1 Chemical mechanism and intermediary of DXR

During exploration of the chemical mechanism of DXR mediated conversion of DXP to MEP, three general mode of actions were proposed, an α-ketol rearrangement (the route A in Fig. 2), a retroaldol-aldol sequence (the route B in Fig. 2), and sequential 1,2-hydride and 1,2-methyl shifts (the route C in Fig. 2) [17–21]. The last mechanism was deduced from the action of ketol-acid isomeroreductase (KARI), one of the key enzymes in the branched-chain amino acid biosynthesis catalyzing the conversion of α-hydroxy-β-ketoacids into 2,3-dihydroxyacids through isomerization and then reduction, namely the alkyl (methyl or ethyl) migration in the leading step and the NADPH reduction in the second [22]. This prediction has been readily denied because it is conflicting to the conclusions drawn from previous isotopic labeling experiments [23,24]. The first mechanism (α-ketol rearrangement) was suggested sooner after the setup of DXP formation from condensation of pyruvate and D-GAP because the early $^{2}H/^{13}C$ incorporation assays had disclosed that an intramolecular carbon scaffold isomerization occurs in the course of DXP transformation to IPP/DMAPP, in which the straight-chain DXP is altered into branched-chain MEP [17,23,24]. This proposed rearrangement begins with deprotonation of the hydroxyl at C-3 of DXP, which triggers shift of the phosphate-bearing C2 moiety, namely 2-hydroxy-1-phosphoethyl subunit to the C-2 carbonyl, generating the intermediate 2-methylerythrose-4-phosphate (MEsP, **5**). The following reduction of the compound by NADPH gives the product MEP. Although no evidence supports the proposal, the possible intermediary of compound **5** has determined because it could be recognized by DXR and reduced to MEP by NADPH [20]. The retroaldol-aldol sequence is initialized by oxidation of the 4-hydroxyl of DXP, which results in the cleavage the C3–C4 bond of

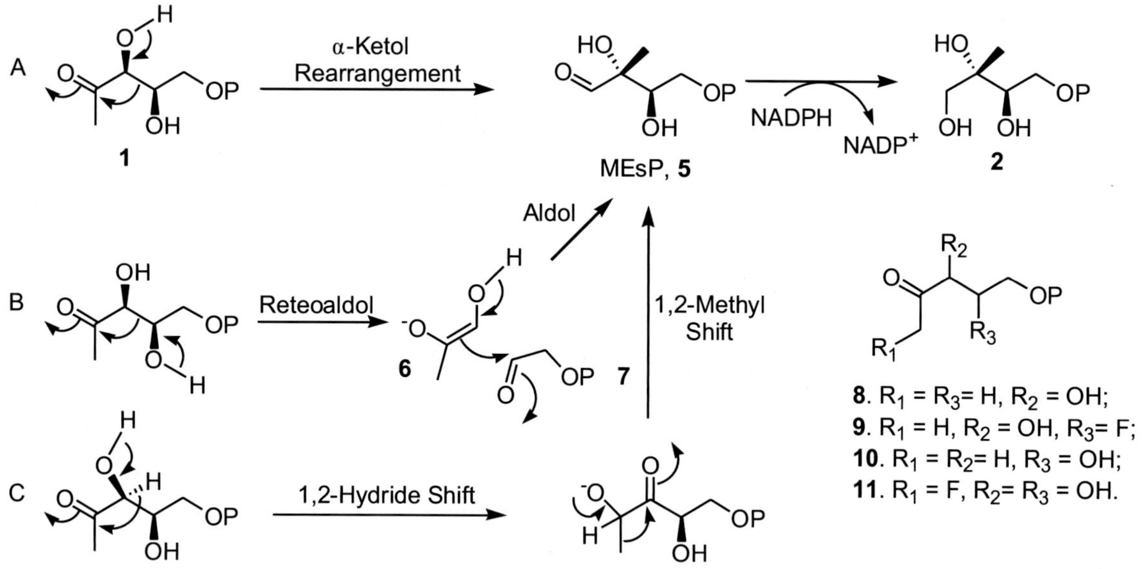

FIG. 2 Proposed DXR mechanisms and some DXP analogs (**8–11**). (A) α-Ketol rearrangement; (B) retroaldol-aldol sequence; (C) sequential 1,2-hydride and 1,2-methyl shifts.

DXP in a retro aldol manner, affording two fragments, the enediolate of hydroxyacetone **6** and glycolaldehyde phosphate **7**, which then reconnect in an aldol reaction via C2 of **6** attacking the carbonyl group of **7** to produce a new C—C bond and form the same aldehyde intermediate **5** as in the upper route. The subsequent NADPH reduction of **5** furnishes MEP [20]. Enzymes catalyzing reactions in a retroaldol-aldol manner in the place of an α-ketol rearrangement have also been found in literature, for example ribulose-5-phosphate 4-epimerase that mediates the interconversion of L-ribulose-5-phosphate and D-xylulose-5-phosphate by inverting the configuration at C-4 follows the retroaldol-aldol mechanism [25,26].

To distinguish between the first two mechanisms, several DXP analogs were prepared and tested on DXR. The results indicated that 4-deoxy-DXP (**8**), 4-fluoro-DXP (**9**) (Fig. 2), and C-4 epimer of DXP are all not substrates of DXS but its inhibitors [20,27,28]. These observations seemed to support the retroaldol-aldol mechanism because the C-4 hydroxyl of DXP is key to it but would not take a direct role in the α-ketol rearrangement. The data, however, obtained from 3-deoxy-DXP (**10**) was contradictory because it was not a substrate of the enzyme either but an inhibitor [20,28]. Therefore, it is difficult to exclude any of the mechanisms based on only the evidence. Stronger proof supporting the retroaldol-aldol pathway was deduced from the data obtained from the C-1 fluorinated DXP analog, namely 1-fluoro-DXP (**11**). Poulter and Liu groups independently synthesized the compound and assayed its competence as a substrate of DXR [27,29]. Because the isomerization step is rate-limiting in the overall reaction of DXR [29], it has been predicted **11** may behave differently from DXP in an α-ketol rearrangement since the electron-withdrawing effect of the fluorine substituent would influence the formation of the transition state. But the kinetic data of **11** show that it is a slightly weaker substrate than DXP with a similar K_m and fourfold lower k_{cat} [27] or almost a same substrate as DXP with comparable k_{cat} (37 and $21\,s^{-1}$, respectively) and single-turnover rate ($k_{max} = 61\,s^{-1}$ for both) [29]. This result prompted the authors to support the retroaldol-aldol sequence in the DXR reaction. The problem is, however, the interaction between the analog **11** and DXR may be largely different from that between DXP and DXR because introduction of a fluorine atom changes the basic parameters of DXP such as its geometry and electrostatics at the transition state, and the data obtained in this way may not reflect the real behavior of DXP. Further explorations are still necessary to differentiate between the two routes of rearrangement.

The other direction to solve the problem is to check the intermediacy of DXR reaction by directly locating the possible intermediates. Experiments designed to trap MEsP were carried out even earlier than the suggestion of retroaldol-aldol mechanism since it was also proposed to be the intermediate in the α-ketol rearrangement [18,19]. Koppisch and colleagues followed the DXR reaction by proton NMR spectroscopy but failed to detect any signals produced by aldehyde and aldehyde hydrate protons, based on which the authors estimated that MEsP constituted less than 0.2% of the sum of MEP and DXP. Further tries to trap MEsP with $NaBH_4$ reduction or *N*-methylnitrosohydrozine derivatization gave negative results as well [30]. Hoeffler and coworkers traced the DXR reaction by ^{13}C NMR utilizing [1-^{13}C]DXP as a substrate to improve the sensitivity of the detection. Unfortunately, they could not get direct evidence for the formation of MEsP in both normal DXR reaction or the reaction without addition of NADPH [20]. Based on the results, the authors concluded that either the isomerization and NADPH reduction proceed synergistically and MEsP is never really formed or it is a transient intermediate strictly confined by the enzyme till NADPH reduction. A reasonable deduction for the former conclusion is that the exogenous MEsP can be reduced by DXR under assistance of NADPH [20] is only a side reaction of the enzyme. This situation is quite similar to that of KRAI catalysis in which the hypothetical intermediate, 3-hydroxy-3-methyl-2-ketobutyrate, resulted from the methyl migration of the substrate acetolactate has not been directly detected [22,31].

Attempts to verify the intermediacy of **6** and **7** (Fig. 2) were performed upon the retroaldol-aldol mechanism was raised [20]. Paralleling the detection of MEsP by ^{13}C NMR using [1-^{13}C]DXP as a substrate, Hoeffler et al. also tried to identify formation of the putative intermediates **6**, no signal correlating with [3-^{13}C]hydroxyacetone appeared. Moreover, they checked as well the activity of compounds **6** and **7** on DXR at concentrations up to 1 mM in the presence of NADPH, no drop of UV signal at 340 nm was observed, showing no MEP formed. In a related experiment, Fox et al. incubated [^{32}P]glycolaldehyde phosphate **7**, hydroxyacetone **6**, and NADPH with DXR but also failed to trap ^{32}P-labeled DXP or MEP [29]. These result manifests that **6** and **7** do not seem to be recognized by DXR and DXR does not have aldolase activity.

A later investigation performed by Lauw and coworkers extensively explored the intermediary of DXR reaction by various stable isotope experiments carried out under conditions of maximal stringency [32]. The results showed when a mixture of [1-^{13}C]- and [3-^{13}C]MEP were used as substrates and a large amount of DXR was added, no fragment exchange was detected by ^{13}C NMR in the reverse direction after the reaction had been followed over thousands of cycles. In addition, exogenous **6** and [1,2-$^{13}C_2$]glycolaldehyde phosphate **7**, the two putative intermediates expected from the retro-aldol cleavage, were not recognized by DXR because no [3,4-$^{13}C_2$]MEP could be obtained, even in the presence of very high concentrations the enzyme and very high concentrations of **6** (>0.24 M). Similarly, the authors also found that when [1,3,4-$^{13}C_3$]MEP reacted with **6**, isotope washout was not detectable. These data seem to confirm that DXR does not have any aldolase activity and support the α-ketol mechanism; this phenomenon, however, would be possible in the context of the retro-aldol/aldol mechanism given the two putative fragments are tightly restrained in the active site of the enzyme during isomerization.

Introduction of isotopes does not alter the basic properties of a compound, but can change its reactivity, the rate difference between the unlabeled and labeled substrate is called kinetic isotope effects (KIEs). Therefore, determination of KIEs is a powerful tool in mechanism clarification. Secondary KIEs are especially useful for verifying changes in hybridization at a carbon atom [33]. In the case of DXR catalysis, hybridization changes at carbon atoms are involved in both of the putative mechanisms, thus measurement of the α-secondary KIEs employing unlabeled and [3-^2H]- or [4-^2H]-DXP as substrates can provide insights of the reaction process. If the DXR reaction goes through the α-ketol rearrangement, C-3 of DXP rehybridizes from sp^3 to sp^2, leading to a normal KIE (>1), while C-4 remains its sp^3 hybridization, yielding a unit KIE (=1). In contrast, if the alternative mechanism works, both C-3 and C-4 of DXP change hybridization from sp^3 to sp^2 and show normal KIEs accordingly. A prerequisition for discrimination of the two mechanisms on KIEs is that the isomerization step in DXR reaction should be at least partially rate limiting, which has been set up for DXRs in early research [21,29].

Two groups finished stereoselective syntheses of DXP, [3-^2H]-, and [4-^2H]-DXP and used them to measure the α-secondary KIEs of the DXR reaction [9,34]. The data acquired by Wong and Cox displayed inverse $^{D}V_{max}$ and $^{D}(V_{max}/K_m)$ for both [3-^2H]DXP (0.56 and 0.92) and [4-^2H]DXP (0.62 and 0.86), respectively. Based on the results, the authors trusted that the retroaldol-aldol sequence is operative in DXR reaction other than the α-ketol rearrangement. Meanwhile, they also suggested that the aldol step, which contains rehybridization from sp^2 to sp^3 at both C-3 and C-4 positions should be rate limiting because of the inverse KIEs [34]. These data were argued by Munos and collaborators that (i) the ^{D}V values measured by Wong and Cox exceed the range of α-secondary KIEs; (ii) the $^{D}(V_{max}/K_m)$ ought to be normal for the two ^2H-labeled DXPs because both C-3 and C-4 are sp^3 hybridization in the reactant state; (iii) the unlabeled or labeled substrate DXP acquired from chiral organic synthesis could be inevitably contaminated by a small amount of C-4 epimer that has been elucidated as an inhibitor of the enzyme [28], and the KIE values based on these compounds would be confounding [9]. On the basis of the above analysis, Munos et al. redetermined the KIEs employing the equilibrium perturbation method developed by Cleland [35] to circumvent the issue confronted by Wong and Cox in their KIE measurement by direct comparison of reaction rates. The normal α-secondary KIEs for both labeled substrates (1.04 for [3-^2H]DXP and 1.11 for [4-^2H]DXP) strongly favor the retroaldol-aldol mechanism.

In the following investigations, Manning and coworkers established a highly precise ^1H-detected 2D [^{13}C,^1H]-HSQC NMR method to determine ^{13}C KIEs of several proteins including DXR of *Mycobacterium tuberculosis* [10]. This is a competitive assay in which the light (^{12}C) and heavy (^{13}C) isotopologues react under identical conditions within the same sample, leading to the superior accuracy and precision of the protocol, which is key in transition state analysis. The KIEs can be calculated using the equation given below:

$$R/R_0 = (1 - F)^{(1/\mathrm{KIE})-1}$$

where $R = [^{13}C\text{-DXP}]_t/[\text{DXP}]_t$ and R_0 is the ratio at time $t = 0$ min. F is the fraction of conversion of the substrate. Utilizing this approach, Manning et al. determined [2-^{13}C]-, [3-^{13}C]-, and [4-^{13}C]-DXP KIEs of 1.0031, 1.0303, and 1.0148, respectively. The latter two data are in consistence with the primary ^{13}C KIEs connected with the cleavage of C3–C4 bond during the retroaldol step. In contrast, the 2-^{13}C KIE that is 10-fold less than that of 3-^{13}C is identical with a secondary ^{13}C KIE, cleanly displaying that the DXR reaction is limited by the retroaldol step generating the intermediates **6** and **7**. In addition, these KIE data disfavor the α-ketol rearrangement because all the three positions should exhibit primary ^{13}C KIEs if it works.

So far the retroaldol-aldol mechanism has been unambiguously verified for DXR, but the supposed intermediates in retroaldol-aldol sequence, namely **6** and **7** produced in retroaldol step and **5** generated in aldol step are still pending because none of them has been captured directly or indirectly in any DXR experiments. Furthermore, the observation that DXR cannot convert exogenous **6** and **7** into MEP or DXP needs a reasonable interpretation because in the context of the retroaldol-aldol mechanism DXR ought to possess aldolase activity.

2 Substrate-binding mode

The crystal structure of *E. coli* DXR in a ternary complex with the inhibitor fosmidomycin **3** and Mn^{2+} has shown the N-formyl hydroxylamine group in **3** provides two oxygen ligands for the Mn^{2+}, which is present in a distorted octahedral coordination sphere [36]. In a subsequent study, Yajima and collaborators [37] reported the three- dimensional structure of *E. coli* DXR in complex with NADPH, **3**, and Mn^{2+} in which **3** coordinates to Mn^{2+} in the identical way with that in the ternary complex. The fosmidomycin-Mn^{2+} chelating mode is therefore deemed to mimic the substrate binding of the carbonyl oxygen and the C3 hydroxyl group in DXP with the enzyme-bound divalent metal ion M^{2+} (Mg^{2+} or Mn^{2+}, C2-C3 binding mode) because DXP can be superimposed onto the inhibitor [36]. These observations were soon confirmed by a following research in which the crystal structures of *M. tuberculosis* DXR with ligands in different combinations including the quaternary complexes of the substrate DXP, Mn^{2+} and NADPH and **3**, $Mn^{2+}$and NADPH were studied and compared. The results show that the two quaternary complexes are highly similar overall [38]. As a consequence, the C2-C3 binding mode of DXP has been generally accepted and used to explain the proceedings of the enzymatic reaction: Initializing deprotonation of the C-4 hydroxyl group of the substrate DXP leads to cleavage of the C3–C4 bond and generation of the retroaldol intermediates **6**, which is stabilized by coordination with M^{2+}, and **7**, which is free from chelation. The ensuing reunion of **6** and **7** with C2 of **6** attacking the aldehyde group of **7** in the aldolization reaction results in the formation of a new C—C bond between carbon atoms C2 and C4 (of DXP) and produces the intermediate **5**. Subsequent reduction of **5** by NADPH gives the product MEP (Fig. 2) [4,5,7,8,12].

Although the crystal structures of DXR complexed with NADPH, M^{2+} and DXP/**3** have furnished many insights into the architecture of the active site of the enzyme and disclosed the similar chelating manners of DXP and the inhibitor to the DXR bound M^{2+}, which represents the C2-C3 binding mode of DXP, this information may not accurately reflect the actual DXP binding mode because the crystals of the complexes were prepared normally under mildly acidic conditions (pH 5.0–5.8) [36–38]. This weakly acidic condition for crystallization differs largely from the functional pH range of the enzyme, which is around pH 7–8 [29,30] and at a pH lower than 6.0, DXR from cyanobacterium *Synechocystis* sp. only shows 50%–60% activity [39]. Moreover, under weakly acidic conditions the protonation state of the clustered acidic residues in the active site of the enzyme would be distinct from that in the enzyme under physiological conditions, degenerating its metal-binding affinity and blocking the formation of the octahedral geometry of the active site residues, which is essential for M^{2+} binding [13,21,40].

More importantly, after perusal of the C2–C3 substrate binding mode, one can realize that this format does not properly interpret the conversion process of the substrate DXP to the product MEP in light of the determined retroaldol-aldol mechanism for DXR: (i) the retro-aldol rearrangement is initiated by deprotonation of the C4 hydroxyl group of DXP, which is assisted by electrostatic interaction between the M^{2+} and the OH group [21,41], but the C2–C3 substrate binding mode could not provide such electrostatic force because the C4 hydroxyl does not chelate with M^{2+}; (ii) the intermediate **6** produced in the retro-aldol step might be stabilized by bidentate coordination with M^{2+}, but its nucleophilicity could be reduced because of the electrostatic interaction. This disfavors the ensuing aldolization reaction in which **6** acts as a nucleophile. Meanwhile, intermediate **7** does not coordinate with the enzyme bound M^{2+}, so its carbonyl group is unactivated. These factors could increase the barrier of the subsequent aldol step; (iii) it has been proved that the intermediates **6** and **7** are tightly restrained by the enzyme [20,32,41], in the C2–C3 binding mode, however, only **6** shows the strongly bound state via coordination with M^{2+}, the aldehyde **7** does not exhibit any strict limitation at all. Therefore, at least **7** derivatizing from isotope-labeled

DXP or MEP ought to be interexchangeable in this binding mode. The deduction contradicts the reported results [32]; (iv) In this binding mode, extraneous **6** and **7** should be converted by DXR to DXP under the assistance of $NADP^+$ or to MEP in the presence of NADPH, because they would bind respectively to the enzyme active sites and then recombine to generate the aldolization product **5**, just like that exogenous **5** can be recognized by DXR and reduced by NADPH [20]. This deduction is also opposite to the published data that showed DXR has not exhibited any aldolase activity (Fig. 2) [20,29,32].

Besides, there were also studies focused on the significance of the C3 and C4 hydroxyl groups of DXP for the chelation of M^{2+}. By preparation and examination of compounds **8** and **10** (Fig. 2), Hoeffler and colleagues verified that the two hydroxyls are apparently essential in the retro aldol reaction in which they are possibly the sites for coordination of a divalent cation such as Mg^{2+} or Mn^{2+}, most likely behaving as Lewis acids prompting the deprotonation step in isomerization reaction [20]. In an attempt to get more details of the DXR mechanism, Wong et al. prepared 3- and 4-fluorinated substrate analogs and tested their competence as substrates upon incubation with DXR. The results indicated that the two compounds are both noncompetitive inhibitors of DXR other than its substrates, confirming that both hydroxyls at C-3 and C-4 of DXP are key to DXR catalysis [27]. These data imply that a C3–C4 DXP binding mode may be an alternative route to explain the interaction between DXR and its substrate DXP.

To discriminate which of the two binding modes works in the reaction catalyzed by DXR, we carried out a couple of ^{18}O incorporation experiments to acquire more insights. The design of the assay was based on the perusal of this intriguing rearrangement-reduction cascade: three carbonyl compounds including substrate DXP, the retro-aldol intermediate **7**, and the aldol product **5** and one potential carbonyl compound, the retro-aldol intermediate **6** that is the enediolate of hydroxyacetone are involved. Among these compounds, **5** and **7** are aldehydes (Fig. 2), which can exchange their carbonyl oxygens with bulky solvent more quickly than DXP and **6** that are ketones. Furthermore, each of the compound possesses an electron-withdrawing substituent at its α-position, which can accelerate the exchange [42]. In light of this analysis, we predicted ^{18}O should be incorporated into the final product MEP via solvent exchange if DXP is incubated with DXR in presence of Mg^{2+} and NADPH in buffered $H_2^{18}O$ [41]. The results revealed that 67% ^{18}O-enrichment could be detected in C2 of MEP that originates from C2 (carbonyl) of DXP, but no incorporation in C1 and C3, which originate respectively from C3 and C4 of DXP. Comparison of the ^{18}O exchange rate of DXP with its conversion rate indicated that the high ^{18}O-enrichment in C2 of MEP could be resulted from both prebinding and postbinding exchange, implying that the carbonyl of DXP was free from coordination with the DXR-bound metal ion during conversion. The lack of isotope exchange at C1 and C3 of MEP displayed that the C3 and C4 hydroxyls of DXP could chelate the DXR-bound metal ion, leading to the carbonyls of the intermediates **5** and **7** coordinating with the metal ion as well, which prevented the carbonyls from exchanging with the bulky solvent. These data prompted the authors to conclude, in combination with the early investigations, which have emphasized the essentiality of C3 and C4 hydroxyls of DXP for metal ion chelation [20,27], that DXP is more likely to bind the metal ion through a C3–C4 binding mode rather than the generally suggested C2–C3 binding mode [10,36,37].

Employing the newly suggested C3–C4 DXP binding model, the transformation of DXP to MEP can be properly explained in terms of the accepted retroaldol-aldol mechanism as follows (Fig. 3). (i) Chelation of the adjacent hydroxyls of DXP with M^{2+} promotes their acidity and eases oxidation of the C4 hydroxy, which could be essential to initiate the retro-aldol rearrangement; (ii) the respective coordination of the intermediates **6** and **7** with M^{2+} as monodentate ligands tightly confines them accordingly [8,20,32]; (iii) although the fragment **6** is only partially stabilized as a monodentate ligand, its C2 may possess more

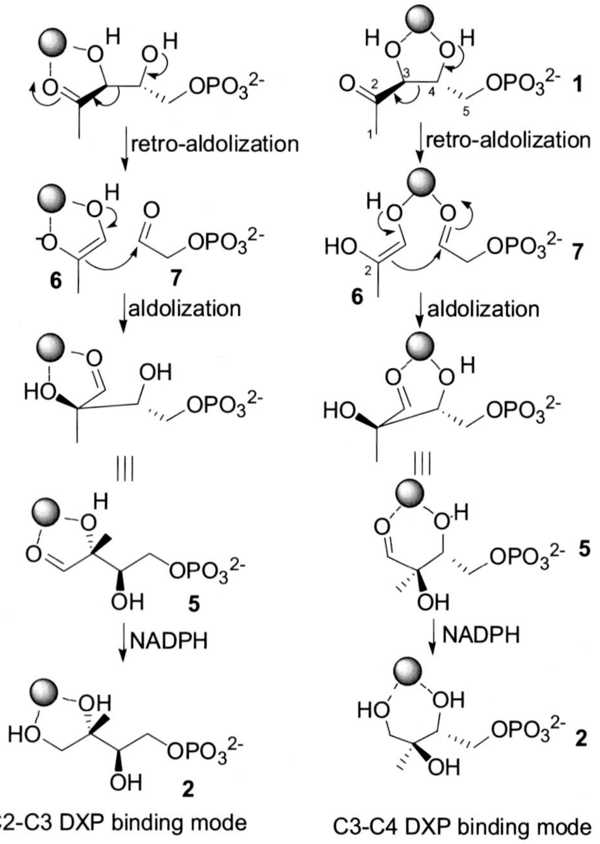

FIG. 3 The conversion of DXP to MEP in C2–C3/C3–C4 DXP binding modes.

carbanionic property [43] than it would as a bidentate ligand in the alternative binding mode, making it a better nucleophile. Furthermore, upon coordination with M^{2+}, fragment **7** may be induced to generate a partial positive charge on its carbonyl C atom, increasing its electrophilicity. More importantly, the separate chelation of **6** and **7** with M^{2+} could create a pseudo-single molecule and thus make the ensuing aldolization pseudo-intramolecular. These active factors could be beneficial for the aldol step; (iv) the restriction of the carbonyl of **7** would fully suppress its hydration, prohibiting ^{18}O introduction at this position; (v) In the aldol step, the formation of the coordinator between **5** and M^{2+} could on the one hand activate the carbonyl of **5**, facilitating hydride transfer from NADPH, it could orient the intermediate on the other. Moreover, this coordination also stops the ^{18}O isotope exchange, which would occur in the carbonyl of **5** if there is no such an action.

Perhaps more significantly, the C3–C4 DXP binding mode provides us with a powerful model to account for the discrepancies about DXR catalysis within the accepted retro-aldol/aldol sequence. The major opposite opinions on the retro-aldol/aldol sequence of DXR lie in (i) DXR cannot recognize exogenous retro-aldol intermediates **6** and **7**, which leads to the view that the protein does not possess any aldolase activity [20,29]; (ii) the retro-aldol intermediates **6** and **7** from differently isotpe-labeled substrates cannot exchange with each other, showing that a retroldol-aldol reaction sequence for DXR would have to proceed with very stringent fragment containment. But how fragment **7** is tightly confined remains unclear in the C2-C3 DXP binding mode [32]. These observations can be explicitly clarified in the context of the retro-aldol/aldol sequence using the C3–C4 DXP binding mode [41]. The two intermediates **6** and **7** generated from the retroaldol step both chelate with the metal cation, forming a pseudo-single molecule transition state and making the subsequent aldolization reaction pseudo-intramolecular. That is to say the aldolase activity of DXR can only be observed under the unique circumstance that the pseudo-single molecule transition state could form. Exogenous **6** and **7** cannot be reunited by aldolization reaction to form **5** [20,29] because the two fragments and the DXR-bound M^{2+} actually constitute a termolecular reaction system in which the simultaneous coordination of **6** and **7** with M^{2+} to form a pseudo-single molecule would be infrequent. That the fragments are not exchanged between the different DXP molecules and how **7** is stringently contained [32] can also be explained by the formation of the pseudo-single molecule transition state preceding the aldolization reaction. Externally added **5** can be converted by DXR to MEP in the presence of NADPH or DXP in the presence of $NADP^+$ [20] because the intact molecule of **5** could readily chelate M^{2+} and form an active complex DXR-M^{2+}-**5** (see catalytic cycle (Fig. 4)).

3 DXR catalytic cycle

Early studies on DXR have showed that the enzyme is divalent cation dependent and its activity is maximal with Mn^{2+} or Co^{2+}, whereas Mg^{2+} is slightly less efficacious. Because the intracellular concentration of Mg^{2+} is much higher than that of Co^{2+} or Mn^{2+}, it is believed that Mg^{2+} is the normal cofactor of DXR in vivo. No effect is observed

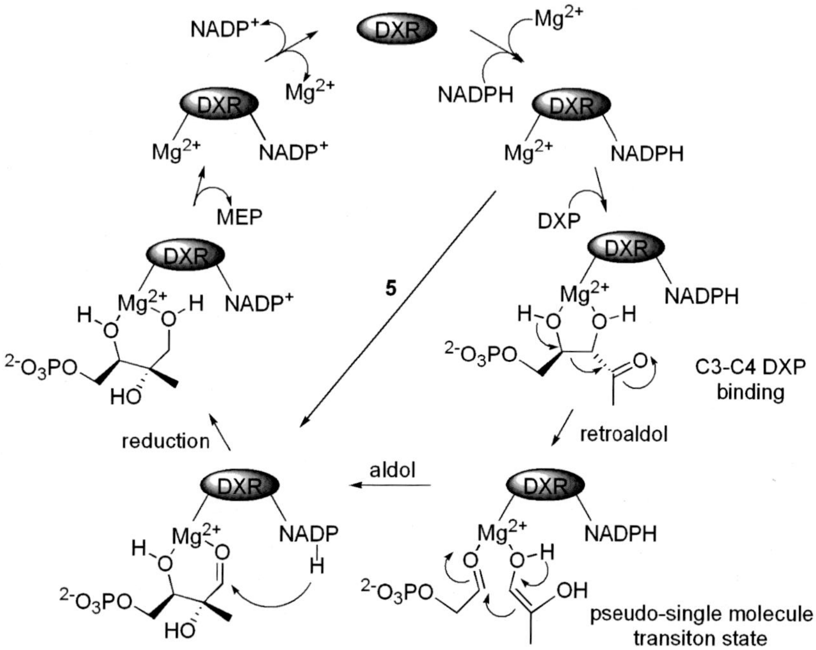

FIG. 4 The proposed DXR catalytic cycle for the conversion of DXP to MEP.

with other metal ions such as Ca^{2+}, Cu^{2+}, Fe^{2+}, Ni^{2+}, and Zn^{2+} [18,19,21,30]. Further investigations on *E. coli* DXR implied an ordered mechanism for the enzyme with NADPH and the divalent cation binding prior to substrate DXP and after reduction, the final product MEP is released before the discharge of $NADP^+$ [20,30]. Interestingly, the experiments carried out on *M. tuberculosis* DXR suggested a random binding of NADPH and DXP and an ordered product release, with $NADP^+$ preceding MEP [21,44]. Research performed independently by several groups also displayed that all the reaction segments, including the retroaldol intermediates enolate of hydroxylacetone **6** and glycolaldehyde phosphate **7** and the aldol product MEsP **5** remain tightly combined to DXR before NADPH reduction [20,32,41]. Explorations on the stereochemistry of the NADPH reduction step of DXR reaction indicated that the enzyme is a class B dehydrogenase that delivers the pro*S* hydride from NADPH [13,21,45,46]. Additional research verified that the hydrogen at C3 of DXP becomes the pro*S* hydrogen at C1 of MEP and the C1 pro*R* hydrogen of MEP is from NADPH [45,47]. Based on these results, it was established that the hydride attack is from the *Re* face of the intermediate aldehyde **5**. In accordance to these results and the C3–C4 DXP binding mode, a catalytic cycle for the isomerization/reduction steps mediated by DXR was proposed using *E. coli* DXR as a model (Fig. 4) [41]: NADPH and the metal cation Mg^{2+} bind first to the enzyme, followed by combination of DXP through coordination of its C3 and C4 hydroxyls with the DXR-bound Mg^{2+}. Subsequent oxidation of the C4 hydroxyl triggers the retro-aldol cleavage of the bond between C3 and C4, producing the two putative fragments **6** and **7** that coordinate respectively to the metal cation, which leads to the strict restraint of them. In the ensuing aldolization reaction, the two intermediates reconnect stereoselectively, affording two new chiral centers in the proposed aldehyde phosphate product **5** whose carbonyl oxygen atom and 3-hydroxyl chelate the Mg^{2+}, which causes the tight confinement of the intermediate as well. The reduction of **5** by hydride addition from the C4 pro-*S* of NADPH to the *Re* face of the aldehyde generates the final product with the correct stereochemistry [21,45–47]. The subsequent yield of MEP and further ejection of $NADP^+$ and Mg^{2+} end up the catalytic cycle. Because the DXR reaction is freely reversible, the catalytic cycle of the reverse reaction would occur in the opposite direction, although the equilibrium largely favors the formation of MEP [20,32].

4 DXR inhibitors

4.1 Analogs of fosmidomycin and FR900098

Fosmidomycin (FSM, **3**) and FR900098 (**4**), two close structural analogs of DXP but each holds a phosphonate moiety instead of the phosphate moiety of the DXR substrate, were isolated from *Streptomyces lavendulae* and *S. rubellomurinus*, respectively, and showed activity against many Gram-negative bacteria [48,49]. They were evaluated for treatment of urinary tract infections in 1980s but was abandoned at Phase II clinical trial due to their less than ideal pharmacokinetics [48–50]. A few years later, they were demonstrated to inhibit the biosynthesis of bacterial isoprenoids, but their target remained yet unclear [51]. Sooner after the identification of the alternative pathway for the boisynthesis of terpenoids and characterization of DXR, these phosphonic acid-containing antibiotics were proven to specific inhibitors of the protein [52,53]. The real breakthrough was the experiments performed by Jomaa and coworkers, which showed that **3** and **4** can suppress the MEP terpenoid biosynthetic pathway of *Plasmodium* through inhibiting its DXR and demonstrated that they could cure *Plasmodium vinckei* infected mice [11]. The finding triggered substantial research interest on the two compounds and plenty of investigations have been carried out to explore their drugability. FSM and FR900098 have shown safety, well-tolerance, and favorable mechanism of action as an antimalarial agent, but recrudescence of patients was a frequent problem [54,55]. Moreover, both the compounds possess high polarity because the phosphonate moiety exists mainly in a dianion form at physiological pH, which results in the suboptimal pharmacokinetic properties such as a plasma half-life of less than 2 h and a gastrointestinal absorption rate of only about 20%–40% [56,57].

The drawbacks of **3** and **4** have largely crippled their market introduction, but their huge potential has prompted chemists and pharmacologists to improve their drug-like properties through structural modification. From the view of the chemical structures of the two compounds and studies on their structure-activity relationship [4,58], their structural modification could be centered on four aspects: (i) synthesis of prodrugs of **3/4** by masking the polar phosphonate and hydroxamate moieties using lipophilic groups; (ii) replacement or modification of the phosphonate motif by (bio)isosteres; (iii) replacement or modification of the hydroxamate moiety; and (iv) modification of the aliphatic C3 spacer. Some important advances are summarized below.

i. Preparation of prodrugs of **3/4** and their reverse hydroxamate analogs: The major shortcomings of **3** and **4** are their high polarity resulting from the charged phosphonate moiety, decreasing their bioavailability, oral efficacy, and preventing penetration across very highly lipophilic cell walls. To promote both cellular uptake and bioavailability, a prodrug strategy in which the phosphonate part is masked by lipophilic esters has adopted and demonstrated efficient at improving the inhibitory activities of the compounds. The nonpolar groups are expected to be hydrolyzed by nonspecific esterases upon uptake into the cell, releasing the active components and

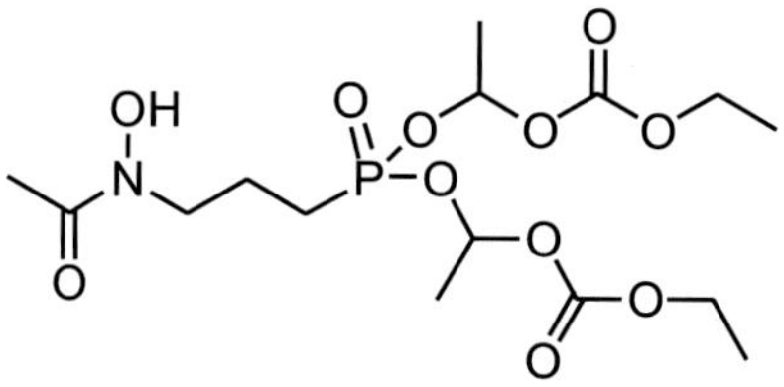

12. Mt H37Rv MIC = 400 μg/mL
Ec k12 > 200
Ec tolc = 200

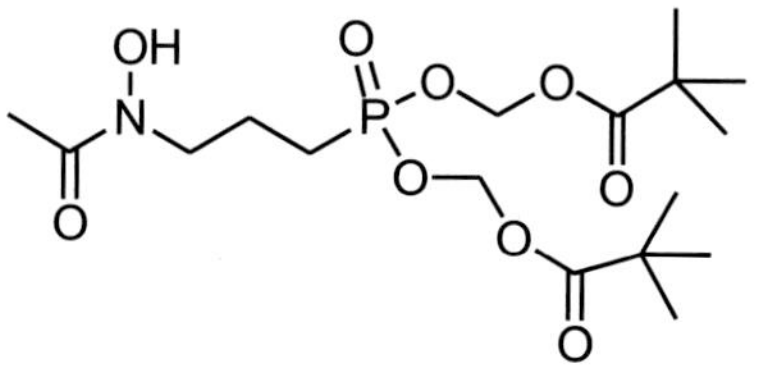

13. Mt H37Rv MIC = 50-100 μg/mL
Ec k12 > 200
Ec tolc = 100

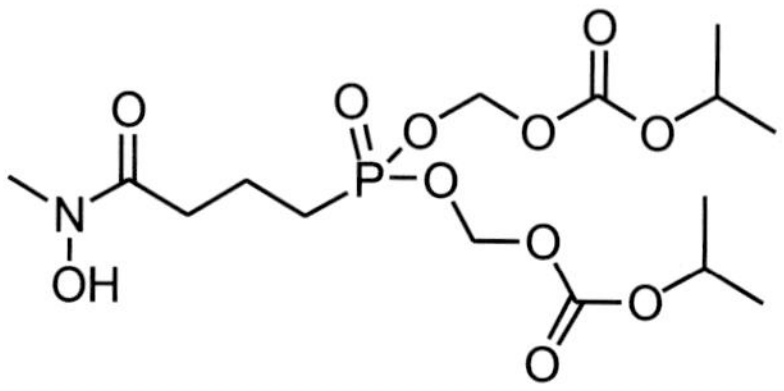

14 . Mt MIC > 50 μM
Pf IC_{50} = 0.11 μM

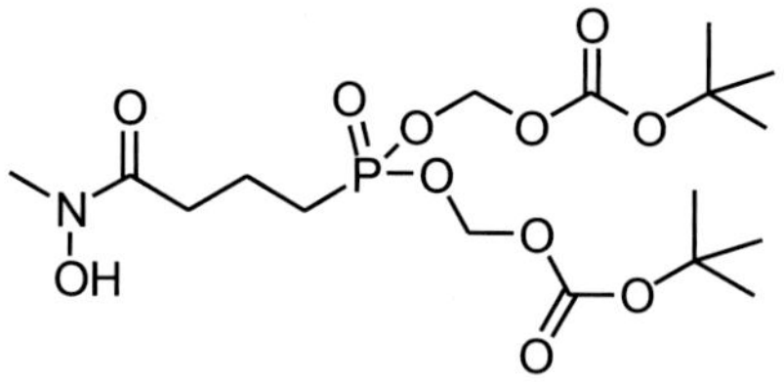

15. Mt MIC = 120-240 μM
Pf IC_{50} = 0.27 μM

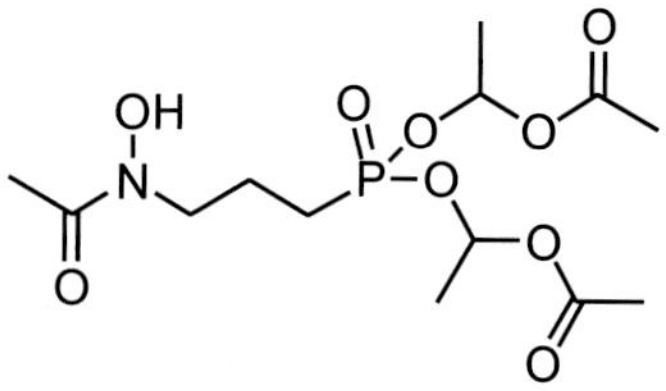

16. Pf IC_{50} = 174 nM

FIG. 5 Representative prodrugs of **3/4** and their reverse hydroxamate analogs. Mt is the short name of *Mycobacterium tuberculosis*; Ec is *E. coli*; Pf is *Plasmodium falciparum*.

inhibiting DXR inside bacterial to cut off the biosynthesis of essential isoprenoids and eventually lead to destroying the bacterial cell [59,60]. To this end, acyloxyalkyl and alkoxycarbonyloxyethyl esters were prepared and tested and the results showed bis-(alkoxycarbonyloxy)ethyl ester **12–15** and acetyloxyethyl ester **16** are representative compounds (Fig. 5) [60–63].

Recently, Van Calenbergh and co-workers prepared a series of FSM based prodrugs and evaluated their activity. The first class includes several alkoxyalkyl and acyloxybenzyl prodrugs and the latter ones that are represented by compound **17** in Fig. 6 proved to be the most attractive and exhibit enhanced antimalarial and antitubercular effect compared with FSM [62].

The second series comprises of *N*-alkoxy analogs of the L-leucine ethyl ester phosphonodiamidate derivatives of a fosmidomycin surrogate with an inversed hydroxamate group. Unfortunately, none of these compounds showed improved activity on either *P. falciparum* or *M. tuberculosis* in comparison with the parent compound (Fig. 6, **18**) [64]. The authors also synthesized some amino acid based prodrugs of an FSM surrogate and determined their activity. Conversion of the phosphonate moiety into tyrosine-derived esters (**19**) increases the in vitro activity against asexual blood stages of *Plasmodium falciparum*, while phosphonodiamidate (**20**) prodrugs display promising antitubercular activities [65]. In last series of compounds, the authors prepared double prodrugs of an FSM surrogate of which both the phosphonate and the hydroxamate motifs were modified with nonpolar groups. The results showed that *N*-benzyl substituted carbamate prodrug **21** was the most active antimalarial analog and 2-nitrothiophene prodrug **22** displayed promising antitubercular activities (Fig. 6) [66].

ii. Replacement or modification of the phosphonate motif by (bio)isosteres: A lot of research has shown that substitution of the phosphonate by (bio)isosteric moieties (e.g., carboxylate and sulfamate) could not lead to increased activity and is generally not a very promising approach [58,67].

iii. Replacement or modification of the hydroxamate moiety: Just like the phosphonate group in FSM and FR900098, the hydroxamate moiety of the two compounds are also key to their activity because it functions as a chelating group to block the coordination between the substrate DXP and enzyme-bound bivalent metal ion. Replacement of the hydroxamate motif by other chelating groups led to limited progress against bacterial DXR [58,67]. The reverse hydroxamate analogs of FSM and FR900098 (compounds **23** and **24**), however, possess comparable DXR inhibitory activity with the parent compounds (Fig. 7) [68,69].

A latest study reported the preparation and activity evaluation of a couple of FSM analogs with the hydroxamate chelating group being replaced with a 3,4-dihydroxyphenyl group [70]. Although all compounds (represented by compounds **25** and **26**) prepared showed almost no activity against DXR of *P. falciparum* (PfDXR) at 20 μM, they displayed very encouraging activity with IC_{50} values of 5.6–16.4 μM against *P. falciparum* parasites and IC_{50} values

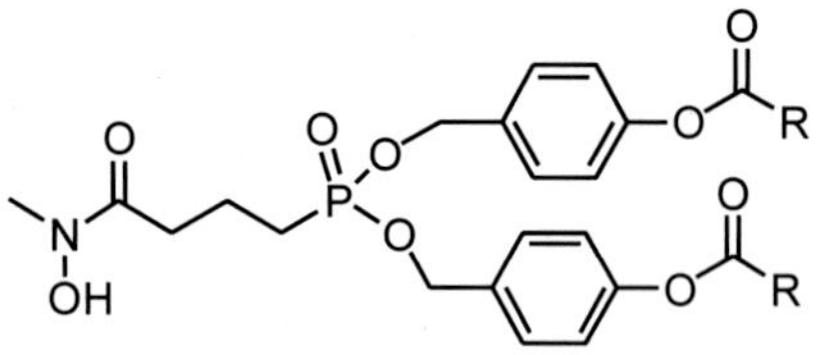

17. R = Ph
Pf-K1 IC_{50} = 0.03 μM
Mt-H37Ra IC_{50} = 0.42 μM
MRC-5 IC_{50} = 0.61 μM

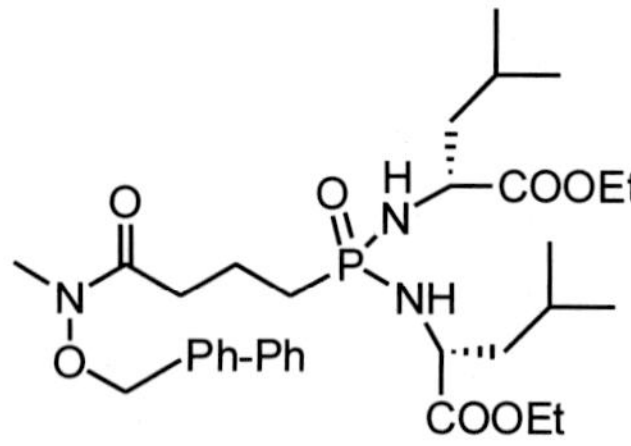

18. Pf-K1 IC_{50} = 5.25 μM
Mt-H37Ra IC_{50} = 35.2 μM
MRC-5 IC_{50} = 5.94 μM

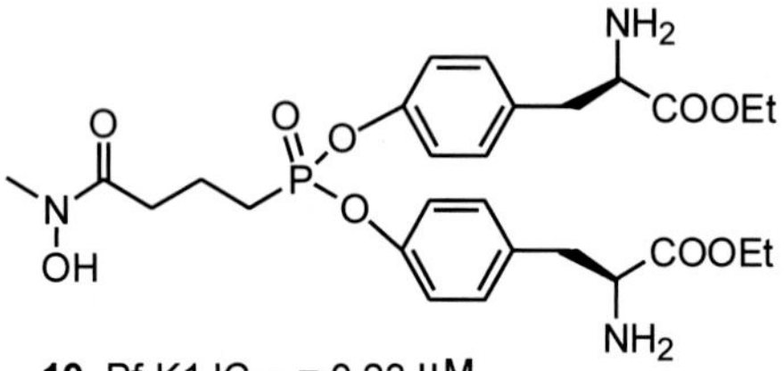

19. Pf-K1 IC_{50} = 0.23 μM
Mt-H37Rv MIC > 50 μM
MRC-5 MIC > 64 μM

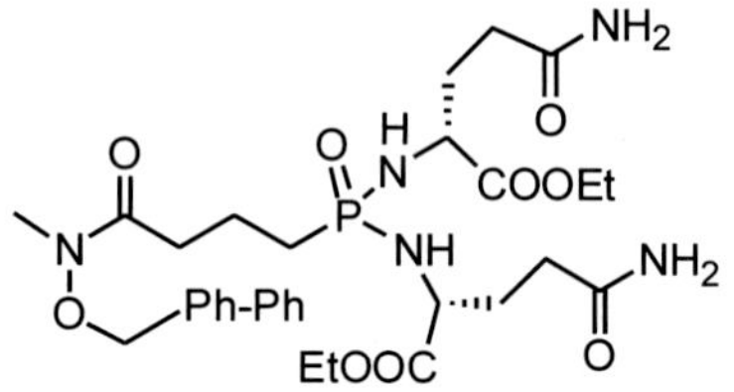

20. Pf-K1 IC_{50} = 0.96 μM
Mt-H37Rv MIC > 50 μM
MRC-5 MIC > 64 μM

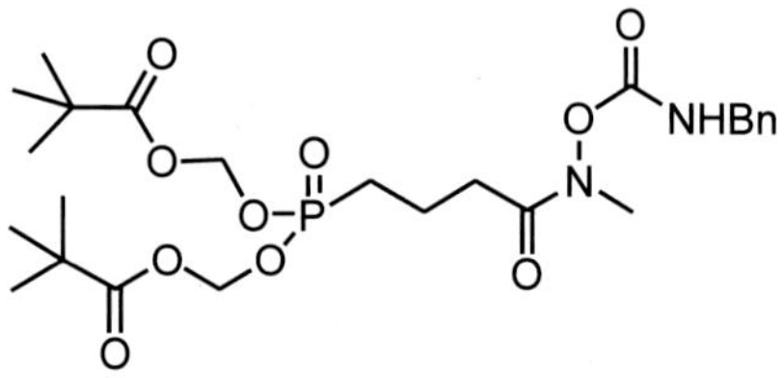

21. Pf-K1 IC_{50} = 0.64 μM
Mt-H37Rv MIC > 50 μM
MRC-5 CC_{50} = 27.9 μM

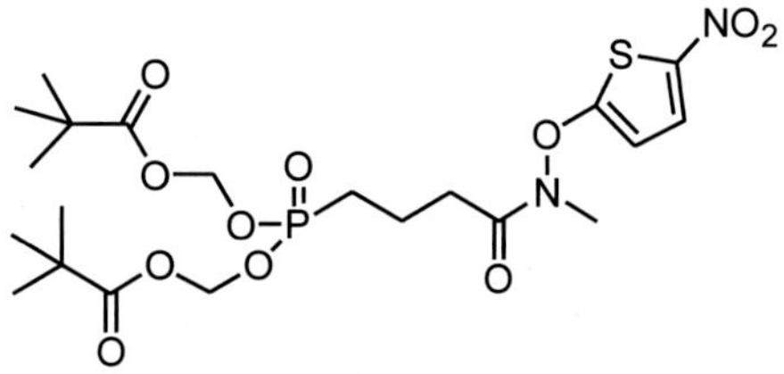

22. Pf-K1 IC_{50} = 4.69 μM
Mt-H37Rv MIC = 12.5 μM
MRC-5 CC_{50} = 20.5 μM

FIG. 6 Recently published prodrugs of FSM and FR900098.

of 5.2–10.2 μM against *Trypanosoma brucei brucei* (Fig. 7). The dramatic difference of the effect of these compounds between protein level and cell level may be because they could be hydrolyzed to their phosphoramidic acid analogs. Data derived from docking experiments of the ligands in the PfDXR receptor cavity support their potential as PfDXR inhibitors.

iv. Modification of the aliphatic C3 spacer: Although the propyl linker between the phosphonate anchor and the metal-chelating hydroxamate group of FSM and FR900098 is very stringent for them and any change of its length can cause significant loss of their inhibitory activity, substitution of its carbons with hydrophobic substituents, replacement of each of its methenes with heteroatoms such as O and S at different positions (α, β, or γ to the phosphonate group), and combination of the two most promising approaches have been proven to be effective ways to prepare better DXR inhibitors and also antibacterials. Moreover, this strategy has provided a new solution to create structural diversity of DXR inhibitors with more potent activity. The representative compounds **27**–**33** are listed in Fig. 8 [58,67].

Based on these encouraging advances, quite many new FSM/FR900098 analogs with modification on the C3 moiety have been generated recently. Sooriyaarachchi et al. synthesized a set of hydroxamate analogs of FR900098 with their β-position of the C3 linker being replaced by various arylpropyl substituents [71]. Some new compounds show almost equal potency on PfDXR to FSM and FR900098 and little lower growth inhibitory effect on the multidrug-resistant *P. falciparum* K1 strain (**34** and **35**) (Fig. 8). The data also exhibit that *meta*-substitution on the aromatic ring favors the enhancement of activity. Using the thio analog **29** as a lead, Lienau et al. prepared a panel of new thioethers and found that some of them showed DXR inhibitory effect and antiplasmodial in vitro activity in the nanomolar range [72]. Although the potency

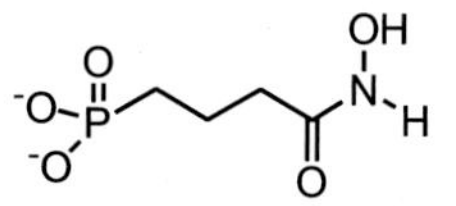

23. R = H, reverse Fosmidomycin
EcDXR IC_{50} = 170 nM

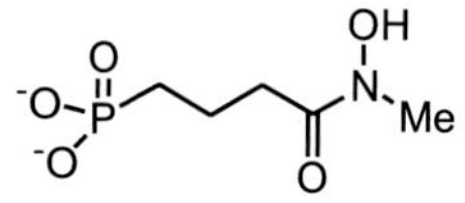

24. R = Me, reverse FR900098
EcDXR IC_{50} = 48 nM

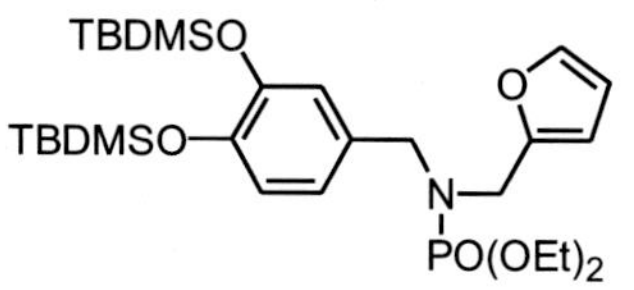

25. Pf IC_{50} = 5.6 μM
Tbb IC_{50} = 5.2 μM

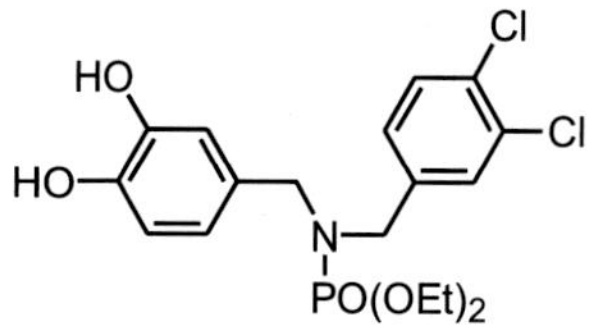

26. Pf IC_{50} = 16.4 μM

FIG. 7 The analogs of FSM and FR900098 with variation on the hydroxamate moiety. Tbb: *Trypanosoma brucei brucei*.

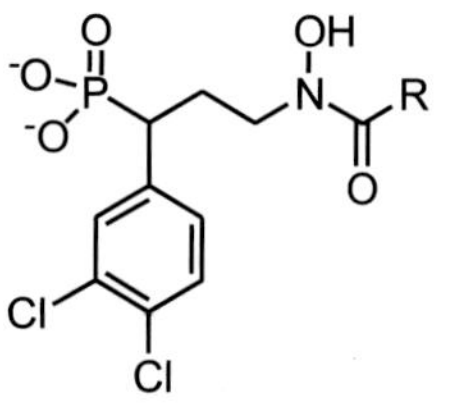

27. R = H, EcDXR IC_{50} = 59 nM
PfD2d IC_{50} = 28 nM
28. R = Me, EcDXR IC_{50} = 119 nM
PfD2d IC_{50} = 90 nM

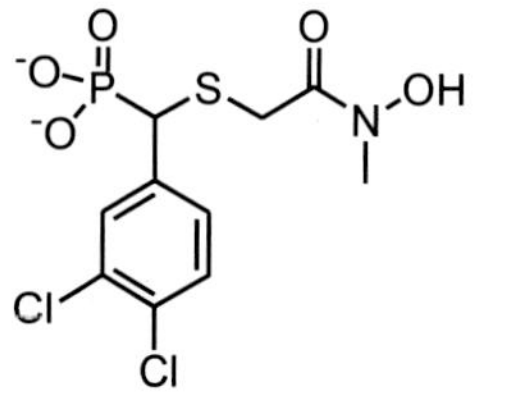

29. EcDXR IC_{50} = 5.9 nM
PfDXR IC_{50} = 4.5 nM
MtDXR IC_{50} = 9.2 nM
Pf3D7 IC_{50} = 91 nM
PfDd2 IC_{50} = 85 nM

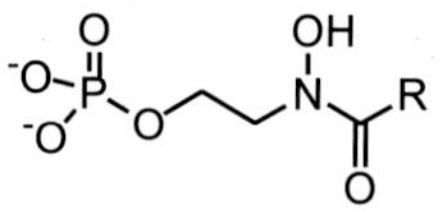

30. R = H, EcDXR IC_{50} = 0.34 μM
MtDXR IC_{50} = 2.1 μM
31. R = Me, EcDXR IC_{50} = 77 nM
PfDXR IC_{50} = 279 nM

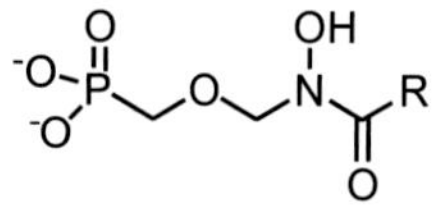

32. R = H, EcDXR IC_{50} = 1.11 μM
PfDXR IC_{50} = 4.9 μM
33. R = Me, EcDXR IC_{50} = 87 nM
PfDXR IC_{50} = 360 nM

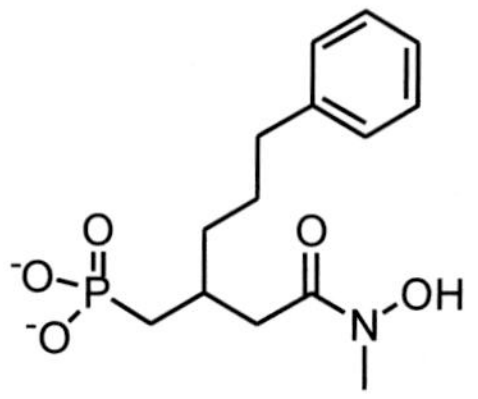

34. EcDXR IC_{50} = 840 nM
PfDXR IC_{50} = 79 nM
Pf K1 IC_{50} = 2.7 μM

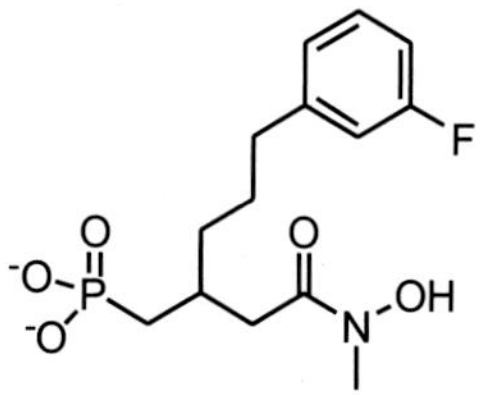

35 EcDXR IC_{50} = 2.0 μM
PfDXR IC_{50} = 67 nM
Pf K1 IC_{50} = 5.7 μM

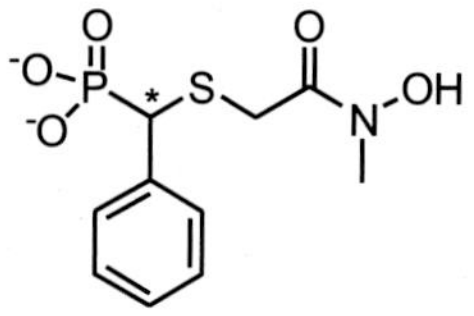

36. racemate Pf3D7 IC_{50} = 190 nM
PfDd2 IC_{50} = 260 nM
(*S*)-isomer Pf3D7 IC_{50} = 87 nM
PfDd2 IC_{50} = 140 nM
(*R*)-isomer Pf3D7 IC_{50} > 130 μM
PfDd2 IC_{50} > 130 μM

FIG. 8 Typical C3-spacer modified FSM/FR900098 analogs **27–33** and newly published structures **34–36**.

of the best new inhibitor on DXR is at least three times as weak as that of **29**, the antiplasmodial efficiency of it is almost same as compound **29**. The experiments also revealed the oxidation state of the thio motif is closely related to the DXR inhibitory activity and oxidation of the thio to sulfone decreased the activity by 2 to 3 orders of magnitude. More importantly, the authors further discussed the enantiospecificity of one of the new inhibitors **36** and by combination with the data obtained for the model compound **29** [73], they concluded that in both the DXR level and the parasite level, the (*S*)-enantiomer displayed around double potency by comparison with the racemate and was more pronounced than that of its (*R*)-enantiomer by at least three orders of magnitude (Fig. 8). In the latest paper, 17 known α-phenylsubstituted reverse derivatives of FSM against *Yersinia pestis* DXR and *Y. pestis* A1122 strain were tested. This series of compounds varied by replacement of the middle methylene with O or S, adding electron-withdrawing or -donating groups on the different positions of α-phenyl ring. The results showed some of these compounds exhibited raised effect on DXR over FSM with IC_{50} values in the nanomolar range and decreased minimal inhibitory concentrations over FSM on the bacterial strain [74].

The saturation of C3 linker is a key factor on DXR inhibitory activity as well. Compound **37**, an α,β-unsaturated analog of FR900098, is the first synthesized and tested item of this kind, which shows quite encouraging effect against MtDXR and its phosphonate prodrug **38** displays much better antimycobacterial 7H9 activity than FSM (Fig. 9) [75]. The further assays performed by Edwards et al. uncovered that the lipophilic prodrug **38** possesses nanomolar activity against various quinine resistant asexual parasites and is both safe and effective in malaria-infected mice [76]. In the following research, Wang and co-workers synthesized a set of α,β-unsaturated analogs of FSM and evaluated their inhibition on *P. falciparum*. The most potent inhibitor **40**, exhibits on-target, strong inhibition against the growth of *P. falciparum* (IC_{50} = 13 nM) and also exceptional in vivo efficacy on *P. berghei* mouse model of malaria (Fig. 9) [77]. The latest report carried out by Edwards et al. determined the activity of compounds **37**–**40** against *Staphylococcus schleiferi* DXR and some *Staphylococcus* spp. Compounds **37** and **39** do not show excellent DXR inhibitory effect, but the prodrug **38** and **40** display 6 and 25 times higher activity than FSM against *S. schleiferi*, rspectively and comparable effect with FSM against *Staphylococcus pseudintermedius*. All four inhibitors show no suppression on *Staphylococcus aureus* [78]. A recent study revealed that compound **38** is also a natural product formed in *S. lavendulae* that is the host of FSM and FR900098 as well [79].

4.2 Other type inhibitors

Besides FSM/FR900098 and their analogs, researchers have also tried to locate DXR inhibitors of different types. Kaiser and co-workers have ever attempted to seek DXR inhibitors from Mediterranean plants and found that the leaf extracts of *Cercis siliquastrum* exhibit strong inhibitory activity against the protein, but no specific compound has been elucidated [80]. Several lipophilic

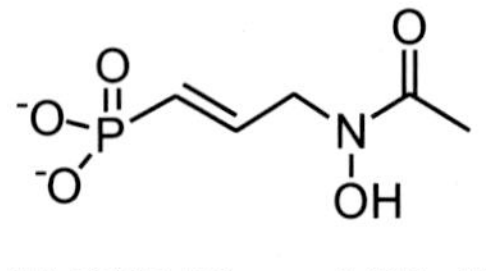

37. Mt DXR IC_{50} = 1.07 μM
Mt 7H9 MIC > 200 μg/mL
Ss DXR IC_{50} = 3.31 μM

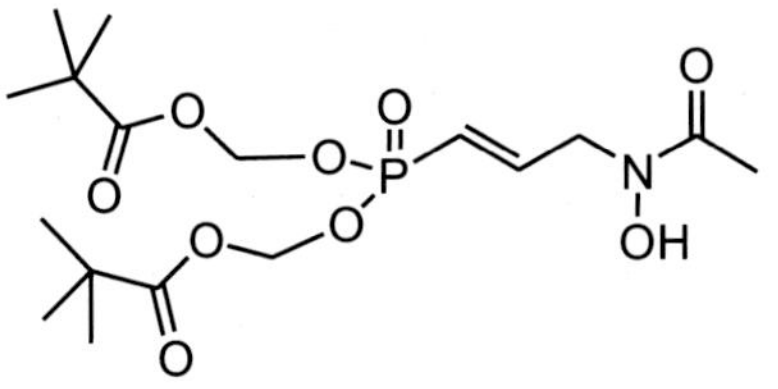

38. Mt7H9 MIC = 9.4 μg/mL
Pf3D7 IC_{50} = 18.3 nM
Pf7G8 IC_{50} = 46.7 nM
Ss IC_{50} = 100 nM
Sp IC_{50} = 260 nM

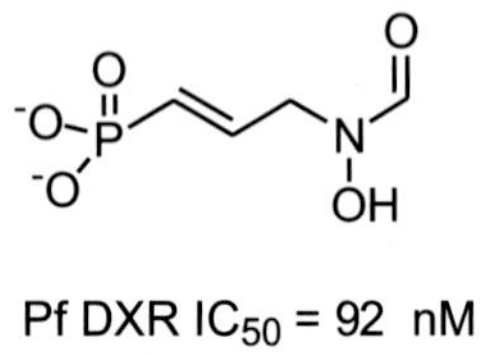

39. Pf DXR IC_{50} = 92 nM
Pf IC_{50} = 19 nM
Ss DXR IC_{50} = 0.41 μM

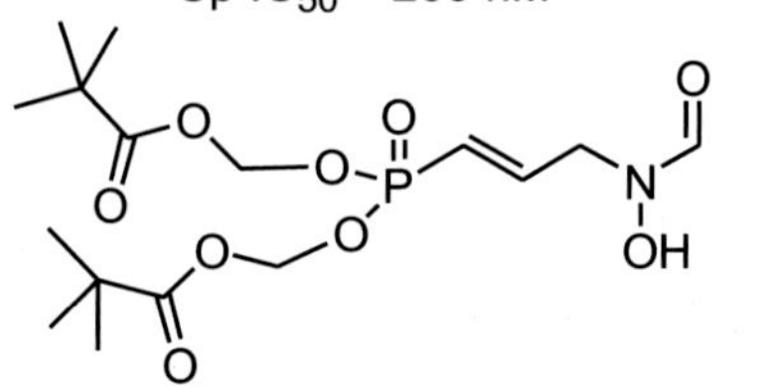

40. Pf IC_{50} = 13 nM
Ss IC_{50} = 30 nM
Sp IC_{50} = 210 nM

FIG. 9 C3-spacer unsaturated analogs of FSM and FR900098. Ss: *S. schleiferi*; Sp: *S. pseudintermedius*.

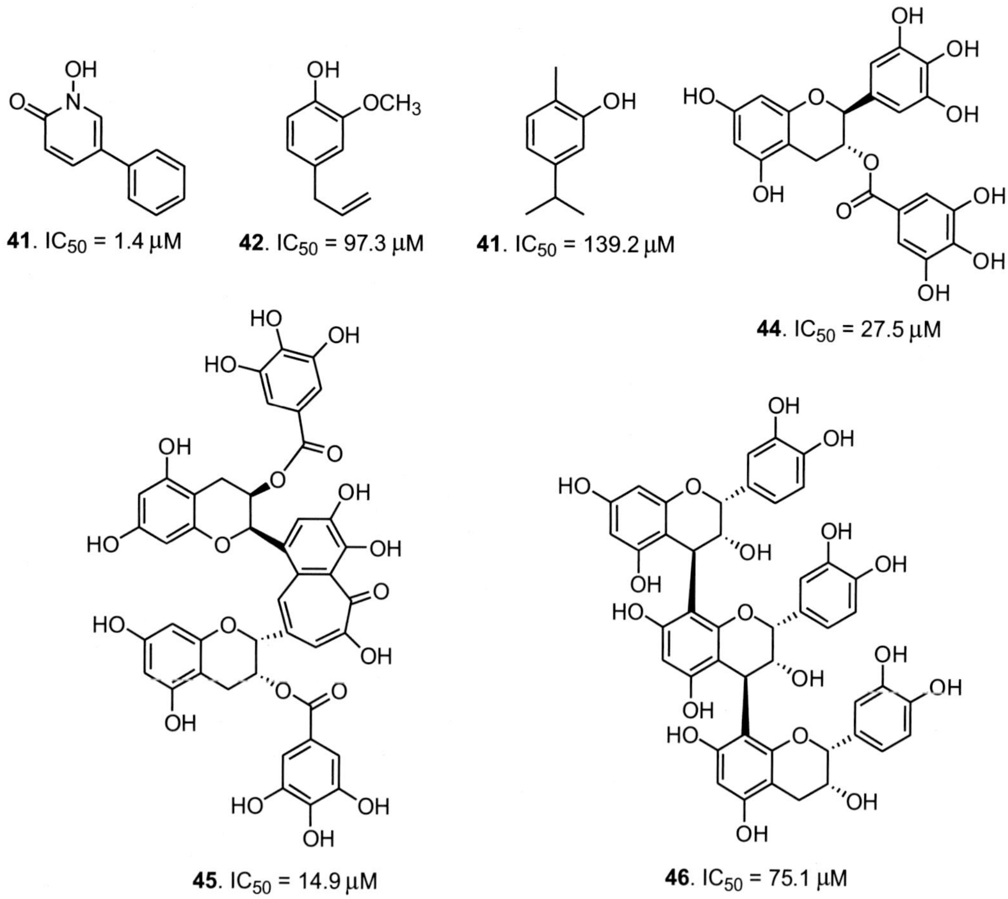

FIG. 10 Non-FSM/FR900098 type DXR inhibitors.

compounds with structural similarity to FSM/FR900098 were recently synthesized and their DXR inhibitory effects were tested. The representative compound (**41** in Fig. 10) exhibits an IC_{50} of 1.4 μM on *E. coli* DXR and a broad spectrum of activity against a set of Gram-negative and -positive bacteria with minimal inhibition concentrations ranging from 20 to 100 μM [81]. It is believed that compound **41** may exert its DXR inhibitory activity through competing Mg^{2+} with the enzyme. The authors group has searched DXR inhibitors from various plant essential oils (PEOs) as well because it has been demonstrated that PEOs possess many sorts of bioactivities such as antimicrobial and antiviral properties [82,83]. The results show that compounds eugenol **42** and carvacrol **43** (Fig. 10) display weak inhibition against *E. coli* DXR with IC_{50} values being about 97.3 and 139.2 μM, respectively [84,85].

Phenolic compounds have been proven to possess many kinds of bioactivities including antibacterial effect [86,87]. To look for phenolic inhibitors of DXR, Tritsch, etc. carried out assays to determine the DXR suppressive effects of several flavonoids and found that most of the tested compounds effectively inhibited DXR activity with encouraging IC_{50} values in the micromolar range (1.5–2.2 μM). But further investigation revealed that the supplement of 0.01% Triton X100 in the assays led to a substantial decrease of the inhibition, showing these compounds are nonspecific inhibitors of DXR [88]. The authors group has also studied the inhibitory activities of some phenolics and the results show catechin compounds gallocatechin gallate (**44**, IC_{50} 27.5 μM) and theaflavin-3,3′-digallate (**45**, IC_{50} 14.9 μM) exhibit moderate activity against *E. coli* DXR, while proanthocyanidin compound procyanid C1 displays weak effect on the protein (**46**, IC_{50} 75.1 μM) (Fig. 10). In contrast to the flavonoids that are aggregating inhibitors of DXR [88], these phenolic compounds have been clarified as specific inhibitors of DXR because the addition of the detergent Triton X100 has no influence on the inhibition [89–92].

In summary, the studies on many aspects of DXR have been fruitful to date except two issues. One is in the investigation of DXR chemical mechanism, namely the capture of the retro-aldol intermediates **6** and **7** (Fig. 2) that have never been detected in any experiments. The other is the development of more potent DXR inhibitors with better druggable properties, focusing on the β-thio analogs **29** and the stereochemistry of the α-substituent.

References

[1] Spurgeon SL, Porter JW. In: Porter JW, Spurgeon SL, editors. Biosynthesis of isoprenoid compounds, vol. 1. New York: Wiley; 1981. p. 1–46.

[2] Bloch K. Sterol molecule: structure, biosynthesis, and function. Steroids 1992;57:378–83.

[3] Qureshi N, Porter JW. In: Porter JW, Spurgeon SL, editors. Biosynthesis of isoprenoid compounds, vol. 1. New York: Wiley; 1981. p. 47.

[4] Rohmer M. Methylerythritol phosphate pathway. In: Mander L, Liu HW, editors. Comprehensive natural products II, chemistry and biology, vol. 1. Elsevier; 2010. p. 517–55.

[5] Frank A, Groll M. The methylerythritol phosphate pathway to isoprenoids. Chem Rev 2017;117:5675–703.

[6] Hemmerlin A, Harwood JL, Bach TJ. A raison d'être for two distinct pathways in the early steps of plant isoprenoid biosynthesis? Prog Lipid Res 2012;51:95–148.

[7] Wang X, Dowd CS. The methylerythritol phosphate pathway: promising drug targets in the fight against tuberculosis. ACS Infect Dis 2018;4:278–90.

[8] Singh N, Cheve G, Avery MA, et al. Targeting the MEP pathway for novel antimalarial, antibacterial and herbicidal drug discovery: inhibition of DXR enzyme. Curr Pharm Design 2007;13:1161–77.

[9] Munos JW, Pu X, Mansoorabadi SO, Kim HJ, Liu HW. A secondary kinetic isotope effect study of the 1-deoxy-D-xylulose-5-phosphate reductoisomerase-catalyzed reaction: evidence for a retroaldol-aldol rearrangement. J Am Chem Soc 2009;131:2048–9.

[10] Manning KA, Sathyamoorthy B, Eletsky A, et al. Highly precise measurement of kinetic isotope effects using ^{1}H-detected 2D [^{13}C,^{1}H]-HSQC NMR spectroscopy. J Am Chem Soc 2012;134:20589–92.

[11] Jomaa H, Wiesner J, Sanderbrand S, et al. Inhibitors of the nonmevalonate pathway of isoprenoid biosynthesis as antimalarial drugs. Science 1999;285:1573–6.

[12] Murkin AS, Manning KA, Kholodar SA. Mechanism and inhibition of 1-deoxy-D-xylulose-5-phosphate reductoisomerase. Bioorg Chem 2014;57:171–85.

[13] Proteau PJ. 1-Deoxy-D-xylulose 5-phosphate reductoisomerase: an overview. Bioorg Chem 2004;32:483–93.

[14] Sangari FJ, Pérez-Gil J, Carretero-Paulet L, et al. A new family of enzymes catalyzing the first committed step of the methylerythritol 4-phosphate (MEP) pathway for isoprenoid biosynthesis in bacteria. Proc Natl Acad Sci U S A 2010;107:14081–6.

[15] Carretero-Paulet L, Lipska A, Pérez-Gil J, et al. Evolutionary diversification and characterization of the eubacterial gene family encoding DXR type II, an alternative isoprenoid biosynthetic enzyme. BMC Evol Biol 2013;13:180.

[16] Pérez-Gil J, Calisto BM, Behrendt C, et al. Crystal structure of *Brucella abortus* deoxyxylulose-5-phosphate reductoisomerase-like (DRL) enzyme involved in isoprenoid biosynthesis. J Biol Chem 2012;287:15803–9.

[17] Rohmer M, Seemann M, Horbach S, et al. Glyceraldehyde 3-phosphate and pyruvate as precursors of isoprenic units in an alternative nonmevalonate pathway for terpenoid biosynthesis. J Am Chem Soc 1996;118:2564–6.

[18] Kuzuyama T, Takahashi S, Watanabe H, et al. Direct formation of 2-C-methyl-D-erythritol 4-phosphate from 1-deoxy-D-Xylulose 5-phosphate by 1-deoxy-D-Xylulose 5-phosphate reductoisomerase, a new enzyme in the non-mevalonate pathway to isopentenyl diphosphate. Tetrahedron Lett 1998;39:4509–12.

[19] Takahashi S, Kuzuyama T, Watanabe H, et al. A 1-deoxy-D-xylulose 5-phosphate reductoisomerase catalyzing the formation of 2-C-methyl-D-erythritol 4-phosphate in an alternative nonmevalonate pathway for terpenoid biosynthesis. Proc Natl Acad Sci U S A 1998;95:9879–84.

[20] Hoeffler JF, Tritsch D, Grosdemange-Billiard C, et al. Isoprenoid biosynthesis via the methylerythritol phosphate pathway. Mechanistic investigations of the 1-deoxy-D-xylulose-5-phosphate reductoisomerase. Eur J Biochem 2002;269:4446–57.

[21] Argyrou A, Blanchard JS. Kinetic and chemical mechanism of mycobacterium tuberculosis 1-deoxy-D-xylulose-5-phosphate isomeroreductase. Biochemistry 2004;43:4375–84.

[22] Dumas R, Biou V, Halgand F, et al. Enzymology, structure, and dynamics of acetohydroxy acid isomeroreductase. Acc Chem Res 2001;34:399–408.

[23] Arigoni D, Sagner S, Latzel C, et al. Terpenoid biosynthesis from 1-deoxy-D-xylulose in higher plants by intramolecular skeletal rearrangement. Proc Natl Acad Sci U S A 1997;94:10600–5.

[24] Eisenreich W, Schwarz M, Cartayrade A, et al. The deoxyxylulose phosphate pathway of terpenoid biosynthesis in plants and microorganisms. Chem Biol 1998;5:R221–33.

[25] Johnson AE, Tanner ME. Epimerization via carbon-carbon bond cleavage. L-Ribulose-5-phosphate 4-epimerase as a masked class II aldolase. Biochemistry 1998;37:5746–54.

[26] Lee LV, Vu MV, Cleland WW. ^{13}C and deuterium isotope effects suggest an aldol cleavage mechanism for L-ribulose-5-phosphate 4-epimerase. Biochemistry 2000;39:4808–20.

[27] Wong A, Munos JW, Devasthali V, et al. Study of 1-deoxy-D-xylulose-5-phosphate reductoisomerase: synthesis and evaluation of fluorinated substrate analogues. Org Lett 2004;6:3625–8.

[28] Phaosiri C, Proteau PJ. Substrate analogs for the investigation of deoxyxylulose 5-phosphate reductoisomerase inhibition: synthesis and evaluation. Bioorg Med Chem Lett 2004;14:5309–12.

[29] Fox DT, Poulter CD. Mechanistic studies with 2-methyl-D-erythritol 4-phosphate synthase from *Escherichia coli*. Biochemistry 2005; 44:8360–8.

[30] Koppisch AT, Fox DT, Blagg BS, et al. *E. coli* MEP synthase: steady-state kinetic analysis and substrate binding. Biochemistry 2002;41:236–43.

[31] Franco TMA, Blanchard JS. Bacterial branched-chain amino acid biosynthesis: structures, mechanisms, and drugability. Biochemistry 2017;56:5849–66.

[32] Lauw S, Illarionova V, Bacher A, et al. Biosynthesis of isoprenoids-studies on the mechanism of 2-C-methyl-D-erythritol-4-phosphate synthase. FEBS J 2008;275:4060–73.

[33] Carpenter BK. Determination of organic reaction mechanisms. Chichester: Wiley; 1984.

[34] Wong U, Cox RJ. The chemical mechanism of D-1-deoxyxylulose -5-phosphate reductoisomerase from *Escherichia coli*. Angew Chem Int Ed 2007;46:4926–9.

[35] Schimerlik MI, Rife JE, Cleland WW. Equilibrium perturbation by isotope substitution. Biochemistry 1975;14:5347–54.

[36] Steinbacher S, Kaiser J, Eisenreich W, et al. Structural basis of fosmidomycin action revealed by the complex with 2-C-methyl-D-erythritol 4-phosphate synthase (IspC). J Biol Chem 2003;278:18401–7.

[37] Yajima S, Hara K, Iino D, et al. Structure of 1-deoxy-D-xylulose 5-phosphate reductoisomerase in a quaternary complex with a magnesium ion, NADPH and the antimalarial drug fosmidomycin. Acta Crystallogr 2007;F63:466–70.

[38] Henriksson LM, Unge T, Carlsson J, et al. Structures of mycobacterium tuberculosis 1-deoxy-D-xylulose-5-phosphate reductoisomerase provide new insights into catalysis. J Biol Chem 2007;282:19905–16.

[39] Yin X, Proteau JP. Characterization of native and histidine-tagged deoxyxylulose 5-phosphate reductoisomerase from the cyanobacterium *Synechocystis* sp. PCC6803. Biochim Biophys Acta 2003;1652:75–81.

[40] Sweeney AM, Lange R, Fernandes RPM, et al. The crystal structure of *E. coli* 1-deoxy-D-xylulose-5-phosphate reductoisomerase in a ternary complex with the antimalarial compound fosmidomycin and NADPH reveals a tight-binding closed enzyme conformation. J Mol Biol 2005;345:115–27.

[41] Li H, Tian J, Sun W, et al. Mechanistic insights into 1-deoxy-D-xylulose 5-phosphate reductoisomerase, a key enzyme of the MEP terpenoid biosynthetic pathway. FEBS J 2013;280:5896–905.

[42] Bell RP. The reversible hydration of carbonyl compounds. Adv Phys Org Chem 1966;4:1–29.

[43] Deupree JD, Wood WA. L-Ribulose-5-phosphate 4-epimerase from *Aerobacter aerogenes*: evidence for a role of divalent metal ions in the epimerization reaction. J Biol Chem 1972;247:3093–7.

[44] Liu J, Murkin AS. Pre-steady-state kinetic analysis of 1-deoxy-D-xylulose-5-phosphate reductoisomerase from *M. tuberculosis* reveals partially rate-limiting product release by parallel pathways. Biochemistry 2012;51:5307–19.

[45] Proteau PJ, Woo Y-H, Williamson RT, et al. Stereochemistry of the reduction step mediated by recombinant 1-deoxy-D-xylulose 5-phosphate isomeroreductase. Org Lett 1999;1:921–3.

[46] Radykewicz T, Rohdich F, Wungsintaweekul J, et al. Biosynthesis of terpenoids: 1-deoxy-D-xylulose-5-phosphate reductoisomerase from *Escherichia coli* is a class B dehydrogenase. FEBS Lett 2000; 465:157–60.

[47] Arigoni D, Giner J-L, Sagner S, et al. Stereochemical course of the reduction step in the formation of 2-C-methylerythritol from the terpene precursor 1-deoxyxylulose in higher plants. Chem Commun 1999;1127–8.

[48] Mine Y, Kamimura T, Nonoyama S, et al. In vitro and in vivo antibacterial activities of FR-31564, a new phosphonic acid antibiotic. J Antibiot 1980;33:36–43.

[49] Neu HC, Kamimura T. In vitro and in vivo antibacterial activity of FR-31564, a phosphonic acid antimicrobial agent. Antimicrob Agents Chemother 1981;19:1013–23.

[50] Okuhara M, Kuroda Y, Goto T, et al. Studies on new phosphonic acid antibiotics. III. Isolation and characterization of FR-31564, FR-32863 and FR-33289. J Antibiot 1980;33:24–8.

[51] Shigi Y. Inhibition of bacterial isoprenoid synthesis by fosmidomycin, a phosphonic acid-containing antibiotic. J Antimicrob Chemother 1989;24:131–45.

[52] Kuzuyama T, Shimizu T, Takahashi S, et al. Fosmidomycin, a specific inhibitor of 1-deoxy-D-xylulose 5-phosphate reductoisomerase in the nonmevalonate pathway for terpenoid biosynthesis. Tetrahedron Lett 1998;39:7913–6.

[53] Zeidler J, Schwender J, Mueller C, et al. Inhibition of the non-mevalonate 1-deoxy-D-xylulose-5-phosphate pathway of plant isoprenoid biosynthesis by fosmidomycin. Z Naturforsch C 1998;53:980–6.

[54] Borrmann S, Issifou S, Esser G, et al. Fosmidomycin-clindamycin for the treatment of *Plasmodium falciparum* malaria. J Infect Dis 2004;190:1534–40.

[55] Borrmann S, Lundgren I, Oyakhirome S, et al. Fosmidomycin plus clindamycin for treatment of pediatric patients aged 1 to 14 years with *Plasmodium falciparum* malaria. Antimicrob Agents Chemother 2006;50:2713–8.

[56] Murakawa T, Sakamoto H, Fukada S, et al. Pharmacokinetics of fosmidomycin, a new phosphonic acid antibiotic. Antimicrob Agents Chemother 1982;21:224–30.

[57] Na-Bangchang K, Ruengweerayut R, Karbwang J, et al. Pharmacokinetics and pharmacodynamics of fosmidomycin monotherapy and combination therapy with clindamycin in the treatment of multidrug resistant falciparum malaria. Malar J 2007;6:70.

[58] Jackson ER, Dowd CS. Inhibition of 1-deoxy-D-Xylulose-5-phosphate reductoisomerase (Dxr): a review of the synthesis and biological evaluation of recent inhibitors. Curr Top Med Chem 2012;12:706–28.

[59] Schultz C. Prodrugs of biologically active phosphate esters. Bioorg Med Chem 2003;11:885–98.

[60] Ortmann R, Wiesner J, Reichenberg A, et al. Acyloxyalkyl ester prodrugs of FR900098 with improved in vivo anti-malarial activity. Bioorg Med Chem Lett 2003;13:2163–6.

[61] Wiesner J, Ortmann R, Jomaa H, et al. Double ester prodrugs of FR900098 display enhanced in-vitro antimalarial activity. Arch Pharm (Weinheim) 2007;340:667–9.

[62] Courtens C, Risseeuw M, Caljon G, et al. Acyloxybenzyl and alkoxyalkyl prodrugs of a fosmidomycin surrogate as antimalarial and antitubercular agents. ACS Med Chem Lett 2018;9:986–9.

[63] Uh E, Jackson ER, San JG, et al. Antibacterial and antitubercular activity of FSM, FR900098, and their lipophilic analogs. Bioorg Med Chem Lett 2011;21:6973–6.

[64] Courtens C, Risseeuw M, Caljon G, et al. Phosphonodiamidate prodrugs of N-alkoxy analogs of a fosmidomycin surrogate as antimalarial and antitubercular agents. Bioorg Med Chem Lett 2019;29:1051–3.

[65] Courtens C, Risseeuw M, Caljon G, et al. Amino acid based prodrugs of a fosmidomycin surrogate as antimalarial and antitubercular agents. Bioorg Med Chem 2019;27:729–47.

[66] Courtens C, Risseeuw M, Caljon G, et al. Double prodrugs of a fosmidomycin surrogate as antimalarial and antitubercular agents. Bioorg Med Chem Lett 2019;29:1232–5.

[67] Lienau C, Konzuch S, Gräwert T, et al. Inhibition of the non-mevalonate isoprenoid pathway by reverse hydroxamate analogues of fosmidomycin. Inhibition of the non-mevalonate isoprenoid pathway by reverse hydroxamate analogues of fosmidomycin. Proc Chem 2015;14:108–16.

[68] Behrendt CT, Kunfermann A, Illarionova V, et al. Reverse fosmidomycin derivatives against the antimalarial drug target IspC (Dxr). J Med Chem 2011;54:6796–802.

[69] Kuntz L, Tritsch D, Grosdemange-Billard C, et al. Isoprenoid biosynthesis as a target for antibacterial and antiparasitic drugs: phosphonohydroxamic acids as inhibitors of deoxyxylulose phosphate reductoisomerase. Biochem J 2005;386:127–35.

[70] Adeyemi CM, Hoppe HC, Isaacs M, et al. Synthesis and anti-parasitic activity of N-benzylated phosphoramidate Mg^{2+}-chelating ligands. Bioorg Chem 2020;105, 104280.

[71] Sooriyaarachchi S, Chofor R, Risseeuw MDP, et al. Targeting an aromatic hotspot in *P. falciparum* 1-deoxy-D-xylulose-5-phosphate reductoisomerase with β-arylpropyl analogues of fosmidomycin. ChemMedChem 2016;1:2024–36.

[72] Lienau C, Graewert T, Alves Avelar LA, et al. Novel reverse thia-analogs of fosmidomycin: synthesis and antiplasmodial activity. Euro J Med Chem 2019;181, 111555.

[73] Kunfermann A, Lienau C, Illarionov B, et al. IspC as target for anti-infective drug discovery: synthesis, enantiomeric separation, and structural biology of fosmidomycin thia isosters. J Med Chem 2013;56:8151–62.

[74] Ball HS, Girma M, Zainab M, et al. Inhibition of the *Yersinia pestis* MEP pathway of isoprenoid biosynthesis by α-phenyl-substituted reverse FSM analogues. ACS Omega 2020;5:5170–5.
[75] Jackson ER, San Jose G, Brothers RC, et al. The effect of chain length and unsaturation on Mtb DXR inhibition and antitubercular killing activity of FR900098 analogs. Bioorg Med Chem Lett 2014;24:649–53.
[76] Edwards RL, Brothers RC, Wang X, et al. MEPicides: potent antimalarial prodrugs targeting isoprenoid biosynthesis. Sci Rep 2017;7:8400.
[77] Wang X, Edwards RL, Ball H, et al. MEPicides: α,β-unsaturated fosmidomycin analogues as DXR Inhibitors against malaria. J Med Chem 2018;61:8847–58.
[78] Edwards RL, Heueck I, Lee SG, et al. Potent, specific MEPicides for treatment of zoonotic staphylococci. PLoS Pathogen 2020;16, e1007806.
[79] Parkinson EI, Erb A, Eliot AC, et al. Fosmidomycin biosynthesis diverges from related phosphonate natural products. Nat Chem Biol 2019;15:1049–56.
[80] Kaiser J, Yassin M, Prakash S, et al. Anti-malarial drug targets: screening for inhibitors of 2C-methyl-D-erythritol 4-phosphate synthase (IspC protein) in Mediterranean plants. Phytomedicine 2007;14:242–9.
[81] Deng L, Sundriyal S, Rubio V, Shi Z-Z, Song Y. Coordination chemistry based approach to lipophilic inhibitors of 1-deoxy-D-xylulose-5-phosphate reductoisomerase. J Med Chem 2009;52:6539–42.
[82] Burt S. EOs: their antibacterial properties and potential applications in foods-a review. Int J Food Microbiol 2004;94:223–53.
[83] Hammer KA, Carson CF. Antibacterial and antifungal activities of EOs. In: Thormar H, editor. Lipids and essential oils as antimicrobial agents. United Kingdom: Wiley; 2011. p. 255–306.
[84] Hui X, Yan G, Tian F-L, Li H, Gao W-Y. Antimicrobial mechanism of the major active essential oil compounds and their structure-activity relationship. Med Chem Res 2017;26:442–9.
[85] Yan G, Zhu B-R, Tian F-L, Hui X, Li H, Li Y-M, Gao W-Y. Inhibitory activity of plant essential oils against *E. coli* 1-deoxy-D-xylulose-5-phosphate reductoisomerase. Molecules 2019;24:2518.
[86] Wu T, He M, Zang X, Zhou Y, Qiu T, Pan S, Xu X. A structure-activity relationship study of flavonoids as inhibitors of *E. coli* by membrane interaction effect. Biochim Biophys Acta 2013; 1828:2751–6.
[87] Tsou LK, Lara-Tejero M, RoseFigura J. Antibacterial flavonoids from medicinal plants covalently inactivate type III protein secretion substrates. J Am Chem Soc 2016;138(7):2209–18.
[88] Tritsch D, Zinglé C, Rohmer M, Grosdemange-Billiard C. Flavonoids: true or promiscuous inhibitors of enzyme? The case of deoxyxylulose phosphate reductoisomerase. Bioorg Chem 2015;59:140–4.
[89] Hui X, Hua S-H, Wu Q-Q, Li H, Gao W-Y. Antimicrobial mechanism of epigallocatechin gallate and gallocatechin gallate: they target 1-deoxy-D-xylulose 5-phosphatereductoisomerase, the key enzyme of the MEP terpenoid biosynthetic pathway. Arch Biochem Biophys 2017;622:1–8.
[90] Hui X, Yue Q, Zhang D-D, Li H, Yang S-Q, Gao W-Y. Antimicrobial mechanism of theaflavins: they target 1-deoxy-D-xylulose5-phosphate reductoisomerase, the key enzyme of the MEP terpenoid biosynthetic pathway. Sci Rep 2016;6:38945.
[91] Hui X, Liu H, Tian F-L, Li F-F, Li H, Gao W-Y. Inhibition of green tea and the catechins against 1-deoxy-D-xylulose 5-phosphate reductoisomerase, the key enzyme of the MEP terpenoid biosynthetic pathway. Fitoterapia 2016;113:80–4.
[92] Hui X, Zhu B-R, Wu L-L, Gao W-Y, Li Y-M, Jia Q, Li H. Inhibitory activity of proanthocyanidins against *E. coli* 1-deoxy-D-xylulose-5-phosphate reductoisomerase. Nat Prod Commun 2021;16:1–8.

Section E

Nickel enzymes

Chapter 5.1

Urease

Ilaria D'Agostino* and Simone Carradori
Department of Pharmacy, University "G. d'Annunzio" of Chieti-Pescara, Chieti, Italy

1 Introduction

Urease, less known as urea amidohydrolase (BRENDA word map is reported in Fig. 1A), is a nickel-containing metalloenzyme (E.C. 3.5.1.5) catalyzing the hydrolysis of urea into ammonia and carbamate, which spontaneously decomposes into ammonia and bicarbonate ion (Fig. 1B).

Ureases are commonly found in a wide range of organisms belonging to all the kingdoms except for that of Animalia since their crucial role in accelerating the rate of urea/ammonia conversion by a factor of at least 10^{14} with respect to un-catalyzed elimination or hydrolysis reactions (which could take 3.6 years) [5], resulting to be the most efficient enzyme known to date [4,6]. This is fundamental in plants to prevent the accumulation of urea from arginine degradation and to implement the mineralization step of the nitrogen cycle. Also, the resulting pH changes serve several microbial pathogens, i.e., *Helicobacter pylori*, to survive and colonize specific host compartments, such as the acid human stomach. Thus, the interest of the scientific community in this enzyme as a pharmacological target is largely justified and, also, demonstrated by the numerous crystallographic structures (at least 79 entries including both the enzyme alone or in complex with a ligand) [7] reported in the Protein Data Bank (PDB) and in the enormous, higher and higher, number of publications related to this topic (Fig. 1C).

Notable curiosities for urease make it one of the Guinness world record enzymes in the history of science. In fact, urease, and in particular that from Jack bean, *Canavalia ensiformis*, was the first enzyme to be crystallized, in the far 1926, and proven to have a protein nature [8]. Moreover, for the first case, scientists recognized the role of nickel in catalytic processes, being urease a nickel-dependent enzyme [9].

However, although the large and detailed information we have obtained since that year, the domain-specificity of the urease structure (i.e., homotrimers in bacteria or heterotrimers in plants) [10] and the very short lifetime of the urease-urea adduct (approximately 20 μs) [6] have hampered the revealing of the catalytic mechanism and urea-binding mode and the issue has been debated for a long time [11], albeit one of the most reliable proposed mechanism is depicted in Fig. 1D [4].

2 Structure and function

According to phylogenetic analyses, ureases from different sources are quite divergent from a common ancestral protein, showing a primary sequence identity (about 55%) although they differ in their quaternary structures [12].

However, almost all [13,14] the urease isoforms share a common active site architecture, containing two ions of nickel (Ni^{2+}) coordinated by a group of four histidines, one aspartate, and a carbamoylated lysine (Fig. 2A).

The two Ni^{2+} ions are bridged by the carbamoylated lysine and a hydroxide ion that acts as a nucleophile in the catalysis. One of the two Ni^{2+} ions, namely Ni (1), is coordinated by two histidines, while the second one, namely Ni (2), by the other two histidines and an aspartate residue. Before the binding to urea, also three terminal water molecules coordinate the nickel ions, forming, along with the hydroxide ion, a hydrogen-bonded water tetrahedral cluster (Fig. 2B) [18].

For the urea binding, the carbonyl oxygen of urea interacts with Ni(1) provoking a conformational change of a mobile flap, a conformationally variable helix-turn-helix motif bearing a conserved cysteine, that covers and closes the active site all the time of the reaction catalysis, acting as a gate. Then, one of the two urea nitrogen atoms binds to Ni (2), establishing a bidentate bond with the enzyme and making the urea carbon more susceptible to the nucleophilic attack by the close hydroxide anion/water molecule. Then, the urea carbon atom undergoes to nucleophilic attack and a proton is transferred from the hydroxide ion to the distal urea nitrogen atom, yielding the urea in a tetrahedral transition state. The following release of

* Current Affiliation: Department of Pharmacy, University of Pisa, Pisa, Italy

Metalloenzymes. https://doi.org/10.1016/B978-0-12-823974-2.00035-8

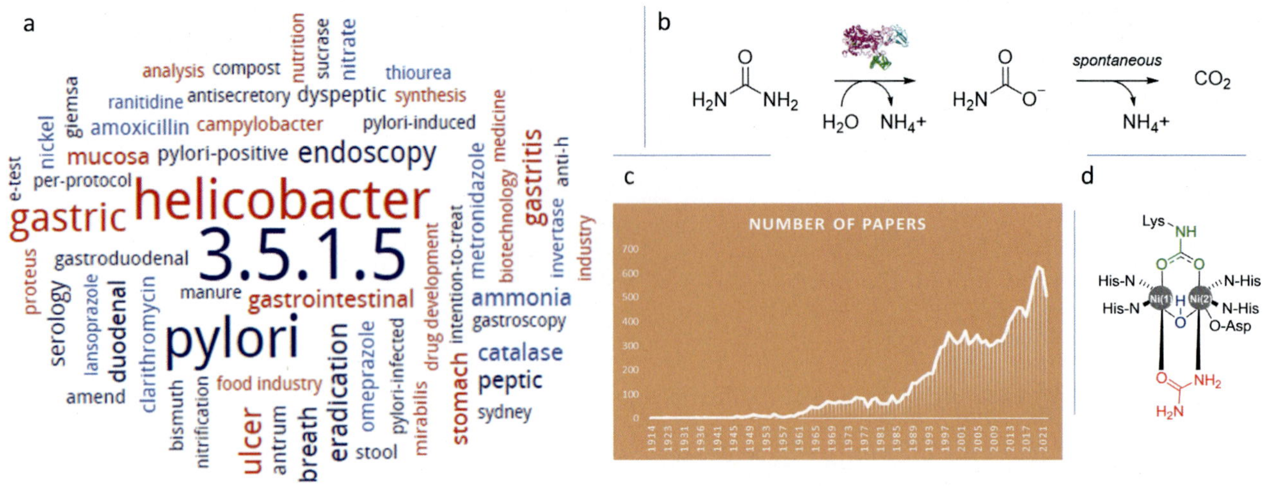

FIG. 1 Overview of urease: (A) word map, (B) catalyzed reaction, (C) number of urease-related publications per year, (D) proposed urea-binding mechanism of urease. (A) The word map for urease is obtained through the BRENDA database [1,2] and modified. The image contains the most relevant topic-related words in the enzyme publication context (PubMed titles and abstracts). The font size of each word is proportional to its relevance, while the color depends on the category. (B) Reaction catalyzed by urease: the substrate is urea and the product is carbamate, which decomposes into carbon dioxide. The enzyme is represented as a colored cartoon. (C) Curve of the number of publications (*y*-axis) per year (*x*-axis) for the period 1914–2021 obtained through the PubMed database [3] with the keyword "urease". (D) Representation of the binding pose of urea into the urease. Urea is shown in *red*, hydroxide in *blue*, and carbamate in *green*; nickel atoms are presented as *gray* balls [4].

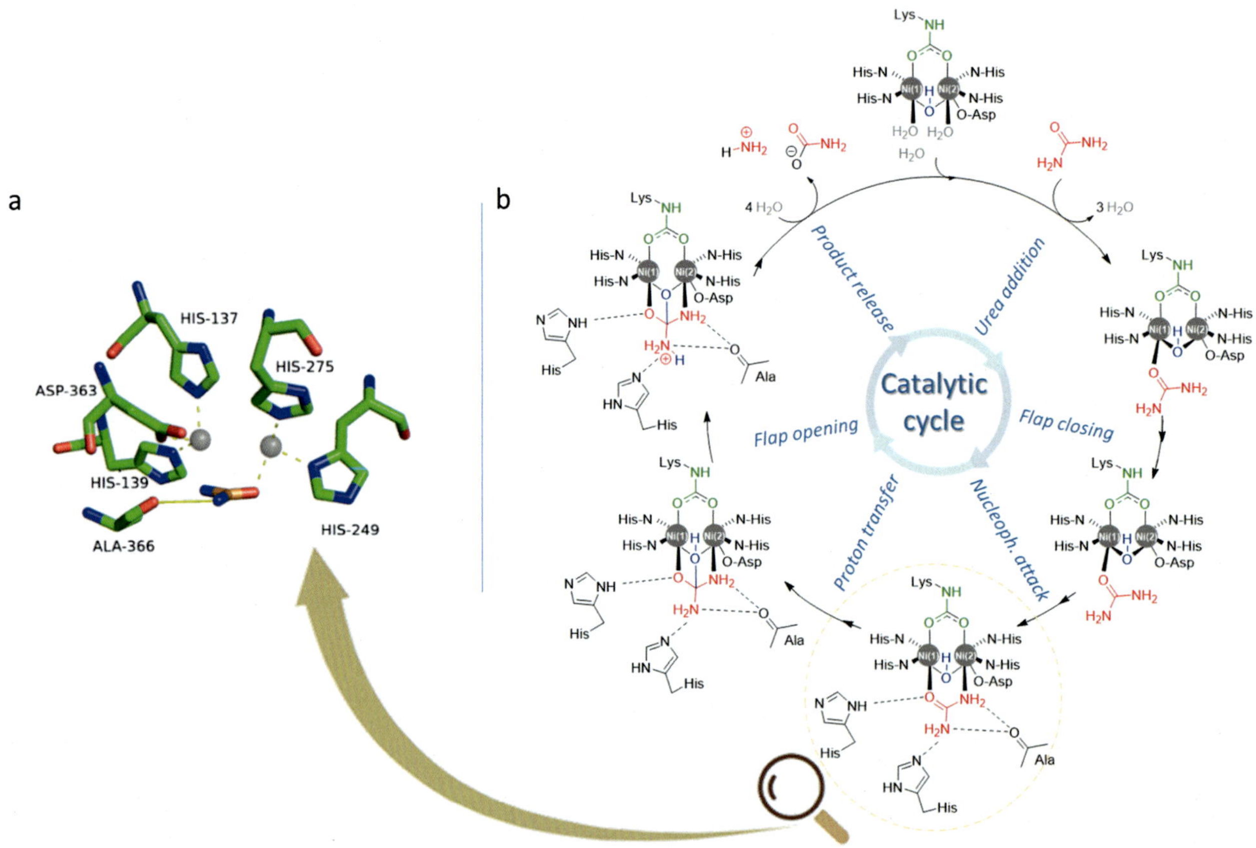

FIG. 2 (A) Binding pose of urea in the active site and (B) the proposed catalytic cycle of urea. (A) Zoom on the binding site of urea in *Sporosarcina pasteurii* urease. The complex is obtained by PDB ID: 6QDY [4]. Interaction residues are shown in green sticks and labeled, urea in orange sticks, and Ni^{2+} ions are presented as gray spheres. Hydrogen bondings and metal coordination are shown in yellow solid and dotted lines, respectively. The figure was prepared through Protein-Ligand Interaction Profiler (PLIP) web service [15] and PyMOL Molecular Graphics System, Version 2.5 Schrödinger, LLC [16]. (B) Reaction mechanism of urease according to Benini and coworkers [17]. The mechanism is based on the addition of the substrate urea, nucleophilic attack, proton transfer, and the release of carbamate and ammonia products through the mobile flap opening and closing.

ammonia yields the carbamate ion, which spontaneously hydrolyses into carbon dioxide (Fig. 2B).

As before mentioned, although the amino acid sequence, the catalytic site, and the mechanism of the whole ureases family are highly conserved [19,20], the overall quaternary structures, the chain topologies, and their oligomeric assembly are variable, even if some features are shared.

In each enzyme, several binding sites are present, one for every functional unit or monomer. One monomer is formed by one or more polypeptide chains according to the source of the enzyme. Generally, ureases belonging to plants and bacteria are characterized by a trimeric structure [21]. Monomers of plant and fungal ureases are composed of just one polypeptide chain (α), while those of bacterial enzymes are formed by three (α, β, and γ) different polypeptide chains, apart from the *Helicobacter* genus ureases that bear two subunits (α and β). Plant ureases are usually dimers of trimers $(\alpha_3)_2$, i.e., Jack bean, while bacterial ones are trimers ($[\alpha\beta\gamma]_3$) and urease from *H. pylori* assembles into a dodecamer (tetramer-of-trimers) composed of dimers ($([\alpha\beta]_3)_4$ [22] (Fig. 3).

To achieve the optimal catalytic activity, two conditions are strongly required: the insertion of the nickel ions in the active site, after they entered the cell through a nickel permease, and the carbamoylation of the lysine residue. To respect them, bacteria encode four accessory proteins (UreD, UreF, UreG, and UreE); the genes for these latter are grouped with UreA, UreB, and UreC genes encoding the urease subunits. The bacterial urease assembly is very complex, and the maturation of the active enzyme is accomplished through a preactivation complex that can be formed through two proposed pathways, as illustrated in Fig. 4A.

The first activation model (Figs. 4A, top, B) proposes that UreF binds UreG, facilitating the GTP uptake and transferring the metal ion to UreG. Then, $UreD_2F_2$ binds Ni^{2+}-UreG, and at this time GTP undergoes hydrolysis thanks to the $KHCO_3/NH_4HCO_3$ catalysis, generating the active enzyme. Instead, in the second proposal (Fig. 4A, bottom), the urease oligomer in apoprotein form (ApoU, $(UreABC)_3$) is first bound by UreD and this event marks the beginning of the urease activation process. Then, the binding to UreF and subsequently UreG is correlated to the conferring of GTPase activity. The consequent hydrolysis of GTP to GDP releases the energy required for the Ni^{2+} ions delivery from the chaperone UreE to the $(UreABC\text{-}UreDFG)_3$ complex [28].

Conversely, plant and fungal genomes do not encode for UreE but only for histidine-rich UreG that performs both the bacterial UreG and UreE functions [29].

3 Physiological roles and involvement in diseases

Catalyzing the urea hydrolysis, ureases play crucial roles in several physiological pathways in living organisms. In general, they provide several organisms with a fundamental source of nitrogen for their growth or they act as actual vir ulence factors in a number of pathogens.

Animals produce urea as a nontoxic form of ammonia, through the Krebs-Henseleit cycle, also known as the urea cycle, and do not need ureases to restore nitrogen atoms. In particular, in humans, urea undergoes hydrolysis in urine, blood serum, sweat, and exocrine gland secretions, and also gut microbiota seems to contribute to this process in a small percentage. However, the excess of ammonia can lead to an increase in pH and so stimulates the growth of opportunistic microbial pathogens and the development of their related diseases [30].

Plants, fungi, and bacteria use ureases to recycle urea nitrogen, which counts for 46% of the urea molecule since urea represents a repository of nitrogen for cell survival and growth. However, in plants, the only enzyme involved in urea production is arginase, an Mn^{2+} enzyme that hydrolyzes arginine residues. As confirmed by the KEGG

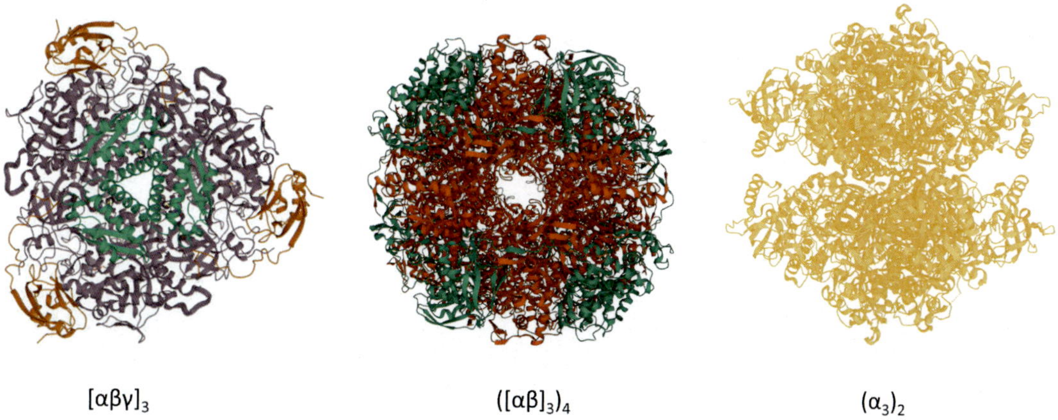

FIG. 3 3D structures of representative ureases. Crystal structures of different functional oligomers-composed ureases from *S. pasteurii* (PDB ID: 2UBP) [17], *H. pylori* (PDB ID: 1E9Z) [23], and *C. ensiformis* (PDB ID: 3LA4, Jack bean urease) [24]. *Images were prepared by using the online Mol* Viewer tool. Sehnal D, Bittrich S, Deshpande M, Svobodová R, Berka K, Bazgier V, et al. Mol* viewer: modern web app for 3D visualization and analysis of large biomolecular structures. Nucleic Acids Res 2021;49(W1):W431–7.*

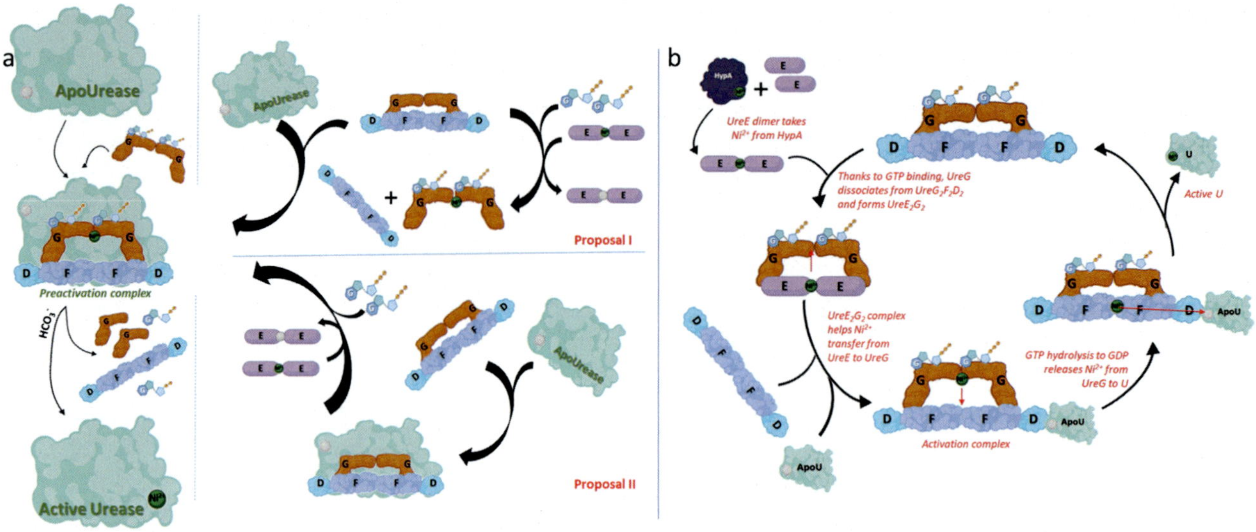

FIG. 4 (A, left) Activation mechanisms of *H. pylori* urease through the preactivation complex, generated through a specific pathway (A, right): two proposed models, 1 and 2, for urease preactivation and (B) detailed mechanism. In general, nickel-bearing UreG dimer is formed thanks to the nickel ion transfer from UreE dimer and generates a complex with $UreF_2D_2$ and ApoUrease. The preactivation complex induces conformational changes in the urease enzyme and bicarbonate anions stimulate the GTPase activity of UreG, leading to the delivery of nickel ions into the urease active site (Fig. 4A, left). The two proposed mechanisms differ in the nickel release from the differently composed complexes. In the first proposal (Fig. 4A, top), the GTP binding to UreG dimer makes it more prone to dissociate from the $UreG_2F_2D_2$ complex. Then, the GTP-bound $UreG_2$ receives the nickel ion from $UreE_2$, and at this point, the new complex is able to start the urease activation process by binding ApoU. Instead, according to the second proposed model, the UreE dimer transfers the nickel ion to the already assembled $UreG_2F_2D_2$ApoU [25–27] (Fig. 4A, bottom). Representation of proteins is for illustrative purposes and not comprehensive or related to their actual 3D structure. *The image was created with BioRender.com, Biorender. 2022. Available from: https://app.biorender.com/. Accessed 3 November 2022.*

database, urease is involved in arginine biosynthesis since arginine is converted into ornithine by arginase-releasing urea, in the degradation of the herbicide atrazine, due to the conversion of allophanate into urea by urease carboxylase, and in purine metabolism where urea derives from allantoate and ureidoglycolate at the end of the enzymatic cascade [31].

In plants, urease has a role in germination, seed sizing, and nickel depletion, which provokes the consequent inactivity of urease, the accumulation of a large amount of urea, and the appearance of necrotic spots on leaves in higher plants. Moreover, plant ureases play a protective role by exhibiting antifungal and insecticidal properties, likely due to the production of toxic ammonia [29,32,33].

Furthermore, as regards the application in agriculture, we need to consider that most of the currently employed nitrogen fertilizers contain urea because of its high nitrogen content (46%), low-cost, high-water solubility, and high foliar uptake along with the expression of specific urea transporters in plants to actively absorb urea from the soil. However, urea suffers from a massive release of ammonia, due to both the volatilization issue and the abundance of ureases in the soil, which hosts several urease-positive microorganisms or extracellular ureases bound to clays and organic colloids [30,34].

Another relevant aspect should be considered in terms of eco-suitability and environment-taking care: a large amount of ammonia in the soil due to the employment of urea-based fertilizers and the soil ureases activity negatively impacts the availability of minerals since the decrease in pH can result in carbonates precipitation. Hence, in this context, the employment and development of urease inhibitors combined with nitrogen fertilizers are strongly required [30].

Most interestingly for medicinal chemistry purposes, the role of ureases in bacteria and fungi is more defined and is implicated in the pathogenesis of several infectious diseases. This is often related to the consequent alkalization effect, fundamental to making the host cell more suitable to be infected and colonized by the pathogen. As an example, *H. pylori*, is a neutrophilic bacterium that is able to survive and persist in the harsh conditions of the acidic gastric environment by maintaining its cytoplasmic pH close to neutral through ureolytic activity, allowing the infection and, thus, gastritis, peptic ulcers, and gastric cancer. Furthermore, *H. pylori* urease causes a stimulation of the host inflammatory reaction and damage to tight junctions, resulting in a serious cytotoxic effect on the host. Notably, the catalytic activity of *H. pylori* urease is linked to the increase in blood platelet aggregation, resulting in gastritis but also cardiovascular diseases [19].

Staphylococcus aureus, *Escherichia coli*, and *Mycobacterium tuberculosis* employ urease for pH homeostasis and as a source of nitrogen. Moreover, the ability of ammonia to block phagosome-lysosome fusion could be an additional virulence factor for *M. tuberculosis*, being reflected in a decrease in the host immune response and infection propagation (urinary infection, hemorrhagic colitis, and tuberculosis). Urease is also responsible for the survival of *Yersinia enterocolitica*, providing nitrogen from the urea present in the soil and water [11]. Its ureolytic activity also stimulates the inflammatory response of the host provoking the development of reactive arthritis [19]. Urease from *Streptococcus thermophilus* is implicated in several amino acid biosynthetic pathways, such as those of aspartate, glutamine, and arginine. In this bacterium, besides the traditional pH modulation, urease also increases the activity of two enzymes, β-galactosidase, and lactate dehydrogenase [35].

Among the urease-positive pathogens, also *Proteus mirabilis* and *Proteus* spp. are worthy of interest since the urease-mediated pH increase from 6.5 to 9.0 causes supersaturation and precipitation of Ca^{2+} and Mg^{2+}, thus serious damage to the glycosaminoglycan layer responsible for urinary infections. The infection of these bacterial strains can cause urinary stones, promote pyelonephritis associated with kidney necrosis, due to urine alkalinization, and urinary catheter obstruction, known to be caused by calcium, phosphor, and magnesium encrustation and crystallization of carbonate apatite [30].

A different trend is recorded for urease-positive oral bacteria such as *Streptococcus sanguinis, Actinomyces naeslundii, Haemophilus parainfluenzae*, and *Streptococcus salivarius*. In fact, ammonia-mediated alkalinization neutralizes acids in the oral cavity, favoring oral health and preventing oral diseases such as dental caries, tooth decay, and plaque formation [36].

Among the urease-positive microorganisms, there are several fungal pathogens, such as *Cryptococcus neoformans*, *Coccidioides immitis*, *Sporothrix schenckii*, *Histoplasma capsulatum,* and *Aspergillus* spp. In particular, in *C. neoformans* and *C. posadasii* ureases hydrolyze urea in the epithelial lining fluid of lungs and the produced ammonia prevents immune function [11].

Pieces of evidence recognize urease as an important virulence factor in several bacterial and fungal phenotypes [35,37] and, as a proof-of-concept, WHO's priority risk pathogens list comprises 10 urease-positive species out of 12 [38].

Although the catalytic activity of urease is accomplished when the enzyme is activated through the whole process above-described, pieces of evidence seem to confirm the ability of the inactive enzyme or its individual subunits to exert biological effects. Moreover, urease has additional functions besides ureolysis, i.e., proinflammatory activity, immunomodulation, extragastric illnesses, and activation of platelets, neutrophils, and endothelial cells. Thus, the role of urease in the pathogenesis of several diseases is not dependent only on its ureolytic activity, albeit it is the most studied to date [30].

4 Insight into the dual urease-carbonic anhydrase enzyme system in *H. pylori*

Unmissable in this list of ureases functions to spend a few words about the involvement of *H. pylori* urease in a dual enzyme system with carbonic anhydrases (CAs). As widely described in Chapter 3.1., CAs are a superfamily of metalloenzymes, almost all Zn^{2+}-containing, involved in the reversible hydration of carbon dioxide to bicarbonate and proton. These ubiquitary enzymes are also found in *H. pylori* strains and, in particular, two isoforms belonging to the α- and the β- classes, *Hp*αCA and *Hp*βCA, have been fully characterized [39–41] and described as a potential pharmacological target for the development of inhibitors as antibacterial agents [42–45]. The dual urease-CA system (Fig. 5) acts as a buffering system in order to maintain an almost neutral cell pH and to ensure the survival of the bacterium since it could not tolerate the acid pH (about 2.0) of the stomach mucus layer.

In brief, urea enters the bacterial cell through an outer membrane porin and the inner membrane urea channel (UreI). In the cytoplasm, it is converted into ammonia and carbon dioxide in two (a urease-catalyzed and an uncatalyzed) steps. The produced carbon dioxide is captured by the intracellular *Hp*βCA and bicarbonate anion is released along with a proton that enters the equilibrium with urease-generated ammonia, yielding ammonium ion. Meanwhile, periplasmatic *Hp*αCA catalyzes the CO_2/HCO_3^- interconversion by maintaining the pH at 6 by using the protons diffused from the acidic gastric lumen through the outer membrane. The overall process is regulated by UreI, whose channel is closed at neutral pH, not allowing the entry of urea in the periplasm, and open at pH below 6.0 to promote the activation of the dual urease-CA system.

5 Urease as a diagnostic tool for *H. pylori* infections

As above-mentioned, human DNA does not encode for ureases, thus, in case of infection from a pathogen with abundant urease activity, this can be used as a biomarker for the diagnosis of the infectious disease. The exemplar case for such an application is that of *H. pylori* infection, which is not easy to detect and eradicate. In fact, two different tests were developed based on the *H. pylori* urease activity: the rapid urease test (RUT) and the urea breath test (UBT). The first tool can be used on every gastric specimen, such as mucus, biopsy, or brushings, and detects the changes in pH caused by urease activity in presence of urea through pH indicators. The UBT is the most commonly used in clinics and consists of the employment of labeled carbon atom (^{13}C or ^{14}C)—bearing urea that is first ingested by the patient and, then, the isotope enrichment of CO_2 in the exhaled air is detected by specific techniques, such as

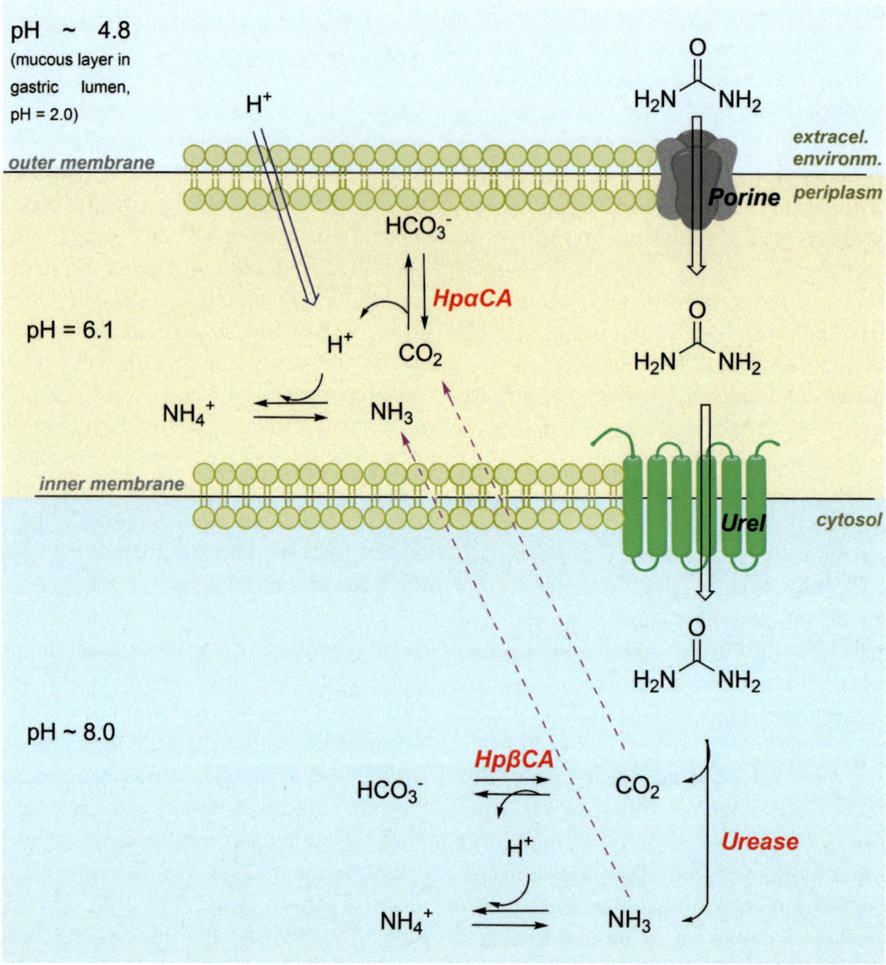

FIG. 5 Schematic representation of the role of the dual urease-CA enzyme system in periplasmic pH maintenance in *H. pylori*. Urea enters the *H. pylori* cell through the outer membrane porins. In acid environment, the urea channel (UreI) is open and urea is its substrate, crossing the inner membrane and moving up to the *H. pylori* cytoplasm, where it is metabolized by urease into CO_2 and NH_3, which diffuse rapidly into the periplasm, being both gases. In both cytosol and periplasm, the former is hydrated to H_2CO_3 by β- and α-CA isoforms, respectively, and dissociates to H^+ and HCO_3^-, while NH_3 is protonated by the CA-released proton into NH_4^+. The NH_3/NH_4^+ system buffers the periplasm. Purple dotted arrows indicate passive diffusion, blue and black arrows indicate several types of active and passive diffusion. Representation of membranes, porin, and UreI is for illustrative purposes and not comprehensive or related to the actual 3D structure of the proteins. *The image was prepared through PerkinElmer Informatics ChemDraw 21.*

isotope ratio mass spectrometry or scintillation according to the employed carbon isotope [46].

6 Urease as pharmacological target: Design and development of inhibitors

In light of all the analyzed physio-pathological roles that urease plays, the rising interest in its modulation is fully justified and several project lines focused on the discovery and development of innovative and safe urease inhibitors (UIs).

A preliminary but basic consideration should be performed as regards the employment of Jack bean urease in the enzymatic inhibition assays. Interestingly, although most of the research efforts in this frame are directed toward the development of UIs for the treatment of *H. pylori* infections and several protocols of expression and purification of *H. pylori* urease were reported [47,48], only a few research groups used to test in vitro their designed inhibitors on the *H. pylori* enzyme [49,50] and the assay-of-choice still remains that on Jack bean urease. In fact, the literature is rich in studies and research and development programs based on enzymatic assays on the Jack bean enzyme. The scientific community seems to justify this approximation since the high degree of identity, especially in the active site, among ureases and the reproducibility and the fast obtainment of results due to the commercial availability of the Jack bean enzyme and its related assay kits. However, this could represent a criticism in the advanced phases of new *H. pylori* UIs design since several discrepancies in kinetic properties and substrate affinities were reported [51–53].

In the last decades, wide libraries of UIs, synthetic and natural, organic and inorganic, have been designed, synthesized, and tested. These compounds can be classified according to their manner of inhibition, thus noncovalent (competitive, noncompetitive, and mixed type) and covalent UIs, or their mode of binding the enzyme. The latter distinction involves three main classes [54]:

(1) The substrate-like inhibitors, which are structural analogs of the natural urease substrates thiourea and hydroxyurea. These UIs usually bind directly to the urease active site with a similar binding pose/manner to that of the substrate urea, i.e., by coordinating nickel ions, with a reversible mechanism;
(2) Binders to the active site of the enzyme via reaction with the nickel ions, with a different binding mode with respect to urea and a nonanalog structure to the natural substrates. An example is acetohydroxamic acid, which binds the two Ni centers in a sort of bridge thanks to its negatively charged oxygen and coordinates nickel ions through its carbonyl oxygen;
(3) Different mechanisms by nonsubstrate-like structures, such as the influence of the urease catalytic mechanism, i.e., phosphorodiamidates activated by urease into potent irreversible inhibitors or cysteine-reactive inhibitors, which covalently bind to the key cysteine residue of the urease mobile flap, preventing the access of the substrate urea to the active site.

Generally, UIs are grouped into inhibitors bearing the urea fragment and its isosteres (thiourea, semicarbazones, barbiturates, etc.), those with different chemical scaffolds (coumarins, chalcones, etc.), and metal complexes.

Although the enormous efforts in the development of anti-urease agents, none is actually in clinical trials (the search was performed on the NIH ClinicalTrials.gov database, data not shown), and no patent describing new UIs with medical purpose has been applied in the past 2 years (the search was carried out in November 2022 with the following official scientific databases Espacenet and GooglePatents, data not shown).

In this paragraph, representative compounds for each class of UIs, their enzymatic and biological data and the interaction hotspots with ureases will be described.

6.1 Urease inhibitors and related compounds in the DrugBank

Several urease-related compounds are reported in the DrugBank Database [55] and are illustrated in Fig. 6.

Acetohydroxamic acid reversibly inhibits bacterial ureases in the urines leading to a decrease in pH and ammonia concentration and is used in clinics in coadministration with antibiotics to treat chronic urea-splitting urinary infections.

Lactulose, the synthetic derivative of lactose, consisting of a disaccharide of galactose and fructose, is hydrolyzed by several saccharolytic bacterial species that populate the large intestine. The resulting lactic acid, formic and acetic acids are very volatile and contribute to the intraluminal gas formation, the peristaltic gut motility, and the osmotic increase in the water content of stool, with a laxative effect. The consequent colon acidification is useful in the treatment of portal-systemic encephalopathy since

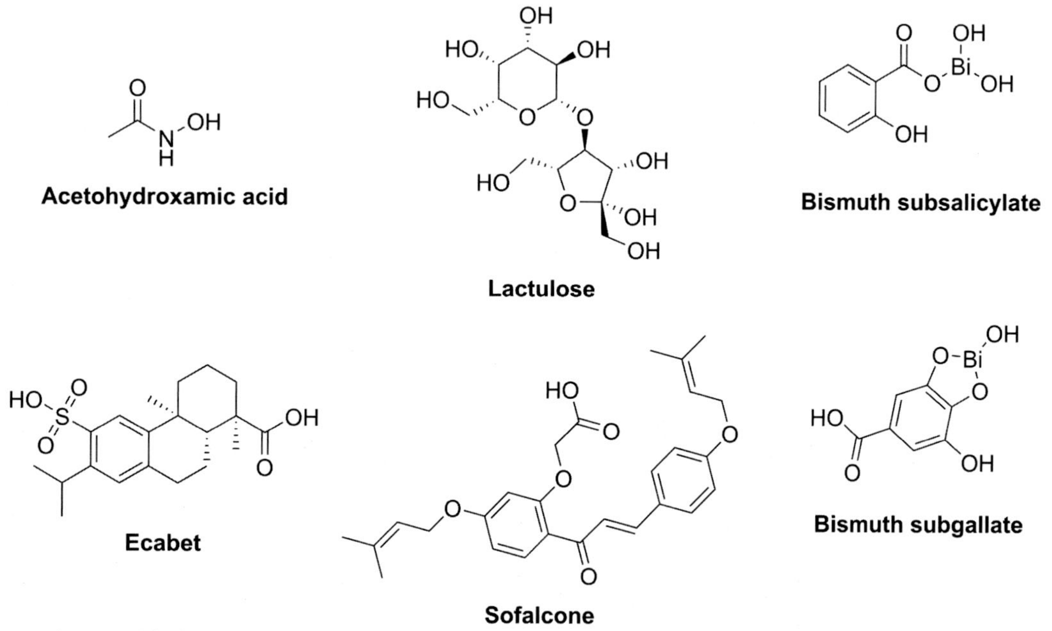

FIG. 6 Compounds reported in the DrugBank as UIs searching for "urease" keyword.

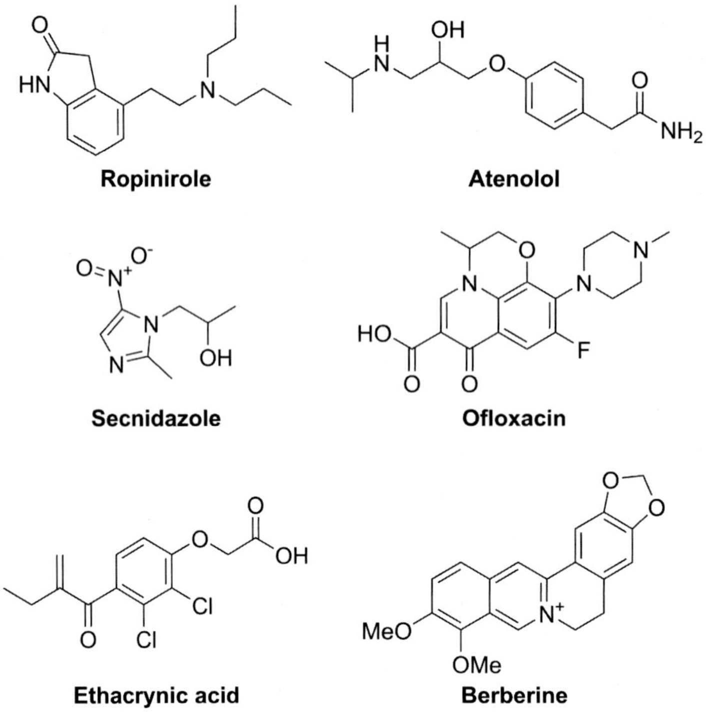

FIG. 7 Representative examples of repurposed drugs as UIs.

the accumulated ammonia is converted into ammonium ions, not able to diffuse through the systemic circulation. Moreover, lactulose is believed to eliminate the urease-producing bacteria responsible for the formation of ammonia.

Bismuth subsalycilate and **subgallate** have been employed in the treatment of *H. pylori* infection for a long time. Indeed, bismuth complexes exert antimicrobial effects against various gastrointestinal tract bacterial species, such as *E. coli*, *Vibrio cholerae*, *H. pylori*, and some enteric viruses, such as *Rotaviruses* with a not determined specific mode of mechanism(s) of action. Bismuth polypharmacology includes also the complexation of bacterial walls and periplasmic membranes, the inhibition of bacterial enzymes like urease, catalase, and lipase along with protein and ATP synthesis. Moreover, it seems to decrease the adherence of *H. pylori* to gastric epithelial cells.

Ecabet is a drug used in dry eye syndrome, also recognized anti-*H. pylori* agent by inhibiting the bacterial NADPH oxidase and urease, preventing the adhesion to the gastric mucosa and the survival in the stomach.

Sofalcone, a mucosal protective compound, is a known anti-*H. pylori* agent through a direct bactericidal effect acting as an inhibitor of the urease activity and reducing the pathogen adhesion to gastric epithelial cells [55].

Additionally, in the era of the repositioning of known drugs for out-of-label applications, some molecules present in the DrugBank Database were investigated for their anti-urease activity and some of them are illustrated in Fig. 7 [55,56].

The anti-Parkinson **ropinirole** (Fig. 7) was subject to a recent patent application as an inhibitor of Jack bean urease with a micromolar IC_{50} (11.7 μM) and a mixed-type of inhibition mechanism, likely interacting with both the active and allosteric sites of the enzyme [57]. Also the beta-blocker **atenolol** resulted in a mild inhibition of urease, showing an IC_{50} of 64.36 μM [58]. However, while the antibiotic **secnidazole** exerted a moderate anti-urease activity with an IC_{50} value of 156 μM [59], quinolones, such as **ofloxacin**, showed very interesting inhibitory profiles [60] and several potent derivatives series were developed and discussed later. No exciting results for the diuretic agent **ethacrynic acid**, which was found to moderately inhibit the urease activity with an IC_{50} of 1.21 mM against Jack beans urease [61], and the antifungal and antidiarrheal alkaloid **berberine**, which showed an $IC_{50} = 5.09$ mM against Jack beans urease and no activity versus *H. pylori* urease [62].

6.2 UIs endowed with urea fragments and isosters

Thioureas. The trisubstituted urea **1** (Fig. 8) was found to inhibit Jack bean urease in vitro with an IC_{50} of 11.73 μM. Docking studies revealed this atenolol derivative well suited to the catalytic cavity and the establishment of ionic bonds between the sulfur and the oxygen atoms and the two nickel ions [58].

Thioureas **2** and **3** (Fig. 8), bearing palmitoyl and aroyl substituents, respectively, were reported to strongly inhibit

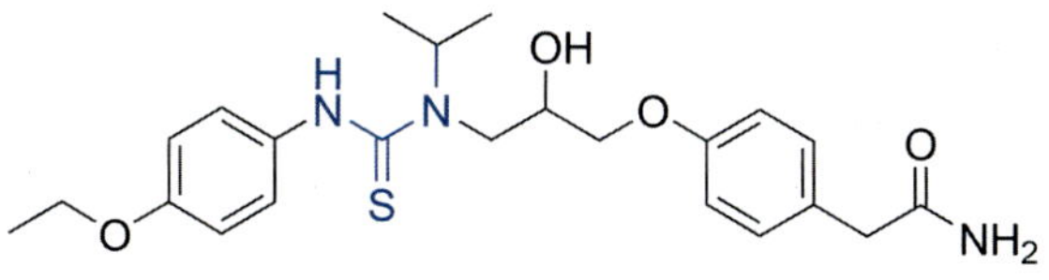

1

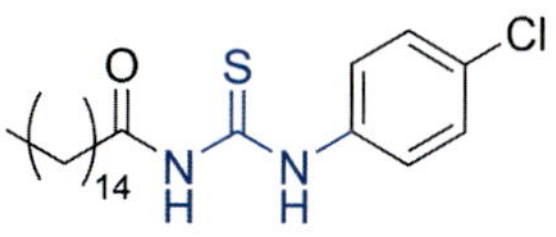

2

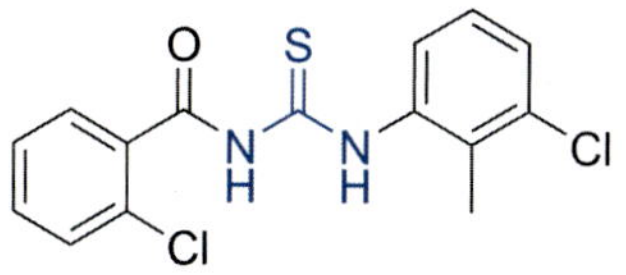

3

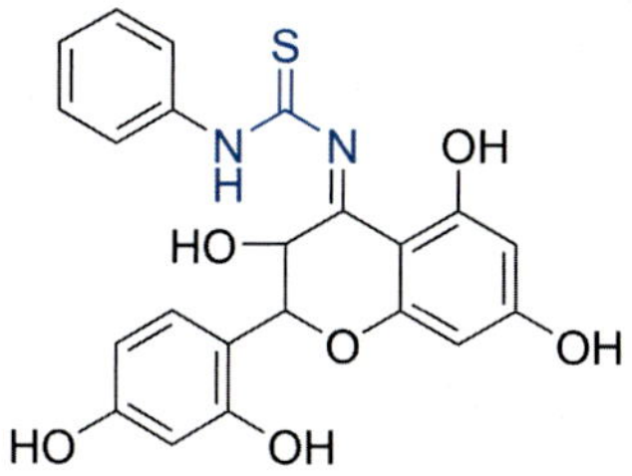

4

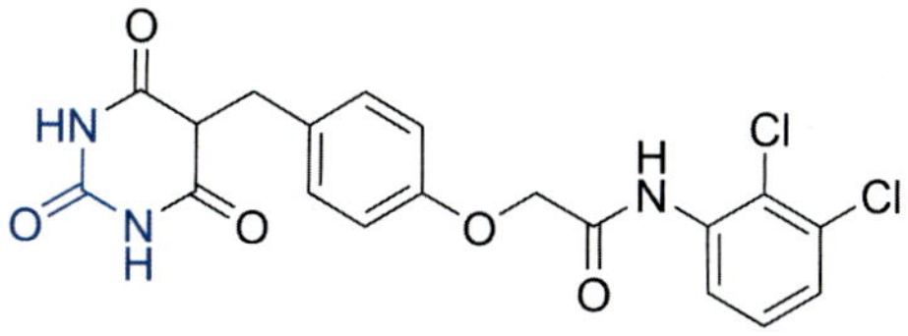

5

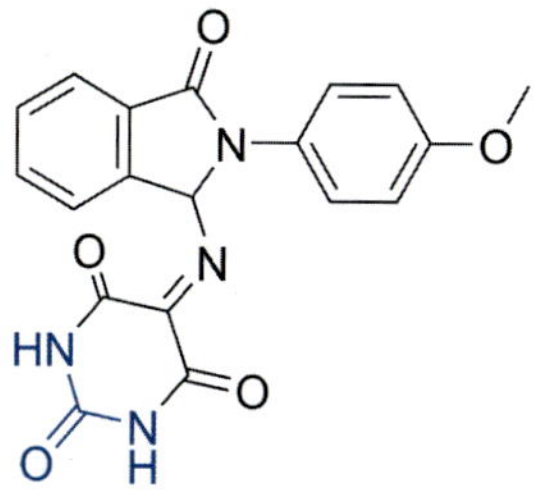

6

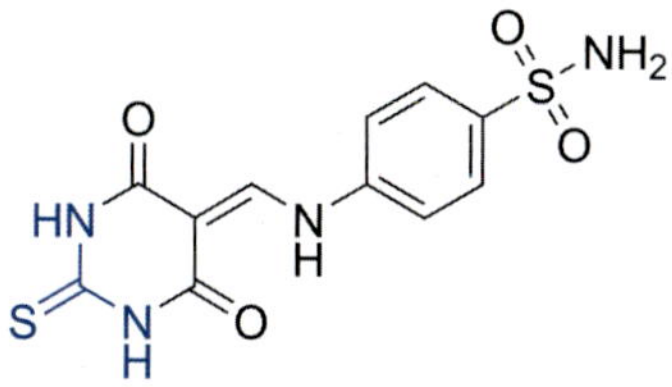

7

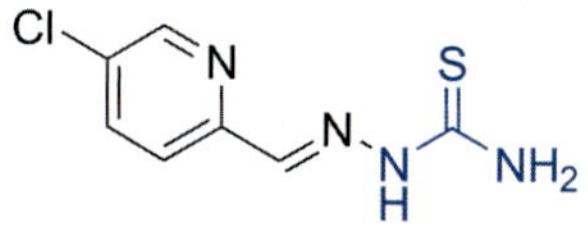

8

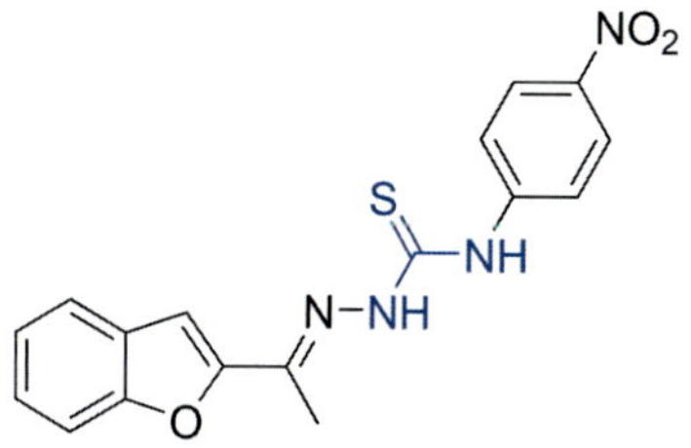

9

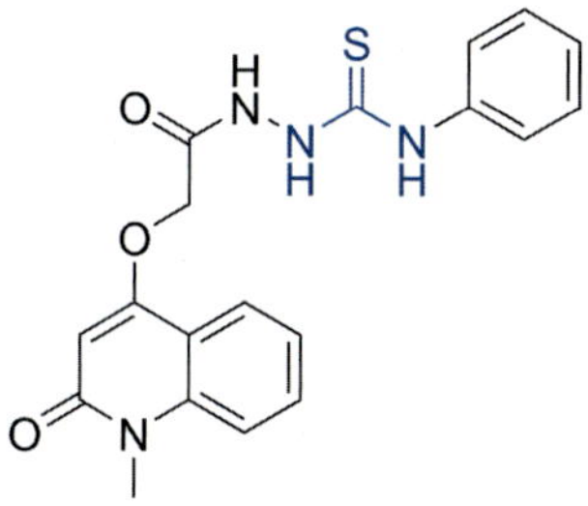

10

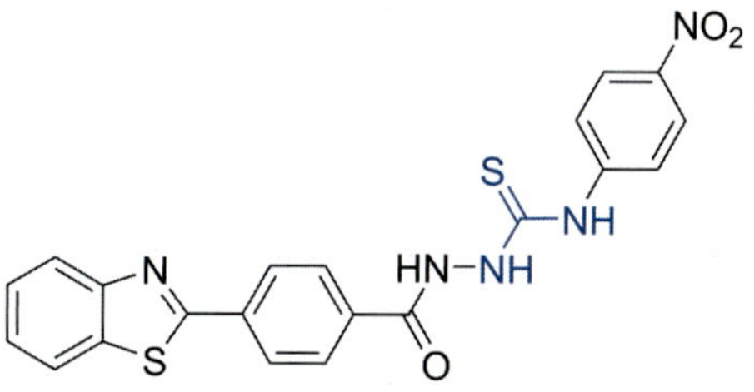

11

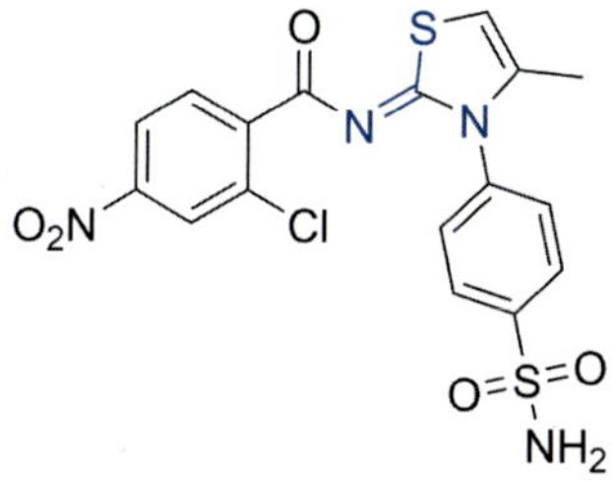

12

FIG. 8 Representative UIs endowed with the urea or thiourea fragment (*blue*).

urease with IC_{50} values of 17 and 1.9nM. Kinetic studies revealed a noncompetitive inhibition, while in silico simulations revealed unexpectedly a non-Ni-binding mode of action for **2** and a strong hydrogen bonding network with the active site for **3**. Moreover, molecular dynamics (MD) confirmed **3** to be firmly bound to the urease binding site [63].

Interestingly, derivative **4** (Fig. 8), endowed with a thiourea-conjugated flavone core, showed a micromolar inhibitory activity against Jack bean urease ($IC_{50}=7.37\,\mu M$) along with peculiar antioxidant activity. The arylthiourea group was found to enhance both the urease inhibition and the antioxidant behavior of morin, the parent natural flavonoid compound, and also, to confer a good inhibition of *H. pylori* growth. Moreover, docking simulations highlighted the establishment of eight hydrogen bonds in the active site [64].

Barbiturates. Barbituric derivative **5** (Fig. 8) inhibits urease in a submicromolar range ($IC_{50}=0.69\,\mu M$), fitting well with the active cavity and interacting with the active site flap residues, as resulted in the docking and molecular dynamics simulations [65]. A comparable activity ($IC_{50}=0.82\,\mu M$) was found for compound **6** (Fig. 8), characterized by the presence of two chemical features recognized to inhibit ureases, the barbituric acid portion and the isoindoline ring. **6** seemed to establish three strong hydrogen bonds with the active site [66].

Sulphonamide-bearing thiobarbiturate **7** (Fig. 8), endowed with an IC_{50} value of 3.76 μM on bean Jack urease, was found to coordinate the urease nickel ion through the thiourea portion of the heterocycle [67].

Semicarbazones. A small compound, pyridine substituted thiosemicarbazone **8** (Fig. 8), was recently reported as a micromolar UI with an IC_{50} = 1.07 μM [68].

In compound **9** (Fig. 8), a benzofuran-linked thiosemicarbazone, the presence of a nitrophenyl ring emerged as a great benefit for the potency in urease inhibition (IC_{50} = 77 nM). Also for **9**, docking simulations suggested the relevant role of the sulfur atom, being the core of the binding to urease through the coordination of the two catalytic nickel ions [69].

Hydrazinecarbothioamide **10** (Fig. 8) emerged for its inhibitory activity in the low micromolar range (IC_{50} = 1.83 μM) and antibacterial activity against P. mirabilis strain (MIC = 27 μM). The thiourea moiety of **10** was also found to establish two hydrogen bonds with urease in a chelation fashion [70]. The same thiourea-containing moiety is present in compound **11** (Fig. 8), a competitive inhibitor (IC_{50} = 1.4 μM). **11** was found to bind tightly with the binding cavity of Jack bean urease, also thanks to the benzothiazole ring, interacting with the catalytic nickel ion [71].

Iminothiazolines. A recently reported representative of 2-iminothiazoline derivatives is compound **12** (Fig. 8), also bearing a benzenesulfonamide group. **12** is a competitive inhibitor and exhibited an anti-urease activity at a nanomolar concentration (IC_{50} = 58 nM). Through in silico studies, the main interacting residues of the Jack bean urease binding site were identified in two aspartates, one histidine, two alanines, and others [72].

6.3 Non-urea-based inhibitors

Hydrazones. Another flavone-based compound (**13**, Fig. 9) was recently reported as UI.

In particular, **13** is characterized by the benzoyl hydrazone portion and the presence of a fluorine atom on the flavone core. These two moieties seem to confer to the compound a potent activity against urease, recording an IC_{50} of 1.21 μM. In particular, the benzoyl hydrazide function forms a π-π interaction with a tryptophan residue [73].

The same nitrogen-containing function is also present in compound **14**, (Fig. 9), which emerged from an innovative fragment-based dynamic combinatorial chemistry and amplification protocol for its promising anti-urease activity in the low-micromolar range (IC_{50} = 10.55 μM) and non-toxic profile. Kinetic studies disclosed a mixed-type inhibition for **14** and docking simulations revealed its accommodation in the center of the active pocket of Jack bean urease. The hydrazone forms a thermodynamically favorable binding with the enzyme through relevant interactions, such as the nickel chelation by the two hydroxyl groups and an intense hydrogen bonding network with binding site histidine and aspartate residues. Also, the imidazole ring is able to generate a π-alkyl interaction, while the phenolic ring establishes π-alkyl hydrophobic interaction with a binding site alanine. A stacked docking showed the extension of the nitroimidazole ring into the active site and the formation of the key interaction with an alanine residue. However, the presence of an indole ring in place of the nitroimidazole in compound **15** (Fig. 9) seems to improve the inhibitory efficiency, reaching a notable submicromolar IC_{50} (0.60 μM) [74].

Benzohydrazides. In enzymatic assays, benzohydrazide **16** (Fig. 9) emerged as a nanomolar mixed-type inhibitor of urease, showing an IC_{50} of 0.87 μM. The ring B methoxy substituent seemed to positively impact the ligand-enzyme interaction since its oxygen atom establishes relevant hydrogen bonds. Another interesting piece of information from docking simulations is the accommodation of **16** in the binding pocket and its tight contact with nickel ions [75].

Hydroxamic acids. An IC_{50} value of 83 nM was determined for the mixed-type inhibitor **17** (Fig. 9), a small molecule endowed with a hydroxamic acid function. The compound was tested as a racemate since docking studies on *H. pylori* urease showed a similar binding pose for the R and S enantiomers, both fitting well in the active site pocket [76].

The introduction of a benzyl ether, unfortunately, led to an increase in the IC_{50} value (0.15 μM) in hydroxamate **18** (Fig. 9). However, the compound resulted in a moderate moderate human and rabbit plasma protein binding, laying the foundation for further studies [77].

Sulfo derivatives. A sulfamate compound (**19**, Fig. 9), previously reported as an inhibitor of steroid sulfatase in JEG-3 placental cell lysate, was widely described for its anti-urease activity. Its nanomolar inhibition of urease (IC_{50} = 62 nM) drove the authors to investigate the key interaction in the active site. In particular, **19** seemed to compete with urea for the binding. The hindered adamantane and phenyl rings establish multiple alkyl and π-alkyl interactions, while the sulfamate oxygen atom and the amide nitrogen are involved in the conventional hydrogen bonds with one of the two enzyme arginine residues. Computational analyses also revealed that the presence of *o*-bromo substituent on the phenyl ring is related to the conformational changes and the binding of compound within the urease active pocket. Moreover, HYDE affinity studies showed a high affinity toward urease and Molecular Dynamics confirmed the significant stability of the compound-enzyme interaction. Notably, predicted pharma cokinetic properties suggested a satisfactory aqueous solubility and a good blood-brain barrier permeation, corroborating its valuable drug-likeness as a drug candidate [78].

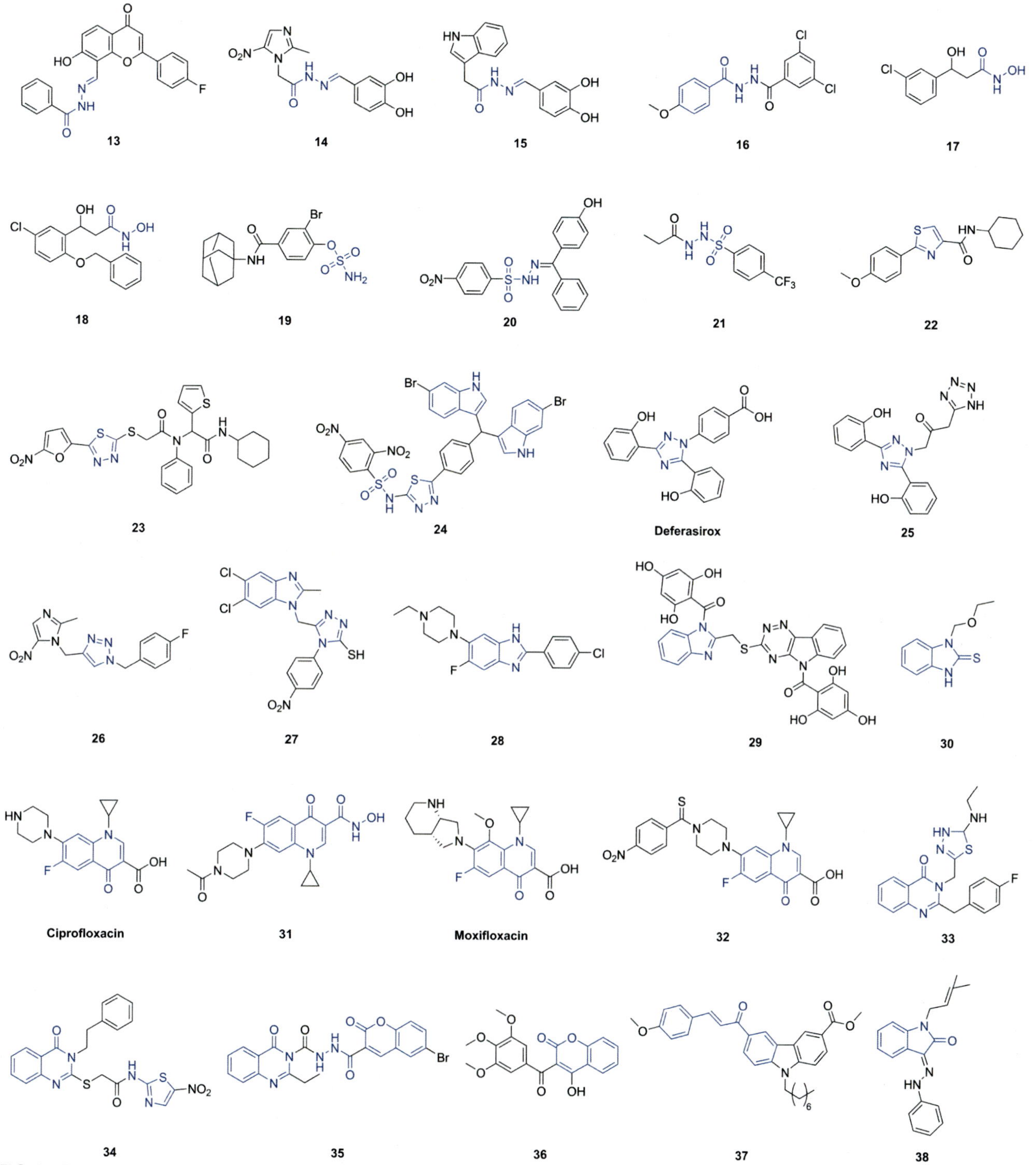

FIG. 9 Representative UIs endowed with non-urea analog fragments but widely studied pharmacophoric features (*blue*).

The modified *p*-nitrobenzene sulfonohydrazide **20** (Fig. 9), exhibited an interesting activity in the in vitro inhibition assays, showing an $IC_{50} = 3.90\,\mu M$. Computational studies revealed a specific aromatic-cation interaction with a catalytic histidine residue and, also, nickel coordination by the hydroxyl group in the benzophenone portion [79]. Moreover, an acyl sulfonohydrazide (**21**, Fig. 9) was recently reported as low-micromolar UI, with an IC_{50} value of $2.50\,\mu M$, and good antibacterial and antifungal activities were observed. The presence of the trifluoro group in the *para*-position of the phenyl ring forms a strong hydrogen bond with the urease binding pocket [80].

Thiazoles. Among the latest 21 reported thiazole-based UIs, compound **22** (Fig. 9) emerged for its inhibitory activity in the low-micromolar range ($IC_{50} = 1.82\,\mu M$). The X-ray single-crystal diffraction and Hirshfeld surfaces analysis highlighted the establishment of N-H/O hydrogen bonds and C-H/O intermolecular interaction [69].

Thiadiazoles. In the sulfur- and nitrogen-containing heterocycles, thiadiazoles shine for potency in inhibiting urease enzymes. Compound **23** (Fig. 9) showed a submicromolar activity ($IC_{50} = 0.94\,\mu M$) as a noncompetitive inhibitor, probably binding to an allosteric site of the enzyme or enzyme-substrate complex through a rich network of hydrophobic interactions [81].

A compound containing several key features for the urease affinity is thiadiazole **24** (Fig. 9). Indeed, besides the thiadiazole ring, also a sulfonamide function and two indoles are present in the chemical structure. The diindolylmethane **24** is a natural alkaloid characterized by an IC_{50} value of 0.50 μM against Jack beans urease. Docking studies revealed the relevant hydrogen bonds established by the nitro groups at the ortho and para positions of the benzenesulfonamide portion [82].

Triazoles. With an IC_{50} value of 1.27 μM, compound **25** (Fig. 9) emerged as one of the most active derivatives of **deferasirox** (Fig. 9), bearing a 1,2,4-triazole. This ring makes **25** well-anchored at the catalytic site. The compound extends to the entrance of the binding pocket, limiting the mobility of the flap [83].

A 1,2,3-triazole is, instead, present in compound **26** (Fig. 9), which inhibits urease with an IC_{50} value of 1.98 μM in a noncompetitive manner by stabilizing the active site flap. In fact, the fluorine atom on the benzyl group pointed toward the center of the double nickel ions and several interactions were established to anchor the compound to the helix-turn-helix motif in the active site cavity [84].

Benzimidazoles. The benzimidazole-triazole hybrid **27** (Fig. 9) bearing a nitro group was found to be a nanomolar inhibitor of urease ($IC_{50} = 29\,nM$). The compound core establishes relevant interactions with the active site, in particular, a hydrogen bond with an arginine residue and a π-π stacking interaction with a histidine, while the nitro function creates a salt bridge with an aspartate [85].

A benzimidazole ring is also present in compound **28** (Fig. 9), which showed an IC_{50} of 3.36 μM. According to computational studies, **28** establishes Van der Waals contacts in the active cavity and hydrogen bondings. Remarkably, the fluorine atom of **28** interacts with a histidine residue [85].

A more complex structure is that of benzimidazole **29** (Fig. 9), in which a triazinoindole is linked. This compound, with a submicromolar activity ($IC_{50} = 0.20\,\mu M$) strongly interacts with the enzyme, also thanks to the tri-hydroxy substitution pattern at the 2,4,6-position of both heterocycles [86]. Also, a small compound (**30**, Fig. 9), a benzimidazole-2-thione, was recently reported to inhibit *H. pylori* and Jack bean ureases with IC_{50} values of 31 and 62 nM, respectively [87].

Fluoroquinolones. As above-mentioned, promising inhibition data were disclosed by testing some antibacterial fluoroquinolones and, in particular, interesting activities on *H. pylori* and *P. mirabilis* ureases came also from **ciprofloxacin** (Fig. 9), which showed $IC_{50} = 3.5\,\mu M$ [88]. Molecular modeling studies revealed the key interaction between the carboxylic acid and the nickel atom; otherwise, when this moiety is substituted by the hydroxamic acid function (**31**, Fig. 9), the affinity to Ni^{2+} increases and improves the affinity and inhibition to the enzyme, inhibiting urease with $IC_{50} = 2.22\,\mu M$ [88]. In this context, a notable inhibitory profile was recorded for **moxifloxacin** (Fig. 9), as demonstrated by the IC_{50} value of 0.66 μg/mL measured for the silver nanoparticles charged with the compound [89].

The derivatization of ciprofloxacin to increase the enzyme affinity led to compound **32** (Fig. 9), bearing a thioamide group on the piperazinyl moiety of the parent compound. In this way, the activity increased till to an IC_{50} of 2.05 μM in the urease inhibition, even if a decrease in the antibacterial activity was observed, with MIC values of 0.781, 0.390, 1.562, and 0.097 μg/mL on *S. aureus*, *Staphylococcus epidermidis*, *P. aeruginosa*, and *E. coli* species, respectively (compared to MICs = 0.024, 0.048, 0.048, and < 0.003 μg/mL exerted by ciprofloxacin). However, an anti-ureolytic assay demonstrated that the compound is able to reduce the urease activity of some representative urease-positive pathogens, such as *C. neoformans* and *P. vulgaris,* recording IC_{50} values of 5.59 and 5.72 μg/mL, respectively. An in-depth investigation of the binding pose of compound **32** was performed through an induced-fit docking and molecular dynamics, resulting in a pronounced interaction with the active site and, in particular, the mobile flap residues (two histidines and one arginine) through nickel coordination by the quinolone ring, respectively [90].

Quinazolinones. Compound **33** (Fig. 9) has a quinazolinone scaffold linked through a methylene group to a dihydrothiazolidine ring. The compound showed an IC_{50} of 0.17 μg/mL and the docking simulation highlighted the establishment of a number of hydrogen bonds and π-stacking interactions by the thiadiazole cycle. While, the quinazolinone skeleton, being another important feature for urease inhibition, forms additional hydrogen bonds and the fluorobenzyl group fits the enzyme hydrophobic region [91].

Compound **34** (Fig. 9), a quinazolinone linked to a 2-aminothiazole, exhibited a potent uncompetitive inhibitory anti-urease activity ($IC_{50} = 2.22\,\mu M$). Molecular docking simulations revealed a good fitting into the binding site thanks to an interaction network rich in π-cation and π-π contacts and H-bondings. **34** also showed IC_{50} values of 129.4 and 172.4 μg/mL against *C. neoformans* and *P. vulgaris*, respectively, in the ureolytic assays [92].

An $IC_{50} = 1.26$ μg/mL was determined for the coumarin-containing quinazoline **35** (Fig. 9). The compound resulted to be fixed to the active site through the CO/H-N and CO/H-O-H/N hydrogen bonds between the compound and histidine residues and, remarkably, the quinazolinone nitrogen forms a hydrogen bond with the urease key cysteine [93].

Coumarins. 4-Hydroxycoumarin **36** (Fig. 9) exhibited a micromolar activity ($IC_{50} = 11.30$ μM) and seems to form a stable complex with the two nickel atoms through the oxygen of its methoxy groups [94].

Chalcones. The carbazole-chalcone hybrid **37** (Fig. 9), with an IC_{50} of 6.88 μM, is a moderate UI. The compound resulted to be cleverly installed on the active site interface and able to mediate the interaction with the nickel ions [95].

Indoles. The oxoindoline **38** (Fig. 9) is a potent urease inhibitor, as demonstrated by the IC_{50} value of 0.71 μM on *H. pylori* urease, and antibacterial agent, showing a MIC = 0.48 μM on a pathogen strain. With a good ADMET profile, the compound seems to be a good candidate for deeper investigation. 3D-QSAR studies highlighted the relevant role of larger electron-donating groups in almost all positions, while electron-withdrawing groups were only suitable for narrow stretching directions [96].

Other compounds. A promising anti-urease and antioxidant activity ($IC_{50} = 11$ and 5.10 μM, respectively) was found for the chlorogenic derivative **39** (Fig. 10). For this compound, the incorporation of hydroxy groups and chloro- and nitro-substituted phenyl rings were demonstrated to enhance the urease inhibitory activity [97].

A rhodamine G6 derivative **40** (Fig. 10) showed a submicromolar anti-urease activity ($IC_{50} = 0.11$ μM) and the phenyl substitutions were found to significatively affect the activity [98].

The plant-derived **coptisine** (Fig. 10) exerts the most potent inhibitory activity toward ruminal bacterial urease ($IC_{50} = 41.53$ nM) and a moderate inhibition of Jack bean urease ($IC_{50} = 2.45$ μM). The compound interacts with the cysteine SH group in the flap, the nickel ions, and key residues in the active site. Furthermore, it is able to slow down the release of ammonia and the decomposition of urea during rumen microbiota fermentation in vitro [99].

The mixed-type UI **41** (Fig. 10) showed IC_{50} values of 74 nM and 0.51 μM in cell-free and intact cell *H. pylori* urease inhibition assays. In-depth biological investigations revealed that the antibacterial activity of the disulfide **41** (MIC = 0.98 μg/mL) was comparable to that of amoxicillin and the cell viability was major than 90% in three different cell lines, ensuring its safety. Unexpectedly, **41** seems to bind the key cysteine residue with a noncovalent interaction [100].

Some analogs of the tetrahedral transition state of the substrate urea have been developed as UIs. A representative example is represented by bis(aminomethyl)phosphinic acid **42** (Fig. 10), endowed with high-nanomolar activity ($K_I = 108$ nM) and uncompetitive binding mode [101].

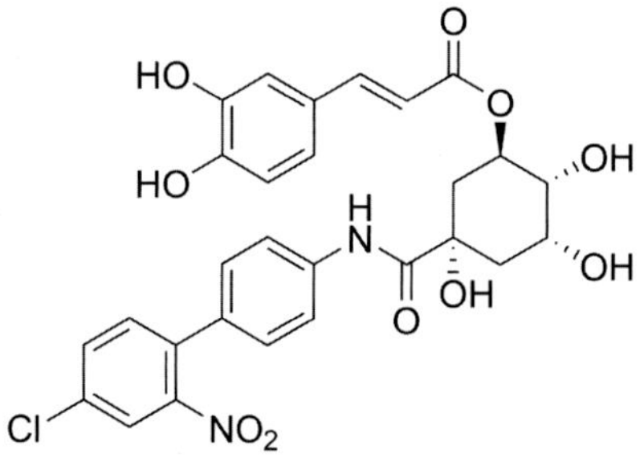

39

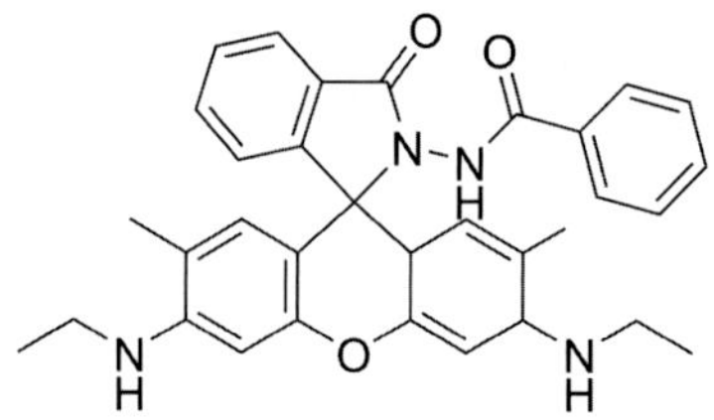

40

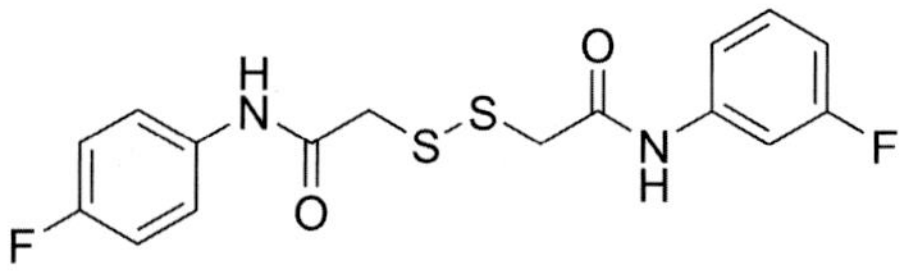

41

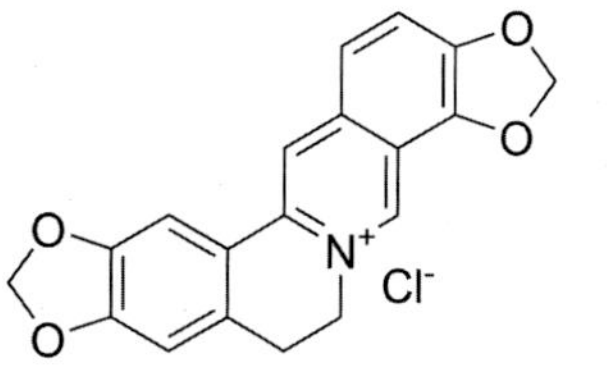

Coptisine

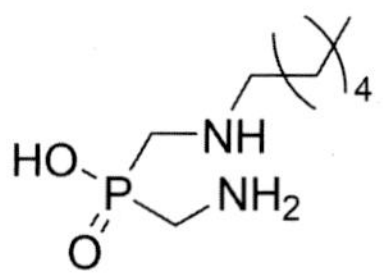

42

FIG. 10 Other UIs not endowed with urea-analog fragments.

6.4 Covalent inhibitors

This wide class of UIs is also known as cysteine-reactive inhibitors since their mechanism of action involves covalent binding to a specific cysteine thiol group. This reaction is very efficient in inhibiting the catalytic activity of the enzyme since this amino acid belongs to the group of residues forming the mobile flap that opens and closes the enzyme active site. Also, it has a high druggability degree, being fully conserved among bacterial ureases. Several covalent UIs, such as Michael acceptors, cinnamic derivatives, selenium- and phosphorus-containing compounds, have been described and different binding modes have been observed.

The advantages of covalent UIs are due to their strong binding to the enzyme and consequently, high potency, extended duration of action, and lower dose are noticed. However, covalent inhibitors are usually toxic and immunogenic and there is no full knowledge of the destiny and the degradation of the modified proteins. Therefore, for this reason, this class of UIs remains poorly explored and only a few examples have been reported.

Recently, an extraordinary inhibitory activity was detected for a bisphosphoramide compound (**43**, Fig. 11), which showed an IC_{50} value of $1.91*10^{-10}$ nM [102].

Both the aniline functions and the P=O moieties seem to play a key role in the affinity to the enzyme, resulting in connecting the compound to the entrance and the interior of the binding cavity through hydrogen bonding, hydrophobic contacts, and π-interactions, without interacting directly with nickel ions. Since their extended structures, it is able to occupy the entire pocket. However, the aromatic amines are not relevant features for this class of compounds since a similar activity profile was provided also by aliphatic amines [102].

The same mechanism of action on *H. pylori* and *S. pasteurii* ureases was found ($K_i = 2.11$ and 226 nM, respectively) for the organoselenium drug ebselen (Fig. 11), an anti-inflammatory, antioxidant, and cytoprotective agent. In fact, also in this case, a nickel complexation by its carbonyl atom and a cysteine sulfur-selenium bond were highlighted in computational studies [103].

Furthermore, the hydroquinone class (general structure **44**, Fig. 11) has been recently reported as covalent UIs through an irreversible inactivation process that involves a radical-based autocatalytic mechanism resulting in the block of the flap in the open conformation, as elucidated by macromolecular X-ray crystallography and kinetic studies [104].

6.5 Metals and coordination complexes

Urease is generally inhibited by heavy metal ions through the formation of bonds with thiol groups on the active site of the enzyme and, in some cases (Ag^+ and Hg^{2+}), forming insoluble sulfides. The efficiency scoring for metals is: $Ag^+ \approx Hg^{2+} > Cu^{2+} > Ni^{2+} > Cd^{2+} > Zn^{2+} > Co^{2+} > Fe^{3+} > Pb^{2+} > Mn^{2+}$ [54].

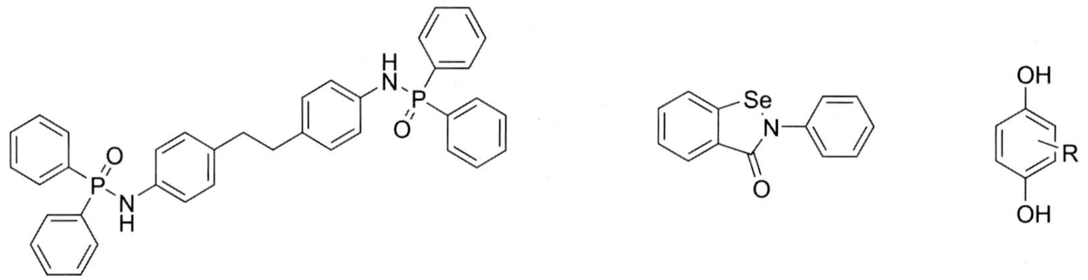

FIG. 11 Representative covalent UIs.

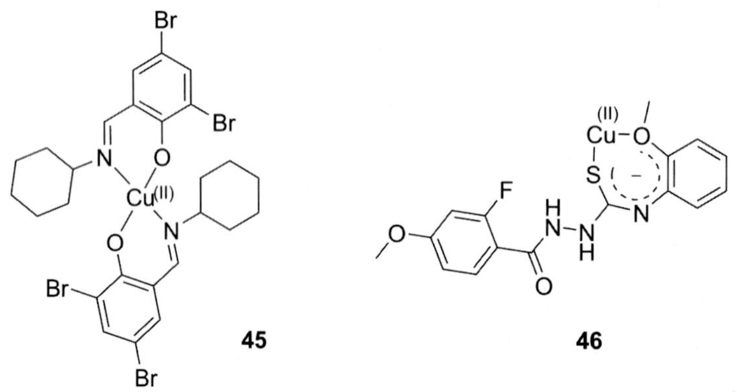

FIG. 12 Two representative metal complexes as UIs.

Metal complexes of small molecules were found able to inhibit ureases by substituting one of the ligands with specific amino acid side chains of the enzyme. Some of these UIs have structures complementary to that of the binding sites of the enzyme. A copper-Schiff base complex (**45**, Fig. 12) showed an IC_{50} of 0.46 μM on Jack bean urease.

The phenolic oxygen forms relevant hydrogen bonds and several hydrophobic contacts were established by the complex [105].

Lower IC_{50} (2.63 μM) was found for compound **46** (Fig. 12), a mixed-type UI containing copper (II) and able to establish three hydrogen bonds stabilizing the metallo-complex inside the active site [106].

References

[1] Chang A, Jeske L, Ulbrich S, Hofmann J, Koblitz J, Schomburg I, et al. BRENDA, the ELIXIR core data resource in 2021: new developments and updates. Nucleic Acids Res 2021;49(D1):D498–508.

[2] BRENDA Enzyme Database., 2022. Available from: https://www.brenda-enzymes.org/. [Accessed 15 October 2022].

[3] NIII PubMed., 2022. Available from: https://pubmed.ncbi.nlm.nih.gov/. [Accessed 15 October 2022].

[4] Mazzei L, Cianci M, Benini S, Ciurli S. The structure of the elusive urease–urea complex unveils the mechanism of a paradigmatic nickel-dependent enzyme. Angew Chem Int Ed 2019;58 (22):7415–9.

[5] Krajewska B, Ureases I. Functional, catalytic and kinetic properties: a review. J Mol Catal B Enzym 2009;59(1–3):9–21.

[6] Callahan BP, Yuan Y, Wolfenden R. The burden borne by urease. J Am Chem Soc 2005;127(31):10828–9.

[7] RCSB PDB., 2022. Available from: https://www.rcsb.org/. [Accessed 15 October 2022].

[8] Dixon NE, Gazzola C, Blakeley RL, Zerner B. Jack bean urease (EC 3.5.1.5). A metalloenzyme. A simple biological role for nickel? J Am Chem Soc 1975;97(14):4131–3.

[9] Sumner JB. The isolation and crystallization of the enzyme urease: preliminary paper. J Biol Chem 1926;69(2):435–41.

[10] Joseph PS, Musa DA, Egwim EC, Uthman A. Function of urease in plants with reference to legumes: a review. In: Jimenez-Lopez JC, Clemente A, editors. Legumes research, vol. 2. London: IntechOpen; 2022. Available from: https://www.intechopen.com/chapters/81535.

[11] Mazzei L, Musiani F, Ciurli S. The structure-based reaction mechanism of urease, a nickel dependent enzyme: tale of a long debate. J Biol Inorg Chem 2020;25(6):829–45.

[12] Ligabue-Braun R, Andreis FC, Verli H, Carlini CR. 3-to-1: unraveling structural transitions in ureases. Naturwissenschaften 2013;100(5):459–67.

[13] Carter EL, Tronrud DE, Taber SR, Karplus PA, Hausinger RP. Iron-containing urease in a pathogenic bacterium. Proc Natl Acad Sci U S A 2011;108(32):13095–9.

[14] Follmer C, Carlini CR, Yoneama ML, Dias JF. PIXE analysis of urease isoenzymes isolated from Canavalia ensiformis (jack bean) seeds. Nucl Inst Methods Phys Res B 2002;189(1–4):482–6.

[15] Adasme MF, Linnemann KL, Bolz SN, Kaiser F, Salentin S, Haupt VJ, et al. PLIP 2021: expanding the scope of the protein–ligand interaction profiler to DNA and RNA. Nucleic Acids Res 2021;49(W1):W530–4.

[16] Delano WL. PyMOL: an open-source molecular graphics tool; 2002.

[17] Benini S, Rypniewski WR, Wilson KS, Miletti S, Ciurli S, Mangani S. A new proposal for urease mechanism based on the crystal structures of the native and inhibited enzyme from Bacillus pasteurii: why urea hydrolysis costs two nickels. Structure 1999;7(2):205–16.

[18] Kappaun K, Piovesan AR, Carlini CR, Ligabue-Braun R. Ureases: historical aspects, catalytic, and non-catalytic properties—a review. J Adv Res 2018;13:3.

[19] Konieczna I, Żarnowiec P, Kwinkowski M, Kolesińska B, Frączyk J, Kamiński Z, et al. Bacterial urease and its role in long-lasting human diseases. Curr Protein Pept Sci 2012;13(8):789.

[20] Mobley HL, Hu LT, Foxal PA. *Helicobacter pylori* urease: properties and role in pathogenesis. Scand J Gastroenterol Suppl 1991;187:39–46.

[21] Mazzei L, Musiani F, Ciurli S. Chapter 5. Urease. In: Zamble D, Rowińska-Żyrek M, Kozlowski H, editors. The biological chemistry of nickel. Royal Society of Chemistry; 2017. p. 60–97.

[22] Carlini CR, Ligabue-Braun R. Ureases as multifunctional toxic proteins: a review. Toxicon 2016;110:90–109.

[23] Ha NC, Oh ST, Sung JY, Cha KA, Lee MH, Oh BH. Supramolecular assembly and acid resistance of *Helicobacter pylori* urease. Nat Struct Biol 2001;8(6):505–9.

[24] Balasubramanian A, Ponnuraj K. Crystal structure of the first plant urease from jack bean: 83 years of journey from its first crystal to molecular structure. J Mol Biol 2010;400(3):274–83.

[25] Farrugia MA, Macomber L, Hausinger RP. Biosynthesis of the urease Metallocenter. J Biol Chem 2013;288(19):13178.

[26] Fong YH, Wong HC, Yuen MH, Lau PH, Chen YW, Wong KB. Structure of UreG/UreF/UreH complex reveals how urease accessory proteins facilitate maturation of *Helicobacter pylori* urease. PLoS Biol 2013;11(10), e1001678.

[27] Nim YS, Wong KB. The maturation pathway of nickel urease. Inorganics (Basel) 2019;7(7):85.

[28] Masetti M, Falchi F, Gioia D, Recanatini M, Ciurli S, Musiani F. Targeting the protein tunnels of the urease accessory complex: a theoretical investigation. Molecules 2020;25(12):2911.

[29] Polacco JC, Mazzafera P, Tezotto T. Opinion: nickel and urease in plants: still many knowledge gaps. Plant Sci 2013;199–200:79–90.

[30] Loharch S, Berlicki Ł. Rational development of bacterial ureases inhibitors. Chem Rec 2022;22(8), e202200026.

[31] KEGG ENZYME., 2022. Available from: https://www.genome.jp/entry/3.5.1.5. [Accessed 3 November 2022].

[32] Becker-Ritt AB, Martinelli AHS, Mitidieri S, Feder V, Wassermann GE, Santi L, et al. Antifungal activity of plant and bacterial ureases. Toxicon 2007;50(7):971–83.

[33] de Jesús R-JT, Ojeda-Barrios DL, Blanco-Macías F, Valdez-Cepeda RD, Parra-Quezada R. Urease and nickel in plant physiology. Rev Chapingo Ser Hortic 2016;22(2):69–82.

[34] Cameron KC, Di HJ, Moir JL. Nitrogen losses from the soil/plant system: a review. Ann Appl Biol 2013;162(2):145–73.

[35] Mora D, Arioli S. Microbial urease in health and disease. PLoS Pathog 2014;10(12), e1004472.

[36] Morou-Bermudez E, Elias-Boneta A, Billings RJ, Burne RA, Garcia-Rivas V, Brignoni-Nazario V, et al. Urease activity in dental plaque and saliva of children during a three-year study period and its relationship with other caries risk factors. Arch Oral Biol 2011;56 (11):1282.

[37] Fiori-Duarte AT, Rodrigues RP, Kitagawa RR, Kawano DF. Insights into the design of inhibitors of the urease enzyme—a major target for the treatment of *Helicobacter pylori* infections. Curr Med Chem 2019;27(23):3967–82.

[38] Svane S, Sigurdarson JJ, Finkenwirth F, Eitinger T, Karring H. Inhibition of urease activity by different compounds provides insight into the modulation and association of bacterial nickel import and ureolysis. Sci Rep 2020;10(1).

[39] Nishimori I, Minakuchi T, Morimoto K, Sano S, Onishi S, Takeuchi H, et al. Carbonic anhydrase inhibitors: DNA cloning and inhibition studies of the α-carbonic anhydrase from *Helicobacter pylori*, a new target for developing sulfonamide and sulfamate gastric drugs. J Med Chem 2006;49(6):2117–26.

[40] Morishita S, Nishimori I, Minakuchi T, Onishi S, Takeuchi H, Sugiura T, et al. Cloning, polymorphism, and inhibition of β-carbonic anhydrase of *Helicobacter pylori*. J Gastroenterol 2008;43(11):849–57.

[41] Ronci M, del Prete S, Puca V, Carradori S, Carginale V, Muraro R, et al. Identification and characterization of the α-CA in the outer membrane vesicles produced by *Helicobacter pylori*. J Enzyme Inhib Med Chem 2019;34(1):189–95.

[42] Rahman MM, Tikhomirova A, Modak JK, Hutton ML, Supuran CT, Roujeinikova A. Antibacterial activity of ethoxzolamide against *Helicobacter pylori* strains SS1 and 26695. Gut Pathog 2020;12(1):1–7.

[43] Nishimori I, Minakuchi T, Morimoto K, Sano S, Onishi S, Takeuchi H, et al. Inhibition of the alpha- and beta-carbonic anhydrases from the gastric pathogen *Helycobacter pylori* with anions. J Enzyme Inhib Med Chem 2013;28(2):388–91.

[44] Angeli A, Ferraroni M, Supuran CT. Famotidine, an antiulcer agent, strongly inhibits *Helicobacter pylori* and human carbonic anhydrases. ACS Med Chem Lett 2018;9(10):1035–8.

[45] Nishimori I, Minakuchi T, Kohsaki T, Onishi S, Takeuchi H, Vullo D, et al. Carbonic anhydrase inhibitors: the β-carbonic anhydrase from *Helicobacter pylori* is a new target for sulfonamide and sulfamate inhibitors. Bioorg Med Chem Lett 2007;17(13):3585–94.

[46] Graham DY, Miftahussurur M. *Helicobacter pylori* urease for diagnosis of *Helicobacter pylori* infection: a mini review. J Adv Res 2018;13:51.

[47] Rokita E, Makristathis A, Hirschl AM, Rotter ML. Purification of surface-associated urease from *Helicobacter pylori*. J Chromatogr B Biomed Sci Appl 2000;737(1–2):203–12.

[48] Dunn BE, Campbell GP, Perez-Perez GI, Blaser MJ. Purification and characterization of urease from *Helicobacter pylori*. J Biol Chem 1990;265(16):9464–9.

[49] Cunha ES, Chen X, Sanz-Gaitero M, Mills DJ, Luecke H. Cryo-EM structure of *Helicobacter pylori* urease with an inhibitor in the active site at 2.0 Å resolution. Nat Commun 2021;12(1):1–8.

[50] Woo HJ, Yang JY, Lee P, Kim JB, Kim SH. Zerumbone inhibits *Helicobacter pylori* urease activity. Molecules 2021;26:2663.

[51] Lu Q, Zhang Z, Xu Y, Chen Y, Li C. Sanguinarine, a major alkaloid from Zanthoxylum nitidum (Roxb.) DC., inhibits urease of *Helicobacter pylori* and jack bean: susceptibility and mechanism. J Ethnopharmacol 2022;295, 115388.

[52] Odaki S, Morikawa T, Tsuchiya M, Imamura L, Kobash K. Inhibition of *Helicobacter pylori* urease activity by hydroxamic acid derivatives. Biol Pharm Bull 1994;17(10):1329–32.

[53] Cesareo SD, Langton SR. Kinetic properties of *Helicobacter pylori* urease compared with jack bean urease. FEMS Microbiol Lett 1992;78(1):15–21.

[54] Habala L, Devínsky F, Egger AE. Review: metal complexes as urease inhibitors. J Coord Chem 2018;71(7):907–40.

[55] Wishart DS, Knox C, Guo AC, Shrivastava S, Hassanali M, Stothard P, Chang Z, Woolsey J. DrugBank: a comprehensive resource for in silico drug discovery and exploration. Nucleic Acids Res 2006;34(Database issue):D668–72.

[56] Aniceto N, Bonifácio VDB, Guedes RC, Martinho N. Exploring the chemical space of urease inhibitors to extract meaningful trends and drivers of activity. J Chem Inf Model 2022;62(15):3535–50.

[57] Khan JAJ, Choudhary MI, MAA AL-G, Huwait E, Wahab A, Javaid S. 4-[2-(Dipropylamino)ethyl]-1,3-dihydro-2H-indol-2-one (ropinirole) a new inhibitor of Jack bean urease enzyme: an example of drug repurposing. US20170252323; 2017.

[58] Wahid S, Jahangir S, Versiani MA, Khan KM, Salar U, Ashraf M, et al. Atenolol thiourea hybrid as potent urease inhibitors: design, biology-oriented drug synthesis, inhibitory activity screening, and molecular docking studies. Bioorg Chem 2020;94, 103359.

[59] Huang XS, Liu K, Yin Y, Li WM, Ran W, Duan M, et al. The synthesis, structure and activity evaluation of Secnidazole derivatives as *Helicobacter pylori* urease inhibitors. Curr Bioact Compd 2011;7 (4):268–80.

[60] Ramadan MA, Tawfik AF, el-Kersh TA, Shibl AM. In vitro activity of subinhibitory concentrations of quinolones on urea-splitting bacteria: effect on urease activity and on cell surface hydrophobicity. J Infect Dis 1995;171(2):483–6.

[61] Janser I, Vortolomei CM, Meka RK, Walsh CA, Janser RFJ. Ethacrynic acid as a lead structure for the development of potent urease inhibitors. C R Chim 2013;16(7):660–4.

[62] Tan L, Li C, Chen H, Mo Z, Zhou J, Liu Y, et al. Epiberberine, a natural protoberberine alkaloid, inhibits urease of *Helicobacter pylori* and jack bean: susceptibility and mechanism. Eur J Pharm Sci 2017;110:77–86.

[63] Rasheed S, Aziz M, Saeed A, Ejaz SA, Channar PA, Zargar S, et al. Analysis of 1-Aroyl-3-[3-chloro-2-methylphenyl] thiourea hybrids as potent urease inhibitors: synthesis, biochemical evaluation and computational approach. Int J Mol Sci 2022;23:11646.

[64] Kataria R, Khatkar A. Molecular docking, synthesis, kinetics study, structure-activity relationship and ADMET analysis of morin analogous as *Helicobacter pylori* urease inhibitors. BMC Chem 2019;13(3):1–17.

[65] Sedaghati S, Azizian H, Montazer MN, Mohammadi-Khanaposhtani M, Asadi M, Moradkhani F, et al. Novel (thio)barbituric-phenoxy-N-phenylacetamide derivatives as potent urease inhibitors: synthesis, in vitro urease inhibition, and in silico evaluations. Struct Chem 2021;32(1):37–48.

[66] Kazemzadeh H, Hamidian E, Hosseini FS, Abdi M, Niasari Naslaji F, Talebi M, et al. Isoindolin-1-ones fused to barbiturates: from design and molecular docking to synthesis and urease inhibitory evaluation. ACS Omega 2022;7(23):19401–11.

[67] Rauf A, Ahmed F, Qureshi AM, Aziz-Ur-Rehman KA, Qadir MI, et al. Synthesis and urease inhibition studies of barbituric and thiobarbituric acid derived sulphonamides. J Chin Chem Soc 2011;58 (4):528–37.

[68] Naseer A, Osra FA, Awan AN, Imran A, Hameed A, Ali Shah SA, et al. Exploring novel pyridine carboxamide derivatives as urease inhibitors: synthesis, molecular docking, kinetic studies and ADME profile. Pharmaceuticals 2022;15:1288.

[69] Hameed A, Khan KM, Zehra ST, Ahmed R, Shafiq Z, Bakht SM, et al. Synthesis, biological evaluation and molecular docking of N-phenyl thiosemicarbazones as urease inhibitors. Bioorg Chem 2015;61:51–7.

[70] Elshaier YAMM, Aly AA, Abdel-Aziz M, Fathy HM, Brown AB, Bräse S, et al. Synthesis and identification of new N,N-disubstituted thiourea, and thiazolidinone scaffolds based on quinolone moiety as urease inhibitor. Molecules 2022;27(20):7126.

[71] Taha M, Ismail NH, Imran S, Wadood A, Rahim F, Saad SM, et al. Synthesis, molecular docking and α-glucosidase inhibition of 5-aryl-2-(6′-nitrobenzofuran-2′-yl)-1,3,4-oxadiazoles. Bioorg Chem 2016;66:117–23.

[72] Channar PA, Saeed A, Afzal S, Hussain D, Kalesse M, Shehzadi SA, et al. Hydrazine clubbed 1,3-thiazoles as potent urease inhibitors: design, synthesis and molecular docking studies. Mol Divers 2020;25(2):1–13.

[73] Liu H, Wang Y, Lv M, Luo Y, Liu BM, Huang Y, et al. Flavonoid analogues as urease inhibitors: synthesis, biological evaluation, molecular docking studies and in-silico ADME evaluation. Bioorg Chem 2020;105, 104370.

[74] Wu Y, Zhao S, Liu C, Hu L. Development of urease inhibitors by fragment-based dynamic combinatorial chemistry. ChemMedChem 2022;17(19), e202200307.

[75] Abbas A, Ali B, Kanwal, Khan KM, Iqbal J, Rahman S, et al. Synthesis and in vitro urease inhibitory activity of benzohydrazide derivatives, in silico and kinetic studies. Bioorg Chem 2019;82:163–77.

[76] Xiao ZP, Peng ZY, Dong JJ, Deng RC, Wang XD, Ouyang H, et al. Synthesis, molecular docking and kinetic properties of β-hydroxy-β-phenylpropionyl-hydroxamic acids as *Helicobacter pylori* urease inhibitors. Eur J Med Chem 2013;68:212–21.

[77] Shi WK, Deng RC, Wang PF, Yue QQ, Liu Q, Ding KL, et al. 3-Arylpropionylhydroxamic acid derivatives as *Helicobacter pylori* urease inhibitors: synthesis, molecular docking and biological evaluation. Bioorg Med Chem 2016;24(19):4519–27.

[78] Zaib S, Tayyab Younas M, Zaraei SO, Khan I, Anbar HS, El-Gamal MI. Discovery of urease inhibitory effect of sulfamate derivatives: biological and computational studies. Bioorg Chem 2022;119, 105545.

[79] Arshia BF, Almandil NB, Lodhi MA, Khan KM, Hameed A, et al. Synthesis and urease inhibitory potential of benzophenone sulfonamide hybrid in vitro and in silico. Bioorg Med Chem 2019;27 (6):1009–22.

[80] Khan S, Iqbal S, Shah M, Rehman W, Hussain R, Rasheed L, et al. Synthesis, in vitro anti-microbial analysis and molecular docking study of aliphatic hydrazide-based benzene sulphonamide derivatives as potent inhibitors of α-glucosidase and urease. Molecules 2022;27(20):7129.

[81] Asadi M, Iraji A, Sherafati M, Nazari Montazer M, Ansari S, Mohammadi Khanaposhtani M, et al. Synthesis and in vitro urease inhibitory activity of 5-nitrofuran-2-yl-thiadiazole linked to different cyclohexyl-2-(phenylamino)acetamides, in silico and kinetic studies. Bioorg Chem 2022;120, 105592.

[82] Taha M, Rahim F, Khan AA, Anouar EH, Ahmed N, Shah SAA, Ibrahim M, Zakari ZA. Synthesis of diindolylmethane (DIM) bearing thiadiazole derivatives as a potent urease inhibitor. Sci Rep 2020;10(1):7969.

[83] Salehi Ashani R, Azizian H, Sadeghi Alavijeh N, Fathi Vavsari V, Mahernia S, Sheysi N, et al. Synthesis, biological evaluation and molecular docking of Deferasirox and substituted 1,2,4-triazole derivatives as novel potent urease inhibitors: proposing repositioning candidate. Chem Biodivers 2020;17(5), e1900710.

[84] Rezaei EB, Abedinifar F, Azizian H, Montazer MN, Asadi M, Hosseini S, et al. Design, synthesis, and evaluation of metronidazole-1,2,3-triazole derivatives as potent urease inhibitors. Chem Pap 2021;75(8):4217–26.

[85] Menteşe E, Emirik M, Sökmen BB. Design, molecular docking and synthesis of novel 5,6-dichloro-2-methyl-1H-benzimidazole derivatives as potential urease enzyme inhibitors. Bioorg Chem 2019;86:151–8.

[86] Mumtaz S, Iqbal S, Shah M, Hussain R, Rahim F, Rehman W, et al. New Triazinoindole bearing benzimidazole/benzoxazole hybrids analogs as potent inhibitors of urease: synthesis, in vitro analysis and molecular docking studies. Molecules 2022;27(19):6580.

[87] Mohammed SO, el Ashry ESH, Khalid A, Amer MR, Metwaly AM, Eissa IH, et al. Expression, purification, and comparative inhibition of *Helicobacter pylori* urease by regio-selectively alkylated benzimidazole 2-thione derivatives. Molecules 2022;27:865.

[88] Abdullah MA, Abuo-Rahma GE, Abdelhafez EM, Hassan HA, Abd El-Baky RM. Design, synthesis, molecular docking, anti-Proteus mirabilis and urease inhibition of new fluoroquinolone carboxylic acid derivatives. Bioorg Chem 2017;70:1–11.

[89] Nisar M, Khan SA, Shah MR, Khan A, Farooq U, Uddin G, et al. Moxifloxacin-capped noble metal nanoparticles as potential urease inhibitors. New J Chem 2015;39(10):8080–6.

[90] Pedrood K, Azizian H, Montazer MN, Moazzam A, Asadi M, Montazeri H, et al. Design and synthesis of new N-thioacylated ciprofloxacin derivatives as urease inhibitors with potential antibacterial activity. Sci Rep 2022;12(1).

[91] Menteşe E, Akyüz G, Emirik M, Baltaş N. Synthesis, in vitro urease inhibition and molecular docking studies of some novel quinazolin-4 (3H)-one derivatives containing triazole, thiadiazole and thiosemicarbazide functionalities. Bioorg Chem 2019;83:289–96.

[92] Sohrabi M, Nazari Montazer M, Farid SM, Tanideh N, Dianatpour M, Moazzam A, Zomorodian K, Yazdanpanah S, Asadi M, Hosseini S, Biglar M, Larijani B, Amanlou M, Barazandeh Tehrani M, Iraji A, Mahdavi M. Design and synthesis of novel nitrothiazolacetamide conjugated to different thioquinazolinone derivatives as anti-urease agents. Sci Rep 2022;12(1):2003.

[93] Kahveci B, Menteşe E, Akkaya E, Yilmaz F, Doğan IS, Özel A. Synthesis of some novel 1,2,4-triazol-3-one derivatives bearing the salicyl moiety and their anticonvulsant activities. Arch Pharm (Weinheim) 2014;347(6):449–55.

[94] Rashid U, Rahim F, Taha M, Arshad M, Ullah H, Mahmood T, et al. Synthesis of 2-acylated and sulfonated 4-hydroxycoumarins: in vitro urease inhibition and molecular docking studies. Bioorg Chem 2016;66:111–6.

[95] Kazmi M, Khan I, Khan A, Halim SA, Saeed A, Mehsud S, et al. Developing new hybrid scaffold for urease inhibition based on carbazole-chalcone conjugates: synthesis, assessment of therapeutic potential and computational docking analysis. Bioorg Med Chem 2019;27(22), 115123.

[96] Yang YS, Su MM, Zhang XP, Liu QX, He ZX, Xu C, et al. Developing potential *Helicobacter pylori* urease inhibitors from novel oxoindoline derivatives: synthesis, biological evaluation and in silico study. Bioorg Med Chem Lett 2018;28(19):3182–6.

[97] Kataria R, Khatkar A. In-silico designing, ADMET analysis, synthesis and biological evaluation of novel derivatives of Diosmin against urease protein and *Helicobacter pylori* bacterium. Curr Top Med Chem 2019;19(29):2658–75.

[98] Vanjare BD, Mahajan PG, Dige NC, Raza H, Hassan M, Seo SY, et al. Synthesis of novel xanthene based analogues: their optical properties, jack bean urease inhibition and molecular modelling studies. Spectrochim Acta A Mol Biomol Spectrosc 2020;241, 118667.

[99] He Y, Zhang X, Li M, Zheng N, Zhao S, Wang J. Coptisine: a natural plant inhibitor of ruminal bacterial urease screened by molecular docking. Sci Total Environ 2022;808, 151946.

[100] Liu ML, Li WY, Fang HL, Ye YX, Li SY, Song WQ, et al. Synthesis and biological evaluation of Dithiobisacetamides as novel urease inhibitors. ChemMedChem 2022;17(2), e202100618.
[101] Macegoniuk K, Dziełak A, Mucha A, Berlicki Ł. Bis(aminomethyl) phosphinic acid, a highly promising scaffold for the development of bacterial urease inhibitors. ACS Med Chem Lett 2015;6(2):146.
[102] Gholivand K, Pooyan M, Mohammadpanah F, Pirastefar F, Junk PC, Wang J, et al. Synthesis, crystal structure and biological evaluation of new phosphoramide derivatives as urease inhibitors using docking, QSAR and kinetic studies. Bioorg Chem 2019;86:482–93.
[103] Macegoniuk K, Grela E, Palus J, Rudzińska E, Grabowiecka A, Biernat M, et al. 1,2-Benzisoselenazol-3(2H)-one derivatives as a new class of bacterial urease inhibitors. J Med Chem 2016;59(17):8125–33.
[104] Mazzei L, Cianci M, Ciurli S. Inhibition of urease by hydroquinones: a structural and kinetic study. Chem Eur J 2022; e202201770.
[105] Chen W, Li Y, Cui Y, Zhang X, Zhu HL, Zeng Q. Synthesis, molecular docking and biological evaluation of Schiff base transition metal complexes as potential urease inhibitors. Eur J Med Chem 2010;45(10):4473–8.
[106] Saleem M, Hanif M, Rafiq M, Hassan M, Tahir T. Synthesis, characterization, optical properties, molecular modeling and urease inhibition analysis of organic ligands and their metal complexes. J Fluoresc 2022;1–12.

Chapter 5.2

Methyl-coenzyme M reductase

Alessandro Bonardi
Department of NEUROFARBA, Section of Pharmaceutical and Nutraceutical Sciences, Pharmaceutical and Nutraceutical Section, University of Florence, Firenze, Italy

1 Methyl-coenzyme M reductase: An important biocatalyst complex in archaea metabolism

Methyl-coenzyme M reductase (MCR) is an important nickel-enzymatic complex that catalyzes the reversible reduction of a methyl group in methane [1–9]. MCR is crucial in the metabolism of methanogenic, anaerobic methanotrophic (ANME) archaea, which are the unique microorganisms responsible for the complete biological production or degradation of methane, leading them to colonize the earth's globe at the beginning of earth life [1–9]. Nowadays, archaea are found in the great part of the anaerobic niches including animal microbiomes, anoxic soils, and marine/freshwater habitats [10–14].

The anaerobic respiration of archaea, which entails the reduction of simple oxidized carbon compounds to generate methane as an end product, is known as methanogenesis, while methanotrophy is the reversible process that uses CH_4 as an energy source. In detail, methanogenic metabolisms are grouped into three historical categories, depending on the used substrate (Fig. 1) [14–16]. In detail, (1) hydrogenotrophic methanogenesis reduces CO_2 or RCH_3 into CH_4 using prevalently H_2 as an electron donor and also formate, alcohols, CO and iron [17,18]; (2) methylotrophic and methoxydotrophic methanogenesis consists of the dismutation of the activated methyl compounds (CoM-SCH_3), respectively obtained from methanol/trimethylamine and methoxylated aromatic as substrates, where the methyl group is reduced to CH_4 and the coenzyme-M (CoM-SH) moiety is oxidized with coenzyme-B (CoB) to the disulfur CoM-S-S-CoB [19,20]; finally, (3) acetoclastic methanogenesis dismutates the activated acetyl compounds (acetyl-CoA), obtained by the acetate or alkanes (e.g., ethane, propane, butane), oxidizing the carboxyl group to CO_2 and reducing the methyl portion in CH_4. Although this is the less bioenergetically favorable mechanism, 2/3 of the whole biological methane is produced by acetate [21].

As reported in Fig. 1, while several distinctions are notable between the three methanogenic metabolisms, the final key methane-generating step is permanently catalyzed by MCR [9]. Again, this enzymatic complex plays a key role in the first step of methanotrophic metabolism, which oxidizes methane in CoM-SCH_3 to finally produce CO_2 [9]. Instead, the alkanotrophic metabolism uses a peculiar MCR, which is crucial for the activation of the alkane substrates to alkyl-S-CoM, known as alkyl-coenzyme M reductase (ACR) [4].

The production of methane by ruminant husbandry (in particular sheep and cows), which is the major cause of greenhouse gas emissions, has been increasingly considered a massive menace to global climate warming. In the past decades, a wide number of investigations have been done to mitigate methane emission in ruminant animals. Given the important role of MCR in the production of methane, the recent discovery of MCR inhibitors resulted in an important outcome to decrease gas emissions by animals and control the greenhouse effect phenomenon [1–12].

2 Phylogenetic and cellular localization of MCRs

Several studies confirmed the exclusive presence of the MCR (or ACR) or MCR-like complex in the overwhelming majority of Archaea. In detail, MCR is expressed in methanogenic (Class I: *Methanobacteriales*, *Methanococcales*, and *Methanopyrales*; Class II: *Methanosarcinales*, *Methanomicrobiales*, and *Methanocellales*) [1,2,9], methanotrophic (ANME1-3) [22–24] and short/long-chain alkane oxidizer archaea [1,2,6,25–35], supporting that they are evolutionarily linked [29,32,36–38]. The *mcr* genes have not yet been found outside the domain of archaea [8]. Indeed, some bacteria (e.g., *Rhodopseudomonas palustris*) that are involved in the biosynthesis of methane are in symbiosis with archaea, furnishing them only the redox species for the methanogenic reaction [39].

Immune-labeling experiments with *Methanothermobacter* [40,41] and *Methanosarcina* [42] have demonstrated that MCR is a membrane-associated complex with cytoplasm localization [43,44]. Despite that, there are no genes

Metalloenzymes. https://doi.org/10.1016/B978-0-12-823974-2.00034-6

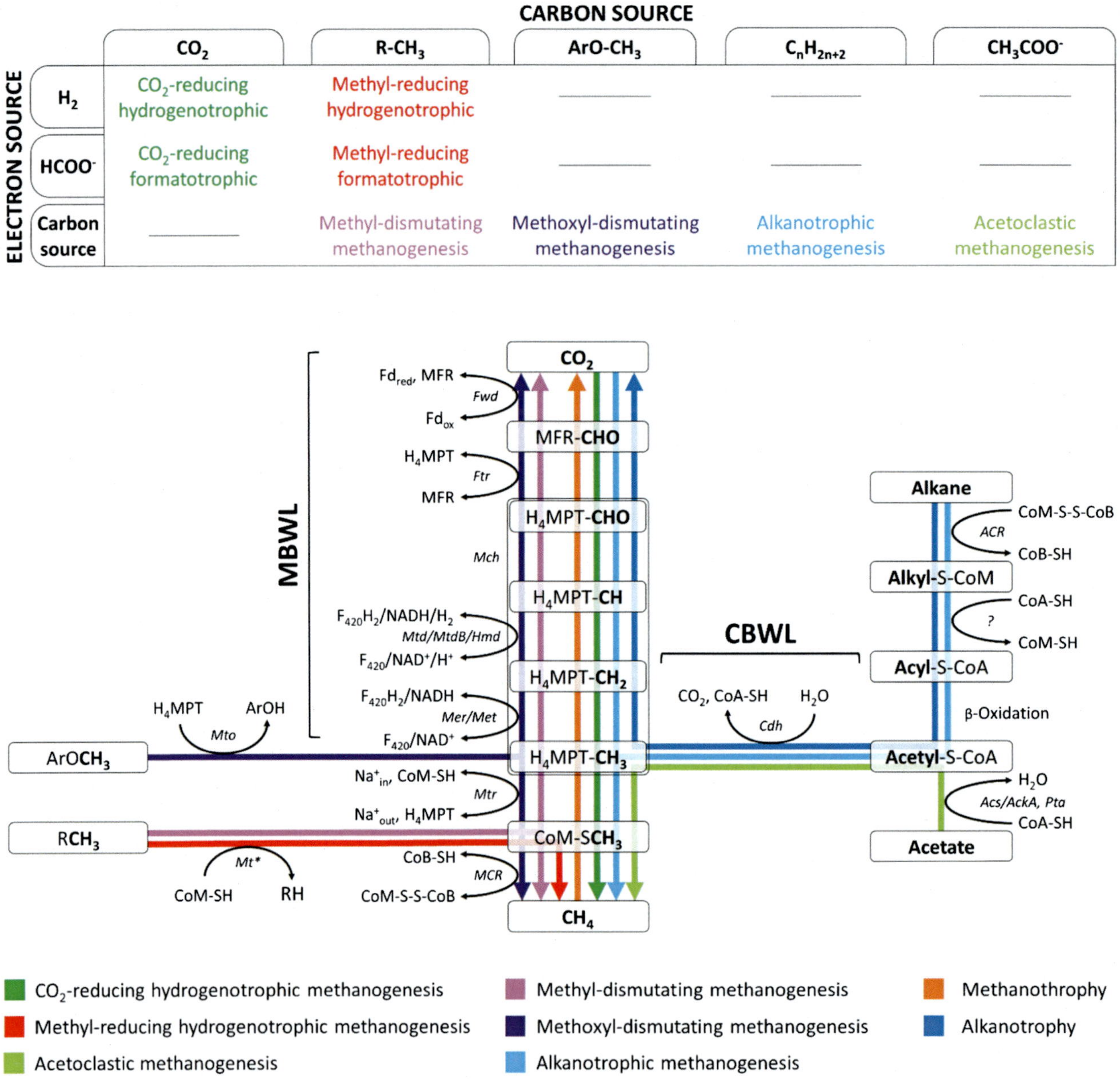

FIG. 1 Archaea metabolisms: the three types of methanogenesis [(1) hydrogenotrophic methanogenesis (dark green and red arrows), (2) methylotrophic/methoxydotrophic methanogenesis (pink and purple arrows), (3) acetoclastic/alkanotrophic methanogenesis (light green and cyan arrows)], methanotrophy (orange arrow) and alkanotrophy (blue arrow). Above, the archaea metabolisms are classified as electron source and carbon source; below, all the enzymes involved in the synthetic pathways, are reported. The beginning of CO_2 reduction starts with the Methyl Branch of the Wood-Ljungdahl pathway (MBWL; all colored arrows). In detail, the CO_2 is reduced by the ferredoxin (Fd) of the formyl-methanofuran dehydrogenase complex (Fwd) and bound to the methanofuran (MFR) into formyl-MFR (MFR-CHO). Then, the formyl moiety is transferred by the formyl-methanofuran-tetrahydromethanopterin formyl-transferase (Ftr) on tetrahydromethanopterin (H_4MPT) obtaining the complex H_4MPT-CHO. Subsequently, the formyl group is reduced into methenyl (H_4MPT-CH), methylene (H_4MPT-CH_2), and methyl (H_4MPT-CH_3) by methenyl-tetrahydromethanopterin cyclohydrolase (Mch), methylene-tetrahydromethanopterin dehydrogenase, F_{420} dependent (Mtd), and methylene-tetrahydromethanopterin reductase (Mer). The intermediate H_4MPT-CH_3 is also obtained from the Carbonyl Branch of the Wood-Ljungdahl pathway (CBWL; light green, cyan and blue arrows) that transforms acetyl-CoA, derived from the activation of alkanes in alkyl-CoM by Acr followed by a β-oxidation or from activation of acetate by acetyl-CoA synthase (Acs), using the CO_2 dehydrogenase complex (Cdh). Again, the methoxydotrophic methanogenesis starts relocating the methyl group by methoxylated aromatic substrates on H_4MPT with the methoxyltransferase (Mto) to give (H_4MPT-CH_3). The methanogenesis pathways proceed with the tetrahydromethanopterin S-methyl-transferase complex (Mtr) that couples the exergonic transfer ($\Delta G^{o\prime} = -30$ kJ/mol) of the methyl group from H_4MPT-CH_3 to CoM-SH with the export of Na^+ across the membrane, producing an electrochemical gradient that is used by the ATP synthase. The final step catalyzed by Mcr provides the reduction of CoM-SCH_3 into CoM-S-S-CoB and CH_4 [9].

associated with MCR synthesis predicted to encode proteins that are membrane anchors. Therefore the close localization of the MCR complex with the cell membrane is probably due to another undiscovered mechanism [8].

3 Conformations of MCR and oxidation states of the coenzyme F430 nickel atom

Among the archaea, the MCR from *Methanothermobacter marburgensis* was the most studied. Only after growing up under 80% H_2 and 20% CO_2 conditions and then being exposed to a 100% H_2 atmosphere, *M. marburgensis* cells mainly contain the active MCR with the electron paramagnetic resonance (EPR) signal designated as MCR_{red1} [45,46]. Instead, the cells exposed to an environment of 80% N_2 and 20% CO_2 express an inactive MCR with the EPR signal MCR_{ox1} [46,47]. Again, harvesting directly the cells the MCR resulted inactive with an EPR defined as MCR_{silent} [46,47]. Probably an explanation of that could be found in the different intracellular concentrations of substrates and products CoM-SCH_3, CoM-SH, CoB-SH, and CoM-S-S-CoB. In fact, the standard redox potential ($E^{o\prime}$) of the couple CoM-S-S-CoB/CoM-SH + CoB-SH is −140 mV [48]. In 100% H_2 and 0% CO_2 conditions, zero concentration of CoM-S-S-CoB and higher intracellular concentrations of CoM-SH and CoB-SH are expected. Moreover, without CO_2, CoM-SCH_3 cannot be produced by methanogenesis (Fig. 1). In that conditions, the couple CoM-S-S-CoB/CoM-SH+CoB-SH redox potential will be much more negative than −140 mV [48]. Conversely, when the cells are under 100% CO_2 and 0% H_2 atmosphere, the thiol CoM-SH and CoB-SH concentrations are zero, while the concentration of disulfide CoM-S-S-CoB is higher because cannot be reduced without H_2. Thus, the redox potential of the couple CoM-S−S-CoB/CoM-SH+CoB-SH will be more positive than −140 mV. In the last case, when the cells are directly harvested, there is an intermediary condition with a redox potential of the couple CoM-S-S-CoB/CoM-SH+CoB-SH near −140 mV. Therefore, the MCR of *M. marburgensis* is in the active state only when the couple CoM-S-S-CoB/CoM-SH +CoB-SH redox potential is more negative than the $E^{0\prime} = -140$ mV [48].

In detail, MCR assumes different states, that are influenced by the environment of the prosthetic group (nickel tetrahydrocorphinoid or coenzyme F_{430}), depending on the three different oxidation states of the nickel atom, Ni(I), Ni(II), and Ni(III), distinguishable by EPR and UV-Vis spectra [46,49–51]. The several states adopted by MCR could be classified as: (1) enzymatically active states MCR_{red1} and MCR_{red2} that contains a Ni(I) species; (2) the inactive states MRC_{silent} (Fig. 2A) and $MCR_{ox1\text{-}silent}$ (Fig. 2B) species with a Ni(II) ion; (3) the inactive MCR_{ox1} with a Ni(III) species [8].

There are three other sub-states of the MCR_{red1} state, united by a Ni(I) in a pentacoordinate geometry with a Gln147, which are (1) MCR_{red1a} without coenzymes, (2) MCR_{red1m} with CoM-SCH_3, and (3) MCR_{red1c} with CoM-SH. The addition of CoB-SH to MCR_{red1c} induces conformational changes in the enzymatic complex structure that approach CoM-SH to the nickel center, causing a partial conversion into the MCR_{red2} state. This latter conformation is an equilibrium between two distinct species: (1) the Ni(I) state side-on coordinated with the CoM-SH thiol group and (2) Ni(III) state hydride complex or a nickel complex in

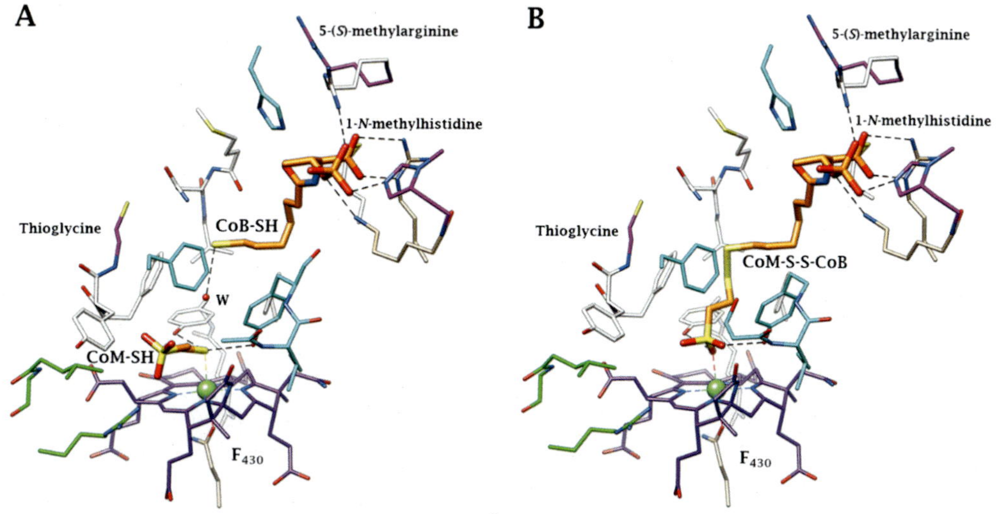

FIG. 2 X-ray crystal structure of (A) $MCR_{ox1\text{-}silent}$ (PDB ID: 1HBN, 1.16 Å) and (B) MCR_{silent} (PDB ID: 1HBM, 1.80 Å) states [56]. CoM-SH, CoB-SH, and CoM-S-S-CoB are colored in orange, while coenzyme F_{430} and posttranslational modified residues are painted in purple and magenta, respectively. Water molecules and nickel ions are represented as red and green spheres, in that order. H-bonds are depicted as black dashed lines.

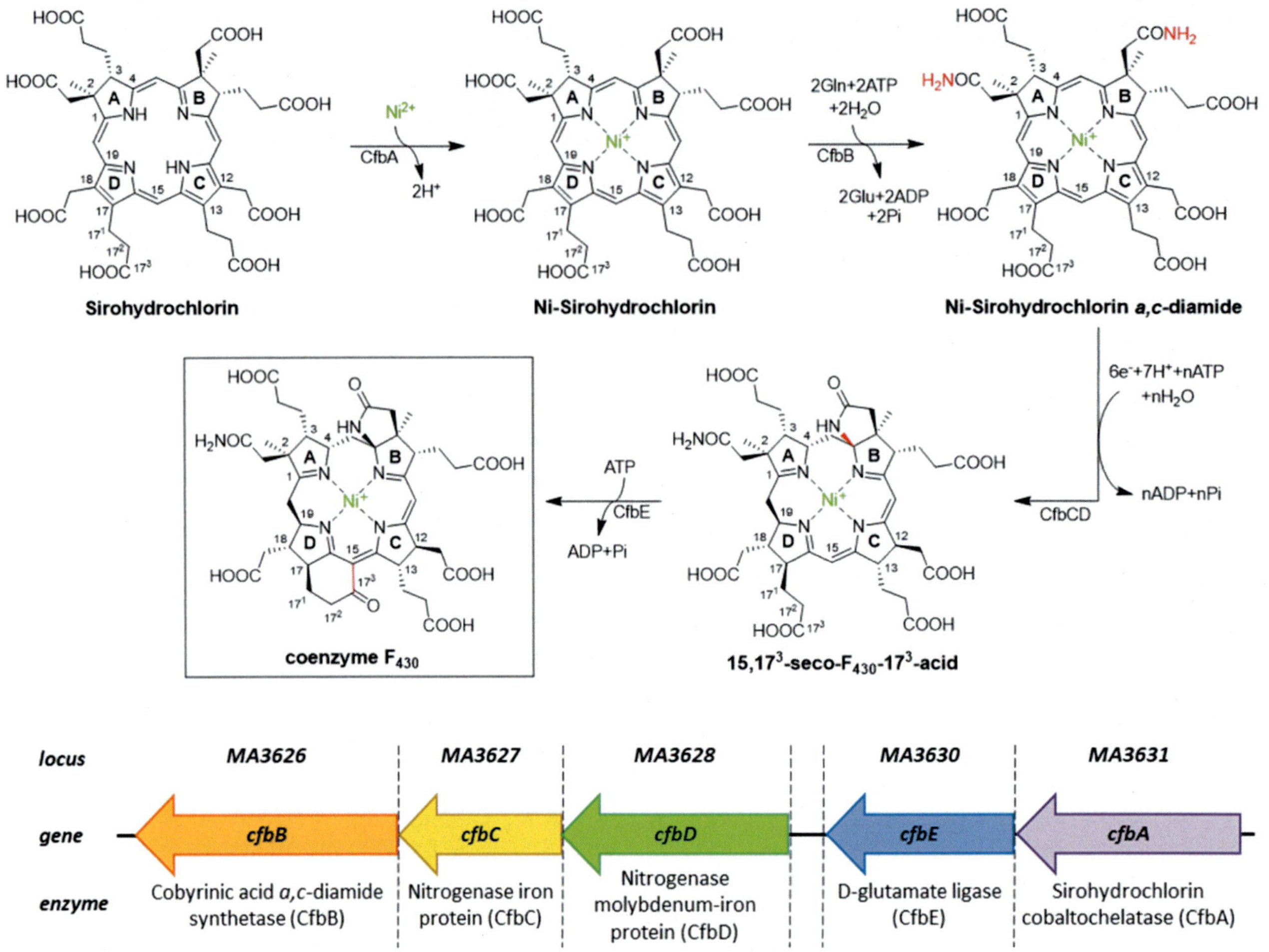

FIG. 3 Biosynthetic pathway of coenzyme F_{430} and coenzyme F_{430} biosynthesis (*cfb*) gene cluster, codifying the enzymes CfbA, CfbB, CfbC, CfbD, and CfbE [57].

H-bond distance with the acidic proton of CoM-SH thiol group [52,53].

Instead, the enzymatically inactive state MCR_{ox1} is an equilibrium between the (1) Ni(III) state axially coordinated with the CoM-SH thiolate and (2) Ni(II) coupled with a thinyl radical species [54]. Under aerobic conditions, MCR_{ox1} is converted in the inactive $MCR_{ox1\text{-}silent}$ state [47], while both $MCR_{ox1\text{-}silent}$ and MCR_{red1} states are transformed into the inactive MCR_{silent} conformation under low H_2 concentrations and in presence of CoM-S-S-CoB (Fig. 2) [50–56]. MCR_{ox2} and MCR_{ox3} are the other two enzymatically inactive oxidized states of MCR with a Ni(III), similar to MCR_{ox1}. These confirmations are experimentally obtained irreversibly by treating the MCR_{red2} with Na_2SO_3 and O_2 [50].

4 Coenzyme F_{430}

The biosynthesis of the nickel tetrahydrocorphinoid scaffold of the coenzyme F_{430} starts from the condensation of four porphobilinogen monomers, catalyzed by porphobilinogen deaminase, to produce the sirohydrochlorin core after several steps (Fig. 3). The synthetic pathway from sirohydrochlorin to coenzyme F_{430} is regulated by the coenzyme F_{430} biosynthesis (*cfb*) gene cluster that codifies the peculiar archaeal enzymes CfbA, CfbB, CfbC, CfbD, and CfbE [57].

An LC-HR/mass spectrometry study identified other nine variants of coenzyme F_{430} from different methanogens and methanotrophic archaea (Fig. 4) [58]. The reactions required to obtain all the F_{430}-modified coenzymes cannot occur spontaneously, except for derivative F_{430}-4, which could be an oxidative product of the native F_{430} type. Nowadays, the variant F_{430}-2, which includes a methylthio group on the tetrahydrocorphinoid scaffold, was identified by X-ray crystallography in ANME-1 archaea and not in ANME-2 or ANME-3 [59,60]. However, the MS studies showed the presence of F_{430}-2 type also in samples that contain exclusively extract cells of ANME-2 archaea [58]. F_{430}-2 has not yet been found in methanogenic archaea, suggesting that the methylthio group is essential only for the oxidation of methane [58].

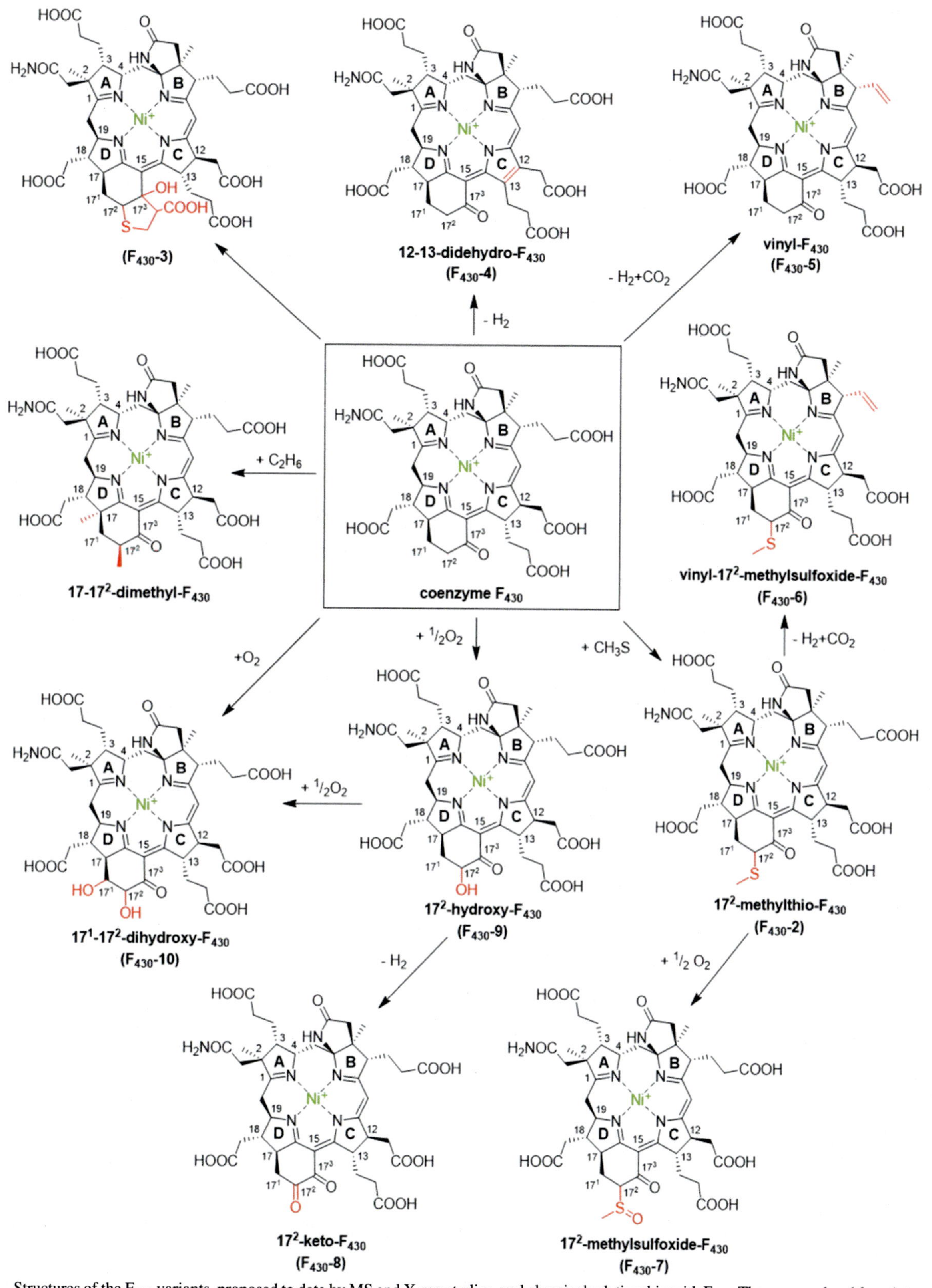

FIG. 4 Structures of the F_{430} variants, proposed to date by MS and X-ray studies, and chemical relationship with F_{430}. The proposed and found structural modifications are colored in red [58,62].

While MS analysis found that the F_{430}-3 type was not correlated to the MCR protein, variants F_{430}-5 and F_{430}-6 derived from the oxidation of the propionic chain of the pyrrolic ring B of coenzyme F_{430} and F_{430}-2, respectively, and were identified in both methanogenic and methanotrophic archaea [58].

Again, the oxidized coenzymes F_{430}-7 (sulfoxide group), F_{430}-8 (keto group), F_{430}-9 (hydroxyl group), and F_{430}-10 (two hydroxyl groups) were detected in both microorganisms. In the case of F_{430}-8, F_{430}-9, and F_{430}-10, it was purposed that the modifications regard positions 17^2 and 17^3 [58].

It is important to stress that only the crystal structures of the coenzymes F_{430} and F_{430}-2 within MCR are available to date [59–61]. The structures of the other variants were predicted and purposed by MS analysis using the calculated molecular weight [58]. Recently, the 17-17^2-dimethyl-F_{430} structure (Fig. 4) was identified within an alkyl-coenzyme M reductase from an ethane oxidizer archaea by crystallography studies [62].

5 The MCR isoforms

Two isoforms MCR I and MCR II were first discovered in methanogen *M. marburgensis* [63] and are encoded by the gene clusters *mcr* and *mrt*, respectively [64]. These isozymes, commonly expressed in methanogens, are important in two different archaeal growth phases in presence of H_2 and CO_2. In particular, MCR I is expressed under a high intracellular concentration of H_2 while MCR II is synthesized when H_2 is repleted [64–66]. Moreover, MCR I and II differ in their pH optimum and the K_M for CoB and CoM-SCH_3 [61,67].

Recently, a phylogenetic analysis of MCRs identified the isoform MCR III (exclusively from *Methanococcales*), which is highly structurally similar to MCR I and MCR II in terms of the coenzyme F_{430} and substrates binding mode, active site architectures and overall protein folding [61]. The main differences regarding the electrostatic surface potentials (that allow their separation by an anion exchange chromatography), the loop folding and the interaction of the C-terminal end of the γ subunits with α and β subunits, responsible for different the kinetic properties [61].

In the case of methanogens, belonging to *Methanomassiliicoccales*, *Methanosarcinales*, *Methanocellales*, and methanotrophic archaea (ANME1-3) only one type of MCR is expressed, while in the butane oxidizer archaea were found four *mcr* gene sets [8,34].

6 Structural features of MCRs

As observed by X-ray crystal structures (Figs. 5–7), the functional MCRs are a dimer of heterotrimers ($\alpha_2\beta_2\gamma_2$), where each monomer ($\alpha\beta\gamma$ and $\alpha'\beta'\gamma'$) is composed by the tight association of three different subunits α (McrA), β (McrB), and γ (McrG). In detail, the assembling of the $\alpha\beta\gamma\alpha'$ and $\alpha'\beta'\gamma'\alpha$ subunits, respectively defined each active site: a long hydrophobic channel of about 50 Å that connects the surface of the complex with a narrow pocket, containing the nickel tetrahydrocorphinoid (coenzyme F_{430}) not covalently bound to the protein [56]. In particular, another 33 Å-long hydrophobic tunnel is established by α' and α subunits for the only exit/entrance of the methane [8].

The literature reported that only acetate-grown *Methanosarcina thermophila* MCR (~130 kDa) is assembled in a monomer of heterotrimer ($\alpha_1\beta_1\gamma_1$) and binds one F_{430}, while the other MCRs (~300 kDa) are always dimer of heterotrimers ($\alpha_2\beta_2\gamma_2$) [68]. Moreover, a dissociation of the dimeric MCR into two heterotrimeric monomers seems improbable thanks to the intertwined hexameric structure. Again, the X-ray crystallography from an ANME-1 archaea MCR confirmed a dimeric structure of heterotrimer (Fig. 6), suggesting that methanotrophic and methanogenic MCRs share the same substrates [59]. Nonetheless, there are some differences between methanogenic and methanotrophic MCRs linked to the ring of F_{430} and different posttranscriptional modifications (PTMs) [59,69].

The genes that encoded for the three subunits are assembled in operons *mcrBGCDA* (for MCR I and III) and *mrtBGDA* (MCR II) [70–72]. While the subunit McrD and McrC are not important for the activity of the enzymatic complex, they seem to be required for the correct posttranslational assembly and the activation of the enzyme, respectively [13,73,74]. In particular, in the "order assembly model", the first expressed subunit McrB forms a primary complex with McrD (McrBD) associating McrG and McrA when they are translated. After facilitating the PTMs and the coordination of the coenzyme F_{430} to the protein, subunit McrD could be lost or weakly associated [13].

6.1 X-ray crystal structures of MCR from methanogens

The X-ray crystal structures of methanogens MCR I from *M. marburgensis* (PDB: 5A0Y; Fig. 5A), MCR II from *M. wolfeii* (PDB: 5A8W; Fig. 5B), and MCR III from *M. thermolithotrophicus* (PDB: 5N1Q; Fig. 5C) are in $MCR_{ox1\text{-}silent}$ state and resulted in a very overall similar structure. Among the three MCR isoforms, the residues close to the prosthetic group, CoB-SH and CoM-SH and their side chain conformations are strictly conserved: only the mutation Ile/Val380γ (MCR I/MCR II or MCR III) was observed. Among the enzymes of the methanogenic pathways, this high conservation is only peculiar to MCR, highlighting the complexity of the catalytic mechanism. Thus, MCR is evolved in a highly optimized polypeptide for catalyzing the MCR reaction, sensitive to minimal

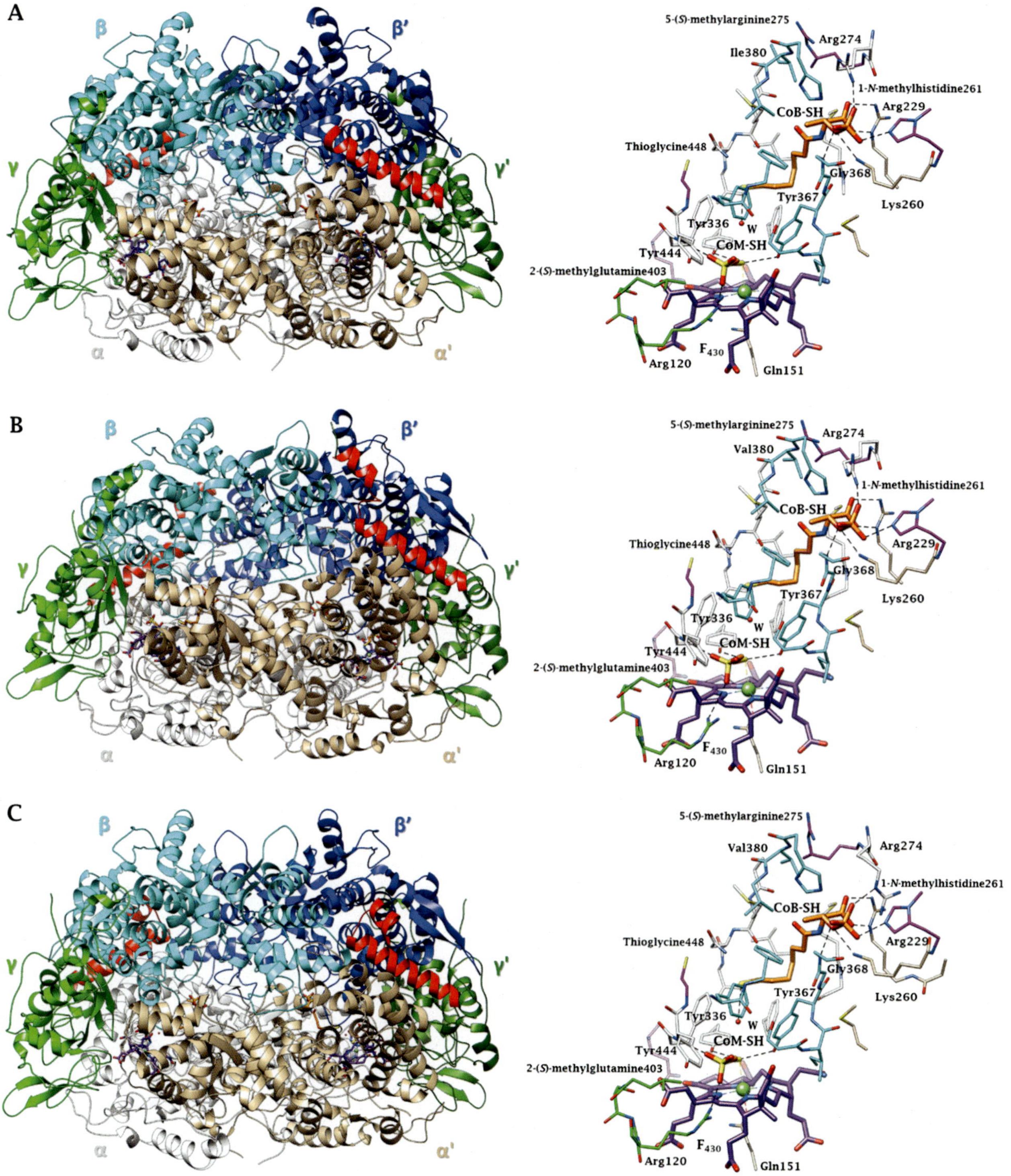

FIG. 5 X-ray crystal structure and zoom of the active site of (A) MCR I from *M. marburgensis* (PDB: 5A0Y; 1.10 Å), (B) MCR II from *M. wolfeii* (PDB: 5A8W; 1.80 Å), and (C) MCR III from *M. thermolithotrophicus* (PDB: 5N1Q; 1.90 Å). The most variable elements between the three isoforms are colored in red. CoM-SH is stabilized by four hydrogen bonds undertaken by two S=O moieties and the thiol group with Tyr444α, Arg120γ, Tyr336α, and Tyr367β, respectively, while CoB-SH engages six H-bonds with 1-*N*-methylhistidine261α, Arg274α, Arg229α′, Lys260α′, and Gly368γ. CoM-SH, CoB-SH, and CoM-S-S-CoB are colored in orange, while coenzyme F_{430} and posttranslational modified residues are painted in purple and magenta, respectively. Water molecules and nickel ions are represented as red and green spheres, in that order. H-bonds are depicted as black dashed lines.

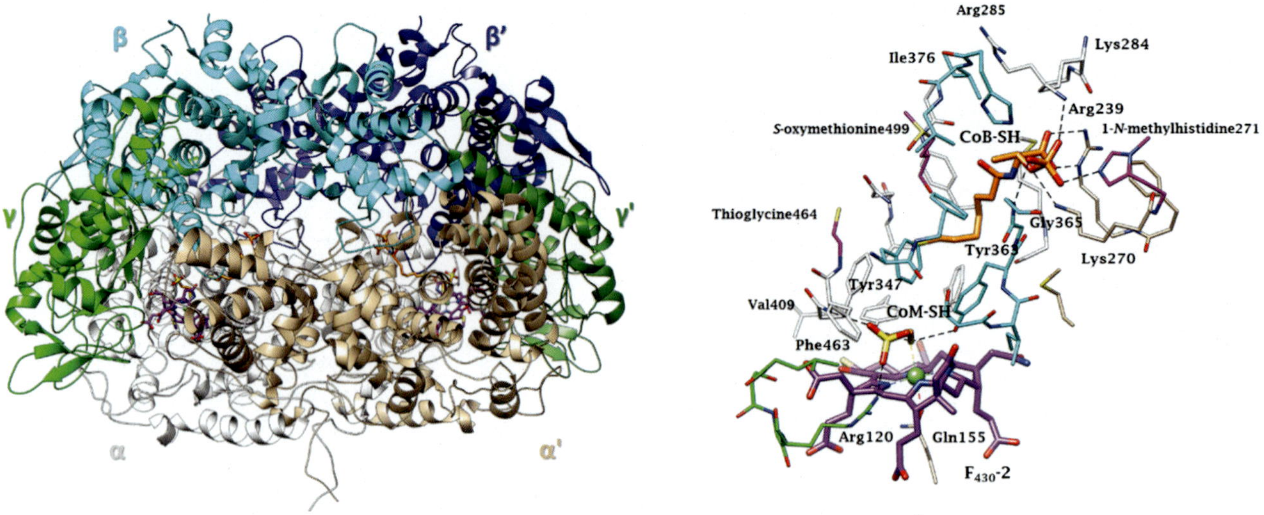

FIG. 6 X-ray crystal structure and zoom of the active site of MCR I from ANME-1 archaea (PDB: 3SQG; 2.10 Å). CoM-SH is stabilized by four hydrogen bonds undertaken by two S=O moieties and the thiol group with Phe463α, Arg120γ, Tyr347α, and Tyr363β, respectively, while CoB-SH engages six H-bonds with 1-*N*-methylhistidine271α, Arg285α, Arg239α′, Lys270α′, and Gly365γ. CoM-SH, CoB-SH, and CoM-S-S-CoB are colored in orange, while coenzyme F_{430}-2 and posttranslational modified residues are painted in purple and magenta, respectively. Water molecules and nickel ions are represented as red and green spheres, in that order. H-bonds are depicted as black dashed lines.

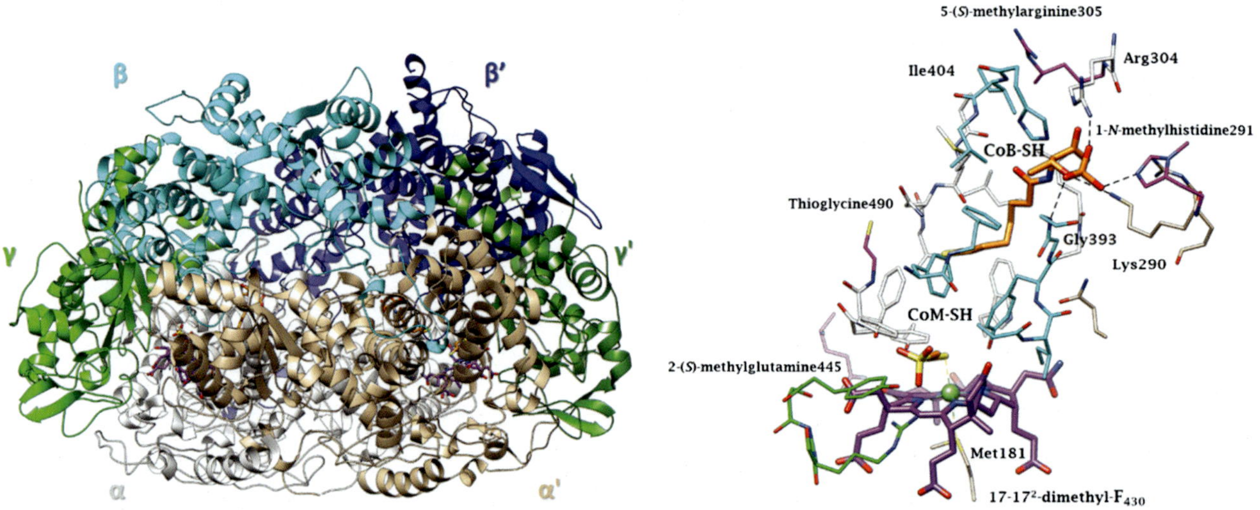

FIG. 7 X-ray crystal structure and zoom of the active site of MCR I from ANME-1 archaea (PDB: 3SQG; 2.10 Å). CoM-SH is not stabilized by hydrogen bond interactions, while CoB-SH engages four H-bonds with 1-*N*-methylhistidine291α, Arg304α, Lys290α′ and Gly393γ. CoM-SH, CoB-SH, and CoM-S-S-CoB are colored in orange, while coenzyme 17-17^2-dimethyl-F_{430} and posttranslational modified residues are painted in purple and magenta, respectively. Water molecules and nickel ions are represented as red and green spheres, in that order. H-bonds are depicted as black dashed lines.

amino acid exchanges [61]. In all three isoforms, the nickel ion of F_{430} is coordinated to the protein structure by the side-chain C=O of Gln151α′.

The structural differences between the three MCR types are principally found in the solvent-exposed loops α40-61, α357-368, β281-291, and γ181-188 [61]. In particular, a lot of basic amino acids on the protein surface are present at the entrance of the active site, probably to better attract the negatively charged substrate CoM-SH, CoM-SH, and CoM-S-CH_3, and are more abundant in MCR III [61]. Again, in MCR I the C-terminal Glu285, found on the eighth helix of the γ subunit (γ8), undertakes a unique salt bridge with the Arg236 of the β subunit. Instead, the presence of an additional helix (γ9) in MCR II and III is fundamental to increasing the interaction with the protein core. In MCR II, the helix γ9 engages a wide network of interactions with the β and β′ subunits and two stable salt bridges between Asp257γ and Arg261γ with Arg253β and Glu234β, respectively. In MCR III the presence of P252 rotates the helix γ9 of about 180 degrees if compared to that of MCR II.

This rotation brought the MCR III helix γ9 to interact with the β and α subunits (and not with β') through several hydrogen bonds and two conserved salt bridges (Arg240γ-Asp284β and Arg243γ-Asp371α/Asp374α). Furthermore, the Arg243γ also interacts with the terminal carboxy group of residue L261γ.

6.2 X-ray crystal structures of MCR from nonmethanogens

The 3D-solved structure of methanotrophic MCR was obtained in the $MCR_{ox1\text{-}silent}$ state from ANME-1 archaea (PDB: 3SQG; Fig. 6) collected by a Black Sea mat with a submersible [59,60]. The presence of CoM-SH and CoB-SH within the two active sites of the heterohexameric protein ($\alpha_2\beta_2\gamma_2$) confirms that methanotrophic and methanogenic archaea have identical substrates [75].

ANME-1 MCR has a highly similar structure to methanogenic MCR but differs for a wide number of cysteine residues in the α subunits [75]. Thioglycine448α and 1-N-methylhistidine261α' (numeration is referred to as the methanogens MCRs) are the unique posttranslational amino acid conserved with respect to methanogenic MCRs, while the other modifications are heterogeneous and are strictly dependent on the enzyme function and biological context [59]. In particular, the active site mutated residues of the ANME-1 MCR against the methanogens MCRs resulted in Phe238α'/Ala228α', Lys284α/Arg274α, Arg285α/5-(*S*)-methylarginine275α, Phe334α/Leu323α, Val419α/2-(*S*)-methylglutamine403α, Phe463α/Tyr447α, *S*-methylcysteine499α/Cys483α, respectively. Moreover, the prosthetic group represented by the 17^2-methylthio-F_{430} (defined as F_{430}-2), a coenzyme F_{430}, which includes a methylthio group on the tetrahydrocorphinoid scaffold (Fig. 6C), was exclusively identified in the crystallography of ANME-1 MCR (Fig. 6) [58,59]. Analog to the three methanogen archaea MCR isoforms, the prosthetic group is coordinated with the nickel ion to the protein structure by the C=O side chain of a Gln155α'.

6.2.1 X-ray crystal structures of MCR-like protein from alkane-oxidizer archaea

The alkyl coenzyme M reductases (ACRs) are MCR-like proteins that oxidize short-chain (e.g., ethane, propane, and butane) and long-chain alkanes as an energy source in some peculiar archaea. Nowadays, only enrichment cultures of ethane or butane-oxidizing archaea are available and metagenomes studies deduced the primary structure of their MCRs.

A very recent study solved the X-ray crystal structure of an MCR-like protein from the ethane oxidizer *Ethanoperedens thermophilum* in the $MCR_{ox1\text{-}silent}$ state (PDB: 7B1S; Fig. 7), obtaining the first structural information regarding an ethyl coenzyme M reductase (ECR) [62]. The treatment of an *E. thermophilum* culture with the classical MCR inhibitor 2-bromoethane sulfonate (BES) showed a decrement in ethane consumption, while the same sensibility was shown by a control culture of ANME archaea, corroborating the hypothesis that ECR performs the activation of ethane into ethyl-CoM [8,76,77]. ECR is a heterohexameric protein ($\alpha_2\beta_2\gamma_2$) such as the MCRs belonging to methanogenic and methanotrophic archaea and CoM-SH and CoB-SH lodge in the active site in an analog manner [62]. The prosthetic group has a 17-17^2-dimethyl-F_{430} structure essential for the ethane oxidation, uniquely identified in the ECR and anchored to the protein structure by the axial coordination of the nickel ion with the sulfur atom of Met181α', highly conserved in thermophilic ethanotrophs [62].

A four-helix bundle dilatation, specific loop extensions and posttranslational methylated residues, such as *S*-methylcysteine354α', 3-(*S*)-methylisoleucine377α', 2-(*S*)-methylglutamine445α', and N_2-methylhistidine491α', resulted in the establishment of a 33 Å-long hydrophobic tunnel, in which an inner diameter of 1.02 Å permits the exclusive entrance of ethane in the active site, preventing the access to the longer alkanes [62]. Moreover, the roomier active site of ECR adapts the catalytic reaction to a substrate with two carbon atoms, impairing the correct positioning of methane, and it is not enough larger to accommodate longer alkanes than ethane [62]. The same thing was observed in the MCR I from methanogen *M. marburgensis*, which can generate ethane by the ethyl-CoM and CoB-SH with low affinity because the volume of the MCR active site prevents a correct placing of larger alkanes than methane [78–80].

The posttranslational residue N_2-methylhistidine491α' in ECR is tyrosine or phenylalanine in all other MCRs (Tyr446α' in MCR I from *M. marburgensis*, Tyr448α' in MCR II from *M. wolfeii*, Tyr449α' in MCR III from *M. thermolithotrophicus*, and Tyr465α' in ANME1-MCR) [62].

The ECR active site residues, which differ from the discussed methanogenic MCRs are Arg308α'/Gly278α', Gln262α/Met233α, Trp373α/Phe333α, Tyr120γ/Leu120γ, and Tyr156γ/Ala157γ, while the differences between ECR against ANME1-MCR are in Gln263α'/Met243α', Arg258α'/Phe238α', Ala259α'/Arg239α', 5(*S*)-methylarginine305α/Arg285α, Leu362α/Phe334α, Met525α/*S*-oxymethionine499α and Tyr120γ/Leu117γ residues. Modification in the loop α365-376/α326-336/α337-347 (ECR/methanogenic MCRs/ANME1-MCR) is responsible for a larger active site of ECR than the MCRs.

In silico studies were able to build the ACR homology model from a butanotrophic archaea, showing a high identity with the primary structure of MCRs and ECRs, translating into an overall similar 3D structure [29,34].

7 Posttranscriptional modifications

The posttranscriptional modifications (PTMs) are assumed to play a key role in regulating MCR and ACR functions. However, the functions and biosynthesis of PTMs in MCR enzymes remain widely unclear [7]. Currently, seven PTMs were identified in methanogenic MCRs (1-*N*-methylhistidine, thioglycine, *S*-methylcysteine, 2-(*S*)-methylglutamine, 5-(*S*)-methylarginine, 6-hydroxytryptophan, didehydroaspartate), four in ANME-MCRs (1-*N*-methylhistidine, thioglycine, 7-hydroxytryptophan, *S*-oxymethionine) and six in ACRs (1-*N*-methylhistidine, *S*-methylcysteine, 2-(*S*)-methylglutamine, 5-(*S*)-methylarginine, 3-(*S*)-methylisoleucine, N_2-methylhistidine) (Fig. 8).

In particular, 1-*N*-methylhistidine is the unique posttranscriptional modified residue conserved in all the MCR types. The methylation might serve to position the imidazole ring for the correct coordination of CoB-SH [81]. Thioglycine, present in methanogenic and ANME-MCRs, could be important in (1) enhancing MCRs stability and facilitating catalysis, (2) reducing the sulfhydryl group pK_a and assisting the CoB-SH deprotonation, and (3) facilitating the catalysis as an electron carrier during the reaction [55,82,83]. *S*-methylcysteine and 5-(*S*)-methylarginine might be crucial in the adaption to mesophilic conditions of methanogenic MCRs and ACRs and play several roles in the cataliysis, assembly or stability in that order [84,85]. The peculiar methanogenic posttranscriptional modified residue didehydroaspartate, identified in MCR I and II, and 6-hydroxytryptophan, found in MCR III, might have the equivalent role in increasing the catalytic efficiency of the enzymes [61,80]. In ANMEs, 7-hydroxytryptophan is assumed to compensate for the arginine methylation absence [59]. Further studies are necessary to better understand the role of *S*-oxymethionine in ANME and 2-(*S*)-methylglutamine in methanogens. In ECR, 2-(*S*)-methylglutamine, *S*-methylcysteine, 3-(*S*)-methylisoleucine and N_2-methylhistidine seem to be important for the 33 Å-long hydrophobic tunnel stability of the ethane [62].

8 Catalytic mechanism of MCRs

The catalytic mechanism of MCRs is not completely clear. Nowadays, three reaction mechanisms were proposed by extensive mechanistic studies where different reaction

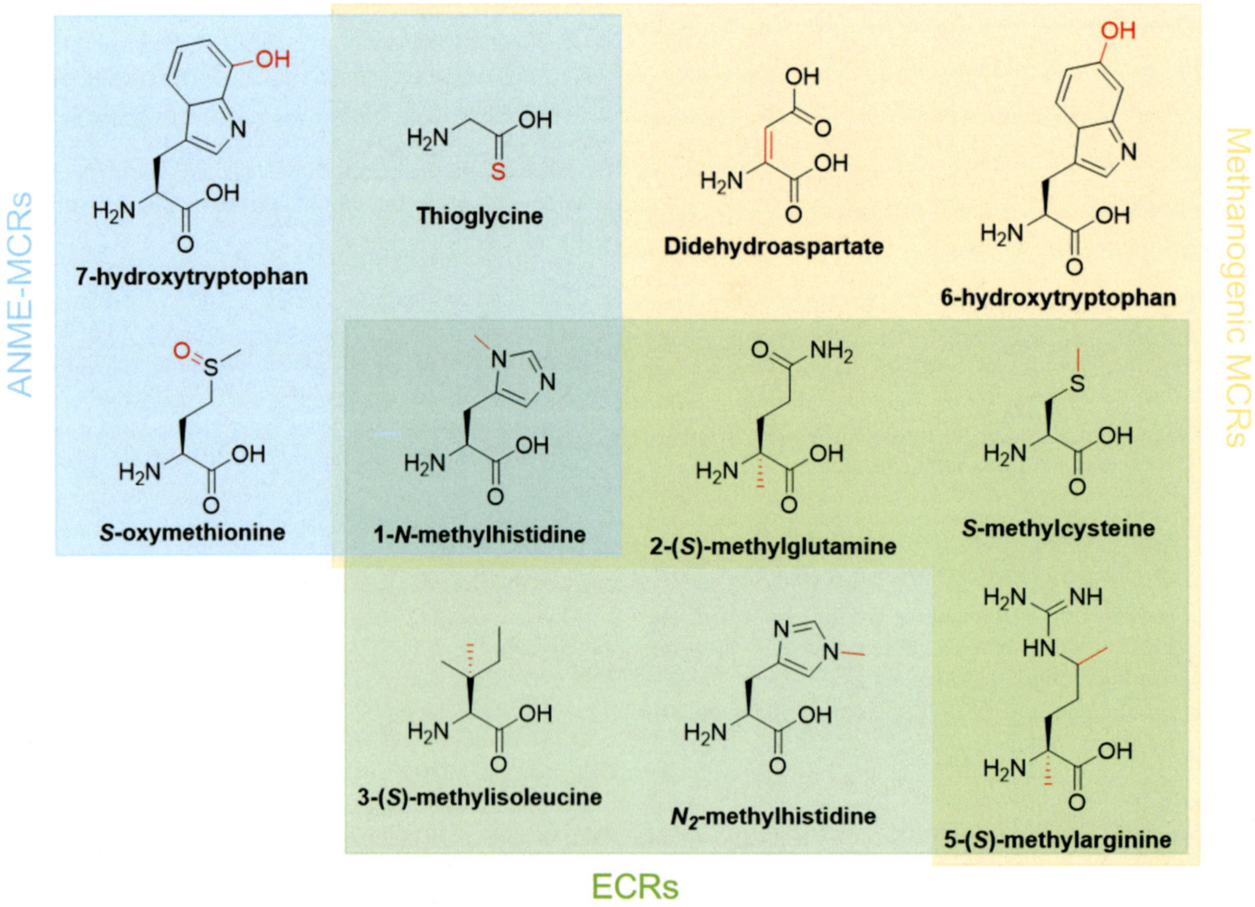

FIG. 8 Posttranslational modifications identified to date in methanogenic (orange), methanotrophic (blue) and alkanotrophic (green) MCRs.

intermediates are involved, such as a CH_3-Ni(III) species in mechanism I and a methyl radical species in mechanisms II and III (Fig. 9) [8,86–88].

In mechanism I is proposed the nucleophilic S_N2-type attack of CoM-SCH_3 by the Ni(I) species of F_{430} generates the transient CH_3-Ni(III) intermediate, followed by the C–S bond cleavage. The electron transfer by the CoM-S^- to the coenzyme triggers the hydrogen abstraction from the CoB-SH, producing methane, the disulfide CoM-S-S-CoB and the Ni(I) species. The mechanism I was proposed as a consequence of the F_{430} derivatives characterization and MCR_{red1} reactions with activated alkyl substrates [54,89–95]. The experiment used the pentamethylester of coenzyme F_{430} (F_{430}M) for its high solubility in noncoordinating organic solvents. In particular, Ni(I) species of F_{430}M (2 equiv.) react with electrophilic methyl donors (1 equiv.), such as methyl tosylate, methyl iodide and trimethylsulfonium compound, affording Ni (II) species (2 equiv.) and methane (1 equiv.). The formed intermediate CH_3-Ni (II) of F_{430}M was identified at low temperatures with NMR studies.

Likewise, the reactions of methyl bromide, methyl bromide and 3-bromopropionate the MCR_{red1} state form CH_3-/alkyl-Ni(III) complexes that are solved by X-ray structural/absorption analysis.

The formation of the CH_3-Ni(III) intermediate is supported by its ability to react with CoM-SH and CoB-SH, affording their thioethers and the regeneration of the MCR_{red1} state, and with Ti(III) citrate to regenerate the MCR_{red1} state producing methane [54,89–95]. Despite this, the CH_3-Ni(III) species has not been observed in the reaction between the MCR_{red1} state and CoM-SCH_3 and, as predicted by the hybrid DFT analysis, its formation due to the nucleophilic attack of the Ni(I) species on the methyl group of CoM-SCH_3 is endergonic and not practicable for the energy barrier of 21.8 $kcal\,mol^{-1}$ [96–99].

To solve this issue, mechanism II was proposed. In the first step, the Ni(I) species causes the C–S bond homolysis of CoM-SCH_3, obtaining a transient methyl radical intermediate. The next hydrogen atom abstraction from CoB-SH generates methane, the CoM-S-Ni(II) complex and the thiyl radical of CoB-SH. This radical intermediate of CoB triggers an electron transfer that initially produces the disulfide anion-Ni(II) complex and finally the original Ni(I) species and the disulfide CoM-S-S-CoB. Theoretical studies and a recent experimental investigation on the MCR_{red1} state reaction with substrate analogs containing thioether or thiol moieties support mechanism II. Furthermore, the hybrid DFT analysis calculates that the energy of the first and second steps in mechanism II is 15.1 and 5.9 $kcal\,mol^{-1}$, in that order, resulting in a more energetically favorable first rate-determining step than the analog step in mechanism I [96–100].

The third mechanism suggests the C–S bond long-term homolytic cleavage of the CoM-SCH_3 that reduces the generated thiyl radical to only afford a methyl radical species. The CoB-SH furnishes the hydrogen atom during the abstraction in the formation of methane and the complex CoB-S-S-CoM-SO_3-Ni(I) is generated by a long-distance electron transfer. The formation of the CH_3S-CoM-SO_3-Ni(I) species is confirmed by spectroscopic, kinetic and computational studies, supporting mechanism III. This reaction mechanism could be reasonable, considering the sulfur atoms distance of CoB-SH and CoM-SH. Moreover, hybrid DFT calculations have estimated mechanisms I and II, which start to coordinate the Ni(I) species with the CoM-SCH_3 carbon/sulfur atom, to be energetically unfavorable [101]. Once catalyzed methane reduction, MCR needs ATP-dependent reductive reactivation [8].

9 Catalytic features of MCRs

Currently, only the catalytic properties of MCR I from the Methanobacteriales *M. marburgensis* are consistently known, but it could be possible that the properties of the other methanogenic MCRs significantly differ. Indeed, catalytic activity and substrate specificity could be different even in phylogenetically closed related enzymes. The differences between the MCR I and II from *M. marburgensis*, sharing the 70% sequence identity, reflect various catalytic and affinity profiles. While is very difficult to obtain the MCR purified isoform I and II in the MCR_{red1} state in the presence of O_2 traces from *M. marburgensis*, it is possible to easily achieve the MCR_{ox1} and $MCR_{ox1\text{-}silent}$ states, thanks to the Ni(I) species protection from the oxidation by CoM-SCH_3 or CoM-SH substrates [102]. Thus, the purified MCR_{red1} I is obtained by the MCR_{ox1} I reduction in an oxygen-free chamber under an atmosphere condition of 95% N_2 and 5% H_2 in presence of Ti(III) and CoM-SCH_3 at 60°C. In the case of MCR $_{red1}$ I, the apparent K_M for CoM-SCH_3 and CoB-SH in the reduction to methane and CoM-S-S-CoB are 0.7 ± 0.2 and 0.2 ± 0.1 mM, respectively with an apparent V_{max} of 100 $\mu mol\,min^{-1}$ mg protein^{-1} [47,103]. Instead, the apparent K_M of MCR $_{red1}$ II for the CoM-SCH_3 and CoB-SH substrates are 1.4 ± 0.2 and 0.5 ± 0.2 mM, in that order. Furthermore, the MCR I and II catalytic activity are optimized for a pH of 7.0–7.5 and 7.5–8.0, correspondingly [47,103].

In particular, MCR I can also use the substrate ethyl-S-CoM with an apparent K_M of about 20 mM and specific activity of 0.1 $\mu mol\,min^{-1}$ mg protein^{-1} for methane reduction, but it cannot catalyze the reaction with propyl-S-CoM, allyl-S-CoM and CoM-SCF_3 [102]. Again, the MCR from *M. thermoautotrophicus* can produce methane from CoM-SCH_2F, CoM-$SCHF_2$, CoM-$SeCH_3$ and 3-methylthiopropionate, with very low rates [79,104,105].

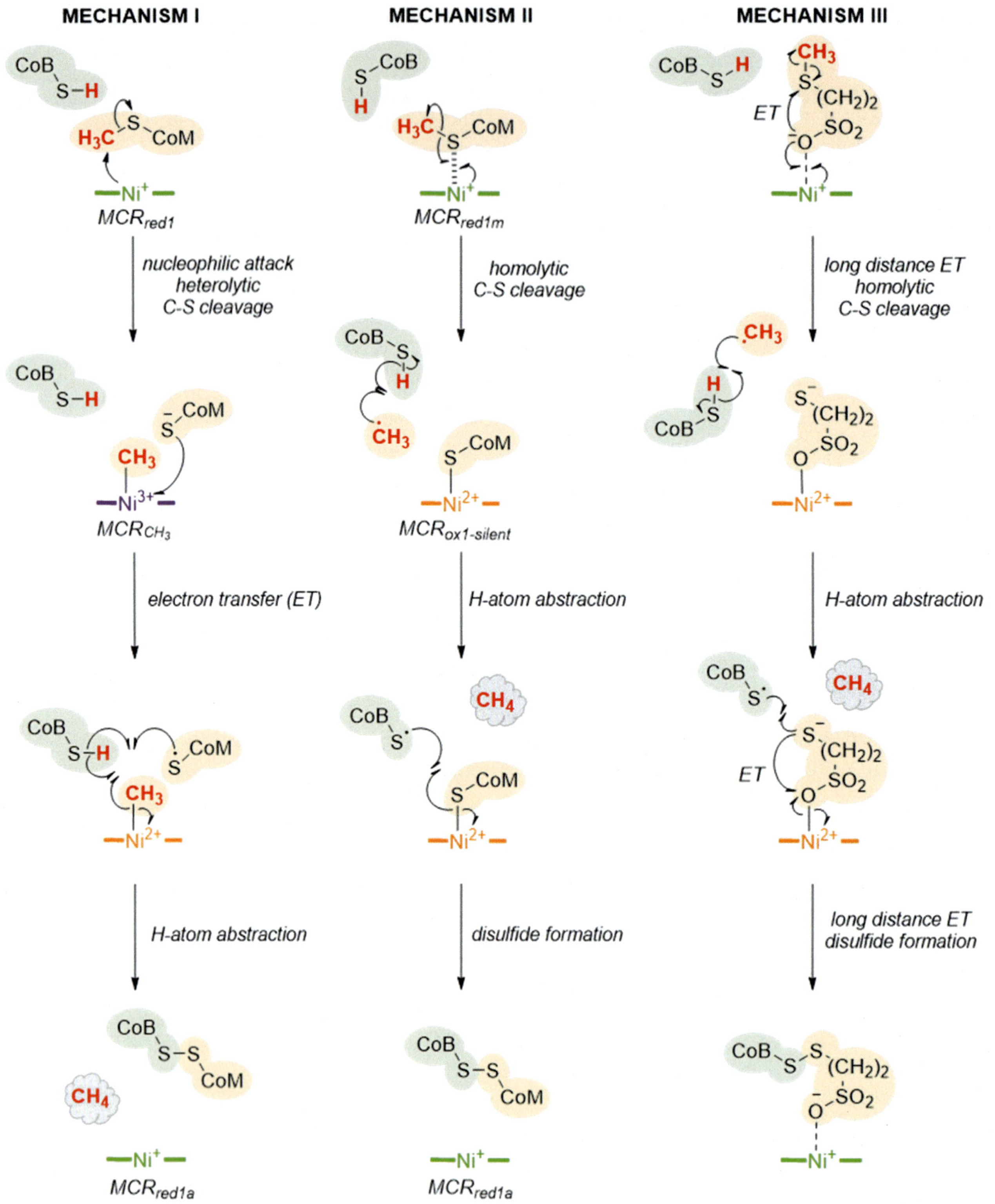

FIG. 9 Schematic representation of the three proposed catalytic mechanisms. The coenzyme F_{430} in Ni(I), Ni(II) and Ni(III) state is colored in green, orange and violet, respectively. The atoms involved in the formation of the methane are labeled in red. The atoms derived from CoB/CoM are surrounded by a green/yellow cloud. The electron transfers are indicated by the black arrows.

Besides CoB-SH (7-mercaptoheptanoyl-threoninephosphate), Co6B-SH (6-mercaptoheptanoyl-threoninephosphate) could act as an electron donor. However, the reduction rates of CoM-SCH_3 with Co6B-SH are less than 1% of the amount obtained with CoB-SH [106,107]. Despite that, the slow analog Co6B-SH resulted in a more important tool for deciphering the MCR catalytic mechanism [100].

Moreover, during the MCR-catalyzed reaction of ethyl-S-CoM, an inversion of the stereo-configuration of the substrate was observed in *M. barkeri* and *M. marburgensis*, using a substrate isotopically chiral form or a deuterated substrate, respectively [78,108].

MCR I can catalyze the oxidation of methane, the back reaction, with a specific activity of about 10 nmol min^{-1} mg $protein^{-1}$ at 60°C and 1 bar (~1 mM of CH_4), corresponding to 0.01% of the forward reaction specific activity [109]. The amount of methane oxidation improved linearly until the highest methane concentration reached experimentally (2 mM), highlighting that the MCR apparent K_M for methane is above 2 mM. These catalytic features of MCR I from *M. marburgensis* are comparable with the ANME-MCRs kinetic properties, enzymes involved in the direct anaerobic methane oxidation in the very slowly growing ANME organisms [109].

10 MCR inhibitors

During the purification of MCR_{red1} in presence of CoM-SH, the apparent K_M for CoM-SCH_3 is about 5 vs 0.7 mM (K_M value in absence of CoM-SH), while the apparent V_{max} resulted in a very lower value than 100 μmol min^{-1} mg $protein^{-1}$. This evidence shows that CoM-SH acts as a competitive inhibitor of MCR with an apparent K_I of 4 mM [102]. Other reversible inhibitors that directly bind the Ni(I) ion of F_{430} are propyl-S-CoM (~K_I=2 mM), allyl-S-CoM (~K_I=0.1 mM), and 2-azidoethane-S-CoM (~K_I=1 μM). Generally, their competitive inhibition is explained also in presence of different analogs and homologs of CoB-SH [74,93,102,106,107,110].

The 2-bromoethanesulfonate (BES) is the first irreversible inhibitor of MCR discovered in 1978. Competing with CoM-SH, BES (~K_I=2 μM) alkylates Ni(I) and produces a labile alkyl-Ni(III) intermediate that collapses to afford ethylene. [102,111,112]. Instead, the irreversible inhibitor 3-bromopropanesulfonate (BPS) with an apparent K_I of 0.1 μM, displaces CoM-SH, binds the Ni(I)ion of F_{430} and generates stable alkyl-Ni(III) derivative [110,113,114]. Also brominated carboxylic acids with a carbon chain length of 5–16 atoms are able to generate a stable alkyl-Ni(III) species [93]. Again, polychlorinated hydrocarbons in C1 and C2 positions act as irreversible MCR inhibitors that oxidize the Ni(I) ion [115]. Another important MCR inhibitor classified as a dual warhead inhibitor is 3-nitrooxypropanol (3-NOP), which, together with BES, is used in the ecological studies of the methane cycle. The action mechanism of 3-NOP consists of the Ni(I) species oxidation that in turn is translated into the reduction of the inhibitor nitro-ester moiety to afford the 1,3-propanediol and nitrite. The nitrite causes the oxidation of a second Ni(I) species in the MCR active site. Nowadays, 3-NOP is tested to inhibit the methanogenesis in the rumen of farmed sheep and cows [116–118]. The not charged 3-NOP is able to permeate the methanogenic cells without necessitating a transporter. Conversely, BES (~K_I=2 μM) uses the transporter of CoM-SH to enter the methanogenic cells, while the most potent inhibitor BPS in vitro (~K_I =0.1 μM) results in less activity in vivo than BES because it is not transported inside the cells. Nitrite inactivates MCR while nitrate does not affect this enzymatic complex [118]. On the other hand, 2-nitroethanol, 2-nitro-L-propanol and nitroethane are nitro-compounds that inhibit methanogenesis in vivo at mM concentrations, with unknown mechanisms [119]. Many other compounds inhibit methanogenesis in vivo and may be acting on MCR. Among these, the most interesting compound is ethylene because, if the Ni(I) species can add to the alkene double bond, that would be evidence of the great one-electron reactivity of the Ni(I) ion [120,121].

Some derivatives explain a CoB-SH-dependent inhibition. The inactivation rate of BPS and BES is correlated to CoB-SH. Indeed, BES acts as an MCR inhibitor at very low concentrations if CoB-SH is present. Other inhibitors whose inactivation depends on CoB-SH are cyano-S-CoM, CoM-SCF_3 and seleno-S-CoM [102]. These discoveries are very important because demonstrate that the CoB-SH binding improves the nucleophilicity of the nickel atom of F_{430}.

11 Conclusions

The treatment of farming animals with MCR inhibitors is very important to reduce methane emissions. This is a fundamental step in the control of the greenhouse effect phenomenon metamorphosing intensive husbandry into more eco-sustainable practices [117,118]. Among the several inhibitors of MCR, 3-NOP resulted in the most potent tool in vivo, classified as a dual warhead inhibitor that permeates the archaeal cells without a transporter. BES is transported in the cell and inhibits methanogenesis while BPS is less active in vivo [116–118]. Another important application of MCR could be in the energetic field for the production of methane by archaea. Currently, activators of MCR have not been discovered yet.

References

[1] Hua ZS, Wang YL, Evans PN, Qu YN, Goh KM, Rao YZ, Qi YL, Li YX, Huang MJ, Jiao JY, Chen YT, Mao YP, Shu WS, Hozzein W,

Hedlund BP, Tyson GW, Zhang T, Li WJ. Insights into the ecological roles and evolution of methyl-coenzyme M reductase-containing hot spring Archaea. Nat Commun 2019;10:4574.

[2] Wang Y, Wegener G, Hou J, Wang F, Xiao X. Expanding anaerobic alkane metabolism in the domain of Archaea. Nat Microbiol 2019;4:595–602.

[3] Gendron A, Allen KD. Overview of diverse methyl/alkyl-coenzyme m reductases and considerations for their potential heterologous expression. Front Microbiol 2022;13, 867342.

[4] Lemaire ON, Wagner T. A structural view of alkyl-coenzyme M reductases, the first step of alkane anaerobic oxidation catalyzed by archaea. Biochemistry 2022;61:805–21.

[5] Miyazaki Y, Oohora K, Hayashi T. Focusing on a nickel hydrocorphinoid in a protein matrix: methane generation by methyl-coenzyme M reductase with F430 cofactor and its models. Chem Soc Rev 2022;51:1629–39.

[6] Wang Y, Wegener G, Ruff SE, Wang F. Methyl/alkyl-coenzyme M reductase-based anaerobic alkane oxidation in archaea. Environ Microbiol 2021;23:530–41.

[7] Chen H, Gan Q, Fan C. Methyl-coenzyme M reductase and its post-translational modifications. Front Microbiol 2020;11, 578356.

[8] Thauer RK. Methyl (alkyl)-coenzyme M reductases: nickel F-430-containing enzymes involved in anaerobic methane formation and in anaerobic oxidation of methane or of short chain alkanes. Biochemistry 2019;58:5198–220.

[9] Garcia PS, Gribaldo S, Borrel G. Diversity and evolution of methane-related pathways in archaea. Annu Rev Microbiol 2022; 76:727–55.

[10] Thomas CM, Desmond-Le Quéméner E, Gribaldo S, Borrel G. Factors shaping the abundance and diversity of the gut archaeome across the animal kingdom. Nat Commun 2022;13:3358.

[11] Borrel G, Brugère JF, Gribaldo S, Schmitz RA, Moissl-Eichinger C. The host-associated archaeome. Nat Rev Microbiol 2020;18: 622–36.

[12] Moissl-Eichinger C, Pausan M, Taffner J, Berg G, Bang C, Schmitz RA. Archaea are interactive components of complex microbiomes. Trends Microbiol 2018;26:70–85.

[13] Lyu Z, Chou CW, Shi H, Wang L, Ghebreab R, Phillips D, Yan Y, Duin EC, Whitman WB. Assembly of methyl coenzyme M reductase in the methanogenic archaeon Methanococcus maripaludis. J Bacteriol 2018;200. e00746–17.

[14] Liu Y, Whitman WB. Metabolic, phylogenetic, and ecological diversity of the methanogenic archaea. Ann N Y Acad Sci 2008; 1125:171–89.

[15] Yan Z, Ferry JG. Electron bifurcation and confurcation in methanogenesis and reverse methanogenesis. Front Microbiol 2018;9:1322.

[16] Costa KC, Wong PM, Wang T, Lie TJ, Dodsworth JA, Swanson I, Burn JA, Hackett M, Leigh JA. Protein complexing in a methanogen suggests electron bifurcation and electron delivery from formate to heterodisulfide reductase. Proc Natl Acad Sci U S A 2010; 107:11050–5.

[17] Ferry JG. CO in methanogenesis. Ann Microbiol 2010;60:1–12.

[18] Dolfing J, Jiang B, Henstra AM, Stams AJ, Plugge CM. Syntrophic growth on formate: a new microbial niche in anoxic environments. Appl Environ Microbiol 2008;74:6126–31.

[19] Kurth JM, Nobu MK, Tamaki H, de Jonge N, Berger S, Jetten MSM, Yamamoto K, Mayumi D, Sakata S, Bai L, Cheng L, Nielsen JL, Kamagata Y, Wagner T, Welte CU. Methanogenic archaea use a bacteria-like methyltransferase system to demethoxylate aromatic compounds. ISME J 2021;15:3549–65.

[20] Mayumi D, Mochimaru H, Tamaki H, Yamamoto K, Yoshioka H, Suzuki Y, Kamagata Y, Sakata S. Methane production from coal by a single methanogen. Science 2016;354:222–5.

[21] Lyu Z, Shao N, Akinyemi T, Whitman WB. Methanogenesis. Curr Biol 2018;28:R727–32.

[22] Chadwick GL, Skennerton CT, Laso-Pérez R, Leu AO, Speth DR, Yu H, Morgan-Lang C, Hatzenpichler R, Goudeau D, Malmstrom R, Brazelton WJ, Woyke T, Hallam SJ, Tyson GW, Wegener G, Boetius A, Orphan VJ. Comparative genomics reveals electron transfer and syntrophic mechanisms differentiating methanotrophic and methanogenic archaea. PLoS Biol 2022;20, e3001508.

[23] Haroon MF, Hu S, Shi Y, Imelfort M, Keller J, Hugenholtz P, Yuan Z, Tyson GW. Anaerobic oxidation of methane coupled to nitrate reduction in a novel archaeal lineage. Nature 2013;500:567–70.

[24] Knittel K, Boetius A. Anaerobic oxidation of methane: progress with an unknown process. Annu Rev Microbiol 2009;63:311–34.

[25] Zhou Z, Zhang CJ, Liu PF, Fu L, Laso-Pérez R, Yang L, Bai LP, Li J, Yang M, Lin JZ, Wang WD, Wegener G, Li M, Cheng L. Non-syntrophic methanogenic hydrocarbon degradation by an archaeal species. Nature 2022;601:257–62.

[26] Zhao R, Biddle JF. Helarchaeota and co-occurring sulfate-reducing bacteria in subseafloor sediments from the Costa Rica margin. ISME Commun 2021;1:1–11.

[27] Wang Y, Wegener G, Williams TA, Xie R, Hou J, Tian C, Zhang Y, Wang F, Xiao X. A methylotrophic origin of methanogenesis and early divergence of anaerobic multicarbon alkane metabolism. Sci Adv 2021;7:eabj1453.

[28] Hahn CJ, Laso-Pérez R, Vulcano F, Vaziourakis KM, Stokke R, Steen IH, Teske A, Boetius A, Liebeke M, Amann R, Knittel K, Wegener G. "Candidatus ethanoperedens," a thermophilic genus of archaea mediating the anaerobic oxidation of ethane. mBio 2020;11. e00600–20.

[29] Chen SC, Musat N, Lechtenfeld OJ, Paschke H, Schmidt M, Said N, Popp D, Calabrese F, Stryhanyuk H, Jaekel U, Zhu YG, Joye SB, Richnow HH, Widdel F, Musat F. Anaerobic oxidation of ethane by archaea from a marine hydrocarbon seep. Nature 2019; 568:108–11.

[30] Borrel G, Adam PS, McKay LJ, Chen LX, Sierra-García IN, Sieber CMK, Letourneur Q, Ghozlane A, Andersen GL, Li WJ, Hallam SJ, Muyzer G, de Oliveira VM, Inskeep WP, Banfield JF, Gribaldo S. Wide diversity of methane and short-chain alkane metabolisms in uncultured archaea. Nat Microbiol 2019;4:603–13.

[31] Laso-Pérez R, Hahn C, van Vliet DM, Tegetmeyer HE, Schubotz F, Smit NT, Pape T, Sahling H, Bohrmann G, Boetius A, Knittel K, Wegener G. Anaerobic Degradation of non-methane alkanes by "Candidatus Methanoliparia" in hydrocarbon seeps of the Gulf of Mexico. mBio 2019;10:e01814–9.

[32] Boyd JA, Jungbluth SP, Leu AO, Evans PN, Woodcroft BJ, Chadwick GL, Orphan VJ, Amend JP, Rappé MS, Tyson GW. Divergent methyl-coenzyme M reductase genes in a deep-subseafloor Archaeoglobi. ISME J 2019;13:1269–79.

[33] Seitz KW, Dombrowski N, Eme L, Spang A, Lombard J, Sieber JR, Teske AP, Ettema TJG, Baker BJ. Asgard archaea capable of anaerobic hydrocarbon cycling. Nat Commun 2019;10:1822.

[34] Laso-Pérez R, Wegener G, Knittel K, Widdel F, Harding KJ, Krukenberg V, Meier DV, Richter M, Tegetmeyer HE, Riedel D,

Richnow HH, Adrian L, Reemtsma T, Lechtenfeld OJ, Musat F. Thermophilic archaea activate butane via alkyl-coenzyme M formation. Nature 2016;539:396–401.

[35] Fournier GP, Gogarten JP. Evolution of acetoclastic methanogenesis in Methanosarcina via horizontal gene transfer from cellulolytic Clostridia. J Bacteriol 2008;190:1124–7.

[36] Vanwonterghem I, Evans PN, Parks DH, Jensen PD, Woodcroft BJ, Hugenholtz P, Tyson GW. Methylotrophic methanogenesis discovered in the archaeal phylum Verstraetearchaeota. Nat Microbiol 2016;1:16170.

[37] Knief C. Diversity and habitat preferences of cultivated and uncultivated aerobic methanotrophic bacteria evaluated based on pmoA as molecular marker. Front Microbiol 2015;6:1346.

[38] Ferry JG, House CH. The stepwise evolution of early life driven by energy conservation. Mol Biol Evol 2006;23:1286–92.

[39] Zheng Y, Harris DF, Yu Z, Fu Y, Poudel S, Ledbetter RN, Fixen KR, Yang ZY, Boyd ES, Lidstrom ME, Seefeldt LC, Harwood CS. A pathway for biological methane production using bacterial iron-only nitrogenase. Nat Microbiol 2018;3:281–6.

[40] Aldrich HC, Beimborn DB, Bokranz M, Schönheit P. Immunocytochemical localization of methyl-coenzyme-M reductase in methanobacterium-thermoautotrophicum. Arch Microbiol 1987; 147:190–4.

[41] Ossmer R, Mund T, Hartzell PL, Konheiser U, Kohring GW, Klein A, Wolfe RS, Gottschalk G, Mayer F. Immunocytochemical localization of component C of the methylreductase system in Methanococcus voltae and Methanobacterium thermoautotrophicum. Proc Natl Acad Sci U S A 1986;83:5789–92.

[42] Mayer F, Rohde M, Salzmann M, Jussofie A, Gottschalk G. The methanoreductosome: a high-molecular-weight enzyme complex in the methanogenic bacterium strain Gö1 that contains components of the methylreductase system. J Bacteriol 1988;170:1438–44.

[43] Wrede C, Walbaum U, Ducki A, Heieren I, Hoppert M. Localization of methyl-coenzyme M reductase as metabolic marker for diverse methanogenic archaea. Archaea 2013;2013, 920241.

[44] Sauer FD, Erfle JD, Mahadevan S. Methane production by the membranous fraction of Methanobacterium thermoautotrophicum. Biochem J 1980;190:177–82.

[45] Rospert S, Böcher R, Albracht SP, Thauer RK. Methyl-coenzyme M reductase preparations with high specific activity from H2-preincubated cells of Methanobacterium thermoautotrophicum. FEBS Lett 1991;291:371–5.

[46] Albracht SPJ, Ankel-Fuchs D, Bocher R, Ellermann J, Moll J, Vanderzwaan JW, Thauer RK. 5 New electron-paramagnetic-res signals assigned to nickel in methylcoenzyme M-reductase from Methanobacterium-thermoautotrophicum, Strain Marburg. Biochim Biophys Acta Protein Struct Mol Enzymol 1988;955:86–102.

[47] Goubeaud M, Schreiner G, Thauer RK. Purified methyl-coenzyme-M reductase is activated when the enzyme-bound coenzyme F430 is reduced to the nickel(I) oxidation state by titanium(III) citrate. Eur J Biochem 1997;243:110–4.

[48] Tietze M, Beuchle A, Lamla I, Orth N, Dehler M, Greiner G, Beifuss U. Redox potentials of methanophenazine and CoB-S-S-CoM, factors involved in electron transport in Methanogenic archaea. Chembiochem 2003;4:333–5.

[49] Duin EC, Signor L, Piskorski R, Mahlert F, Clay MD, Goenrich M, Thauer RK, Jaun B, Johnson MK. Spectroscopic investigation of the nickel-containing porphinoid cofactor F-430. Comparison of the free cofactor in the + 1, + 2 and + 3 oxidation states with the cofactor bound to methyl-coenzyme M reductase in the silent, red and ox forms. J Biol Inorg Chem 2004;9:563–76.

[50] Mahlert F, Bauer C, Jaun B, Thauer RK, Duin EC. The nickel enzyme methyl-coenzyme M reductase from methanogenic archaea: In vitro induction of the nickel-based MCR-ox EPR signals from MCR-red2. J Biol Inorg Chem 2002;7:500–13.

[51] Mahlert F, Grabarse W, Kahnt J, Thauer RK, Duin EC. The nickel enzyme methyl-coenzyme M reductase from methanogenic archaea: in vitro interconversions among the EPR detectable MCR-red1 and MCR-red2 states. J Biol Inorg Chem 2002;7:101–12.

[52] Finazzo C, Harmer J, Bauer C, Jaun B, Duin EC, Mahlert F, Goenrich M, Thauer RK, Van Doorslaer S, Schweiger A. Coenzyme B induced coordination of coenzyme M via its thiol group to Ni(I) of F430 in active methyl-coenzyme M reductase. J Am Chem Soc 2003;125:4988–9.

[53] Albracht SPJ, Ankel-Fuchs D, Böcher R, Ellermann J, Moll J, van der Zwaan JW, Thauer RK. Five new EPR signals assigned to nickel in methyl-coenzyme M reductase from Methanobacterium thermoautotrophicum, strain Marburg. Biochim Biophys Acta, Protein Struct Mol Enzymol 1988;955:86–102.

[54] Yang N, Reiher M, Wang M, Harmer J, Duin EC. Formation of a nickel-methyl species in methyl-coenzyme m reductase, an enzyme catalyzing methane formation. J Am Chem Soc 2007;129:11028–9.

[55] Grabarse W, Mahlert F, Duin EC, Goubeaud M, Shima S, Thauer RK, Lamzin V, Ermler U. On the mechanism of biological methane formation: structural evidence for conformational changes in methyl-coenzyme M reductase upon substrate binding. J Mol Biol 2001;309:315–30.

[56] Ermler U, Grabarse W, Shima S, Goubeaud M, Thauer RK. Crystal structure of methyl-coenzyme M reductase: the key enzyme of biological methane formation. Science 1997;278:1457–62.

[57] Zheng K, Ngo PD, Owens VL, Yang XP, Mansoorabadi SO. The biosynthetic pathway of coenzyme F430 in methanogenic and methanotrophic archaea. Science 2016;354:339–42.

[58] Allen KD, Wegener G, White RH. Discovery of multiple modified F(430) coenzymes in methanogens and anaerobic methanotrophic archaea suggests possible new roles for F(430) in nature. Appl Environ Microbiol 2014;80:6403–12.

[59] Shima S, Krueger M, Weinert T, Demmer U, Kahnt J, Thauer RK, Ermler U. Structure of a methyl-coenzyme M reductase from Black Sea mats that oxidize methane anaerobically. Nature 2011;481: 98–101.

[60] Krüger M, Meyerdierks A, Glöckner FO, Amann R, Widdel F, Kube M, Reinhardt R, Kahnt J, Böcher R, Thauer RK, Shima S. A conspicuous nickel protein in microbial mats that oxidize methane anaerobically. Nature 2003;426:878–81.

[61] Wagner T, Wegner CE, Kahnt J, Ermler U, Shima S. Phylogenetic and structural comparisons of the three types of methyl coenzyme M reductase from methanococcales and methanobacteriales. J Bacteriol 2017;199. e00197–17.

[62] Hahn CJ, Lemaire ON, Kahnt J, Engilberge S, Wegener G, Wagner T. Crystal structure of a key enzyme for anaerobic ethane activation. Science 2021;373:118–21.

[63] Rospert S, Linder D, Ellermann J, Thauer RK. Two genetically distinct methyl-coenzyme M reductases in Methanobacterium thermoautotrophicum strain Marburg and delta H. Eur J Biochem 1990;194:871–7.

[64] Pihl TD, Sharma S, Reeve JN. Growth phase-dependent transcription of the genes that encode the 2 methyl-coenzyme-M reductase isoenzymes and N-5-methyltetrahydromethanopterin-coenzyme-M methyltransferase in Methanobacterium thermoautotrophicum Delta-H. J Bacteriol 1994;176:6384–91.

[65] Bonacker LG, Baudner S, Thauer RK. Differential expression of the 2 methyl-coenzyme M reductases in Methanobacterium thermoautotrophicum as determined immunochemically via isoenzyme-specific antisera. Eur J Biochem 1992;206:87–92.

[66] Rospert S, Linder D, Ellermann J, Thauer RK. 2 genetically distinct methyl-coenzyme M reductases in Methanobacterium thermoautotrophicum Strain Marburg and Delta-H. Eur J Biochem 1990; 194:871–7.

[67] Bonacker LG, Baudner S, Morschel E, Bocher R, Thauer RK. Properties of the 2 isoenzymes of methyl-coenzyme-M reductase in Methanobacterium thermoautotrophicum. Eur J Biochem 1993;217: 587–95.

[68] Jablonski PE, Ferry JG. Purification and properties of methyl coenzyme-M methylreductase from acetate grown Methanosarcina thermophila. J Bacteriol 1991;173:2481–7.

[69] Selmer T, Kahnt J, Goubeaud M, Shima S, Grabarse W, Ermler U, Thauer RK. The biosynthesis of methylated amino acids in the active site region of methyl-coenzyme M reductase. J Biol Chem 2000;275:3755–60.

[70] Bokranz M, Klein A. Nucleotide-sequence of the methyl coenzyme-M reductase gene-cluster from Methanosarcina barkeri. Nucleic Acids Res 1987;15:4350–1.

[71] Cram DS, Sherf BA, Libby RT, Mattaliano RJ, Ramachandran KL, Reeve JN. Structure and expression of the genes, McrBDCGA, which encode the subunits of component-C of methyl coenzyme-M reductase in Methanococcus vannielii. Proc Natl Acad Sci U S A 1987;84:3992–6.

[72] Allmansberger R, Bollschweiler C, Konheiser U, Müller B, Muth E, Pasti G, Klein A. Arrangement and expression of methyl coenzyme M reductase genes in Methanococcus voltae. Syst Appl Microbiol 1986;7:13–7.

[73] Prakash D, Wu Y, Suh S, Duin EC. Elucidating the process of activation of methyl-coenzyme M reductase. J Bacteriol 2014; 196:2491–8.

[74] Ellermann J, Rospert S, Thauer RK, Bokranz M, Klein A, Voges M, Berkessel A. Methyl-coenzyme-M reductase from Methanobacterium thermoautotrophicum (Strain Marburg)—purity, activity and novel inhibitors. Eur J Biochem 1989;184:63–8.

[75] Mayr S, Latkoczy C, Krüger M, Günther D, Shima S, Thauer RK, Widdel F, Jaun B. Structure of an F430 variant from archaea associated with anaerobic oxidation of methane. J Am Chem Soc 2008;130:10758–67.

[76] Dong X, Rattray JE, Campbell DC, Webb J, Chakraborty A, Adebayo O, Matthews S, Li C, Fowler M, Morrison NM, MacDonald A, Groves RA, Lewis IA, Wang SH, Mayumi D, Greening C, Hubert CRJ. Thermogenic hydrocarbon biodegradation by diverse depth-stratified microbial populations at a Scotian Basin cold seep. Nat Commun 2020;11:5825.

[77] Holler T, Widdel F, Knittel K, Amann R, Kellermann MY, Hinrichs KU, Teske A, Boetius A, Wegener G. Thermophilic anaerobic oxidation of methane by marine microbial consortia. ISME J 2011; 5:1946–56.

[78] Scheller S, Goenrich M, Thauer RK, Jaun B. Methyl-coenzyme M reductase from methanogenic archaea: isotope effects on label exchange and ethane formation with the homologous substrate ethyl-coenzyme M. J Am Chem Soc 2013;135:14985–95.

[79] Gunsalus RP, Romesser JA, Wolfe RS. Preparation of coenzyme M analogues and their activity in the methyl coenzyme M reductase system of Methanobacterium thermoautotrophicum. Biochemistry 1978;17:2374–7.

[80] Wagner T, Kahnt J, Ermler U, Shima S. Didehydroaspartate modification in methyl-coenzyme M reductase catalyzing methane formation. Angew Chem Int Ed Engl 2016;55:10630–3.

[81] Grabarse W, Mahlert F, Shima S, Thauer RK, Ermler U. Comparison of three methyl-coenzyme M reductases from phylogenetically distant organisms: unusual amino acid modification, conservation and adaptation. J Mol Biol 2000;303:329–44.

[82] Horng YC, Becker DF, Ragsdale SW. Mechanistic studies of methane biogenesis by methyl-coenzyme M reductase: evidence that coenzyme B participates in cleaving the C-S bond of methyl-coenzyme M. Biochemistry 2001;40:12875–85.

[83] Nayak DD, Mahanta N, Mitchell DA, Metcalf WW. Post-translational thioamidation of methyl-coenzyme M reductase, a key enzyme in methanogenic and methanotrophic archaea. Elife 2017;6, e29218.

[84] Nayak DD, Liu A, Agrawal N, Rodriguez-Carerro R, Dong SH, Mitchell DA, Nair SK, Metcalf WW. Functional interactions between posttranslationally modified amino acids of methyl-coenzyme M reductase in Methanosarcina acetivorans. PLoS Biol 2020;18, e3000507.

[85] Lyu Z, Shao N, Chou CW, Shi H, Patel R, Duin EC, Whitman WB. Posttranslational methylation of arginine in methyl coenzyme M reductase has a profound impact on both methanogenesis and growth of Methanococcus maripaludis. J Bacteriol 2020;202. e00654–19.

[86] Patwardhan A, Sarangi R, Ginovska B, Raugei S, Ragsdale SW. Nickel-sulfonate mode of substrate binding for forward and reverse reactions of methyl-SCoM reductase suggest a radical mechanism involving long-range electron transfer. J Am Chem Soc 2021;143:5481–96.

[87] Ragsdale SW, Raugei S, Ginovska B, Wongnate T. In: Zamble D, Rowinska-Zyrek M, Kozlowski H, editors. The biological chemistry of nickel. Cambridge, UK: Royal Society of Chemistry; 2017. p. 149–69.

[88] Ermler U. On the mechanism of methyl-coenzyme M reductase. Dalton Trans 2005;21:3451–8.

[89] Cedervall PE, Dey M, Li X, Sarangi R, Hedman B, Ragsdale SW, Wilmot CM. Structural analysis of a Ni-methyl species in methyl-coenzyme M reductase from Methanothermobacter marburgensis. J Am Chem Soc 2011;133:5626–8.

[90] Kunz RC, Dey M, Ragsdale SW. Characterization of the thioether product formed from the thiolytic cleavage of the alkyl-nickel bond in methyl-coenzyme M reductase. Biochemistry 2008;47:2661–7.

[91] Sarangi R, Dey M, Ragsdale SW. Geometric and electronic structures of the Ni(I) and methyl-Ni(III) intermediates of methyl-coenzyme M reductase. Biochemistry 2009;48:3146–56.

[92] Dey M, Telser J, Kunz RC, Lees NS, Ragsdale SW, Hoffman BM. Biochemical and spectroscopic studies of the electronic structure and reactivity of a methyl-Ni species formed on methyl-coenzyme M reductase. J Am Chem Soc 2007;129:11030–2.

[93] Dey M, Kunz RC, Lyons DM, Ragsdale SW. Characterization of alkyl-nickel adducts generated by reaction of methyl-coenzyme m reductase with brominated acids. Biochemistry 2007;46:11969–78.
[94] Lin S-K, Jaun B. Coenzyme F430 from methanogenic bacteria: detection of a paramagnetic methylnickel(II) derivative of the pentamethyl ester by 2H-NMR spectroscopy. Helv Chim Acta 1991;74:1725–38.
[95] Jaun B, Pfaltz A. Coenzyme F430 from methanogenic bacteria: methane formation by reductive carbon–sulphur bond cleavage of methyl sulphonium ions catalysed by F430 pentamethyl ester. J Chem Soc Chem Commun 1988;4:293–4.
[96] Chen SL, Blomberg MR, Siegbahn PE. How is methane formed and oxidized reversibly when catalyzed by Ni-containing methyl-coenzyme M reductase? Chemistry 2012;18:6309–15.
[97] Chen SL, Pelmenschikov V, Blomberg MR, Siegbahn PE. Is there a Ni-methyl intermediate in the mechanism of methyl-coenzyme M reductase? J Am Chem Soc 2009;131:9912–3.
[98] Pelmenschikov V, Siegbahn PE. Nickel superoxide dismutase reaction mechanism studied by hybrid density functional methods. J Am Chem Soc 2006;128:7466–75.
[99] Pelmenschikov V, Blomberg MR, Siegbahn PE, Crabtree RH. A mechanism from quantum chemical studies for methane formation in methanogenesis. J Am Chem Soc 2002;124:4039–49.
[100] Wongnate T, Sliwa D, Ginovska B, Smith D, Wolf MW, Lehnert N, Raugei S, Ragsdale SW. The radical mechanism of biological methane synthesis by methyl-coenzyme M reductase. Science 2016;352:953–8.
[101] Chen SL, Blomberg MR, Siegbahn PE. An investigation of possible competing mechanisms for Ni-containing methyl-coenzyme M reductase. Phys Chem Chem Phys 2014;16:14029–35.
[102] Goenrich M, Mahlert F, Duin EC, Bauer C, Jaun B, Thauer RK. Probing the reactivity of Ni in the active site of methyl-coenzyme M reductase with substrate analogues. J Biol Inorg Chem 2004;9: 691–705.
[103] Bonacker LG, Baudner S, Mörschel E, Böcher R, Thauer RK. Properties of the two isoenzymes of methyl-coenzyme M reductase in Methanobacterium thermoautotrophicum. Eur J Biochem 1993; 217:587–95.
[104] Wackett LP, Honek JF, Begley TP, Wallace V, Orme-Johnson WH, Walsh CT. Substrate analogues as mechanistic probes of methyl-S-coenzyme M reductase. Biochemistry 1987;26:6012–8.
[105] Balch WE, Wolfe RS. Specificity and biological distribution of coenzyme M (2-mercaptoethanesulfonic acid). J Bacteriol 1979;137:256–63.
[106] Goenrich M, Duin EC, Mahlert F, Thauer RK. Temperature dependence of methyl-coenzyme M reductase activity and of the formation of the methyl-coenzyme M reductase red2 state induced by coenzyme B. J Biol Inorg Chem 2005;10:333–42.
[107] Ellermann J, Kobelt A, Pfaltz A, Thauer RK. On the role of N-7-mercaptoheptanoyl-O-phospho-L-threonine (component B) in the enzymatic reduction of methyl-coenzyme M to methane. FEBS Lett 1987;220:358–62.
[108] Ahn Y, Krzycki J, Floss HG. Steric course of reduction of ethyl-coenzyme M to ethane catalyzed by methylcoenzyme M reductase from Methanosarcina barkeri. J Am Chem Soc 1991;113:4700–1.
[109] Scheller S, Goenrich M, Boecher R, Thauer RK, Jaun B. The key nickel enzyme of methanogenesis catalyses the anaerobic oxidation of methane. Nature 2010;465:606–8.
[110] Cedervall PE, Dey M, Pearson AR, Ragsdale SW, Wilmot CM. Structural insight into methyl-coenzyme M reductase chemistry using coenzyme B analogues. Biochemistry 2010;49:7683–93.
[111] Li X, Telser J, Kunz RC, Hoffman BM, Gerfen G, Ragsdale SW. Observation of organometallic and radical intermediates formed during the reaction of methyl-coenzyme M reductase with bromoethanesulfonate. Biochemistry 2010;49:6866–76.
[112] Holliger C, Kengen SW, Schraa G, Stams AJ, Zehnder AJ. Methyl-coenzyme M reductase of Methanobacterium thermoautotrophicum delta H catalyzes the reductive dechlorination of 1,2-dichloroethane to ethylene and chloroethane. J Bacteriol 1992;174:4435–43.
[113] Kunz RC, Horng YC, Ragsdale SW. Spectroscopic and kinetic studies of the reaction of bromopropanesulfonate with methyl-coenzyme M reductase. J Biol Chem 2006;281:34663–76.
[114] Hinderberger D, Piskorski RP, Goenrich M, Thauer RK, Schweiger A, Harmer J, Jaun B. A nickel-alkyl bond in an inactivated state of the enzyme catalyzing methane formation. Angew Chem Int Ed Engl 2006;45:3602–7.
[115] Yu ZT, Smith GB. Inhibition of methanogenesis by C-1- and C-2-polychlorinated aliphatic hydrocarbons. Environ Toxicol Chem 2000;19:2212–7.
[116] Hristov AN, Oh J, Giallongo F, Frederick TW, Harper MT, Weeks HL, Branco AF, Moate PJ, Deighton MH, Williams SR, Kindermann M, Duval S. An inhibitor persistently decreased enteric methane emission from dairy cows with no negative effect on milk production. Proc Natl Acad Sci U S A 2015;112:10663–8.
[117] Yu G, Beauchemin KA, Dong R. A review of 3-nitrooxypropanol for enteric methane mitigation from ruminant livestock. Animals (Basel) 2021;11:3540.
[118] Duin EC, Wagner T, Shima S, Prakash D, Cronin B, Yáñez-Ruiz DR, Duval S, Rümbeli R, Stemmler RT, Thauer RK, Kindermann M. Mode of action uncovered for the specific reduction of methane emissions from ruminants by the small molecule 3-nitrooxypropanol. Proc Natl Acad Sci U S A 2016;113:6172–7.
[119] Zhang ZW, Cao ZJ, Wang YL, Wang YJ, Yang HJ, Li SL. Nitrocompounds as potential methanogenic inhibitors in ruminant animals: a review. Anim Feed Sci Technol 2018;236:107–14.
[120] Lan XW, Wang NX, Xing YL. Recent advances in radical difunctionalization of simple alkenes. Eur J Org Chem 2017;39:5821–51.
[121] Schink B. Inhibition of methanogenesis by ethylene and other unsaturated hydrocarbons. FEMS Microbiol Lett 1985;31:63–8.

Section F

Iron enzymes (heme-containing)

Chapter 6.1

Cyclooxygenase

Maria Novella Romanelli
NEUROFARBA—Department of Neurosciences, Psychology, Drug Research and Child Health, Section of Pharmaceutical and Nutraceutical Sciences, University of Florence, Italy

1 Introduction

Non-Steroidal Antiinflammatory Drugs (NSAIDs) are compounds used in acute and chronic inflammation, as analgesic, antipyretic and antithrombotic agents. NSAIDs are among the most commonly used medications in the world, also owing to the over the counter availability of some of them without a prescription. NSAIDs inhibit cyclooxygenases (COXs), also called Prostaglandin Endoperoxide H Synthase (PGHS), a ubiquitous enzyme, which catalyze the oxidation of arachidonic acid (AA) to Prostaglandin H (PGH_2) [1,2]. The clinical application of many NSAIDs started well before the discovery of their molecular target: aspirin, the first NSAID, is used since the end of the nineteenth century, indomethacin and ibuprofen from the mid-1900s, but the COX enzyme was isolated and purified only in 1976 [3,4], and the existence of two isoforms (called COX1 and COX2) was reported in 1991 [5,6]. The history of NSAIDs and COX enzymes has already been described in some excellent reviews [7,8]. Owing to the huge amount of literature on these enzymes and their modulators, this chapter will briefly summarize the most important points, and the reader is referred for an in-depth analysis to the cited references.

2 Structure and function of the enzyme

COXs are homodimers. Each monomer is composed of three domains (Fig. 1): an epidermal growth factor (EGF) domain, a membrane-binding domain and the catalytic domain that comprises the COX and Peroxidase (POX) active sites. The function of the EGF domain is still unknown; it partly forms and stabilizes the dimers interface [7,9]. The membrane-spanning domain consists in four α-helices (indicated as A–D), which provide a hydrophobic surface inserting into one leaflet of the membrane, thus anchoring the enzyme. Above the A–D helices there is a cavity called "lobby" at the entrance of the catalytic site. At the top of the lobby residues Arg120, Tyr355 and Glu524 form a constriction that must open to give access to the catalytic site. The cyclooxygenase active site is L-shaped, about 25 Å long and 8 Å wide, and hydrophobic [10]. Tyr 385 and Ser530 are found at the bending of the L-shaped channel, while the distal part of the active site is composed mainly of hydrophobic residues [9,10]. The active site is divided into COX and POX sites; the heme prosthetic group is located at the base of the latter, and the iron ion is coordinated to His388.

The catalytic mechanism (Fig. 2) starts with the oxidation of the heme iron in the POX site that causes an electron to be transferred from Tyr385. The formed tyrosyl radical (**I**) pulls out the pro-S C13 hydrogen atom of AA, producing the radical moiety **II** (step 1). Isomerization and addition of one oxygen molecule on C11 give the radical peroxide **III** (step 2), which attacks carbon 9, yielding the cyclic endoperoxide **IV** (step 3). A new bond is then formed between C8 and C12 (step 4) giving the dioxabicycloheptane moiety of **V** and shifting the unpaired electron to C15. The addition of a second oxygen molecule leads to a new peroxyl radical **VI** (step 5), which is reduced to the hydroperoxide PGG_2 by the transfer of a hydrogen atom from Tyr385; this step regenerates the tyrosyl radical **I** for a new catalysis. After this cyclooxygenase run, a peroxide reaction (step 7) reduces the C15 hydroperoxide moiety to alcohol, producing PGH_2 [11].

COX1 and COX2 share a high degree of sequence identity (about 60%) and their three-dimensional structures are nearly superimposable. Important differences are found in the active sites of the two isozymes: the active site of COX2 is larger (about 25%) than that of COX1, making COX2 able to oxygenates substrates other than arachidonic acid, that are not processed by COX1. The main differences between COX1 and COX2 active sites is the replacement of an isoleucine residue (I523) with a valine immediately above the constriction: this open a side pocket where selective inhibitors can insert. Two other substitution at position 513 and 434 provide a H-bond forming group (Arg for His513) and additional room (Val for Ile434) [12].

The X-ray structure of the AA-COX1 complex (Fig. 3, left) [13] shows a salt bridge between the carboxylate ion and Arg120 and H-bond between the same group and

Metalloenzymes. https://doi.org/10.1016/B978-0-12-823974-2.00010-3

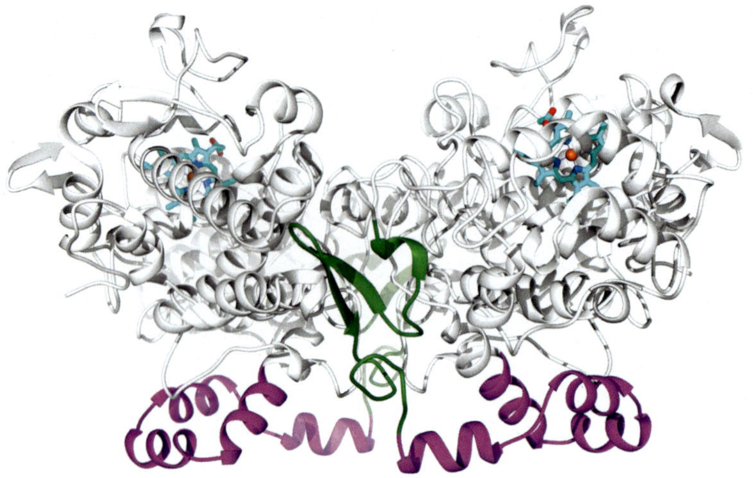

FIG. 1 Structure of COX1 dimer (PDB code 1CQE). The protein is shown in *white ribbon*, the epidermal growth factor domain in *magenta*, the membrane binding domain in *green*, the heme in *cyan*, and the iron atom in *orange*.

Tyr355. At the bending site Tyr 385 is close to the 13 pro-S hydrogen, ready for the catalytic mechanism. On the contrary, the X-ray structures of the complex of AA with COX2 show two possible binding modes [14,15]: the orientation of the "productive" conformation is similar to that seen in COX1, but the AA carboxylate forms only one H-bond with Tyr355 and Arg120 is engaged in a salt bridge with Glu254. Alternatively, AA interacts in an "inverted conformation" (Fig. 3, right), in which the carboxylate makes H-bond with Ser530 and Tyr385 in the bending site. Besides these polar interactions, the binding of AA in both isoforms is stabilized through a large number of hydrophobic contacts. Inhibitors bind in the same site as the substrate but usually occupy the proximal part and the bending region.

COX enzymes are homodimers that behave as functional heterodimers [16]. It has been demonstrated that in the dimer there is only one affinity site for heme, and that complete enzyme inhibition is achieved with one NSAID molecule per dimer [17]. Each COX dimer is composed of an allosteric (E_{allo}) and a catalytic (E_{cat}) monomer that function cooperatively. Molecules can differently affect COX

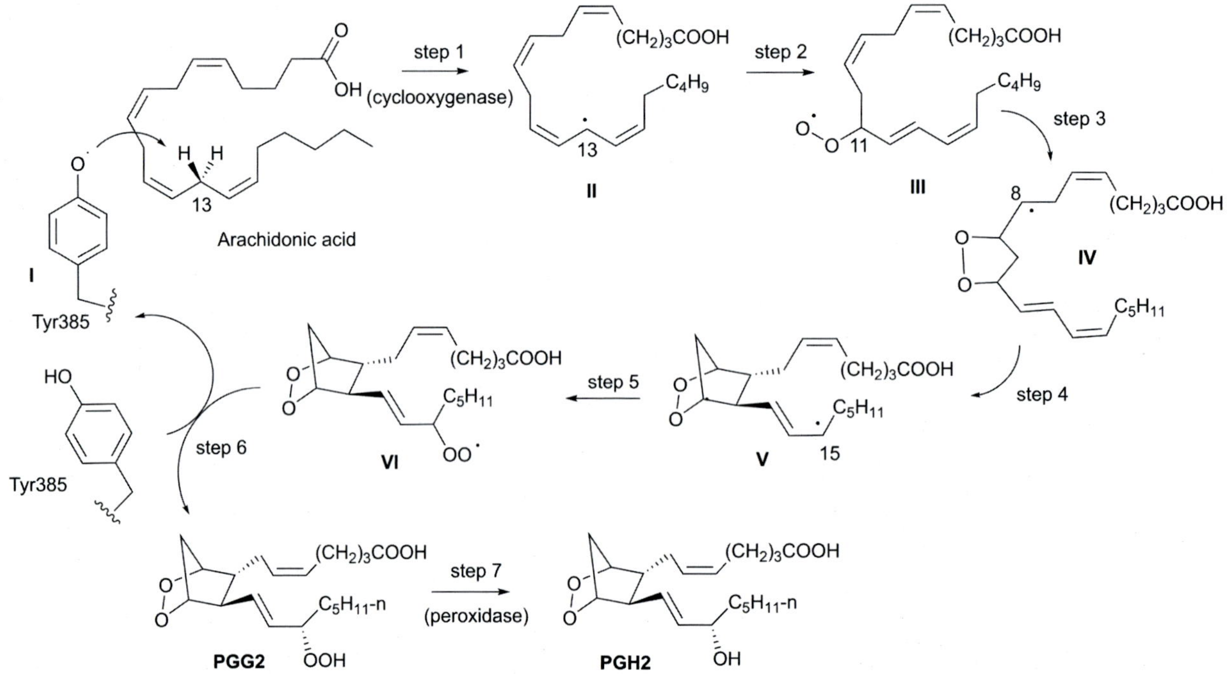

FIG. 2 The conversion of arachidonic acid into PGG_2 and PGH_2.

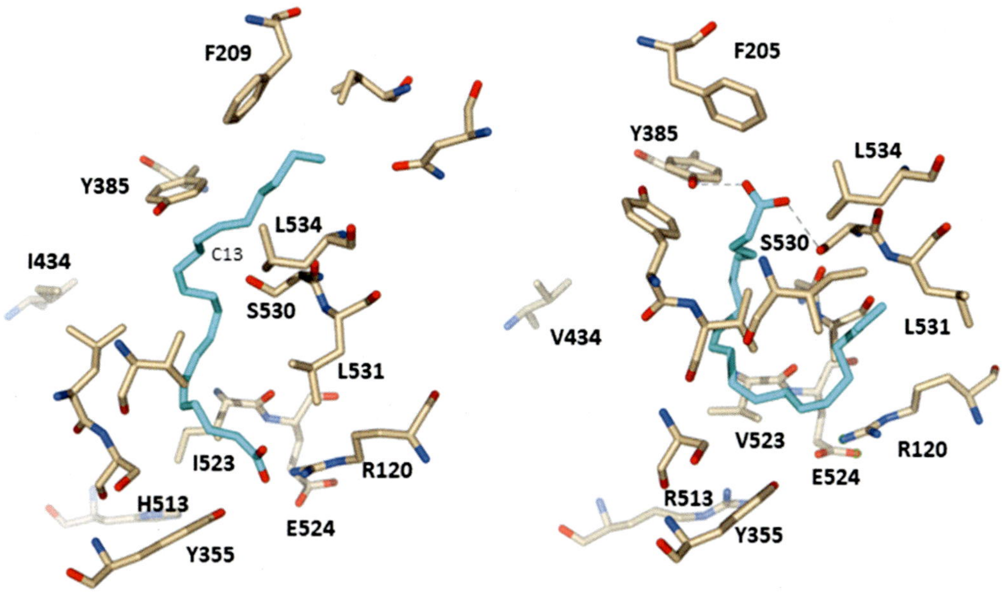

FIG. 3 The active site of COX1 (left, PDB code 1DIY) and COX2 (right, PDB code 1CVU) showing, respectively a productive and nonproductive (inverted) conformation of the substrate. The protein residues are colored in *sand* and AA in *cyan*. A label indicates the C13 atom of AA in the COX1 site.

activity by interacting with the two monomers (see Section 4).

3 Physiological and pathological role

COXs convert AA into PGH_2, which is the substrate of other enzymatic systems producing prostaglandins PGD_2, PGE_2, $PGF_{2\alpha}$, and PGI_2 and thromboxane TXA_2. These prostanoids are found almost in every tissue, thus highlighting the importance of COX enzymes. Prostanoids are involved in the maintenance of physiological processes such as the protection of stomach mucosa from damage by gastric acid, the control of blood clotting, the control of blood flow and glomerular filtration rate in the kidney, the increase of the tone or the contractile activity of uterus. Prostanoid biosynthesis is significantly increased in inflamed tissues. Both COX isoforms are engaged, although the main source is COX2, which is markedly upregulated during inflammation. Prostanoids regulate vascular permeability, the expression of cellular adhesion molecules, the release of cytokines and chemokines, the peripheral and central sensitization. PGE_2 and PGI_2 are the predominant pro-inflammatory and hyperalgesic agents, but other prostanoids are also involved [18]. PGE_2 mediates the febrile response in the CNS [19].

The COX1 isoform is constitutively expressed in almost all tissues; the expression of COX2 is generally at low levels, but it is constitutive in brain and kidney, and found also in colon and lung. The expression of COX2 is however rapidly increased in response to inflammatory stimuli such as cytokines, oncogenes, and tumor inducers [20,21]. Although COX2 is usually referred to as the inducible COX isoform, also the expression of COX1 can be upregulated in several types of cancer [22].

Pharmacological or genetic inhibition of COX activity have helped the elucidation of the physiological functions of both isozymes (reviewed in Ref. [23]). COX1 is responsible for platelet aggregation through the production of TXA_2 (platelets do not express COX2), and for the initiation of parturition through the production of $PGF_{2}\alpha$. COX2 controls ovulation and implantation through the production of PGE_2 and PGI_2. Processes that depend on both isoforms are inflammation, wound healing, gastric ulceration and carcinogenesis.

Other differences between the two isoforms regard their cellular localization: both isozymes are found in the endoplasmic reticulum and nuclear envelope, but COX2 is more concentrated in the latter and it is found also in the Golgi apparatus [24–26]. In addition, there is evidence that COX2 can function at lower AA concentrations than COX1 [27].

An important difference regards the substrate specificity of the two isoforms (see also Section 4). COX2 is capable of metabolizing ester and amide derivatives of AA that are poor substrates for COX1. 2-Arachidonoylglycerol (2-AG) and andanamide (*N*-arachidonoylethanolamine, AEA), two endocannabinoids (eCBs) whose analgesic activity is mediated by the interaction with the CB receptors, are oxidized by COX2 to PGH_2-gliceryl ester (PGH_2-G) and PGH_2-ethanolamide (PGH_2-EA), which can be further processed to other PG-G or PG-EA [28]. The most studied is PGE_2-gliceryl ester (PGE_2-G) that exerts hyperalgesic activity with mechanisms either dependent or independent on the EP receptors. Thus, COX2 inhibition prevents the formation of such compounds and at the same time increases eCBs levels resulting in analgesia (reviewed in Ref. [29]). The hydrolysis of eCBs releases arachidonic acid that is processed by COX isoforms to prostanoids; there is evidence that 2-AG is the primary source of arachidonic acid in the brain, while in other districts the main pathway is the phospholipase A2 (PLA2)-mediated hydrolysis of membrane lipids [30]. The oxidation of eCBs and of AA differs in the sensitivity to peroxide tone, the former being more sensible than the latter [31].

Prostanoids are cytoprotective in the gastrointestinal tract, and the ulcerogenic side effect of NSAID has been linked to the inhibition of COX1 activity in this tissue. While COX1 is constitutively expressed, COX2 can be induced by several stimuli, including inflammation, growth factors and cytokines. The finding that only COX2 is overexpressed in inflamed tissues initially led to the conclusion that targeting this isoform would be ideal for antiinflammatory drugs since such selective inhibition could spare the physiological activity of COX1. Later it has been shown that COX2 is constitutively expressed in various tissues [32], and that COX1 and COX2 are differently linked to terminal prostanoid synthases [33]. In particular, COX2 is the isozyme controlling the production of PGI_2 in vascular endothelial cells while COX1 controls the synthesis of TXA_2 in platelets; therefore, selective inhibition of COX2 disrupts the homeostatic control of blood clotting carried out by the balance between thromboxane and prostacyclin. This is one of the reasons of the cardiotoxicity of COX2 inhibitors, which led to the withdrawal of some of them (rofecoxib, valdecoxib) from the market, and the non-approval of others (etoricoxib, parecoxib) by FDA. Indeed, cardiotoxicity may be due also to other kind of on-target and off-target effects (reviewed in Refs. [34,35]); for rofecoxib, either an interaction with hERG channel and the production of toxic metabolites have been proposed to account for its adverse effects [36,37]. Moreover, reducing prostaglandins biosynthesis in the kidneys has also hemodynamic consequences. Indeed, the risk of cardiovascular events and gastrointestinal toxicity is common to several NSAIDs, eithers COX2-selective or traditional drugs, although to different extent [38].

Several lines of evidences prove that COX enzymes are linked to tumors. The most studied isoform is COX2, whose overexpression is found in different kind of cancer, among which colorectal, breast, pancreatic and lung carcinoma. PGE_2 and other prostanoids produced through COX2 activity cause a variety of processes that favor tumor growth (inflammation, decreased apoptosis, angiogenesis, metastasis) [39–41]. Transcription factors such as NF-κB stimulate the expression of pro-inflammatory cytokines and COX2; chronic inflammation may predispose to some cancers and stimulate their progression [42]. However, also overexpression of COX1 has been found in several tumors (ovarian, renal, head and neck, breast cancer) [41,43]. Moreover, COX1 and COX2 deficient mice showed about 75% reduction of intestinal polyps [44], suggesting that both isoform contribute to cancer progression [41].

4 Classes of modulators

Molecules interacting with COX can be substrates, activators or inhibitors. The physiological substrate for COX1 is arachidonic acid, which is oxidized mainly to PGH_2 but also 11-hydroxyeicosatetraenoic acid (11-HETE) and 15-HETE are formed in small amounts. COX2, as stated before, can also oxidize 2-AG and AEA. Besides these substrates, COX1 and COX2 also oxidize polyunsaturated fatty acids (PUFAs), with COX2 being in several cases more efficient that COX1 [45,46]. Linoleic acid [18:2 (n-6)] and gamma-linolenic acid [18:3 (n-6)] are converted to the mono-oxygenated products, while dihomo-γ-linolenic acid [20:3 (n-6)] and eicosapentaenoic acid [EPA, 20:5 (n-3)] are converted to PGH_1 and PGH_3, respectively. PUFAs are oxidized less efficiently than AA, competing for the binding site occupation, possibly behaving as antagonist as it happens for EPA on COX1, and for linoleic acid at high doses [47]. The acetylenic analog of AA, can also behave as inhibitors (reviewed in Ref. [48]).On the contrary, other fatty acids stimulate the oxidation of AA with an allosteric mechanism, the most efficient being the saturated derivative palmitic acid [49,50]. Moreover, *S*-(13-methyl)arachidonic acid is a substrate-selective activator, with no effect on the AA oxygenation but increasing the 2-AG oxidation on both COX1 and COX2 isoforms [51].

COX inhibitors fall into different structural categories: salicylates, pyrazolones, fenamates, arylpropionates, arylacetates, oxicams, and diarylheterocycles (Table 1). Pyrazolone derivatives such as antipyrine are historically important [54] but now scarcely used, while phenylbutazone is mainly used now in veterinary medicine. The structures of the compounds listed in Table 1 are reported in Figs. 4, 7–10; the list is not exhaustive and does not contain veterinary drugs.

From the mechanistic point of view COX inhibitors can be reversible or irreversible. Aspirin is the only inhibitor, which forms covalent bonds with the enzyme. The S530 residue is acetylated by this compound, resulting in an irreversible inhibition of COX. Both acetylated isoforms are unable to produce PGH_2 from arachidonic acid; however, COX-2 can still oxidize the substrate to 15-HETE predominantly in the *15-R* configuration [55,56]. 15-HETE is the precursor of the so-called aspirin-triggered lipoxins endowed with antiinflammatory activity [57].

All the other inhibitors bind to the enzyme establishing noncovalent interactions, but according to their kinetic mode(s) of inhibition they can be classified into rapidly reversible, slowly reversible and tight-binding inhibitors (Table 2). The last two classes are time-dependent

TABLE 1 Approved COX inhibitors,[a] divided in structural classes, and their therapeutic indications.

Category	Drugs	Therapeutic indications[b]
Salicylates	Aspirin (1899)	Pain, fever, inflammation, cardiac risk
	Diflunisal (1982)	For symptomatic treatment of mild to moderate pain accompanied by inflammation, osteoarthritis, and rheumatoid arthritis
	Mesalamine (1987)	Ulcerative colitis
Pyrazolones	*Antypirine* (1885)	Pain
	Phenylbutazone (1952)	Backache, ankylosing spondylitis
Fenamates	Mefenamic acid (1967)	Rheumatoid arthritis, osteoarthritis, dysmenorrhea, mild to moderate pain, inflammation, fever
	Meclofenamic acid (1980)	Pain, primary dysmenorrhea, rheumatoid arthritis, osteoarthritis
	Niflumic acid[c]	Treatment of rheumatoid arthritis, and joint and muscular pain
Arylpropionates	Ibuprofen (1974)	Pain, fever, inflammation
	Flurbiprofen (1986)	Symptomatic treatment of rheumatoid arthritis, osteoarthritis, ankylosing spondylitis
	Ketoprofen (1986)	Rheumatoid arthritis, osteoarthritis, ankylosing spondylitis, primary dysmenorrhea, pain
	Fenoprofen (1976)	Rheumatoid arthritis, osteoarthritis, pain
	Naproxen (1976)	Rheumatoid arthritis, osteoarthritis, ankylosing spondylitis, polyarticular juvenile idiopathic arthritis, tendinitis, bursitis, acute gout, primary dysmenorrhea, pain
	Ketorolac (1989)	For short-term management of acute pain
Arylacetates	Diclofenac (1988)	Pain and inflammation from varying sources including inflammatory conditions such as osteoarthritis, rheumatoid arthritis, and ankylosing spondylitis
	Indometacin (1965), Sulindac (1978)	Rheumatoid arthritis, ankylosing spondylitis, osteoarthritis, acute painful shoulder (bursitis and/or tendinitis), acute gouty arthritis
	Etodolac (1991)	Osteoarthritis, rheumatoid arthritis, pain
	Nepafenac (2005)[d]	Pain and inflammation associated with cataract surgery
	Amfenac[e]	Treatment of pain and inflammation associated with rheumatoid and osteoarthritis and low back pain, as well as the treatment of pain and inflammation following surgery, injury or tooth extraction.
	Tolmetin (1976) *Lumiracoxib*[c]	rheumatoid arthritis, osteoarthritisFor the acute and chronic treatment of the signs and symptoms of osteoarthritis of the knee in adults
	Tolmetin (1976)	Rheumatoid arthritis, osteoarthritis
Oxicams	Piroxicam (1982)	Osteoarthritis, rheumatoid arthritis
	Meloxicam (2000)	Arthritis, osteoarthritis, juvenile rheumatoid arthritis
Diarylheterocycles	Celecoxib (1998)	Osteoarthritis, rheumatoid arthritis
	Rofecoxib (1999)	Osteoarthritis, rheumatoid arthritis, acute pain in adults, primary dysmenorrhea
	Valdecoxib (2001)	Osteoarthritis, dysmenorrhea
	Parecoxib[c]	Short term perioperative pain control
	Etoricoxib[c]	Osteoarthritis, rheumatoid arthritis, ankylosing spondylitis, acute pain conditions, gout, postoperative dental surgery pain
	Mofezolac[e]	Postoperative and posttraumatic pain, acute upper respiratory tract pain, osteoarthritis, and lumbago

Continued

TABLE 1 Approved COX inhibitors, divided in structural classes, and their therapeutic indications—cont'd

Category	Drugs	Therapeutic indications
Miscellanea	*Nimesulide*	Acute pain, osteoarthritis, primary dysmenorrhea
	Oxaprozin (1992)	Inflammation and joint pain associated with rheumatoid arthritis and osteoarthritis
	Nabumetone (1991)[f]	Osteoarthritis, rheumatoid arthritis
	Acetaminophen (1950)	Pain, fever

In round brackets is reported the year of approval by FDA; compounds not or no longer available in United States but available somewhere else are shown in italic. The chemical structures are reported in Figs. 4, 7–10.
[a]*Compounds available worldwide [53].*
[b]*Data taken from DrugBank Online (https://go.drugbank.com/) and from Inxight Drugs (https://drugs.ncats.io/).*
[c]*Approved by EMA.*
[d]*Prodrug of Amfenac.*
[e]*Approved in Japan.*
[f]*Prodrug metabolized to 6-methoxy-2-naphtylacetic acid (6MNA).*

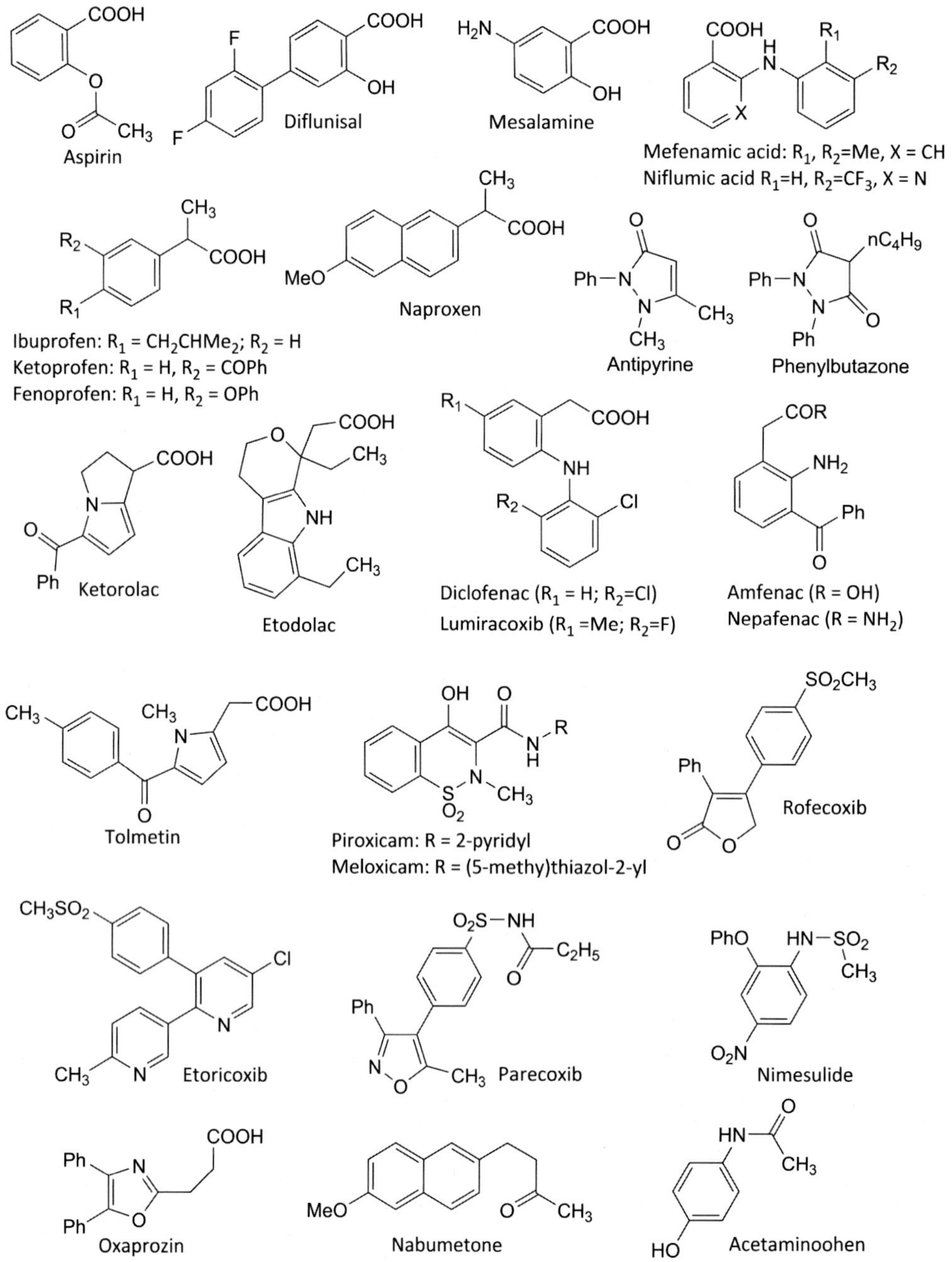

FIG. 4 Structure of some compounds listed in Table 1.

TABLE 2 Classification of noncovalent COX inhibitors according to their binding kinetic.

Rapidly reversible	Slowly reversible	Tight binders: Two-steps mechanism	Tight binders: Three steps mechanism
Celecoxib (COX1) Flurbiprofen methyl ester (COX1) Ibuprofen (COX1, COX2) Indomethacin methyl ester (COX1) Mefenamic acid Naproxen (COX1)	Etodolac (COX1) Mefenamic acid (COX1, COX2) Meloxicam (COX1) Naproxen (COX2) 6-NMA (COX1, COX2)	Diclofenac Etodolac (COX2) Flurbiprofen Indomethacin Indomethacin methyl ester (COX2) Meclofenamic acid Piroxicam (COX2) Lumiracoxib (COX-2)	Celecoxib (COX2) SC299 (COX1, COX2)

In brackets the tested isoform, when available.
Data taken from Blobaum AL, Marnett LJ. Structural and functional basis of cyclooxygenase inhibition. J Med Chem 2007;50:1425–41. Gierse JK, Hauser SD, Creely DP, Koboldt C, Rangwala SH, Isakson PC, Seibert K. Expression and selective inhibition of the constitutive and inducible forms of human cyclo-oxygenase. Biochem J 1995;305:479–84. Esser R, Berry C, Du Z, Dawson J, Fox A, Fujimoto RA, Haston W, Kimble EF, Koehler J, Peppard J, Quadros E, Quintavalla J, Toscano K, Urban L, van Duzer J, Zhang X, Zhou S, Marshall PJ. Preclinical pharmacology of lumiracoxib: a novel selective inhibitor of cyclooxygenase-2. Br J Pharmacol 2005;144:538–50. Rome LH, Lands WE. Structural requirements for time-dependent inhibition of prostaglandin biosynthesis by anti-inflammatory drugs. Proc Natl Acad Sci 1975;72:4863–65. Walker MC, Kurumbail RG, Kiefer JR, Moreland KT, Koboldt CM, Isakson PC, Seibert K, Gierse JK. A three-step kinetic mechanism for selective inhibition of cyclo-oxygenase-2 by diarylheterocyclic inhibitors. Biochem J 2001;357:709–18.

inhibitors, and bind with a two-steps or three steps mechanism; the dissociation from the enzyme is so slow that such inhibitors appear to be functionally irreversible (reviewed in Ref. [7]). The same compounds may have different binding kinetic on COX1 and COX2, generally leading to COX2 selectivity [58]. The assessment of selectivity can be done on the purified enzymes or using the whole blood assay, which measure the production of TXB_2 from platelet COX1 and PGE_2 from monocyte COX2 after LPS stimulation [59–65]. The methods give different results, as the experimental conditions (enzyme source and purification, enzymes and substrate concentration) and the dynamics of enzyme-inhibitor interaction can affect the outcomes [66]; Fig. 5 shows the selectivity ratios obtained with the whole human blood assay and reported in selected publications [64,65,67]. In general the "coxibs" are COX2 selective compounds while aspirin is considered as COX1 selective; the members of the other categories are nonselective or have a various degree of preference toward a single isoform, which may vary according to the text.

Coxibs' selectivity is usually explained taking into account the structural differences within the binding site, namely the I523V substitution that opens the side pocket where the H513R replacement inserts a residue for strong H-bonding. The X-ray structures of diarylheterocycle-COX2 complexes show such binding mode: the aromatic ring carrying the sulfone (rofecoxib) or sulfonamide group (SC558, celecoxib) is inserted into the side pocket and the substituent is engaged in H-bond with H90 and R513 [68–70]. However, this binding mode alone is not enough to explain selectivity, as celecoxib bind in the same pose also in COX1, where a rotation of the I523 side chain makes room to accommodate the benzenesulfonamide moiety [71]. The COX2 selectivity of diaryl-heterocycles has been linked to the different binding kinetic, being these compounds reversible inhibitors on COX1 and tight binders on COX2 [58,72]. A comparison of the X-ray structures of celecoxib in COX1 and COX2 (Fig. 6) shows that the SO_2NH_2 group establishes H-bonds but not with H513 in COX1, while R513 in celecoxib-COX2 complex is strongly engaged in the interaction with the inhibitor. The strong H-bond formed by COX2 inhibitors with R513 is proposed to be critical for time dependency [73,74].

COX-2 inhibition can be also substrate-selective. In fact, some arylpropionic acids with R configuration (R-profens: R-flurbiprofen, R-ibuprofen, R-naproxen), mefenamic acid and other fenamates (tolfenamic and flufenamic acid) are able to inhibit the oxidation of eCBs at doses much lower than those required to oxidize AA [75–78]. Such property is characteristic of reversible inhibitors while tight binding inhibitors block the oxidation of both kinds of substrate with similar potency. Such behavior has been explained with the ability of reversible inhibitors to bind to E_{allo} and to induce a conformational change that does not allow 2-AG oxidation, while to prevent AA oxidation reversible inhibitors must occupy both monomers. For tight binders only the occupancy of one monomer is sufficient to inhibit oxidation of both substrates [76].

5 Design of inhibitors

The most part of the nonselective compounds reported in Table 1 have been developed before knowing their mechanism of action; in general, it is difficult to go back to their design, which could have been based on some empirical observations or intuitions (see, for instance Ref. [79]).

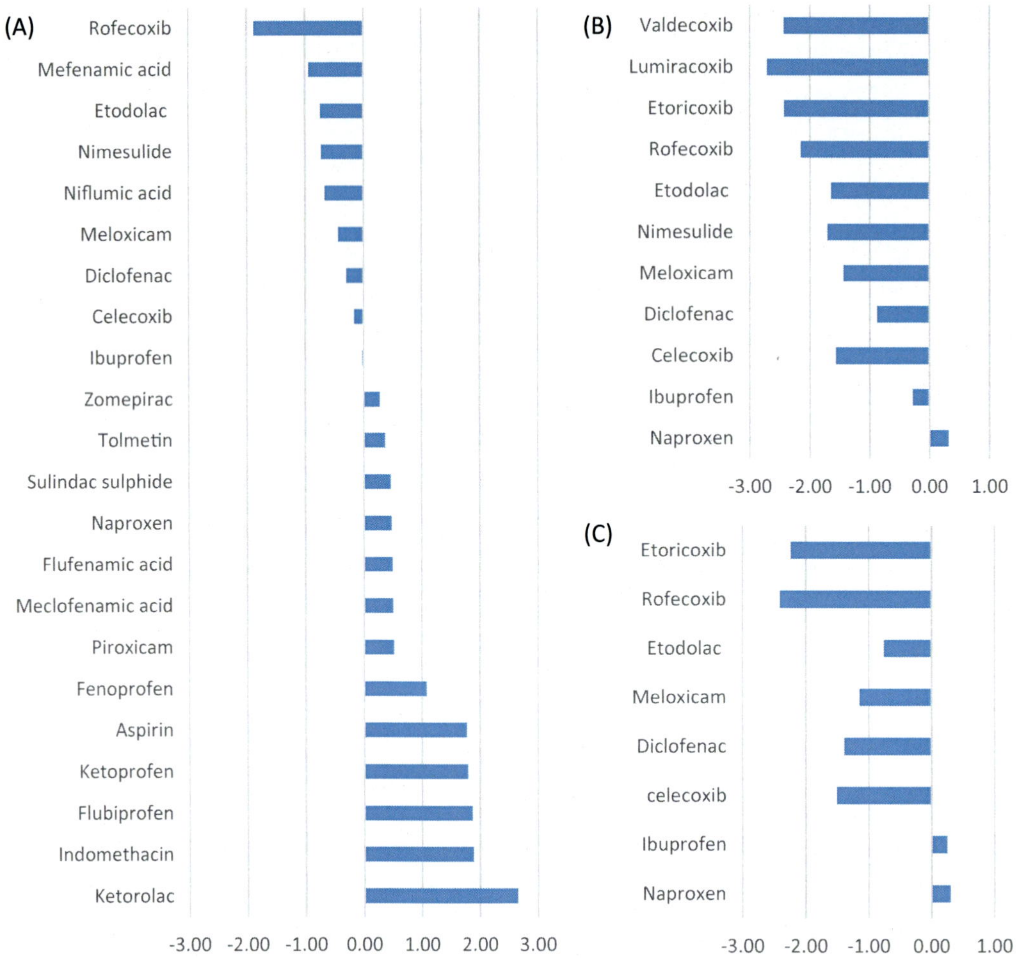

FIG. 5 Selectivity of inhibitors, expressed as $\log(IC_{50(COX2)}/IC_{50(COX1)})$. *(Data are taken from (A) Warner TD, Giuliano F, Vojnovic I, Bukasa A, Mitchell JA, Vane JR. Nonsteroid drug selectivities for cyclo-oxygenase-1 rather than cyclo-oxygenase-2 are associated with human gastrointestinal toxicity: a full in vitro analysis. Proc Natl Acad Sci 1999;96:7563–68, (B) Esser R, Berry C, Du Z, Dawson J, Fox A, Fujimoto RA, Haston W, Kimble EF, Koehler J, Peppard J, Quadros E, Quintavalla J, Toscano K, Urban L, van Duzer J, Zhang X, Zhou S, Marshall PJ. Preclinical pharmacology of lumiracoxib: a novel selective inhibitor of cyclooxygenase-2. Br J Pharmacol 2005;144:538–50, and (C) García Rodríguez LA, Tacconelli S, Patrignani P. Role of dose potency in the prediction of risk of myocardial infarction associated with nonsteroidal anti-inflammatory drugs in the general population. J Am Coll Cardiol 2008;52:1628–36.)*

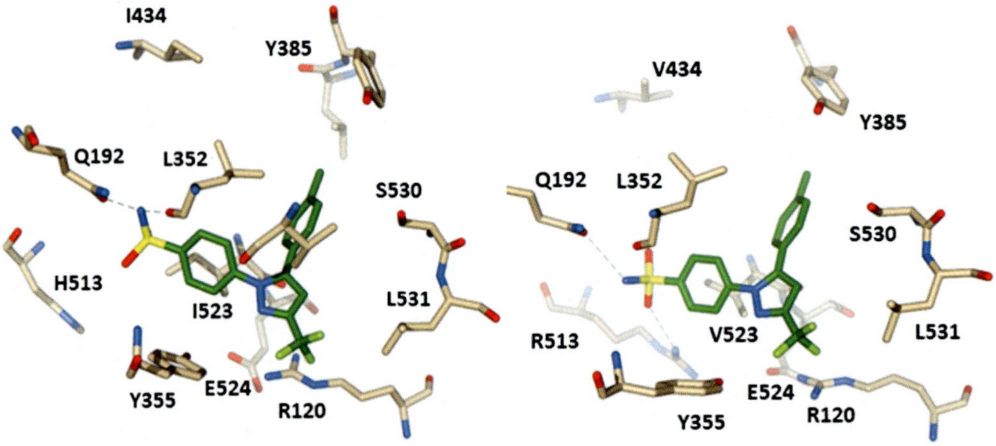

FIG. 6 Celecoxib *(green)* in the active site of COX1 (left, PDB code 3kk6) and COX2 (right, PDB code 3ln1). H-bonds are shown as *black dotted lines*.

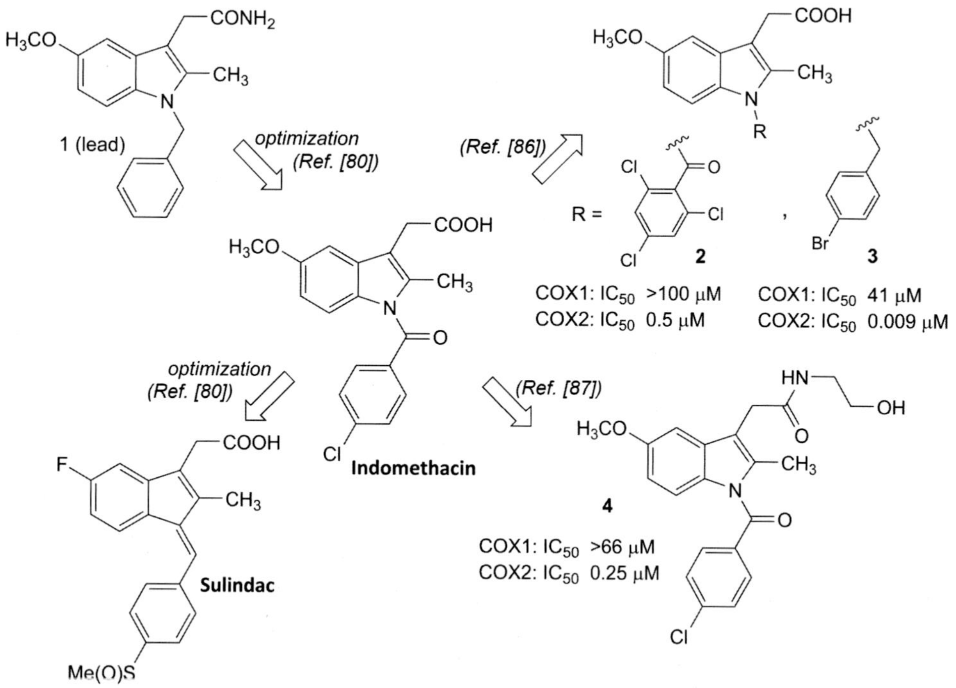

FIG. 7 Discovery of indometacin, sulindac, and COX2 selective indometacin analogs.

As an exception, the development of indomethacin and sulindac made at Merck, Sharp & Dohme has been described in details by Shen and Winter [80]. This research originated from the intensive debate, in the middle part of 1900, on serotonin as possible mediator of inflammation and was considerably facilitated by a high number of indole-containing antiserotoninergic agents that could be tested in vivo using the granuloma inhibition test in rats. Indeed, some compounds showed antiinflammatory activity comparable to phenylbutazone, supporting the initiation of the program even if, as found some years later, the initial hypothesis was not fulfilled. From the initial lead molecule (**1**, Fig. 7) indomethacin was obtained after the synthesis and in vivo testing of more than 2000 compounds. Sulindac derived from the application on indomethacin of known medicinal chemistry strategies such as isosteric replacement (indole-indane) and the reduction of conformational flexibility (the double bond in place of the amidic linkage). The methylsulfinyl group was inserted to improve solubility, and the replacement of the 5-methoxy with 5-F gave a compound with activity comparable to indomethacin but much less irritating properties at the gastrointestinal level; this improvement was ascribed to the prodrug properties of sulindac, whose antiinflammatory activity is due to the reduced metabolite (sulfide). Obviously, in vivo tests complicated the derivation of structure-activity relationships, even if facilitated the discovery of the prodrug sulindac; moreover, nor the mechanism of action nor the existence of two isoforms of the target enzyme was known at that time, but it is likely that the in vivo activity reflected the inhibition of both COX1 and COX2.

Many of the COX-2 selective inhibitors have a diarylheterocyclic structure, a skeleton probably derived from phenylbutazone; several derivatives were described much before the identification of COX2 [81]. The first compound reported as COX2 selective was the thiophene derivative Dup697 (Fig. 8) [82], which was not developed due to a unacceptably long half-life in humans [83]. Despite the intense research on diarylheterocyclic, only celecoxib is now available in United States for human use, while in Europe also etoricoxib has been approved. Researchers at Searle have reported the medicinal chemistry efforts leading to celecoxib (Fig. 8) [52]. The pharmacophore consists of a central unsaturated ring carrying two vicinal aryl groups, one of which has an H-bond acceptor group. They selected the 1,5-diphenylpyrazole as core structure, and found that high COX2 activity is achieved replacing the methyl sulfone group in para position on the 1-phenyl ring of the initial lead SC58125 with SO_2NH_2. Substituents on the positions 3 and 4 on the pyrazole ring and on the other phenyl moiety were then extensively investigated. It was found that the para substitution of the 5-phenyl ring greatly affected potency and selectivity: an electron withdrawing group was detrimental for activity on both isoforms, while and electron donating group could increase both COX1 and COX2 potency, reducing selectivity. However, bulky substituents in the para position were not tolerated on COX2. Position 3 in the pyrazole ring could accept several different

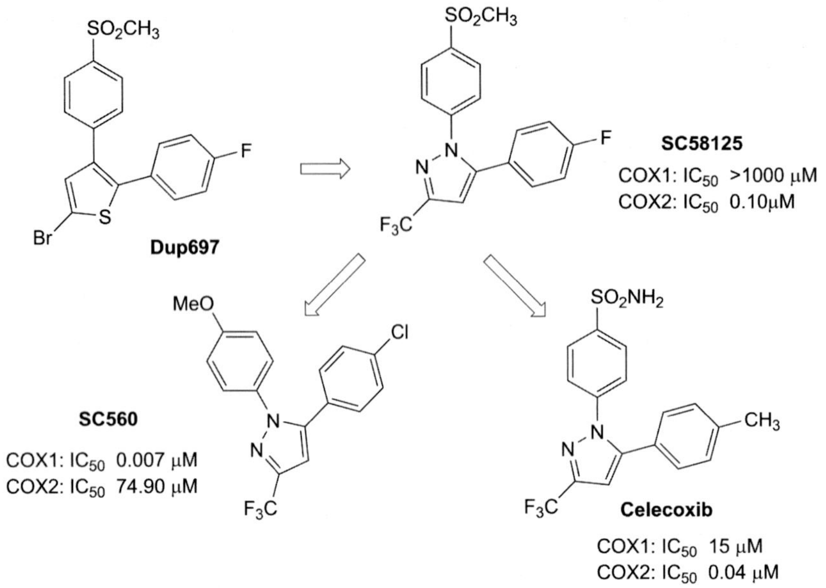

FIG. 8 Discovery of celecoxib [52].

groups, fluoroalkyl moieties being the best; on the contrary substituents in position 4 often increased COX1 activity. SC58635 (celecoxib) was selected for its potency, selectivity and pharmacokinetic (bioavailability, half-life) properties.

Etoricoxib (Fig. 4) belongs to the bispyridine group, in which the central 5-membered heterocycle and one of the vicinal phenyl rings have been replaced with pyiridine moieties. The activity and the pharmacokinetic profile of these compounds have been modulated by placing suitable substituents on both pyridine rings and keeping $MeSO_2$ as H-bond forming group to obtain higher COX2 selectivity [84]. Lumiracoxib (Fig. 4) is the only example of COX2 selective compound belonging to the arylacetic acid class of inhibitors. This compound differs from diclofenac in one of the ortho-halogens of the aniline moiety (a F instead of a Cl) and in the presence of a methyl group in the 5′-position of the arylacetic portion. The X-ray structure of lumiracoxib with COX2 showed an "inverted" binding pose similar to that of diclofenac and to the nonproductive conformation of AA (the carboxylate close to Ser503 and not to Arg120), with the additional Me group inserted into a lipophilic pocket created by the rotation of Leu384. The impossibility that a similar movement could occur in COX1 explains the selectivity. Lumiracoxib is also a substrate-selective inhibitor [85].

Selectivity can be achieved also by modifying nonselective inhibitors. After the discovery of the second isoform, the observation that the active site is larger in COX2 than in COX1 suggested that the increase the steric bulkiness of some substituents in indomethacin could result into COX2 selectivity. This indeed was the case when the 4-chlorobenzoyl group was replaced with 2,4,6-trichlorobenzoyl or with 4-bromobenzyl moieties (compounds **2** and **3**, respectively; Fig. 7) [86]. Another observation that led to COX2 selective inhibitor regarded the interaction of carboxylates with Arg120. All the COX1 X-ray structures with a ligand carrying a carboxylate group (AA and most of the nonselective inhibitors) show an ion-ion interaction between this moiety and the positively charged Arg120 residue, and the importance of such interaction is highlighted also by site-directed mutagenesis studies. On the contrary, this salt bridge does not occur with COX2. Moreover, COX2 but not COX1 is able to oxidize eCB (AA esters or amides). Indeed, transformation of indomethacin into esters and amides reduced or abolished activity on COX1 while improved that on COX2, resulting in selective compounds (indomethacin ethanolamide **4** is shown as example in Fig. 7) [87]. These strategies worked also on other nonselective inhibitors: by increasing the steric bulk on the distal phenyl ring flurbiprofen was converted into COX2 selective compounds (**5**, Fig. 9) [88]. The transformation of meclofenamic acid into esters or amides increased COX-2 selectivity, as exemplified by compound **6** [87], while on zomepirac the replacement of the carboxylic acid moiety with a 6-pyrazyn-3-one group gave **7** as COX2 selective compound [89].

As demonstrated in the development of celecoxib, suitable modification of the substituents on the phenyl rings can provide COX1 selective compounds, exemplified by SC560 (Fig. 8), lacking the sulfonamide group on the 1-phenyl ring. Analogously, the removal of this group on valdecoxib (Fig. 10) produced a two orders of magnitude decrease of the IC_{50} value on COX-1 (compound **8**) introducing selectivity for this isoform [90]; the selectivity was then improved by modifying the second aryl moiety obtaining **9**. The high potency of compounds SC560,

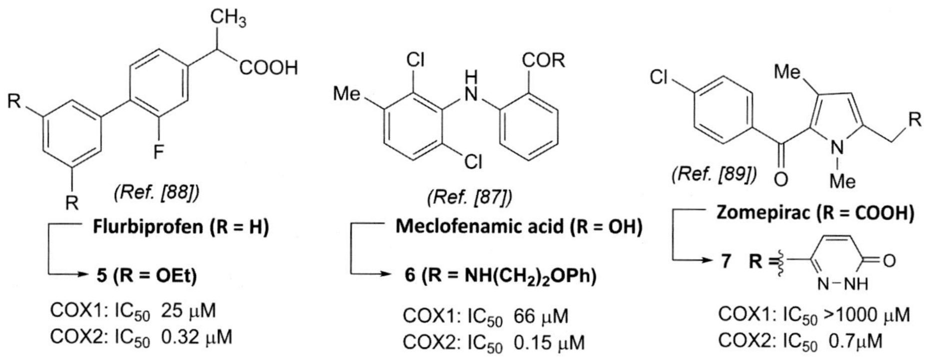

FIG. 9 COX-2 selectivity achieved from modifications of the nonselective compounds flurbiprofen, meclofenamic acid and zomepirac.

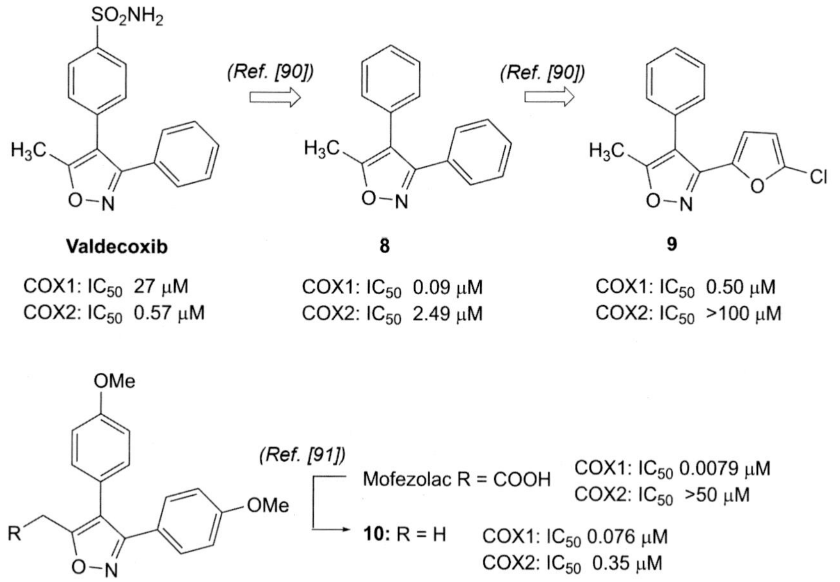

FIG. 10 From COX2 to COX1 selectivity in the diarylheterocycle class.

8 and **9** shows that a carboxylic acid function is not essential for activity on COX-1, although it may improve potency and selectivity. For instance, Mofezolac is about 10 times more potent and much more selective than its methyl analog **10** [91]; this may be due to the ion-ion interaction with Arg120 [92].

Also modification on sulindac sulfide can lead to COX1 selective compounds (Fig. 11). On studying the effect of the 2-methyl group of indomethacin and sulindac on the kinetics of inhibition of these two nonselective molecules, the group of Marnett found that the removal of this moiety transforms the parent compounds from tight binders into weak inhibitors [93,94]. In case of sulindac, the removal of the 2-methyl group changed also the stereochemistry of the double bond, yielding a weak but COX1 preferring compound (**11**), which was further modified into **12** and **13** showing COX1 selectivity [95].

The search for new inhibitors is still active in the academia, mainly in the field of COX2 inhibitors, as proved by recent literature surveys [96–98]. The search for COX1 inhibitors is less actively pursued [99]. For both isoforms, the availability of a high number of X-ray structures in complex with inhibitors can help the design through a structure-based approach [7,9], but also other techniques can be exploited. For instance, the group of Wuest has used kinetic target-guided synthesis (also called in-situ click chemistry) to obtain COX2 selective inhibitors [100]. Different libraries of azide-containing pyrazoles have been reacted with libraries of alkyne derivatives to perform Huisgen cycloaddition inside the protein (Fig. 12A). Two potent and selective COX2 inhibitors were synthesized using the COX2 binding site as reaction vessel. The protein was able to select from the libraries 1-phenylpyrazoles containing the $MeSO_2$ group able to form the pivotal H-bond in the COX2 binding site subpocket and the alkyne reagent with suitable electronic and steric characteristics to fit the binding site. Compounds **14** and **15** showed in vitro potency and selectivity similar to celecoxib, but were much more

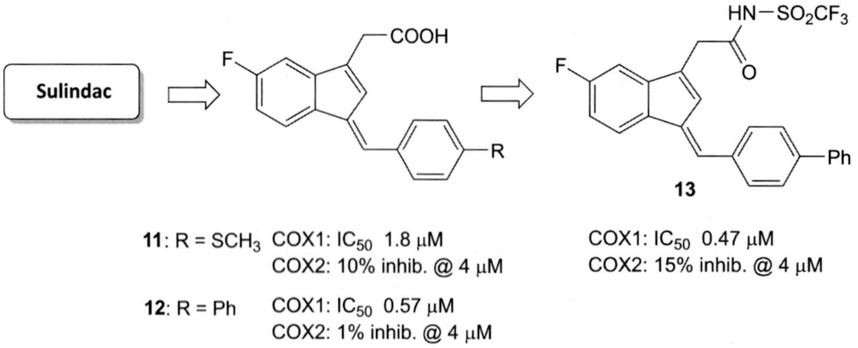

FIG. 11 COX-1 selective compounds from sulindac modifications [93,94].

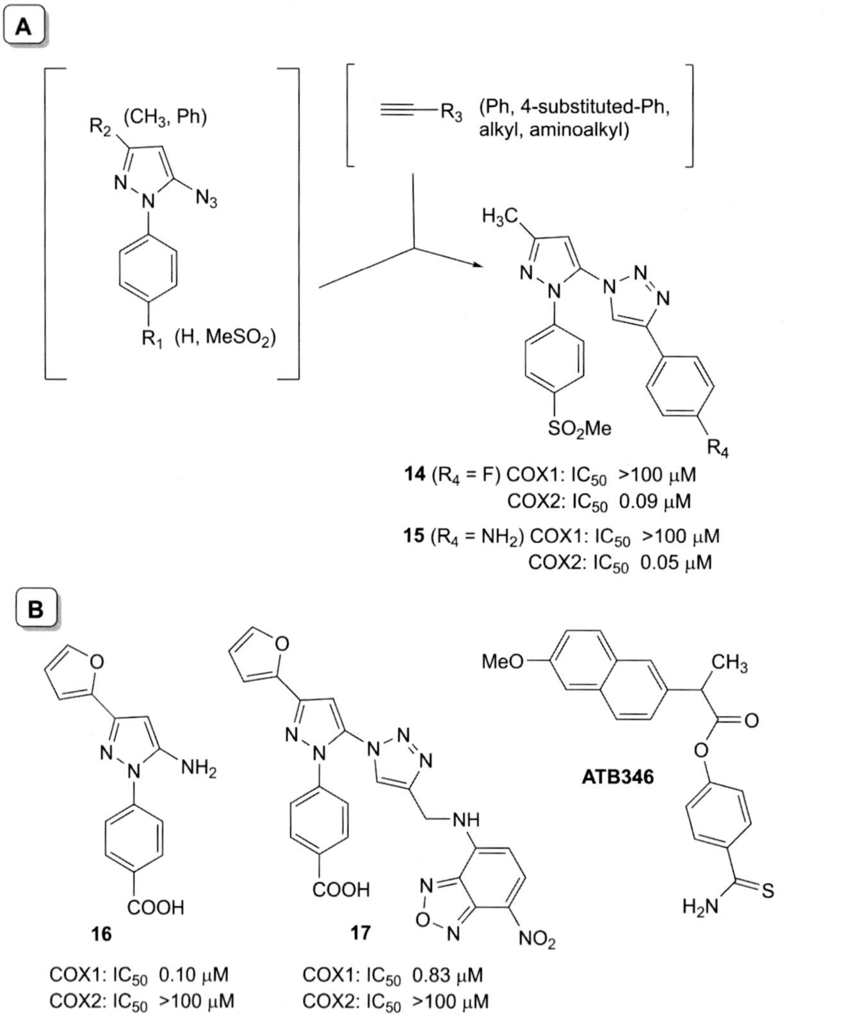

FIG. 12 (A) Design of COX2 selective inhibitors using in situ click chemistry. (B) Structure of compounds **16**, **17**, and ATB346.

potent in vivo in the carrageenan-induced rat paw edema assay. More recently the same group demonstrated that careful interpretation of the SAR on diarylheterocycles carrying a pyrazole moiety could lead to COX1 selective compounds. In fact, by shifting the position of the second phenyl ring from C5 (as in SC560) to C3 and replacing it with a furane moiety they obtained a series of COX1 selective compounds among which compound **16** (Fig. 12B), carrying a carboxylic acid group, was the most potent and selective inhibitor; it is intriguing that molecular docking studies predict for **16** a sort of "inverted conformation" in which the H-bond is formed between the carboxylate moiety and Tyr385, but not a salt bridge with Arg120. The attachment of a nitro-benzoxadiazole moiety to **16** gave compound **17**, which was used as fluorescence imaging probe for labeling COX1 in ovarian cancer cells [101].

In recent time the concept of polypharmacology has emerged, leading to the design of agents able to modulate multiple targets or disease pathways, which should result particularly useful in complex diseases in which the modulation of only one target proved to be not effective [102]. Indeed, cancer and inflammation can be considered complex diseases; by applying this approach, multitarget ligands directed to enzymes involved in the AA cascade have been prepared with the hope to improve antiinflammatory activity [103]. With a similar reasoning, the structure of known NSAIDs has been modified in order to interact with COX and other signaling pathways [43]; an example is ATB346 (Fig. 12B) described in Section 6.

6 Clinically used agents and compounds in clinical development

All the compounds reported in Table 1 are clinically available worldwide for the treatment of acute or chronic conditions where pain and inflammation are present. Ketorolac is used for short-term management of moderate to severe pain, while aspirin's main use is presently as antithrombotic.

The H_2S-releasing drug ATB346 is an example of multitarget ligand, since it is able to inhibit COX enzymes and to release H_2S. H_2S is a signaling molecule that produces many important effects including antiinflammatory and prohealing actions, and that contributes to the gastric mucosa defense [104]. In addition, long-time exposure to H_2S through H_2S-releasing drugs results in cancer cell death through increased apoptosis [105].

ATB346 is the ester of naproxen with 4-carbamothioylphenol. Preclinical studies showed that in the mouse air pouch model this compound suppressed COX2 activity and inhibited leukocyte infiltration more effectively than naproxen; the antiinflammatory effect was obtained without inducing gastrointestinal lesion in rats [106]. ATB346 reduced human melanoma cells proliferation in vitro, inhibited NF-κB activation and increased apoptosis, and showed efficacy also in a mouse model of melanoma in vivo [107]. Antibe therapeutics has sponsored several clinical trials on ATB346. The phase 1 studies showed that the compound was safe and well tolerated at doses from 25 to 2000mg; the rate of adverse effects was similar to placebo. The inhibition of COX activity appeared very rapidly (within 1 h) and lasted at least 24h. In a phase 2 clinical trial a once daily 250mg dose of ATB-346 was evaluated for its pain-relieving properties in patients with osteoarthritis of the knee; the performance of the drug was evaluated through the Western Ontario-McMaster University Arthritis Index (WOMAC) pain score. After 4days treatment, ATB346 (250mg once daily) significantly decreased the pain score with an efficacy similar to naproxen or celecoxib as found in previous studies; pain relief was further improved after 10days treatment [108]. In a Phase 2B study ATB346 (250mg once daily for 2weeks) was tested on 244 healthy volunteers to control the formation of gastric lesions; the effect of ARB346 was compared to naproxen at a dose (550mg twice daily), which produced the same extent of COX inhibition. ATB346 was much better tolerated: only 3% of the subjects taking this drug developed ulcers, compared to 44% in the naproxen group, and also the incidence of gastrointestinal side effects was lower. This study demonstrated for the H_2S-releasing ATB346, a marked reduction of gastrointestinal toxicity compared to naproxen [109]. ATB346 showed efficacy also in a dermal model of acute inflammation, triggered by ultraviolet-killed *Escherichia coli*, in healthy human volunteers. Compared to untreated controls, ATB346 significantly inhibited neutrophil infiltration into the site of injection of bacteria; the efficacy was similar to naproxen [110]. Therefore ATB-346 is a new antiinflammatory drug, with reduced gastrointestinal adverse effects with respect to conventional NSAIDs.

Many clinically used COX inhibitors (aspirin, celecoxib, sulindac, naproxen) have been and are being tested in clinical trials for cancer as mono- or combination therapy. The anticancer activity associated with these drugs is however not completely dependent on COX inhibition: the reduction of prostanoids biosynthesis (mainly PGE_2) decreases inflammation, tumor cell proliferation, angiogenesis, and increases apoptosis (see Section 3), but at the same time other COX-independent mechanisms seem involved (reviewed in Ref. [43]). To highlight the importance of this possible therapeutic application, the clinical trials of NSAIDs in cancer have been the topic of several recent reviews, to which the reader is referred for details [42,111–114].

Since long time aspirin has been shown to reduce the risk of colorectal cancer [115]. Aspirin is a selective COX1 inhibitor, which is mainly used to reduce platelet aggregation; it has recently been shown that deletion of COX1 gene in platelet decreased the number and size of tumors in the small intestine in $Apc^{Min/+}$ mice, due to the reduction of TXA_2 production, which lowers COX2 induction [116]. It is possible that the effect of aspirin is due to its activity on both intestine and platelets. In 2016 the US Preventive Services Task Force (USPSTF) recommended low-dose aspirin for primary prevention of cardiovascular diseases and of colorectal cancer in adults aged 50–59 years, indicating that also people aged 60–69 showing certain conditions are likely to benefit [117]. More recently, despite several evidences on the effectiveness of this drug [115], the same authority concluded that the proof on the reduction of colorectal cancer incidence or mortality is inadequate [118], the main concern being the risks of gastrointestinal bleeding associated to a long usage.

References

[1] Vane JR. Inhibition of prostaglandin synthesis as a mechanism of action for aspirin-like drugs. Nat New Biol 1971;231:232–5.

[2] Smith JB, Willis AL. Aspirin selectively inhibits prostaglandin production in human platelets. Nat New Biol 1971;231:235–7.

[3] Miyamoto T, Ogino N, Yamamoto S, Hayaishi O. Purification of prostaglandin endoperoxide synthetase from bovine vesicular gland microsomes. J Biol Chem 1976;251:2629–36.

[4] Hemler M, Lands WE. Purification of the cyclooxygenase that forms prostaglandins. Demonstration of two forms of iron in the holoenzyme. J Biol Chem 1976;251:5575–9.

[5] Kujubu D, Fletcher B, Varnum B, Lim R, Herschman H. TIS10, a phorbol ester tumor promoter-inducible mRNA from Swiss 3T3 cells, encodes a novel prostaglandin synthase/cyclooxygenase homologue. J Biol Chem 1991;266:12866–72.

[6] Xie WL, Chipman JG, Robertson DL, Erikson RL, Simmons DL. Expression of a mitogen-responsive gene encoding prostaglandin synthase is regulated by mRNA splicing. Proc Natl Acad Sci 1991;88:2692–6.

[7] Blobaum AL, Marnett LJ. Structural and functional basis of cyclooxygenase inhibition. J Med Chem 2007;50:1425–41.

[8] Simmons DL, Botting RM, Hla T. Cyclooxygenase isozymes: the biology of prostaglandin synthesis and inhibition. Pharmacol Rev 2004;56:387–437.

[9] Rouzer CA, Marnett LJ. Structural and chemical biology of the interaction of cyclooxygenase with substrates and non-steroidal anti-inflammatory drugs. Chem Rev 2020;120:7592–641.

[10] Picot D, Loll PJ, Garavito RM. The X-ray crystal structure of the membrane protein prostaglandin H2 synthase-1. Nature 1994;367:243–9.

[11] Rouzer CA, Marnett LJ. Mechanism of free radical oxygenation of polyunsaturated fatty acids by cyclooxygenases. Chem Rev 2003;103:2239–304.

[12] For the sake of comparison, the numbering of COX1 is adopted also for COX2.

[13] Malkowski MG, Ginell SL, Smith WL, Garavito RM. The productive conformation of arachidonic acid bound to prostaglandin synthase. Science 2000;289:1933–7.

[14] Vecchio AJ, Simmons DM, Malkowski MG. Structural basis of fatty acid substrate binding to cyclooxygenase-2. J Biol Chem 2010;285:22152–63.

[15] Kiefer JR, Pawlitz JL, Moreland KT, Stegeman RA, Hood WF, Gierse JK, Stevens AM, Goodwin DC, Rowlinson SW, Marnett LJ, Stallings WC, Kurumbail RG. Structural insights into the stereochemistry of the cyclooxygenase reaction. Nature 2000;405:97–101.

[16] Smith WL, Malkowski MG. Interactions of fatty acids, nonsteroidal anti-inflammatory drugs, and coxibs with the catalytic and allosteric subunits of cyclooxygenases-1 and -2. J Biol Chem 2019;294:1697–705.

[17] Kulmacz RJ, Lands WE. Stoichiometry and kinetics of the interaction of prostaglandin H synthase with anti-inflammatory agents. J Biol Chem 1985;260:12572–8.

[18] Smyth EM, Grosser T, Wang M, Yu Y, FitzGerald GA. Prostanoids in health and disease. J Lipid Res 2009;50:S423–8.

[19] Engblom D, Saha S, Engström L, Westman M, Audoly LP, Jakobsson P-J, Blomqvist A. Microsomal prostaglandin E synthase-1 is the central switch during immune-induced pyresis. Nat Neurosci 2003;6:1137–8.

[20] Claria J. Cyclooxygenase-2 biology. Curr Pharm Des 2003;9:2177–90.

[21] Ji XK, Madhurapantula SV, He G, Wang KY, Song CH, Zhang JY, Wang KJ. Genetic variant of cyclooxygenase-2 in gastric cancer: more inflammation and susceptibility. World J Gastroenterol 2021;27:4653–66.

[22] Perrone MG, Scilimati A, Simone L, Vitale P. Selective COX-1 inhibition: a therapeutic target to be reconsidered. Curr Med Chem 2010;17:3769–805.

[23] Smith WL, Langenbach R. Why there are two cyclooxygenase isozymes. J Clin Invest 2001;107:1491–5.

[24] Morita I, Schindler M, Regier MK, Otto JC, Hori T, DeWitt DL, Smith WL. Different intracellular locations for prostaglandin endoperoxide H synthase-1 and 2. J Biol Chem 1995;270:10902–8.

[25] Spencer AG, Woods JW, Arakawa T, Singer II, Smith WL. Subcellular localization of prostaglandin endoperoxide H synthases-1 and -2 by immunoelectron microscopy. J Biol Chem 1998;273:9886–93.

[26] Yuan C, Smith WL. A cyclooxygenase-2-dependent prostaglandin E2 biosynthetic system in the Golgi apparatus. J Biol Chem 2015;290:5606–20.

[27] Morita I. Distinct functions of COX-1 and COX-2. Prostaglandins Other Lipid Mediat 2002;68–69:165–75.

[28] Vila A, Rosengarth A, Piomelli D, Cravatt B, Marnett LJ. Hydrolysis of prostaglandin glycerol esters by the endocannabinoid-hydrolyzing enzymes, monoacylglycerol lipase and fatty acid amide hydrolase. Biochemistry 2007;46:9578–85.

[29] Buisseret B, Alhouayek M, Guillemot-Legris O, Muccioli GG. Endocannabinoid and prostanoid crosstalk in pain. Trends Mol Med 2019;25:882–96.

[30] Nomura DK, Morrison BE, Blankman JL, Long JZ, Kinsey SG, Marcondes MCG, Ward AM, Hahn YK, Lichtman AH, Conti B, Cravatt BF. Endocannabinoid hydrolysis generates brain prostaglandins that promote neuroinflammation. Science 2011;334:809–13.

[31] Musee J, Marnett LJ. Prostaglandin H synthase-2-catalyzed oxygenation of 2-arachidonoylglycerol is more sensitive to peroxide tone than oxygenation of arachidonic acid. J Biol Chem 2012;287:37383–94.

[32] Zidar N, Odar K, Glavac D, Jerse M, Zupanc T, Stajer D. Cyclooxygenase in normal human tissues—is COX-1 really a constitutive isoform, and COX-2 an inducible isoform? J Cell Mol Med 2009;13:3753–63.

[33] Ueno N, Takegoshi Y, Kamei D, Kudo I, Murakami M. Coupling between cyclooxygenases and terminal prostanoid synthases. Biochem Biophys Res Commun 2005;338:70–6.

[34] Khan S, Andrews KL, Chin-Dusting JPF. Cyclo-oxygenase (COX) inhibitors and cardiovascular risk: are non-steroidal anti-inflammatory drugs really anti-inflammatory? Int J Mol Sci 2019;20:4262.

[35] Arora M, Choudhary S, Singh PK, Sapra B, Silakari O. Structural investigation on the selective COX-2 inhibitors mediated cardiotoxicity: a review. Life Sci 2020;251:117631.

[36] Park SJ, Buschmann H, Bolm C. Bioactive sulfoximines: syntheses and properties of Vioxx® analogs. Bioorg Med Chem Lett 2011;21:4888–90.

[37] Preston Mason R, Walter MF, McNulty HP, Lockwood SF, Byun J, Day CA, Jacob RF. Rofecoxib increases susceptibility of human LDL and membrane lipids to oxidative damage: a mechanism of cardiotoxicity. J Cardiovasc Pharmacol 2006;47:S7–S14.

[38] Coxib and traditional NSAID Trialists' (CNT) Collaboration, Bhala N, Emberson J, Merhi A, Abramson S, Arber N, Baron J, Bombardier C, Cannon C, Farkouh M, FitzGerald G, Goss P, Halls H, Hawk E, Hawkey C, Hennekens C, Hochberg M, Holland L, Kearney P, Laine L, Lanas A, Lance P, Laupacis A, Oates J, Patrono C, Schnitzer T, Solomon S, Tugwell P, Wilson K, Wittes J, Baigent C. Vascular and upper gastrointestinal effects of non-steroidal anti-inflammatory drugs: meta-analyses of individual participant data from randomised trials. Lancet 2013;382:769–79.

[39] Jara-Gutiérrez Á, Baladrón V. The role of prostaglandins in different types of cancer. Cells 2021;10:1487.

[40] Hashemi Goradel N, Najafi M, Salehi E, Farhood B, Mortezaee K. - Cyclooxygenase-2 in cancer: a review. J Cell Physiol 2019;234:5683–99.

[41] Pannunzio A, Coluccia M. Cyclooxygenase-1 (COX-1) and COX-1 inhibitors in cancer: a review of oncology and medicinal chemistry literature. Pharmaceuticals 2018;11:101.

[42] Maniewska J, Jeżewska D. Non-steroidal anti-inflammatory drugs in colorectal cancer chemoprevention. Cancers 2021;13:594.

[43] Ramos-Inza S, Ruberte AC, Sanmartín C, Sharma AK, Plano D. NSAIDs: old acquaintance in the pipeline for cancer treatment and prevention—structural modulation, mechanisms of action, and bright future. J Med Chem 2021;64:16380–421.

[44] Langenbach R, Loftin CD, Lee C, Tiano H. Cyclooxygenase-deficient mice: a summary of their characteristics and susceptibilities to inflammation and carcinogenesis. Ann N Y Acad Sci 1999;889:52–61.

[45] Laneuville O, Breuer DK, Xu N, Huang ZH, Gage DA, Watson JT, Lagarde M, DeWitt DL, Smith WL. Fatty acid substrate specificities of human prostaglandin-endoperoxide H synthase-1 and −2: formation of 12-hydroxy-(9Z,13E/Z,15Z)-octadecatrienoic acids from α-linolenic acid. J Biol Chem 1995;270:19330–6.

[46] Thuresson ED, Malkowski MG, Lakkides KM, Rieke CJ, Mulichak AM, Ginell SL, Garavito RM, Smith WL. Mutational and X-ray crystallographic analysis of the interaction of dihomo-γ-linolenic acid with prostaglandin endoperoxide H synthases. J Biol Chem 2001;276:10358–65.

[47] Wada M, DeLong CJ, Hong YH, Rieke CJ, Song I, Sidhu RS, Yuan C, Warnock M, Schmaier AH, Yokoyama C, Smyth EM, Wilson SJ, FitzGerald GA, Garavito RM, Sui DX, Regan JW, Smith WL. Enzymes and receptors of prostaglandin pathways with arachidonic acid-derived versus eicosapentaenoic acid-derived substrates and products. J Biol Chem 2007;282:22254–66.

[48] Flower RJ. Drugs which inhibit prostaglandin biosynthesis. Pharmacol Rev 1974;26:33–67.

[49] Dong L, Vecchio AJ, Sharma NP, Jurban BJ, Malkowski MG, Smith WL. Human cyclooxygenase-2 is a sequence homodimer that functions as a conformational heterodimer. J Biol Chem 2011;286:19035–46.

[50] Yuan C, Sidhu RS, Kuklev DV, Kado Y, Wada M, Song I, Smith WL. Cyclooxygenase allosterism, fatty acid-mediated cross-talk between monomers of cyclooxygenase homodimers. J Biol Chem 2009;284:10046–55.

[51] Kudalkar SN, Nikas SP, Kingsley PJ, Xu S, Galligan JJ, Rouzer CA, Banerjee S, Ji L, Eno MR, Makriyannis A, Marnett LJ. 13-Methylarachidonic acid is a positive allosteric modulator of endocannabinoid oxygenation by cyclooxygenase. J Biol Chem 2015;290:7897–909.

[52] Penning TD, Talley JJ, Bertenshaw SR, Carter JS, Collins PW, Docter S, Graneto MJ, Lee LF, Malecha JW, Miyashiro JM, Rogers RS, Rogier DJ, Yu SS, Anderson GD, Burton EG, Cogburn JN, Gregory SA, Koboldt CM, Perkins WE, Seibert K, Veenhuizen AW, Zhang YY, Isakson PC. Synthesis and biological evaluation of the 1,5-diarylpyrazole class of cyclooxygenase-2 inhibitors: identification of 4-[5-(4-methylphenyl)-3-(trifluoromethyl)-1H-pyrazol-1-yl]benzenesulfonamide (SC-58635, celecoxib). J Med Chem 1997;40:1347–65.

[53] LiverTox: Clinical and research information on drug-induced liver injury [Internet]. Bethesda, MD: National Institute of Diabetes and Digestive and Kidney Diseases; 2012-. Nonsteroidal Antiinflammatory Drugs (NSAIDs) [Updated 2020 Mar 18]. Available from: https://www.ncbi.nlm.nih.gov/books/NBK548614/.

[54] Brune K. The early history of non-opioid analgesics. Acute Pain 1997;1:33–40.

[55] Lecomte M, Laneuville O, Ji C, DeWitt DL, Smith WL. Acetylation of human prostaglandin endoperoxide synthase-2 (cyclooxygenase-2) by aspirin. J Biol Chem 1994;269:13207–15.

[56] Holtzman MJ, Turk J, Shornick LP. Identification of a pharmacologically distinct prostaglandin H synthase in cultured epithelial cells. J Biol Chem 1992;267:21438–45.

[57] Poorani R, Bhatt AN, Dwarakanath BS, Das UN. COX-2, aspirin and metabolism of arachidonic, eicosapentaenoic and docosahexaenoic acids and their physiological and clinical significance. Eur J Pharmacol 2016;785:116–32.

[58] Copeland RA, Williams JM, Giannaras J, Nurnberg S, Covington M, Pinto D, Pick S, Trzaskos JM. Mechanism of selective inhibition of the inducible isoform of prostaglandin G/H synthase. Proc Natl Acad Sci 1994;91:11202–6.

[59] Gierse JK, Hauser SD, Creely DP, Koboldt C, Rangwala SH, Isakson PC, Seibert K. Expression and selective inhibition of the constitutive and inducible forms of human cyclo-oxygenase. Biochem J 1995;305:479–84.

[60] Gierse JK, Koboldt CM, Walker MC, Seibert K, Isakson PC. Kinetic basis for selective inhibition of cyclo-oxygenases. Biochem J 1999;339(Pt 3):607–14.

[61] Laneuville O, Breuer DK, Dewitt DL, Hla T, Funk CD, Smith WL. Differential inhibition of human prostaglandin endoperoxide H synthases-1 and -2 by nonsteroidal anti-inflammatory drugs. J Pharmacol Exp Ther 1994;271:927–34.

[62] Patrignani P, Panara MR, Sciulli MG, Santini G, Renda G, Patrono C. Differential inhibition of human prostaglandin endoperoxide synthase-1 and -2 by nonsteroidal anti-inflammatory drugs. J Physiol Pharmacol 1997;48:623–31.

[63] Cryer B, Feldman M. Cyclooxygenase-1 and cyclooxygenase-2 selectivity of widely used nonsteroidal anti-inflammatory drugs. Am J Med 1998;104:413–21.

[64] Warner TD, Giuliano F, Vojnovic I, Bukasa A, Mitchell JA, Vane JR. Nonsteroid drug selectivities for cyclo-oxygenase-1 rather than cyclo-oxygenase-2 are associated with human gastrointestinal toxicity: a full *in vitro* analysis. Proc Natl Acad Sci 1999;96:7563–8.

[65] García Rodríguez LA, Tacconelli S, Patrignani P. Role of dose potency in the prediction of risk of myocardial infarction associated with nonsteroidal anti-inflammatory drugs in the general population. J Am Coll Cardiol 2008;52:1628–36.

[66] Marnett LJ, Kalgutkar AS. Cyclooxygenase 2 inhibitors: discovery, selectivity and the future. Trends Pharmacol Sci 1999;20:465–9.

[67] Esser R, Berry C, Du Z, Dawson J, Fox A, Fujimoto RA, Haston W, Kimble EF, Koehler J, Peppard J, Quadros E, Quintavalla J, Toscano K, Urban L, van Duzer J, Zhang X, Zhou S, Marshall PJ. Preclinical pharmacology of lumiracoxib: a novel selective inhibitor of cyclooxygenase-2. Br J Pharmacol 2005;144:538–50.

[68] Kurumbail RG, Stevens AM, Gierse JK, McDonald JJ, Stegeman RA, Pak JY, Gildehaus D, Iyashiro JM, Penning TD, Seibert K, Isakson PC, Stallings WC. Structural basis for selective inhibition of cyclooxygenase-2 by anti-inflammatory agents. Nature 1996;384:644–8.

[69] Xing L, Hamper BC, Fletcher TR, Wendling JM, Carter J, Gierse JK, Liao S. Structure-based parallel medicinal chemistry approach to improve metabolic stability of benzopyran COX-2 inhibitors. Bioorg Med Chem Lett 2011;21:993–6.

[70] Orlando BJ, Malkowski MG. Crystal structure of rofecoxib bound to human cyclooxygenase-2. Acta Crystallogr Sect F 2016;72:772–6.

[71] Rimon G, Sidhu RS, Lauver DA, Lee JY, Sharma NP, Yuan C, Frieler RA, Trievel RC, Lucchesi BR, Smith WL. Coxibs interfere with the action of aspirin by binding tightly to one monomer of cyclooxygenase-1. Proc Natl Acad Sci U S A 2010;107:28–33.

[72] Ouellet M, Percival MD. Effect of inhibitor time-dependency on selectivity towards cyclooxygenase isoforms. Biochem J 1995;306:247–51.

[73] Wong E, Bayly C, Waterman HL, Riendeau D, Mancini JA. Conversion of prostaglandin G/H synthase-1 into an enzyme sensitive to PGHS-2-selective inhibitors by a double His513→ Arg and Ile523→ Val mutation. J Biol Chem 1997;272:9280–6.

[74] Khan YS, Gutiérrez-de-Terán H, Åqvist J. Molecular mechanisms in the selectivity of nonsteroidal anti-inflammatory drugs. Biochemistry 2018;57:1236–48.

[75] Prusakiewicz JJ, Duggan KC, Rouzer CA, Marnett LJ. Differential sensitivity and mechanism of inhibition of COX-2 oxygenation of arachidonic acid and 2-arachidonoylglycerol by ibuprofen and mefenamic acid. Biochemistry 2009;48:7353–5.

[76] Duggan KC, Hermanson DJ, Musee J, Prusakiewicz JJ, Scheib JL, Carter BD, Banerjee S, Oates JA, Marnett LJ. (R)-Profens are substrate-selective inhibitors of endocannabinoid oxygenation by COX-2. Nat Chem Biol 2011;7:803–9.

[77] Windsor MA, Hermanson DJ, Kingsley PJ, Xu S, Crews BC, Ho W, Keenan CM, Banerjee S, Sharkey KA, Marnett LJ. Substrate-selective inhibition of cyclooxygenase-2: development and evaluation of achiral profen probes. ACS Med Chem Lett 2012;3:759–63.

[78] Orlando BJ, Malkowski MG. Substrate-selective inhibition of cyclooxygeanse-2 by fenamic acid derivatives is dependent on peroxide tone. J Biol Chem 2016;291:15069–81.

[79] Harrison IT, Lewis B, Nelson P, Rooks W, Roszkowski A, Tomolonis A, Fried JH. Nonsteroidal antiinflammatory agents. I. 6-Substituted 2-naphthylacetic acids. J Med Chem 1970;13:203–5.

[80] Shen T-Y, Winter CA. Chemical and biological studies on indomethacin, sulindac and their analogs. In: Simmonds AB, editor. Advances in drug research. Academic Press; 1977. p. 89–245.

[81] Talley JJ. Selective inhibitors of cyclooxygenase-2 (COT-2). In: King FD, Oxford AW, editors. Progress in medicinal chemistry. Elsevier; 1999. p. 201–34.

[82] Gans KR, Galbraith W, Roman RJ, Haber SB, Kerr JS, Schmidt WK, Smith C, Hewes WE, Ackerman NR. Anti-inflammatory and safety profile of DuP 697, a novel orally effective prostaglandin synthesis inhibitor. J Pharmacol Exp Ther 1990;254:180–7.

[83] Pinto DJP, Copeland RA, Covington MB, Pitts WJ, Batt DG, Orwat MJ, Lam GN, Joshi A, Chan Y-C, Wang S, Trzaskos JM, Magolda RL, Kornhauser DM. Chemistry and pharmacokinetics of diarylthiophenes and terphenyls as selective COX-2 inhibitors. Bioorg Med Chem Lett 1996;6:2907–12.

[84] Friesen RW, Brideau C, Chan CC, Charleson S, Deschênes D, Dubé D, Ethier D, Fortin R, Gauthier JY, Girard Y, Gordon R, Greig GM, Riendeau D, Savoie C, Wang Z, Wong E, Visco D, Xu LJ, Young RN. 2-Pyridinyl-3-(4-methylsulfonyl)phenylpyridines: selective and orally active cyclooxygenase-2 inhibitors. Bioorg Med Chem Lett 1998;8:2777–82.

[85] Windsor MA, Valk PL, Xu S, Banerjee S, Marnett LJ. Exploring the molecular determinants of substrate-selective inhibition of cyclooxygenase-2 by lumiracoxib. Bioorg Med Chem Lett 2013;23:5860–4.

[86] Black WC, Bayly C, Belley M, Chan CC, Charleson S, Denis D, Gauthier JY, Gordon R, Guay D, Kargman S, Lau CK, Leblanc Y, Mancini J, Ouellet M, Percival D, Roy P, Skorey K, Tagari P, Vickers P, Wong E, Xu L, Prasit P. From indomethacin to a selective COX-2 inhibitor: development of indolalkanoic acids as potent and selective cyclooxygenase-2 inhibitors. Bioorg Med Chem Lett 1996;6:725–30.

[87] Kalgutkar AS, Crews BC, Rowlinson SW, Marnett AB, Kozak KR, Remmel RP, Marnett LJ. Biochemically based design of cyclooxygenase-2 (COX-2) inhibitors: facile conversion of nonsteroidal antiinflammatory drugs to potent and highly selective COX-2 inhibitors. Proc Natl Acad Sci 2000;97:925–30.

[88] Bayly CI, Black WC, Léger S, Ouimet N, Ouellet M, Percival MD. Structure-based design of COX-2 selectivity into flurbiprofen. Bioorg Med Chem Lett 1999;9:307–12.

[89] Luong C, Miller A, Barnett J, Chow J, Ramesha C, Browner MF. Flexibility of the NSAID binding site in the structure of human cyclooxygenase-2. Nat Struct Biol 1996;3:927–33.

[90] Di Nunno L, Vitale P, Scilimati A, Tacconelli S, Patrignani P. Novel synthesis of 3,4-diarylisoxazole analogues of valdecoxib: reversal cyclooxygenase-2 selectivity by sulfonamide group removal. J Med Chem 2004;47:4881–90.

[91] Pati ML, Vitale P, Ferorelli S, Iaselli M, Miciaccia M, Boccarelli A, Di Mauro GD, Fortuna CG, Souza Domingos TF, da Silva LCRP, de Pádula M, Cabral LM, Sathler PC, Vacca A, Scilimati A, Perrone MG. Translational impact of novel widely pharmacological characterized mofezolac-derived COX-1 inhibitors combined with bortezomib on human multiple myeloma cell lines viability. Eur J Med Chem 2019;164:59–76.

[92] Cingolani G, Panella A, Perrone MG, Vitale P, Di Mauro G, Fortuna CG, Armen RS, Ferorelli S, Smith WL, Scilimati A. Structural basis for selective inhibition of Cyclooxygenase-1 (COX-1) by diarylisoxazoles mofezolac and 3-(5-chlorofuran-2-yl)-5-methyl-4-phenylisoxazole (P6). Eur J Med Chem 2017;138:661–8.

[93] Prusakiewicz JJ, Felts AS, Mackenzie BS, Marnett LJ. Molecular basis of the time-dependent inhibition of cyclooxygenases by indomethacin. Biochemistry 2004;43:15439–45.

[94] Walters MJ, Blobaum AL, Kingsley PJ, Felts AS, Sulikowski GA, Marnett LJ. The influence of double bond geometry in the inhibition of cyclooxygenases by sulindac derivatives. Bioorg Med Chem Lett 2009;19:3271–4.

[95] Liedtke AJ, Crews BC, Daniel CM, Blobaum AL, Kingsley PJ, Ghebreselasie K, Marnett LJ. Cyclooxygenase-1-selective inhibitors

based on the (E)-2′-des-methyl-sulindac sulfide scaffold. J Med Chem 2012;55:2287–300.
[96] Ahmadi M, Bekeschus S, Weltmann K-D, von Woedtke T, Wende K. Non-steroidal anti-inflammatory drugs: recent advances in the use of synthetic COX-2 inhibitors. RSC Med Chem 2022;13:471–96.
[97] Ju Z, Li M, Xu J, Howell DC, Li Z, Chen F-E. Recent development on COX-2 inhibitors as promising anti-inflammatory agents: the past 10 years. Acta Pharm Sin B 2022;12:2790–807.
[98] Mahboubi Rabbani SMI, Zarghi A. Selective COX-2 inhibitors as anticancer agents: a patent review (2014-2018). Expert Opin Ther Pat 2019;29:407–27.
[99] Vitale P, Panella A, Scilimati A, Perrone MG. COX-1 inhibitors: beyond structure toward therapy. Med Res Rev 2016;36:641–71.
[100] Bhardwaj A, Kaur J, Wuest M, Wuest F. In situ click chemistry generation of cyclooxygenase-2 inhibitors. Nat Commun 2017;8:1.
[101] Kaur J, Bhardwaj A, Wuest F. Development of fluorescence imaging probes for labeling COX-1 in live ovarian cancer cells. ACS Med Chem Lett 2021;12:798–804.
[102] Kabir A, Muth A. Polypharmacology: the science of multi-targeting molecules. Pharmacol Res 2022;176:106055.
[103] Sala A, Proschak E, Steinhilber D, Rovati GE. Two-pronged approach to anti-inflammatory therapy through the modulation of the arachidonic acid cascade. Biochem Pharmacol 2018;158:161–73.
[104] Ianaro A, Cirino G, Wallace JL. Hydrogen sulfide-releasing anti-inflammatory drugs for chemoprevention and treatment of cancer. Pharmacol Res 2016;111:652–8.
[105] Cao X, Ding L, Xie Z-z, Yang Y, Whiteman M, Moore PK, Bian J-S. A review of hydrogen sulfide synthesis, metabolism, and measurement: is modulation of hydrogen sulfide a novel therapeutic for cancer? Antioxid Redox Signal 2019;31:1–38.
[106] Wallace JL, Caliendo G, Santagada V, Cirino G. Markedly reduced toxicity of a hydrogen sulphide-releasing derivative of naproxen (ATB-346). Br J Pharmacol 2010;159:1236–46.
[107] De Cicco P, Panza E, Ercolano G, Armogida C, Sessa G, Pirozzi G, Cirino G, Wallace JL, Ianaro A. ATB-346, a novel hydrogen sulfide-releasing anti-inflammatory drug, induces apoptosis of human melanoma cells and inhibits melanoma development in vivo. Pharmacol Res 2016;114:67–73.
[108] Wallace JL, Vaughan DJ, Dicay M, MacNaughton WK, de Nucci G. Hydrogen sulfide-releasing therapeutics: translation to the clinic. Antioxid Redox Signal 2018;28:1533–40.
[109] Wallace JL, Nagy P, Feener TD, Allain T, Ditrói T, Vaughan DJ, Muscara MN, de Nucci G, Buret AG. A proof-of-concept, Phase 2 clinical trial of the gastrointestinal safety of a hydrogen sulfide-releasing anti-inflammatory drug. Br J Pharmacol 2020; 177:769–77.
[110] Glanville JRW, Jalali P, Flint JD, Patel AA, Maini AA, Wallace JL, Hosin AA, Gilroy DW. Potent anti-inflammatory effects of an H2S-releasing naproxen (ATB-346) in a human model of inflammation. FASEB J 2021;35:e21913.
[111] Ganduri V, Rajasekaran K, Duraiyarasan S, Adefuye MA, Manjunathaet N. Colorectal carcinoma, cyclooxygenases, and COX inhibitors. Cureus 2022;14:e28579.
[112] Stiller C-O, Hjemdahl P. Lessons from 20 years with COX-2 inhibitors: importance of dose–response considerations and fair play in comparative trials. J Intern Med 2022;292:557–74.
[113] Saxena P, Sharma PK, Purohit P. A journey of celecoxib from pain to cancer. Prostaglandins Other Lipid Mediat 2020; 147:106379.
[114] Li J, Hao Q, Cao W, Vadgama JV, Wu Y. Celecoxib in breast cancer prevention and therapy. Cancer Manag Res 2018;10:4653–67.
[115] Menter DG, Bresalier RS. An aspirin a day: new pharmacological developments and cancer chemoprevention. Annu Rev Pharmacol Toxicol 2023;63.
[116] Bruno A, Contursi A, Tacconelli S, Sacco A, Hofling U, Mucci M, Lamolinara A, Del Pizzo F, Ballerini P, Di Gregorio P, Yu Y, Patrignani P. The specific deletion of cyclooxygenase-1 in megakaryocytes/platelets reduces intestinal polyposis in ApcMin/+ mice. Pharmacol Res 2022;185:106506.
[117] US Preventive Services Task Force, Bibbins-Domingo K. Aspirin use for the primary prevention of cardiovascular disease and colorectal cancer: U.S. Preventive Services Task Force recommendation statement. Ann Intern Med 2016;164:836–45.
[118] US Preventive Services Task Force. Aspirin use to prevent cardiovascular disease: US Preventive Services Task Force recommendation statement. JAMA 2022;327:1577–84.

Chapter 6.2

Cytochrome P450 (inhibitors for the metabolism of drugs)

Atilla Akdemir
Istinye University, Faculty of Pharmacy, Istanbul, Turkey

1 Introduction

Cytochrome P450 monooxygenases (CYP or P450) are heme-containing enzymes that are widely found in archaea, bacteria and eukaryotes [1]. The "450" refers to the characteristic UV/VIS Soret peak at approximately 450 nm and is besides P450 enzymes also observed in other hemoproteins such as Nitric Oxide Synthase (NOS), Chloroperoxidases, and protein H450 [2]. P450s generally acts in close collaboration with redox domains such as Cytochrome P450 reductase (EC 1.6.2.4; CPR) that supply electrons to the P450 heme group.

In humans, there are 57 P450 enzymes (Table 1) that are all membrane-bound proteins of which 50 are found in the endoplasmic reticulum (microsomal types) and 7 in mitochondria (mitochondrial types) [3,4].

The human enzymes are involved in the oxidation of sterols, fatty acids, eicosanoids, vitamins, and xenobiotics. In addition, prokaryotic isoforms are involved in the biosynthesis of erythromycin and mycinamicin and fungal isoforms are involved in the biosynthesis of ergosterol. The variety of chemical reactions performed by the heme group is remarkable and the substrate selectivity is mainly dictated by differences in the active site. Substrate-specificity is mainly determined by the amino acids lining the active site.

The drug metabolizing P450 enzymes are of great importance in pharmacology and pharmacotherapy as they are also involved in the metabolism of drugs and hence are key factors in drug-drug and drug-food interactions. As such pharmacokinetic boosters, which are compounds that are mainly aiming to increase the bioavailability of co-administered target drugs, are of increasing importance.

In this chapter, the focus will be mainly on the most important drug metabolizing P450 enzyme 3A4 as well as its inhibitors with a specific emphasis on pharmacokinetic boosters. First the general structure of P450 enzymes will be discussed, followed by the (patho)physiology of P450 3A4. Finally, the chapter will conclude with P450 3A4 enzyme inhibitors as pharmacokinetic boosters.

2 Structure and function

All P450 enzymes have a conserved three-dimensional fold. Structural insight into the human membrane-bound P450 became available with the crystallization of *Pseudomonas putida* P450 in complex with its substrate camphor ($P450_{cam}$ aka P450 101A1) [5–7]. This structure served as a template to generate homology models of the other P450 enzymes [3]. Almost a decade later a second bacterial P450 enzyme from *Priesta megaterium* was crystallized ($P450_{BM3}$ aka P450 102A1); however, this protein did not contain any substrates in its active site [8].

One of the most important crystal structure was the structure of modified rabbit P450 2C5 as it showed that the enzyme could be mutated and the membrane-binding α-helix could be removed from the structure while keeping its catalytic activity [2]. As such, the way was open for mammalian P450 enzymes to be investigated with protein crystallization studies as it was relatively easy to crystallize the enzymes without the N-terminal membrane binding α-helical domain. Afterward, many more prokaryotic, eukaryotic, and human cocrystal structures of P450 enzymes were obtained.

2.1 Overall structure of P450 enzymes

The mammalian P450 enzymes share a common architecture with one another and with prokaryotic P450 enzymes (Fig. 1, Table 2) [1,9]. The enzymes have an N-terminal transmembrane α-helix with a length of approximately 20–25 residues. This α-helix is not present in many cocrystal structures of mammalian P450 enzymes and may include targeting sequences for either the endoplasmic reticulum or mitochondria [1]. The enzyme mainly contains α-helixes that are numbered from A to L starting from the N-terminal (Fig. 1, Table 2). In addition, several β-sheets are present in the structure. Six of these β-strands are located spatially close to the N-terminal transmembrane domain (β1–β6), while the remaining two β-strands (β7–β8) are

Metalloenzymes. https://doi.org/10.1016/B978-0-12-823974-2.00012-7

TABLE 1 A list of the 57 human P450 enzymes known to date and their main functions [3].

Xenobiotics/drugs		Sterols		Fatty acids	Eicosanoids	Vitamins	Others
1A1	**2C19**	1B1	19A1	**2J2**	**4F2**	2R1	2A7
1A2	**2D6**	7A1	21A2	2U1	4F3	24A1	2S1
2A6	2E1	7B1	27A1	4A11	4F8	26A1	2W1
2A13	2F1	8B1	39A1	4B1	5A1	26B1	4A22
2B6	**3A4**	11A1	46A1	4F11	8A1	26C1	4F22
2C8	3A5	11B1	51A1	4F12		27B1	4X1
2C9	3A7	11B2		4V2		27C1	4Z1
2C18	3A43	17A1					20A1

Enzymes that have been shown to contribute to drug metabolism are indicated in bold.

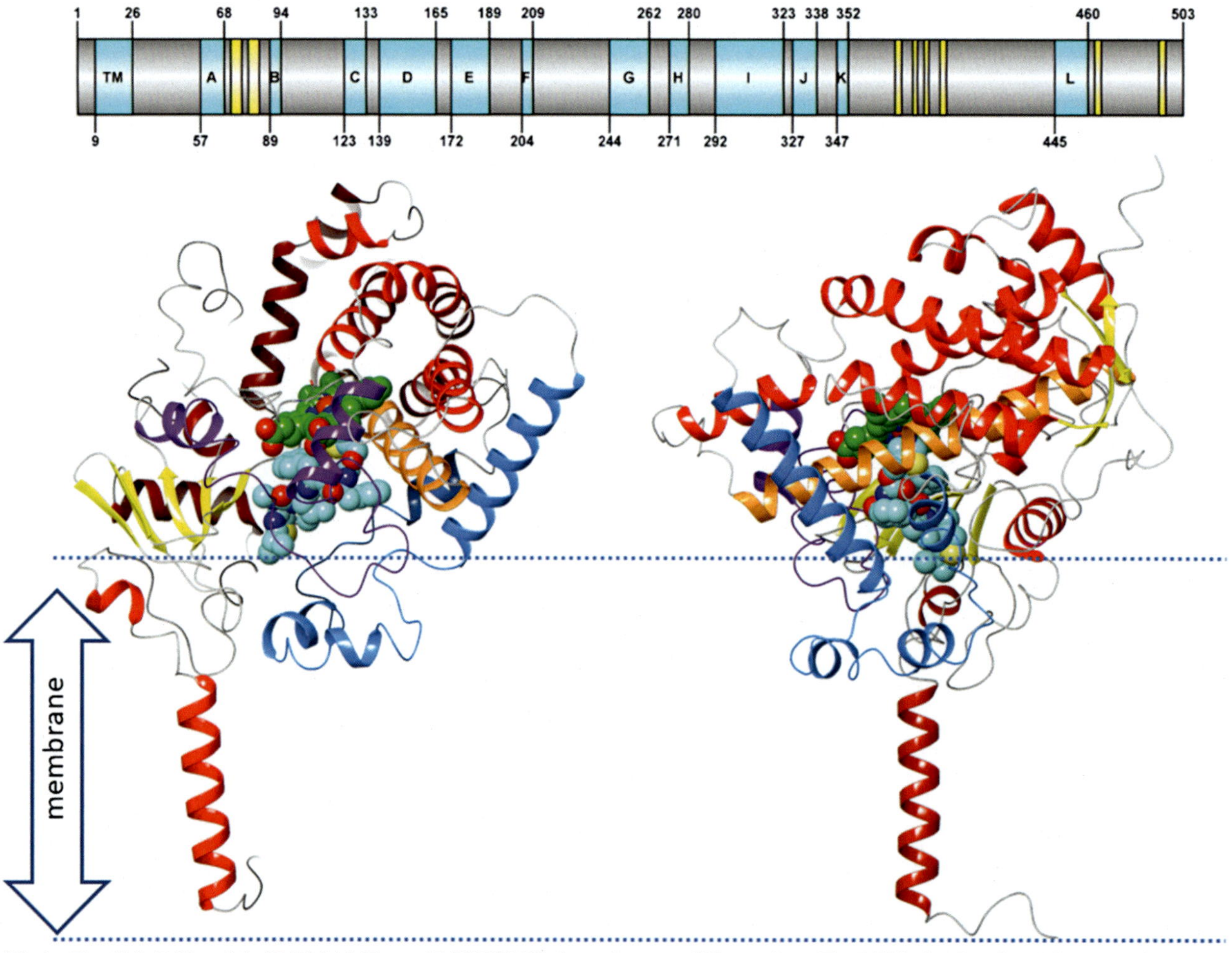

FIG. 1 The AlphaFold model of P450 3A4 (Entry: AF-P08684-F1) shown from two different views. The BC block (αB, αC, and the connecting loop) is indicated in *purple*. The FG block (αF, αG, and the connecting loop) is indicated in *blue*. αL is indicated in *brown*. The remaining α-helices are indicated in *red*. The β-sheets are indicated in *yellow*. The heme group is indicated in *green CPK representation*. Ritonavir (*turquoise CPK*; obtained from PDB entry: 3NXU) is bound to the active site. The approximate location of the membrane is indicated in *blue dashed lines*.

TABLE 2 The positions of the α-helices and β-sheets in human P450 3A4.

Helix—sheet	P450 3A4	Helix—sheet	P450 3A4
αA	Phe57-Tyr68	αK	Glu354-Leu366
αB	Met89-Leu94	αL	Met445-Leu460
αC	Asp123-Leu133	β1	Val71-Asp76[a]
αD	Ser139-Gly165	β2	Gln79-Ile84[a]
αE	Leu172-Phe189	β3	Leu373-Val376[a]
αF	Val204-Lys209[b]	β4	Val381-Ile383[c]
αG	Glu244-Glu262	β5	Met386-Ile388[c]
αH	Phe271-Asn280	β6	Val393-Ile396[a]
αI	Asp292-Thr323	β7	Phe463-Lys466[d]
αJ	Val327-VAl338	β8	Val490-Ser495[d]

[a]β-sheet cluster 1A. [b]Regions Asn198-Phe103 and Leu210-Asp214 are not folded as α-helix, while they are in most other enzymes. [c]β-sheet cluster 1B. [d]β-sheet cluster 2.

further away from where the membrane would be located (Fig. 1, Table 2). To illustrate this, the cocrystal structure of P450 3A4 in complex with ritonavir (PDB entry: 3nxu) and the P450 3A4 AlphaFold model (Entry: AF-P08684-F1) are superposed and the location of the N-terminal transmembrane α-helix, the location of the active site and secondary structure are indicated (Fig. 1).

The enzyme is not only bound to the membrane via the first α-helix but also immersed into the membrane, especially through the F/G-loop and the β-sheets (β1–β6) near the membrane (Fig. 1).

The proximal side of the heme group (Cys442 side) is close to the enzyme's surface and is available for interaction with P450 reductase [10]. Here mainly basic residues from helix K and M are located in such a way that contacts with redox partners of the enzyme are possible.

The heme-group is located between helices I (D292-T323) and L (M445-N462). A cysteine residue (Cys442) just before the L helix forms a bond with the heme iron atom. This cysteine is part of the heme-binding decapeptide loop (FXXGX$_b$XXCXG; X$_b$ indicating a basic residue and C the axial Cys residue to the heme group), which motif is conserved in all P450 enzymes [1]. The heme group is positioned approximately perpendicular to the surface of the membrane and the substrate binding site (distal side of heme) is facing toward the membrane. The active site can be reached by hydrophobic substrates via the membrane through an access tunnel located near the F/G-loop and parts of the F and G helices that are in close proximity to the membrane (Fig. 1). There is also a tunnel with its entrance near the B/C loop through which water molecules and metabolized products may enter and leave the active site (Fig. 1). As such, hydrophobic substrates may be concentrated in the membrane and gain easy access to the active site and subsequently converted into more water soluble products that leave toward the cytosol [11].

2.2 The heme prosthetic group

The heme group is the heart of the reaction center and is basically responsible for the generation of the reactive $Fe^{4+}{=}O$ radical that reacts with the substrate S to form an oxidized metabolite [12]. In addition to hydroxylations, cytochrome P450s also catalyze a broad variety of other chemical reactions including deamination, desulfuration, dehalogenation, epoxidation, N-, S-, and O-dealkylation, N-oxidation, peroxidation and sulfoxidation [1].

$$S - H + \left(Fe^{4+} = O\right)\bullet \Rightarrow S\bullet + Fe^{4+} - OH \Rightarrow SOH + Fe^{3+}$$

The heme group is a complex between the tetradentate porphyrin ring in complex with an iron ion, which is mainly in the Fe^{2+} or Fe^{3+} ionization states. Both Fe^{2+} and Fe^{3+} are thermodynamically stable and can form octahedral complexes of which four interactions are being provided by the heme and one by the cysteine. The sixth interaction site can be occupied by water, oxygen species, the ligand, or in some cases, it can be unoccupied.

The iron chemistry is especially important since it forms the actual catalytic site that actively takes part in the reactions performed by P450 enzymes. The two iron oxidation states can readily interconvert, and this is used in biological systems for electron transfer reactions and acid/base reactions. Fe^{2+} is considered a borderline soft/hard acid that prefers to interact with nitrogen and sulfur atoms over oxygen atoms. In contrast, Fe^{3+} is a hard acid that prefers to interact with oxygen atoms. In addition, both iron ions can be present in either the low-spin (all electrons paired) or high-spin states (unpaired electrons), but Fe^{3+} mainly favors the low-spin state. The state influences the ionic radii of the iron ions. For example, conversion of Fe^{2+} from low-spin to high-spin increases (for example after desolvation of the iron to form deoxy-Fe^{2+}) the ionic radius from 62 pm to 78 pm. For Fe^{3+}, the conversion from low-spin to high-spin increases the ionic radius from 55 pm to 64.5 pm. The larger radii of the high-spin ions result in the outward motion of the $Fe^{2+/3+}$ ions from the heme plane toward the Cysteine residue, thus low-spin to high-spin as well as high-spin to low-spin changes result in a change in conformation of the heme group and probably trigger conformational changes of the enzyme.

2.3 Binding of ligands to the active site and the catalytic cycle

In the resting state of the enzyme a water molecule is bound to the Fe^{3+} ion (Fig. 2). This ion is in the low spin state and has a smaller ion radius of approximately 55 pm. As such, this ion lies in the heme plane. Binding of substrates to the active site may displace the water molecule from the Fe^{3+} ion [13]. In this case, Fe^{3+} adopts the high-spin state with 78 pm radius and as such distort the heme group. The change of Fe^{3+} state from low-spin to high-spin involves a change in the UV-visible absorbance spectrum (the Soret band; λ_{max} ~390 nm). This type of substrate binding is termed type I.

Afterward, two electrons from the redox partner (generally CPR) and molecular oxygen is being transferred to the deoxy-Fe^{3+} ion to form complex II (Fig. 2). Subsequent addition of a hydrogen atom results in the loss of a water molecule and the reactive Fe^{4+}=O radical (complex III, Fig. 2) is being formed. This radical then reacts with the substrate to produce a hydroxylated product and a deoxy-Fe^{3+} ion is generated again. Finally, a water molecule binds to this iron atom and the system is ready for a new catalytic cycle.

Alternatively, the ligand may not only displace the water molecule from the Fe^{3+} ion but also form an interaction between its nitrogen atom and the low-spin Fe^{3+} ion. In this case, the catalytic cycle is not started and the ligand acts as an inhibitor. This type of binding does not cause a change in spin state of the ion and is termed type II (λ_{max} ~430 nm). Finally, the substrate may bind to the active site without displacement of the water molecule [14].

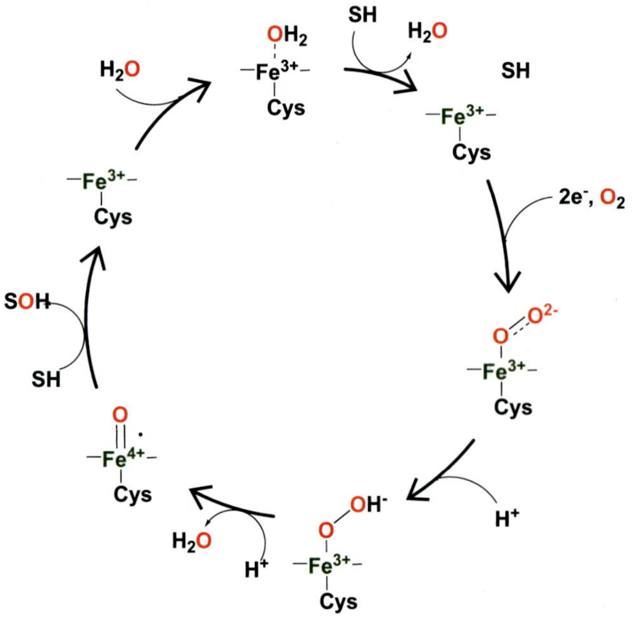

FIG. 2 The catalytic cycle of the heme group.

2.4 Substrate specificity

The P450 3A4 enzyme has a broad substrate specificity and thus can metabolize a wide range of structurally different compounds.

In part, this is caused by strong capacity of the active site to change size and form and thus adjust itself to structurally different substrates (Fig. 3) [3,4]. Besides these induced-fit effects, the enzyme has been shown to be able to easily adopt different conformational states even in the absence of substrates [4]. These distinct conformations of the enzyme may be able to recognize different substrates. This phenomenon is known as conformational selection [3,4].

In addition, six regions have been identified in or near the active site as that are responsible for substrate specificity and reaction kinetics, the so-called Substrate Recognition Sites (SRSs), which seem to be located generally at the same regions in eukaryotic and prokaryotic P450 enzymes [15–20]. However, residues that do not belong to the SRS's may also be important in selectivity and kinetics [21].

Besides binding to the active site, inhibitors may also bind to an allosteric site and interfere with the enzyme's function, thus showing noncompetitive inhibition.

3 Physiology and pathophysiology

The drug-metabolizing P450 enzymes are very important in physiology, pharmacology and toxicology as they dictate the bioavailability and elimination of drugs and as such their extent of biological effects. These enzymes have low substrate selectivity and as such can metabolize many different drugs, xenobiotics, and endogenous compounds (including hormones, cholesterol, vitamin D, and arachidonic acid). As a result, inhibitors of drug-metabolizing P450s generally have a profound effect on the pharmacokinetics of their substrates and as such may have a strong impact on physiology.

3.1 The drug-metabolizing P450 enzymes

The most important P450 enzymes in the metabolism of commonly used drugs have been identified (Table 1) [2,22–25]. For the majority (~75%) of the 200 most prescribed drugs the metabolism was mainly performed by P450 enzymes. Amongst the P450 enzymes, the most common five enzymes (involved in the metabolization of ~90% of drugs) have been identified as P450 1A2, 2C9, 2C19, 2D6, and 3A4. Remarkably, the P450 3A4 enzyme was involved in the metabolization of ~46% of the drugs.

Besides these most common five enzymes, also other P450s that are primarily involved in (other) biosynthetic/metabolic reactions can be involved in the metabolism of drugs including P450 2J2 and 4F2. In addition, some P450 enzymes have been shown to have a dominant role

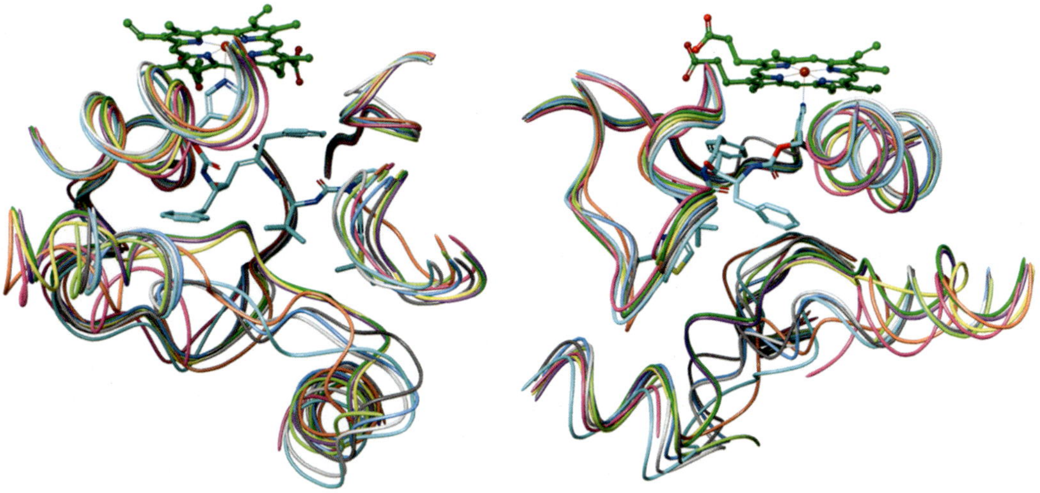

FIG. 3 The plasticity of the P450 3A4 active site.

in the bioactivation of chemicals, i.e., P450 1A1, 1A2, 1B1, 2A6, 2E1, and 3A4 [25].

3.2 The P450 3A4 enzyme and P450 3A family

The P450 3A4 enzyme (UniProt entry: P08684) is a member of the larger P450 3A family (EC: 1.14.14.1), which also includes P450 3A5 (UniProt entry: P20815), 3A7 (UniProt entry: P24462) and 3A43 (UniProt entry: Q9HB55). The sequence identify amongst these enzymes is at least 70%, with the highest similarity between 3A4, 3A5, and 3A7 (at least 81% identity) and the lowest with 3A43 (71%–76% identity toward 3A4/5/7) [26]. The most common isoforms of the 3A family in adults are P450 3A4 and 3A5. The P450 3A7 and 3A43 isoforms are predominantly expressed in the fetus; however, P450 3A7 has been identified in human liver of some patients [27].

The P450 3A4 enzyme is located in the endoplasmic reticulum mainly expressed in the liver, kidney, and small intestines. It is the most expressed enzyme of the P450 3A family and most common enzyme of this family in the human liver. Its expression levels among humans shows 10–100-fold differences [27–29]. These differences are mainly caused by receptor-mediated enzyme induction by xenobiotics and drugs as well as inhibition by endogenous/exogenous compounds (Table 3) [26]. In addition, genetic polymorphisms, age, sex, and (liver)diseases also influence P450 3A4 enzyme activity [26].

The expression of enzymes of the P450 3A family are regulated by nuclear factors, interleukins, tumor necrosis factor α (TNFα), hormones, and bile acids (Table 3) [26]. In addition, the enzymes are being induced by drugs such as phenytoin, carbamazepine, rifampin, and efavirenz and natural compounds such as extracts from St. John's Wort. These inducers mainly mediate their effects by receptors (PXR, CAR, VDR, FXR, PPARα, GR, ERα, LXR) (Table 3) [26]. Finally, several drugs as well as natural compounds (constituents of grapefruit juice) can inhibit these enzymes.

The P450 3A5 enzyme has approximately 85% sequence identity to 3A4 and shows overlapping substrate specificity [30,31]. After 3A4, it is the most common enzyme of the P450 3A family in the liver. However, it is only present in approximately 20% of the livers [30,32]. P450 3A5 is also expressed in the small intestines, kidneys, lungs, and prostate. Similar to 3A4, it shows 10–100-fold differences in expression levels.

The other two isozymes of the 3A family, P450 3A7 (P24462) and 3A43 (Q9HB55), are mainly expressed in the fetus. P450 3A7 is believed to show overlapping substrate specificity to 3A4 [30].

3.3 Physiological effect of the inhibition or induction of drug-metabolizing P450 enzymes

The P450 3A4 enzyme is very important in the metabolism and pharmacokinetics of many drugs due to its expression profile (most abundant drug-metabolizing enzyme in liver and present in small intestines on apical domain of enterocytes) and broad substrate specificity. It is not only involved in the metabolism of many drugs, xenobiotics and endogenous compounds, but also prone to both inhibition and induction. As such, coadministration of drugs that are substrates of P450 3A4 together with P450 3A4 inhibitors or inducers may lead to significant increase or decrease in the plasma concentrations of the drug. This leads not only to unwanted drug-drug or drug-food interactions but also to beneficial "pharmacokinetic enhancing" drug-drug interactions.

TABLE 3 The expression and activity of P450 3A enzyme family is regulated by various endogenous and exogenous compounds [1,26].

Inhibitors	Inducers	Regulators of P450 expression/ activation	Receptors involved
Itraconazole, ketoconazole	St. John's Wort	Hepatocyte nuclear factor 4α (HFN4α)	Pregnane X receptor (PXR)
Clarithromycin, erythromycin	Phenytoin	Interleukin 1β, 6	Constitutive androstane receptor (CAR)
Nefazodone	Carbamazepine	Tumor necrosis factor α (TNFα)	Vitamin D receptor (VDR)
Ritonavir	Rifampin	Thyroid hormone (TH), parathyroid hormone (PTH)	Farnesoid X receptor (FXR)
Grape fruit juice	Efavirenz	Growth hormone (GH)	Peroxisome proliferator-activated receptor α (PPARα)
	Barbiturates	Bile acids	Glucocorticoid receptor (GR)
	Glucocorticoids		Estrogen receptor α (ERα)
	Phenobarbital		Liver X receptor (LXR)

3.4 Unwanted drug-drug or drug-food interactions

Less common adverse effects of several drugs may become more frequent or more severe when the drugs are coadministered with inhibitors of drug-metabolizing P450 enzymes [33]. For example, coadministration of the cholesterol lowering statins (HMG-CoA reductase inhibitors) with P450 3A4 inhibitors can cause rhabdomyolysis, which under normal conditions is an uncommon adverse effect of statins. Similarly, *Torsades de pointes* is more commonly observed when terfenadine, astemizole, cisapride, or pimozide are coadministered with P450 3A4 inhibitors.

Inducers of drug-metabolizing P450 enzymes increase the expression of the enzymes mainly via nuclear receptor mediated responses and as such increase the metabolism of the target drug and therefore its plasma concentrations.

3.5 Pharmacokinetic enhancers

These compounds, also known as pharmacokinetic boosters, inhibit the either phase I or phase II metabolism or the P-glycoprotein mediated excretion of target drugs. Thereby, they increase the plasma levels of the target drug and result in the requirement of lower doses. Lower doses of drugs may be beneficial not only due to less costs but also to potentially less dose-related side effects.

4 Classes of inhibitors/activators and their design

To date, there are two pharmacokinetic enhancers of antiretroviral drugs in clinical use. These analogous compounds, cobicistat and ritonavir (Fig. 4), are peptidomimetics and potently inhibit the drug metabolizing P450 3A4 enzyme, with cobicistat being more selective than ritonavir.

Ritonavir was initially developed as a HIV protease inhibitor and entered clinical use in 1996. It was originally used at a dose of 600 mg twice daily for the treatment of HIV patients, but various side effects were observed, including diarrhea, nausea, vomiting, hyperlipidemia, hyperglycemia, hepatitis, abdominal pain, paresthesia, rash, and fatigue [34–36]. In addition, mutations in HIV caused the emerging of drug resistance against ritonavir [37–39].

The compound is mainly metabolized by P450 3A4 and to a lesser extent by P450 2D6. More interestingly, the compound potently inhibits P450 3A4 and also 2C8, 2C9, 2C19, and 2D6 and induces P450 1A2, 3A4, 2B6, 2C9, and 2C19 [36,40,41]. In addition, ritonavir inhibits P-glycoproteins (p-Gp) [41]. Especially, due to its potent inhibition of P450 3A4 and p-Gp, ritonavir is mainly used as a pharmacokinetic enhancer of protease inhibitors lopinavir, atazanavir, darunavir, elvitegravir, and fosamprenavir (prodrug of amprenavir) in antiretroviral therapy at doses of 100–200 mg per day (Fig. 5) [34,36]. This lower dose of ritonavir minimizes its side effects and also increases the bioavailability and therefore lowers the doses of the coadministered protease inhibitors. This is favorable since protease inhibitors in general have poor oral bioavailability due to

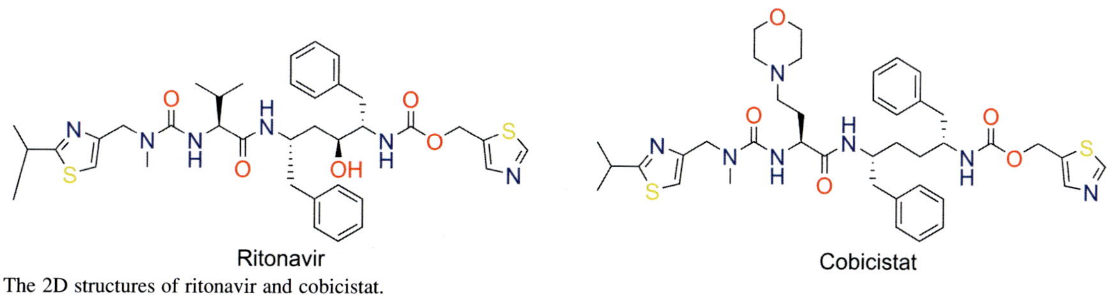

FIG. 4 The 2D structures of ritonavir and cobicistat.

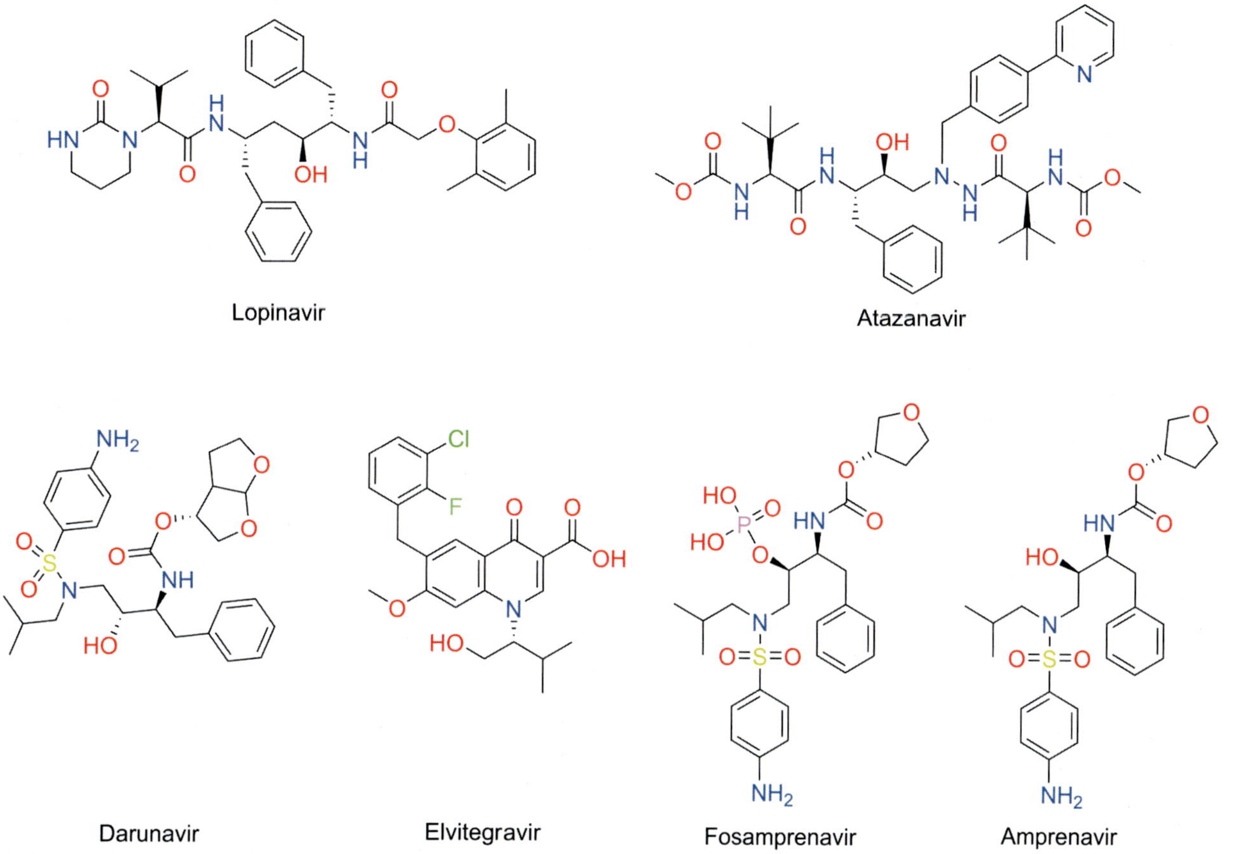

FIG. 5 The 2D structures of inhibitors lopinavir, atazanavir, darunavir, elvitegravir, fosamprenavir, and amprenavir.

extensive metabolism by P450 3A4 and excretion by p-Gp and therefore high pill burden [42].

Cobicistat (Fig. 4), which was developed with the aim to reduce the ritonavir-related side effects, does not show protease inhibition and has no antiretroviral effects [35]. The compound is a potent inhibitor of P450 3A4, a potent blocker of p-Gp, a weak inhibitor of P450 2D6, and does not show significant inhibition or induction of the other P450 enzymes [34,35,41,43,44]. The compound has been approved as a pharmacokinetic enhancers in 2012 as part of once-daily formulation also including HIV-1 integrase strand transfer inhibitor elvitegravir, nucleoside reverse transcriptase inhibitor emtricitabine and the cyclic nucleotide diester analog of adenosine monophosphate tenofovir (Figs. 5 and 6). The cobicistat dose is 150 mg per day.

Several companies such as Sequoia, Tibotec, and Concert Pharmaceuticals/GSK have been developing new P450-targetting pharmacokinetic enhancers such as SPI-452, TMC-558445, and CTP-518 (deuterated atazanavir) (Fig. 7), however, none of these compounds have been approved yet [34].

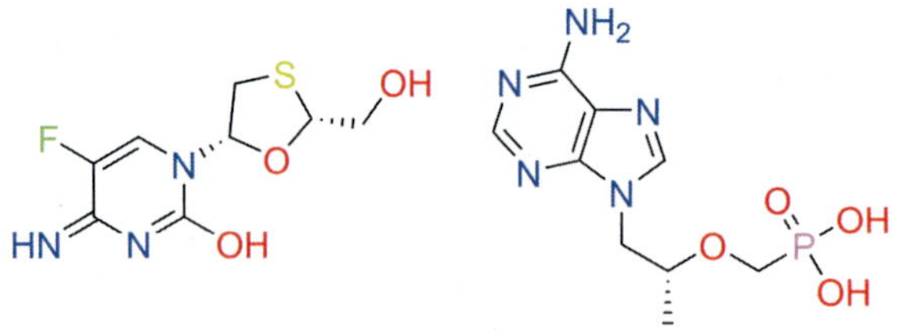

Emtricitabine Tenofovir

FIG. 6 The 2D structures of emtricitabine and tenofovir.

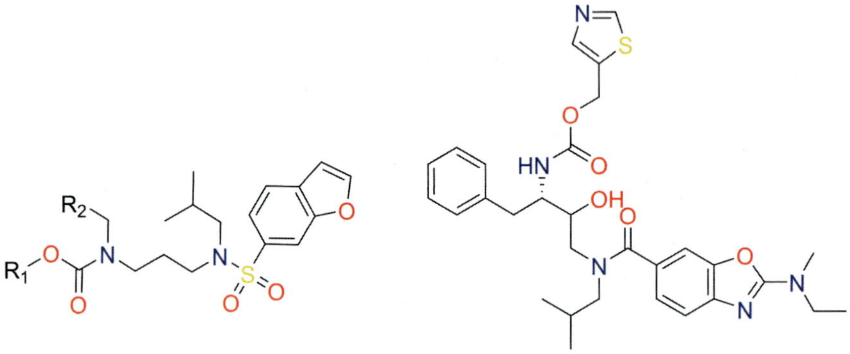

SPI-452 TMC-558445

FIG. 7 The 2D structures of SPI-452 and TMC-558445.

5 Conclusion

Drug-metabolizing P450 enzymes control the bioavailability and thus physiological effects of drugs by controlling their pharmacokinetics. Especially, P450 3A4 is an important drug-metabolizing enzyme as it is expressed in high concentrations in hepatocytes and enterocytes and due to its plastic and malleable active site shows a broad substrate recognition profile. Thus orally administered drugs are likely to encounter this enzyme on their way into the systemic circulation. The coadministration of P450 3A4 inhibitors together with target drugs can therefore increase the bioavailability of the latter and thus lower the required doses, the risk of serious side effects, and costs of pharmacotherapy.

References

[1] Danielson PB. The cytochrome P450 superfamily: biochemistry, evolution and drug metabolism in humans. Curr Drug Metab 2002;3(6):561–97.

[2] Williams PA, et al. Mammalian microsomal cytochrome P450 monooxygenase: structural adaptations for membrane binding and functional diversity. Mol Cell 2000;5(1):121–31.

[3] Guengerich FP, Waterman MR, Egli M. Recent structural insights into cytochrome P450 function. Trends Pharmacol Sci 2016;37(8):625–40.

[4] Guengerich FP. Human cytochrome P450 enzymes. In: Ortiz de Montellano PR, editor. Cytochrome P450: Structure, mechanism, and biochemistry. Cham: Springer International Publishing; 2015. p. 523–785.

[5] Barnes HJ, Arlotto MP, Waterman MR. Expression and enzymatic activity of recombinant cytochrome P450 17 alpha-hydroxylase in *Escherichia coli*. Proc Natl Acad Sci U S A 1991;88(13):5597–601.

[6] Katagiri M, Ganguli BN, Gunsalus IC. A soluble cytochrome P-450 functional in methylene hydroxylation. J Biol Chem 1968;243(12):3543–6.

[7] Poulos TL, Perez M, Wagner GC. Preliminary crystallographic data on cytochrome P-450CAM. J Biol Chem 1982;257(17):10427–9.

[8] Ravichandran KG, et al. Crystal structure of hemoprotein domain of P450BM-3, a prototype for microsomal P450's. Science 1993;261(5122):731–6.

[9] Hasemann CA, et al. Structure and function of cytochromes P450: a comparative analysis of three crystal structures. Structure 1995;3(1):41–62.

[10] Estrada DF, Laurence JS, Scott EE. Substrate-modulated cytochrome P450 17A1 and cytochrome b5 interactions revealed by NMR. J Biol Chem 2013;288(23):17008–18.

[11] Šrejber M, et al. Membrane-attached mammalian cytochromes P450: an overview of the membrane's effects on structure, drug binding, and interactions with redox partners. J Inorg Biochem 2018;183:117–36.

[12] Guengerich FP. Cataloging the repertoire of nature's blowtorch, P450. Chem Biol 2009;16(12):1215–6.

[13] Schenkman JB, Remmer H, Estabrook RW. Spectral studies of drug interaction with hepatic microsomal cytochrome. Mol Pharmacol 1967;3(2):113–23.

[14] Mast N, et al. Binding of a cyano- and fluoro-containing drug bicalutamide to cytochrome P450 46A1: unusual features and spectral response. J Biol Chem 2013;288(7):4613–24.

[15] Lindberg RL, Negishi M. Alteration of mouse cytochrome P450coh substrate specificity by mutation of a single amino-acid residue. Nature 1989;339(6226):632–4.

[16] Lindberg RL, Negishi M. Modulation of specificity and activity in mammalian cytochrome P-450. Methods Enzymol 1991;202:741–52.

[17] Wang H, et al. Structure-function relationships of human liver cytochromes P450 3A: aflatoxin B1 metabolism as a probe. Biochemistry 1998;37(36):12536–45.

[18] Gotoh O. Substrate recognition sites in cytochrome P450 family 2 (CYP2) proteins inferred from comparative analyses of amino acid and coding nucleotide sequences. J Biol Chem 1992;267(1):83–90.

[19] Raucy JL, Allen SW. Recent advances in P450 research. Pharmacogenomics J 2001;1(3):178–86.

[20] Sirim D, et al. Prediction and analysis of the modular structure of cytochrome P450 monooxygenases. BMC Struct Biol 2010;10:34.

[21] Podust LM, Poulos TL, Waterman MR. Crystal structure of cytochrome P450 14alpha-sterol demethylase (CYP51) from *Mycobacterium tuberculosis* in complex with azole inhibitors. Proc Natl Acad Sci U S A 2001;98(6):3068–73.

[22] Wienkers LC, Heath TG. Predicting in vivo drug interactions from in vitro drug discovery data. Nat Rev Drug Discov 2005; 4(10):825–33.

[23] Eisenmann ED, et al. Boosting the oral bioavailability of anticancer drugs through intentional drug-drug interactions. Basic Clin Pharmacol Toxicol 2022;130(Suppl. 1):23–35.

[24] Williams JA, et al. Drug-drug interactions for UDP-glucuronosyltransferase substrates: a pharmacokinetic explanation for typically observed low exposure (AUCi/AUC) ratios. Drug Metab Dispos 2004;32(11):1201–8.

[25] Rendic S, Guengerich FP. Survey of human oxidoreductases and cytochrome P450 enzymes involved in the metabolism of xenobiotic and natural chemicals. Chem Res Toxicol 2015;28(1):38–42.

[26] Fujino C, Sanoh S, Katsura T. Variation in expression of cytochrome P450 3A isoforms and toxicological effects: endo- and exogenous substances as regulatory factors and substrates. Biol Pharm Bull 2021;44(11):1617–34.

[27] Ohtsuki S, et al. Simultaneous absolute protein quantification of transporters, cytochromes P450, and UDP-glucuronosyltransferases as a novel approach for the characterization of individual human liver: comparison with mRNA levels and activities. Drug Metab Dispos 2012;40(1):83–92.

[28] Achour B, Barber J, Rostami-Hodjegan A. Expression of hepatic drug-metabolizing cytochrome p450 enzymes and their intercorrelations: a meta-analysis. Drug Metab Dispos 2014;42(8):1349–56.

[29] Huang W, et al. Evidence of significant contribution from CYP3A5 to hepatic drug metabolism. Drug Metab Dispos 2004;32(12):1434–45.

[30] Daly AK. Significance of the minor cytochrome P450 3A isoforms. Clin Pharmacokinet 2006;45(1):13–31.

[31] Williams JA, et al. Comparative metabolic capabilities of CYP3A4, CYP3A5, and CYP3A7. Drug Metab Dispos 2002;30(8):883–91.

[32] Kuehl P, et al. Sequence diversity in CYP3A promoters and characterization of the genetic basis of polymorphic CYP3A5 expression. Nat Genet 2001;27(4):383–91.

[33] Dresser GK, Spence JD, Bailey DG. Pharmacokinetic-pharmacodynamic consequences and clinical relevance of cytochrome P450 3A4 inhibition. Clin Pharmacokinet 2000;38(1):41–57.

[34] Krauß J, Bracher F. Pharmacokinetic enhancers (boosters)—escort for drugs against degrading enzymes and beyond. Sci Pharm 2018;86 (4):43.

[35] Renjifo B, et al. Pharmacokinetic enhancement in HIV antiretroviral therapy: a comparison of ritonavir and cobicistat. AIDS Rev 2015; 17(1):37–46.

[36] Larson KB, et al. Pharmacokinetic enhancers in HIV therapeutics. Clin Pharmacokinet 2014;53(10):865–72.

[37] Danner SA, et al, European-Australian Collaborative Ritonavir Study Group. A short-term study of the safety, pharmacokinetics, and efficacy of ritonavir, an inhibitor of HIV-1 protease. N Engl J Med 1995;333(23):1528–33.

[38] Hsu A, et al. Multiple-dose pharmacokinetics of ritonavir in human immunodeficiency virus-infected subjects. Antimicrob Agents Chemother 1997;41(5):898–905.

[39] Schmit JC, et al. Resistance-related mutations in the HIV-1 protease gene of patients treated for 1 year with the protease inhibitor ritonavir (ABT-538). AIDS 1996;10(9):995–9.

[40] Shah BM, et al. Cobicistat: a new boost for the treatment of human immunodeficiency virus infection. Pharmacotherapy 2013;33(10): 1107–16.

[41] Tseng A, et al. Cobicistat versus ritonavir: similar pharmacokinetic enhancers but some important differences. Ann Pharmacother 2017;51(11):1008–22.

[42] Lv Z, Chu Y, Wang Y. HIV protease inhibitors: a review of molecular selectivity and toxicity. HIV AIDS (Auckl) 2015;7:95–104.

[43] Lepist EI, et al. Cobicistat boosts the intestinal absorption of transport substrates, including HIV protease inhibitors and GS-7340, in vitro. Antimicrob Agents Chemother 2012;56(10):5409–13.

[44] Mathias AA, et al. Pharmacokinetics and pharmacodynamics of GS-9350: a novel pharmacokinetic enhancer without anti-HIV activity. Clin Pharmacol Ther 2010;87(3):322–9.

Chapter 6.3

Aromatase

Özlen Güzel-Akdemir
Istanbul University, Faculty of Pharmacy, Department of Pharmaceutical Chemistry, Istanbul, Turkey

1 Introduction

In humans, there are four major estrogens, i.e., estrone (E1), estradiol (E2), estriol (E3), and estetrol (E4) (Fig. 1). E2 is the most potent and most prevalent estrogen. E3 is mainly present in pregnant women, while E4 seems to be important in the fetus. Estrogens are the female sex hormones that are involved in the development of the female secondary sex characteristics, genital tract, and breasts (Fig. 2). However, estrogens are also found in men, although in lower concentrations. They are also involved in the maturation and maintenance of bone mineral density, fat storage, cholesterol levels, and muscle strength (Fig. 2). As such, these hormones are involved in the pathophysiology of several diseases such as hormone-receptor positive breast cancer and conditions such as gynecomastia and infertility.

The human aromatase enzyme (also known as P450 19A1) is crucial in the synthesis of estrogens from androgens and is therefore important in estrogen physiology. As such, it is a well-accepted drug target and aromatase inhibitors have clinical use in the treatment of hormone-receptor positive breast cancer in postmenopausal women, in the treatment of gynecomastia in men and have potential use in the pharmacotherapy against infertility.

Currently, there are several orally active aromatase inhibitors in clinical use, such as exemestane, anastrozole, and letrozole. Exemestane has a steroid structure and forms a covalent bond with the aromatase active site (a suicide inhibitor) and irreversibly inhibits P450 19A1. Anastrozole and letrozole are reversible inhibitors that interact with the heme-iron via their triazole nitrogen atoms.

This chapter will discuss the structure and function of human aromatase. In addition, the clinical use as well as the most frequent side effects of aromatase inhibitors will be discussed.

2 Structure and function

Human aromatase (UniProt accession code: P11511; EC:1.14.14.14) is a heme-containing enzyme that belongs to the cytochrome P450 (CYP) family. This enzyme, which is also known as P450 19A1, forms a heterodimer with NADPH cytochrome P450 reductase (CPR) and is a membrane-bound protein that is located in the endoplasmic reticulum membrane [1,2]. The enzyme is responsible for the most crucial step in the conversion of androgens into estrogens and is therefore important in estrogen physiology.

The aromatase enzyme has a similar fold as the other mammalian as well as the prokaryotic P450 enzymes. A detailed description of their common structure as well as the mechanism of catalysis has been given in Chapter 6.2.

Several cocrystal structures of human aromatase have been solved to date, including the structures in complex with the endogenous substrate androstenedione (PDB entry: 3S79). The endogenous substrate forms a hydrogen bond between one of its carbonyl group and the backbone NH group of Met374 (Fig. 3). The carbonyl group on the A ring of the substrate forms an aromatic hydrogen bond with the side chain of Trp224. In addition, extensive hydrophobic interactions are present between androstenedione and the hydrophobic residues of the active site.

Aromatase converts the androgens androstenedione and testosterone into the estrogens E1 and E2. To this end, the methyl group between the A and B rings of androgens (carbon number 19) is stepwise hydroxylated by aromatase (steps i and ii; Fig. 4). Afterward, a water molecule is released from the intermediate to result in an aldehyde (step iii; Fig. 4). This step also converts the carbonyl group on position 3 into a hydroxyl group and introduces a double bond between carbon atoms 2 and 3. Finally, a formic acid molecule is released from the intermediate to yield estrone. During the final step, the A ring of the hormone is converted into an aromatic ring, hence the resulting in the name of the enzyme.

3 Physiology and pathophysiology

The biosynthesis of estrogens E1 and E2 starts with the conversion of cholesterol into pregnenolone by the *cholesterol side chain cleavage* enzyme P450 11A1 (aka. P450 Scc) (Fig. 5). Subsequently, pregnenolone is converted into another progestogen progesterone by the 3β-hydroxysteroid dehydrogenase enzyme (HSD3B1). Finally, progesterone is

Metalloenzymes. https://doi.org/10.1016/B978-0-12-823974-2.00013-9

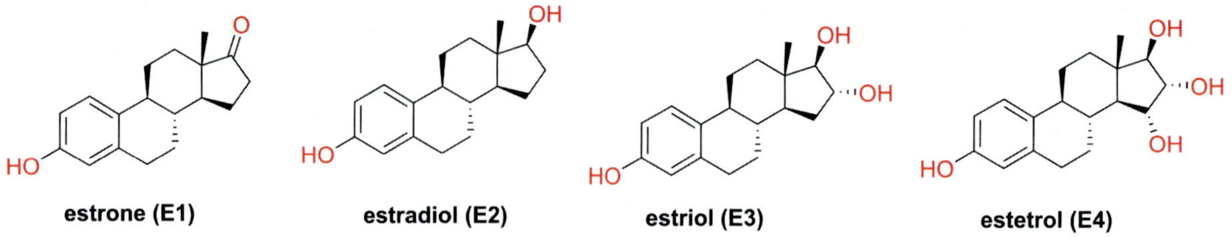

FIG. 1 The four major estrogens in humans.

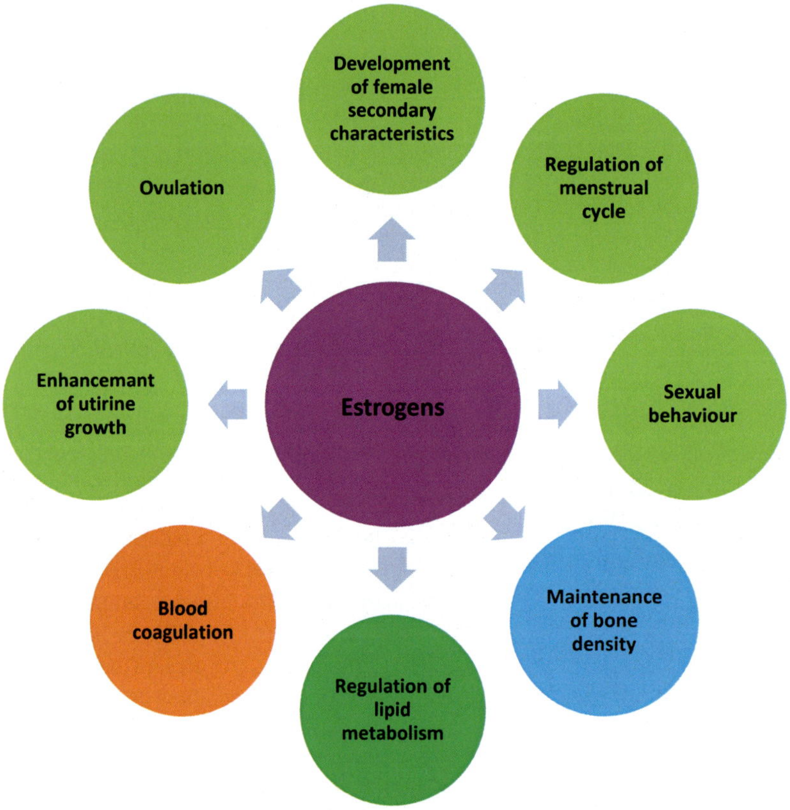

FIG. 2 The most important physiological effects of estrogens.

converted into 17α-hydroxy-progesterone by 17α-hydroxylase (P450 17). Alternatively, pregnenolone can also be converted to 17α-hydroxypregnenolone by P450 17 and then to 17α-hydroxy-progesterone by the 17,20 lyase enzyme. Pregnenolone, 17α-hydroxypregnenolone, progesterone, and 17α-hydroxyprogesterone are all 21 carbon containing progestogens. 17α-Hydroxyprogesterone is converted into androstenedione by the P450 17 enzyme. This hormone can be converted into E1 by aromatase or into testosterone by 17β-hydroxysteroid dehydrogenase type 1 (HSD17B1). Testosterone is also a substrate of aromatase and is converted into E2. These estrogens may be subsequently converted into E3 and E4 by P450 3A4 and 15α- and 16α-hydroxylases. As such, the aromatase enzyme is the source of estrogens E1 and E2 (Fig. 5) and indirectly also of E3 and E4.

Estrogen biosynthesis mainly occurs in the ovaries of premenopausal women and in the placenta during pregnancy. Other minor sites of estrogen synthesis are the pancreas, liver, adrenal glands, testis, brain, breasts, bone, skin, and adipose tissue. However, in postmenopausal women, the synthesis in the ovaries and placenta are strongly decreases and the synthesis in the secondary sites, including the breasts, become more important.

The aromatase enzyme is basically the source of estrogens in the human body and in situations where high estrogen levels are not preferred the enzyme can be inhibited. This includes adjuvant therapy in hormone-receptor positive breast cancer in postmenopausal women, gynecomastia in men, and aromatase excess syndrome [3,4]. Approximately two thirds of all breast cancers are estrogen receptor positive and as such are amenable to

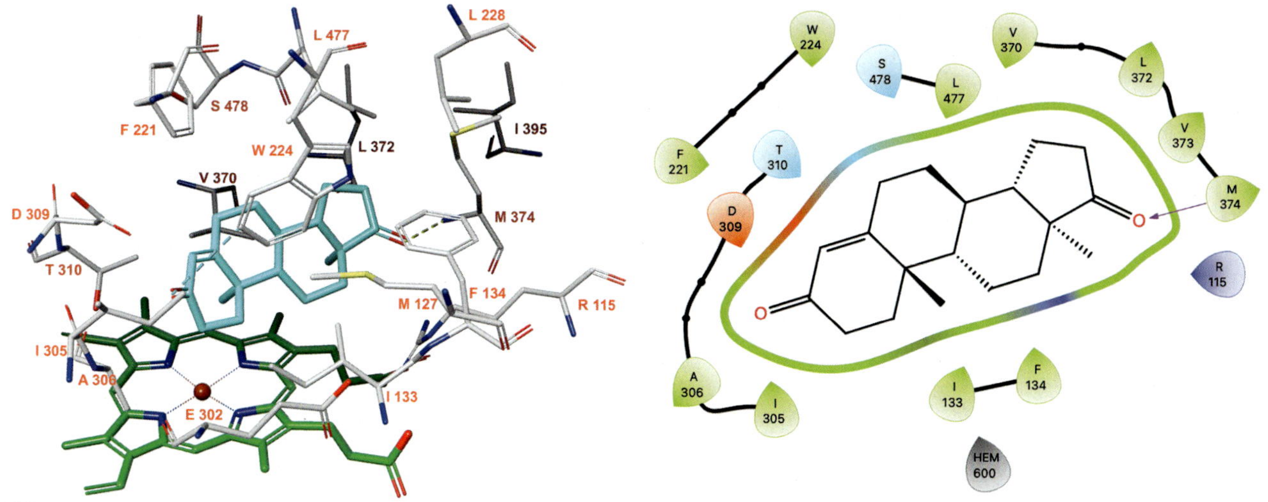

FIG. 3 The crystal structures and binding interactions of human placental aromatase in complex with the substrate androstenedione (PDB entry: 3S79). The heme group is indicated in *green sticks*. The iron is indicated as a *brown sphere*. The ligand is indicated in *turquoise sticks*. Hydrogen bonds are indicated in *dashed yellow* (left panel) or *plain purple lines* (right panel). Aromatic hydrogen bonds are indicated in *dashed turquoise lines*. Anionic residues are indicated in *red*, cationic residues are indicated in *purple*, polar residues are indicated in *blue*, and hydrophobic residues are indicated in *green*.

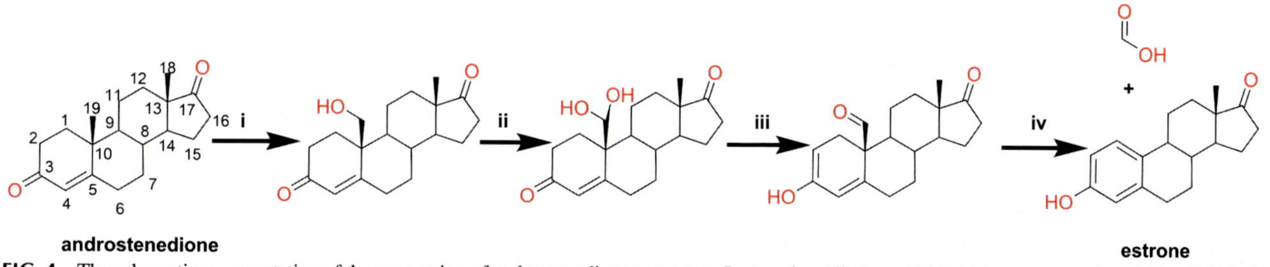

FIG. 4 The schematic representation of the conversion of androstenedione to estrone. In steps **i** and **ii**, O_2 and NADPH are converted to H_2O and $NADP^+$ and the methyl group (C19) is double hydroxylated. In step **iii**, a water molecule is being released and the carbonyl group is converted into a hydroxyl group. In the last step **iv**, HCO_2H is being released, ring A is aromatized and estrone is obtained. All steps are being performed by aromatase.

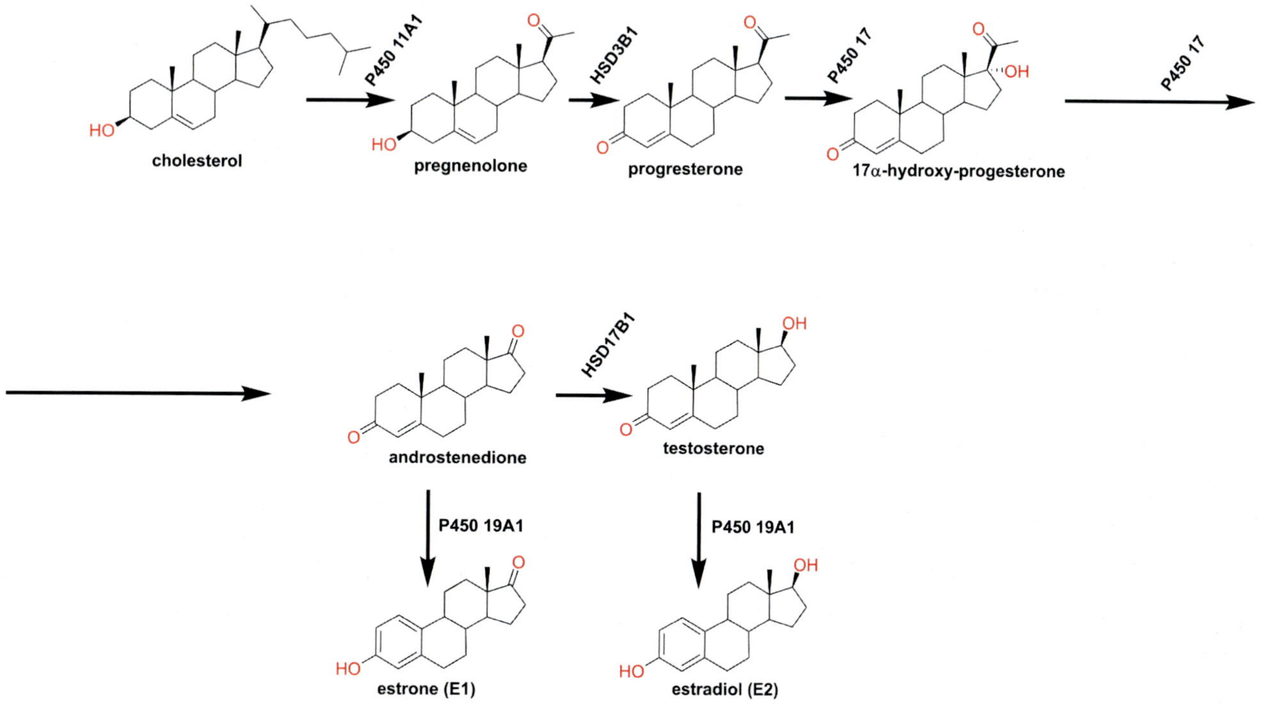

FIG. 5 A simplified overview of the synthesis of E1 and E2.

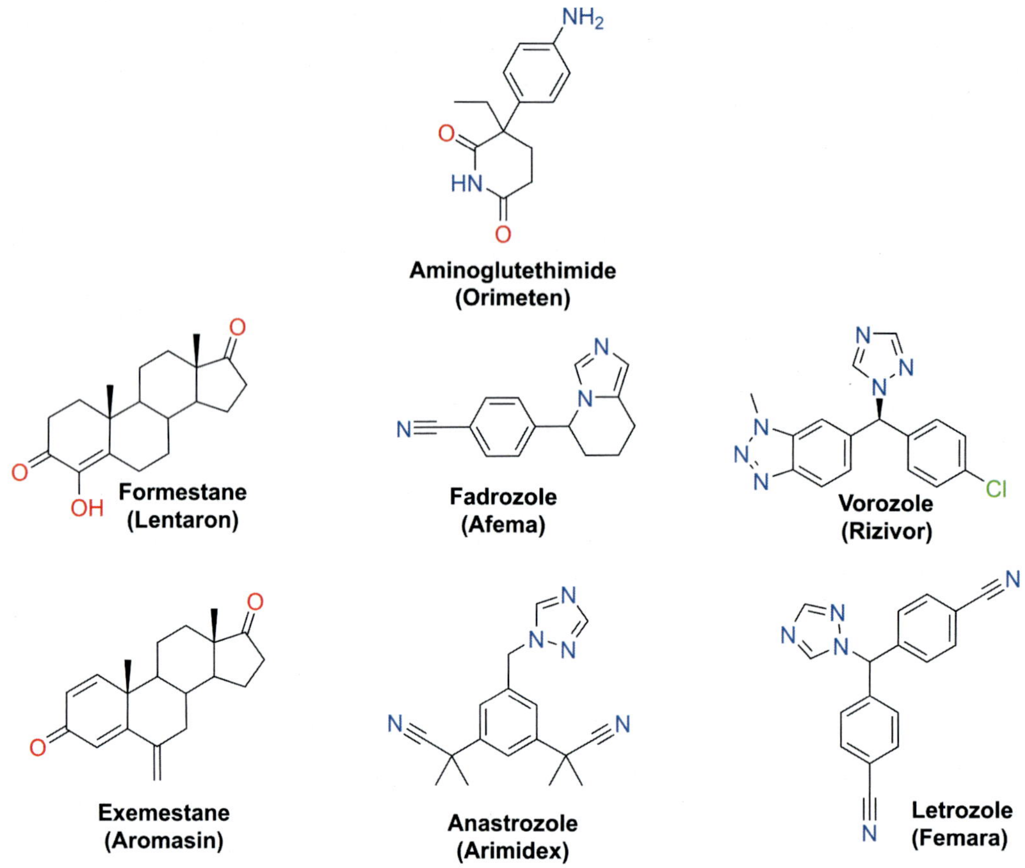

FIG. 6 The 2D structures of the first-, second-, and third-generation aromatase inhibitors.

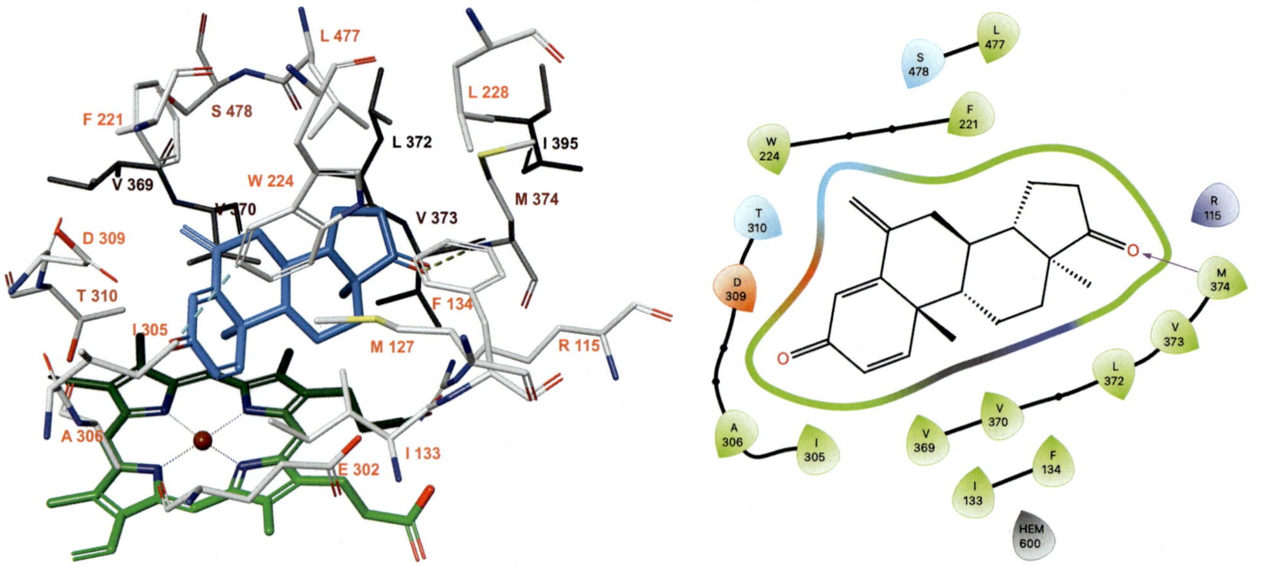

FIG. 7 The crystal structure and binding interactions of human aromatase in complex with exemestane (PDB entry: 3S7S). The heme group is indicated in *green sticks*. The iron is indicated as a *brown sphere*. The ligand is indicated in *blue sticks*. Hydrogen bonds are indicated in *dashed yellow* (left panel) or *plain purple lines* (right panel). Aromatic hydrogen bonds are indicated in *dashed turquoise lines*. Anionic residues are indicated in *red*, cationic residues are indicated in *purple*, polar residues are indicated in *blue*, hydrophobic residues are indicated in *green*.

treatment with adjuvant hormonal therapy. Aromatase inhibitors may also be used in the pharmacotherapy of infertility and endometrial carcinoma [5–8].

Aromatase excess syndrome is characterized by increased levels of estrogens in both females and males. In males, this may result in heterosexual precocity (development of sexual characteristics of females), including gynecomastia, and short stature as an adult. In females, it may result in isosexual precocity in females (development of phenotypically appropriate sexual characteristics but at early age), such as macromastia, and also irregular menstrual cycles and short stature.

4 Classes of inhibitors/activators

The first-generation nonselective aromatase inhibitor was aminoglutethimide (Fig. 6) [9]. Besides aromatase, this compound also inhibits the P450 11A1, 11B1, and 21A2 enzymes and thus results in a decrease in the production of adrenal glucocorticoids, mineralocorticoids, estrogens, and androgens [9]. The drug was first used as a mild anticonvulsant and later it was used in the treatment of breast and prostate cancer. However, the lack of selectivity results in a wide range of side effects, including high cholesterol levels, hepatotoxicity, lethargy, depression, apathy, personality changes, sleep disturbances, and adrenal insufficiency.

The lack of selectivity and broad range of side effects of aminoglutethimide required the development of second-generation aromatase inhibitors. Formestane has a steroidal structure and is an androstenedione analog (Fig. 6) [10]. The compound is an irreversible inhibitor and it is suggested that the compounds free hydroxyl group may be involved in covalently binding to the aromatase active site [9,11]. Due to its poor bioavailability it is administered to patients intramuscularly on a biweekly basis and therefore it is not currently not preferred.

Fadrozole and vorozole (withdrawn from testing due to efficacy issues) are second-generation nonsteroidal aromatase inhibitors (Fig. 6). The compounds have a diazole and triazole moiety in their structure, which are expected to interact with the heme-iron of the aromatase active site [9]. These compounds are superseded by third-generation aromatase inhibitors exemestane, anastrozole, and letrozole due to the lack of bioavailability and/or potency (Fig. 6).

5 Clinically used agents or compounds in clinical development

The third-generation aromatase inhibitors are currently used in the pharmacotherapy of hormone-receptor positive breast cancer in postmenopausal women. Exemestane is an androstenedione analog. The compound is an irreversible inhibitor of aromatase and is converted by the enzyme into a reactive metabolite that forms a covalent bond with the active site [2,9,12–16]. The active site is permanently blocked and the enzyme is degraded by the protease system. In contrast to formestane, this compound has better bioavailability, and the drug is available for oral administration (25 mg/day) [14]. The compound is easily distributed throughout the body. The drug or its metabolites are mainly excreted via feces and urine. Several P450 enzymes (1A1/2, 4A11, 3A4/5, 2B6, 2A6, 2C8, 2C9, 2C19) as well as aldo-ketoreductases are involved in the biotransformation of exemestane into metabolites of which several show bioactivity. The adverse effects are generally mild to moderate and include hot flashes, fatigue and musculoskeletal side effects (osteoporosis, fractures, arthralgias) [3,4,17]. Exemestane has also some androgenic effects such as weight gain.

The cocrystal structure of human aromatase in complex with exemestane (PDB entry: 3S7S) reveals important binding interactions (Fig. 7). Interestingly, the binding pose and the ligand-enzyme binding interactions are very similar as observed for androstenedione (Fig. 3). Exemestane has an additional double bond compared to androstenedione and this double bond is in close proximity to the heme group and Ile133.

Anastrozole (1 mg/day) and letrozole (2.5 mg/day) are triazole derivatives and reversibly inhibit the enzyme by forming competitive interactions with the heme-iron atom [9]. They show an increased selectivity toward aromatase compared to the first- and second-generation aromatase inhibitors and better bioavailability [16,18,19]. When both second-generation competitive inhibitors were compared in various studies, letrozole demonstrated higher potency compared to anastrazole [18,19]. Letrozole is mainly metabolized by P450 2A6 and 3A4 and UGT2B7 [20,21]. Anastrozole is mainly metabolized by P450 3A4 and UGT1A4, but P450 2C8 and 3A8 are also involved [22]. The bioavailability of both compounds is high and therefore both compounds are administered orally. The adverse effects of anastrozole and letrozole are similar to exemestane.

References

[1] Amarneh B, et al. Functional domains of human aromatase cytochrome P450 characterized by linear alignment and site-directed mutagenesis. Mol Endocrinol 1993;7(12):1617–24.

[2] Chumsri S, et al. Aromatase, aromatase inhibitors, and breast cancer. J Steroid Biochem Mol Biol 2011;125(1–2):13–22.

[3] Gaillard S, Stearns V. Aromatase inhibitor-associated bone and musculoskeletal effects: new evidence defining etiology and strategies for management. Breast Cancer Res 2011;13(2):205.

[4] Garreau JR, et al. Side effects of aromatase inhibitors versus tamoxifen: the patients' perspective. Am J Surg 2006;192(4):496–8.

[5] Yang AM, et al. Letrozole for female infertility. Front Endocrinol (Lausanne) 2021;12:676133.

[6] Yang C, Li P, Li Z. Clinical application of aromatase inhibitors to treat male infertility. Hum Reprod Update 2021;28(1):30–50.

[7] Gao C, et al. The therapeutic significance of aromatase inhibitors in endometrial carcinoma. Gynecol Oncol 2014;134(1):190–5.

[8] Schlegel PN. Aromatase inhibitors for male infertility. Fertil Steril 2012;98(6):1359–62.

[9] Brueggemeier RW. Overview of the pharmacology of the aromatase inactivator exemestane. Breast Cancer Res Treat 2002;74(2):177–85.

[10] Wiseman LR, McTavish D. Formestane. A review of its pharmacodynamic and pharmacokinetic properties and therapeutic potential in the management of breast cancer and prostatic cancer. Drugs 1993;45(1): 66–84.

[11] Brodie AM, et al. The effect of an aromatase inhibitor, 4-hydroxy-4-androstene-3,17-dione, on estrogen-dependent processes in reproduction and breast cancer. Endocrinology 1977;100(6):1684–95.

[12] Hong Y, Chen S. Aromatase inhibitors: structural features and biochemical characterization. Ann N Y Acad Sci 2006;1089:237–51.

[13] Wang X, Chen S. Aromatase destabilizer: novel action of exemestane, a food and drug administration-approved aromatase inhibitor. Cancer Res 2006;66(21):10281–6.

[14] Sobral AF, et al. Unravelling exemestane: from biology to clinical prospects. J Steroid Biochem Mol Biol 2016;163:1–11.

[15] Giudici D, et al. 6-Methylenandrosta-1,4-diene-3,17-dione (FCE 24304): a new irreversible aromatase inhibitor. J Steroid Biochem 1988;30(1–6):391–4.

[16] Nabholtz JM, et al. Comparative review of anastrozole, letrozole and exemestane in the management of early breast cancer. Expert Opin Pharmacother 2009;10(9):1435–47.

[17] Tenti S, et al. Aromatase inhibitors-induced musculoskeletal disorders: current knowledge on clinical and molecular aspects. Int J Mol Sci 2020;21(16).

[18] Bhatnagar AS. The discovery and mechanism of action of letrozole. Breast Cancer Res Treat 2007;105(Suppl. 1):7–17.

[19] Bhatnagar AS. The early days of letrozole. Breast Cancer Res Treat 2007;105(Suppl. 1):3–5.

[20] Jeong S, et al. Inhibition of drug metabolizing cytochrome P450s by the aromatase inhibitor drug letrozole and its major oxidative metabolite 4,4′-methanol-bisbenzonitrile in vitro. Cancer Chemother Pharmacol 2009;64(5):867–75.

[21] Precht JC, et al. The letrozole phase 1 metabolite carbinol as a novel probe drug for UGT2B7. Drug Metab Dispos 2013;41(11): 1906–13.

[22] Kamdem LK, et al. In vitro and in vivo oxidative metabolism and glucuronidation of anastrozole. Br J Clin Pharmacol 2010;70(6): 854–69.

Section G

Iron enzymes, nonheme containing

Chapter 7.1

Nonheme mono- and dioxygenases

Marta Ferraroni
Dipartimento di Chimica "Ugo Schiff", Università di Firenze, Firenze, Italy

1 Introduction

Oxygenases catalyze the insertion of oxygen atom(s) into an organic substrate, using molecular oxygen as oxygen donor. According to the number of oxygen atoms transferred to their substrate, they can be categorized as monoxygenases or dioxygenases. They catalyze a vast array of chemical transformations including hydroxylation, chlorination, and epimerization as well as both cyclization and ring cleavage of organic substrates [1]. In order to carry out this type of reactions, oxygenases need to activate molecular oxygen upon donation of electrons to overcome its spin-forbidden reaction with the organic substrate. They either take all four electrons needed for the reaction from the substrate or two from the substrate and two from an external donor (cosubstrate) [2].

Thus, oxygenases can be grouped according to the respective electron-source into cofactor independent enzymes and O_2 activating enzymes that depend on cofactors such as pterin, 2-oxoglutarate or an electron providing Rieske cluster (Rieske-type dioxygenases).

Almost all the oxygenases are iron-dependent enzymes, in which the iron ion is covalently bound to porphyrin (heme enzymes) or is coordinated to side chains of enzyme residues, generally histidine, glutamate, or aspartate (nonheme enzymes). In the latter, the iron can be mononuclear or binuclear as in the bacterial di-iron monoxygenases [3]. These enzymes and the Rieske-type dioxygenases are bacterial multicomponent enzymes which catalyze the hydroxylation of aromatic intermediates, the first step of the oxidative degradation of aromatic compounds. Since they are exclusively found in bacteria, they fall outside the scope of this book and will not be discussed in this chapter.

The majority of the mononuclear nonheme iron-dependent oxygenases utilize Fe(II) and are characterized by a metal-binding motif that has been named the "2-histidine-1-carboxylate facial triad," in which the metal is bound to two histidine residues and one carboxylic acid side chain of either a glutamate or an aspartate residue, arranged at one face of an octahedron. However in several enzymes, the carboxylate is replaced by a third histidine [4].

This chapter will focus on the different families of mononuclear nonheme iron-dependent enzymes of pharmacological interest and will discuss more thoroughly a quite peculiar dioxygenase, the 4-hydroxyphenylpyruvate dioxygenase.

2 Pterin-dependent monooxygenases

The pterin-dependent monooxygenases constitute a small family which catalyze the oxygenation of the aromatic amino acids phenylalanine, tyrosine, and tryptophan with the formation of tyrosine, 3,4-dihydroxyphenylalanine, and 5-hydroxytryptophan, respectively. The enzymes are found in mammal liver and in the central nervous system, as well as in a few bacteria [5]. Phenylalanine hydroxylase (PAH) catalyzes the first step of the catabolic degradation of L-Phe, whereas tyrosine (TH) and tryptophan hydroxylase participate in the biosynthesis of the neurotransmitters catecholamine and serotonin, respectively [6].

The crystallographic structure of truncated forms of the rat TH [7] and human PAH [8] were determined in 1997. The structure of the full-length human PAH both unbound and complexed with tetrahydrobiopterin (BH4) has been recently solved (Fig. 1) [9]. The eukaryotic pterin monooxygenases are all homotetramers. Each subunit is formed by three domains: an N-terminal catalytic, a central regulatory domain, and a C-terminal oligomerization domain. The catalytic domains of the pterin-dependent hydroxylases are sufficiently similar to assume that all three enzymes have essentially the same catalytic mechanism. They are characterized by the 2-His-1-carboxylate facial triad motif anchoring the catalytic mononuclear ferrous iron ion at the active site and utilize BH4 as the cosubstrate. It has been suggested that O_2 is activated by both the ferrous ion and the bound reduced tetrahydrobiopterin, after which the side-product 4α-hydroxybiopterin is formed in addition to the reactive Fe(IV)=O intermediate. This intermediate then hydroxylates the substrate, yielding the product and ferrous enzyme [10].

Phenylalanine hydroxylase is the most important and thoroughly studied pterin monooxygenase since mutations

Metalloenzymes. https://doi.org/10.1016/B978-0-12-823974-2.00028-0

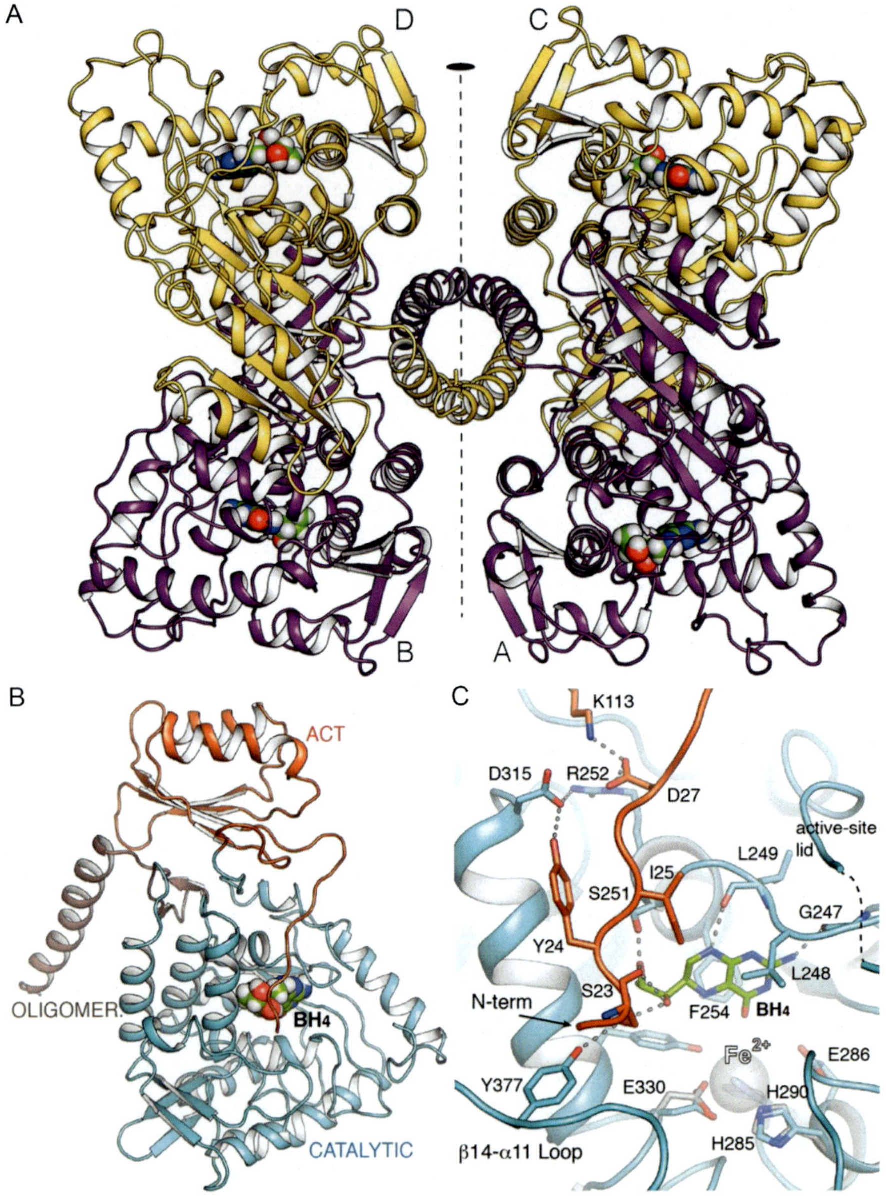

FIG. 1 Crystal structure of human PAH. (A) Ribbon representation of human PAH in complex with BH_4 at the four active sites. The tetramer is formed as a dimer of dimers by a twofold symmetry (depicted). BH_4 cofactor is drawn as spheres. (B) The human PAH monomer with each domain colored differently. (C) Detailed view of the active site of human PAH in complex with BH_4 (*green* sticks). Relevant residues are represented as capped sticks and labeled. *Reproduced with permission from Flydal MI, Alcorlo-Pagés M, Johannessen FG, Martínez-Caballero S, Skjærven L, Fernandez-Leiro R, Martinez A, Hermoso JA. Structure of full-length human phenylalanine hydroxylase in complex with tetrahydrobiopterin. Proc Natl Acad Sci U S A 2019;116(23):11229-11234.*

in its sequence are responsible for phenylketonuria (PKU), a misfolding genetic disorder where loss of enzymatic function is caused mainly by folding defects leading to decreased stability of the protein. Dysfunction of phenylalanine hydroxylase leads to accumulation of phenyalanine in blood and tissues and the subsequent disturbance in brain neurotransmitters causing neurological symptoms, such as mental retardation, purposeless movements, and depression [11]. Phenylketonuria treatment is based on a low phenylalanine diet and, for milder patients, supplementation with BH_4, the cofactor of the enzyme, as this compound acts as a pharmacological chaperone partly recovering the lost enzymatic activity [12]. Sapropterin dihydrochloride, a synthetic form of the natural cofactor BH_4, is since 2007 a FDA-approved drug under the name of Kuvan.

On the other hand, a number of compounds with pharmacological chaperone potential for PKU mutants have been discovered. The stabilizing effect of these compounds has been established in vitro, in cells, and in animal models [13]. One compound (named IV) was identified as promising alternative to BH_4 in the treatment of PKU. Compound IV (5,6-dimethyl-3-(4-methyl-2-pyridinyl)-2-thioxo-2,3-dihydrothieno[2,3-d]pyrimidin-4(1H)-one) was described to stabilize the tetrameric functional form of the enzyme in vitro and appeared to act as a canonical pharmacological chaperone, displacing the folding equilibrium toward the native form [14]. Compound IV has been crystallized in complex with PAH [15]. The crystal structure shows a direct coordination to the iron atom in PAH, in a position overlapping with both the cofactor and substrate-binding site, which explains its behavior as a weak competitive PAH inhibitor.

3 Ring cleaving dioxygenases

Many aerobic degradation pathways of aromatic compounds by microorganisms converge to catecholic substrates which can undergo an ortho cleavage, catalyzed by intradiol dioxygenases, or a meta cleavage, catalyzed by extradiol dioxygenases [16].

Although they perform similar reactions the two family are not structurally related and possess a different catalytic mechanism. Extradiol ring cleaving dioxygenases are structurally characterized by βαβββ modules and share a common metal-binding motif, the "2-His-1-carboxylate facial triad" [17]. The other three coordination positions provide sites for O_2 and the organic substrate or water ligands in the enzyme's resting state. On the other hand, the intradiol dioxygenases use Fe(III) centers to activate substrate for reaction with O_2 and a different metal coordination composed by two tyrosines and two histidines. The substrate binds as a dianion displacing one water and one tyrosine ligand. The general topology of catechol intradiol dioxygenases comprises two catalytic domains separated by a common "α-helical zipper" motif that consists of six N-terminal helices from each subunit [18,19].

However, in several bacterial degradation pathways, intermediates may be converted to para-diols or hydroxylated aromatic carboxylic acids. The oxidation of these noncatecholic substrates is catalyzed by a third class of less studied dioxygenases [20]. This class is part of the vast superfamily of cupins, one of the most functionally diverse protein classes named on the basis of a conserved β-barrel fold (also known as double-stranded β-helix or jelly-roll) [21]. In fact, they all share a common architecture composed of a motif of 6–8 antiparallel β-strands located within a conserved β-barrel structure. Members of this superfamily are quite functionally diverse and include nonenzymatic proteins and a wide variety of enzymes, like epimerases, isomerases, carboxylases, and many oxygenases. Most cupin dioxygenases either comprise a single cupin domain (monocupins) or have a duplicated domain structure (bicupins) [20].

Cupins are characterized by two motifs, originally designated as $G(X)_5\mathbf{H}X\mathbf{H}(X)_{3,4}\mathbf{E}(X)_6G$ (motif1) and $G(X)_5PXG(X)_2\mathbf{H}(X)_3N$ (motif2), although much less conserved than initially suggested. The two His residues and the Glu residue in motif 1, together with the His residue in motif 2, can act as ligands for the binding of the active site metal. In the cupin ring cleaving dioxygenases the active site is centered on an Fe(II) ion which is generally coordinated to three histidines (two from motif1 and the other from motif2) like in gentisate, 1-hydroxy-2-naphthoate and salycilate 1,2-dioxygenase or to two His and one Glu (each from motif1) facial triad, like in homogentisate 1,2-dioxygenase and 3-hydroxyanthranilate 3,4-dioxygenase [20]. The last two enzymes are found in bacteria, fungi, as well as in mammals. In humans, homogentisate 1,2-dioxygenase catalyzes the cleavage of the homogentisate ring, produced by 4-hydroxyphenylpyruvate dioxygenase in the degradation pathway of tyrosine, to yield maleylacetoacetate. Disfunction of the gene encoding for the enzyme causes Alkaptonuria (see below). On the other hand, 3-hydroxyanthranilate 3,4-dioxygenase (HAD) catalyzes the oxidative ring opening of 3-hydroxyanthranilate in the so-called kynurenine pathway of tryptophan metabolism.

Both the bacterial and the yeast HADs are homodimeric and some contain a noncatalytic rubredoxin-like site characterized by a FeS_4 center on the protein surface that appears to act as a metal storage site for the enzyme (Fig. 2) [22]. In contrast to the bacterial HADs, the human protein is a

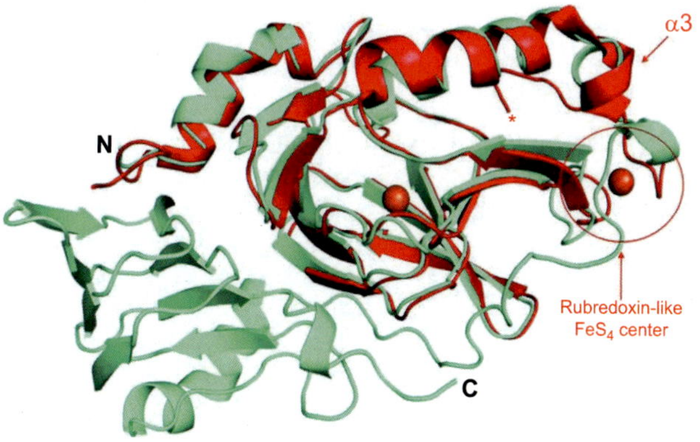

FIG. 2 Superposition of human HAD (*cyan*) and *R. metallodurans* HAD (*red*). Fe atoms are shown as *orange* spheres. The N- and C-termini of human HAD are labeled. The C-terminus of *R. metallodurans* HAD is labeled with a *red* asterisk. The locations of the rubredoxin-like FeS4 center in *R. metallodurans* HAD is indicated. *Reproduced with permission of the International Union of Crystallography from Pidugu LS, Neu H, Wong TL, Pozharski E, Molloy JL, Michel SL, Toth EA. Crystal structures of human 3-hydroxyanthranilate 3,4-dioxygenase with native and non-native metals bound in the active site. Acta Crystallogr D Struct Biol 2017;73(Pt 4): 340-348.*

monomeric bicupin lacking the rubredoxin-like center, with only the N-terminal cupin domain containing an active site [23]. Despite these structural differences, the active-site residues considered relevant for the catalytic reaction appear to be fully conserved in the prokaryotic and eukaryotic proteins. However, in contrast to the 2His-1Glu facial triad of the extradiol dioxygenases, the glutamate binds in a bidentate manner [23].

The degradation of 3-hydroxyanthranilate by HAD produces an unstable product 2-amino-3-carboxymuconic semialdehyde, which spontaneously cyclize to quinolinic acid (Fig. 3), a precursor for the de novo biosynthesis of NAD^+ for use in fundamental metabolic processes [24]. Quinolinic acid is a molecule endowed with several physiological implications. In the mammalian central nervous system, quinolinic acid binds to the *N*-methyl-d-aspartate receptor, which controls synaptic plasticity and memory functions, and thus has potent neurotoxic effects.

Quinolinic acid has been implicated in depressive and important neurodegenerative disorders, including Huntington's disease, Parkinson's disease, Alzheimer's disease, epilepsy, hepatic encephalopathy, and AIDS-related dementia [25–27]. Moreover, high levels of quinolinic acid have been detected in several inflammatory diseases and infections, as well as in both serum and cerebrospinal fluid of patients with amyotrophic lateral sclerosis (ALS) [28]. Recently, Edaravone (3-methyl-1-phenyl-2-pyrazolin-5-one), a molecule approved for the treatment of ALS, has been found to induce the suppression of quinolinic acid production through direct competitive HAD inhibition, supporting the idea of HAD involvement in the neurodegenerative disease [29].

Hence, HAD has long been recognized as a potential drug target. Nevertheless, inhibition studies on this enzyme have never taken off although it is a well characterized enzyme. Only a class of HAD inhibitors based on the 2-aminonicotinic acid 1-oxide nucleus, chemically stable and active in vivo, has been reported [30]. 4-halo-hydroxyanthranilates have also been identified as selective and potent mechanism-based inhibitors of HAD and are potentially useful drug candidates for controlling quinolinate levels [31].

4 2-Oxoglutarate-dependent dioxygenases

2-Oxoglutarate (2OG)-dependent dioxygenases catalyze a wide range of oxidative reactions. In animals, these comprise hydroxylations and *N*-demethylations, proceeding via hydroxylation; in plants and microorganisms they catalyze also cyclizations, rearrangements, desaturations, and halogenations [32–34].

Most 2OG-oxygenases employ Fe (II) as a cofactor and 2OG and O_2 as co-substrates and couples substrate oxidation to the decarboxylation of 2OG to produce succinate and CO_2.

2OG-oxygenase structures are characterized by the double-stranded β-helix core fold characteristic of cupins. The Fe(II) ion is normally bound by three residues forming the highly conserved HXD/E...H triad with the proximal histidine on β-II and the distal histidine on β-VII. The third metal-chelating residue is normally an aspartate and three water molecules generally complete the metal coordination sphere. There is variation in the oligomerization states of 2OG-oxygenases, with both monomeric and dimeric forms being common.

Most 2OG-dioxygenases employ a common reaction mechanism that starts with 2OG binding to the active site Fe(II) that promotes the binding of O_2. 2OG decarboxylation occur via a cyclic peroxy intermediate to produce succinate, carbon dioxide, and a Fe(IV)=O intermediate which then reacts with the primary substrate to cause hydroxylation (see below).

In general, the biological functions of 2OG-oxygenases in plants and microbes principally involve biosynthesis of secondary metabolites and signaling molecules. In animals, 2OG-oxygenases are involved in physiologically important processes, including collagen biosynthesis, lipid metabolism, nucleic acid repair, chromatin/transcription factor modification, and hypoxia sensing.

4-Hydroxyphenylpyruvate dioxygenase is classified as a 2OG-dependent dioxygenase based on the reaction it catalyzes, although it presents significant differences respect to the other members of the family. In humans, 4-hydroxyphenylpyruvate dioxygenase is involved in the tyrosine degradation pathway and represents an interesting

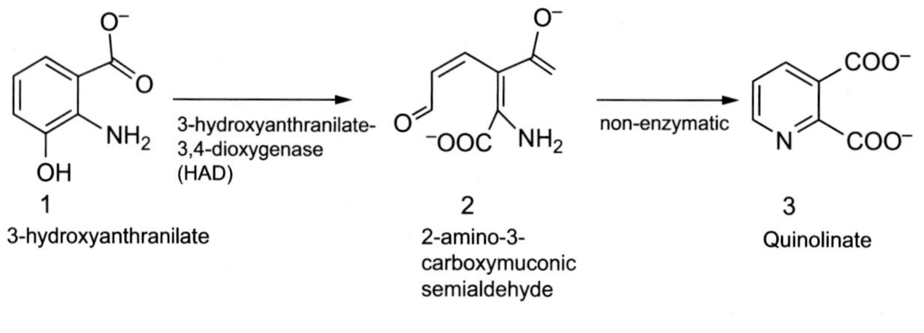

FIG. 3 The reaction catalyzed by HAD. *Reproduced with permission from Sanz I, Altomare A, Mondanelli G, Protti M, Valsecchi V, Mercolini L, Volpi C, Regazzoni L. Chromatographic measurement of 3-hydroxyanthranilate 3,4-dioxygenase activity reveals that edaravone can mitigate the formation of quinolinic acid through a direct enzyme inhibition. J Pharm Biomed Anal 2022;219:114948. Copyright 2017 American Chemical Society.*

pharmacological target for the treatment of different disorders all related to abnormalities in the tyrosine catabolism.

5 4-Hydroxyphenylpyruvate dioxygenase

4-Hydroxyphenylpyruvate dioxygenase (HPPD; EC 1.13.11.27) is a nonheme Fe(II)-dependent metalloenzyme that catalyzes the conversion of 4-hydroxyphenylpyruvate (HPPA) to 2,5-dihydroxyphenylacetate (also known as homogentisate, HGA), the second step of the L-tyrosine catabolic pathway that yields acetoacetate and fumarate (Fig. 4) [35–38]. The transformation catalyzed by HPPD is a complex reaction, which includes decarboxylation of the 2-oxo group of 4-hydroxyphenylpyruvate, accompanied by hydroxylation of the aromatic ring and 1,2-migration of the carboxymethyl group in a single catalytic cycle (Fig. 5).

HPPD belongs to the 2OG-dependent subgroup of mononuclear nonheme dioxygenases, which utilize 2-oxoglutarate as a cosubstrate to generate a high-energy iron species that is subsequently used to effect substrate oxidation [32]. Within this large family of enzymes, HPPD is an exception as it catalyzes the incorporation of both atoms of molecular oxygen into a single substrate and since the 2OG co-substrate is provided from the pyruvate substituent of HPPA. Therefore, it is not surprising that the sequences of all HPPDs reveal only low homology to other 2OG-dependent dioxygenases [39]. HPPD shares this

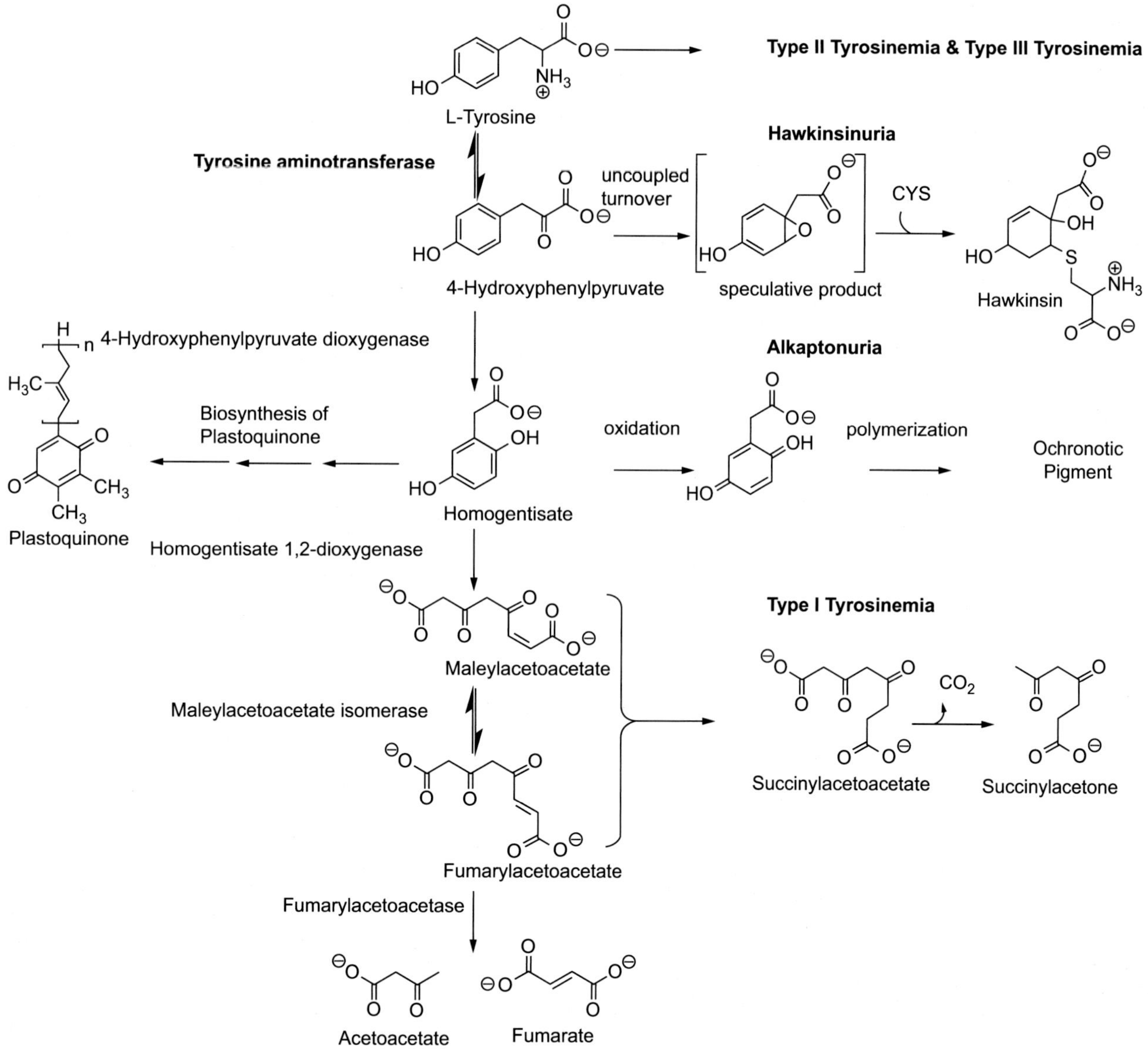

FIG. 4 L-tyrosine catabolic pathway. *Reprinted with permission from Moran GR, 4-Hydroxyphenylpyruvate dioxygenase. Arch Biochem Biophys 2005;433(1):117-128. Copyright 2005 from Elsevier.*

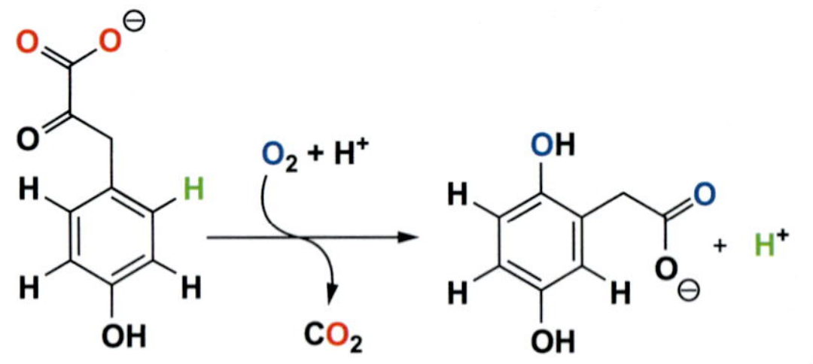

FIG. 5 The reaction catalyzed by HPPD. *Reprinted with permission from Moran GR, 4-Hydroxyphenylpyruvate dioxygenase. Arch Biochem Biophys 2005;433(1):117-128. Copyright 2005 from Elsevier.*

peculiarity only with another enzyme of the 2OG-dependent dioxygenase family, the hydroxymandelate synthase [38].

HPPD is found in nearly all aerobic organisms except for some gram-positive bacteria, although it has different functions in various organisms. In animal, HPPD plays a central role in the catabolism of tyrosine and modulates blood tyrosine levels. In humans, HPPD is mainly expressed in liver and kidney [40], and variants have been directly or indirectly involved in a number of metabolic disorders (see below). A single subunit of the human enzyme is composed of 392 amino acid residues, and the mature enzyme is a homodimer of identical subunits with an MW of 43,000 [41]. The human HPPD gene is over 30kb long, split into 14 exons [42], and it has been located to chromosome 12, region q243ter [40].

In plants, HGA formed by the action of HPPD is a key intermediate for the biosynthesis of redox cofactors such as plastoquinone and tocopherols, which are necessary compounds for the photosynthesis. Tocopherols function as antioxidants that protect the plastid membranes from lipid oxidation. Plastoquinone is a required cofactor for phytoene desaturase, a key enzyme for the biosynthesis of carotenoids, which are essential photoprotectants. Inhibition of HPPD results in the characteristic bleaching effect in plants and eventually death as chlorophyll molecules are destroyed by excessive light energy. Thus, the enzyme is a molecular target for potent herbicides [43].

6 Crystal structures of 4-hydroxyphenylpyruvate dioxygenase

The Protein Data Bank (PDB) collects crystal structures of HPPD from different organisms (*Pseudomonas fluorescens, Streptomyces avermitilis, Zea mays, Arabidopsis thaliana, Rat norvegicus,* and *Homo sapiens*) (Table 1) [44–47]. These depositions refer to the native forms of HPPD from these organisms and to the structures of complexes with the substrate HPPA [49] and with several inhibitors such as NTBC, sulcotrione, mesotrione, benziquitrione (triketones) [48], DAS645, DAS869 (pyrazoles) [46] and others. The structures of inhibitor complexes are mainly of *Arabidopsis thaliana* HPPD (*At*HPPD). Two entries referring to human HPPD structures were deposited in the PDB database (PDB code 3ISQ, 5EC3), but the structures have not yet been published.

Despite the high structural similarity shared between the HPPDs from various organisms, they differ in their oligomeric state. In fact, the HPPD from mammals, plants, and *Streptomyces avermitilis* [47] are organized as homodimers, whereas the *Pseudomonas fluorescens* HPPD is a homotetramer [44].

Each monomeric unit of the HPPD structure consists of two beta-barrel like structural domains: the flexible and diversified N-terminus and the conserved C-terminus which contains the active site, each composed by two βαβββα modules (Fig. 6A and B). Despite the mechanistic relation to 2OG-dependent dioxygenases, HPPD shares these structural features with one class of extradiol ring-cleaving dioxygenases which includes catechol-2,3-dioxygenase [54]. The N-terminal domain apparently has no direct catalytic function, even if a single mutation at position 33 in the human enzyme was thought to be responsible of Hawkinsinuria (see below) [55].

The most significant difference at the monomeric level between plant and microbial HPPDs is an insertion preceding the C-terminal helix of a long and flexible loop of fifteen residues that adopts multiple conformations in the presence or absence of ligands [46,47].

The active site of HPPD is remarkably well conserved and consists of a catalytic Fe (II) ion coordinated by two histidine and a glutamate, the "2-His-1-carboxylate facial triad" (Glu394, His226, His308 in *At*HPPD). The spacing motif between these three ligands in the primary structure is $HX_{\sim 80}HX_{\sim 80}E$, more similar to the $HX_{\sim 65}HX_{\sim 50}E$ spacing motif of the extradiol dioxygenases than to the $HX(D/E)X_{50\text{-}210}H$ of the 2OG-dependent dioxygenases. This, together with the topological similarity, led to suggest that HPPD evolution converged with that of the 2OG-dependent enzymes from an extradiol dioxygenase progenitor [37].

In addition to the residues of the facial triad motif, three water molecules coordinate the ferrous ion in the wild-type enzyme structures of the plant HPPDs, i.e., *A. thaliana* and *Z. mays* [46], giving an overall octahedral coordination. In the active site of *P. fluorescens* HPPD [44], an acetate ion complete the iron coordination sphere, while for *R. norvegicus* [47] and *S. avermitilis* HPPDs [45] only the structures of inhibitor complexes (with NBCT and DAS869, respectively) have been reported. In the human HPPD models, deposited in the PDB, two solvent molecules were placed in the iron coordination sphere, although the correspondent electron density suggests the presence of an acetate ion, as in the *P. fluorescens* HPPD, or a similar anion. In several published structures, the nature of the metal ion in the HPPD active site is uncertain since the

TABLE 1 Structures of native HPPDs from various source and their complexes deposited in the PDB.

Organism	Description	PDB code	Reference	Resolution (Å)
Pseudomonas fluorescens				
	Native	1CJX	1999 [44]	2.4
Streptomyces avermitilis				
	Complex with NTBC	1T47	2004 [45]	2.5
Rat norvegicus				
	Complex with DAS869	1SQI	2004 [46]	2.15
Arabidopsis thaliana				
	Native	1SQD	2004 [46]	1.8
	Complex with DAS645	1TG5	2004 [46]	1.9
	Complex with DAS869	1TFZ	2004 [46]	1.8
	Native	1SP9	2004 [47]	3.0
	Complex with sulcotrione	6ISD	2019 [48]	2.4
	Complex with mesotrione	5YWG	2019 [48]	2.6
	Complex with NTBC	6J63	2019 [48]	2.6
	Complex with HPPA	5XGK	2019 [49]	2.8
	Complex with Benquitrione	5YY6	2019 [49]	2.4
	Complex with 4ae (Y17107)	6JX9	2019 [50]	1.8
	Complex with methyl-benquitrione	5YWK	2021 [51]	2.8
	Complex with 60	6M6D	2021 [51]	1.8
	Complex with 9bj (Y16542)	6LGT	2021 [52]	1.75
	Mutant in complex with HPA	7E0X	2021 [53]	1.9
Zea mais				
	Native	1SP8	2004 [47]	2.0
Homo sapiens	Co(II) derivative	3ISQ 5EC3	To be published	1.75

enzyme buffer in the crystallization was supplemented by cobaltous chloride [46,49]. A mutational study of human HPPD designed to understand the contribution of the facial triad to the enzyme function revealed that one Glu and one His residue but not two His residues are sufficient for catalytic function [56].

The HPPDs from prokaryotes and eukaryotes show a limited sequence identity of about 30%. However, the iron center is surrounded by an almost strictly conserved environment that is clearly dominated by hydrophobic residues [47]. In fact, sequence alignments of HPPDs from a wide variety of sources show that 44 residues out of more than 350 residues are strictly conserved and that 41% of the conserved residues are clustered around the metal-binding site [35].

7 Catalytic mechanism of 4-hydroxyphenylpyruvate dioxygenase

The HPPD mechanism has been extensively investigated using a combination of structural, spectroscopic, kinetic, and computational studies.

The physiological reaction of 4-HPPD involves 2-oxoglutarate decarboxylation, phenyl ring hydroxylation, and sidechain rearrangement (Fig. 5). In other words, the pyruvic side chain is transformed into an acetic acid group, which migrates to the adjacent carbon of the ring, while the original position of the pyruvate binding is hydroxylated. In this complex transformation catalyzed by HPPD, the two oxygen atoms from molecular oxygen are incorporated into HGA, the 2-hydroxyl and one of the carboxylate oxygens.

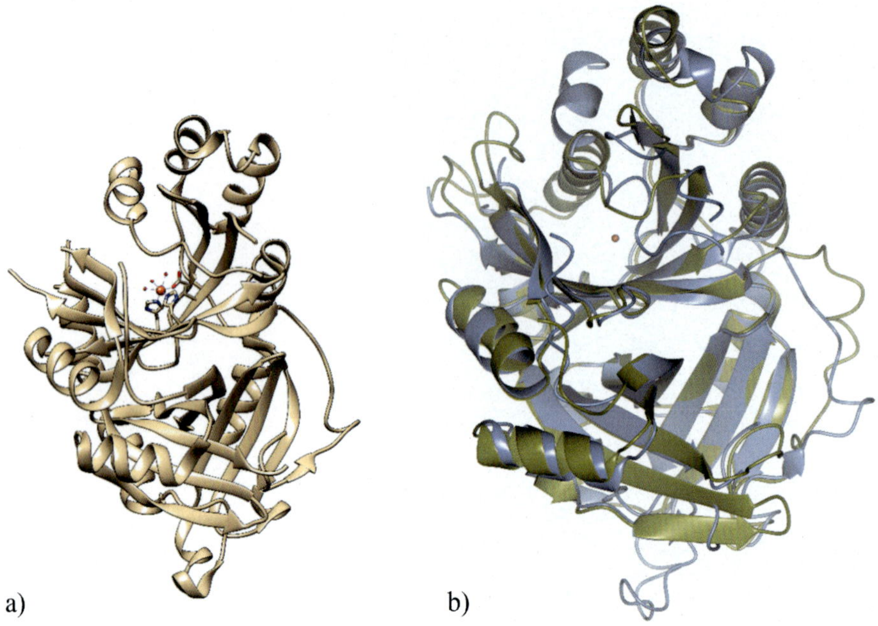

FIG. 6 (A) Crystallographic structure of one *At*HPPD subunit. The Fe(II) ion and the coordinating residues are also shown; (B) superposition of one subunit of *At*HPPD with that of human HPPD. The Fe(II) ion is represented as an *orange* sphere.

Steady-state kinetic experiments indicate an ordered Bi Bi mechanism in which HPPA is the first substrate to bind and carbon dioxide is the first product to dissociate [57].

According to the most accepted mechanism (Fig. 7), the catalysis is initiated by the bidentate coordination of HPPA with the ferrous ion in the HPPD active site. It is recognized that HPPA has an activating effector role for the reaction of HPPD ferrous ion with molecular oxygen [58]. Then the dioxygen nucleophilic attack to the keto carbon of HPPA generates a peracid intermediate which undergoes decarboxylation by heterolytic cleavage with release of CO_2. A key intermediate HPPD-Fe(IV)=O-HPA (4-hydroxyphenylacetic acid) is produced after the decarboxylation. C1 hydroxylation of the aromatic ring then occurs via an electrophilic attack by the Fe(IV)=O species with subsequent dearomatization. A 1,2 shift of the acetate side chain then leads to the formation of a 4-hydroxycyclohexandienone, which then undergoes a tautomerization involving a proton transfer from the sp3 carbon ring to the keto oxygen, with the formation of the final product HGA. Some biochemical studies also support the hypothesis that an arene oxide intermediate forms during the aromatic C1 hydroxylation prior to ortho-migration [59].

Concerning the residues involved in substrate binding and catalysis, QM/MM studies together with mutagenic and kinetic experiments led to the suggestion that residues Gln293, Gln307, and Gln379 could play important roles in the binding of HPPA and in the first nucleophilic attack during the catalytic cycle [59], interacting the first two with the 4-hydroxyl group of HPPA and the third with the carboxylate moiety. It was hypothesized that in the subsequent step, the intermediates move and start to interact, via the 4-hydroxyl group, with Ser267 and Asn282, and these interactions presumably play a key role in directing the electrophilic attack on the aromatic ring of HPA. Hence, the first part of the catalysis (dioxygen association and decarboxylation) and the second part (phenol ring hydroxylation and side chain migration) occur in two different zones of the active site connected by an amide-rich region. This amide-rich zone, consisting of four residues Gln307, Gln293, Asn282, and Ser267, is proposed to rotate the HPA intermediate, using the hydroxyl group as a pivot, to a final position where the benzylic carbon is at a favorable distance from the oxo atom and HGA is generated. Mutation of these residues was found to block the transport of HPA and suppress HGA conversion, yielding HPA as alternative product. Moreover, in the structure of the *At*HPPD mutant Ser267Trp, the HPA intermediate was captured in the active site of HPPD, supporting the decisive role of the amide-rich zone during the later stage of the HPPD catalysis [53].

The X-ray structure of the complex of *At*HPPD with the natural substrate HPPA has been recently determined [49]. In the active site, the pyruvate moiety of HPPA coordinates the metal ion in a bidentate mode, resulting in an octahedral coordination geometry involving the facial triad and a water molecule (Fig. 8). Differently from previous studies, the structure shows that the phenolic hydroxyl of HPPA forms a hydrogen bond with the side chain of Asn423. In addition, the benzene ring of HPPA forms a T-π stacking interaction with Phe381 and a weak hydrophobic interaction with other

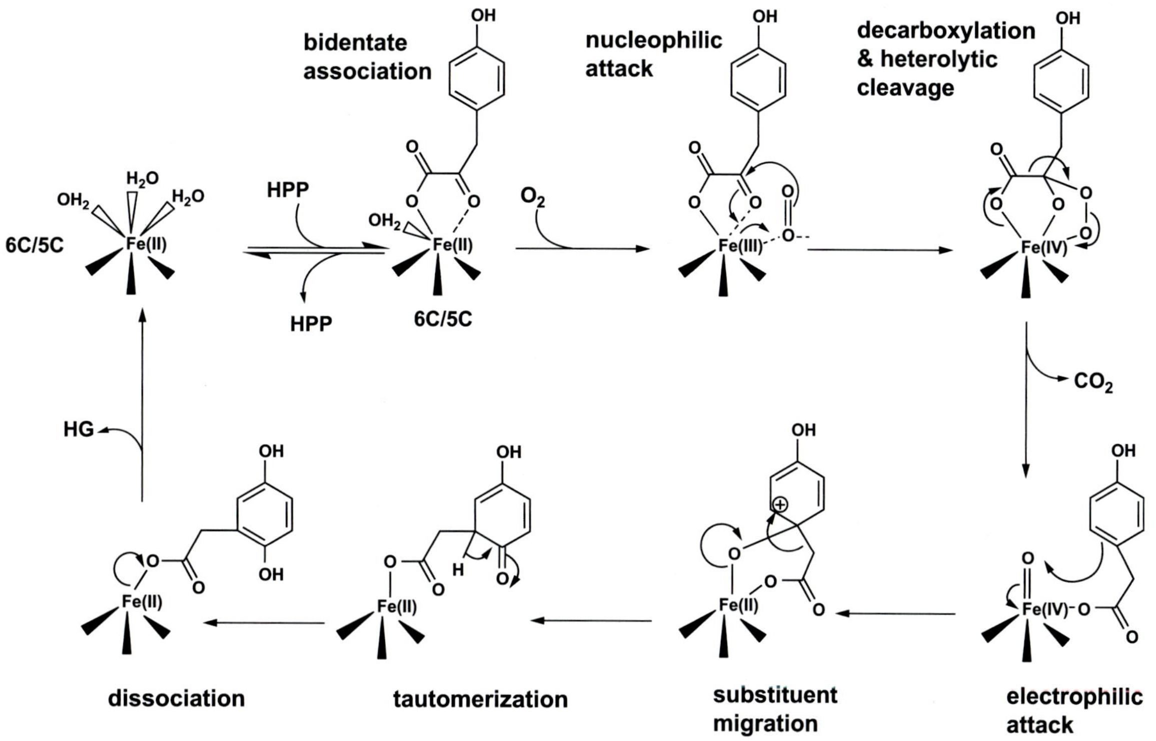

FIG. 7 Catalytic mechanism of HPPD. *Reprinted with permission from Moran GR, 4-Hydroxyphenylpyruvate dioxygenase. Arch Biochem Biophys 2005;433(1):117-128. Copyright 2005 from Elsevier.*

surrounding residues such as Leu368 and Leu427. Significant conformational changes were revealed upon HPPA binding, of which the movement of residue Gln293 may be regarded as the most prominent feature. In fact, Gln293 undergoes a significant rotation forming a H-bond with Gln307, which triggers the generation of the H-bond network of Ser267-Asn282-Gln307-Gln293. Moreover, after the binding of HPPA, the formation of a H-bond between Asn423 and the hydroxyl group of HPPA led to the disappearance of the H-bond between Gln379 and Asn423 present in the native structure. Site-directed mutational studies of the residues interacting with the substrate

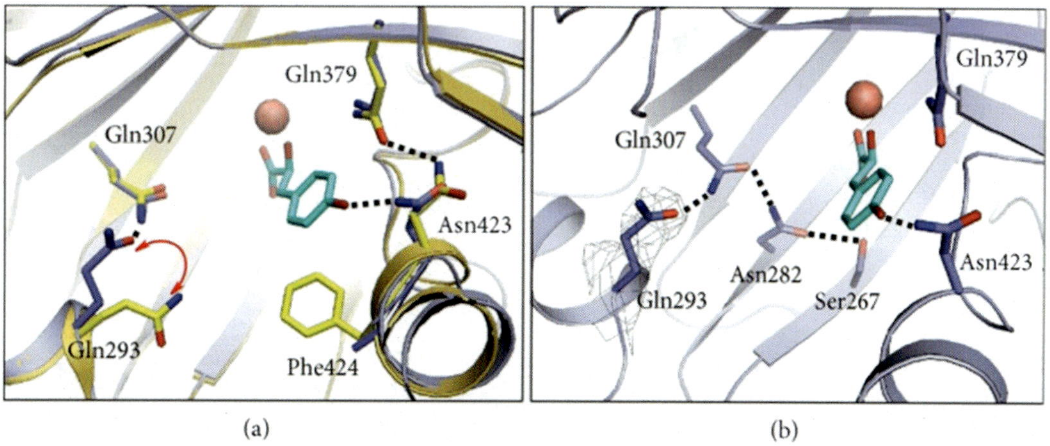

FIG. 8 (A) Superposition of the active sites of native *At*HPPD (*yellow*) and *At*HPPD-HPPA complex (*light blue*). (B) The H-bond network in the active site of *At*HPPD-HPPA complex. The electron densities (2Fo-Fc) map corresponding to Gln293 is contoured at 1.0 σ. *Reprinted with permission from Lin H-Y, Chen X, Chen J-N, Wang D-W, Wu, Lin S-Y, Zhan C-G, Wu J-W, Yang W-C, Yang G-F. Crystal structure of 4-hydroxyphenylpyruvate dioxygenase in complex with substrate reveals a new starting point for herbicide discovery. Research, 2019;11:2019. Copyright © 2019 Exclusive licensee Science and Technology Review Publishing House. Distributed under a Creative Commons Attribution License (CC BY 4.0).*

(Asn423Ala and Phe381Ala) and those forming the H-bond network that stabilizes the HPPA conformation (Gln293Ala, Gln307Ala, and Ser267Ala, Asn282Ala) confirmed their involvement in the catalytic mechanism [49].

8 Classes of HPPD inhibitors: Triketones, pyrazoles, and isoxazoles

The design and development of HPPD inhibitors found a principal application in the field of herbicides for weed control, since HPPD inhibition prevents the production of HGA, precluding the synthesis of tocopherols and plastoquinone that results in bleaching of the plants due to diminished chlorophyll levels.

To date, three main classes of inhibitors of HPPD are known: triketones, pyrazoles, and isoxazoles. Some of them were developed based on naturally occurring allelopathic compounds produced by a number of plants and lichens which are HPPD inhibitors, such as leptospermone and usnic acid (Fig. 9) [60]. In the early eighties, Pyrazolinate and Pyrazoxyfen were developed for weed control in rice. The two compounds are pro-drugs and release the free hydroxypyrazole, which binds to the HPPD enzyme. In 1982, the discovery of the triketone-type HPPD inhibitors, which are related to the natural product leptospermone, brought to the launch on the market of two prominent herbicides used in fields worldwide sulcotrione and later mesotrione. In the late 1980s, an interesting new lead structure was identified that, through optimization, led to isoxaflutole, the first isoxazole HPPD inhibitor patented in 1991. The isoxazoles are also prodrugs that owe their HPPD inhibitory activity to the diketonitrile group that is formed upon ring opening [61].

Since then several other inhibitors have been synthesized and some of them were registered as pesticide and have reached the market for the treatment of corn, rice, wheat, and maize, such as tefuryltrione, tembotrione, bicyclopyrone, benzobycyclon, topramezone, and pyrasulfotole (Fig. 10) [36,60].

The crystallographic structures of *At*HPPD complexes with the two pyrazole inhibitors DAS645 and DAS869 were first determined, which disclosed the HPPD inhibitor mode of binding and inhibition mechanism [46]. The inhibitors are coordinated to the metal ion in the active site in a bidentate mode by the 1,3-diketone moiety, displacing two water molecules in the metal coordination sphere (Fig. 11). No hydrogen bonds or ionic interactions involving the inhibitors were found for complex stabilization. The benzoyl group of the inhibitors is sandwiched between Phe424 and Phe381 forming π-stacking interactions. The N-terbutyl substituent on the DAS869 pyrazole ring interacts with Pro280. The DAS645 inhibitor has a bulkier 3-(2,4-dichlorophenyl) substituent on the pyrazole ring, compared to DAS869. For that reason, the side chain of Phe424 rotates away, because of the steric hindrance, to make space for the substituent. The inhibitor DAS645 is selective for the plant enzyme (IC_{50} 12 nM) versus mammalian HPPD, i.e., no significant inhibition was found against Rat norvegicus HPPD (*Rn*HPPD). In fact, in *Rn*HPPD Phe424 cannot rearrange in order to avoid collision with DAS645 since it is positioned on the C-terminal α-helix that is less flexible in *Rn*HPPD than in *At*HPPD.

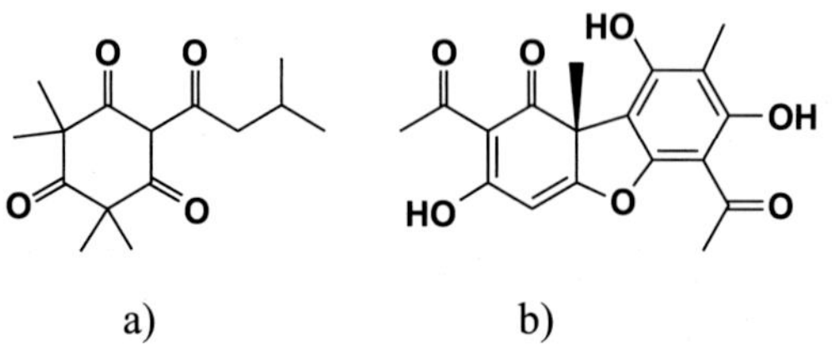

FIG. 9 Natural inhibitors of HPPD: (A) Leptospermone is produced by oil producing plants such as the Australian bottle brush; (B) Usnic acid is produced by lichens.

Later, the crystallographic structures of *At*HPPD with the triketone-based inhibitor NTBC (Fig. 12), the first effective drug approved by FDA to treat hereditary type I tyrosinemia (see below), and the two commercial herbicides sulcotrione and mesotrione were also solved [48]. As observed in the previous structures, the two carbonyl groups of the triketone inhibitors are coordinated to the metal ion and the phenyl ring forms π-stacking with Phe381 and Phe424. Two important differences were identified around the HPPD active site upon binding of the triketone inhibitor: (i) the β-strand 250–253 becomes a loop and rotate roughly 30 degrees (this conformational change stabilizes the *At*HPPD N-terminal loop, a five-residue fragment which is disordered in wild-type *At*HPPD) and (ii) the side chain of Phe428 rotate in order to avoid steric clashes with the inhibitor. This conformational change of Phe428 was supposed to be involved in the generally slow-binding inhibition of HPPD, and it was confirmed by site-directed mutagenesis and computational simulations [48]. However, it is worth to note that the Phe428 movement is a general feature of HPPD inhibition, and it is observed also in the structure of the pyrazole complexes.

Many studies have reported that the modification of the aromatic part of triketone derivatives is an effective way to obtain new inhibitors with improved potency. In recent years, fenquinotrione, the first commercial triketone HPPD inhibitor with a bicyclic core skeletal structure, demonstrated excellent potential as a herbicide [62]. Therefore, also HPPD inhibitors based on bicyclic skeletal structures have been widely exploited. For examples, triketone-based hybrid compounds were synthesized replacing the benzene ring of the benzoyl cycloexane-1,3-dione scaffold with a quinazoline-2,4-dione [51,63,64], quinoline [65], quinoxaline [66], aminopyridine [67], benzoxazinone [68], and a phenoxymethyl moieties [69,70]. New molecular scaffolds were also designed and synthesized by hybridizing the pyrazol-1,3 dione with isoindoline-1,3-dione [50], quinazoline-2,4-dione [52], quinazolone [71], and

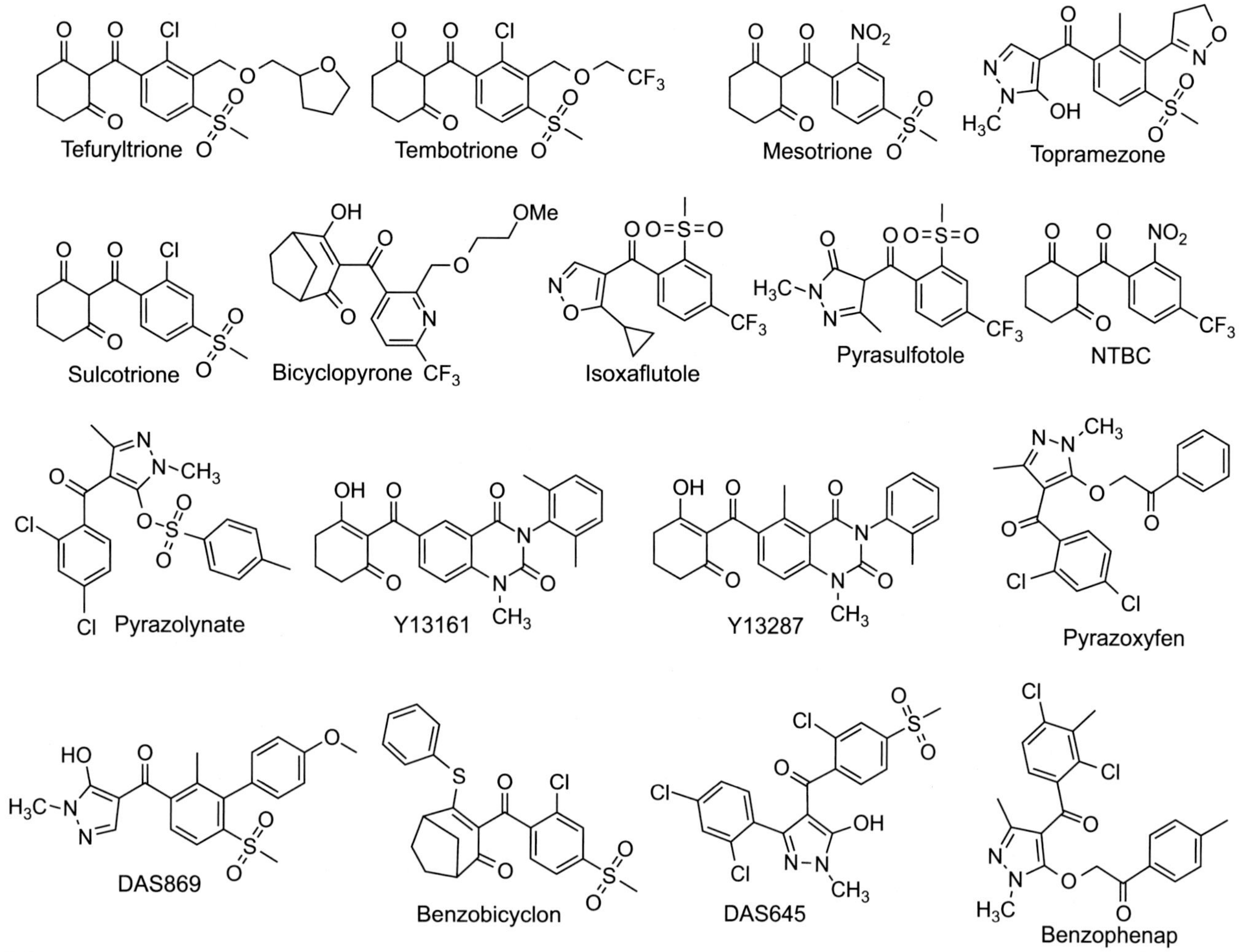

FIG. 10 Chemical structures of common commercial HPPD inhibitors. Some of them are used as ligands in current 3D structures. *Reprinted with permission from Ndikuryayo F, Moosavi B, Yang W-C, Yang G-F, 4-Hydroxyphenylpyruvate dioxygenase inhibitors: from chemical biology to agrochemicals. J Agric Food Chem 2017;65 (39):8523-8537. Copyright 2017 American Chemical Society.*

benzimidazolone [72] fragments. Otherwise, starting from the (2-benzoylethen-1-ol) triketone substructure, compounds containing 1,2-benzothiazine derivatives were also synthesized [73].

These modification brought to the discovery of compounds showing very high inhibition activity. Recently, the first subnanomolar-range *At*HPPD inhibitor was identified (Ki = 0.86 nM) [51], optimizing the hydrophobic interactions between side chains of the core structure of quinazoline-2,4-dione in triketone-hybrid derivatives and a hydrophobic pocket at the active site entrance of *At*HPPD.

9 Disorders associated with tyrosine metabolism

Genetic disruption of the enzymes involved in the tyrosine catabolism is linked to various human metabolic diseases. They arise when gene mutations render specific enzymes of the pathway inactive or catalytically altered. In particular, alteration in the structure and activity of HPPD is causally related to two of these different metabolic disorders: type III tyrosinemia and Hawkinsinuria (Fig. 4).

Type III tyrosinemia is a rare disorder characterized by elevated serum tyrosine levels and massive excretion of tyrosine derivatives into urine. Patients with this disorder have neurological symptoms and mental retardation. On the other hand, chronic acidosis and growth failure in infancy are common in Hawkinsinuria, characterized by the excretion of a sulfur amino acid identified as (2-amino-3-{[2-(carboxymethyl)-2,5-dihydroxy-1-cyclohex-3-enyl]sulfanyl}propanoic acid), which is known as "hawkinsin," from the surname of the first patient described in the literature [74]. Symptoms that occur in Hawkinsinuria are transient and improve within the first year of life [75].

Tyrosinemia-III is an autosomal recessive disorder and in patients both alleles encode for an HPPD catalytically inactive variant. Conversely, Hawkinsinuria is inherited

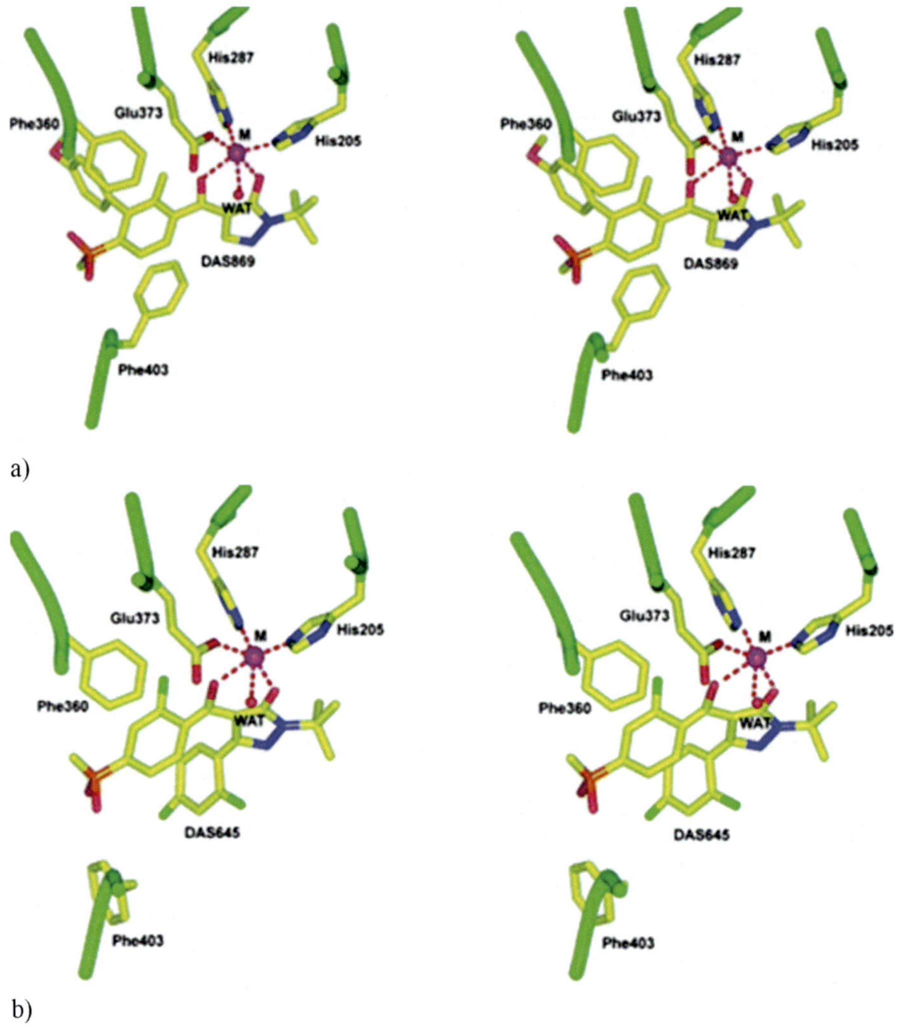

FIG. 11 Stereodiagrams of the active sites of the (A) *At*HPPD-DA869 and (B) *At*HPPD-DA645 structures. Two water molecules are displaced by the 1,3-diketone moiety from the inhibitor in the ligand-bound structures. Two phenylalanine residues also form a π-stacking interaction with the inhibitor. The phenyl ring of Phe403 rotates away to avoid steric clash when DAS645 binds to the *At*HPPD active site (the *At*HPPD residue numbering is shifted by -21 with respect to numbering in the other figures and main text). *Reprinted with permission from Yang C, Pflugrath JW, Camper DL, Foster ML, Pernich DJ, Walsh TA. Structural basis for herbicidal inhibitor selectivity revealed by comparison of crystal structures of plant and mammalian 4-hydroxyphenylpyruvate dioxygenases. Biochemistry, 2004;43 (32):10414-10423. Copyright 2004 American Chemical Society.*

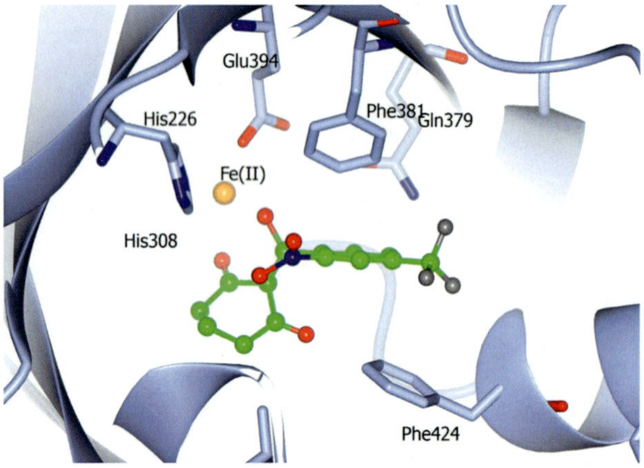

FIG. 12 Active site of the *At*HPPD-NTBC complex.

as an autosomal dominant disorder. In this case, the HPPD variant converts the substrate but fails to produce HGA, yielding an alternative reactive product known to form a covalent adduct with cellular thiols. In fact, the HPPD variant is capable of decarboxylation and oxidation but cannot catalyze the final rearrangement step, producing a reactive epoxide which will then dissociate from the enzyme and react with cytoplasmic components such as glutathione (to form hawkinsin) or water (to form 4-hydroxycyclohexylacetic acid) [76].

Tomoeda et al. [77] identified a homozygous missense mutation predicting an Ala to Val change at codon 268 (Ala268Val) in the HPPD gene in a patient with tyrosinemia type III. Moreover, in the HPPD gene of two patients with Hawkinsinuria, a heterozygous missense mutation, predicting an Ala to Thr change at codon 33 (Ala33Thr), was also found. This mutation is in a region of the gene that would result in an N-terminal defect of the protein; however, it was subsequently shown that the Ala33Thr mutation is a frequent polymorphism and not the causative mutation [78].

Later, a case emerged in which an individual suffering of Hawkinsinuria presented a previously unreported missense mutation encoding an Asn to Ser switch at position 241, a well-conserved residue in all known HPPDs. Furthermore, since both the rat and the Streptomyces avermitilis Asn241Ser enzyme variants yield quinolacetic acid as the primary product it was speculated that this compound or one of its by-products can cause the pathology of Hawkinsinuria and the Asn241Ser mutation was therefore identified as causative for Hawkinsinuria [79].

Other three hereditary human diseases, namely type I tyrosinemia, type II tyrosinemia, and alkaptonuria, occur due to misregulation of previous and later steps in tyrosine catabolism.

Tyrosinemia type II (also known as Richner-Hanhart Syndrome) is an extremely rare hereditary disease that is caused by an inborn defect in the gene encoding tyrosine aminotransferase (TAT), which blocks the transamination reaction converting tyrosine to 4-hydroxyphenylpyruvate, the first step of the tyrosine pathway. Tyrosinemia type I patients suffer from keratitis and palmar and plantar hyperkeratosis, accompanied by drastically elevated serum and urine levels of tyrosine and its metabolites. Mild or severe mental retardation may also be present [80,81]. To date, only 143 cases and 33 disease-associated variant of the TAT gene are reported in the literature [82].

Alkaptonuria is the oldest known inherited disease [83]. It was discovered by Sir Archibald Garrod, a founding father of inherited metabolic disorders. In 1899, he first postulated that Alkaptonuria was due to a chemical aberration which he believed was congenital.

Alkaptonuria is a rare autosomal recessive genetic disorder, which results from low activity of homogentisate 1,2 dioxygenase (HGD) due to mutations in the gene encoding for the enzyme. In healthy individuals, HGD converts HGA, produced by HPPD activity, to maleylacetoacetic acid. In Alkaptonuria patients, it was found that HGA first undergo spontaneous oxidation into 1,4-benzoquinone-2-acetic acid, and then polymerization into a compound of undefined composition and properties, with the production of a melanin-like polymer which accumulate in the collagen. This process of deposition is called ochronosis. Ochronotic pigment deposits bind to connective tissues of various organs leading to the destruction of joints, valves, and vessels. Hence, homogentisic aciduria, ochronosis, and arthritis-like damage constitute the major symptoms characterizing Alkaptonuria [84].

Tyrosinemia type I, the most serious of these diseases, is a rare autosomal recessive genetic disease caused by an inactive form of fumarylacetoacetate hydrolase, the last enzyme in the tyrosine catabolism pathway, which cause an accumulation of fumarylacetoacetone [80,85]. This beta-keto acid readily saturates and/or decarboxylates to form succinylacetoacetate and succinylacetone. Then, the keto groups of these molecules can react with cellular thiols [86]. As such, Tyrosinemia type I patients suffer of severe liver and kidney dysfunction likely due to mutagenic effects and influences on the cell cycle by accumulated metabolites. The pathology has an average estimated incidence of 1 in 100,000 births worldwide, but the highest incidence is recorded in specific geographical areas such as Quebec [87]. Symptomatology appears in the first few months of life and causes renal tubular dysfunction, neurologic crises, rickets and liver failure, with the long-term risk of development of hepatocellular carcinoma. Untreated children often do not survive past the age of 10 [88]. Liver transplantation was the only medical option to prevent liver cancer until Lindstedt et al. [89] demonstrated that the levels of excreted markers for the condition could be brought to normal levels using NTBC, an effective inhibitor of HPPD, as a drug. Thus, they proposed that NTBC could be used to treat Tyrosinemia type I, since it prevents substrate from reaching fumarylacetoacetate hydrolase, impedes the accumulation of fumarylacetoacetate and its eventual conversion to succinylacetone.

10 Therapeutical uses of NTBC and other human HPPD inhibitors

NTBC, 2-[2-nitro-4-(trifluoromethyl)benzoyl]-1,3-cyclohexanedione, [90] represents an excellent example of repurposing of an active ingredient, well before the concept of "drug repositioning" [91,92] was conceived.

NTBC was one of the early discovered HPPD inhibitors, but it was not launched on the market as a herbicide due to prolonged retention in human metabolism and (at the time) incomplete toxicity profiles [93,94]. Later, NTBC has become the paradigm triketone HPPD inhibitor with numerous known and potential therapeutic functions. In fact, NTBC is the first effective drug approved by the United States FDA and used for over 25 years to clinically treat hereditary type I tyrosinemia [95,96].

In 1995 NTBC was designated an Orphan drug by the US Office for Orphan Product Development and attained FDA (2002) and European approval (2005) under the drug name Nitisinone (Ordafin), for use "under exceptional circumstances" as a lifesaving agent in infant patients of hereditary type I tyrosinemia. Orfadin is currently a proprietary product of Swedish Orphan Biovitrum AB (Sobi) [97].

As an inhibitor of HPPD, NTBC can reduce the formation of HGA. Based on this action, it was expected that treatment of Alkaptonuria patients with NTBC may also reduce the consequences of this disorder. Hence, as an approved drug for type I tyrosynemia, Nitisinone was also originally used as an off-label drug for the treatment of Alkaptonuria [97]. Later, four clinical trials have been undertaken to explore the use of NTBC for treating Alkaptonuria (NCT00107783; NCT01390077; NCT01828463, SONIA1; and NCT01916382, SONIA2). Each study showed that NTBC is well tolerated and effective in reducing urinary excretion of HGA, decreasing ochronosis and improving clinical signs, indicating a slower disease progression [98–100]. Nevertheless, these studies emphasized that patients treated with nitisinone, as a consequence of chronic high plasma levels of tyrosine, may develop ophthalmologic complications like keratopathy, liver failure, convulsions, and cognitive difficulties [101,102]. These symptoms, which mimics in part those of Tyrosinemia type II and III, are due to HPPD inhibition which has the side effect of driving the reversible reaction catalyzed by tyrosine aminotransferase toward the production of tyrosine, increasing the blood tyrosine concentration.

In September 2020, nisitinone has been officially approved by EMA for the treatment of adults affected by Alkaptonuria [103] (https://www.ema.europa.eu/en/medicines/human/EPAR/orfadin).

Nitisinone has also been indicated as a therapeutic candidate for the treatment of Hawkinsinuria. In fact, it has been shown that also the Asn241Ser HPPD variant responsible for the symptoms of Hawkinsinuria is susceptible to NTBC inhibition as the wild-type HPPD and should block the production of quinolacetic acid in patients with this genetic defect [74].

At present, it is unknown if this unique drug will be appropriate to prevent problems for a long-term period of time, suggesting the need for the discovery and development of new treatments. Nevertheless, although the herbicide field has attracted significant efforts in the last three decades, the development of more effective drugs with less severe side effects than NTBC is just at an initial stage.

In two recent studies, NTBC, mesotrione, sulcotrione, tembotrione, and other commercial compounds with similar structures to NTBC were analyzed as human HPPD inhibitors by computational method, in vitro inhibition, and in vivo toxicologic experiments [98,104]. Mesotrione, tembotrione, and another commercial compound were found to be promising in terms of IC_{50}, LD_{50}, and also tyrosine accumulation [98]. In silico experiments also found that mesotrione has higher affinity for hHPPD in terms of binding free energy compared to NTBC, sulcotrione, and tembotrione [104]. In this study, Ser226, Asn241, Gln265, Phe336, Phe359, and Phe364 (Ser267, Asn282, Gln307, Phe381, Phe419, and Phe424 in *At*HPPD) were identified as the residues involved in the interaction between NTBC and hHPPD. Moreover, also Tyr221 (which has not a counterpart in *At*HPPD) and Leu224 (Leu265 in *At*HPPD) were found as key residues in determining the selective binding affinity to hHPPD, whose mutation into Ala caused a significant decrease of NTBC-binding ability.

Analyzing the molecular docking of NTBC into the active site of hHPPD homology model, Ndikuryayo et al. [105] found an empty hydrophobic region surrounded by Phe347 (Phe392 in *At*HPPD) residue. Thus, they hypothesized that a π-π stacking interaction between Phe347 and inhibitor may increase the hydrophobic interaction with this hHPPD pocket, and thus improve the inhibitory activity (Fig. 13). Guided by this rationale, they prepared a series of new 2-(3-(benzyloxy)benzoyl)-3-hydroxycyclohex-2-en-1-one analogs, bearing a benzyloxy substituent at the meta-position of the benzene ring of triketone moiety. As expected, a number of compounds exhibited improved activity, and compound d23 (IC_{50} 0.047 mM), the most

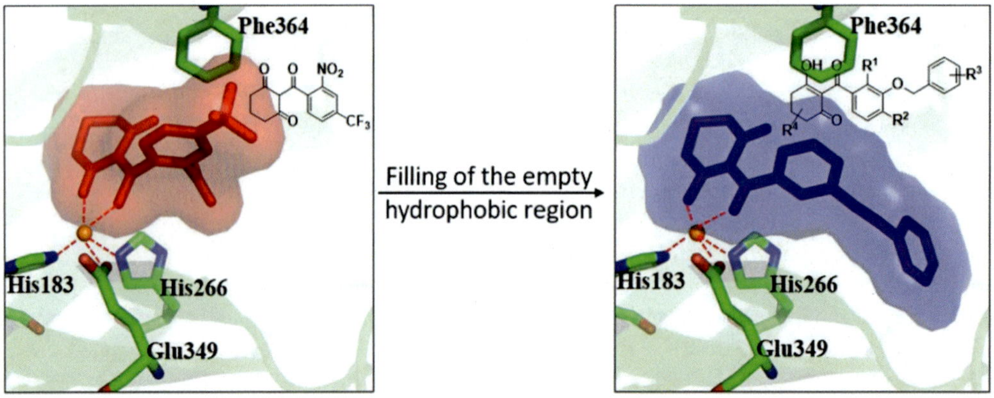

FIG. 13 Design strategy of the 2-(3-(benzyloxy)benzoyl)-3-hydroxycyclohex-2-en-1-one analogs as new human HPPD inhibitors. *Reprinted with permission from Ndikuryayo F, Kang WM, Wu FX, Yang WC, Yang GF. Hydrophobicity-oriented drug design (HODD) of new human 4-hydroxyphenylpyruvate dioxygenase inhibitors. Eur J Med Chem 2019;166:22-31. Copyright 2019 from Elsevier.*

active candidate, showed about twofold higher potency than NTBC (IC_{50} 0.085 mM) and lower binding energy.

Also, a series of hybrids of pyrazole-benzimidazolone and -quinazolone were rationally designed, synthesized, and evaluated as hHPPD inhibitors, leading to the discovery of several novel inhibitors with excellent inhibitory potency [71,72].

Nitisinone has also been proposed for the treatment of oculocutaneous albinism type 1B (OCA-1B), a genetic disease that is related to a decreased production of melanin due to defect of the TYR gene which produces tyrosinase, an essential enzyme in the melanogenesis, catalyzing the hydroxylation of tyrosine to dihydroxyphenylalanine (DOPA) [106]. In OCA-1B patients, a reduced tyrosinase activity leads to hypopigmentation of skin, hair, and eyes with characteristic ocular abnormalities. Onojafe et al. first demonstrated that administering nitisinone in transgenic murine OCA-1B model elevates plasma tyrosine levels and increases coat and ocular pigmentation. They also suggested that elevated tyrosine concentrations might stabilize the tyrosinase enzyme enough to increase its enzymatic activity [107].

More recently, the same research group reported a human clinical trial designed to determine if nitisinone is effective also in humans, and if an increase in pigmentation can improve visual function. The results of this pilot study were that nitisinone did not increase melanin content in iris but increased hair and skin pigmentation in patients with OCA-1B [108].

Currently, HPPD inhibitors are also evaluated as insecticide to kill blood-feeding arthropods that are vectors of infectious diseases such as dengue, Zika, Chagas disease, and malaria, as an alternative less toxic and more environmentally friendly than the conventional neurotoxic insecticides [109–111].

References

[1] Peck SC, van der Donk WA. Go it alone: four-electron oxidations by mononuclear non-heme iron enzymes. J Biol Inorg Chem 2017;22:381–94.

[2] Wang Y, Li J, Liu A. Oxygen activation by mononuclear nonheme iron dioxygenases involved in the degradation of aromatics. J Biol Inorg Chem 2017;22:395–405.

[3] Ullrich R, Hofrichter M. Enzymatic hydroxylation of aromatic compounds. Cell Mol Life Sci 2007;64:271–93.

[4] Buongiorno D, Straganz GD. Structure and function of atypically coordinated enzymatic mononuclear non-heme-Fe(II) centers. Coord Chem Rev 2013;257(2):541–63.

[5] Torres Pazmiño DE, Winkler M, Glieder A, Fraaije MW. Monooxygenases as biocatalysts: classification, mechanistic aspects and biotechnological applications. J Biotechnol 2010;146(1-2):9–24.

[6] Roberts KM, Fitzpatrick PF. Mechanisms of tryptophan and tyrosine hydroxylase. IUBMB Life 2013;65(4):350–7.

[7] Goodwill KE, Sabatier C, Marks C, Raag R, Fitzpatrick PF, Stevens RC. Crystal structure of tyrosine hydroxylase at 2.3 A and its implications for inherited neurodegenerative diseases. Nat Struct Biol 1997;4:578–85.

[8] Erlandsen H, Fusetti F, Martínez A, Hough E, Flatmark T, Stevens RC. Crystal structure of the catalytic domain of human phenylalanine hydroxylase reveals the structural basis for phenylketonuria. Nat Struct Biol 1997;4:995–1000.

[9] Flydal MI, Alcorlo-Pagés M, Johannessen FG, Martínez-Caballero S, Skjærven L, Fernandez-Leiro R, Martinez A, Hermoso JA. Structure of full-length human phenylalanine hydroxylase in complex with tetrahydrobiopterin. Proc Natl Acad Sci U S A 2019;116(23):11229–34.

[10] Fitzpatrick PF. Tetrahydropterin-dependent amino acid hydroxylases. Annu Rev Biochem 1999;68:355–81.

[11] Flydal MI, Martinez A. Phenylalanine hydroxylase: function, structure, and regulation. IUBMB Life 2013;65(4):341–9.

[12] Underhaug J, Aubi O, Martinez A. Phenylalanine hydroxylase misfolding and pharmacological chaperones. Curr Top Med Chem 2012;12(22):2534–45.

[13] Hole M, Jorge-Finnigan A, Underhaug J, Teigen K, Martinez A. Pharmacological chaperones that protect tetrahydrobiopterin dependent aromatic amino acid hydroxylases through different mechanisms. Curr Drug Targets 2016;17(13):1515–26.

[14] Pey AL, Ying M, Cremades N, Velázquez-Campoy A, Scherer T, Thöny B, Sancho J, Martínez A. Identification of pharmacological chaperones as potential therapeutic agents to treat phenylketonuria. J Clin Investig 2008;118:2858–67.

[15] Torreblanca R, Lira-Navarrete E, Sancho J, Hurtado-Guerrero R. Structural and mechanistic basis of the interaction between a pharmacological chaperone and human phenylalanine hydroxylase. Chembiochem 2012;13:1266–9.

[16] Vaillancourt FH, Bolin JT, Eltis LD. The ins and outs of ring-cleaving dioxygenases. Crit Rev Biochem Mol Biol 2006;41 (4):241–67.

[17] Ohlendorf DH, Orville AM, Lipscomb JD. Structure of protocatechuate 3,4-dioxygenase from Pseudomonas aeruginosa at 2.15 A resolution. J Mol Biol 1994;244(5):586–608.

[18] Ferraroni M, Kolomytseva M, Scozzafava A, Golovleva L, Briganti F. X-ray structures of 4-chlorocatechol 1,2-dioxygenase adducts with substituted catechols: new perspectives in the molecular basis of intradiol ring cleaving dioxygenases specificity. J Struct Biol 2013;181(3):274–82.

[19] Ferraroni M, Solyanikova IP, Kolomytseva MP, Scozzafava A, Golovleva L, Briganti F. Crystal structure of 4-chlorocatechol 1,2-dioxygenase from the chlorophenol-utilizing gram-positive Rhodococcus opacus 1CP. J Biol Chem 2004;279(26): 27646–55.

[20] Fetzner S. Ring-cleaving dioxygenases with a cupin fold. Appl Environ Microbiol 2012;78(8):2505–14.

[21] Dunwell JM, Purvis A, Khuri S. Cupins: the most functionally diverse protein superfamily? Phytochemistry 2004;65(1):7–17.

[22] Zhang Y, Colabroy KL, Begley TP, Ealick SE. Structural studies on 3-hydroxyanthranilate-3,4-dioxygenase: the catalytic mechanism of a complex oxidation involved in NAD biosynthesis. Biochemistry 2005;44(21):7632–43.

[23] Pidugu LS, Neu H, Wong TL, Pozharski E, Molloy JL, Michel SL, Toth EA. Crystal structures of human 3-hydroxyanthranilate

3,4-dioxygenase with native and non-native metals bound in the active site. Acta Crystallogr D Struct Biol 2017;73(Pt 4):340–8.

[24] Colabroy KL, Begley TP. The pyridine ring of NAD is formed by a nonenzymatic pericyclic reaction. J Am Chem Soc 2005;127:840–1.

[25] Hestad K, Alexander J, Rootwelt H, Aaseth JO. The role of tryptophan dysmetabolism and quinolinic acid in depressive and neurodegenerative diseases. Biomolecules 2022;12(7):998.

[26] Stepanova P, Srinivasan V, Lindholm D, Voutilainen MH. Cerebral dopamine neurotrophic factor (CDNF) protects against quinolinic acid-induced toxicity in in vitro and in vivo models of Huntington's disease. Sci Rep 2020;10(1):19045.

[27] Schwarcz R, Okuno E, White RJ, Bird ED, Whetsell Jr WO. 3-Hydroxyanthranilate oxygenase activity is increased in the brains of Huntington disease victims. Proc Natl Acad Sci U S A 1988;85 (11):4079–81.

[28] Lee JM, Tan V, Lovejoy D, Braidy N, Rowe DB, Brew BJ, Guillemin GJ. Involvement of quinolinic acid in the neuropathogenesis of amyotrophic lateral sclerosis. Neuropharmacology 2017;112:346–64.

[29] Sanz I, Altomare A, Mondanelli G, Protti M, Valsecchi V, Mercolini L, Volpi C, Regazzoni L. Chromatographic measurement of 3-hydroxyanthranilate 3,4-dioxygenase activity reveals that edaravone can mitigate the formation of quinolinic acid through a direct enzyme inhibition. J Pharm Biomed Anal 2022;219, 114948.

[30] Vallerini GP, Amori L, Beato C, Tararina M, Wang XD, Schwarcz R, Costantino G. 2-Aminonicotinic acid 1-oxides are chemically stable inhibitors of quinolinic acid synthesis in the mammalian brain: a step toward new antiexcitotoxic agents. J Med Chem 2013;56 (23):9482–95.

[31] Colabroy KL, Zhai H, Li T, Ge Y, Zhang Y, Liu A, Ealick SE, McLafferty FW, Begley TP. The mechanism of inactivation of 3-hydroxyanthranilate-3,4-dioxygenase by 4-chloro-3-hydroxyanthranilate. Biochemistry 2005;44(21):7623–31.

[32] Islam MS, Leissing TM, Chowdhury R, Hopkinson RJ, Schofield CJ. 2-Oxoglutarate-dependent oxygenases. Annu Rev Biochem 2018;87:585–620.

[33] McDonough MA, Loenarz C, Chowdhury R, Clifton IJ, Schofield CJ. Structural studies on human 2-oxoglutarate dependent oxygenases. Curr Opin Struct Biol 2010;20(6):659–72.

[34] Rose NR, McDonough MA, King ON, Kawamura A, Schofield CJ. Inhibition of 2-oxoglutarate dependent oxygenases. Chem Soc Rev 2011;40:4364–97.

[35] Santucci A, Bernardini G, Braconi D, Petricci E, Manetti F. 4-Hydroxyphenylpyruvate dioxygenase and its inhibition in plants and animals: small molecules as herbicides and agents for the treatment of human inherited diseases. J Med Chem 2017;60: 4101–25.

[36] Ndikuryayo F, Moosavi B, Yang W-C, Yang G-F. 4-Hydroxyphenylpyruvate dioxygenase inhibitors: from chemical biology to agrochemicals. J Agric Food Chem 2017;65(39): 8523–37.

[37] Moran GR. 4-Hydroxyphenylpyruvate dioxygenase. Arch Biochem Biophys 2005;433(1):117–28.

[38] Moran GR. 4-Hydroxyphenylpyruvate dioxygenase and hydroxymandelate synthase: exemplars of the a-keto acid dependent oxygenase. Arch Biochem Biophys 2014;544:58–68.

[39] Prescott AG. A dilemma of dioxygenases (or where biochemistry and molecular biology fail to meet). J Exp Bot 1993;44:849–61.

[40] Ruetschi U, Rymo L, Lindstedt S. Human 4-hydroxyphenylpyruvate dioxygenase gene (HPD). Genomics 1997;44:292–9.

[41] Endo F, Awata H, Tanoue A, Ishiguro M, Eda Y, Titani K, Matsuda I. Primary structure deduced from complementary DNA sequence and expression in cultured cells of mammalian 4-hydroxyphenylpyruvic acid dioxygenase. Evidence for that the enzyme is a homodimer of identical subunits homologous to rat liver-specific alloantigen F. J Biol Chem 1992;267:24235–40.

[42] Awata H, Endo F, Matsuda I. Structure of the human 4-hydroxyphenylpyruvic acid dioxygenase gene (HPD). Genomics 1994;23:534–9.

[43] Schenck CA, Maeda HA. Tyrosine biosynthesis, metabolism, and catabolism in plants. Phytochemistry 2018;149:82–102.

[44] Serre L, Sailland A, Sy D, Boudec P, Rolland A, Pebay-Peyroula E, Cohen-Addad C. Crystal structure of *Pseudomonas fluorescens* 4-hydroxyphenylpyruvate dioxygenase: an enzyme involved in the tyrosine degradation pathway. Structure 1999;7(8):977–88.

[45] Brownlee JM, Johnson-Winters K, Harrison DHT, Moran GR. Structure of the ferrous form of (4-hydroxyphenyl)pyruvate dioxygenase from *Streptomyces avermitilis* in complex with the therapeutic herbicide. NTBC-Biochemistry 2004;43(21):6370–7.

[46] Yang C, Pflugrath JW, Camper DL, Foster ML, Pernich DJ, Walsh TA. Structural basis for herbicidal inhibitor selectivity revealed by comparison of crystal structures of plant and mammalian 4-hydroxyphenylpyruvate dioxygenases. Biochemistry 2004;43 (32):10414–23.

[47] Fritze IM, Linden L, Freigang J, Auerbach G, Huber R, Steinbacher S. The crystal structures of *Zea mays* and *Arabidopsis* 4-hydroxyphenylpyruvate dioxygenase. Plant Physiol 2004;134 (4):1388–400.

[48] Lin HY, Yang JF, Wang DW, Hao GF, Dong JQ, Wang YX, Yang WC, Wu JW, Zhan CG, Yang GF. Molecular insights into the mechanism of 4-hydroxyphenylpyruvate dioxygenase inhibition: enzyme kinetics, X-ray crystallography and computational simulations. FEBS J 2019;286:975–90.

[49] Lin H-Y, Chen X, Chen J-N, Wang D-W, Wu LS-Y, Zhan C-G, Wu J-W, Yang W-C, Yang G-F. Crystal structure of 4-hydroxyphenylpyruvate dioxygenase in complex with substrate reveals a new starting point for herbicide discovery. Research 2019;11:2019.

[50] He B, Dong J, Lin H-Y, Wang M-Y, Li X-K, Zheng B-F, Chen Q, Hao G-F, Yang W-C, Yang G-F. pyrazole-isoindoline-1,3-dione hybrid: a promising scaffold for 4-Hydroxyphenylpyruvate dioxygenase inhibitors. J Agric Food Chem 2019;67(39):10844–52.

[51] Qu R-Y, Nan J-X, Yan Y-C, Chen Q, Ndikuryayo F, Wei X-F, Yang W-C, Lin H-Y, Yang G-F. Structure-guided discovery of silicon-containing subnanomolar inhibitor of hydroxyphenylpyruvate dioxygenase as a potential herbicide. J Agric Food Chem 2021;69: 459–73.

[52] He B, Wu F-X, Yu L-K, Wu L, Chen Q, Hao G-F, Yang W-C, Lin H-Y, Yang G-F. Discovery of novel pyrazole–hybrids as 4-Hydroxyphenylpyruvate dioxygenase inhibitors. J Agric Food Chem 2020;68(18):5059–67.

[53] Lin H-Y, Chen X, Dong J, Yang J-F, Xiao H, Ye Y, Li L-H, Zhan C-G, Yang W-C, Yang G-F. Rational redesign of enzyme via the combination of quantum mechanics/molecular mechanics, molecular dynamics, and structural biology study. J Am Chem Soc 2021;143(38):15674–87.

[54] Kita A, Kita S, Fujisama I, Inaka K, Ishida T, Horrike K, Nozaki M, Miki K. An archetypical extradiol-cleaving catecholic dioxygenase: the crystal structure of catechol 2,3-dioxygenase (metapyrocatechase) from Pseudomonas putida mt-2. Structure 1999;7:25–34.

[55] Feng AN, Huang CW, Lin CH, Chang YL, Ni MY, Lee HJ. Role of the N-terminus in human 4-hydroxyphenylpyruvate dioxygenase activity. J Biochem 2020;167(3):315–22.

[56] Huang C-W, Liu H-C, Shen C-P, Chen Y-T, Lee S-J, Lloyd MD, Lee H-J. The different catalytic roles of the metal- binding ligands in human 4-hydroxyphenylpyruvate dioxygenase. Biochem J 2016;473(9):179–1189.

[57] Rundgren M. Steady state kinetics of 4-hydroxyphenylpyruvate dioxygenase from human liver (III). J Biol Chem 1977;252(14): 5094–9.

[58] Johnson-Winters K, Purpero VM, Kavana M, Nelson T, Moran GR. (4-Hydroxyphenyl)pyruvate dioxygenase from streptomyces avermitilis: the basis for ordered substrate addition. Biochemistry 2003;42(7):2072–80.

[59] Raspail C, Graindorge M, Moreau Y, Crouzy S, Lefèbvre B, et al. 4-hydroxyphenylpyruvate dioxygenase catalysis: identification of catalytic residues and production of a hydroxylated intermediate shared with a structurally unrelated enzyme. J Biol Chem 2011;286(29): 26061–70.

[60] Beaudegnies R, Edmunds AJ, Fraser TE, Hall RG, Hawkes TR, Mitchell G, Schaetzer J, Wendeborn S, Wibley J. Herbicidal 4-hydroxyphenylpyruvate dioxygenase inhibitors—a review of the triketone chemistry story from a Syngenta perspective. Bioorg Med Chem 2009;17(12):4134–52.

[61] Pallett KE, Cramp SM, Little JP, Veerasekaran P, Crudace AJ, Slater AE. Isoxaflutole: the background of its discovery and the basis of its herbicidal properties. Pest Manag Sci 2001;57:133–42.

[62] Yamamoto S, Tanetani Y, Uchiyama C, Nagamatsu A, Kobayashi M, Ikeda M, Kawai K. Mechanism of action and selectivity of a novel herbicide, fenquinotrione. J Pestic Sci 2021;46(3):249–57.

[63] Wang D, Lin H, Cao R, Ming Z, Chen T, Hao G, Yang W, Yang G. Design, synthesis and herbicidal activity of novel quinazoline-2,4-diones as 4-hydroxyphenylpyruvate dioxygenase inhibitors. Pest Manag Sci 2015;71:1122–32.

[64] Wang D, Lin H, Cao R, Yang S, Chen Q, Hao G, Yang W, Yang G. Synthesis and herbicidal evaluation of triketone containing quinazoline-2,4-diones. J Agric Food Chem 2014;62:11786–96.

[65] Wang D, Lin H, Cao R, Chen T, Wu F, Hao G, Yang W, Yang G, Chen Q. Synthesis and herbicidal activity of triketone quinoline hybrids as novel 4-hydroxyphenylpyruvate dioxygenase inhibitors. J Agric Food Chem 2015;63:5587–96.

[66] Hu W, Gao S, Zhao LX, Guo KL, Wang JY, Gao YC, Shao XX, Fu Y, Ye F. Design, synthesis and biological activity of novel triketone-containing quinoxaline as HPPD inhibitor. Pest Manag Sci 2022;78 (3):938–46.

[67] Nan J-X, Yang F-J, Lin H-Y, Yan Y-C, Zhou S-M, Wei X-F, Chen Q, Yang W-C, Qu R-Y, Yang G-F. Synthesis and herbicidal activity of triketone-aminopyridines as potent p-Hydroxyphenylpyruvate Dioxygenase inhibitors. J Agric Food Chem 2021;69(20):5734–45.

[68] Li HB, Li L, Li JX, Han TF, He JL, Zhu YQ. Novel HPPD inhibitors: triketone 2H-benzo[b][1,4]oxazin-3(4H)-one analogs. Pest Manag Sci 2018;74(3):579–89.

[69] Wang D-W, Lin H-Y, He B, Wu F-X, Chen T, Chen Q, Yang W-C, Yang G-F. An efficient one-pot synthesis of 2-(aryloxyacetyl) cyclohexane-1,3-diones as herbicidal 4-hydroxyphenylpyruvate dioxygenase inhibitors. J Agric Food Chem 2016;64:8986–93.

[70] Wang M-M, Huang H, Shu L, Liu J-M, Zhang J-Q, Yan Y-L, Zhang D-Y. Synthesis and herbicidal activities of aryloxyacetic acid derivatives as HPPD inhibitors. Beilstein J Org Chem 2020;16:233–47.

[71] Xu YL, Lin HY, Cao RJ, Ming ZZ, Yang WC, Yang GF. Pyrazolone-quinazolone hybrids: a novel class of human 4-hydroxyphenylpyruvate dioxygenase inhibitors. Bioorg Med Chem 2014;22(19):5194–211.

[72] Xu YL, Lin HY, Ruan X, Yang SG, Hao GF, Yang WC, Yang GF. Synthesis and bioevaluation of pyrazole-benzimidazolone hybrids as novel human 4-Hydroxyphenylpyruvate dioxygenase inhibitors. Eur J Med Chem 2015;92:427–38.

[73] Lei K, Hua XW, Tao YY, Liu Y, Liu N, Ma Y, Li YH, Xu XH, Kong CH. Discovery of (2-benzoyethen-1-ol)-containing 1,2-benzothiazine derivatives as novel 4-hydroxyphenylpyruvate dioxygenase (HPPD) inhibiting-based herbicide lead compounds. Bioorg Med Chem 2016;24(2):92–103.

[74] Brownlee JM, Heinz B, Bates J, Moran GR. Product analysis and inhibition studies of a causative Asn to Ser variant of 4-hydroxyphenylpyruvate dioxygenase suggest a simple route to the treatment of Hawkinsinuria. Biochemistry 2010;49(33):7218–26.

[75] Lehnert W, Stogmann W, Engelke U, Wevers RA, van den Berg GB. Long-term follow up of a new case of hawkinsinuria. Eur J Pediatr 1999;158:578–82.

[76] Zhao D, Tian Y, Li X, Ni M, Zhu X, Jia L. Variant analysis of HPD genes from two families showing elevated tyrosine upon newborn screening by tandem mass spectrometry (MS/MS). J Pediatr Endocrinol Metab 2020;33(4):563–7.

[77] Tomoeda K, Awata H, Matsuura T, Matsuda I, Ploechl E, Milovac T, Boneh A, Scott CR, Danks DM, Endo F. Mutations in the 4-hydroxyphenylpyruvic acid dioxygenase gene are responsible for tyrosinemia type III and hawkinsinuria. Mol Genet Metab 2000;71: 506–10.

[78] Ruetschi U, Cerone R, Perez-Cerda C, Schiaffino MC, Standing S, Ugarte M, Holme E. Mutations in the 4-hydroxyphenylpyruvate dioxygenase gene (HPD) in patients with tyrosinemia type III. Hum Genet 2000;106:654–62.

[79] Item CB, Mihalek I, Lichtarge O, Jalan A, Vodopiutz J, Muhl A, Bodamer OA. Manifestation of hawkinsinuria in a patient compound heterozygous for hawkinsinuria and tyrosinemia III. Mol Genet Metab 2007;91:379–83.

[80] Scott CR. The genetic tyrosinemias. Am J Med Genet C: Semin Med Genet 2006;142C(2):121–6.

[81] Hühn R, Stoermer H, Klingele B, Bausch E, Fois A, Farnetani M, Di Rocco M, Boué J, Kirk JM, Coleman R, Scherer G. Novel and recurrent tyrosine aminotransferase gene mutations in tyrosinemia type II. Hum Genet 1998;102(3):305–13.

[82] Peña-Quintana L, Scherer G, Curbelo-Estévez ML, Jiménez-Acosta F, Hartmann B, La Roche F, Meavilla-Olivas S, Pérez-Cerdá C, García-Segarra N, Giguère Y, Huppke P, Mitchell GA, Mönch E, Trump D, Vianey-Saban C, Trimble ER, Vitoria-Miñana I, Reyes-Suárez D, Ramírez-Lorenzo T, Tugores A. Tyrosinemia type II: mutation update, 11 novel mutations and description of 5 independent subjects with a novel founder mutation. Clin Genet 2017;92(3):306–17.

[83] Garrod EA. The incidence of alkaptonuria: a study in chemical individuality. Lancet 1902;1616–20.

[84] Braconi D, Millucci L, Bernardini G, Santucci A. Oxidative stress and mechanisms of ochronosis in alkaptonuria. Free Radic Biol Med 2015;88(Pt A):70–80.
[85] Russo PA, Mitchell GA, Tanguay RM. Tyrosinemia: a review. Pediatr Dev Pathol 2001;4(3):212–21.
[86] Jorquera R, Tanguay RM. Fumarylacetoacetate, the metabolite accumulating in hereditary tyrosinemia, activates the ERK pathway and induces mitotic abnormalities and genomic instability. Hum Mol Genet 2001;10(17):1741–52.
[87] De Braekeleer M, Larochelle J. Genetic epidemiology of hereditary tyrosinemia in Quebec and in Saguenay-Lac-St-Jean. Am J Hum Genet 1990;47(2):302–7.
[88] Russo P, O'Regan S. Visceral pathology of hereditary tyrosinemia type I. Am J Hum Genet 1990;47(2):317–24.
[89] Lindstedt S, Holme E, Lock EA, Hjalmarson O, Strandvik B. Treatment of hereditary tyrosinaemia type I by inhibition of 4-hydroxyphenylpyruvate dioxygenase. Lancet 1992;340:813–7.
[90] Michaely WJ, Kratz GW. US Patent 4 780 127; 1988.
[91] Pushpakom S, Iorio F, Eyers PA, Escott KJ, Hopper S, Wells A, Doig A, Guilliams T, Latimer J, McNamee C, Norris A, Sanseau P, Cavalla D, Pirmohamed M. Drug repurposing: progress, challenges and recommendations. Nat Rev Drug Discov 2019;18(1):41–58.
[92] Nosengo N. Can you teach old drugs new tricks? Nature 2016;534:314–6.
[93] Lock EA, Ellis MK, Gaskin P, Robinson M, Auton TR, Provan WM, Smith LL, Prisbylla MP, Mutter LC, Lee DL. From toxicological problem to therapeutic use: the discovery of the mode of action of 2-(2-nitro-4-trifluoromethylbenzoyl)-1,3-cyclohexanedione (NTBC), its toxicology and development as a drug. J Inherit Metab Dis 1998;21(5):498–506.
[94] Hall MG, Wilks MF, Provan WM, Eksborg S, Lumholtz B. Pharmacokinetics and pharmacodynamics of NTBC (2-(2-nitro-4-fluoromethylbenzoyl)-1,3-cyclohexanedione) and mesotrione, inhibitors of 4-hydroxyphenyl pyruvate dioxygenase (HPPD) following a single dose to healthy male volunteers. Br J Clin Pharmacol 2001;52:169–77.
[95] Yang DY. 4-Hydroxyphenylpyruvate dioxygenase as a drug discovery target. Drug Target Perspect 2003;16:493–6.
[96] Lindblad B. On the enzymic defects in hereditary tyrosinemia. Proc Natl Acad Sci U S A 1977;74:4641–5.
[97] Lock E, Ranganath LR, Timmis O. The role of nitisinone in tyrosine pathway disorders. Curr Rheumatol Rep 2014;16(11):457.
[98] Laschi M, Bernardini G, Dreassi E, Millucci L, Geminiani M, Braconi D, Marzocchi B, Botta M, Manetti F, Santucci A. Inhibition of para-hydroxyphenylpyruvate dioxygenase by analogues of the herbicide nitisinone as a strategy to decrease homogentisic acid levels, the causative agent of alkaptonuria. ChemMedChem 2016;11(7):674–8.
[99] Ranganath LR, Milan AM, Hughes AT, Dutton JJ, Fitzgerald R, Briggs MC, et al. Suitability of nitisinone in alkaptonuria 1 (SONIA 1): an international, multicentre, randomised, open-label, no-treatment controlled, parallel-group, dose-response study to investigate the effect of once daily nitisinone on 24-h urinary homogentisic acid excretion in patients with alkaptonuria after 4 weeks of treatment. Ann Rheum Dis 2016;75(2):362–7.
[100] Ranganath LR, Psarelli EE, Arnoux JB, Braconi D, Briggs M, Broijersen A, et al. Efficacy and safety of once-daily nitisinone for patients with alkaptonuria (SONIA 2): an international, multicentre, open-label, randomized controlled trial. Lancet Diabetes Endocrinol 2020;8:762–72.
[101] Masurel-Paulet A, Poggi-Bach J, Rolland MO, Bernard O, Guffon N, Dobbelaere D, Sarles J, de Baulny HO, Touati G. NTBC treatment in tyrosinaemia type I: long-term outcome in French patients. J Inherit Metab Dis 2008;31(1):81–7.
[102] Bendadi F, de Koning TJ, Visser G, Prinsen HC, de Sain MG, Verhoeven-Duif N, Sinnema G, van Spronsen FJ, van Hasselt PM. Impaired cognitive functioning in patients with tyrosinemia type I receiving nitisinone. J Pediatr 2014;164(2):398–401.
[103] Wolffenbuttel BHR, Heiner-Fokkema MR, van Spronsen FJ. Preventive use of nitisinone in alkaptonuria. Orphanet J Rare Dis 2021;16:343.
[104] Liu Y-X, Zhao L-X, Ye T, Gao S, Li J-Z, Ye F, Fu Y. Identification of key residues determining the binding specificity of human 4-hydroxyphenylpyruvate dioxygenase. Eur J Pharm Sci 2020;154, 105504.
[105] Ndikuryayo F, Kang WM, Wu FX, Yang WC, Yang GF. Hydrophobicity-oriented drug design (HODD) of new human 4-hydroxyphenylpyruvate dioxygenase inhibitors. Eur J Med Chem 2019;166:22–31.
[106] Liu S, Kuht HJ, Moon EH, Maconachie GDE, Thomas MG. Current and emerging treatments for albinism. Surv Ophthalmol 2021;66 (2):362–77.
[107] Onojafe IF, Adams DR, Simeonov DR, Zhang J, Chan CC, Bernardini IM, Sergeev YV, Dolinska MB, Alur RP, Brilliant MH, Gahl WA, Brooks BP. Nitisinone improves eye and skin pigmentation defects in a mouse model of oculocutaneous albinism. J Clin Invest 2011;121(10):3914–23.
[108] Adams DR, Menezes S, Jauregui R, Valivullah ZM, Power B, Abraham M, Jeffrey BG, Garced A, Alur RP, Cunningham D, Wiggs E, Merideth MA, Chiang PW, Bernstein S, Ito S, Wakamatsu K, Jack RM, Introne WJ, Gahl WA, Brooks BP. One-year pilot study on the effects of nitisinone on melanin in patients with OCA-1B. JCI Insight 2019;4(2), e124387.
[109] Vergaray Ramirez MA, Sterkel M, Martins AJ, Bp Lima J, Oliveira PL. On the use of inhibitors of 4-hydroxyphenylpyruvate dioxygenase as a vector-selective insecticide in the control of mosquitoes. Pest Manag Sci 2022;78(2):692–702.
[110] Sterkel M, Perdomo HD, Guizzo MG, Barletta AB, Nunes RD, Dias FA, Sorgine MH, Oliveira PL. Tyrosine detoxification is an essential trait in the life history of blood-feeding arthropods. Curr Biol 2016;26(16):2188–93.
[111] Sterkel M, Haines LR, Casas-Sánchez A, Owino Adung'a V, Vionette-Amaral RJ, Quek S, Rose C, Silva Dos Santos M, García Escude N, Ismail HM, Paine MI, Barribeau SM, Wagstaff S, MacRae JI, Masiga D, Yakob L, Oliveira PL, Acosta-Serrano Á. Repurposing the orphan drug nitisinone to control the transmission of African trypanosomiasis. PLoS Biol 2021;19(1), e3000796.

Chapter 7.2

Indoleamine 2,3-dioxygenase*

Michele Coluccia, Daniela Secci, and Paolo Guglielmi
Department of Drug Chemistry and Technologies, Sapienza University of Rome, Rome, Italy

1 Introduction

L-Tryptophan (Trp or W) is the heaviest proteinogenic amino acid (AA) in humans and one of the nine essential amino acids for mammals, which are the ones that cannot be endogenously synthesized by our metabolism and that have to be supplied through the diet. In addition to protein synthesis, it is involved in a wide variety of biochemical pathways, including the metabolic synthesis of neurotransmitters, like serotonin (5-hydroxytryptamine or 5-HT) and melatonin, and the construction of fundamental coenzymes, such as nicotinamide adenine dinucleotide (NAD); hence, alterations of Trp delicate homeostasis may result in several pathologic conditions. Though the Trp indole group plays an important role in stabilizing peptides structures and promoting inter-protein interactions, the presence of tryptophan in proteins is scarce (1%–2%, while other AAs' frequency is 5% on average) [1–3]. Of the total tryptophan daily intake, which should be approximately 4 mg/kg, only a small percentage is involved in protein synthesis while most of the free Trp (90% of the total, which is in equilibrium with the remaining 10%, bound to albumin [4]) is driven into three main metabolic pathways: the kynurenine pathway, where the indole group is broken down and whose first and rate-limiting step is catalyzed by indoleamine 2,3-dioxygenase (IDO, the topic of this chapter); the serotonin pathway, that maintains undamaged the indole ring and mainly occurs in the gastrointestinal tract (where approximately 95% of mammalian serotonin is synthesized), and also leads to the production of melatonin in the pineal gland; the indole pyruvate pathway, relying on the microbial degradation of Trp. The kynurenine pathway alone is responsible for the metabolization of 95%–99% of free tryptophan, while 1%–5% of the amino acid is directed into the serotonin pathway [5–8]. As just said, the main way through which tryptophan is degraded and converted into active metabolites is the kynurenine metabolic route (Fig. 1): the first step can be catalyzed by either the indoleamine 2,3 dioxygenase or the tryptophan 2,3-dioxygenase (TDO) and consists in the oxygenation of L-tryptophan to *N*′-formyl-kynurenine [9]. Although these two analogous heme-containing enzymes catalyze the same oxygenation reaction, they do not share structural similarity, are expressed in different tissues and their expressions respond to diverse stimuli [10]. Nonetheless, some similarities between the two proteins can still be found and will be further discussed in the paragraph about TDO. *N*-formyl-kynurenine is an unstable intermediate and is rapidly converted into L-kynurenine (KYN) either by *N*-formyl-kynurenine formamidase or by spontaneous conversion.

At this stage, kynurenine can undergo three different destinies, catalyzed by three different enzymes: (i) kynureninase catalyzes the conversion of KYN into anthranilic acid; (ii) kynurenine aminotransferase (KAT) catalyzes the production of kynurenic acid (KA); and (iii) kynurenine 3-monooxygenase (KMO), catalyzes the production of 3-hydroxykynurenine (or 3-HK) [9]. These intermediates as well as the others in the metabolic route possess important biological effects, which will be now briefly described. Kynurenic acid seems to interact as an antagonist with all ionotropic glutamate receptors, like NMDA and AMPA receptors, in the central nervous system, and with α7-nicotinic cholinoceptor (α7NR) as a negative allosteric modulator. This blockage would result in an antiepileptic effect deriving from the suppression of the epileptiform burst in the hippocampal region [11]. Furthermore, besides its neuroprotective effect, it has been reported the involvement of KA in inflammatory diseases, in which it may play an immunosuppressive role thanks to its interaction with aryl hydrocarbon receptor (AhR) and relative pathway and G-protein-coupled receptor 35 (GPR35), but also cancer, mental, and neurodegenerative disorders [12–14].

Another downstream product of the metabolization of L-kynurenine is 3-hydroxykynurenine, produced thanks to the catalytic action of the enzyme KMO. Kynurenine monooxygenase is a NADP-dependent and flavin adenine dinucleotide (FAD)-containing enzyme located in the outer mitochondrial membrane and catalyzes the insertion of a hydroxyl group in the ortho position of KYN [15]. 3-HK possesses important redox properties and can behave both

*Due to a formal error this chapter has been included among the non-heme containing iron enzyme, although this is a heme-containing enzyme.

Metalloenzymes. https://doi.org/10.1016/B978-0-12-823974-2.00024-3

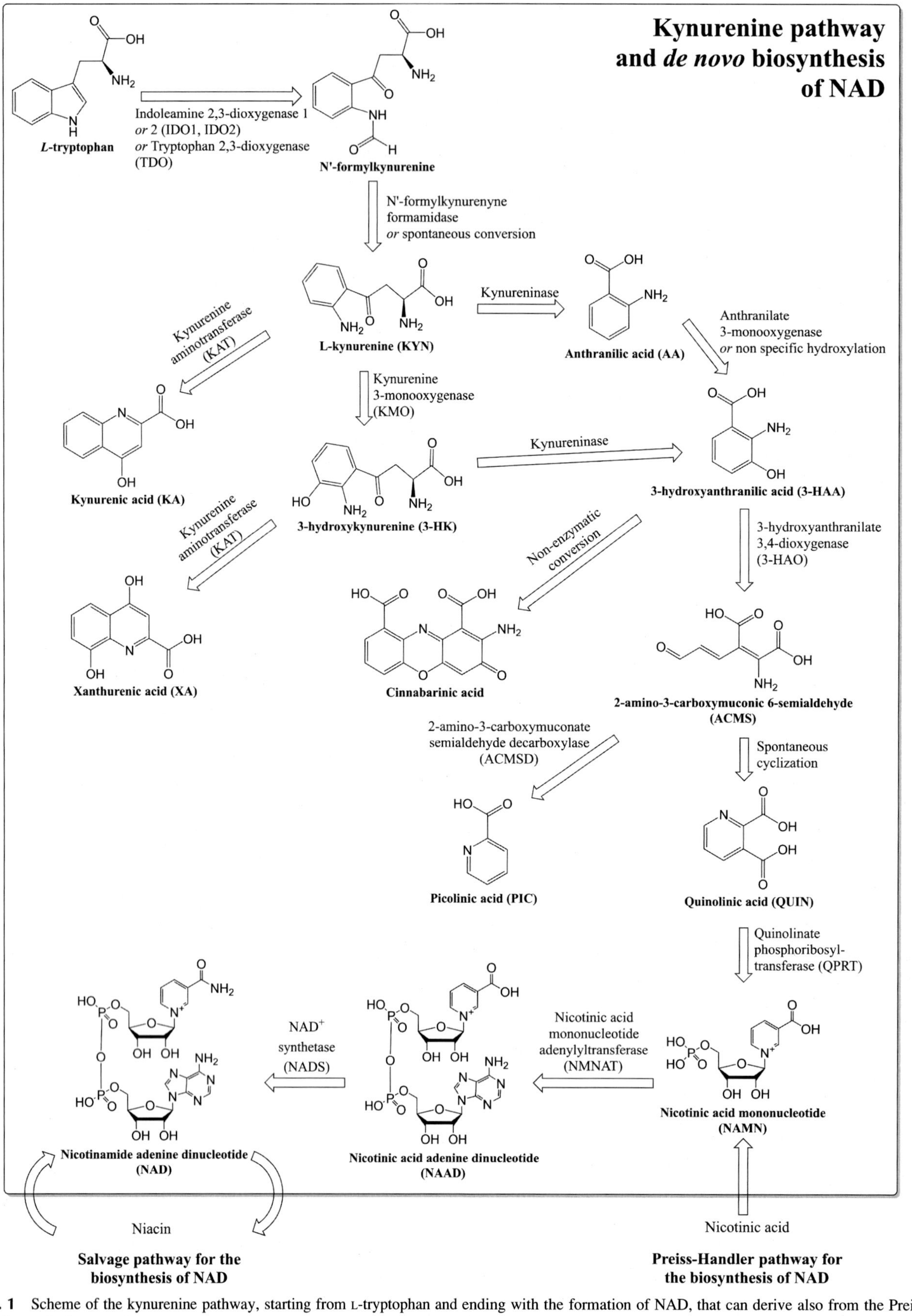

FIG. 1 Scheme of the kynurenine pathway, starting from L-tryptophan and ending with the formation of NAD, that can derive also from the Preiss-Handler and salvage pathways.

as an antioxidant and as a prooxidant by the donation of electrons and the formation of reactive oxygen species (ROS) [16]. Because of this characteristic, the part played by this molecule in several conditions is still not completely clear. The Janus-face nature of 3-hydroxykynurenine strongly influences its role in our organisms and 3-HK appears to be involved in several physiological mechanisms as well as in diverse central nervous system pathologies, like Huntington's, Parkinson's, amyotrophic lateral sclerosis, but also non-neurological diseases like inflammatory conditions and cancer [17–23]. Furthermore, 3-HK, together with other kynurenines like KYN, can be found in the eye's lens of diverse animals, including primates and humans, where it interacts with amino acid residues of lens proteins and acts as a UV light filter, thanks to its absorbance of 365 nm wavelength radiation [17,24,25]. Herein, on the other hand, 3-HK may interact with methionine and tryptophan lens proteins, like α-crystallin, and induce the yellowing of the lens and the formation of cataracts [24]. 3-HK can be converted by the enzyme kynureninase into 3-hydroxyanthranilic acid (3-HAA), the direct precursor of quinolinic acid (QUIN). Another possible fate for 3-HK is the conversion into xanthurenic acid (XA) by kynurenine aminotransaminase (the same enzyme that catalysed the conversion of L-KYN into KA). XA has shown to possess antioxidant and vasorelaxant activities, but it is also active as a modulator of glutamatergic transmission, causing a decrease in glutamate concentration through the inhibition of vesicular glutamate transport and activating group II metabotropic glutamate receptors [26,27]. Anthranilic acid derives from the action of kynureninase on L-kynurenine and shares the same metabolic fate of 3-hydroxykynurenine, which is the production of 3-HAA and then QUIN, a fundamental precursor in the de novo biosynthesis of NAD, which mainly occurs in lack of niacin, the other crucial precursor of NAD in the salvage pathway [28]. Kynureninase, which is an aminotransferase, is a pyridoxal-5′-phosphate (PLP)-dependent enzyme with a homodimeric structure and in the kynurenine pathway catalyzes both the conversion of 3-HK into 3-HAA and of KYN into anthranilic acid, releasing alanine as a product. In both of the reactions that this enzyme catalyzes, KYN to anthranilic acid and 3-HK to 3-HAA, the amino acidic part of the molecules, deriving from L-Trp structure, is converted into a carboxylic group [29–31]. Experimental studies have shown how kynureninase is overexpressed in inflammatory diseases like psoriasis, while the expression of this enzyme results lower in cancer cells if compared with other kynurenine-related enzymes, like IDO, suggesting a complicated mechanism that still has to be fully comprehended [32–34]. Anthranilic acid, similarly to 3-HK, can be converted into 3-hydroxyanthranilic acid, in a reaction catalyzed by anthranilate 3-monooxygenase [9], even if some authors describe a nonspecific hydroxylation of the anthranilic acid [35,36]. 3-HAA, like 3-HK, is characterized by the presence of both aromatic amine and phenolic groups, which confer to it an interesting redox activity, reflected by the roles it seems to play in various conditions [37]. This intermediate exerts an antioxidant and neuroprotective ability, especially in the central nervous system where it seems to determine cytoprotective and anti-inflammatory effects in the glia, through the induction of enzymes such as hemeoxygenase-1 [38,39]. Furthermore, lower levels of 3-HAA (and higher levels of anthranilic acid), if compared with healthy individuals, have been observed in people suffering from various diseases, like osteoporosis, Huntington's disease, depression, stroke, and more [40].

The effect of 3-hydroxyanthranilic acid on T cells has been investigated as well and this metabolite has been seen to induce T-reg cells activation and activated T cells apoptosis via glutathione reserves depletion and consequent exposure of the cells to oxidative stress. Another mechanism that could be involved in T cells suppression is the direct inhibition of dendritic cells activation, which are the primary responsible for the activation of naïve T cells. Moreover, this molecule seems to exert the highest proapoptotic effect on activated T cells among L-tryptophan metabolites [41–43]. These findings suggest the possible use of 3-hydroxyanthranilate as immunosuppressive drug for the treatment of autoimmune diseases and transplant rejection, even if more data are needed for further evaluation. 3-Hydroxy anthranilate is converted by the iron-containing enzyme (non-heme) 3-hydroxyanthranilate 3,4-dioxygenase (3-HAO) to 2-amino-3-carboxymuconic 6-semialdehyde (ACMS), which eventually cyclizes spontaneously into quinolinic acid, which will be employed for the production of NAD by quinolinate phosphoribosyltransferase (QPRT). Furthermore, 3-HAA can nonenzymatically be converted through oxidation in cinnabarinic acid, which plays a role as immunomodulator and neuroprotector compound [44]. Human 3-HAO, mainly found in the liver, kidney, and brain, is a monomeric enzyme that contains iron at its active site and the catalytic mechanism involves the formation of an oxygen radical through the transfer of an electron from the Fe atom to oxygen, thus promoting the addition of the two O atoms to the 3-HAA structure to form QUIN [45,46]. The proposed mechanism for this enzyme suggests that the formation of the pyridine ring of the quinolinic acid happens after the deprotonation of the phenol group of the 3-HAA and the subsequent first oxygen insertion, bond migration via Criegee rearrangement that would end in the ring expansion and opening, addition of the second oxygen, formation of ACMS as unstable dioxygenated intermediate and, eventually, the closure of the ring after the nucleophilic attack of the amine group, resulting in the formation of quinolinic acid [46–49]. In addition to QUIN, picolinic acid (PIC), an isomer of the nicotinic acid (carboxyl group at position 2), is another L-Trp metabolite directly deriving from 3-hydroxyanthranilate. After the production of ACMS, it can either nonenzymatically be converted into QUIN or it can be converted into picolinic acid thanks to 2-amino-3-carboxymuconate semialdehyde decarboxylase (ACMSD). This metabolite appears to possess a neuroprotective and immunomodulatory role, through the enhancement of macrophage functions, and seems to be involved in bone formation with an osteoprotective and osteogenic effect; indeed, PIC has been investigated as a potential treatment for osteoporosis [50–52]. Among all kynurenine metabolites, quinolinic acid is probably one of the most important, because of its involvement in numerous biological activities, first of all, the de novo biosynthesis of NAD; over the years it has emerged the neurotoxicity of this metabolite and the role it plays in the pathogenesis and development of several pathologies. In the brain, the macrophages (and the activated resident microglia with a significantly smaller role) are the primary source of QUIN, while astrocytes appear not to be able to produce this molecule. The mechanisms through which QUIN exert its main neurotoxic effect are different, but the main one is surely the interaction with NMDA glutamate receptors in the central nervous system, especially in the hippocampus, neocortex, and striatum, where the neurons seem to be more sensitive to QUIN, either because of a different NMDA receptors configuration and because of a reduced presence of the enzyme that catabolizes QUIN, QPRT (interestingly, the effect on NDMA receptors by QUIN is often opposed to the neuroprotective and inhibitory one exerted by kynurenic acid, KA [53]). Furthermore, QUIN increases glutamate release and inhibits its uptake by astrocytes, causing augmented glutamate concentration and consequent neurotoxicity. The effects on brain cells caused by QUIN-NMDA interaction (QUIN activity on NMDA receptors is comparable with the one of glutamate and aspartate) are various and comprehend cytoskeleton destabilization and microtubule disruption, lipid peroxidation, and oxidative stress, with these two last effects that can be exerted also with an NMDA-independent mechanism, relying on the interaction with ferrous ions. If QUIN biosynthesis represents a risk per se, but, while in physiological conditions it is normally consumed and directed into NAD pathway and modulates local events, under pathological conditions it can play an important neurodegenerative, pro-inflammatory and pro-oxidant effect [54–57]. Because of the excitotoxic activity of QUIN, high levels of quinolate have been found in patients suffering from major depression [58], Alzheimer's (where quinolinic acid is associated with the formation of plaques, surrounded by macrophages and microglia [59]), Huntington's [60], and Parkinson's disease and in the latter QUIN and its iron complexes seem to actively participate in the degeneration of the dopaminergic neurons in the substantia nigra [53]. Furthermore, QUIN activity seems to be implicated also in many other conditions, such as amyotrophic lateral sclerosis and HIV-associated neurocognitive disorders [61–63]. Additionally, similarly to 3-HK and 3-HAA, QUIN is able to inhibit T cell proliferation and exogenous administration of quinolinic acid has been shown to enhance human colon cancer cells proliferation and tumor growth in mice, suggesting an implication of QUIN in tumor pathogenesis [64,65].

At the end of the kynurenine pathway, quinolinic acid is metabolized by quinolinate phosphoribosyl transferase (QPRT). This enzyme, characterized by a hexameric structure in humans, catalyzes the phosphoribosyl moiety transfer between QUIN and 5-phosphoribosyl-1-pyrophosphate (PRPP) and the subsequent decarboxylation, to obtain pyrophosphate, carbon dioxide, and nicotinic acid mononucleotide (NAMN), one of the last intermediates before the ultimate formation of NAD [66,67]. NAMN is an intermediate in the Preiss-Handler salvage pathway that originates from nicotinic acid and will be converted into nicotinic acid adenine dinucleotide (NAAD) by nicotinic acid mononucleotide adenylyltransferase (NMNAT). Eventually, NAD^+ synthetase (NADS) will convert NAAD into NAD [68,69]. NAD (and NADH) plays a critical role in our organism, being a fundamental cofactor for countless enzymes and being involved in the maintenance of the redox status of the cell. As a consequence, due to the impact of this molecule in so many biological processes, NAD concentration and availability in the human body are extremely important and alterations of NAD levels have been observed to be implied in carcinogenesis, aging processes, neurodegenerative and inflammatory diseases, and many more [70–74].

As emerged from the brief description of the kynurenine pathway and its products, the intermediates of the metabolization of L-tryptophan interact with a huge variety of biological targets and exert an effect in many different pathologies and conditions. Neurological disorders, such as depression, bipolar disorder, schizophrenia, and anxiety, together with neurodegenerative conditions, such as Parkinson's, Alzheimer's, and Huntington's diseases, but also autoimmune diseases, inflammation, obesity, cardiovascular pathologies, and cancer (the fingerprint of the involvement of kynurenine-related metabolites seems to be observable in almost any kind of tumor) [28,36,75–79]. All things considered, it is conspicuous the extreme importance of the kynurenine pathway in our organism and how the study of its mechanism and functions may be relevant for the discovery of new treatments for numerous pathologies.

Hence, in the next paragraphs of this chapter, we will focus on the structure, catalytic mechanism and physiopathological role of indoleamine 2,3-dioxygenase, the opening gate of tryptophan metabolism and the most important enzyme of the kynurenine pathway.

2 Discovery of IDO

The history of the kynurenine pathway dates to 100 years ago, with the discovery by Matsuoka and Yoshimatsu of kynurenine in 1925 [80]. The relationship with tryptophan metabolism was clear right from the beginning because KYN had been discovered in the urine of high L-Trp diet-fed rabbits. The structure of kynurenine was discovered in 1943 by Butenandt [81]. In 1936, the existence of TDO enzyme was discovered by Kotake and Masayama [82] in rat livers (even though the name of the enzyme hadn't yet been fixed) and in 1959 it was understood that this dioxygenase enzyme contained a heme group [83]; but the presence of another enzyme, IDO, would have been unveiled only thirty years later. Prof. Osamu Hayaishi, in the post-war and destroyed Japan, was donated some precious grams of tryptophan by prof. Kotake. Hayaishi then took scorched soil from the back of the university (animal systems were not practicable in the poor conditions of Osaka University and so he decided to use microorganisms) and mixed it with water and the amino acid in a test tube and with only water in another tube. Only in the test tube with Trp, a cloudiness formed in the following days. This allowed the scientist to isolate the *Pseudomonas* bacterium that had grown with only Trp as source of carbon and nitrogen. Starting from this almost accidental episode, with low equipment and chemicals, Hayaishi started studying tryptophan metabolism and, in the successive years, he was able to verify the insertion of an oxygen molecule in the structure of the amino acid and the eventual formation of *N*-formylkynurenine from tryptophan [84]. At that time, IDO1 had not yet been discovered and the incorporation of oxygen gas in Trp was attributed to a non-identified tryptophan peroxidase (also called oxidase or pyrrolase) [85–87]. In 1967, Yamamoto and Hayaishi discovered IDO, observing that, on the contrary to Kotake's enzyme, which only catalyzed the metabolization of L-tryptophan, the newly discovered "tryptophan pyrrolase" could also oxidize the D isomer of the amino acid. The research group continued to investigate the nature and the characteristics of this enzyme [88,89] and eventually, in 1978, the tryptophane pyrrolase was named indoleamine 2,3-dioxygenase [90].

3 Gene, structure and catalytic mechanism of IDO

Indoleamine 2,3-dioxygenase (IDO or INDO) is a heme-containing cytosolic enzyme that catalyzes the conversion of L-Trp into L-kynurenine: the first and rate-limiting step of the kynurenine pathway, which is where more than 95% of L-tryptophan amount in our organism is directed. There are two known isoforms of IDO, IDO1, and IDO2. Both enzymes catalyze the same reaction, but their activity rate and their expressions are different. Another enzyme, tryptophan 2,3-dioxygenase (TDO) is also able to catalyze the same reaction, but its structure is completely different, as it is its expression in human tissues. Both IDO2 and TDO will be described in a successive paragraph. IDO1 can oxidize not only L-Trp but also D-Trp and other indoleamine derivatives [91].

IDO1 expression relies on *IDO1* gene, located on the short arm of chromosome 8 (8p11) and discovered in 1993 by Burkin and his colleagues [92]. *IDO1* is a single-copy gene, contains 10 exons and comprehends a total of 15423 base pairs, with the final peptide product composed of 403 amino acids (1531 nucleotides in the transcript). *IDO1* possesses two important interferon-stimulated response elements (ISREs), which are responsible for the extremely important induction of IDO1 expression mediated by IFN-γ [93,94]. Contiguously to *IDO1* gene, on chromosome 8p11, is located *IDO2* gene (or *INDOL1*). Both genes are associated with various genetic variants, found in different ethnicities and with different functionality, and that may result in diverse effects on the organism. Indeed, a study conducted in 2009 on a population of 48 Caucasians and 48 African-Americans found that, in the latter group, around 4% of the individuals carried a nonfunctional *IDO1* allele, and none of the Caucasians involved presented alleles with impaired functionality [95].

For what does regard the structure of IDO1, its crystal was firstly analyzed by X-ray diffraction in 2006 by Sugimoto [96], who analyzed the structure of recombinant human IDO1 complexed with 4-phenyimidazole, a ligand inhibitor, and cyanide. It was only in 2017 that the X-ray crystal of IDO1 in complex with L-tryptophan, its natural substrate, was obtained, thanks to the work of Lewis-Ballester and co-workers [97]. Indoleamine 2,3-dioxygenase 1 weighs approximately 45 kDa and is composed of 403 amino acids [98]. IDO1 structure comprises two domains, large (C-terminal, residues 155–403) and small (N-terminal, residues 1–154) (Fig. 2) [99].

The former is an all-helical domain and comprehends thirteen α-helices and two 3_{10} helices. Four helices (called G, I, Q, and S) are positioned parallel to the plane that contains the porphyrinic structure of the heme group, with the iron ion in the center. Helix Q has a fundamental importance because it contains the imidazole group of His346, which thanks to its nitrogen atom is the endogenous ligand for the fifth coordination position (the so-called "proximal side") of the iron (Fig. 3A). No histidine has been found in the distal pocket of the enzyme: the active site lacks polar residues and, as will be better described later, His does not play a role in the catalytic mechanism (no base-catalyzed abstraction of indole proton) [100]. The sixth coordination position of the iron is maintained free because is deputed to the catalytic mechanism (Fig. 3A). A salt bridge between Arg343 and Asp274 contributes to the stability of the

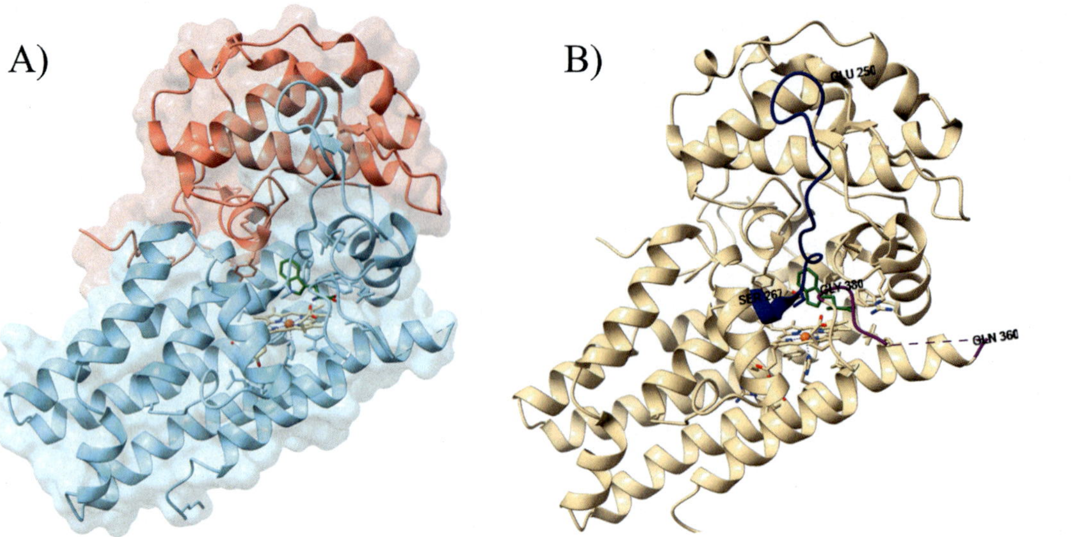

FIG. 2 (A) IDO1 is composed of two domains: an N-terminal small one (*red*) and a C-terminal large one (*blue*). The active site is located near the interface between the two domains. (B) The two main loops of the enzyme are highlighted: *blue*, long loop from Glu250 to Ser267; *purple*, the JK loop from Gln360 to Gly380, it contains the two phosphorylation sites (PDB: 6e46).

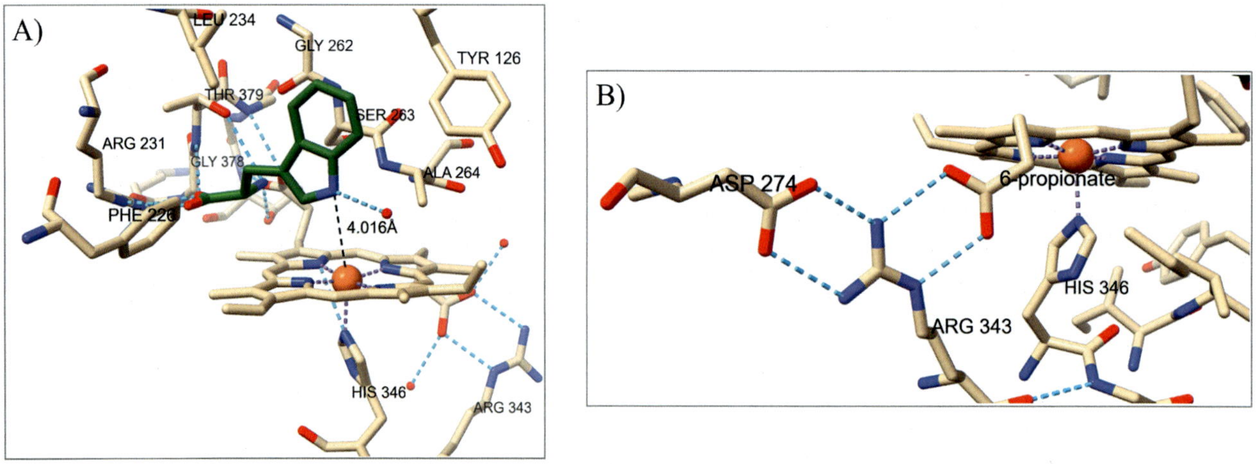

FIG. 3 (A) The active site of IDO1, the main residues involved in the catalytic cycle and in the stabilization of the substrate are shown. (B) The salt bridge between Arg343 and Asp274, which contributes to the stability of the heme group (PDB: 6E46).

prosthetic group (Fig. 3B). The heme group is located in a binding pocket (or pocket A) created by the four previously reported helices and by the helices K, L, and N. K and L helices are also responsible for heme-proteins interactions and for the connection of the two domains of the enzyme. Overall, the heme pocket results from the combination of the two domains and a loop. The main residues of pocket A are Tyr126, Cys129, Val130, Phe164, Gly262, and Ala264. The roof of the heme pocket is made up of the small domain of IDO1 and by a long loop, which contains the residues from 250 to 267; furthermore, the sequence from 260 to 265, which also connects the two domains, is highly conserved [101]. The entrance to the catalytic pocket is permitted by the presence of hydrophobic pocket B (Phe163, Phe226, Phe227, Arg231, Ile354, and Leu 384). In pocket A, the indole ring of the side chain of L-Trp is perpendicular to the heme group, while in pocket B, the amino acidic portion of tryptophan lies parallel to the heme and is stabilized by the residues, mainly Arg231 and Thr379, as will be described later. The large domain, besides the heme (and catalytic) pocket, also contains a flexible ring consisting of the residues from glutamine 360 to glycine 380, which acts as a shuttle of substrate/product. The small domain is composed of six α-helices, two β-sheets, and three 3_{10} helices and contains two tyrosine-based inhibitory motifs. As emerges from the description of the numerous interactions between the two parts of the enzyme, the surface area between the two domains of IDO1 is considerably wide (3100 Å2), and the interconnection is mainly based on hydrophobic interactions, hydrogen bonds, and salt bridges [96,98,102–104]. In addition to the two main pockets (A and B), other two subpockets have been identified: C, which comprehends a small surface near the entrance of the active site; and D, in the proximal heme side. Both C and D seem to have a role only in the interaction with enzyme inhibitors and become accessible only after conformational changes of IDO1 [105]. The heme contains two propionate groups, the proximal (6-propionate) and the distal one (7-propionate), depending on the side of the plane on which they are located (6- and 7-propionate groups point downward and upward from the heme plane, respectively) (Fig. 3B). The proximal propionic group and the carbonyl group of Leu388 interact with two water molecules (called wa1 and wa2), inducing the partial formation of the anionic form of His346, which is bonded to a water molecule (wa1) itself. The presence of water molecules in the active site of IDO1 is extremely important. The analysis of IDO1 structure indicates a strongly hydrophobic environment (the structure contains plenty of large hydrophobic residues, such as phenylalanine and tyrosine) in the catalytic pocket which has to accommodate the highly hydrophobic L-tryptophan, with Ser167 being the only hydrophilic enzyme in the active site.

The presence of water molecules in the active site of IDO1 is extremely important (Fig. 4A and B). The analysis of IDO1 structure indicates a strongly hydrophobic environment (the structure contains plenty of large hydrophobic residues, such as phenylalanine and tyrosine) in the catalytic pocket, which has to accommodate the highly hydrophobic L-tryptophan, with Ser167 being the only hydrophilic enzyme in the active site. Since this residue seems not to be relevant for the binding of the oxygen molecule or the stabilization of the ferrous-oxy complex, as shown by mutagenesis studies, it has been suggested that, as it happens in other enzymes (e.g., cytochrome P450), water molecules mediate the formation of stabilizing hydrogen bonds [106,107]. On the other hand, Ser167 appears to have a key function in the recognition of L-tryptophan, even if to a greater extent way than the distal tyrosine (Tyr126) [108,109]. Spectroscopic analyses have shown that the binding and stabilization of L-Trp critically depend on the H-bond interaction of the substrate with two residues: Arg231 and Thr379. The arginine long side chain stabilizes the carboxylic group of Trp, while the 7-propionic group of heme interacts with the ammonium group (NH_3^+) on the C_α atom. Thr379 is involved in the coordination and stabilization of the ammonium group [109]. Another important portion of the IDO1 structure is a highly dynamic and flexible loop, called JK-Loop, in the larger domain (residues from 360 to 380). JK-Loop, in its N-terminal fragment, contains two phosphorylation sites, that suggest the function of this loop to be related with signal transduction or post-translational modifications, and not with the regulation of enzyme activity (elimination of JK-Loop residues does not significantly alter IDO1 efficiency). On the other hand, the C-terminal fragment can adopt different conformations, depending on the absence or presence of the substrate Trp, thanks to the interaction with Thr379 and Arg231 [97,110]. As emerges from the description of the enzyme structure, IDO1 has a complex organization, and the active site is not the only site of interaction of the protein with small molecules, but diverse inhibitory substrate-binding sites seem to be present [101,111]. A later paragraph of this chapter will be dedicated to IDO1 inhibitors.

IDO1 catalyzes the formation of *N′*-formylkynurenine (that will then be converted into L-kynurenine) from L-Trp. The reaction consists of the deoxygenation and ring opening of the indole group of tryptophan and the exact mechanism through which IDO1 exerts its catalytic action has been debated and has not been completely understood. The presence of the nitrogen atom of the proximal His346 in the active site of the enzyme erroneously brought the researchers (Hamilton proposed this mechanism in 1969 [112]) to think that a base-catalyzed deprotonation of the Trp indole was involved in the catalytic mechanism, but a lot of evidence has now refuted this theory. First,

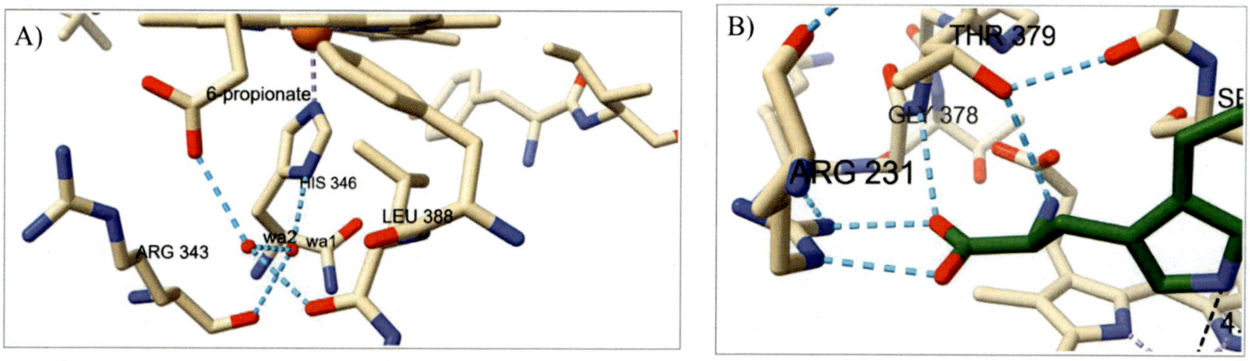

FIG. 4 (A) Leu388, with its carbonylic group, together with the 6-propionate group of the heme, interacts with two water molecules (wa1 and wa2), inducing the partial formation of anionic His346. (B) H-bon stabilization of L-tryptophan by Arg231 and Thr379. The ammonium group of Arg stabilizes the carboxylic group of the substrate, while Thr interacts with the amino acidic ammonium group of Trp (PDB: 6E46).

mutagenesis studies have verified that this resides, although important, doesn't play a fundamental role in the catalytic cycle [113]. Second, the deprotonation is not energetically favorable and is not consistent with the reaction chemistry of indoles and does not happen in other oxygen-dependent heme enzymes (such as cytochrome P450). Moreover, the indole oxygenation is observed also when 1-methyl tryptophan is employed as the enzyme substrate, resulting in the formation of *N*-formyl-methylkynurenine and eventually demonstrating that the deprotonation of indole nitrogen does not occur in IDO1 [114]. An alternative to the base-catalyzed mechanism for the first step of the catalytic cycle of IDO1 is the electrophilic addition (Fig. 5): the lone pair of the nitrogen atom would initiate an electrophilic addition to the iron-bound O_2 molecule, even though oxygen molecule does not possess excellent electrophilic properties. Another proposed mechanism for the first step of the cycle is a radical addition [107]. Oxygen molecule, in fact, is bound to ferrous heme and seems to be present as ferric-superoxide species. The radical pathway consists of the formation of ferryl heme. Computational works seem to encourage the radical mechanism. An interesting reflection is that the heterolytic cleavage normally requires a sufficient charge separation: in this case, electrostatic stabilization by active site residues is required. But, IDO1 active site does not present hydrophilic residues, apart from Arg231 involved in Trp binding. This supports the homolytic radical mechanism.

However, the exact mechanism has not yet been understood. Both the electrophilic and radical addition mechanisms would result in the formation of an epoxide intermediate. Mass spectrometry analyses [115], molecular dynamics study, and QM/MM simulations [116,117] of the reaction revealed the formation of a stable amino acetal intermediate, deriving from an initial and metastable 2,3-alkyl epoxide species, which seems to originate from the direct attack of the oxygen to C=C bond and first insertion of a single oxygen atom between C2 and C3 in the indole ring of tryptophan. The remaining ferryl oxygen can attack the epoxide and bring to the formation of amino acetal intermediate and then *N*-formylkynurenine and ferrous heme [85]. The insertion of the two oxygen atoms through a sequential and two-step mechanism has been supported by resonance Raman analysis [118–121]. It is interesting to note that, on the contrary to what happens in tryptophan 2,3-dioxygenase (TDO), a precise order for the binding of Trp to the active site and oxygen to the heme is not required for the formation of the oxy-ferrous adduct. Despite this, the binding order of the two ligands may play a role in the self-inhibition of the enzyme activity. In fact, tryptophan binding velocity depends on its concentration: at high concentrations of L-Trp, the amino acid would bind

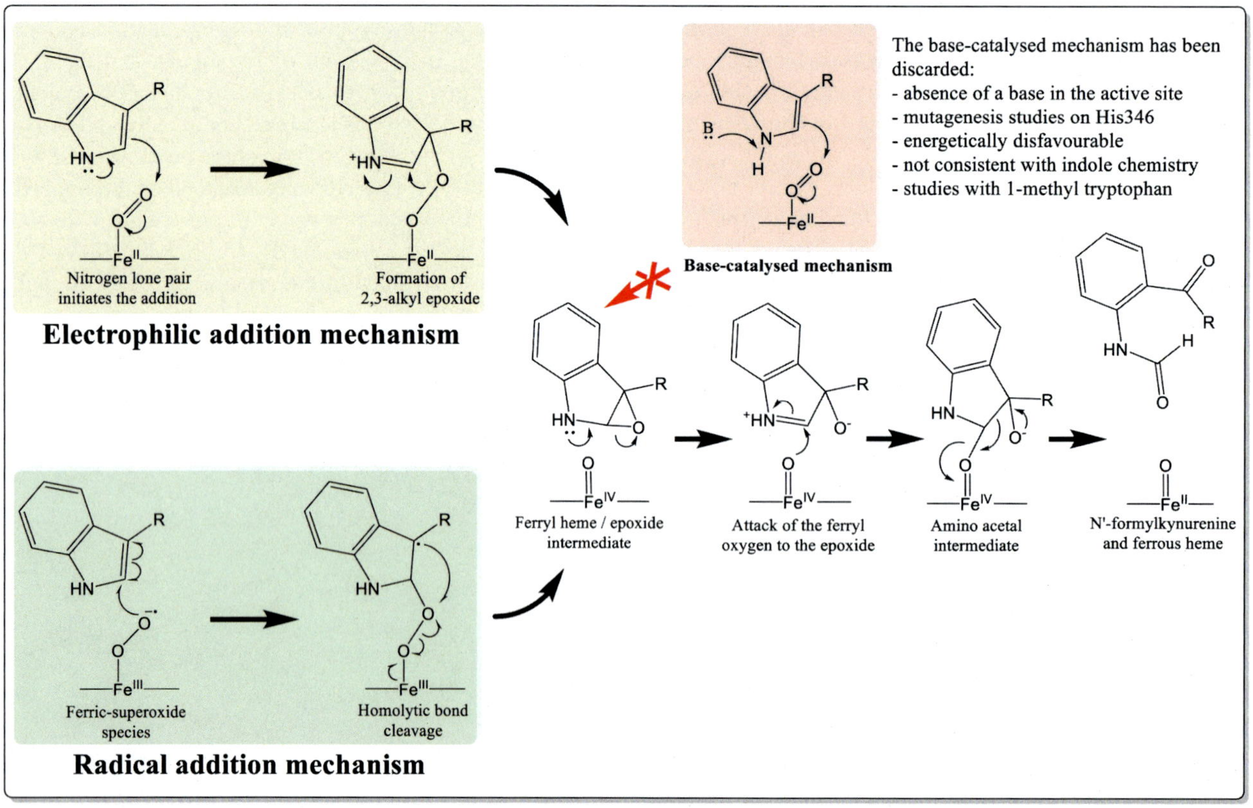

FIG. 5 The two proposed mechanisms for the catalytic cycle of IDO1: the electrophilic addition and the radical addition. The base-catalyzed mechanism theory has now been discarded due to numerous experimental and computational data.

first and consequently block the access to the oxygen molecule and increase the heme reduction potential (hence hampering O_2 binding); while at low Trp concentrations, O_2 would bind before the amino acid and the catalytic complex can be formed more easily. Another proposed mechanism for the L-Trp-driven self-inhibition of IDO1 is the presence of an inhibitory site [10,122].

4 IDO1 expression in tissues and expression regulation

IDO1 is constitutively expressed in a few cells of the human organism: pulmonary endothelial cells, epididymis, epithelial cells of the feminine genital tract, and antigen-presenting cells (APCs), like dendritic cells (DCs), macrophages, B lymphocytes, NK cells, granulocytes, and activated monocytes. IDO1 expression has been observed also in the gastrointestinal tract, specifically in the epithelial and endothelial cells, fibroblasts, and mesenchymal stem cells but also in immune cells normally present in this anatomic portion of the organism [123–125]. Another important tissue where IDO1 is constitutively expressed is the placenta, especially in endothelial cells in the areas close to the feto-maternal interface [126].

IDO1 expression is finely regulated by numerous factors and molecules, and interferon-γ (IFN-γ, type II interferon) is the most important among those. *IDO1* gene has been the first IFN-regulated gene to be characterized and it contains numerous IFN-activated sites and IFN-stimulated response elements, or ISREs, located on the upstream region of the gene and gamma activation sequences (GAS1 and 2), responding to signal transducer and activator of transcription 1 (STAT1) and interferon-regulatory factors (IRFs). IFN-γ-mediated induction of IDO1 is mediated by the IFN signal transduction pathway: the interferon binds to its receptor activate (Janus kinases, or Jak, 1 and 2), inducing the phosphorylation and subsequent dimerization of STAT1, which is the final actor of the route. STAT1 enters the nucleus and interacts with IFN-promoted genes. The response to IFN-γ seems to be related to the immune escape in inflammatory conditions, even if the mechanism, that should prevent the organism to suffer from an exaggerated immune response, is exploited by many kinds of tumors to escape the immune surveillance [127]. IFN-γ is not the only inducer of *IDO1*. IFN-α and IFN-β (type I interferons) too can strongly induce IDO1 expression by immune cells, especially in dendritic cells, where the induction by the two types, I and II, of interferons exert an equipotent and synergistic effect. On the other hand, the effect of IFN-γ on other cells can be up to 100 times more effective than α and β IFNs. Other molecules that can induce the expression of IDO1 are: transforming growth factor β (TGF-β), tumor necrosis factor α (TNF-α), IL-6, the anti-inflammatory interleukin IL-10, IL-27. Pathogen- and damage-associated molecular patterns (PAMPs and DAMPs), like lipopolysaccharide (LPS) and specific CpG motif-rich oligodeoxynucleotides (ODN), which seem to modulate and induce IDO1 expression through the interaction with toll-like receptors, like TLR 4 and 9, whose activation induces the release of type I interferon by DCs. IDO1 regulation can also be influenced by the interactions between immune cells: the binding of inhibitory T cells co-receptor CTLA-4 (on T reg cells) with B7 ligands CD80 and CD86 expressed on the membrane of dendritic cells causes and IFN-mediated induction of IDO [125–131].

From the description of the principal factors able to modulate IDO1 activity and expression, has emerged the fact that this enzyme, and consequently, the whole kynurenine pathway, are strongly connected with the immune system and plays a decisive role in immune response and immune tolerance. In the next paragraph, it will be described the role of IDO1 in physiological and pathological pathways, and many of those are strongly connected with immune response, like pregnancy, infections, and cancer.

5 Physiological functions and involvement in diseases of IDO

Surely one of the most important roles played by IDO1 is the regulation of the immune response, as suggested by the extremely complex expression and regulation of this enzyme that takes place in immune cells. Munn and coworkers, in a seminal work published in 1998 [132], verified that mice trophoblast, which develops into the placenta, could suppress the maternal T cell response that tends to consider the allogeneic fetus as an allograft. The involvement of IDO in this immune tolerance mechanism, necessary to avoid fetal rejection, was studied by treating pregnant mice with 1-methyltryptophan, an IDO inhibitor. Conversely, to the nontreated mice, the ones whose tryptophan catabolism had been inhibited showed a significantly increased occurrence of fetal rejection. The fact that the rejection had occurred 4–5 days after the implantation, brought the researchers to the hypothesis that maternal T cells were involved in the fetal loss, rather than antibodies, which would have required a longer time to be produced by B cells. This limiting effect of the immune response, which also occurs in inflammation, makes IDO a critically important enzyme in the regulation of immune homeostasis, essential not only during pregnancy or organ transplants, which necessitate the tolerance to the self, but also in many other conditions, like cancer [133].

The immune modulatory action of IDO1 is exerted through different mechanisms (Fig. 6) [134]. The first one, since the importance of this enzyme in the

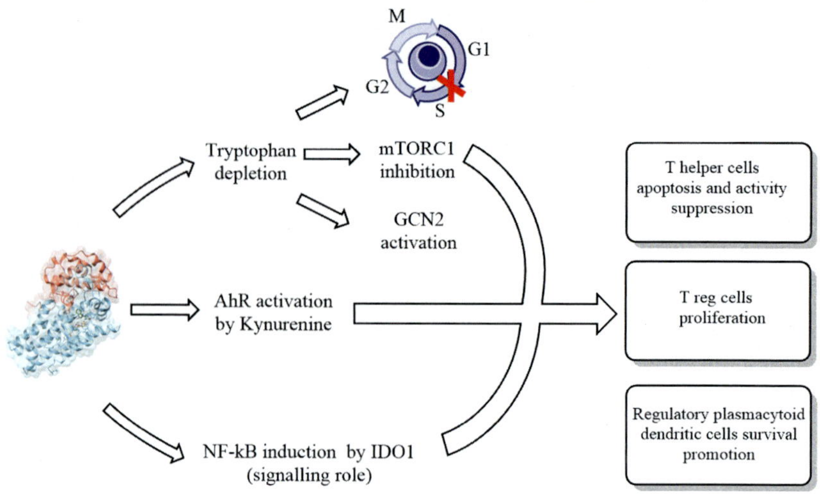

FIG. 6 The three main mechanisms through which IDO1 exerts its immune-modulatory effect: (i) Trp depletion and consequent T cell cycle arrest in G1, mTORC1 inhibition, and GCN2 activation; (ii) kynurenine deriving from the reaction catalyzed by IDO1 activates AhR; (iii) IDO1 can act as a signaling protein and activates NF-κB.

metabolization of most of the ingested tryptophan, is tryptophan depletion. The scarce presence of this amino acid determines the arrest of T cell cycle, precisely: the cells enter the G1 phase but are unable to move into the S phase. This makes T cells extremely vulnerable to Fas-ligand mediated cell death [135]; moreover, apoptosis is provoked also through the inhibition of the mechanistic target of rapamycin complex 1 (mTORC1), together with the induction of a stress response that activates the general control non-depressible-2 (GCN2), able to sense the low tryptophan concentration. The Trp concentration under which the proliferation of T cells is significantly inhibited is 0.5 μM [136]. Second, kynurenine has been seen to activate the aryl hydrocarbon receptor (AhR), which is a cytosolic transcription factor pivotal in the differentiation of T cells into $Foxp3^+$ regulatory cells (T reg cells), instead of proinflammatory T cells; the pathway triggered by the activation of AhR, which includes IL-6 and STAT3 involvement, also brings to the induction of T helper cells apoptosis [124,137].

A third mechanism through which IDO1 exerts a regulation of the immune response relies on the nuclear factor kappa-light-chain-enhancer of activated B cells (or NF-κ B). IDO1 seems to act as a signaling protein, in a rare case of functional eclectism, able to induce NF-κB, which, in a synergistic action with TGF-β, promotes the survival of regulatory plasmacytoid dendritic cells. This kind of cell can promote the formation of T reg cells and suppress the activity of T cells [133].

For what does regard the role of IDO in pregnancy, as described above the enzyme is expressed at the maternal-fetal interface, specifically in syncytiotrophoblast of placental villi, decidual immune, and stroma cells, but also in the endometrium epithelial cells and the vascular endothelial cells of chorion and endometrium. As demonstrated by Munn and by other studies, IDO can reduce the risk of T cell-dependent and antibody-independent maternal fetal rejection, based on the activation of maternal complement [138]. IDO involvement in pregnancy, in addition to its fundamental action in favor of mother-to-fetus tolerance, appears to be significant also in embryo implantation. Interestingly, in preeclampsia, a pregnancy disorder characterized by maternal hypertension, proteinuria, and even generalized seizures, the kynurenine/tryptophan ratio is similar to non-pregnant women, while healthy pregnant women possess an increased value of the ratio, suggesting a reduced tryptophan degradation and minor IDO activity [126,139–141].

The expression of IDO1 in the uteroplacental region of pregnant women may also act as a tool for the regulation of the immune response in case of infections of the female reproductive tract, like *Chlamydia* or other pathogens (Fig. 7) [142]. The involvement of IDO in the regulation of the response to infections has been extensively investigated. Since the first years of research on IDO, it was already clear the involvement of IDO in the immune and inflammatory response to pathogen infections, as demonstrated by a work of 1978 by Yoshida and Hayaishi, who injected mice with bacterial lipopolysaccharide in and observed a significant increase in the enzyme activity [143]. IFN-γ is abundantly produced as a response to viruses, bacteria, or parasites infections, and the result is a great induction of the IDO expression in APCs and the consequent depletion of tryptophan reserves and increase in kynurenine-related metabolites, especially in stimulated macrophages [144]. The immune-suppressive effect and the simultaneous depletion of Trp, fundamental for the pathogens, make the understanding of the role of IDO in infections a hard task to achieve. In fact, in viral infections, the increase in proinflammatory IFN-γ causes the induction of IDO1, resulting in the attenuation of the immune response and a state of infectious tolerance: apoptosis and reduced

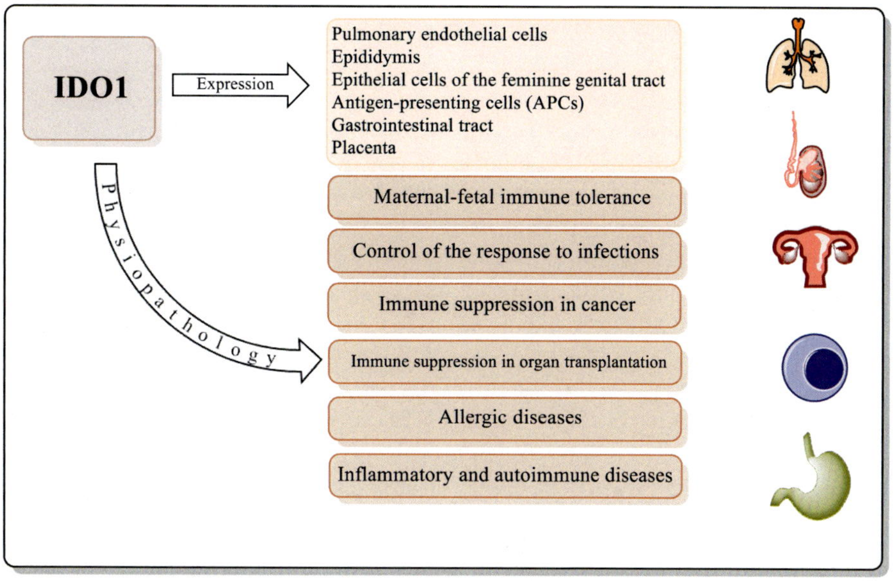

FIG. 7 Scheme of the expression and the physio-pathological conditions in which IDO1 play an important role.

proliferation of T helper cells, expansion of T reg cell population [145]. For example, a superior IDO activity has been observed in the lung parenchyma during influenza virus infection [146]. The induction of IDO by IFN and other proinflammatory cytokines has been hypothesized to play an important role also in the Sars-Cov-2 infection. In fact, a higher kynurenine/tryptophan ratio has been observed in severe COVID-19 patients [147], even if the role of IDO1 and IDO2 have not been completely understood yet [148,149]. Low levels of tryptophan in patients infected by HIV (human immunodeficiency virus type 1) have been reported more than 30 years ago and it became clear that the diminished plasma concentration of this amino acid was due to the induction of the extra-hepatic catabolism of Trp exerted by IDO1 [150]. Interferon-induction is not the only mechanism exploited by HIV; indeed, it has been observed that HIV viral proteins Nef and Tat are able to directly induce the expression of IDO in macrophages [151]. Interestingly, it was demonstrated that, in monocytic dendritic cells, Tat is only able to induce IDO1 expression and not IDO2, and that the mechanisms through which Tat and IFN-γ induce IDO expression are different. Precisely, Tat seems to work with an NF-κB pathway involving TLR4 [152]. But why is IDO activated in HIV infections? The augmented activation of IDO in HIV-infected patients may be a defense mechanism of our organism against the virus-elicited immune and inflammatory reaction, in an attempt of reducing tissue damage; furthermore, HIV viral replication preferentially occurs in cycling $CD4^+$ cells and the virus would find a reduced number of cells for the infection due to the immune suppressive effect on $CD4^+$ population exerted by IDO and mediated by GCN2. This mechanism, that, theoretically, could be beneficial to the host, eventually results in a dysregulation between immune suppression and activation, leading to chronic infection, replication of viral reservoirs and opportunistic infections [153–156]. A further proof of the involvement of IDO in the pathologic mechanism of HIV infection is the fact that the antiretroviral treatment has been shown to reduce IDO activity in treated patients [157]. HIV is not the only virus whose infection of the human organism is promoted by the immune suppressive activity of IDO. An upregulation of IDO and an overexpression of CTLA-4, which is involved in the regulation of IDO expression have been observed in patients with chronic hepatitis C, caused by the infection by hepatitis C virus (HCV) [158]. Similarly to what happens in HIV infections, the role of IDO in HCV infection is not completely defined and beneficial and detrimental effects seem to be exerted by this enzyme. In the onset of HCV infection, IDO participates in the host innate immune response and seems to slow the virus replication in hepatocytes, but, with the development of the infection into a chronic state, IFN-γ upregulation of IDO and the subsequent effect on T cells favor the induction of a tolerance state toward the virus [159]. Chronic infection by HCV is often associated with hepatocarcinoma and other oncoviruses exploit the immune escape mechanism offered by IDO, like Epstein-Barr virus (EBV), and ubiquitous virus (95% of the world population is infected by EBV) whose related cancers represent 1.8% of all cancer-related deaths in the world [160], and human papillomavirus (HPV), which is critically related with cervical cancer. IDO expression in cervical epithelium from women with HPV infection, squamous intraepithelial lesions or cervical cancer has been reported to be superior if compared with normal cervical epithelium [161].

The same dual nature that is observed in viral infections can be seen in bacterial and fungal infections, where IDO plays an immunosuppressive role but, at the same time, it depletes Trp reserves, fundamental to the infectious pathogen and produces kynurenine metabolites, that possess antimicrobial effects (like 3-hydroxykynurenine and picolinic acid) [162].

Pregnancy and infections are not the only conditions in which the immune system needs to be particularly regulated and alterations of the fine immune homeostasis may lead to detrimental effects on the organism. Cancer cells strongly rely on the escape from the immune system to proliferate in the human tissues and, because of the involvement in the regulation of the immune cells, kynurenine metabolites have been linked to many types of tumors [163]. Cancer cells are able to modulate the immune system through different mechanisms, like the downregulation of class 1 major histocompatibility complex (MHC class I), recruitment of T reg cells, and the induction of immune suppressive molecules. One of the first clues that tryptophan metabolism was somehow involved in the response of the organism to cancer came in 1955, with the discovery of high Trp levels in the urine of patients suffering from bladder cancer [164]. Since this first observation and with the discoveries made in the field of the kynurenine pathway, augmented IDO expression and activity have frequently been observed in cancer patients because of the high expression of proinflammatory signals that characterizes the tumor microenvironment: IFN-γ, TGF-β, PAMP, and DAMP, numerous kinds of cytokines, protein kinase C (PKC), and many more, as described above. Furthermore, the downregulation of the tumor suppressor gene *Bin1* has been seen to correlate with the increase in IDO1 expression in cancer cells, thanks to a STAT1 and NF-κB-mediated route. IDO1 heightened expression can be related to cancer progression, decreased survival, and poor outcome of chemotherapy. Despite the clear involvement of this enzyme and the kynurenine pathway in the immune escape mechanism, IDO's role has not been completely understood yet: if its immune regulator effect is clear, the consequences of the tryptophan depletion on the inhibition of cancer cells proliferation must be precisely evaluated [165–168]. Different immune cells, like macrophages, neutrophils, and granulocytes, are involved in the IDO-mediated tumor immune escape; nonetheless, dendritic cells seem to play a fundamental role in this context. B7 costimulatory molecules, expressed on DCs, are in fact essential in the activation of T reg cells. IDO acts as a downstream effector in the immune tolerance mechanism mediated by CTLA-4, present on T reg cells [169]. The complicated network of relationships between IDO, immune cells, and cancer cells can be observed also from the perspective of the tumor microenvironment (TME), which is fundamental in tumorigenesis. The wide variety of cells and molecules that constitutes TME help the proliferation of cancer cells, the formation of metastasis, and the switch toward an aggressive phenotype. The high expression of IDO in TME leads, among numerous other effects, to effector T cells apoptosis, and T reg proliferation due to increased concentration of kynurenine, which binds AhR; CTLA-4-mediated induction of tolerogenic dendritic cells; NK cells disfunction; immunosuppression by tumor-related macrophages (via AhR pathway) [170,171].

Furthermore, IDO participates also in the angiogenesis in tumors, mediating IL-6/STAT3 and VEGF signaling [172]. All things considered, IDO1 contributes to the immune escape of cancer cells with several mechanisms, in an intricated labyrinth of interconnections and relationships resulting in a synergistic tolerogenic effect [173]. The involvement of IDO in cancer has been verified in many tumors, like, among others, endometrial and ovarian cancer [174], cervical cancer [175], esophageal cancer [176], glioblastoma [177], colorectal carcinoma [178,179], breast cancer [180], hematological malignancies [181,182], and brain tumor, where the kynurenine metabolites, especially the neurotoxic quinolinic acid and the neuroprotective picolinic acid, play an important role as well [183,184].

The immunosuppressive effect of IDO1, unfavorable in the immune escape of cancer cells, may be helpful to the organism in a condition where the response of the immune cells to the nonself is not only undesirable but fought: organ transplantation. Immunosuppressive therapies allow transplant patients to experience long-term life of the allografts, but new therapies are always needed to improve the health conditions of these individuals. In 2014, in a study conducted on stable pediatric kidney transplant patients, it has been observed that urine kynurenine/tryptophan ratio and IDO levels were higher than in healthy children, suggesting a significative role of the kynurenine pathway in the regulation of the immune response [185]. These kinds of results, along with the knowledge of the IDO pathway in immune suppression, make IDO a potential target for post-transplant treatment [186]. Immune regulation is fundamental also in allergic diseases, caused by the exaggerated response of the organism to an allergen. Clinical experiments have suggested that IDO may induce tolerance to the allergens in allergic disorders, through the stimulation of T reg cells [187]. Lastly, the same immune suppressive mechanism may result particularly useful in the treatment of inflammatory and autoimmune diseases, like endometriosis [188], autoimmune diabetes [189], rheumatoid arthritis, and lupus erythematosus [190].

As it has been described in this paragraph, IDO1 is involved in the regulation of the immune response in various conditions, such as pregnancy, infections by viruses or other pathogens, cancer, allergies, autoimmune diseases, and others. IDO-mediated immune suppression is strictly linked to the fine regulation of IDO expression and activity exerted by IFN-γ and other pro-inflammatory cytokines. The often-dual nature of indoleamine 2,3-dioxygenase and the tangled network of interactions between signaling molecules, immune cells, and IDO itself make the understanding of this complicated system a hard task for researchers. Despite this, the advances made in the disentangling of the IDO pathway open fascinating perspectives on new therapeutic approaches and better comprehension of many physiopathologic conditions.

6 Tryptophan 2,3-dioxygenase (TDO) and indoleamine 2,3-dioxygenase 2 (IDO2)

The oxygenation of L-tryptophan into *N*-formylkynurenine can be catalyzed by two enzymes other than IDO1: indoleamine 2,3-dioxygenase 2 (IDO2) and tryptophan 2,3-dioxygenase (TDO).

The discovery of IDO2 happened much more recently than IDO1, in 1967. Its gene was identified in 2007 by Murray and co-workers [93], and three distinct research groups independently demonstrated that this gene encoded for an enzyme involved in the catabolism of tryptophan [191–193]. *IDO2* gene has a structure similar to *IDO1* and is adjacent to *IDO1* gene, on chromosome 8. It has been hypothesized that the two genes originated through gene duplication in a mammalian ancestor. Other animals, like lower vertebrates, only possess one *IDO* gene, with more similarity to *IDO2* than *IDO1*, suggesting the former to be the closest of the two to the ancestor *IDO* gene. Interestingly, human IDO2 is characterized by two highly frequent single nucleotide polymorphisms (SNPs), which correspond to a reduction of more than 90% of the enzyme activity and premature stop codon with a complete loss of activity, respectively. These two nonfunctional alleles are widely present in the world population [192]. IDO2 is composed of 420 amino acids and has a molecular weight of approximately 47 kDa and a 43% homology with IDO1 for what does regard the primary structure, while little of the amino acid sequence is shared with TDO, with which IDO1 and 2 are not structurally related. While IDO1 is mainly responsible for the extra-hepatic metabolism of tryptophan, IDO2 is expressed in the liver and kidneys, but also in the testis, epididymis, ovaries, and uterus. The regulation of IDO2 by IFN-γ, a cytokine so essential in the induction of IDO1 expression, has been evaluated in different cancer cell lines, and contrasting results have been obtained. IDO2 activity seems to be lower than IDO1's and L-Trp possesses a higher affinity for IDO1 than IDO2. This suggests that the role of IDO2 in Trp metabolism is neglectable and lacks a relevant enzymatic activity [194–196]. IDO2 is also expressed in antigen-presenting cells, like DCs, and this has suggested the involvement of IDO2, like IDO1, in the regulation of the immune response. Some evidence so far reveals an immunomodulatory role for IDO2. Even if this effect appears to involve APCs and T cell populations, it seems that the mechanisms of this immune regulation are somehow different from IDO1's. IDO2 expression has also been observed in some types of cancer, such as basal cell skin carcinoma, gastric, colon, renal, and pancreatic cancer, including pancreatic ductal adenocarcinoma (PDAC), brain tumors, and also nonsmall-cell lung cancer (NSCLC), but the involvement of this enzyme in the tumorigenesis still has to be clarified [197–200].

Tryptophan 2,3-dioxygenase (TDO) was discovered in 1936 by Kotake and Masayama. In mammals, TDO is limited to the liver, with a low expression in the brain, where it seems to play a role in neurodegenerative diseases [82]. Like IDO1 and 2, TDO is a heme protein; nevertheless, the primary structures of TDO and IDO1 are not similar and their homology derives from the function (functional homology), the similarity of the three-dimensional structures and the conservation of specific residues in the catalytic sites. An important difference between the two enzymes is substrate specificity. While IDO can metabolize diverse molecules other than L-Trp (like D-Trp, tryptamine, and serotonin among others), TDO only catalyzes the deoxygenation of L-tryptophan few 5- and 6-substituted Trp derivatives [201–203]. There is a growing amount of evidence that TDO is involved in brain and neurodegenerative diseases, like schizophrenia, Parkinson's, and Alzheimer's diseases, due to its expression in the brain; hence, the modulation of TDO activity at this level may be beneficial to the treatment of numerous conditions [204,205]. For what does regard the involvement of TDO in cancer, TDO may favor cancer metastasis and progression acting on the immune response to tumoral cells, in addition, it also seems to play a role in angiogenesis. These effects, as for IDO1, make TDO an important predictive biomarker of progression, survival, and resistance to immunotherapy and an important target in the ongoing research for the development of new cancer treatments [206–210].

7 IDO inhibitors

7.1 From first discoveries to Indoximod

In the light of the key roles that IDO enzymes cover in different pathological conditions, they have been, and still are, attractive immunotherapy targets [166]. Accordingly, many inhibitors have been designed and some of them are currently evaluated in clinical trials; these compounds are mainly IDO1 inhibitors investigated for anticancer purposes. Due to the particular history of the kynurenine pathway discovery and IDO1 identification, it is quite difficult to establish what were the first reported inhibitors of this enzyme. In the attempt to untangle this aspect, we identified as starting point the work of Schor and Frieden, which in 1958 described the ability of serotonin, carbobenzoxy L-tryptophan, and L-epinephrine to weakly inhibit ($K_I \approx 3 \times 10^{-3}$ M), from a rat liver extract, the enzyme at that time still named tryptophan peroxidase (Fig. 8A) [211]. Three years later the same research group assessed

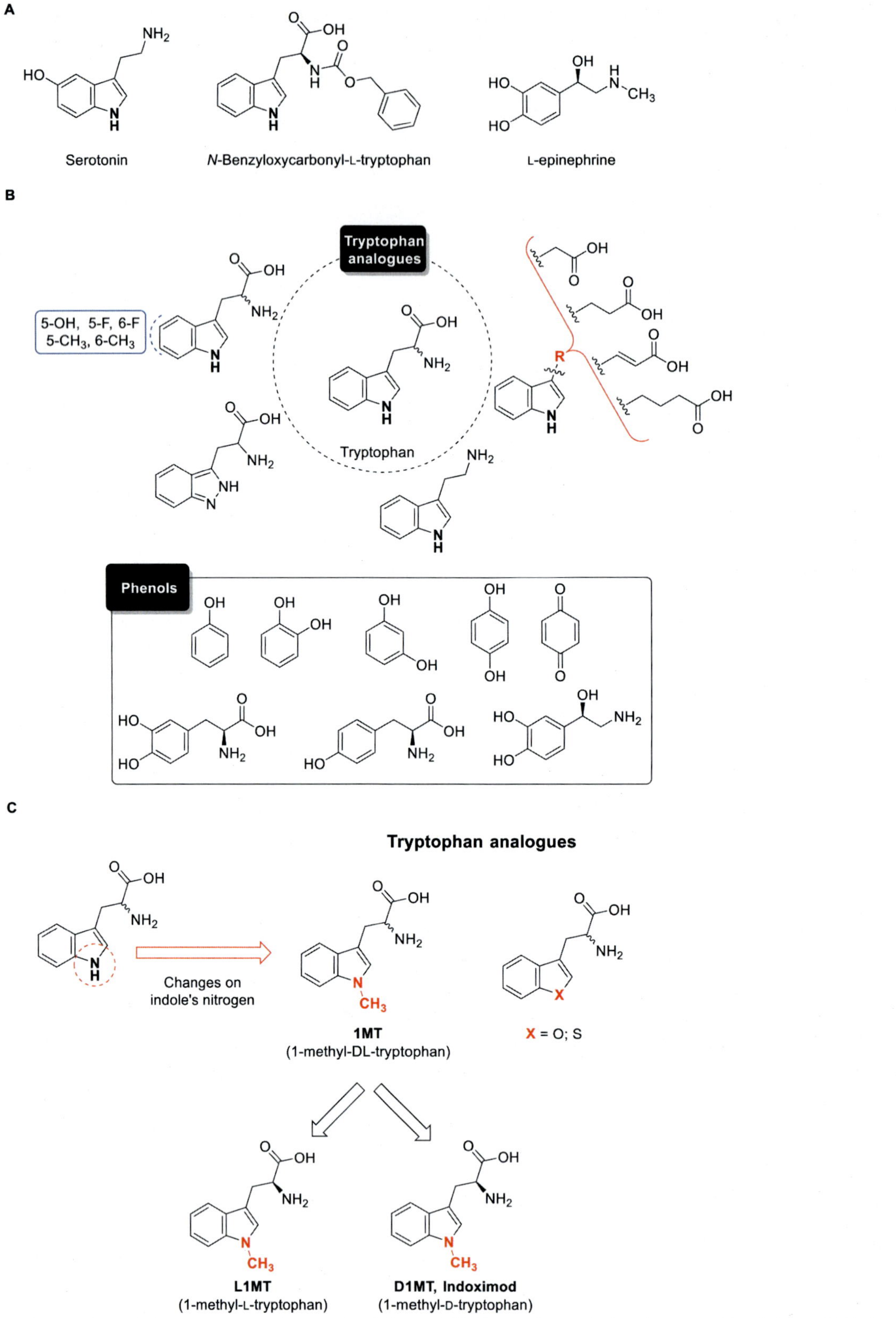

FIG. 8 (A) Structures of serotonin, carbobenzoxy L-tryptophan, L-epinephrine and (B) Trp analogs and phenols proposed by Schor and Frieden. (C) Development of 1MT and Indoximod.

against IDO1 isolated from rat liver a series of compounds including serotonin, indole derivatives, phenols, and tryptophananalogs [212]. The last ones were designed by taking advantage of a common approach exploited in medicinal chemistry to develop enzyme inhibitors, based on the design of substrate analogs, even though this strategy could let to not selective compounds. Interestingly, some of these derivatives effectively blocked the conversion of tryptophan to kynurenine. In particular, Trp analogs bearing side-chain variants (i.e., indole-3-acrylic acid) functioned as competitive inhibitors while the substitution on benzene ring of the indole ring (i.e., 5-F, 5-Me, etc.) led to noncompetitive inhibitors (Fig. 8B).

This line of research, regarding molecules designed on the basis of the Trp scaffold, underwent a "dormant period" that lasted up to 1991 when Cady and Sono reported three Trp analogs exhibiting micromolar K_I values in the range 7–70 μM against rabbit small intestinal IDO (Fig. 8C) [213]. The discovery of the ability of D,L racemic mixture of 1-methyltryptophan to weakly inhibits IDO1 (**1MT**, K_I IDO = 34 μM) represented an important breakthrough, this inhibitor having been (and still being) largely employed as reference IDO inhibitor in preclinical studies. The first studies aimed at evaluating the antitumoral effects of 1MT demonstrated that it moderately retarded the growth of cancer in different mouse models [214,215]. The step forward in the understanding of its therapeutic potential came from its combination with chemotherapy. 1MT was investigated for its antitumoral activity in association with DNA-damaging antitumoral drugs (cyclophosphamide, cisplatin, doxorubicin, and paclitaxel) eliciting the regression of tumors otherwise resistant to classic therapies [216]. These results were not associated to drug-drug interactions but to the ability of 1MT to elicit T cell attacks in the presence of chemotherapy [217].

Successively, the chiral resolution of the racemic mixture and the isolation of the single enantiomers led to the assessment of them against IDO enzymes, figuring out an unanswered key question concerning their inhibitory activity (Fig. 8C) [218]. Interestingly, L1MT functioned as a competitive inhibitor (K_I IDO = 19 μM) or, as successively described by one of the authors (i.e., Prendergast), as a weak substrate of IDO [128,219]. This inhibition mechanism was unraveled by means of a meticulous study requiring fine control of O_2 levels, demonstrating that L1MT is competitive with respect to L-tryptophan and noncompetitive with respect to O_2 [220]. The D-isomer (Fig. 8C, D1MT, Indoximod) instead was much less effective against IDO1, its K_I being higher than 100 μM; similar results were achieved in biological assays based on the evaluation of kynurenine production by intracellular IDO enzyme expressed in HeLa cells. Actually, studies revealed that D1MT efficiently inhibits IDO2 [192] and is more effective than L1MT in reversing the suppression of T cells and enhancing anticancer activity in chemo-immunotherapy regimens [218]. D1MT was further investigated to understand the largely debated complex cellular mechanism [221,222]. Recently, Brincks and co-workers demonstrated that D1MT influences the effect of IDO beyond and distinct from direct enzymatic inhibition of the enzyme. Indeed, this compound reactivates the mammalian target of rapamycin complex 1 (mTORC1), involved in monitoring the level of nutrients, like amino acids such as tryptophan eliciting direct effects on T cells ($CD8^+$ and $CD4^+$ T cells) [223].

In the light of the preclinical results, D1MT (from here named Indoximod) has been largely investigated in clinical trials. Indoximod has been examined in eighteen studies (Phase 1 or Phase 2), against different tumor types and in association with some currently employed clinical antitumoral drugs (ClinicalTrials.gov). For what concern the current status of the clinical trials, twelve result to be completed, three terminated, two have been withdrawn, and two are recruiting. The safety of Indoximod as single agent was assessed in a phase I trial, demonstrating a good oral bioavailability along with mild toxicity and almost absence of myelosuppression (NCT00567931). Moreover, 2000 mg orally twice per day (BID, *bis in die*) was found as the maximum safe dosage of Indoximod [224]. In a successive phase 1 trial, Indoximod (1200 mg BID) was combined with docetaxel at the recommended dose level of 75 mg/m^2, in patients with metastatic solid tumors [225]. This association was well tolerated, with main adverse events due to treatment (treatment emergent adverse events, TEAEs) being fatigue, anemia, hyperglycemia, infection, and nausea [225]. The trial NCT02835729 assessed the safety and preliminary effectiveness of Indoximod in combination with induction chemotherapy (Idarubicin 12 mg/m^2/d × 3 days with cytarabine 100 mg/m^2/d × 7 days) in patients with newly diagnosed acute myeloid leukemia [226]. After the induction, the patients were treated with consolidation therapy based on the administration of up to 4 cycles of high-dose cytarabine (HiDAC) while continuing Indoximod. As observed in other clinical trials, the combination of Indoximod with induction chemotherapy was well tolerated, and a high percentage (79%) of the patients receiving their scheduled Indoximod doses achieved remission. In a phase 1/2 clinical trial, Indoximod has been associated with temozolomide, bevacizumab and stereotactic radiosurgery (Indoximod + temozolomide, Indoximod + temozolomide + bevacizumab; Indoximod + temozolomide + stereotactic radiosurgery) for the treatment of temozolomide-refractory primary high-grade gliomas in adults, still proving Indoximod safety also in conjunction with the other treatments (NCT02052648). Moreover, encouraging survival outcomes was observed when Indoximod was combined with temozolomide alone or temozolomide + stereotactic radiosurgery. In the phase II trial NCT02073123, Indoximod was evaluated in combination with checkpoint inhibitors pembrolizumab (the most employed in the trial), nivolumab, or ipilimumab for the treatment of advanced melanoma [227]. Indoximod was administered at the recommended dose of 1200 mg BID, while checkpoint inhibitors were dosed per US Food and Drug Administration (FDA)-approved label. The combination Indoximod-pembrolizumab was the most common one, and 89 of the patients from the phase II cohort with nonocular melanoma received this co-administration leading to objective response rate (ORR), as indicated in the RECIST guidelines [228], of 51% with confirmed complete response of 20% and disease control rate of 70%. The combination was well tolerated with side effects similar to the ones observed for the single drug pembrolizumab.

Some trials, however, did not have the expected results. As a matter of fact, two Phase 1/2 trials involving patients with metastatic breast cancer did not report evidence of usefulness by the treatment [229]. The trial NCT01042535 assessed the combination of Indoximod with vaccine Ad. p53 (Adenovirus p53 has been used to obtain vaccine against p53). The primary endpoints of this study were to investigate maximum tolerate doses (phase I) and objective response (phase 2), respectively. As regard the phase 1 results, the maximum tolerated dose of Indoximod in combination with Ad.p53-DC vaccine was 1600 mg twice daily. However, the same combination did not increase the objective response rate over the trial threshold of 20%. Similarly, the phase 2 trial NCT01792050 evaluating the treatment of ERBB2-negative metastatic breast cancer treated with Indoximod plus a taxane and did not improve progression-free survival compared with the taxane alone.

Further investigations have to be performed in order to better understand the real therapeutical potential and employment of this drug.

7.2 Navoximod

The beginning of Navoximod history took place simultaneously with the development of the first Trp analogs. In 1979, taking advantage of previous research concerning inhibitors of brain tryptophan 5-monooxygenase, Takeuchi and Umezawa evaluated 2,5-dihydro-L-phenylalanine as putative inhibitor of both IDO1 (obtained by extracts of rabbit small intestine) and TDO (recovered by supernatant fraction of mice's liver homogenates) [230,231]. This compound moderately inhibited IDO1 ($K_I = 0.23$ mM) and TDO ($K_I = 0.70$ mM) in a competitive manner with respect to L-Trp. With the aim to investigate novel chemical entities able to inhibit IDO and TDO, the same research group evaluated a series of novel compounds including indole-based derivatives and β-carbolines (Fig. 9) [232]. The indole-based compounds were ineffective against IDO1 while showing moderate activity against TDO; on the contrary, some of the β-carbolines effectively inhibited these enzymes, so describing for the first time the ability of this scaffold to inhibit IDO1 and TDO [232]. In particular, among the assessed compounds, the most active one was norharman that inhibited IDO1 and TDO with K_I values of 0.12 mM and 0.29 mM, respectively (Fig. 9). Preliminary kinetic studies categorized norharman as uncompetitive inhibitor of IDO and competitive inhibitor of TDO. The mechanism of action of norharman was untangled five years later, when Sono and Cady demonstrated through enzyme kinetic studies and spectroscopic investigation the ability of this and 4-phenyl-1*H*-imidazole (Fig. 9, **4PI**) to similarly bind to IDO enzyme as heme binders [233].

Moreover, the study demonstrated that both the investigated compounds had noncompetitive inhibition mechanisms with respect to L-Trp and D-Trp, thus contradicting the previously reported observations [232]. Nevertheless, some differences were assessed between the two inhibitors. Norharman directly binds as a nitrogen donor ligand to the heme iron of the ferrous (active) enzyme, competing with O_2. Instead, 4PI seemed to compete for heme iron in the ferric enzyme thus preventing the reductive re-activation of the dioxygenase. The binding of 4PI to the IDO1 heme iron was then confirmed by the cocrystal structure of 4PI/IDO1. Indeed, in 2006, a pioneering study performed by Sugimoto and co-workers led to the first crystal structure of IDO1 complexed with 4PI confirming the ability of this weak (IC_{50} IDO1 = 48 μM) noncompetitive IDO1 inhibitor to coordinate heme moiety through the imidazole nitrogen, placing its phenyl group to a hydrophobic domain pocket (Fig. 10) [96]. Interestingly, in the crystal structure were also included two molecules of *N*-cyclohexyl-2-aminoethanesulfonic acid, the buffer employed for crystallization which occupied an adjacent hydrophobic domain.

This discovery led to scientific campaigns aimed at ameliorating the pharmacodynamic and pharmacokinetic (PK) properties of 4PI. In particular, with an intensive SAR study performed in different steps, Kumar and colleagues found some important changes on the 4PI scaffold, leading to the discovery of NLG919 (Fig. 11, Navoximod) one of the IDO1 inhibitors investigated in clinical trials [236–238]. Fig. 11 reports some of the highlights in the phases that led to the discovery of Navoximod. Initial studies performed on 4PI scaffold evaluated the effects of small hydrophilic and hydrophobic moieties located on the 4PI phenyl ring to affect IDO inhibitory activity at enzymatic (IC_{50}) as well as cellular (EC_{50}) assays. Interestingly, the presence of 2′-OH group on the phenyl ring of 4PI resulted in a 10-fold increase of enzymatic inhibition, ascribed to the hydrogen bonding interactions established by hydroxyl group with Ser167; further improvement was obtained by the insertion of fluorine and chlorine at the positions 3′ and 5′ respectively (Fig. 11, **1-2**). Albeit ameliorated in their IDO inhibition, these derivatives exhibited a short half-life likely due to phenol metabolization. With the aim to establish

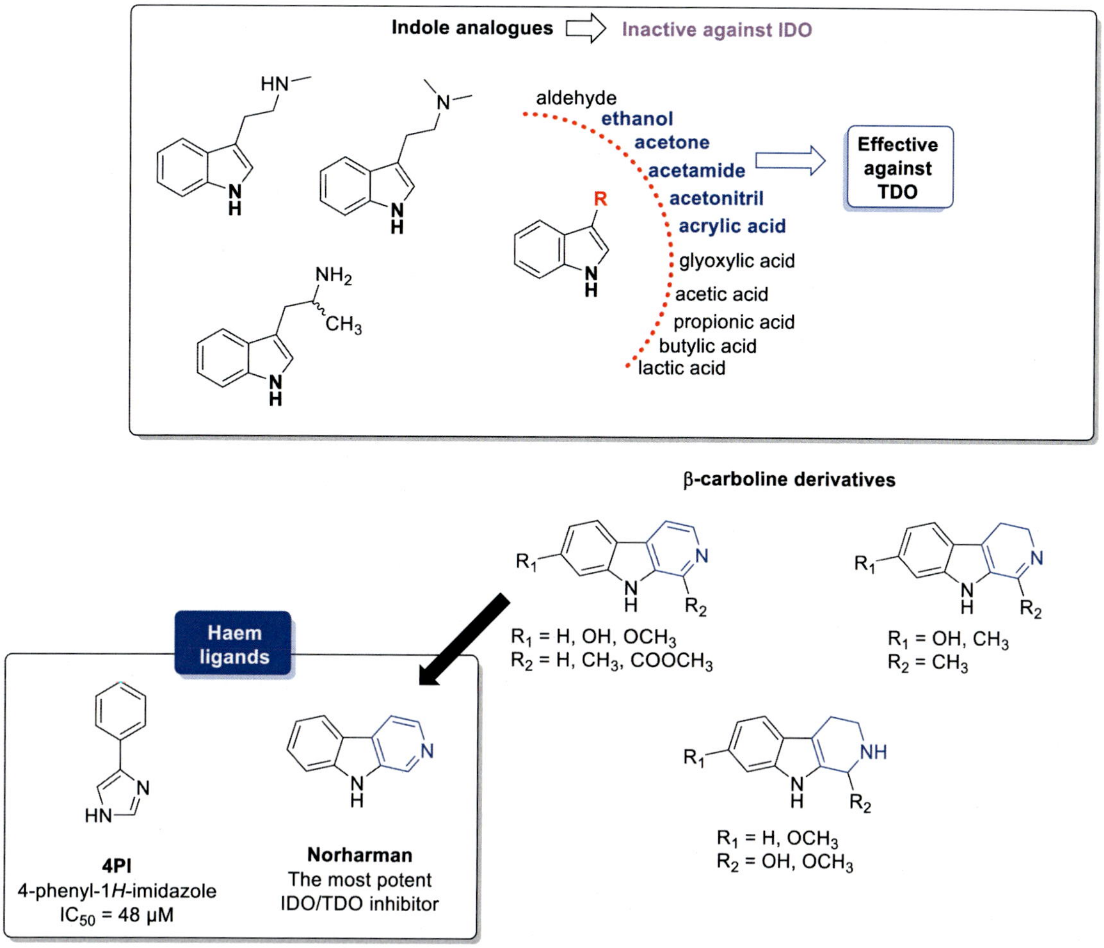

FIG. 9 Structures of indole-based compounds (Takeuchi and Umezawa), β-carbolines and 4PI.

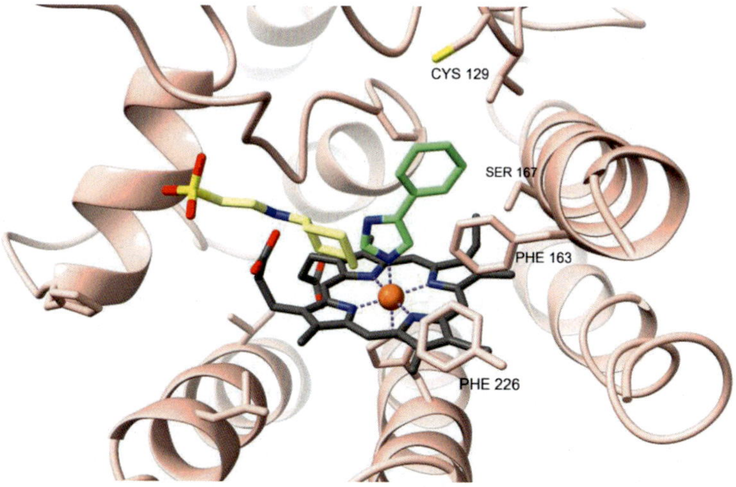

FIG. 10 Crystal structure of 4-PI (*green*) bound to heme iron of IDO (PDB: 2D0T) [96]. Protein images have been created starting by the PDB files elaborated with ChimeraX 1.5 software [234,235].

interaction with the hydrophobic pocket formed by Phe163 and Phe226 (Fig. 10), the hydroxyl was functionalized, resulting in OH capping, leading to ether derivatives (Fig. 11, **3**). However, among the evaluated substituents only the *O*-ethylene linker terminating with a cyclohexyl group exhibited outcomes comparable with phenol **1**. The successive modifications were addressed to increase membrane permeability and cellular potency by reducing the number of rotatable bonds. In this regard, molecular rigidification led to the discovery of the imidazo[5,1-*a*]isoindole

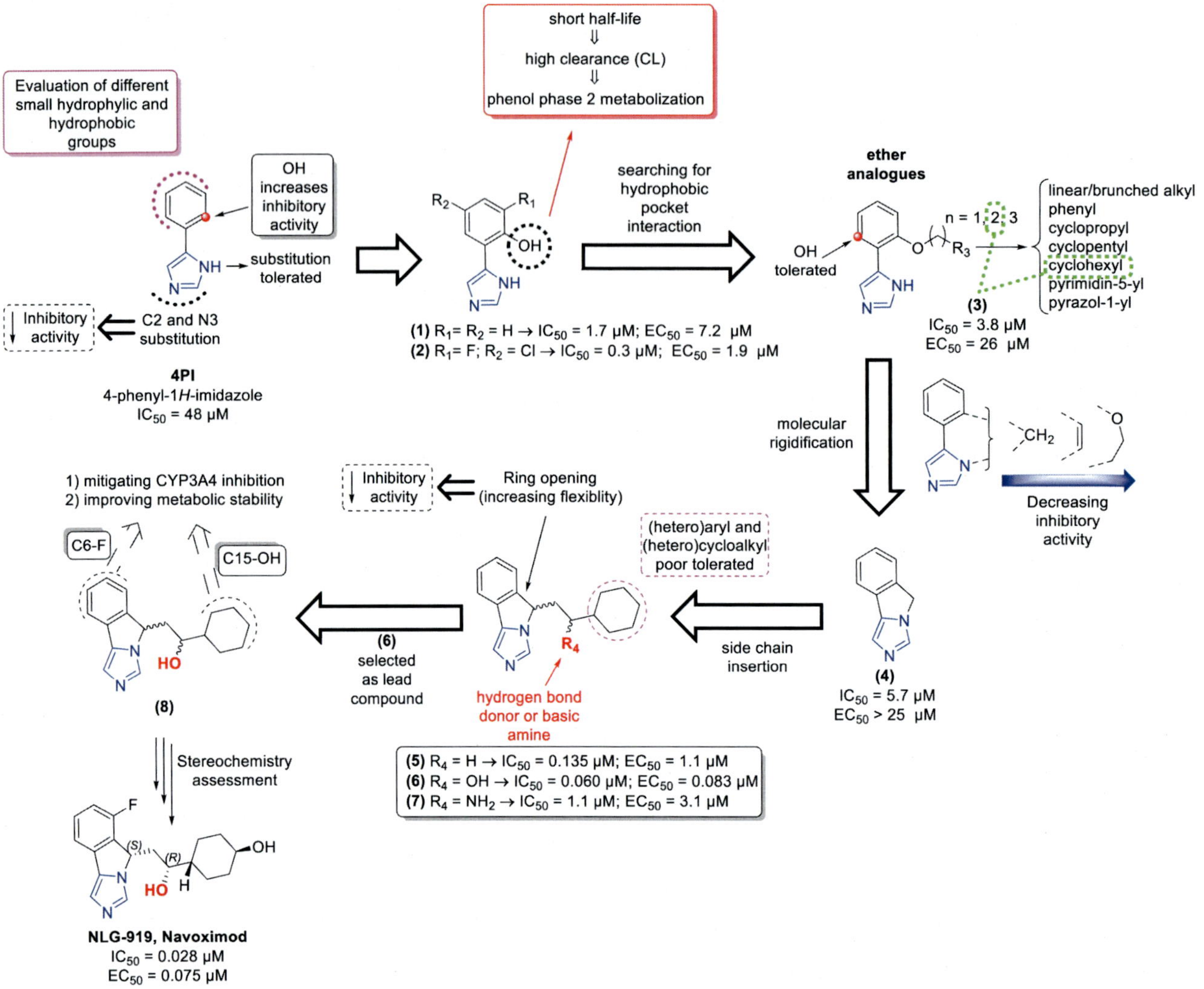

FIG. 11 Structure-activity relationship (SAR) studies leading to Navoximod discovery.

ring, with the other attempts leading to weaker inhibitors (Fig. 11, **4**). Keeping in mind the good results demonstrated by the insertion of ethylcyclohexyl ether side chain, a similar approach was attempted at the 5-position of the imidazoisoindole ring, affording derivative **5** exhibiting improved potency at enzymatic as well as cellular assays (Fig. 11).

Moreover, docking studies hypothesized that the methylene group binding the cyclohexyl group could be in close proximity to heme propionate side chain. Besides, the authors evaluated the introduction of a hydrogen bond donor or a basic amine to form a hydrogen bond. Interestingly, the presence of OH group led to a dramatic increase in potency providing the analogue **6**, which was selected as lead compound for successive studies. In particular, the last efforts were directed at the improvement of PK properties by mitigating CYP3A4 inhibition and enhancing metabolic stability. The authors assessed the effects of small halogens on the phenyl ring of the imidazo[5,1-*a*]isoindole core, obtaining the best results for 6-F. Furthermore, taking into account the first-pass metabolism of cyclohexyl moiety [239], the C15 position (considered a hot spot for metabolization) was blocked to avoid high clearance. Moreover, docking studies showed that cyclohexyl ring was in proximity with Ser235 and Arg231, prompting the insertion of a polar group in this position.

As matter of fact, substitution of C15-position with hydroxyl group led to the best balance of potency, metabolic stability, off-target liabilities, and physiochemical properties. Final inquiry concerned the understanding of the impact that stereochemistry exerts on drug properties, achieving the best results for **NLG-919** (currently **Navoximod**). Crystal structure of Navoximod in complex with IDO1 corroborated the coordination of the imidazoisoindole with heme iron, also establishing hydrophobic interactions in the cavity created by IDO1 residues

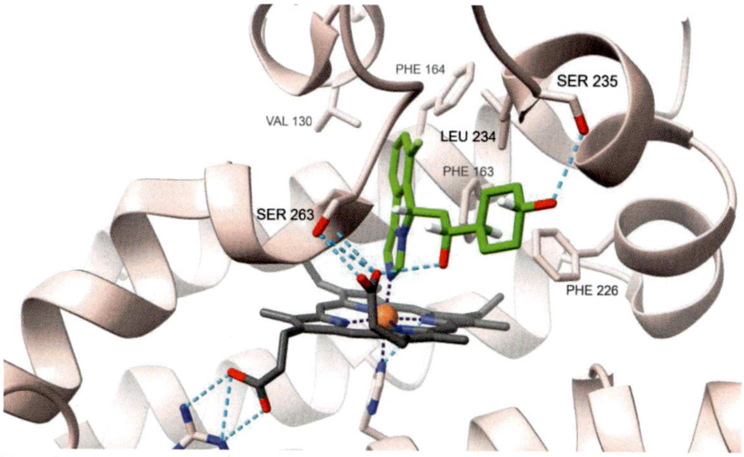

FIG. 12 Crystal structure of Navoximod (*green*) bound to heme iron of IDO (PDB: 6O3I) [96].

Val130, Phe163, Phe164, and Leu234 (Fig. 12) [237]. The stereochemistry of hydroxyl group on the side chain linker allows the establishment of a hydrogen bond with the heme propionate, while the terminal hydroxyl moiety located on the cyclohexyl group is able to form a hydrogen bond interaction with Ser235. All these interactions account for the tight bind of this inhibitor with IDO1 enzyme [237].

From the pharmacological point of view, Navoximod displayed moderate metabolization by human and rat microsomes and no or very weak inhibitory effect for different CYP isoforms (CYP1A2, CYP2B6, CYP2C8, CYP2C9, CYP2C19, and CYP2D6), while slightly affected CYP3A4. Caco-2 assay established its high permeability; in addition, it did not result in a potential substrate for P-glycoprotein. The pharmacokinetic profile of NLG-919 was investigated in rats, mice, and dogs by analyzing $t_{1/2}$ and clearance after single intravenous or oral administration, and it exhibited a moderate bioavailability for the latter [238]. The antitumor activity of Navoximod was investigated against Pan02 tumor model in C57Bl/6 mice, demonstrating tumor growth reduction and survival enhancement; furthermore, in combination with cyclophosphamide, improvement of antitumor effects was observed. Further studies performed at preclinical level on sarcoma models, demonstrated that Navoximod provokes a considerable decrease in plasmatic Kyn to Trp ratio along with a reduction of tumoral Kyn [99,240]. However, in sarcoma-bearing mice, Navoximod used alone or in combination with an anti-PDL1, did not exhibit anti-tumoral effects, and also unaffected the tumor immune cell infiltrate. On the contrary, different outcomes came from experiments involving 4T1 tumor cell engrafted mouse model treated with doxorubicin, inducing immunogenic cell death and the upregulation of IDO1 and Navoximod [241]. Navoximod combined with DOX reversed $CD8^+$ T cell suppression and significantly inhibited the tumor growth with respect to the single agent, accounting for synergistic effect. To date, three clinical trials involving Navoximod have been realized: two have been completed (NCT02048709, NCT02471846) while the more recent one, aimed at evaluating the combination of stereotactic body radiation therapy (SBRT) with Navoximod and NLG802 (a prodrug of Indoximod), has been withdrawn (NCT05469490, no information available). In the NCT02048709 Phase 1 trial, Navoximod was evaluated for safety, pharmacokinetics, and pharmacodynamics. Moreover, a preliminary antitumor activity in patients with recurrent/advanced solid tumors was assessed [242]. This drug was well tolerated at doses up to 800mg BID on a cycle of 21/28 days or 600mg continuous dosing. As regard to clinical activity, no objective responses were observed, the best outcomes have been stable disease (36% of patients), while the majority (45%) had progression of illness. Another Phase I/II clinical trial starting in 2015, investigated the combination treatment of Navoximod with Atezolizumab (programmed cell death ligand 1, PDL-1, inhibitor) in advanced solid tumors (NCT02471846). Similar to the previous study, the combination of Navoximod and atezolizumab exhibited acceptable outcomes concerning safety, tolerability, and pharmacokinetics for patients with advanced cancer. Albeit activity was observed in various tumor types (melanoma, pancreatic, prostate, ovarian, head and neck squamous cell carcinoma, cervical, neural sheath, nonsmall cell lung cancer, triple-negative breast cancer, renal cell carcinoma, and urothelial bladder cancer), clear evidence of benefit coming from adding Navoximod to atezolizumab was not encountered [243].

7.3 Epacadostat

The history of Epacadostat (INCB024360), the IDO1 inhibitor most investigated in clinical trials, starts in 2007 when a high-throughput screening performed by the pharmaceutical company Incyte Holdings Corp led to the discovery of a series of *N*-hydroxyamidinoheterocycles able to modulate indoleamine 2,3-dioxygenase activity [244]. In particular, the efforts were directed to the identification of a novel lead compound exploitable to afford potent and selective IDO1 inhibitors endowed with suitable pharmacokinetic properties for successive in vivo studies. Among the screened compounds, 4-amino-*N*-benzyl-*N'*-hydroxy-1,2,5-oxadiazole-3-carboximidamide (Fig. 13, compound **9**) was identified as a micromolar inhibitor of IDO (IC_{50} IDO1 = 1.5 μM); moreover, in-depth analysis demonstrated a competitive inhibition mechanism based on the direct binding to the ferrous heme active site and the aptitude to potently inhibit IDO1 also at cellular level (IC_{50} HeLa = 1.0 μM). The successive removal of methylene group between phenyl ring and carboximidamide group, led to 4-amino-*N'*-hydroxy-*N*-phenyl-1,2,5-oxadiazole-3-carboximidamides showing a slight decrease in the IDO1 enzymatic inhibitory activity while increasing its inhibition in HeLa cells assay (Fig. 13, molecule **10**). On the basis of these results, SAR study was performed on this scaffold by evaluating different substituents on the phenyl ring, halogens resulting as the most effective ones in increasing potency at both enzymatic as well as cellular level. In particular, for the monosubstituted analogues, the best results were obtained by placing the halogen in the *meta* position; eventually, the insertion of a second halogen in the *para* position is well tolerated and afforded the most active compound of the series (Fig. 13, derivatives **11–14**). The hydroxyamidine structural motif was challenged resulting fundamental for IDO inhibition, its removal leading to ineffective compounds [244]. The derivative **15**, displaying improved human intrinsic clearance (assessed in in vitro experiments), was further investigated demonstrating no affinity for TDO ($IC_{50} > 10$ μM) and good permeability (Caco-2 assay). In in vivo experiments, compound **15** subcutaneously administered to naive C57BL/6 mice, provoked reduction of kynurenine levels by ~50%–60% thus confirming in vivo blockage of ID01. The same compound was also assessed in C57BL/6 mice bearing GMCSF-secretingB16 tumors, displaying dose-dependent inhibition of tumor growth [244]. However, compound **15** suffered from poor oral bioavailability and, for this reason,

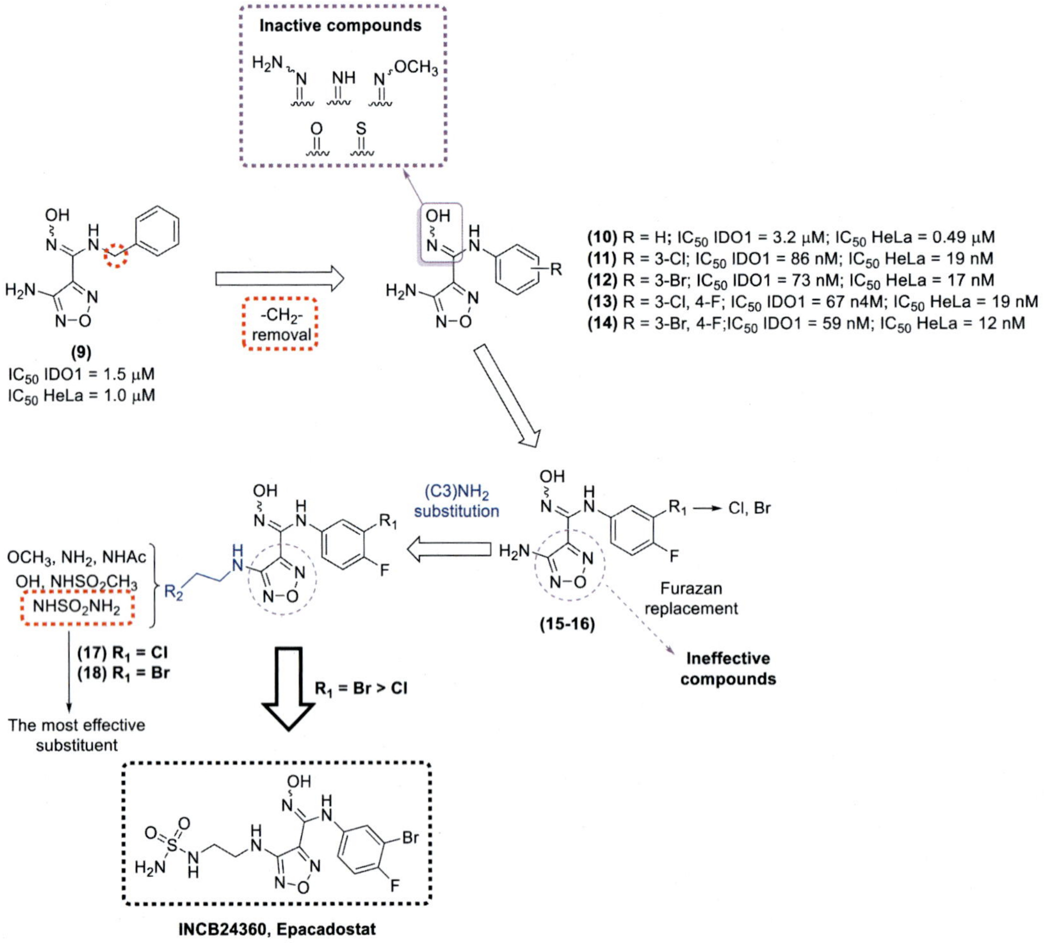

FIG. 13 SAR studies leading to Epacadostat discovery.

successive studies were performed also including the 3-Br, 4-F analogue endowed with better outcomes at the cellular assay (Fig. 13, compound **16**). The replacement of the furazan ring with other heterocycles was investigated, demonstrating the importance of this nucleus for potent IDO1 inhibition [245]. With the aim to reduce the binding with the proximal glucuronidase active site that through glucuronidation affects the drug clearance, substitution at the nitrogen on the C3 position was attempted [245].

These substituents consisted of an ethyl linker bearing a polar group at the end of the chain; the most effective moiety was the sulfamide group (Fig. 13, compounds **17** and **18**). A final round of SAR focused on the substituent located at the *meta* position of the phenyl ring, being both the chloro- (**17**) and bromo- (**18**) analogs endowed with comparable enzymatic (against IDO1) and cellular potency as well as pharmacokinetics properties. However, a head-to-head efficacy study was performed in a CT26 tumor growth model in immunocompetent mice to which it has orally been administered the two drugs. The *meta*-bromo analog demonstrated superior activity and was fully profiled through in vitro and in vivo studies. In vitro studies demonstrated that the *meta*-bromo derivative (i.e., Epacadostat) is a potent and selective IDO1 inhibitor (HeLa $IC_{50}=7.4\,nM$), being devoid of activity against TDO and IDO2. Pharmacokinetics studies performed on rats, dogs, and cynomolgus monkeys demonstrated good oral bioavailabilities. Moreover, albeit the high polar surface area and the large number of hydrogen bond donors, the compound exhibited good permeability (Caco-2 assay). This characteristic, along with oral bioavailability, is attributable to the two intramolecular hydrogen bonds observed in the crystal structure of Epacadostat.

As assessed above, during SAR studies was observed the prominent role covered by the hydroxyamidine motif and its removal or capping caused a dramatic loss of IDO1 inhibition. Based on this evidence, along with enzymatic kinetics and spectroscopic results, the authors suggested as putative inhibition mechanism the establishment of dative bond with the hydroxyamidine oxygen to the heme iron in the ferrous state. In 2017, the crystal structure of IDO1 in complex with Epacadostat was reported and the inhibitor binding mode unraveled [97]. This crystal structure confirmed the previous hypothesis, the drug coordinating the heme iron through the oxygen of the hydroxyamidine group (Fig. 14). The binding of Epacadostat is stabilized through a H-bond with A264; the benzene ring is situated in a hydrophobic task lined by F163, L234, F164, V130, Y126, and S167 residues. The Br and F atoms are placed next to the C129, thus allowing the establishment of fluorine-sulfur contact that stabilizes protein-inhibitor interactions [246]. Lastly, the side arm of Epacadostat bearing the sulfamide group is able to establish interaction with R231, while the furazan ring is stabilized by F163, L234, and F226.

As assessed above, Epacadostat is the IDO1 inhibitor mostly investigated in clinical trials. Currently, of the 60 clinical trials that have been reported for this drug, 18 are completed, 14 are terminated, one figures as "unknown," nine are active but not recruiting, five are recruiting and 13 have been withdrawn. The first among human clinical trials aimed at assessing the maximum tolerated dose, safety, pharmacokinetics, pharmacodynamics, and antitumor activity of Epacadostat started in 2010 (NCT01195311). It enrolled 52 patients with advanced solid malignancies, treated with escalating Epacadostat doses (from 50 mg up to 700 mg BID). In general, the drug was well tolerated; the most common adverse events were fatigue, nausea, decreased appetite, vomiting, constipation,

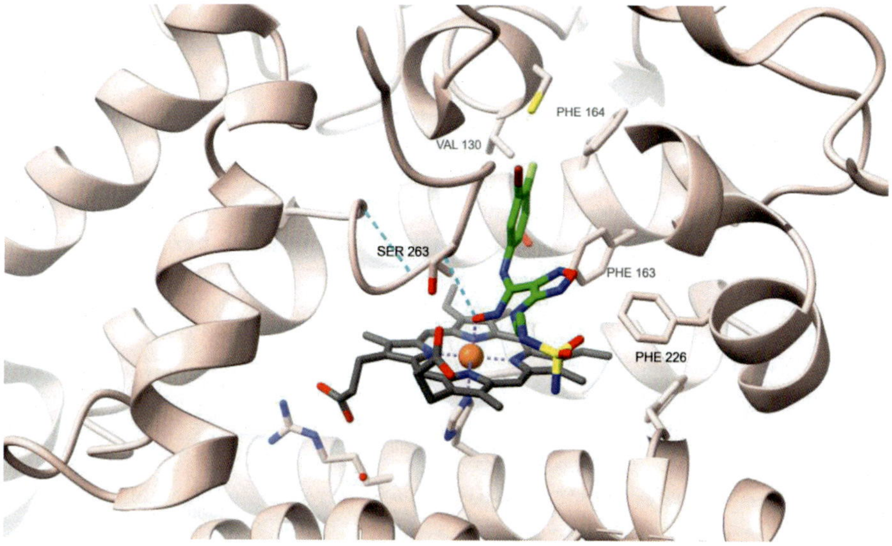

FIG. 14 Crystal structure of Epacadostat (*green*) bound to heme iron of IDO (PDB: 5wn8) [97].

abdominal pain, diarrhea, dyspnea, back pain, and cough. However, a dose-limiting toxicity occurred at 300 mg BID (grade 3, radiation pneumonitis) and 400 mg BID (grade 3, fatigue). As regard the antitumor activity, Epacadostat did not produce objective responses with only 7 patients experiencing stable disease lasting 16 weeks [247]. Another study examined the effects of Epacadostat as single agent for myelodysplastic syndromes treatment (NCT01822691) [248]. This phase 2 trial was realized with fifteen patients that prior had received azacytidine and that successively has been treated with 600 mg of orally Epacadostat twice a day for 16 weeks. Albeit the treatment was well tolerated, the best outcome was stable disease in 80% patients while the others 20% experienced progressive disease. Half of the clinical trials occurred with a combination of Epacadostat with Pembrolizumab, a monoclonal antibody working as an inhibitor of programmed death protein 1 (PD-1). This coadministration has been employed for different cancer types. In the phase 1/2 trial NCT02178722, have been evaluated safety and efficacy of Epacadostat orally administered in combination with intravenous pembrolizumab, in patients with advanced tumors. The outcomes of phase 1 defined Epacadostat 100 mg BID plus pembrolizumab 200 mg every 3 weeks as the dosage for successive phase 2, which included patients with advanced or recurrent melanoma (MEL), nonsmall cell lung cancer (NSCLC), renal cell carcinoma (RCC), urothelial carcinoma (UC), triple-negative breast cancer, squamous cell carcinoma of head and neck (SCCHN), ovarian cancer, diffuse large B-cell lymphoma, or microsatellite instability—high colorectal cancer. During the study, the majority of patients (84%) experienced treatment-related adverse events (TRAEs), mainly fatigue, rash, arthralgia, pruritus, and nausea and 11% of patients discontinued the study because of them. Nevertheless, objective responses occurred in 55% of the patients with melanoma and in patients with non-small-cell lung cancer, renal cell carcinoma, endometrial adenocarcinoma, urothelial carcinoma, and squamous cell carcinoma of the head and neck treated, thus demonstrating encouraging antitumor activity [249,250]. Similar results were observed also for the trials NCT02327078 and NCT01604889, treating advanced melanoma and combining Epacadostat with Nivolumab and Ipilimumab (two (Cytotoxic T-Lymphocyte Antigen 4, CTLA-4, inhibitors), respectively [251,252]. In the former, the overall response rate was 62% with 18% of complete response and 36% of partial response. Albeit some patients discontinued treatment due to TRAEs, the co-administration of the two drugs showed promising antitumor activity. The final outcomes have not yet published. The other trial (NCT01604889) combining Epacadostat with Ipilimumab was initially scheduled as phase 1/2 trial. On the contrary, this trial was terminated early, for the sponsor's decision (Incyte Corporation) to focus on other treatments. This study suggested that Epacadostat $\leq$50 mg BID in combination with Ipilimumab 3 mg/kg could have an acceptable safety profile, along with the potential to enhance clinical activity in patients with unresectable or metastatic melanoma [251]. On the basis of these results, it was attempted to run a randomized phase III trial, by comparing pembrolizumab (fixed dose of 200 mg every 3 weeks) with either Epacadostat (100 mg BID) or placebo [253]. Strikingly, the findings of this trial displayed no significant differences between the patients' groups for progression-free survival or overall survival [254]. These results represented a real setback for the research field reducing the enthusiasm for IDO-1 inhibitors. Successively, a series of clinical trials were early terminated because of these outcomes or for ineffectiveness of treatments (NCT02298153, NCT03361228, NCT01685255, NCT03471286, NCT03823131, NCT02959437, NCT03347123, NCT02575807, NCT03277352, NCT03348904, NCT01604889, NCT02559492, NCT03707457; the trial NCT03463161 was terminated due to investigator conflict of interest). Moreover, the studies NCT03322540 and NCT03322566, both focused on patients with metastatic nonsmall cell lung cancer treated with Pembrolizumab plus Epacadostat, failed to meet the set endpoint of improvement in overall response rate. Among the other six phase 3 clinical trials, three result as completed (NCT02752074, NCT03374488, and NCT03361865), two are active but not recruiting (NCT03260894, NCT03358472), one is terminated (NCT03348904) and one has been withdrawn prior to enrolling any patient (NCT03342352).

The reasons behind the failure of these trials have been widely debated. Muller and colleagues poured the attention on the presence of other enzymes (TDO and/or IDO2) that may compensate the inhibition of IDO1 in tumor cells [255]. Another issue concerns the dose employed in the trials, the enzymatic inhibition of IDO1 being dose-dependent. Due to due toxicity problems related to concentrations higher than 100 mg BID of Epacadostat, this concentration has been often employed in the studies. However, the low concentration could be insufficient to ensure an inhibition rate able to fully counteract the immunosuppressive effects of IDO1 in the tumor [253]. Moreover, also the drugs selected for the therapeutic approach have been addressed as the putative reason for the trial failure because more effective combination can be obtained with DNA-damaging chemotherapy agents [253]. All these aspects should be kept into account for future clinical trial studies.

7.4 IPD (EOS200271, PF-06840003)

In 2017, the screening of a library including ~180,000 compounds performed at iTeos Therapeutics led to the discovery of a weak IDO1 inhibitor (Fig. 15, molecule **19**) [256]. This

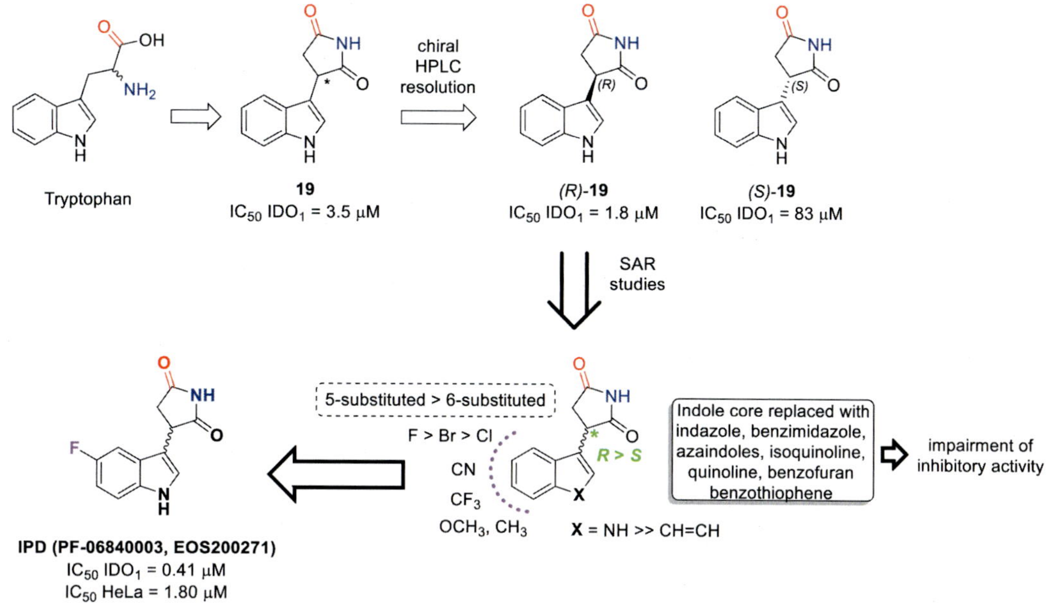

FIG. 15 SAR investigation leading to IPD discovery.

compound, an analogue of tryptophan with the aminoacidic moiety cyclized in the succinimide group, was judged as a suitable starting point for subsequent optimization. The chiral HPLC resolution followed by the enzymatic assessment of the single enantiomers demonstrated as the first eluted, corresponding to the (*R*)-enantiomer, inhibited IDO1 stronger than the second eluted one (Fig. 15: (*R*)-**19**, IC_{50} IDO1 = 1.8 μM; (*S*)-**19**, IC_{50} IDO1 = 83 μM). The active enantiomer was further profiled by in vitro experiments displaying slight protein binding, good stability in human microsomes, and excellent permeability in the Caco-2 assay (without active efflux). Moreover, assessment against five cytochrome P450 isoforms (1A2, 2C19, 2C9, 2D6, and 3A4) demonstrated their scarce inhibition by (*R*)-**19** with $IC_{50} > 50$ μM. In in vivo mice model, (*R*)-**19** demonstrated moderate clearance and good oral bioavailability. However, this compound lacked potency; so, with the aim to increase drug effectiveness, successive SAR studies were performed by evaluating a series of changes on the indole nucleus. Among the different substituents placed on the indole's benzene ring, the halogens exhibited the best outcomes, especially when placed at the C5 position. The insertion of two halogens at the positions C5/C6 was tolerated as well.

On the contrary, nitrile, methyl, and methoxy moieties placed either alone or with halogens at the C5/C6 led to impairment of inhibitory activity. The best effect was obtained by introducing a fluorine atom at the C5 position, leading to a submicromolar IDO1 inhibitor (**IPD**, IC_{50} IDO1 = 0.41 μM). Similarly to what was observed for compound **19**, the chiral resolution and evaluation of the single enantiomers led to the identification of the (*R*) one as the eutomer ((*R*)-IPD, IC_{50} IDO1 = 0.20 μM). Indole nucleus replacement and/or expansion was also attempted by creating a series of molecules that, however, were less effective than the parent compound.

The binding mode of (*R*)-IPD was determined through X-ray and spectroscopic methods, indicating IPD as a tryptophan noncompetitive and nonheme-binding IDO-1 inhibitor [256]. Crystal structure of (*R*)-IPD-IDO1 complex showed that the succinimide ring closely and parallelly lies to the heme moiety, while the indole ring locates in a lipophilic pocket. Interestingly, there is no direct interaction between (*R*)-IPD and the iron, while the inhibitor participates in the establishment of hydrogen bonding network enforcing the interaction with the enzyme (Fig. 16). In particular, the side chain of Ser167 establishes interaction with indole NH, while the succinimide NH hydrogen bonds interact with heme carboxylic acid; finally, the two carbonyls are in contact with residues of Ala264 and Thr379. In addition to the hydrogen bonds, there are several aromatic groups (Tyr126, Phe163, and Phe164) properly distributed to afford π-π interactions with the indole ring that is in proximity with other lipophilic residues (Leu234, Val130, and Cys129) (Fig. 16).

Successive in-depth analysis demonstrated for (*R*)-IPD excellent inhibitory activity in an IDO-1 human whole blood assay. Nevertheless, its rapid racemization occurring both in vitro and in vivo was observed, justifying the advancement of racemate in spite of a single enantiomer for further clinical investigation. In particular, among the good pharmacokinetic properties demonstrated by IPD, the compound showed the ability to cross blood brain barrier (BBB) thus allowing central nervous system (CNS) penetration in rat. In the light of these results, this

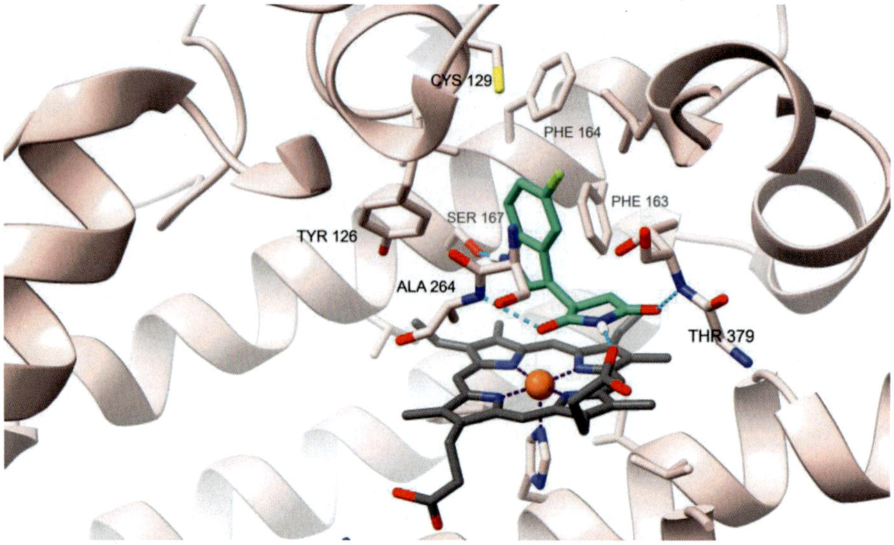

FIG. 16 Crystal structure of IPD (*green*) bound to the IDO1 active site (PDB: 5WHR) [256].

compound was investigated in a dose escalation (125 mg or 250 mg once-daily, 250 mg or 500 mg twice-daily) clinical trial for malignant gliomas (NCT02764151) [257]. This trial demonstrates that IPD was usually well tolerated at the dose of 500 mg twice-daily also evidencing pharmacodynamic effect, with almost the half (47%) of the patients experiencing disease control. Nevertheless, the sponsor of the study decided to no longer pursue this investigation and prematurely terminated the trial before starting phase 2 [257].

7.5 Linrodostat (BMS-986205)

BMS-986205 (also known as Linrodostat) is an IDO1 inhibitor based on a quinolinic core and clinically developed by Bristol-Myers Squibb Corporate (Fig. 16). Linrodostat is a potent and selective inhibitor of IDO1, its in vitro activity falling in the sub-nanomolar range (IC_{50} HeLa = 0.5 nM) [258]. Unfortunately, information about the discovery of Linrodostat is still not available. Linrodostat is endowed with a particular inhibition mechanism that differs from the ones seen so far. Nelp and colleagues reported for the first time this behavior, describing this compound as an irreversible IDO1 inhibitor that binds its apo-form, i.e., the enzyme devoid of heme moiety [258]. The authors exploited different experiments to demonstrate the ease with which IDO1 loses its heme cofactor, also in function of the temperature, this mirroring the temperature dependence of Linrodostat inhibition. Moreover, the inhibitory activity did not show a linear relationship with inhibitor concentration, thus enforcing the opinion that apo-IDO1 formed from intrinsic heme loss is the target of Linrodostat. Cellular assays aimed at justifying the high potency found in cellular assay, correlated to the binding of inhibitor to the apo-IDO1 that was found at least 85% inside the cell and that can be activated through addiction of exogenous heme [259]. Moreover, the lability of IDO1-heme interaction was found to be strongly influenced by the redox status of iron in the heme cofactor, with the ferrous heme binding more tightly than the ferric one. The crystal structure of the complex obtained with a Linrodostat analog (BMS-116, bearing *p*-CN group in spite of *p*-Cl one) and IDO1 confirmed the binding of the inhibitor in a manner that displaced the heme cofactor, albeit the overall structure of IDO1 remained largely unperturbed with respect to heme-containing IDO1 [258]. The quinoline of BMS-116 occupies an additional pocket made available by side-chain rearrangements of Phe-270, Phe-214, His-346, and Arg-343 that are stabilized by edge-to-face π-interactions with Phe-270 and a hydrogen bond with Arg-343 (Fig. 17).

One year later, another group proposed a different point of view performing further crystallographic analysis [260]. The authors crystallized the IDO1 complex devoid of inhibitor, at first; then, soaked these crystals with Linrodostat as a function of time. A large-scale screening process led to the isolation of three unique forms of the hIDO1-Linrodostat complex whose crystal structures were solved. Albeit human IDO1 work as a monomer in free solution, all the three crystal structures were solved in the dimeric form, which were called C0/C2, C0/C3, and C1/C3 (Fig. 18A–D) [260].

C0 represents the inhibitor-free structure displaying an overall protein fold similar to that observed in other complexes albeit some slight differences regarding a few protein loops and water molecules occurred (Fig. 18A) [97]. In the C1 structure, the inhibitor stays perpendicular to the heme moiety, establishing π-stacking interactions with the heme, F226, and F163 (Fig. 18B). However, this distribution

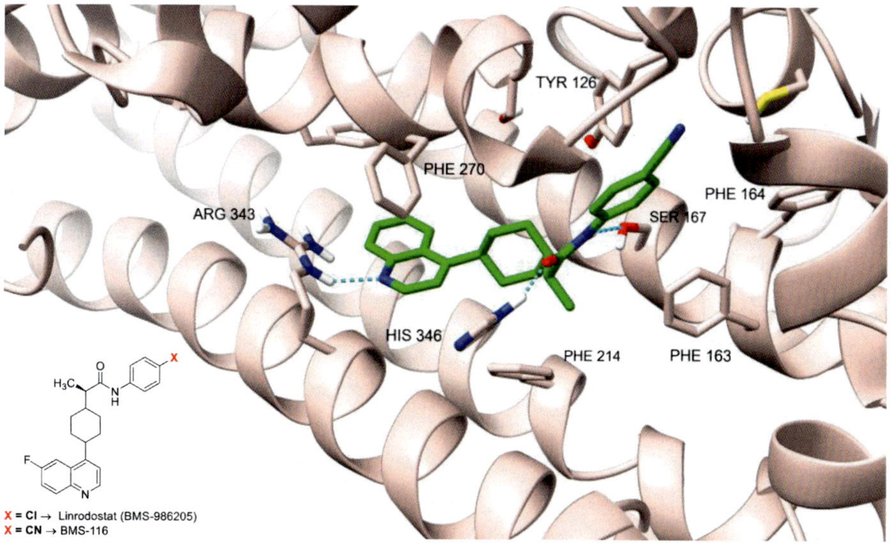

FIG. 17 Crystal structure of BMS-116 (*green*) bound to the IDO1 active site (PDB: 5WHR) [256].

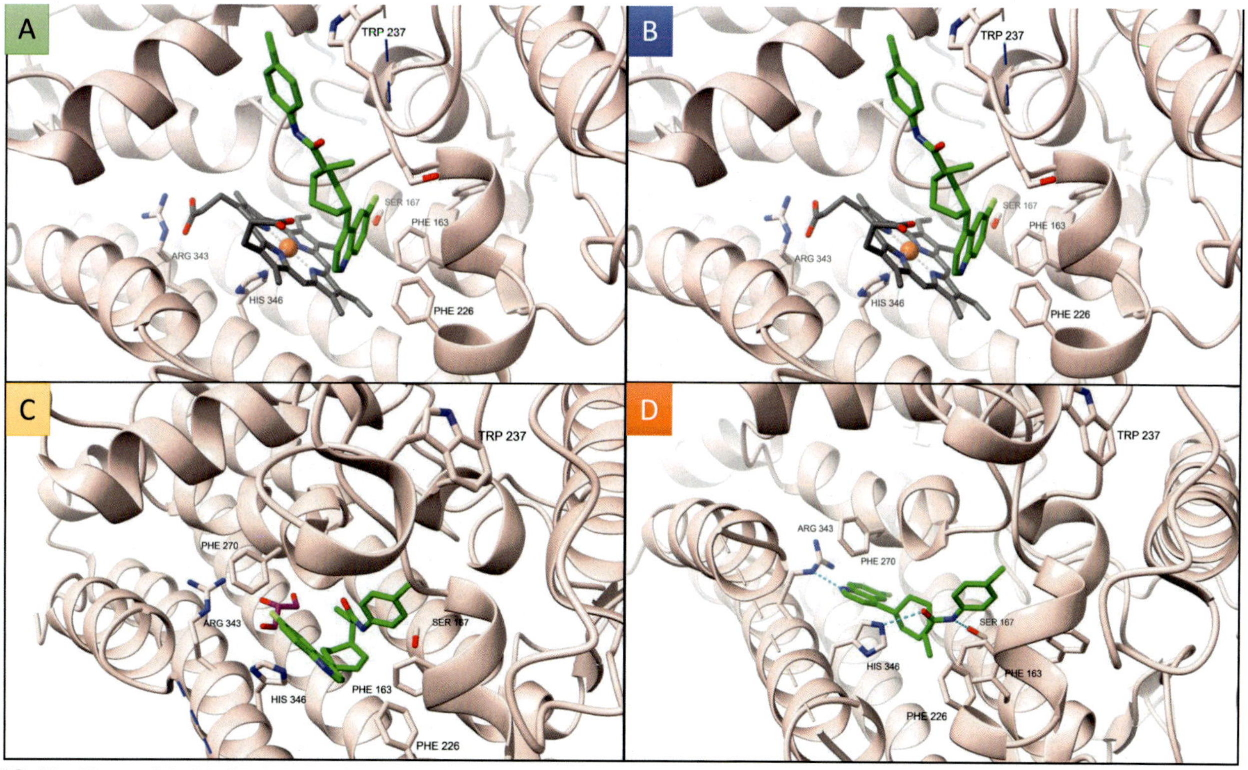

FIG. 18 Phases of the binding between IDO1 and Linrodostat. (A) C0 structure, including also glycerol molecule as part of the solvent employed for crystallization (*violet*). (B) C1 structure. (C) C2 structure. (D) C3 structure. Linrodostat is reported in *green* (PDBs: 6DPQ, 6DPR, 6MQ6).

provokes the rotation of W237 side chain favored by the phenyl propanamide moiety, which led to the partial unfolding of a protein hairpin and consecutive weakening of the H346-iron coordination bond, thus causing heme release. The spacious hydrophobic site early occupied by heme is then occupied by the inhibitor through a large-scale movement, leading to the C2 structure (Fig. 18C). Interestingly, in this position, the inhibitor is in a kinked conformation with the quinoline located at the space equivalent to that occupied by the heme, while the phenyl group establishes π-stacking with Y126 and F163. The conformational strain of the inhibitors at the C2 structure led to a further structural transition (C3) with a "bent" conformation of Linrodostat with quinoline ring stabilized by π-stacking and H-bonding with F270 and R343, respectively (Fig. 18D). In this condition, the propanamide moiety establishes H-bonds with S167 and H346 via the amine and carbonyl groups, respectively. This last structure accounts for the most extensive interactions of Linrodostat with the protein and is the minimum energy state of the complex [260].

In the light of the above, the binding of Linrodostat can be conceived as consecutive steps that, starting from C0, pass through the first interaction in C1 with successive active site cracks opening and release of heme to afford C2. C2 contains high-energy kinked inhibitor, whose torsional energy release led to the crystal-inhibitor complex found in C3. So, while the study of Nelp suggested that Linrodostat can target only the apo-form of hIDO1 to form C3, this work assessed that Linrodostat is able to promote heme displacement from IDO enzyme. Currently, the debate is still open albeit some authors described Linrodostat as an apo-form IDO1 inhibitor on the basis of kinetic studies and spectroscopic determinations also keeping into account experiments involving IDO in physiological conditions supporting the existence of IDO1 in apo-form more than the holo-form [104,105,261,262]. However, further insights are needed to unravel this debate.

Clinicaltrial.org reports 22 clinical studies for Linrodostat including 13 completed, 4 active but not recruiting, three withdrawn and one recruiting. In 2017 was reported the first outcomes concerning the employment of Linrodostat in a phase 1/2 trial, alone or in combination with Nivolumab for the treatment of patients with advanced cancers (NCT02658890) [263]. The therapy was administered following a dose escalation schedule with Linrodostat administered at concentrations up to 200 mg (25–200 mg) orally once daily for 2 weeks, followed by Linrodostat plus Nivolumab (240 mg) intravenous administered every 2 weeks. This study demonstrated that Linrodostat, in combination with Nivolumab, is well tolerated at doses up to 200 mg. Interestingly, the reduction of serum kynurenine took place at the lower concentration (25 mg once a day) thus demonstrating the effectiveness of this drug. Moreover, at the higher concentrations, intratumoral kynurenine reduction occurred [263]. In the expansion of this clinical trial, information about safety across all tumor cohorts and efficacy in the immuno-oncology naive advanced bladder cancer cohort was updated. However, the dose escalation regimen underwent slight modification in the patients receiving Linrodostat 100 or 200 mg once daily plus intravenous Nivolumab at two different dosages (240 mg every 2 weeks or 480 mg every 4 weeks) [264]. Fifty-seven percent of the patients suffered from TRAEs, mainly fatigue and nausea, with little percentage discontinuing the treatment (4%); 3 patients (<1%) died because of TRAEs (myocarditis, Stevens-Johnson syndrome, and hepatic failure). The frequency and severity of TRAEs diminished at the lowering of the Linrodostat dose (100 mg once daily). As regard antitumor effects, the advanced bladder cancer cohort exhibited an objective response rate and disease control rate of 37% and 56%, respectively. These results paved the way for a successive phase 3 clinical trial (NCT03661320) aimed at investigating the effectiveness of chemotherapy (monotherapy with cis-platin) with respect to chemotherapy plus nivolumab with or without Linrodostat in patients with muscle-invasive bladder cancer [265]. Results regarding this trial are still missing.

Recently, some information about the trial NCT04106414 has been reported [266]. This has been a randomized phase 2 study performed on patients with endometrial cancer that have received 1–4 prior lines of chemotherapy. The treatment consisted of Nivolumab (480 mg, intravenous every 4 weeks) with or without Linrodostat (100 mg orally daily). Unfortunately, in the cohort receiving Nivolumab no response was observed, while in the group receiving the combination of the two drugs the ORR was very little (8.3%) and, consequently, the trial was closed due to lack of observed clinical efficacy.

8 Conclusions

Since its discovery, indoleamine-2,3-dioxigenase attracted interest for its mechanism of action, at first, and successively for the roles that play in physiological as well as pathological functions, making IDO a critically important enzyme in the regulation of immune homeostasis. Among the molecular processes involving IDO enzymes, the escape from immune system exploited by tumor cells to proliferate has been widely investigated. IDO1 contributes to the immune escape of cancer cells with several mechanisms, in an intricated labyrinth of interconnections and relationships resulting in a synergistic tolerogenic effect.

In the past years, extensive efforts have been focused on the design of compounds able to affect these enzymes, whose modulation can be exploited for the development of novel therapeutic approaches. The discovery of drugs like Indoximod, Novoximod, and Epacadostat (among others), along with the first results coming from clinical trials, prompted a general enthusiasm that was successively soften because of limited effectiveness observed in some clinical trials. Nevertheless, some concerns about the design of the studies and the selection of the drugs combined with IDO inhibitors could justify the insufficient results obtained paving the way to novel investigations aimed at understanding the real potential of these drugs.

References

[1] Barik S. The uniqueness of tryptophan in biology: properties, metabolism, interactions and localization in proteins. Int J Mol Sci 2020;21:1–22. https://doi.org/10.3390/ijms21228776.

[2] le Floc'h N, Otten W, Merlot E. Tryptophan metabolism, from nutrition to potential therapeutic applications. Amino Acids 2011;41:1195–205. https://doi.org/10.1007/s00726-010-0752-7.

[3] Palego L, Betti L, Rossi A, Giannaccini G. Tryptophan biochemistry: structural, nutritional, metabolic, and medical aspects in humans. J Amino Acids 2016;2016. https://doi.org/10.1155/2016/8952520.

[4] Guillemin GJ, Vincent S, Chen Y. Kynurenine pathway metabolites in humans: disease and healthy states. Int J Tryptophan Res 2009;2:1–19.

[5] Badawy AAB. Kynurenine pathway of tryptophan metabolism: regulatory and functional aspects. Int J Tryptophan Res 2017;10. https://doi.org/10.1177/1178646917691938.

[6] Dougherty DM, Richard DM, Dawes MA, Mathias CW, Acheson A, Hill-Kapturczak N. L-Tryptophan: basic metabolic functions, behavioral research and therapeutic indications. Int J Tryptophan Res 2009;2:45–60.

[7] Hubková B, Valko-Rokytovská M, Čižmárová B, Zábavníková M, Mareková M, Birková A. Tryptophan: its metabolism along the kynurenine, serotonin, and indole pathway in malignant melanoma. Int J Mol Sci 2022;23. https://doi.org/10.3390/ijms23169160.

[8] Platten M, Nollen EAA, Röhrig UF, Fallarino F, Opitz CA. Tryptophan metabolism as a common therapeutic target in cancer, neurodegeneration and beyond. Nat Rev Drug Discov 2019;18:379–401. https://doi.org/10.1038/s41573-019-0016-5.

[9] Hughes TD, Güner OF, Iradukunda EC, Phillips RS, Bowen JP. The kynurenine pathway and kynurenine 3-monooxygenase inhibitors. Molecules 2022;27. https://doi.org/10.3390/molecules27010273.

[10] Nienhaus K, Nienhaus GU. Different mechanisms of catalytic complex formation in two L-tryptophan processing dioxygenases. Front Mol Biosci 2018;4. https://doi.org/10.3389/fmolb.2017.00094.

[11] Stone TW, Stoy N, Darlington LG. An expanding range of targets for kynurenine metabolites of tryptophan. Trends Pharmacol Sci 2013;34:136–43. https://doi.org/10.1016/j.tips.2012.09.006.

[12] Ostapiuk A, Urbanska EM. Kynurenic acid in neurodegenerative disorders—unique neuroprotection or double-edged sword? CNS Neurosci Ther 2022;28:19–35. https://doi.org/10.1111/cns.13768.

[13] Moroni F, Cozzi A, Sili M, Mannaioni G. Kynurenic acid: a metabolite with multiple actions and multiple targets in brain and periphery. J Neural Transm 2012;119:133–9. https://doi.org/10.1007/s00702-011-0763-x.

[14] Wirthgen E, Hoeflich A, Rebl A, Günther J. Kynurenic acid: the Janus-faced role of an immunomodulatory tryptophan metabolite and its link to pathological conditions. Front Immunol 2018;8. https://doi.org/10.3389/fimmu.2017.01957.

[15] Smith JR, Jamie JF, Guillemin GJ. Kynurenine-3-monooxygenase: a review of structure, mechanism, and inhibitors. Drug Discov Today 2016;21:315–24. https://doi.org/10.1016/j.drudis.2015.11.001.

[16] Colín-González AL, Maya-López M, Pedraza-Chaverrí J, Ali SF, Chavarría A, Santamaría A. The Janus faces of 3-hydroxykynurenine: dual redox modulatory activity and lack of neurotoxicity in the rat striatum. Brain Res 2014;1589:1–14. https://doi.org/10.1016/j.brainres.2014.09.034.

[17] Colín-González AL, Maldonado PD, Santamaría A. 3-Hydroxykynurenine: an intriguing molecule exerting dual actions in the central nervous system. Neurotoxicology 2013;34:189–204. https://doi.org/10.1016/j.neuro.2012.11.007.

[18] Parrott JM, O'Connor JC. Kynurenine 3-monooxygenase: an influential mediator of neuropathology. Front Psychiatry 2015;6. https://doi.org/10.3389/fpsyt.2015.00116.

[19] Lu Y, Shao M, Wu T. Kynurenine-3-monooxygenase: a new direction for the treatment in different diseases. Food Sci Nutr 2020;8:711–9. https://doi.org/10.1002/fsn3.1418.

[20] Chen Y, Zhang J, Yang Y, Xiang K, Li H, Sun D, et al. Kynurenine-3-monooxygenase (KMO): from its biological functions to therapeutic effect in diseases progression. J Cell Physiol 2022. https://doi.org/10.1002/jcp.30876.

[21] Vázquez Cervantes GI, Pineda B, Ramírez Ortega D, Salazar A, González Esquivel DF, Rembao D, et al. Kynurenine monooxygenase expression and activity in human astrocytomas. Cells 2021;10. https://doi.org/10.3390/cells10082028.

[22] Capucciati A, Galliano M, Bubacco L, Zecca L, Casella L, Monzani E, et al. Neuronal proteins as targets of 3-hydroxykynurenine: implications in neurodegenerative diseases. ACS Chem Neurosci 2019;10:3731–9. https://doi.org/10.1021/acschemneuro.9b00265.

[23] Guidetti P, Bates GP, Graham RK, Hayden MR, Leavitt BR, MacDonald ME, et al. Elevated brain 3-hydroxykynurenine and quinolinate levels in Huntington disease mice. Neurobiol Dis 2006;23:190–7. https://doi.org/10.1016/j.nbd.2006.02.011.

[24] Korlimbinis A, Hains PG, Truscott RJW, Aquilina JA. 3-Hydroxykynurenine oxidizes α-crystallin: potential role in cataractogenesis. Biochemistry 2006;45:1852–60. https://doi.org/10.1021/bi051737.

[25] Staniszewska MM, Nagaraj RH. 3-Hydroxykynurenine-mediated modification of human lens proteins: structure determination of a major modification using a monoclonal antibody. J Biol Chem 2005;280:22154–64. https://doi.org/10.1074/jbc.M501419200.

[26] Sathyasaikumar KV, Tararina M, Wu HQ, Neale SA, Weisz F, Salt TE, et al. Xanthurenic acid formation from 3-hydroxykynurenine in the mammalian brain: neurochemical characterization and physiological effects. Neuroscience 2017;367:85–97. https://doi.org/10.1016/j.neuroscience.2017.10.006.

[27] Taleb O, Maammar M, Klein C, Maitre M, Mensah-Nyagan AG. A role for xanthurenic acid in the control of brain dopaminergic activity. Int J Mol Sci 2021;22. https://doi.org/10.3390/ijms22136974.

[28] Fathi M, Vakili K, Yaghoobpoor S, Tavasol A, Jazi K, Hajibeygi R, et al. Dynamic changes in metabolites of the kynurenine pathway in Alzheimer's disease, Parkinson's disease, and Huntington's disease: a systematic review and meta-analysis. Front Immunol 2022;13. https://doi.org/10.3389/fimmu.2022.997240.

[29] Phillips RS. Structure and mechanism of kynureninase. Arch Biochem Biophys 2014;544:69–74. https://doi.org/10.1016/j.abb.2013.10.020.

[30] Lima S, Khristoforov R, Momany C, Phillips RS. Crystal structure of Homo sapiens kynureninase. Biochemistry 2007;46:2735–44. https://doi.org/10.1021/bi0616697.

[31] Phillips RS. Structure, mechanism, and substrate specificity of kynureninase. Biochim Biophys Acta, Proteins Proteomics 2011;1814:1481–8. https://doi.org/10.1016/j.bbapap.2010.12.003.

[32] Wang M, Wang Y, Zhang M, Duan Q, Chen C, Sun Q, et al. Kynureninase contributes to the pathogenesis of psoriasis through pro-

inflammatory effect. J Cell Physiol 2022;237:1044–56. https://doi.org/10.1002/jcp.30587.

[33] Liu Y, Feng X, Lai J, Yi W, Yang J, Du T, et al. A novel role of kynureninase in the growth control of breast cancer cells and its relationships with breast cancer. J Cell Mol Med 2019;23:6700–7. https://doi.org/10.1111/jcmm.14547.

[34] Harden JL, Lewis SM, Lish SR, Suárez-Fariñas M, Gareau D, Lentini T, et al. The tryptophan metabolism enzyme L-kynureninase is a novel inflammatory factor in psoriasis and other inflammatory diseases. J Allergy Clin Immunol 2016;137:1830–40. https://doi.org/10.1016/j.jaci.2015.09.055.

[35] Fujigaki S, Saito K, Takemura M, Fujii H, Wada H, Noma A, et al. Species differences in L-tryptophan-kynurenine pathway metabolism: quantification of anthranilic acid and its related enzymes. Arch Biochem Biophys 1998;358:329–35. https://doi.org/10.1006/abbi.1998.0861.

[36] Savitz J. The kynurenine pathway: a finger in every pie. Mol Psychiatry 2020;25:131–47. https://doi.org/10.1038/s41380-019-0414-4.

[37] Chadha R, Mahal HS, Mukherjee T, Kapoor S. Evidence for a possible role of 3-hydroxyanthranilic acid as an antioxidant. J Phys Org Chem 2009;22:349–54. https://doi.org/10.1002/poc.1485.

[38] Krause D, Suh HS, Tarassishin L, Cui QL, Durafourt BA, Choi N, et al. The tryptophan metabolite 3-hydroxyanthranilic acid plays anti-inflammatory and neuroprotective roles during inflammation: role of hemeoxygenase-1. Am J Pathol 2011;179:1360–72. https://doi.org/10.1016/j.ajpath.2011.05.048.

[39] Pérez-González A, Alvarez-Idaboy JR, Galano A. Dual antioxidant/pro-oxidant behavior of the tryptophan metabolite 3-hydroxyanthranilic acid: A theoretical investigation of reaction mechanisms and kinetics. New J Chem 2017;41:3829–45. https://doi.org/10.1039/c6nj03980d.

[40] Darlington LG, Forrest M, Mackay GM, Smith RA, Smith AJ, Stoy N, et al. On the biological importance of the 3-hydroxyanthranilic acid: anthranilic acid ratio. Int J Tryptophan Res 2010;3:51–9. https://doi.org/10.4137/ijtr.s4282.

[41] Lee WS, Lee SM, Kim MK, Park SG, Choi IW, Choi I, et al. The tryptophan metabolite 3-hydroxyanthranilic acid suppresses T cell responses by inhibiting dendritic cell activation. Int Immunopharmacol 2013;17:721–6. https://doi.org/10.1016/j.intimp.2013.08.018.

[42] Lee SM, Lee YS, Choi JH, Park SG, Choi IW, Joo YD, et al. Tryptophan metabolite 3-hydroxyanthranilic acid selectively induces activated T cell death via intracellular GSH depletion. Immunol Lett 2010;132:53–60. https://doi.org/10.1016/j.imlet.2010.05.008.

[43] Piscianz E, Cuzzoni E, de Iudicibus S, Valencic E, Decorti G, Tommasini A. Differential action of 3-hydroxyanthranilic acid on viability and activation of stimulated lymphocytes. Int Immunopharmacol 2011;11:2242–5. https://doi.org/10.1016/j.intimp.2011.09.009.

[44] Costantino G. Inhibitors of quinolinic acid synthesis: new weapons in the study of neuroinflammatory diseases. Future Med Chem 2014;6:841–3. https://doi.org/10.4155/fmc.14.35.

[45] Dang Y, Xia C, Brown OR. Effects of oxygen on 3-hydroxyanthranilate oxidase of the kynurenine pathway. Free Radic Biol Med 1998;25(9):1033–43. https://doi.org/10.1016/S0891-5849(98)00136-1.

[46] Wang Y, Liu KF, Yang Y, Davis I, Liu A. Observing 3-hydroxyanthranilate-3,4-dioxygenase in action through a crystalline lens. Proc Natl Acad Sci U S A 2020;117:19720–30. https://doi.org/10.1073/pnas.2005327117.

[47] Zhang Y, Colabroy KL, Begley TP, Ealick SE. Structural studies on 3-hydroxyanthranilate-3,4-dioxygenase: the catalytic mechanism of a complex oxidation involved in NAD biosynthesis. Biochemistry 2005;44:7632–43. https://doi.org/10.1021/bi047353l.

[48] Colabroy KL, Zhai H, Li T, Ge Y, Zhang Y, Liu A, et al. The mechanism of inactivation of 3-hydroxyanthranilate-3,4-dioxygenase by 4-chloro-3-hydroxyanthranilate. Biochemistry 2005;44:7623–31. https://doi.org/10.1021/bi0473455.

[49] Dilović I, Gliubich F, Malpeli G, Zanotti G, Matković-Čalogović D. Crystal structure of bovine 3-hydroxyanthranilate 3,4-dioxygenase. Biopolymers 2009;91:1189–95. https://doi.org/10.1002/bip.21167.

[50] Ding K, McGee-Lawrence ME, Kaiser H, Sharma AK, Pierce JL, Irsik DL, et al. Picolinic acid, a tryptophan oxidation product, does not impact bone mineral density but increases marrow adiposity. Exp Gerontol 2020;133. https://doi.org/10.1016/j.exger.2020.110885.

[51] Duque G, Vidal C, Li W, al Saedi A, Khalil M, Lim CK, et al. Picolinic acid, a catabolite of tryptophan, has an anabolic effect on bone in vivo. J Bone Miner Res 2020;35:2275–88. https://doi.org/10.1002/jbmr.4125.

[52] Grant RS, Coggan SE, Smythe GA. The physiological action of picolinic acid in the human brain. Int J Tryptophan Res 2009;2:71–9.

[53] Hestad K, Alexander J, Rootwelt H, Aaseth JO. The role of tryptophan dysmetabolism and quinolinic acid in depressive and neurodegenerative diseases. Biomolecules 2022;12. https://doi.org/10.3390/biom12070998.

[54] Guillemin GJ. Quinolinic acid, the inescapable neurotoxin. FEBS J 2012;279:1356–65. https://doi.org/10.1111/j.1742-4658.2012.08485.x.

[55] Lugo-Huitrón R, Ugalde Muñiz P, Pineda B, Pedraza-Chaverrí J, Ríos C, Pérez-De La Cruz V. Quinolinic acid: an endogenous neurotoxin with multiple targets. Oxidative Med Cell Longev 2013. https://doi.org/10.1155/2013/104024.

[56] la Cruz VPD, Carrillo-Mora P, Santamaría A. Quinolinic acid, an endogenous molecule combining excitotoxicity, oxidative stress and other toxic mechanisms. Int J Tryptophan Res 2013;5:1–8. https://doi.org/10.4137/IJTR.S8158.

[57] Chiarugi A, Meli E, Moroni F. Similarities and differences in the neuronal death processes activated by 3OH-kynurenine and quinolinic acid. J Neurochem 2001;77:1310–8. https://doi.org/10.1046/j.1471-4159.2001.00335.x.

[58] Öztürk M, Yalın Sapmaz Ş, Kandemir H, Taneli F, Aydemir Ö. The role of the kynurenine pathway and quinolinic acid in adolescent major depressive disorder. Int J Clin Pract 2021;75. https://doi.org/10.1111/ijcp.13739.

[59] Guillemin GJ, Brew BJ. Implications of the kynurenine pathway and quinolinic acid in Alzheimer's disease. Redox Rep 2002;7:199–206. https://doi.org/10.1179/135100002125000550.

[60] Schwarcz R, Guidetti P, Sathyasaikumar KV, Muchowski PJ. Of mice, rats and men: revisiting the quinolinic acid hypothesis of Huntington's disease. Prog Neurobiol 2010;90:230–45. https://doi.org/10.1016/j.pneurobio.2009.04.005.

[61] Kandanearatchi A, Brew BJ. The kynurenine pathway and quinolinic acid: pivotal roles in HIV associated neurocognitive disorders. FEBS J 2012;279:1366–74. https://doi.org/10.1111/j.1742-4658.2012.08500.x.

[62] Lee JM, Tan V, Lovejoy D, Braidy N, Rowe DB, Brew BJ, et al. Involvement of quinolinic acid in the neuropathogenesis of

amyotrophic lateral sclerosis. Neuropharmacology 2017;112:346–64. https://doi.org/10.1016/j.neuropharm.2016.05.011.

[63] Guillemin GJ, Meininger V, Brew BJ. Implications for the kynurenine pathway and quinolinic acid in amyotrophic lateral sclerosis. Neurodegener Dis 2006;2:166–76. https://doi.org/10.1159/000089622.

[64] Thaker AI, Rao MS, Bishnupuri KS, Kerr TA, Foster L, Marinshaw JM, et al. IDO1 metabolites activate β-catenin signaling to promote cancer cell proliferation and colon tumorigenesis in mice. Gastroenterology 2013;145. https://doi.org/10.1053/j.gastro.2013.05.002.

[65] Nkandeu DS, Basson C, Joubert AM, Serem JC, Bipath P, Nyakudya T, et al. The involvement of a chemokine receptor antagonist CTCE-9908 and kynurenine metabolites in cancer development. Cell Biochem Funct 2022;40:608–22. https://doi.org/10.1002/cbf.3731.

[66] Kang GB, Kim MK, Youn HS, An JY, Lee JG, Park KR, et al. Crystallization and preliminary X-ray crystallographic analysis of human quinolinate phosphoribosyltransferase. Acta Crystallogr Sect F Struct Biol Cryst Commun 2011;67:38–40. https://doi.org/10.1107/S1744309110041011.

[67] Liu H, Woznica K, Catton G, Crawford A, Botting N, Naismith JH. Structural and kinetic characterization of quinolinate phosphoribosyltransferase (hQPRTase) from Homo sapiens. J Mol Biol 2007;373:755–63. https://doi.org/10.1016/j.jmb.2007.08.043.

[68] Cambronne XA, Kraus WL. Compartmentalization of NAD+ synthesis and functions in mammalian cells. Trends Biochem Sci 2020;45:858–73. https://doi.org/10.1016/j.tibs.2020.05.010.

[69] Grant R, Nguyen S, Guillemin G. Kynurenine pathway metabolism is involved in the maintenance of the intracellular NAD+ concentration in human primary astrocytes. Int J Tryptophan Res 2010;3:151–6.

[70] Kincaid JWR, Berger NA. NAD metabolism in aging and cancer. Exp Biol Med 2020;245:1594–614. https://doi.org/10.1177/1535370220929287.

[71] Navas LE, Carnero A. NAD+ metabolism, stemness, the immune response, and cancer. Signal Transduct Target Ther 2021;6. https://doi.org/10.1038/s41392-020-00354-w.

[72] Castro-Portuguez R, Sutphin GL. Kynurenine pathway, NAD+ synthesis, and mitochondrial function: targeting tryptophan metabolism to promote longevity and healthspan. Exp Gerontol 2020;132. https://doi.org/10.1016/j.exger.2020.110841.

[73] Massudi H, Grant R, Guillemin GJ, Braidy N. NAD+ metabolism and oxidative stress: the golden nucleotide on a crown of thorns. Redox Rep 2012;17:28–46. https://doi.org/10.1179/1351000212Y.0000000001.

[74] Amjad S, Nisar S, Bhat AA, Shah AR, Frenneaux MP, Fakhro K, et al. Role of NAD+ in regulating cellular and metabolic signaling pathways. Mol Metab 2021;49. https://doi.org/10.1016/j.molmet.2021.101195.

[75] Ala M. The footprint of kynurenine pathway in every cancer: a new target for chemotherapy. Eur J Pharmacol 2021;896. https://doi.org/10.1016/j.ejphar.2021.173921.

[76] Pires AS, Sundaram G, Heng B, Krishnamurthy S, Brew BJ, Guillemin GJ. Recent advances in clinical trials targeting the kynurenine pathway. Pharmacol Ther 2022;236. https://doi.org/10.1016/j.pharmthera.2021.108055.

[77] Gouasmi R, Ferraro-Peyret C, Nancey S, Coste I, Renno T, Chaveroux C, et al. The kynurenine pathway and cancer: why keep it simple when you can make it complicated. Cancers (Basel) 2022;14. https://doi.org/10.3390/cancers14112793.

[78] Davis I, Liu A. What is the tryptophan kynurenine pathway and why is it important to neurotherapeutics? Expert Rev Neurother 2015;15:719–21. https://doi.org/10.1586/14737175.2015.1049999.

[79] Chung KT, Gadupudi GS. Possible roles of excess tryptophan metabolites in cancer. Environ Mol Mutagen 2011;52:81–104. https://doi.org/10.1002/em.20588.

[80] Matsuoka Z, Yoshimatsu N. Über eine neue Substanz, die aus Tryptophan im Tierkörper gebildet wird. Hoppe Seylers Z Physiol Chem 1925;143:206–10. https://doi.org/10.1515/bchm2.1925.143.4-6.206.

[81] Butenandt A, Weidel W, Weichert R, von Derjugin W, Kynurenin Ü. Physiologie, Konstitutionsermittlung und Synthese. Hoppe Seylers Z Physiol Chem 1943;279:27–43. https://doi.org/10.1515/bchm2.1943.279.1-2.27.

[82] Kotake Y. Studien über den intermediären Stoffwechsel des Tryptophans XVIII—XXIV. Hoppe Seylers Z Physiol Chem 1936;243:237–65. https://doi.org/10.1515/bchm2.1936.243.6.237.

[83] Tanaka T, Knox WE. The nature and mechanism of the tryptophan pyrrolase (peroxidase-oxidase) reaction of pseudomonas and of rat liver. J Biol Chem 1959;234:1162–70. https://doi.org/10.1016/S0021-9258(18)98149-4.

[84] Hayaishi O, Rothberg S, Mehler AH, Saito Y. Studies on oxygenases: enzymatic formation of kynurenine from tryptophan. J Biol Chem 1957;229:889–96. https://doi.org/10.1016/S0021-9258(19)63693-8.

[85] Raven EL. A short history of heme dioxygenases: rise, fall and rise again. J Biol Inorg Chem 2017;22:175–83. https://doi.org/10.1007/s00775-016-1412-5.

[86] Hayaishi O. From oxygenase to sleep. J Biol Chem 2008;283:19165–75. https://doi.org/10.1074/jbc.X800002200.

[87] Narumiya S, Shimizu T, Yamamoto S. In Memoriam: Osamu Hayaishi (1920 – 2015). J Lipid Res 2016;57:517–22. https://doi.org/10.1194/jlr.e067579.

[88] Hirata F, Hayaishi O. New degradative routes of 5-hydroxytryptophan and serotonin. Biochem Biophys Res Commun 1972;47:1112–9. https://doi.org/10.1016/0006-291X(72)90949-7.

[89] Hirata F, Hayaishi O, Tokuyama T, Senoh S. In vitro and in vivo formation of two new metabolites of melatonin. J Biol Chem 1974;249:1311–3. https://doi.org/10.1016/S0021-9258(19)42976-1.

[90] Shimizu T, Nomiyama S, Hirata F, Hayaishi O. Indoleamine 2,3-dioxygenase. Purification and some properties. J Biol Chem 1978;253:4700–6. https://doi.org/10.1016/S0021-9258(17)30447-7.

[91] Batabyal D, Yeh SR. Human tryptophan dioxygenase: a comparison to indoleamine 2,3-dioxygenase. J Am Chem Soc 2007;129:15690–701. https://doi.org/10.1021/ja076186k.

[92] Burkin DJ, Kimbro KS, Barr BL, Jones C, Taylor MW, Gupta SL. Localization of human indoleamine 2,3-dioxygenase (IDO) gene to the pericentromeric region of human chromosome 8. Genomics 1993;17:262–3.

[93] Murray MF. The human indoleamine 2,3-dioxygenase gene and related human genes. Curr Drug Metab 2007;8:197–200.

[94] Tone S, Kadoya A, Maeda H, Minatogawa Y, Kido R. Assignment of the human indoleamine 2,3-dioxygenase gene to chromosome 8 using the polymerase chain reaction. Hum Genet 1994;93:201–3.

[95] Arefayene M, Philips S, Cao D, Mamidipalli S, Desta Z, Flockhart DA, et al. Identification of genetic variants in the human indoleamine 2,3-dioxygenase (IDO1) gene, which have altered enzyme activity. Pharmacogenet Genomics 2009;19:464–76. https://doi.org/10.1097/FPC.0b013e32832c005a.

[96] Sugimoto H, Oda SI, Otsuki T, Hino T, Yoshida T, Shiro Y. Crystal structure of human indoleamine 2,3-dioxygenase: catalytic mechanism of O2 incorporation by a heme-containing dioxygenase. Proc Natl Acad Sci U S A 2006;103:2611–6. https://doi.org/10.1073/pnas.0508996103.

[97] Lewis-Ballester A, Pham KN, Batabyal D, Karkashon S, Bonanno JB, Poulos TL, et al. Structural insights into substrate and inhibitor binding sites in human indoleamine 2,3-dioxygenase. Nat Commun 2017;8. https://doi.org/10.1038/s41467-017-01725-8.

[98] Lancellotti S, Novarese L, De Cristofaro R. Biochemical properties of indoleamine 2,3-dioxygenase: from structure to optimized design of inhibitors. Curr Med Chem 2011;18:2205–14.

[99] Tang K, Wu YH, Song Y, Yu B. Indoleamine 2,3-dioxygenase 1 (IDO1) inhibitors in clinical trials for cancer immunotherapy. J Hematol Oncol 2021;14. https://doi.org/10.1186/s13045-021-01080-8.

[100] Chauhan N, Basran J, Rafice SA, Efimov I, Millett ES, Mowat CG, et al. How is the distal pocket of a heme protein optimized for binding of tryptophan? FEBS J 2012;279:4501–9. https://doi.org/10.1111/febs.12036.

[101] Nickel E, Nienhaus K, Lu C, Yeh SR, Nienhaus GU. Ligand and substrate migration in human indoleamine 2,3-dioxygenase. J Biol Chem 2009;284:31548–54. https://doi.org/10.1074/jbc.M109.039859.

[102] Duan H-C, Peng L-X, Hu Y-C, Luo Q, Liu X-Y, Sun X, et al. The advances of the structure and function of indoleamine 2, 3- dioxygenase 1 and its inhibitors. Curr Protein Pept Sci 2020;21:1027–39. https://doi.org/10.2174/1389203721666200526122304.

[103] Singh R, Salunke DB. Diverse chemical space of indoleamine-2,3-dioxygenase 1 (Idol) inhibitors. Eur J Med Chem 2021;211. https://doi.org/10.1016/j.ejmech.2020.113071.

[104] Kassab SE, Mowafy S. Structural basis of selective human indoleamine-2,3-dioxygenase 1 (hIDO1) inhibition. ChemMedChem 2021;16:3149–64. https://doi.org/10.1002/cmdc.202100253.

[105] Röhrig UF, Michielin O, Zoete V. Structure and plasticity of indoleamine 2,3-dioxygenase 1 (IDO1). J Med Chem 2021;64:17690–705. https://doi.org/10.1021/acs.jmedchem.1c01665.

[106] Chauhan N, Basran J, Efimov I, Svistunenko DA, Seward HE, Moody PCE, et al. The role of serine 167 in human indoleamine 2,3-dioxygenase: a comparison with tryptophan 2,3-dioxygenase. Biochemistry 2008;47:4761–9. https://doi.org/10.1021/bi702405a.

[107] Efimov I, Basran J, Thackray SJ, Handa S, Mowat CG, Raven EL. Structure and reaction mechanism in the heme dioxygenases. Biochemistry 2011;50:2717–24. https://doi.org/10.1021/bi101732n.

[108] Yuasa HJ. High l-Trp affinity of indoleamine 2,3-dioxygenase 1 is attributed to two residues located in the distal heme pocket. FEBS J 2016;3651–61. https://doi.org/10.1111/febs.13834.

[109] Nienhaus K, Nickel E, Nienhaus GU. Substrate binding in human indoleamine 2,3-dioxygenase 1: a spectroscopic analysis. Biochim Biophys Acta, Proteins Proteomics 2017;1865:453–63. https://doi.org/10.1016/j.bbapap.2017.02.008.

[110] Álvarez L, Lewis-Ballester A, Roitberg A, Estrin DA, Yeh SR, Marti MA, et al. Structural study of a flexible active site loop in human indoleamine 2,3-dioxygenase and its functional implications. Biochemistry 2016;55:2785–93. https://doi.org/10.1021/acs.biochem.6b00077.

[111] Lewis-Ballester A, Karkashon S, Batabyal D, Poulos TL, Yeh SR. Inhibition mechanisms of human indoleamine 2,3 dioxygenase 1. J Am Chem Soc 2018;140:8518–25. https://doi.org/10.1021/jacs.8b03691.

[112] Hamilton GA. Mechanisms of two- and four-electron oxidations catalyzed by some metalloenzymes. Adv Enzymol Relat Areas Mol Biol 1969;32:55–96. https://doi.org/10.1002/9780470122778.ch3.

[113] Thackray SJ, Bruckmann C, Anderson JLR, Campbell LP, Xiao R, Zhao L, et al. Histidine 55 of tryptophan 2,3-dioxygenase is not an active site base but regulates catalysis by controlling substrate binding. Biochemistry 2008;47:10677–84. https://doi.org/10.1021/bi801202a.

[114] Chauhan N, Thackray SJ, Rafice SA, Eaton G, Lee M, Efimov I, et al. Reassessment of the reaction mechanism in the heme dioxygenases. J Am Chem Soc 2009;131:4186–7. https://doi.org/10.1021/ja808326g.

[115] Basran J, Efimov I, Chauhan N, Thackray SJ, Krupa JL, Eaton G, et al. The mechanism of formation of N -formylkynurenine by heme dioxygenases. J Am Chem Soc 2011;133:16251–7. https://doi.org/10.1021/ja207066z.

[116] Capece L, Lewis-Ballester A, Batabyal D, di Russo N, Yeh SR, Estrin DA, et al. The first step of the dioxygenation reaction carried out by tryptophan dioxygenase and indoleamine 2,3-dioxygenase as revealed by quantum mechanical/molecular mechanical studies. J Biol Inorg Chem 2010;15:811–23. https://doi.org/10.1007/s00775-010-0646-x.

[117] Capece L, Lewis-Ballester A, Yeh SR, Estrin DA, Marti MA. Complete reaction mechanism of indoleamine 2,3-dioxygenase as revealed by QM/MM simulations. J Phys Chem B 2012;116:1401–13. https://doi.org/10.1021/jp2082825.

[118] Lewis-Ballester A, Batabyal D, Egawa T, Lu C, Lin Y, Marti MA, et al. Evidence for a ferryl intermediate in a heme-based dioxygenase. Proc Natl Acad Sci U S A 2009;106:17371–6.

[119] Basran J, Booth ES, Lee M, Handa S, Raven EL. Analysis of reaction intermediates in tryptophan 2,3-dioxygenase: a comparison with indoleamine 2,3-dioxygenase. Biochemistry 2016;55:6743–50. https://doi.org/10.1021/acs.biochem.6b01005.

[120] Booth ES, Basran J, Lee M, Handa S, Raven EL. Substrate oxidation by indoleamine 2,3-dioxygenase: evidence for a common reaction mechanism. J Biol Chem 2015;290:30924–30. https://doi.org/10.1074/jbc.M115.695684.

[121] Geng J, Liu A. Heme-dependent dioxygenases in tryptophan oxidation. Arch Biochem Biophys 2014;544:18–26. https://doi.org/10.1016/j.abb.2013.11.009.

[122] Kolawole AO, Hixon BP, Dameron LS, Chrisman IM, Smirnov V, v. Catalytic activity of human indoleamine 2,3-dioxygenase (hIDO1) at low oxygen. Arch Biochem Biophys 2015;570:47–57. https://doi.org/10.1016/j.abb.2015.02.014.

[123] Pallotta MT, Rossini S, Suvieri C, Coletti A, Orabona C, Macchiarulo A, et al. Indoleamine 2,3-dioxygenase 1 (IDO1): an up-to-date overview of an eclectic immunoregulatory enzyme. FEBS J 2021. https://doi.org/10.1111/febs.16086.

[124] Bello C, Heinisch PP, Mihalj M, Carrel T, Luedi MM. Indoleamine-2,3-dioxygenase as a perioperative marker of the immune system. Front Physiol 2021;12. https://doi.org/10.3389/fphys.2021.766511.

[125] Acovic A, Gazdic M, Jovicic N, Harrell CR, Fellabaum C, Arsenijevic N, et al. Role of indoleamine 2,3-dioxygenase in pathology of the gastrointestinal tract. Ther Adv Gastroenterol 2018;11. https://doi.org/10.1177/1756284818815334.

[126] Sedlmayr P, Blaschitz A. Placental expression of indoleamine 2,3-dioxygenase. Wien Med Wochenschr 2012;162:214–9. https://doi.org/10.1007/s10354-012-0082-3.

[127] Heidari F, Ramezani A, Erfani N, Razmkhah M. Indoleamine 2, 3-dioxygenase: a professional immunomodulator and its potential

functions in immune related diseases. Int Rev Immunol 2022;41:346–63. https://doi.org/10.1080/08830185.2020.1836176.
[128] Prendergast GC, Smith C, Thomas S, Mandik-Nayak L, Laury-Kleintop L, Metz R, et al. Indoleamine 2,3-dioxygenase pathways of pathogenic inflammation and immune escape in cancer. Cancer Immunol Immunother 2014;63:721–35. https://doi.org/10.1007/s00262-014-1549-4.
[129] Puccetti P. On watching the watchers: IDO and type I/II IFN. Eur J Immunol 2007;37:876–9. https://doi.org/10.1002/eji.200737184.
[130] Grohmann U, Orabona C, Fallarino F, Vacca C, Calcinaro F, Falorni A, et al. CTLA-4-Ig regulates tryptophan catabolism in vivo. Nat Immunol 2002;3:1097–101. https://doi.org/10.1038/ni846.
[131] Huang L, Baban B, Johnson BA, Mellor AL. Dendritic cells, indoleamine 2,3 dioxygenase and acquired immune privilege. Int Rev Immunol 2010;29:133–55. https://doi.org/10.3109/08830180903349669.
[132] Munn DH, Zhou M, Attwood JT, Bondarev I, Conway SJ, Marshall B, et al. Prevention of allogeneic fetal rejection by tryptophan catabolism. Science 1979;1998(281):1191–3. https://doi.org/10.1126/science.281.5380.1191.
[133] Pallotta MT, Orabona C, Volpi C, Vacca C, Belladonna ML, Bianchi R, et al. Indoleamine 2,3-dioxygenase is a signaling protein in long-term tolerance by dendritic cells. Nat Immunol 2011;12:870–8. https://doi.org/10.1038/ni.2077.
[134] Bilir C, Sarisozen C. Indoleamine 2,3-dioxygenase (IDO): only an enzyme or a checkpoint controller? J Oncol Sci 2017;3:52–6. https://doi.org/10.1016/j.jons.2017.04.001.
[135] Lee GK, Park HJ, MacLeod M, Chandler P, Munn DH, Mellor AL. Tryptophan deprivation sensitizes activated T cells to apoptosis prior to cell division. Immunology 2002;107:452–60. https://doi.org/10.1046/j.1365-2567.2002.01526.x.
[136] Munn DH, Shafizadeh E, Attwood JT, Bondarev I, Pashine A, Mellor AL. Inhibition of T cell proliferation by macrophage tryptophan catabolism. J Exp Med 1999;189:1363–72.
[137] Mbongue JC, Nicholas DA, Torrez TW, Kim NS, Firek AF, Langridge WHR. The role of indoleamine 2, 3-dioxygenase in immune suppression and autoimmunity. Vaccines (Basel) 2015;3:703–29. https://doi.org/10.3390/vaccines3030703.
[138] Mellor AL, Chandler P, Lee GK, Johnson T, Keskin DB, Lee J, et al. Indoleamine 2,3-dioxygenase, immunosuppression and pregnancy. J Reprod Immunol 2002;57:143–50. https://doi.org/10.1016/s0165-0378(02)00040-2.
[139] Chang RQ, Li DJ, Li MQ. The role of indoleamine-2,3-dioxygenase in normal and pathological pregnancies. Am J Reprod Immunol 2018;79. https://doi.org/10.1111/aji.12786.
[140] Alegre E, Dìaz A, Lopez AS, Uriz M, Murillo O, Melero I, et al. Indoleamine 2,3-dioxygenase: from tolerance during pregnancy to cancer. Mol Cell Oncol 2005;24:20–7.
[141] Sedlmayr P. Indoleamine 2,3-dioxygenase in materno-fetal interaction. Curr Drug Metab 2007;8:205–8. https://doi.org/10.2174/138920007780362491.
[142] Sedlmayr P, Blaschitz A, Stocker R. The role of placental tryptophan catabolism. Front Immunol 2014;5. https://doi.org/10.3389/fimmu.2014.00230.
[143] Yoshida R, Hayaishi O. Induction of pulmonary indoleamine 2,3-dioxygenase by intraperitoneal injection of bacterial lipopolysaccharide. Proc Natl Acad Sci U S A 1978;78:3998–4000. https://doi.org/10.1073/pnas.75.8.3998.
[144] Murakami Y, Hoshi M, Imamura Y, Arioka Y, Yamamoto Y, Saito K. Remarkable role of indoleamine 2,3-dioxygenase and tryptophan metabolites in infectious diseases: potential role in macrophage-mediated inflammatory diseases. Mediat Inflamm 2013;2013, 391984. https://doi.org/10.1155/2013/391984.
[145] Greco FA, Coletti A, Camaioni E, Carotti A, Marinozzi M, Gioiello A, et al. The Janus-faced nature of IDO1 in infectious diseases: challenges and therapeutic opportunities. Future Med Chem 2016;8:39–54. https://doi.org/10.4155/fmc.15.165.
[146] Fox JM, Sage LK, Huang L, Barber J, Klonowski KD, Mellor AL, et al. Inhibition of indoleamine 2,3-dioxygenase enhances the T-cell response to influenza virus infection. J Gen Virol 2013;94:1451–61. https://doi.org/10.1099/vir.0.053124-0.
[147] Cihan M, Doğan Ö, Ceran Serdar C, Altunçekiç Yıldırım A, Kurt C, Serdar MA. Kynurenine pathway in Coronavirus disease (COVID-19): Potential role in prognosis. J Clin Lab Anal 2022;36. https://doi.org/10.1002/jcla.24257.
[148] Chilosi M, Doglioni C, Ravaglia C, Martignoni G, Salvagno GL, Pizzolo G, et al. Unbalanced IDO1/IDO2 endothelial expression and skewed keynurenine pathway in the pathogenesis of COVID-19 and post-COVID-19 pneumonia. Biomedicines 2022;10. https://doi.org/10.3390/biomedicines10061332.
[149] Guo L, Schurink B, Roos E, Nossent EJ, Duitman JW, Vlaar APJ, et al. Indoleamine 2,3-dioxygenase (IDO)-1 and IDO-2 activity and severe course of COVID-19. J Pathol 2022;256:256–61. https://doi.org/10.1002/path.5842.
[150] Werner ER, Fuchs D, Hausend A, Jaeger H, Reibnegger G, Werner-Felmayer G, et al. Tryptophan degradation in patients infected by human immunodeficiency. Virus 1988;369.
[151] Smith DG, Guillemi GJ. Quinolinic acid is produced by macrophages stimulated by platelet activating factor, Nef and Tat. J Neurovirol 2001;7:56–60. https://doi.org/10.1080/135502801300069692.
[152] Bahraoui E, Serrero M, Planès R. HIV-1 Tat – TLR4/MD2 interaction drives the expression of IDO-1 in monocytes derived dendritic cells through NF-κB dependent pathway. Sci Rep 2020;10. https://doi.org/10.1038/s41598-020-64847-y.
[153] Boasso A, Shearer GM. How does indoleamine 2,3-dioxygenase contribute to HIV-mediated immune dysregulation. Curr Drug Metab 2007;8:217–23.
[154] Chen J, Xun J, Yang J, Ji Y, Liu L, Qi T, et al. Plasma indoleamine 2,3-dioxygenase activity is associated with the size of the human immunodeficiency virus reservoir in patients receiving antiretroviral therapy. Clin Infect Dis 2019;68:1274–81. https://doi.org/10.1093/cid/ciy676.
[155] Manches O, Munn D, Fallahi A, Lifson J, Chaperot L, Plumas J, et al. HIV-activated human plasmacytoid DCs induce Tregs through an indoleamine 2,3-dioxygenase–dependent mechanism. J Clin Invest 2008;118:3431–9.
[156] Adu-Gyamfi CG, Savulescu D, George JA, Suchard MS. Indoleamine 2, 3-dioxygenase-mediated tryptophan catabolism: a leading star or supporting act in the tuberculosis and HIV Pas-de-Deux? Front Cell Infect Microbiol 2019;9. https://doi.org/10.3389/fcimb.2019.00372.
[157] Chen J, Shao J, Cai R, Shen Y, Zhang R, Liu L, et al. Anti-retroviral therapy decreases but does not normalize indoleamine 2,3-dioxygenase activity in HIV-infected patients. PLoS One 2014;9. https://doi.org/10.1371/journal.pone.0100446.

[158] Larrea E, Riezu-Boj JI, Gil-Guerrero L, Casares N, Aldabe R, Sarobe P, et al. Upregulation of indoleamine 2,3-dioxygenase in hepatitis C virus infection. J Virol 2007;81:3662–6. https://doi.org/10.1128/jvi.02248-06.
[159] Lepiller Q, Soulier E, Li Q, Lambotin M, Barths J, Fuchs D, et al. Antiviral and immunoregulatory effects of indoleamine-2,3-dioxygenase in hepatitis C virus infection. J Innate Immun 2015;7:530–44. https://doi.org/10.1159/000375161.
[160] Sawada L, Vallinoto ACR, Brasil-Costa I. Regulation of the immune checkpoint indoleamine 2,3-dioxygenase expression by epstein–barr virus. Biomolecules 2021;11. https://doi.org/10.3390/biom11121792.
[161] Venancio PA, Consolaro MEL, Derchain SF, Boccardo E, Villa LL, Maria-Engler SS, et al. Indoleamine 2,3-dioxygenase and tryptophan 2,3-dioxygenase expression in HPV infection, SILs, and cervical cancer. Cancer Cytopathol 2019;127:586–97. https://doi.org/10.1002/cncy.22172.
[162] Narui K, Noguchi N, Saito A, Kakimi K, Motomura N, Kubo K, et al. Anti-infectious activity of tryptophan metabolites in the L-tryptophan-L-kynurenine pathway. Biol Pharm Bull 2009;32:41–4.
[163] Günther J, Fallarino F, Fuchs D, Wirthgen E. Editorial: Immunomodulatory roles of tryptophan metabolites in inflammation and cancer. Front Immunol 2020;11. https://doi.org/10.3389/fimmu.2020.01497.
[164] Boyland E, Williams DC. The estimation of tryptophan metabolites in the urine of patients with cancer of the bladder. Biochem J 1955;60:v.
[165] Godin-Ethier J, Hanafi LA, Piccirillo CA, Lapointe R. Indoleamine 2,3-dioxygenase expression in human cancers: clinical and immunologic perspectives. Clin Cancer Res 2011;17:6985–91. https://doi.org/10.1158/1078-0432.CCR-11-1331.
[166] Li F, Zhang R, Li S, Liu J. Corrigendum to "IDO1: An important immunotherapy target in cancer treatment" [International Immunopharmacology 47 (2017) 70–77](S156757691730125X)(10.1016/j.intimp.2017.03.024). Int Immunopharmacol 2017;49:231. https://doi.org/10.1016/j.intimp.2017.04.020.
[167] Curti A, Trabanelli S, Salvestrini V, Baccarani M, Lemoli RM. The role of indoleamine 2,3-dioxygenase in the induction of immune tolerance: focus on hematology. Blood 2009;113:2394–401. https://doi.org/10.1182/blood-2008-07-144485.
[168] Moon YW, Hajjar J, Hwu P, Naing A. Targeting the indoleamine 2,3-dioxygenase pathway in cancer. J Immunother Cancer 2015;3. https://doi.org/10.1186/s40425-015-0094-9.
[169] Katz JB, Muller AJ, Prendergast GC. Indoleamine 2,3-dioxygenase in T-cell tolerance and tumoral immune escape. Immunol Rev 2008;222:206–21. https://doi.org/10.1111/j.1600-065X.2008.00610.x.
[170] Huang X, Zhang F, Wang X, Liu K. The role of indoleamine 2, 3-dioxygenase 1 in regulating tumor microenvironment. Cancers (Basel) 2022;14. https://doi.org/10.3390/cancers14112756.
[171] Cheong JE, Sun L. Targeting the IDO1/TDO2–KYN–AhR pathway for cancer immunotherapy – challenges and opportunities. Trends Pharmacol Sci 2018;39:307–25. https://doi.org/10.1016/j.tips.2017.11.007.
[172] Song X, Si Q, Qi R, Liu W, Li M, Guo M, et al. Indoleamine 2,3-dioxygenase 1: a promising therapeutic target in malignant tumor. Front Immunol 2021;12. https://doi.org/10.3389/fimmu.2021.800630.
[173] Ricciuti B, Leonardi GC, Puccetti P, Fallarino F, Bianconi V, Sahebkar A, et al. Targeting indoleamine-2,3-dioxygenase in cancer: scientific rationale and clinical evidence. Pharmacol Ther 2019;196:105–16. https://doi.org/10.1016/j.pharmthera.2018.12.004.
[174] Passarelli A, Pisano C, Cecere SC, di Napoli M, Rossetti S, Tambaro R, et al. Targeting immunometabolism mediated by the IDO1 pathway: a new mechanism of immune resistance in endometrial cancer. Front Immunol 2022;13. https://doi.org/10.3389/fimmu.2022.953115.
[175] Sato N, Saga Y, Mizukami H, Wang D, Takahashi S, Nonaka H, et al. Downregulation of indoleamine-2,3-dioxygenase in cervical cancer cells suppresses tumor growth by promoting natural killer cell accumulation. Oncol Rep 2012;28:1574–8. https://doi.org/10.3892/or.2012.1984.
[176] He K, Shen F, Zhou F. Prognostic value of indoleamine 2, 3-dioxygenase expression in esophageal cancer: a systematic review and meta-analysis. Asian J Surg 2022. https://doi.org/10.1016/j.asjsur.2022.06.122.
[177] Hosseinalizadeh H, Mahmoodpour M, Samadani AA, Roudkenar MH. The immunosuppressive role of indoleamine 2, 3-dioxygenase in glioblastoma: mechanism of action and immunotherapeutic strategies. Med Oncol 2022;39. https://doi.org/10.1007/s12032-022-01724-w.
[178] Colombo G, Gelardi ELM, Balestrero FC, Moro M, Travelli C, Genazzani AA. Insight into nicotinamide adenine dinucleotide homeostasis as a targetable metabolic pathway in colorectal cancer. Front Pharmacol 2021;12. https://doi.org/10.3389/fphar.2021.758320.
[179] Brandacher G, Perathoner A, Ladurner R, Schneeberger S, Obrist P, Winkler C, et al. Prognostic value of indoleamine 2,3-dioxygenase expression in colorectal cancer: effect on tumor-infiltrating T cells. Clin Cancer Res 2006;12:1144–51. https://doi.org/10.1158/1078-0432.CCR-05-1966.
[180] Soliman H, Rawal B, Fulp J, Lee JH, Lopez A, Bui MM, et al. Analysis of indoleamine 2-3 dioxygenase (IDO1) expression in breast cancer tissue by immunohistochemistry. Cancer Immunol Immunother 2013;62:829–37. https://doi.org/10.1007/s00262-013-1393-y.
[181] Masaki A, Ishida T, Maeda Y, Ito A, Suzuki S, Narita T, et al. Clinical significance of tryptophan catabolism in Hodgkin lymphoma. Cancer Sci 2018;109:74–83. https://doi.org/10.1111/cas.13432.
[182] Masaki A, Ishida T, Maeda Y, Suzuki S, Ito A, Takino H, et al. Prognostic significance of tryptophan catabolism in adult t-cell leukemia/lymphoma. Clin Cancer Res 2015;21:2830–9. https://doi.org/10.1158/1078-0432.CCR-14-2275.
[183] Adams S, Braidy N, Bessesde A, Brew BJ, Grant R, Teo C, et al. The kynurenine pathway in brain tumor pathogenesis. Cancer Res 2012;72:5649–57. https://doi.org/10.1158/0008-5472.CAN-12-0549.
[184] Platten M, Friedrich M, Wainwright DA, Panitz V, Opitz CA. Tryptophan metabolism in brain tumors — IDO and beyond. Curr Opin Immunol 2021;70:57–66. https://doi.org/10.1016/j.coi.2021.03.005.
[185] Al Khasawneh E, Gupta S, Tuli SY, Shahlaee AH, Garrett TJ, Schechtman KB, et al. Stable pediatric kidney transplant recipients run higher urine indoleamine 2, 3 dioxygenase (IDO) levels than healthy children. Pediatr Transplant 2014;18:254–7. https://doi.org/10.1111/petr.12232.

[186] Palafox D, Llorente L, Alberú J, Torres-Machorro A, Camorlinga N, Rodríguez C, et al. The role of indoleamine 2,3 dioxygenase in the induction of immune tolerance in organ transplantation. Transplant Rev 2010;24:160–5. https://doi.org/10.1016/j.trre.2010.04.003.
[187] Esmaeili SA, Hajavi J. The role of indoleamine 2,3-dioxygenase in allergic disorders. Mol Biol Rep 2022;49:3297–306. https://doi.org/10.1007/s11033-021-07067-5.
[188] Yang HL, Li MQ. Indoleamine 2,3-dioxygenase in endometriosis. Reprod Dev Med 2019;3:110–6. https://doi.org/10.4103/2096-2924.262391.
[189] Ueno A, Cho S, Cheng L, Wang J, Hou S, Nakano H, et al. Transient upregulation of indoleamine 2,3-dioxygenase in dendritic cells by human chorionic gonadotropin downregulates autoimmune diabetes. Diabetes 2007;56:1686–93.
[190] Furuzawa-Carballeda J, Lima G, Jakez-Ocampo J, Llorente L. Indoleamine 2,3-dioxygenase-expressing peripheral cells in rheumatoid arthritis and systemic lupus erythematosus: a cross-sectional study. Eur J Clin Investig 2011;41:1037–46. https://doi.org/10.1111/j.1365-2362.2011.02491.x.
[191] Ball HJ, Sanchez-Perez A, Weiser S, Austin CJD, Astelbauer F, Miu J, et al. Characterization of an indoleamine 2,3-dioxygenase-like protein found in humans and mice. Gene 2007;396:203–13. https://doi.org/10.1016/j.gene.2007.04.010.
[192] Metz R, DuHadaway JB, Kamasani U, Laury-Kleintop L, Muller AJ, Prendergast GC. Novel tryptophan catabolic enzyme IDO2 is the preferred biochemical target of the antitumor indoleamine 2,3-dioxygenase inhibitory compound D-1-methyl-tryptophan. Cancer Res 2007;67:7082–7. https://doi.org/10.1158/0008-5472.CAN-07-1872.
[193] Yuasa HJ, Takubo M, Takahashi A, Hasegawa T, Noma H, Suzuki T. Evolution of vertebrate indoleamine 2,3-dioxygenases. J Mol Evol 2007;65:705–14. https://doi.org/10.1007/s00239-007-9049-1.
[194] Ball HJ, Yuasa HJ, Austin CJD, Weiser S, Hunt NH. Indoleamine 2,3-dioxygenase-2; a new enzyme in the kynurenine pathway. Int J Biochem Cell Biol 2009;41:467–71. https://doi.org/10.1016/j.biocel.2008.01.005.
[195] Fatokun AA, Hunt NH, Ball HJ. Indoleamine 2,3-dioxygenase 2 (IDO2) and the kynurenine pathway: characteristics and potential roles in health and disease. Amino Acids 2013;45:1319–29. https://doi.org/10.1007/s00726-013-1602-1.
[196] Pantouris G, Serys M, Yuasa HJ, Ball HJ, Mowat CG. Human indoleamine 2,3-dioxygenase-2 has substrate specificity and inhibition characteristics distinct from those of indoleamine 2,3-dioxygenase-1. Amino Acids 2014;46:2155–63. https://doi.org/10.1007/s00726-014-1766-3.
[197] Merlo LMF, Mandik-Nayak L. IDO2: a pathogenic mediator of inflammatory autoimmunity. Clin Med Insights Pathol 2016;9s1:21–8. https://doi.org/10.4137/CPath.S39930.
[198] Prendergast GC, Metz R, Muller AJ, Merlo LMF, Mandik-Nayak L. IDO2 in immunomodulation and autoimmune disease. Front Immunol 2014;5. https://doi.org/10.3389/fimmu.2014.00585.
[199] Mondanelli G, Mandarano M, Belladonna ML, Suvieri C, Pelliccia C, Bellezza G, et al. Current challenges for IDO2 as target in cancer immunotherapy. Front Immunol 2021;12. https://doi.org/10.3389/fimmu.2021.679953.
[200] Li P, Xu W, Liu F, Zhu H, Zhang L, Ding Z, et al. The emerging roles of IDO2 in cancer and its potential as a therapeutic target. Biomed Pharmacother 2021;137. https://doi.org/10.1016/j.biopha.2021.111295.
[201] Thackray SJ, Mowat CG, Chapman SK. Exploring the mechanism of tryptophan 2,3-dioxygenase. Biochem Soc Trans 2008;36:1120–3. https://doi.org/10.1042/BST0361120.
[202] Kozlova A, Frédérick R. Current state on tryptophan 2,3-dioxygenase inhibitors: a patent review. Expert Opin Ther Pat 2019;29:11–23. https://doi.org/10.1080/13543776.2019.1556638.
[203] Naismith JH. Tryptophan oxygenation: mechanistic considerations. Biochem Soc Trans 2012;40:509–14. https://doi.org/10.1042/BST20120073.
[204] Boros FA, Vécsei L. Tryptophan 2,3-dioxygenase, a novel therapeutic target for Parkinson's disease. Expert Opin Ther Targets 2021;25:877–88. https://doi.org/10.1080/14728222.2021.1999928.
[205] Yu CP, Pan ZZ, Luo DY. TDO as a therapeutic target in brain diseases. Metab Brain Dis 2016;31:737–47. https://doi.org/10.1007/s11011-016-9824-z.
[206] Hoffmann D, Dvorakova T, Stroobant V, Bouzin C, Daumerie A, Solvay M, et al. Tryptophan 2,3-dioxygenase expression identified in human hepatocellular carcinoma cells and in intratumoral pericytes of most cancers. Cancer Immunol Res 2020;8:19–31. https://doi.org/10.1158/2326-6066.CIR-19-0040.
[207] Hu Y, Liu Z, Tang H. Tryptophan 2,3-dioxygenase may be a potential prognostic biomarker and immunotherapy target in cancer: a meta-analysis and bioinformatics analysis. Front Oncol 2022;12. https://doi.org/10.3389/fonc.2022.977640.
[208] Hjortsø MD, Larsen SK, Kongsted P, Met Ö, Frøsig TM, Andersen GH, et al. Tryptophan 2,3-dioxygenase (TDO)-reactive T cells differ in their functional characteristics in health and cancer. Oncoimmunology 2015;4, 968480. https://doi.org/10.4161/21624011.2014.968480.
[209] Sumitomo M, Takahara K, Zennami K, Nagakawa T, Maeda Y, Shiogama K, et al. Tryptophan 2,3-dioxygenase in tumor cells is associated with resistance to immunotherapy in renal cell carcinoma. Cancer Sci 2021;112:1038–47. https://doi.org/10.1111/cas.14797.
[210] Yu CP, Song YL, Zhu ZM, Huang B, Xiao YQ, Luo DY. Targeting TDO in cancer immunotherapy. Med Oncol 2017;34. https://doi.org/10.1007/s12032-017-0933-2.
[211] Schor JM, Frieden E. Induction of tryptophan peroxidase of rat liver by insulin and alloxan. J Biol Chem 1958;233:612–8. https://doi.org/10.1016/s0021-9258(18)64714-3.
[212] Frieden E, Westmark GW, Schor JM. Inhibition of tryptophan pyrrolase by serotonin, epinephrine and tryptophan analogs. Arch Biochem Biophys 1961;92:176–82. https://doi.org/10.1016/0003-9861(61)90233-8.
[213] Cady SG, Sono M. 1-Methyl-dl-tryptophan, β-(3-benzofuranyl)-dl-alanine (the oxygen analog of tryptophan), and β-[3-benzo(b)thienyl]-dl-alanine (the sulfur analog of tryptophan) are competitive inhibitors for indoleamine 2,3-dioxygenase. Arch Biochem Biophys 1991;291:326–33. https://doi.org/10.1016/0003-9861(91)90142-6.
[214] Friberg M, Jennings R, Alsarraj M, Dessureault S, Cantor A, Extermann M, et al. Indoleamine 2,3-dioxygenase contributes to tumor cell evasion of T cell-mediated rejection. Int J Cancer 2002;101:151–5. https://doi.org/10.1002/ijc.10645.
[215] Uyttenhove C, Pilotte L, Théate I, Stroobant V, Colau D, Parmentier N, et al. Evidence for a tumoral immune resistance mechanism based on tryptophan degradation by indoleamine 2,3-dioxygenase. Nat Med 2003;9:1269–74. https://doi.org/10.1038/nm934.
[216] Muller AJ, DuHadaway JB, Donover PS, Sutanto-Ward E, Prendergast GC. Inhibition of indoleamine 2,3-dioxygenase, an immunoregulatory target of the cancer suppression gene Bin1, potentiates

cancer chemotherapy. Nat Med 2005;11:312–9. https://doi.org/10.1038/nm1196.

[217] Fox E, Oliver T, Rowe M, Thomas S, Zakharia Y, Gilman PB, et al. Indoximod: an immunometabolic adjuvant that empowers T cell activity in cancer. Front Oncol 2018;8. https://doi.org/10.3389/fonc.2018.00370.

[218] Hou DY, Muller AJ, Sharma MD, DuHadaway J, Banerjee T, Johnson M, et al. Inhibition of indoleamine 2,3-dioxygenase in dendritic cells by stereoisomers of 1-methyl-tryptophan correlates with antitumor responses. Cancer Res 2007;67:792–801. https://doi.org/10.1158/0008-5472.CAN-06-2925.

[219] Prendergast GC, Smith C, Thomas S, Mandik-Nayak L, Laury-Kleintop L, Metz R, et al. IDO in inflammatory programming and immune suppression in cancer. In: Tumor-induced immune suppression: Mechanisms and therapeutic reversal. Springer New York; 2014. p. 311–46. https://doi.org/10.1007/978-1-4899-8056-4_11.

[220] Dolušić E, Larrieu P, Blanc S, Sapunaric F, Norberg B, Moineaux L, et al. Indol-2-yl ethanones as novel indoleamine 2,3-dioxygenase (IDO) inhibitors. Bioorg Med Chem 2011;19:1550–61. https://doi.org/10.1016/j.bmc.2010.12.032.

[221] Munn DH, Bronte V. Immune suppressive mechanisms in the tumor microenvironment. Curr Opin Immunol 2016;39:1–6. https://doi.org/10.1016/j.coi.2015.10.009.

[222] Munn DH, Sharma MD, Johnson TS, Rodriguez P. IDO, PTEN-expressing Tregs and control of antigen-presentation in the murine tumor microenvironment. Cancer Immunol Immunother 2017;66:1049–58. https://doi.org/10.1007/s00262-017-2010-2.

[223] Brincks EL, Adams J, Wang L, Turner B, Marcinowicz A, Ke J, et al. Indoximod opposes the immunosuppressive effects mediated by IDO and TDO via modulation of AhR function and activation of mTORC1. Oncotarget 2020;11:2438–61.

[224] Soliman HH, Minton SE, Han HS, Ismail-Khan R, Neuger A, Khambati F, et al. A phase I study of indoximod in patients with advanced malignancies. Oncotarget 2016;7:22928–38.

[225] Soliman HH, Jackson E, Neuger T, Dees EC, Harvey RD, Han H, et al. A first in man phase I trial of the oral immunomodulator, indoximod, combined with docetaxel in patients with metastatic solid tumors. Oncotarget 2014;5:8136–46. https://doi.org/10.18632/oncotarget.2357.

[226] Emadi A, Duong VH, Pantin J, Imran M, Koka R, Singh Z, et al. Indoximod combined with standard induction chemotherapy is well tolerated and induces a high rate of complete remission with MRD-negativity in patients with newly diagnosed AML: results from a phase 1 trial. Blood 2018;132:332. https://doi.org/10.1182/blood-2018-99-117433.

[227] Zakharia Y, McWilliams RR, Rixe O, Drabick J, Shaheen MF, Grossmann KF, et al. Phase II trial of the IDO pathway inhibitor indoximod plus pembrolizumab for the treatment of patients with advanced melanoma. J Immunother Cancer 2021;9. https://doi.org/10.1136/jitc-2020-002057.

[228] Eisenhauer EA, Therasse P, Bogaerts J, Schwartz LH, Sargent D, Ford R, et al. New response evaluation criteria in solid tumours: revised RECIST guideline (version 1.1). Eur J Cancer 2009;45:228–47. https://doi.org/10.1016/j.ejca.2008.10.026.

[229] Soliman H, Khambati F, Han HS, Ismail-Khan R, Bui MM, Sullivan DM, et al. A phase-1/2 study of adenovirus-p53 transduced dendritic cell vaccine in combination with indoximod in metastatic solid tumors and invasive breast cancer. Oncotarget 2018;9:10110–7.

[230] Okabayashi K, Morishima H, Hamada M, Takeuchi T, Umezawa H. A tryptophan hydroxylase inhibitor produced by a streptomycete: 2,5-dihydro-L-phenylalanine. J Antibiot (Tokyo) 1977;30:675–7. https://doi.org/10.7164/antibiotics.30.675.

[231] Watanabe Y, Fujiwara M, Hayaishi O, Takeuchi T, Umezawa H. 2,5-Dihydro-L-phenylalanine: a competitive inhibitor of indoleamine 2,3-dioxygenase and tryptophan 2,3-dioxygenase. Biochem Biophys Res Commun 1978;85:273–9. https://doi.org/10.1016/S0006-291X(78)80039-4.

[232] Eguchi N, Watanabe Y, Kawanishi K, Hashimoto Y, Hayaiship' O. Inhibition of lndoleamine 2,3-dioxygenase and tryptophan 2,3-dioxygenase by β-carboline and lndole derivatives. Arch Biochem Biophys 1984;232:602–9.

[233] Sono M, Cady SG. Enzyme kinetic and spectroscopic studies of inhibitor and effector interactions with indoleamine 2,3-dioxygenase. 1. Norharman and 4-phenylimidazole binding to the enzyme as inhibitors and heme ligands. Biochemistry 1989;28:5392–9.

[234] Goddard TD, Huang CC, Meng EC, Pettersen EF, Couch GS, Morris JH, et al. UCSF ChimeraX: meeting modern challenges in visualization and analysis. Protein Sci 2018;27:14–25. https://doi.org/10.1002/pro.3235.

[235] Pettersen EF, Goddard TD, Huang CC, Meng EC, Couch GS, Croll TI, et al. UCSF ChimeraX: structure visualization for researchers, educators, and developers. Protein Sci 2021;30:70–82. https://doi.org/10.1002/pro.3943.

[236] Kumar S, Jaller D, Patel B, LaLonde JM, DuHadaway JB, Malachowski WP, et al. Structure based development of phenylimidazole-derived inhibitors of indoleamine 2,3-dioxygenase. J Med Chem 2008;51:4968–77. https://doi.org/10.1021/jm800512z.

[237] Peng YH, Ueng SH, Tseng CT, Hung MS, Song JS, Wu JS, et al. Important hydrogen bond networks in indoleamine 2,3-dioxygenase 1 (IDO1) inhibitor design revealed by crystal structures of imidazoleisoindole derivatives with IDO1. J Med Chem 2016;59:282–93. https://doi.org/10.1021/acs.jmedchem.5b01390.

[238] Kumar S, Waldo JP, Jaipuri FA, Marcinowicz A, van Allen C, Adams J, et al. Discovery of clinical candidate (1 R,4 r)-4-(R)-2-(S)-6-fluoro-5 H-imidazo[5,1-A(isoindol-5-yl)-1-hydroxyethyl] cyclohexan-1-ol (navoximod), a potent and selective inhibitor of indoleamine 2,3-dioxygenase 1. J Med Chem 2019;62:6705–33. https://doi.org/10.1021/acs.jmedchem.9b00662.

[239] May HE, Boose R, Reed DJ. Hydroxylation of the carcinostatic 1-(2-chloroethyl)-3-cyclohexyl-1-nitrosourea (CCNU) by rat liver microsomes. Biochem Biophys Res Commun 1974;57:426–33. https://doi.org/10.1016/0006-291X(74)90948-6.

[240] Nafia I, Toulmonde M, Bortolotto D, Chaibi A, Bodet D, Rey C, et al. IDO targeting in sarcoma: biological and clinical implications. Front Immunol 2020;11. https://doi.org/10.3389/fimmu.2020.00274.

[241] Gao J, Deng F, Jia W. Inhibition of indoleamine 2,3-dioxygenase enhances the therapeutic efficacy of immunogenic chemotherapeutics in breast cancer. J Breast Cancer 2019;22:196–209. https://doi.org/10.4048/jbc.2019.22.e23.

[242] Nayak-Kapoor A, Hao Z, Sadek R, Dobbins R, Marshall L, Vahanian NN, et al. Phase Ia study of the indoleamine 2,3-dioxygenase 1 (IDO1) inhibitor navoximod (GDC-0919) in patients with recurrent advanced solid tumors. J Immunother Cancer 2018;6. https://doi.org/10.1186/s40425-018-0351-9.

[243] Jung KH, LoRusso P, Burris H, Gordon M, Bang YJ, Hellmann MD, et al. Phase I study of the indoleamine 2,3-dioxygenase 1 (IDO1) inhibitor navoximod (GDC-0919) administered with PD-L1 inhibitor

(atezolizumab) in advanced solid tumors. Clin Cancer Res 2019;25:3220–8. https://doi.org/10.1158/1078-0432.CCR-18-2740.

[244] Yue EW, Douty B, Wayland B, Bower M, Liu X, Leffet L, et al. Discovery of potent competitive inhibitors of indoleamine 2,3-dioxygenase with in vivo pharmacodynamic activity and efficacy in a mouse melanoma model. J Med Chem 2009;52:7364–7. https://doi.org/10.1021/jm900518f.

[245] Yue EW, Sparks R, Polam P, Modi D, Douty B, Wayland B, et al. INCB24360 (epacadostat), a highly potent and selective indoleamine-2,3-dioxygenase 1 (IDO1) inhibitor for immuno-oncology. ACS Med Chem Lett 2017;8:486–91. https://doi.org/10.1021/acsmedchemlett.6b00391.

[246] Bauer MR, Jones RN, Baud MGJ, Wilcken R, Boeckler FM, Fersht AR, et al. Harnessing fluorine-sulfur contacts and multipolar interactions for the design of p53 mutant Y220C rescue drugs. ACS Chem Biol 2016;11:2265–74. https://doi.org/10.1021/acschembio.6b00315.

[247] Beatty GL, O'Dwyer PJ, Clark J, Shi JG, Bowman KJ, Scherle PA, et al. First-in-human phase I study of the oral inhibitor of indoleamine 2,3-dioxygenase-1 epacadostat (INCB024360) in patients with advanced solid malignancies. Clin Cancer Res 2017;23:3269–76. https://doi.org/10.1158/1078-0432.CCR-16-2272.

[248] Komrokji RS, Wei S, Mailloux AW, Zhang L, Padron E, Sallman D, et al. A phase II study to determine the safety and efficacy of the oral inhibitor of indoleamine 2,3-dioxygenase (IDO) enzyme INCB024360 in patients with myelodysplastic syndromes. Clin Lymphoma Myeloma Leuk 2019;19:157–61. https://doi.org/10.1016/j.clml.2018.12.005.

[249] Mitchell TC, Hamid O, Smith DC, Bauer TM, Wasser JS, Olszanski AJ, et al. Epacadostat plus pembrolizumab in patients with advanced solid tumors: phase I results from a multicenter, open-label phase I/II trial (ECHO-202/KEYNOTE-037). J Clin Oncol 2018;36:3223–30. https://doi.org/10.1200/JCO.2018.78.9602.

[250] Hamid O, Bauer TM, Spira AI, Smith DC, Olszanski AJ, Tarhini AA, et al. Safety of epacadostat 100 mg bid plus pembrolizumab 200 mg Q3W in advanced solid tumors: phase 2 data from ECHO-202/KEYNOTE-037. J Clin Oncol 2017;35:3012. https://doi.org/10.1200/JCO.2017.35.15_suppl.3012.

[251] Gibney GT, Hamid O, Lutzky J, Olszanski AJ, Mitchell TC, Gajewski TF, et al. Phase 1/2 study of epacadostat in combination with ipilimumab in patients with unresectable or metastatic melanoma. J Immunother Cancer 2019;7. https://doi.org/10.1186/s40425-019-0562-8.

[252] Daud A, Saleh MN, Hu J, Bleeker JS, Riese MJ, Meier R, et al. Epacadostat plus nivolumab for advanced melanoma: Updated phase 2 results of the ECHO-204 study. J Clin Oncol 2018;36:9511. https://doi.org/10.1200/JCO.2018.36.15_suppl.9511.

[253] van den Eynde BJ, van Baren N, Baurain J-F. Is there a clinical future for IDO1 inhibitors after the failure of epacadostat in melanoma? Annu Rev Cancer Biol 2020;4:241–56. https://doi.org/10.1146/annurev-cancerbio-030419.

[254] Long GV, Dummer R, Hamid O, Gajewski TF, Caglevic C, Dalle S, et al. Epacadostat plus pembrolizumab versus placebo plus pembrolizumab in patients with unresectable or metastatic melanoma (ECHO-301/KEYNOTE-252): a phase 3, randomised, double-blind study. Lancet Oncol 2019;20:1083–97. https://doi.org/10.1016/S1470-2045(19)30274-8.

[255] Muller AJ, Manfredi MG, Zakharia Y, Prendergast GC. Inhibiting IDO pathways to treat cancer: lessons from the ECHO-301 trial and beyond. Semin Immunopathol 2019;41:41–8. https://doi.org/10.1007/s00281-018-0702-0.

[256] Crosignani S, Bingham P, Bottemanne P, Cannelle H, Cauwenberghs S, Cordonnier M, et al. Discovery of a novel and selective indoleamine 2,3-dioxygenase (IDO-1) inhibitor 3-(5-fluoro-1H-indol-3-yl)pyrrolidine-2,5-dione (EOS200271/PF-06840003) and its characterization as a potential clinical candidate. J Med Chem 2017;60:9617–29. https://doi.org/10.1021/acs.jmedchem.7b00974.

[257] Reardon DA, Desjardins A, Rixe O, Cloughesy T, Alekar S, Williams JH, et al. A phase 1 study of PF-06840003, an oral indoleamine 2,3-dioxygenase 1 (IDO1) inhibitor in patients with recurrent malignant glioma. Investig New Drugs 2020;38:1784–95. https://doi.org/10.1007/s10637-020-00950-1.

[258] Nelp MT, Kates PA, Hunt JT, Newitt JA, Balog A, Maley D, et al. Immune-modulating enzyme indoleamine 2,3-dioxygenase is effectively inhibited by targeting its apo-form. Proc Natl Acad Sci U S A 2018;115:3249–54. https://doi.org/10.1073/pnas.1719190115.

[259] Thomas SR, Salahifar H, Mashima R, Hunt NH, Richardson DR, Stocker R. Antioxidants inhibit indoleamine 2,3-dioxygenase in IFN-activated human macrophages: posttranslational regulation by pyrrolidine dithiocarbamate. J Immunol 2001;166:6332–40.

[260] Pham KN, Yeh SR. Mapping the binding trajectory of a suicide inhibitor in human indoleamine 2,3-dioxygenase 1. J Am Chem Soc 2018;140:14538–41. https://doi.org/10.1021/jacs.8b07994.

[261] Ortiz-Meoz RF, Wang L, Matico R, Rutkowska-Klute A, de la Rosa M, Bedard S, et al. Characterization of apo-form selective inhibition of indoleamine 2,3-dioxygenase**. ChemBioChem 2021;22:516–22. https://doi.org/10.1002/cbic.202000298.

[262] Röhrig UF, Reynaud A, Majjigapu SR, Vogel P, Pojer F, Zoete V. Inhibition mechanisms of indoleamine 2,3-dioxygenase 1 (IDO1). J Med Chem 2019;62:8784–95. https://doi.org/10.1021/acs.jmedchem.9b00942.

[263] Siu LL, Gelmon K, Chu Q, Pachynski R, Alese O, Basciano P, et al. Abstract CT116: BMS-986205, an optimized indoleamine 2,3-dioxygenase 1 (IDO1) inhibitor, is well tolerated with potent pharmacodynamic (PD) activity, alone and in combination with nivolumab (nivo) in advanced cancers in a phase 1/2a trial. Cancer Res 2017;77:CT116. https://doi.org/10.1158/1538-7445.AM2017-CT116.

[264] Luke JJ, Tabernero J, Joshua A, Desai J, Varga AI, Moreno V, et al. BMS-986205, an indoleamine 2, 3-dioxygenase 1 inhibitor (IDO1i), in combination with nivolumab (nivo): updated safety across all tumor cohorts and efficacy in advanced bladder cancer (advBC). J Clin Oncol 2019;37:358. https://doi.org/10.1200/JCO.2019.37.7_suppl.358.

[265] Sonpavde G, Necchi A, Gupta S, Steinberg GD, Gschwend JE, van der Heijden MS, et al. ENERGIZE: a phase III study of neoadjuvant chemotherapy alone or with nivolumab with/without linrodostat mesylate for muscle-invasive bladder cancer. Future Oncol 2019;16:4359–68. https://doi.org/10.2217/fon-2019-0611.

[266] Kyi C, Rubinstein MM, Shah P, Zhou Q, Iasonos A, Liu Y, et al. A phase II trial of IDO-inhibitor, BMS-986205 (IDO), and PD-1 inhibitor, nivolumab (NIVO), in recurrent or persistent endometrial cancer (EC; CA017-056). J Clin Oncol 2022;40:5589. https://doi.org/10.1200/JCO.2022.40.16_suppl.5589.

Section H

Copper enzymes

Chapter 8.1

Superoxide dismutases inhibitors

Azadeh Hekmat[a], **Ali Akbar Saboury**[b], **and Luciano Saso**[c]
[a]*Department of Biology, Science and Research Branch, Islamic Azad University, Tehran, Iran,* [b]*Institute of Biochemistry and Biophysics, University of Tehran, Tehran, Iran,* [c]*Department of Physiology and Pharmacology "vittorio erspamer", Sapienza University, Rome, Italy*

1 Introduction

Aerobic organisms benefit from high-yield energy production acquired through the electron transport chain (ETC) and the controlled conversion of molecular oxygen to water [1]. Nevertheless, electron leakage from ETC causes the accumulation of reactive intermediates, i.e., reactive oxygen species (ROS), which can produce significant cellular damage [2]. ROS contain singlet oxygen, superoxide anion radicals, hydroxyl radicals ($HO^{\bullet}$), and hydrogen peroxide (H_2O_2). The excessive redox active species can induce damage in RNA, lipids, proteins, and DNA, repress the activity of cellular enzymes, and induce cell death through activation of caspase and kinase cascades [3]. In 1954, reactive species were first demonstrated to be present in biological materials by Commoner et al. [4]. Superoxide was formally found in 1934 when Edward W. Neuman prepared a potassium oxide sample and measured its magnetic susceptibility. To maintain redox homeostasis, aerobic organisms have developed effective defense systems of nonenzymatic and enzymatic antioxidants. The superoxide dismutase (SOD) family is specialized in removing superoxide anion radicals derived from extracellular stimulants, containing oxidative insults and ionizing radiation, together with those primarily produced within the mitochondrial matrix as by-products of oxygen metabolism through the electron transport chain [5]. SODs (EC1.15.1.1) are enzymes that function to catalytically convert superoxide anion ($O_2^{\bullet-}$) to hydrogen peroxide (H_2O_2) and oxygen (O_2):

$$2O_2^{\bullet-} + 2H^+ \rightarrow H_2O_2 + O_2 \quad (1)$$

H_2O_2 is then converted by other enzymes such as peroxidases (GPx) and catalase (CAT) into harmless product water. The first SOD was discovered from bovine erythrocytes in 1969 at Duke University [6]. An antioxidant enzyme SOD was introduced into the market in the 1990s. However, it did not perform up to expectations. Subsequently, its use was limited to drug applications in animals and nondrug applications in humans (such as in agriculture, food, chemical, and cosmetic industries).

2 Structure and catalytic mechanism of SOD isoforms

SOD is a highly conserved enzyme composed of proteins and metal cofactors. According to the metal cofactor they harbor, SODs can be classified into four groups: manganese SOD (Mn-SOD), copper-zinc SOD (Cu/Zn-SOD), nickel SOD (Ni-SOD), and iron SOD (Fe-SOD). All four groups can be found in prokaryotic organisms; however, eukaryotes only express Mn-SODs and Cu/Zn-SODs. Fe-SOD can be also found only in chloroplasts. In mammals, distinct isoforms of SOD have been identified and characterized: Cu/Zn-SOD (SOD1), Mn-SOD (SOD2), and extracellular superoxide dismutase (EcSOD or SOD3) (Fig. 1). The structure and mechanism of these SOD are briefly introduced below.

2.1 Copper-zinc superoxide dismutase (Cu/Zn-SOD)

The first crystal structures of Cu/Zn-SOD were published in 1992 [7]. Commonly, Cu/Zn-SOD is homodimeric and is present in various locations in distinct organisms (Table 1). It can be found in the periplasm of gram-negative bacteria (sodC), cytoplasm, and chloroplast of plants, intermembrane space of mitochondria and various compartments for example nucleus, cytosol (SOD1), peroxisome, lysosome, and extracellular space (SOD3, EC SOD) in animals [21,22]. The structure of human Cu/Zn-SOD (PDB ID: 1PU0) is constituted by two subunits linked by histidine, each of which has an active site composed of Zn and Cu (Fig. 2). Cu/Zn-SODs are hypothesized to be the most modern SOD family due to the fact that Cu/Zn-SOD is highly abundant in animals and plants. Furthermore, the bioavailability of Zn and Cu enhances as oceanic and atmospheric O_2 levels increase [23]. Each subunit is the classical β-barrel made of eight antiparallel β-strands and connected by three loop regions. Two of these loops (called the zinc-binding loop and the electrostatic loop) together with a section of the β-barrel form the walls of a channel

Metalloenzymes. https://doi.org/10.1016/B978-0-12-823974-2.00004-8

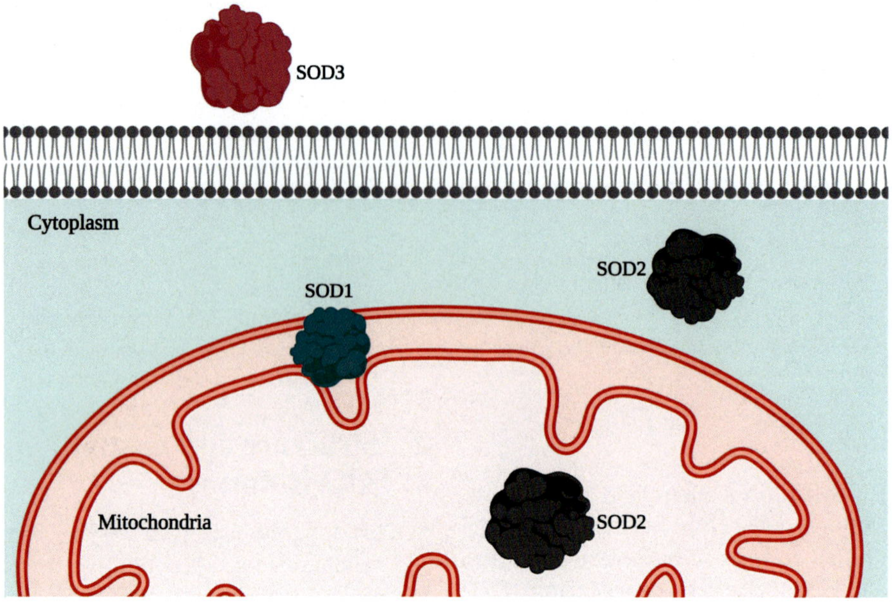

FIG. 1 Mammalian isoforms of SOD.

TABLE 1 Nomenclature and characteristics of Cu/Zn-SOD across various species.

Species	Gene	Protein	Localization	Refs.
Escherichia coli	*sodC*	SodC	Periplasmic space	[8]
Saccharomyces cerevisiae	*SOD1*	SOD1	Cytoplasm and mitochondrial intermembrane space	[9]
Caenorhabditis elegans	*sod-1* *sod-4* *sod-5*	Sod-1 Sod-4 Sod-5	Cytoplasm Extracellular space and plasma membrane Cytoplasm	[10] [11] [12]
Drosophila	Sod1 Sod3	SOD1 SOD3	Cytoplasm and mitochondrial intermembrane space Extracellular space and plasma membrane	[13] [14,15]
Rodent	Sod1 Sod3	SOD1 SOD3/ ecSOD	Cytoplasm, mitochondrial intermembrane space, nucleus, and peroxisomes Extracellular matrix and endothelial surface	[16] [17]
Human	*SOD1* *SOD3*	SOD1 SOD3/ EcSOD	Cytoplasm and mitochondrial intermembrane space and nucleus Extracellular matrix and endothelial surface	[18] [19,20]

from the enzyme surface to the active site. The Zn and Cu ions are bound at the base of the active site cavity. In the active site, Cu^{2+} acts as the catalytic site, whereas Zn^{2+} preserves the stability of the enzyme [21,24]. In the catalytic process, Cu^{2+} should be exposed to the solvent outside the enzyme to facilitate its binding with $O_2^{\bullet-}$ to initiate the reaction. Through the catalytic reaction, copper is alternatively reduced and oxidized in continuous encounters with $O_2^{\bullet-}$, changing between +2 and +1 valence states. Cu^{2+} Zn^{2+} SOD first adsorbs and binds $O_2^{\bullet-}$ and then facilitates its oxidation. In this process, Cu^{2+} Zn^{2+} SOD turns into Cu^{+} Zn^{2+} SOD with minor variations in the structure while oxidizing $O_2^{\bullet-}$ to O_2. Later, Cu^{+} Zn^{2+} SOD combines with another molecule $O_2^{\bullet-}$, reducing it to H_2O_2 and returning to the original form of Cu^{2+} Zn^{2+} SOD. Metal ions can be eliminated by dialysis at pH 3.8 against a metal chelator for instance ethylenediaminetetraacetic acid (EDTA) [25,26]. When the zinc ion is removed, the second reduction reaction (Eq. 3) in the dismutation reaction of Cu^{2+} apoSOD becomes pH dependent.

$$Cu^{+2}Zn^{+2}SOD + O_2^{\bullet-} \rightarrow Cu^{+}Zn^{2+}SOD + O_2 \quad (2)$$

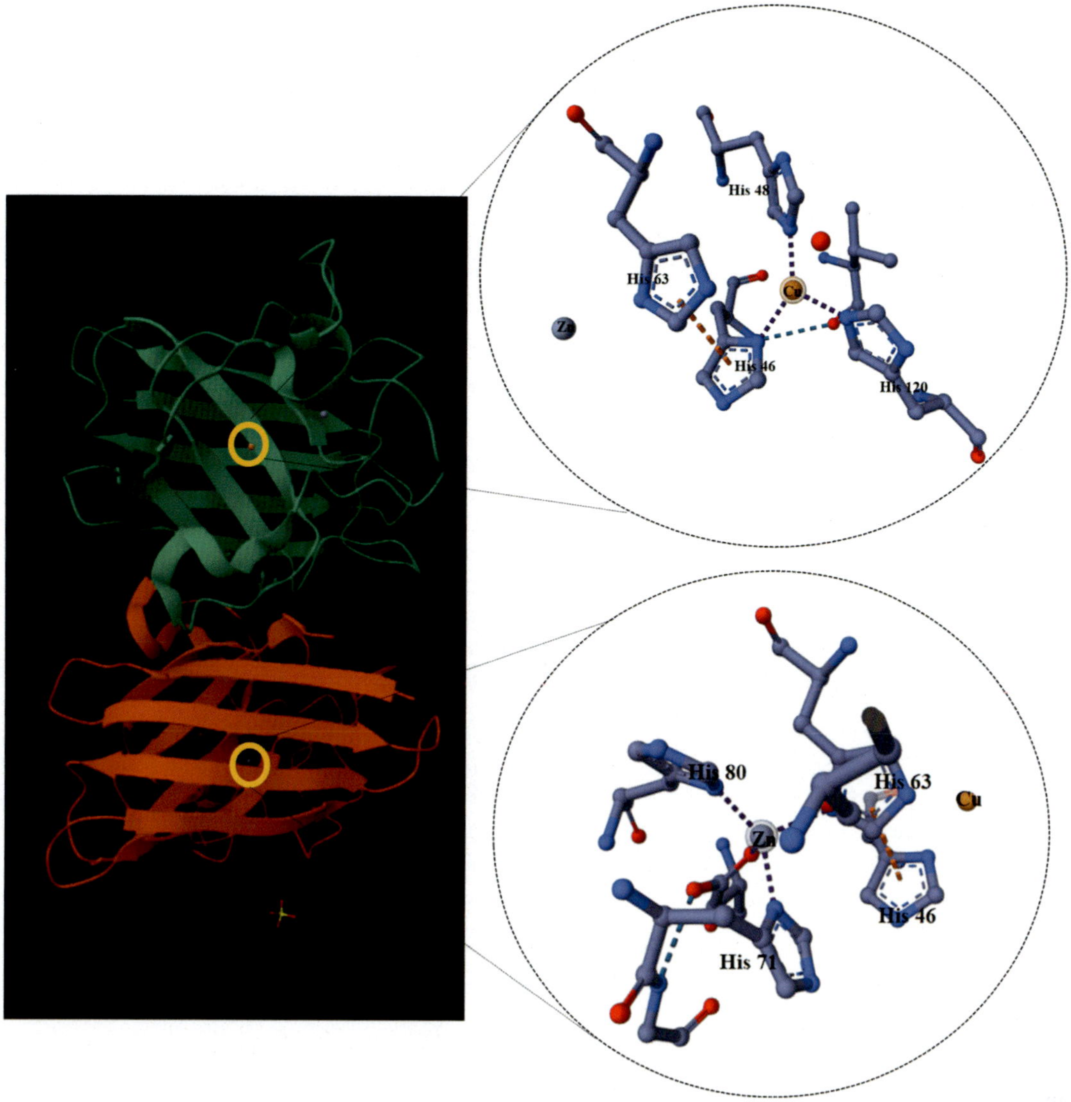

FIG. 2 Schematic illustration of Cu/Zn-SOD. The structure and active sites of human Cu/Zn-SOD (PDB ID: 1PU0). *(From RCSB Protein Data Bank (RCSB PDB).)*

$$Cu^{+}Zn^{+2}SOD + O_2^{\bullet -}(+2H^{+}) \rightarrow Cu^{2+}Zn^{2+}SOD + H_2O_2 \quad (3)$$

The catalytic reaction rate is associated with the electrostatic attraction of the negatively charged $O_2^{\bullet -}$ into the positively charged active center. Positively charged arginine near the active site significantly improves electrostatic guidance [24]. Dismutation reaction catalyzed by Cu/Zn-SOD, as a result of an electrostatic loop, is one of the fastest enzymatic reactions (its maximum speed has been approximated to $2 \times 10^9\ M^{-1}\ s^{-1}$) [27].

The monomeric subunit of human SOD3 is synthesized as a 32kDa monomeric protein that displays higher order dimers, tetramers, and octamers cross-linked through disulfide bridges between cysteine residues in the C-terminal region. Mature SOD3 can be separated into three regions, the amino-terminal features an asparagine at position 89 discovered by mass spectrometry to be a singular glycosylated residue greatly enhancing protein solubility and has been demonstrated as required for secretion but not activity [19,28]. The second domain bears 60% homology to SOD1 and contains the conserved active site and ion binding folds for the singular Zn and Cu ions required for catalysis [29].

2.2 Manganese superoxide dismutase (Mn-SOD)

Mn-SOD was first discovered in prokaryotes by Keele et al. in 1970 [30]. Mn-SODs are typically found in prokaryotes and mitochondria of the eukaryotes (Table 2). Across various phyla of archaea, eubacteria, and eukaryotes, the active center structure of the Mn-SOD family has both

TABLE 2 Nomenclature and characteristics of Mn-SOD across various species

Species	Gene	Protein	Localization	Refs.
Escherichia coli	*sodA*	SodA	Cytoplasm	[30]
Saccharomyces cerevisiae	*SOD2*	SOD2	Mitochondrial matrix	[31]
Caenorhabditis elegans	*sod-2* *sod-3*	Sod-2 Sod-3	Mitochondrial matrix Mitochondrial matrix	[32]
Drosophila	Sod2	SOD2	Mitochondrial matrix	[33]
Rodent	Sod2	SOD2	Mitochondrial matrix	[34]
Human	*SOD2*	SOD2	Mitochondrial matrix	[35]

dimer and tetramer structures with considerable sequence homology and well-conserved protein folds. In eukaryotic cells, the Mn-SOD is typically in the tetramer form; however, Mn-SOD is in the dimer form for prokaryotes. The human Mn-SOD (PDB ID: 1N0J) is constituted by one manganese active site in each subunit and each manganese is coordinated with four residues (three histidine residues and one aspartic acid residue) and a single-oxygen-containing molecule (H_2O or OH^-) (Fig. 3). The catalytic mechanism of Mn-SOD includes a cycle between Mn^{3+} and Mn^{2+}. Nevertheless, owing to the complexity of the structure, the catalytic mechanism of Mn-SOD requires more steps than other SODs and involves a series of proton transfers (Eqs. 4–7).

$$Mn^{+3}SOD(OH^-) + O_2^{\bullet -}(+H^+) \rightarrow Mn^{+2}SOD(H_2O) + O_2 \quad (4)$$

$$Mn^{+2}SOD(H_2O) + O_2^{\bullet -}(+H^+) \rightarrow Mn^{+3}SOD(OH^-) + H_2O_2 \quad (5)$$

$$Mn^{+2}SOD(H_2O) + O_2^{\bullet -}(+H^+) \rightarrow Mn^{+3}SOD(H_2O) + O_2^{2-} \quad (6)$$

$$Mn^{+3}SOD(H_2O) + O_2^{-2}(+H^+) \rightarrow Mn^{+3}SOD(OH^-) + H_2O_2 \quad (7)$$

Mn ion is caved in the lumen of the active site and forms a hydrogen bond network with the surrounding side chain residues (Tyr34 and Gln143) and with two single-oxygen-containing molecules at the opening of the cavity. The role of the hydrogen bond network is possible to promote proton transfer in the process of $O_2^{\bullet -}$ reduction to H_2O_2 and preserve the catalytic activity and stability of Mn-SOD [21,24,26].

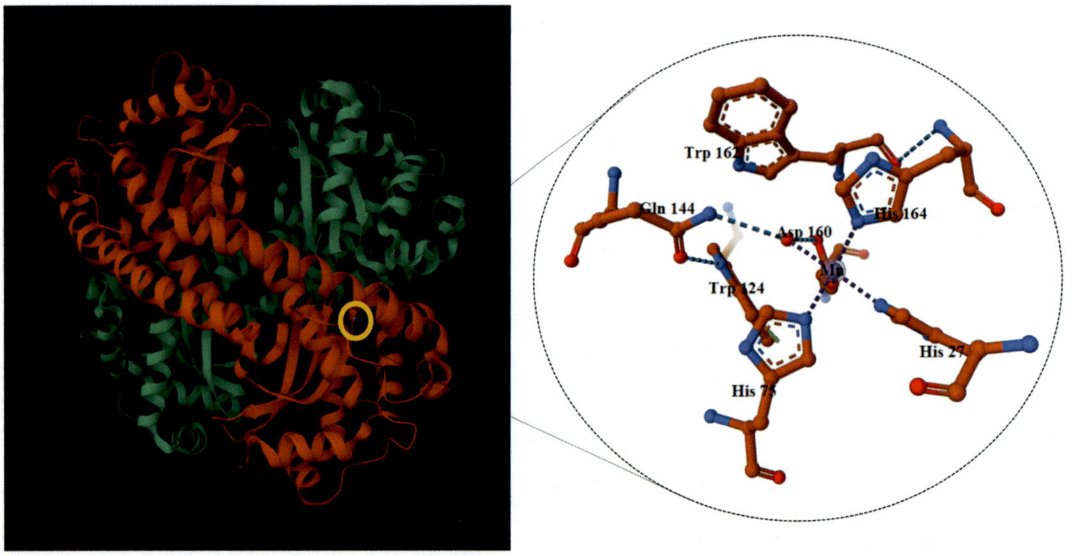

FIG. 3 Schematic illustration of Mn-SOD. The structure and active sites of human Mn-SOD (PDB ID: 1N0J). *(From RCSB Protein Data Bank (RCSB PDB).)*

2.3 Iron superoxide dismutase (Fe-SOD)

Fe-SOD was first discovered in prokaryotes by Yost and Fridovich in 1973 [36]. Fe-SOD is found in prokaryotes (SodB) [36], protozoans [37], and chloroplasts of algae and in plants [38] two families of Fe-SODs have been found: the plastidial localized and the chloroplastic localized. Its structure is similar to that of Mn-SOD. Fe-SOD is a dimer, and each monomer has an iron-centered active site that comprises a single iron binding an aspartic acid, three histidines and a water molecule (Fig. 4). Even though the active sites of Fe-SOD and Mn-SOD are similar in structure, the substitution of their metal ions with each other results in reduced activity. The overall catalytic mechanism of Fe-SOD relies on the conversion between Fe^{2+} and Fe^{3+} in a ping-pong manner [21,24,26].

2.4 Nickel superoxide dismutase (Ni-SOD)

Ni-SOD was discovered in 1996 [39] and is typically in marine bacteria and algae. The Ni-SOD has a homohexamer structure, with a nickel ion-binding hook formed by the combination of Ni^{2+} and some amino acid residues in the center of each monomer. Unlike other SODs, in addition to histidine and aspartic acid, Ni-SOD has numerous cysteine residues linked to the active center Ni^{2+} via S–Ni^{2+} bonds (Fig. 5). Nevertheless, the metal-bound cysteines are highly oxidation-sensitive and susceptible to damage. Current research has exposed that all Ni-related redox enzymes comprise cysteine or redox nonharmless ligands. How Ni^{2+}-SOD evades the oxidative damage of S–Ni^{2+} bonds and the catalytic mechanism of Ni-SOD are still being explored. Nevertheless, the mechanism of catalytic

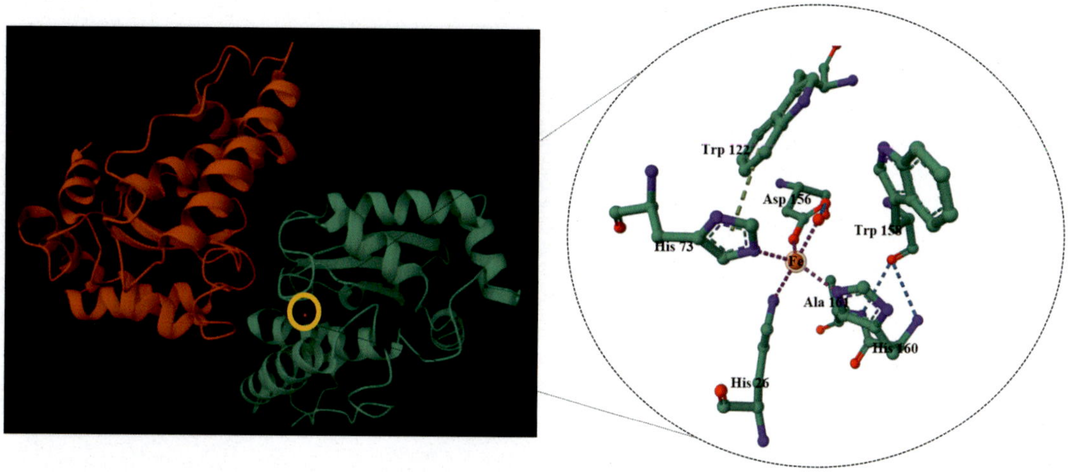

FIG. 4 Schematic illustration of Fe-SOD. The structure and active sites of *E. col* Fe-SOD (PDB ID: 1ISA). *(From RCSB Protein Data Bank (RCSB PDB).)*

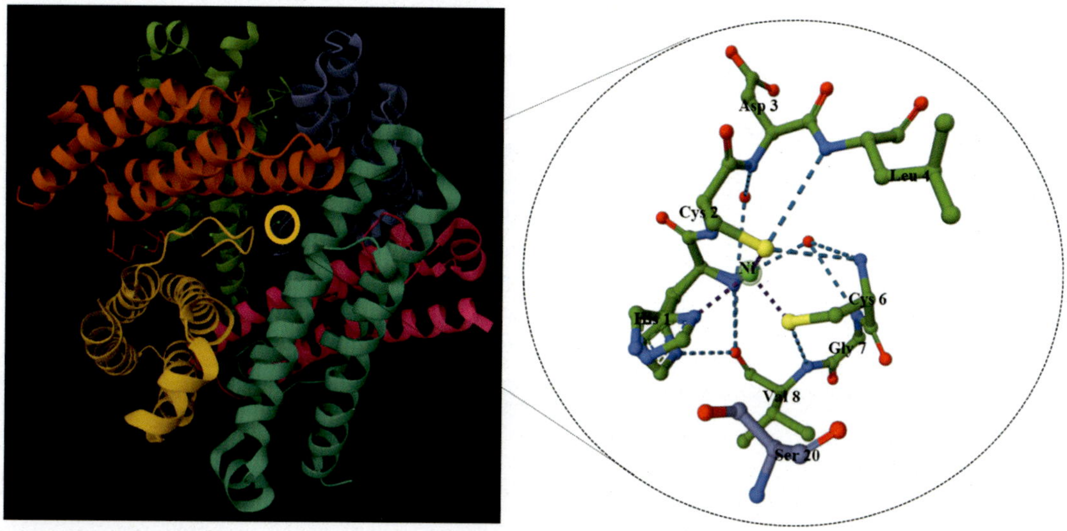

FIG. 5 Schematic illustration of Ni-SOD. The structure and active sites of *Streptomyces coelicolor* Ni-SOD (PDB ID: 1T6U). *(From RCSB Protein Data Bank (RCSB PDB).)*

disproportionation is certainly associated with the transformation between Ni^{2+} and Ni^{3+} [21,24,26].

3 The roles of SODs in human diseases

Alterations in the catalytic activity or expression level of SODs have been found in a variety of pathological conditions, including some of the most common age-dependent diseases; for example, cancer, cardiovascular diseases, and neurodegenerative diseases. In the following section, most common diseases in which SOD plays a crucial role were discussed.

3.1 SOD in cancer

Several studies have discovered the critical role of oxidative stress in carcinogenesis. Oxidative stress has been strongly involved in several aspects of cancer biology. Higher ROS levels can promote DNA mutations, consequently playing a potential causal function in tumorigenesis [40]. After transformation and during tumor maintenance, cancer cells must cope with the toxic effects of high oxidative stress and possibly modify their redox homeostasis to maintain fast growth. It has been shown that increasing ROS production is an adaptive process that cancer cells utilize to generate variations in signaling pathways beneficial for their normal function [41]. Thus, SOD, being a crucial cellular antioxidant, is highly responsible for the elimination of $O_2^{\bullet-}$. It has been proposed that SOD can regulate cancer progression and consequently can be utilized as a unique target for cancer treatment [41]. It has been shown that SOD1 can be utilized as a novel therapeutic target for the treatment of multiple myeloma [42]. In contrast, the migratory and invasive activity of pancreatic cancer is encouraged by SOD through activation of the H_2O_2/ERK/NF-κB axis [43]. Oncomine analysis of neoplastic *vs* normal tissues revealed that SOD3 expression levels were drastically downregulated across a majority of cancers including breast cancer, sarcoma, lung cancer, and head and neck cancer [44]. In breast cancer cells, overexpression of SOD3 inhibited *in vitro* clonogenic survival, proliferation, and invasion of triple-negative breast cancer cells partially through suppressing heparanase-mediated fragmentation of cell surface proteoglycans and by diminishing vascular endothelial growth factor (VEGF) bioavailability [44]. SOD3 expression is significantly downregulated in primary human lung cancer with further reduction occurring between stages I and IV [45].

3.2 SOD in neurodegenerative diseases

Oxidative stress has been confirmed to be involved in the pathophysiology of various neurodegenerative diseases. The most commonly explored connection between human diseases and SOD is amyotrophic lateral sclerosis (ALS), a late-onset fatal neurodegenerative disorder that primarily affects motor neurons [46]. In 1993, mutations in the SOD1 gene were first reported as a cause of familial ALS [47]. Since then, numerous mutations spread throughout the SOD1 gene have been described to result in ALS [48,49]. In addition, the affected regions of patients having Alzheimer's disease (AD) and Parkinson's disease (PD) have reduced activity of antioxidant enzymes specifically SOD [50]. It has been also experimentally confirmed that SOD-2 overexpression could reduce hippocampal superoxide and consequently inhibits memory deficits in a mouse model of AD [51]. SOD supplementation revealed enhancement in mice models of AD [52].

3.3 SOD in diabetes

It has been shown that increased oxidative stress performs a key role in the etiology of diabetes and its complications [53]. In diabetes, persistent hyperglycemia stimulates the production of ROS from different sources [54]. Accordingly, diabetes typically leads to enhanced ROS formation and weakened antioxidant defenses [54,55]. Under hyperglycemic conditions, endothelial cells produce elevated levels of $O_2^{\bullet-}$ [56] Overproduction of $O_2^{\bullet-}$ can inhibit glyceraldehyde-3-phosphate dehydrogenase (GAPDH), which is an essential enzyme of the glycolytic pathway [57,58]. This leads to the glucose accumulation and other intermediate metabolites of this pathway and shifts to other alternative pathways of glucose metabolism as well as enhanced production of advanced glycation end products [58]. It has been shown that extracellular SOD can act as a therapeutic agent to defend the progression of diabetic nephropathy [59].

3.4 SOD3 in cardiovascular diseases

The key role of SOD3 in modulating $O_2^{\bullet-}$ levels is in the vasculature, thus SOD3 has been linked to pathological conditions that involve vascular dysfunction such as diabetes, hypertension, and atherosclerosis. Oxidative stress, particularly the production of $O_2^{\bullet-}$, can stimulate endothelial dysfunction, a vital early step in atherosclerosis [60]. Furthermore, excess $O_2^{\bullet-}$ itself can directly antagonize $NO^{\bullet}$ by a direct chemical interaction. The endothelium-derived $NO^{\bullet}$ controls the extent of vascular smooth muscle relaxation, inhibits platelet aggregation, and attenuates neutrophil adherence to the endothelium [61]. SOD3, by opposing the inactivation of vascular $NO^{\bullet}$, can therefore be vital for the maintenance of blood vessel tone and protection against plaque formation. In patients with coronary artery disease, the activity of SOD3 was also found to be reduced, whereas SOD1 and SOD2 remain unaffected [61].

3.5 SOD in inflammatory diseases

Neutrophils perform an essential function in the pathogenesis of inflammation. Activated neutrophils adhere to vascular endothelium and transmigrate to the extravascular space, releasing protease enzymes and large amounts of chemokines as well as ROS [61]. Proteases and ROS can damage normal tissue and extracellular matrix proteins. $O_2^{\bullet -}$ serves to activate endothelial cells and enhance neutrophil infiltration [61]. Neutrophil apoptosis can also be an essential step in the resolution of inflammation. In individuals with down syndrome, neutrophil apoptosis improves, and Cu/Zn-SOD is overexpressed [62]. SOD1, SOD2, and SOD3 have been described as potential inhibitors of inflammation by numerous reporters [63,64]

4 SOD inhibitors

As mentioned in the previous section, SOD is associated with various diseases, such as aging, neurodegenerative diseases, tumorigenesis, and diabetes. Thus, it is essential to discover inhibitors for this enzyme and knowledge about their mechanism of action.

4.1 Selected inhibitors of Cu/Zn-SOD

4.1.1 Hydrogen peroxide

H_2O_2 can inhibit both SOD1 and SOD3. This mechanism facilitates Cu/Zn-SOD activity adjustment by an excessive concentration of H_2O_2, which can be a result of modified H_2O_2 decomposition by GPx and CAT. By reversibly binding to the histidine residues in the active site of the enzyme, H_2O_2 alters the integrity of the active site of Cu/Zn-SOD, delaying the regeneration of the bond between Cu^{2+} and the bridging histidine, vital for conserving the Cu/Zn-SOD activity [19,65].

4.1.2 Choline tetrathiomolybdate (ATN-224) and N,N′-diethyldithiocarbamate

N,N′-diethyldithiocarbamate (DDC), a chelating agent, is the first product of disulfiram biotransformation. Choline tetrathiomolybdate (ATN-224) is a second-generation tetrathiomolybdate analog. The mechanism of Cu/Zn-SOD inhibition by DDC was proposed in 1976 [66]. The inhibitory effect of both ATN-224 and DDC on Cu/Zn-SOD stems from the capability of these compounds to bind with metals, that is, with Cu ion in the active site of the enzyme [67]. Additionally, disulfiram toxicity could be associated with the prooxidative action of DDC [68]. As it is the Cu ion, not Zn ion, which is reduced in the dismutation reaction, Cu/Zn-SOD activity is considerably reduced. The mechanism of Cu/Zn-SOD inhibition of both compounds is yet to be comprehensively analyzed, even though it has already been utilized in numerous research. In the case of DDC, the mechanism is catalyzed with first-order kinetics and needs at least two molecules of DDC. It has been proven that DDC is potentially useful as a supplementary factor in antifungal treatment [68]. On the other hand, ATN-224 can be utilized as a prooxidative drug in cancer treatment [67].

4.1.3 2-Methoxyestradiol and other estrogen derivatives

The structure of 2-ME is shown in Fig. 2I. 2-ME cannot bind with estrogen receptors and has anticancer properties, owing to its antiangiogenic, antimetastatic, and antiproliferative action [69,70]. There are a few observations for 2-ME being an inhibitor of Cu/Zn-SOD, and the potential mechanism has not been discovered yet. It seems that the anticancer properties of this drug originate from its ability to initiate the production of superoxide, causing ROS-mediated damage and leading to cell apoptosis. However, still it is hard to determine whether 2-ME inhibits Cu/Zn-SOD. Other inhibitors for Cu/Zn-SOD are listed in Table 3. It seems that despite their prooxidative action, Cu/Zn-SOD inhibitors can have a potential application in human diseases.

TABLE 3 Some inhibitors for Cu/Zn-SOD activity.

Inhibitor	Mechanism of inhabitation	Refs.
4-Aminoantipyrine (4-AAP)	Bind to Cu/Zn-SOD in the region connecting it's both subunits Increase the amount of α-helix and change the structure of the enzyme	[71]
2-Mercaptobenzimidazole (MBI)	Bind to Cu/Zn-SOD in the region connecting its both subunits Increase the amount of *β*-sheet and turn and change the structure of the enzyme	[72]
Phenylglyoxal	Modify arginine residues through binding with its guanidine group	[73]
2,3-Butanedione	Modify arginine residues through binding with its guanidine group	[73,74]
Cadmium (Cd)	Enzyme structure misfolding and partial denaturation as a result of Cd substitution with Zn in the active site of the enzyme	[75,76]
Sodium nitroprusside (SNP)	SNP-$O^{\bullet -}$ adducts bind to the enzyme and lower its activity	[77]
Hemin	Not known	[78]

5 Future perspectives

SODs are the only enzymes that interact with superoxide specifically and can control the levels of ROS. They also serve as crucial regulators of signaling. While SOD catalyzes the production of H_2O_2, which is also an effective oxidant, some research applied SOD in combination with H_2O_2 metabolizing antioxidants for example catalase. Nowadays, SODs have received much attention in efforts to minimize oxygen radical-induced damage to normal tissues. Recent research reveals the potential therapeutic applications of SOD in the prevention/control of several diseases. SOD can be also an effective antioxidant therapy for managing the detrimental consequences of inflammatory diseases. Although different SOD-based compounds have been verified, from plant and animal extracts and SOD recombinant forms to SOD mimetics and SOD gene therapy. However, the literature data on enzyme inhibitors are rare. Consequently, according to the crucial role of SOD in human health, it is necessary to synthesize biocompatible SOD inhibitors.

References

[1] Alberts B, Johnson A, Lewis J, Raff M, Roberts K, Walter P. Molecular biology of the cell. 4th ed. New York: Garland Science; 2002.
[2] Kirkman HN, Gaetani GF. Trends Biochem Sci 2007;32:44.
[3] Galasso M, Gambino S, Romanelli MG, Donadelli M, Scupoli MT. Free Rad Biol Med 2021;172:264.
[4] Commoner B, Townsend J, Pake GE. Nature 1954;174:689.
[5] Miao L, Clair DKS. Free Rad Biol Med 2009;47:344.
[6] McCord JM, Fridovich I. J Biol Chem 1969;244:6049.
[7] Parge HE, Hallewell RA, Tainer JA. Proc Natl Acad Sci USA 1992;89:6109.
[8] Imlay K, Imlay JA. J Bacteriol 1996;178:2564.
[9] Sturtz LA, Diekert K, Jensen LT, Lill R, Culotta VC. J Biol Chem 2001;276:38084.
[10] Larsen PL. Proc Natl Acad Sci USA 1993;90:8905.
[11] Fujii M, Ishii N, Joguchi A, Yasuda K, Ayusawa D. DNA Res 1998;5:25.
[12] Jensen LT, Culotta VC. J Biol Chem 2005;280:41373.
[13] Campbell SD, Hilliker AJ, Phillips JP. Genetics 1986;112:205.
[14] Jung I, Kim T-Y, Kim-Ha J. FEBS Lett 1973;585:2011.
[15] Blackney MJ, Cox R, Shepherd D, Parker JD. Biosci Rep 2014;34.
[16] Chang L-Y, Slot JW, Geuze HJ, Crapo JD. J Cell Biol 1988;107:2169.
[17] Folz RJ, Guan J, Seldin MF, Oury TD, Enghild JJ, Crapo JD. Am J Resp Cell Mol Biol 1997;17:393.
[18] Levanon D, Lieman-Hurwitz J, Dafni N, Wigderson M, Sherman L, Bernstein Y, Laver-Rudich Z, Danciger E, Stein O, Groner Y. EMBO J 1985;4:77.
[19] Antonyuk SV, Strange RW, Marklund SL, Hasnain SS. J Mol Biol 2009;388:310.
[20] Strålin P, Karlsson K, Johansson BO, Marklund SL. Arterioscl Thromb Vasc Biol 1995;15:2032.
[21] Culotta VC, Yang M, O'Halloran TV. Biochim Biophys Acta 2006;1763:747.
[22] Huang J-K, Wen L, Ma H, Huang Z-X, Lin C-T. J Agric and Food Chem 2005;53:6319.
[23] Saito MA, Sigman DM, Morel FM. Inorg Chim Acta 2003;356:308.
[24] Zhao H, Zhang R, Yan X, Fan K. J Mater Chem B 2021;9:6939.
[25] Li H-T, Jiao M, Chen J, Liang Y. Acta Biochim Biophys Sin 2010;42:183.
[26] Wang Y, Branicky R, Noë A, Hekimi S. J Cell Biol 1915;217:2018.
[27] Shin DS, DiDonato M, Barondeau DP, Hura GL, Hitomi C, Berglund JA, Getzoff ED, Cary SC, Tainer JA. J Mol Biol 2009;385:1534.
[28] Ota F, Kizuka Y, Kitazume S, Adachi T, Taniguchi N. FEBS Lett 2016;590:3357.
[29] Tibell L, Aasa R, Marklund S. Arch Biochem Biophys 1993;304:429.
[30] Keele BB, McCord J, Fridovich I. J Biol Chem 1970;245:6176.
[31] Ravindranath S, Fridovich I. J Biol Chem 1975;250:6107.
[32] Hunter T, Bannister WH, Hunter GJ. J Biol Chem 1997;272:28652.
[33] Duttaroy A, Parkes T, Emtage P, Kirby K, Boulianne GL, Wang X, Hilliker AJ, Phillips JP. DNA Cell Biol 1997;16:391.
[34] Jones PL, Kucera G, Gordon H, Boss JM. Gene 1995;153:155.
[35] Wan XS, Devalaraja MN, ClairR DKST. DNA Cell Biol 1994;13:1127.
[36] Yost FJ, Fridovich I. J Biol Chem 1973;248:4905.
[37] Anju A, Jeswin J, Thomas P, Paulton M, Vijayan K. Fish Shellfish Immunol 2013;34:946.
[38] Wang F, Wu Q, Zhang Z, Chen S, Zhou R. Protein J 2013;32:259.
[39] Youn H-D, Kim E-J, Roe J-H, Hah YC, Kang S-O. Biochem J 1996;318:889.
[40] Panieri E, Santoro M. Cell Death Dis 2016;7, e2253.
[41] Hekmat A, Saboury AA. Protein Kinase Inhibitors. Elsevier; 2022. p. 23.
[42] Salem K, McCormick ML, Wendlandt E, Zhan F, Goel A. Redox Biol 2015;4:23.
[43] Li T-M, Chen G-W, Su C-C, Lin J-G, Yeh C-C, Cheng K-C, Chung J-G. Anticancer Res 2005;25:971.
[44] Griess B, Tom E, Domann F, Teoh-Fitzgerald M. Free Rad Biol Med 2017;112:464.
[45] Teoh-Fitzgerald ML, Fitzgerald MP, Jensen TJ, Futscher BW, Domann FE. Mol Cancer Res 2012;10:40.
[46] Robberecht W, Philips T. Nat Rev Neurosci 2013;14:248.
[47] Rosen DR, Siddique T, Patterson D, Figlewicz DA, Sapp P, Hentati A, Donaldson D, Goto J, O'Regan JP, Deng H-X. Nature 1993;362:59.
[48] Andersen PM, Al-Chalabi A. Nat Rev Neurol 2011;7:603.
[49] Taylor JP, Brown RH, Cleveland DW. Nature 2016;539:197.
[50] Choi J, Rees HD, Weintraub ST, Levey AI, Chin L-S, Li L. J Biol Chem 2005;280:11648.
[51] Massaad CA, Washington TM, Pautler RG, Klann E. Proc Natl Acad Sci 2009;106:13576.
[52] Persichilli S, Gervasoni J, Di Napoli A, Fuso A, Nicolia V, Giardina B, Scarpa S, Desiderio C, Cavallaro RA. J Alzheimer's Dis 2015;44:1323.
[53] Ceriello A. Metabolism 2000;49:27.
[54] Yan L-J. J Diab Res 2014;2014.
[55] Kayama Y, Raaz U, Jagger A, Adam M, Schellinger IN, Sakamoto M, Suzuki H, Toyama K, Spin JM, Tsao PS. Int J Mol Sci 2015;16:25234.
[56] Kar S, Kavdia M. Free Rad Biol Med 2013;63:161.
[57] Du X, Matsumura T, Edelstein D, Rossetti L, Zsengellér Z, Szabó C, Brownlee M. J Clin Invest 2003;112:1049.
[58] Lazarev VF, Guzhova IV, Margulis BA. Pharmaceutics 2020;12:416.

[59] Kuo C-W, Shen C-J, Tung Y-T, Chen H-L, Chen Y-H, Chang W-H, Cheng K-C, Yang S-H, Chen C-M. Life Sci 2015;135:77.
[60] Fukai T, Folz RJ, Landmesser U, Harrison DG. Cardiovas Res 2002;55:239.
[61] Tousoulis D, Kampoli A-M, Papageorgiou CTN, Stefanadis C. Curr Vasc Pharmacol 2012;10:4.
[62] Arbuzova S, Hutchin T, Cuckle H. Bioessays 2002;24:681.
[63] Porfire AS, Leucuţa SE, Kiss B, Loghin F, Pârvu AE. Pharmacol Rep 2014;66:670.
[64] A. Joseph, Y. Li, H.-c. Koo, J.M. Davis, S. Pollack, J.A. Kazzaz, Free Rad Biol Med, 45, 2008, 1143.
[65] Kurahashi T, Miyazaki A, Suwan S, Isobe M. J Am Chem Soc 2001;123:9268.
[66] Heikkila RE, Cabbat FS, Cohen G. J Biol Chem 1976;251:2182.
[67] Che M, Wang R, Li X, Wang H-Y, Zheng XS. Drug Discov Today 2016;21:143.
[68] Bink A, Vandenbosch D, Coenye T, Nelis H, Cammue BP, Thevissen K. Antimicrob Agents Chemother 2011;55:4033.
[69] Lakhani NJ, Sarkar MA, Venitz J, Figg WD. Pharmacotherapy 2003;23:165.
[70] Mueck A, Seeger H. Steroids 2010;75:625.
[71] Teng Y, Liu R. J Hazardous Mater 2013;262:318.
[72] Teng Y, Zou L, Huang M, Chen Y. PloS One 2014;9, e106003.
[73] Borders C, Johansen JT. Carlsberg Res Commun 1980;45:185.
[74] Borders Jr CL, Saunders JE, Blech D, Fridovich I. Biochem J 1985;230:771.
[75] Sun H, Cui E, Liu R. Environ Sci Pollut Res 2015;22:18267.
[76] Huang YH, Shih CM, Huang CJ, Lin CM, Chou CM, Tsai ML, Liu TP, Chiu JF, Chen CT. J Cell Biochem 2006;98:577.
[77] Misra HP. J Biol Chem 1984;259:12678.
[78] Percival SS, Harris ED. Biochem J 1991;274:153.

Chapter 8.2

Tyrosinase enzyme and its inhibitors: An update of the literature

Simone Carradori[a], Francesco Melfi[a], Josip Rešetar[b], and Rahime Şimşek[c]

[a]*Department of Pharmacy, University "G. d'Annunzio" of Chieti-Pescara, Chieti, Italy,* [b]*Faculty of Pharmacy and Biochemistry, University of Zagreb, Zagreb, Croatia,* [c]*Faculty of Pharmacy, Department of Pharmaceutical Chemistry, Hacettepe University, Ankara, Turkey*

1 Introduction

Tyrosinase (EC 1.14.18.1, TYR), also known as polyphenol oxidase, is a copper-containing oxidation enzyme, broadly distributed in plants, animals, and microorganisms [1]. This enzyme plays an essential role in the production of melanin, by being involved in two different reactions of melanin synthesis: monophenolase and diphenolase activities. Its activity is implied also in disagreeable browning of fruits, vegetables, and beverages, a paramount issue in food industry [2]. In insects, the oxidase function is reflected in the mechanisms of cuticle sclerotization, wound healing, defensive encapsulation, and melanization of foreign organisms [3,4]. Furthermore, tyrosinase also exists in many bacteria and plays a role in the bacterial melanin biosynthesis, vital for the protection role from ultraviolet (UV) radiation. It is worth noting that bacterial melanin can neutralize antibiotics and chelate metal ions, allowing it to improve bacterial survival, especially under environmental stress [5,6]. In agreement with that, tyrosinase has been recognized as a therapeutic target for the development of anti-browning, antibacterial, skin whitening, insecticide, and therapeutic agents. Taking advantage of its great potential application in food, agricultural, cosmetic, and pharmaceutical industries, several synthetic tyrosinase inhibitors have been reported in recent years. According to a report by Grand View Research, Inc., it is predicted that the global skin whitening market will arrive at $13.7 billion by 2025 [7]. Moreover, the skin whitening practice is common in some ethnic groups, such as Asia, Africa, and Middle East, as the result of a complex interplay of social, cultural, psychological, and political factors, since time immemorial. In these cultures, lighter skin complexion is often associated with beauty and health, and according to it, darker skin tone is linked to lower social conditions, indeed, since 1980s, the interest in skin whitening has exponentially grown [8].

2 Structure and function of the enzyme

Tyrosinase is a structural-complicated multisubunit binuclear copper-containing metalloenzyme and in humans it is expressed by melanocytes as a transmembrane protein [9]. Structurally, it contains four conserved regions—*N*-terminal signal peptide, intra-melanosomal domain, single transmembrane α-helix, and flexible *C*-terminal cytoplasmic domain [10]. TYR is anchored inside of the melanosomal membrane via a single transmembrane α-helix, while a small *C*-terminal tail oriented to the cytoplasm has an important role in sorting and diverting the protein to the melanosomal membrane. A large intramelanosomal domain is divided into a cysteine-rich subdomain, whose function is still unknown, and a catalytically active tyrosinase-like subdomain with two copper ion-binding sites (Fig. 1).

The main function of TYR is the catalytic addition of oxygen to phenol compounds. TYR's active site holds a binuclear copper center (CuA and CuB), where each Cu ion is coordinated with three histidine residues. It can catalyze both the hydroxylation of monophenols (monophenolase activity) and the oxidation of *o*-diphenols (diphenolase activity). Tyrosinase is recognized as the ancestor protein of hemocyanin because it can be found in very primitive organisms [11]. Studies have shown that they are highly homologous, and there is also a close similarity between the two active sites and catalytic states. Three different states of tyrosinase are involved in catalytic mechanisms, namely, deoxy-form (E deoxy), met-form (E met), and oxy-form (E oxy) [1,12,13]. The deoxy-form is characterized by two Cu(I) with a $3d^{10}$ electron configuration. The coordination form of Cu(I) is as independent as that of metal copper, so the bridging ligand is absent [14]. The oxy-form consists of two Cu(II) with a $3d^{9}$ electron configuration, every copper atom is characterized by one weaker axial and two strong equatorial histidine residues. The oxygen

Metalloenzymes. https://doi.org/10.1016/B978-0-12-823974-2.00003-6

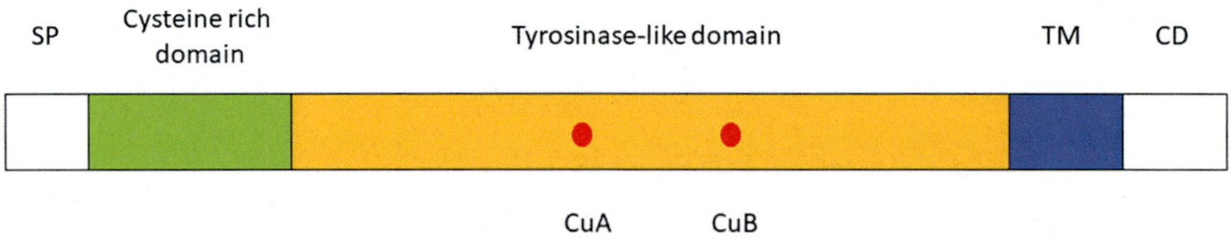

FIG. 1 Organization of tyrosinase domains. *SP*, signal peptide; *TM*, transmembrane domain; *CD*, cytosolic domain.

is positioned in the equatorial plane of the copper ions by means of peroxide [15]. The met-form consists of two Cu(II) ions too, but the bridging ligand is a hydroxo ligand, other than peroxide [16]. When the oxygen is added, the tyrosinase changes from oxy-form to met-form. In the catalytic cycles, tyrosinase can catalyze the oxidation of monophenols and diphenols, and clearly, oxygen is implicated in both catalytic circles (Fig. 2) [17]. In the monophenolase cycle, the monophenol is connected to the vertical position of a copper ion in the oxy-form's active site, which turns into hydroxylated monophenol, making a tetragonal intermediate [18,19]. This step's conversion rate is the monophenolase activity [20]. Afterward, the hydroxylated monophenol is oxidized to the *o*-benzoquinone. Therefore, oxy-form is reduced to deoxy-form. Oxygen interacts with deoxy-form, with the aim to produce oxy-form again. In the diphenolase cycle, oxy-form can interact with *o*-diphenol and oxidize it to *o*-benzoquinone, which is reduced to met-form subsequently. The diphenolase activity is the rate of *o*-diphenol oxidized to *o*-benzoquinone. Studies have shown that both the monophenol and *o*-diphenol compete for the central binding site of met-form, but the combination of monophenol and met-form will generate the inactive dead-end compound (E metM), which will abandon the diphenolase catalytic circle. The *o*-diphenol can be oxidized to *o*-benzoquinone accompanied by the state change of tyrosinase from met-form to deoxy-form. The deoxy-form can be transferred to oxy-form with the help of oxygen. Above all, the catalytic mechanisms of tyrosinase allow us to understand its crucial role in melanin biosynthesis.

Although several tyrosinases have been isolated from different natural sources in the past years, but only few tyrosinase structures have been understood. Among them, mushroom tyrosinase from *Agaricus bisporus* (mTYR) is the most cited and well-characterized one because it is highly similar to other tyrosinases regarding enzymatic core region [21,22]. The design studies for human tyrosinase inhibitors have always considered mushroom tyrosinase in the screening assays for human-directed applications molecules. Moreover, mushroom *Agaricus bisporus* tyrosinase has always been available at low cost from commercial sources (often as a mixture of isoenzymes), while the human enzyme (hTYR) is still costly and difficult to obtain in a proper form. The two types of the enzyme are different, in fact mammalian TYRs have only 22%–24% overall identity with mTYR in a region of 48%–49% sequence coverage [23]. Additionally, mTYR is a soluble tetrameric enzyme placed in the cytosol, while hTYR is a highly glycosylated monomeric protein tied on the melanosome membrane [24]. Lastly, the active sites of mushroom and human tyrosinases carry structural differences which affect substrate or inhibitor binding, despite the same capacity to interact with oxygen. It is known that the active site consists of two copper ions, and each of them can engage in strong interactions with three histidine residues. To be specific, in the mTYR, one copper ion coordinates with nitrogen atoms of His61, His85, and His94, while the other copper ion coordinates with His259, His263, and His296. Both copper ions are interlinked with an internal water molecule. Phe90 and Phe292 seem to restrict the side-chain conformations of six histidine residues [25,26].

In addition, the human enzyme contains a unique cysteine-rich subdomain placed in the cytosol of melanosomes (called EGF—epidermal growth factor—domain), a transmembrane hydrophobic domain, and seven asparagine glycosylation sites; these characteristics are not present in mTYR. Also in the active sites, there are variations. Even if the dicopper center surrounded by six histidines is conserved, the environment of the second sphere of coordination can differ. Some salient residues located near the hTYR active site that were expected to make important interactions with hTYR inhibitors [27–29], such as His304, Lys306, Arg308, Thr343, Thr352, Ile368, Ser375, and Ser380, are completely absent from abPPO3 and abPPO4 (the two most abundant isoforms of mTYR). There are also other important variations between mushroom and human TYRs; for example, a covalent thioether bond occurs between a cysteine and a copper-chelating histidine in mTYR: Cys83 and His85 in abPPO3 [21] Cys80 and His82 in abPPO4 [30], ratified by the corresponding crystal structures too. Another important divergence is the absence in abPPO3 and abPPO4 of Ser380 close to dicopper center, characteristic of hTYR, on the contrary, it is conserved in hTYRP1 as Ser394.

This amino acid may be involved in substrate activation, controlling hTYR action [10]. Consequently, Ser380 may also have a crucial influence in hTYR inhibitor binding. Therefore, many mTYR inhibitors gave unsatisfactory results when tested against hTYR. Regarded as reference molecules used as standards for TYR inhibition and employed especially in dermo-cosmetic industry, kojic acid

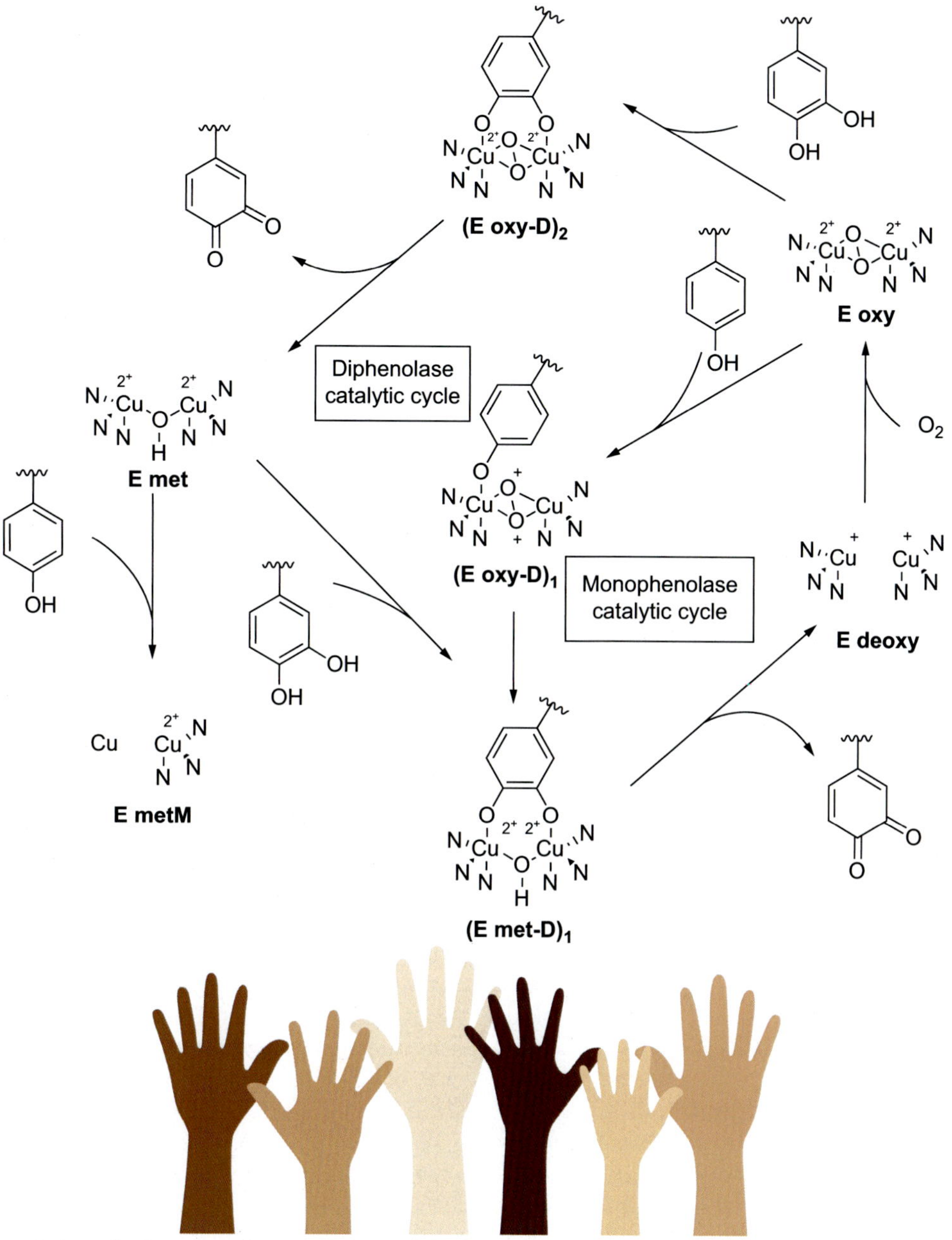

FIG. 2 Two-step catalytic mechanism of tyrosinase.

(**1**), hydroquinone (**2**), or arbutin (**3**) (Fig. 3) showed minimal effect against the human enzyme, with inhibition values 100-fold or 1000-fold higher than the results from mTYR-based assays. Rucinol (**4**) (Fig. 3), that showed a significant anti-hTYR activity (IC_{50}=21 μM), also inhibited mTYR with more efficacy (IC_{50}=0.6 μM) [8]. Kojic acid is a fungal metabolite produced by many species of *Aspergillus*, *Acetobacter*, and *Penicillium*, it is consumed daily by people as a part of diet, due to its presence in several foods, such as miso (soybean paste), shoyu (soy sauce), and sake (the alcoholic Japanese beverage) [31]. Nowadays, this molecule is used both in cosmetic and food industries because it is considered as a whitening agent. Moreover, it is known that kojic acid can inhibit melanin synthesis due to its chelating function. The study conducted by Karakaja [32] reported that IC_{50} value for kojic acid in the inhibition of mushroom tyrosinase was equal to 418.2 μM. Since 1960s, hydroquinone and its derivatives (arbutin, for example) have been used for skin hyperpigmentation treatments, as topical formulations. Boo's study has reported tyrosinase inhibitory effects of hydroquinone (97.2% of monophenolase activity inhibition at 3 mM) and arbutin (82.0% of monophenolase activity inhibition at 3 mM) on mushroom tyrosinase [33].

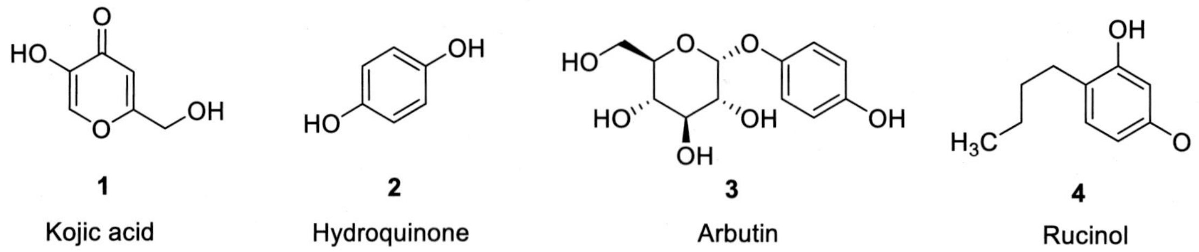

FIG. 3 Classical tyrosinase inhibitor's structures.

In Raper-Mason's melanogenesis pathway, the first two reaction steps include the oxidation of L-tyrosine to L-3,4-dihydroxyphenylalanine (L-DOPA), and then dehydrogenation of L-DOPA to L-dopaquinone, conducted both by tyrosine [34]. After L-dopaquinone has been formed, the pathway can progress spontaneously to the formation of final melanin. However, data strongly suggested the importance of post-tyrosinase regulation in mammals, which beyond TYR includes two tyrosinase-related proteins (TYRP1 and TYRP2) [35]. Mutations or impairment in any of these melanogenic metal-containing glycoproteins can result in serious health disorders.

3 Physiological/pathological role

Human tyrosinases' pivotal physiologic role is reflected in the synthesis of melanin precursor L-dopaquinone from amino acid L-tyrosine. The skin color is conditioned by the presence of four substances: brown melanin, blue reduced hemoglobin, yellow carotenoids, and red oxygenated hemoglobin [36]. Melanin is the larger quantity pigment, depending on its amount in the epidermis, it influences the color from lighter tan to a deeper brown or black. Studies have implied that melanin has an important role in skin homeostasis, such as absorbing or scattering UV radiation, metal chelation, scavenging free radicals, and electromagnetic induction [37]. Once the skin is exposed to UV radiation, in the basal layer of the epidermis, melanins are produced in melanosomes by melanocytes and moved to the adjacent keratinocytes, where they are stored in granules. The epidermal cells divide the unnucleated keratinocytes to release melanin, preventing, in this way, UV radiation from damaging cellular DNA [38,39]. The melanin mainly includes eumelanin and pheomelanin, which are both mostly present in the skin, the hair, the eye (iris and choroids) and the inner ear [40]. Eumelanin is a brown/black insoluble polymer, while pheomelanin is a red/yellow soluble polymer [41]. The melanin biosynthesis involves diverse complex enzymatic reactions, such as tyrosinase, tyrosinase-related protein 1 (TYRP-1), and tyrosinase-related protein 2 (TYRP-2) [42–45]. All of them are metal-containing glycoproteins with a single transmembrane α-helix which present at least 40% amino acid sequence identity and 70% similarity circa. TYRP-1 and TYRP-2 both contain two zinc ions in the active site [10]. In the case of a small amount of tyrosinase, dopaquinone can react with cysteine or glutathione to form cysteinyldopa or glutathionedopa. Following a series of oxidoreduction reactions, the benzothiazine intermediate is generated. Pheomelanin is finally formed after a series of polymerization reactions. Due to excessive tyrosinase, dopaquinone can self-cyclize to produce leukodopachrome, converting in dopachrome. Once decarboxylation is completed, dopachrome converts to 5,6-dihydroxyindole (DHI), which is further oxidized to generate indole-5,6-quinone by the tyrosinase-mediated oxidation. Dopachrome can change into 5,6-dihydroxyindole carboxylic acid (DHICA) with the help of TYRP-2. Indole-5,6-quinone carboxylic acid is oxidized from both DHICA and leukodopachrome. Eumelanin is formed through polymerization of indole-5,6-quinone and indole-5,6-quinone carboxylic acid. Studies have shown that tyrosinase is an oxidoreductase and TYRP-2 is an isomerase. On the contrary, the function of TYRP-1 is not completely explained yet. Moreover, evidence confirmed that TYRP-1 in mice is an oxidoreductase enzyme, which can catalyze the oxidation of DHICA [46]. Nevertheless, Boissy et al. reported that TYRP-1 of humans is not able to catalyze DHICA oxidation [47]. In 2017, Lai et al. obtained the crystal structure of human TYRP-1 (PDB ID: 5M8L) for the first time, providing new insights to explain that TYRP-1 did not show any tyrosinase redox activity [48]. The production of melanin is activated by regulating different intracellular signaling pathways, including the Wnt/b-catenin pathway [49], NO pathway [50], MAPK pathway [51], PI3K/Akt pathway [52], and MC1R pathway [53]. MITF (microphthalmia-associated transcription factor) is the wholesale regulator of these five signal pathways [54–56]. They establish the signal transmission maps of melanogenesis, which plays a vital role in the control of the gene expression of TYR, TYRP-1, and TYRP-2 [43]. The unchecked expression of tyrosinase is linked to several diseases. When tyrosinase activity decreases, melanin biosynthesis blocks, leading to oculocutaneous albinism, vitiligo, and other de/hypopigmentary diseases [57,58]. Nowadays, more than 300 mutations in the TYR gene are well known. If expressed, they usually result in TYR catalytic inactivity and are associated with oculocutaneous albinism type 1, autosomal recessive condition, and melanoma. In people with oculocutaneous albinism type 1A (OCA1A) mutations completely abolish TYR activity and no melanin pigment is formed. In people with oculocutaneous albinism type 1B (OCA1B), mutations render partial enzyme activity allowing some accumulation of melanin pigment over time. Additionally, mutations within the gene encoding for tyrosinase-related protein 1 (*TYRP1*) are found in oculocutaneous albinism type 3 (OCA3) [59].

On the contrary, excessive melanin causes hyperpigmentation diseases and malignant melanoma. Although hyperpigmentation diseases such as freckles, melasma, post-inflammatory melanoderma, and age spots are not fatal, but they have a huge impact on the psychology and, therefore, patients' quality of life [60]. In addition, studies have shown that tyrosinase is also involved in the neuromelanin synthesis in the substantia nigra (SN) of the brain, having neuroprotective effects, being able to eliminate potential neurotoxic substances such as dopaquinone [61–63]. Furthermore, tyrosinase seems to be related to both Alzheimer's disease and Parkinson's disease (PD), [64]. Evidence has suggested that there is a close relationship between PD and malignant melanoma. In fact, the risk of melanoma for PD patients is 2-3 times higher in comparison with healthy people [65,66]. A study report involving about 160,000 people has shown that a family history of melanoma in a first-degree relative is linked to a higher risk of PD [67]. The synthesis of neuromelanin in the substantia nigra of dopaminergic neurons is extremely similar in melanocytes. In humans, neuromelanin accumulates and amasses with age, being the main risk factor for PD [68]. When neuromelanin is accumulated above a specific threshold, it can lead to DA neuron death and characteristic PD phenotypes [69]. Even though the contribution of neuromelanin to PD pathogenesis is still not completely understood, the control of intracellular neuromelanin levels via brain tyrosinase activity might be a promising way to delay PD initiation. The overexpression of TYR is frequently found in many skin hyperpigmentary diseases, and melasma is one of the most frequent. These diseases seriously affect patients' quality of life, which contributed to the development of cosmeceuticals that target uncontrolled melanocytes and block elementary steps in melanogenesis. Since TYR is essential for the first steps in melanogenesis, the introduction and investigation of tyrosinase inhibitors as safe and efficacious cosmeceuticals represented a milestone in the treatment melasma and disorders of pigmentation [70]. Studies have revealed that the APOE4 variant as the largest monogenetic risk factor for Alzheimer's disease displayed favorable outcomes in melanoma progression, while the APOE2 variant which was the protective factor exhibited poor outcomes [71].

It is estimated that approximately 207,390 people in the United States were diagnosed with melanoma in 2021, which is considered one of the most common cancers in the world [72]. The amount of tyrosine in melanoma cells is generally higher than that of normal melanocytes, and this contributes to survival, proliferation, and metastasis for the tumor. Fu et al. discovered that the restriction of tyrosine and phenylalanine can inhibit cell proliferation of murine melanoma in vitro and in vivo [73]. Therefore, reducing melanin production, by inhibiting tyrosinase activity, could be a potential therapeutic strategy for the treatment of melanoma.

4 Tyrosinase as pharmacological target: design and development of inhibitors

The mTYR-based assays in the literature are ubiquitous because of the difficulty to produce hTYR. The maturation of human tyrosinase is indeed a laborious process, which provides for heavy and heterogeneous post-translational modifications, a transmembrane anchor, and a trafficking through the endoplasmic reticulum and Golgi, controlled by six or seven specific N-glycans and by copper uptake. Nonetheless, in literature, there are works that reported the purification of hTYR from human melanotic melanoma metastases, [74] but the native form of the protein is open to question, and the method does not work for regular hTYR production, as it needs the removal of biological tissues from patients by surgery. Other sources of human tyrosinase can be *E. coli*, insect cells and larvae, and human HEK cells [8]. A limited number of works have focused on the anti-TYR activity measure of natural products using a variety of hTYR-overexpressing human cell lines, among which melanogenic human malignant melanoma cells (e.g., G-361, HBL, HMV-II, SKMEL), normal adult or newborn melanocytes (HEMa and HEMn, respectively), and human non-melanogenic cells transfected with a hTYR construct (HEK-293-TYR), which provide for cell-free crude preparations of hTYR for inhibition tests screening.

Peng's group [24] reported, in a review, a series of tyrosinase inhibitors and their structure-activity relationship (SAR). Molecules reported in Fig. 4 are those that presented the best inhibition activity on mTYR. Among which there are (i) kojic acid derivatives, characterized by a central triazole and an aryl portion, or an ester function connected to a substituted benzene or amino acid/peptide function; (ii) thiourea compounds, with one functionalized amine group, while the other one free; (iii) thiosemicarbazone derivatives, characterized by the presence of aryl ring; and (iv) 4-substituted benzaldehydes, whose compound with the best activity presents a terminal dimethoxy-phosphate group.

Another substantial group of the identified compounds are phenol derivatives, such as 4-*tert*-butylphenol (**16**) [75] or 6-hydroxyindole (**27**) [76] (Fig. 5). It is not surprising that L-tyrosine and L-DOPA structural analogs,

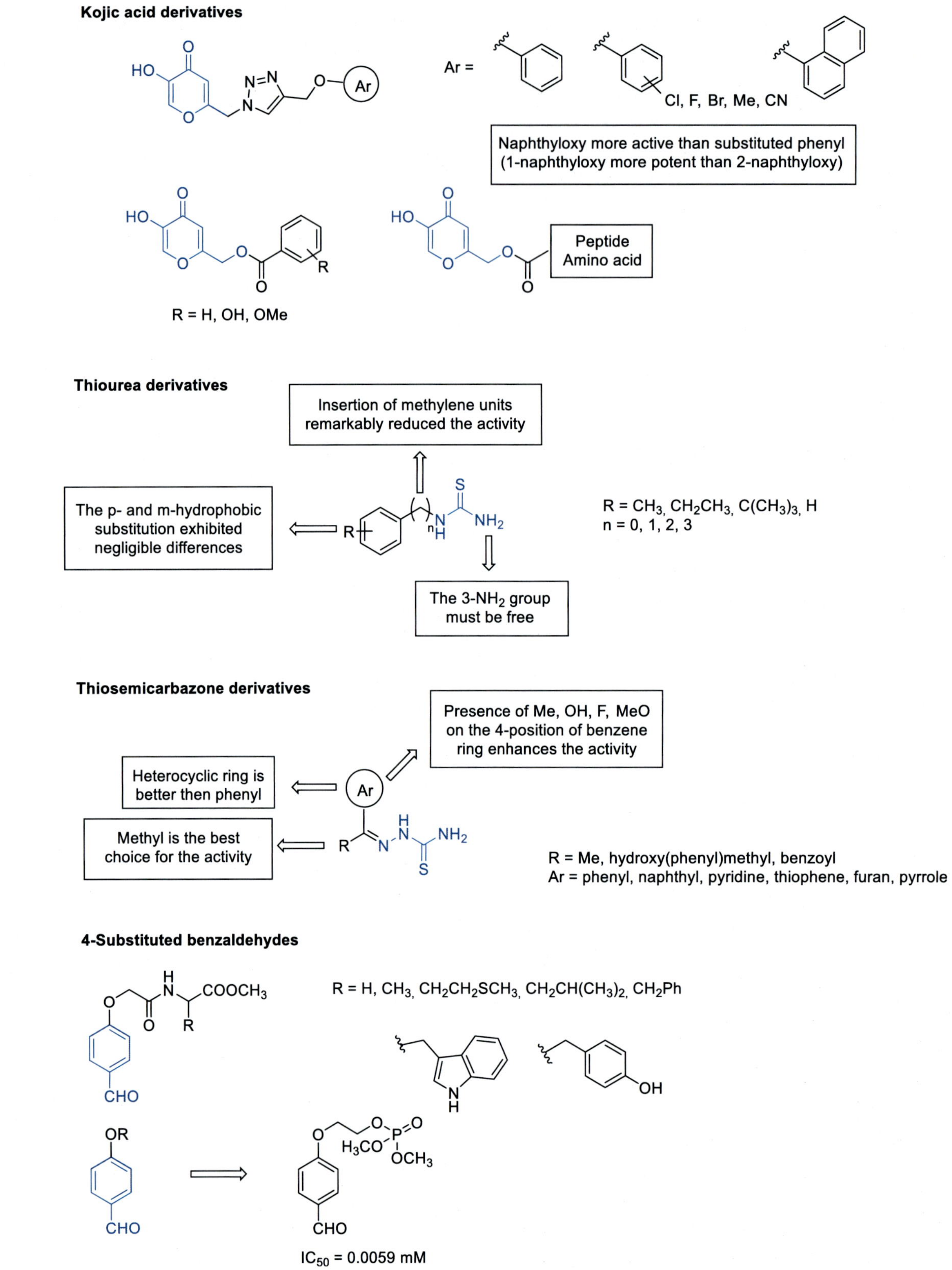

FIG. 4 Structure-activity relationship analyses of compounds endowed with the best mTYR inhibitory activity.

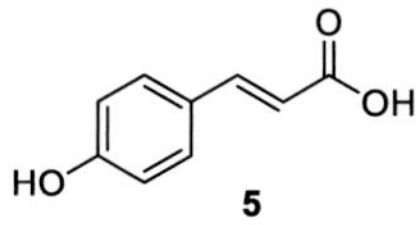

5
IC_{50} = 2-6 μM

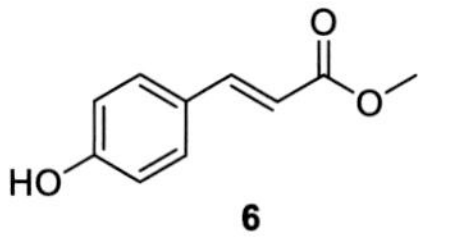

6
IC_{50} = 30 μM

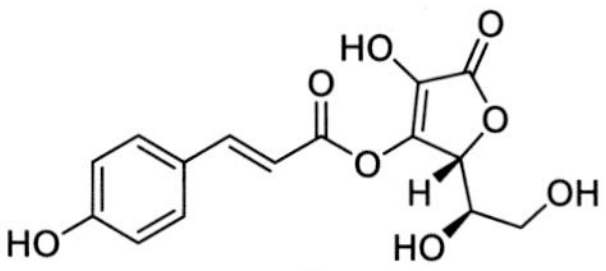

7
IC_{50} = 10 μM

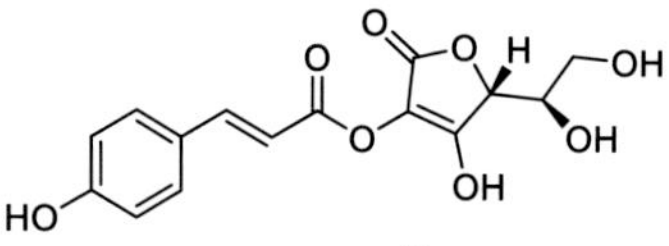

8
IC_{50} = 40 μM

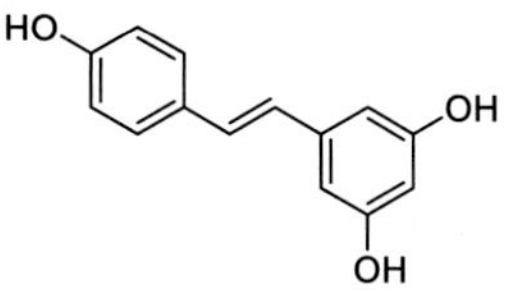

9
IC_{50} = 1.7-9 μM

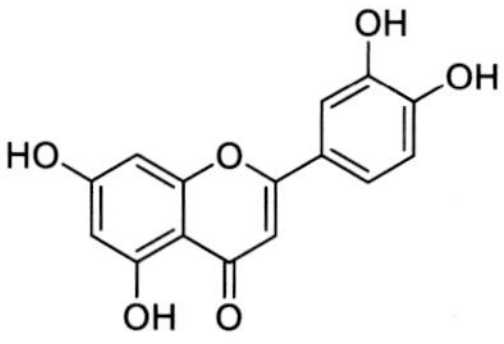

10
IC_{50} = 60 μM

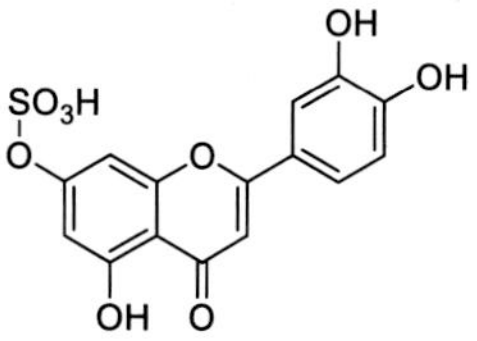

11
IC_{50} = 16 μM

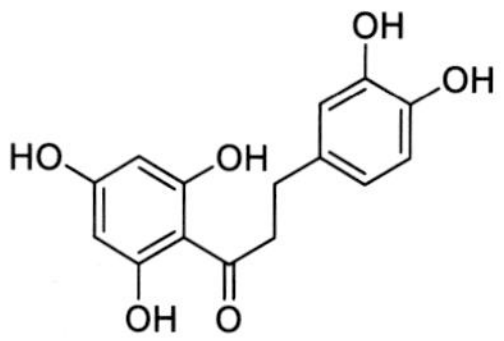

12
IC_{50} = 32 μM

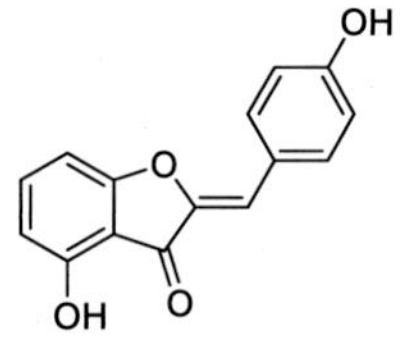

13
IC_{50} = 32 μM

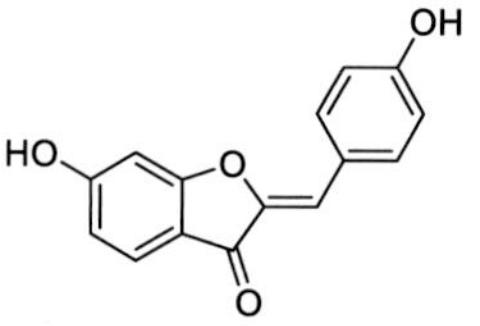

14
IC_{50} = 38 μM

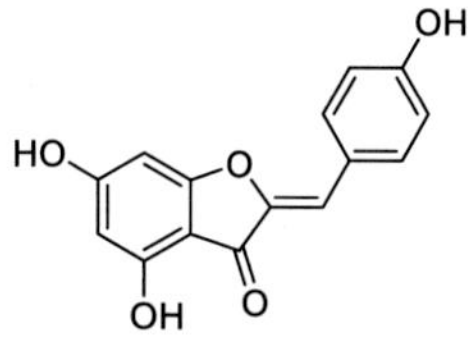

15
IC_{50} = 38 μM

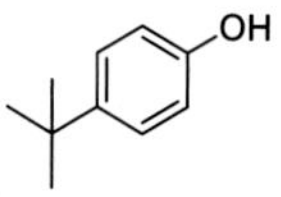

16
IC_{50} = 10 μM

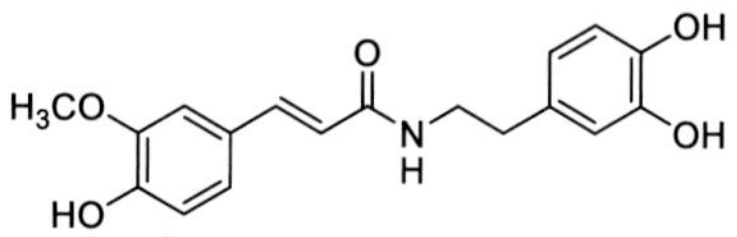

17
IC_{50} = 50 μM

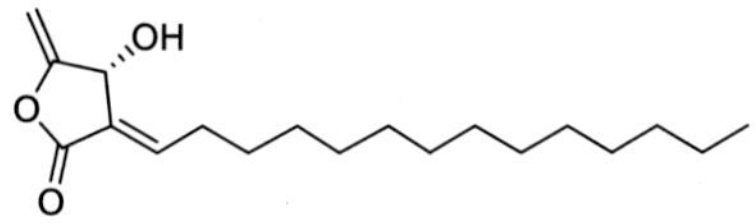

18
IC_{50} = 1 mM

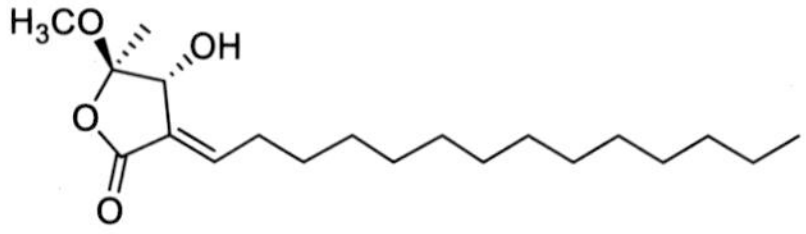

19
IC_{50} = 1 mM

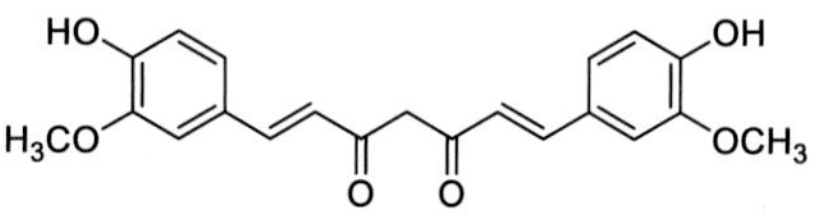

20
IC_{50} = 5 mM

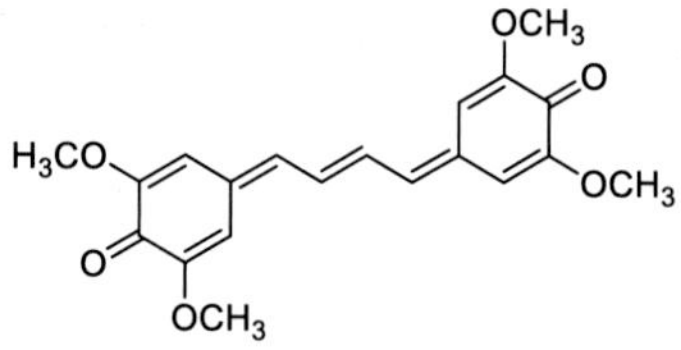

21
IC_{50} = 2.5 mM

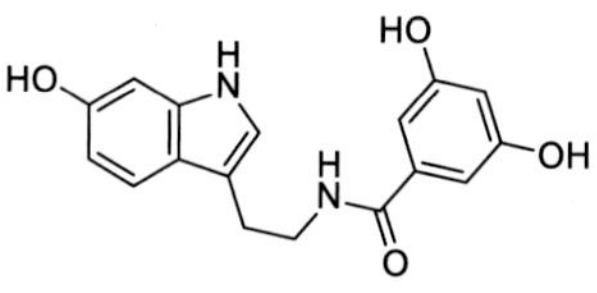

22
IC_{50} = 9.1 mM

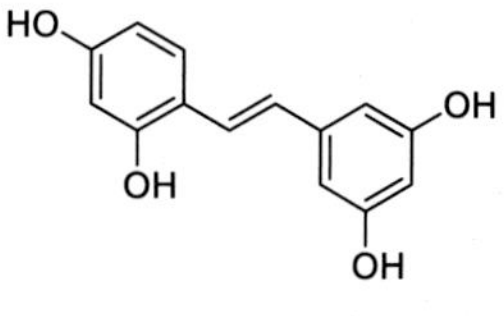

23
IC_{50} = 0.09-7.6 mM

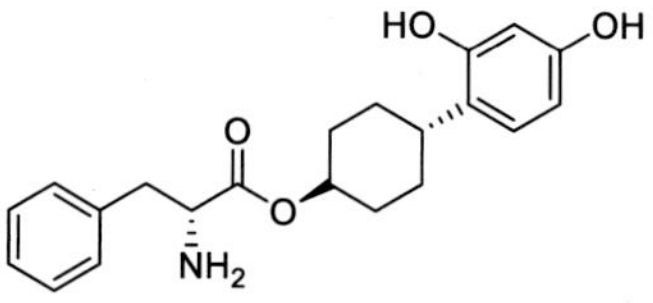

24
IC_{50} = 23 mM

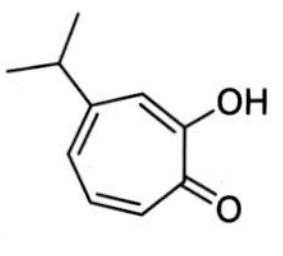

25
IC_{50} = 9 mM

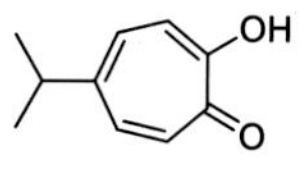

26
IC_{50} = 1.2 mM

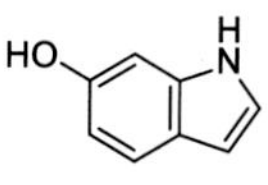

27
IC_{50} = 22 mM

FIG. 5 Other compounds tested as hTYR inhibitors.

derived from *p*-coumaric acid (**5**-**8**) [77], showed some affinity for hTYR. Interestingly, also *p*-coumaric acid (**5**) resulted far more active against crude hTYR preparations than against mTYR (IC_{50}=2−6 μM vs >100 μM). [77–79] Resveratrol (**9**) showed good IC_{50} values (1.7–9 μM), but its resorcinol-based analog oxyresveratrol (**23**) resulted more active (IC_{50}=0.09–7.6 μM), probably because of the presence of resorcinol moieties [80,81]. Flavonoid derivatives were also occasionally identified, belonging to flavone (**10**, **11**) [82], dihydrochalcone (**12**) [83], and aurone (**13**−**15**) subclasses [84], with modest activities. A larger 3,4-catechol, *N*-feruloyldopamine (**17**), was also tested [85]. Other extended linear molecules were reported in the Roulier's study, such as linderanolide B (**18**) and subamolide A (**19**), both with IC_{50}=1 μM [86], curcumin (**20**) [87], kinobeon A (**21**) [88], *N*-(3,5-dihydroxybenzoyl)-6-hydroxytryptamine (**22**), [76] or trans-4-(2,4-dihydroxyphenyl)cyclohexyl-(2*R*)-2-amino-3-phenylpropanoate (**24**) [89] (IC_{50}=2−85 μM), putting forward that the presence of long chains, like internal linker or hydrophobic tail, does not prevent hTYR inhibition (Fig. 5). Phenols or catechols posturing a distinctive substituent at the para position of one OH group (i.e., 4-phenols and 3,4-catechols) are the reminiscence of the structures of natural substrates L-tyrosine and L-DOPA. Given the nonspecific nature of hTYR catalytic activity, in the conditions of the tests usually performed, it resulted that at least some of them are substrates and not inhibitors. Yet, the oxidation of depigmenting agents by hTYR can generate reactive quinones, by triggering the formation of potentially toxic polymers or conjugates, like hydroquinone does. The work from Yoshimori et al. about thujaplicin isomers (**25** and **26**) [90] explained that the tropolone scaffold embodies the typical copper-chelating group that cannot be oxidable, ensuring an actual inhibitor behavior.

Globally, two hTYR constructs (from insect *S. frugiperda* Sf9 cells and human HEK cells) were used for most anti-hTYR assays reported. Among the compounds seen as classical TYR inhibitors, 3-hydroxycoumarin (**27**), L-mimosine (**29**), and phenylthiourea (**30**) presented a significant influence on the catalytic activity of hTYR (K_i=3.4 μM, 10.3 μM, and 1.7 μM, respectively), whereas kojic acid, hydroquinone, or arbutin showed almost to have an inactive profile. Molecules that presented acceptable values of K_i are reported in Fig. 6. Aurone (**37**) (K_i=0.35 μM) and thiamidol (**38**) (K_i=0.25 μM) are the best inhibitors of hTYR, among several less potent derivatives [8].

Compounds **37** and **38** were found to interact with a single copper atom, through the OH group (N−OH or a phenol group); it seems that they can also interact with residues Val377 and Ser380 residues, placed next to the dicopper center (Fig. 7). Crucial interactions were also found with other residues, such as Ile368 or Thr352. Moreover, thiamidol can accept two hydrogen bonds between Ser375 and the amide carbonyl and between Asn364 and the sulfur atom of the thiazole ring. In conclusion, both molecules seem to interact with the 302−310 loop of hTYR (especially with His304, Lys306, and Arg308 residues), a flexible part of the protein placed far from the dicopper center. Thiamidol showed its fully retained ability to act in a MelanoDerm pigmented 3D tissue model (IC_{50}=0.9 μM vs 1.1 μM against isolated hTYR). Contrariwise, aurone (**37**) resulted less efficient in preventing melanogenesis in MNT-1 human melanoma cells (IC_{50}=16.6 μM against MNT-1 lysates and 85.3 μM against whole cells, vs K_i=0.35 μM against isolated hTYR).

A vast study of high-efficiency screening of a 50,000 compounds' library against recombinant hTYR has been recently conducted [28]. The screening suggested that thiazolyl-resorcinol derivatives are the most promising class of human tyrosinase inhibitors for further development. As cited previously, thiamidol is a competitive inhibitor that exhibited a great potential of hTYR inhibition, especially if it is compared to classical compounds used in mTYR inhibition assays. Moreover, other resorcinol derivatives demonstrated promising results in the study conducted by Ghani, such as rucinol (compound **4**, Fig. 3) (K_i=9 μM), 4-phenylethylresorcinol (K_i=24 μM), and 4-hexylresorcinol (K_i=39 μM) (compounds **46** and **47**—Fig. 9). Several new amine and amide derivatives of 4-(2-aminothiazol-4-yl) benzene-1,3-diol were synthesized by alkylation or acylation [91]. The results confirmed the importance of resorcinol in the molecular structure, in fact compounds with no hydroxy group, or a fluorine in 3′-position presented bigger IC_{50} values, therefore ineffective for enzyme inhibition. In the amide series, the thiazole ring proved to be essential for a potent inhibition of hTYR, and according to it, when thiazole is substituted with a benzene ring or imidazole, the activity decreased dramatically. In all the target compounds, both thiazole and resorcinol rings showed an essential role for a potent inhibition of the human enzyme. Interestingly, the type of substituents at 2-amino group held less importance since several different chemical groups could be accommodated at this position. Tests suggested that substituents of small sizes are most potent and selective enzyme inhibitors. Moreover, methylation of the thiazole 2-amino group led to lower activities in the amine series. However, this effect was much more evident in the amide series. Alkylation of any of the aromatic rings conferred the compounds inactive because they impose steric effects that hinder proper enzyme binding and inhibition. If resorcinol was linked to thiazole in 2-position and amine/amide portion was connected to thiazole in 4-position, there was a decrease of activity (Fig. 8).

The β-phenyl-α,β-unsaturated carbonyl scaffold is known to inhibit tyrosinase activity [92]. In the study conducted by Ghani [91], two thiooxazolidinone derivatives

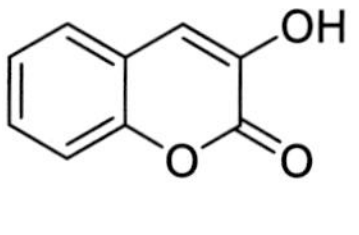

28
K_i = 3.4 μM

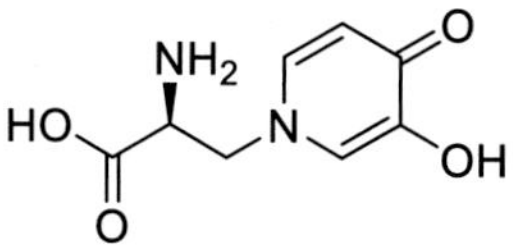

29
K_i = 10.3 μM

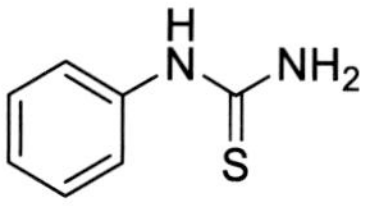

30
K_i = 1.7 μM

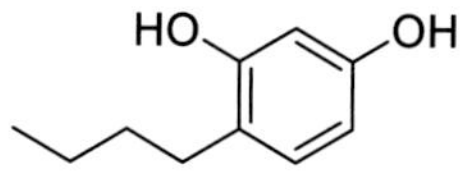

31
K_i = 9.1 μM

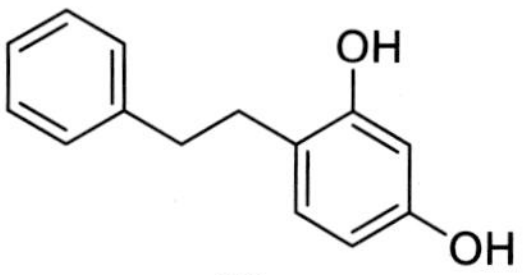

32
K_i = 24 μM

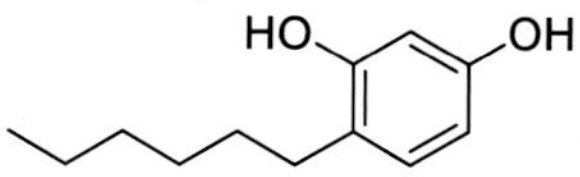

33
K_i = 39 μM

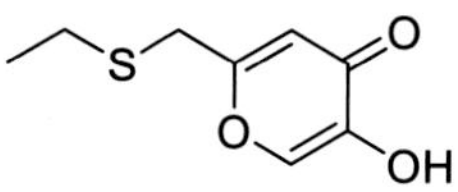

34
K_i = 19 μM

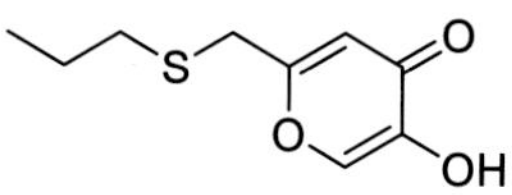

35
K_i = 27 μM

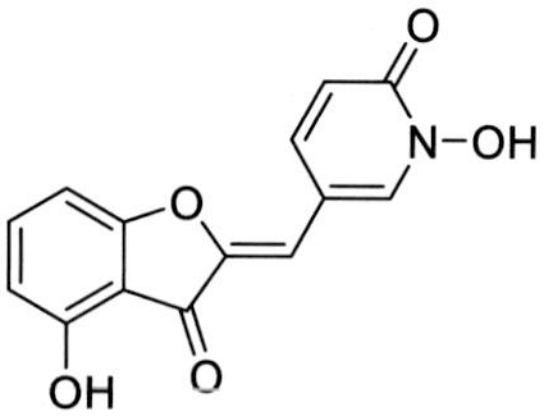

36
K_i = 1.02 μM

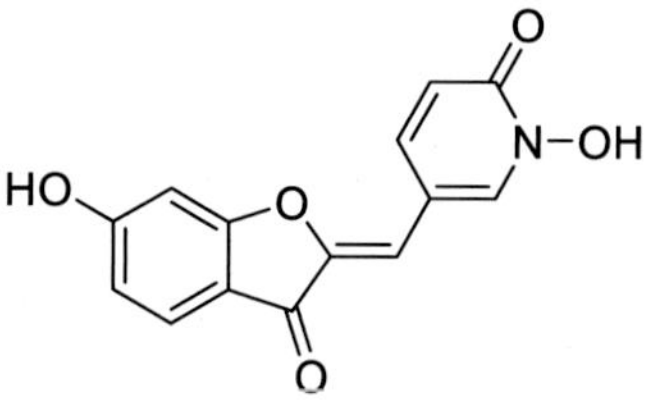

37
K_i = 0.35 μM

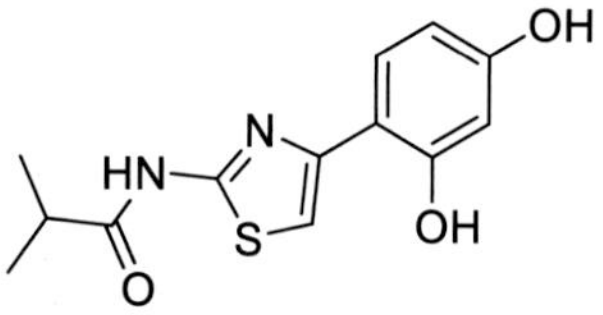

38
K_i = 0.25 μM

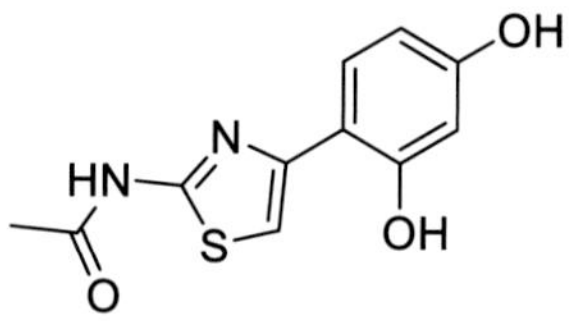

39
IC_{50}= 3.2 μM

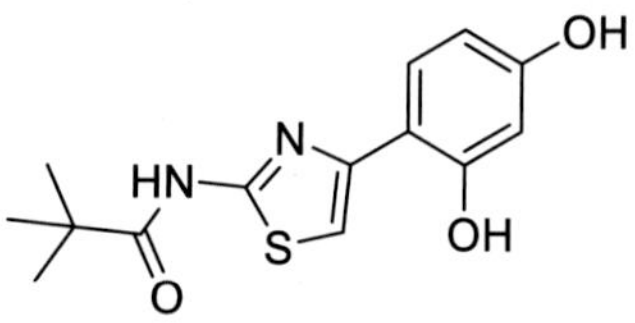

40
IC_{50}= 1.4 μM

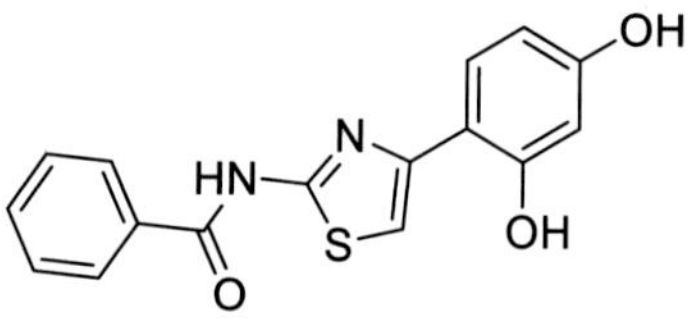

41
IC_{50}= 2.5 μM

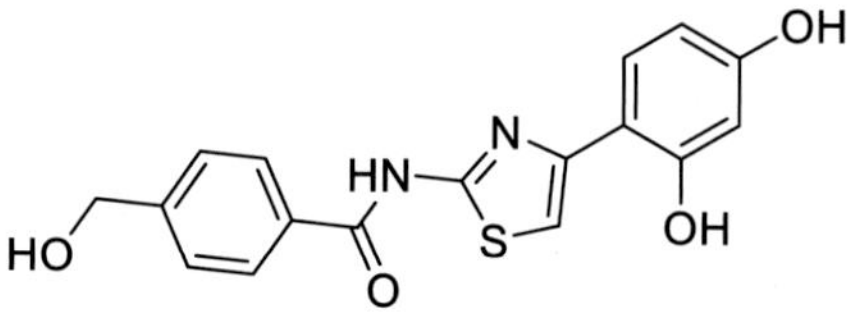

42
IC_{50}= 1.6 μM

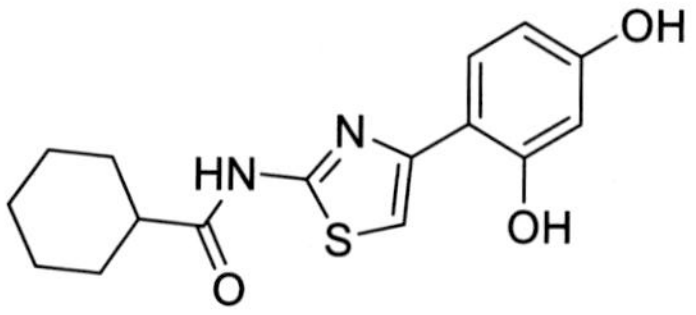

43
IC_{50}= 4.6 μM

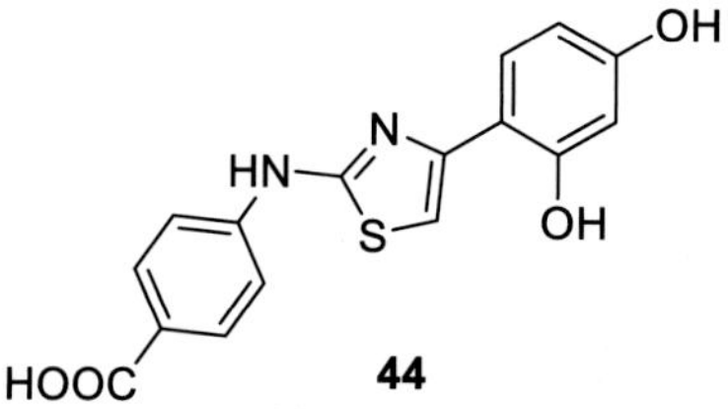

44
IC_{50}= 2.6 μM

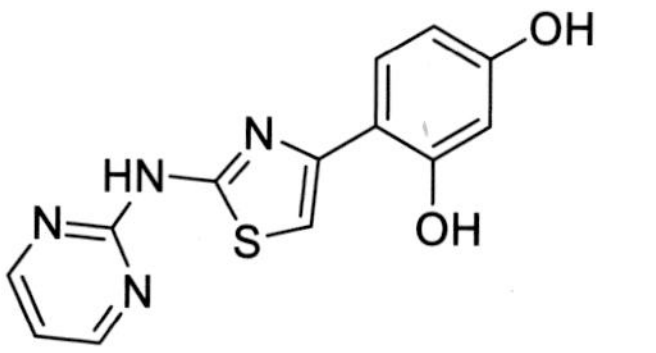

45
IC_{50}= 3.2 μM

FIG. 6 Compounds tested as hTYR inhibitors.

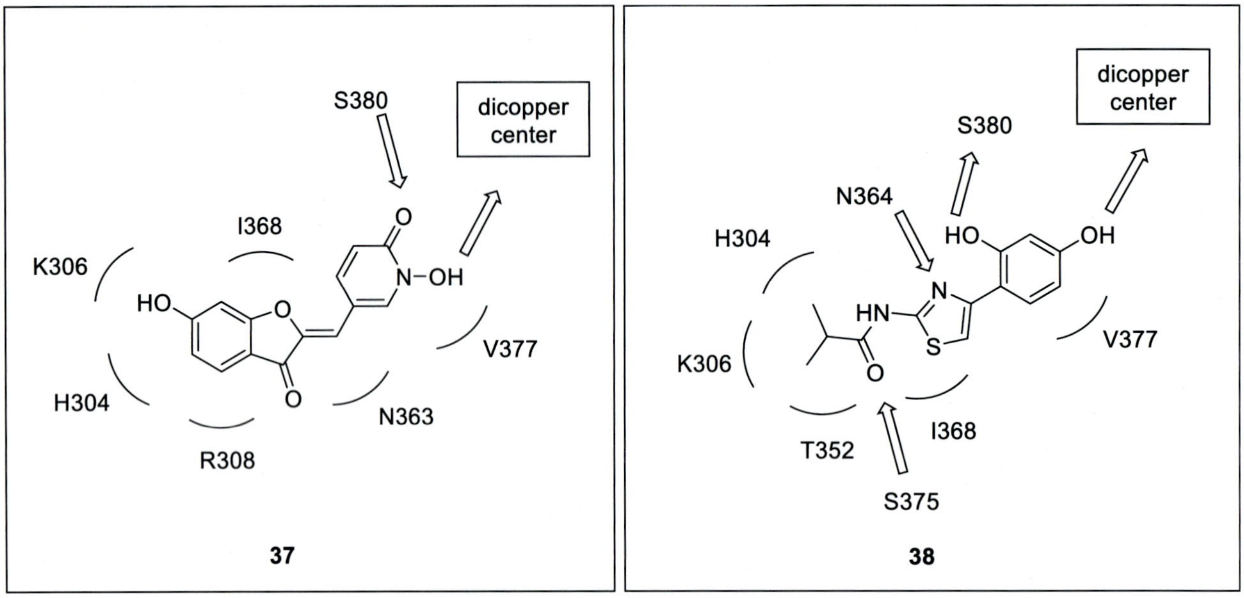

FIG. 7 Interactions between human tyrosinase and compounds 37 and 38.

were studied in detail, which led to promising results. Compound **48** (IC_{50}=4.7 μM) (Fig. 9), characterized by 2,4-dihydroxyphenyl group, demonstrated high potency of mTYR inhibition owing to its two hydroxy groups at 2- and 4-positions (resorcinol portion) of the β-phenyl group. Computational data showed that these hydroxy groups engage Asn260 and Met280 residues for hydrogen bonding interactions. Hydroxy groups at other positions proved to abolish the potency. Comparison of the structural features and activity of compound **48** with compound **49** (IC_{50}=11.18 μM) (Fig. 9) showed significant differences in their activities and binding modes since the latter is devoid of a hydroxyl group at 4-position of the phenyl group.

Docking studies highlighted differences between the binding modes of the compounds in the crystal structure of mTYR and a homology model of hTYR active sites. In the mTYR active site, the two hydroxy groups of compound **48** form hydrogen bonds with Asn260 and Met280 and β-phenyl ring is stabilized by hydrophobic interactions with Val283 and Ala286 residues. The binding mode of **48** in the hTYR active site is different, in fact, it coordinates

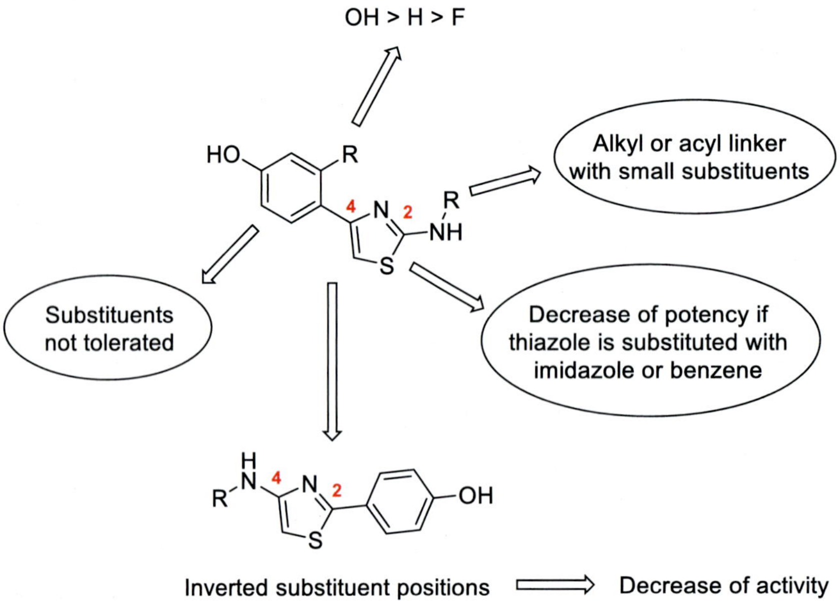

FIG. 8 SAR among thiamidol derivatives.

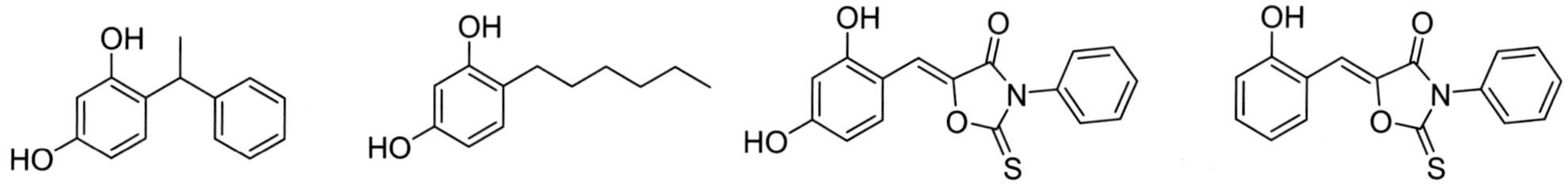

46 $K_i = 24$ M
47 $K_i = 39$ M
48 $IC_{50} = 4.7$ mM
49 $IC_{50} = 11.2$ mM

FIG. 9 Resorcinol/phenol derivatives tested as hTYR inhibitors.

with the metal center through one of its hydroxyl groups on the β-phenyl ring. Compound **49** is mainly stabilized in the mTYR active by hydrophobic forces, however, in the hTYR active site it is involved in two hydrogen bonds with Lys306 and Val377 residues via the hydroxy group on the β-phenyl ring and the carbonyl group of its thiooxazolidinone ring. In addition, both compounds dose-dependently inhibited melanogenesis in B16–F10 melanoma cells when stimulated by α-MSH and 3-isobutyl-1-methylxanthine without signs of cytotoxicity up to a 20-μM concentration [93].

5 Clinically used agents or compounds in clinical development: An update of the literature

Inhibiting tyrosinase activity is the most effective approach to counteract cutaneous hyperpigmentation by means of reduced melanin production. Most of the published compounds in the literature were assessed against mTYR, thus resulting in poor inhibitors of the human enzyme and lacking high translational efficacy when incorporated in commercial products. In addition, hyperpigmentation can be refractory to routinary treatment, and relapses can occur frequently. A new arsenal of cosmeceuticals was proposed to be perceived as more natural and less toxic with respect to hydroquinone. Kojic acid, arbutin, vitamin C, nicotinamide, ferulic acid, retinol and its derivatives, antioxidants as cysteamine, silymarin and glutathione, thiamidol, and resorcinol were proposed alone or in association to be more efficacious and safer. In addition, hydrolyzed plant extracts such as *Psoralea corylifolia*, *Glycyrrhiza glabra*, and *Polypodium leucotomos* were tested. Among repurposed drugs, methimazole, tranexamic acid, and flutamide possessed interesting characteristics in in vivo studies. Despite the results of large randomized controlled trials or follow-up analyses are still missing, small in vivo and in vitro studies supported their therapeutic potential [94,95].

Hydroquinone should be administered for at least three months at concentrations ranging 2% to 5% to obtain results after 5 weeks. Treatments are usually well-tolerated if discontinued within one year. Irritation (mild erythema) can be reduced with the application of a triple combination cream (hydroquinone, dexamethasone/fluocinolone acetonide/hydrocortisone, and tretinoin) [96]. Despite no episodes of carcinogenicity in humans were registered, animal studies confirmed the association with cancer.

Kojic acid was first commercialized in Japan to prevent discoloration of meat and vegetables and then as a whitening agent in cosmetic preparations at 1% (0.1%–2%). Higher doses or transdermal penetration can influence thyroid function by iodine uptake inhibition, whereas mutagenicity studies are controversial [97].

Arbutin, and α-arbutin, can be converted into hydroquinone after dermal penetration by skin microbe metabolism or enzymatic hydrolysis. It is characterized by low stability to heat. Hydroquinone is associated with corticosteroids or anti-inflammatory agents. It is less toxic than hydroquinone (till 0.5% in body lotions and 2% in facial creams) especially if nanotechnologically formulated [98,99].

In the past years, thiamidol was the new and promising tyrosinase inhibitor proposed to effective for the depigmenting of facial melasma with respect to hydroquinone. A recent randomized, evaluator-blinded, controlled study put in comparison the efficacy and safety of thiamidol *versus* hydroquinone enrolling fifty female patients with facial melasma (mean age=43 years and belonging to phototypes III or IV). The application consisted of a double layer of 0.2% thiamidol twice a day or 4% hydroquinone cream overnight for three months. The outcomes were measured based on the Modified Melasma Area Severity Index, Melasma Quality of Life Index, skin pigmentation and color, and Global Aesthetic Improvement Scale. The most interesting results were related to mild side effects and an increment in the Global Aesthetic Improvement Scale score in the thiamidol group with respect to hydroquinone [100]. Kolbe's group further extended work on thiamidol by conducting clinical studies to define its effects on post-inflammatory hyperpigmentation in human volunteers [101]. Previous work by the same group had already demonstrated that for 4–12 weeks treatment, thiamidol was effective for reducing facial hyperpigmentation and age spots at doses ranging from 0.1% to 0.2% [102]. If approved, this derivative would be the first azole drug for the treatment of skin hyperpigmentation.

References

[1] Sánchez-Ferrer A, Rodríguez-López JN, García-Cánovas F, García-Carmona F. Tyrosinase: a comprehensive review of its mechanism. Biochim Biophys Acta 1995;1247(1):1–11.

[2] McEvily AJ, Iyengar R, Otwell WS. Inhibition of enzymatic browning in foods and beverages. Crit Rev Food Sci Nutr 1992;32 (3):253–73.

[3] Bai Q, Yu J, Su M, Bai R, Katumata G, Katumata M, Chen X. Antioxidant function of solanesol and its inhibitory effect on tyrosinase. Sheng Wu Yi Xue Gong Cheng Xue Za Zhi 2014;31 (4):833–6. 841.

[4] Hu X, Yu MH, Yan GR, Wang HY, Hou AJ, Lei C. Isoprenylated phenolic compounds with tyrosinase inhibition from *Morus nigra*. J Asian Nat Prod Res 2018;20(5):488–93.

[5] García-Rivera J, Casadevall A. Melanization of *Cryptococcus neoformans* reduces its susceptibility to the antimicrobial effects of silver nitrate. Med Mycol 2001;39(4):353–7.

[6] Nosanchuk JD, Casadevall A. The contribution of melanin to microbial pathogenesis. Cell Microbiol 2003;5(4):203–23.

[7] https://www.grandviewresearch.com/press-relate/global-skin-lightnening-products-market (Accessed 29 Nov 2022).

[8] Roulier B, Pérès B, Haudecoeur R. Advances in the design of genuine human tyrosinase inhibitors for targeting melanogenesis and related pigmentations. J Med Chem 2020;63:13428–43.

[9] De Luca L, Germanò MP, Fais A, Pintus F, Buemi MR, Vittorio S, Mirabile S, Rapisarda A, Gitto R. Discovery of a new potent inhibitor of mushroom tyrosinase (*Agaricus bisporus*) containing 4-(4-hydroxyphenyl)piperazin-1-yl moiety. Bioorg Med Chem 2020;28(11), 115497.

[10] Lai X, Wichers HJ, Soler-Lopez M, Dijkstra BW. Structure and function of human tyrosinase and tyrosinase-related proteins. Chemistry 2018;24(1):47–55.

[11] Della Longa S, Ascone I, Bianconi A, Bonfigli A, Castellano AC, Zarivi O, Miranda M. The dinuclear copper site structure of *Agaricus bisporus* tyrosinase in solution probed by X-ray absorption spectroscopy. J Biol Chem 1996;271(35):21025–30.

[12] Matoba Y, Kumagai T, Yamamoto A, Yoshitsu H, Sugiyama M. Crystallographic evidence that the dinuclear copper center of tyrosinase is flexible during catalysis. J Biol Chem 2006;281 (13):8981–90.

[13] Sanjust E, Cecchini G, Sollai F, Curreli N, Rescigno A. 3-hydroxykynurenine as a substrate/activator for mushroom tyrosinase. Arch Biochem Biophys 2003;412(2):272–8.

[14] Beltramini M, Salvato B, Santamaria M, Lerch K. The reaction of CN with the binuclear copper site of *Neurospora* tyrosinase: its relevance for a comparison between tyrosinase and hemocyanin active sites. Biochim Biophys Acta 1990;1040(3):365–72.

[15] Cramer CJ, Włoch M, Piecuch P, Puzzarini C, Gagliardi L. Theoretical models on the Cu2O2 torture track: mechanistic implications for oxytyrosinase and small-molecule analogues. J Phys Chem A 2006;110(5):1991–2004.

[16] Espín JC, Wichers HJ. Slow-binding inhibition of mushroom (*Agaricus bisporus*) tyrosinase isoforms by tropolone. J Agric Food Chem 1999;47(7):2638–44.

[17] Fenoll LG, Rodríguez-López JN, García-Sevilla F, García-Ruiz PA, Varón R, García-Cánovas F, Tudela J. Analysis and interpretation of the action mechanism of mushroom tyrosinase on monophenols and diphenols generating highly unstable o-quinones. Biochim Biophys Acta 2001;1548(1):1–22.

[18] Burton SG. Biocatalysis with polyphenol oxidase—a review. Catal Today 1994;1994:459–87.

[19] Wilcox DE, Porras AG, Hwang YT, Lerch K, Winkler ME, Solomon EI. Substrate-analog binding to the coupled binuclear copper active-site in tyrosinase. J Am Chem Soc 1985;107:4015–27.

[20] Naish-Byfield S, Riley PA. Oxidation of monohydric phenol substrates by tyrosinase. An oximetric study. Biochem J 1992;288(Pt 1):63–7.

[21] Ismaya WT, Rozeboom HJ, Weijn A, Mes JJ, Fusetti F, Wichers HJ, Dijkstra BW. Crystal structure of *Agaricus bisporus* mushroom tyrosinase: identity of the tetramer subunits and interaction with tropolone. Biochemistry 2011;50(24):5477–86.

[22] Seo SY, Sharma VK, Sharma N. Mushroom tyrosinase: recent prospects. J Agric Food Chem 2003;51(10):2837–53.

[23] Fogal S, Carotti M, Giaretta L, Lanciai F, Nogara L, Bubacco L, Bergantino E. Human tyrosinase produced in insect cells: a landmark for the screening of new drugs addressing its activity. Mol Biotechnol 2015;57(1):45–57.

[24] Peng Z, Wang G, Zeng QH, Li Y, Liu H, Wang JJ, Zhao Y. A systematic review of synthetic tyrosinase inhibitors and their structure-activity relationship. Crit Rev Food Sci Nutr 2022;62(15):4053–94.

[25] Ando H, Kondoh H, Ichihashi M, Hearing VJ. Approaches to identify inhibitors of melanin biosynthesis via the quality control of tyrosinase. J Invest Dermatol 2007;127(4):751–61.

[26] Mendes E, Perry Mde J, Francisco AP. Design and discovery of mushroom tyrosinase inhibitors and their therapeutic applications. Expert Opin Drug Discovery 2014;9(5):533–54.

[27] Haudecoeur R, Carotti M, Gouron A, Maresca M, Buitrago E, Hardré R, Bergantino E, Jamet H, Belle C, Réglier M, Bubacco L, Boumendjel A. 2-Hydroxypyridine-*N*-oxide-embedded aurones as potent human tyrosinase inhibitors. ACS Med Chem Lett 2016;8 (1):55–60.

[28] Mann T, Gerwat W, Batzer J, Eggers K, Scherner C, Wenck H, Stäb F, Hearing VJ, Röhm KH, Kolbe L. Inhibition of human tyrosinase requires molecular motifs distinctively different from mushroom tyrosinase. J Invest Dermatol 2018;138(7):1601–8.

[29] Mann T, Scherner C, Röhm KH, Kolbe L. Structure-activity relationships of thiazolyl resorcinols, potent and selective inhibitors of human tyrosinase. Int J Mol Sci 2018;19(3):690.

[30] Pretzler M, Bijelic A, Rompel A. Heterologous expression and characterization of functional mushroom tyrosinase (AbPPO4). Sci Rep 2017;7(1):1810.

[31] Burdock GA, Soni MG, Carabin IG. Evaluation of health aspects of kojic acid in food. Regul Toxicol Pharmacol 2001;33(1):80–101.

[32] Karakaya G, Türe A, Ercan A, Öncül S, Aytemir MD. Synthesis, computational molecular docking analysis and effectiveness on tyrosinase inhibition of kojic acid derivatives. Bioorg Chem 2019;88, 102950.

[33] Boo YC. Arbutin as a skin depigmenting agent with antimelanogenic and antioxidant properties. Antioxidants 2021;10(7):1129.

[34] McLarin MA, Leung IKH. Substrate specificity of polyphenol oxidase. Crit Rev Biochem Mol Biol 2020;55(3):274–308.

[35] Solano F. On the metal cofactor in the tyrosinase family. Int J Mol Sci 2018;19(2):633.

[36] Kameyama K, Sakai C, Kuge S, Nishiyama S, Tomita Y, Ito S, Wakamatsu K, Hearing VJ. The expression of tyrosinase, tyrosinase-related proteins 1 and 2 (TRP1 and TRP2), the silver protein, and a melanogenic inhibitor in human melanoma cells of differing melanogenic activities. Pigment Cell Res 1995;8 (2):97–104.

[37] Hu D, Clarke JA, Eliason CM, Qiu R, Li Q, Shawkey MD, Zhao C, D'Alba L, Jiang J, Xu X. A bony-crested Jurassic dinosaur with evidence of iridescent plumage highlights complexity in early paravian evolution. Nat Commun 2018;9(1):217.

[38] Tsatmali M, Ancans J, Thody AJ. Melanocyte function and its control by melanocortin peptides. J Histochem Cytochem 2002;50 (2):125–33.

[39] Thakur R, Batheja P, Kaushik D, Michniak B. Chapter 4. Structural and biochemical changes in aging skin and their impact on skin permeability barrier. In: Dayan N, editor. Skin aging handbook. Norwich, NY: William Andrew Publishing; 2009. p. 55–90.

[40] Barozzi S, Ginocchio D, Socci M, Alpini D, Cesarani A. Audiovestibular disorders as autoimmune reaction in patients with melanoma. Med Hypotheses 2015;85(3):336–8.

[41] Pillaiyar T, Manickam M, Jung SH. Downregulation of melanogenesis: drug discovery and therapeutic options. Drug Discov Today 2017;22(2):282–98.

[42] Hearing VJ, Jiménez M. Mammalian tyrosinase—the critical regulatory control point in melanocyte pigmentation. Int J Biochem 1987;19(12):1141–7.

[43] Pillaiyar T, Namasivayam V, Manickam M, Jung SH. Inhibitors of melanogenesis: an updated review. J Med Chem 2018;61 (17):7395–418.

[44] Pillaiyar T, Manickam M, Jung SH. Inhibitors of melanogenesis: a patent review (2009–2014). Expert Opin Ther Pat 2015;25 (7):775–88.

[45] Ullah S, Son S, Yun HY, Kim DH, Chun P, Moon HR. Tyrosinase inhibitors: a patent review (2011-2015). Expert Opin Ther Pat 2016;26(3):347–62.

[46] Kobayashi T, Urabe K, Winder A, Jiménez-Cervantes C, Imokawa G, Brewington T, Solano F, García-Borrón JC, Hearing VJ. Tyrosinase related protein 1 (TRP1) functions as a DHICA oxidase in melanin biosynthesis. EMBO J 1994;13(24):5818–25.

[47] Boissy RE, Sakai C, Zhao H, Kobayashi T, Hearing VJ. Human tyrosinase related protein-1 (TRP-1) does not function as a DHICA oxidase activity in contrast to murine TRP-1. Exp Dermatol 1998;7 (4):198–204.

[48] Lai X, Wichers HJ, Soler-Lopez M, Dijkstra BW. Structure of human tyrosinase related protein 1 reveals a binuclear zinc active site important for melanogenesis. Angew Chem Int Ed Eng 2017;56 (33):9812–5.

[49] Kovacs D, Migliano E, Muscardin L, Silipo V, Catricalà C, Picardo M, Bellei B. The role of Wnt/β-catenin signaling pathway in melanoma epithelial-to-mesenchymal-like switching: evidences from patients-derived cell lines. Oncotarget 2016;7(28):43295–314.

[50] Park HY, Kosmadaki M, Yaar M, Gilchrest BA. Cellular mechanisms regulating human melanogenesis. Cell Mol Life Sci 2009;66(9):1493–506.

[51] Wang Y, Viennet C, Robin S, Berthon JY, He L, Humbert P. Precise role of dermal fibroblasts on melanocyte pigmentation. J Dermatol Sci 2017;88(2):159–66.

[52] Tsang TF, Ye Y, Tai WC, Chou GX, Leung AK, Yu ZL, Hsiao WL. Inhibition of the p38 and PKA signaling pathways is associated with the anti-melanogenic activity of Qian-wang-hong-bai-san, a Chinese herbal formula, in B16 cells. J Ethnopharmacol 2012;141(2):622–8.

[53] Newton RA, Roberts DW, Leonard JH, Sturm RA. Human melanocytes expressing MC1R variant alleles show impaired activation of multiple signaling pathways. Peptides 2007;28(12):2387–96.

[54] Tachibana M. Cochlear melanocytes and MITF signaling. J Investig Dermatol Symp Proc 2001;6(1):95–8.

[55] Levy C, Khaled M, Fisher DE. MITF: master regulator of melanocyte development and melanoma oncogene. Trends Mol Med 2006;12(9):406–14.

[56] Ye Y, Chu JH, Wang H, Xu H, Chou GX, Leung AK, Fong WF, Yu ZL. Involvement of p38 MAPK signaling pathway in the anti-melanogenic effect of San-bai-tang, a Chinese herbal formula, in B16 cells. J Ethnopharmacol 2010;132(2):533–5.

[57] Qu Y, Zhan Q, Du S, Ding Y, Fang B, Du W, Wu Q, Yu H, Li L, Huang W. Catalysis-based specific detection and inhibition of tyrosinase and their application. J Pharm Anal 2020;10(5):414–25.

[58] Liu S, Kuht HJ, Moon EH, Maconachie GDE, Thomas MG. Current and emerging treatments for albinism. Surv Ophthalmol 2021;66 (2):362–77.

[59] Maranduca MA, Branisteanu D, Serban DN, Branisteanu DC, Stoleriu G, Manolache N, Serban IL. Synthesis and physiological implications of melanic pigments. Oncol Lett 2019;17(5):4183–7.

[60] Pillaiyar T, Manickam M, Namasivayam V. Skin whitening agents: medicinal chemistry perspective of tyrosinase inhibitors. J Enzyme Inhib Med Chem 2017;32(1):403–25.

[61] Zucca FA, Basso E, Cupaioli FA, Ferrari E, Sulzer D, Casella L, Zecca L. Neuromelanin of the human substantia nigra: an update. Neurotox Res 2014;25(1):13–23.

[62] Pan T, Li X, Jankovic J. The association between Parkinson's disease and melanoma. Int J Cancer 2011;128(10):2251–60.

[63] Ikemoto K, Nagatsu I, Ito S, King RA, Nishimura A, Nagatsu T. Does tyrosinase exist in neuromelanin-pigmented neurons in the human substantia nigra? Neurosci Lett 1998;253 (3):198–200.

[64] Hasegawa T. Tyrosinase-expressing neuronal cell line as in vitro model of Parkinson's disease. Int J Mol Sci 2010;11(3):1082–9.

[65] Rugbjerg K, Friis S, Lassen CF, Ritz B, Olsen JH. Malignant melanoma, breast cancer and other cancers in patients with Parkinson's disease. Int J Cancer 2012;131(8):1904–11.

[66] Bertoni JM, Arlette JP, Fernandez HH, Fitzer-Attas C, Frei K, Hassan MN, Isaacson SH, Lew MF, Molho E, Ondo WG, Phillips TJ, Singer C, Sutton JP, Wolf Jr JE. North American Parkinson's and Melanoma Survey Investigators. Increased melanoma risk in Parkinson disease: a prospective clinicopathological study. Arch Neurol 2010;67(3):347–52.

[67] Gao X, Simon KC, Han J, Schwarzschild MA, Ascherio A. Family history of melanoma and Parkinson disease risk. Neurology 2009;73 (16):1286–91.

[68] Kamaraj B, Purohit R. Mutational analysis of oculocutaneous albinism: a compact review. Biomed Res Int 2014;2014, 905472.

[69] Carballo-Carbajal I, Laguna A, Romero-Giménez J, Cuadros T, Bové J, Martinez-Vicente M, Parent A, Gonzalez-Sepulveda M, Peñuelas N, Torra A, Rodríguez-Galván B, Ballabio A, Hasegawa T, Bortolozzi A, Gelpi E, Vila M. Brain tyrosinase overexpression implicates age-dependent neuromelanin production in Parkinson's disease pathogenesis. Nat Commun 2019;10(1):973.

[70] Nagatsu T, Nakashima A, Watanabe H, Ito S, Wakamatsu K. Neuromelanin in Parkinson's disease: tyrosine hydroxylase and tyrosinase. Int J Mol Sci 2022;23(8):4176.

[71] Ostendorf BN, Bilanovic J, Adaku N, Tafreshian KN, Tavora B, Vaughan RD, Tavazoie SF. Common germline variants of the human APOE gene modulate melanoma progression and survival. Nat Med 2020;26(7):1048–53.

[72] Taylor JS. Biomolecules. The dark side of sunlight and melanoma. Science 2015;347(6224):824.

[73] Fu YM, Yu ZX, Ferrans VJ, Meadows GG. Tyrosine and phenylalanine restriction induces G0/G1 cell cycle arrest in murine melanoma in vitro and in vivo. Nutr Cancer 1997;29(2):104–13.

[74] Nishioka K. Particulate tyrosinase of human malignant melanoma. Eur J Biochem 1978;85:137–46.

[75] Yang F, Boissy RE. Effects of 4-tertiary butylphenol on the tyrosinase activity in human melanocytes. Pigment Cell Res 1999;12 (4):237–45.

[76] Yamazaki Y, Kawano Y. N-(3,5-dihydroxybenzoyl)-6-hydroxytryptamine as a novel human tyrosinase inhibitor that inactivates the enzyme in cooperation with l-3,4-dihydroxyphenylalanine. Chem Pharm Bull 2010;58(11):1536–40.

[77] Kwak JY, Park S, Seok JK, Liu KH, Boo YC. Ascorbyl coumarates as multifunctional cosmeceutical agents that inhibit melanogenesis and enhance collagen synthesis. Arch Dermatol Res 2015;307 (7):635–43.

[78] Kim M, Park J, Song K, Kim HG, Koh JS, Boo YC. Screening of plant extracts for human tyrosinase inhibiting effects. Int J Cosmet Sci 2012;34(2):202–8.

[79] An SM, Koh JS, Boo YC. p-coumaric acid not only inhibits human tyrosinase activity in vitro but also melanogenesis in cells exposed to UVB. Phytother Res 2010;24(8):1175–80.

[80] Park J, Park JH, Suh HJ, Lee IC, Koh J, Boo YC. Effects of resveratrol, oxyresveratrol, and their acetylated derivatives on cellular melanogenesis. Arch Dermatol Res 2014;306(5):475–87.

[81] Park S, Seok JK, Kwak JY, Choi YH, Hong SS, Suh HJ, Park W, Boo YC. Anti-melanogenic effects of resveratryl triglycolate, a novel hybrid compound derived by esterification of resveratrol with glycolic acid. Arch Dermatol Res 2016;308(5):325–34.

[82] Kwak JY, Seok JK, Suh HJ, Choi YH, Hong SS, Kim DS, Boo YC. Antimelanogenic effects of luteolin 7-sulfate isolated from *Phyllospadix iwatensis* Makino. Br J Dermatol 2016;175(3):501–11.

[83] Lin YP, Hsu FL, Chen CS, Chern JW, Lee MH. Constituents from the Formosan apple reduce tyrosinase activity in human epidermal melanocytes. Phytochemistry 2007;68(8):1189–99.

[84] Okombi S, Rival D, Bonnet S, Mariotte AM, Perrier E, Boumendjel A. Discovery of benzylidenebenzofuran-3(2H)-one (aurones) as inhibitors of tyrosinase derived from human melanocytes. J Med Chem 2006;49(1):329–33.

[85] Leoty-Okombi S, Bonnet S, Rival D, Degrave V, Lin X, Vogelgesang B, André-Frei V. In vitro melanogenesis inhibitory effects of N-feruloyldopamine. J Cosmet Sci 2013;64(2):133–44.

[86] Wang HM, Chen CY, Wen ZH. Identifying melanogenesis inhibitors from *Cinnamomum subavenium* with in vitro and in vivo screening systems by targeting the human tyrosinase. Exp Dermatol 2011;20 (3):242–8.

[87] Tu CX, Lin M, Lu SS, Qi XY, Zhang RX, Zhang YY. Curcumin inhibits melanogenesis in human melanocytes. Phytother Res 2012;26(2):174–9.

[88] Kanehira T, Takekoshi S, Nagata H, Osamura RY, Homma T. Kinobeon A as a potent tyrosinase inhibitor from cell culture of safflower: in vitro comparisons of kinobeon A with other putative inhibitors. Planta Med 2003;69(5):457–9.

[89] Pfizer Product INC. Resorcinol derivatives. PCT WO02/24613; 2002.

[90] Yoshimori A, Oyama T, Takahashi S, Abe H, Kamiya T, Abe T, Tanuma S. Structure-activity relationships of the thujaplicins for inhibition of human tyrosinase. Bioorg Med Chem 2014;22 (21):6193–200.

[91] Ghani U. Azole inhibitors of mushroom and human tyrosinases: current advances and prospects of drug development for melanogenic dermatological disorders. Eur J Med Chem 2022;239, 114525.

[92] Kim SJ, Yang J, Lee S, Park C, Kang D, Akter J, Ullah S, Kim YJ, Chun P, Moon HR. The tyrosinase inhibitory effects of isoxazolone derivatives with a (Z)-β-phenyl-α, β-unsaturated carbonyl scaffold. Bioorg Med Chem 2018;26(14):3882–9.

[93] Choi I, Park Y, Ryu IY, Jung HJ, Ullah S, Choi H, Park C, Kang D, Lee S, Chun P, Young Chung H, Moon HR. In silico and in vitro insights into tyrosinase inhibitors with a 2-thioxooxazoline-4-one template. Comput Struct Biotechnol J 2020;19:37–50.

[94] Searle T, Al-Niaimi F, Ali FR. The top 10 cosmeceuticals for facial hyperpigmentation. Dermatol Ther 2020;33(6), e14095.

[95] Charoo NA. Hyperpigmentation: looking beyond hydroquinone. J Cosmet Dermatol 2022;21(10):4133–45.

[96] Searle T, Al-Niaimi F, Ali FR. Hydroquinone: myths and reality. Clin Exp Dermatol 2021;46(4):636–40.

[97] Brtko J. Biological functions of kojic acid and its derivatives in medicine, cosmetics, and food industry: Insights into health aspects. Arch Pharm 2022;355(10), e2200215.

[98] Saeedi M, Khezri K, Seyed Zakaryaei A, Mohammadamini H. A comprehensive review of the therapeutic potential of α-arbutin. Phytother Res 2021;35(8):4136–54.

[99] Radmard A, Saeedi M, Morteza-Semnani K, Hashemi SMH, Nokhodchi A. An eco-friendly and green formulation in lipid nanotechnology for delivery of a hydrophilic agent to the skin in the treatment and management of hyperpigmentation complaints: arbutin niosome (Arbusome). Colloids Surf B: Biointerfaces 2021;201, 111616.

[100] Lima PB, Dias JAF, Cassiano DP, Esposito ACC, Miot LDB, Bagatin E, Miot HA. Efficacy and safety of topical isobutylamido thiazolyl resorcinol (Thiamidol) vs. 4% hydroquinone cream for facial melasma: an evaluator-blinded, randomized controlled trial. J Eur Acad Dermatol Venereol 2021;35(9):1881–7.

[101] Roggenkamp D, Dlova N, Mann T, Batzer J, Riedel J, Kausch M, Zoric I, Kolbe L. Effective reduction of post-inflammatory hyperpigmentation with the tyrosinase inhibitor isobutylamido-thiazolyl-resorcinol (Thiamidol). Int J Cosmet Sci 2021;43 (3):292–301.

[102] Arrowitz C, Schoelermann AM, Mann T, Jiang LI, Weber T, Kolbe L. Effective tyrosinase inhibition by thiamidol results in significant improvement of mild to moderate melasma. J Invest Dermatol 2019;139(8):1691–8.

Section I

Cadmium enzymes CAs

Chapter 9

CDCA1, a versatile member of the ζ-class of carbonic anhydrase family

Vincenzo Alterio*, Emma Langella*, Davide Esposito, Martina Buonanno, Simona Maria Monti, and Giuseppina De Simone

Institute of Biostructures and Bioimaging-CNR, Naples, Italy

1 Introduction

Among the eight different classes of carbonic anhydrases (CAs) existing in nature [1–8], the ζ-class is certainly the most intriguing both for structural and functional features [9–13]. It was first defined by Morel's group upon the isolation of the CDCA1 enzyme from the *Thalassiosira weissflogii* marine diatom [14] and, subsequently, demonstrated to be ubiquitous in the environment, being present in several diatom species such as *Thalassiosira oceanica*, *Thalassiosira rotula*, *Ditylum brightwellii*, *Phaeodactylum tricornutum*, *Skeletonema costatum*, *Chaetoceros calcitrans*, *Nitzschia* cf. *pusilla*, and *Asterionellopsis glacialis* [14–17]. Despite this wide distribution, CDCA1 remains to date the only member of ζ-class extensively characterized from a biochemical, structural, and functional point of view. This enzyme is distinguished by three peculiar features: the first is its cambialistic nature, since it is able to use both Cd^{2+} and Zn^{2+} ions for catalysis, spontaneously exchanging them into the catalytic site according to their bioavailability [14,15,18]. The second is its catalytic versatility, being able to catalyze the hydration of CO_2 as all known CAs, but also the conversion of CS_2 to H_2S with COS as intermediate [13]. Finally, CDCA1 expression in *T. weissflogii* is highly modulated by pH/pCO_2 and Cd/Zn availabilities in seawater [15,19]. It has been suggested that altogether these properties confer to *T. weissflogii*, and probably to the other diatoms which express CDCA1-like proteins, the capability to adapt to the external environment, being able to survive in limiting conditions [13].

In this chapter, the main catalytic, functional, and structural features of CDCA1 will be summarized and critically discussed in the light of the hypothesis that the enzyme may participate in a novel carbon uptake pathway in diatoms.

2 Biochemical features, CO_2 hydration activity and its modulation

CDCA1 is a 616-residue protein consisting of three nearly identical repeats (R1, R2, and R3), which share about 85% sequence identity one each other (Fig. 1) [14]. Genes coding for similar proteins were also identified in other diatom species, highlighting a wide distribution of this protein in nature [17]. Interestingly, a sequence highly homologous to the single repeats of the *cdca1* gene is present in *Thalassiosira pseudonana* genome (Fig. 1), thus suggesting that a single repeat is sufficient for catalytic activity [14].

The kinetic properties of this enzyme in catalyzing the CO_2 hydration reaction were first characterized in 2008 by Xu and coworkers using mass spectrometry [18] and then confirmed in 2010 by means of stopped-flow assay by Supuran's group [20]. These studies demonstrated that the single repeats and the full-length protein exhibited high catalytic activity with either cadmium or zinc ion in the active site, with the zinc-bound form slightly more efficient. As already observed for other known CAs [21], in both metal-bound forms, CDCA1 showed a pH-dependent catalytic activity with enhanced catalytic efficiency at higher pH [18].

Inhibition studies of the CO_2 hydration reaction were carried out on CDCA1 single repeats R1/R3 either with zinc or cadmium ion within the catalytic site, using different classes of compounds [12,20]. These studies revealed that sulfonamide and sulfamate derivatives properly work as inhibitors, but generally with a lower efficacy for CDCA1 with respect to what observed for CAs belonging to the α-class. Authors ascribed this behavior to the rather big molecular size of investigated sulfonamide and sulfamate compounds that can be better accommodated in the large active site of α-CAs rather than in the small catalytic cavity of the CDCA1 repeats [12]. Additionally, when zinc is present

*These authors contributed equally to the work.

Metalloenzymes. https://doi.org/10.1016/B978-0-12-823974-2.00025-5

```
R1-CDCA1      -----------------------------------------GWQAEIVTEFSLLNEMVDV 45
R2-CDCA1      ---------------------------SISPAQIAEALQGRGWDAEIVTDASMAGQLVDV 255
R3-CDCA1      ---------------------------SITPPQIVSALRGRGWKASIVKASTMSSELKRV 465
TpCDCA        MCMHVDLQVAMSSILSKLTGKDDTSAPPLTPKDIVAALQSRGWEAEIISASSISQDMVEV 60
                                                       ** * *            *

R1-CDCA1      DPQGILKCVDGRGSDNTQFCGPKMPGGIYAIAHNRGVTTLEGLKQITKEVASKGHVPSVH 105
R2-CDCA1      RPEGILKCVDGRGSDNTRMGGPKMPGGIYAIAHNRGVTSIEGLKQITKEVASKGHLPSVH 315
R3-CDCA1      DPQGILKCVDGRGSDNTQFGGPKMPGGIYAIAHNRGVTTLEGLKDITREVASKGHVPSVH 525
TpCDCA        DPAGILKCVDGRGSDNTRMAGPKMPGGIYAIAHNRGTTSVDGLKEITKEVASKGHVPSVH 120
               * **************   **************** *   *** ** *******  ***

R1-CDCA1      GDHSSDMLGCGFFKLWVTGRFDDMGYPRPQFDADQGAKAVENAGGVIEMHHGSHAEKVVY 165
R2-CDCA1      GDHSSDMLGCGFFKLWVTGRFDDMGYPRPQFDADQGANAVKDAGGIIEMHHGSHTEKVVY 375
R3-CDCA1      GDHSSDMLGCGFFKLWLTGRFDDMGYPRPEFDADQGALAVRAAGGVIEMHHGSHEEKVVY 585
TpCDCA        GDHSADMLGCGFFRLWVTGEFDSMGYPRPEFDADQGAAAVKESGGVIEMHHGSHTEKVVY 180
              **** ******** ** ** ** ****** ******* **   ** ******** *****

R1-CDCA1      INLVENKTLEPDEDDQRFIVDGWAAGKFGLDVPKFLIAAAATVEMLGGPKKAKIVIP    222
R2-CDCA1      INLLANKTLEPNENDQRFIVDGWAADKFGLDVPKFLIAAAATVEMLGGPKNAKIVVP    432
R3-CDCA1      INLVSGMTLEPNEHDQRFIVDGWAASKFGLDVVKFLVAAAATVEMLGGPKKAKIVIP    642
TpCDCA        INLVENKTLEPDENDQRFIVDGWAAIKFNLDVVKFLVAAAATVEMLGGPRIAKIVVA    237
              ***    **** * *********** ** *** *** ************  ****
```

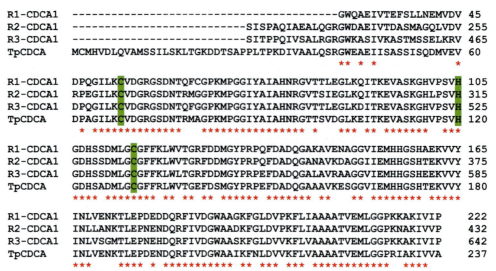

FIG. 1 Multiple sequence alignment of the three CDCA1 single repeats and CDCA from *T. pseudonana*. Numbering of CDCA1 repeats refers to that used in Ref. [18]. Strictly conserved residues are indicated by *asterisks*, and metal-binding residues are highlighted in *green*.

in the active site, the enzyme shows greater affinity for sulfonamides/sulfamates compared to the cadmium containing enzyme [12] likely due to an intrinsic different affinity of sulfonamide derivatives toward the two metal ions [22,23].

CDCA1 repeats are also sensitive to anion inhibitors, with inhibition constants from the micromolar to millimolar range [20]. Also in this case, the zinc and the cadmium containing repeats showed some differences in the inhibition profile. Indeed, the best anion inhibitors of R1 containing cadmium (here referred as Cd-R1) were thiocyanate, sulfamate, and sulfamide, with K_Is of 10–89 μM, whereas the best anion inhibitors of the same enzyme with zinc in the active site (here referred as Zn-R1) were sulfamate and sulfamide with K_Is of 60–72 μM. It is worth noting that these enzymes were only weakly inhibited by chloride, bromide, or sulfate, the main anion components of sea water, with inhibition constants in the range of 0.24–0.85 mM [20].

More recently, Supuran and coworkers investigated also the possibility to activate Zn-R1 and Cd-R1 using a panel of amino acid and amine derivatives [24]. Surprisingly, although the Cd-containing enzyme was completely insensitive to activation, the zinc-containing one was activated by all tested compounds with activation constants ranging between 0.092 and 37.9 μM. The most effective activators were L-adrenaline, 1-(2-aminoethyl)-piperazine, and 4-(2-aminoethyl)-morpholine. Further studies are necessary to understand the molecular mechanisms controlling the activation reactions and to correlate results according to the metal ion within the active site [24].

3 Structural features: Overall fold, substrate binding pocket, and access route

CDCA1 has been extensively characterized from a structural point of view [12,13,18]; indeed, the crystal structures of the three repeats were determined [12,18] and a model of the full-length protein was obtained by docking approaches [12]. In particular, in agreement with their high sequence identity, the three repeats show a very similar structure with ellipsoidal shape, characterized by nine β-strands and seven α-helices (Fig. 2A). Seven of the nine β-strands are located in the center of the structure creating two contiguous β-sheets. The active site is positioned in a cleft containing the metal ion at the bottom, coordinated by three conserved protein residues (two cysteines and one histidine) and one water molecule. In the case of the cadmium-containing enzyme, a second water molecule completes the coordination sphere [12,18] (Fig. 2B).

Structural data allowed also to explain the facile metal exchange in CDCA1, revealing a stable opening of the metal-coordinating site in the absence of metal. Interestingly, although the three repeats show a completely different fold with respect to the other CA classes previously characterized [2–4,8,21], they can be considered as structural and functional mimics of β-CAs, particularly in the region of the active site [18].

Fig. 3 shows the docking model of the full-length CDCA1. It is characterized by an asymmetric structure with two covalently linked interfaces (R1-R2 and R2-R3) and a small noncovalent R1-R3 interface. The three active sites, present in each repeat, are far from each other and completely accessible to the substrate [12].

More recently, our group determined the structure of the zinc-containing R3 repeat (Zn-R3) in complex with CO_2 substrate, thus allowing for the first time the identification of the substrate-binding site of a ζ-class CA enzyme [13]. As shown in Fig. 4, CO_2 binds within the active site without altering the metal ion coordination geometry but only displacing a water molecule, which is hydrogen bonded to the Zn-bound solvent molecule. It is located in

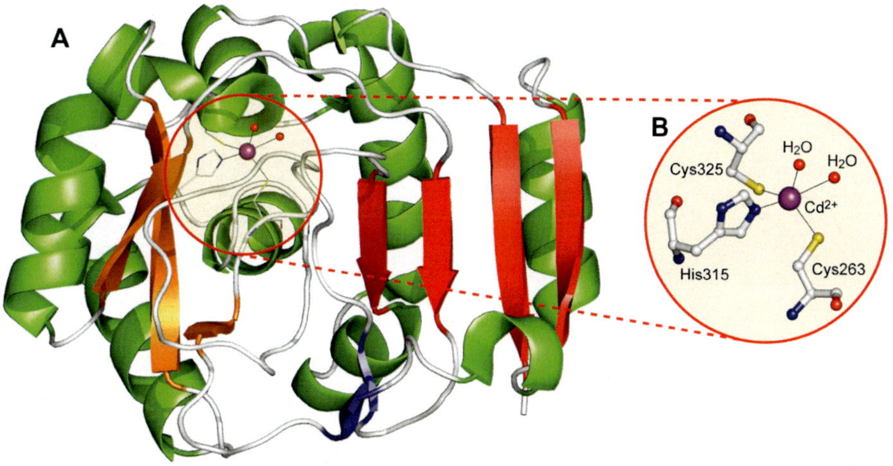

FIG. 2 (A) Cartoon representation and (B) active site detail of CDCA1-R2 structure (PDB code 3BOB) [18]. The two contiguous β-sheets in the center of the structure are shown in *orange* and *red*, while a third β-sheet and helices are shown in *blue* and *green*, respectively. Cd^{2+} ion is displayed as a *purple sphere*, and its coordinating residues are shown in *stick* representation.

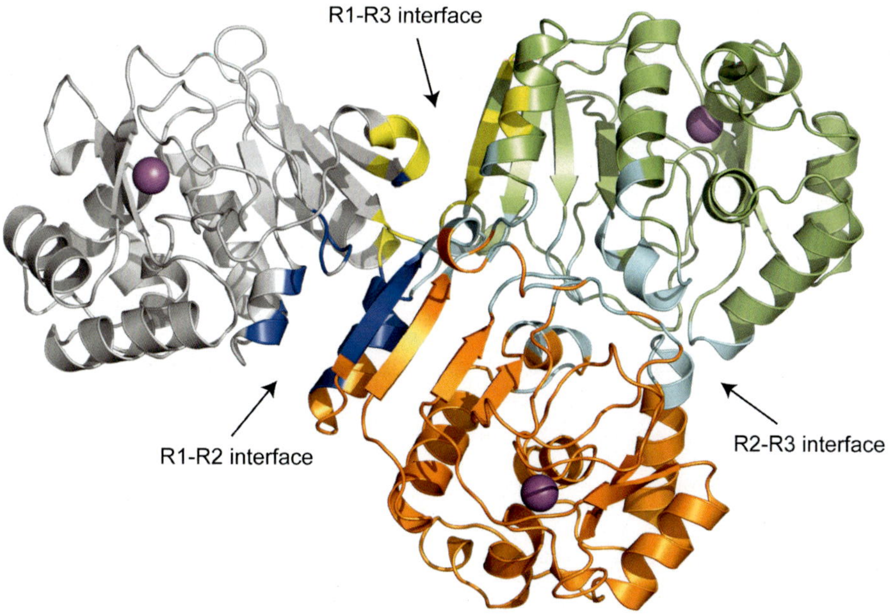

FIG. 3 Cartoon representation of the CDCA1 full length docking model. R1, R2, and R3 repeats are in *gray*, *orange*, and *green*, respectively. Interface regions are indicated by *arrows* and displayed in *blue* (R1-R2), *cyan* (R2-R3), and *yellow* (R1-R3). Cadmium ions are represented as *magenta spheres*.

a hydrophobic pocket bordered by residues Val474, Phe537, Phe603, Thr627, and Leu631 (Fig. 4C) [13].

Starting from this structure and by using MD simulations, the access route of the CO_2 substrate to the active site was identified. It consists of a narrow, long, and highly hydrophobic tunnel (Fig. 5).

This was a quite surprising result compared to what was observed for other canonical CA-classes in which the substrate entry route had been characterized. Indeed, in the case of α-CAs, the CO_2 substrate arrives at the active site passing through a quite large cavity consisting of both hydrophobic and hydrophilic residues (Fig. 6A), whereas in the case of β-CAs, a funnel shaped cavity, containing again both hydrophobic and hydrophilic residues, was identified as possible CO_2 migration pathway (Fig. 6B) [13,26–29].

4 From structure to function: ζ-CAs show CS_2 hydrolase activity

As described in the previous paragraph, the presence of a highly hydrophobic tunnel as substrate entry route is a unique feature of CDCA1 with respect to other canonical CA-classes so far investigated. Notably, a similar substrate

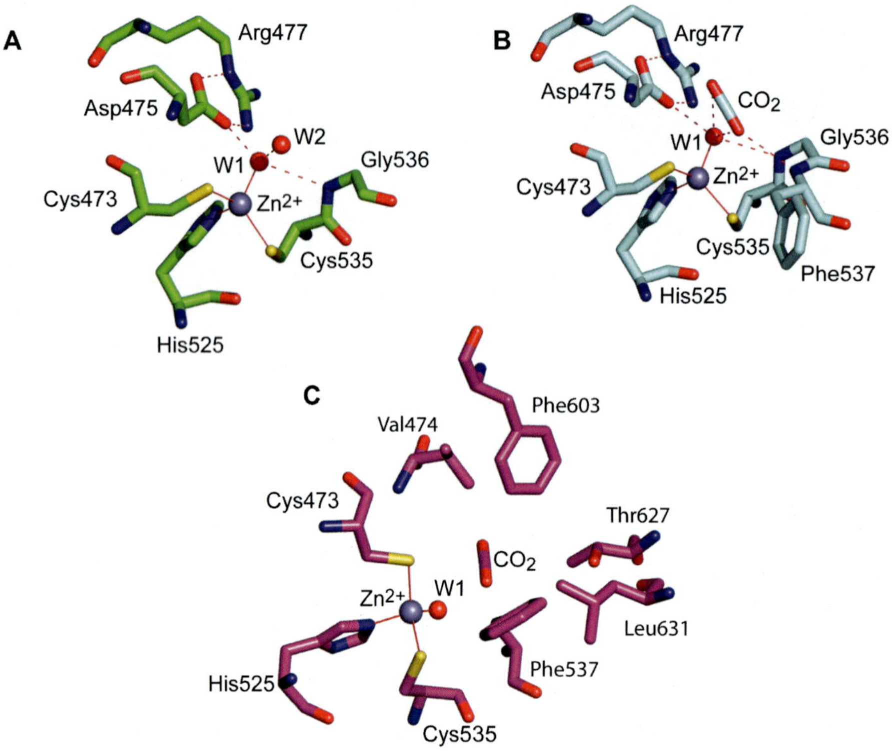

FIG. 4 Stick representation of the active sites of (A) Zn-R3 and (B) Zn-R3/CO_2 complex; (C) CO_2-binding pocket in Zn-R3/CO_2 active site. Zinc ion and water molecules are represented as *spheres*. Continuous lines show zinc ion coordination, whereas dashed lines indicate potential hydrogen bonds.

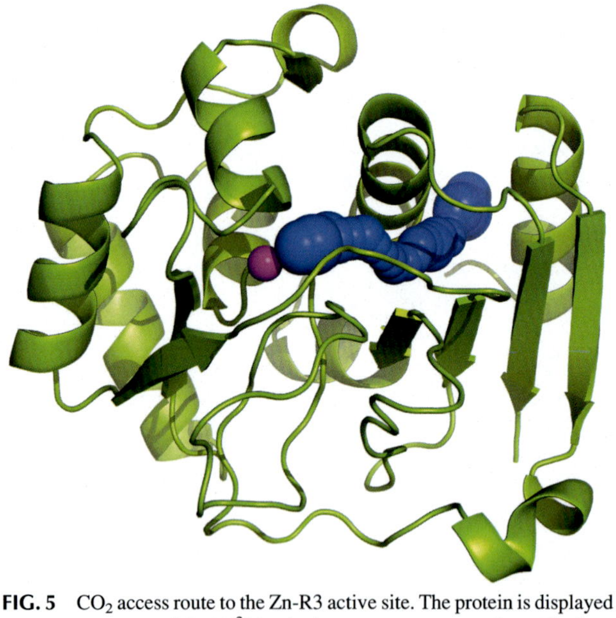

FIG. 5 CO_2 access route to the Zn-R3 active site. The protein is displayed as *green cartoon*, and the Zn^{2+} ion is shown as a *magenta sphere*. The long hydrophobic tunnel leading to the active site is represented by a sequence of *blue spheres*.

entry route has been observed in a functionally related enzyme, namely the CS_2 hydrolase from *Acidianus* A1-3 [30]. Notably, this enzyme is not able to catalyze the CO_2 hydration reaction, but only the conversion of CS_2 to H_2S with COS as intermediate ($CS_2 + H_2O \rightarrow COS + H_2S$ and $COS + H_2O \rightarrow CO_2 + H_2S$). Based on this structural similarity, the investigation of CDCA1 capability to hydrate CS_2 to H_2S by means of dedicated enzymatic assays was carried out. Obtained results confirmed the starting hypothesis, showing that, when zinc is present in the active site, CDCA1 can catalyze the CS_2 hydrolysis reaction. Accordingly, CDCA1 emerges as the only enzyme known so far able to use CS_2 and CO_2 as substrates [13]. Interestingly, as the CO_2 hydration activity, also this additional catalytic activity, was inhibited by the classical CA inhibitor acetazolamide [13].

5 Conclusions and future perspectives

To date, CDCA1 is the only member of the ζ-CA class extensively characterized. Studies reviewed in this chapter clearly show that this enzyme has a much greater catalytic versatility than other currently known CAs. Indeed, CDCA1 can indiscriminately use zinc or cadmium in its active site to catalyze the CO_2 hydration reaction and, in its zinc-bound form, it can also catalyze the conversion of CS_2 to H_2S. The physiological implications of this versatility are all to be investigated. Some authors suggested that the capability of the enzyme to bind in the active site cadmium, an element previously known only for its toxicity, is necessary to support the catalytic needs of diatoms in the metal-poor

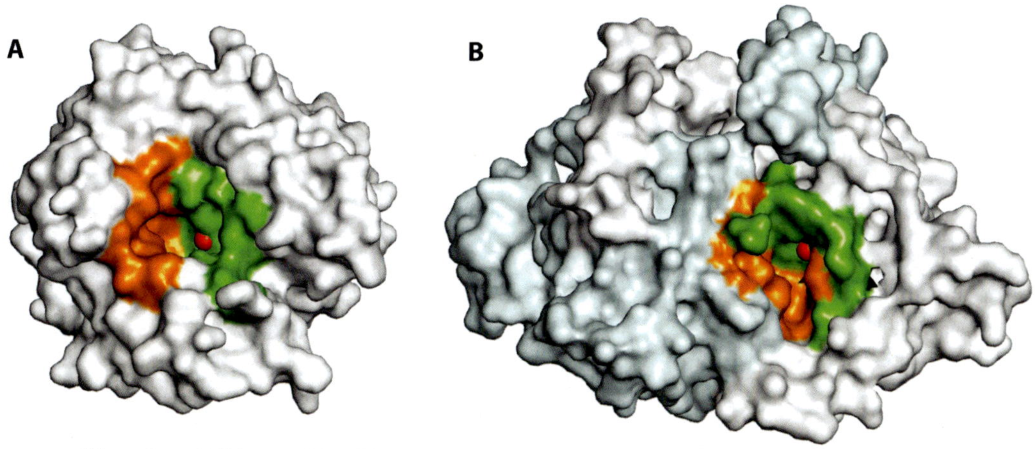

FIG. 6 Solvent accessible surface of (A) human CA II (PDB code 1CA2) [25] and (B) *Saccharomyces cerevisiae* β-CA Nce103 (PDB code 3EYX) [26]. Putative CO_2 access routes are highlighted in *orange* (hydrophobic region) and *green* (hydrophilic region). The catalytic zinc ion is shown as a *red sphere*. β-CA dimer chains A and B are reported in *white* and *cyan*, respectively.

environment of the oceans [10,18,31]. Similarly, the CS_2 hydration catalytic activity has also been interpreted in light of the adaptation of diatoms to the external environment [13]. Indeed, since CS_2 is present in marine environments as a breakdown product from organic matter, mainly dimethyl sulfide [32,33], the capability of CDCA1 to convert CS_2 may represent an alternative source of carbon acquisition for diatoms. However, these hypotheses need to be further investigated, leaving ample room for further functional studies on CDCA1 as well as on other members of the ζ-CA class.

References

[1] Langella E, Di Fiore A, Alterio V, Monti SM, De Simone G, D'Ambrosio K. α-CAs from photosynthetic organisms. Int J Mol Sci 2022;23(19):12045.

[2] Kikutani S, Nakajima K, Nagasato C, Tsuji Y, Miyatake A, Matsuda Y. Thylakoid luminal Θ-carbonic anhydrase critical for growth and photosynthesis in the marine diatom *Phaeodactylum tricornutum*. Proc Natl Acad Sci U S A 2016;113(35):9828–33.

[3] De Simone G, Di Fiore A, Capasso C, Supuran CT. The zinc coordination pattern in the η-carbonic anhydrase from *Plasmodium falciparum* is different from all other carbonic anhydrase genetic families. Bioorg Med Chem Lett 2015;25(7):1385–9.

[4] Jensen EL, Clement R, Kosta A, Maberly SC, Gontero B. A new widespread subclass of carbonic anhydrase in marine phytoplankton. ISME J 2019;13(8):2094–106.

[5] Del Prete S, Vullo D, Fisher GM, Andrews KT, Poulsen SA, Capasso C, et al. Discovery of a new family of carbonic anhydrases in the malaria pathogen *Plasmodium falciparum*—the η-carbonic anhydrases. Bioorg Med Chem Lett 2014;24(18):4389–96.

[6] Ferry JG. The γ-class of carbonic anhydrases. Biochim Biophys Acta Proteins Proteomics 2010;1804(2):374–81.

[7] Zimmerman S, Ferry J. The β and γ classes of carbonic anhydrase. Curr Pharm Des 2008;14(7):716–21.

[8] Alterio V, Di Fiore A, D'Ambrosio K, Supuran CT, De Simone G. Multiple binding modes of inhibitors to carbonic anhydrases: how to design specific drugs targeting 15 different isoforms? Chem Rev 2012;112(8):4421–68.

[9] Langella E, De Simone G, Esposito D, Alterio V, Monti SM. ζ-carbonic anhydrases. In: Supuran CT, Nocentini A, editors. Carbonic anhydrases biochemistry and pharmacology of an evergreen pharmaceutical target. Elsevier; 2019. p. 131–7.

[10] Alterio V, Langella E, De Simone G, Monti SM. Cadmium-containing carbonic anhydrase CDCA1 in marine diatom *Thalassiosira weissflogii*. Mar Drugs 2015;13(4):1688–97.

[11] Monti SM, De Simone G, Supuran CT, Alterio V. CDCA1 from *Thalassiosira weissflogii* as representative member of ζ-class CAs: general features and biotechnological applications. In: Supuran CT, De Simone G, editors. Carbonic anhydrases as biocatalysts. Elsevier; 2015. p. 351–9.

[12] Alterio V, Langella E, Viparelli F, Vullo D, Ascione G, Dathan NA, et al. Structural and inhibition insights into carbonic anhydrase CDCA1 from the marine diatom *Thalassiosira weissflogii*. Biochimie 2012;94(5):1232–41.

[13] Alterio V, Langella E, Buonanno M, Esposito D, Nocentini A, Berrino E, et al. Zeta-carbonic anhydrases show CS2 hydrolase activity: a new metabolic carbon acquisition pathway in diatoms? Comput Struct Biotechnol J 2021;19:3427–36.

[14] Lane TW, Saito MA, George GN, Pickering IJ, Prince RC, Morell FMM. A cadmium enzyme from a marine diatom. Nature 2005;435:42.

[15] Park H, McGinn PJ, Morel FMM. Expression of cadmium carbonic anhydrase of diatoms in seawater. Aquat Microb Ecol 2008;51(2):183–93.

[16] McGinn PJ, Morel FMM. Expression and regulation of carbonic anhydrases in the marine diatom *Thalassiosira pseudonana* and in natural phytoplankton assemblages from Great Bay, New Jersey. Physiol Plant 2008;133(1):78–91.

[17] Park H, Song B, Morel FMM. Diversity of the cadmium-containing carbonic anhydrase in marine diatoms and natural waters. Environ Microbiol 2007;9(2):403–13.

[18] Xu Y, Feng L, Jeffrey PD, Shi Y, Morel FMM. Structure and metal exchange in the cadmium carbonic anhydrase of marine diatoms. Nature 2008;452:56–61.

[19] Xu Y, Supuran CT, Morel FMM. Cadmium-carbonic anhydrase. In: Encyclopedia of inorganic and bioinorganic chemistry. John Wiley & Sons, Ltd; 2011. p. 1–5.

[20] Viparelli F, Monti SM, De Simone G, Innocenti A, Scozzafava A, Xu Y, et al. Inhibition of the R1 fragment of the cadmium-containing ζ-class carbonic anhydrase from the diatom *Thalassiosira weissflogii* with anions. Bioorg Med Chem Lett 2010;20(16):4745–8.

[21] Supuran CT, De Simone G, editors. Carbonic anhydrases as biocatalysts. 1st ed. Amsterdam: Elsevier; 2015. 373 p.

[22] Avdeef A, Hartenstein F, Chemotti AR, Brownlb JA. Cadmium binding by biological ligands. 5. Solution studies of cadmium and zinc binding by sulfhydryl ligands N, N'-dimethyl-N, N'-bis(2-mercaptoethyl)ethylenediamine and (2-mercaptoethyl)amine. Inorg Chem 1992;31(18):3701–5.

[23] Rosati AM, Traversa U. Mechanisms of inhibitory effects of zinc and cadmium ions on agonist binding to adenosine A1 receptors in rat brain. Biochem Pharmacol 1999;58(4):623–32.

[24] Angeli A, Buonanno M, Donald WA, Monti SM, Supuran CT. The zinc- but not cadmium-containing ζ-carbonic from the diatom *Thalassiosira weissflogii* is potently activated by amines and amino acids. Bioorg Chem 2018;80:261–5.

[25] Eriksson AE, Jones TA, Liljas A. Refined structure of human carbonic anhydrase II at 2.0 Å resolution. Proteins Struct Funct Bioinform 1988;4(4):274–82.

[26] Teng YB, Jiang YL, He YX, He WW, Lian FM, Chen Y, et al. Structural insights into the substrate tunnel of *Saccharomyces cerevisiae* carbonic anhydrase Nce103. BMC Struct Biol 2009;9(1):1–8.

[27] Maupin CM, Castillo N, Taraphder S, Tu C, McKenna R, Silverman DN, et al. Chemical rescue of enzymes: proton transfer in mutants of human carbonic anhydrase II. J Am Chem Soc 2011;133(16): 6223–34.

[28] De Simone G, Alterio V, Supuran CT. Exploiting the hydrophobic and hydrophilic binding sites for designing carbonic anhydrase inhibitors. Expert Opin Drug Discov 2013;8(7):793–810.

[29] Liang JY, Lipscomb WN. Binding of substrate CO2 to the active site of human carbonic anhydrase II: a molecular dynamics study. Proc Natl Acad Sci 1990;87(10):3675–9.

[30] Smeulders MJ, Barends TRM, Pol A, Scherer A, Zandvoort MH, Udvarhelyi A, et al. Evolution of a new enzyme for carbon disulphide conversion by an acidothermophilic archaeon. Nature 2011;478 (7369):412–6.

[31] Morel FMM, Milligan AJ, Saito MA. Marine bioinorganic chemistry: the role of trace metals in the oceanic cycles of major nutrients. In: Treatise on geochemistry. Pergamon; 2003. p. 113–43.

[32] Smeulders MJ, Pol A, Venselaar H, Barends TRM, Hermans J, Jetten MSM, et al. Bacterial CS2 hydrolases from *Acidithiobacillus thiooxidans* strains are homologous to the archaeal catenane CS2 hydrolase. J Bacteriol 2013;195(18):4046–56.

[33] Watts SF. The mass budgets of carbonyl sulfide, dimethyl sulfide, carbon disulfide and hydrogen sulfide. Atmos Environ 2000;34(5): 761–79.

Section J

Molybdenum enzymes

Chapter 10

Molybdenum enzymes

Simone Giovannuzzi
NEUROFARBA Department, Pharmaceutical and Nutraceutical Section, University of Florence, Firenze, Italy

1 Introduction

Molybdenum (Mo, atomic number 42) is one of the transition metals of the 5th row (4d) of the periodic table with biological activity. It was discovered as a molybdic acid ($MoO_3{*}H_2O$) by Carl Wilhelm Scheele and, subsequently, was first isolated as a dark metallic powder by Peter Jacob Hjelm in 1781, but it does not occur naturally as a free metal on earth [1]. It was found in nature in different oxidation states ranging from zero to six and forms complexes with organic and inorganic ligands with coordination numbers between four and eight. Molybdenum does not exist naturally in the metallic state but occurs in association with other elements. The predominant form of molybdenum occurring in soil and natural waters is the molybdate anion, MoO_4^{2-}, and it is the only known form that cells can take up from the environment. Despite that scarcity, molybdenum is essential to most organisms [2–4], from archaea and bacteria to higher plants and mammals, being found in the active site of enzymes that catalyze oxidation-reduction reactions involving carbon, nitrogen and sulfur atoms of key metabolites [5–10]; in all these living organisms, it is present at low concentrations. Up to now, more than 50 different Mo-dependent enzymes have been found in all kingdom's life [11]. Most of them are prokaryotic and, except for the multinuclear iron Fe-Mo cofactor in nitrogenase [12–14], they all bind molybdenum through a pterin-based prosthetic group forming the molybdenum cofactor, called Moco. In eukaryotes, Moco (Fig. 1) is composed of a fully reduced pterin backbone with a C6-substituted four-carbon side chain forming a third pyran ring that hosts a terminal phosphate and the unique dithiolene group, which binds molybdenum [15]. Five Mo-enzymes were found in eukaryotes that belong to two families: (I) Nitrate reductase, sulfite oxidase (SO), and the amidoxime-reducing component (mARC) are members of the SO family, on the other side, (II) xanthine dehydrogenase (XDH) or oxidase (XO) and aldehyde oxidase (AOX) are members of the XO family [3]. Interesting, in the SO family as shown in Fig. 1, Moco is covalently bound to the enzyme via an invariant cysteine residue forming the third S-ligand in the molybdenum coordination sphere [3].

Noteworthy, even if eukaryotic organisms use this element, a lot of unicellular eukaryotes, including some parasites, Saccharomyces, and other yeasts, have lost the ability to use molybdenum [16,17]. Considering humans, the highest levels of molybdenum were found in specific parts of the human body, such as kidney, liver, small intestine, and adrenals [18]. In serum, the concentration is about $0.6\,ng\,mL^{-1}$ [19], but it depends on dietary intake [20]. Molybdenum results in an essential element [21], but, at the same time, it is potentially toxic for humans [22]. The discovery of molybdenum in different enzymes such as nitrogenase, nitrate reductase (NR), and xanthine oxidase (XO) has demonstrated the biological importance of molybdenum as a cofactor in the active site of such enzymes.

2 The molybdenum cofactor (Moco)

In all kingdoms of life, Moco is synthesized following a highly conserved metabolic pathway, which could be divided into four different and subsequently steps [23], that take place within the mitochondrial and cytosol (Fig. 2). At least six genes are needed for Moco biosynthesis in eukaryotes. The nomenclature that is used for these genes depends on the organisms, in fact, human genes use the MOCS nomenclature (MO cofactor synthesis), while plants follow the CNX nomenclature (cofactor for nitrate reductase and xanthine dehydrogenase) [24]. Moco biosynthesis needs a molybdenum availability in cells in the form of oxyanion molybdate, MoO_4^{2-}, being mediated by specific molybdate transporters [25].

2.1 Molybdenum uptake

All organisms, which depend on molybdenum, need to acquire the oxyanion molybdate as the only source of molybdenum from the external medium. In prokaryotes, oxyanion molybdate enters into the cells through the action of proteins belonging to the ATP-binding cassette (ABC) transporters family [26–28]. In contrast to bacteria, molybdate transport in eukaryotes is poorly understood, and the first eukaryotic molybdate transporters were

Metalloenzymes. https://doi.org/10.1016/B978-0-12-823974-2.00005-X

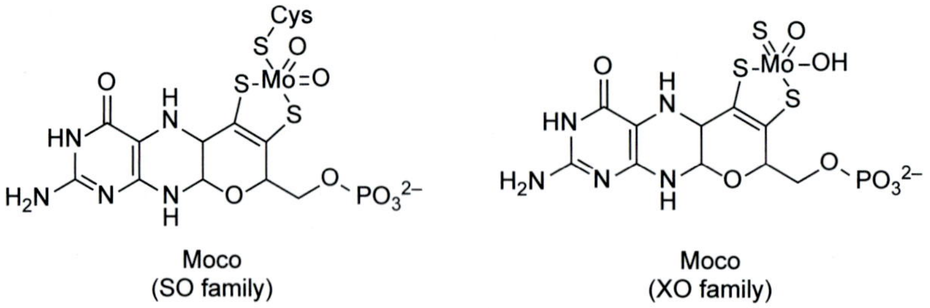

FIG. 1 The chemical structure of Moco was found in sulfite oxidase (SO), with cysteine ligation, and in xanthine oxidase (XO), with a terminal sulfide ligand.

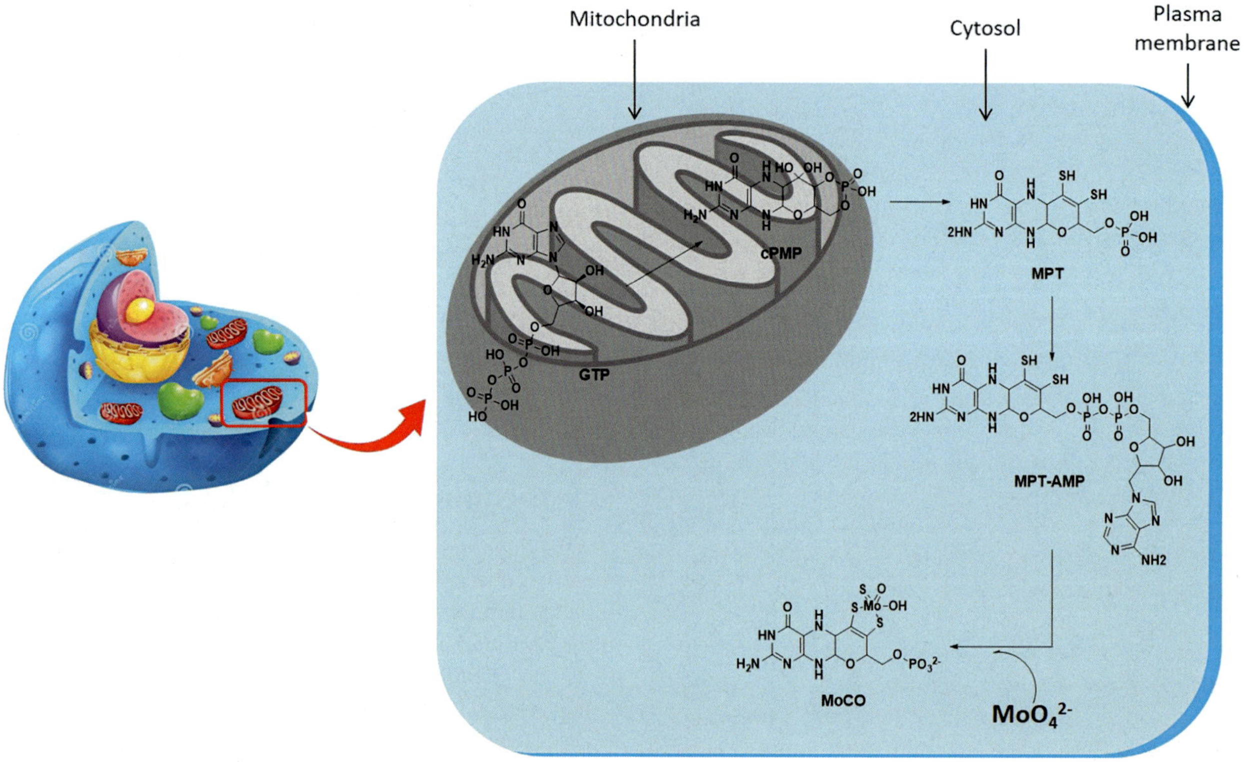

FIG. 2 Schematic representation of the metabolic steps involved in the Mo cofactor biosynthesis pathway. *GTP*, guanosine triphosphate; *cPMP*, cyclic pyranopterin monophosphate; *MTP*, molybdopterin; *MTP-AMP*, adenylated molybdopterin.

identified only recently [29–31]. In eukaryotes, transporters mediating specific and high-affinity molybdate transport belong to the MOT1 or MOT2 families [29,31,32]. The first eukaryotic transporter identified involved in molybdate transport belongs to the MOT1 family (Molybdate Transporter type 1) and, simultaneously, it was discovered in the alga *Chlamydomonas reinhardtii* and the plant *Arabidopsis thaliana* [33]. MOT1 shows both a high-specificity and high-affinity transport for molybdate. Particularly, in *A. thaliana* were found two different MOT1 members, called AtMOT1 and AtMOT2, and both are involved in the molybdenum transport: AtMOT1 seems to carry out a crucial role in the molybdenum uptake from the soil, but, up to now, its physiological role still needs to be clarified [34], whereas AtMOT2 is a vacuolar molybdenum transporter and it is involved in the molybdenum storage and molybdenum homeostasis in the cytosol. AtMOT1 was identified in two different subcellular localizations: plasma membrane and mitochondria [33,34]. The MOT1 family resulted important in legume plants, which should be useful for symbiotic nitrogen fixation [35–37]. A pivotal role was reported in *Medicago truncatula*, a small annual legume plant: MtMOT1 family is important for Mo supply to the nitrogen-fixing tissue, specifically, isozyme MtMOT1.2 releases from the vascular apparatus to the nodules, where the nitrogen fixation takes places [38], while MtMOT1.3 result responsible for the movement of Mo between nodule apoplast and nodule cells [39]. Recently, physiological studies in *Chlamydomonas* suggested the presence of a second transporter, which was recently identified as the first member of the MOT2 family of molybdate transporters. Specifically, in *C. reinhardtii,* CrMOT2 shares only around 12% of homology with MOT1 proteins. MOT2 was found not only in plants but also in animals and humans [30]. During heterologous expression studies with *Saccharomyces*, expressing the human member of the MOT2 family (HsMOT2), a molybdate uptake activity was visible and comparable to *Chlamydomonas* cells expressing MOT1 or MOT2 [30]. This study suggested the role of human MOT2 in Mo transport, being the first human protein that has been related to Mo homeostasis. However, the MOT2 function as a molybdate transporter has only been verified in *C. reinhardtii*, for this reason, more experiments are needed to confirm this role in other eukaryotes, such as humans.

2.2 The MocO biosynthesis pathway

For all Mo-enzymes except nitrogenase, Mo is activated and chelated into a prosthetic group named Moco. This cofactor consists of Mo covalently bound to the metal-binding molybdopterin (MPT), which is highly conserved in eukaryotes, eubacteria, and archaebacteria [11]. Moco is synthesized by a conserved biosynthetic pathway that can be divided into four steps inside the cells, between the mitochondria and the cytosol, as reported in Fig. 2 [4,11]. The first step, called GTP-cyclization, occurs in the mitochondria matrix [40], and it starts with the conversion of GTP into cPMP in a complex rearrangement reaction catalyzed in humans by two proteins, MOCS1A and MOCS1AB, expressed from the MOCS1 gene [41,42]; in plants, this reaction requires two gene products, CNX2 and CNX3 [43]. cPMP consists in a sulfur-free pyranopterin containing an uncommon geminal diol group, resulting in the most stable intermediate within the Moco biosynthetic pathway. Once cPMP is synthesized, it is exported to the cytosol by the mitochondrial inner membrane transporter ATM3 [44], then, cPMP incorporates two sulfur atoms in the so-called dithiolene reaction step, to form MPT. This reaction is mediated by the MTP synthase complex, a hetero-tetrameric complex, which is formed by two monomers (plants/humans) of CNX6/MOCS2B (large subunit) and two monomers of CNX7/MOCS2A (small subunit) [32,45,46]. In plants, CNX6 and -7 are encoded by an independent gene, while, in humans, MOCS2A and -2B belong to the same bicistronic gene MOCS2 [47]. To complete this second step, after the dithiolene reaction, a resulfuration is required and mediated by the proteins CNX5 and MOCS3, respectively, in plants and humans, to obtain MPT [48,49]. The insertion of Mo cannot take place immediately after MPT synthesis, since this molecule requires a prior activation, an MPT adenylation. In plants, this reaction is mediated by CNX1 [50], but in humans, it is Gephyrin, a multifunctional protein involved in the clustering of inhibitory neuroreceptors and consists in an N-terminal G-domain (MPT adenylyltransferase, GEPH-G) and C-terminal *E*-domain (Mo insertase, GEPH-E) [51,52]. These proteins are involved not only in the adenylation but also in the Mo insertion, the fourth and final step, thanks to the presence of these two domains in each protein. The adenylation of MPT generates MTP-AMP, the last intermediate of the pathway. In *C. reinhardtii*, the CNX1G and CNX1E domains are encoded by two independent genes and the chimeric fusions containing these two domains with different orientations can synthesize the Mo cofactor [53,54]. In *A. thaliana*, the three-dimensional structure of CNX1G showed that the MPT adenylation step is Mg^{2+}- and ATP-dependent, and it leads to MPT-AMP that is bound to the CNX1G domain, then, MPT-AMP intermediate is transferred to the *E*-domain where the deadenylation reaction takes place in a Zn^{2+}/Mn^{2+}-dependent manner [55–58]. Finally, the cytosolic molybdate, entered inside the cell through specific molybdate transporters [56], will be released from the vacuoles in the shape of MoO_4^{2-}, can also bind to the E-domain triggering the AMP hydrolysis and yielding the active Moco [25].

In Fig. 3, it was shown the metabolism of molybdenum in higher plant cells, from the just described biosynthesis to all the Moco-enzyme pathways.

2.3 Molybdenum cofactor stability

Upon formation, Moco requires a quick insertion into the apoenzymes or into MCP (Moco Carrier Protein) [59] because its free form is highly labile [59]. It can bind directly to SO, NR, or mARC apoenzymes, or it can undertake a final sulfuration in order to bind to the apoenzymes AO or XOR/XD, as reported in Fig. 4.

The free Moco is sensitive to oxidation; moreover, in vitro, it exhibits a half-life between 15 min and 1 h [59,60]. To counteract this little stability, eukaryotic organisms developed some mechanisms for Moco protection and storage that are achieved by employing Moco binding proteins (MBP) [59–61]; it can improve the half-life of Moco until 24 h [60]. In *C. reinhardtii,* MCP seems to be able to bind and protect Moco, increasing its half-time and, on the other side, it can facilitate the transfer of the Moco to the apoenzymes [62]. The MCP crystal structure reveals the presence of a Rossmann fold in each monomer, so, the putative-binding site has been proposed; in addition, in silico docking studies together with mutagenesis studies suggested a probable Moco-binding site because mutations in conserved residues resulted in a reduced capacity to bind and protect Moco [62]. Considering higher plant genomes, no similar MCP enzymes were found; however, members of a lysine decarboxylase-like proteins family (MoBP) share a

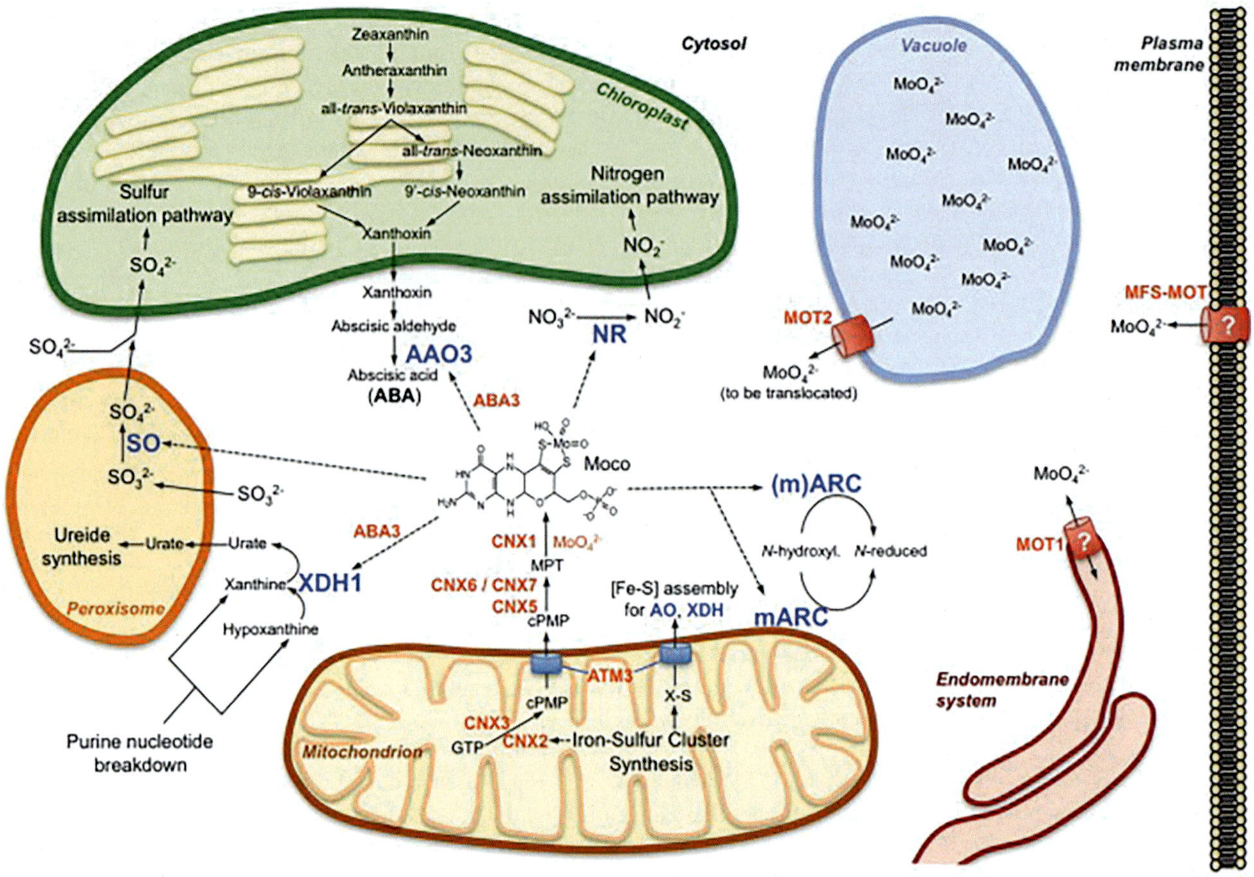

FIG. 3 Molybdenum metabolism in higher plant cells [46].

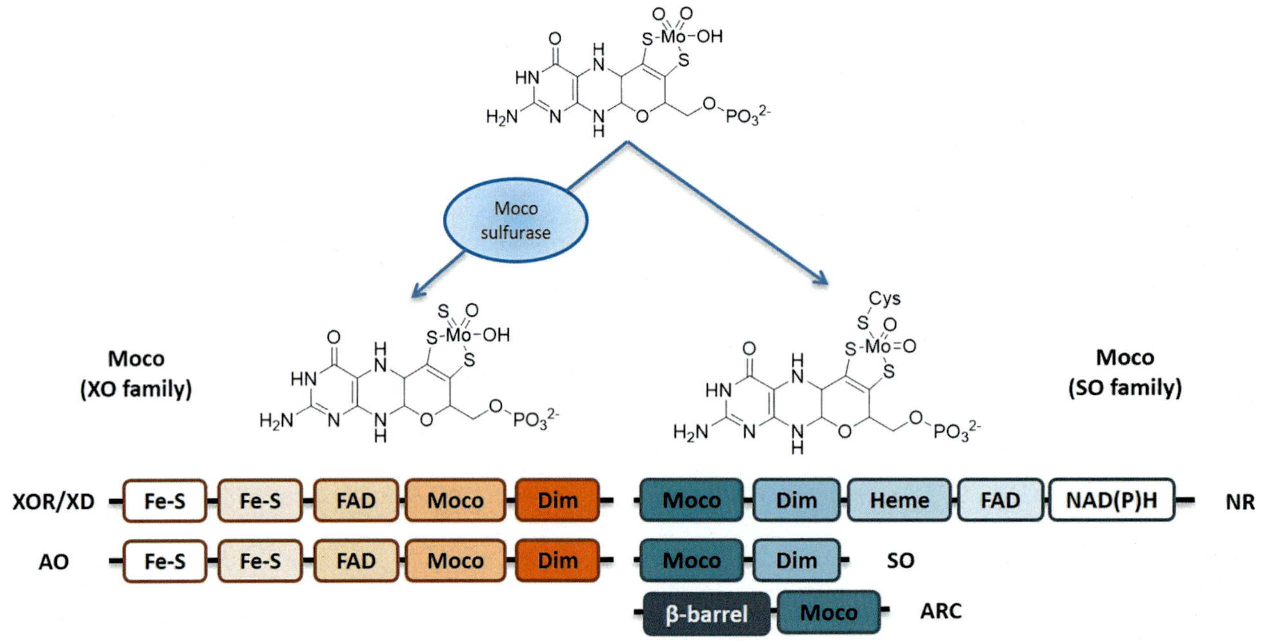

FIG. 4 Eukaryotic Moco-containing enzyme. Domain structure of the eukaryotic Mo-enzymes of the xanthine oxidoreductase and sulfite oxidase families. Moco's final conversion needed for its transference to the corresponding Mo-enzyme family is shown. *XOR/XD*, xanthine oxidoreductase/xanthine dehydrogenase; *AO*, aldehyde oxidase; *NR*, nitrate reductase; *SO*, sulfite oxidase; *mARC*, mitochondrial amidoxime reducing component; *Dim*, dimerization domain [32].

similar structure with MCP; in fact, these enzymes are able to bind Moco in *A. thaliana* [63,64]. Generally, MoBPs can interact with the CNX1 domain and the Mo-containing enzymes.

3 Moco enzymes

Molybdenum enzymes have long been recognized to be important in eukaryotes, dating to 1954 and the near-simultaneous demonstration of molybdenum in xanthine oxidase [65], aldehyde oxidase [66], and nitrate reductase [67]; molybdenum was subsequently identified as a component of sulfite oxidase in 1971 [68]. The last Mo-enzyme was recently discovered in 2006, and it was named mARC, mitochondrial amidoxime reducing component [69,70]. These enzymes are involved in key metabolic processes such as nitrate reduction, sulfite detoxification, and purine catabolism that play fundamental roles in the global carbon, nitrogen and sulfur cycles. Except for nitrate reductase, which is only found in autotroph organisms such as plants, fungi and algae, all other four Mo-enzymes are expressed in humans [8]. Historically, Moco-containing enzymes in eukaryotes can be divided into two families, the sulfite oxidase (SO) and the xanthine oxidase (XO) families, which are defined by the identity of the ligands bound to the Mo center (Fig. 4) [71]. As reported in Fig. 4, members of the xanthine oxidase family possess an $LMo^{VI}OS(OH)$ coordination sphere, where L is a pyranopterin cofactor that coordinates the metal via an enedithiolate side chain; in contrast, the sulfite oxidases and nitrate reductase have an $LMo^{I-V}O_2$(S-Cys) coordination sphere, with the cysteine ligand contributed by the polypeptide. Both types of center possess a square-pyramidal coordination geometry.

3.1 Xanthine oxidoreductase

Xanthine oxidoreductase (XOR) is a member of a highly conserved family of molybdo-flavoenzymes that are widely distributed in all living organisms and are hypothesized to derive from a common ancestral progenitor [72]. The catabolism of hypoxanthine and xanthine to uric acid is ensured by the xanthine dehydrogenase activity (XDH, EC 1.17.1.4), but only mammals possess the xanthine oxidase (XO, EC 1.17.3.2), whose activity in milk was already described at the end of the 19th century [73]. We will use the term xanthine oxidoreductase (XOR) for both forms of the enzyme. In mammals, XDH may be converted into XO as a post-translational regulation of XOR activity. This conversion occurs under a variety of pathophysiological conditions, such as the presence of inflammation, during hypoxia/reoxygenation, and ischemia/reperfusion injury, or in presence of liver damage by a viral infection or toxic substances. While XDH prevails intracellularly, XO is the prevalent form in body fluids, such as milk and plasma, where XOR may be secreted or released from dead cells and converted into XO. This transition may take place either irreversibly, by partial proteolysis, or reversibly, by the chemical or enzymatic oxidation of thiol groups [71,73].

3.1.1 Introduction to the structure

The first X-ray crystal structures of xanthine oxidoreductase were reported from bovine milk, in both its dehydrogenase and oxidase forms (utilizing NAD^+ and O_2, respectively, as substrates) by Enroth et al. in 2000 [74]. Mammalian XOR protein is a homodimer of approximately 300 kDa [75], in which each subunit has three domains: from the N-terminus, one domain with a [2Fe-2S] ferredoxin-like cluster, a second domain that possesses FAD (this domain is absent in a little number of bacterias, such as the *Desulfovibrio gigas* [76,77]), and the large, C-terminal domains that bind the active site molybdenum cofactor [74]. The three domains showed proper and different characteristics, in fact, only the NADH oxidase activity takes place in the FAD domain, while the largest domain contains the substrate pocket for XDH (xanthine dehydrogenase), XO (xanthine oxidase), and nitrite reductase activity of XOR (Fig. 5).

Two aspects of the structure of the molybdenum center as inferred from the initial protein crystallography ultimately required revision. First, the portion Mo=S rather than Mo=O was initially considered to occupy the apical position in the molybdenum coordination sphere [76]. This statement was reverted based on a similar magnetic circular dichroism (MCD) between a catalytically relevant Mo(V) species and a model compound possessing an apical Mo=O, and it was concluded that the portion Mo=O should be apical in the enzyme [77]. Then, this conclusion has been confirmed in the crystal structures of other members of this enzyme family, particularly the bacterial CO dehydrogenase [80] and quinoline 2-oxidoreductase [81,82]; it is now generally known that the portion Mo=O is apical in all enzymes of the xanthine oxidase family. Second, the equatorial oxygen ligand was initially thought to be water (H_2O) rather than hydroxide (OH^-), but a subsequent high-resolution X-ray absorption spectroscopy (XAS) has shown that, at a distance of 1.98 Å, the ligand is a hydroxide [83,84].

The two iron-sulfur centers of xanthine oxidase have long been known to be distinguishable based on their EPR signals [85]: The Fe/S I signal has $g_{1,2,3} = 2.022$, 1.932, 1.894, with unexceptional relaxation properties for a [2Fe-2S] cluster, while Fe/S II has $g_{1,2,3} = 2.110$, 1.991, 1.902, with unusually relaxation properties [86,87]. This assignment was possible thanks to site-directed mutagenesis studies with a heterologously expressed rat xanthine oxidoreductase [88], and it was consistent with the known position of the Fe/S I cluster closed to the Moco in the crystal structure [71,89].

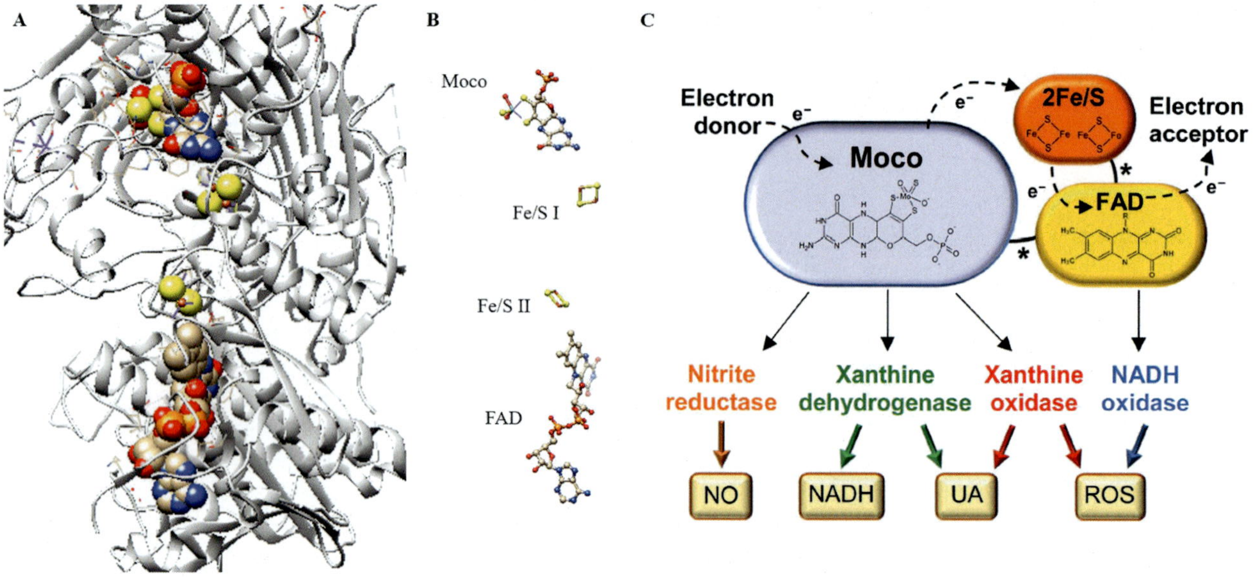

FIG. 5 (A) Section of the monomer of bovine xanthine oxidoreductase (PDB file: 1FO4) with highlighted functional groups in each domain; the minimal backbone in *light gray*. (B) Structures of functional groups showed in the xanthine oxidoreductase, a Moco group, Fe/S I and II, and the FAD co-factor. Distances: 14.7 Å between Moco and Fe/S I, 11.4 Å between Fe/S I and II, and 7.6 Å between Fe/S II and FAD. (C) XOR has two identical subunits, each composed of three domains connected by unstructured hinge regions (indicated by asterisks): the 20-kDa N-terminal domain (*orange*) has two nonidentical iron-sulfur clusters (2Fe/S), the 40-kDa intermediate domain (*yellow*) has a flavin adenine dinucleotide (FAD) cofactor and the 85-kDa C-terminal domain (lilac) has a molybdopterin cofactor containing a molybdenum atom (Moco). The electron (e^-) flux moves from the Moco site, where oxidation occurs, through the two iron-sulfur redox centers toward the FAD site, where the electron acceptor is reduced. The products of XOR activities are uric acid (UA) and reduced nicotinamide adenine dinucleotide (NADH) from xanthine dehydrogenase (XDH), UA, superoxide ion, and hydrogen peroxide (ROS) from xanthine oxidase (XO), nitric oxide (NO) from nitrate and nitrite reductase, and ROS from NADH oxidase [78,79].

3.1.2 *Reaction mechanism*

Xanthine oxidoreductase (XOR) catalyzes the last two reactions of uric acid formation in human purine catabolism. It oxidizes hypoxanthine to xanthine as well as xanthine to uric acid at the molybdopterin center through a concomitant reduction of NAD^+ to NADH at the FAD cofactor domain [3,5,7]. Mechanistically, each enzyme reaction is an oxidative hydroxylation and two electrons are passed from the substrate to the enzyme in each step where the metal passed from Mo(VI) to Mo(IV) [3,5,7]. The enzyme is subsequently reoxidized by NAD^+ or molecular oxygen in a reaction that occurs at the FAD (after electron transfer from the molybdenum center via the iron-sulfur centers) [90]. Normally, NAD^+ is the electron acceptor, forming NADH, but under certain conditions in mammals, the conformational change of the enzyme protein is induced and the enzyme is transformed to xanthine oxidase (XO), of which the principal electron acceptor is molecular oxygen.

The hydroxylation reactions of purine occur at the Moco, composed of molybdenum that binds the pterin ring through two sulfur atoms. A further sulfur atom and two oxygen atoms are coordinated to the molybdenum and are exposed to solvent [90]. One of the oxygen atoms is derived from a water molecule and is subsequently incorporated into the substrate [91]. The general equation of the dehydrogenase reaction can be written as follows:

$$RH + OH^- \rightarrow ROH + H^+ + 2e^-$$

where R may represent various nitrogen-containing cyclic molecules, such as hypoxanthine and xanthine, which are the physiological substrates of the enzyme, as well as aldehydes. As reported in the described reaction, the enzyme receives $H^+ + 2e^-$ from the substrate and this is called the reductive half-reaction. On the other side, the transfer of an electron from the enzyme to NAD^+ or O_2 is called the oxidative half-reaction. First, two electrons are transferred, in the form of hydride (H^-), from the substrate to a sulfur ligand of the molybdenum atom ($Mo(VI){=}S \rightarrow Mo(IV)$-SH), reducing Mo(VI) to Mo(IV); however, Mo(V) is observed only transiently and its involvement has been demonstrated only recently [71,91,92]. In the purine catabolic pathway, the first hydroxylation of hypoxanthine (6-hydroxypurine) initially takes place at the 2-position, obtaining xanthine (2,6-dihydroxypurine). The next hydroxylation occurs at the 8-position, giving urate (2,6,8-trihydroxypurine); this sequence was predicted from spectral observation [93,94]. Electrons transferred to molybdenum are quickly distributed to other reaction centers [95,96], as demonstrated by pulse radiolysis analysis [96]; the slowest step is the dissociation of urate as shown thanks to a stopped-flow assay [97], as reported in Fig. 6.

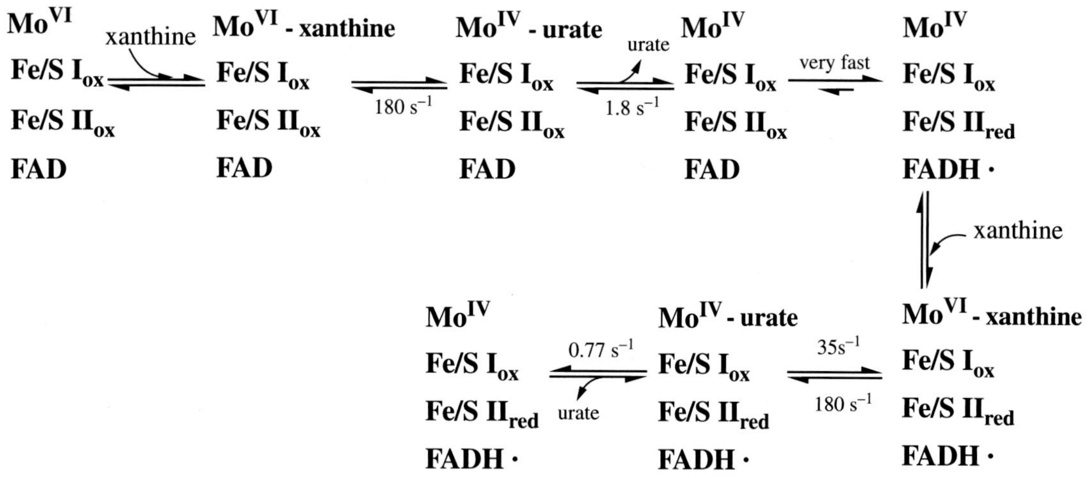

FIG. 6 Schematic diagram of the reaction of molybdenum during xanthine hydroxylation. The results of a fast reaction between xanthine and chicken xanthine dehydrogenase were followed using a stopped-flow method at 4°C. Dissociation of urate (1.8/s) is the slowest step and is rate limiting for the overall process [97].

The structure of the active site of the molybdenum center is reported in Fig. 7, and it is generally accepted that XOR transfers the hydroxyl group and not the Mo=O ligand of molybdenum to the substrate [99]. The free electron pairs of oxygen can attack the electrophilic carbon to obtain the hydroxylated product, then the Mo=S group accept the hydride ion from the reactive carbon atom, affording Mo-SH and a reduced molybdenum center. The geometry of the molybdenum-coordinated atoms was shown through X-ray crystal structure analysis of the complex with FYX-051, an inhibitor, which can slowly react with the substrate and forms a quite stable reaction intermediate with the Moco. Regarding the role of charged residues in the active center, the research group of Yamaguchi et al. in 2007 found that the enzyme activity was decreased significantly upon mutation of Glu803 and Arg881 [98], in fact, proposed binding modes of substrates hypoxanthine and xanthine suggested that the nucleophilic

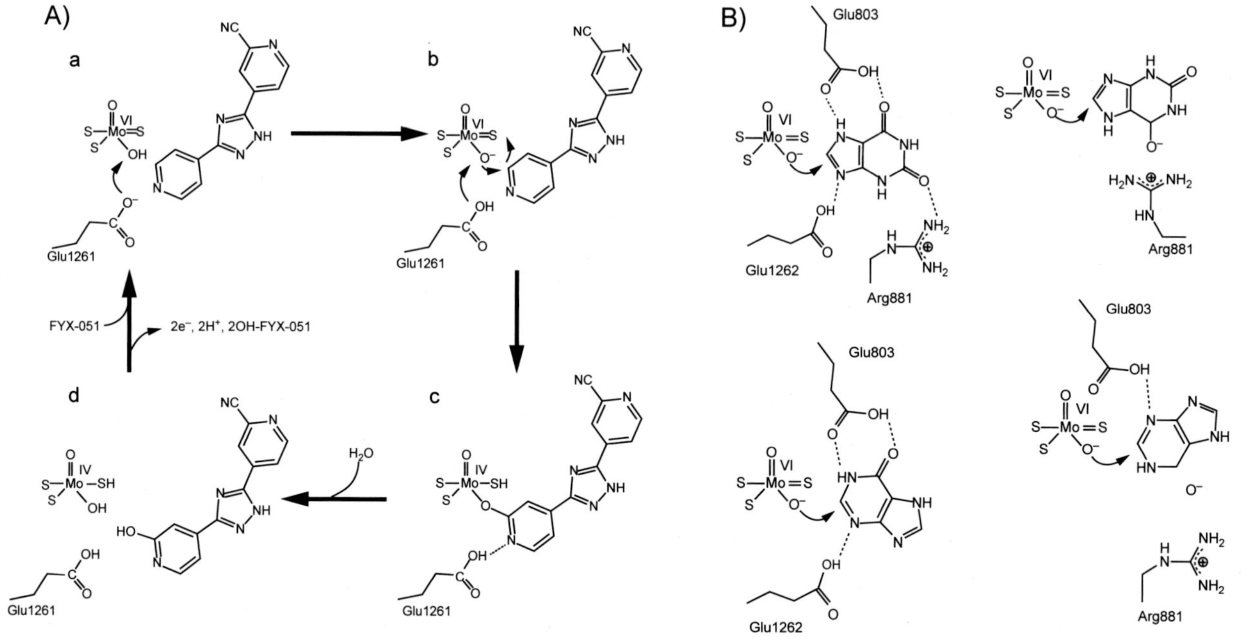

FIG. 7 (A) Structure of mammalian molybdopterin and hydroxylation Mechanism of Artificial Slow Substrate FYX-051. Glu1261 works as a base, abstracting the proton from Mo-OH (a). The generated Mo-O^- nucleophilically attacks the carbon center to be hydroxylated, with concomitant hydride transfer (b). The protonated Glu1261 forms a hydrogen bond to the N1-nitrogen of the substrate, and this facilitates the nucleophilic attack (b). The reduced Mo(IV) coordinated with the product via the newly introduced hydroxyl group (c). The intermediate breaks down by hydroxide displacement of the product (d) [92]. (B) Proposed model of xanthine binding mode based on the analysis of mutant enzymes [98].

reaction is facilitated by hydrogen bonds between Glu803 and Arg881 and the substrate.

These binding modes are consistent with the metabolic sequence, hydroxylation at the 2-position of hypoxanthine precedes that at the 8-position, in fact, as shown in Fig. 7, the interaction between the keto group (C=O) in 2-position and Arg881 result important for efficient hydroxylation at the 8-position. This was demonstrated by the report that mutation of the corresponding arginine to glutamate in *Aspergillus nidulans* XOR resulted in a loss of xanthine hydroxylation activity [100]. Finally, a water molecule, located between atoms 3N and 9N of xanthine, assists in the release of the urate product [101].

3.1.3 Physiological role

Xanthine oxidoreductase (XOR) is a widely distributed enzyme that has been studied for more than 100 years, largely because of its presence in cows' milk, from which it is readily available on a large scale. Despite many decades of research into XOR, the human enzyme has only relatively recently been characterized, purified from breast milk using an affinity purification on heparin, essentially as for cows' milk XOR [102].

Xanthine oxidoreductase is highly expressed in some tissues, such as the breast, kidney, gut, and liver, and the normal subcellular localization is the cytosol, but it was also found in peroxisomes [103]; however, XOR can be found in extracellular compartments, such as milk and blood. Particularly, serum XOR derives from the physiological hepatic cell turnover, which induces the release of liver enzymes from dead cells into the circulation, in fact, levels of serum XOR can increase in presence of several liver pathologies [104]. The production of human XOR is regulated at both transcriptional and posttranslational levels. The basal activity of the human XOR promoter is minimal if compared with other mammals, probably due to repressor elements identified in noncoding regions of XOR [105]. The expression of the human XOR gene (hXOR) is usually subjected to downregulation in almost every tissue except for the epithelial cells of lactating breast, gastrointestinal tract, kidney, and liver [106].

In the breast, in all mammals, milk-fat globule secretion during lactation is mediated by the clustering of the transmembrane protein, named butyrophilin1A1 (Btn). XOR contributes to apocrine lipid secretion by inducing apical membrane reorganization, thus allowing the Btn clustering and the membrane docking of milk-fat droplets [107,108]. In milk, XOR can produce hydrogen peroxide (H_2O_2) and nitric oxide (NO), which lactoperoxidase utilizes to form hypothiocyanite and nitrogen dioxide to counteract the growth of bacteria, thus protecting the breast from mastitis. Nevertheless, this bactericidal action in milk leaves unharmed the commensal flora of the neonatal oral cavity, stomach, and intestine, thus regulating the intestinal microbiome. For all these reasons, Xanthine oxidoreductase shows essential features for the regular growth of the mammal newborn [109,110].

In the gut, expression of XOR is high in intestinal epithelial cells, where a microbicidal role has been proposed. The frequent turnover of enterocytes ensures an abundant presence of XOR in the intestinal lumen where XOR-derived reactive oxygen species (ROS) and reactive nitrogen species (RNS) exert a protective activity against some infections saving the commensal microbiome [111].

In the kidney, most purine catabolism occurs; in this district, XOR is responsible for uricosuria and consequently for uricemia levels, which in turn contributes to supporting blood pressure by upregulating cyclooxygenase-2 (COX-2) expression and consequently the renin/angiotensin pathway. XOR oxidant products can also contribute to keeping sterile the urinary tract [112].

In the liver, XOR carries out all its activities and, in addition to purines, metabolizes a lot of endogenous and exogenous substrates, including drugs. The uric acid produced by XOR influences the hepatic metabolism of glucose and lipids and increases gluconeogenesis and fat accumulation [113].

Finally, serum XOR mainly derives from the physiological hepatic cell turnover, which induces the release of liver enzymes from dead cells [114]. Circulating XOR can bind endothelial cells, thus promoting endothelial activation during inflammation, modulating vascular tone, and consequently contributing to blood pressure regulation [115]. However, the situation is far from clear, in fact, circulating XOR, with its capacity to generate ROS, can be viewed as potentially pathogenic. It can bind to glycosaminoglycans on the surface of vascular endothelial cells [116,117] and, thus concentrated, initiate oxidative damage in distal organs [118] (Fig. 8).

3.1.4 Xanthinuria type I

Deficiency of XOR results in the accumulation of xanthine in urine (xanthinuria) with hypoxanthine being also elevated. Although inherited XOR deficiency was first reported in 1954 [119], a detailed analysis of mutation sites of XOR was first reported in 1997 [120] and, subsequently, there have been several reports on XOR protein mutations associated with xanthinuria. The incidence of XOR deficiency has been reported to be 1/69,000, but SNP analysis suggested a higher frequency of mutation in XOR, possibly because most mutations do not cause dysfunction, being asymptomatic [121,122]. Hereditary type I xanthinuria lacks only XOR activity, while type II and molybdenum cofactor deficiency (MoCD) lack two or more Mo enzyme activities, respectively. Type I xanthinuria patients show mutations in the XDH gene 2p22, particularly, mutation

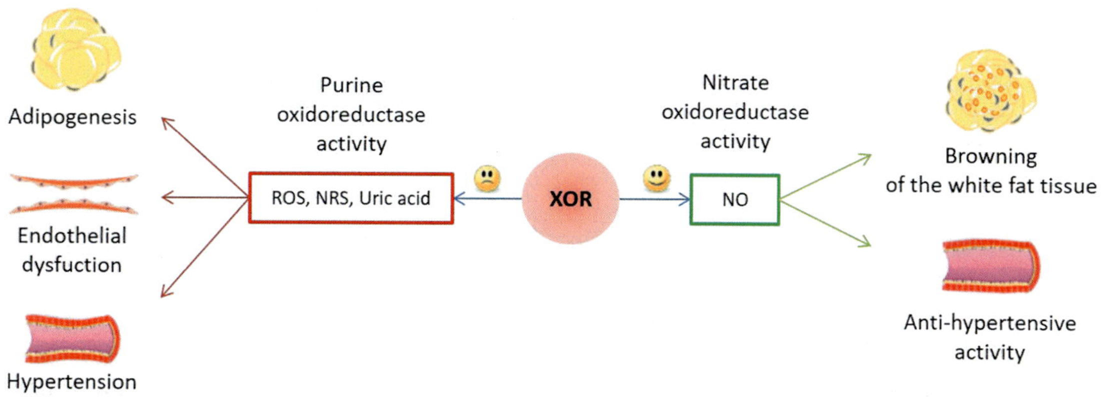

FIG. 8 Effects of xanthine oxidoreductase (XOR) activities on the vasculature and adipose tissue. The negative effects of XOR-purine oxidoreductase activity are adipogenesis, inflammation and hypertension. Positive effects of XOR-nitrate reductase activity are browning of the white fat tissue and anti-hypertensive activity.

of Arg149Cys at the Fe/S I cluster motif may influence the formation of the cluster [123], resulting in loss of electron transfer, and the mutation of Thr910 (located at a distance of 7.3 Å from Mo=S in the molybdenum center) to a bulky methionine or lysine residue seems likely to result in the loss of Moco or its sulfur atom, which is essential for the activity [124]. Alternatively, insertion of the lysine residue may change the electrostatic environment in the active center cavity. The affected individuals by xanthinuria type I may develop urinary tract calculi, acute renal failure or myositis due to tissue deposition of xanthine, while some subjects remain asymptomatic.

3.1.5 *Xanthinuria type II*

Xanthinuria type II is characterized by the simultaneous loss of two Mo-enzyme activities, XOR and AOX [120]. The molecular basis of this dual loss of Mo-enzyme function is due to a mutation in the MCSU gene located on chromosome 18q12.2, encoding for a two-domain protein catalyzing the sulfuration of Moco in enzymes of the XO family [125]; other mutants, such as Ala156 to Pro [126] and Arg776 to Cys [127], were reported to cause xanthinuria type II. Patients with Xanthinuria type II develop symptoms, which are caused by the deposition of xanthine in the urinary tract. This often results in hematuria or renal colic and acute renal failure or chronic complications related to urolithiasis [124]. Furthermore, in a little number of cases, muscle pains caused by xanthine deposition have been reported [127]. In one case with hereditary xanthinuria type II the association with a mental delay, autism, cortical renal cysts, osteopenia, hair and teeth defects, and various behavioral symptoms was observed [128]. Both types result in low plasma uric acid levels and elevated plasma xanthine concentrations, but patients with type I retain their ability to metabolize allopurinol (via the activity of AOX), while those with type II xanthinuria cannot.

3.1.6 *Xanthinuria type III*

Type III xanthinuria involves a triple deficiency of xanthine oxidoreductase, aldehyde oxidase, and sulfite oxidase, due to a defect in the synthesis of the molybdopterin/pyranopterin cofactor common of all three enzymes; it is the so-called molybdenum cofactor deficiency. This extremely serious condition involves several problems associated with the central nervous system and, except in very rare cases, it usually presents with neonatal feeding problems, neonatal seizures, increased or decreased muscle tone, ocular lens dislocation, severe intellectual disability, and death in early childhood. Type III xanthinuria can arise from functional mutations in any of 3 genes: MOCS1 (encoding 2 enzymes for synthesis of the precursor via a bicistronic transcript), MOCS2 (encoding molybdopterin synthase), or GPHN (encoding gephyrin), located at 6p21.2, 5q11.2, and 14q23.3, respectively [129].

3.1.7 *Hyperuricemia*

Hyperuricemia is characterized by an accumulation of uric acid levels in the blood (over 7 mg/dL). The major cause of the disease is seen in an imbalance between the rates of production and excretion of uric acid. Elevated uric acid can also be seen in accelerated purine degradation, in high cell turnover states (hemolysis and rhabdomyolysis), and in decreased excretion (renal insufficiency and metabolic acidosis) [130]. A decreased renal excretion of uric acid takes to primary hyperuricemia, while causes of secondary hyperuricemia are more complex and include several factors: increased XOR activity, increased purine release due to chronic inflammation, increased purine synthesis, and high dietary intake of purines [130]. Longstanding hyperuricemia can lead to gout (deposition of urate crystals in the joints and periarticular structures) and nephrolithiasis; furthermore, it has also been implicated as an indicator for diseases like metabolic syndrome, diabetes mellitus, cardiovascular disease, and chronic renal disease [130].

3.1.8 Xanthine oxidoreductase inhibitors (XOI)

Xanthine oxidoreductase is the target of drugs for the treatment of hyperuricemia and gout [131]. Inhibition at the Moco center is important from a therapeutic viewpoint because many proteins containing flavin or iron-sulfur cofactors exist in the human body, and if they were also inhibited, there might be several side effects. Theoretically, as inhibitors of xanthine oxidoreductase that target the Moco center, they can also inhibit the formation of ROS generated by the enzyme, so they might also be effective in the treatment of other diseases, particularly, in the prevention of cardiovascular events, as reported by Bredemeier et al. in 2018 [132] in their meta-analysis. However, there is contradictory evidence regarding the possible cardiovascular protective effect exerted by XOI. Xanthine oxidoreductase inhibitors belong to two classes: (I) purine analogs, such as allopurinol, oxypurinol, or tisopurine and (II) others, such as febuxostat and topiroxostat.

Allopurinol ($C_5H_4N_4O$, MW: 136.11), approved in 1966 by FDA, is the archetypal and longest established XO inhibitor in clinical use. It was initially synthesized as an attempt to produce new antineoplastic agents, but it was found to have inhibitory activity on XO, reducing both urinary and serum uric acid levels. It is a weak acid with a dissociation constant (pK_a) of 9.4. It is rapidly converted to **oxypurinol**, or so-called alloxanthine, by aldehyde oxidoreductase or xanthine oxidase. Allopurinol is considered a kind of suicide inhibitor because it is almost completely metabolized to oxypurinol by its target within 2h of oral administration, whereas oxypurinol is slowly excreted by the kidneys over 18–30h. For this reason, oxypurinol is believed responsible for the majority of allopurinol's effects [131]. Common adverse effects of allopurinol are a gastrointestinal disturbance, hypersensitivity reactions and skin rash [133].

Tisopurine or thiopurinol (5H-Pyrazolo[3,4-d] pyrimidine thione-4), a thio-analog of allopurinol, has a molecular formula of $C_5H_4N_4S$ (MW: 152.18), and it is applicable for treatment of gout [134]. It is known as an antiparasitic and antithrombotic agent having very similar pharmaceutical and biomedical behaviors to allopurinol counterparts, in fact, potential inhibitor activity against xanthine oxidase was reported by Robin et al. in 1985 [135].

Febuxostat, IUPAC name 2-(3-cyano-4-[2-methylpropoxyl]phenyl)-4-methylthiazole-5-carboxylic acid, is a thiazolecarboxylic acid derivative, selective for inhibition of both the oxidized and reduced forms of xanthine oxidase and does not resemble a purine or pyrimidine [136]. Febuxostat has a selective affinity for both the oxidized and reduced forms of xanthine oxidase, with an in vitro inhibition (K_i) value of <1 nM [137], but it is generally recommended only for people who cannot take allopurinol due to the increased risk of death. Common side effects include liver problems, nausea, joint pain and a rash.

Topiroxostat, IUPAC name [4-[5-(4-Pyridinyl)-1H-1,2,4-triazol-3-yl]-2-pyridinecarbonitrile] is a non-purine XOR inhibitor approved in Japan in 2013 for the treatment of hyperuricemia; unfortunately, there are limited international experiences with this agent. Topiroxostat behaves initially as a competitive-type inhibitor to xanthine oxidoreductase before forming a strong covalent linkage to molybdenum via oxygen in the hydroxylation reaction intermediate [138]. It also displays a potent noncovalent competitive-type inhibition of XOR with a K_i value of 5.7 nM [139]. The structures of the commercially available drugs are reported in Fig. 9.

Because purine-based analogs can be converted by hypoxanthine-guanine phosphoribosyltransferase and orotate phosphoribosyltransferase to nucleotide analogs, which can interfere with normal purine metabolism and pyrimidine synthesis in the human body, which can lead

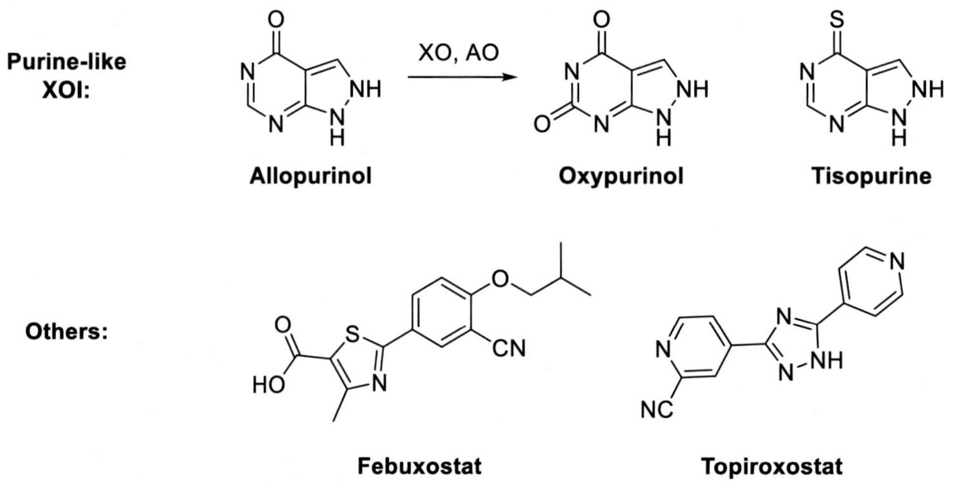

FIG. 9 Molecular structures of purine-like and other drugs as commercially XOR inhibitors.

to severe life-threatening adverse effects, like fulminant hepatitis, Stevens-Johanson syndrome, and chronic renal failure [140], the global research switched toward non-purine analogs.

3,4-Dihydroxy-5-nitrobenzaldehyde, or DHNB, is a new and powerful time-dependent XOI with a mechanism of action still under investigation, but that seems to be similar to allopurinol. Some studies have suggested that DHNB has low toxicity and enhances therapeutic efficacy, even at low doses, in co-administration with allopurinol. Moreover, DHNB has a direct antioxidant capacity and reduces the production of free radicals and ROS, to limit cell damage. However, while it appears to be extremely safe and effective in mice, no human data are available yet [141].

BOF-4272 is a classic example of a potent anti-hyperuricemic agent in animal models, but, in humans, its efficacy differs from person to person due to the variation in hepatic metabolism which limits its clinical use [142].

Various research groups reported potent XO inhibitors of natural origin. Among them, flavonoids were found to show excellent inhibitory potential against the XO enzyme [143–145]. **Quercetin** is one of them that even reaches clinical studies for the treatment of hyperuricemic conditions. The results indicated that daily supplementation of quercetin (500mg), for 4weeks significantly reduces elevated plasma uric acid levels in healthy males without affecting blood pressure, fasting glucose and urinary excretion [146]. Its XO inhibitory potential was first reported by Nagao et al. in 1999 [147] and, recently, a Phase I clinical study completed in 2014, reported that its chronic consumption (500mg tablet) does not affect the blood pressure and chemical composition of blood and urine, suggesting that the use of Quercetin should be safe [148].

KUX-1151, developed by Kessei pharmaceutical Co., is a dual inhibitor of XO and URAT1. It can reduce the production of uric acid as well as excrete out its excess amount from the body. The compound has entered the clinical trial Phase II.

LC350189 is a non-purine-based selective XO inhibitor, developed by LG Life Sciences, Korea. The in vitro efficacy of the compound is comparable to Febuxostat, but it resulted in more safety after the clinical trial phase I, which was terminated in 2013.

3.2 Aldehyde oxidase

The vertebrate aldehyde oxidases (AOX) are closely related to the xanthine oxidoreductases described above, sharing very similar overall protein architectures, overlapping substrate specificities, and the requirement for a catalytically essential Mo=S group in the molybdenum coordination sphere. AOX enzymes (EC 1.2.3.1) originate from a duplication of the *XDH* gene in eukaryotes before the origin of multicellularity [149], in fact, even AO is a cytosolic molybdoflavoprotein that utilizes molybdenum (Mo) and flavin adenine dinucleotide (FAD) for its catalytic activity in oxidizing and/or reducing substrates [150]. The idea of the gene's duplication is supported by the fact that vertebrate *AOX* and *XDH* genes are characterized by very similar exon structures and by strict conservation of the exon/intron junctions. Structurally, AOX and XOR are very similar in their overall topology and share around 50% of their amino acid sequence [72]. As XOR, the AOX family exists as a homodimer in its catalytically active form with two identical subunits of about 150kDa. Each monomeric subunit consists of three conserved domains that are connected by two linking regions: the small 25-kDa N-terminal domain contains two non-equivalent and spectroscopically distinct iron-sulfur (2Fe-2S) centers, a 45-kDa central domain accommodates a FAD-binding site and an 85-kDa carboxyl-terminal domain harbors the Moco [151,152]. As for the mechanisms of catalysis, XDH can use both NAD^+ and molecular oxygen as the final acceptors of the reducing equivalents generated during substrate oxidation depending on the enzyme being under the dehydrogenase or oxidase form, but, in contrast, the classic reactions catalyzed by AOXs produce electrons, which reduce only molecular oxygen and generate superoxide anions or hydrogen peroxide. A typical AOX catalytic cycle is represented in Fig. 10.

Mammals are characterized by a maximum of four *AOX* genes coding for a corresponding number of AOX isoenzymes. The two extremes are represented by humans, which express a single AOX1 protein, and rodents, such as mice

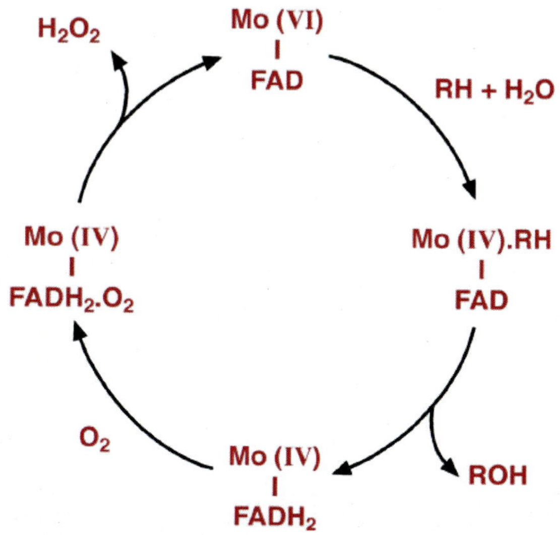

FIG. 10 The catalytic cycle of the AOX enzymes is schematically represented. A generic type of substrate, RH, is oxidized to the corresponding product, ROH with concomitant reduction of Mo(VI) to Mo(IV). The electrons generated are subsequently transferred to FAD, with the production of $FADH_2$, and molecular O_2 which is the final electron acceptor giving rise to H_2O_2 [72].

and rats, which have the four AOX1, AOX2, AOX3, and AOX4 isoenzymes [153]. The presence of a single AOX isoform in humans and between 1 and 4 AOX isoenzymes in different mammalian species is due to species-specific gene deletion or pseudogenization events, which may have been associated with the necessity to acquire new physiological functions.

The first data on the structure of mammalian AOXs derive from the crystallization of the native mouse liver (PDB ID: 3ZYV; 2.9 Å resolution) [154,155], after that, very recently, the crystal structures of human AOX1, in its substrate-free form (PDB ID: 4UHW; 2.6 Å resolution), as well as in complex with the substrate, phthalazine, and the inhibitor, thioridazine (PDB ID: 4UHX; 2.7 Å resolution), were solved.

3.2.1 Physiological role

Except for the involvement of human AOX1 in xenobiotic metabolism and inactivation of environmental pollutants characterized by potential toxicity, the physiological functions, substrates, and products of mammalian AOXs are still largely unknown or just hypothesized; however, a careful analysis of the data available on the tissue- and cell distribution of human AOX1 and mouse AOX1, AOX2, AOX3, and AOX4 is a prerequisite to formulate hypotheses on the role played by AOXs in homeostatic processes other than xenobiotic metabolism.

The richest source of human AOX1 protein is represented by the liver, although elevated amounts of the protein were found in respiratory, digestive, urogenital, and endocrine organs. Particularly, in the respiratory tract, the epithelial component of the trachea and bronchi contains high levels of protein, as well as the epithelial cells of the small and large intestines [72]. Large amounts of the AOX1 protein were even found in the adrenal gland cortex, and this observation is consistent with a potential role of AOX1 in the steroid hormone biosynthetic pathway. It is also involved in various metabolic pathways such as neurotransmitters, the conversion of the hydroquinone-precursor aldehyde into gentisate, the catabolism of valine, leucine or isoleucine as well as vitamins (nicotinamide and pyridoxal) [156], as well as a drug-metabolizing enzyme against different xenobiotics, such as the antitumor agents, methotrexate [157], and 6-mercaptopurine [158], the antidepressant, citalopram [159], and other compounds of medical relevance.

The largest amounts of mouse AOX1 mRNA are present in the inner ear and the seminal vesicles, whereas a little presence was observed in the liver. It was also detectable in the central nervous system [160], especially in the epithelial component of the choroid plexus, where AOX1 may play a role in the absorption/secretion of the cerebrospinal fluid. Interestingly, the major sources of human and mouse AOX1 are not overlapping with the notable exception of the liver.

Similar to mouse AOX1, mouse AOX3 expression is also relatively restricted, as significant amounts of the corresponding isoenzyme are found only in the oviduct, testis, lung, and liver, where AOX3 and AOX1 are coexpressed [161]. In mouse liver, both are synthesized by the hepatocyte, but it is unknown if the same hepatocytic population is responsible for the synthesis of these two isozymes.

Mouse AOX4 mRNA is very abundant in the fertilized ovum, the inner ear, the tongue, and beyond the oral cavity, while new studies confirmed the presence of AOX4 in the digestive tube and skin [162]. However, recent results indicate that, by far, the richest source of AOX4 is represented by a largely overlooked organ present in most vertebrates, but not in humans and many other primates, the Harderian gland. In this organ, approximately 2% of all the cytosolic proteins consist of AOX4. The Harderian gland is a large exocrine gland located behind the eye bulb and characterized by the secretion of a lipid-rich fluid, which is believed to serve for the lubrication of the eye surface and the fur [163,164], where it acts as a thermic isolator to contribute to the regulation of the body temperature.

AOX2 is the mouse AOX characterized by the most restricted pattern of expression. Indeed the corresponding transcript is detectable only in the nasal cavity [161]. In this location, high levels of protein are expressed in the Bowman's gland, which is the principal exocrine gland present in the submucosal layer, and it is responsible for the production of the mucous fluid excreted in the nasal cavity. It is speculated that AOX2 may play a role in olfaction and it could explain the much more developed olfactory function of rodents and other mammalian species than humans, where the *AOX2* gene is inactivated.

Considering higher plants, there are between two and four genes encoding aldehyde oxidases [151], enzymes that catalyze the final steps in the biosynthesis of important plant hormones. For example, in *A. thaliana* there are four genes, AOX1–4 [165,166], with AOX3 specifically involved in abscisic acid biosynthesis, whereas in *Pisum sativum*, there are three genes encoding aldehyde oxidases, Psaox1–3 [167].

3.2.2 AOX inhibitors

AO is subject to inhibition by xenobiotics reversibly and irreversibly. Reversible inhibitors of AO span a wide range of chemical space with various sizes and functional groups covering many classes of drugs. In 2004, Obach et al. [168] investigated more than 200 drugs and other xenobiotics for their potential to inhibit AOX in cytosolic fractions. The results of this study revealed that around 36 compounds of certain drug classes were potent AO inhibitors. These included estrogenic compounds, phenothiazines, tricyclic antidepressants, tricyclic atypical antipsychotic agents, and dihydropyridine calcium channel blockers. Some compounds that could not be classified into a particular class also inhibited AO [168]. Up to now, raloxifene, of the estrogen compound class, is the most potent inhibitor, showing an IC_{50} of 2.9–5.7nM. A closer look into the structure-inhibitor relationship suggests that the phenolic groups in raloxifene are important for inhibitory activity. Likewise, the 3-aroyl-p-ethoxypiperidinyl substituent is also important for potency. Molecular docking studies using mAO3 by Coehlo et al. with raloxifene suggest two different binding modes for this molecule: (I) one of the phenol moieties enters into the pocket and interacts via the hydrogen bond with Glu1266 and aromatic stacking interaction with Phe919; (II) the piperidinyl group replaces the phenol and the molecule is held together by van der Waals interactions between the piperidinyl group of the inhibitor and Phe919 in the pocket [154]. Obach et al. have shown that several phenothiazine drugs inhibit AO with an IC_{50} ranging from 33nM to 1.6μM. Perphenazine was identified as the most potent inhibitor among all the phenothiazine tested and displayed an IC50 of 33nM [168]. The crystal structure of hAO complexed with phthalazine and thioridazine has revealed the molecular details of the binding site for the phenothiazine class [169]. The complex has shown that thioridazine binds near the interface in a groove at the enzyme surface which is 34Å away from the active site. The binding site in hAO contains a proline residue (Pro576) that confers extra flexibility to the loop in the binding pocket and allows inhibitor access. The authors proposed that this is a phenothiazine binding site and that possibly all inhibitors belonging to the phenothiazine class will bind to this site and inhibit the enzyme in a noncompetitive manner [169].

A wide range of natural components including polyphenols and flavonoids are known to interfere with AO-catalyzed reactions. The effects of 12 flavonoids from three subclasses of flavon-3-ol, flavan-3-ol, and flavanone on the oxidation of vanillin and phenanthridine by AO were studied by Pirouzpanah et al. and Hamzeh-Mivehroud et al. Among the flavonoids studied by the two groups, quercetin, myricetin, and genistein were the most potent inhibitors of AO, whereas naringin displayed negligible inhibition [170,171]. In a similar manner, Siah and co-workers investigated the inhibitory effects of several phenolic compounds from three subclasses of aurone, flavanone, and phenolic lactone compounds on the activity of AO [172]; among the phenolic compounds tested, ellagic acid was the most potent agent with higher inhibitory. In Fig. 11, are shown different AO inhibitors belonging to different but just mentioned classes.

Only for knowledge, a few time-dependent inhibitors (TDIs) have been found and recognized, such as hydralazine, quinazoline, and cyanide [173–175].

3.3 Sulfite oxidase

Sulfite oxidase is a Mo-enzyme ubiquitous among animals, plants, and humans, and it catalyzes the physiologically vital oxidation of sulfite to sulfate, the terminal reaction in the oxidative degradation of the sulfur-containing amino acids cysteine and methionine. In mammals, it involves ferricytochrome *c* ((cyt c)$_{ox}$) as the physiological electron acceptor:

$$SO_3^{2-} + H_2O + 2(\text{cyt c})_{ox} \rightarrow SO_4^{2-} + 2(\text{cyt c})_{red} + 2H^+.$$

In vertebrates, SO is localized in the mitochondrial intermembrane space where electrons derived from sulfite oxidation are directly passed to the physiological electron acceptor cytochrome c [176].

3.3.1 Structure and function

The eukaryotic sulfite oxidases differ fundamentally from the xanthine- and aldehyde-utilizing enzymes described above in two regards: (I) their molybdenum centers possess two Mo=O groups and a cysteinyl ligand to the metal, LMo(*VI*)O_2(S-Cys), and (II) the chemistry catalyzed is a simpler oxygen atom transfer reaction rather than a carbon center hydroxylation involving cleavage of a C—H bond. The first X-ray crystal structures were derived from a vertebrate, a chicken [177], and a plant, *Arabidopsis thaliana* [178]. Considering first the chicken enzyme, each subunit of the homodimer consists of an N-terminal heme-containing domain with structural homologies to vertebrate cytochromes b5, followed by a 12–15 residue long tether connecting the heme domain to the molybdenum-containing domain, which shows a mixed α/β fold unique to this family of proteins, and, finally, a C-terminal domain that constitutes the dimer interface and possesses a β-sandwich fold. As in the xanthine oxidase family of molybdenum enzymes, the redox-active centers of one subunit are well-separated from each other, furthermore, the two heme domains do not occupy the same orientation relative to their respective molybdenum subunits [177]. In contrast to all vertebrate sulfite oxidases, the *A. thaliana* enzyme lacks a heme domain [179]. Considering the active site of chicken, in its reduced form, in addition to an LMo(*IV*)O(OH)(S-Cys185) molybdenum center (with an elongated equatorial Mo—O bond), a group of highly conserved residues consisting of Arg138, Arg190, Trp204, Tyr322 and Arg450 surrounds a bound product sulfate molecule and clearly defines the substrate binding site. The sulfate is positioned near the equatorial oxygen of the molybdenum coordination sphere [177]. The corresponding active site residues of the (oxidized) plant enzyme are Arg 51, Arg 103, Trp 117, Tyr 241, and Arg 374 [178]. Subsequent crystallographic studies of the recombinant chicken enzyme in the absence of substrate have demonstrated that Arg450

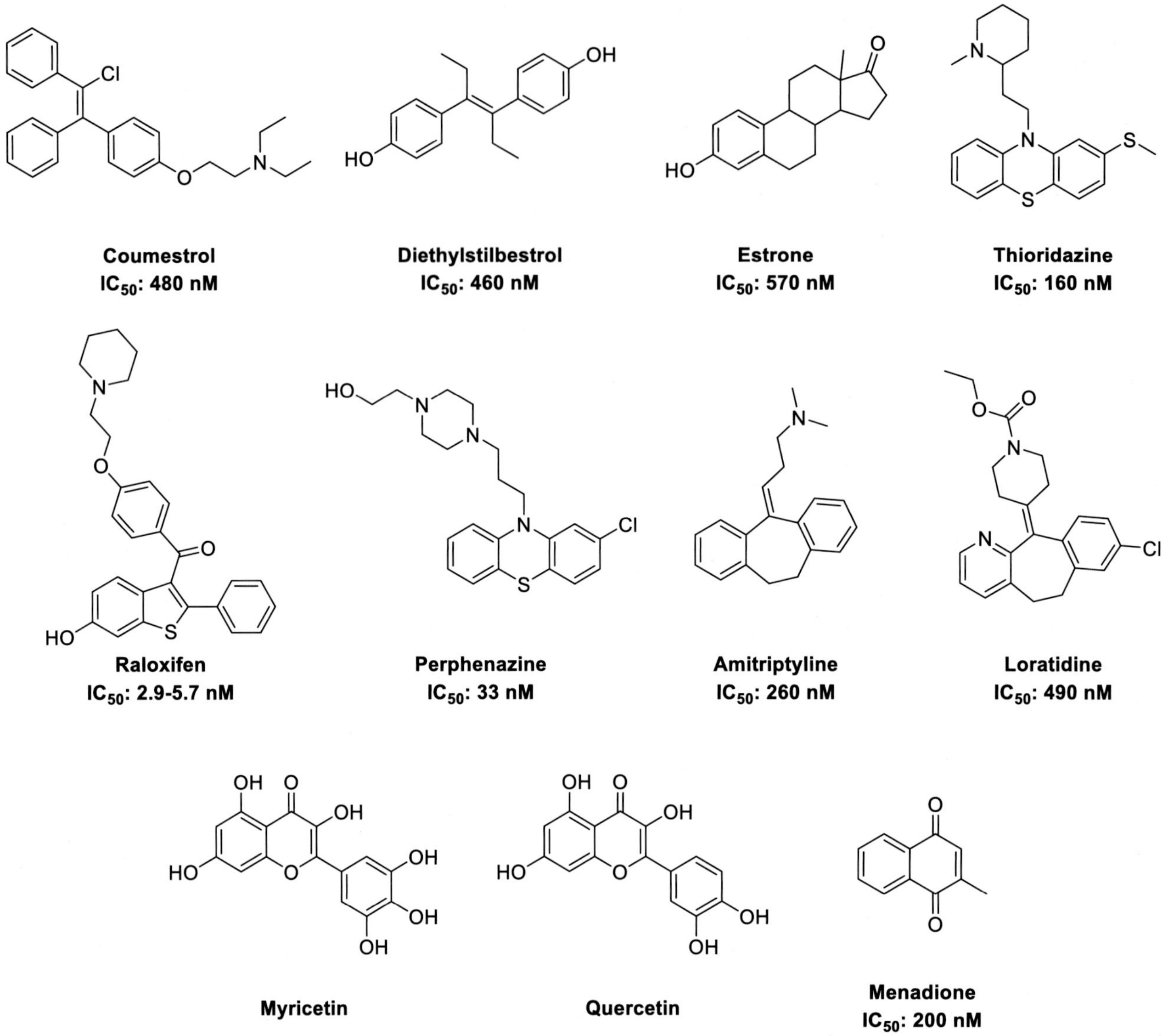

FIG. 11 AO reversible inhibitors belong to different classes.

does indeed swing away from the substrate binding site in the absence of bound sulfate [180]. without sulfate, chloride occupies the binding site, interacting with Arg190 and Trp204. Unexpectedly, in the protein R138Q mutant, Arg450 faces into the substrate binding cavity, even in the absence of sulfate: this appears to represent a "blocked" form of the enzyme. Considering the active site of the SO, it is ideally suited for binding anions such as the substrate sulfite and the product sulfate, in fact, all four oxygen atoms of the sulfate are involved in a dense hydrogen bond network. This coordination geometry might allow a nucleophilic attack of the sulfite lone-pair electrons to the equatorial oxo group of Moco, as has been proposed by Hille et al. in 1994 [181]. According to the postulated mechanism, a bidentate intermediate would be formed after the nucleophilic attack, with the fifth ligand being coordinated with Mo and the substrate. After that, sulfate would be released from the active site and a hydroxide ion from the solvent would be coordinated to the Moco. This completes the reductive half-cycle of the reaction, in which Mo is reduced from (VI) to (IV), while sulfite is oxidized to sulfate. In the oxidative half cycle, two electrons are transferred sequentially from the Mo center to the heme in the cytochrome *b*5 domain and from there are passed to cytochrome *c*. The electron transfer and the concomitant deprotonation of the Mo-bound hydroxyl group are necessary to reestablish the initial Mo(VI) state [181]. A surprising result of this structure analysis is the long distance (about 32 Å) between the Moco and the heme Fe within the same monomer. Based on current models for long-range electron transfer [182], the observed rates for electron transfer and turnover of the substrate are quite fast [183], considering the large intermetallic distance. This implies that the two redox centers should be connected by a very efficient electron-transfer pathway. One possible pathway involves Arg-138, which forms a hydrogen bond with the water/hydroxo ligand. From Arg 138, the electron could be transferred via the main chain atoms of β strand β8 to Leu-134, which is located at the interface between domains I and II and is in van der Waals contact to Phe-58, which itself is close to the heme; the pterin ring does not seem to be part of this electron transfer pathway, since it is not oriented toward the heme. Observing similar Moco enzymes, such as aldehyde oxidoreductase from *Desulfovibrio gigas* and formate dehydrogenase from *Escherichia coli* [184,185], another possibility considers the different conformation of the cytochrome *b*5 domain in solution, which could be oriented toward the Moco. However, additional experiments are necessary to confirm this last vision.

3.3.2 Sulfite oxidase deficiency

Sulfite oxidase deficiency is a sulfur metabolic disorder that results in profound birth defects, severe neonatal neurological problems and early death, without effective therapies known [186]. The inborn error is characterized by dislocation of ocular lenses, mental retardation, and, in severe cases, attenuated growth of the brain [187]. These severe neurological symptoms result from either point mutations in the SO protein itself (so-called SO deficiency, in which only SO activity is affected) or the inability to properly produce the pyranopterin-dithiolate cofactor, which results in deficiencies in all Mo-containing enzymes (so-called Moco deficiency) [48]. The biochemical basis of this pathology is unclear yet. Fatal brain damage may be due to the accumulation of a toxic metabolite, possibly SO_3^{2-}, which is a strong nucleophile that can react with a wide variety of cell components. It was reported that sulfite reacts with protein disulfides to form sulfonated cysteine derivatives and since the integrity of disulfide bonds is crucial to the tertiary structure and thus protein function, the disruption of protein structure may result in altered cellular activities leading to biochemical lesions [188,189]. On the other side, a deficiency in the reaction product (sulfate, SO_4^{2-}) may disturb normal fetal and neonatal development of the brain [190].

3.4 Nitrate reductase

Nitrate reductase (NR; EC 1.7.1.1–3) catalyzes the following reaction, the first and rate-limiting steps of nitrate assimilation in plants, algae, and fungi [191]:

$$\begin{aligned} &NO_3^- + LMo(IV)O(OH)(S-Cys) + NADH \\ &\quad \rightarrow NO_2^- + LMo(VI)O_2(S-Cys) + NAD^+ + OH^-. \end{aligned}$$

This enzyme is a member of the sulfite oxidase family of molybdenum enzymes and has the same square-pyramidal $LMo(VI)O_2(SCys)$ Moco in the oxidized state. NR occurs in three different forms: NADH-specific forms are frequently present in higher plants and algae, NADPH-specific forms are unique in fungi, and NAD(*P*)H-bispecific forms are found in all aforementioned organisms, being most common in fungi [192]. The catalytic cycle of NR can be divided into three different steps: (I) a reductive half-reaction in which NAD(*P*)H reduces FAD, (II) electron transfer via the intermediate cytochrome *b*5 domain, and (III) an oxidative half-reaction in which the Moco transfers its electrons toward nitrate (NO_3^-), thereby forming nitrite (NO_2^-) and hydroxide (OH^-) [193]. NR is an unstable protein with a half-life of a few hours and its degradation is influenced by environmental conditions [194].

3.4.1 Introduction to the structure

Nitrate reductase from *A. thaliana*, as well as from other higher plants, is a homodimer of 2 × 110kDa. Each subunit contains a large N-terminal portion (~59kDa) that possesses the Moco, a small central domain possessing a b-type cytochrome (14kDa), and a C-terminal domain containing FAD and the NADH-binding site (24kDa) [195]. In the course of turnover, NADH introduces reducing equivalents into the enzyme at its FAD site, and these are subsequently transferred via the heme to the Moco, where nitrate is reduced [191]. The molybdenum- and heme-containing domains have significant sequence and structural homology to the corresponding portions of the eukaryotic sulfite oxidases and cytochrome *b*5, respectively. The substrate-binding site also includes Met427 (a Val in the sulfite oxidases), Asn272 (a Tyr in the sulfite oxidases), and Thr425 (an Arg in the sulfite oxidases) that impart substrate specificity. Up to now, the orientation of the molybdenum, heme, and flavin domains concerning one another in the holoenzyme is unknown.

3.4.2 Regulation of NR

NR catalytic flux or the total nitrate-reducing capacity of a plant system depends on: (I) the availability of the substrates in the cytoplasm, (II) the level of functional NR, and (III) the activity level of the functional NR. Each process is regulated either directly or indirectly, and the overall level of nitrate reduction capacity is controlled concerning the overall plant metabolic level by metabolic sensors and signal transduction pathways. In normal plants with optimum growing conditions and sufficient nitrate, the nitrate-reducing capacity is about two times greater than the plant needs and NR activity levels cycle daily with low activity in the dark [196]; nitrate essentially acts as a hormone in plants by inducing functional NR, furthermore, even the expression of nitrate transporter genes is also induced by nitrate [197]. Presumably, other regulatory factors such as light, tissue specificity, Gln/Glu balance, water and carbohydrate status, the photosynthesis rate of the plant, and other limiting conditions are included between the promoters of NR and related nitrate response genes [198,199].

3.4.3 NR inhibition

Azide (as NaN_3), cyanate (as NaOCN), and cyanide (as KCN) are inhibitors of NR at micromolar concentrations in both plants and green algae, but their inhibition mechanisms are not clear yet [200,201]. Previous studies have shown that cyanide enhances the inactivation of NR in the presence of NAD(*P*)H [202]. Inactive NR can be reactivated either by an oxidant such as ferricyanide or by blue light photoirradiation [203]. Cyanide was proposed to stabilize the inactive state through the binding to the pterin-Moco at the active site [204]. The mechanism of inhibition of NR by azide is also not clear. During past decades, it has been demonstrated that azide acts as a competitive inhibitor for nitrate [201,202] and that azide inhibits the photoreactivation of NR [205]. Azide can also inhibit the photoreduction of Cyt b by flavin in endoplasmic reticulum NR preparations from *Neurospora crassa* leading to the inactivation of the enzyme [206].

3.5 mARC

Investigation of the aerobic reduction of amidoxime structures led to the discovery of the fourth mammalian molybdenum enzyme mitochondrial amidoxime-reducing component (mARC) by Havemeyer et al. in 2006 [69]. mARC belongs to the SO family of molybdenum-containing enzymes [207]. It is part of an N-reductive enzyme system consisting of mARC, cytochrome *b*5 type B, and NADH cytochrome *b*5 reductase [69,208]. Mitochondrial amidoxime-reducing component proteins represent the simplest mammalian molybdenum enzymes; nevertheless, heterologous studies in *E. coli* allowed for purification as monomers [209]. Unfortunately, no crystal structures of these proteins are available yet. The human genome harbors two mARC genes referred to as mARC1 and mARC2, which are organized in a tandem arrangement on chromosome 1 (1q41) [70] and show an overall sequence identity/similarity of 66/80% [209]. Human MARC1 (NM_022746.3) encodes an mARC-1 protein consisting of 337 amino acids, while the human MARC2 encodes an mARC-2 protein consisting of 335 amino acids. Based on computational analysis, both proteins contain an amino-terminal mitochondrial signal sequence, a predicted β-barrel domain, and an MOSC (MOS C-terminal) domain near the carboxyl-terminus [210]; this last domain is present in the MOS (Moco sulfurases) enzymes that are involved in the transfer of a sulfide ligand to Moco, and the resulting sulfurated Moco is essential for the activity of XO family enzymes. Considering other species, even the *Arabidopsis genome*, such as in many eukaryotes, contains two genes that code for mARC proteins, but both show strong similarities in the amino acid and at the nucleotide levels. The alga *Chlamydomonas reinhardtii*, however, contains only one mARC gene that was named crARC [24]. Interestingly, in vitro crARC shows a Zn^{2+}-dependent activity [207], which remains to be confirmed in other mARC enzymes. Proteins belonging to the plant mARC family are present in trees, herbaceous mono- and di-cotyledonous plant species, as well as several algae [211].

3.5.1 Localization and distribution

These molybdoproteins are widely expressed in different tissues. Human mARC-1 shows the highest mRNA expression levels in adipocytes, pancreas, liver, and kidney. The highest expression levels of human mARC-2 are found in the kidney, the thyroid gland, liver, and small intestine [212]. Interestingly, in most human cell lines and tissues, only mARC-1 is highly abundant. Considering other mammals, several studies with porcine and murine tissue allowed to a similar tissue localization, in fact, high protein levels of mARC-1 were found in mitochondrial subcellular fractions of liver, kidney, and pancreas, whereas mARC-2 was abundant in liver, kidney, thyroid gland, lung, small intestine, and pancreas, which is in good agreement with high N-reductive activity in mitochondrial fractions obtained from these tissues. All tissues investigated so far showed N-reductive activity and expression of at least one mARC isoform [213]. A recent study with human cell lines revealed that the mARC-1 protein is associated with the mitochondrial membrane with an $N_{(in)}$-$C_{(out)}$ membrane orientation. The catalytic domain is localized at the C-terminus and exposed to the cytosol, which is consistent with the localization and orientation of cytochrome *b*5 type B and NADH cytochrome *b*5 reductase [214]. This submitochondrial localization is in agreement with a maximum of N-reductive activity detectable in the outer mitochondrial membrane [215]. Once, it was reported a dual-localization, mitochondria, and peroxisomes, in mice and rats for mARC-2, but the absence of cytochrome *b*5 and its reductase on peroxisomes must be considered [216]. Thus, it is probable that a peroxisomal localized mARC has a different function than N-reductive activity, for example, cell response to oxidative stress or lipid metabolism [217,218]. However, the responsible mechanisms for the targeting of peroxisomal membrane proteins are still rather poorly understood.

3.5.2 Proposed physiological role

Although there is no doubt about the N-reductive activity of mARC, the physiological function of mARC is not fully understood. It is generally accepted that mARC enzymes play a major role in the reduction of NHCs (N-hydroxylated compounds) and some NHCs are prodrugs [208]. Up to now, all tested NHC substrates were converted by mARC in vitro into the corresponding amine compounds. Amidines are known functional groups in drugs and they are used for the treatment of several diseases [219]. However, amidines are easily protonated under physiological conditions due to their strong basicity and, for this reason, they are poorly absorbed by gastrointestinal tracts showing insufficient oral absorption. This problem can be solved by the use of a prodrug strategy because amidoxime prodrugs improve the bioavailability of the amidine moiety since the electronegative charges resulting from oxygenation at a physiological pH prevent these compounds to be protonated. One of the NHC prodrugs converted to its active form by human ARCO is ximelagatran (Astra Zeneca), the first approved oral drug for direct thrombin inhibition [220]. Another one is MesupronVR (WX-671) the oral prodrug of the urokinase inhibitor WX-UK1 [221]. Interestingly, the N-oxides are only reduced by mARC1 and not by mARC2 [222]. These findings indicate that there could be some structural differences in the catalytic centers of mARC1 and mARC2, which would allow mARC1 and mARC2 to participate in different metabolic reactions.

Several mARC proteins have been involved in the reduction of many N-hydroxylated nucleobases and nucleosides, which are potentially toxic or mutagenic components. The mammalian mARCs have been involved in the reduction of N-hydroxycytosine, HAP, N-hydroxycytidine, and N-hydroxyadenosine [213]. This hypothesis of a mARC-mediated role in detoxification reactions is also supported by a study that describes a significant and progressively changed down-regulated expression of human mARC-2 in colon tumors in samples of human tissue [223].

Several studies have shown a close connection between mARC and lipid metabolism. In the adipose tissue of rodents, NHC reduction activity was found to be high [224]. The knockdown of murine mARC2 expression in adipocytes has a significant effect on the fatty acid composition and impairs lipid synthesis [225]. As a consequence, the mARC2 knockout mouse is able to survive, and the phenotype of the male mice is characterized by decreased fat levels in the entire body, while female mARC2 knockout mice exhibit an increased circulating glucose level. Additionally, the caloric diet restriction of obese patients caused a decreased expression of mARC2 in liver but it did not affect mARC1 expression [226]. However, how mARC regulates lipogenesis still remains unresolved, and more experiments are needed to uncover the mechanism of mARC action.

Expression of mARC-2 is up-regulated in diabetic kidneys of Goto-Kakizaki rats, a prototype for the study of non-overweight type 2 diabetes, and by glucose in human and rat renal cells [217]. Therefore, it might be possible that there is a connection between only one specific mARC form and diabetes mellitus.

4 Molybdenum cofactor deficiencies (MoCD)

Molybdenum cofactor deficiency (MoCD) is a severe autosomal recessive inborn error of metabolism first described in 1978. MoCD is characterized by the simultaneous loss of all Mo-enzyme activities due to a mutational block in the biosynthesis of Moco. Human MoCD is a rare autosomal recessive disorder, which mostly affects neonates and is characterized by progressive brain injury leading to early childhood death [227]. Patients are typically normal at birth but within hours to weeks develop intractable seizures, such as severe neurological abnormalities, lens dislocation of the eyes, and major dysmorphic features of the head. MoCD is caused by biallelic pathogenic variants in either MOCS1, MOCS2, or GPHN, essential genes in the synthesis of Moco as described above in the proper chapter. MOCS1 is responsible for around two-thirds of cases (MoCD Type A), followed by MOCS2 (MoCD Type B) and then GPHN (MoCD Type C) (Fig. 12).

Due to the high prevalence of type A MoCD, a *mocs1*-knockout mouse was generated by homologous recombination with a targeting vector. As observed in humans, biochemical characterization of *mocs1*$^{-/-}$ mice revealed markedly elevated urinary sulfite and xanthine levels, while uric acid was undetectable. Furthermore, neither MPT nor Moco could be detected in homozygous mice and, as a result, Mo-enzyme activities were absent. So, in terms of biomarkers and disease progression (survival), *mocs1*$^{-/-}$ mice can be considered a suitable animal model to study the molecular basis and treatment of human MoCD. To probe that *mocs1*$^{-/-}$ mice are still able to convert cPMP into MPT and to determine the required dose, purified cPMP was

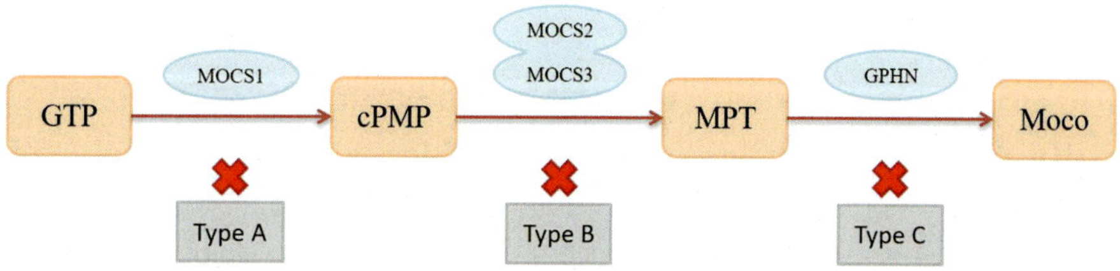

FIG. 12 Biosynthesis of Moco with relative MoCD typology: GTP is converted to cPMP by MOCS1 (Type A). cPMP is converted to MPT by MOCS2 and MOCS3 (Type B). The final step is the conversion of MPT to Moco by GPHN (Type C).

inoculated to crude liver extracts of *mocs1*$^{-/-}$ mice and in vitro Moco synthesis was observed as a function of cPMP added. cPMP-treated Mocs1-deficient mice developed normally, gained weight, and reached adulthood, were fertile and not distinguishable from their wild-type littermates [228]. In conclusion, a first experimental and causal treatment approach using cPMP substitution has been established for type A MoCD [228]. Considering human treatments, using sulfite, SSC, xanthine, and uric acid as biomarkers to monitor treatment efficacy, the administration of cPMP took to a normalization of MoCD biomarker [229]. Depending on the time of treatment initiation, patients can reach an almost normal neurodevelopmental outcome [230]. In contrast, MoCD-type B patients are unable to synthesize MPT and therefore a substitution therapy with either MPT or mature Moco would be required. Neither MPT nor Moco have been stably isolated in a protein-free form yet and are therefore not available for substitution therapies [11]. MoCD-type C has only been reported in two cases. Given the overall increased severity of a gephyrin loss of function in both, an animal model as well as a patient, we assume that most cases remain undiagnosed due to their early neonatal death [41].

In 2019, FDA approved the first treatment for Molybdenum Cofactor Deficiency Type A called Nulibry (fosdenopterin), an intravenous medication that replaces the missing cPMP. The patients treated with Nulibry had a survival rate of 84% at 3 years, compared to 55% for the untreated patients. The most common side effects included complications related to the intravenous line, fever, respiratory infections, vomiting, gastroenteritis, and diarrhea.

References

[1] Greenwood NN, Earnshaw A. Chemistry of the elements. Oxford: Pergamon Press; 1984.

[2] Zhang Y, Gladyshev VN. Molybdoproteomes and evolution of molybdenum utilization. J Mol Biol 2008;379:881–99.

[3] Hille R. The mononuclear molybdenum enzymes. Chem Rev 1996;96:2757–816.

[4] Schwarz G, Mendel RR. Molybdenum cofactor biosynthesis and molybdenum enzymes. Annu Rev Plant Biol 2006;57:623–47.

[5] Hille R. The molybdenum oxotransferases and related enzymes. Dalton Trans 2013;42:3029–42.

[6] Majumdar A. Bioinorganic modelling chemistry of carbon monoxide dehydrogenases: description of model complexes, current status and possible future scopes. Dalton Trans 2014;43:12135–45.

[7] Hille R, Hall J, Basu P. The mononuclear molybdenum enzymes. Chem Rev 2014;114:3963–4038.

[8] Mendel RR, Kruse T. Cell biology of molybdenum in plants and humans. Biochim Biophys Acta 2012;1823:1568–79.

[9] Mendel RR. Cell biology of molybdenum in plants. Plant Cell Rep 2011;30:1787–97.

[10] Leimkuhler S, Lobbi-Nivol C. Bacterial molybdoenzymes: old enzymes for new purposes. FEMS Microbiol Rev 2016;40:1–18.

[11] Schwarz G, Mendel R, Ribbe M. Molybdenum cofactors, enzymes and pathways. Nature 2009;460:839–47.

[12] Lawson DM, Smith BE. Molybdenum nitrogenases: a crystallographic and mechanistic view. Met Ions Biol Syst 2002;39:75–119.

[13] Rubio LM, Ludden PW. Biosynthesis of the iron-molybdenum cofactor of nitrogenase. Annu Rev Microbiol 2008;62:93–111.

[14] Allen RM, Roll JT, Rangaraj P, Shah VK, Roberts GP, Ludden PW. Incorporation of molybdenum into the iron-molybdenum cofactor of nitrogenase. J Biol Chem 1999;274:15869–74.

[15] Yang J, Enemark JH, Kirk ML. Metal-dithiolene bonding contributions to pyranopterin molybdenum enzyme reactivity. Inorganics 2020;8:19.

[16] Zhang Y, Rump S, Gladyshev VN. Comparative genomics and evolution of molybdenum utilization. Coord Chem Rev 2011;255:1206–17.

[17] Peng T, Xu Y, Zhang Y. Comparative genomics of molybdenum utilization in prokaryotes and eukaryotes. BMC Genomics 2018;19:691.

[18] Burguera JL, Burguera M. Molybdenum in human whole blood of adult residents of the Merida state (Venezuela). J Trace Elem Med Biol 2007;21:178–83.

[19] Versieck J, Hoste J, Barbier F, Vanballenberghe L, Derudder J, Cornelis R. Determination of molybdenum in human serum by neutron activation analysis. Clin Chim Acta 1978;87:135–40.

[20] Turnlund JR, Keyes WR. Plasma molybdenum reflects dietary molybdenum intake. J Nutr Biochem 2004;15:90–5.

[21] Underwood EI. Trace elements in human and animal nutrition. 4th ed. New York: Academic Press; 1977. p. 109–31.

[22] Vyskocil A, Viau C. Assessment of molybdenum toxicity in humans. J Appl Toxicol 1999;19:185–92.

[23] Schwarz G, Belaidi AA. Molybdenum in human health and disease. Met Ions Life Sci 2013;13:415–50.

[24] Llamas A, Tejada-Jimenez M, Fernandez E, Galvan A. Molybdenum metabolism in the alga *Chlamydomonas* stands at the crossroad of those in Arabidopsis and humans. Metallomics 2011;3:578–90.

[25] Mendel RR, Leimkuhler S. The biosynthesis of the molybdenum cofactors. J Biol Inorg Chem 2015;20:337–47.

[26] Bevers LE, Schwarz G, Hagen WR. A molecular basis for tungstate selectivity in prokaryotic ABC transport systems. J Bacteriol 2011;193:4999–5001.

[27] Grunden AM, Shanmugam KT. Molybdate transport and regulation in bacteria. Arch Microbiol 1997;168:345–54.

[28] Thiel T, Pratte B, Zahalak M. Transport of molybdate in the cyanobacterium *Anabaena variabilis* ATCC 29413. Arch Microbiol 2002;179:50.

[29] Tejada-Jimenez M, Llamas A, Sanz-Luque E, Galvan A, Fernandez E. A high-affinity molybdate transporter in eukaryotes. Proc Natl Acad Sci U S A 2007;104:20126–30.

[30] Tejada-Jiménez M, Galván A, Fernández E. Algae and humans share a molybdate transporter. Proc Natl Acad Sci U S A 2011;108:6420–5.

[31] Self WT, Grunden AM, Hasona A, Shanmugam KT. Molybdate transport. Res Microbiol 2001;152:311–21.

[32] Tejada-Jimenez M, Chamizo-Ampudia A, Calatrava V, Galvan A, Fernandez E, Llamas A. From the eukaryotic molybdenum cofactor biosynthesis to the moonlighting enzyme mARC. Molecules 2018;23:3287.

[33] Tomatsu H, Takano J, Takahashi H, Watanabe-Takahashi A, Shibagaki N, Fujiwara T. An Arabidopsis thaliana high-affinity molybdate transporter required for efficient uptake of molybdate from soil. Proc Natl Acad Sci U S A 2007;104:18807–12.

[34] Baxter I, Muthukumar B, Park HC, Buchner P, Lahner B, Danku J, Zhao K, Lee J, Hawkesford MJ, Guerinot ML, Salt DE. Variation in molybdenum content across broadly distributed populations of

Arabidopsis thaliana is controlled by a mitochondrial molybdenum transporter (MOT1). PLoS Genet 2008;4:1000004.
[35] Duan G, Hakoyama T, Kamiya T, Miwa H, Lombardo F, Sato S, Tabata S, Chen Z, Watanabe T, Shinano T. LjMOT1, a high-affinity molybdate transporter from Lotus japonicus, is essential for molybdate uptake, but not for the delivery to nodules. Plant J 2017;90:1108–19.
[36] Anderson AJ, Spencer D. Molybdenum in nitrogen metabolism of legumes and non-legumes. Aust J Sci Res 1950;3:414–30.
[37] Bambara S, Ndakidemi PA. The potential roles of lime and molybdenum on the growth, nitrogen fixation and assimilation of metabolites in nodulated legume: a special reference to Phaseolus vulgaris L. African J Biotech 2010;8:2482–9.
[38] Gil-Diez P, Tejada-Jimenez M, Leon-Mediavilla J, Wen J, Mysore KS, Imperial J, Gonzalez-Guerrero M. MtMOT1.2 is responsible for molybdate supply to Medicago truncatula nodules. Plant Cell Environ 2018;42:310–20.
[39] Tejada-Jimenez M, Gil-Diez P, Leon-Mediavilla J, Wen J, Mysore KS, Imperial J, Gonzalez-Guerrero M. Medicago truncatula molybdate transporter type 1 (MtMOT1.3) is a plasma membrane molybdenum transporter required for nitrogenase activity in root nodules under molybdenum deficiency. New Phytol 2017;216:1223–35.
[40] Teschner J, Lachmann N, Schulze J, Geisler M, Selbach K. A novel role for Arabidopsis mitochondrial ABC transporter ATM3 in molybdenum cofactor biosynthesis. Plant Cell 2010;22:468–80.
[41] Reiss J, Hahnewald R. Molybdenum cofactor deficiency: mutations in GPHN, MOCS1, and MOCS2. Hum Mutat 2010;32:10–8.
[42] Reiss J, Johnson JL. Mutations in the molybdenum cofactor biosynthetic genes MOCS1, MOCS2, and GEPH. Hum Mutat 2003;21:569–76.
[43] Hoff T, Schnorr KM, Meyer C, Caboche M. Isolation of two Arabidopsis cDNAs involved in early steps of molybdenum cofactor biosynthesis by functional complementation of Escherichia coli mutants. J Biol Chem 1995;270:6100–7.
[44] Kruse I, Maclean AE, Hill L, Balk J. Genetic dissection of cyclic pyranopterin monophosphate biosynthesis in plant mitochondria. Biochem J 2018;475:495–509.
[45] Kaufholdt D, Gehl C, Geisler M, Jeske O, Voedisch S, Ratke C, Bollhoner B, Mendel RR, Hansch R. Visualization and quantification of protein interactions in the biosynthetic pathway of molybdenum cofactor in Arabidopsis thaliana. J Exp Bot 2013;64:2005–16.
[46] Bittner F. Molybdenum metabolism in plants and crosstalk to iron. Front Plant Sci 2014;5:28.
[47] Stallmeyer B, Drugeon G, Reiss J, Haenni AL, Mendel RR. Human molybdopterin synthase gene: identification of a bicistronic transcript with overlapping reading frames. Am J Hum Genet 1999;64:698–705.
[48] Matthies A, Rajagopalan KV, Mendel RR, Leimkuhler S. Evidence for the physiological role of a rhodanese-like protein for the biosynthesis of the molybdenum cofactor in humans. Proc Natl Acad Sci U S A 2004;101:5946–51.
[49] Wuebbens MM, Rajagopalan KV. Mechanistic and mutational studies of Escherichia coli molybdopterin synthase clarify the final step of molybdopterin biosynthesis. J Biol Chem 2003;278:14523–32.
[50] Schwarz G, Schulze J, Bittner F, Eilers T, Kuper J, Bollmann G, Nerlich A, Brinkmann H, Mendel RR. The molybdenum cofactor biosynthetic protein Cnx1 complements molybdate-repairable mutants, transfers molybdenum to the metal binding pterin, and is associated with the cytoskeleton. Plant Cell 2000;12:2455–72.
[51] Smolinsky B, Eichler SA, Buchmeier S, Meier JC, Schwarz G. Splice-specific functions of gephyrin in molybdenum cofactor biosynthesis. J Biol Chem 2008;283:17370–9.
[52] Stallmeyer B, Schwarz G, Schulze J, Nerlich A, Reiss J, Mendel RR. The neurotransmitter receptor-anchoring protein gephyrin reconstitutes molybdenum cofactor biosynthesis in bacteria, plants, and mammalian cells. Proc Natl Acad Sci U S A 1999;96:1333–8.
[53] Llamas A, Tejada-Jimenez M, Gonzalez-Ballester D, Higuera JJ, Schwarz G, Galvan A, Fernandez E. Chlamydomonas reinhardtii CNX1E reconstitutes molybdenum cofactor biosynthesis in Escherichia coli mutants. Eukaryot Cell 2007;6:1063–7.
[54] Li W, Fingrut DR, Maxwell DP. Characterization of a mutant of Chlamydomonas reinhardtii deficient in the molybdenum cofactor. Physiol Plant 2009;136:336–50.
[55] Llamas A, Mendel RR, Schwarz G. Synthesis of adenylated molybdopterin: an essential step for molybdenum insertion. J Biol Chem 2004;279:55241–6.
[56] Llamas A, Otte T, Multhaup G, Mendel RR, Schwarz G. The mechanism of nucleotide-assisted molybdenum insertion into molybdopterin. A novel route toward metal cofactor assembly. J Biol Chem 2006;281:18343–50.
[57] Krausze J, Probst C, Curth U, Reichelt J, Saha S, Schafflick D, Heinz DW, Mendel RR, Kruse T. Dimerization of the plant molybdenum insertase Cnx1E is required for synthesis of the molybdenum cofactor. Biochem J 2017;474:163–78.
[58] Probst C, Yang J, Krausze J, Hercher TW, Richers CP, Spatzal T, Khadanand KC, Giles LJ, Rees DC, Mendel RR, Kirk ML, Kruse T. Mechanism of molybdate insertion into pterin-based molybdenum cofactors. Nat Chem 2021;13:758–65.
[59] Witte CP, Igeno MI, Mendel R, Schwarz G, Fernandez E. The Chlamydomonas reinhardtii MoCo carrier protein is multimeric and stabilizes molybdopterin cofactor in a molybdate charged form. FEBS Lett 1998;431:205–9.
[60] Ataya FS, Witte CP, Galvan A, Igeno MI, Fernandez E. Mcp1 encodes the molybdenum cofactor carrier protein in Chlamydomonas reinhardtii and participates in protection, binding, and storage functions of the cofactor. J Biol Chem 2003;278:10885–90.
[61] Aguilar M, Kalakoutskii K, Cardenas J, Fernandez E. Direct transfer of molybdopterin cofactor to aponitrate reductase from a carrier protein in Chlamydomonas reinhardtii. FEBS Lett 1992;307:162–3.
[62] Fischer K, Llamas A, Tejada-Jimenez M, Schrader N, Kuper J, Ataya FS, Galvan A, Mendel RR, Fernandez E, Schwarz G. Function and structure of the molybdenum cofactor carrier protein from Chlamydomonas reinhardtii. J Biol Chem 2006;281:30186–94.
[63] Kruse T, Gehl C, Geisler M, Lehrke M, Ringel P, Hallier S, Hansch R, Mendel RR. Identification and biochemical characterization of molybdenum cofactor-binding proteins from Arabidopsis thaliana. J Biol Chem 2010;285:6623–35.
[64] Kuper J, Palmer T, Mendel RR, Schwarz G. Mutations in the molybdenum cofactor biosynthetic protein Cnx1G from Arabidopsis thaliana define functions for molybdopterin binding, molybdenum insertion, and molybdenum cofactor stabilization. Proc Natl Acad Sci U S A 2000;97:6475–80.
[65] De Renzo EC, Heytler PG, Kaleita E. Further evidence that molybdenum is a cofactor of xanthine oxidase. Arch Biochem Biophys 1954;49:242–4.
[66] Mahler HR, Mackler B, Green DE, Bock RM. Studies on metalloflavoproteins. III. Aldehyde oxidase: a molybdoflavoprotein. J Biol Chem 1954;210:465–80.

[67] Nicholas DJD, Nason A, McElroy WD. Molybdenum and nitrate reductase. I. Effect of molybdenum deficiency on the neurospora enzyme. J Biol Chem 1954;207:341–51.

[68] Cohen HJ, Fridovich I, Rajagopalan KV. Hepatic sulfite oxidase. A functional role for molybdenum. J Biol Chem 1971;246:374–82.

[69] Havemeyer A, Bittner F, Wollers S, Mendel R, Kunze T. Identification of the missing component in the mitochondrial benzamidoxime prodrug-converting system as a novel molybdenum enzyme. J Biol Chem 2006;281:34796–802.

[70] Gruenewald S, Wahl B, Bittner F, Hungeling H, Kanzow S, Kotthaus J, Schwering U, Mendel RR, Clement B. The fourth molybdenum-containing enzyme mARC: cloning and involvement in the activation of N-hydroxylated prodrugs. J Med Chem 2008;51:8173–7.

[71] Hille R, Nishino T, Bittner F. Molybdenum enzymes in higher organisms. Coord Chem Rev 2011;255:1179–205.

[72] Terao M, Romao MJ, Leimkühler S, Bolis M, Fratelli M, Coelho C, Santos-Silva T, Garattini E. Structure and function of mammalian aldehyde oxidases. Arch Toxicol 2016;90:753–80.

[73] Massey V, Harris CM. Milk xanthine oxidoreductase: the first one hundred years. Biochem Soc Trans 1997;25:750–5.

[74] Enroth C, Eger BT, Okamoto K, Nishino T, Nishino T, Pai EF. Crystal structures of bovine milk xanthine dehydrogenase and xanthine oxidase: structure-based mechanism of conversion. Proc Natl Acad Sci U S A 2000;97:10723–8.

[75] McManaman JL, Bain DL. Structural and conformational analysis of the oxidase to dehydrogenase conversion of xanthine oxidoreductase. J Biol Chem 2002;277:21261–8.

[76] Morais-Silva FO, Rezende AM, Pimentel C, Santos CI, Clemente C, Varela-Raposo A, Resende DM, da Silva SM, de Oliveira LM, Matos M, Costa DA, Flores O, Ruiz JC, Rodrigues-Pousada C. Genome sequence of the model sulfate reducer Desulfovibrio gigas: a comparative analysis within the Desulfovibrio genus. Microbiology 2014;3:513–30.

[77] Jones RM, Inscore FE, Hille R, Kirk ML. Freeze-quench magnetic circular dichroism spectroscopic study of the "very rapid" intermediate in xanthine oxidase. Inorg Chem 1999;38:4963–70.

[78] Battelli MG, Bortolotti M, Bolognesi A, Polito L. Pro-aging effects of xanthine oxidoreductase products. Antioxidants 2020;9:839.

[79] Bortolotti M, Polito L, Battelli MG, Bolognesi A. Xanthine oxidoreductase: one enzyme for multiple physiological tasks. Redox Biol 2021;41, 101882.

[80] Dobbek H, Gremer L, Kiefersauer R, Huber R, Meyer O. Catalysis at a dinuclear [CuSMo(==O)OH] cluster in a CO dehydrogenase resolved at 1.1-a resolution. Proc Natl Acad Sci U S A 2002;99: 15971–6.

[81] Bonin I, Martins BM, Purvanov V, Fetzner S, Huber R, Dobbek H. Active site geometry and substrate recognition of the molybdenum hydroxylase quinoline 2-oxidoreductase. Structure 2004;12: 1425–35.

[82] Canne C, Stephan I, Finsterbusch J, Lingens F, Kappl R, Fetzner S, Hüttermann J. Comparative EPR and redox studies of three prokaryotic enzymes of the xanthine oxidase family: quinoline 2-oxidoreductase, quinaldine 4-oxidase, and isoquinoline 1-oxidoreductase. Biochemistry 1997;36:9780–90.

[83] Doonan CJ, Stockert A, Hille R, George GN. Nature of the catalytically labile oxygen at the active site of xanthine oxidase. J Am Chem Soc 2005;127:4518–22.

[84] Hille R. Molybdenum-containing hydroxylases. Arch Biochem Biophys 2005;433:107–16.

[85] Orme-Johnson WH, Beinert H. Heterogeneity of paramagnetic species in two iron-sulfur proteins: *Clostridium pasteurianum* ferredoxin and milk xanthine oxidase. Biochem Biophys Res Commun 1969;36:337–44.

[86] Canne C, Lowe DJ, Fetzner S, Adams B, Smith AT, Kappl R, Bray RC, Hüttermann J. Kinetics and interactions of molybdenum and iron-sulfur centers in bacterial enzymes of the xanthine oxidase family: mechanistic implications. Biochemistry 1999;38:14077–87.

[87] Nishino T, Okamoto K. The role of the [2Fe-2s] cluster centers in xanthine oxidoreductase. J Inorg Biochem 2000;82:43–9.

[88] Iwasaki T, Okamoto K, Nishino T, Mizushima J, Hori H. Sequence motif-specific assignment of two [2Fe-2S] clusters in rat xanthine oxidoreductase studied by site-directed mutagenesis. J Biochem 2000;127:771–8.

[89] Liu M, Wang JA, Klysubun W, Wang GG, Sattayaporn S, Li F, Cai YW, Zhang F, Yu J, Yang Y. Interfacial electronic structure engineering on molybdenum sulfide for robust dual-pH hydrogen evolution. Nat Commun 2021;12:5260.

[90] Okamoto K, Matsumoto K, Hille R, Eger BT, Pai EF, Nishino T. The crystal structure of xanthine oxidoreductase during catalysis: implications for reaction mechanism and enzyme inhibition. Proc Natl Acad Sci U S A 2004;101:7931–6.

[91] Maiti NC, Tomita T, Kitagawa T, Okamoto K, Nishino T. Resonance Raman studies on xanthine oxidase: observation of Mo(VI)-ligand vibrations. J Biol Inorg Chem 2003;8:327–33.

[92] Okamoto K, Kusano T, Nishino T. Chemical nature and reaction mechanisms of the molybdenum cofactor of xanthine oxidoreductase. Curr Pharm Des 2013;19:2606–14.

[93] Bergmann F, Dikstein S. Studies on uric acid and related compounds. J Biol Chem 1956;223:765–80.

[94] Jezewska MM. Xanthine accumulation during hypoxanthine oxidation by milk xanthine oxidase. Eur J Biochem 1973;36:385–90.

[95] Olson JS, Ballou DP, Palmer G, Massey V. The mechanism of action of xanthine oxidase. J Biol Chem 1974;249:4363–82.

[96] Kobayashi K, Miki M, Okamoto K, Nishino T. Electron transfer process in milk xanthine dehydrogenase as studied by pulse radiolysis. J Biol Chem 1993;268:24642–6.

[97] Schopfer LM, Massey V, Nishino T. Rapid reaction studies on the reduction and oxidation of chicken liver xanthine dehydrogenase by the xanthine/urate and NADH/NAD couples. J Biol Chem 1988;263:13528–38.

[98] Yamaguchi Y, Matsumura T, Ichida K, Okamoto K, Nishino T. Human xanthine oxidase changes its substrate specificity to aldehyde oxidase type upon mutation of amino acid residues in the active site: roles of active site residues in binding and activation of purine substrate. J Biochem 2007;141:513–24.

[99] Hille R, Nishino T. Flavoprotein structure and mechanism. 4. Xanthine oxidase and xanthine dehydrogenase. FASEB J 1995;9:995–1003.

[100] Glatigny A, Hof P, Romao MJ, Huber R, Scazzocchio C. Altered specificity mutations define residues essential for substrate positioning in xanthine dehydrogenase. J Mol Biol 1998;278:431–8.

[101] Bayse CA. Density-functional theory models of xanthine oxidoreductase activity: comparison of substrate tautomerization and protonation. Dalton Trans 2009;13:2306–14.

[102] Godber BL, Schwarz G, Mendel RR, Lowe DJ, Bray RC, Eisenthal R, Harrison R. Molecular characterization of human xanthine oxidoreductase: the enzyme is grossly deficient in molybdenum and substantially deficient in iron-sulphur centres. Biochem J 2005;388:501–8.

[103] Angermüller S, Bruder G, Volkl A, Wesch H, Fahimi HD. Localization of xanthine oxidase in crystalline cores of peroxisomes. A cytochemical and biochemical study. Eur J Cell Biol 1987;45:137–44.

[104] Battelli MG, Polito L, Bortolotti M, Bolognesi A. Xanthine oxidoreductase in drug metabolism: beyond a role as a detoxifying enzyme. Curr Med Chem 2016;23:4027–36.

[105] Xu P, LaVallee P, Hoidal JR. Repressed expression of the human xanthine oxidoreductase gene. E-box and TATA-like elements restrict ground state transcriptional activity. J Biol Chem 2000;275:5918–26.

[106] Battelli MG, Polito L, Bortolotti M, Bolognesi A. Xanthine oxidoreductase in cancer: more than a differentiation marker. Cancer Med 2016;5:546–57.

[107] Mcmanaman JL, Palme CA, Wright RM, Neville MC. Functional regulation of xanthine oxidoreductase expression and localization in the mouse mammary gland: evidence of a role in lipid secretion. J Physiol 2022;542:567–79.

[108] Monks J, Dzieciatkowska M, Bales ES, Orlicky DJ, Wright RM, McManaman JL. Xanthine oxidoreductase mediates membrane docking of milk-fat droplets but is not essential for apocrine lipid secretion. J Physiol 2016;594:5899–921.

[109] Silanikove N, Shapiro F, Shamay A, Leitner G. Role of xanthine oxidase, lactoperoxidase, and NO in the innate immune system of mammary secretion during active involution in dairy cows: manipulation with casein hydrolyzates. Free Radic Biol Med 2005;38:1139–11351.

[110] Al-Shehri SS, Duley JA, Bansal N. Xanthine oxidase-lactoperoxidase system and innate immunity: biochemical actions and physiological roles. Redox Biol 2020;34, 101524.

[111] Martin HM, Hancock JT, Salisbury V, Harrison R. Role of xanthine oxidoreductase as an antimicrobial agent. Infect Immun 2004;72:4933–9.

[112] Chung HY, Song SH, Kim HJ, Ikeno Y, Yu BP. Modulation of renal xanthine oxidoreductase in aging: gene expression and reactive oxygen species generation. J Nutr Health Aging 1999;3:19–23.

[113] Lima WG, Martins-Santos ME, Chaves VE. Uric acid as a modulator of glucose and lipid metabolism. Biochimie 2015;116:17–23.

[114] Harrison R. Structure and function of xanthine oxidoreductase: where are we now? Free Radic Biol Med 2002;33:774–97.

[115] Neogi T, George J, Rekhraj S, Struthers AD, Choi H, Terkeltaub RA. Are either or both hyperuricemia and xanthine oxidase directly toxic to the vasculature? A critical appraisal. Arthritis Rheum 2012;64:327–38.

[116] Adachi T, Fukushima T, Usami Y, Hirano K. Binding of human xanthine oxidase to sulphated glycosaminoglycans on the endothelial cell surface. Biochem J 1993;289:523–7.

[117] Radi R, Rubbo H, Bush K, Freeman BA. Xanthine oxidase binding to glycosaminoglycans: kinetics and superoxide dismutase interactions of immobilised xanthine oxidase-heparin complexes. Arch Biochem Biophys 1997;339:125–35.

[118] Yokoyama Y, Beckman JS, Beckman TK, Wheat JK, Cash TG, Freeman BA, Parks DA. Circulating xanthine oxidase: potential mediator of ischemic injury. Am J Phys 1990;258:565–70.

[119] Dent CE, Philpot GR. Xanthinuria, an inborn error (or deviation) of metabolism. Lancet 1954;266:182–5.

[120] Ichida K, Amaya Y, Kamatani N, Nishino T, Hosoya T, Sakai O. Identification of two mutations in human xanthine dehydrogenase gene responsible for classical type I xanthinuria. J Clin Invest 1997;99:2391–7.

[121] Nakamura M, Yuichiro Y, Jorn Oliver S, Tomohiro M, Schwab KO, Takeshi N, Tatsuo H, Ichida K. Identification of a xanthinuria type I case with mutations of xanthine dehydrogenase in an Afghan child. Clin Chim Acta 2012;414:158–60.

[122] Levartovsky D, Lagziel A, Sperling O, Liberman U, Yaron M, Hosoya T, Ichida K, Peretz H. XDH gene mutation is the underlying cause of classical xanthinuria: a second report. Kidney Int 2000;57:2215–20.

[123] Sakamoto N, Yamamoto T, Moriwaki Y, Teranishi T, Toyoda M, Onishi Y, Kuroda S, Sakaguchi K, Fujisawa T, Maeda M. Identification of a new point mutation in the human xanthine dehydrogenase gene responsible for a case of classical type I xanthinuria. Hum Genet 2001;108:279–83.

[124] Ichida K, Amaya Y, Okamoto K, Nishino T. Mutations associated with functional disorder of xanthine oxidoreductase and hereditary xanthinuria in humans. Int J Mol Sci 2012;13:15475–95.

[125] Watanabe T, Ihara N, Itoh T, Fujita T, Sugimoto Y. Deletion mutation in Drosophila ma-l homologous, putative molybdopterin cofactor sulfurase gene is associated with bovine xanthinuria type II. J Biol Chem 2000;275:21789–92.

[126] Yamamoto T, Moriwaki Y, Takahashi S, Tsutsumi Z, Tuneyoshi K, Matsui K, Cheng J, Hada T. Identification of a new point mutation in the human molybdenum cofactor sulferase gene that is responsible for xanthinuria type II. Metabolism 2003;52:1501–4.

[127] Peretz H, Naamati MS, Levartovsky D, Lagziel A, Shani E, Horn I, Shalev H, Landau D. Identification and characterization of the first mutation (Arg776Cys) in the C-terminal domain of the human molybdenum cofactor sulfurase (HMCS) associated with type II classical xanthinuria. Mol Genet Metab 2007;91:23–9.

[128] Zannolli R, Micheli V, Mazzei MA, Sacco P, Piomboni P, Bruni E, Miracco C, de Santi MM, Terrosi-Vagnoli P, Volterrani L, Pellegrini L, Livi W, Lucani B, Gonnelli S, Burlina AB, Jacomelli G, Macucci F, Pucci L, Fimiani M, Swift JA, Zappella M, Morgese G. Hereditary xanthinuria type II associated with mental delay, autism, cortical renal cysts, nephrocalcinosis, osteopenia, and hair and teeth defects. J Med Genet 2003;40:121.

[129] Robert M, Kliegman MD. Textbook of pediatrics. Disorders of purine and pyrimidine metabolism. vol. 21. Philadelphia, PA: Elsevier; 2020. p. 817–27.

[130] Barkas F, Elisaf M, Liberopoulos E, Kalaitzidis R, Liamis G. Uric acid and incident chronic kidney disease in dyslipidemic individuals. Curr Med Res Opin 2018;34:1193–9.

[131] Day RO, Graham GG, Hicks M, McLachlan AJ, Stocker SL, Williams KM. Clinical pharmacokinetics and pharmacodynamics of allopurinol and oxypurinol. Clin Pharmacokinet 2007;46:623–44.

[132] Bredemeier M, Lopes LM, Eisenreich MA, Hickmann S, Bongiorno GK, d'Avila R, Morsch ALB, da Silva-Stein F, Campos GGD. Xanthine oxidase inhibitors for prevention of cardiovascular events: a systematic review and meta-analysis of randomized controlled trials. BMC Cardiovasc Disord 2018;18:24.

[133] Schmidt HHHW, Ghezzi P, Cuadrado A. Reactive oxygen species. Network pharmacology and therapeutic applications. vol. 264. Springer; 2021. p. 205–28.

[134] El Fal M, Ramli Y, Essassi EM, Saadi M, El Ammari L. Crystal structure of 1-ethyl-pyrazolo[3,4-d]pyrimidine-4(5H)-thione. Acta Crystallogr 2014;70:1005–6.

[135] Robins RK, Revankar GR, O'Brien DE, Springer RH, Albert TNA, Senga K. Purine analog inhibitors of xanthine oxidase structure-activity relationships and proposed binding of the molybdenum cofactor. J Heterocyclic Chem 1985;22:601–34.

[136] Ernst ME, Fravel MA. Febuxostat: a selective xanthine-oxidase/xanthine-dehydrogenase inhibitor for the management of hyperuricemia in adults with gout. Clin Ther 2009;31:2503–18.

[137] Takano Y, Hase-Aoki K, Horiuchi H, Zhao L, Kasahara Y, Kondo S, Becker MA. Selectivity of febuxostat, a novel non-purine inhibitor of xanthine oxidase/xanthine dehydrogenase. Life Sci 2005;76:1835–47.

[138] Chen C, Lu JM, Yao Q. Hyperuricemia-related diseases and xanthine oxidoreductase (xor) inhibitors: an overview. Med Sci Monit 2016;22:2501–12.

[139] Matsumoto K, Okamoto K, Ashizawa N, Nishino T. Fyx-051: a novel and potent hybrid-type inhibitor of xanthine oxidoreductase. J Pharmacol Exp Ther 2011;336:95–103.

[140] Kumar R, Joshi G, Kler H. Toward an understanding of structural insights of xanthine and aldehyde oxidases: an overview of their inhibitors and role in various diseases. Med Res Rev 2018;38:1073–125.

[141] Lü JM, Yao Q, Chen C. 3,4-Dihydroxy-5-nitrobenzaldehyde (DHNB) is a potent inhibitor of xanthine oxidase: a potential therapeutic agent for the treatment of hyperuricemia and gout. Biochem Pharmacol 2013;86:1328–37.

[142] Seiji S, Kunihiko T, Takeshi N. A novel xanthine dehydrogenase inhibitor (BOF-4272). In: Purine and pyrimidine metabolism, vol. 309. New York: Springer; 1991. p. 135–8.

[143] Tung YT, Hsu CA, Chen CS. Phytochemicals from *Acacia confusa* heartwood extracts reduce serum uric acid levels in oxonate-induced mice: their potential use as xanthine oxidase inhibitors. J Agric Food Chem 2010;58:9936–41.

[144] Lin S, Zhang G, Liao Y. Dietary flavonoids as xanthine oxidase inhibitors: structure-affinity and structure-activity relationships. J Agric Food Chem 2015;63:7784–94.

[145] Takahama U, Koga Y, Hirota S, Yamauchi R. Inhibition of xanthine oxidase activity by an oxathiolanone derivative of quercetin. Food Chem 2011;126:1808–11.

[146] Shi Y, Williamson G. Quercetin lowers plasma uric acid in pre-hyperuricaemic males: a randomised, double-blinded, placebo-controlled, cross-over trial. Br J Nutr 2016;115:800–6.

[147] Nagao A, Seki M, Kobayashi H. Inhibition of xanthine oxidase by flavonoids. Biosci Biotechnol Biochem 1999;63:1787–90.

[148] Clinicaltrials.gov. NCT04161872. Nutraceutical on hyperuricemia, 2020. Available from: https://clinicaltrials.gov/ct2/show/NCT04161872?term=Quercetin&cond=Hyperuricemia&draw=2&rank=1. Retrieved on 25-April-2020.

[149] Rodríguez-Trelles F, Tarrío R, Ayala FJ. Convergent neofunctionalization by positive Darwinian selection after ancient recurrent duplications of the xanthine dehydrogenase gene. Proc Natl Acad Sci U S A 2003;100:13413–7.

[150] Romão MJ, Coelho C, Santos-Silva T, Foti A, Terao M, Garattini E, Leimkühler S. Structural basis for the role of mammalian aldehyde oxidases in the metabolism of drugs and xenobiotics. Curr Opin Chem Biol 2017;37:39–47.

[151] Garattini E, Fratelli M, Terao M. Mammalian aldehyde oxidases: genetics, evolution and biochemistry. Cell Mol Life Sci 2008;65:1019–48.

[152] Garattini E, Terao M. The role of aldehyde oxidase in drug metabolism. Expert Opin Drug Metab Toxicol 2012;8:487–503.

[153] Kurosaki M, Bolis M, Fratelli M. Structure and evolution of vertebrate aldehyde oxidases: from gene duplication to gene suppression. Cell Mol Life Sci 2013;70:1807–30.

[154] Coelho C, Mahro M, Trincao J. The first mammalian aldehyde oxidase crystal structure: insights into substrate specificity. J Biol Chem 2012;287:40690–702.

[155] Mahro M, Coelho C, Trincao J. Characterization and crystallization of mouse aldehyde oxidase 3: from mouse liver to Escherichia coli heterologous protein expression. Drug Metab Dispos 2011;39:1939–45.

[156] Garattini E, Mendel R, Romão MJ, Wright R, Terao M. Mammalian molybdo-flavoenzymes, an expanding family of proteins: structure, genetics, regulation, function and pathophysiology. Biochem J 2003;372:15–32.

[157] Jordan CG, Rashidi MR, Laljee H, Clarke SE, Brown JE, Beedham C. Aldehyde oxidase-catalysed oxidation of methotrexate in the liver of Guinea-pig, rabbit and man. J Pharm Pharmacol 1999;51:411–8.

[158] Rashidi MR, Beedham C, Smith JS, Davaran S. In vitro study of 6-mercaptopurine oxidation catalysed by aldehyde oxidase and xanthine oxidase. Drug Metab Pharmacokinet 2007;22:299–306.

[159] Rochat B, Kosel M, Boss G, Testa B, Gillet M, Baumann P. Stereoselective biotransformation of the selective serotonin reuptake inhibitor citalopram and its demethylated metabolites by monoamine oxidases in human liver. Biochem Pharmacol 1998;56:15–23.

[160] Bendotti C, Prosperini E, Kurosaki M, Garattini E, Terao M. Selective localization of mouse aldehyde oxidase mRNA in the choroid plexus and motor neurons. Neuroreport 1997;8:2343–9.

[161] Terao M, Kurosaki M, Saltini G. Cloning of the cDNAs coding for two novel molybdo-flavoproteins showing high similarity with aldehyde oxidase and xanthine oxidoreductase. J Biol Chem 2000;275:30690–700.

[162] Terao M, Kurosaki M, Barzago MM. Role of the molybdoflavoenzyme aldehyde oxidase homolog 2 in the biosynthesis of retinoic acid: generation and characterization of a knockout mouse. Mol Cell Biol 2009;29:357–77.

[163] Buzzell GR. The harderian gland: perspectives. Microsc Res Tech 1996;34:2–5.

[164] Hardeland R, Pandi-Perumal SR, Cardinali DP. Melatonin. Int J Biochem Cell Biol 2006;38:313–6.

[165] Seo M, Peeters AJ, Koiwai H, Oritani T, Marion-Poll A, Zeevaart JA, Koornneef M, Kamiya Y, Koshiba T. The Arabidopsis aldehyde oxidase 3 (AAO3) gene product catalyzes the final step in abscisic acid biosynthesis in leaves. Proc Natl Acad Sci U S A 2000;97:12908–13.

[166] Hoff T, Frandsen GI, Rocher A, Mundy J. Biochemical and genetic characterization of three molybdenum cofactor hydroxylases in Arabidopsis thaliana. Biochim Biophys Acta 1998;1398:397–402.

[167] Zdunek-Zastocka E. Molecular cloning, characterization and expression analysis of three aldehyde oxidase genes from Pisum sativum L. Plant Physiol Biochem 2008;46:19–28.

[168] Obach RS, Huynh P, Allen MC, Beedham C. Human liver aldehyde oxidase: inhibition by 239 drugs. J Clin Pharmacol 2004;44:7–19.

[169] Coelho C, Foti A, Hartmann T, Santos-Silva T, Leimkühler S, Romão MJ. Structural insights into xenobiotic and inhibitor binding to human aldehyde oxidase. Nat Chem Biol 2015;11:779–83.

[170] Pirouzpanah S, Hanaee J, Razavieh SV, Rashidi MR. Inhibitory effects of flavonoids on aldehyde oxidase activity. J Enzyme Inhib Med Chem 2009;24:14–21.

[171] Hamzeh-Mivehroud M, Rahmani S, Feizi MA, Dastmalchi S, Rashidi MR. In vitro and in silico studies to explore structural features of flavonoids for aldehyde oxidase inhibition. Arch Pharm 2014;347:738–47.

[172] Siah M, Farzaei MH, Ashrafi-Kooshk MR, Adibi H, Arab SS, Rashidi MR, Khodarahmi R. Inhibition of Guinea pig aldehyde oxidase activity by different flavonoid compounds: an in vitro study. Bioorg Chem 2016;64:74–84.
[173] Johnson C, Stubley-Beedham C, Stell JG. Hydralazine: a potent inhibitor of aldehyde oxidase activity in vitro and in vivo. Biochem Pharmacol 1985;34:4251–6.
[174] Mccormack JJ, Allen BA, Hodnett CN. Oxidation of quinazoline and quinoxaline by xanthine oxidase and aldehyde oxidase. J Heterocyclic Chem 1978;15:1249–54.
[175] Wahl RC, Rajagopalan KV. Evidence for the inorganic nature of the cyanolyzable sulfur of molybdenum hydroxylases. J Biol Chem 1982;257:1354–9.
[176] Cohen HJ, Betcher-Lange S, Kessler DL, Rajagopalan KV. Hepatic sulfite oxidase. J Biol Chem 1972;247:7759–66.
[177] Kisker C, Schindelin H, Pacheco A, Wehbi WA, Garrett RM, Rajagopalan KV, Enemark JH, Rees DC. Molecular basis of sulfite oxidase deficiency from the structure of sulfite oxidase. Cell 1997;91:973–83.
[178] Schrader N, Fischer K, Theis K, Mendel RR, Schwarz G, Kisker C. The crystal structure of plant sulfite oxidase provides insights into sulfite oxidation in plants and animals. Structure 2003;11:1251–63.
[179] Eilers T, Schwarz G, Brinkmann H, Witt C, Richter T, Nieder J, Koch B, Hille R, Hänsch R, Mendel RR. Identification and biochemical characterization of Arabidopsis thaliana sulfite oxidase. A new player in plant sulfur metabolism. J Biol Chem 2001;276:46989–94.
[180] Karakas E, Wilson HL, Graf TN, Xiang S, Jaramillo-Busquets S, Rajagopalan KV, Kisker C. Structural insights into sulfite oxidase deficiency. J Biol Chem 2005;280:33506–15.
[181] Hille R. The reaction mechanism of oxomolybdenum enzymes. Biochim Biophys Acta 1994;1184:143–69.
[182] Gray HB, Winkler JR. Electron transfer in proteins. Annu Rev Biochem 1996;65:537–61.
[183] Sullivan EP, Hazzard JT, Tollin G, Enemark JH. Electron transfer in sulfite oxidase: effects of pH and anions on transient kinetics. Biochemistry 1993;32:12465–70.
[184] Romao MJ, Archer M, Moura I, LeGall J, Engh R, Schneider M, Hof P, Huber R. Crystal structure of the xanthine oxidase-related aldehyde oxido-reductase from *Desulfovibrio gigas*. Science 1995;270:1170–6.
[185] Boyington JC, Gladyshev VN, Khangulov SV, Stadtman TC, Sun PD. Crystal structure of formate dehydrogenase H: catalysis involving Mo, molybdopterin, selenocysteine and an Fe_4S_4 cluster. Science 1997;275:1305–8.
[186] Johnson JL. Prenatal diagnosis of molybdenum cofactor deficiency and isolated sulfite oxidase deficiency. Prenat Diagn 2003;23:6–8.
[187] Dublin AB, Hald JK, Wootton-Gorges SL. Isolated sulfite oxidase deficiency: MR imaging features. AJNR Am J Neuroradiol 2002;23:484–5.
[188] Bailey JL, Cole RD. Studies on the reaction of sulfite with proteins. J Biol Chem 1959;234:1733–9.
[189] Menzel DB, Leung KH. Covalent reactions in the toxicity of SO_2 and sulfite. Adv Exp Med Biol 1986;197:477–92.
[190] Schindelin H, Kisker C, Rajagopalan KV. Advances in protein chemistry. San Diego: Academic Press Inc; 2001. p. 47–94.
[191] Campbell WH. Structure and function of eukaryotic NAD(P)H: nitrate reductase. Cell Mol Life Sci 2001;58:194–204.
[192] Campbell WH, Kinghorn KR. Functional domains of assimilatory nitrate reductases and nitrite reductases. Trends Biochem Sci 1990;15:315–9.
[193] Skipper L, Campbell WH, Mertens JA, Lowe DJ. Pre-steady-state kinetic analysis of recombinant Arabidopsis NADH: nitrate reductase: rate-limiting processes in catalysis. J Biol Chem 2001;276:26995–7002.
[194] Zalogin TR, Pick U. Inhibition of nitrate reductase by azide in microalgae results in triglycerides accumulation. Algal Res 2014;3:17–23.
[195] Kubo Y, Ogura N, Nakagawa H. Limited proteolysis of the nitrate reductase from spinach leaves. J Biol Chem 1988;263:19684–9.
[196] Scheible WR, Gonzalez-Fontes A, Morcuende R, Lauerer M, Geiger M. Tobacco mutants with a decreased number of functional nia genes compensate by modifying the diurnal regulation of transcription, post-translational modification and turnover of nitrate reductase. Planta 1997;203:304–19.
[197] Tsay YF, Schroeder JI, Feldmann KA, Crawford NM. The herbicide sensitivity gene CHL1 of Arabidopsis encodes a nitrate-inducible nitrate transporter. Cell 1993;72:705–13.
[198] Crawford NM. Nitrate: nutrient and signal for plant growth. Plant Cell 1995;7:859–68.
[199] Redinbaugh MG, Campbell WH. Higher plant responses to environmental nitrate. Physiol Plant 1991;82:640–50.
[200] Amy NK, Garrett RH. Purification and characterization of the nitrate reductase from the diatom *Thalassiosira pseudonana*. Plant Physiol 1974;54:629–37.
[201] Solomonson LP, Vennesland B. Properties of a nitrate reductase of *Chlorella*. Biochim Biophys Acta 1972;267:544–57.
[202] Mikami B, Ida S. Reversible inactivation of ferredoxin-nitrate reductase from the cyanobacterium *Plectonema boryanum*. The role of superoxide anion and cyanide. Plant Cell Physiol 1986;27:1013–21.
[203] Fritz BJ, Ninnemann H. Photoreactivation by triplet flavin and photoinactivation by singlet oxygen of *neurospora crassa* nitrate reductase. Photochem Photobiol 1985;41:39–45.
[204] Garrett RH, Greenbaum P. The inhibition of the *Neurospora crassa* nitrate reductase complex by metal-binding agents. Biochim Biophys Acta 1973;302:24–32.
[205] Roldan JM, Butler WL. Photoactivation of nitrate reductase from *Neurospora crassa*. Photochem Photobiol 1980;32:375–81.
[206] Borgeson CE, Bowman BJ. Blue light-reducible cytochromes in membrane fractions from *Neurospora crassa*. Plant Physiol 1985;78:433–7.
[207] Chamizo-Ampudia A, Galvan A, Fernandez E, Llamas A. The *Chlamydomonas reinhardtii* molybdenum cofactor enzyme crARC has a Zn-dependent activity and protein partners similar to those of its human homologue. Eukaryot Cell 2011;10:1270–82.
[208] Havemeyer A, Lang J, Clement B. The fourth mammalian molybdenum enzyme mARC: current state of research. Drug Metab Rev 2011;43:524–39.
[209] Wahl B, Reichmann D, Niks D, Krompholz N, Havemeyer A, Clement B, Messerschmidt T, Rothkegel M, Biester H, Hille R, Mendel RR, Bittner F. Biochemical and spectroscopic characterization of the human mitochondrial amidoxime reducing components hmARC-1 and hmARC-2 suggest the existence of a new molybdenum enzyme family in eukaryotes. J Biol Chem 2010;285:37847–59.
[210] Anantharaman V, Aravind L. MOSC domains: ancient, predicted sulfur-carrier domains, present in diverse metal sulfur cluster biosynthesis proteins including molybdenum cofactor sulfurases. *FEMS Microbiol* Lett 2002;207:55–61.

[211] Tejada-Jimenez M, Chamizo-Ampudia A, Galvan A, Fernandez E, Llamas A. Molybdenum metabolism in plants. Metallomics 2013;5:1191–203.

[212] Rixen S, Havemeyer A, Tyl-Bielicka A, Pysniak K, Gajewska M, Kulecka M, Ostrowski J, Mikula M, Clement B. Mitochondrial amidoxime-reducing component 2 (MARC2) has a significant role in N-reductive activity and energy metabolism. J Biol Chem 2019;294:17593–602.

[213] Krompholz N, Krischkowski C, Reichmann D, Garbe-Schönberg D, Mendel R, Bittner F, Clement B, Havemeyer A. The mitochondrial amidoxime reducing component (mARC) is involved in detoxification of N-hydroxylated base analogues. Chem Res Toxicol 2012;25:2443–50.

[214] D'Arrigo A, Manera E, Longhi R, Borgese N. The specific subcellular localization of two isoforms of cytochrome b5 suggests novel targeting pathways. J Biol Chem 1993;268:2802–8.

[215] Kotthaus J, Wahl B, Havemeyer A, Kotthaus J, Schade D, Garbe-Schönberg D, Mendel R, Bittner F, Clement B. Reduction of N(ω)-hydroxy-L-arginine by the mitochondrial amidoxime reducing component (mARC). Biochemist 2011;433:383–91.

[216] Islinger M, Lüers GH, Li KW, Loos M, Völkl A. Rat liver peroxisomes after fibrate treatment. A survey using quantitative mass spectrometry. J Biol Chem 2007;282:23055–69.

[217] Malik AN, Rossios C, Al-Kafaji G, Shah A, Page RA. Glucose regulation of CDK7, a putative thiol related gene, in experimental diabetic nephropathy. Biochem Biophys Res Commun 2007;357:237–44.

[218] Pieuchot L, Jedd G. Peroxisome assembly and functional diversity in eukaryotic microorganisms. Annu Rev Microbiol 2012;66:237–63.

[219] Peterlin-Masic L, Cesar J, Zega A. Metabolism-directed optimization of antithrombotics: the prodrug principle. Curr Pharm Des 2006;12:73–91.

[220] Gustafsson D, Elg M. The pharmacodynamics and pharmacokinetics of the oral direct thrombin inhibitor ximelagatran and its active metabolite melagatran: a mini-review. Thromb Res 2003;109:9–15.

[221] Froriep D, Clement B, Bittner F, Mendel RR, Reichmann D. Activation of the anti-cancer agent upamostat by the mARC enzyme system. Xenobiotica 2013;43:780–4.

[222] Jakobs HH, Froriep D, Havemeyer A, Mendel RR, Bittner F. The mitochondrial amidoxime reducing component (mARC): involvement in metabolic reduction of N-oxides, oximes and N-hydroxyamidinohydrazones. Chem Med Chem 2014;9:2381–7.

[223] Mikula M, Rubel T, Karczmarski J, Goryca K, Dadlez M, Ostrowski J. Integrating proteomic and transcriptomic high-throughput surveys for search of new biomarkers of colon tumors. Funct Integr Genom 2011;11:215–24.

[224] Andersson S, Hofmann Y, Nordling A, Li XQ, Nivelius S. Characterization and partial purification of the rat and human enzyme systems active in the reduction of N-hydroxymelagatran and benzamidoxime. Drug Metab Dispos 2005;33:570–8.

[225] Neve EP, Nordling A, Andersson TB, Hellman U, Diczfalusy U. Amidoxime reductase system containing cytochrome b5 type B (CYB5B) and MOSC2 is of importance for lipid synthesis in adipocyte mitochondria. J Biol Chem 2012;287:6307–17.

[226] Llamas A, Chamizo-Ampudia A, Tejada-Jimenez M, Galvan A, Fernandex E. The molybdenum cofactor enzyme mARC: moonlighting or promiscuous enzyme? Biofactors 2017;43:486–94.

[227] Schwarz G. Molybdenum cofactor biosynthesis and deficiency. Cell Mol Life Sci 2005;62:2792–810.

[228] Schwarz G, Santamaria-Araujo JA, Wolf S, Lee HJ, Adham IM, Gröne HJ, Schwegler H, Sass JO, Otte T, Hänzelmann P, Mendel RR, Engel W, Reiss J. Rescue of lethal molybdenum cofactor deficiency by a biosynthetic precursor from Escherichia coli. Hum Mol Genet 2004;13:1249–55.

[229] Veldman A, Santamaria-Araujo JA, Sollazzo S, Pitt J, Gianello R, Yaplito-Lee J, Wong F, Ramsden CA, Reiss J, Cook I, Fairweather J, Schwarz G. Successful treatment of molybdenum cofactor deficiency type a with cPMP. Pediatrics 2010;125: 1249–54.

[230] Hitzert MM, Bos AF, Bergman KA, Veldman A, Schwarz G, Santamaria-Araujo JA, Heiner-Fokkema R, Sival DA, Lunsing RJ, Arjune S, Kosterink JG, van Spronsen FJ. Favorable outcome in a newborn with molybdenum cofactor type a deficiency treated with cPMP. Pediatrics 2012;130:1005–10.

Section K

Tungsten-containing enzymes

Chapter 11

Tungsten-containing enzymes

Niccolò Paoletti

Department of NEUROFARBA, Section of Pharmaceutical and Nutraceutical Sciences, Pharmaceutical and Nutraceutical Section, University of Florence, Firenze, Italy

1 Introduction

Tungsten or wolframium (W) is a heavy metal of the sixth group of the periodic table of elements with electronic configurations [Xe] $4f^{14}$ $5d^4$ $6s^2$ and atomic number 74 [1,2]. It has an atomic radius of 1.40 Å and an ionic radius of 0.68 Å. It has various oxidation states (from +0 to +6), but only the oxidation states +4, +5, and +6 are biologically important [3 7]. Tungsten is relatively scarce in nature, and it ranks 54 in the abundance of the elements on earth [3,4,8]. But W is valuable because it has the highest fusion point of all metals, great strength, and good conductivity [4]. Furthermore, it is the heaviest element with a biological role [2,9].

Tungsten chemical properties are very similar to molybdenum (Mo) as shown in Table 1 [1–4,8]. Both have a chemical versatility that is useful to biological systems: they are redox-active under physiological conditions (ranging between oxidation states VI and IV); because the V valence state is also accessible, they can transduce two-electron and one-electron oxidation-reduction systems; and they can catalyze reactions such as the hydroxylation of carbon centers under more moderate conditions than are required by other systems. The physiological roles of these enzymes are fundamental, including the catalysis of key steps in carbon, nitrogen, and sulfur metabolism [9,10]. The role of molybdenum in biology has been known for decades and molybdoenzymes are ubiquitous. However, it is only recently that a biological role for tungsten has been established in prokaryotes, although not yet in eukaryotes [4].

For these reasons, for many years, tungsten was considered only a biological antagonist of molybdenum and was used to study the properties and functions of molybdenum in Mo enzymes.

This was because tungsten can replace molybdenum in Mo-enzymes due to the similar properties, forming analogs that are catalytically inactive (or with very low activity) [3,8].

Tungsten has never been investigated for its possible role in biological function until recent years. In 1930, H. Bortels reported a putative function for nitrogen fixation in some *Azotobacter* species by tungsten, but only under certain conditions [2]. Until the 1950s, tungsten was considered only as a selective inhibitor of Molybdoenzymes and was used for studying the properties and functions of molybdenum in enzymes such as nitrate reductase [2,3,11] formate dehydrogenase [2,12] xanthine oxidase, aldehyde oxidase [2,13], and sulfite oxidase [2,3,14].

Since 1970, the influence of W on the growth of a wide variety of microorganisms has been investigated, and in many cases, the effects on various enzyme activities have also been reported [4]. The results of these studies are reported in Table 2 and show that W stimulates the growth of many types of prokaryotes, but a positive influence of W on the growth of a eukaryote has not been demonstrated [4,15–40].

Only in 1990 after the isolation of some hyperthermophilic archaea growing at very high temperatures (80°C or even up to 113°C) that require the addition of tungstate for growth, become accepted that tungsten can also be a positively acting element for certain enzymes, even being essential for some prokaryotes [2,25,41,42].

The best-characterized organisms regarding their metabolism of tungsten are certain species of hyperthermophilic archaea (*Pyrococcus furiosus* and *Thermococcus litoralis*), methanogens (*Methanobacterium thermoautotrophicum* and *Methanobacterium wolfei*), gram-positive bacteria (*Clostridium thermoaceticum*, *C. formicoaceticum*, and *Eubacterium acidaminophilum*), gram-negative anaerobes (*Desulfovibrio gigas* and *Pelobacter acetylenicus*), and gram-negative aerobes (*Methylobacterium* sp. RXM). Of these, only the hyperthermophilic archaea appear to be obligately tungsten-dependent [4].

1.1 Abundance and chemical forms of tungsten

W and Mo are relatively scarce in nature. W is usually found as oxo-rich tungstate minerals [W(VI)], either as scheelite ($CaWO_4$) or wolframite ([Fe/Mn]WO_4), whereas the more reduced tungstenite [WS_2] is very rare because WS_2 is readily solubilized according to Eq. (1) [4].

Metalloenzymes. https://doi.org/10.1016/B978-0-12-823974-2.00007-3

TABLE 1 Physical and chemical properties of tungsten and molybdenum.

	Tungsten (W)	Molybdenum (Mo)
Atomic number	74	42
Atomic weight	183.85	95.94
Atomic radius	1.40Å	1.40Å
Ionic radius	0.68Å	0.68Å
Electronic configuration	[Xe] $4f^{14}$ $5d^4$ $6s^2$	[Kr] $4d^5$ $5s^1$
Electronegativity	1.4	1.3
M=O bond length	1.76Å	1.76Å
Melting point	3410°C	2610°C
Density	19g/cm^3	10g/cm^3
Stable isotopes	182, 183b, 184, 186	92, 94, 95, 96, 97, 98

TABLE 2 Bacteria grow in response to the administration of tungsten into the culture medium.

Genus	Species	Growth substrate	W supply	Reported effect	Enzymes involved	Ref.
Methanogenic archaea	*Methanococcus vannielii*	Formate	100µM +1µM Se	Growth was dramatically enhanced	FDH	[15,16]
		H_2/CO_2	100µM +1µM Se	No effected reported		
	Methanococcus formicicum	H_2/CO_2	W≫Mo	Growth decreases	FDH inactive	[4]
	Methanogenium tatii	Formate		Required W for growth	FDH	[17]
		H_2/CO_2			FMDH	
	Methanogenium liminatans	H_2/CO_2	1–2µM	Growth increase	FMDH	[18]
	Methanoplantus endosymbiosus	Formate	0.1mM	Optimal concentrations for growth	FDH	[19]
		H_2/CO_2			FMDH	
	Methanobacterium wolfei	H_2/CO_2	1µM	Growth increase	FMDH	[20,21]
	Methanobacterium thermoautotrophilum	H_2/CO_2	1µM	Growth increase	FMDH	[22,23]
	Methanocorpusculum parvum	Formate	0.5µM	Optimal concentrations for growth	FDH (Mo and Se-dependent)	[24]
			1µM	Growth was inhibited, this effect could be alleviated by the addition of Mo and Se		
		H_2/CO_2	1µM	Optimal concentrations for growth	FMDH	

TABLE 2 Bacteria grow in response to the administration of tungsten into the culture medium—cont'd

Genus	Species	Growth substrate	W supply	Reported effect	Enzymes involved	Ref.
Hyperthermophiles	*Pyrococcus furiosus*	Sugars	10 μM	Growth increase 5- to 10-fold	AOR, GAPOR, FOR	[25,26]
	Thermococcus litoralis	Peptides	10 μM	Growth increase	AOR, GAPOR, FOR	[4]
	Thermotoga maritima	Carbohydrates	10 μM	Growth increase		[27]
Gram-positive bacteria	*Clostridium thermoaceticum*	Sugars	10 μM	Required W for growth	FDH, CAR	[28–32]
	Eubacterium acidaminophilum	Amino acids	0.1 μM	Growth increase	FDH, CAR	[33]
		Amino acids	0.1 μM Mo	Growth decreases of 70%	FDH (Mo)	
	Clostridium formicoceticum		10 μM	Growth increase	FDH, CAR	[34]
Sulfate-reducing bacteria	*Desulfovibrio gigas*		0.1 μM	Required bot W and Mo for growth	ADH	[35]
Nitrogen-fixing bacteria	*Azotobacter vinelandii*	N_2 or NO_3^-	0.1–10 mM	Growth decreases		[36–38]
		NH_4^+		No effected reported		
Aerobic and facultatively aerobic bacteria	*Proteus mirabilis*			Concentration-dependent decrease		[39]
	Escherichia coli		1 mM	Growth decreases		[40]

$$WS_2 + 4H_2O \leftrightarrows WO_4^{2-} + 2H_2S + 4H^+ + 2e^- \quad (1)$$

The W content of most rock formations is similar to Mo, but in surface waters, the solubility and mobility of W are low compared to Mo. So, only in four types of natural water tungsten is present in significant amounts:

- groundwaters directly associated with W-containing ore deposits.
- Alkaline, nitrogenous fissure—vein thermal waters of crystalline rocks.
- Alkaline waters of lakes in arid zones.
- Hot-spring water.
- Hydrothermal vents.

Deep-sea hydrothermal vents represent unique environments in which to consider W accumulation and availability [4], and it is no coincidence that most of the hyperthermophilic organisms, which possess tungstoenzymes, were discovered in marine hydrothermal vent waters where W is more available [1,9].

2 Tungsten an ancestral precursor of molybdenum

To date more than 50 different Mo-dependent enzymes have been found in all kingdom's life, in fact, Mo-containing enzymes are ubiquitous in nature; but, strangely, only relatively few organisms utilize tungsten. It is very puzzling if we consider the chemical similarities between tungsten and molybdenum and the fact that both metals are coordinated by the same organic cofactor [1,9,10].

So, could the modern scarcity of tungstoenzymes reflect the early earth scenario? It's possible that W-enzymes were the predecessor of their Mo counterparts? Could the "pollution "of the atmosphere by oxygen force organisms to use molybdenum (available as highly water-soluble MoO_4)? [1,9].

This is a plausible scenario if considering tungsten's higher availability under euxinic conditions and its chemical features (instead of the similarities between the two metals). Under euxinic conditions (sulfidic and anoxic conditions), tungsten forms relatively soluble salts (WS_4^{2-}), and it was therefore probably more available in the euxinic ocean than molybdenum (which would have been present as the water-insoluble MoS_2) [1,9]. The same reasoning explains the higher tungsten availability in today's marine hydrothermal vent waters [1,3,4,9,10]. Furthermore, tungsten compounds exhibit lower reduction potential than molybdenum compounds, so they were, therefore, more useful to early life forms, which probably had a low intracellular redox pose. Moreover, tungsten compounds have higher bond strengths and enhanced thermal stability compared to isostructural molybdenum counterparts but are more sensitive to dioxygen [1,3,8,10,43,44]. These differences support the idea that tungsten would have been a better choice for anaerobic low-reduction potential reactions carried out under euxinic conditions at higher operating temperatures. For this reason, tungsten was likely an essential element for the earliest life forms [1,10,45].

However, as a result of the cooling of the earth's crust and the development of photosynthetic organisms that have made the atmosphere aerobic, the O_2 sensitivity of tungsten compounds became a liability and the water solubility of molybdenum oxides an asset. Thus, the surviving organisms could have exploited the chemical similarities between the two metals to evolve enzymes that enabled them to continue catalyzing the same old reactions and new reactions imposed by the new environment [1,3,4,9,10].

This hypothesis is supported by the biological distribution of Mo/W-enzymes today, where the molybdenum-containing enzymes are found in all aerobic organisms, whereas tungsten occurs only in obligate, typically thermophilic, anaerobes which grow in conditions like the euxinic ones where the earliest life forms developed. Some mesophilic anaerobes are transitional in that they can utilize either molybdenum or tungsten depending on availability and growth conditions [1,3,4,9,10].

3 Tungsten-containing enzymes

Organisms use tungsten to catalyze oxidation-reduction reactions, most of which involve oxygen atom transfer to/from a carbon, nitrogen, and sulfur atom of key metabolites [8,10,45–49]. They are redox-active under physiological conditions and their oxidation state can range from 6^+, 5^+, and 4^+, this versatility allows tungsten-containing enzymes to catalyze either two-electron ($M^{6+} \leftrightarrow M^{4+}$) or one-electron ($M^{6+} \leftrightarrow M^{5+}, M^{5+} \leftrightarrow M^{4+}$) oxidation-reduction reactions [1].

To date, all tungsten enzymes described contain a pyranopterin-dithiolene cofactor in which the metal is coordinated by the dithiolene moiety [9]. The active sites of all well-characterized tungsten-containing enzymes are mononuclear, with a single equivalent of the metal coordinated by the cis-dithiolene group of two molecules of the same pyranopterin cofactor found in molybdoenzymes [1,8,10,50]. The tungsten coordination sphere is completed with oxygen and/or sulfur and/or selenium atoms in a diversity of arrangements (Fig. 1) [1,10,51].

3.1 The pyranopterin cofactor

The cofactor was first discovered by Johnson and coworkers, who studied its oxidized inactive form [52,53]. This was initially formulated in the ring-open form (Fig. 1A) [9]. In recent years, the crystal structures of several proteins of this class have been reported and these structures show that the metal atom is coordinated by theene-1,2-dithiolate unit, as originally proposed by Rajagopalan [52–55]. However, all these structures show one notable difference from the original proposal; the special pterin adopts a tricyclic structure (Fig. 1B) rather than the bicyclic structure (Fig. 1A). The third ring is a pyran ring formed by an *O,N*-acetal linkage via the hydroxyl group of the side chain [54].

The cofactor is often referred to in the literature as "molybdenum cofactor," abbreviated Moco, because it was originally believed to be present only in molybdenum enzymes [54–57]. The metal-free apo-form was termed "molybdopterin" (MPT) (Fig. 1C) [54]. For some enzymes from bacterial sources, the phosphate group of Fig. 1C is linked to a nucleotide derived from bases such as guanine and cytosine. The nomenclature of these systems with two phosphate groups provides that the first letter, "M," represented molybdopterin; the second letter designated the base; and the third letter was "D" for dinucleotide. Thus, the compound shown in Fig. 1D was called molybdopterin guanosine dinucleotide (MGD), although the pterin part of the "dinucleotide" does not actually have sugar attached to the base. Similar acronyms have been put forward for related dinucleotides involving other bases: cytosine is MCD; adenine is MAD; the cofactor containing the nucleotide with the purine base hypoxanthine has been termed MHD [54]. However, with the isolation of new proteins containing a tungsten ion bound to molybdopterin the situation has changed dramatically [8,54,57,58]. Although the term molybdopterin is still often used, but the alternative name pyranopterin is less confusing [57].

3.1.1 *Structure of the pyranopterin cofactor*

MPT is a tricyclic ligand that possesses a dithiolene moiety fused to a heterocyclic pterin ring system by a bridging

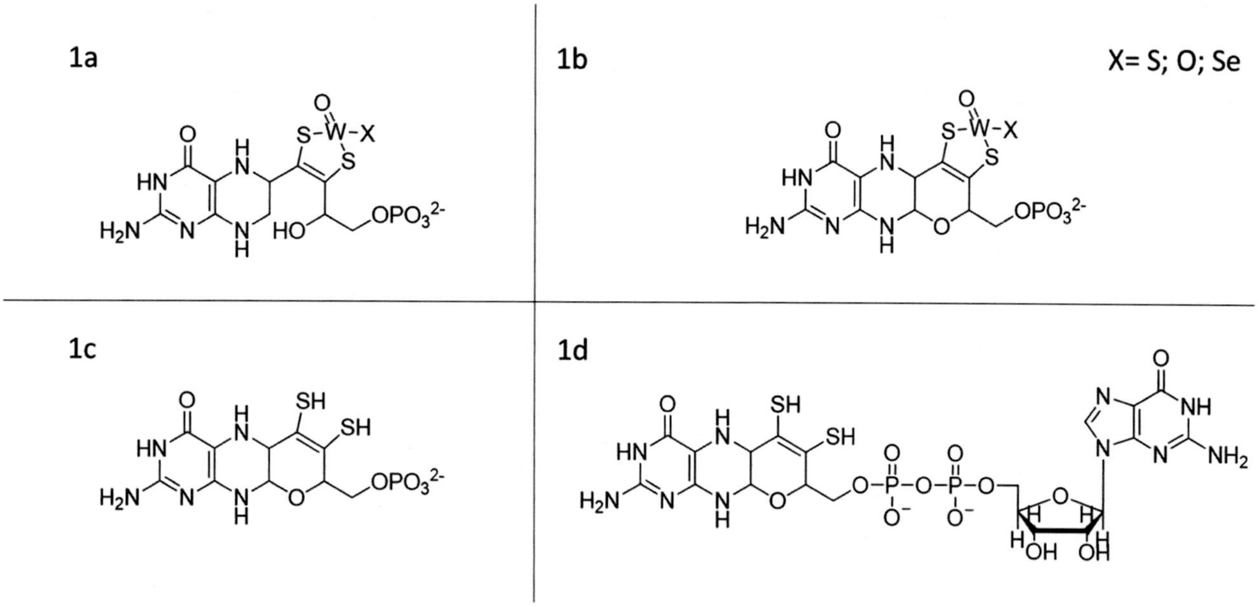

FIG. 1 The chemical structure proposed for the pyranopterin cofactor. (A) First structure proposed by Rajagopalan. (B) The tricyclic structure found in some W-enzymes. (C) The metal-free apo-form (MPT). (D) The molybdopterin guanosine dinucleotide form (MGD).

pyran ring. An alternative form of MPT may also exist as a pyran ring-opened structure in some enzymes [59–61]. MPT represents one of the most electronically flexible ligands in biology [62]. This electronic flexibility derives from the fact that both the dithiolene and pterin components of MPT are redox-active [63]. The MPT is known to be extensively hydrogen bonded to the enzymes but is not covalently attached to the protein. Surprisingly, different degrees of MPT nonplanarity related to protein function have been observed, and the nature of MPT conformations has been shown to correlate with a given family of enzymes [63].

To date, in all higher organisms studied, Moco is synthesized by a conserved biosynthesis pathway that can be divided into four steps:

1. Conversion of GTP to cyclic pyranopterin monophosphate
2. Synthesis of molybdopterin
3. Adenylation of molybdopterin
4. Metal Insertion into molybdopterin [64].

Biosynthesis has already been carefully explained in Chapter 10, so it will not be further explored in this chapter.

3.1.2 *Physiological role of the pyranopterin cofactor*

The primary role of the pterin cofactor is to position the catalytic metal (Mo or W) in the active site [57]. However, the pyranopterin cofactor is not considered an "innocent scaffold," [62,65–68] but has also been suggested to facilitate intramolecular electron transfer, acting as a "wire" to conduct electrons to other redox-active centers that are found in these enzymes [1]. Moreover, because the pyranopterin cofactor has several potential structural isoforms and oxidation states [69,70] each adopting a different geometry, it has recently been suggested that each enzyme holds a binding pocket that selectively controls the pyranopterin conformation, mainly in the dihydro and tetrahydro reduced forms, to regulate the oxidation state of the metal center and facilitate the transfer of electrons [71]. Also, two crystal structures have been reported in which the pyran ring of the cofactor is in the open form, that of the respiratory *Escherichia coli* and that of *Aromatoleum aromaticum* ethylbenzene reductase [59–61]. Although it can be suggested that the opening and closing of the pyran ring might provide protons for the reaction [1].

3.2 Classification

Mo and W-enzymes are divided into four macro groups:

- The xanthine oxidase family
- The sulfite oxidase family
- The DMSO reductase family
- The tungsten family [1]

The tungsten family is further subdivided into three families:

(1) **AOR** family that comprises the aldehyde ferredoxin oxidoreductase (AOR) a homodimeric enzyme that oxidizes a broad range of both aliphatic and aromatic aldehydes and shows the highest catalytic efficiency

with the aldehyde derivatives of the common amino acids, such as acetaldehyde (from alanine), isovaleraldehyde (from valine), and phenylacetaldehyde (from phenylalanine) [25,72]. AOR is thought to play a key role in peptide fermentation [73]. Formaldehyde ferredoxin oxidoreductase (FOR) is a homotetrameric enzyme that has a very limited substrate specificity and only oxidizes small (C1-C3) aliphatic aldehydes [74]. Glyceraldehyde-3-phosphate ferredoxin oxidoreductase (GAPOR) is a monomeric enzyme specific for the oxidation of glyceraldehyde 3-phosphate (GAP) [8,75]. GAPOR is thought to play a key role in the unusual glycolytic pathway in the fermentative hyperthermophiles [8,75,76]. All three enzymes are unable to use nicotinamide nucleotides as electron carriers. Furthermore, AOR and FOR do not oxidize aldehyde phosphates and the presence of zinc in GAPOR is not accidental. AOR formed a dimer and maintained the substrate range, FOR acquired much more limited specificity perhaps in part by forming a tetrameric structure, while GAPOR remained monomeric and acquired its absolute specificity for GAP in part by incorporation zinc, which specifically orients the negatively charged phosphate to the catalytic site [8]. Carboxylic acid reductase (CAR) is the fourth enzyme of this family. CAR was first identified by its ability to catalyze the reduction of unactivated carboxylic acids; however, it also catalyzes the reverse reaction, the oxidation of aldehyde [8]. The last tungstoenzyme that is included in the AOR family is aldehyde dehydrogenase (ADH) [8].

(2) **F(*M*)DH** is the second family and involves two types of enzymes, both of which utilize CO_2 as the substrate. One is formate dehydrogenase (FDH), which was the first W-containing enzyme to be purified. It is a complex tetrameric enzyme of two different subunits that catalyzes the first step in the conversion of CO_2 into acetate (the production of formate), using NADPH as a physiological electron donor [8,77]. The other member of the second class of tungstoenzyme is *N*-formylmethanofuran dehydrogenase (FMDH). This enzyme catalyzes the first step in the conversion of CO_2 to methane in methanogens, using methanofuran as substrate [8].

(3) **AH** is the last family, and it includes a single enzyme, acetylene hydratase (AH) that catalyzes the hydration of acetylene to acetaldehyde. It catalyzes hydration in contrast to the oxidoreductase-type reaction catalyzed by all other tungstoenzymes [8,10,78].

However, the classification of tungstoenzymes has been a matter of some controversy. Some authors group all the tungsten-containing enzymes into a single family (as described above), characterized by the presence of a tungsten-bis-pyranopterin system. Others prefer to classify the tungsten-containing aldehyde: ferredoxin oxidoreductase separately, in a distinct family, with the other tungstoenzymes in the molybdenum-containing dimethyl sulfoxide reductase family. This latter classification is based on the fact that the vast majority of prokaryotic molybdo- and tungstoenzymes have an active site with the metal coordinated by two molecules of pyranopterin esterified to guanosine monophosphate, forming a pyranopterin guanosine dinucleotide (which is not the case of aldehyde: ferredoxin oxidoreductase, whose pyranopterin cofactor molecules are in mononucleotide form) [1].

4 Aldehyde ferredoxin oxidoreductase (AOR)

Aldehyde ferredoxin oxidoreductase catalyzes the oxidation of a range of aliphatic and aromatic aldehydes reducing ferredoxin, the proposed physiological electron acceptor, but not NAD(P) [9].

AOR has been identified in various microorganisms including *Pyrococcus furiosus* [78], *Thermococcus* strain ES-1 [72,79] *Pyrococcus endeavori* (previously named *Pyrococcus* ES-4) [80], *Thermococcus litoralis* [4] *Pyrobaculum aerophilum* [80], *Moorella thermoacetica* [81], *Clostridium* formicoaceticum [34], *Eubacterium acidaminophilum* [82], and *Desulfovibrio gigas* [83].

The first X-ray crystal structure of ferredoxin oxidoreductase aldehyde was reported by *Pyrococcus furiosus* and is also the best characterized. However, thanks to the kinetic analysis of the AOR from *Thermococcus* strain ES-I has shown that the most efficient aldehyde substrates are acetaldehyde, phenylacetaldehyde, indoleacetaldehyde, and isovaleraldehyde, which correspond to the oxidation products of amino acids. This led to the proposal that the hyperthermophiles AORs function in vivo to oxidize aldehydes generated from two-keto acids, which are key intermediates in both peptide and carbohydrate catabolism [9,72,79].

4.1 General features

AOR is a homodimer composed of two subunits of 605 residuals (66 kDa) that are bridged by an iron atom (Fig. 2A). Moreover, two other different types of metal sites are found in the AOR protein dimer; a [4Fe:4S] cluster and a tungsten site coordinated by two molybdopterin molecules, which are contained within each subunit. The [4Fe:4S] cluster and the tungsten cofactor within a subunit are very close together, but the mononuclear iron center is positioned at 25 Å by these groups. The [4Fe:4S] clusters and tungsten cofactors in different dimer subunits are separated by 50 Å [72,78].

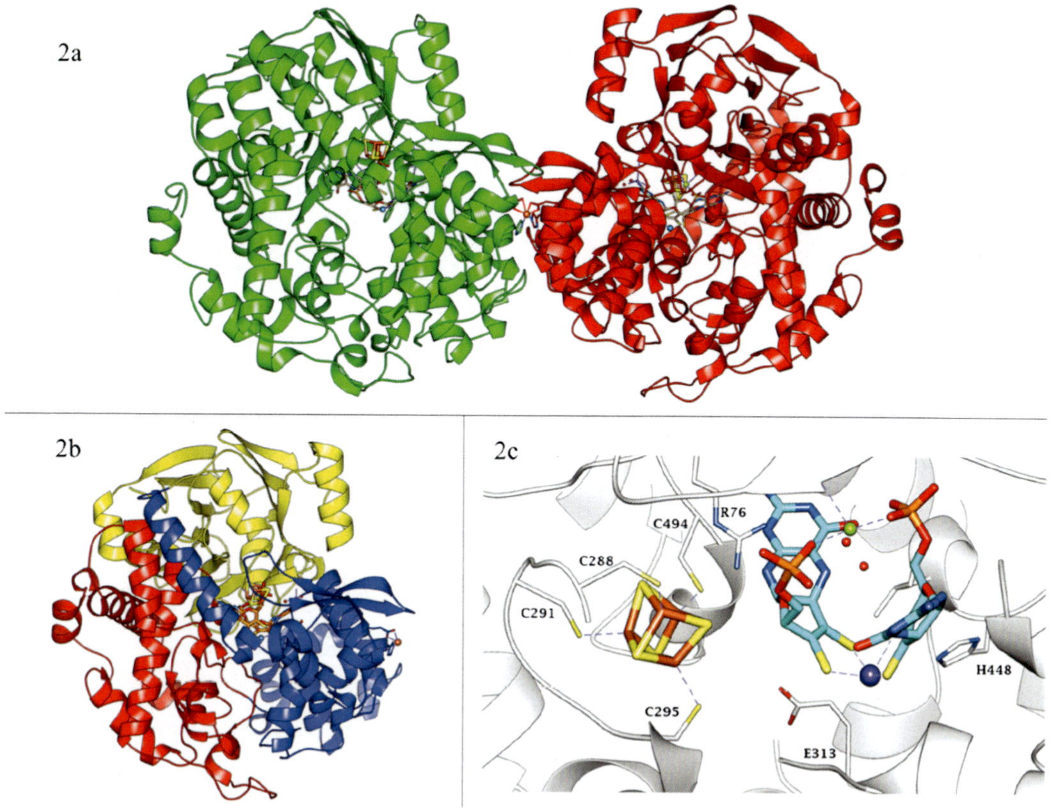

FIG. 2 The structure of the aldehyde ferredoxin oxidoreductase from *Pyrococcus furiosus.* (A) The homodimer structure with at the center the mononuclear iron *(orange)* coordinated by Glu332 and His383 from each subunit. (B) Representation of the domain's subdivision in each subunit: domain 1 *yellow*, domain 2 *blue*, and domain 3 *red*. (C) Detail of the tungsten center where we can see the W atom *(violet)* coordinated by the dithiolene sulfurs from each cofactor; the magnesium ion *(green)* bridge between the phosphate groups of the cofactors, the four cysteines that coordinate the [4Fe:4S] cluster and Glu313 and His448 that participate in proton transfers.

4.2 Structure of the metal sites

In the mononuclear site, iron is symmetrically coordinated by Glu332 and His383 from each subunit in a tetrahedral fashion (Fig. 2A). The mononuclear iron is approximately 25 Å away from the other metal centers, and a redox role for this iron was considered unlikely, so it may be involved in stabilizing the protein dimer at high temperatures [78,80]. However, the hyperthermophilic AOR of *Pyrobaculum aerophilum* is a homodimer, although it does not contain the binding motifs for the mononuclear iron site. Therefore, also the mononuclear iron in *Pyrococcus furiosus* AOR may not be essential for the dimerization of the enzyme at elevated temperatures [80].

The tungsten atom, located at the center of each subunit, is coordinated by two molecules of pyranopterin by the dithiolene sulfurs. The side chains of residues Glu313 and His448 near the substrate binding site of this center could participate in proton transfers associated with oxidation-reduction reactions. In addition to interactions between the dithiolene sulfurs and tungsten, the two MPT ligands are also linked through their phosphate groups, which coordinate the same magnesium ion. Two water molecules in this center form an intricate, asymmetric hydrogen bonding network within the tungsten cofactor (Fig. 2C) [78].

The X-ray structure reveals that the [4Fe:4S] cluster is positioned approximately 10 Å from the tungsten atom and buried 6 Å below the van der Waals surface of the protein. This arrangement is consistent with the postulated role of the [4Fe:4S] cluster as an intermediary for electron transfer between the tungsten cofactor and ferredoxin, the physiological electron acceptor of AOR [78]. As we can see in Fig. 2C, four cysteine ligands Cys288, Cys291, Cys295, and Cys494 coordinate the [4Fe:4S] cluster. Where the first three Cys residues are part of a characteristic iron-sulfur cluster binding sequence. The [4Fe:4S] cluster is linked to one of the two pyranopterins by two distinct sets of interactions:

- the side chain of Arg76 that bridges, thanks to a hydrogen bond, inorganic sulfur of the [4Fe:4S] cluster and two sites on the pyranopterin; and
- the sulfur of Cys494 (one of the four [4Fe:4S] cluster ligand) which is positioned to accept a hydrogen bond from the pterin ring nitrogen N-8.

These interactions could provide electron transfer pathways between the two centers [78].

4.3 Protein structure

The binding sites for the tungsten cofactor and [4Fe:4S] cluster are located at the interfaces of the three domains that are formed by each AOR subunit. Domain 1 (yellow in Fig. 2B) consists of 12 β strands arranged in two 6-stranded P3 sheets, predominantly antiparallel. Residues Asn93 and Ala183, from domain 1, coordinate the magnesium ion through their carbonyl oxygens. Within this domain, there are also the Arg182 residue, which forms a salt bridge to the pyranopterin phosphate group closest to the [4Fe:4S] cluster, and the Arg76 that binds hydrogen to both, pyranopterin oxygen and a sulfur inorganic [4Fe:4S] [78]. The secondary structure of domains 2 (blue) and 3 (red) consists primarily of α helices, with few β sheet regions. Both contain, respectively, Asp-X-X- Gly-Leu-(Cys or Asp)-X sequences, where the Asp carboxylate group and Leu main chain carbonyl oxygen are arranged to bind the primary amine group of their respective MPT ligand. Asp343 and Cys494 interact with the secondary ring nitrogen, N-8, near the ether linkage, whereas Thr344 and Leu495 form a hydrogen bond with a secondary ring nitrogen on the pterin [84,85]. In addition to the structural role, domains 1 and 2 can regulate substrate access to the AOR. In fact, between the two domains, it is possible to define a long hydrophobic channel that leads from the surface of the protein to the tungsten site where the nonpterin coordination sites on the tungsten atom are located [78,86].

4.4 Activators and inhibitors

AOR was extremely sensitive to inactivation by O_2. Moreover, its activity is inhibited at different concentrations of these substances: p-chloro-mercuribenzoate (10 M), ZnCl2 (0.25 mM), iodoacetate (0.5 mM), sodium arsenite (0.5 mM), and 2,2-bipyridyl (10 mM). To date, however, there are no drugs in use that act on this target [72].

5 Formaldehyde ferredoxin oxidoreductase (FOR)

The enzyme formaldehyde ferredoxin oxidoreductase of the AOR family oxidizes only short-chain (at most C4) unsubstituted aldehydes, but even with these substrates the catalytic efficiencies are very low, and the physiological reaction of FOR is not known [87]. FOR, like AOR, utilizes ferredoxin as a physiological electron acceptor [88]. It has been purified from *Thermococcus litoralis* [74] and *Pyrococcus furiosus* [89].

5.1 General features

FOR and AOR are closely related as their subunits are similar in size and exhibit a high sequence identity. Residues of functional importance in AOR are conserved in FOR as the four cysteine residues coordinate the single cluster [4Fe:4S] and the first of two pterin-binding motifs. However, these two proteins differ in the state of oligomerization, as FOR exists as a homotetramer (Fig. 3A), while AOR is a dimer (Fig. 2A). FOR, also

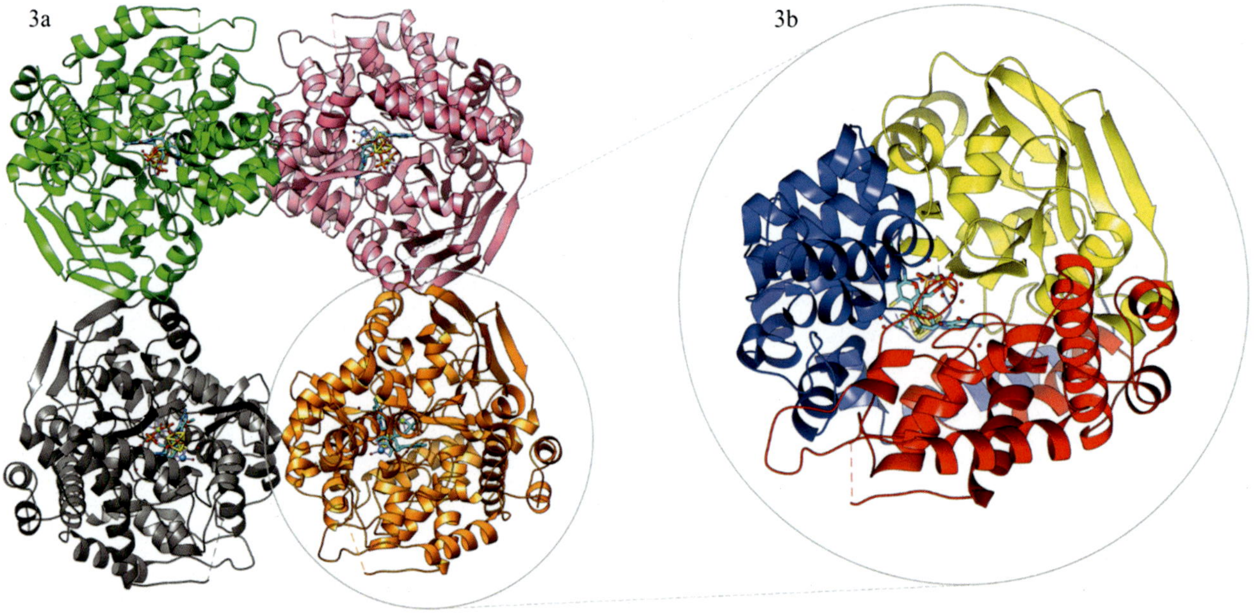

FIG. 3 The structure of the formaldehyde ferredoxin oxidoreductase. (A) The homotetramer structure with a channel of 27 Å diameter in the center. (B) Detail of the domain's subdivision in each monomer: domain 1 *yellow*, domain 2 *blue*, and domain 3 *red*.

contains calcium (one Ca^{2+} ion for subunit) but does not have the iron metal coordination residues [90].

5.2 Structure of the metal sites

As for AOR, the tungstopterin moiety of FOR consists of two tricyclic pterin molecules and a tungsten atom, coordinated by all four dithiolenic sulfur atoms. The two pterin molecules are further linked together by a magnesium ion, which bridges the phosphate groups of each pterin. In addition to the four sulfur atoms, another ligand was found to coordinate the W atom, and it is assumed to be an oxygen atom [90]. The pterins in each subunit are sandwiched between domain 1 on one side and domains 2 and 3 on the other. Two phosphate groups, one from each pterin, form salt bridges with the sidechains of Lys75, Arg180, and Lys438, and hydrogen bonds with numerous amino groups from the main chain. Except for the Lys438 sidechain, all the protein groups involved in the binding of the phosphate groups are from domain 1, while only domains 2 and 3 are involved in the interactions with the ring systems of the two pterin atoms. The ring systems of the two pterins each interact with a pterin-binding motif corresponding to residues Asp333-Asp338 and Glu486-Cys491 [90].

The calcium ion was found to provide additional interactions between the first pterin and the protein. This site is 7.8 Å from a tungsten atom and is octahedrally coordinated by the carbonyl oxygen atoms of the first pterin and residues Gly179, the side-chain carboxylate groups of Glu304 and Asp306, and two water molecules. One of the water molecules is also hydrogen bonded to the phosphate group of the second pterin. This site is inaccessible to the substrate and is not in the putative electron transfer pathway, so, its function is likely to be structural [90].

The [4Fe:4S] cluster is located 10 Å from the tungsten atom and is buried 6 Å below the protein surface. As anticipated, it is coordinated by the S atoms of Cys284, Cys287, Cys291, and Cys491. The cluster is linked also to the second pterin through several hydrogen bond interactions. The mononuclear iron site found in AOR at the center of the dimerization interface is not present in FOR [90].

5.3 Protein structure

As shown in Fig. 3A, FOR is a tetramer with a channel of 27 Å diameter in the center, so each subunit contacts only two of the other three subunits in the tetramer.

The FOR monomer is approximately spherically shaped with a diameter of 60 Å. The polypeptide chain (as was observed for AOR) consists of three domains: the ellipsoid-shaped domain 1 that forms the base of the protein, while domains 2 and 3 are on top of the first domain (Fig. 3B). The pterin and the [4Fe:4S] cluster are located at the interface of the domains [90]. The overall secondary structure of FOR is dominated by α-helices. The channel that leads from the surface of the protein to the tungsten site would be partly filled in FOR than in AOR, as the side chains of the Arg481 and Arg492 residues replace the smaller residues Thr485 and Leu495, respectively, in AOR, furthermore, the insertion of a residue between Ile480 and Gly483 further reduces the cavity volume in FOR shifting the main chain atoms toward the center of the cavity. These variations may contribute to the differences in substrate specificity between FOR and AOR [90].

5.4 Reaction mechanism

Thr240, Asp306, Tyr307, Glu308, Tyr416, and His437 have very similar conformations in AOR and FOR and are conserved in AOR-like enzymes that are involved in the catalytic mechanism. Many reaction mechanisms have been proposed for formaldehyde ferredoxin oxidoreductase; however, the reaction in Fig. 4 appears to be the most likely. Tyr416 is proposed to participate in the enzymatic mechanism by donating a hydrogen bond to the carbonyl oxygen atom of the substrate, thus serving to activate the aldehyde group to undergo the nucleophilic attack of the W═O group. During the nucleophilic attack, the tungsten ion stabilizes the negative charge developed on the formaldehyde oxygen. So, the two negatively charged oxygens are coordinated to tungsten with similar bond distances, and one of the two protons attached to the diol carbon is taken by Glu308. This step is not a simple proton transfer between a C—H bond and a carboxylate, because concomitantly with the proton transfer, two electrons are transferred from the

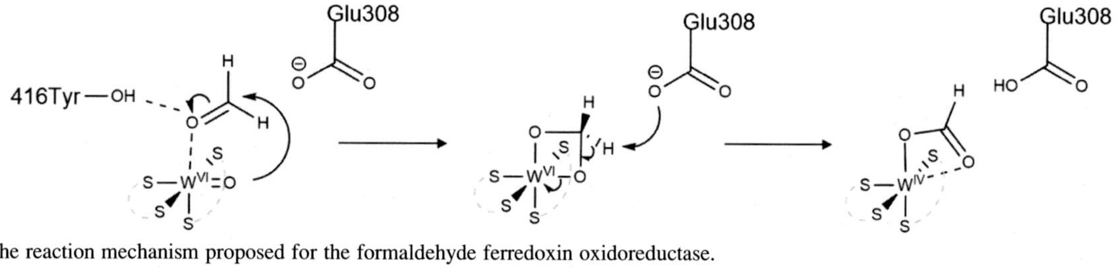

FIG. 4 The reaction mechanism proposed for the formaldehyde ferredoxin oxidoreductase.

substrate to W^{VI} to afford W^{IV}. Following substrate oxidation, electrons from the reduced tungsten site are proposed to be transferred to the [4Fe:4S] cluster of FOR, and then to the [4Fe:4S] of ferredoxin. Electron transfer from W to the cluster of FOR could be facilitated by the second pterin of the tungsten bis(pterin) site, through a hydrogen bond between the pterin N8 atom and the S atom of Cys491. Other possible pathways for coupling these two redox centers are through the side chains of Arg180 or Lys75, which form hydrogen bonds or salt bridges with both the second pterin and the inorganic sulfur atoms of the cluster [88,90].

6 Glyceraldehyde-3-phosphate ferredoxin oxidoreductase (GAPOR)

The third tungstoenzyme of the AOR family is glyceraldehyde-3-phosphate ferredoxin oxidoreductase (GAPOR), which, to date, has been purified only from *Pyrococcus furiosus*. GAPOR oxidizes only glyceraldehyde 3-phosphate (GAP) and is thought to play a key role in the unusual glycolytic pathway in fermentative hyperthermophiles [8]. This enzyme is the least characterized of the three; however, its N-terminal amino acid sequence and its size and W and Fe contents strongly suggest that GAPOR is very closely related in structure. Unlike AOR and FOR, GAPOR is monomeric, moderately sensitive to oxygen, and has 1 W and 6 Fe atoms/mol; it also contains 2 Zn atoms/mol, an element not found in AOR or FOR [75].

The Tungsten center appears to be coordinated by no more than three sulfur ligands. This suggests that it may be coordinated by the dithiolene side chain of a single pterin cofactor or that two pterin cofactors are attached with one weakly coordinated through only one dithiolenic sulfur in reduced forms [4].

GAPOR did not oxidize nonphosphorylated aldehydes such as formaldehyde, acetaldehyde, crotonaldehyde, or benzaldehyde, but its specific for glyceraldehyde 3-phosphate (GAP). Its specificity is due to the incorporation of zinc, which is close to the tungsten atom within the hydrophobic channel and could serve to bind the GAP phosphate group to place the aldehyde group close to the W atom. However, further studies will be required to validate this postulate [75].

7 Carboxylic acid reductase (CAR)

Carboxylic acid reductase was the second tungsten-containing enzyme to be identified. Its named CAR for the ability to reduce unactivated carboxylic acids, but it also catalyzes aldehydes oxidation (the reverse reaction) and for this reason, it has been included in the AOR family [31]. CAR was first obtained and purified in 1989 from the thermophilic acetogen *Clostridium thermoaceticum* [31], and 2 years later, also from the mesophilic acetogen *C. formicoaceticum* [34]. They have a broad spectrum of substrate specificity for carboxylic acids and both aliphatic and aromatic aldehydes can be oxidized [34]. The affinity of CAR for aldehydes is much higher than for acids suggesting that the oxidation of the aldehyde is the physiological role. However, its function in vivo is not yet clear, and the natural electron carriers are not known [4]. Although *C. thermoaceticum* and *C. formicoaceticum* CARs have the same function and similar properties, their structures are very different:

- CAR from *C. formicoaceticum* is a homodimer of two Alfa subunits with an M_r of 67 kDa, each one also contains 1 W atom, 5–6 atoms of Fe, 1 mol of pterin (that appear to be in the nonnucleotide form), and 8 acid-labile sulfide atoms [4,34].
- In *C. thermoaceticum*, two forms of CAR have been identified. CAR I is a heterodimer $\alpha\beta$ with an M_r of 86 kDa (64 kDa for α and 14 kDa for β) that contains 1 W atom, 29 atoms of Fe, 1 mol of pterin in the mononucleotide form, and 25 acid-labile sulfide atoms. CAR II, on the other hand, is a large trimeric complex (about 300 kDa) with a possible $\alpha_3\beta_3\gamma$ stoichiometry. α and β subunits have the same M_r shown in CAR, I while γ has a M_r of 43 kDa. CAR II contains about three times the amount of tungsten, iron, pterin, and acid-labile sulfide present in CAR I, and there are also 2 mol of FAD in the γ subunit. According to this consideration, CAR II, but not CAR I, was able to oxidize aldehydes with NADP, as well as reduce $NADP^+$ using reduced viologen as the electron donor. So, thus all three reactions below reported (Eqs. 2–4) are possible in the ternary complex, while the dimer can perform only the first reaction [31,34,81,91].

$$RCHO + 2V^{2+} + OH^- \leftrightharpoons RCOO^- + 2H^+ + 2V^{+\bullet} \quad (2)$$

$$RCHO + NADP^+ + OH^- \leftrightharpoons RCOO^- + NADPH + H^+ \quad (3)$$

$$NADPH + 2V^{2+} \leftrightharpoons NADP^+ + 2V^{+\bullet} + H^+ \quad (4)$$

8 Aldehyde dehydrogenase (ADH)

The last enzyme of the AOR family is represented by aldehyde dehydrogenase, and it is the first enzyme containing tungsten isolated in gram-negative bacteria, in detail from the sulfate-reducing bacterium, *Desulfovibrio gigas* [83]. Unlike CAR, the activity of ADH is limited to the oxidation of aldehydes, where acetaldehyde, propionaldehyde, and benzaldehyde are excellent substrates, but also a broad spectrum of aliphatic and aromatic aldehydes can be oxidated. ADH is a homodimer and each subunit, of 62 kDa, contains approximately 0.6 W and 5 Fe atoms. The form of the pterin-type cofactor has not been identified, and the physiological electron carrier is also unknown. However, two potential redox vectors have been identified in *D. gigas* that could function as natural acceptor electrons, flavodoxin and ferredoxin. Given the observation derived from the other enzymes of this class, ferredoxin could be the physiological electron carrier for ADH [83].

ADH, as shown for all the enzymes of the AOR family, is extremely sensitive to O_2; moreover, its activity is strongly inhibited by the presence of cyanide, arsenite, and iodoacetate [83].

9 Formate dehydrogenase (FDH)

Formate dehydrogenases catalyze the reversible two-electron oxidation of formate to carbon dioxide [92], according to Eq. (5):

$$HCOO^- \leftrightharpoons CO_2 + H^+ + 2^{e-} \quad (5)$$

FDH is the first tungstoenzyme to be discovered, from *C. thermoaceticum* [2], since that time it is one of the best-characterized tungsten-containing enzymes, and it has also been found in many other bacteria such as *C. formicoaceticum* [93], *C. acidiurici* [33], *C. carboxidovorans* [94], *D. gigas* [95], *D. alaskensis* [96], and *Methylobacterium extorquens* (the only example of a W-dependent enzyme in aerobic bacteria) [29].

FDHs are a large group of heterogeneous proteins, diffuse in all life kingdoms, that have different cellular localization (cytoplasmic, periplasmic, and membrane-bound), and are involved in a multiplicity of pathways, in both biosynthetic and energy metabolism. In fact, prokaryotes connect FDH directly, or through some electron carriers, to other enzymes to exploit the redox potential generated by FDH in multiple energy and metabolic pathways [97].

9.1 General features

FDH is a broad enzyme class and can be divided into two main groups based on metal content: (i) metal-independent FDH class, most common in aerobic bacteria and (ii) metal-containing FDH class, found only in prokaryotes. The latter is divided into Mo-FDH and W-FDH, and each is further divided into two other classes based on dependence on NAD^+ [97]. We will focus only on the W-FDH class. In all these enzymes, the tungsten atom is coordinated with the four dithiolate sulfur atoms from the two cofactors, present as guanine dinucleotide (MGD), a residue of cysteine or selenocysteine, and a terminal sulfido ligand that is essential for catalytic activity. Furthermore, histidine and arginine residue are highly conserved near the active site [98].

The simplest W-FDH is from *C. carboxidovorans*, it is monomeric (80.7 kDa) and contains only one W-center and a single [4Fe:4S] cluster [94]. Going up in complexity we find W-FDH from *Methylobacterium extorquens*. These enzymes, NAD-dependent, are localized to the periplasm rather than cytoplasm and is a heterodimer of two subunits ($\alpha_1\beta_1$) of 107 and 61 kDa, respectively [29]. The most complex are the enzymes of the Mo-FDH class so we will not go into their details [98]. In the next chapters, we will analyze in detail the FDH structure from *D. gigas*, the best characterized.

9.2 Structure of the metal sites

The structure of the active site follows the typical characteristics shown above. The tungsten atom is coordinated by two pterin cofactors (MGD1001 and MGD1002), MGD1002 is oxidized with two planar sulfur bonds, while MGD1001 appears to be reduced. SeCys158 and one hydroxyl/sulfide ligand complete the coordination. Near this center, there are His159, involved in the reaction mechanism, and Arg407, which is responsible for substrate orientation [99]. The active site is buried 25 Å from the protein surface, so formate reaches that center through the tunnel formed by many histidine, lysine, and arginine. Among that Arg407, which provides a positive charge and a hydrogen bond to orient the substrate molecule for catalysis, while Hys159 establishes a π interaction with the Se atom of Se-Cys158. Perpendicular to the "formate cleft," there is a channel formed by side chains of proton-able amino acids (Glu409, Asp132, Glu134, and Glu579) and a chain of water molecules that brings out the H^+ molecule generated by the reaction [99]. The other product of the reaction, CO_2 is carried out by a further channel formed by the residues Val412, His159, Tyr428, Arg172, Arg734, Trp458, Trp459, Trp693, and Trp730. All four [4Fe:4S] clusters are 9–10 Å apart [100], providing an easy electron pathway to the physiological electron donor (monoheme cytochrome). The electrons reach the first [4Fe:4S] clusters, from MGD1002, thanks to the nitrogen of the Lys56 [99].

9.3 Protein structure

D. gigas-FDH is a heterodimer (αβ) of 110 and 24 kDa subunits (Fig. 5).The large one (977 amino acids) can be divided into four domains, which contain the W-center and one of the four [4Fe:4S] clusters. The N terminus residues of the large subunit embrace the small subunit, strengthening the interaction between both subunits. The first domain (domain 1) carries the characteristic cysteine motif (-CXXCXnCXmC-) that binds the [4Fe:4S] center. Domain 2 of 205 amino acids is defined by three polypeptide traits, one of these protrudes over the cleft that gives access to the buried active site [99]. Domain 3 is defined by

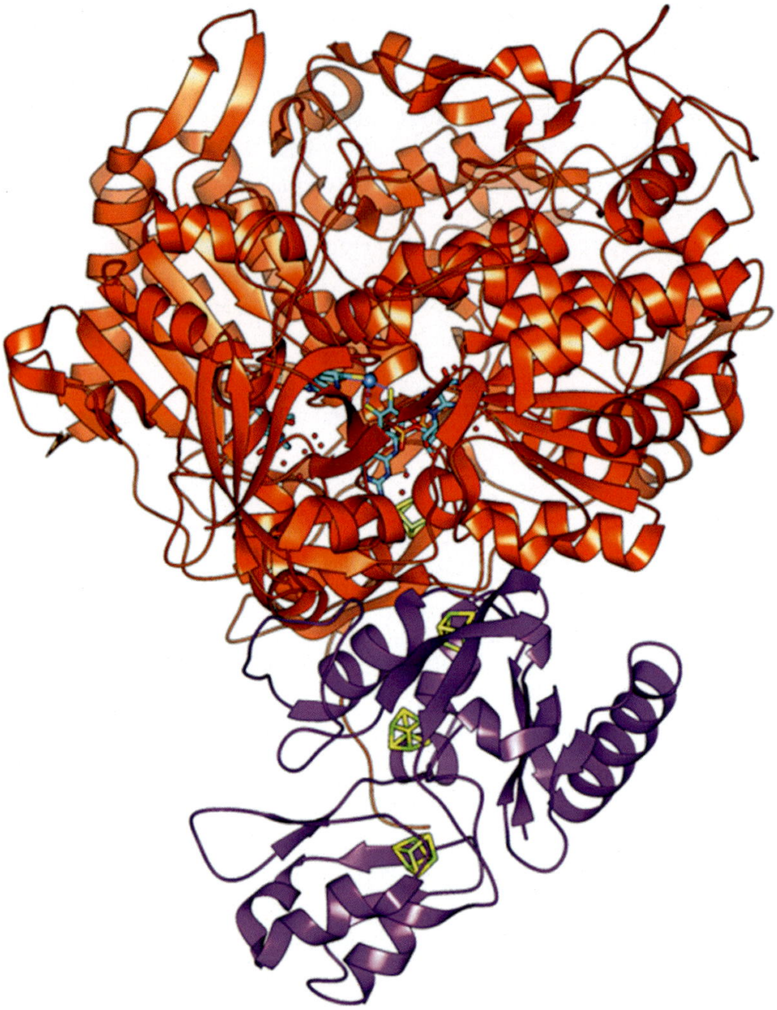

FIG. 5 The formate dehydrogenase's structure from *D. gigas*. In orange, the α subunit contains the W center and one of the four [4Fe:4S] clusters. In the small subunit β *(violet)*, we can see the other three clusters that bridge the tungstocenter with the protein surface.

a continuous polypeptide (residues 157–409) and presents a Ca^{2+}-binding site, which does not have catalytic utility. Domain IV, defined by the last 241 amino acids, can be subdivided into domain IVa (residues 735–795) and IVb (residues 796–977). In domain IVb, there is a disulfide bridge, between Cys817 and Cys844, located in a hydrophobic pocket on the surface. This bridge is far from the electron transport chain, so it is unlikely to play a role in catalysis; furthermore, in the active form, it will be broken to open the formate entry cleft. Thus, its protective role in mildly oxidized (toxic) environments has been proposed [99].

The small subunit (214 amino acids) is located opposite from the cleft leading to the active site and interacts with the large subunit by several hydrogen bonds, salt bridges, bound waters, and hydrophobic interactions, stabilizing the interface between the subunits of the heterodimer. The small subunit contains three domains: Domain A which holds two [4Fe:4S] clusters, Domain B that have the last one [4Fe:4S] cluster, and Domain C [99] (Fig. 5).

9.4 Reaction mechanism

For FDH four different mechanisms of action have been proposed; however, the first is based on the first resolved structure of *Escherichia coli*, where the sixth coordination ligand of the metal center was assumed to be a hydroxylic group. So, the subsequent identification of a terminal sulfo group in the metal coordination led to a discard of this mechanism [101].

- The second mechanism suggests that when the formate enters in the active site replaces the SeCys140 interaction with tungsten. So, the now uncoordinated SeCys140, stabilized by interaction with Arg333, takes the formate Cα hydrogen, together with the two-electron transfer to the W atom (W^{6+} to W^{4+}). Then the proton is subsequentially transferred to His141 (the final proton acceptor) and the carbon dioxide is released. The catalytic cycle would be closed with the oxidation of W^{4+} to W^{6+}, by electron transfer to the Fe/S center, and re-binding of the selenocysteine to the tungsten center [97].

- The third mechanism proposed involves also the selenocysteine dissociation from the tungsten center, but, in this case, through a sulfur shift. In this mechanism, the hexacoordination of the tungsten atom represents the inactive form of the enzyme, and it would be activated when the formate reaches the active site. When the formate enters the active site, oriented by the positively charged arginine residue, the repulsive environment generated would trigger the insertion of the sulfur atom into the W-SeCys bond, to produce a W-S-SeCys portion. In this process, the tungsten is formally reduced to W^{4+} and a new binding position is created that can, now, coordinate the formate molecule. Subsequently, the S-Secys bond is cleaved and the formed selenol anion, stabilized by a hydrogen bond with the histidine residue, can abstract the Cα proton from the formate. The catalytic cycle is closed by releasing carbon dioxide, oxidizing W^{4+} a W^{6+}, transferring intramolecular electrons to the FE/S center, and deprotonating the residue of selenocysteine. At this point, the tungsten center can bind a new formate molecule and start a new catalytic cycle or reorient the selenocysteine-containing loop to return to the inactive form [97].
- In the last mechanism that was proposed, the formate oxidation would occur through an initial hydride transfer from formate to the tungsten atom, followed by proton transfer from tungsten to the selenocysteine (the final proton acceptor) [97].

10 *N*-formylmethanofuran dehydrogenase (FMDH)

N-formylmethanofuran dehydrogenase is the last enzyme of this family. It was found only in methanogens where it catalyzes the first step in the conversion of CO_2 to methane, the other substrate is represented by methanofuran (MFR), and the physiological electron donor is not known. The reaction is herein reported (Eq. 6):

$$CO_2 + MFR^+ + H^+ + 2e^- \leftrightharpoons CHO - MFR + H_2O \quad (6)$$

FMDH has been found only in *Methanobacterium thermoautotrophicum* and *Methanobacterium wolfei* [102]. Bot produces two FMDHs: one containing W (FMDH II) and the other with Mo (FMDH I), all of these contain the GMP derivative of the pterin cofactor [103]. W-FMDH II from *Methanobacterium wolfei* is only induced by tungstate, whereas in *M. thermoautotrophicum*, the tungsten isoenzyme is constitutively synthesized [22].

10.1 General structure

The *Methanobacterium thermoautotrophicum* FMDH is a dimer of heterohexamer composed of six different subunits, which are shown in Fig. 6: FwdA (63 kDa), FwdB (47 kDa), FwdC (29 kDa), FwdD (14 kDa), FwdF (39 kDa), FwdG (8.6 kDa), while *Methanobacterium wolfei* FMDH has only FwdA and FwdB [84]. The subunit FwdA has a binuclear metal center composed of two Zinc atoms that are coordinated to four histidines and an N6-carboxylysine. One is also ligated to the catalytically essential Asp385. FwdB is structurally related to domains I, II, and III of FDH and has inside the W-center and a [4Fe:4S] cluster. As in FDH, tungsten is coordinated by four dithiolene thiolates of two tungstopterin guanine dinucleotide molecules, by the thiolate of Cys118, and by an inorganic sulfide ligand. Moreover, FwdD is structurally related to domain IV. Each FwdF has a [4Fe:4S] cluster that serves as an entry point for electrons, these follow a transport chain formed by a sequence of clusters (23 clusters overall), extend into the G subunit, and lead the electron to the tungsten centers for the CO_2 reduction (the W-center in FwdB). It is possible that this large network of clusters does not only perform a function of transposing electrons to the active centers but can also function to store reduction equivalents similarly to multi-cytochromes and multiheme enzymes [101].

The deeply buried tungsten center (in FwdB) relates to the protein surface by a hydrophobic tunnel 40 Å long, accessible to CO_2 but not to formate. Here the first step of the reaction takes place, where CO_2 is reduced to formate. In a subsequent step, formate is covalently bound to the amino group of MFR to generate formyl-MFR, this happens in FwdA. A further tunnel consisting of a wide solvent-filled cavity allows the transport of formate and formic acid between the active sites of FwdA and FwdB. Arg 288 from FwdB and Lys64 from FwdA control gate opening and closing, to arrest the formate until MFR binding [101].

11 Acetylene hydratase (AH)

Acetylene hydratase is the only member of the third class of tungstoenzymes, and it catalyzes the hydration of acetylene to acetaldehyde, according to the reaction (Eq. 7).

$$HC \equiv CH + H_2O \leftrightharpoons CH_3CHO \quad (7)$$

It was purified only from acetylene utilizing anaerobe *Pelobacter acetylenicus* [78]. AH catalyzes hydration (not a redox reaction) in contrast to the oxidoreductase-type reaction catalyzed by all other tungstoenzymes. However, AH activity depends on the presence of a strong reducing agent, such as Ti(III) citrate or Na^+ dithionite. AH is

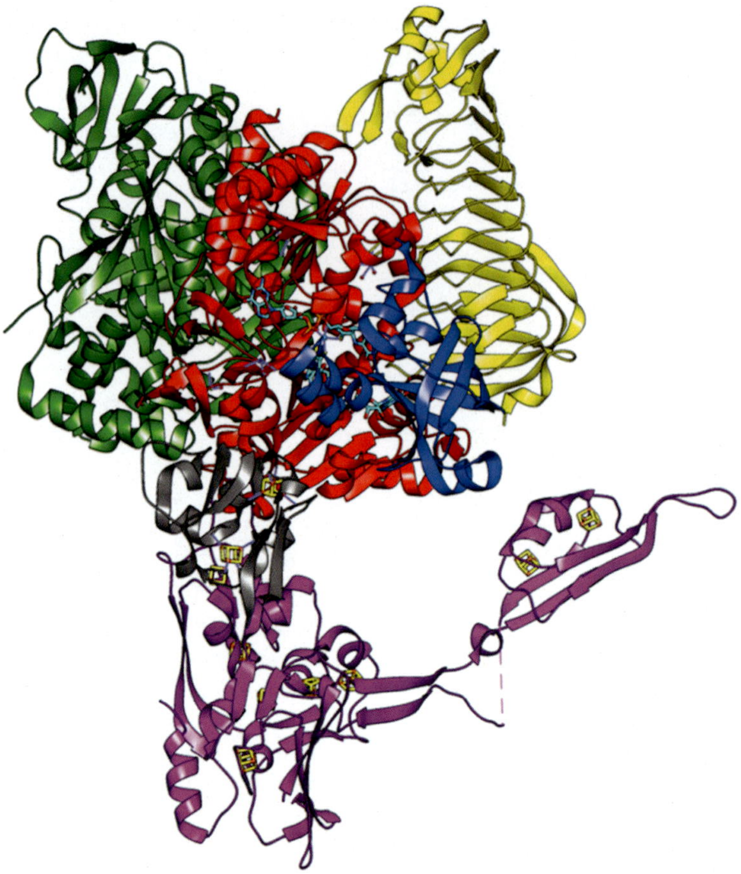

FIG. 6 *N*-formylmethanofuran dehydrogenase's monomer from *Methanobacterium thermoautotrophicum*. The FwdA *(green)* contains the binuclear Zinc center; FwdB *(red)* has the W-center and a [4Fe:4S] cluster; FwdF *(violet)* and FwdG *(gray)* contain the chain of clusters that connect the protein surface with the CO_2 reduction center; the last FwdC and FwdD are, respectively, *yellow* and *blue*.

extremely sensitive to oxygen, and its activity is irreversibly lost with exposure to air [85].

11.1 Structure

AH, shown in Fig. 7, is a monomer of 730 amino acids (83 kDa), the tertiary structure is subdivided into four domains: Domain I has the [4Fe:4S] cluster, ligated by four cysteine residues, Cys-9, Cys-12, Cys-16, and Cys-46. Domains II and III provide hydrogen bonds, required to bind one of the MGD cofactors. Domain IV completes the coordination of the MGD cofactors [104].

11.2 Structure of the active sites

The W center is in the reduced W(IV) state and is coordinated by the four sulfur atoms of the dithiolene moieties of the two pyranopterin guanine dinucleotide cofactors and by one sulfur atom of Cys141. The sixth ligand is an oxygen atom at 2.04 Å from the W atom, which could be OH^- or H_2O.

Acetylene can access the active site through a channel 17 Å deep, positioned close to the only [4Fe:4S] cluster. This funnel ends in a ring of six bulky hydrophobic residues (Ile14, Ile113, Ile142, Trp179, Trp293, and Trp472), originating from domains I, II, and III, and they form a small hydrophobic pocket with perfect dimensions for binding acetylene [105]. Asp13, near the catalytic center, forms a tight hydrogen bond to the oxygen ligand of the W atom and is assumed to be catalytically important to activate the oxygen atom for the addition to the C≡C triple bond. Residue Lys48, between the [4Fe:4S] cluster and the MGD cofactor, extremely important for electron transport in other tungstoenzymes, has no function in this case as ACH does not involve net electron transfer [104–106].

11.3 Reaction mechanism

Many reaction mechanisms have been proposed based on the nature of Asp 13 (protonated or deprotonated) and the oxygen ligand (OH^- or H_2O). If we assume that Asp13 is protonated, two different mechanistic scenarios are possible:

- A hydroxy ligand would constitute a strong nucleophile and would yield a vinyl anion with acetylene of sufficient basicity to deprotonate Asp-13 and form the

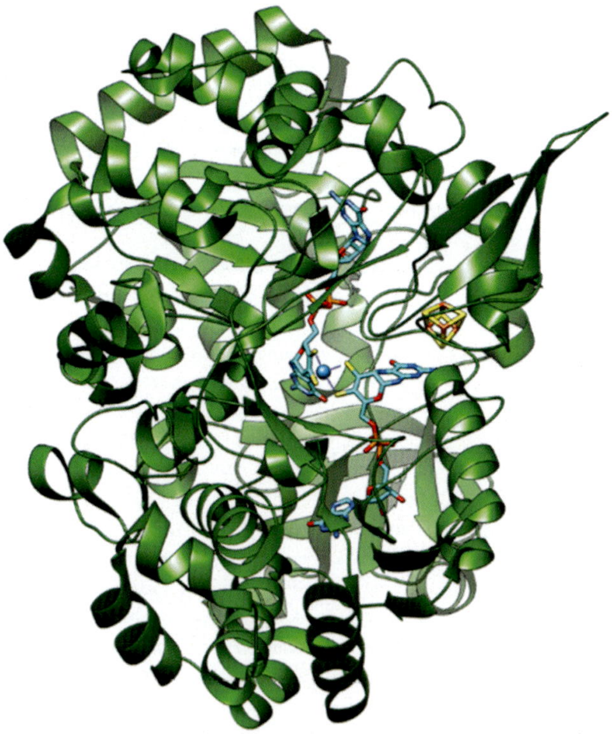

FIG. 7 The acetylene hydratase's structure from *Pelobacter acetylenicus*.

vinyl alcohol. Then another water molecule enters and regenerates the hydroxy ligand for the next reaction cycle.

- Alternatively, the H_2O molecule, with a partially positive net charge for the proximity of the protonated Asp13, could directly attack the triple bond in a Markovnikov-type addition reaction.
- If Asp13 is deprotonated acetylene can form a complex with the W ion by displacing the water ligand. The water molecule that was displaced attacks the acetylene complex forming a vinyl anion, and a water proton is transferred to the Asp13 carboxylate group. In a subsequent step, the resulting vinyl anion is protonated by Asp13 yielding the corresponding vinyl alcohol. At the end, the tautomerization of the vinyl alcohol produces acetaldehyde.
- In the last mechanism, the water molecule coordinated to the W center is activated to Lewis's acid and Asp13 is assumed to be in anionic form. This activated water then donates one of its protons to the anionic Asp13, forming the W-bound hydroxide and protonated Asp13. The W-bound hydroxide then attacks the C atom of acetylene together with the transfer of a proton from Asp13 to the other carbon atom, resulting in the formation of a vinyl alcohol intermediate complex. The final step corresponds to the tautomerization of a vinyl alcohol intermediate to the acetaldehyde via intermolecular assistance of two water molecules [104–107].

12 Tungstoenzymes and human health

As shown, enzymes containing tungsten are diverse and widespread in the microbial world, and some of these were found in the human gut microbiome. These enzymes can catalyze the conversion of toxic aldehydes to the corresponding acid. The human gut is rich in aldehydes produced by microorganisms or assumed in the diet, which are chemically reactive and potentially toxic to human health and intestinal microbes. Furfural and HMF generated during baking are reported to be a mutagen. Cooking oils also contain many aliphatic aldehydes, some of which are toxic. The removal of toxic aldehyde by generating acid can also produce intracellularly a strong reductant which can be used by the organism to drive thermodynamically unfavorable metabolic reactions. Moreover, this W-dependent process may therefore be beneficial for humans by not only removing toxic aldehydes but also adding to the pool of beneficial organic acids.

The prevalence of tungstoenzymes was found in the anaerobic large intestine, given their susceptibility to oxidation. Studies in patients with intestinal disorders, such as Crohn's disease, which have changes in the anaerobicity of the colon, have shown a different response to the intake of oligofructose-enriched prebiotics. In both cases, sick and healthy, there is an increase in the concentration of acetaldehyde, but in the healthy, it is converted into acetate while in the sick, given the intestinal aerobic condition, the activity of tungstoenzymes is zero and the levels of acetaldehyde remain high. These results, therefore, suggest that the oxidation of aldehydes to their corresponding organic acids in the gut is specific to the presence of anaerobic microbes, and this process is at least in part W dependent involving W-containing enzymes [108].

13 Conclusion

We have explored many different classes of tungsten-containing enzymes, going into detail about their distribution, physiological function, structure, and reaction mechanism. We have seen how these enzymes could have an important role in the euxinic era, thanks to the chemical-physical features of tungsten. And finally, watching human health, although tungsten-containing enzymes have been found only in prokaryotes that usually live in extreme conditions, we have shown a possible role of these enzymes in the detoxification of aldehydes in the human gut. To date, the interest in these enzymes is rising but have not been identified their possible therapeutic applications, and no drugs were produced yet.

References

[1] Luisa BM, Isabel M, Moura JJG. Molybdenum and tungsten enzymes: biochemistry. In: Molybdenum and tungsten-containing enzymes: an overview. The Royal Society of Chemistry; 2016. p. 1–80 [chapter 1].

[2] Andreesen JR, Makdessi K. Tungsten, the surprisingly positively acting heavy metal element for prokaryotes. Ann N Y Acad Sci 2008 Mar;1125:215–29.

[3] L'vov NP, Nosikov AN, Antipov AN. Tungsten-containing enzymes. Biochemistry (Mosc) 2002 Feb;67(2):196–200.

[4] Kletzin A, Adams MW. Tungsten in biological systems. FEMS Microbiol Rev 1996 Mar;18(1):5–63.

[5] Rajagopalan KV, Johnson JL. The pterin molybdenum cofactors. J Biol Chem 1992 May 25;267(15):10199–202.

[6] Burgmayer SJN, Stiefel EI. Molybdenum enzymes, cofactors, and systems: the chemical uniqueness of molybdenum. J Chem Educ 1985;62(11):943.

[7] Edward IS, Dimitri C, William EN. Molybdenum enzymes, cofactors, and model systems. ACS 1993;535.

[8] Johnson MK, Rees DC, Adams MW. Tungstoenzymes. Chem Rev 1996 Nov 7;96(7):2817–40.

[9] Pushie MJ, Cotelesage JJ, George GN. Molybdenum and tungsten oxygen transferases—and functional diversity within a common active site motif. Metallomics 2014 Jan;6(1):15–24.

[10] Hille R. Molybdenum and tungsten in biology. Trends Biochem Sci 2002 Jul;27(7):360–7.

[11] Nicholas DJ, Nason A. Molybdenum and nitrate reductase. II. Molybdenum as a constituent of nitrate reductase. J Biol Chem 1954;207:353–60.

[12] Pinsent J. The need for selenite and molybdate in the formation of formic dehydrogenase by members of the coli-aerogenes group of bacteria. Biochem J 1954;57:10–6.

[13] Higgins ES, Richert DA, Westerfeld WW. Molybdenum deficiency and tungstate inhibition studies. J Nutr 1956;59:539–59.

[14] Johnson JL, Rajagopalan KV, Cohen HJ. Molecular basis of the biological function of molybdenum. Effect of tungsten on xanthine oxidase and sulfite oxidase in the rat. J Biol Chem 1974;249:859–66.

[15] Jones JB, Stadtman TC. Methanococcus vannielii: culture and effects of selenium and tungsten on growth. J Bacteriol 1977 Jun;130(3):1404–6.

[16] Jones JB, Stadtman TC. Selenium-dependent and selenium-independent formate dehydrogenases of Methanococcus vannielii. Separation of the two forms and characterization of the purified selenium-independent form. J Biol Chem 1981 Jan 25;256 (2):656–63.

[17] Zabel HP, Winter J, König H. Isolation and characterization of a new coccoid methanogen, Methanogenium tatii sp. nov. from a solfataric field on mount Tatio. Arch Microbiol 1984;137:308–15.

[18] Zellner G, Sleytr UB, Messner P, Kneifel H, Winter J. Methanogenium liminatans spec. Nov., a new coccoid, mesophilic methanogen able to oxidize secondary alcohols. Arch Microbiol 1990;153:287–93.

[19] Zellner G, Alien C, Stackebrandt E, Conway de Macario E, Winter J. Isolation and characterization of Methanocorpusculum pare,urn, gen. nov., spec. nov., a new tungsten requiring, coccoid methanogen. Arch Microbiol 1987;147:13–20.

[20] Schmitz RA, Albracht SP, Thauer RK. Properties of the tungsten-substituted molybdenum formylmethanofuran dehydrogenase from Methanobacterium wolfei. FEBS Lett 1992 Aug 31;309(1):78–81.

[21] Schmitz RA, Richter M, Linder D, Thauer RK. A tungsten-containing active formylmethanofuran dehydrogenase in the thermophilic archaeon Methanobacterium wolfei. Eur J Biochem 1992 Jul 15;207(2):559–65.

[22] Bertram PA, Schmitz RA, Linder D, Thauer RK. Tungstate can substitute for molybdate in sustaining growth of Methanobacterium thermoautotrophicum. Identification and characterization of a tungsten isoenzyme of formylmethanofuran dehydrogenase. Arch Microbiol 1994;161(3):220–8.

[23] Bertram PA, Karrasch M, Schmitz RA, Böcher R, Albracht SP, Thauer RK. Formylmethanofuran dehydrogenases from methanogenic Archaea. Substrate specificity, EPR properties and reversible inactivation by cyanide of the molybdenum or tungsten iron-sulfur proteins. Eur J Biochem 1994 Mar 1;220(2):477–84.

[24] Zellner G, Winter J. Growth promoting effect of tungsten on methanogens and incorporation of tungsten185 into cells. FEMS Microbiol Lett 1987;40:81–7.

[25] Mukund S, Adams MW. The novel tungsten-iron-sulfur protein of the hyperthermophilic archaebacterium, Pyrococcus furiosus, is an aldehyde ferredoxin oxidoreductase. Evidence for its participation in a unique glycolytic pathway. J Biol Chem 1991 Aug 5;266(22):14208–16.

[26] Bryant FO, Adams MW. Characterization of hydrogenase from the hyperthermophilic archaebacterium, *Pyrococcus furiosus*. J Biol Chem 1989 Mar 25;264(9):5070–9.

[27] Juszczak A, Aono S, Adams MW. The extremely thermophilic eubacterium, Thermotoga maritima, contains a novel iron-hydrogenase whose cellular activity is dependent upon tungsten. J Biol Chem 1991 Jul 25;266(21):13834–41.

[28] Andreesen JR, Ljungdahl LG. Formate dehydrogenase of Clostridium thermoaceticum: incorporation of selenium-75, and the effects of selenite, molybdate, and tungstate on the enzyme. J Bacteriol 1973 Nov;116(2):867–73.

[29] Laukel M, Chistoserdova L, Lidstrom ME, Vorholt JA. The tungsten-containing formate dehydrogenase from Methylobacterium extorquens AM1: purification and properties. Eur J Biochem 2003 Jan;270(2):325–33.

[30] Andreesen JR, El Ghazzawi E, Gottschalk G. The effect of ferrous ions, tungstate and selenite on the level of formate dehydrogenase in Clostridium formicoaceticum and formate synthesis from CO2 during pyruvate fermentation. Arch Microbiol 1974;96:103–18.

[31] White H, Strobl G, Feicht R, Simon H. Carboxylic acid reductase: a new tungsten enzyme catalyses the reduction of non-activated carboxylic acids to aldehydes. Eur J Biochem 1989 Sep 1;184(1):89–96.

[32] White H, Simon H. The role of tungstate and/or molybdate in the formation of aldehyde oxidoreductase in Clostridium thermoaceticum and other acetogens; immunological distances of such enzymes. Arch Microbiol 1992;158(2):81–4.

[33] Wagner R, Andreesen JR. Differentiation between Clostridium acidiurici and Clostridium cylindrosporum on the basis of specific metal requirements for formate dehydrogenase formation. Arch Microbiol 1977 Sep 28;114(3):219–24.

[34] White H, Feicht R, Huber C, Lottspeich F, Simon H. Purification and some properties of the tungsten-containing carboxylic acid reductase from Clostridium formicoaceticum. Biol Chem Hoppe Seyler 1991 Nov;372(11):999–1005.

[35] Bertram PA, Karrasch M, Schmitz RA, Bocher R, Albracht SPJ, Thauer RK. Formylmethanofuran dehydrogenases from methanogenic Archaea: substrate specificity, EPR properties and reversible inactivation by cyanide of the molybdenum or tungsten iron-sulfur proteins. Eur J Biochem 1994;220:477–84.

[36] Keeler RF, Varner JE. Tungstate as an antagonist of molybdate in Azotobacter cinelandii. Arch Biochem Biophys 1957;70:585–90.

[37] Takahashi H, Nason A. Tungstate as competitive inhibitor of molybdate in nitrate assimilation and in N2 fixation by Azotobacter. Biochim Biophys Acta 1957 Feb;23(2):433–5.

[38] Bulen WA. Effect of tungstate on the uptake and function of molybdate in azotobacter agilis. J Bacteriol 1961 Jul;82(1): 130–4.

[39] Oltmann LF, Claassen VP, Kastelein P, Reijnders WN, Stouthamer AH. Influence of tungstate on the formation and activities of four reductases in proteus mirabilis: identification of two new molybdo-enzymes: chlorate reductase and tetrathionate reductase. FEBS Lett 1979 Oct 1;106(1):43–6.

[40] Scott RH, Sperl GT, DeMoss JA. In vitro incorporation of molybdate into demolybdoproteins in Escherichia coli. J Bacteriol 1979 Feb;137(2):719–26.

[41] Schmitz RA, Albracht SP, Thauer RK. A molybdenum and a tungsten isoenzyme of formylmethanofuran dehydrogenase in the thermophilic archaeon Methanobacterium wolfei. Eur J Biochem 1992;209:1013–8.

[42] Adams MWW. Novel iron-sulfur clusters in metalloenzymes and redox proteins from extremely thermophilic bacteria. Adv Inorg Chem 1992;38:341–96.

[43] Dellien I, Hall FM, Hepler L. Chromium, molybdenum, and tungsten: thermodynamic properties, chemical equilibriums, and standard potentials. Chem Rev 1976;76:283.

[44] Callis GE, Wentworth RA. Tungsten vs. Molybdenum in models for biological systems. Bioinorg Chem 1977;7(1):57–70.

[45] Hille R. The mononuclear molybdenum enzymes. Chem Rev 1996 Nov 7;96(7):2757–816.

[46] Hille R, Nishino T, Bittner F. Molybdenum enzymes in higher organisms. Coord Chem Rev 2011 May 1;255(9–10):1179–205.

[47] Mendel RR, Kruse T. Cell biology of molybdenum in plants and humans. Biochim Biophys Acta 2012 Sep;1823(9):1568–79.

[48] Hille R. The molybdenum oxotransferases and related enzymes. Dalton Trans 2013 Mar 7;42(9):3029–42.

[49] Hille R, Hall J, Basu P. The mononuclear molybdenum enzymes. Chem Rev 2014 Apr 9;114(7):3963–4038.

[50] Roy R, Adams MW. Tungsten-dependent aldehyde oxidoreductase: a new family of enzymes containing the pterin cofactor. Met Ions Biol Syst 2002;39:673–97.

[51] Mendel R, Schwarz G. Molybdoenzymes and molybdenum cofactor in plants. Crit Rev Plant Sci 1999;18:33–69.

[52] Johnson JL, Rajagopalan KV. Structural and metabolic relationship between the molybdenum cofactor and urothione. Proc Natl Acad Sci U S A 1982 Nov;79(22):6856–60.

[53] Johnson JL, Hainline BE, Rajagopalan KV. Characterization of the molybdenum cofactor of sulfite oxidase, xanthine, oxidase, and nitrate reductase. Identification of a pteridine as a structural component. J Biol Chem 1980 Mar 10;255(5):1783–6.

[54] Fischer B, Enemark JH, Basu P. A chemical approach to systematically designate the pyranopterin centers of molybdenum and tungsten enzymes and synthetic models. J Inorg Biochem 1998 Oct;72(1–2):13–21.

[55] Johnson JL, Hainline BE, Rajagopalan KV, Arison BH. The pterin component of the molybdenum cofactor. Structural characterization of two fluorescent derivatives. J Biol Chem 1984 May 10;259(9): 5414–22.

[56] Hinton SM, Dean D. Biogenesis of molybdenum cofactors. Crit Rev Microbiol 1990;17(3):169–88.

[57] Romão MJ. Molybdenum and tungsten enzymes: a crystallographic and mechanistic overview. Dalton Trans 2009 Jun 7;21:4053–68.

[58] Mukund S, Adams MW. Characterization of a tungsten-iron-sulfur protein exhibiting novel spectroscopic and redox properties from the hyperthermophilic archaebacterium Pyrococcus furiosus. J Biol Chem 1990 Jul 15;265(20):11508–16.

[59] Bertero MG, Rothery RA, Palak M, Hou C, Lim D, Blasco F, Weiner JH, Strynadka NC. Insights into the respiratory electron transfer pathway from the structure of nitrate reductase a. Nat Struct Biol 2003 Sep;10(9):681–7.

[60] Kloer DP, Hagel C, Heider J, Schulz GE. Crystal structure of ethylbenzene dehydrogenase from Aromatoleum aromaticum. Structure 2006 Sep;14(9):1377–88.

[61] Jormakka M, Richardson D, Byrne B, Iwata S. Architecture of NarGH reveals a structural classification of Mo-bisMGD enzymes. Structure 2004 Jan;12(1):95–104.

[62] Rothery RA, Weiner JH. Shifting the metallocentric molybdoenzyme paradigm: the importance of pyranopterin coordination. J Biol Inorg Chem 2015 Mar;20(2):349–72.

[63] Kirk ML, Khadanand KC. Molybdenum and tungsten cofactors and the reactions they catalyze. Met Ions Life Sci 2020 Mar 23;20.

[64] Mendel RR. The molybdenum cofactor. J Biol Chem 2013 May 10;288(19):13165–72.

[65] Matz KG, Mtei RP, Leung B, Burgmayer SJ, Kirk ML. Noninnocent dithiolene ligands: a new oxomolybdenum complex possessing a donor-acceptor dithiolene ligand. J Am Chem Soc 2010 Jun 16;132(23):7830–1.

[66] Matz KG, Mtei RP, Rothstein R, Kirk ML, Burgmayer SJ. Study of molybdenum(4+) quinoxalyldithiolenes as models for the noninnocent pyranopterin in the molybdenum cofactor. Inorg Chem 2011 Oct 17;50(20):9804–15.

[67] Williams BR, Fu Y, Yap GP, Burgmayer SJ. Structure and reversible pyran formation in molybdenum pyranopterin dithiolene models of the molybdenum cofactor. J Am Chem Soc 2012 Dec 5;134(48): 19584–7.

[68] Dong C, Yang J, Leimkühler S, Kirk ML. Pyranopterin dithiolene distortions relevant to electron transfer in xanthine oxidase/dehydrogenase. Inorg Chem 2014 Jul 21;53(14):7077–9.

[69] Sugimoto H, Tsukube H. Chemical analogues relevant to molybdenum and tungsten enzyme reaction centres toward structural dynamics and reaction diversity. Chem Soc Rev 2008 Dec;37(12): 2609–19.

[70] Basu P, Burgmayer SJ. Pterin chemistry and its relationship to the molybdenum cofactor. Coord Chem Rev 2011 May;255(9–10): 1016–38.

[71] Rothery RA, Stein B, Solomonson M, Kirk ML, Weiner JH. Pyranopterin conformation defines the function of molybdenum and tungsten enzymes. Proc Natl Acad Sci U S A 2012 Sep 11;109(37): 14773–8.

[72] Heider J, Ma K, Adams MW. Purification, characterization, and metabolic function of tungsten-containing aldehyde ferredoxin oxidoreductase from the hyperthermophilic and proteolytic archaeon Thermococcus strain ES-1. J Bacteriol 1995 Aug;177(16):4757–64.

[73] Adams MW, Kletzin A. Oxidoreductase-type enzymes and redox proteins involved in fermentative metabolisms of hyperthermophilic Archaea. Adv Protein Chem 1996;48:101–80.

[74] Mukund S, Adams MW. Characterization of a novel tungsten-containing formaldehyde ferredoxin oxidoreductase from the hyperthermophilic archaeon, Thermococcus litoralis. A role for tungsten in peptide catabolism. J Biol Chem 1993 Jun 25;268(18):13592–600.

[75] Mukund S, Adams MW. Glyceraldehyde-3-phosphate ferredoxin oxidoreductase, a novel tungsten-containing enzyme with a potential glycolytic role in the hyperthermophilic archaeon Pyrococcus furiosus. J Biol Chem 1995 Apr 14;270(15):8389–92.

[76] Kengen SW, De Bok FA, Van Loo ND, Dijkema C, Stams AJ, De Vos WM. Evidence for the operation of a novel Embden-Meyerhof pathway that involves ADP-dependent kinases during sugar fermentation by Pyrococcus furiosus. J Biol Chem 1994 Jul 1;269(26): 17537–41.

[77] Yamamoto I, Saiki T, Liu SM, Ljungdahl LG. Purification and properties of NADP-dependent formate dehydrogenase from Clostridium thermoaceticum, a tungsten-selenium-iron protein. J Biol Chem 1983 Feb 10;258(3):1826–32.

[78] Chan MK, Mukund S, Kletzin A, Adams MW, Rees DC. Structure of a hyperthermophilic tungstopterin enzyme, aldehyde ferredoxin oxidoreductase. Science 1995 Mar 10;267(5203):1463–9.

[79] Ma K, Loessner H, Heider J, Johnson MK, Adams MW. Effects of elemental sulfur on the metabolism of the deep-sea hyperthermophilic archaeon Thermococcus strain ES-1: characterization of a sulfur-regulated, non-heme iron alcohol dehydrogenase. J Bacteriol 1995 Aug;177(16):4748–56.

[80] Hagedoorn PL, Chen T, Schröder I, Piersma SR, De Vries S, Hagen WR. Purification and characterization of the tungsten enzyme aldehyde:ferredoxin oxidoreductase from the hyperthermophilic denitrifier Pyrobaculum aerophilum. J Biol Inorg Chem 2005 May;10(3):259–69.

[81] Strobl G, Feicht R, White H, Lottspeich F, Simon H. The tungsten-containing aldehyde oxidoreductase from Clostridium thermoaceticum and its complex with a viologen-accepting NADPH oxidoreductase. Biol Chem Hoppe Seyler 1992 Mar;373(3):123–32.

[82] Rauh D, Graentzdoerffer A, Granderath K, Andreesen JR, Pich A. Tungsten-containing aldehyde oxidoreductase of eubacterium acidaminophilum. Eur J Biochem 2004 Jan;271(1):212–9.

[83] Hensgens CM, Hagen WR, Hansen TA. Purification and characterization of a benzylviologen-linked, tungsten-containing aldehyde oxidoreductase from Desulfovibrio gigas. J Bacteriol 1995 Nov;177(21):6195–200.

[84] Hochheimer A, Hedderich R, Thauer RK. The formylmethanofuran dehydrogenase isoenzymes in Methanobacterium wolfei and Methanobacterium thermoautotrophicum: induction of the molybdenum isoenzyme by molybdate and constitutive synthesis of the tungsten isoenzyme. Arch Microbiol 1998 Oct;170(5):389–93.

[85] Meckenstock RU, Krieger R, Ensign S, Kroneck PM, Schink B. Acetylene hydratase of Pelobacter acetylenicus. Molecular and spectroscopic properties of the tungsten iron-sulfur enzyme. Eur J Biochem 1999 Aug;264(1):176–82.

[86] Kleywegt GJ, Jones TA. Detection, delineation, measurement and display of cavities in macromolecular structures. Acta Crystallogr D Biol Crystallogr 1994 Mar 1;50(Pt 2):178–85.

[87] Roy R, Mukund S, Schut GJ, Dunn DM, Weiss R, Adams MW. Purification and molecular characterization of the tungsten-containing formaldehyde ferredoxin oxidoreductase from the hyperthermophilic archaeon Pyrococcus furiosus: the third of a putative five-member tungstoenzyme family. J Bacteriol 1999 Feb;181(4): 1171–80.

[88] Liao RZ, Yu JG, Himo F. Tungsten-dependent formaldehyde ferredoxin oxidoreductase: reaction mechanism from quantum chemical calculations. J Inorg Biochem 2011 Jul;105(7):927–36.

[89] Kletzin A, Mukund S, Kelley-Crouse TL, Chan MK, Rees DC, Adams MW. Molecular characterization of the genes encoding the tungsten-containing aldehyde ferredoxin oxidoreductase from Pyrococcus furiosus and formaldehyde ferredoxin oxidoreductase from Thermococcus litoralis. J Bacteriol 1995 Aug;177(16):4817–9.

[90] Hu Y, Faham S, Roy R, Adams MW, Rees DC. Formaldehyde ferredoxin oxidoreductase from Pyrococcus furiosus: the 1.85 a resolution crystal structure and its mechanistic implications. J Mol Biol 1999 Feb 26;286(3):899–914.

[91] Huber C, Skopan H, Feicht R, et al. Pterin cofactor, substrate specificity, and observations on the kinetics of the reversible tungsten-containing aldehyde oxidoreductase from Clostridium thermoaceticum. Arch Microbiol 1995;164:110–8.

[92] Groysman S, Holm RH. Synthesis and structures of bis(dithiolene) tungsten (IV,VI) thiolate and selenolate complexes: approaches to the active sites of molybdenum and tungsten formate dehydrogenases. Inorg Chem 2007 May 14;46(10):4090–102.

[93] Andreesen JR, El Ghazzawi E, Gottschalk G. The effect of ferrous ions, tungstate and selenite on the level of formate dehydrogenase in Clostridium formicoaceticum and formate synthesis from CO_2 during pyruvate fermentation. Arch Mikrobiol 1974 Mar 4;96 (2):103–18.

[94] Alissandratos A, Kim HK, Matthews H, Hennessy JE, Philbrook A, Easton CJ. Clostridium carboxidivorans strain P7T recombinant formate dehydrogenase catalyzes reduction of CO(2) to formate. Appl Environ Microbiol 2013 Jan;79(2):741–4.

[95] Almendra MJ, Brondino CD, Gavel O, Pereira AS, Tavares P, Bursakov S, Duarte R, Caldeira J, Moura JJ, Moura I. Purification and characterization of a tungsten-containing formate dehydrogenase from Desulfovibrio gigas. Biochemistry 1999 Dec 7;38(49): 16366–72.

[96] Brondino CD, Passeggi MC, Caldeira J, Almendra MJ, Feio MJ, Moura JJ, Moura I. Incorporation of either molybdenum or tungsten into formate dehydrogenase from Desulfovibrio alaskensis NCIMB 13491; EPR assignment of the proximal iron-sulfur cluster to the pterin cofactor in formate dehydrogenases from sulfate-reducing bacteria. J Biol Inorg Chem 2004 Mar;9(2):145–51.

[97] Maia LB, Moura JJ, Moura I. Molybdenum and tungsten-dependent formate dehydrogenases. J Biol Inorg Chem 2015 Mar;20(2):287–309.

[98] Niks D, Hille R. Molybdenum- and tungsten-containing formate dehydrogenases and formylmethanofuran dehydrogenases: structure, mechanism, and cofactor insertion. Protein Sci 2019 Jan;28(1):111–22.

[99] Raaijmakers H, Macieira S, Dias JM, Teixeira S, Bursakov S, Huber R, Moura JJ, Moura I, Romão MJ. Gene sequence and the 1.8 a crystal structure of the tungsten-containing formate dehydrogenase from Desulfovibrio gigas. Structure 2002 Sep;10(9):1261–72.

[100] Raaijmakers H, Teixeira S, Dias JM, Almendra MJ, Brondino CD, Moura I, Moura JJ, Romão MJ. Tungsten-containing formate dehydrogenase from Desulfovibrio gigas: metal identification and preliminary structural data by multi-wavelength crystallography. J Biol Inorg Chem 2001 Apr;6(4):398–404.

[101] Wagner T, Ermler U, Shima S. The methanogenic CO_2 reducing-and-fixing enzyme is bifunctional and contains 46 [4Fe-4S] clusters. Science 2016 Oct 7;354(6308):114–7.

[102] Bertram PA, Thauer RK. Thermodynamics of the formylmethanofuran dehydrogenase reaction in Methanobacterium thermoautotrophicum. Eur J Biochem 1994 Dec 15;226(3):811–8.

[103] Hochheimer A, Schmitz RA, Thauer RK, Hedderich R. The tungsten formylmethanofuran dehydrogenase from Methanobacterium thermoautotrophicum contains sequence motifs characteristic for

enzymes containing molybdopterin dinucleotide. Eur J Biochem 1995 Dec 15;234(3):910–20.

[104] Boll M, Einsle O, Ermler U, Kroneck PM, Ullmann GM. Structure and function of the unusual tungsten enzymes acetylene hydratase and class II benzoyl-coenzyme a reductase. J Mol Microbiol Biotechnol 2016;26(1–3):119–37.

[105] Seiffert GB, Ullmann GM, Messerschmidt A, Schink B, Kroneck PM, Einsle O. Structure of the non-redox-active tungsten/[4Fe:4S] enzyme acetylene hydratase. Proc Natl Acad Sci U S A 2007 Feb 27;104(9):3073–7.

[106] Kroneck PM. Acetylene hydratase: a non-redox enzyme with tungsten and iron-sulfur centers at the active site. J Biol Inorg Chem 2016 Mar;21(1):29–38.

[107] Habib U, Riaz M, Hofmann M. Unraveling the way acetaldehyde is formed from acetylene: a study based on DFT. ACS Omega 2021 Mar 2;6(10):6924–33.

[108] Schut GJ, Thorgersen MP, Poole FL, Haja DK, Putumbaka S, Adams MWW. Tungsten enzymes play a role in detoxifying food and antimicrobial aldehydes in the human gut microbiome. Proc Natl Acad Sci U S A 2021 Oct 26;118(43), e2109008118.

Index

Note: Page numbers followed by *f* indicate figures, *t* indicate tables, and *s* indicate schemes.

B

D

E

F

O

P

Q

Printed in the United States
by Baker & Taylor Publisher Services